Polarized Light and Optical Systems

Optical Sciences and Applications of Light

Series Editor
James C. Wyant
University of Arizona

Polarized Light and Optical Systems, *Russell A. Chipman, Wai-Sze Tiffany Lam, and Garam Young*

Fundamentals of Optomechanics, *Paul Yoder and Daniel Yukobratovich*

Optics Manufacturing: Components and Systems, *Christoph Gerhard*

Femtosecond Laser Shaping: From Laboratory to Industry, *Marcos Dantus*

Photonics Modelling and Design, *Slawomir Sujecki*

Handbook of Optomechanical Engineering, Second Edition, *Anees Ahmad*

Lens Design: A Practical Guide, *Haiyin Sun*

Nanofabrication: Principles to Laboratory Practice, *Andrew Sarangan*

Blackbody Radiation: A History of Thermal Radiation Computational Aids and Numerical Methods, *Sean M. Stewart and R. Barry Johnson*

High-Speed 3D Imaging with Digital Fringe Projection Techniques, *Song Zhang*

Introduction to Optical Metrology, *Rajpal S. Sirohi*

Charged Particle Optics Theory: An Introduction, *Timothy R. Groves*

Nonlinear Optics: Principles and Applications, *Karsten Rottwitt and Peter Tidemand-Lichtenberg*

Numerical Methods in Photonics, *Andrei V. Lavrinenko, Jesper Lægsgaard, Niels Gregersen, Frank Schmidt, and Thomas Søndergaard*

Please visit our website www.crcpress.com for a full list of titles

Polarized Light and Optical Systems

Russell A. Chipman
Wai-Sze Tiffany Lam
Garam Young

CRC Press
Taylor & Francis Group
Boca Raton London New York

CRC Press is an imprint of the
Taylor & Francis Group, an **informa** business

MATLAB® is a trademark of The MathWorks, Inc. and is used with permission. The MathWorks does not warrant the accuracy of the text or exercises in this book. This book's use or discussion of MATLAB® software or related products does not constitute endorsement or sponsorship by The MathWorks of a particular pedagogical approach or particular use of the MATLAB® software.

CRC Press
Taylor & Francis Group
6000 Broken Sound Parkway NW, Suite 300
Boca Raton, FL 33487-2742

© 2019 by Taylor & Francis Group, LLC
CRC Press is an imprint of Taylor & Francis Group, an Informa business

No claim to original U.S. Government works

Printed in Canada on acid-free paper

International Standard Book Number-13: 978-1-4987-0056-6 (Hardback)

This book contains information obtained from authentic and highly regarded sources. Reasonable efforts have been made to publish reliable data and information, but the author and publisher cannot assume responsibility for the validity of all materials or the consequences of their use. The authors and publishers have attempted to trace the copyright holders of all material reproduced in this publication and apologize to copyright holders if permission to publish in this form has not been obtained. If any copyright material has not been acknowledged please write and let us know so we may rectify in any future reprint.

Except as permitted under U.S. Copyright Law, no part of this book may be reprinted, reproduced, transmitted, or utilized in any form by any electronic, mechanical, or other means, now known or hereafter invented, including photocopying, microfilming, and recording, or in any information storage or retrieval system, without written permission from the publishers.

For permission to photocopy or use material electronically from this work, please access www.copyright.com (http://www.copyright.com/) or contact the Copyright Clearance Center, Inc. (CCC), 222 Rosewood Drive, Danvers, MA 01923, 978-750-8400. CCC is a not-for-profit organization that provides licenses and registration for a variety of users. For organizations that have been granted a photocopy license by the CCC, a separate system of payment has been arranged.

Trademark Notice: Product or corporate names may be trademarks or registered trademarks, and are used only for identification and explanation without intent to infringe.

Library of Congress Cataloging-in-Publication Data

Names: Chipman, Russell A., author. | Lam, Wai-Sze Tiffany, author. | Young, Garam, author.
Title: Polarized light and optical systems / Russell A. Chipman, Wai-Sze Tiffany Lam, and Garam Young.
Description: Boca Raton : Taylor & Francis, CRC Press, 2019. | Series: Optical sciences and applications of light | Includes bibliographical references and index.
Identifiers: LCCN 2017049570| ISBN 9781498700566 (hardback : alk. paper) | ISBN 9781498700573 (ebook)
Subjects: LCSH: Optical instruments--Design and construction. | Polarization (Light) | Optics--Mathematics.
Classification: LCC TS510 .C44 2018 | DDC 535.5/2--dc23
LC record available at https://lccn.loc.gov/2017049570

Visit the Taylor & Francis Web site at
http://www.taylorandfrancis.com

and the CRC Press Web site at
http://www.crcpress.com

Contents

Authors xxi
Acknowledgments xxiii
Preface xxv
How This Book Came to Be xxix
Suggested Curricula xxxi
Guided Tour of the Chapters xxxiii
Learning Features xlvii
List of Abbreviations xlix

CHAPTER 1

Introduction and Overview 1
    1.1   Polarized Light 2
    1.2   Polarization States and the Poincaré Sphere 2
    1.3   Polarization Elements and Polarization Properties 5
    1.4   Polarimetry and Ellipsometry 8
    1.5   Anisotropic Materials 11
    1.6   Typical Polarization Problems in Optical Systems 12
            1.6.1   Angle Dependence of Polarizers 12
            1.6.2   Wavelength and Angle Dependence of Retarders 13
            1.6.3   Stress Birefringence in Lenses 14
            1.6.4   Liquid Crystal Displays and Projectors 16
    1.7   Optical Design 17
            1.7.1   Polarization Ray Tracing 20
            1.7.2   Polarization Aberrations of Lenses 22
            1.7.3   High Numerical Aperture Wavefronts 25
    1.8   Comment on Historical Treatments 26
    1.9   Reference Books on Polarized Light 27
    1.10 Problem Sets 27
    References 29

CHAPTER 2

Polarized Light 31

- 2.1 The Description of Polarized Light 31
- 2.2 The Polarization Vector 32
- 2.3 Properties of the Polarization Vector 34
- 2.4 Propagation in Isotropic Media 36
- 2.5 Magnetic Field, Flux, and Polarized Flux 36
- 2.6 Jones Vectors 37
- 2.7 Evolution of Overall Phase 41
- 2.8 Rotation of Jones Vectors 42
- 2.9 Linearly Polarized Light 42
- 2.10 Circularly Polarized Light 43
- 2.11 Elliptically Polarized Light 45
- 2.12 Orthogonal Jones Vectors 48
- 2.13 Change of Basis 49
- 2.14 Addition of Jones Vectors 49
- 2.15 Polarized Flux Components 50
- 2.16 Converting Polarization Vectors into Jones Vectors 51
- 2.17 Decreasing Phase Sign Convention 54
- 2.18 Increasing Phase Sign Convention 55
- 2.19 Polarization State of Sources 56
- 2.20 Problem Sets 59
- References 61

CHAPTER 3

Stokes Parameters and the Poincaré Sphere 63

- 3.1 The Description of Polychromatic Light 63
- 3.2 Phenomenological Definition of the Stokes Parameters 64
- 3.3 Unpolarized Light 65
- 3.4 Partially Polarized Light and the Degree of Polarization 66
- 3.5 Spectral Bandwidth 70
- 3.6 Rotation of the Polarization Ellipse 72
- 3.7 Linearly Polarized Stokes Parameters 73
- 3.8 Elliptical Polarization Parameters 73
- 3.9 Orthogonal Polarization States 74
- 3.10 Stokes Parameter and Jones Vector Sign Conventions 75
- 3.11 Polarized Fluxes and Conversions between Stokes Parameters and Jones Vectors ... 75
- 3.12 The Stokes Parameters' Non-Orthogonal Coordinate System 78
- 3.13 The Poincaré Sphere 80
- 3.14 Flat Mappings of the Poincaré Sphere 83
- 3.15 Summary and Conclusion 85
- 3.16 Problem Sets 86
- References 88

CHAPTER 4

Interference of Polarized Light 91

- 4.1 Introduction 91
- 4.2 Combining Light Waves 92
- 4.3 Interferometers 93
- 4.4 Interference of Nearly Parallel Monochromatic Plane Waves 95
- 4.5 Interference of Plane Waves at Large Angles 100
- 4.6 Polarization Considerations in Holography 102
- 4.7 The Addition of Polarized Beams 103
    - 4.7.1 Addition of Polarized Light of Two Different Frequencies 103
    - 4.7.2 Addition of Polychromatic Beams 105
    - 4.7.3 A Gaussian Wave Packet Example 109
- 4.8 Conclusion 113
- 4.9 Problem Sets 113
- References 116

CHAPTER 5

Jones Matrices and Polarization Properties 117

- 5.1 Introduction 117
- 5.2 Dichroic and Birefringent Materials 118
- 5.3 Diattenuation and Retardance 118
    - 5.3.1 Diattenuation 119
    - 5.3.2 Retardance 120
- 5.4 Jones Matrices 123
    - 5.4.1 Eigenpolarizations 124
    - 5.4.2 Jones Matrix Notation 127
    - 5.4.3 Rotation of Jones Matrices 127
- 5.5 Polarizer and Diattenuator Jones Matrices 128
    - 5.5.1 Polarizer Jones Matrices 128
    - 5.5.2 Linear Diattenuator Jones Matrices 130
- 5.6 Retarder Jones Matrices 133
    - 5.6.1 Linear Retarder Jones Matrices 133
    - 5.6.2 Circular Retarder Jones Matrices 137
    - 5.6.3 Vortex Retarders 137
- 5.7 General Diattenuators and Retarders 138
    - 5.7.1 Linear Diattenuators 139
    - 5.7.2 Elliptical Diattenuators 140
    - 5.7.3 Elliptical Retarders 141
- 5.8 Non-Polarizing Jones Matrices for Amplitude and Phase Change 143
- 5.9 Matrix Properties of Jones Matrices 144
    - 5.9.1 Hermitian Matrices: Diattenuation 144
    - 5.9.2 Unitary Matrices and Unitary Transformations: Retarder 145
    - 5.9.3 Polar Decomposition: Separating Retardance from Diattenuation 149

5.10	Increasing Phase Sign Convention	153
5.11	Conclusion	153
5.12	Problem Sets	155
References		161

CHAPTER 6

Mueller Matrices 163

6.1	Introduction	163
6.2	The Mueller Matrix	164
6.3	Sequences of Polarization Elements	165
6.4	Non-Polarizing Mueller Matrices	165
6.5	Rotating Polarization Elements about the Light Direction	166
6.6	Retarder Mueller Matrices	168
6.7	Polarizer and Diattenuator Mueller Matrices	174
	6.7.1 Basic Polarizers	174
	6.7.2 Transmittance and Diattenuation	177
	6.7.3 Polarizance	180
	6.7.4 Diattenuators	180
6.8	Poincaré Sphere Operations	182
	6.8.1 The Operation of Retarders on the Poincaré Sphere	182
	6.8.2 The Operation of a Rotating Linear Retarder	186
	6.8.3 The Operation of Polarizers and Diattenuators	187
	6.8.4 Indicating Polarization Properties	188
6.9	Weak Polarization Elements	190
6.10	Non-Depolarizing Mueller Matrices	192
6.11	Depolarization	193
	6.11.1 The Depolarization Index and the Average Degree of Polarization	195
	6.11.2 Degree of Polarization Surfaces and Maps	196
	6.11.3 Testing for Physically Realizable Mueller Matrices	198
	6.11.4 Weak Depolarizing Elements	200
	6.11.5 The Addition of Mueller Matrices	201
6.12	Relating Jones and Mueller Matrices	202
	6.12.1 Transforming Jones Matrices into Mueller Matrices Using Tensor Product	202
	6.12.2 Conversion of Jones Matrices to Mueller Matrices Using Pauli Matrices	207
	6.12.3 Transforming Mueller Matrices into Jones Matrices	207
6.13	Ray Tracing with Mueller Matrices	211
	6.13.1 Mueller Matrices for Refraction	212
	6.13.2 Mueller Matrices for Reflection	212
6.14	The Origins of the Mueller Matrix	213
6.15	Problem Sets	214
References		218

CHAPTER 7

Polarimetry 219

7.1	Introduction	219
7.2	What Does the Polarimeter See?	220

7.3	Polarimeters		221
	7.3.1	Light-Measuring Polarimeters	221
	7.3.2	Sample-Measuring Polarimeters	221
	7.3.3	Complete and Incomplete Polarimeters	222
	7.3.4	Polarization Generators and Analyzers	222
7.4	Mathematics of Polarimetric Measurement and Data Reduction		222
	7.4.1	Stokes Polarimetry	222
	7.4.2	Measuring Mueller Matrix Elements	225
	7.4.3	Mueller Data Reduction Matrix	226
	7.4.4	Null Space and the Pseudoinverse	230
7.5	Classes of Polarimeters		240
	7.5.1	Time-Sequential Polarimeters	240
	7.5.2	Modulated Polarimeters	240
	7.5.3	Division of Amplitude	241
	7.5.4	Division of Aperture	241
	7.5.5	Imaging Polarimeters	241
7.6	Stokes Polarimeter Configurations		242
	7.6.1	Simultaneous Polarimetric Measurement	242
		7.6.1.1 Division-of-Aperture Polarimetry	242
		7.6.1.2 Division-of-Focal-Plane Polarimetry	242
		7.6.1.3 Division-of-Amplitude Polarimetry	246
	7.6.2	Rotating Element Polarimetry	249
		7.6.2.1 Rotating Analyzer Polarimeters	249
		7.6.2.2 Rotating Analyzer Plus Fixed Analyzer Polarimeter	250
		7.6.2.3 Rotating Retarder and Fixed Analyzer Polarimeters	251
	7.6.3	Variable Retarder and Fixed Polarizer Polarimeter	254
	7.6.4	Photoelastic Modulator Polarimeters	255
	7.6.5	The MSPI and MAIA Imaging Polarimeters	258
	7.6.6	Example Atmospheric Polarization Images	259
7.7	Sample-Measuring Polarimeters		262
	7.7.1	Polariscopes	263
		7.7.1.1 Linear Polariscope	263
		7.7.1.2 Circular Polariscope	266
		7.7.1.3 Interference Colors	267
		7.7.1.4 Polariscope with Tint Plate	269
		7.7.1.5 Conoscope	270
	7.7.2	Mueller Polarimetry Configurations	272
		7.7.2.1 Dual Rotating Retarder Polarimeter	274
		7.7.2.2 Polarimetry Near Retroreflection	275
7.8	Interpreting Mueller Matrix Images		276
7.9	Calibrating Polarimeters		279
7.10	Artifacts in Polarimetric Images		280
	7.10.1	Pixel Misalignment	281
7.11	Optimizing Polarimeters		282
7.12	Problem Sets		286
Acknowledgments			291
References			291

CHAPTER 8

Fresnel Equations ... 295

- 8.1 Introduction ... 295
- 8.2 Propagation of Light ... 296
  - 8.2.1 Plane Waves and Rays ... 296
  - 8.2.2 Plane of Incidence ... 296
  - 8.2.3 Homogeneous and Isotropic Interfaces ... 297
  - 8.2.4 Light Propagation in Media ... 297
- 8.3 Fresnel Equations ... 298
  - 8.3.1 s- and p-Polarization Components ... 298
  - 8.3.2 Amplitude Coefficients ... 299
  - 8.3.3 The Fresnel Equations ... 300
  - 8.3.4 Intensity Coefficients ... 301
  - 8.3.5 Normal Incidence ... 304
  - 8.3.6 Brewster's Angle ... 305
  - 8.3.7 Critical Angle ... 306
  - 8.3.8 Intensity and Phase Change with Incident Angle ... 307
  - 8.3.9 Jones Matrices with Fresnel Coefficients ... 308
- 8.4 Fresnel Refraction and Reflection ... 308
  - 8.4.1 Dielectric Refraction ... 308
  - 8.4.2 External Reflection ... 309
  - 8.4.3 Internal Reflection ... 311
  - 8.4.4 Metal Reflection ... 313
    - 8.4.4.1 Normal Incidence Reflectance ... 315
    - 8.4.4.2 Retardance and Diattenuation of Metal at Non-Normal Incidence ... 315
- 8.5 Approximate Representations of Fresnel Coefficients ... 316
  - 8.5.1 Taylor Series for the Fresnel Coefficients ... 317
- 8.6 Conclusion ... 318
- 8.7 Problem Sets ... 318
- References ... 320

CHAPTER 9

Polarization Ray Tracing Calculus ... 323

- 9.1 Definition of Polarization Ray Tracing Matrix, **P** ... 324
- 9.2 Formalism of Polarization Ray Tracing Matrix Using Orthogonal Transformation ... 325
- 9.3 Retarder Polarization Ray Tracing Matrix Examples ... 331
- 9.4 Diattenuation Calculation Using Singular Value Decomposition ... 334
- 9.5 Example—Interferometer with a Polarizing Beam Splitter ... 337
  - 9.5.1 Ray Tracing the Reference Path ... 338
  - 9.5.2 Ray Tracing through the Test Path ... 340
  - 9.5.3 Ray Tracing through the Analyzer ... 341
  - 9.5.4 Cumulative **P** Matrix for Both Paths ... 342
- 9.6 The Addition Form of Polarization Ray Tracing Matrices ... 344
  - 9.6.1 Combining **P** Matrices for the Interferometer Example ... 346
- 9.7 Example—A Hollow Corner Cube ... 347
- 9.8 Conclusion ... 351

| 9.9 | Problem Sets | 352 |

References 357

CHAPTER 10

Optical Ray Tracing 359

- 10.1 Introduction 359
- 10.2 Goals for Ray Tracing 360
- 10.3 Specification of Optical Systems 364
  - 10.3.1 Surface Equations 366
  - 10.3.2 Apertures 366
  - 10.3.3 Optical Interfaces 367
  - 10.3.4 Dummy Surfaces 367
- 10.4 Specifications of Light Beams 368
- 10.5 System Descriptions 369
  - 10.5.1 Object Plane 369
  - 10.5.2 Aperture Stop 369
  - 10.5.3 Entrance and Exit Pupils 370
  - 10.5.4 Importance of the Exit Pupil 370
  - 10.5.5 Marginal and Chief Rays 372
  - 10.5.6 Numerical Aperture and Lagrange Invariant 373
  - 10.5.7 Etendué Ξ 374
  - 10.5.8 Polarized Light 375
- 10.6 Ray Tracing 376
  - 10.6.1 Ray Intercept 377
  - 10.6.2 Multiplicity of Ray Intercepts with a Surface 378
  - 10.6.3 Optical Path Length 378
  - 10.6.4 Reflection and Refraction 380
  - 10.6.5 Polarization Ray Tracing 381
  - 10.6.6 s- and p-Components 382
  - 10.6.7 Amplitude Coefficients and Interface Jones Matrix 383
  - 10.6.8 Polarization Ray Tracing Matrix 385
- 10.7 Wavefront Analysis 393
  - 10.7.1 Normalized Coordinates 393
  - 10.7.2 Wavefront Aberration Function 393
  - 10.7.3 Polarization Aberration Function 394
  - 10.7.4 Evaluation of the Aberration Function 395
  - 10.7.5 Seidel Wavefront Aberration Expansion 398
  - 10.7.6 Zernike Polynomials 401
  - 10.7.7 Wavefront Quality 405
  - 10.7.8 Polarization Quality 406
- 10.8 Non-Sequential Ray Trace 407
- 10.9 Coherent and Incoherent Ray Tracing 407
  - 10.9.1 Polarization Ray Tracing with Mueller Matrices 409
- 10.10 The Use of Polarization Ray Tracing 410
- 10.11 Brief History of Polarization Ray Tracing 411
- 10.12 Summary and Conclusion 412
- 10.13 Problem Sets 413
- 10.14 Appendix: Cell Phone Lens Prescription 416

References 420

CHAPTER 11

The Jones Pupil and Local Coordinate Systems 423

- 11.1 Introduction: Local Coordinates for Entrance and Exit Pupils 423
- 11.2 Local Coordinates 424
- 11.3 Dipole Coordinates 426
- 11.4 Double Pole Coordinates 430
- 11.5 High Numerical Aperture Wavefronts 436
- 11.6 Converting **P** Pupils to Jones Pupils 437
- 11.7 Example: Cell Phone Lens Aberrations 439
- 11.8 Wavefront Aberration Function Difference between Dipole and Double Pole Coordinates 440
- 11.9 Conclusion 441
- 11.10 Problem Sets 442
- References 444

CHAPTER 12

Fresnel Aberrations 447

- 12.1 Introduction 447
- 12.2 Uncoated Single-Element Lens 448
- 12.3 Fold Mirror 455
- 12.4 Combination of Fold Mirror Systems 461
- 12.5 Cassegrain Telescope 469
- 12.6 Fresnel Rhomb 474
- 12.7 Conclusion 475
- 12.8 Problem Sets 475
- References 477

CHAPTER 13

Thin Films 479

- 13.1 Introduction 479
- 13.2 Single-Layer Thin Films 480
  - 13.2.1 Antireflection Coatings 482
  - 13.2.2 Ideal Single-Layer Antireflection Coating 485
  - 13.2.3 Metal Beam Splitters 485
- 13.3 Multilayer Thin Films 486
  - 13.3.1 Algorithms 487
  - 13.3.2 Quarter and Half Wave Films 488
  - 13.3.3 Reflection-Enhancing Coatings 489
  - 13.3.4 Polarizing Beam Splitters 492
- 13.4 Contributions to Wavefront Aberrations 497
- 13.5 Phase Discontinuities 499
- 13.6 Conclusion 501
- 13.7 Appendix: Derivation of Single-Layer Equations 502
- 13.8 Problem Sets 504
- References 505

CHAPTER 14

Jones Matrix Data Reduction with Pauli Matrices ... 507

- 14.1 Introduction ... 507
- 14.2 Pauli Matrices and Jones Matrices ... 509
  - 14.2.1 Pauli Matrix Identities ... 509
  - 14.2.2 Expansion in a Sum of Pauli Matrices ... 510
  - 14.2.3 Pauli Sign Convention ... 511
  - 14.2.4 Pauli Coefficients of a Polarization Element Rotated about the Optical Axis ... 511
  - 14.2.5 Eigenvalues and Eigenvectors and Matrix Functions for the Pauli Sum Form ... 513
  - 14.2.6 Canonical Summation Form ... 514
- 14.3 Sequences of Polarization Elements ... 515
- 14.4 Exponentiation and Logarithms of Matrices ... 518
  - 14.4.1 Exponentiation of Matrices ... 518
  - 14.4.2 Logarithms of Matrices ... 519
  - 14.4.3 Retarder Matrices ... 520
  - 14.4.4 Diattenuator Matrices ... 521
  - 14.4.5 Polarization Properties of Homogeneous Jones Matrices ... 523
- 14.5 Elliptical Retarders and the Retarder Space ... 529
- 14.6 Polarization Properties of Inhomogeneous Jones Matrices ... 531
- 14.7 Diattenuation Space and Inhomogeneous Polarization Elements ... 533
  - 14.7.1 Superposing the Diattenuation and Retardance Spaces ... 534
- 14.8 Weak Polarization Elements ... 535
- 14.9 Summary and Conclusion ... 536
- 14.10 Problem Sets ... 537
- References ... 540

CHAPTER 15

Paraxial Polarization Aberrations ... 543

- 15.1 Introduction ... 543
- 15.2 Polarization Aberrations ... 545
  - 15.2.1 Interaction of Weakly Polarizing Jones Matrices ... 546
  - 15.2.2 Polarization of a Sequence of Weakly Polarizing Ray Intercepts ... 548
- 15.3 Paraxial Polarization Aberrations ... 550
  - 15.3.1 Paraxial Angle and Plane of Incidence ... 550
  - 15.3.2 Paraxial Diattenuation and Retardance ... 553
  - 15.3.3 Diattenuation Defocus ... 553
  - 15.3.4 Diattenuation Defocus and Retardance Defocus ... 555
  - 15.3.5 Diattenuation and Retardance across the Field of View ... 556
  - 15.3.6 Polarization Tilt and Piston ... 557
  - 15.3.7 Binodal Polarization ... 558
  - 15.3.8 Summation of Paraxial Polarization Aberrations over Surfaces ... 558
- 15.4 Paraxial Polarization Analysis of a Seven-Element Lens System ... 560
- 15.5 Higher-Order Polarization Aberrations ... 567
  - 15.5.1 Electric Field Aberrations ... 568
  - 15.5.2 Orientors ... 572

	15.5.3	Diattenuation and Retardance	578
15.6	Polarization Aberration Measurements		580
15.7	Summary and Conclusion		583
15.8	Appendix		583
	15.8.1	Paraxial Optics	583
	15.8.2	Setting Up the Optical System	585
	15.8.3	The Paraxial Ray Trace	586
	15.8.4	Reduced Thicknesses and Angles	587
	15.8.5	Paraxial Skew Rays	588
15.7	Problem Sets		589
References			592

CHAPTER 16

Image Formation with Polarization Aberration 593

16.1	Introduction	593
16.2	Discrete Fourier Transformation	594
16.3	Jones Exit Pupil and Jones Pupil Function	598
16.4	Amplitude Response Matrix (**ARM**)	601
16.5	Mueller Point Spread Matrix (**MPSM**)	603
16.6	The Scale of the ARM and **MPSM**	605
16.7	Polarization Structure of Images	607
16.8	Optical Transfer Matrix (**OTM**)	608
16.9	Example—Polarized Pupil with Unpolarized Object	610
16.10	Example—Solid Corner Cube Retroreflector	614
16.11	Example—Critical Angle Corner Cube Retroreflector	618
16.12	Discussion and Conclusion	622
16.13	Problem Sets	623
References		627

CHAPTER 17

Parallel Transport and the Calculation of Retardance 629

17.1	Introduction		629
	17.1.1	Purpose of the Proper Retardance Calculation	631
17.2	Geometrical Transformations		631
	17.2.1	Rotation of Local Coordinates: Polarimeter Viewpoint	631
	17.2.2	Non-Polarizing Optical Systems	633
	17.2.3	Parallel Transport of Vectors	634
	17.2.4	Parallel Transport of Vectors with Reflection	636
	17.2.5	Parallel Transport Matrix, **Q**	636
17.3	Canonical Local Coordinates		640
17.4	Proper Retardance Calculations		642
	17.4.1	Definition of the Proper Retardance	642
17.5	Separating Geometric Transformations from **P**		642
	17.5.1	The Proper Retardance Algorithm for **P**, Method 1	643
	17.5.2	The Proper Retardance Algorithm for **P**, Method 2	644
	17.5.3	Retardance Range	645

Contents

17.6	Examples	645
	17.6.1 Ideal Reflection at Normal Incidence	646
	17.6.2 An Aluminum-Coated Three-Fold Mirror System Example	647
17.7	Conclusion	649
17.8	Problem Sets	649
	References	651

CHAPTER 18

A Skew Aberration 653

18.1	Introduction	653
18.2	Definition of Skew Aberration	654
18.3	Skew Aberration Algorithm	655
18.4	Lens Example—U.S. Patent 2,896,506	658
18.5	Skew Aberration in Paraxial Ray Trace	660
18.6	Example of Paraxial Skew Aberration	662
18.7	Skew Aberration's Effect on PSF	663
18.8	PSM for U.S. Patent 2,896,506	665
18.9	Statistics—CODE V Patent Library	666
18.10	Conclusion	667
18.11	Problem Sets	667
	References	668

CHAPTER 19

Birefringent Ray Trace 669

19.1	Ray Tracing in Birefringent Materials	669
19.2	Description of Electromagnetic Waves in Anisotropic Media	672
19.3	Defining Birefringent Materials	673
19.4	Eigenmodes of Birefringent Materials	679
19.5	Reflections and Refractions at Birefringent Interface	681
19.6	Data Structure for Ray Doubling	694
19.7	Polarization Ray Tracing Matrices for Birefringent Interfaces	695
	19.7.1 Case I: Isotropic-to-Isotropic Intercept	698
	19.7.2 Case II: Isotropic-to-Birefringent Interface	700
	19.7.3 Case III: Birefringent-to-Isotropic Interface	701
	19.7.4 Case IV: Birefringent-to-Birefringent Interface	703
19.8	Example: Ray Splitting through Three Biaxial Crystal Blocks	706
19.9	Example: Reflections Inside a Biaxial Cube	707
19.10	Conclusion	710
19.11	Problem Sets	711
	References	713

CHAPTER 20

Beam Combination with Polarization Ray Tracing Matrices 715

20.1	Introduction	715
20.2	Wavefronts and Ray Grids	716

	20.3	Co-Propagating Wavefront Combination	718
	20.4	Non-Co-Propagating Wavefront Combination................................	728
	20.5	Combining Irregular Ray Grids	730
		20.5.1 General Steps to Combine Misaligned Ray Data	730
		20.5.2 Inverse-Distance Weighted Interpolation	732
	20.6	Conclusion...	736
	20.7	Problem Sets...	737
	References		739

CHAPTER 21

Uniaxial Materials and Components 741

	21.1	Optical Design Issues in Uniaxial Materials	741
	21.2	Descriptions of Uniaxial Materials...	743
	21.3	Eigenmodes of Uniaxial Materials	745
	21.4	Reflections and Refractions at a Uniaxial Interface....................	746
	21.5	Index Ellipsoid, Optical Indicatrix, and K- and S-Surfaces	749
	21.6	Aberrations of Crystal Waveplates...	765
		21.6.1 A-Plate Aberrations	767
		21.6.2 C-Plate Aberrations	769
	21.7	Image Formation through an A-Plate	771
	21.8	Walk-Off Plate	777
	21.9	Crystal Prisms	778
	21.10	Problem Sets...	779
	References		784

CHAPTER 22

Crystal Polarizers 785

	22.1	Introduction to Crystal Polarizers...	785
	22.2	Materials for Crystal Polarizers...	786
	22.3	Glan–Taylor Polarizer	787
		22.3.1 Limited FOV	787
		22.3.2 Multiple Potential Ray Paths..	789
		22.3.3 Multiple Polarized Wavefronts	793
		22.3.4 Polarized Wavefronts Exiting from the Polarizer.........	796
	22.4	Aberrations of the Glan–Taylor Polarizer	797
	22.5	Pairs of Glan–Taylor Polarizers	799
	22.6	Conclusion...	804
	22.7	Problem Sets...	805
	References		808

CHAPTER 23

Diffractive Optical Elements 811

	23.1	Introduction..	811
	23.2	The Grating Equation...	814

23.3 Ray Tracing DOEs 818
   23.3.1 Reflection Diffractive Gratings 818
   23.3.2 Wire Grid Polarizers 822
   23.3.3 Diffractive Retarders 825
   23.3.4 Diffractive Subwavelength Antireflection Coatings 826
23.4 Summary of the RCWA Algorithm 829
23.5 Problem Sets 832
Acknowledgments 834
References 834

CHAPTER 24

Liquid Crystal Cells 837

24.1 Introduction 837
24.2 Liquid Crystals 838
   24.2.1 Dielectric Anisotropy 840
24.3 Liquid Crystal Cells 841
   24.3.1 Construction of Liquid Crystal Cells 842
   24.3.2 Restoring Forces 843
   24.3.3 Liquid Crystal Display: High Contrast Ratio Intensity Modulation 845
24.4 Configurations of Liquid Crystal Cells 846
   24.4.1 The Fréedericksz Cell 846
   24.4.2 90° Twisted Nematic Cell 847
   24.4.3 Super Twisted Nematic Cell 849
   24.4.4 Vertically Aligned Nematic Cell 850
   24.4.5 In-Plane Switching Cell 852
   24.4.6 Liquid Crystal on Silicon Cells 854
   24.4.7 Blue Phase LC Cells 855
24.5 Polarization Models 856
   24.5.1 Extended Jones Matrix Model 856
   24.5.2 Single Pass with Polarization Ray Tracing Matrices 857
   24.5.3 Multilayer Interference Models 859
   24.5.4 Calculation for Liquid Crystal Cell ZLI-1646 859
24.6 Issues in the Construction of LC Cells 861
   24.6.1 Spacers 861
   24.6.2 Disclinations 861
   24.6.3 Pretilt 862
   24.6.4 Oscillating Square Wave Voltage 863
24.7 Limitations on LC Cell Performance 864
   24.7.1 LC Cell Speed 865
   24.7.2 Spectral Variation of Exiting Polarization State 867
   24.7.3 Variation of Retardance with Angle of Incidence 867
   24.7.4 Compensating LC Cells' Polarization Aberrations with Biaxial Films 868
   24.7.5 Polarizer Leakage 870
   24.7.6 Depolarization 871
24.8 Testing Liquid Crystal Cells 872
   24.8.1 Twisted Nematic Cell Example 873
   24.8.2 IPS Tests 874

	24.8.3	VAN Cell	875
	24.8.4	MVA Cell Test	875
	24.8.5	Sheet Retarder Defect	876
	24.8.6	Misalignment between Analyzer and Exiting Polarization State	877
24.9	Problem Sets		877
Acknowledgment			878
References			878

CHAPTER 25

Stress-Induced Birefringence 879

25.1	Introduction to Stress Birefringence	879
25.2	Stress Birefringence in Optical Systems	881
25.3	Theory of Stress-Induced Birefringence	881
25.4	Ray Tracing in Stress Birefringent Components	883
25.5	Ray Tracing through Stress Birefringence Components with Spatially Varying Stress	889
	25.5.1 Storage of System Shape	890
	25.5.2 Refraction and Reflections	891
	25.5.3 Stress Data Format	891
	25.5.4 Polarization Ray Tracing Matrix for Spatially Varying Biaxial Stress	892
	25.5.5 Examples of Spatially Varying Stress Function	895
25.6	Effects of Stress Birefringence on Optical System Performance	898
	25.6.1 Observing Stress Birefringence Using Polariscope	898
	25.6.2 Simulations of Injection-Molded Lens	901
	25.6.3 Simulation of a Plastic DVD Lens	903
25.7	Conclusion	905
25.8	Problem Sets	906
Acknowledgments		907
References		907

CHAPTER 26

Multi-Order Retarders and the Mystery of Discontinuities 909

26.1	Introduction	909
26.2	Mystery of Retardance Discontinuity	910
26.3	Retardance Unwrapping for Homogeneous Retarder Systems Using a Simple Dispersion Model	912
	26.3.1 Dispersion Model	912
	26.3.2 Retardance of the Homogeneous Retarder System	912
	26.3.3 Homogeneous Retarder's Trajectory and Retardance Unwrapping in Retarder Space	914
26.4	Discontinuities in Unwrapped Retardance Values for Compound Retarder Systems with Arbitrary Alignment	916
	26.4.1 Compound Retarder Jones Matrix Decomposition	917
	26.4.2 Compound Retarder's Trajectory in Retarder Space	919
	26.4.3 Multiple Modes Exit the Compound Retarder System	920
	26.4.4 Compound Retarder Example at 45°	922
26.5	Conclusion	924

26.6 Appendix 925
26.7 Problem Sets 925
References 927

CHAPTER 27

Summary and Conclusions 929

27.1 Difficult Issues 929
27.2 Polarization Ray Tracing Complications 930
    27.2.1 Optical System Description Complications 930
    27.2.2 Elliptical Polarization Properties of Ray Paths 931
    27.2.3 Optical Path Length and Phase 931
    27.2.4 Definition of Retardance 932
    27.2.5 Retardance and Skew Aberration 932
    27.2.6 Multi-Order Retardance 933
    27.2.7 Birefringent Ray Tracing Complications 934
    27.2.8 Coherence Simulation 935
    27.2.9 Scattering 935
    27.2.10 Depolarization 936
27.3 Polarization Ray Tracing Concepts and Methods 937
    27.3.1 Jones Matrices and Jones Pupil 937
    27.3.2 **P** Matrix and Local Coordinates 937
    27.3.3 Generalization of PSF and OTF 937
    27.3.4 Ray Doubling, Ray Trees, and Data Structures 938
    27.3.5 Mode Combination 939
    27.3.6 Alternative Simulation Methods 940
27.4 Polarization Aberration Mitigation 940
    27.4.1 Analyzing Polarization Ray Tracing Output 941
27.5 Comparison of Polarization Ray Tracing and Polarization Aberrations 942
    27.5.1 Aluminum Coating and Polarization Aberration Expression 943
    27.5.2 Polarization Ray Trace and the Jones Pupil 945
    27.5.3 Aberration Expression for the Jones Pupil 946
    27.5.4 Diattenuation and Retardance Contributions 949
    27.5.5 Design Rules Based on Polarization Aberrations 950
        27.5.5.1 Diattenuation at the Center of the Pupil 951
        27.5.5.2 Retardance at the Center of the Pupil 951
        27.5.5.3 Linear Variation of Diattenuation 952
        27.5.5.4 Linear Variation of Retardance, the PSF Shear between the XX- and YY-Components 952
        27.5.5.5 The Polarization-Dependent Astigmatism 952
        27.5.5.6 The Fraction of Light in the Ghost PSF in XY- and YX-Components 953
    27.5.6 Amplitude Response Matrix 954
    27.5.7 Mueller Matrix Point Spread Matrices 955
    27.5.8 Location of the PSF Image Components 958
References 959

Index 961

Authors

Russell A. Chipman, PhD, is professor of optical sciences at the University of Arizona and a visiting professor at the Center for Optics Research and Education (CORE), Utsunomiya University, Japan. He teaches courses in polarized light, polarimetry, and polarization optical design at both universities. Prof. Chipman received his BS in physics from Massachusetts Institute of Technology (MIT) and his MS and PhD in optical sciences from the University of Arizona. He is a fellow of The Optical Society (OSA) and The International Society for Optics and Photonics (SPIE). He received SPIE's 2007 G.G. Stokes Award for research in Polarimetry and OSA's Joseph Fraunhofer Award/Robert Burley Award for Optical Engineering in 2015. He is a co-investigator on NASA/JPL's Multi-Angle Imager for Aerosols, a polarimeter scheduled for launch into Earth's orbit around 2021 for monitoring aerosols and pollution in metropolitan areas. He is also developing UV and IR polarimeter breadboards and analysis methods for other NASA exoplanet and remote sensing missions. He has recently focused on developing the Polaris-M polarization ray tracing code, which analyzes optical systems with anisotropic materials, electro-optic modulators, diffractive optical elements, polarized scattered light, and many other effects. His hobbies include hiking, Japanese language, rabbits, and music.

Wai-Sze Tiffany Lam, PhD, was born and raised in Hong Kong. She is currently an optical scientist in Facebook's Oculus Research. She received her BS in optical engineering and her MS and PhD in optical sciences from the University of Arizona. In her research, she developed robust optical modeling and polarization simulation for birefringent and optically active optical components, components with stress birefringence, the aberrations in crystal retarders and polarizers, and the modeling of liquid crystal cells. Many of these algorithms form the basis of the commercial ray tracing code, Polaris-M, marketed by Airy Optics.

Garam Young, PhD, graduated with a BS in physics from Seoul National University in Korea and received her doctorate from the University of Arizona's College of Optical Sciences, also earning Valedictorian and Outstanding Graduate Student honors. She then developed polarization features and optimization features for CODE V and LightTools with Synopsys in Pasadena, and she currently works as an optical and illumination engineer in the San Francisco Bay area. Her husband and daughter keep her busy at home.

Acknowledgments

To our families who have provided so much support during the writing of this book: Laure, Peter, Kin Lung, Tek Yin, Wai Kwan, Stefano, and Sofia.

With special appreciation to the College of Optical Sciences of the University of Arizona, the Jet Propulsion Laboratory, the Center for Optics Research and Education at Utsunomiya University, Oculus Research, and Nalux Co., Limited.

This book would not be possible without the assistance of so many colleagues including the following: Lloyd Hillman, Steve McClain, Jim McGuire, James B. Breckinridge, Stacey Sueoka, Christine Bradley, Brian Daugherty, Scott Tyo, Anna-Britt Mahler, Scott McEldowney, Michihisa Onishi, Hannah Nobel, Paula Smith, Matthew Smith, Kyle Hawkins, Meredith Kupinski, Lisa Li, Dennis Goldstein, Shih-Yau Lu, Karen Twietmeyer, David Chenault, John Gonglewski, Yukitoshi Otani, Ashley Gasque, Larry Pezzaniti, Robert Galter, Nasrat Raouf, Glenn Boreman, David Diner, Greg Smith, Rolland Shack, Dan Reiley, Angus Macleod, Cindy Gardner, Stanley Pau, David Voeltz, Ab Davis, Joseph Shaw, Kira Hart, James C. Wyant, Amy Phillips, Tom Brown, John Stacey, Suchandra Banerjee, Brian DeBoo, David Salyer, Toyohiko Yatagai, Julie Gillis, Alba Peinado, Jeff Davis, Juliana Richter, Jack Jewell, Alex Erstad, Chanda Bartlett-Walker, Jaden Bankhead, Kazuhiko Oka, Wei-Liang Hsu, Adriana Stohn, Eustace Dereniak, Chikako Sugaya, Justin Wolfe, John Greivenkamp, Momoka Sugimura, Erica Mohr, Alex Schluntz, Charles LaCasse, Jason Auxier, Karlton Crabtree, Israel Vaughn, Pierre Gerligand, Jose Sasian, Ami Gupta, Ann Elsner, Juan Manuel Lopez, Kurt Denninghoff, Toru Yoshizawa, Kyle Ferrio, Tom Bruegge, Bryan Stone, James Harvey, Brian Cairns, Charles Davis, Adel Joobeur, Robert Shannon, Robert Dezmelyk, Matt Dubin, Quinn Jarecki, Masafumi Seigo, Tom Milster, James Hadaway, Dejian Fu, Steven Burns, James Trollinger Jr., Beth Sorinson, Mike Hayford, Jennifer Parsons, Johnathan Drewes, Bob Breault, Rodney Fuller, Peter Maymon, Alan Huang, Jacob Krause, Kasia Sieluzycka, Jurgen Jahns, Phillip Anthony, Aristide Dogariu, Michaela May, Jon Herlocker, Robert Pricone, Charlie Hornback, Krista Drummond, Barry Cense, Lena Wolfe, Neil Beaudry, Virginia Land, Noah Gilbert, Helen Fan, Eugene Waluschka, Phil McCulloch, Thomas Germer, Thiago Jota, Morgan Harlan, Tracy Gin, Cedar Andre, Dan Smith, Victoria Chan, Lirong Wang, Christian Brosseau,

Andre Alanin, Graham Myhre, Mona Haggard, Eugene W. Cross, Ed West, Shinya Okubo, Matt Novak, Andrew Stauer, Conrad Wells, Michael Prise, Caterina Ubacch, David Elmore, Oersted Stavroudis, Weilin Liu, Tyson Ririe, Tom Burleson, Long Yang, Sukumar Murali, Julia Craven, Goldie Goldstein, Adoum Mahamat, Ravi Kinnera, Livia Zarnescu, Robert Rodgers, Randy Gove, Gordon Knight, Randall Hodgeson, Dan Brown, Nick Craft, Stephen Kupiac, Graeme Duthie, John Caulfield Jr., Joseph Shamir, Ken Cardell, and Bill Galloway.

Preface

Polarized Light and Optical Systems addresses the need for a polarized light class for undergraduate and graduate students using a curriculum that has been developed and refined for a decade at the University of Arizona's College of Optical Sciences. This book is also intended as a reference for the optical engineer and optical designer in building polarimeters, designing polarization critical optical systems, and manipulating polarized light for a myriad of purposes.

Polarization is central to the operation of liquid crystal displays, 3D movies, advanced remote sensing satellites, microlithography systems, and numerous other products. The sophistication of optical systems that utilize polarized light has rapidly advanced as have the tools for simulation and design. The development of more accurate and complex polarizers, waveplates, polarizing beam splitters, and thin films provides designers and scientists with new choices, as well as many simulation challenges.

Unlike bees and ants, humans are essentially polarization blind. Humans miss the rich and subtle polarization information in the sky, water, and the rest of nature. Similarly, we can't see how polarized light is changing and evolving as it propagates through windshields, eyeglasses, and all nature of optical systems. Therefore, students often have difficulties understanding the importance of polarization to optical system design, metrology, image formation, atmospheric optics, and the propagation of light in tissue. *Polarized Light and Optical Systems* provides a guide to the optics behind these technologies by tying together the fundamentals of polarized light with the practice of the optical engineer and designer.

Polarization involves as many as 16 degrees of freedom: linear, circular and elliptical polarization, diattenuation, retardance, and depolarization. These degrees of freedom are not visible to us and thus may seem abstract. In discussions with students, we occasionally hear that polarization is regarded as *complicated*, *difficult*, and *mostly misunderstood*. Unfortunately, this often occurs because many polarization concepts are taught in a rushed and overly simplified treatment.

Polarized Light and Optical Systems contains detailed discussions to clarify several topics the authors have found confusing; topics often avoided the following:

- *Sign conventions* for the electromagnetic field and Fresnel equations
- The treatment of polarized light in local coordinates in the transverse plane with the *Jones calculus*
- Many subtle issues of *phase*

- The three degrees of freedom of *retardance* and the three degrees of freedom of *diattenuation*, which are defined in a new simpler way using matrix exponentials
- The non-orthogonal coordinate systems of the *Stokes parameters* and *Mueller matrix*
- Applying the *Jones or Mueller calculus* to ray tracing optical systems in three dimensions, in particular for stray light or tissue optics

To address these issues, a new pedagogical approach was developed and tested in the classroom. This approach starts with a three-dimensional approach to light propagation. Light propagates in arbitrary directions in optical systems and the mathematics needs to reflect this. But Jones matrices and Mueller matrices only describe propagation along a z-axis. Reorienting this z-axis, making it a local coordinate, introduces some significant problems, particularly for the description of reflection. Here, a three-dimensional polarization ray tracing matrix is taught, which simplifies the analysis of optical systems and polarization components. Then, Jones vectors and Jones matrices are treated as a useful special case. This three-dimensional approach may sound more complex, but it actually makes polarization calculations more straightforward. We have come to love this 3D method because *it solves a long-standing coordinate-related paradox* regarding Jones matrices and normal incidence reflection!

In many introductory optics classes, polarization is introduced early but then rarely discussed again. The Fresnel equations are often introduced, but their consequences are neglected. Thus, a detailed treatment is provided describing how the Fresnel equations affect light passage through lenses and mirrors, and image formation. By learning diffraction and image formation with polarized light, students come to understand how the Fresnel equations can change the structure of the point spread function and how the polarization state can vary within the image of a point object.

In optical engineering, the polarization elements have usually been treated as a separate subsystem. The mathematics of polarization, the Jones calculus and Mueller calculus, has been kept separate from the mathematics for first-order optics, aberration theory, lens design, and mostly from interference and diffraction. Earlier treatments of polarization mostly focused on the Jones calculus, Mueller calculus, and polarization elements. *Polarized Light and Optical Systems* treats polarization elements as optical elements and optical elements as polarization elements. The properties of lens and mirror systems vary with wavelength, angle, and position—these are the aberrations. Similarly, the polarization properties of polarization elements vary with wavelength, angle, and position—these are the *polarization aberrations*. Just as the optical designer needs to perform detailed accounting of optical path lengths in traditional optical design, a similar detailed accounting of polarization properties is performed by *polarization ray tracing*.

Most optical design programs now provide polarization ray tracing calculations, so now many users have a greater need to understand the subtlety of the propagation of polarized light in optical systems, to succeed with polarization ray tracing software, and to communicate the results clearly. Until now, polarization ray tracing has never been part of the optics curriculum. To address that, we provide the materials for the instructor to base a class on polarization in optical systems, not polarization as a subsystem. This book addresses these needs by teaching polarization elements, sequences of polarization elements, polarimetry, Fresnel equations, and anisotropic materials, the basics.

Polarization elements are never ideal. The theories and analysis methods presented provide insights on the polarization effects of common optical elements such as lenses, fold mirrors, and prisms. Hence, *Polarized Light and Optical Systems* devotes considerable space to their polarization aberrations: the variation of retardance of waveplates with angle of incidence and wavelength, and the angular dependence of wire grid, sheet, and Glan–Taylor polarizers.

To complement the book's treatment of optical systems, an overview of ray tracing algorithms and paraxial optics is provided, with enough materials to familiarize the scientists and engineers from outside optics with the basic concepts of ray tracing algorithms. Many of the example systems in this book are calculated by our in-house research polarization ray tracing software *Polaris-M*, which is based on the *3D polarization ray tracing matrix*.

We find these concepts fun and hope that our sense of enchantment is communicated to our readers. Geometry is one of the most pleasurable fields of mathematics, and the combination of polarization with optical systems provides a wealth of great geometrical problems and insights. It is a big step for the optical designer to move from the surfaces of the scalar wavefront aberration function (one degree of freedom, optical path length) to the eight dimensions of the Jones pupil and its high dimensional shapes. But once one becomes accustomed to diattenuation aberrations and retardance aberrations, there is great beauty and symmetry in generalizing the Seidel and Zernike aberrations into eight-dimensional space. A step-by-step guide to this eight-dimensional Jones matrix space is provided in our treatments of polarization aberrations in the structure of the Jones calculus.

Russell A. Chipman
Wai-Sze Tiffany Lam
Garam Young

MATLAB® is a registered trademark of The MathWorks, Inc. For product information, please contact:

The MathWorks, Inc.
3 Apple Hill Drive
Natick, MA 01760-2098 USA
Tel: 508 647 7000
Fax: 508-647-7001
E-mail: info@mathworks.com
Web: www.mathworks.com

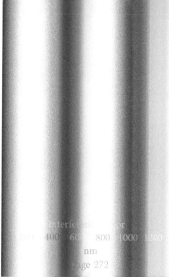

How This Book Came to Be

The first author, Russell Chipman, has been teaching polarization for over 30 years while simultaneously conducting a wide-ranging research program into polarization in optical design, polarimetry, and polarizing devices. Over the years, research took priority over book writing, but course materials were steadily developed and taught.

Starting in 2006, Garam Young wrote a dissertation on polarization ray tracing, during the course of which many underlying issues in polarization ray tracing surrounding phase, retardance, and skew aberration were uncovered and deep issues were clarified. These advances led to support from the Science Foundation Arizona to write a research polarization ray tracing program, Polaris-M, to demonstrate these new polarization ray tracing methods. Tiffany Lam, together with Steve McClain, took the responsibility for the anisotropic materials ray tracing algorithm development and testing, generating many highly instructive polarization ray tracing examples and developing special treatment for birefringent ray trace.

Polarized Light and Optical Systems is the culmination of the first author's research into polarization aberrations that began in 1982 in graduate school in optics at the University of Arizona under the direction of Jim Wyant and Jim Breckinridge of the Jet Propulsion Laboratory. Between Chipman's teaching materials and research experience, Young's polarization ray tracing dissertation, and Lam's anisotropic ray trace dissertation, the pieces fell into place for this ambitious project. This is not just a book about polarized light and the polarization calculus, but one that will lead you toward a modern view of optical systems, where everything is a polarization element.

Suggested Curricula

Through experimentation over years of teaching this subject, a logical order for the chapters emerged. The sequence is organized for a natural flow in the classroom. The chapters are organized from simpler to more complex, working from more fundamental to more applied concepts. Different sequences of chapters can be selected depending on the class goals. Below are three suggested curricula, highlighted in pink for (1) an undergraduate polarized light course, in blue for (2) a graduate polarization optics course, and in purple for (3) an advanced course in polarization optical design.

Undergraduate Course: Polarized Light

1. Introduction and Overview
2. Polarized Light
3. Stokes Parameters and the Poincaré Sphere
4. Interference of Polarized Light
5. Jones Matrices and Polarization Properties
6. Mueller Matrices
7. Polarimetry
8. Fresnel Equations
9. Polarization Ray Tracing Calculus
10. Optical Ray Tracing
11. The Jones Pupil and Local Coordinate Systems
12. Fresnel Aberrations
13. Thin Films
14. Jones Matrix Data Reduction with Pauli Matrices
15. Paraxial Polarization Aberrations
16. Image Formation with Polarization Aberration
17. Parallel Transport and the Calculation of Retardance
18. A Skew Aberration
19. Birefringent Ray Trace
20. Beam Combination with Polarization Ray Tracing Matrices
21. Uniaxial Materials and Components
22. Crystal Polarizers
23. Diffractive Optical Elements
24. Liquid Crystal Cells
25. Stress-Induced Birefringence
26. Multi-Order Retarders and the Mystery of Discontinuities
27. Summary and Conclusions

Graduate Course: Polarization Optics

1. Introduction and Overview
2. Polarized Light
3. Stokes parameters and the Poincaré Sphere
4. Interference of Polarized Light
5. Jones Matrices and Polarization Properties
6. Mueller Matrices
7. Polarimetry
8. Fresnel Equations
9. Polarization Ray Tracing Calculus
10. Optical Ray Tracing
11. The Jones Pupil and Local Coordinate Systems
12. Fresnel Aberrations
13. Thin Films
14. Jones Matrix Data Reduction with Pauli Matrices
15. Paraxial Polarization Aberrations
16. Image Formation with Polarization Aberration
17. Parallel Transport and the Calculation of Retardance
18. A Skew Aberration
19. Birefringent Ray Trace
20. Beam Combination with Polarization Ray Tracing Matrices
21. Uniaxial Materials and Components
22. Crystal Polarizers
23. Diffractive Optical Elements
24. Liquid Crystal Cells
25. Stress Induced Birefringence
26. Multi-Order Retarders and the Mystery of Discontinuities
27. Summary and Conclusions

Advanced Course: Polarization Optical Design

1. Introduction and Overview
2. Polarized Light
3. Stokes Parameters and the Poincaré Sphere
4. Interference of Polarized Light
5. Jones Matrices and Polarization Properties
6. Mueller Matrices
7. Polarimetry
8. Fresnel Equations
9. Polarization Ray Tracing Calculus
10. Optical Ray Tracing
11. The Jones Pupil and Local Coordinate Systems
12. Fresnel Aberrations
13. Thin Films
14. Jones Matrix Data Reduction with Pauli Matrices
15. Paraxial Polarization Aberrations
16. Image Formation with Polarization Aberration
17. Parallel Transport and the Calculation of Retardance
18. A Skew Aberration
19. Birefringent Ray Trace
20. Beam Combination with Polarization Ray Tracing Matrices
21. Uniaxial Materials and Components
22. Crystal Polarizers
23. Diffractive Optical Elements
24. Liquid Crystal Cells
25. Stress-Induced Birefringence
26. Multi-Order Retarders and the Mystery of Discontinuities
27. Summary and Conclusions

Guided Tour of the Chapters

Chapter 1: Introduction and Overview surveys polarized light and polarization optics and introduces several polarization issues in optical systems. The electromagnetic nature of light is introduced. The polarization elements such as polarizers, retarders, depolarizers, and associated properties such as birefringence are explained. A series of polarization issues are treated graphically: the *Maltese cross* pattern of uncoated lenses, the field of view dependence of polarizers and retarders, polarization aberrations due to thin film optical coatings, stress birefringence, and the angle dependence of liquid crystals.

Linear polarization over a hemispherical wavefront.

Chapter 2: Polarized Light covers the mathematical treatment of monochromatic light and plane waves in two dimensions with *Jones vectors* and in three dimensions with the *polarization vector*. The *polarization ellipse* is explored in detail. Basic vector mathematics is reviewed along the way. The chapter concludes with a discussion of spherical wavefronts and the polarization of light sources.

3D view of the electric and magnetic fields for circularly polarized light.

xxxiii

Chapter 3: Stokes Parameters and the Poincaré Sphere treats polychromatic light, partially polarized light, and incoherent light. The *Stokes parameters* have an unusual non-orthogonal coordinate system that works wonderfully for problems in radiometry and remote sensing. The three-dimensional representation of the Stokes parameters, the Poincaré sphere simplifies the analysis of many problems with polarization elements.

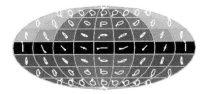

The Poincaré sphere in a Mercator projection.

Chapter 4: Interference of Polarized Light studies *polarization fringes* as well as intensity fringes in interference. Polarization issues can compromise interferometers and prevent good holograms from being recorded. Interference of partially polarized and polychromatic light leads naturally to the Stokes parameters description of light.

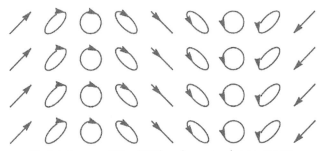

The interference of 0° and 90° laser beams produces a pattern with constant intensity and periodic polarization fringes.

Chapter 5: Jones Matrices and Polarization Properties develops the powerful model of polarization elements as matrices. Each polarization element and polarization property has a corresponding family of Jones matrices. The propagation of light through a series of polarization elements is considered, with valuable examples for the optical engineer. The Jones matrix is considered for the reflection of light at normal incidence from mirrors; this *Jones matrix presents a paradox that will be resolved with the polarization ray tracing matrix*. The concept of Jones matrices where the incident and exiting beams are not parallel, very important in optical design, is developed.

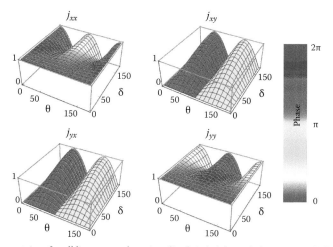

The Jones matrices for all linear retarders. Amplitude is height and phase is encoded as color.

Chapter 6: Mueller Matrices provides a powerful method for performing *incoherent light calculations for polarization elements*: polarizers, diattenuators, retarders, as well as *depolarizers*. Reflection and refraction Mueller matrices are introduced to extend the method to problems containing optical elements. *Depolarization* is a rich phenomenon with nine degrees of freedom.

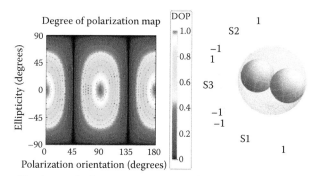

The degree of polarization map in 2D and 3D for an aperture shared between horizontal and vertical polarizers.

Chapter 7: Polarimetry applies the Mueller matrix methods to the measurement of Stokes parameters and Mueller matrices. *Stokes polarimeters* have many applications in remote sensing including characterizing aerosols in the atmosphere and finding man-made objects in clutter. Mueller matrices are used to test polarization elements and photonic devices, and as ellipsometers to measure film thicknesses and refractive indices.

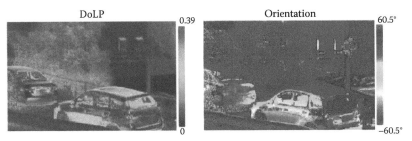

The degree of polarization and its orientation in the image of a car.

Chapter 8: Fresnel Equations describes polarization changes that occur at dielectric interfaces, total internal reflection, and reflection from metal mirrors. Incident light is analyzed into *s*- and *p*-polarized components that are studied separately as eigenpolarizations. Very large polarization effects occur at the critical angle where the derivative of the retardance becomes infinite.

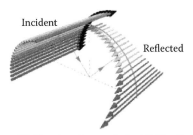

The incident and reflected field vectors for *p*-polarized light reflecting at a glass surface.

Chapter 9: Polarization Ray Tracing Calculus develops the 3 × 3 polarization ray tracing matrix and associated algorithms. This calculus systematizes polarization ray tracing with a three-dimensional polarization ray tracing matrix, the **P** matrix, *a generalization of the Jones matrix into three dimensions*. A major advantage of the **P** matrix is its definition in global coordinates; it solves deep problems with Jones matrices and local coordinates due to singularities and non-uniqueness, a theme developed throughout this book. As a result, anyone who ray traces an optical system with **P** will get the same matrix, unlike a Jones or Mueller matrix calculation where the answer depends on the sequence of local coordinates selected. Algorithms are provided to calculate diattenuation and retardance using **P**. *The Jones matrix for reflection at normal incidence from a mirror is the same as the Jones matrix for a half wave retarder for transmission; this paradox is resolved.* Ray tracing through a polarization interferometer (shown) provides an important example of the calculus.

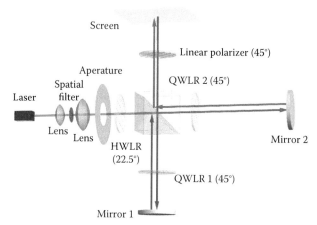

A polarization interferometer to be analyzed with the polarization ray tracing calculus.

Chapter 10: Optical Ray Tracing presents algorithms for ray tracing optical systems and calculating the wavefront and polarization aberration functions. Polarization ray tracing algorithms account for the polarization effects of coated and uncoated interfaces, for example, Fresnel coefficients, during the ray trace. Polarization ray tracing matrices are used to calculate the transmittance (apodization), diattenuation, and retardance properties of rays. A grid of rays across a wavefront form the basis for determination of the polarization aberration function. A cell phone lens with biaspheric surfaces is included as an example for the ray tracing concepts. The chapter concludes with a historical review of polarization ray tracing.

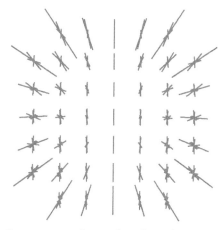

The diattenuation components for an off-axis beam through a cell phone lens, surface by surface, overlapped in the exit pupil.

Chapter 11: The Jones Pupil and Local Coordinate Systems analyzes the difficult issues of converting from ray trace results defined on spherical surfaces in three dimensions into flat surface representations as a Jones pupil. The Jones pupil is commonly used in industry for the representation of polarization aberration. In order to use Jones pupils properly, the subtleties of local coordinate systems are explained and optimal methods are presented. Two principal local coordinate systems are developed: *dipole coordinates* and *double pole coordinates*. For high numerical aperture wavefronts, double pole coordinate systems become more convenient since this coordinate system more closely approximates the natural behavior of lenses. Double pole coordinates also contain a fascinating doubly degenerate singular point. The cell phone lens example is continued to illustrate how **P** arrays are converted into the Jones pupil.

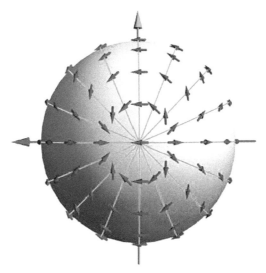

The 720° polarization rotation around the double singularity in the double pole polarization pattern associated with lenses.

Chapter 12: Fresnel Aberrations The Fresnel equations are applied to several example optical systems and the resulting polarization aberrations are surprising. An uncoated lens between crossed polarizers leaks light in the *Maltese cross* pattern. The metal coatings in a Cassegrain telescope introduce a small amount of astigmatism into the on-axis beam! The telescope's point spread function between crossed polarizers is dark in the center with four islands of light in a square. One clever application of the Fresnel equations is the Fresnel rhomb, a total internal reflection-based quarter wave retarder.

The polarization state across an f/1 beam reflecting from an aluminum fold mirror, when the incident polarization is oriented between 45° between s and p.

Guided Tour of the Chapters

Chapter 13: Thin Films covers several of the most important classes of optical thin films and their polarization properties:

- Antireflection coatings
- Enhanced reflection coatings
- Metal beam splitting coatings
- Polarization beam splitting coatings

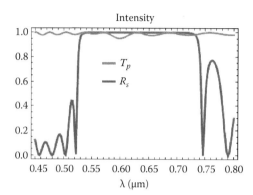

The intensity *p*-transmission and *s*-reflectance for a MacNeille polarizing beam splitting cube coating.

Chapter 14: Jones Matrix Data Reduction with Pauli Matrices deals with the interpretation of Jones pupils as diattenuation and retardance aberration functions. The Jones matrix chapter presents the *forward problem* of calculating Jones matrices from the polarization properties, diattenuation and retardance. Here, *Pauli matrices* and *matrix exponentials* are used to convert Jones matrices into diattenuation and retardance components by finding a canonical form for Jones matrices.

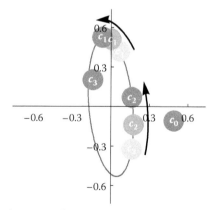

A polarization element can be expressed as a sum of the identity matrix and three Pauli matrices with complex coefficients. When the element is rotated, the two linear coefficients trace ellipses while the identity coefficient and circular coefficient remain fixed.

Polarized Light and Optical Systems

Chapter 15: Paraxial Polarization Aberrations examines the form of polarization aberrations in radially symmetric systems, starting from the angle of incidence function, integrating typical diattenuation and aberration functions, leading to the second-order polarization aberration patterns, which resemble defocus, tilt, and piston. The aberrations associated with the paraxial ray trace are examined as a launching point for a full aberration expansion.

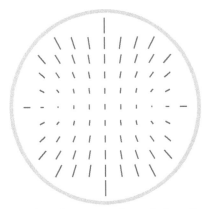

The bimodal distribution of retardance across the pupil of an off-axis field in a telescope has two zeros with 180° of fast axis rotation around each zero.

Chapter 16: Image Formation with Polarization Aberration treats diffraction and the calculation of the point spread function and optical transfer function in the presence of polarization aberrations. Variations in the polarization state exiting an optical system cause interesting variations in the image and its polarization structure. The polarization aberrations of corner cubes are very large due to *skew aberration* and the retardance arising from total internal reflection.

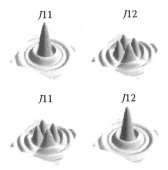

The point spread function of a Cassegrain telescope between x and x, y and x, x and y, and y and y oriented polarizers yields a four-lobed point spread function, dark in the middle for the crossed polarizer combinations.

Chapter 17: Parallel Transport and the Calculation of Retardance describes *retarders* and the calculation of retardance. Retardance is a particularly subtle and sometimes paradoxical concept, typically described as an optical path difference between two orthogonal polarization states. However, when a system has more than two interfering beams, other conceptual issues arise, issues that complicate measuring and interpreting the properties of birefringent films for displays. A paradox occurs in the *calculation of retardance in three dimensions*, when the incident ray, the intermediate ray segments, and the exiting ray are not parallel to each other. It is shown how, with detailed knowledge of the ray path through an optical system, the retardance paradox is resolved simply.

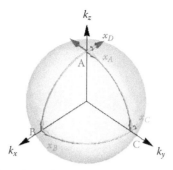

The calculation of retardance depends on the area of spherical polygons formed on a unit propagation sphere formed from all the individual propagation vectors for each ray segment.

Chapter 18: A Skew Aberration applies the resolution of the retardance paradox to rotations of polarization state that *occur in non-polarizing optical systems*. A new type of polarization aberration, skew aberration, is explained using the *Pancharatnam/Berry phase*. A unique characteristic of skew aberration is its presence in ideal non-polarizing optical systems. This effect is significant for high-NA optical systems with large fields of view; microlithography systems are good examples of such optical systems. The effects of skew aberration on polarization point spread functions and optical transfer functions are derived.

The exiting polarization for off-axis fields acquires a polarization rotation across the pupil due to the parallel transport of polarization states.

Chapter 19: Birefringent Ray Trace presents ray tracing methods for anisotropic materials and addresses the handling of *ray doubling*. Polarization ray tracing through anisotropic materials requires tracking a large number of parameters for all split rays:

- Propagation direction **k**
- Poynting vector **S**
- Mode refractive indices n
- Complex Fresnel coefficients a
- Electric field orientations $\hat{\mathbf{E}}$

The anisotropic algorithm handles *double refraction and reflection*, coated anisotropic interfaces, evanescent rays, including total internal reflection and inhibited reflection. The book's website www.polarizedlight.org contains animations of light beams through biaxial materials as well as the evolution of polarization states.

The 16 transmitted ray paths through a sequence of three anisotropic crystals, each with a different sequence of modes and different optical path lengths. The 16 rays exit parallel; each exits at a different location in the shape of a double parallelogram.

Chapter 20: Beam Combination with Polarization Ray Tracing Matrices analyzes interactions of multiple wavefronts with similar and different propagation directions. As part of the ray tracing processes, the interaction of multiple exiting beams must be simulated to analyze *birefringent devices* and interferometers. One of the many issues combining multiple wavefronts is the relative positions of ray grids and the necessity for interpolation.

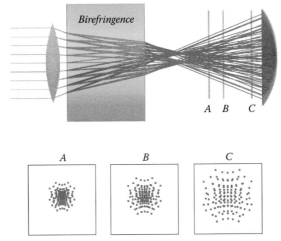

The ray grids for two modes (red and blue) through a thick waveplate focusing in different places have different aberrations. The simulation needs to add the corresponding electric fields to simulate interference, diffraction, and image formation.

Chapter 21: Uniaxial Materials and Components explores light propagation and ray tracing in common uniaxial devices. The *index ellipsoid* helps explain wavefront propagation through birefringent interface. In uniaxial materials, the extraordinary mode's birefringent aberrations are complicated due to an angularly varying refractive index. It is important to understand such aberrations in analyzing waveplates.

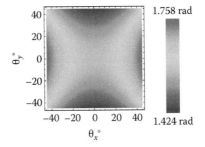

The variation of retardance through a waveplate as a function of direction has a toroidal shape.

Guided Tour of the Chapters

Chapter 22: Crystal Polarizers presents new analyses on common, but misunderstood, optical components—the crystal polarizers, including Glan–Taylor and Glan–Thompson. This treatment results in new insights into the polarizer's field of view, the apodization of its beams, and their aberrations. The numerous minor beams generated from crystal polarizers are described and their paths are explained. The analysis becomes more complicated, but fascinating for pairs of parallel and crossed crystal polarizers with incident spherical wavefronts.

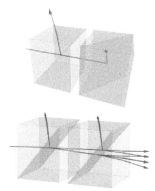

On-axis ray does not make it through a pair of crossed Glan–Taylor polarizers, but for a 4° off-axis ray, five mode pairs transmit. This limits the polarizer's field of view to 3°.

Chapter 23: Diffractive Optical Elements Diffractive optical elements are simulated using *rigorous coupled wave analysis*. The polarization properties of reflection gratings, *wire grid polarizers*, and subwavelength structures for antireflection coatings are explored. Incorporating diffractive optical elements accurately in polarization ray tracing by integration of their amplitude coefficients (Fresnel coefficients) is explained.

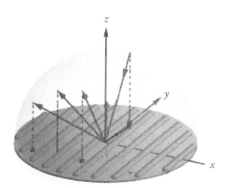

For out-of-plane illumination on a diffraction grating, the propagation vectors for the diffraction orders lie on the surface of a cone, but the projection of the vectors onto the xy plane remains equally spaced.

Chapter 24: Liquid Crystal Cells Liquid crystal cells manipulate the polarization of light by rotating liquid crystal molecules to create *electrically controllable retarders*, polarization controllers, and spatial light modulators. The most common and historically important configurations are analyzed. One of the major issues in liquid crystal cells is the variation of retardance with angle. These angular polarization aberrations are compared between designs and studied as a basis for pairing liquid crystal cells with *field correcting biaxial multilayer films* to make high-performance liquid crystal displays. To attain their position of dominance in the display market, liquid crystal technology overcame many obstacles, including absorption, scattering, low contrast, switching time, uniformity, limited viewing angle, disclinations, and polarization aberration.

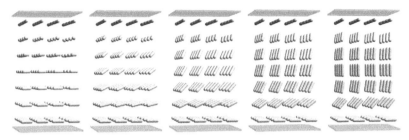

The distribution of directors across a twisted nematic liquid crystal cell from 0 V (left) to 5 V (right).

Chapter 25: Stress-Induced Birefringence Stressed-induced birefringence is a widespread problem in optics that frequently occurs in *injection-molded plastic optics* due to molding processes and in glass lenses as a result of poor opto-mechanical mounting techniques. Stress, internal forces in optical elements, changes the distances between atoms, generating a *spatially varying birefringence*. Stress can arise during glass forming, during the injection molding of plastic lenses, or from the mounting of optical elements. This chapter's algorithms simulate the propagation of polarized light through optical elements with stress birefringence. Common data structures for storing stress information in CAD files are discussed. Methods are included for interpreting colored *polariscope images* of stress birefringence.

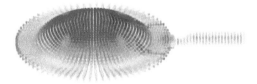

The spatial distribution of stresses (blue) and strain (red) in an injection-molded DVD lens.

Chapter 26: Multi-Order Retarders and the Mystery of Discontinuities Multi-order retarders are retarders with retardance greater than one wave. The retardance of a *compound retarder* (formed from several birefringent plates) can, for example, vary continuously from 1½ waves to 2½ waves of retardance without ever passing through 2 waves of retardance! Retardance can simultaneously assume multiple values when the fast axes of birefringent components are not parallel or perpendicular to each other. Measured data verify this complex but fascinating issue.

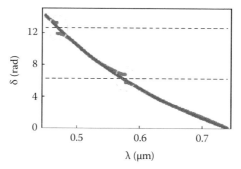

The retardance spectrum of a two-plate retarder with the axes aligned is continuous (red), but when one plate is slightly rotated, discontinuities occur around values of $2n\pi$ (blue), circled in green.

Chapter 27: Summary and Conclusions brings all the book's issues into perspective to get the big picture. Critical issues such as optical path length, phase, retardance, and coordinate systems are reviewed and tolerancing of polarization is discussed. *Polarized Light and Optical Systems* wraps up by discussing the issues in communicating polarization effects and aberrations. How can the optical designers and engineers best communicate this complex information with colleagues, vendors, and production peers?

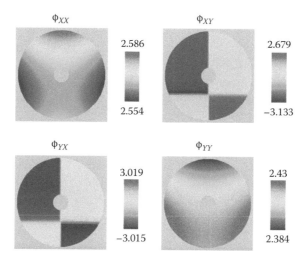

Phase part of a Jones pupil showing the polarization aberration *linear retardance tilt*, which causes the x-and y-polarized parts of a point spread function to slightly separate.

Learning Features

This course has been developed over several years of teaching polarization to undergraduate and graduate students at the University of Arizona by the author, Russell Chipman. In addition, much of the curriculum have been taught in many SPIE and OSA short courses. Watching students learn, observing their difficulties, and obtaining feedback from classes have motivated several features of the book, including the integration of optical design into the treatment.

Color Figures

Polarization and geometrical optics are highly geometrical subjects and many of the concepts can be effectively communicated more effectively through three-dimensional color figures than equations. Color conveys more detail and speeds comprehension. Preparing so many figures required an intense effort, and one of our authors, Tiffany Lam, took the lead in the graphics development.

Poincaré Sphere

The Poincaré sphere is very useful for understanding and explaining polarized light. To really appreciate and apply the sphere, a three-dimensional Poincaré sphere is essential. Therefore, a Poincaré sphere is available from the book's website: www.polarizedlight.org. We suggest printing this Poincaré sphere on heavy stock paper, then cutting and folding it up and taping it into a sphere.

Viewgraphs

Color figures work very well in the classroom to engage students and maintain their interest. A set of teaching viewgraphs containing the majority of figures are available to faculty who adopt the text.

Worked Examples

Integrated throughout the text are numerical examples closely related to the problem sets. These worked examples explain basic concepts such as coordinate rotation and provide useful numerical values for important examples. The student who works through these examples can speed their development of mathematical problem solving skills and gain an expanded understanding of the fundamental physical processes. The worked examples provide step-by-step instructions on linear algebra methods when they are first introduced to aid mastery of matrix manipulations.

Problem Sets and Solution Manual

Problem sets are included with each chapter to help students master the concepts and test their problem solving skills. Generally, the first few problems should only take a few minutes and then the difficulty of the problems increases. A partial solution set is available to faculty from the publisher.

Three-Dimensional Polarization Ray Tracing Calculus

The 3 × 3 matrices that operate on light propagating in any direction find their full explanation here with integration into the algorithms of optical design. Garam Young brought this matrix concept to maturity with her dissertation and presents many examples here that demonstrate its advantages over the Jones calculus. Tiffany Lam further extended the polarization ray tracing calculus to birefringent components in her dissertation, including uniaxial, biaxial, and stress birefringent optical components.

Reference Tables

Organized tables are included throughout for the key properties of polarized light, Jones matrices, Mueller matrices, Stokes parameters, and material properties.

List of Abbreviations

a, a_p, a_s	amplitude coefficients, s- and p-coefficients, such as Fresnel, thin film
A	area
A_E	area of entrance pupil
a_i	aberration coefficient
A_x, A_y, A_z	real amplitudes of the field along the coordinate axes
A	analyzer vector, top row of Mueller matrix
A(ρ)	Jones matrix for a polarization-independent change of amplitude
$a(x,y)$	amplitude function
Abs	absolute value
$AoP(\mathbf{S})$	angle of polarization of Stokes parameters **S**
aperture(x,y)	aperture function
ARM	amplitude response matrix
Arg	argument of a complex number
B and **B**	magnetic induction and magnetic induction vector
BP	blue phase
BRDF	bidirectional reflectance distribution function
c	speed of light
C	contrast ratio
C_0, C_1, C_2	stress optic coefficient
C	stress optic tensor
C_q	curvature of surface q
CA	crystal axis
CR(δ)	circular retarder, retardance δ
d	liquid crystal cell gap
d	film thickness
$D, \mathcal{D}$	diattenuation
d	vector for dipole axis
D and **D**	displacement field and displacement field vector
D	diagonal matrix, matrix of singular values
D	dyad matrix
Det	determinant

xlix

List of Abbreviations

DFT	discrete Fourier transform
D_H, D_{45}, D_R	diattenuation components
$DoCP(\mathbf{S})$	degree of circular polarization of Stokes parameters $\mathbf{S}$
DOE	diffractive optical element
$DoLP(\mathbf{S})$	degree of linear polarization of Stokes parameters $\mathbf{S}$
$DoP(\mathbf{S})$	degree of polarization of Stokes parameters $\mathbf{S}$
e	ellipticity, ratio of the minor to the major axis of ellipse
e	extraordinary mode label
E	extinction ratio of a polarizer or diattenuator
E	extraordinary principal axis label
E	Young's modulus
$\mathbf{E}$	Jones vector
$\mathbf{E}$	polarization vector
$E(\mathbf{r}, t)$	electric field of a monochromatic plane wave
$\mathbf{E}_q$	electric field vector at qth ray intercept
E_x, E_y, E_z	complex amplitudes of the field along the three coordinate axes
$\mathbf{ED}(d_H, d_{45}, d_R)$	elliptical diattenuator, diattenuation components d_H, d_{45}, d_L
$\mathbf{E}(\varepsilon, \psi)$	elliptical polarizer with ellipticity ε oriented at angle ψ
EPD	entrance pupil diameter
$\mathbf{ER}(\delta_H, \delta_{45}, \delta_R)$	elliptical retarder, retardance components $\delta_H, \delta_{45}, \delta_L$
f or *fast*	fast mode label
F	fast principal axis label
FEM	finite element modeling
FOV	field of view
$\mathbf{G}$ and g	gyrotropic tensor and gyrotropic constant
$\mathbf{g}_{In,i}, \mathbf{g}_{Exit,i}$	double pole coordinates defined in entrance and exit pupil
H	Lagrange invariant
$\mathbf{H}$	Jones vector for horizontally polarized light
H and $\mathbf{H}$	magnetic field and magnetic field vector
$\mathbf{H}$	hermitian matrix
$\mathbf{H}$	object coordinates
$H(\mathbf{r}, t)$	magnetic field of a monochromatic plane wave
i, j	summation indices
i	isotropic mode label
i_c	chief ray angle of incidence
i_m	marginal ray angle of incidence
I	flux, light intensity, fraction of flux in particular state
I	first Stokes parameter, S_0, flux
I	inhomogeneity of a Mueller matrix
$\mathbf{I}$	identity matrix
$\mathcal{I}m$	imaginary component of complex value
inc	incident ray label
I_{max}, I_{min}	maximum and minimum intensity transmittances
IPS	in-plane switching
IR	inhibited refraction
$\mathbf{J}$	Jones matrix
$\mathbf{J}_{pupil}$	Jones matrix pupil
JonesPupil(x,y)	full description of the wavefront and polarization at exit pupil
k	wavenumber
$\mathbf{k}$	propagation vector
$\hat{\mathbf{k}}_q$	normalized propagation vector at qth ray intercept

List of Abbreviations

ℓ	physical ray path
l or *left*	left circularly polarized mode label
L	Jones vector for left circularly polarized light
L, L_2	condition number of a matrix
L	left principal index label
$\hat{\mathbf{L}}(\theta)$	linearly polarized light oriented at θ
LC	liquid crystal
LCoS	liquid crystal on silicon
LD(t_1, t_2, θ)	linear diattenuator (partial polarizer), amplitude transmittances t_1, t_2, axis at θ
LCD	liquid crystal display
LED	light-emitting diode
LR(δ, θ)	linear retarder, retardance δ, oriented with fast axis at θ
LP(·)	Jones matrix for a linear polarizer
M	subscript to indicate quantities in increasing phase convention
M	magnification
M	medium principal axis label
M	Mueller matrix
$\vec{\mathbf{M}}$	flattened Mueller vector, 16 × 1
MMBRDF	Mueller matrix bidirectional reflectance distribution function
MPSM	Mueller point spread matrix
MVA	multi-domain vertical-aligned LC cell
MTF	modulation transfer function
MTM	modulation transfer matrix
m_{ij}, M_{ij}	Mueller matrix elements, $m_{00}, m_{01},\ldots,m_{33}$
Δn	birefringence
n	refractive index
$\mathbf{N}_i$	a null space vector
NA	numerical aperture
o	ordinary mode label
O	ordinary principal axis label
O, O^{-1}	orthogonal transformation matrices
$\mathbf{O}_{n,e}^m$	orientor basis functions
OA	optic axis
OPL	optical path length
OTM	optical transfer matrix
P	flux, irradiance, polarized fluxes, $P_H, P_V, P_{45}, \ldots$
p	p-component, in the plane of incidence
p_1, p_2	strain optic coefficient
$\hat{\mathbf{p}}$	p-component basis vector
P	vector of flux measurements
P	3D polarization ray tracing matrix
$\breve{\mathbf{P}}$	3D polarization ray tracing matrix for addition with **k**'s singular values set to zero
PBS	polarizing beam splitter
PDL	polarization-dependent loss, the extinction ratio specified in decibels
POI	plane of incidence
PSA	polarization state analyzer
PSF	point spread function
PSG	polarization state generator
PSM	point spread matrix
PTM	phase transfer matrix

$p_q(x, y, z)$	one–zero aperture function
PVA	patterned vertical-aligned LC cell
$\|\mathbf{W}\|_p$	p-norm of matrix or vector $\mathbf{W}$
q	index for surface number, element number
Q	total number of surfaces, elements, and so on
Q	second Stokes parameter, S_1, 0°–90° flux
$\mathbf{Q}$	3D parallel transport matrix
r, r_p, r_s	amplitude reflection coefficients
R	radius of curvature
r or *right*	right circularly polarized mode label
R_s, R_p	intensity reflection coefficients
$\mathbf{r}$	position vector
R	right principal index label
r, r_q	ray intercept coordinate, ray intercept at surface q
$\mathbf{R}$	Jones vector for right circularly polarized light
$\mathbf{R}$, $\mathbf{Rot}$	rotation matrix
$\mathbf{R}(\alpha)$	Jones matrix rotation operator
$\mathbf{R}_M(\cdot)$	rotation matrix for Stokes parameters and Mueller matrices
RMS	root mean square
$\mathcal{R}e(\cdot)$	real part
RCWA	rigorous coupled wave analysis
s	s-component, perpendicular to plane of incidence
$\hat{\mathbf{s}}$	s-component basis vector
$\mathbf{S}$	Poynting vector
$\mathbf{S}$	stress tensor
s or *slow*	slow mode label
S	slow principal axis label
S_0, S_1, S_2, S_3	Stokes parameters
$1, s_1, s_2, s_3$	normalized Stokes parameters
$\mathbf{S}$	Stokes parameters
sup	the supremum, limiting maximum value
STN	super twisted nematic LC cell
SVD	singular value decomposition
SWG	subwavelength grating
t	time, thickness
t, t_p, t_s	amplitude transmission coefficients
T	transmittance, exiting flux divided by incident flux
$t_{q-1,q}$	length of ray segment, surface $q-1$ to q
TE	transverse electric
TIR	total internal reflection
TM	transverse magnetic
TN	twisted nematic LC cell
T_s, T_p	intensity transmission coefficients
Tr	trace of a matrix
T_{max}	maximum transmittance over polarization states
T_{min}	minimum transmittance over polarization states
u	energy density
U	third Stokes parameter, S_2, 45°–135° flux
u	marginal ray angle
$\bar{u}_q$	chief ray angle after surface
$\mathbf{U}$	Stokes parameters for unpolarized light, (1, 0, 0, 0)

U, **V**	unitary matrix
V	fourth Stokes parameter, S_3, right–left flux
V	fringe visibility
V	velocity
V	Jones vector for vertically polarized light
$\mathbf{V}_{n,e}^{m}(\rho,\phi)$	vector Zernike polynomial
VAN	vertically aligned mode
W	polarimetric measurement matrix
W^{-1}	polarimetric data reduction matrix
$\mathbf{W}_P^{-1}$	pseudoinverse of **W**
W^T	transpose of matrix **W**
$W(x, y)$	wavefront aberration function
x_E, y_E	entrance pupil coordinates
$\hat{\mathbf{x}}_{Loc}, \hat{\mathbf{y}}_{Loc}$	local x and y coordinates
$\overline{y}_q$	chief ray height at surface q
$z(x, y)$	sag (height) function for a surface
x, y, z	*Cartesian coordinate axes*
135	Jones vector for 135° linearly polarized light
45	Jones vector for 45° linearly polarized light
α	optical rotatory power
$\alpha_{s,t,q}, \alpha_p$	Fresnel s and p coefficients
β	phase thickness of film
β_m	diffraction angle for mth diffraction order
γ	walk-off angle
γ	strain tensor coefficient
$\mathbf{\Gamma}$	strain tensor
δ	retardance
$\delta_H, \delta_{45}, \delta_R, \delta_L$	retardance components
$\delta_{i,j}$	Kronecker delta function
$\delta_{principal}$	principal retardance
$\delta_{unwrapped}$	unwrapped retardance
Δn	birefringence
$\Delta\lambda$	spectral bandwidth
ΔOPL	optical path difference
Δr	lateral shear
Δt	optical ray path
ε	ellipticity of polarization ellipse
ε_0	permittivity of free space
$\tilde{\boldsymbol{\varepsilon}}, \boldsymbol{\varepsilon}$	dielectric tensor
η	latitude on the Poincaré sphere
η	characteristic admittance of a thin film layer or substrate
$\eta, \hat{\boldsymbol{\eta}}$	surface normal, points away from incident medium
η	impermeability tensor
θ	an angle, angle of incidence, rotation angle, polarizer or retarder angle
$\theta_i, \theta_{in}, \theta_{inc}$	incident angle
θ_B	Brewster's angle
θ_B	grating blaze angle
θ_C	critical angle
$\overline{\theta}_q$	chief ray angle of incidence at surface q
Θ_1^m	polarized basis vector

κ	imaginary part of refractive index, absorption coefficient
κ	conic constant
λ	wavelength
λ_B	blaze wavelength
$\mathbf{\Lambda}_i$	singular values
μ_i	singular values
$\boldsymbol{\mu}$	magnetic permeability tensor
Ξ	etendué
ξ, ξ_q, ξ_r	eigenvalues
v_p and v_r	phase velocity and ray velocity
v	Poisson's ratio
ρ	amplitude of complex number
ρ	normalized pupil coordinate
ρ, ρ_p, ρ_s	magnitude of amplitude coefficient
$\vec{\rho}$	normalized pupil coordinates
σ	normal stress
$\boldsymbol{\sigma}_0$	identity matrix
$\boldsymbol{\sigma}_1, \boldsymbol{\sigma}_2, \boldsymbol{\sigma}_3$	Pauli matrices
Σ	phase expressed in waves
τ	shear stress
τ_q	reduced thickness after surface q
ϕ	phase of light, phase of a complex number
φ	pupil angle measured counterclockwise from the x-axis
Φ	plane of incidence
Φ_q	power of surface q
$\boldsymbol{\Phi}(\phi)$	Jones matrix for an overall change of phase
ψ	angle of major axis of polarization ellipse
ψ	fast axis orientation of retarder
ψ_o and ψ_e	critical angle of ordinary and extraordinary modes
ω	angular frequency in radians per second
$\bar{\omega}_q$	reduced chief ray angle at surface q
Ω	solid angle, steradians
$\boldsymbol{\Omega}$	strain optic tensor
$\bullet$	dot product, matrix multiplication
$\dagger$	adjoint of vector, the complex conjugate of the transpose

1

Introduction and Overview

For many optical systems, selecting good combinations of *polarization elements* may be difficult. Similarly, understanding and controlling the optical system's polarization properties can present substantial challenges and, for some systems, requires man-years of dedicated *polarization engineering*. Such systems can be called *polarization critical optical systems* because they present polarization challenges and have specifications that are difficult to meet. Liquid crystal displays and optics for microlithography are just two examples of polarization critical systems. Polarization engineering is the task of designing, fabricating, testing, and often mass producing with high yield such polarization critical optical systems.

Of course, the polarization properties of many other optical systems are small and not always significant for their operation. For example, many lenses make relatively small changes to the polarization state, of the order of a percent or less. Even if the polarization effects are small, they may still be interesting and it may be necessary to ensure that some polarization specification is met.

This book is dedicated to understanding these polarization effects or *polarization aberrations*, large and small. One of the principal techniques will be *polarization ray tracing*. Ray tracing is a set of algorithms for calculating the paths of light rays through optical systems. Polarization ray tracing adds calculations to follow the evolution of the polarization state and will display information on how the distribution of polarization and the polarization properties are for the ray paths.

 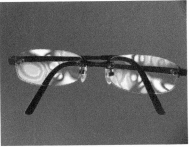

Figure 1.1 (Left top) A leptocephalus eel larva seen in visible light is nearly transparent to avoid predators. (Left bottom) When viewed with polarizing filter, the eel is more visible due to polarization change associated with birefringence in parts of its body. Thus, polarization vision is useful in finding such prey. (Photo from NOAA.[1]) (Right) Eyeglasses under crossed-circular polarizers.

1.1 Polarized Light

Light is a *transverse electromagnetic wave*, a moving *electric* and *magnetic field*. The light's electric and magnetic fields oscillate in a direction transverse to the direction of light propagation. Light is generated when charges, electrons and protons, accelerate and oscillate. Then, the light's forces cause charges to oscillate in return. *Polarization* refers to the properties of the light in this *transverse plane*, describing whether the light is *polarized* or *unpolarized* and the orientation of this polarization. The polarization state of light can be controlled with *polarization elements*. The polarization state also changes at lenses, mirrors, optical coatings, diffraction gratings, crystals, liquid crystals, and many other interfaces and materials. The book's goal is to develop efficient and general methods for understanding the changes of polarization of light with polarization elements, through optical systems, and in the natural environment.

The *human eye* is not sensitive to the polarization of light. We sense the brightness and color of light but cannot tell if light is polarized or unpolarized, or how the polarization is oriented. Since humans are polarization blind, we are unaware of the many polarization effects around us, indoors and outdoors, as shown in Figures 1.1 and 1.2. Our eyes cannot tell that the light from rainbows and the glare reflecting from the road are highly polarized, nor can they sense the polarized light from liquid crystal displays. We do not see how polarized light is scrambled by the stress in our eyeglasses (Figure 1.1) and automobile windows (Figure 1.2). *Polarizers* and *polarimeters* can make such polarization effects visible. Many *animals* do see the polarization of light, such as *bees*, *ants*, and *octopus*. They use polarization for a variety of purposes including to *navigate*, *communicate*, and *find prey*.

1.2 Polarization States and the Poincaré Sphere

Monochromatic light is light with a single pure frequency, a single wavelength, and a cosinusoidal *electric field*. Monochromatic light is an idealization, the limit as the wavelength spread from a light source approaches zero. Light from single-frequency lasers is nearly monochromatic, but the light has some small *spectral bandwidth*. Chapter 2 (Polarized Light) presents the mathematical description of light. Here, a few of these concepts are introduced.

The *polarization ellipse* is the figure traced by the tip of the electric field vector for monochromatic light, repeating each period. Monochromatic light must be *polarized*, either *linearly*, *elliptically*, or *circularly polarized*, because it is periodic. When the light's electric field only oscillates in a single direction, such as along the positive and negative *y*-axis as in Figure 1.3, the light is linearly polarized. The direction of *oscillation* is the *plane of polarization* or *angle of polarization*.

1.2 Polarization States and the Poincaré Sphere

Figure 1.2 Moonroof photographed under a sunny day with (top) and without (bottom) a polarizer filter on the camera. The rainbow checkerboard pattern reveals the stress birefringence of the tempered glass.

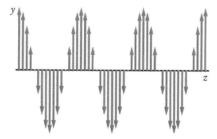

Figure 1.3 Electric field associated with a monochromatic light field linearly polarized in the y-direction.

For linearly polarized light propagating along the z-axis, the light's electric field has x- and y-components that are *in phase*; both components go to zero at the same time, twice per period. Otherwise, the light is elliptically polarized, as in Figure 1.4, and the x- and y-components are out of phase. The tip of the electric field vector traces an ellipse, once per period, creating the polarization ellipse, one of the iconic figures of polarization optics. The polarization ellipse will always be

Polarized Light and Optical Systems

Figure 1.4 (Left) The polarization ellipse for an elliptical polarization state is the ellipse drawn by the tip of the electric field vector during one period. (Middle) The electric field as a function of time is shown with the bounding ellipse. (Right) Usually, just the ellipse is drawn to indicate the polarization state. The location of the arrow indicates the phase of the wave.

drawn looking into the beam. Monochromatic circularly polarized light has a constant electric field amplitude whose orientation uniformly rotates in the transverse plane. Circularly polarized light occurs in two forms, right circularly polarized and left circularly polarized. By convention, right circularly polarized light rotates clockwise and left circularly polarized light rotates counterclockwise in time looking into the beam. Similarly, elliptically polarized light has *right* or *left helicity* depending on the direction of rotation.

Light also consists of a *magnetic field* with a vector pointing perpendicular to the electric field in the transverse plane, oscillating in phase with the electric field. Figure 1.5 shows monochromatic light's electric field (red) and magnetic field (blue) oscillating in space for several polarization states. The interaction of light with matter is dominated by the electric field in most types of light–matter interactions; thus, by convention, the light's polarization state is described by its electric field. The magnetic field is important for calculating the interaction of light with several types of media including magnetic materials, birefringent materials, diffraction gratings, and liquid crystals.

The family of polarization ellipses can be conveniently represented on the surface of a unit sphere, the *Poincaré sphere*, shown in Figure 1.6, shown in front and back views. Right circularly polarized light is at the top of the sphere, the *north pole*. Left circularly polarized light is at the bottom of the sphere, the *south pole*. Linearly polarized states lie on the equator. Note that for a complete circuit around the equator, the polarization state rotates by 180°, which returns *x*-polarized light into *x*-polarized light. The rest of the Poincaré sphere's surface describes *elliptically polarized light*, with a right circular *helicity* in the northern hemisphere and a left circular helicity in the southern hemisphere. The ellipticity is nearly linear close to the equator; the ellipticity approaches circular near the poles. The surface of the Poincaré sphere *continuously* represents all possible polarized ellipses. It is useful to represent the Poincaré sphere on a flat surface as shown in Figure 1.7.

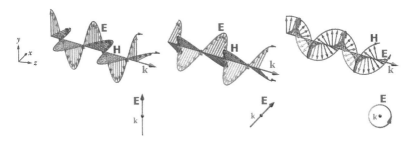

Figure 1.5 Electric (red) and magnetic (blue) fields for (left) 90° polarized, (center) 45° polarized, and (right) left circularly polarized light in space.

1.3 Polarization Elements and Polarization Properties

Figure 1.6 The Poincaré sphere contains a representation of all polarized states on its surface, with linearly polarized states arranged around the equator, and the two circularly polarized states at the poles.

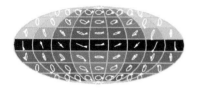

Figure 1.7 Two representations of the Poincaré sphere on a plane: (left) Mollweide projection and (right) equi-rectangular projection, where the entire top row is right circularly polarized and the entire bottom row is left circularly polarized.

Several methods are used to describe polarization states depending on the application. Two element complex vectors, *Jones vectors*, and three element *polarization vectors* that are covered in Chapter 2 are particularly well suited for *optical design*, *diffraction*, and *interferometry*, cases where the phase of the light is essential. The *Stokes parameters* covered in Chapter 3 are well suited for laboratory measurements and descriptions of the *natural light* outdoors; these are polychromatic and incoherent applications where phase has less meaning.

1.3 Polarization Elements and Polarization Properties

A *polarization element* is any optical element used to alter or control the polarization state of light and to transform light between polarization states. Polarization elements are classified into three broad categories—*polarizers*, *retarders*, and *depolarizers*—based on whether they change the amplitudes, phases, or coherence of the light. Mirrors, lenses, thin films, and nearly all optical elements alter polarization to some extent, but are not usually considered as polarization elements because that is not their primary role, but a side effect.

Polarizers transmit a known polarization state independent of the incident polarization state. Most common are linear polarizers that transmit linearly polarized light along their *transmission axis*. A *linear polarizer* is a device that, when placed in an incident unpolarized beam, produces a beam of light where the electric field vector is oscillating primarily in one plane with only a small component in the perpendicular plane. An *ideal polarizer* has a transmission of one for the specified polarization state and transmission of zero for the orthogonal polarization state. Polarizers are an example of *diattenuators*, or *partial polarizers*, which have two transmissions T_{max} and T_{min} for two orthogonal polarization states. Diattenuators can be characterized by their *diattenuation*

$$D = \frac{T_{max} - T_{min}}{T_{max} + T_{min}}, \quad 0 \leq D \leq 1, \tag{1.1}$$

which varies from one for a polarizer to zero for an optical element that transmits all incident polarization states equally. Sheet polarizers, or Polaroid, which come in large plastic sheets, absorb one polarization state transmitting the orthogonal state. Polarizing beam splitters direct two orthogonal polarization components into different directions, as shown in Figure 1.8.

Retarders have two different optical path lengths associated with two special polarization states, the *fast state* and the *slow state*. The slow state is delayed, or *retarded*, with respect to the fast state. The *retardance* is the difference in optical path lengths, which will describe the relative phase change between the two states. Arbitrary incident polarization states divide into the fast state and slow state when entering a retarder, and these two components emerge from the retarder with different optical path lengths, as shown in Figure 1.9. *Linear retarders* divide the light into two linear polarized components, 90° apart, retarding one of the states. A *quarter wave linear retarder* introduces a relative phase delay of a quarter of a wavelength of the light, and is useful for converting linearly polarized light into circularly polarized light. A *half wave linear retarder* delays one linearly polarized component by half a wavelength, and is useful for changing the orientation of linearly polarized light.

Depolarizers scramble the state of polarization and convert polarized light into unpolarized light. *Depolarization* is usually associated with *scattering*, particularly multiple scattering. *Integrating spheres* will readily depolarize a beam of light. Thin slabs of *opal*, a gem consisting of closely packed spheres of quartz, are sold as depolarizers. Most *projection screens* commonly used in classrooms and meeting rooms will effectively depolarize a beam of light. Try illuminating a screen with polarized light and observe that the scattered light cannot be extinguished with a polarizer. Lenses, mirrors, filters, and other typical optical elements exhibit very small amounts of depolarization, typically less than a few tenths of a percent. Hence, in the majority of optical systems, the magnitude of depolarization is small and not significant. Optical surfaces are carefully fabricated and coated to minimize scattering; thus, depolarization is generally very small from high-quality optical surfaces. Figure 1.10 shows a spatially depolarized set of polarization ellipses.

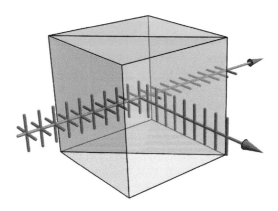

Figure 1.8 A polarizing beam splitter splits orthogonal polarization components into two directions.

1.3 Polarization Elements and Polarization Properties

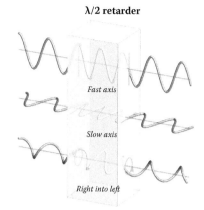

Figure 1.9 Three states of polarized light, propagating from the left in air, enter a half wave linear retarder (birefringent waveplate) with a vertical fast axis (lower refractive index). The two planes are the entrance and exit faces. The vertical polarized mode (top, lower refractive index) has a longer wavelength inside the retarder, so its optical path length is less, 2 waves, than the horizontally polarized mode (center) by ½ wave. (Bottom) Right circularly polarized light divides into the two modes that propagate separately and then combine exiting the retarder. This exiting beam is now left circularly polarized due to the half wave optical path difference (retardance) between the two modes.

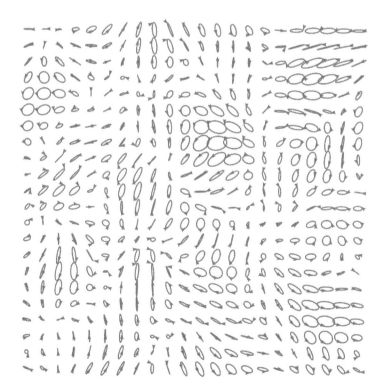

Figure 1.10 A depolarized field with spatially varying polarization states.

1.4 Polarimetry and Ellipsometry

The polarization properties of a beam of light are measured with a *polarimeter*, which is a camera, radiometer, or spectrometer configured to measure the flux through a set of polarizers and calculate the polarization state; these are *light-measuring polarimeters*. Figure 1.11 shows the layout of one type of light-measuring polarimeter, a rotating retarder imaging polarimeter. Figure 1.12 shows a polarization image of the Meinel Optical Sciences building at the University of Arizona. One interesting feature is the polarization of the windows along the front side. Partially polarized skylight reflecting from the windows has its polarization changed different amounts depending on the angle of the windows. Thus, polarimetery can be used to help recover information on the orientations of objects. One application of imaging polarimetry is the measurement of aerosols. Light scattered from small particles undergoes a significant change in polarization state, which can then be used to determine the size and density of aerosol particles. Aerosol polarimeters measure sunlight scattered from the atmosphere to study the properties of atmospheric aerosols. Imaging polarimeters in earth orbit can map the aerosol content of the atmosphere globally.

Sample measuring polarimeters, such as *Mueller matrix polarimeters*, determine a material's polarization properties by illuminating a sample with a sequence of polarization states and measuring the exiting polarization states (Figure 1.13). Mueller matrix polarimeters are used for measuring polarization elements, liquid crystal cells, retinal imaging, and other forms of biological imaging. One form of a sample-measuring polarimeter is the *ellipsometer*, originally developed in the 1880s by Paul Drude, who was seeking a method to measure the refractive index of materials, particularly metals. He realized that if he reflected linearly polarized light with the angle of polarization at 45° from the plane of incidence, he could measure the refractive index of a surface from the orientation of the major axis and ellipticity of the reflected light's polarization. The name *ellipsometry* followed from this characterization of the polarization ellipse.[2] Later, ellipsometric technology evolved to be able to very precisely measure the thickness and refractive index of single-layer thin films. Thus, ellipsometry aided the development of *antireflection coatings*, such as *quarter wave thick magnesium fluoride coatings* in the 1920s. By measuring at multiple angles of incidence and wavelengths, the properties of multilayer thin films are accurately obtained by ellipsometry.[3] Ellipsometry is now an essential technique in the fabrication of integrated circuits and microelectronics, as well as other applications in industrial metrology.

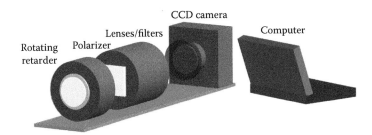

Figure 1.11 A rotating retarder imaging polarimeter measures the polarization state by passing the light through a retarder, a linear polarizer, and into a camera. The retarder steps to several angles and the measurements from several images are used to calculate the degree of polarization, the angle of polarization, and the ellipticity of the light as images.

1.4 Polarimetry and Ellipsometry

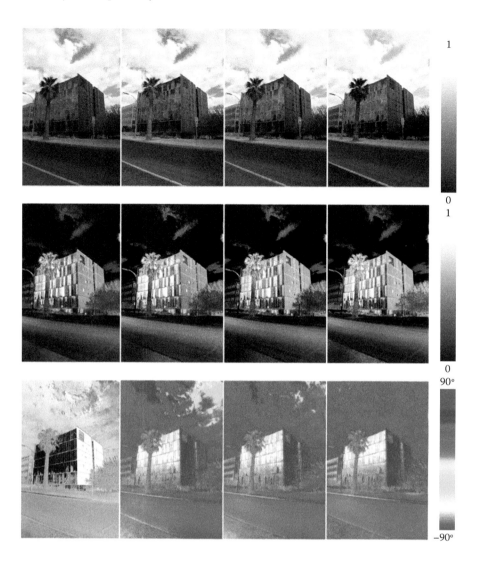

Figure 1.12 Polarization images of the front of the Meinel Optical Sciences building at the University of Arizona. (Top row) Intensity images. (Middle row) Degree of linear polarization images. (Bottom row) Angle of linear polarization images. (Left column) Red/Green/Blue images (RGB). (Second column) 470-nm images. (Third column) 550-nm images. (Right column) 660-nm images. The north side of the Meinel building is all glass panes at a variety of different angles, reflecting skylight with a range of degrees of polarization and angles of polarization (ranging from orange ~−80° to green ~50°. The right side of the building is copper metal with few windows and an AoLP of ~80°. The sky is mostly cloudy with a low DoLP (dark) except for a few blue patches showing white in the middle row. At 880 nm, the cloudy sky mostly has the same AoLP as the blue sky, while at 470 nm, there is considerable variation of AOLP in the cloudy sky. (Taken by Karlton Crabtree and Narantha Balagopal.)

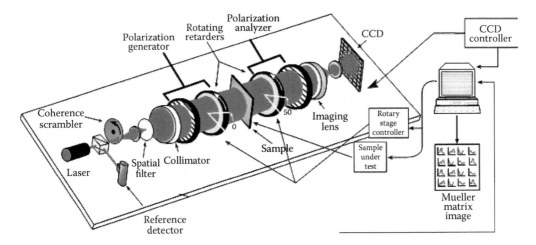

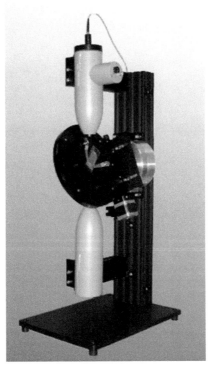

Figure 1.13 (Top) A Mueller matrix imaging polarimeter illuminates and measures a sample with many combinations of illuminating polarizers and analyzing polarizers to measure the Mueller matrix, diattenuation, retardance, and depolarization of samples. This particular Mueller matrix polarimeter illuminates the sample through a linear polarizer and a retarder that steps in angle. The light interacts with a sample and exits with its polarization state changed. The light may reflect from, transmit through, diffract from, or scatter from a sample depending on the measurement. This change is analyzed by a retarder that steps in angle and a linear polarizer before entering a camera or radiometer. (Bottom) A commercial Mueller matrix polarimeter configured for transmission Mueller matrix measurements. This polarimeter measures thirty Mueller matrices per second. (Courtesy of Axometrics, Huntsville, Alabama.) The light source and polarization generator are in the top head and the analyzer and detector are in the lower head.

1.5 Anisotropic Materials

Anisotropic materials have a refractive index that varies with the direction of the light's electric field. Many anisotropic materials are crystals, for example, *calcite* and *quartz*. Within a single crystal calcite, all of the calcium-to-carbon bonds are oriented in one direction, called the *optic axis*. The three carbon oxygen bonds in the carbonate radical are oriented in the perpendicular plane. Light polarized along the optic axis interacts with a different set of chemical bonds compared to light polarized in the orthogonal plane, resulting in two refractive indices for two polarizations associated with one incident direction of propagation. When light refracts into calcite or other anisotropic materials, it refracts into two *modes* with orthogonal polarization, *ordinary* and *extraordinary*, which, in general, propagate in different directions. This is clearly seen in Figure 1.14.

Hence, during polarization ray tracing, each ray entering an *anisotropic crystal* results in two exiting rays that need to be traced to the output of the optical system. As shown in Figure 1.15, when there is a second anisotropic optical element, the rays double again and four rays need to be traced through the remainder of the system. In general, a system containing N anisotropic elements produces 2^N separate rays, all of which need to be traced to simulate the light through the system. Each of these rays takes a separate path and has its own amplitude, polarization, and optical path length. Thus, the light in the exit pupil may be described by 2^N separate *wavefront aberration functions*; each of these *partial waves* can have different *amplitude aberration*, *defocus*, *spherical aberration*, *coma*, *astigmatism*, and so on.

Figure 1.14 Light propagating through calcite divides into two modes, which follow different paths and generates two images.

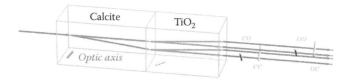

Figure 1.15 Polarization ray trace showing a ray propagates into calcite and splits into two rays. Entering a second anisotropic material, titanium oxide, each ray doubles again. Each incident wavefront generates four exiting wavefronts.

1.6 Typical Polarization Problems in Optical Systems

Polarization provides the basis of operation for many types of optical systems. Examples include *liquid crystal displays* and the *ellipsometers* used in the microlithography industry, polarization instruments that test the composition and thicknesses of the many layers that are deposited during chip fabrication. But polarization is also a problem in many systems and is thus frequently analyzed during the optical design process. A quick survey of the *polarization aberrations* of some simple optical systems will help understand the goals and methods for polarization ray tracing and analysis.

1.6.1 Angle Dependence of Polarizers

One of the issues optical designers face is the dependence of *polarizers* on the angle of incidence of light. Both *dichroic polarizers* (*sheet polarizers*) and *wire grid polarizers* absorb the polarization component that projects onto their *absorption axis*. The *extinction* is high when the angle of incidence varies in two planes: the plane perpendicular to the transmission axes and the parallel plane. But for fundamental geometrical reasons, this high extinction is reduced when the light propagates at other directions. The leakage is worst for propagation in planes at 45° to the extinction axes. Figure 1.16 is a 3D view of two polarizers, one polarizer with its absorption aligned along the y-axis (green) followed by a polarizer aligned along the x-axis. The on-axis direction is represented by a short black line. As an incident beam of light moves off-axis in the diagonal direction as shown in Figure 1.17, the absorption axes are no longer orthogonal and an increasing amount of light leaks through. Thus, if a 45° cone of light is incident, the transmitted intensity appears as in Figure 1.18.

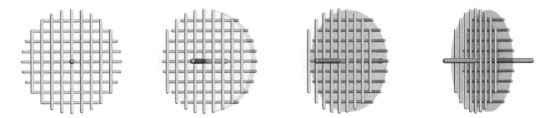

Figure 1.16 A schematic 3D view through two polarizers from several angles in the x–z plane. The transmission axis of the front polarizer is vertical and indicated with green lines. The transmission axis of the back polarizer is horizontal indicated with red lines. The thick black line indicates the z-axis, normal to the two polarizers. (Left) At normal incidence, the front and back transmission axes are perpendicular. Viewing at 15° (second), 30° (third), and 45° (right), the green and red lines remain perpendicular. The extinction of the polarizer pair is good when illuminated in these directions.

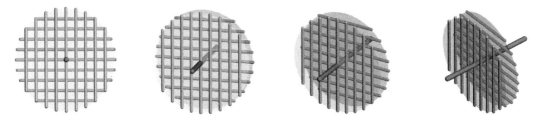

Figure 1.17 A view through the same two polarizers as the light propagation direction varies in the diagonal plane. (Left) At 0° angle of incidence, along the z-axis, the lines are perpendicular. As the angle of incidence in the plane at 45° to x and y increases from (second) 15°, to (third) 30°, to (right) 45°, the projection of the polarizers' transmission axes no longer appears orthogonal and the polarizer pair leaks light accordingly.

1.6 Typical Polarization Problems in Optical Systems

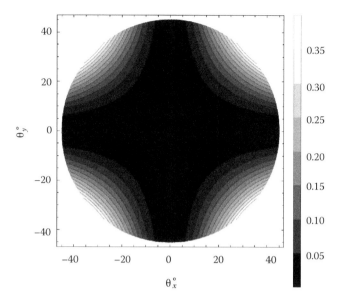

Figure 1.18 Contour plot of the fractional leakage through a crossed pair of polarizers for a 45° cone of light.

1.6.2 Wavelength and Angle Dependence of Retarders

Simple *birefringent waveplates*, the most common retarders, have properties that vary with wavelength and angle of incidence. Since the *ordinary* and *extraordinary* refractive indices of a *uniaxial* crystal vary with wavelength, known as *dispersion*, the *birefringence* is also a function of wavelength. Figure 1.19 shows the wavelength dependence of retardance for a quartz quarter wave retarder and MgF_2 quarter wave retarder.

A retarder's retardance also varies with the angle of incidence. Because the path length through the retarder increases with angle of incidence, and the extraordinary refractive index varies with angle, the retarder's retardance usually has a *toroid-like* variation with angle; a calculation for a quartz quarter wave retarder is shown in Figure 1.20. For y-polarized light, the exiting polarization state is unchanged for light incident along the x–z plane and the y–z plane but changes in all other directions. The variation of retardance is easily seen when the retarders or other birefringent crystals are placed in a *conoscope*, an optical instrument that focuses polarized light through a sample

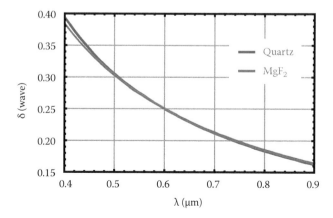

Figure 1.19 The wavelength dependence of birefringent (red) quartz, and (blue) magnesium fluoride waveplate retarders with a quarter wave of retardance at 600 nm.

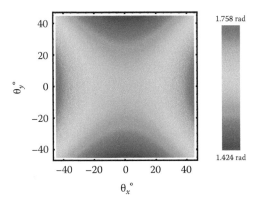

Figure 1.20 The variation of a quarter wave retarder's retardance (1.571 rad) with angle of incidence decreases with angle along the plane containing the optical axis (vertical) and increases with angle along the orthogonal plane (horizontal). Along the diagonals, the retardance is nearly constant with angle of incidence.

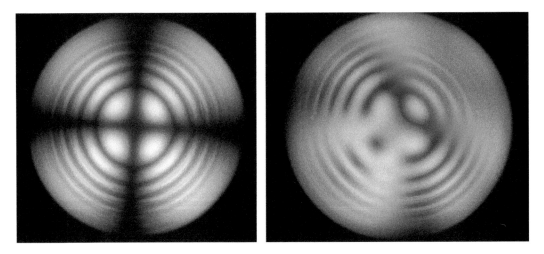

Figure 1.21 Conoscopic images through (left) an A-plate and (right) a C-plate of calcite, showing the variation of exiting polarization with angle and wavelength as colored fringes.

and analyzes the light with a second polarizer. Figure 1.21 (left) shows a conoscopic image of a thick A-plate of calcite in a conoscope and Figure 1.21 (right) shows the image for a thick crystal of calcite with its *optic axis* along the plate normal, a configuration known as a *C-plate*.

1.6.3 Stress Birefringence in Lenses

Another issue with polarized light and lenses, particularly *injection-molded lenses*, is stress birefringence. *Stress birefringence* is a spatially varying birefringence resulting from forces within the lens that compress or stretch the material's atoms, causing birefringence. Stress birefringence can become frozen into a glass blank or molded lens during fabrication as the lens material cools unevenly. Stress can also arise from externally applied stresses, such as forces on the lens from lens mounts or even gravity. Thus, stress birefringence causes unwanted retardance variation. Stress birefringence can be made visible by placing the lens between *crossed polarizers* in a system called a *polariscope*. The stress-induced retardance will change the polarization state of

1.6 Typical Polarization Problems in Optical Systems

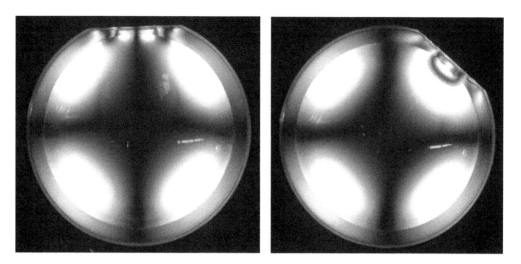

Figure 1.22 An injection-molded lens with stress birefringence (left) between crossed polarizers (at the top) and (right) after rotating the lens by 45°.

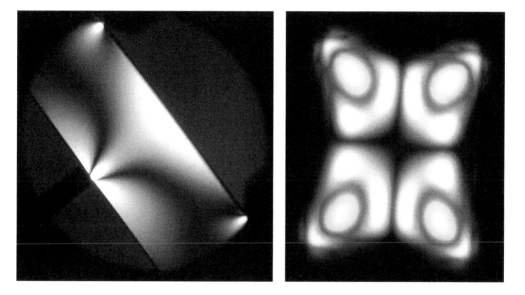

Figure 1.23 (Left) Pressure-induced stress birefringence in the polariscope. A piece of glass supported by two points on the upper right side is being stressed by a screw on the lower left side. (Right) A square piece of glass with large stresses frozen into the glass viewed between crossed polarizers showing the wavelength variation of the stress-induced retardance.

some of the light, and this changed component will pass the second polarizer. By viewing the *polarization leakage*, as shown in Figure 1.22, the stress can be visualized. Stress birefringence can often be reduced by *annealing*, heating a lens close to the glass transition temperature, and then cooling it slowly to reduce internal stresses. Annealing is a routine process for high-quality optical glass. Figure 1.23 (left) shows a piece of glass under mechanical stress in the polariscope. The intensity distribution reveals regions where the polarization state has been modified. Figure 1.23 (right) shows a piece of glass with high stress from rapid cooling. Stress birefringence in lenses reduces their image quality.

1.6.4 Liquid Crystal Displays and Projectors

Liquid crystals are a thick soup of birefringent molecules, typically rod-shaped molecules chosen for their *dipole moment*. These molecules rotate in response to applied electric fields that modulate the retardance of the cell. Figure 1.24 shows the molecular orientation in a typical *twisted nematic liquid crystal cell* for several voltages. When placed between polarizers, the liquid crystal functions as a voltage-controlled *intensity modulator*. To create a liquid crystal display, a thin layer of liquid crystal is placed in a cell between two glass plates. An array of electrodes is fabricated in an addressable array to adjust the electric field to each pixel and packaged with an array of tiny red, green, and blue filters.

Liquid crystal cells, *displays*, and *projectors* present some of the most challenging *polarization aberration* problems. Undesired polarization variation in displays ends up as undesired color variation in the display. The *eye* is very sensitive to color variations; hence, undesired polarization variation must be kept to a minimum. One major cause of color variation in liquid crystal cells is a large variation of retardance with angle of incidence. The retardance magnitude can change with angle, the fast axis can change, and the ellipticity of the fast axis can change. An example of such retardance variation is shown in Figure 1.24 on the right. The ellipses denote the fast state and the

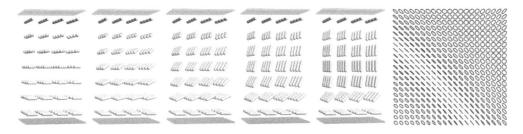

Figure 1.24 Orientation of liquid crystal molecules in a twisted nematic cell for several voltages, increasing from left to right. Light propagates in the vertical direction. Liquid crystal molecules are anchored to top and bottom horizontal planes. At 0 V (far left), the molecules twist in the horizontal plane about the *z*-axis (vertical). As the voltage increases (moving to the right), the molecules begin to tip out of the horizontal plane in the center of the cell, reducing their contribution to the retardance. At high voltages, the molecules at the center are rotated to the vertical plane, while molecules at the top and bottom remain anchored to the substrates. (Far right) Retardance variation (fast axis orientation and retardance magnitude) of the example liquid crystal cell across the field.

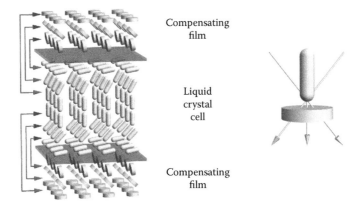

Figure 1.25 In all displays, liquid crystal cells are matched with compensating films to reduce the polarization aberrations. Here, the film on the top compensated the top half of the liquid crystal cell, and the compensating film on the bottom compensates the lower half, greatly improving the color quality of the display by reducing undesired polarization state variation and leakage through polarizers.

magnitude of the ellipses indicates the retardance magnitude. Such variations, if not compensated, cause the color and brightness of displays to vary with angle of incidence, which is very distracting. This angle-dependent retardance is commonly *fixed* by adding another film of birefringent molecules, a *biaxial multilayer film*, typically fabricated with *disk-shaped molecules*, which can effectively compensate the rod-shaped liquid crystal molecules birefringence with angle as shown in Figure 1.25 (right).

1.7 Optical Design

Optical design is the engineering practice of finding good and useful combinations of *optical elements*. The subject is also referred to as *lens design*, because the design of lenses, telescopes, and microscopes and the understanding of their *aberrations* were one of the central research areas as optical design became formalized in the late 1800s.

Optical systems can be divided into *imaging systems* and *other types* of optical systems, such as *illumination systems*. Imaging systems are designed to take input *spherical waves* and transform them into spherical output waves. However, it is not possible with lenses and mirrors to transform the input waves from a finite area of the object into perfectly spherical output waves.* Some deviation from sphericity for the exiting waves is inevitable with combinations of lenses and mirrors. These *deviations* from spherical wavefronts are the *aberrations*. The *wavefronts* are surfaces of constant *phase* and constant *optical path length* from the source. The optical path length can be thought of as the number of wavelengths along a path through an optical system, although its value is usually given in meters. Variations of the optical path length of a small fraction of a wavelength have significant impact on image quality. The priority in *conventional optical design*, by which we mean optical design without consideration of polarization, is the calculation of optical path length by *ray tracing*. Variations of optical path length have a much larger effect on image quality than variations of *amplitude* or *polarization state*. One of the most important tasks in optical design is to minimize the optical system's aberrations over the desired range of wavelengths and object positions by *optimizing* the system. Control of wavefront aberrations is exquisite in many types of optical systems such as lenses for television and movie production and lenses for microlithography.

Consider an example *cell phone lens* (Figure 1.26; U.S. Patent 7,453,654 embodiment #3) and a set of ray paths calculated by ray tracing. Five *collimated* (parallel rays) beams of light are shown entering on the left in *object space* and are represented by *rays*, lines normal to the wavefront. An optical analysis program, *Polaris-M*† in the case of this example, calculates the intersection of each ray with the first surface; these are the *ray intercepts*. Using Snell's law, the ray directions are calculated inside the first lens; these are the *propagation vectors*. Then, the refracted ray is propagated until it intersects the second surface and the length of the ray is calculated between the first and second ray intercept. The product of the ray length and the refractive index is the *optical path length* for the *ray segment*. The process repeats, finding a ray intercept and then refracting the ray, until the ray exits the last surface into *image space*. The *aperture stop* for this lens is located at the first surface, where the different colored rays from each field intersect. More details on the process of ray tracing are found in Chapter 10 (see discussion on geometrical and polarization ray tracing).

To evaluate the lens' *image quality* and *aberrations*, a set of rays are traced through the system to the image plane. The rays for the *on-axis beam* are seen to converge to a small area, almost to a single point, while the rays from the *off-axis beams* do not converge as well. To evaluate the aberrations, the optical path length for a grid of rays is calculated on a spherical surface, the reference sphere, centered on the image point, as shown in Figure 1.27. Figure 1.28 shows the *wavefront*

---

* There are a few exceptions such as Maxwell's fisheye lens and the Luneburg lens with a curved object and curved image surface. These exceptions are not suitable for camera lenses, cell phone lenses, and most imaging applications.
† Polaris-M is a polarization ray tracing program available from Airy Optics, Tucson, AZ.

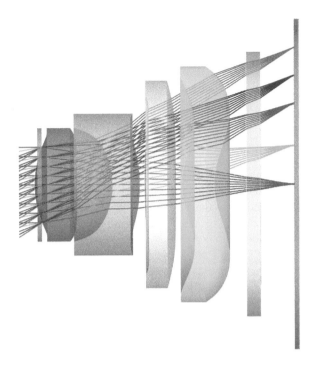

Figure 1.26 An example of a ray trace through a four-element cell phone camera lens. On-axis rays are in red. Rays from the off-axis fields are in green, blue, purple, and brown. A plane parallel IR-rejecting filter, blue, is located to the left of the *image plane* at the right.

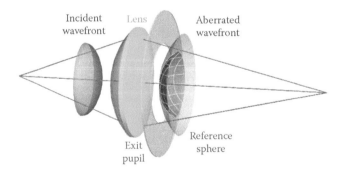

Figure 1.27 The incident wavefront (purple) at an example lens (green) is spherical. The exiting wavefront converges to the image point (right). Intersecting the center of the exit pupil (lavender), a reference sphere (blue) is constructed centered on the image point. The exiting aberrated wavefront (purple, yellow lines) is a surface of constant optical path length from the object and entrance pupil. The separation along the ray paths between the reference sphere and aberrated wavefront is the wavefront aberration. When these surfaces coincide, the wavefront is spherical and aberration free. A "diffraction-limited wavefront" is generally regarded as remaining within one-quarter wavelength of the reference sphere.

aberration expressed in fractions of a wavelength for the on-axis object point. Figure 1.29 shows the wavefront aberration for the green-colored off-axis beam.

The effect of these aberrations on the image is calculated by the methods of *Fourier optics* as described in Chapter 16 (Image Formation with Polarization Aberration). The *image of a point source* is called the *point spread function* or *PSF*, which is calculated by taking a *Fourier transform* of the wavefront at the *exit pupil*. Figure 1.30 shows two representations of the on-axis PSF, an

1.7 Optical Design

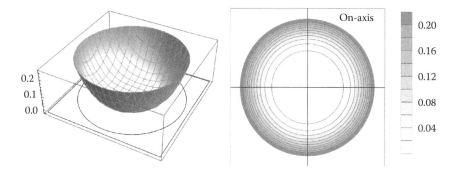

Figure 1.28 The wavefront aberration for the on-axis beam of Figure 1.26 in two presentation formats, an oblique plot (left) and a colored contour plot (right), is shown. About a quarter of a wavelength of *spherical aberration*, a fourth-order bowl-shaped aberration, is visible. An ideal spherical wavefront would have a flat wavefront aberration plot.

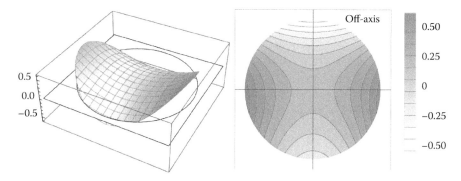

Figure 1.29 The wavefront aberration for the off-axis beam of Figure 1.26 in two presentation formats has about four-tenths of a wave of *coma* and one wave of *astigmatism* and a quarter of a wave of spherical aberration.

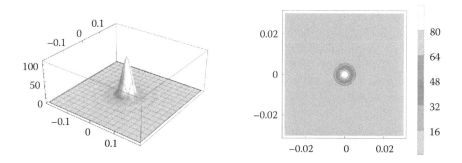

Figure 1.30 The on-axis PSF, the distribution of flux in the image of a point source, in (left) oblique perspective plot and (right) false-colored contour plot. Arbitrary flux units.

oblique projection plot and a colored contour plot. This PSF has a form close to the PSF of an ideal wavefront, known as the *Airy disk*, but is enlarged by the spherical aberration. The peak intensity has been reduced to about 40% of the intensity of the PSF formed without aberration; thus, the image's *Strehl ratio* is 0.4. Figure 1.31 shows two views of the off-axis PSF, where the PSF is much broader and the peak intensity is even further reduced because of the larger aberration.

Polarized Light and Optical Systems

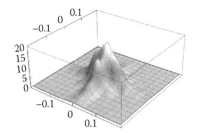 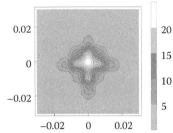

Figure 1.31 The off-axis PSF in (left) oblique perspective plot and (right) false-colored contour plot. Because of the larger aberration for this beam, the peak flux is lower and the distribution of light is much broader (producing lower resolution) than the on-axis PSF of Figure 1.30.

In conventional optical design, the assumptions used to calculate the PSF in these examples are the following: (1) the transmission of all the rays is equal; (2) the output polarization state is constant across the exit pupil. The calculations consider only the effects of optical path length variation, the effect of wavefront aberration. These for Figures 1.28 to 1.31 are the assumptions that this book refers to as the assumptions of *conventional optical design*. The *polarization aberrations* have been neglected, which, since they are often small, is a fine approximation for many systems.

In fact, the *transmission* of rays does vary. Each ray has a different set of angles of incidence leading to a variation of transmittance at each interface. Further, the polarization state of the light is slightly changed upon *refraction* so the polarization state is not uniform in the exit pupil. These amplitude and polarization changes depend on the *antireflection coatings* used on each surface. Thus, to calculate the effect of the coatings with polarization ray tracing, the *coatings* must be specified as well as the *lens shapes* and *refractive indices*. Some coatings will cause much larger amplitude and polarization changes than others. For a system like this cell phone lens, the effect of the coatings on the wavefront aberration and PSF can be quite small. In these cases when the effects of the wavefront aberration are much larger than the effects of the amplitude and polarization aberration, the assumptions of conventional optical design are justified. To find out when these assumptions are justified, it is necessary to perform the appropriate extra polarization calculations, determining the amplitude, optical path length, and *polarization changes at each ray intercept*, and cascading these effects together into a *polarization ray trace*.

From the early 1960s through the mid-1990s, commercial optical design programs were based only on *optical path length* calculations. During this period, the conventional ray tracing assumptions were adequate for a majority of optical design calculations. But by the beginning of the twenty-first century, polarization calculations were needed in many optical design problems to accurately simulate advanced optical systems with high *numerical aperture*, to perform tolerance analyses on such polarization sensitive systems, and to understand the effects of optical coatings on the wavefront aberrations and polarization aberrations. Now, full-featured *optical design programs* allow coatings to be specified on optical surfaces, the output polarization states to be calculated, and the polarization properties of ray paths to be determined.

1.7.1 Polarization Ray Tracing

The objective of *polarization ray tracing* is to calculate the polarization states exiting from optical systems and to determine the polarization properties, the *diattenuation*, *retardance*, and *depolarization*, associated with the ray paths. It is very useful to understand the light paths through optical systems in terms of the equivalent *polarization elements*. What are the polarization properties, the diattenuation, retardance, and depolarization, of the light paths? What would be the equivalent polarization elements, diattenuators and retarders, which reproduce the polarization state changes?

1.7 Optical Design

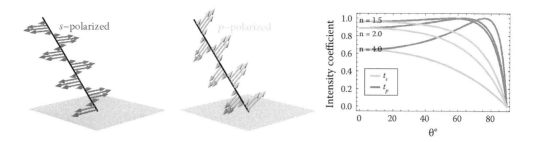

Figure 1.32 (Left) The definition of *s* and *p* components of a light beam at an interface. (Right) The Fresnel coefficients for transmission from air into lenses of different refractive indices for *s* light (green) and *p* light (blue) are a function of the angle of incidence. The difference in transmission is a source of diattenuation or partial polarization.

Polarization effects occur due to *s- and p-component* differences in *reflection* and *refraction*. The *s-* and *p-*components of the incident light are defined in Figure 1.32 (left and middle). Figure 1.32 (right) plots the *intensity transmission coefficients*, the fraction of light transmitted as a function of angle of incidence, calculated from the *Fresnel equations*, for refraction from air into uncoated surfaces with refractive indices of 1.5, 2, and 4. *Antireflection-coated interfaces* have similar curves but with generally improved transmission, closer to one.

The principal polarization ray tracing method is the *polarization matrix propagation method*. A polarization matrix is calculated for each ray intercept and ray segment. Matrix multiplication cascades the polarizing interactions. Finally, a polarization matrix, such as a *Jones matrix* or *Mueller matrix*, is calculated for each ray path from *object space* to *image space*. This information is combined with the optical path length from conventional ray tracing and a variety of additional analyses performed.

The polarization matrix propagation method can determine the output polarization state for all incident polarization states and describe the *diattenuation* and *retardance* for the ray paths. Then, it is useful to understand why the polarization state changed and, if the magnitude is troublesome, what might be done about it. The simplest way to describe the ray paths is with Jones matrices, 2×2 matrices with complex elements, shown in Equation 1.2; here, matrix elements are expressed in both Cartesian complex number form and polar form,

$$\mathbf{J} = \begin{pmatrix} j_{11} & j_{21} \\ j_{12} & j_{22} \end{pmatrix} = \begin{pmatrix} x_{11} + iy_{11} & x_{12} + iy_{12} \\ x_{21} + iy_{21} & x_{22} + iy_{22} \end{pmatrix} = \begin{pmatrix} \rho_{11} e^{i\phi_{11}} & \rho_{12} e^{i\phi_{12}} \\ \rho_{21} e^{i\phi_{21}} & \rho_{22} e^{i\phi_{22}} \end{pmatrix}. \quad (1.2)$$

Stop and contemplate the consequences of this polarization matrix propagation method for the polarization optical designer and the other engineers who need to use and understand his work. Conventional optical design describes the aberration with the wavefront aberration function, a scalar function with one value at each point on the wavefront in the exit pupil. The polarization matrix propagation method replaces this scalar function with a Jones matrix at each point on the wavefront. This function is called the *polarization aberration function* or the *Jones pupil*. Going from a representation with one variable, optical path length, for each ray to a matrix with eight variables at each ray is a very substantial complexification! This book goes one small step at a time, taking several chapters to elaborate on all these degrees of freedom and provide guidance on how to use this additional information. Thus, we will learn to interpret the *polarization aberrations* and understand their effect on image formation and various measurements. It is no wonder that the early optical designers did not include the calculations for uncoated or coated lens surfaces and mirrors in their image quality calculations; it's not easy.

And it gets more complicated!

1.7.2 Polarization Aberrations of Lenses

The polarization aberration of lenses arises due to the effect of the *Fresnel equations* and *thin film equations* at their surfaces. For an on-axis *spherical wave* at a spherical surface, the angle of incidence increases approximately linearly from the center of the pupil and the plane of incidence is radially oriented as shown in Figure 1.33 (left). For most antireflection coatings, the difference between the transmission for light polarized in the *p*-plane (radially) and *s*-plane (tangentially) increases approximately quadratically, as is seen near the origin (left side) of Figure 1.32. Thus, an uncoated lens surface actually acts as a weak *linear polarizer* with a radially oriented transmission axis with an approximately quadratically increasing diattenuation, as shown in Figure 1.33 (right). Figure 1.34 shows the polarization pupil maps, series of *polarization ellipses* sampled around the exit pupil of an uncoated lens when 90° linear, 45° linear, and left circularly polarized beams enter the lens. In the left figure, the beam is brighter at the top and bottom, dimmer on the right and left, and the polarization is rotated toward the radial direction at the edge of the pupil along the diagonals. The middle pattern has the same form but is rotated about the center by 45°. When *circularly polarized light* is incident, the light becomes steadily more elliptical and less circular toward the edge of the pupil, and the ellipse's major axis is oriented radially. If the uncoated lens is placed between *crossed polarizers*, light is extinguished along the two polarizer axes but leaks along the diagonals of the pupil as shown in Figure 1.35, a pattern known as the *Maltese cross*. An interesting pattern is observed when a Maltese cross beam is brought to focus. Because of this polarization

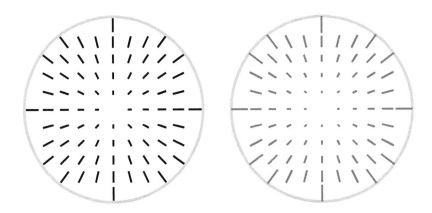

Figure 1.33 (Left) The plane of incidence and angle of incidence functions for an on-axis spherical wave incident at a spherical surface are shown. The angle of incidence is radially oriented and increases linearly from the center. (Right) The diattenuation orientation and magnitude at a lens surface for an on-axis source are shown. The diattenuation is radially oriented and the magnitude increases quadratically. Diattenuation aberrations will be represented with brown lines.

Figure 1.34 The effect of the diattenuation aberration of Figure 1.33 on incident (left) 90° linear, (center) 45° linear, and (right) left circularly polarized beams.

1.7 Optical Design

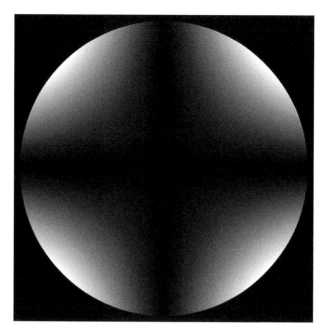

Figure 1.35 When an uncoated lens is observed between crossed (*x*- and *y*-oriented) polarizers, light leaks; this is a Maltese cross pattern.

aberration, the PSF, in the absence of any wavefront aberration, is dark in the center and along the *x*- and *y*-axes, but has four islands of light in one ring and dimmer islands of light further from the center as shown in Figure 1.36.

Most lens surfaces have *antireflection coatings*. For a lens surface with an antireflection layer of a quarter of a wavelength thick MgF_2, the diattenuation is typically reduced to 1/5 of its uncoated value, providing a substantial reduction in *polarization aberration*. Typically, a very small amount of *retardance* is also introduced.

For a multi-element lens, the *diattenuation* and *retardance* contributions accumulate. Both positive and negative lenses introduce diattenuation of the same sign. Consider the pair of *microscope objectives* in Figure 1.37 where collimated light enters the first objective, comes to a focus between the two objectives, and is collimated by the second objective. This pair of low polarization microscope objectives has a *numerical aperture* of 0.55. Figure 1.38 shows the measured polarization

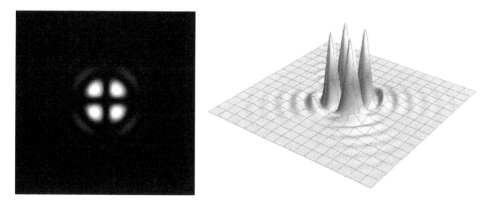

Figure 1.36 The PSF of an uncoated lens between crossed (*x*- and *y*-oriented) polarizers in (left) an intensity plot and (right) an oblique perspective plot.

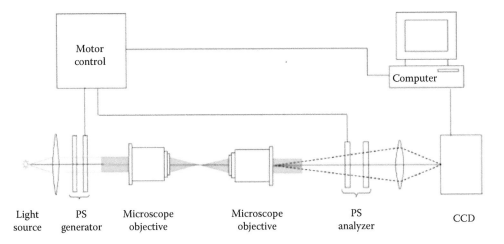

Figure 1.37 The optical layout for a polarization aberration measurement in an imaging polarimeter where collimated light enters one microscope objective, comes to focus, and exits the second objective collimated.

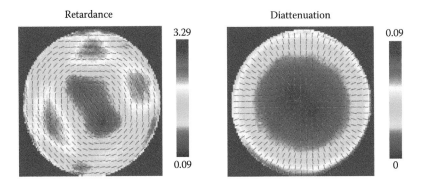

Figure 1.38 Polarization aberration measurement of a pair of microscope objectives for an on-axis beam shows the (left) *diattenuation aberration* distribution, which reaches a maximum of 0.09, and the (right) *retardance aberration* distribution, which reaches a maximum of 3° or 0.09 waves. Both aberrations have small polarization aberration near the center of the pupil where angles of incidence are small. The measurement is performed with a *Mueller matrix polarimeter*.

aberrations of the low polarization microscope objective pair. The diattenuation, which here has a larger effect than the retardance, reaches 0.09 at the edge of the pupil. Even at this level of low polarization aberration, an almost 10% polarizer, when the first lens is illuminated with collimated linearly polarized light, and the exiting light is blocked with an orthogonal polarizer, the pupil-averaged leakage has a Maltese cross pattern.

The presence of *retardance* in an optical system indicates that the system has polarization-dependent *optical path lengths* and thus will have different *interferograms* in different polarization states. The retardance polarization aberration pattern of spherical, parabolic, ellipsoidal, and hyperbolic mirrors with metal coatings seen on-axis and of associated optical systems like Cassegrain telescopes has a tangentially oriented fast axis with a retardance magnitude that increases from the center of the pupil. Figure 1.39 (left) shows the form of retardance aberration of on-axis spherical and conic mirrors. The *phase* advances for polarization states parallel to the lines and decreases for polarization states perpendicular to the lines. Thus, for a spherical wavefront with 90° linearly polarized light, the wavefront becomes deformed like Figure 1.39 (right). For on-axis sources, because of the *retardance*, these *metal mirrors, Cassegrain telescopes*, and similar optics have a different quadratic

 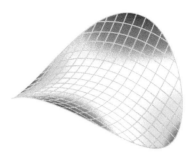

Figure 1.39 (Left) Retardance aberration of spherical or parabolic mirrors illuminated on-axis; lines indicate the orientation of the retardance fast axis and magnitude of the retardance over the pupil. (Right) An astigmatic wavefront results when a spherical wavefront of 90° linearly polarized light interacts with the retardance aberration (left). White indicates the optical path length of the chief ray. Violet shows shorter optical path lengths and green shows longer optical path lengths.

phase variation along the two axes; the mirrors have introduced *astigmatism into the on-axis beam*, something that is not calculated by *conventional ray tracing*! This astigmatism is oriented with the *plane of polarization* of the incident light; when the polarization state is rotated, the astigmatic wavefront aberration rotates with it. Fortunately, common optical systems like Cassegrain telescopes usually have less than a tenth of a wave of this metal coating-induced astigmatism and so this source of astigmatism is not a high priority to optical designers. Nonetheless, this source of astigmatism should be understood because it is easily seen in interferometric tests and begs for an explanation when it appears.

1.7.3 High Numerical Aperture Wavefronts

Numerical aperture characterizes the range of angles over which an optical system can accept or transmit light; F-number or F/# describes the same property. Beams with high numerical aperture, a large cone angle, are valuable in optics because they can focus light into smaller images. Hence, there is a constant push for systems with still higher and higher numerical aperture.

High numerical aperture beams must have polarization variations, because the polarization state, which is transverse to the wavefront, cannot remain uniform in three dimensions; it must curve around the sphere. These intrinsic high numerical aperture polarization state variations are frequently detrimental, broadening the image from the ideal *diffraction-limited* patterns.

Consider a hemispherical light beam, which corresponds to a numerical aperture of one, which subtends a solid angle of 2π steradians. For example, when x-polarized light is incident on such a high numerical aperture lens, the exiting polarization is of the form shown in Figure 1.40. Near the z-axis, the *optical axis* in the center of the beam, the polarization is nearly uniformly polarized. Along the y-axis, the light can remain polarized in the x-direction all the way to the edge of the pupil, since these vectors along x are tangent to the sphere. Along the x-axis, the light must tip upward, with a negative z-component, and downward, with a positive z-component, to remain on the surface of the sphere. The polarization state of the light continues to rotate until, at the right and left sides of the pupil in Figure 1.40 (left), the light becomes polarized in the $\pm z$-direction, since here the light is propagating in the $\pm x$-direction. Around the edge of the pupil, the polarization varies as shown in Figure 1.41. The result of this polarization variation is a PSF that becomes elongated in one direction, much like astigmatism. Figure 1.40 shows only one way that the polarization might vary in a high numerical aperture beam.

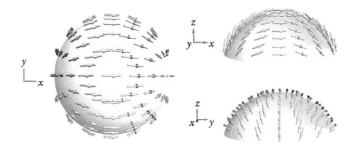

Figure 1.40 A high numerical aperture (NA) spherical wave linearly polarized in the *x*-direction (left) viewed along the *z*-axis, (center) viewed along the *y*-axis, and (right) viewed along the *x*-axis. This polarization is aligned with the double pole coordinates of Section 11.4 with a double pole located on the -z-axis.

Figure 1.41 The distribution of linearly polarized light around the edge of the hemispherical wavefront of Figure 1.40.

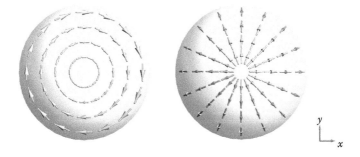

Figure 1.42 The polarization distributions in (left) a tangentially polarized wavefront and (right) a radially polarized wavefront.

There is great interest in *microlithography* and *microscopy* for other polarization distributions with useful imaging properties, particularly the radial and tangentially polarized beams shown in Figure 1.42. Note that these light states cannot be extended to the origin without discontinuity and thus are created with a dark spot in the center.

1.8 Comment on Historical Treatments

The transverse nature of light waves and the properties of polarization have played a central role in the development of optics and physics. A very nice summary of the history of polarized light is contained in Goldstein's *Polarized Light*[4] before Chapter 1. Another summary is Brosseau's *Fundamentals of Polarized Light, A Statistical Optics Approach* in Part 1 ("Historical Survey of Understanding of Polarized Light"). The understanding of polarized light and diffraction advanced rapidly from 1800 through 1830, when consensus finally emerged that light was a *transverse wave*. A compelling account of the scientific controversies during this formative period is found in Buchwald's book, *The Rise of the Wave Theory of Light*.[4] An 1842 book, *Lectures on Polarized Light* by Pereira, available online, documents the sophisticated understanding of polarized light in the first half of the nineteenth century.[5]

Polarized light optics took a giant step forward with the invention of *Polaroid* plastic sheet polarizer. Before the invention of Polaroid, polarizers tended to be small and expensive, such as the *Nichol prism*. The availability of large inexpensive sheet polarizers and retarders helped propel a rapid advance in polarization optics and related fields. The history of *dichroic polarizers* is described by Land and West[6] and by Grabau.[7]

1.9 Reference Books on Polarized Light

The following is a short list of books on polarized light or texts with significant polarized light sections that the authors feel students would find helpful. Goldstein contains a thorough treatment of polarization mathematics, discussion of Fresnel equations, ellipsometry, and many other topics. Können is a good starting point for introductory users such as high school and undergraduate students with a nonmathematical discussion of polarized light. A free version of Können[8] is available online. Mansuripur provides a nonmathematical discussion of many polarization effects at the graduate level.[9]

1. Shurcliff, W. A., *Polarized Light. Production and Use*, Cambridge, MA: Harvard University Press, 1966.
2. Azzam, R. M. A. and Bashara, N. M., *Ellipsometry and Polarized Light*, 2nd edition Amsterdam: Elsevier, 1987.
3. Kliger, D. S. and Lewis, J. W., *Polarized Light in Optics and Spectroscopy*, Elsevier, 1990.
4. Können, G. P., *Polarized Light in Nature*, CUP Archive, 1985.
5. Hecht, E., *Optics*, 4th edition, Addison Wesley Longman, 1998.
6. Brosseau, C., *Fundamentals of Polarized Light: A Statistical Optics Approach*, New York: Wiley, 1998.
7. Born, M. and Wolf, E., *Principles of Optics: Electromagnetic Theory of Propagation, Interference and Diffraction of Light*, 7th expanded edition, Cambridge University Press, 1999.
8. Mansuripur, M., *Classical Optics and Its Applications*, 2nd edition, Cambridge University Press, 2009.
9. Collette, E., *Field Guide to Polarization*, SPIE Press, 2006.
10. Cloude, S., *Polarisation: Applications in Remote Sensing*, Oxford University Press, 2009.
11. Goldstein, D., *Polarized Light,* Revised and expanded 3rd edition, Vol. 83, Boca Raton, FL: CRC Press, 2011.
12. Horváth, G., Polarized Light and Polarization Vision in Animal Sciences, *Springer Series in Vision Research* (2), 2014.

1.10 Problem Sets

1.1 Why do monochromatic waves have periodic electric fields? Why must the electric field of a monochromatic plane wave trace an ellipse in the transverse plane?

1.2 Draw the polarization ellipse for linearly polarized light oriented at 30°. Plot $E_x(t)$ and $E_y(t)$, the x- and y-components in which plane does the magnetic field oscillate?

1.3 In Figure 1.34, how does the major axis of the polarization ellipses vary moving around the edge of the pupil for 90° incident polarization? For 45° polarization? For left circular polarization?

1.4 Using a sheet polarizer, rotate a pair of polarizing sunglasses and estimate the alignment of the transmission axis for the left and right lenses. Is the axis horizontal or vertical? Is the axis exactly the same or can a small difference be observed?

1.5 Illuminate a projection screen with a liquid crystal projector with polarized output.
   a. How is the light from the projector polarized in the red, green, and blue bands?
   b. Examine the light scattered from the screen with a rotating polarizer. Project a red, green, and then blue scene and visually estimate the degree of linear polarization of the scattered light.
   c. Does the polarization depend on the angle of scatter?

1.6 Study the polarization properties of an artificial rainbow with a linear polarizer. Face away from the sun. Create a mist of water with a hose, lawn sprinkler, or mister. Working from a balcony or place where you can look downward into the mist against a dark background provides the best results. View the rainbow through the polarizer. Produce a diagram of the polarization of the rainbow across the arc, showing the orientation of the linear polarization. The degree of polarization is very difficult to estimate visually but is probably near 90%. Does the degree of linear polarization appear to vary for the different colors?

1.7 For a lens spherical surface illuminated from infinity with a collimated beam on-axis, does the angle of incidence increase approximately linearly or quadratically from the center?

1.8 Consider the linearly polarized hemispherical wavefront in Figure 1.40.
   a. In which plane through the origin are all the **E**-fields pointing in the same direction?
   b. At which opposite points in the pupil are the electric fields in opposite directions?
   c. How does the electric field rotate as one moves around the edge of this hemispherical wavefront?

1.9 Where in Figure 1.22 is the stress the greatest?

1.10 A particular diattenuator, or partial polarizer, has a maximum transmittance $T_{max} = 0.7$ and a diattenuation $D = 0.999$. Find T_{min}.

1.11 Contrast ratio is defined as $C = \dfrac{T_{max}}{T_{min}}$. Find an expression for the diattenuation as a function of the contrast ratio. When the diattenuation is 0.999, what is the contrast ratio?

1.12 At 600 nm, a particular quartz plate has three waves of retardance and a MgF_2 plate has four waves or retardance, yielding a one-wave retarder. Referring to Figure 1.19, at what wavelength will the combination be a 3/4 wave retarder? A 5/4 wave retarder?

1.13 Take two sheet polarizers and cross their transmission axes. Take pictures as they rotate about the horizontal axis, vertical axis, and diagonal axes. Describe how the leakage, the lack of extinction varies.

1.14 Consider light passing through two crossed polarizers aligned along $\mathbf{x} = (1, 0, 0)$ and $\mathbf{y} = (0, 1, 0)$ propagating along the direction $\mathbf{k} = (\sin\phi\cos\theta, \sin\phi\sin\theta, \cos\phi)$. By projecting the two polarizers onto the transverse plane, find the apparent angle χ between their absorption axes. The transmission through the polarizer pair is $T = \cos^2\chi$. Perform a Taylor series of T in the angle of incidence θ along the diagonal $\phi = 45°$ to determine the lowest-order polynomial variation in transmittance.

1.15 Malus's law states that when a perfect linear polarizer is placed in a linearly polarized beam, the fraction of light transmitted (F) is given by: $F = \cos^2\theta$, where θ is the angle between the incident polarization and the transmission axis of the polarizer. Consider N polarizers placed one after another (a cascade) in a beam polarized at 0°. The first polarizer is oriented horizontally (0°), and each subsequent polarizer is rotated a fixed amount relative to the previous polarizer such that the last polarizer is always vertically oriented (90°). For example, if $N = 4$, the orientations of the polarizers would be (0°, 30°, 60°, 90°). Using Malus's law, what is the fraction of light transmitted through 2, 4, and 8 polarizers arranged in this fashion? How does the transmission behave as N increases? (If light passes through a linear polarizer at β, the exiting light is linearly polarized at β with an attenuated magnitude.) If the incident light is linearly polarized at α, through the linear polarizer at β, the exiting light is linearly polarized at β with an attenuated magnitude $\cos^2(\alpha - \beta)$.

References

1. S. Johnsen and T. Frank, Polarization Vision, Operation Deep Scope (2005). S. Johnsen using images from E. Widder, NOAA (http://oceanexplorer.noaa.gov/explorations/05deepscope/background/polarization/media/eel.html, accessed on July 15, 2017).
2. R. M. A. Azzam and N. M. Bashara, *Ellipsometry and Polarized Light*, Elsevier Science (1987).
3. I. Ohlidal and D. Franta, Ellipsometry of thin film systems, in *Progress in Optics*, Vol. 41, ed. E. Wolf, Elsevier (2000), pp. 181–282.
4. J. Z. Buchwald, *The Rise of the Wave Theory of Light, Optical Theory and Experiment in the Early Nineteenth Century*, The University of Chicago Press (1989).
5. J. Pereira, On the polarization of light, and its useful applications, *Pharm. J* 2 (1842): 619–637. (https://play.google.com/store/books/details?id=OylbAAAAcAAJ&rdid=book-OylbAAAAcAAJ&rdot=1, accessed October 25, 2016.)
6. E. H. Land and C. D. West, Dichroism and dichroic polarizers, *Colloid Chemistry* 6 (1946): 160–190.
7. M. Grabau, Polarized light enters the world of everyday life, *Journal of Applied Physics* 9.4 (1938): 215–225.
8. G. P. Können, Polarized light in nature, CUP Archive (1985). (http://s3.amazonaws.com/guntherkonnen/documents/249/1985_Pol_Light_in_Nature_book.pdf?1317929665, accessed October 25, 2016.)
9. M. Mansuripur, *Classical Optics and Its Applications*, Cambridge: Cambridge University Press (2002).

2

Polarized Light

2.1 The Description of Polarized Light

This chapter starts with the description of monochromatic plane waves propagating in arbitrary directions and the description of their electric and magnetic fields. Table 2.1 lists several common and useful methods for the description of polarized light.[1–3] In this chapter, first the *polarization vector* is used to describe the polarization state of plane waves. Then, *Jones vectors*, which describe monochromatic plane waves propagating along the z-axis, are treated in detail, analyzing the polarization state and ellipse, phase, polarized flux components, and orthogonality of plane waves. Such a monochromatic wave can be generated with a laser.* Finally, considering the polarization of sources, two models for the polarization of light beams entering optical systems are defined, the dipole spherical wave and the double pole spherical wave. Chapter 3 continues with the description of incoherent light and polychromatic light, light that is not monochromatic, using the Stokes parameters and the Poincaré sphere, a graphical representation of the polarization state.

This book principally uses Jones vectors for plane waves and polarization vectors (three-dimensional electric field vectors) on spherical (or nearly spherical) wavefronts for describing polarized light and uses Stokes parameters where they have advantages.

---

* Laser light is almost monochromatic but always has a small wavelength spread.

Table 2.1 Common Methods for Polarized Light Calculations

Light Representation	Properties	Polarization Element Representation
Jones vector	Monochromatic plane wave along z-axis Two complex elements	*Jones matrix*
Polarization vector	Monochromatic plane wave in arbitrary direction Three complex elements	*Polarization ray tracing matrix*
Stokes parameters	Incoherent light along z-axis Four real elements	*Mueller matrix*

2.2 The Polarization Vector

Light is a transverse electromagnetic wave, an oscillating electric and magnetic field propagating in vacuum at the speed of light.[4–6] The simplest light wave is the monochromatic plane wave with a flat wavefront, representing a collimated beam of light, as shown in Figure 2.1 for a linearly polarized electromagnetic transverse wave. Consider a monochromatic plane wave with wavelength λ, propagating in the direction of unit propagation vector $\mathbf{k}$, with an angular frequency ω in radians per second.* The electric field $\mathbf{E}(\mathbf{r}, t)$ of the monochromatic plane wave in space, $\mathbf{r}$, and time, t, is

$$\mathbf{E}(\mathbf{r},t) = \mathcal{R}e\left[\mathbf{E}\, e^{i\left(\frac{2\pi}{\lambda}\mathbf{k}\cdot\mathbf{r}-\omega t-\phi_o\right)}\right] = \mathcal{R}e\left[\begin{pmatrix} E_x \\ E_y \\ E_z \end{pmatrix} e^{i\left(\frac{2\pi}{\lambda}\mathbf{k}\cdot\mathbf{r}-\omega t-\phi_o\right)}\right] = \mathcal{R}e\left[e^{i\left(\frac{2\pi}{\lambda}\mathbf{k}\cdot\mathbf{r}-\omega t-\phi_o\right)}\begin{pmatrix} A_x e^{-i\phi_x} \\ A_y e^{-i\phi_y} \\ A_z e^{-i\phi_x} \end{pmatrix}\right]. \tag{2.1}$$

The polarization vector $\mathbf{E}$ describes the polarization state. $\mathbf{E}$ is the complex amplitude with units of volts per meter. The three field components of $\mathbf{E}$ are shown in three different forms, from left to right in Equation 2.1, where $(E_x\ E_y\ E_z)$ are the complex amplitude components along the three coordinate axes x, y, and z. A_x, A_y, and A_z are the magnitudes of the complex amplitudes and ϕ_x, ϕ_y, and ϕ_z are their phases. The surfaces of constant phase fronts are shown as blue planes separated by λ in Figure 2.1.

$\mathbf{E}$ is an electric field vector, but since it fully characterizes the polarization ellipse in three dimensions, $\mathbf{E}$ is also called the *polarization vector*. The *propagation vector*[7] $\mathbf{k}$

$$\mathbf{k} = (k_x, k_y, k_z), \quad |\mathbf{k}| = 1, \tag{2.2}$$

is normalized and describes the direction of propagation.† $\mathbf{k}$ points along the light rays associated with the wavefront. The three components (k_x, k_y, k_z) are the *direction cosines*, the cosines of the angles between $\mathbf{k}$ and the x-, y-, and z-axes. The vector $\mathbf{r}$ identifies position in a global coordinate system,

$$\mathbf{r} = (r_x, r_y, r_z). \tag{2.3}$$

---
* ω is 2π times the frequency in hertz.
† In most references, the magnitude of the wavevector $\mathbf{k}$ is $2\pi/\lambda$. In this book, $\mathbf{k}$ is defined as the normalized wavevector; thus, the wavevector is $2\pi/\lambda\, \mathbf{k}$ and $|\mathbf{k}| = 1$.

2.2 The Polarization Vector

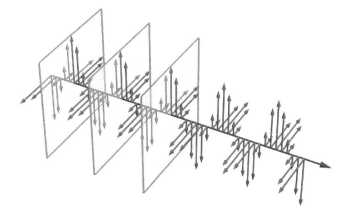

Figure 2.1 The electric field vectors (red) and magnetic field vectors (blue) for a linearly polarized electromagnetic wave. Three wavefronts, surfaces of constant phase are shown, separated by one wavelength.

At the origin of the chosen coordinate system, $\mathbf{r} = (0, 0, 0)$ and $t = 0$, the **E**-field is

$$\mathbf{E} = \begin{pmatrix} E_x \\ E_y \\ E_z \end{pmatrix}. \tag{2.4}$$

For propagation in isotropic media, such as vacuum, air, water, and glass, the polarization vector is orthogonal to the propagation vector, so their dot product is zero,

$$\mathbf{E} \cdot \mathbf{k} = \begin{pmatrix} E_x \\ E_y \\ E_z \end{pmatrix} \cdot \begin{pmatrix} k_x \\ k_y \\ k_z \end{pmatrix} = 0. \tag{2.5}$$

The term $\frac{2\pi}{\lambda} \mathbf{k} \cdot \mathbf{r} - \omega t - \phi_o$ is the *phase* of the plane wave, which is specified for all space and time, and is usually expressed in radians, sometimes in degrees. Phase specifies a location within a periodic phenomenon. ϕ_o is the *absolute phase* of the light, which is the phase of the light at the origin ($\mathbf{r} = (0, 0, 0)$ and $t = 0$). Phase can also be expressed in *waves*, Φ,

$$\Phi = \frac{\mathbf{k} \cdot \mathbf{r}}{\lambda} - \frac{\omega t + \phi_o}{2\pi}. \tag{2.6}$$

Φ is constant on a surface of constant phase, a plane wavefront in this case. The phase can be separated into an integer part and a fractional part. The integer part of Φ provides a method to number wavefronts, such as wavefront $-2, -1, 0, 1, 2, \ldots$. The fractional part of Φ identifies where within a cosinusoidal wave a point in $\mathbf{r}$ and t is located, or for polarized light, where on the polarization ellipse the electric field is located. The phases of the individual x, y, and z components of the field in Equation 2.1 are $\phi_x + \phi_o$, $\phi_y + \phi_o$, and $\phi_z + \phi_o$.

2.3 Properties of the Polarization Vector

E is a *complex vector*, so each complex number can be expressed in polar coordinate form,

$$\mathbf{E} = \begin{pmatrix} E_x \\ E_y \\ E_z \end{pmatrix} = \begin{pmatrix} A_x e^{-i\phi_x} \\ A_y e^{-i\phi_y} \\ A_z e^{-i\phi_z} \end{pmatrix}. \qquad (2.7)$$

A_x, A_y, and A_z are the real amplitudes, the maximum value each component reaches during a period. The phases ϕ_x, ϕ_y, and ϕ_z are provided with minus signs due to the decreasing phase sign convention, $\left(\frac{2\pi}{\lambda}\mathbf{k}\cdot\mathbf{r} - \omega t\right)$, see Section 2.17. Multiplying **E** by the temporal phase $e^{-i\omega t}$ and taking the real part at $\mathbf{r} = (0, 0, 0)$ yields $\mathbf{E}(t)$, the *polarization ellipse*, oriented in three dimensions,

$$\mathbf{E}(t) = \mathcal{R}e\left[e^{-i\omega t} \begin{pmatrix} E_x \\ E_y \\ E_z \end{pmatrix} \right] = \mathcal{R}e\left[e^{-i\omega t} \begin{pmatrix} A_x e^{-i\phi_x} \\ A_y e^{-i\phi_y} \\ A_z e^{-i\phi_z} \end{pmatrix} \right]. \qquad (2.8)$$

Figure 2.2 shows the three-dimensional polarization ellipse for left circularly polarized light.[1] The symbol **E** will be used for both Jones vectors and polarization vectors, since both are electric field amplitudes. The electric and magnetic vectors must oscillate only in the plane perpendicular to the propagation vector **k**, the *transverse plane*, shown in Figure 2.3.

For light to be *linearly polarized*, all three **E**-field components should pass through zero at the same time, twice per period. This requires that the three phases in Equation 2.8 differ from each other by either 0 or π,

$$\phi_x - \phi_y = 0 \text{ or } \pi, \quad \phi_x - \phi_z = 0 \text{ or } \pm \pi. \qquad (2.9)$$

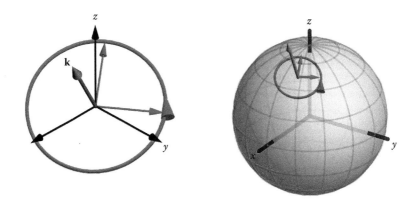

Figure 2.2 (Left) A left circularly polarized polarization ellipse (red) shown in 3D with propagation vector k = (6, 2, 9)/11 (black). (Right) Ellipse drawn on the unit sphere with **k** emerging normal to the sphere. Two basis vectors defining the transverse plane are shown in blue.

2.3 Properties of the Polarization Vector

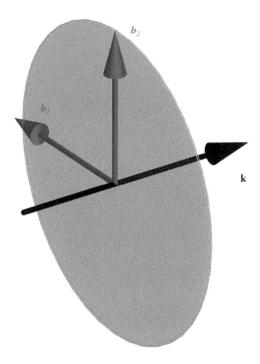

Figure 2.3 With respect to a propagation vector **k**, the transverse plane contains the two orthogonal directions perpendicular to **k** with basis vectors b_1 and b_2, which can be chosen in arbitrary directions.

For *circularly polarized light*, the magnitude of the electric field vector is constant. This can be tested by evaluating the magnitude of the real part of **E** at several times, for example, at $t = 0$ and a quarter period later ($t = \pi/(2\omega)$) when the imaginary part of **E** yields a real E-field. Hence, **E** describes circularly polarized light when these two vectors $\mathbf{E}(0)$ and $\mathbf{E}(\pi/2)$ are orthogonal and equal in magnitude,

$$\mathcal{R}e(\mathbf{E}) \cdot \mathcal{J}m(\mathbf{E}) = 0 \text{ and } |\mathcal{R}e(\mathbf{E})| = |\mathcal{J}m(\mathbf{E})|. \tag{2.10}$$

If the light is not linearly or circularly polarized, it is *elliptically polarized light*. To determine the *helicity* of the polarization state, right or left, one can evaluate the cross product of the electric field vector at time $t = 0$ and $t = \pi/(2\omega) = T/4$. If the cross product is antiparallel to the propagation vector, the light's helicity is *right handed*. If the cross product is parallel to the propagation vector, the light's helicity is *left handed*. Thus, the helicity (handedness) is determined by the sign of

$$[\mathbf{E}(0) \times \mathbf{E}(\pi/(2\omega))] \cdot \mathbf{k}. \tag{2.11}$$

If Equation 2.11 is positive, the electric field is rotating counterclockwise (i.e., left circular), and if it is negative, the electric field is rotating clockwise (i.e., right circular). To determine the ellipticity and orientation of the major axis, consult Section 2.16.

Math Tip 2.1 Adjoint of a Vector

The dagger superscript† indicates the *adjoint* of a vector, the complex conjugate of the transpose. For example, the adjoint of $\mathbf{E} = \begin{pmatrix} A_x e^{-i\phi_x} & A_y e^{-i\phi_y} & A_z e^{-i\phi_z} \end{pmatrix}$ is

$$\mathbf{E}^\dagger = \begin{pmatrix} A_x e^{i\phi_x} & A_y e^{i\phi_y} & A_z e^{i\phi_z} \end{pmatrix}. \tag{2.12}$$

The adjoint provides a shortcut to get the magnitude squared of a complex vector $|\mathbf{E}|^2$,

$$|\mathbf{E}|^2 = \mathbf{E}^\dagger \cdot \mathbf{E} = \begin{pmatrix} A_x e^{i\phi_x} & A_y e^{i\phi_y} & A_z e^{i\phi_z} \end{pmatrix} \cdot \begin{pmatrix} A_x e^{-i\phi_x} \\ A_y e^{-i\phi_y} \\ A_z e^{-i\phi_z} \end{pmatrix} = A_x^2 + A_y^2 + A_z^2. \tag{2.13}$$

2.4 Propagation in Isotropic Media

The description of $\mathbf{E}$ above assumed propagation in vacuum. In glass, water, and other media, the velocity of light, V, is reduced and the medium is characterized by its refractive index n,

$$n = \frac{c}{V}. \tag{2.14}$$

The light is slowed down because as its $\mathbf{E}$- and $\mathbf{H}$-fields propagate through transparent materials, the fields induce motion in the atom's charges at ω, and this charge motion gives rise to light in the same direction at the same frequency but slightly delayed. Thus, the refractive index characterizes the strength of the interaction of light of a particular frequency with a material. A material is isotropic if the strength of this interaction is independent of the direction of $\mathbf{E}$ and $\mathbf{H}$; that is, n is the same for all propagation directions and polarization states. Anisotropic materials such as calcite and quartz are discussed in Chapter 19 (see discussion on anisotropic materials).

2.5 Magnetic Field, Flux, and Polarized Flux

Light is a transverse electromagnetic wave. The light's *magnetic field* $\mathbf{H}(\mathbf{r}, t)$ oscillates perpendicular to and in phase with the electric field. For light propagating in vacuum, the associated magnetic field is

$$\mathbf{H}(\mathbf{r},t) = \mathcal{R}e\left[\mathbf{H} e^{i\left(\frac{2\pi}{\lambda}\mathbf{k}\cdot\mathbf{r} - \omega t - \phi_o\right)}\right] = \mathcal{R}e\left[\begin{pmatrix} H_x \\ H_y \\ H_z \end{pmatrix} e^{i\left(\frac{2\pi}{\lambda}\mathbf{k}\cdot\mathbf{r} - \omega t - \phi_o\right)}\right], \tag{2.15}$$

which is specified in units of amperes per meter. Figure 2.4 shows the electric and magnetic fields as a function of time for several polarization states.

2.6 Jones Vectors

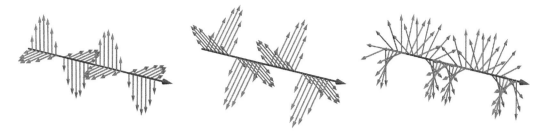

Figure 2.4 Electric (red) and magnetic (blue) fields rendered in three dimensions for three polarization states. (Left) Vertical linearly polarized light. (Middle) 45° linearly polarized light. (Right) Right circularly polarized light.

The *Poynting vector* **S** describes the instantaneous flow of energy of an electromagnetic wave in watts per meter squared,[2]

$$\mathbf{S}(\mathbf{r},t) = \mathbf{E}(\mathbf{r},t) \times \mathbf{H}(\mathbf{r},t). \tag{2.16}$$

The Poynting vector has the same units as irradiance, energy per area per second, or kilogram per second.[3,8] For linearly polarized light, the Poynting vector oscillates twice per period; for circularly polarized light, the Poynting vector is constant. Optical detectors make measurements that average over many periods of the light, because detectors cannot respond at optical frequencies of hundreds of terahertz. Thus, optical detectors measure a time-averaged flow of energy, called the *flux* or *irradiance P*. The *flux* of a beam of light is the time average of the energy crossing a unit area perpendicular to the direction of energy flow per unit time. The flux is often referred to as the *intensity*, but in radiometry, intensity specifically refers to the watts per steradian from a point source. Nonetheless, this use of the term intensity is widespread and well understood, even if it doesn't follow official definitions.[9,10] The irradiance of a light beam, P, is the power per unit area transported by the light's electromagnetic fields and is measured in watts per meter squared. The flux P of our monochromatic plane wave Equation 2.1 is

$$P = \frac{\varepsilon_0 c}{2} \mathbf{E}^\dagger \cdot \mathbf{E} = \frac{\varepsilon_0 c}{2} \begin{pmatrix} A_x e^{i\phi_x} & A_y e^{i\phi_y} & A_z e^{i\phi_z} \end{pmatrix} \cdot \begin{pmatrix} A_x e^{-i\phi_x} \\ A_y e^{-i\phi_y} \\ A_z e^{-i\phi_z} \end{pmatrix} = \frac{\varepsilon_0 c}{2} \left(A_x^2 + A_y^2 + A_z^2 \right). \tag{2.17}$$

Frequently, the constant $\varepsilon_0 c/2$ is dropped and calculations such as Jones vector problem sets are worked in "normalized flux" $\mathbf{E}^\dagger \cdot \mathbf{E}$. The $\varepsilon_0 c/2$ is only necessary to work in MKS units and produces answers in radiometric units.

2.6 Jones Vectors

In many polarization problems, light propagation is often restricted to a single direction. When light propagates through a sequence of sheet polarizers and sheet retarders, the light propagates in a single direction. *Our convention* is to propagate light along the z-axis, from negative toward positive. Such light is frequently modeled as a plane wave with a propagation vector

$$\mathbf{k} = (0,0,1). \tag{2.18}$$

Such light has no **E**-field component along the direction of propagation, since **E** always oscillates in the *transverse plane* of **k** for all isotropic media. In this configuration, the light's z electric field component E_z is zero, so it can be dropped and the monochromatic plane wave's description is thus simplified to only the x and y electric field components. The two-element vector is the *Jones vector*.[11,12] The Jones vector **E** has two complex elements and represents the polarization state of this z-propagating monochromatic light field as

$$\mathbf{E} = \begin{pmatrix} E_x \\ E_y \end{pmatrix} = \begin{pmatrix} A_x e^{-i\phi_x} \\ A_y e^{-i\phi_y} \end{pmatrix}. \tag{2.19}$$

The Jones vector has four degrees of freedom that define the polarization state of the wave, two arbitrary electric field amplitudes, A_x and A_y, and two arbitrary phases, ϕ_x and ϕ_y. The units of the Jones vector are the same as the units of the electric field, volts per meter. A *normalized Jones vector* $\hat{\mathbf{E}}$, indicated with a caret above, has been scaled to a magnitude of one,

$$|\hat{\mathbf{E}}| = \mathbf{E}^\dagger \cdot \mathbf{E} = A_x^2 + A_y^2 = 1. \tag{2.20}$$

Here, "normalized" polarized fluxes mean the flux is one. Table 2.2 lists the normalized Jones vectors for six polarization states. These six states are considered as our *basis polarization states*, those that most frequently occur in defining polarization element properties.

By definition, monochromatic light is periodic with a single frequency. Because of this periodicity, its electric field vector traces a simple figure, the *polarization ellipse*, the most iconic representation of polarized light. Looking into the beam, the polarization ellipse is defined by the tip of the electric field vector, which traces an ellipse as a function of time. To generate the polarization ellipse, the Jones vector is multiplied by the temporal phase $e^{-i\omega t}$ and the time is varied over one period, yielding

$$\mathbf{E}(t) = \mathcal{R}e\left[e^{-i\omega t} \begin{pmatrix} A_x e^{-i\phi_x} \\ A_y e^{-i\phi_y} \end{pmatrix} \right], \tag{2.21}$$

as shown in Figure 2.5. Overall, the electric field exerts a much larger force than the magnetic field in most light–matter interactions; thus, *the electric field, not the magnetic field, is used to define the direction of polarization and the polarization ellipse.*[1] This is the essence of polarization,

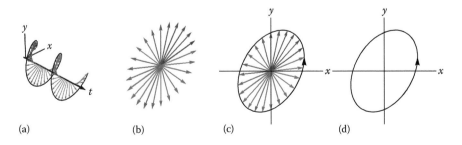

Figure 2.5 (a) The electric field vectors for an elliptical polarization state over one period. (b) The electric field vector as a function of time during one period. (c) The electric field vectors are shown with their bounding ellipse as a function of time. (d) Usually, just the ellipse is drawn to indicate the polarization state. This is the polarization ellipse. The location of the arrow indicates the absolute phase of the wave. This polarization ellipse has a left circularly polarized component because it circulates counterclockwise when looking into the beam.

2.6 Jones Vectors

Table 2.2 Normalized Jones Vectors for the Six Basis Polarization States

Polarization State	Jones Vector
Horizontal, 0°	$\mathbf{H} = \begin{pmatrix} 1 \\ 0 \end{pmatrix}$
Vertical, 90°	$\mathbf{V} = \begin{pmatrix} 0 \\ 1 \end{pmatrix}$
45°	$\mathbf{45} = \dfrac{1}{\sqrt{2}} \begin{pmatrix} 1 \\ 1 \end{pmatrix}$
135°	$\mathbf{135} = \dfrac{1}{\sqrt{2}} \begin{pmatrix} 1 \\ -1 \end{pmatrix}$
Right circular	$\mathbf{R} = \dfrac{1}{\sqrt{2}} \begin{pmatrix} 1 \\ -i \end{pmatrix}$
Left circular	$\mathbf{L} = \dfrac{1}{\sqrt{2}} \begin{pmatrix} 1 \\ i \end{pmatrix}$

The orientation is measured from horizontal *x*-axis and increasing counterclockwise.

the transverse nature of **E**. One important task of the polarization calculi, the Jones calculus, the Mueller calculus, and the polarization ray tracing calculus, is to describe these transverse properties and the associated polarization state transformations in many different types of optical systems.

The two complex Jones vector components, E_x and E_y, can be expressed in polar coordinate form with an amplitude A and phase ϕ, or (on the right) with the x-phase ϕ_x factored out,

$$\mathbf{E} = \begin{pmatrix} A_x e^{-i\phi_x} \\ A_y e^{-i\phi_y} \end{pmatrix} = e^{-i\phi_x} \begin{pmatrix} A_x \\ A_y e^{-i\phi} \end{pmatrix}. \tag{2.22}$$

In this form, the *relative phase* ϕ, the phase difference between the x and y field components, is evident,

$$\phi = \phi_y - \phi_x. \tag{2.23}$$

Example 2.1 A Family of Polarization Ellipses

Figure 2.6 shows the polarization ellipses for a family of Jones vectors

$$\mathbf{E} = \begin{pmatrix} 1 \\ 0.5 e^{-i\phi} \end{pmatrix}. \tag{2.24}$$

The ellipses vary as a function of relative phase ϕ. Here the y-amplitude is set to one-half the x-amplitude.

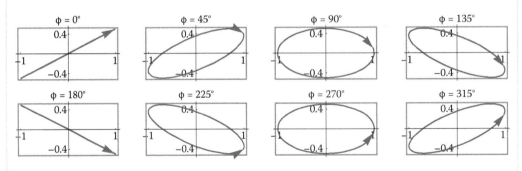

Figure 2.6 Polarization ellipses for $E_x = 1$ and $E_y = \frac{1}{2}$ for several relative phases.

The polarization ellipse fits inside a rectangular box twice the size of its amplitudes A_x and A_y. The relative phase determines how the ellipse fits in the box:

1. When $\phi = 0°$ or $180°$, the state is linear along one of the diagonals.
2. When $\phi = \pm 90°$, the ellipse touches the middle of all four sides and has the maximum ellipticity within this set.
3. For $0° < \phi < 180°$, the ellipse is right elliptically polarized.
4. For $-180° < \phi < 0°$ or $180° < \phi < 360°$, the ellipse is left elliptically polarized.

The location of the ellipse arrow indicates the absolute phase of the polarization state. In this set, E_x was chosen to equal to one at $t = 0$; hence, the arrow is at the $+x$ side. Multiplying the Jones vector by a complex phase $e^{-i\phi}$ advances the arrow to another point on the ellipse.

2.7 Evolution of Overall Phase

Normalizing a Jones vector is the operation of adjusting the amplitude; thus, the *normalized flux* P is one. For the Jones vector of Equation 2.22, the normalized Jones vector $\hat{\mathbf{E}}$ is

$$\hat{\mathbf{E}} = \frac{1}{\sqrt{A_x^2 + A_y^2}} \begin{pmatrix} A_x e^{-i\phi_x} \\ A_y e^{-i\phi_y} \end{pmatrix}, \tag{2.25}$$

so that $|\hat{\mathbf{E}}| = 1$.

Math Tip 2.2 Matrix Vector Multiplication

Matrix multiplication performs a linear transformation on a vector producing a new vector from a linear combination of the original vector's elements. Given an N-element vector $\mathbf{A}$ and an $M \times N$-element matrix $\mathbf{C}$ (M rows and N columns), the elements of the resulting matrix $\mathbf{B}$ are

$$b_j = \sum_{n=1}^{N} c_{j,n} a_n, \tag{2.26}$$

where a, b, and c are components of vector $\mathbf{A}$, matrix $\mathbf{B}$, and vector $\mathbf{C}$. For example, the general equation for 3×3 matrix vector multiplication is

$$\mathbf{C} \cdot \mathbf{A} = \mathbf{B} = \begin{pmatrix} b_1 \\ b_2 \\ b_3 \end{pmatrix} = \begin{pmatrix} c_{11} & c_{12} & c_{13} \\ c_{21} & c_{22} & c_{23} \\ c_{31} & c_{32} & c_{33} \end{pmatrix} \begin{pmatrix} a_1 \\ a_2 \\ a_3 \end{pmatrix} = \begin{pmatrix} c_{11} a_1 + c_{12} a_2 + c_{13} a_3 \\ c_{21} a_1 + c_{22} a_2 + c_{23} a_3 \\ c_{31} a_1 + c_{32} a_2 + c_{33} a_3 \end{pmatrix}. \tag{2.27}$$

2.7 Evolution of Overall Phase

The phase of a polarization state is changed by multiplying a Jones vector by $e^{-i\phi_o}$,

$$e^{-i\phi_o} \mathbf{E} = e^{-i\phi_o} \begin{pmatrix} E_x \\ E_y \end{pmatrix}. \tag{2.28}$$

This operation advances the electric field and moves the polarization ellipse's arrow but the shape of the polarization ellipse is unchanged by the change of overall phase, as shown in Figure 2.7.

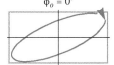
$\phi_o = 0°$

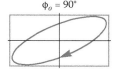
$\phi_o = 90°$

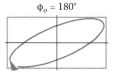
$\phi_o = 180°$

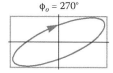
$\phi_o = 270°$

Figure 2.7 Polarization ellipses for $E_x = 1$ and $E_y = e^{-i\pi/4}/2$ for several phases ϕ_o.

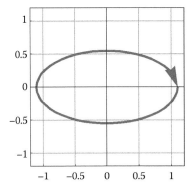 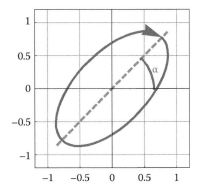

Figure 2.8 An elliptically polarized state shown on the left is rotated counterclockwise by an angle α. The state after rotation is shown on the right.

2.8 Rotation of Jones Vectors

One application of matrix multiplication is performing *rotations* on polarization states as shown in Figure 2.8. The orientation of a Jones vector is rotated by multiplying the vector by the *Cartesian rotation matrix* $\mathbf{R}(\alpha)$,

$$\mathbf{R}(\alpha) = \begin{pmatrix} \cos\alpha & -\sin\alpha \\ \sin\alpha & \cos\alpha \end{pmatrix}. \tag{2.29}$$

This rotation operation rotates the major axis of the ellipse, leaving the ellipticity the same. The phase remains the same with respect to the major axis. For the example of Figure 2.8, the rotation equation is

$$\begin{pmatrix} \cos\alpha & -\sin\alpha \\ \sin\alpha & \cos\alpha \end{pmatrix} \begin{pmatrix} 1 \\ i/2 \end{pmatrix} = \begin{pmatrix} \cos\alpha - i\sin\alpha/2 \\ \sin\alpha + i\cos\alpha/2 \end{pmatrix}. \tag{2.30}$$

2.9 Linearly Polarized Light

The electric field vector associated with *linearly polarized light* oscillates in a single direction between the positive and negative direction. The magnitude of the magnetic field oscillates along the orthogonal direction and in phase with the electric field as shown in Figure 2.9. The magnitude of the electric field goes to zero twice per period. With elliptically and circularly polarized light, the magnitude never goes to zero as is seen in Figure 2.5. These zero magnitudes require the relative phase between the *x*- and *y*-components $\phi_x - \phi_y$ be either 0° or 180°. The Jones vector for horizontal linearly polarized light (0°) is

$$\mathbf{E} = A e^{-i\phi_o} \begin{pmatrix} 1 \\ 0 \end{pmatrix}, \tag{2.31}$$

where A is the amplitude and ϕ_o is the absolute phase. ϕ_o is associated with a negative sign due to the decreasing phase sign convention. The Jones vector for *normalized* horizontal linearly polarized light (with unit amplitude and zero absolute phase) has the symbol $\mathbf{H}$,

$$\mathbf{H} = \begin{pmatrix} 1 \\ 0 \end{pmatrix}, \tag{2.32}$$

2.10 Circularly Polarized Light   43

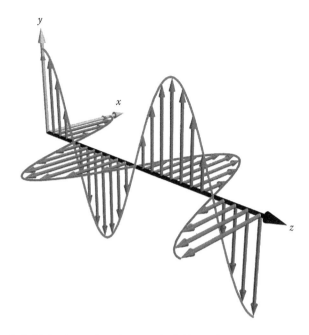

Figure 2.9 The electric field (horizontal) and magnetic field (vertical) for a linearly polarized state.

where **H** is not to be confused with the magnetic field vector. The *Jones vector for normalized linearly polarized light* **LP**(α) polarized at an angle α, measured counterclockwise from the *x*-axis, is obtained via the *Jones vector rotation operation* **R** (α) (Equation 2.29), where

$$\mathbf{LP}(\alpha) = \mathbf{R}(\alpha) \cdot \mathbf{H} = \begin{pmatrix} \cos\alpha & -\sin\alpha \\ \sin\alpha & \cos\alpha \end{pmatrix} \cdot \begin{pmatrix} 1 \\ 0 \end{pmatrix} = \begin{pmatrix} \cos\alpha \\ \sin\alpha \end{pmatrix}. \tag{2.33}$$

The Jones vector for an arbitrary beam of linearly polarized light is

$$\mathbf{E} = A e^{-i\phi_o} \begin{pmatrix} \cos\alpha \\ \sin\alpha \end{pmatrix}, \tag{2.34}$$

where A is the amplitude and ϕ_o is the absolute phase at $t = 0$.

2.10 Circularly Polarized Light

Monochromatic circularly polarized light has a constant electric field amplitude whose orientation uniformly rotates in the transverse plane. Circularly polarized light occurs in two forms, right circularly polarized and left circularly polarized, depending on the direction of rotation or *helicity* of the electric and magnetic field vectors as shown in Figure 2.6. By convention, as shown in Figure 2.10, when looking into the beam toward the negative *z*-direction through time, the electric field vector for right circularly polarized light rotates clockwise.[13] If you align the thumb of your left hand along the direction of propagation, out of the page, then the fingers of your left hand point the direction of motion of the electric field. Thus, right circularly polarized light obeys the *left hand rule*. Similarly, the electric field vector for left circularly polarized light rotates counterclockwise when looking into the beam and obeys the *right hand rule*.[1]

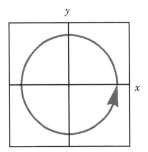
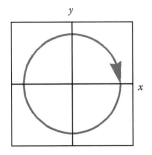

Figure 2.10 (Left) Looking into a left circularly polarized beam, the electric and magnetic fields circulate counterclockwise in time. (Right) Looking into a right circularly polarized beam, the electric and magnetic fields circulate clockwise in time.

Looking into *a right circularly polarized beam, the electric and magnetic fields rotate clockwise in time*. The right circularly polarized Jones vector **R** for a normalized beam is

$$\mathbf{R} = \frac{1}{\sqrt{2}} \begin{pmatrix} 1 \\ -i \end{pmatrix}. \tag{2.35}$$

For a *left circularly polarized beam, the electric and magnetic fields circulate counterclockwise* and the normalized Jones vector **L** is

$$\mathbf{L} = \frac{1}{\sqrt{2}} \begin{pmatrix} 1 \\ i \end{pmatrix}. \tag{2.36}$$

For each state, a *time helix* **E**(x,y,t) and a *space helix* **E**(x,y,z) can be drawn as a three-dimensional space curve. The space helix for right circularly polarized light is found by setting $t = 0$, yielding

$$\mathcal{R}e[\mathbf{E}(z,0)] = A\left[\cos\left(\frac{2\pi}{\lambda}z\right)\hat{\mathbf{x}} + \cos\left(\frac{2\pi}{\lambda}z - \frac{\pi}{2}\right)\hat{\mathbf{y}}\right] = A\left[\cos\left(\frac{2\pi}{\lambda}z\right)\hat{\mathbf{x}} + \sin\left(\frac{2\pi}{\lambda}z\right)\hat{\mathbf{y}}\right]. \tag{2.37}$$

This helix is right handed; when the fingers of the right hand curl in the direction the vector is advancing, the thumb points toward increasing z. The time helix for right circularly polarized light is found by setting $z = 0$, yielding

$$\mathcal{R}e[\mathbf{E}(0,t)] = A\left[\cos\left(\frac{2\pi}{\lambda}z\right)\hat{\mathbf{x}} + \cos\left(\frac{2\pi}{\lambda}z + \frac{\pi}{2}\right)\hat{\mathbf{y}}\right] = A\left[\cos\left(\frac{2\pi}{\lambda}z\right)\hat{\mathbf{x}} - \sin\left(\frac{2\pi}{\lambda}z\right)\hat{\mathbf{y}}\right]. \tag{2.38}$$

This *right circularly polarized time helix is left handed*. Note that *the space helix and the time helix have opposite helicity*, due to the minus sign in $\frac{2\pi}{\lambda}z - \omega t$. The left side of Figure 2.11 shows the time helix for right circularly polarized light and the right side shows the space helix. Note the opposite helicity. Thus, the terms *left and right circular polarization are named after the corresponding space helices*. For left circularly polarized light, the helicity is reversed in the two figures.

2.11 Elliptically Polarized Light

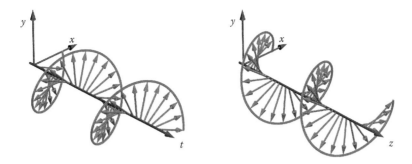

Figure 2.11 The time helix (left) and space helix (right) for right circularly polarized light rotate in opposite directions. The time helix is the space curve drawn through the ends of the electric field vectors in x, y, and t, showing that the time helix is left handed. The space helix in x, y, and z, for right circularly polarized light, is right handed.

2.11 Elliptically Polarized Light

Figure 2.12 graphs the key geometrical features of a polarization ellipse. The *ellipticity* ε of the ellipse is defined as the *length of the minor axis b divided by the length of the major axis a*,

$$\varepsilon = \frac{b}{a}. \tag{2.39}$$

The *orientation of the major axis* ψ is measured counterclockwise from the x-axis. Figure 2.13 shows a family of ellipses of increasing ellipticity.

With polarized light, the ellipses are associated with the oscillation of the electric and magnetic field vectors; both clockwise and counterclockwise ellipses occur. Therefore, the ellipticity of the polarization ellipse is generalized to positive and negative values and varies from −1 for right

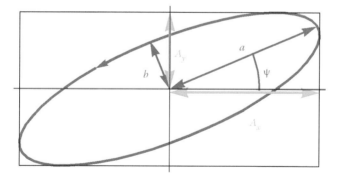

Figure 2.12 a is the length of the major axis of the polarization ellipse; b is the length of the minor axis. ψ is the orientation of the major axis.

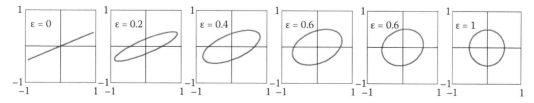

Figure 2.13 Ellipses with ellipticity of $\varepsilon = 0, 0.2, 0.4, 0.6, 0.8, 1$.

circularly polarized light to 1 for left circularly polarized light. Ellipticity of 0 describes linearly polarized light. Similarly, when $\varepsilon < 0$, the light has a right *helicity*; when $\varepsilon > 0$, it has a left helicity. For a horizontal major axis, $\psi = 0$, with major axis amplitude a along x, minor axis b along y, the normalized elliptically polarized Jones vector is[12]

$$\mathbf{E}(\varepsilon, 0) = \frac{1}{\sqrt{a^2 + b^2}} \begin{pmatrix} a \\ ib \end{pmatrix} = \frac{1}{\sqrt{1+\varepsilon^2}} \begin{pmatrix} 1 \\ -i\varepsilon \end{pmatrix}. \tag{2.40}$$

The imaginary i occurs on the y-component because the field components along the major and minor axes are always 90° out of phase for elliptically polarized light. The normalized Jones vector for an arbitrary major axis orientation is obtained by the rotation operation Equation 2.29 through an angle ψ,

$$\mathbf{E}(\varepsilon, \psi) = \frac{1}{\sqrt{1+\varepsilon^2}} \begin{pmatrix} \cos\psi + i\varepsilon\sin\psi \\ -i\varepsilon\cos\psi + \sin\psi \end{pmatrix}. \tag{2.41}$$

The inverse problem determines the polarization ellipse parameters for an arbitrary Jones vector $\mathbf{E}$. With the components of $\mathbf{E}$ specified in polar coordinate form

$$\mathbf{E} = \begin{pmatrix} A_x e^{-i\phi_x} \\ A_y e^{-i\phi_y} \end{pmatrix}, \tag{2.42}$$

the polarization ellipse must fit in a rectangle $2A_x \times 2A_y$ and be tangent to the four sides as shown in Figure 2.15. The major axis must lie along one of the two diagonals as shown in Figure 2.14. The axis lies in the first and third quadrants if the relative phase in Equation 2.23 meets the following conditions:

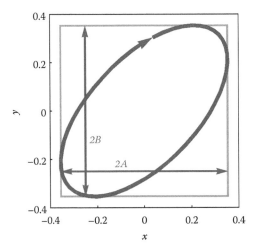

Figure 2.14 The bounding box for a polarization ellipse is a rectangle with sides equal to twice the x- and y-amplitudes.

2.11 Elliptically Polarized Light

$$-\pi/4 < \phi < \pi/4 \quad \text{or} \quad 3\pi/4 < \phi < 5\pi/4. \tag{2.43}$$

Otherwise, the major axis lies in the second and fourth quadrants.
The *major axis orientation*, ψ, is related to the Jones vector as

$$\tan(2\psi) = \frac{2 A_x A_y \cos\phi}{A_x^2 - A_y^2} \quad \text{or} \quad \psi = \frac{1}{2}\arctan\left(\frac{2 A_x A_y \cos\phi}{A_x^2 - A_y^2}\right). \tag{2.44}$$

The *semi-major axis* with a length of one-half the major axis, a, from the origin to the furthest distance, has a complicated expression,

$$a = \sqrt{A_x^2 \cos^2\psi + A_y^2 \sin^2\psi + 2 A_x A_y \cos\psi \sin\psi \cos\phi}. \tag{2.45}$$

Similarly, the *semi-minor axis* is

$$b = \sqrt{A_x^2 \sin^2\psi + A_y^2 \cos^2\psi - 2 A_x A_y \cos\psi \sin\psi \cos\phi}. \tag{2.46}$$

Hence, the equation for the *ellipticity* is quite involved,

$$\varepsilon = \frac{b}{a} = \sqrt{\frac{A_x^2 \sin^2\psi + A_y^2 \cos^2\psi - 2 A_x A_y \cos\psi \sin\psi \cos\phi}{A_x^2 \cos^2\psi + A_y^2 \sin^2\psi + 2 A_x A_y \cos\psi \sin\psi \cos\phi}}. \tag{2.47}$$

Note that a and b obey the relation

$$a^2 + b^2 = A_x^2 + A_y^2. \tag{2.48}$$

Example 2.2 Elliptically Polarized Jones Vector

The Jones vector for a normalized state with major axis orientation $\psi = \pi/4$ and ellipticity $\varepsilon = \frac{1}{2}$ shown in Figure 2.15 is

$$\mathbf{E} = \frac{1}{\sqrt{10}} \begin{pmatrix} 2+i \\ 2-i \end{pmatrix}. \tag{2.49}$$

$\mathbf{E}$ can be multiplied by any arbitrary phase $e^{-i\phi}$.

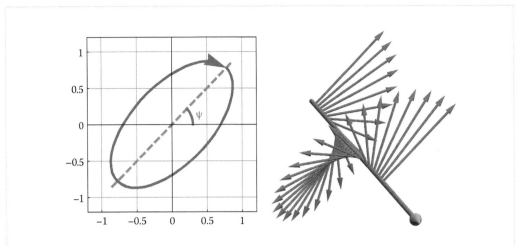

Figure 2.15 The polarization ellipse of Equation 2.49 looking into the beam (left) and a 3D view of the time helix (right).

Figure 2.16 Orthogonal polarization states have major axes 90° apart, opposite helicities, and equal ellipticity magnitudes.

2.12 Orthogonal Jones Vectors

Polarization states are *orthogonal* when their major axes are 90° apart, their helicities are opposite, and the ellipticities have equal magnitude. Three pairs of orthogonal polarization states are shown in Figure 2.16. Horizontal (0°) and vertical (90°) linearly polarized light are orthogonal to each other, as are right and left circularly polarized light. The phases of orthogonal polarization states are not specified and can assume arbitrary values. For two Jones vectors $\mathbf{E}_1$ and $\mathbf{E}_2$ to represent orthogonal polarizations, the dot product of the adjoint of $\mathbf{E}_1$ with $\mathbf{E}_2$ is zero,

$$\mathbf{E}_1^\dagger \cdot \mathbf{E}_2 = \begin{pmatrix} E_{1x}^* & E_{1y}^* \end{pmatrix} \cdot \begin{pmatrix} E_{2x} \\ E_{2y} \end{pmatrix} = 0. \tag{2.50}$$

Example 2.3 Orthogonality of Two Jones Vectors

The Jones vector $\mathbf{F}$ orthogonal to

$$\hat{\mathbf{E}} = \frac{1}{\sqrt{10}} \begin{pmatrix} 2+i \\ 2-i \end{pmatrix} \tag{2.51}$$

2.14 Addition of Jones Vectors

can be found easily in non-normalized form by setting either element of **F** to any numerical value and solving for the other element. Here, F_y has been set to one, and the equation solved for F_x,

$$\hat{\mathbf{E}}^\dagger \cdot \mathbf{F} = \frac{(2-i \quad 2+i)^*}{\sqrt{10}} \begin{pmatrix} F_x \\ 1 \end{pmatrix} = 0, \quad F_x = \frac{-3-4i}{5}. \tag{2.52}$$

Then, **F** can be normalized if desired and an arbitrary phase $e^{-i\phi}$ applied,

$$\hat{\mathbf{F}} = \frac{e^{-i\phi}}{5\sqrt{2}} \begin{pmatrix} -3-4i \\ 5 \end{pmatrix}. \tag{2.53}$$

2.13 Change of Basis

The Jones vector is most commonly expressed in terms of its x- and y-components, but other orthogonal polarization bases can be used. The following matrix transforms a Jones vector from the xy-basis into a basis with the normalized and orthogonal Jones vectors **A** and **B**,

$$\begin{pmatrix} E_A \\ E_B \end{pmatrix} = \begin{pmatrix} A_x^* & A_y^* \\ B_x^* & B_y^* \end{pmatrix} \cdot \begin{pmatrix} E_x \\ E_y \end{pmatrix}. \tag{2.54}$$

This is an example of a *unitary change of basis*, a generalized rotation (the rotation matrix may have real or complex values). An "ordinary rotation" should have a real-valued rotation matrix. If the unitary change of basis is applied to two vectors, then the inner product between the two, the dot product, is preserved; neither the angle between the vectors nor their lengths have changed.

Example 2.4 Circularly Polarized Basis

One useful basis for Jones vectors uses left and right circularly polarized basis states instead of x- and y-basis states, and is obtained from the xy-basis as[14]

$$\begin{pmatrix} E_L \\ E_R \end{pmatrix} = \frac{1}{\sqrt{2}} \begin{pmatrix} 1 & -i \\ 1 & i \end{pmatrix} \cdot \begin{pmatrix} E_x \\ E_y \end{pmatrix}. \tag{2.55}$$

2.14 Addition of Jones Vectors

The combination of two plane waves of the same frequency traveling in the same direction is simulated by the addition of their Jones vectors. Two such monochromatic beams can be input into two different faces of a beam splitter and adjusted so that, after exiting, they propagate in the same direction. If the two Jones vectors following the beam splitters individually have Jones matrices $\mathbf{E}_A$ and $\mathbf{E}_B$, then the combined beam has Jones vector **E**,

$$\mathbf{E}_A + \mathbf{E}_B = \begin{pmatrix} E_{x,A} \\ E_{y,A} \end{pmatrix} + \begin{pmatrix} E_{x,B} \\ E_{y,B} \end{pmatrix} = \mathbf{E}. \tag{2.56}$$

Such beam combination occurs at the output of interferometers. One must be careful when adding Jones vectors to ensure that the phases are specified properly, since a family of polarization ellipses results when either phase is varied.

> **Example 2.5** Addition of Circularly Polarized Beams
>
> The combination of the monochromatic right and left circularly polarized light with equal amplitudes (here set to 1/2) and an adjustable phase ϕ on the right circularly polarized component is
>
> $$\mathbf{E} = \frac{e^{i\phi}}{2}\begin{pmatrix} 1 \\ -i \end{pmatrix} + \frac{1}{2}\begin{pmatrix} 1 \\ i \end{pmatrix} = \frac{e^{i\phi/2}}{2}\begin{pmatrix} e^{i\phi/2} + e^{-i\phi/2} \\ -i\left(e^{i\phi/2} - e^{-i\phi/2}\right) \end{pmatrix} = e^{i\phi/2}\begin{pmatrix} \cos\left(\frac{\phi}{2}\right) \\ \sin\left(\frac{\phi}{2}\right) \end{pmatrix}, \tag{2.57}$$
>
> yielding linearly polarized light. The orientation of the linearly polarized light $\phi/2$ depends on the relative phase between the circularly polarized beams. Thus, the interference of left and right circularly polarized beams of equal amplitude and a variable phase provides one method to generate an adjustable linear polarization orientation.

2.15 Polarized Flux Components

The flux P of a Jones vector $\mathbf{E}$ is calculated using Equation 2.17. The *polarized flux* I_A of $\mathbf{E}$ is the part of the flux in a normalized state $\hat{\mathbf{E}}_A$, the flux transmitted by an ideal polarizer transmitting $\hat{\mathbf{E}}_A$,

$$I_A = |\hat{\mathbf{E}}_A^\dagger \cdot \mathbf{E}|^2 = \left| \begin{pmatrix} E_{A,x}^* & E_{A,y}^* \end{pmatrix} \begin{pmatrix} A_x e^{-i\phi_x} \\ A_y e^{-i\phi_y} \end{pmatrix} \right|^2. \tag{2.58}$$

The fraction of $\mathbf{E}$'s flux in state $\hat{\mathbf{E}}_A$ is I_A/P. The total flux in state $\mathbf{E}$ is the sum of the polarized flux in any two orthogonal polarization states. For example, the sum of the polarized flux in x and the polarized flux in y is

$$\mathbf{E}^\dagger \mathbf{E} = \begin{pmatrix} E_x^* & E_y^* \end{pmatrix} \begin{pmatrix} E_x \\ E_y \end{pmatrix} = \begin{pmatrix} A_x e^{i\phi_x} & A_y e^{i\phi_y} \end{pmatrix} \begin{pmatrix} A_x e^{-i\phi_x} \\ A_y e^{-i\phi_y} \end{pmatrix} = A_x^2 + A_y^2. \tag{2.59}$$

A monochromatic beam in a polarization state $\mathbf{E}$ can be divided by an ideal polarizing beam splitter into two *orthonormal* (orthogonal and normalized) polarization states $\hat{\mathbf{E}}_A$ and $\hat{\mathbf{E}}_B$ with complex amplitudes α and β given by

2.16 Converting Polarization Vectors into Jones Vectors

Table 2.3 The Flux of **E** in the Basis Polarization States

Polarization State	Basis Jones Vector	Polarized Flux I
Horizontal, 0°	$\mathbf{H} = \begin{pmatrix} 1 \\ 0 \end{pmatrix}$	$I_H = \lvert E_x \rvert^2 = A_x^2$
Vertical, 90°	$\mathbf{V} = \begin{pmatrix} 0 \\ 1 \end{pmatrix}$	$I_V = \lvert E_y \rvert^2 = A_y^2$
45°	$\mathbf{45} = \dfrac{1}{\sqrt{2}}\begin{pmatrix} 1 \\ 1 \end{pmatrix}$	$I_{45} = \dfrac{1}{2}\lvert E_x + E_y \rvert^2 = \dfrac{A_x^2 + A_y^2}{2} + A_x A_y \cos\phi$
135°	$\mathbf{135} = \dfrac{1}{\sqrt{2}}\begin{pmatrix} 1 \\ -1 \end{pmatrix}$	$I_{135} = \dfrac{1}{2}\lvert E_x - E_y \rvert^2 = \dfrac{A_x^2 + A_y^2}{2} - A_x A_y \cos\phi$
Right circular	$\mathbf{R} = \dfrac{1}{\sqrt{2}}\begin{pmatrix} 1 \\ -i \end{pmatrix}$	$I_R = \dfrac{1}{2}\lvert E_x + iE_y \rvert^2 = \dfrac{A_x^2 + A_y^2}{2} + A_x A_y \sin\phi$
Left circular	$\mathbf{L} = \dfrac{1}{\sqrt{2}}\begin{pmatrix} 1 \\ i \end{pmatrix}$	$I_L = \dfrac{1}{2}\lvert E_x - iE_y \rvert^2 = \dfrac{A_x^2 + A_y^2}{2} - A_x A_y \sin\phi$

$$\mathbf{E} = (\hat{\mathbf{E}}_A^\dagger \cdot \mathbf{E})\, \mathbf{E}_A + (\hat{\mathbf{E}}_B^\dagger \cdot \mathbf{E})\, \mathbf{E}_B = \alpha\, \mathbf{E}_A + \beta\, \mathbf{E}_B. \tag{2.60}$$

For an arbitrary Jones vector **E**,

$$\mathbf{E} = e^{-i\phi_x} \begin{pmatrix} A_x \\ A_y e^{-i\phi} \end{pmatrix}, \tag{2.61}$$

the component of flux into each of the basis polarization states, I_H, I_V, I_{45}, I_{135}, I_R, and I_L, are listed in Table 2.3. These equations are used in Chapter 3 for conversion between Jones vectors and Stokes parameters.

2.16 Converting Polarization Vectors into Jones Vectors

The polarization ellipse for polarization states defined in three dimensions can be converted into (two-dimensional) Jones vectors by choosing a basis in the transverse plane and calculating the amplitudes along the new basis vectors. Then, the state can be analyzed for ellipticity, flux components, and other metrics by applying the Jones matrix analysis methods of the last few sections. Their selection of two basis states in the transverse plane is arbitrary. For a given **k**, two normalized real-valued vectors ($\hat{\mathbf{v}}_1$ and $\hat{\mathbf{v}}_2$) are selected, which are orthogonal to each other and perpendicular to **k**,

$$\hat{\mathbf{v}}_1 \cdot \hat{\mathbf{v}}_2 = 0,\ \mathbf{k} \cdot \hat{\mathbf{v}}_1 = 0,\ \text{and}\ \mathbf{k} \cdot \hat{\mathbf{v}}_2 = 0. \tag{2.62}$$

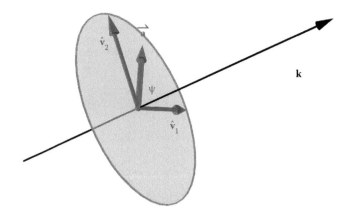

Figure 2.17 The major axis orientation ψ measured from $\hat{\mathbf{v}}_1$ to $\vec{\mathbf{a}}$. The axis vector is shown in red arrow.

Then, any polarization vector **E** transverse to **k** can be written as a superposition of $\hat{\mathbf{v}}_1$ and $\hat{\mathbf{v}}_2$,

$$\begin{aligned}\mathbf{E}(\mathbf{r},t) &= \mathcal{R}e\left[\mathbf{E}e^{i\left(\frac{2\pi}{\lambda}\mathbf{k}\cdot\mathbf{r}-\omega t-\phi\right)}\right] = \mathcal{R}e\left\{\left[(\mathbf{E}\cdot\hat{\mathbf{v}}_1)\hat{\mathbf{v}}_1 + (\mathbf{E}\cdot\hat{\mathbf{v}}_2)\hat{\mathbf{v}}_2\right]e^{i\left(\frac{2\pi}{\lambda}\mathbf{k}\cdot\mathbf{r}-\omega t-\phi\right)}\right\} \\ &= \mathcal{R}e\left[\left(A_{v1}e^{-i\phi_{v1}}\hat{\mathbf{v}}_1 + A_{v2}e^{-i\phi_{v2}}\hat{\mathbf{v}}_2\right)e^{i\left(\frac{2\pi}{\lambda}\mathbf{k}\cdot\mathbf{r}-\omega t-\phi\right)}\right],\end{aligned} \quad (2.63)$$

where A_{vi} are real amplitudes along $\hat{\mathbf{v}}_i$. This defines a Jones vector

$$\mathbf{E} = \begin{pmatrix} A_{v1}e^{-i\phi_{v1}} \\ A_{v2}e^{-i\phi_{v2}} \end{pmatrix}, \quad (2.64)$$

in a $(\hat{\mathbf{v}}_1, \hat{\mathbf{v}}_2)$ basis. The major axis orientation ψ in local coordinates is calculated using Equation 2.44, where ψ is measured from $\hat{\mathbf{v}}_1$ to $\hat{\mathbf{v}}_2$ as shown in Figure 2.17.

The selection of local coordinates is discussed further in Chapter 11 (see discussion on the Jones pupil and local coordinates systems).

Example 2.6 Conversion into Jones Vectors

As an example, a left circularly polarized beam propagating along **k** = (6, 6, 7)/11 will be expressed in three different Jones vector basis sets to appreciate how arbitrary basis sets are and show how they interrelate. The Jones vector basis sets use three different dipole (latitude and longitude) systems, seen in Figure 2.18, to generate the $\hat{\mathbf{v}}_1$ and $\hat{\mathbf{v}}_2$ basis vectors. The polarization vector for this left circularly polarized beam is

$$\mathbf{E} = \frac{1}{22}\begin{pmatrix} 11+7i \\ -11+7i \\ -12 \end{pmatrix}. \quad (2.65)$$

2.16 Converting Polarization Vectors into Jones Vectors

The basis vector along latitude is generated by taking the normalized cross product between **k** and the dipole axis **d** as

$$\hat{\mathbf{v}}_1 = \frac{\mathbf{d} \times \mathbf{k}}{|\mathbf{d} \times \mathbf{k}|}, \quad \hat{\mathbf{v}}_2 = \mathbf{k} \times \hat{\mathbf{v}}_1. \tag{2.66}$$

For the first example, a Jones vector basis uses latitude and longitude with the pole along x; the basis vectors are

$$\hat{\mathbf{v}}_1 = \frac{1}{\sqrt{85}}\begin{pmatrix} 0 \\ -7 \\ 6 \end{pmatrix}, \hat{\mathbf{v}}_2 = \frac{1}{11\sqrt{85}}\begin{pmatrix} 85 \\ -36 \\ -42 \end{pmatrix} \tag{2.67}$$

and the Jones vector $\mathbf{E}_{jx}$ becomes

$$\mathbf{E}_{jx} = \begin{pmatrix} \mathbf{E} \cdot \hat{\mathbf{v}}_1 \\ \mathbf{E} \cdot \hat{\mathbf{v}}_2 \end{pmatrix} = \frac{1}{2\sqrt{85}}\begin{pmatrix} -7+11i \\ -11-7i \end{pmatrix} = \frac{-7+11i}{2\sqrt{85}}\begin{pmatrix} 1 \\ i \end{pmatrix} \approx \frac{e^{2.138i}}{\sqrt{2}}\begin{pmatrix} 1 \\ i \end{pmatrix}. \tag{2.68}$$

Choosing a latitude and longitude basis with the pole along y yields basis vectors

$$\hat{\mathbf{v}}_1 = \frac{1}{\sqrt{85}}\begin{pmatrix} 7 \\ 0 \\ -6 \end{pmatrix}, \hat{\mathbf{v}}_2 = \frac{1}{11\sqrt{85}}\begin{pmatrix} -36 \\ 85 \\ -42 \end{pmatrix} \tag{2.69}$$

and the Jones vector $\mathbf{E}_{jy}$ becomes

$$\mathbf{E}_{jy} = \begin{pmatrix} \mathbf{E} \cdot \hat{\mathbf{v}}_1 \\ \mathbf{E} \cdot \hat{\mathbf{v}}_2 \end{pmatrix} = \frac{1}{2\sqrt{85}}\begin{pmatrix} -7-11i \\ 11-7i \end{pmatrix} = \frac{-7-11i}{2\sqrt{85}}\begin{pmatrix} 1 \\ i \end{pmatrix} \approx \frac{e^{-2.318i}}{\sqrt{2}}\begin{pmatrix} 1 \\ i \end{pmatrix}. \tag{2.70}$$

Finally, using a latitude and longitude basis with the pole along z yields basis vectors

$$\hat{\mathbf{v}}_1 = \frac{1}{\sqrt{2}}\begin{pmatrix} -1 \\ 1 \\ 0 \end{pmatrix}, \hat{\mathbf{v}}_2 = \frac{1}{22}\begin{pmatrix} -11-7i \\ 11-7i \\ 6i \end{pmatrix} \tag{2.71}$$

and the Jones vector $\mathbf{E}_{jz}$ becomes

$$\mathbf{E}_{jz} = \begin{pmatrix} \mathbf{E} \cdot \hat{\mathbf{v}}_1 \\ \mathbf{E} \cdot \hat{\mathbf{v}}_2 \end{pmatrix} = \frac{1}{\sqrt{2}}\begin{pmatrix} 1 \\ i \end{pmatrix}, \tag{2.72}$$

which is the standard Jones vector for left circularly polarized light (Equation 2.36). In the case of circularly polarized light, the only difference between the three Jones vectors $\mathbf{E}_{jx}$, $\mathbf{E}_{jy}$, and $\mathbf{E}_{jz}$ is their absolute phase,

$$\mathbf{E}_{jz} = e^{-2.138\,i}\mathbf{E}_{jx} = e^{2.138\,i}\mathbf{E}_{jy}. \tag{2.73}$$

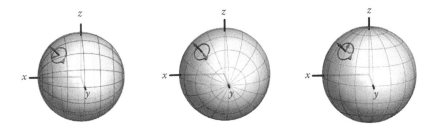

Figure 2.18 A left circularly polarized state (red circle with arrow indicating phase) with propagation vector (black arrow) is shown with basis vectors in three possible local coordinate systems for E_x and E_y. On the left, the first basis vector (green) is along a latitude line and the second basis vector (blue) along a longitude line with the axis along x. In the middle, the same polarization state is shown but with basis vectors chosen along latitude and longitude with the axis along y, and in the right, the axis is along z.

2.17 Decreasing Phase Sign Convention

Electromagnetic waves are commonly written with one of two different *sign conventions*. Either the phase *decreases with time* and *increases in space*, $\frac{2\pi}{\lambda}\mathbf{k}\cdot\mathbf{r} - \omega t - \phi_o$, the convention adopted here, or the phase *increases with time* and *decreases in space*, $\omega t - \frac{2\pi}{\lambda}\mathbf{k}\cdot\mathbf{r} + \phi_o$. Depending on the choice, various plus and minus signs must be adjusted in the mathematical descriptions for circularly and elliptically polarized light. Both conventions, *decreasing phase* and *increasing phase*, are in widespread use; thus, care is necessary when using Jones calculus equations from different sources. The student should learn to identify the sign convention in manuscripts from the context, since the sign convention is often not specified.

The two conventions for phase occur because cosine is an even function,

$$\cos(\theta) = \cos(-\theta). \tag{2.74}$$

Therefore, the choice of phase convention would not appear to matter since

$$\cos\left(\frac{2\pi}{\lambda}z - \omega t - \phi_o\right) = \cos\left(\omega t - \frac{2\pi}{\lambda}z + \phi_o\right), \tag{2.75}$$

and

$$\mathcal{R}e\left[e^{i\left(\frac{2\pi}{\lambda}z - \omega t - \phi_o\right)}\right] = \mathcal{R}e\left[e^{i\left(\omega t - \frac{2\pi}{\lambda}z + \phi_o\right)}\right] \tag{2.76}$$

for all z and t. In our decreasing phase convention, a monochromatic plane wave propagating along z has the form

$$\mathbf{E}(z,t) = \mathcal{R}e \left[\begin{pmatrix} A_x e^{-i\phi_x} \\ A_y e^{-i\phi_y} \\ 0 \end{pmatrix} e^{i\left(\frac{2\pi}{\lambda}z - \omega t - \phi_o\right)} \right] = \begin{pmatrix} A_x \cos\left(\frac{2\pi}{\lambda}z - \omega t - \phi_o - \phi_x\right) \\ A_y \cos\left(\frac{2\pi}{\lambda}z - \omega t - \phi_o - \phi_y\right) \\ 0 \end{pmatrix}. \quad (2.77)$$

Here, a wave is *advanced* through time by *subtracting* from the phase. A wave is delayed or retarded by adding to the phase. In our decreasing phase convention, the equation for a right circularly polarized plane wave of amplitude A and absolute phase ϕ_o is

$$\mathbf{E}(z,t) = \mathcal{R}e \left[e^{i\left(\frac{2\pi}{\lambda}z - \omega t - \phi_o\right)} \frac{A}{\sqrt{2}} \begin{pmatrix} 1 \\ -i \\ 0 \end{pmatrix} \right]. \quad (2.78)$$

Thus, the Jones vector for right circularly polarized light in this book is

$$\mathbf{R} = \frac{A e^{-i\phi_o}}{\sqrt{2}} \begin{pmatrix} 1 \\ -i \end{pmatrix}. \quad (2.79)$$

Similarly, in our decreasing phase convention, left circularly polarized light has the Jones vector

$$\mathbf{L} = \frac{A e^{-i\phi_o}}{\sqrt{2}} \begin{pmatrix} 1 \\ i \end{pmatrix}. \quad (2.80)$$

2.18 Increasing Phase Sign Convention

The other sign convention, the increasing phase sign convention, is not used in this work except in this section. In the increasing phase sign convention, the phase increases with time so the sign of ϕ changes. Subscript M is used in this section to indicate that quantities expressed are using the increasing phase sign convention. The monochromatic plane wave in the increasing phase convention takes the form

$$\mathbf{E}_M(z,t) = \mathcal{R}e \left[\begin{pmatrix} A_x e^{i\phi_x} \\ A_y e^{i\phi_y} \\ 0 \end{pmatrix} e^{i\left(\omega t - \frac{2\pi}{\lambda}z + \phi_o\right)} \right] = \begin{pmatrix} A_x \cos\left(\omega t - \frac{2\pi}{\lambda}z + \phi_o + \phi_x\right) \\ A_y \cos\left(\omega t - \frac{2\pi}{\lambda}z + \phi_o + \phi_y\right) \\ 0 \end{pmatrix}. \quad (2.81)$$

The only change in the plane wave equation is the sign of the exponent. Note that since cosine is an even function, the sign of the argument actually does not matter. Hence, in the increasing phase sign convention, the left and right circularly polarized light Jones vectors are

$$\mathbf{L}_M = \frac{A\,e^{i\phi_o}}{\sqrt{2}} \begin{pmatrix} 1 \\ -i \end{pmatrix}, \quad \mathbf{R}_M = \frac{A\,e^{i\phi_o}}{\sqrt{2}} \begin{pmatrix} 1 \\ i \end{pmatrix}. \tag{2.82}$$

The Jones vector for left circularly polarized light in the increasing phase convention is the Jones vector for right circularly polarized light in the decreasing phase convention and vice versa. This could lead to confusion! The choice of sign convention also affects the signs of Jones matrix elements as is documented in Chapter 5.

2.19 Polarization State of Sources

The polarization state exiting a source or entering an optical system is determined by many factors: whether the light is emitted, scattered, or reflected; the temperature and roughness of the source; the polarization and direction of the illuminating light; and many other factors. The polarization can vary in simple or complex ways; for example, Figure 2.19 contains an example of a complicated polarization variation across a spherical wavefront. Very often, the polarization of the incident light is not well understood, or may vary from measurement to measurement. Two basic models for source polarization will be briefly reviewed to provide an introduction: the *dipole model* and the *double pole model*.

The dipole electromagnetic wave is the most commonly encountered wave after the plane wave because of its simple form. Dipole radiation is generated by charges oscillating along a straight line in simple harmonic oscillation. Dipoles are important in modeling many sources of electromagnetic

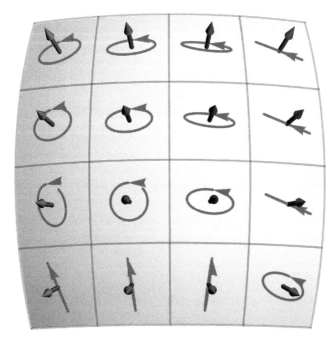

Figure 2.19 An arbitrarily polarized wavefront with **k** vectors (black) and polarization ellipses (red) representing a grid of wavefront patches.

2.19 Polarization State of Sources

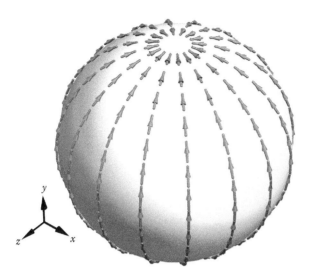

Figure 2.20 The polarization vector for a dipole wavefront looking into the z-axis. The axis of charge oscillation is the vertical y-axis.

waves such as radio antennae and the emission from excited atoms. The dipole wave is a spherical wave that is linearly polarized along lines of longitude.[2,15,16] The maximum amplitude is radiated in the plane perpendicular to the axis of charge oscillation. The dipole source does not radiate along the axis of charge oscillation. Dipole waves are radiated by charges undergoing simple harmonic oscillation along a line, chosen here in Figure 2.20 as the y-axis. Let θ be the longitude, measured from the z-axis, coming out of the page, and ϕ be the latitude. In this polar coordinate system, the normalized propagation vector **k** and the dipole wave's polarization vector **E** are

$$\mathbf{k} = \begin{pmatrix} \cos\phi \sin\theta \\ \sin\phi \\ \cos\phi \cos\theta \end{pmatrix}, \quad \mathbf{E} = \begin{pmatrix} -\sin\phi \sin\theta \\ \cos\phi \\ -\sin\phi \cos\theta \end{pmatrix}. \qquad (2.83)$$

The dipole **E**-field is shown in Figure 2.20 on a 4π steradian spherical wavefront. A dipole field can be expected for the scattered light from an isolated atom illuminated with linearly polarized light, and many other sources.

The *double pole spherical wave* in Figure 2.21 is another very common linearly polarized spherical wave in optics. The double pole spherical wave is generated when a lens is illuminated by collimated linearly polarized light. The incident polarization state is rotated at each lens interface as the light refracts and changes direction. When the light exits the lens, the polarization ellipse has rotated about an axis perpendicular to the incident and exiting rays; the rotation axis is their cross products, shown in Figure 2.22, where vertical linearly polarized light has its propagation direction changed by a lens, represented as the vertical line. Thus, the polarized wavefront exiting the lens is different from the dipole spherical wave. When this rotation is performed for every exiting ray in a spherical wavefront, the double pole polarization pattern results as shown in Figure 2.21. This form of polarized wavefront occurs for both positive and negative lenses. Great circles have been drawn out from the z-axis. Along each of these great circles, the polarization state forms a constant angle with the great circle; this is the result of the rotation operation shown in Figure 2.22. The double pole wave occurs for the ideal non-polarizing lenses. Deviations from this pattern due to the s- and p-Fresnel reflection and transmission coefficients occur, but typically the exiting wavefronts are in

58   Chapter 2: Polarized Light

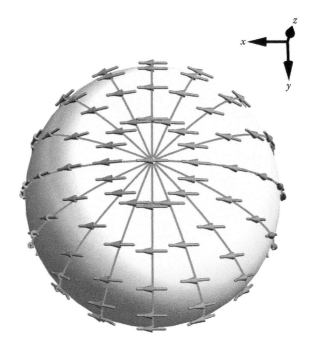

Figure 2.21 A spherical wavefront linearly polarized in the double pole polarization pattern is the basic polarized wavefront that approximates light exiting optical lenses, both positive and negative lenses (top and bottom). The polarization forms a constant angle with great circles through the axis.

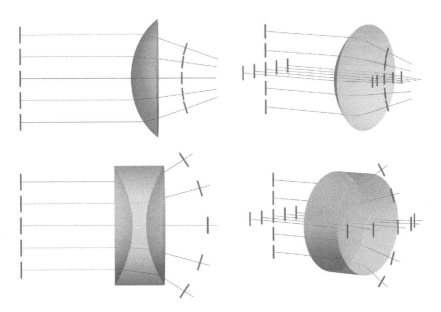

Figure 2.22 Two ideal non-polarizing lenses, (top) positive lens and (bottom) negative lens, rotate the plane of polarization of the incident light (left, red lines) when the light changes direction refracting through and out of the lens.

nearly the double pole form. Thus, the double pole polarized wavefront is a useful and common source model. This polarization pattern is also used as a basis for describing spherical waves and flattening them onto computer screens and pages. This topic is treated in depth in Chapter 11 (see discussion on the Jones pupil and local coordinates systems).

2.20 Problem Sets

2.1 Estimate the Jones vectors for the following polarization ellipses. Find the phases to place the arrow correctly at $t = 0$.

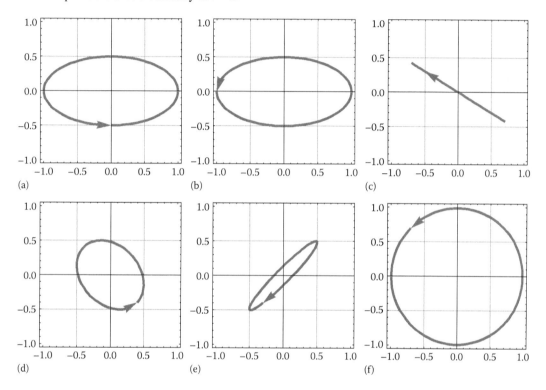

2.2 Which of the following Jones vectors are (a) linearly polarized, (b) circularly polarized, 90° or $\pi/2$ out of phase with equal amplitudes, (c) elliptically polarized, with arbitrary phase relationship?

a. (2, 2)   b. ($i/2$, 1)   c. (i, $-i$)   d. (1, -4)   e. ($2 + 2i$, $-2 + 2i$)
f. ($2 + 2i$, $-3 + 2i$)   g. (0, $1 + i$)   h. (3, $-6i$)   i. ($2 + 3i$, $-3 + 2i$)   j. (2, $-i$)

2.3 Consider the plane wave $\mathbf{E}(\mathbf{r},t) = \mathcal{Re}\left[e^{-i\left(\frac{2\pi}{\lambda}\mathbf{k}\cdot\mathbf{r} - \omega t\right)} \begin{pmatrix} a_x + ib_x \\ a_y + ib_y \\ a_z + ib_z \end{pmatrix} \right]$.

Find the electric field and the Poynting vector at the following times and locations.
a. $t = 0$, $\mathbf{r} = (0, 0, 0)$
b. $t = 0$, $\mathbf{r} = \lambda^2\, \mathbf{k}/(4\pi)$

c. $t = \pi$, $\mathbf{r} = (0, 0, 0)$
d. $t = 4\pi/\omega$, $\mathbf{r} = 8\lambda^2 \mathbf{k}/\pi$

2.4 a. Find the general equation for the normalized Jones vector orthogonal to
$$\begin{pmatrix} E_x \\ E_y \end{pmatrix} = \begin{pmatrix} A_x e^{-i\phi_x} \\ A_y e^{-i\phi_y} \end{pmatrix}.$$
b. Verify the equation with right circularly polarized light.
c. Why is the phase of the orthogonal Jones vector a free parameter that can be chosen arbitrarily?
d. Given a propagation direction $\mathbf{k} = (k_x, k_y, k_z)$ and a polarization vector $\mathbf{F} = \left(A_x e^{-i\phi_x}, A_y e^{-i\phi_y}, A_z e^{-i\phi_z} \right)$, find the polarization vector orthogonal to $\mathbf{F}$, normalized or non-normalized.

2.5 Rotate the following Jones vectors 45° counterclockwise (from +x toward +y).
  a. $\mathbf{v}_1 = (1, i)$ b. $\mathbf{v}_2 = (3, 3)$ c. $\mathbf{v}_3 = (0, -2)$ d. $\mathbf{v}_4 = (1 + i, 1 - i)$

2.6 a. What is the normalized flux of the Jones vector $\mathbf{E}_3 = (w + ix, y + iz)$?
  b. What is the flux of $\mathbf{E}_3$ in W/m² (watts per meter squared)?
  c. What are the units of the Jones vector elements $w + ix$ and $y + iz$?

2.7 a. Find the matrix for a change of basis from the left and right circularly polarized basis states to the xy-basis.
  b. Convert the Jones vector
$$\begin{pmatrix} E_L \\ E_R \end{pmatrix} = \frac{1}{\sqrt{2}} \begin{pmatrix} e^{i\eta} \\ e^{-i\eta} \end{pmatrix}$$
from the circular basis into the xy-basis and identify the type of polarization state.

2.8 Convert the six basis polarization states (Table 2.2) from Jones vectors in the ordinary linear xy-basis into the LR circular basis (Equation 2.55).

2.9 Given a Jones vector $\mathbf{E} = (1, 1)$ in the circular Jones vector basis, perform a change of basis to the xy-basis.

2.10 What are the two directions that light propagating with polarization vector $(4i, 6, 4i)$ might be propagating? This can be determined by taking the cross product between the **E**-field at two different times using the real field representation, not the exponential form.

2.11 Given the polarization vector $\mathbf{E} = (6i, 10, -8i)$:
  a. Show that **E** is circularly polarized.
  b. Find the axis of light propagation. The direction along the axis is undetermined.
  c. Which direction would the light be propagating to be left circularly polarized?

2.12 The equation for a circularly polarized monochromatic plane wave electric field traces a helix for a given point in space or time. Graph this helix for right circularly polarized light with amplitude 1, $\mathbf{E}(z,t) = \mathcal{Re}\left[e^{i\left(\frac{2\pi}{\lambda} k z - \omega t - \phi\right)} \frac{A}{\sqrt{2}} \begin{pmatrix} 1 \\ -i \\ 0 \end{pmatrix} \right]$.
  a. Make plots of the helix in space at $t = 0$, $(E_x(z, 0), E_y(z, 0), z)$, looking toward $-E_z$, front view in three-dimensional perspective plot.
  b. Is the space helix a right-handed or left-handed helix?

c. Next, make the same set of plots of the helix at a fixed point as a function of time, $x = y = z = 0$, $(E_x(0, t), E_y(0, t), t)$.
d. Is the time helix a right-handed or left-handed helix?

2.13 For each of the following Jones vectors

$$\mathbf{E}_1 = \begin{pmatrix} 1 \\ 1-i \end{pmatrix},\ \mathbf{E}_2 = \begin{pmatrix} -i \\ -i \end{pmatrix},\ \mathbf{E}_3 = \begin{pmatrix} -i \\ 1 \end{pmatrix},\ \mathbf{E}_4 = \begin{pmatrix} 5/4 \\ 3i/4 \end{pmatrix},\ \mathbf{E}_5 = \frac{1}{2}\begin{pmatrix} i \\ 1+i\sqrt{2} \end{pmatrix},$$

and $\mathbf{E}_6 = \begin{pmatrix} -i \\ -\pi/3 \end{pmatrix}$:

a. Plot the polarization ellipse and indicate the direction the electric field is rotating.
b. Calculate the phase difference between the x- and y-components, $\delta(\phi) = \phi_x - \phi_y$, orientation of the major ellipse, ψ, and the normalized flux, P. For circularly polarized light, the orientation may be undefined.
c. Calculate the degree of circular polarization (DoCP) defined as $|P_L - P_R|/(P_L + P_R)$.

2.14 a. Compute the normal vector $\hat{\mathbf{w}}$ to complete the right-handed orthonormal basis set $(\hat{\mathbf{u}}, \hat{\mathbf{v}}, \hat{\mathbf{w}})$, where $\hat{\mathbf{u}} = \frac{1}{\sqrt{3}}\begin{pmatrix} 1 \\ 1 \\ 1 \end{pmatrix},\ \hat{\mathbf{v}} = \frac{1}{\sqrt{6}}\begin{pmatrix} 1 \\ 1 \\ -2 \end{pmatrix}$.

b. Write the expression for a rotating unit vector $\hat{s}(t)$ perpendicular to $\hat{\mathbf{u}}$, rotating clockwise about the $\hat{\mathbf{u}}$-axis when looking into $\hat{\mathbf{u}}$, at an angular velocity of ω rad/s, such that $\hat{s}(0) = \hat{\mathbf{v}}$. This is an expression for the electric field at a point associated with right circularly polarized light propagating in the $\hat{\mathbf{u}}$ direction.
c. Calculate the polarization vector $\mathbf{E}$ for this wave.

References

1. W. A. Shurcliff, *Polarized Light. Production and Use*, Cambridge, MA: Harvard University Press (1966).
2. D. Goldstein, *Polarized Light, Revised and Expanded*, Vol. 83, Boca Raton, FL: CRC Press, (2011).
3. C. Brosseau, *Fundamentals of Polarized Light: A Statistical Optics Approach*, New York: Wiley, (1998).
4. M. Born and E. Wolf, *Principles of Optics*, 6th edition, Cambridge, UK: Cambridge University Press, (1980).
5. E. Hecht, *Optics*, Reading, MA: Addison-Wesley, 1987.
6. R. M. A. Azzam and N. M. Bashara, *Ellipsometry and Polarized Light*, 2nd edition, Amsterdam: Elsevier (1987).
7. G. Yun, K. Crabtree, and R. Chipman, Three-dimensional polarization ray-tracing calculus I: Definition and diattenuation, *Appl. Opt.* 50 (2011): 2855–2865.
8. M. Born and E. Wolf, *Principles of Optics: Electromagnetic Theory of Propagation, Interference and Diffraction of Light*, CUP Archive (1999).
9. E. L. Dereniak and G. D. Boreman, *Infrared Detectors and Systems*, New York: Wiley (1996).
10. G. J. Zissis and W. L. Wolfe, *The Infrared Handbook*, Ann Arbor, MI: Infrared Information and Analysis Center (1978).
11. R. C. Jones, A new calculus for the treatment of optical systems: I. Description and discussion of the calculus, *J. Opt. Soc. Am.* 31 (1941): 488–493.
12. R. M. A. Azzam and N. M. Bashara, *Ellipsometry and Polarization Light*, New York: North-Holland (1977), p. 97.

13. W. Swindell. Handedness of polarization after metallic reflection of linearly polarized light, *JOSA* 61.2 (1971): 212–215.
14. J. P. McGuire Jr. and R. A. Chipman, Diffraction image formation in optical systems with polarization aberrations. I: Formulation and example, *JOSA A* 7.9 (1990): 1614–1626.
15. J. D. Jackson, *Classical Electromagnetism*, New York: Wiley (1975).
16. A. Shadowitz, *The Electromagnetic Field*, New York: McGraw-Hill (1975).

3

Stokes Parameters and the Poincaré Sphere

3.1 The Description of Polychromatic Light

Unpolarized light and *partially polarized light* are the dominant forms of light in the universe. Starlight is almost always unpolarized. This unpolarized light is *incoherent*; it does not form stable interference patterns. Monochromatic light, the topic of Chapter 2, which mostly comes from lasers, is the exception. The polarization vectors and Jones vectors from Chapter 2 do not describe partially polarized light.

The *Stokes parameters* are a method for characterizing the polarization properties of light beams, suitable for unpolarized, partially polarized, and polarized beams. The beams described by Stokes parameters can be incoherent or coherent light beams.[1–3] Incoherent light obeys the same basic principles as coherent light during interference and diffraction; however, because of incoherent light's polychromatic nature, the details are more complex. *Polychromatic light* refers to light of multiple wavelengths, light that is not monochromatic. Figure 3.1 (left) shows the electric field amplitude of four monochromatic waves of different frequencies. When these fields are added, the resulting amplitude, shown in Figure 3.1 (right), has a random appearance with peaks of different heights and irregular periods. This is the character of polychromatic light. One difference between monochromatic and polychromatic light is that monochromatic light cannot be unpolarized. Monochromatic light has a single frequency; it is cosinusoidal. Thus, monochromatic light is trapped in a single polarization state for eternity.

In this chapter, the Stokes parameters and their mathematical properties are presented, and partially polarized and unpolarized light are discussed. The Stokes parameter properties of linear, circular, and elliptically polarized light are developed. The transformation between Stokes

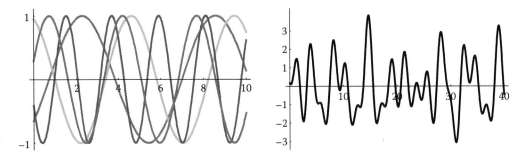

Figure 3.1 Plot of the electric field amplitude versus time for a polychromatic light waveform (right) formed from the addition of four monochromatic waves of different wavelengths (left).

parameters and Jones vectors is presented. The Stokes parameter's non-orthogonal coordinate system is described; this unusual coordinate system correctly handles the superposition and interference of incoherent light. Then, the Poincaré sphere, a graphical representation of the polarization state related to Stokes parameters, is developed. The addition of incoherent light waves and their interference properties are covered in Chapter 4.

The four Stokes parameters are often referred to as *Stokes vectors*. The Stokes parameters are not true vectors since they do not transform or rotate as vectors. The Stokes parameters are added as vectors to simulate the addition of incoherent light beams. The Stokes parameters are also operated on by Mueller matrices using matrix–vector multiplication. Thus, it is acceptable and common to refer to the Stokes parameters as Stokes vectors, but one should remember that they are not true vectors.

3.2 Phenomenological Definition of the Stokes Parameters

The four Stokes parameters,[4] S_0, S_1, S_2, and S_3, are defined in terms of six polarized flux measurements performed with ideal polarizers as

$$\mathbf{S} = \begin{pmatrix} S_0 \\ S_1 \\ S_2 \\ S_3 \end{pmatrix} = \begin{pmatrix} P_H + P_V \\ P_H - P_V \\ P_{45} - P_{135} \\ P_R - P_L \end{pmatrix} = \begin{pmatrix} I \\ Q \\ U \\ V \end{pmatrix}. \tag{3.1}$$

The Stokes parameters are usually written in vector-form **S** with four real-valued elements. S_0 is the total irradiance of the beam. S_1 is the horizontal (0°) polarized flux component (P_H) minus the vertical (90°) flux component (P_V). When $S_1 = 0$, the flux measured through a horizontal linear polarizer and that measured through a vertical linear polarizer are equal. Thus, S_1 measures the excess of horizontal polarization over vertical polarization and is negative if $P_V > P_H$. Similarly, S_2 is the 45° flux (P_{45}) minus the 135° flux (P_{135}). Finally, S_3 measures the difference of the right (P_R) minus left (P_L) circularly polarized flux. The Stokes parameters are frequently labeled I, Q, U, and V, particularly in remote sensing and astronomy.

Table 3.1 lists the Stokes parameters for the six basis polarization states; each has been normalized to a flux (S_0) of one. The normalized Stokes parameters for unpolarized light are appended to the end.

3.3 Unpolarized Light

Table 3.1 Basis Polarization States and Their Stokes Parameters

Type of Polarization	Symbol	Stokes Parameters
Horizontal linearly polarized	H	(1, 1, 0, 0)
Vertical linearly polarized	V	(1, −1, 0, 0)
45° Linearly polarized	45	(1, 0, 1, 0)
135° Linearly polarized	135	(1, 0, −1, 0)
Right circularly polarized	R	(1, 0, 0, 1)
Left circularly polarized	L	(1, 0, 0, −1)
Unpolarized light	U	(1, 0, 0, 0)

Example 3.1 Stokes Parameters

A beam of light is measured to have the following Stokes parameters, $\mathbf{S} = \begin{pmatrix} S_0 \\ S_1 \\ S_2 \\ S_3 \end{pmatrix} = \begin{pmatrix} 6 \\ 4 \\ 2 \\ -4 \end{pmatrix}$.

Find the six fluxes that define the Stokes parameters in Equation 3.1.

Since $S_0 = P_H + P_V$ and $S_1 = P_H - P_V$, $P_H = \dfrac{S_0 + S_1}{2}$ and $P_V = \dfrac{S_0 - S_1}{2}$, and similar equations are obtained for the other four fluxes. Thus, the six flux measurements, expressed as a flux vector **P**, are

$$\mathbf{P} = \begin{pmatrix} P_H \\ P_V \\ P_{45} \\ P_{135} \\ P_R \\ P_L \end{pmatrix} = \begin{pmatrix} 5 \\ 1 \\ 4 \\ 2 \\ 1 \\ 5 \end{pmatrix}.$$

3.3 Unpolarized Light

Unpolarized light has a randomly varying polarization state with no preference for any particular state. Sunlight is unpolarized. The atoms emitting in the sun's photosphere have no preferred orientation and are constantly buffeted by other atoms moving at high velocity. For unpolarized light, any ideal polarizer will transmit one-half the light. If a linear polarizer is rotated, the exiting light has a constant flux.

Following the electric field of unpolarized light, as in Figure 3.2, the tip of the vector randomly moves around the origin, sometimes clockwise, sometimes counterclockwise, sometimes nearly linearly. Each small segment of the arc is nearly elliptical, but the ellipse parameters are continuously evolving. Unpolarized light is generated when a light source emits photons with random distribution of polarization states. Each photon's polarization is uncorrelated with the photons previously emitted by the emitting atom and by its neighbors. The polarization ellipse of the beam is constantly evolving.

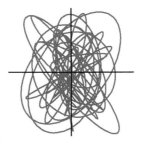

Figure 3.2 An example of electric field patterns for unpolarized light. The tip of the electric field traces patterns with a randomly evolving polarization ellipse. This pattern was generated to simulate *sunlight*, where the vector field changes its local ellipse parameters in less than an optical period.

3.4 Partially Polarized Light and the Degree of Polarization

The *degree of polarization* (*DoP*) is a metric for Stokes parameters that characterizes the randomness of a polarization state. The degree of polarization is defined as

$$DoP(\mathbf{S}) = \frac{\sqrt{S_1^2 + S_2^2 + S_3^2}}{S_0}, \quad 0 \le DoP \le 1. \tag{3.2}$$

A beam with a *DoP* of 1 is polarized. The light is in a single polarization state and the light can be completely blocked by the matched polarizer. For such a completely polarized beam, the Stokes vector elements obey the identity

$$S_0^2 = S_1^2 + S_2^2 + S_3^2. \tag{3.3}$$

Conversely, when the *DoP* is 0, the beam is unpolarized. Then, all ideal polarizers will block one-half of the beam, because the light has no preference for any particular polarization state.

Partially polarized light has a randomly changing polarization state but has a tendency toward a particular polarization state; there is an overall average polarization state. In partially polarized light, some polarization ellipses are more likely than others. Figure 3.3 shows example electric field patterns for circularly polarized (left) and partially circularly polarized beams with decreasing degree of polarization.

The *polarized flux*

$$\sqrt{S_1^2 + S_2^2 + S_3^2} \tag{3.4}$$

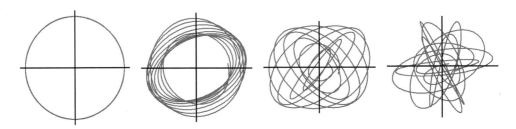

Figure 3.3 Examples of time-varying electric fields with evolving polarization ellipses for partially circularly polarized light with DoCP of 1.0, 0.95, 0.75, and 0.5 (from left to right).

3.4 Partially Polarized Light and the Degree of Polarization

is the amount of flux that is polarized. Thus, the *DoP* is the ratio of the polarized flux to the total flux S_0. With a beam of *linearly polarized light*, a linear polarizer can be oriented to transmit the entire beam, and the orthogonal polarizer, rotated by 90°, blocks the entire beam. *Partially linearly polarized light* has a random distribution but a preference for a particular linearly polarized state. The *degree of linear polarization*, *DoLP*,

$$DoLP(\mathbf{S}) = \frac{\sqrt{S_1^2 + S_2^2}}{S_0}, \tag{3.5}$$

describes the extent to which the polarization ellipse distribution tends to be linear, or equivalently the extent to which the electric field is confined in one plane. When the *DoLP* is equal to 1, the light is linearly polarized. A linear polarizer rotated in this beam will transmit all the light when aligned with the plane of polarization. As the *DoP* or *DoLP* decreases from 1 to 0, the polarization ellipse becomes more random as shown in Figure 3.4 for 45° linearly polarized light.

The *DoLP* also relates to the flux modulation that occurs when a polarizer rotates in a beam, where P_{Max} and P_{Min} are the maximum and minimum transmitted fluxes,

$$DoLP(\mathbf{S}) = \frac{P_{\text{Max}} - P_{\text{Min}}}{P_{\text{Max}} + P_{\text{Min}}} = \frac{\sqrt{S_1^2 + S_2^2}}{S_0}, \tag{3.6}$$

as is shown in Figure 3.5. The orientation ψ of the polarization is the angle of the polarizer for P_{Max}. Both a circularly polarized beam and an unpolarized beam have a *DoLP* of 0.

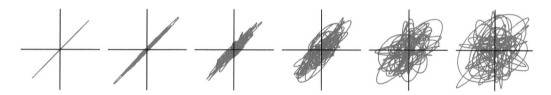

Figure 3.4 Examples of time-varying electric fields with randomly evolving polarization ellipses for partially linearly polarized light oriented at 45° with *DoLP* of 1.0, 0.8, 0.6, 0.4, 0.2, and 0.

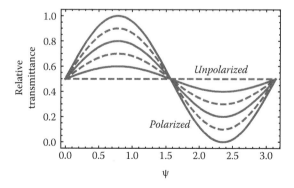

Figure 3.5 The transmittance through a rotating polarizer in a partially linearly polarized beam results in modulation of the flux proportional to the degree of linear polarization. Unpolarized light is unmodulated. The different lines shown correspond to light with *DoLP* of 0 (unpolarized), 0.2, 0.4, 0.6, 0.8, and 1 (polarized). The polarized case is described by Malus' law.

Partially linearly polarized light often arises when unpolarized sunlight scatters from surfaces; typically more light scatters polarized in the plane perpendicular to the plane of incidence (*s*-polarized light as described in Section 1.7.1 and Figure 1.32) than scatters polarized in the plane of incidence (*p*-polarized light). The polarization ellipse of the scattered light is still randomly varying in time and space, but scattering has increased the fraction of the *s*-component. As a result, sunlight or incandescent light scattered from surfaces is usually partially polarized, not unpolarized. When unpolarized light reflects from a smooth water surface, the reflected beam becomes partially polarized. Figure 3.6 shows the *DoP* for the reflected light, which can be calculated using the Fresnel equations (Section 8.3.3).

The *angle of polarization*, *AoP*, for linearly polarized light is the angle of the electric field oscillation measured counterclockwise from the *x*-axis in radians,

$$AoP(\mathbf{S}) = \frac{1}{2}\arctan\frac{S_2}{S_1} = \frac{1}{2}\arctan 2(S_1, S_2). \tag{3.7}$$

The function *arctan2* is the form of the arctan function that takes the numerator (S_2) and the denominator (S_1) as two separate arguments, so that it can return values over twice the range, $-\pi$ to π, of the conventional arctan function by using the signs of the numerator and denominator separately. For elliptically polarized light, *AoP* is the *angle of the major axis* of the polarization ellipse. The *degree of circular polarization*, *DoCP*,

$$DoCP(\mathbf{S}) = \frac{S_3}{S_0}, \tag{3.8}$$

characterizes the fraction of the polarized flux that is circularly polarized and expresses the sign of the helicity. *DoCP* = 1 indicates right circularly polarized light; *DoCP* = −1 indicates left circularly polarized, and *DoCP* = 0 indicates linearly polarized, partially linearly polarized, or unpolarized light. Natural light in the environment generally has a *DoCP* near 0.

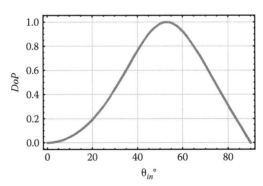

Figure 3.6 The degree of polarization of unpolarized incident light after specularly reflecting from a smooth water surface becomes increasingly linearly polarized until 57°, the Brewster angle, where the reflected light is completely polarized in the horizontal direction. Beyond the Brewster angle, the degree of polarization decreases to grazing incidence where the reflected light is unpolarized, as it is at normal incidence.

3.4 Partially Polarized Light and the Degree of Polarization

Example 3.2 Stokes Image of a Building

Figure 3.7 shows an intensity image, S_0, of the University of Arizona's Optical Sciences building in 660-nm light taken with an imaging polarimeter, GroundMSPI.[5,6] GroundMSPI accurately measures the linearly polarized Stokes parameters (S_0, S_1, S_2). Figure 3.8 shows the Stokes parameter images S_1 and S_2 in grayscale while Figure 3.9 shows the same data in false color.

Figure 3.7 An intensity image of the University of Arizona's Meinel Optical Sciences building. The entire north face of the building is windows tilted at various angles. The top of the football stadium is seen at the upper right. High intensity is white, and low intensity is black.

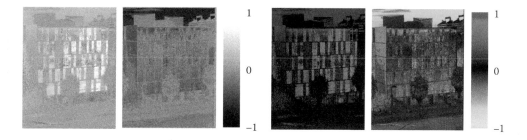

Figure 3.8 The S_1 (first and third panel) and S_2 (second and fourth panel) Stokes parameters for the image of Figure 3.7. In panels 1 and 2, gray is zero, and brighter regions have positive S_1 or S_2 and darker regions have negative S_1 and S_2. In the third and fourth panels, black is zero, red is positive, and green is negative. The Stokes parameters clearly vary with the window angle.

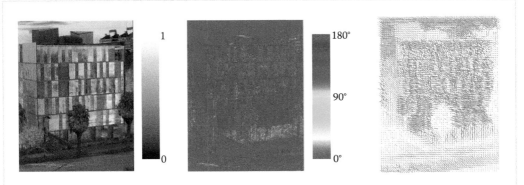

Figure 3.9 The DoP image (left) for the Meinel building image shows large variations. The structures in the image, windows, trees, lamppost, sidewalks, and so on, are easily identified, but the contrast mechanisms are completely different from the intensity image. The AoP image (middle) shows that the polarization orientation for most of the image (red and purple) doesn't vary much, but some regions (blue) show significantly different orientation. Regions of low DoP tend to be much noisier in AoP. In the quiver plot (right), an array of line segments show DoP as line length and AoP as line orientation. The trees and roof structures are seen to have low DoP. Again, the variation of DoP and AoP from window pane to window pane is striking.

3.5 Spectral Bandwidth

Figures 3.2 through 3.4 show unpolarized and partially polarized light as randomly evolving ellipses. This section examines how rapidly the ellipses evolve.

The *spectrum* of a light beam refers to the range of wavelengths in the beam and, in particular, to plots of the spectral density, flux units per nanometer, of the light. The spectrum of visible light can be made visible with a prism, or measured with a spectrometer. *Spectral bandwidth* $\Delta\lambda$ refers to the range of wavelengths in a spectrum and is frequently specified in terms of the *full width at half maximum* of the light. The spectral bandwidth of the light determines how rapidly the polarization ellipse can change in partially polarized light. True monochromatic light has a spectral bandwidth of zero, and the ellipse does not change. Laser light is quasi-monochromatic, the spectral bandwidth is close to, but not equal to, zero. Some ultra-long coherence length lasers achieve a spectral bandwidth $\Delta\lambda/\lambda < 10^{-9}$.

For unpolarized light, different wavelengths can have different polarization states, and as these evolve in and out of phase, the polarization state fluctuates. For sunlight and other broadband visible light, these polarization changes occur so rapidly, with variations on the order of 10^{-14} s, that light detectors cannot follow the rapid fluctuations of polarization state. In Figure 3.10 (left), a simulation of the electric field of unpolarized sunlight is shown. The electric field vector traces a curve that can be described as being elliptical for short segments, but these local ellipse parameters are continuously changing. For sunlight, the elliptical parameters change significantly in less than an optical period, and the ellipse arcs appear random. The rate of change is related to the *spectral bandwidth* of the light. For nearly monochromatic light, the ellipse parameters for unpolarized light change rather slowly. Figure 3.10 (center) shows the evolving electric field for a spectral bandwidth $\Delta\lambda/\lambda = 0.1$ while Figure 3.10 (right) shows the case of $\Delta\lambda/\lambda = 0.05$. Such small spectral bandwidths produce an evolving ellipse where the change in ellipse is small from period to period; the unpolarized beam's ellipse parameters are nearly constant for a few periods. The center and right examples look partially polarized (i.e., not randomized over all states), but only because a short portion of the

3.5 Spectral Bandwidth

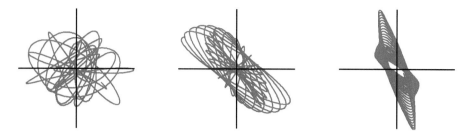

Figure 3.10 Unpolarized light has a randomly evolving polarization ellipse. For sunlight (left), the vector field changes its local ellipse parameters in less than an optical period. For a small spectral bandwidth, the polarization ellipse for unpolarized light evolves more slowly, remaining in a similar ellipse over several periods. Two unpolarized examples with the spectral bandwidth $\Delta\lambda/\lambda = 0.1$ (center) and $\Delta\lambda/\lambda = 0.05$ (right) show typical polarization ellipse evolution.

waveform is shown. Over a long period of time, these functions will fully randomize. Figure 3.11 (four figures from the left) shows the state of Figure 3.10 (right) for four consecutive time periods of equal duration, demonstrating the evolution of the polarization ellipse through many different regions of polarization state. Figure 3.11 (right) shows the state over a longer period, demonstrating the overall randomization of the state.

Polarized light ($DoP = 1$) with a spectral bandwidth also has an electric field with a time-varying polarization ellipse. As the ellipses for different wavelengths add, the ellipse parameters evolve. Figure 3.12 shows examples of the field for circularly polarized light for several $\Delta\lambda/\lambda$. For monochromatic light, $\Delta\lambda/\lambda = 0$, the ellipse (circle) is fixed. As the spectral bandwidth increases, the ellipse changes more rapidly, but always remains close to its basic circular form. In these simulations, every wavelength has the same circular state. These figures are generated by adding circular functions of different frequencies. These fully polarized figures could easily be mistaken for partially polarized light such as Figure 3.3. A Fourier transform of the x- and y-components of the functions in Figure 3.12 would reveal that each spectral component had a circularly polarized Jones vector.

Figure 3.11 The randomly evolving polarization ellipse for the state of Figure 3.10 (right) in four (left, second, third, and fourth) consecutive time periods of equal duration. (Right) The polarization ellipse graphed over a longer period of time shows it approaching a random unpolarized distribution.

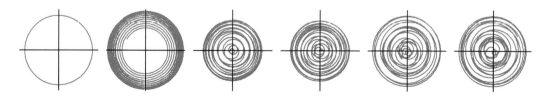

Figure 3.12 Circularly polarized light ($DoP = 1$) of spectral bandwidth 0%, 4%, 8%, 12%, 16%, and 20%.

3.6 Rotation of the Polarization Ellipse

Next, the mathematical operations commonly performed with Stokes parameters are developed. When a polarization state is rotated relative to the coordinate axes by ψ, as in Figure 3.13, the Stokes parameters transform as

$$\mathbf{S}_{ER} = \mathbf{R}_M(\psi) \cdot S = \begin{pmatrix} 1 & 0 & 0 & 0 \\ 0 & \cos 2\psi & -\sin 2\psi & 0 \\ 0 & \sin 2\psi & \cos 2\psi & 0 \\ 0 & 0 & 0 & 1 \end{pmatrix} \begin{pmatrix} S_0 \\ S_1 \\ S_2 \\ S_3 \end{pmatrix} = \begin{pmatrix} S_0 \\ S_1 \cos 2\psi - S_2 \sin 2\psi \\ S_1 \sin 2\psi + S_2 \cos 2\psi \\ S_3 \end{pmatrix}. \tag{3.9}$$

$\mathbf{S}_{ER}$ indicates the Stokes parameters for a rotated ellipse. Hence, Stokes parameters are rotated with respect to their coordinate system by the matrix $\mathbf{R}_M(\psi)$,

$$\mathbf{R}_M(\theta) = \begin{pmatrix} 1 & 0 & 0 & 0 \\ 0 & \cos 2\psi & -\sin 2\psi & 0 \\ 0 & \sin 2\psi & \cos 2\psi & 0 \\ 0 & 0 & 0 & 1 \end{pmatrix}. \tag{3.10}$$

In Section 6.5, this matrix is also shown to be the rotation matrix for Mueller matrices. The 2ψ occurs because the Stokes S_1 and S_2 axes are only 45° apart (see Section 3.12). A rotation of 180°

Example 3.3 Stokes Parameter Rotation by 45°

To rotate the Stokes parameters describing a polarization state by +45°, the transformation is

$$\mathbf{S}_{ER} = \mathbf{R}_M(45°) \cdot S = \begin{pmatrix} 1 & 0 & 0 & 0 \\ 0 & 0 & -1 & 0 \\ 0 & 1 & 0 & 0 \\ 0 & 0 & 0 & 1 \end{pmatrix} \begin{pmatrix} S_0 \\ S_1 \\ S_2 \\ S_3 \end{pmatrix} = \begin{pmatrix} S_0 \\ -S_2 \\ S_1 \\ S_3 \end{pmatrix}. \tag{3.11}$$

The S_1 and S_2 elements are reversed and the S_2 element changes sign. The circular polarization component S_3 is unchanged by rotation.

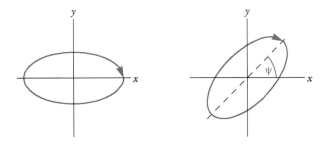

Figure 3.13 Rotation of a polarization state counterclockwise through an angle of ψ.

returns a "Stokes vector" to its original state: 0° horizontally polarized light rotates into 180° horizontally polarized light; thus, $\mathbf{R}_M(180°)$ must equal the identity matrix. By definition, ψ is positive for rotations that initially move an *x*-component toward the +*y*-axis, rotations that are counterclockwise looking into the beam.

3.7 Linearly Polarized Stokes Parameters

The normalized Stokes parameters $\mathbf{LP}_S(\psi)$ for *linearly polarized light* oriented at an angle ψ, measured counterclockwise from zero, are readily determined using the Mueller rotation matrix (Equation 3.9) operating on horizontally polarized Stokes parameters $\mathbf{H}$ as

$$\mathbf{LP}_S(\psi) = \mathbf{R}_M(\psi) \cdot \mathbf{H} = \begin{pmatrix} 1 & 0 & 0 & 0 \\ 0 & \cos 2\psi & -\sin 2\psi & 0 \\ 0 & \sin 2\psi & \cos 2\psi & 0 \\ 0 & 0 & 0 & 1 \end{pmatrix} \begin{pmatrix} 1 \\ 1 \\ 0 \\ 0 \end{pmatrix} = \begin{pmatrix} 1 \\ \cos 2\psi \\ \sin 2\psi \\ 0 \end{pmatrix}. \quad (3.12)$$

Again, notice that the 2ψ dependence is necessary, so that light polarized at 0° and 180° have the same Stokes parameters. Thus, it is seen that the Stokes parameters do not transform as vectors and are not true vectors.

3.8 Elliptical Polarization Parameters

The Stokes parameters $\mathbf{S}$ for a partially polarized beam can be mathematically represented as a sum of a completely polarized "Stokes vector" $\mathbf{S}_P$ and an unpolarized "Stokes vector" $\mathbf{S}_U$, which are uniquely related to $\mathbf{S}$ as follows:

$$\mathbf{S} = \mathbf{S}_P + \mathbf{S}_U = \begin{pmatrix} S_0 \\ S_1 \\ S_2 \\ S_3 \end{pmatrix} = \begin{pmatrix} \sqrt{S_1^2 + S_2^2 + S_3^2} \\ S_1 \\ S_2 \\ S_3 \end{pmatrix} + \left(S_0 - \sqrt{S_1^2 + S_2^2 + S_3^2}\right) \begin{pmatrix} 1 \\ 0 \\ 0 \\ 0 \end{pmatrix}. \quad (3.13)$$

The first vector has a *DoP* of 1 while the second vector is unpolarized (1, 0, 0, 0) with a *DoP* of 0. Thus, *partially polarized light* can be treated mathematically as a superposition of polarized and unpolarized light. Although the Stokes parameters are mathematically separated into polarized and unpolarized parts by Equation 3.13, there is not a corresponding polarization element to separate polarized and unpolarized parts of a beam.

The polarization ellipse corresponding to the fully polarized part of the Stokes parameters has the following parameters:

$$\text{Orientation of major axis, azimuth: } \psi = \frac{1}{2}\arctan\left(\frac{S_2}{S_1}\right), \quad (3.14)$$

$$\text{Ellipticity: } \varepsilon = \frac{b}{a} = \frac{|S_3|}{\sqrt{S_1^2 + S_2^2 + S_3^2} + \sqrt{S_1^2 + S_2^2}}, \quad 0 \le \varepsilon \le 1, \quad (3.15)$$

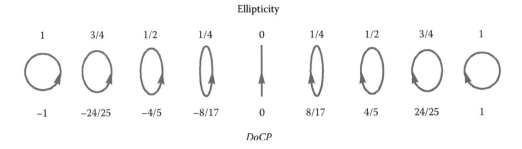

Figure 3.14 Polarization ellipses with varying ellipticity and degree of circular polarization.

$$\text{Eccentricity: } e = \sqrt{1 - \varepsilon^2}, \quad (3.16)$$

$$\text{Degree of circular polarization: } DoCP = S_3/S_0, -1 \leq DoCP \leq 1. \quad (3.17)$$

The *ellipticity* ε is the ratio of the minor (b) to the major axis (a) of the corresponding electric field polarization ellipse and varies from 0 for linearly polarized light to 1 for circularly polarized light. The *polarization ellipse* is alternatively described by its *eccentricity*, which is zero for circularly polarized light, increases as the ellipse becomes thinner (more cigar-shaped), and becomes one for linearly polarized light. Figure 3.14 shows a series of polarization ellipses (always as electric field amplitudes and never as irradiances) for a set of ellipticities and *DoCP*s.

3.9 Orthogonal Polarization States

Polarization states are *orthogonal* when their major axes are 90° apart, their helicities are opposite, and the ellipticities have equal magnitude, as seen in Figure 3.15 and discussed in Section 2.12. For a polarized state ($DoP = 1$), the states with Stokes parameters $\mathbf{S}$ and $\mathbf{S}_{\text{orth}}$,

$$\mathbf{S} = \begin{pmatrix} S_0 \\ S_1 \\ S_2 \\ S_3 \end{pmatrix} \quad \text{and} \quad \mathbf{S}_{\text{orth}} = \begin{pmatrix} S_0 \\ -S_1 \\ -S_2 \\ -S_3 \end{pmatrix}, \quad (3.18)$$

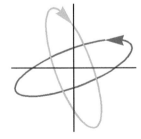

Figure 3.15 An orthogonal pair of polarization ellipses have orthogonal major axes, equal ellipticities, and opposite helicities.

3.11 Polarized Fluxes and Conversions between Stokes Parameters and Jones Vectors

Table 3.2 Basis Polarization States and Their Stokes Parameters

Type of Polarization	Symbol	Stokes Parameters	Jones Vector
Horizontal linearly polarized	H	(1, 1, 0, 0)	(1, 0)
Vertical linearly polarized	V	(1, −1, 0, 0)	(0, 1)
45° Linearly polarized	45	(1, 0, 1, 0)	$(1,1)/\sqrt{2}$
135° Linearly polarized	135	(1, 0, −1, 0)	$(1,-1)/\sqrt{2}$
Right circularly polarized	R	(1, 0, 0, 1)	$(1,-i)/\sqrt{2}$
Left circularly polarized	L	(1, 0, 0, −1)	$(1,i)/\sqrt{2}$
Unpolarized light	U	(1, 0, 0, 0)	Not available

are *orthogonal polarization states*. Orthogonal states are not defined for partially polarized states; for example, unpolarized light does not have an orthogonal polarization state. But the polarized parts of partially polarized beams have orthogonal states.

3.10 Stokes Parameter and Jones Vector Sign Conventions

The conversion between Jones vectors and Stokes parameters depends on the sign convention chosen for circular polarization in the two representations. Using the decreasing phase for the electric fields of monochromatic waves (Table 3.2), left circularly polarized light is $(1,i)/\sqrt{2}$, the y-component is positive complex. This book has adopted the most common sign convention for the Stokes parameters that uses a positive S_3 for a right circular polarized component. Thus, right circular polarization is positive in our Stokes parameters and left circular polarization is positive in our Jones vectors. Appropriate minus signs are included in our equations that convert between Jones vectors and Stokes parameters and later in the equations that convert between Jones matrices and Mueller matrices (Section 6.12). For other sign convention choices, checking a few conversions between elliptical and circular states and elliptical and circular retarders can quickly verify the consistency of sign conventions for other choices.

3.11 Polarized Fluxes and Conversions between Stokes Parameters and Jones Vectors

For a beam of light with Stokes parameters **S**, a certain fraction of the flux will be transmitted by a particular polarizer. That flux component is the *polarized flux* in that particular polarization state. Next, the polarized fluxes for the basis polarization states will be calculated as a step in the conversion between Stokes parameters and Jones vectors. The Jones vector $\mathbf{E} = \left(A_x e^{-\phi_x}, A_y e^{-\phi_y} \right)$ (Section 2.6) describes a monochromatic plane wave propagating along z as

$$\mathbf{E}(\mathbf{r},t) = \mathcal{R}e\left\{ \hat{\mathbf{x}} A_x \exp\left[i\left(\frac{2\pi z}{\lambda} - \omega t - \phi_x\right)\right] + \hat{\mathbf{y}} A_y \exp\left[i\left(\frac{2\pi z}{\lambda} - \omega t - \phi_y\right)\right] \right\}. \quad (3.19)$$

The Jones vector has units of volts per meter while the Stokes parameters have units of watts per meter squared, Sets of Stokes parameters describing fully polarized light are readily converted into equivalent Jones vectors, and vice versa. Equivalent means both the Jones vector and Stokes parameters describe a beam with the same polarization ellipse and flux. The Stokes parameters, however, will not specify the light's *phase*. For a partially polarized beam, the *polarized part of the flux* P_P is

$$P_P = \sqrt{S_1^2 + S_2^2 + S_3^2}. \tag{3.20}$$

The horizontal P_H (0°) and vertical P_V (90°) polarized fluxes of the Jones vector

$$\mathbf{E} = \begin{pmatrix} E_x \\ E_y \end{pmatrix} = \begin{pmatrix} A_x e^{-i\phi_x} \\ A_y e^{-i\phi_y} \end{pmatrix} \tag{3.21}$$

are the components of flux transmitted through ideal *x*- and *y*-oriented polarizers:

$$P_H = \frac{\varepsilon_0 c}{2} A_x^2, \quad P_V = \frac{\varepsilon_0 c}{2} A_y^2. \tag{3.22}$$

P_H and P_V are independent of the phases. Here, c is the speed of light and ε_0 is the permittivity of free space. The factor $\frac{\varepsilon_0 c}{2}$ is the conversion between amplitude squared, given in (volts per meter)^2, and watts/meter^2. The 45° and 135° polarized fluxes P_{45} and P_{135} expressed in terms of the *x* and *y* amplitudes and phases are

$$P_{45} = \frac{\varepsilon_0 c}{4}\left(A_x^2 + A_y^2 + 2A_x A_y \cos(\phi_x - \phi_y)\right), P_{135} = \frac{\varepsilon_0 c}{4}\left(A_x^2 + A_y^2 - 2A_x A_y \cos(\phi_x - \phi_y)\right). \tag{3.23}$$

For 45° polarized light, the *x*- and *y*-components are in phase or equal $\phi_x - \phi_y = 0$, while for 135° polarized light $\phi_x - \phi_y = \pi$. The right and left circularly polarized fluxes P_R and P_L are associated with the *x*- and *y*-phases differing by $\pm\pi/2$,

$$P_R = \frac{\varepsilon_0 c}{4}\left(A_x^2 + A_y^2 + 2A_x A_y \sin(\phi_x - \phi_y)\right), \quad P_L = \frac{\varepsilon_0 c}{4}\left(A_x^2 + A_y^2 - 2A_x A_y \sin(\phi_x - \phi_y)\right). \tag{3.24}$$

To convert a Jones vector into Stokes parameters, these expressions can be applied to the definition of the Stokes parameters in Equation 3.1. A Jones vector **E** transforms into Stokes parameters as follows:

$$\mathbf{S}(\mathbf{E}) = \frac{\varepsilon_0 c}{2} \begin{pmatrix} A_x^2 + A_y^2 \\ A_x^2 - A_y^2 \\ 2A_x A_y \cos(\phi_x - \phi_y) \\ 2A_x A_y \sin(\phi_x - \phi_y) \end{pmatrix}. \tag{3.25}$$

When the relative phase of the *x*- and *y*-components of the electric field is 0 or π, the light is linearly polarized, and the Stokes parameters have non-zero S_1 and/or S_2 components, but zero S_3 components, so $\sin(\phi_x - \phi_y) = 0$. Similarly, when the relative phase is $\pm\pi/2$, $\cos(\phi_x - \phi_y) = 0$, the

3.11 Polarized Fluxes and Conversions between Stokes Parameters and Jones Vectors

polarization ellipse has maximum ellipticity for a given set of amplitudes A_x and A_y. The *absolute phase* of the Jones vector does not change the corresponding Stokes parameters; therefore,

$$\mathbf{S}(e^{-i\phi}\mathbf{E}) = \mathbf{S}(\mathbf{E}). \tag{3.26}$$

Example 3.4 Finding the Flux Components

Figure 3.16 (left) graphs the polarization ellipse for the polarized state with the Jones vector **E**,

$$\mathbf{E} = \sqrt{\frac{2}{\varepsilon_0 c}} e^{0.6i} \begin{pmatrix} i \\ 0.8 + 0.5i \end{pmatrix}. \tag{3.27}$$

The $\sqrt{2/(\varepsilon_0 c)}$ is the unit conversion to express the Jones vector **E** in a normalized form, not in units of volts per meter. The polarized flux component pairs are $P_H = |i|^2 = 1$ and $P_V = |0.8 + 0.5 i|^2 = 0.89$. According to Equations 3.23 and 3.24, $P_{45} = 1.445$ and $P_{135} = 0.445$, and $P_R = 1.745$ and $P_L = 0.145$, yielding Stokes parameters $\mathbf{S} = (1.89, 0.11, 1, 1.6)$. The corresponding orthogonal pairs of states are graphed to scale in Figure 3.16 (second, third, and last figure).

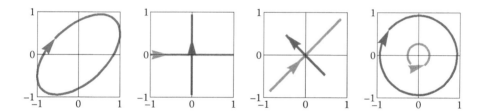

Figure 3.16 An elliptical polarization state represented by its polarization ellipse (left) and its decomposition into horizontal and vertical flux components (second), 45° and 135° components (third), and right and left circular components (right). Note how the arrows on the horizontal and vertical components line up with the elliptical state's arrow. Similarly, the arrows on the 45° and 135° components add as vectors into the elliptical state's arrow. Vectors from the origin to the right and to the left circular component's arrows also add, yielding the location of the elliptical state's arrow.

A fully polarized set of Stokes parameters, or the polarized part of a set of partially polarized Stokes parameters, is equivalent to the Jones vector **E**,

$$\mathbf{E}(\mathbf{S}) = \sqrt{\frac{2}{\varepsilon_0 c}} \begin{pmatrix} \sqrt{\frac{S_0 + S_1}{2}} \\ \sqrt{\frac{S_0 - S_1}{2}} e^{-2i \arctan(S_2, S_3)} \end{pmatrix}. \tag{3.28}$$

This Jones vector can be multiplied by an arbitrary phase $e^{-i\phi}$ since the Stokes parameters do not specify an *absolute phase*.

Example 3.5 Converting a Jones Vector into Stokes Parameters

The polarization ellipse of the Jones vector

$$\mathbf{E} = \sqrt{\frac{2}{\varepsilon_0 c}} \begin{pmatrix} 1.5 \\ 0.5 + 0.5i \end{pmatrix} = \sqrt{\frac{2}{\varepsilon_0 c}} \begin{pmatrix} 1.5 \\ 0.707 e^{i0.785} \end{pmatrix} \quad (3.29)$$

is plotted in Figure 3.17 along with its decomposition into the pairs of basis polarization states. Note the positions of the arrows identifying the phases of the polarized E-field components. When two vectors from the origin to these arrows are added, they equal the vector from the origin to the ellipse in Figure 3.17 (left). **E** has the following polarized flux component pairs: $P_H = 2.25$ and $P_V = 0.5$, $P_{45} = 2.125$ and $P_{135} = 0.625$, and $P_R = 0.625$ and $P_L = 2.125$. The corresponding Stokes parameters are

$$\mathbf{S} = \begin{pmatrix} 1.5^2 + 0.707^2 \\ 1.5^2 - 0.707^2 \\ 2(1.5)(0.707)\cos(-0.785) \\ 2(1.5)(0.707)\sin(-0.785) \end{pmatrix} = \begin{pmatrix} 2.75 \\ 1.75 \\ 1.5 \\ -1.5 \end{pmatrix}. \quad (3.30)$$

The degree of polarization must be 1, since this state also has a Jones vector. The other elliptical polarization parameters are as follows: $DoLP = \sqrt{85}/11 \approx 0.838$, orientation of major axis $\psi \approx 20°$, $DoCP = -6/11$, ellipticity $\varepsilon = -0.30$, and eccentricity $e \approx 0.95$.

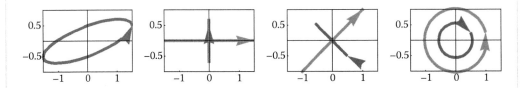

Figure 3.17 An elliptical polarization state represented by its polarization ellipse (left) and its decomposition into horizontal and vertical flux components (second), 45° and 135° components (third), and right and left circular components (right).

3.12 The Stokes Parameters' Non-Orthogonal Coordinate System

The Stokes parameters have an unusual coordinate system, because the S_1, S_2, and S_3 axes do not form an orthogonal coordinate system. This *non-orthogonal coordinate system* is an extremely clever system, which usefully describes the addition of partially polarized light beams.

3.12 The Stokes Parameters' Non-Orthogonal Coordinate System

An orthogonal coordinate system uses basis vectors that are perpendicular to each other, such as the Cartesian basis vector set $\hat{\mathbf{x}}$, $\hat{\mathbf{y}}$, and $\hat{\mathbf{z}}$:

$$\begin{cases} \hat{\mathbf{x}} \times \hat{\mathbf{y}} = 0, \\ \hat{\mathbf{y}} \times \hat{\mathbf{z}} = 0, \text{ and} \\ \hat{\mathbf{z}} \times \hat{\mathbf{x}} = 0. \end{cases} \quad (3.31)$$

Orthogonal coordinate systems simplify most geometrical calculations and are the principal coordinate systems used in physics and optics. The Jones vector uses the standard x–y orthogonal coordinate system. On the other hand, Stokes parameters use basis vectors where plus and minus values represent orthogonal polarizations, not opposite directions. The S_1 and S_2 basis states represent electric fields that are only 45° apart; these two states are halfway between orthogonal in the Jones calculus as shown in Figure 3.18.

There are two reasons for the great utility of this non-orthogonal coordinate system. First, the Stokes parameters do not differentiate between light polarized at 0° and 180° because for incoherent light, there is no effective difference. Linearly polarized light is polarized along a line, such as the 0° and 180° line in the transverse plane. The electric field takes equal excursions in both the +x- and −x-directions. A polarizer oriented at 0° has the same optical effect as a polarizer rotated through 180°. This is why polarizers have a plane of polarization, such as "horizontal," not a single direction such as +x. Since the Stokes parameters do not differentiate between light polarized at 0° and 180°, the Stokes parameters for linearly polarized light $\hat{\mathbf{L}}(\psi)$ repeat after rotations of 180°,

$$\hat{\mathbf{L}}(\psi) = \begin{pmatrix} 1 \\ \cos 2\psi \\ \sin 2\psi \\ 0 \end{pmatrix} = \hat{\mathbf{L}}(\psi + \pi) = \begin{pmatrix} 1 \\ \cos(2(\psi + \pi)) \\ \sin(2(\psi + \pi)) \\ 0 \end{pmatrix}. \quad (3.32)$$

This explains the 2ψ in the Mueller rotation matrix. Second, the plus and minus signs of the Stokes parameter elements represent orthogonal polarizations. Adding orthogonally polarized incoherent beams reduces the degree of polarization of the light. For example, two beams of horizontally and vertically polarized polychromatic light with equal flux per area add to yield unpolarized light.

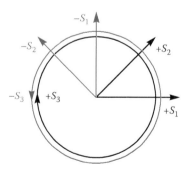

Figure 3.18 The three basis vectors for the Stokes parameters, **H**, **45°**, and **R**, corresponding to the S_1, S_2, and S_3 basis states, are not orthogonal in Cartesian space.

3.13 The Poincaré Sphere

The *Poincaré sphere* is a geometric representation for polarization states and the Stokes parameters that simplifies the analysis of many polarization problems, particularly problems involving retarders.[2] Henri Poincaré introduced the sphere in 1892 in his text *Traité de Lumiéré*.[7] The Poincaré sphere, shown in Figure 3.19, maps all fully polarized states into points covering the surface of a sphere. Three views of a partially transparent (front and back) Poincaré sphere are provided. Figure 3.20 shows additional views of the Poincaré sphere with latitude and longitude lines superposed. Ellipses are drawn looking toward the center of the sphere; hence, the northern (top) hemisphere is right elliptically polarized.

Consider the *normalized Stokes parameters* $\hat{\mathbf{S}}$ obtained by dividing the Stokes parameters $\mathbf{S}$ by the flux S_0,

$$\hat{\mathbf{S}} = \begin{pmatrix} 1 \\ s_1 \\ s_2 \\ s_3 \end{pmatrix} = \frac{\mathbf{S}}{S_0} = \begin{pmatrix} 1 \\ S_1/S_0 \\ S_2/S_0 \\ S_3/S_0 \end{pmatrix}. \tag{3.33}$$

The degree of polarization of $\hat{\mathbf{S}}$, $DoP(\hat{\mathbf{S}})$ is

$$DoP(\hat{\mathbf{S}}) = \sqrt{s_1^2 + s_2^2 + s_3^2}. \tag{3.34}$$

Figure 3.19 The Poincaré sphere is a three-dimensional representation of polarization states. The coordinates at a set of points on the surface have been converted into normalized Stokes parameters and the ellipses plotted on the surface. These transparent views from three different angles show the front and back simultaneously showing how the orientation of the polarization rotated through 180° for a full 360° circuit around the sphere.

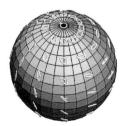

Figure 3.20 The Poincaré sphere, shown here in four views centered above 135° light or $-S_2$ (left), centered just above 0° light or $+S_1$ (second from left), centered just below 45° light or $+S_2$ (second from right), and centered below 90° light or $-S_1$ (right). All fully polarized states are represented on the surface of the sphere. The dark band around the equator represents linearly polarized light, with circularly polarized light at the poles. Polarization states are drawn looking into the sphere, so the top hemisphere is right elliptically polarized.

3.13 The Poincaré Sphere

For all polarized states, the degree of polarization is one; hence, the normalized Stokes components obey the relation

$$\sqrt{s_1^2 + s_2^2 + s_3^2} = 1. \tag{3.35}$$

The Poincaré sphere represents a normalized set of Stokes parameters for each polarization state (s_1, s_2, s_3) as a point in three-dimensional space. The axes of the Poincaré sphere are the Stokes parameter basis states, $\pm S_1, \pm S_2, \pm S_3$, as shown in Figure 3.21. The fully polarized states lie on the surface of the sphere because their *DoP* is equal to 1. Figure 3.22 shows views of the circular polarized regions at the top and bottom of the Poincaré sphere. Around each pole, the light is in elliptical states that are nearly circular and the major axis rotates through 180° as we move in a circle of latitude about the pole.

The origin (0, 0, 0) of the Poincaré sphere represents unpolarized light. Linearly polarized states occupy the unit circle in the s_1–s_2 plane. Right circularly polarized light is at the top of the sphere (0, 0, 1) at the *north pole*. Left circularly polarized light is at the bottom of the sphere (0, 0, −1) at the *south pole*. The rest of the surface of the sphere describes elliptically polarized light, with a right circular helicity in the northern hemisphere and a left circular helicity in the southern hemisphere. The ellipticity of the light approaches 0 when approaching the equator; the ellipticity approaches ±1 when approaching the poles. The surface of the Poincaré sphere *continuously* represents all possible polarized states.

The interior of the Poincaré sphere represents partially polarized light with the distance from the origin indicating the *DoP*. Thus, a sphere of radius ½ centered at the origin describes all the partially polarized states with a *DoP* of ½. Figure 3.23 (top) shows a series of spheres; each contains the polarization states with a fixed *DoP*. A radial line from the center through each sphere describes a state with the same polarization ellipse but varying *DoP*. Along the line segment $(s_1, 0, 0)$,

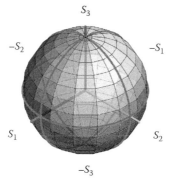

Figure 3.21 The axes associated with the Poincaré sphere are the Stokes parameter basis polarization states.

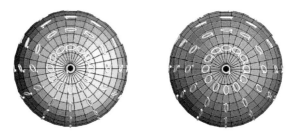

Figure 3.22 The polar regions of the Poincaré sphere represent states around right circularly polarized (left, north pole) and left circularly polarized light (right, south pole).

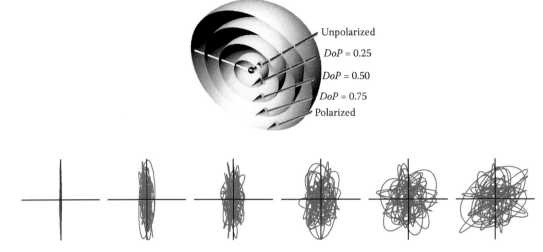

Figure 3.23 (Top) Surfaces of constant degree of polarization are a family of spheres inside the Poincaré sphere centered at the origin. The point at the center of these spheres represents unpolarized light. (Bottom) Random polarization ellipses for polarization states along the white line, varying from (left) vertically polarized to (right) unpolarized.

$-1 \leq s_1 \leq 0$, shown in Figure 3.23 (top), the polarization state varies from fully vertically polarized, strongly vertically polarized, weakly vertically polarized to unpolarized, with examples of the random polarization ellipses plotted in Figure 3.23 (bottom).

Moving around the Poincaré sphere equator from (1, 0, 0), to (0, 1, 0), (–1, 0, 0), (0, –1, 0), and back to the beginning, (1, 0, 0), the polarization states change from horizontal or 0°, to 45°, 90°, 135°, and back to horizontal, 0°, as shown in Figure 3.24 (left). The polarization state with an orientation of 180° is the same as 0°; polarizers oriented at 0° and 180° generate the same polarization state. Thus, moving around the equator of the Poincaré sphere, the orientation of the polarization axis ψ changes at one-half the rate of the *longitude* ζ,

$$\psi = \frac{\zeta}{2}, \qquad (3.36)$$

a consequence of the factors of two in Equation 3.7.

Figure 3.24 (Left) The linearly polarized states on the equator rotate by 180° as we move 360° around the equator. (Middle) Polarization ellipses along the longitude great circle from left circularly polarized light, through horizontal, to right circularly polarized light all have a horizontal major axis. Continuing onto the back side, all the ellipses have a vertical major axis. (Right) Polarization change shown around two circles of latitude. At each latitude, the ellipse's major axis orientation changes while the degree of circular polarization remains constant.

3.14 Flat Mappings of the Poincaré Sphere

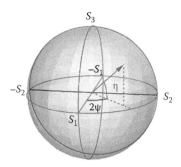

Figure 3.25 Position of a state with major axis orientation ψ and latitude η on the Poincaré sphere.

Along a longitude line, moving from the south (left circularly polarized) pole, across the equator, to the north (right circularly polarized) pole, the ellipticity of the light changes but the *orientation of the major axis* remains constant as shown in the front side of the sphere in Figure 3.24 (middle). The lines of latitude on a sphere, the circles in planes perpendicular to the S_3 axis through the poles, are labeled from −90° at the south pole, to 0° at the equator, to 90° at the north pole, or as we prefer to work in radians, from −π/2 to 0 to π/2. Moving around a circle of latitude, the normalized Stokes parameter s_3 and the ellipticity remain constant, but the orientation of the ellipse's major axis rotates through 180° as shown in Figure 3.24 (right). Each circle of latitude corresponds to the states of a *constant degree of circular polarization*. The degree of circular polarization for a normalized set of Stokes parameters is

$$DoCP(\mathbf{S}) = \frac{S_3}{S_0} = s_3. \tag{3.37}$$

Thus, the latitude η on the Poincaré sphere in radians is proportional to s_3,

$$\sin \eta = s_3 = DoCP. \tag{3.38}$$

Figure 3.25 shows the orientation of the major axis on the Poincaré sphere as positioned at 2ψ in sphere coordinates and at a latitude η.

3.14 Flat Mappings of the Poincaré Sphere

Just as it is useful to represent the earth, not just on spherical globes, but on flat maps, it is similarly helpful to use various flat representations for the surface of the Poincaré sphere. First, the Poincaré sphere can be represented within a rectangle where the x-axis represents the orientation of the polarization axis and the y-axis is *DoCP*, as shown in Figure 3.26a. In this projection, the *equi-rectangular projection*, the entire line across the top represents right circularly polarized light and the entire line across the bottom represents left circularly polarized light. Other standard map transformations can be used to represent the Poincaré sphere. Figure 3.26b depicts the *Mollweide projection* of the sphere; Figure 3.26c is the *sinusoidal projection* and Figure 3.26d is the *interrupted sinusoidal projection*.

Before Henri Poincaré introduced the Poincaré sphere, he first developed an interesting representation of polarized light on the complex plane. This parameterization of the Stokes parameters is generated by placing a plane tangent to the bottom of the Poincaré sphere (at left circular), and projecting polarization states from the sphere onto the complex plane along lines as shown in Figure 3.27. If the radius of the Poincaré sphere is set to ½, conveniently, linearly polarized light lies

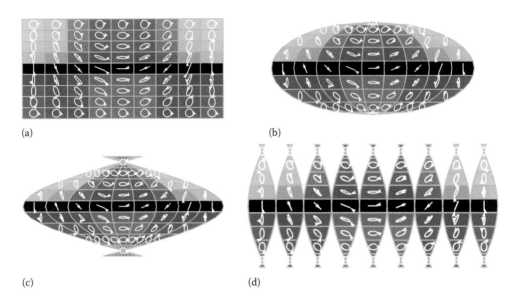

Figure 3.26 Several flat representations of the Poincaré sphere: (a) the equi-rectangular projection, (b) the Mollweide projection, (c) the sinusoidal projection, and (d) the interrupted sinusoidal projection.

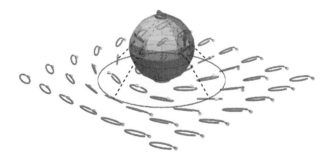

Figure 3.27 The set of all polarized states can be represented on the complex plane by projecting the Poincaré sphere from the right circularly polarized state.

around the unit circle. The resulting parameterization of the normalized Stokes parameters in terms of the complex number $z = x + iy$ is

$$\mathbf{S} = \begin{pmatrix} 1 \\ s_1 \\ s_2 \\ s_3 \end{pmatrix} = \begin{pmatrix} 1 \\ \dfrac{2x}{1+x^2+y^2} \\ \dfrac{2y}{1+x^2+y^2} \\ \dfrac{2}{1+x^2+y^2} - 1 \end{pmatrix}. \qquad (3.39)$$

Figure 3.28 shows the polarization states near the origin of Poincaré's complex plane. Centered circles contain all states of common ellipticity. Radial lines represent constant major

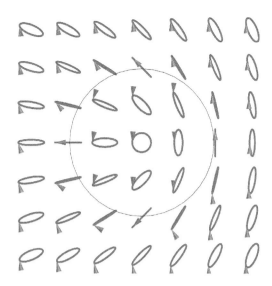

Figure 3.28 Projection of the Poincaré sphere onto the plane such that left circularly polarized light is at the origin, linearly polarized light is on the unit circle, and right circularly polarized light is at infinity.

axis orientation. The parameterization of the complex plane in terms of normalized Stokes vector elements is

$$x + iy = \frac{s_1 + i s_2}{1 + s_3}. \tag{3.40}$$

3.15 Summary and Conclusion

The Stokes parameters are a standard method in optics for characterizing the polarization states of light beams. Jones vectors are most suited for coherent light beams. Laser light is well approximated by Jones vectors and polarization vectors since the spectral bandwidth is nearly monochromatic ($\Delta\lambda \ll 1$ nm) and since laser beams can be well collimated. These laser-generated light waves are comparatively simple to describe. On the other hand, light from the sun, light bulbs, LEDs, and most other sources is far from monochromatic. The superposition of all these waves is more challenging to describe mathematically, as treated in Section 4.7.

Stokes parameters can be used for either coherent or incoherent beams. Incoherent light obeys the same basic principles as coherent light during interference and diffraction; however, because of the incoherent light's polychromatic nature, the details are more complex. For example, monochromatic light cannot be unpolarized. In particular, polarized incoherent beams can add to form unpolarized light whereas mutually coherent beams cannot. Combining the light from one flashlight with a horizontal polarizer and another with a vertical polarizer generates unpolarized light. Combining the light from a horizontally polarized and vertically polarized laser generates a family of polarized states, not unpolarized light; this is treated further in Section 4.4. Thus, incoherent light uses different polarization mathematics compared to the Jones vector, and this mathematics has the remarkable coordinate system of Section 3.12.

Stokes parameters are particularly useful for characterizing the polarization in outdoor and indoor scenes.[8] These Stokes parameter images and spectra are measured with Stokes polarimeters (Chapter 7). Jones vectors are not used to describe natural light.

3.16 Problem Sets

3.1 For the following Stokes parameters, determine the degree of polarization, the degree of linear polarization, the orientation of the major axis of the polarization ellipse, the degree of circular polarization, and the ellipticity.
   a. (1, 0, 1, 0)
   b. (1, 0, 0.5, 0)
   c. (2, 1, 0.5, −0.5)
   d. (3, 1, 0, −1)
   e. (1, 0.2, 0.3, 0.6)
   f. (1, −0.8, 0.1, 0.4)

   If all six beams are incoherently combined, what is the resulting Stokes vector?

3.2 Decompose the following Stokes parameters in Problem 3.1 into completely polarized, $\mathbf{S}_P$, and unpolarized, $\mathbf{S}_U$, components. Plot the polarization ellipses for the $\mathbf{S}_P$ all to the same scale.

3.3 Each row in the table below represents a set of four polarization component measurements chosen from the six measurement types on the heading. Calculate the Stokes vector and fill in the expected measurements for the two missing measurements (marked ☐).

	P_H	P_V	P_{45}	P_{135}	P_R	P_L
a.	10	1	3	☐	6	☐
b.	5	2	5	☐	☐	2
c.	9	☐	9	☐	0	18
d.	6	☐	☐	4	10	7
e.	☐	7	6	6	10	☐
f.	4	☐	6	4	☐	4

3.4 Why do Stokes vector elements for linearly polarized light vary with orientation as 2ψ while Jones vector elements have a ψ dependence?

3.5 Show that for a partially polarized beam with Stokes parameters (S_0, S_1, S_2, S_3), the maximum amount of light that can be transmitted through the matched ideal polarizer is more than the polarized flux $\sqrt{S_1^2 + S_2^2 + S_3^2}$. How much more flux can be transmitted?

3.6 a. Where on the Poincaré sphere are all the states with a $DoCP = 2/3$?
   b. Where are all the states on the Poincaré sphere with the major axis orientation at 30°?
   c. Indicate the states on the Poincaré sphere with a transmittance of 50% through a 45° linear polarizer.
   d. Show the trajectory for 45° linearly polarized light propagating through, first, a quarter wave linear retarder with fast axis at 0°, then a quarter wave linear retarder with fast axis at 45°, and then a quarter wave left circular retarder.

3.16 Problem Sets

3.7 Match the Jones vectors with the Stokes parameters that represent the same polarization state:

a. (1, 1, 0, 0)
b. $\left(\sqrt{2}, 1, 1, 0\right)$
c. (1, 0, 1, 0)
d. $\left(\sqrt{2}, -1, 1, 0\right)$
e. (1, −1, 0, 0)
f. (1, 0, 0, 1)
g. (1, 0, 0, 0)

i. $(1 + i, 0)$
ii. (0, 1)
iii. (i, i)
iv. $(1, i)$
v. $2^{1/4} (\cos 22.5°, \sin 22.5°)$
vi. $(-i, 1)$
vii. $2^{1/4} (\cos 67.5°, -\sin 67.5°)$

3.8 Determine the completely polarized set of Stokes parameters with the specified irradiance P, orientation θ, and ellipticity ε:

P (W/m²)	θ	ε
2	0°	0
10	22.5°	0.1
10	45°	0.25
100	60°	0.4
30	90°	0.5
0.1	113°	1

3.9 Transform the following Jones vectors into Stokes parameters:

a. (1, 1) b. $\left(e^{\frac{i\pi}{4}}, 0\right)$ c. $(1, e^{i\delta})$ d. $(1, i/2)$

3.10 Transform the following Stokes parameters into Jones vectors:

a. (1, 0, 0, 1) b. $\left(1, 1/\sqrt{2}, 1/\sqrt{2}, 0\right)$ c. $\left(1, 0, 1/\sqrt{2}, 1/\sqrt{2}\right)$

d. $\left(\sqrt{14}, 1, 2, 3\right)$ e. $(1, -\cos 2\theta, \sin 2\theta, 0)$ f. $\left(\sqrt{3}, 1, 1, 1\right)$

3.11 Light with the Stokes vector $\mathbf{S} = (11, 6, 6, 7)$ is transmitted through an adjustable polarization rotator that rotates the major axis of the polarization ellipse.
   a. Determine the resulting Stokes vector for the following rotations: 10°, 22.5°, 30°, 45°, 90°, 135°, and 180°.
   b. Draw a Poincaré sphere and label the basis polarization states.
   c. Show the incident state and all the states from part (a). Describe the trajectory on the Poincaré sphere as the rotation angle increases.

3.12 Perform the following operations on the Stokes vector below:

(1, 0, −1, 0)   (1, −0.5, 0, 0)   (1, 0.5, 0.5, 0)   (5, 0, 0, 2)

a. Determine the degree of polarization.
b. Identify the polarization state and note whether it is polarized, partially polarized, or unpolarized.
c. Decompose the Stokes parameters **S** into a polarized **S**$_P$ and unpolarized **S**$_U$ component.
d. Determine the polarized Stokes vector **S**$_O$ (D_O = 1) orthogonal to **S**$_P$.

3.13 An elliptically polarized state has a horizontal major axis, has a degree of polarization of one, and rotates clockwise looking into the beam. A horizontal polarizer passes 70% of the flux and a vertical polarizer passes 30%.
a. Find the normalized Stokes parameters, **S**.
b. What is the ellipticity and degree of circular polarization?
c. Find the orthogonal polarization state.
d. If the major axis of the ellipse is rotated by 60°, find the resulting Stokes vector.

3.14 Light with the Stokes parameters **S** = (27, 22, 14, 7) is transmitted through an adjustable polarization rotator that rotates the major axis of the polarization ellipse.
a. Determine the resulting Stokes parameters for the following rotations: 10°, 22.5°, 30°, 45°, 90°, 135°, and 180°.
b. Draw a Poincare sphere and label the six basis polarization states. Show the incident state and all the states from part (a). Describe the trajectory.
c. For **S**, determine the degree of polarization and plot the polarization ellipse.

3.15 Consider the partially polarized Stokes vector **S** = (1, 0, 0.3, 0).
a. Decompose **S** into the sum of two fully polarized linear states **L**$_1$ and **L**$_2$ with equal flux.
b. Are **L**$_1$ and **L**$_2$ orthogonal Stokes parameters?
c. Is this decomposition of **S** into two the sum of two fully polarized Stokes parameters with equal flux unique? Choose a numerical example if that helps. For an arbitrary Stokes vector, is the decomposition unique? It might help to reason on the Poincaré sphere.
d. For which state or states is the decomposition not unique?

3.16 Why does the Stokes vector use a non-orthogonal basis such that the *x*- and *y*-components occur as positive and negative values on the same basis state, **S**$_1$. In typical vectors, *x* and *y* occur as two orthogonal vector basis states.

3.17 Show that if two partially polarized beams with $DoP_1 > DoP_2$ are added, the resulting degree of polarization cannot be greater than DoP_1, but it can be less than DoP_2. When does it equal DoP_1?

3.18 If two polarized Stokes parameters with equal amplitude are added, where in the Poincaré sphere is the result located? If three polarized Stokes parameters with equal amplitude are added, where in the Poincaré sphere is the result located?

References

1. R. M. A. Azzam and N. M. Bashara, *Ellipsometry and Polarized Light*, 2nd edition, Amsterdam: Elsevier (1987).
2. W. A. Shurcliff, *Polarized Light—Production and Use*, Cambridge, MA: Harvard University Press (1962).

References

3. D. Goldstein, *Polarized Light*, 2nd edition, New York, NY: Marcel Dekker (2003).
4. G. G. Stokes, On the composition and resolution of streams of polarized light from different sources, *Transactions of the Cambridge Philosophical Society* 9 (1851): 399.
5. D. J. Diner, A. Davis, B. Hancock, S. Geier, B. Rheingans, V. Jovanovic, M. Bull, D. M. Rider, R. A. Chipman, A.-B. Mahler, and S. C. McClain, First results from a dual photoelastic-modulator-based polarimetric camera, *Applied Optics* 49 (2010): 2929–2946.
6. D. J. Diner, A. Davis, B. Hancock, G. Gutt, R. A. Chipman, and B. Cairns, Dual-photoelastic-modulator-based polarimetric imaging concept for aerosol remote sensing, *Applied Optics*, 46 (2007): 8428–8445.
7. H. Poincaré, *Traite de la Lumiere*, Paris 2, 165 (1892).
8. G. P. Können, *Polarized Light in Nature*, CUP Archive (1985).

4

Interference of Polarized Light

4.1 Introduction

Interference is the ability of light waves when combined or interfered to produce interference fringes. Monochromatic light easily produces *interference fringes* and *speckle patterns* over a wide variety of situations. Thus, laser light is used for most holographic recording and as the light source for many interferometers. *Incoherent light*, such as sunlight and light from light-emitting diodes, only produces interference fringes in very restricted conditions where the optical path lengths (*OPLs*) of the combined beams differ by less than a few wavelengths. This chapter first treats the interference of coherent monochromatic polarized light. Then, the interference of *polychromatic light* is studied to understand why this beam combination process is well modeled by the *Stokes parameters*.

The performance of an interferometer is critically dependent on the flux, spectrum, wavefront quality, and polarization of the beams; the latter is our focus. The fundamental metric for coherence is the quality of the *interferograms* it can produce, the *fringe visibility*, which is the *contrast* of the fringes. The visibility of interference fringes is highest when the polarization states of the interfering beams are aligned and decreases as the angle between polarization vectors increases. The fringe visibility becomes zero when orthogonal polarizations interfere. There may still be *polarization fringes* present, a modulation of the polarization state. Even though the fringe visibility may be zero, an interference pattern can be recovered by introducing a polarizer oriented at an intermediate angle.

The interference of light has provided essential clues into the *nature of light* since Thomas Young published the results of his early interferometry experiments in 1803.[1] The fact that two equal beams of light could cancel in some areas and generate twice the flux in other areas provided compelling evidence for the *wave nature of light*. Later, Fresnel and Arago[2] discovered that interference fringes were not formed if the two beams had orthogonal polarization states. Therefore, the control of polarization is an important factor in optimizing the performance of interferometers and holographic recording setups.

4.2 Combining Light Waves

Interference refers to the measurable effects that occur when two or more light waves are combined. When the two beams are coherent, they can *constructively* and *destructively interfere*, creating interference patterns that reveal important information of the two beams.

When beams of light are combined in a linear optical medium, their electric fields add. Labeling individual beams with index q, the electric field throughout the region of space where the beams overlap becomes

$$\mathbf{E}(\mathbf{r},t) = \mathbf{E}_1(\mathbf{r},t) + \mathbf{E}_2(\mathbf{r},t) + \mathbf{E}_3(\mathbf{r},t) + \ldots = \sum_q \mathbf{E}_q(\mathbf{r},t). \tag{4.1}$$

When the beams are monochromatic, such as if the beams all derive from the same laser, all beams have a common angular frequency ω and a fixed phase relationship with respect to each other. The interference of the beams is expressed by summing their fields, where each field has an x, y, and z amplitude distribution $A(\mathbf{r})$ and a *phase distribution* $\phi(\mathbf{r})$,

$$\begin{aligned}
\mathbf{E}(\mathbf{r},t) &= \mathcal{R}e\left(e^{-i\omega t} \sum_q \mathbf{E}_q(\mathbf{r}) e^{-i\frac{2\pi}{\lambda}\mathbf{k}_q \cdot \mathbf{r}} \right) = \mathcal{R}e\left(e^{-i\omega t} \sum_q \begin{pmatrix} E_{x,q}(\mathbf{r}) \\ E_{y,q}(\mathbf{r}) \\ E_{z,q}(\mathbf{r}) \end{pmatrix} \right) \\
&= \mathcal{R}e\left(e^{-i\omega t} \sum_q \begin{pmatrix} A_{x,q}(\mathbf{r}) e^{-i\phi_x(\mathbf{r})} \\ A_{y,q}(\mathbf{r}) e^{-i\phi_y(\mathbf{r})} \\ A_{z,q}(\mathbf{r}) e^{-i\phi_z(\mathbf{r})} \end{pmatrix} \right).
\end{aligned} \tag{4.2}$$

Young's double slit interferometer, shown in Figure 4.1, creates an interference pattern by *division of wavefront*, transmitting two beams of light from an incident wavefront through two slits of nominally equal width. The light from each slit spreads out in angle due to diffraction, and interference

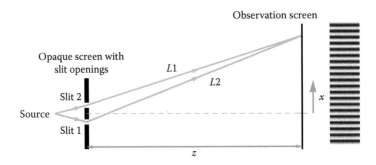

Figure 4.1 Schematic of Young's double slit. The corresponding interference pattern is shown on the right.

4.3 Interferometers   93

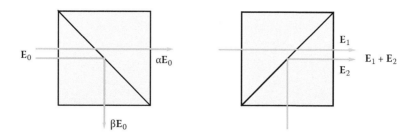

Figure 4.2 (Left) Non-polarizing beam splitter dividing incident light $\mathbf{E}_0$ into two beams $\alpha\mathbf{E}_0$ and $\beta\mathbf{E}_0$. (Right) Beam splitter combining two beams. The beam incident from the left exits with Jones vector $\mathbf{E}_1$. The beam incident from the bottom exits with $\mathbf{E}_2$. For coherent monochromatic beams, the combined beam has Jones vector $\mathbf{E}_1 + \mathbf{E}_2$.

fringes form on the observation screen. When light is normally incident on the two slits, the phase of the light emerging from both slits is equal. At the screen, the phase of the two beams differs due to the different path length ΔL to the screen as a function of angle. Where ΔL is an integral number of wavelengths, constructive interference occurs, the interference pattern is bright, and the two beams are in phase. Where the two beams are π radians out of phase, the interference pattern is dark. Since the slits are assumed to not change the polarization, the two beams have the same polarization, and polarization is not a consideration in the calculation of the interference pattern.

Young's double slit generates interference by division of wavefront, taking two separate pieces from a wavefront and combining them. Most interferometers operate by *division of amplitude*, splitting the amplitude of a wave with a *beam splitter* or *diffractive optical element* such as a grating to create two or more beams. Figure 4.2 shows a *non-polarizing beam splitter* dividing an incident beam into two beams (left) and combining two beams (right). Beam splitters can be designed to divide a beam equally into two beams of equal amplitude, but in general, the division is unequal. In practice, beam splitters also change the polarization of the light; the first beam will have one polarization state and the second beam will have a different polarization state. Beam splitters can be designed to be non-polarizing, where the polarization changes are minimal. Alternatively, *polarizing beam splitters* are common where one polarization component, the *p*-polarized component, is transmitted and the *s*-polarized component is reflected.

4.3 Interferometers

Interferometers divide and combine waves and measure the resulting constructive and destructive interference, often to obtain information about the shape of wavefronts or the shape of optical surfaces. Hundreds of configurations for interferometers have been described and interferometers are employed for a large variety of applications, including optical testing, optical communications, and holography.[3–10] Two representative interferometers are the *Mach–Zehnder interferometer*, shown in Figure 4.3, and the *Twyman–Green interferometer*, shown in Figure 4.4. Both of these interferometers split an incident wavefront into two waves, operate on the two beams separately, and then recombine the wavefronts with a beam splitter to generate an interference pattern. In Figure 4.3, the wavefront quality of a transmitting sample is shown being tested. A laser beam is collimated by a beam expander illuminating beam splitter 1. The transmitted beam, the *test beam*, is transmitted through a sample, reflecting from mirror 1, and then reflecting out of beam splitter 2 toward the camera. The beam reflected from beam splitter 1, the *reference beam*, reflects from mirror 2 and transmits through beam splitter 2. Upon exiting beam splitter 2, a lens images the sample onto a screen, a camera, or a detector where the interference can be observed. If the reference beam is a high-quality plane wave, the aberrations of the sample can be measured with the interference

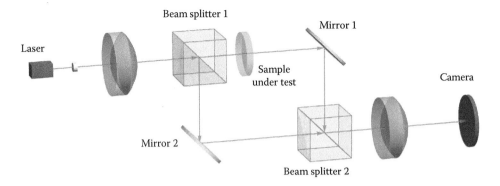

Figure 4.3 A schematic of a Mach–Zehnder interferometer, which utilizes two beam splitters to interfere beams taking two different paths. A transmitting sample under test is shown, but many different configurations of samples are used.

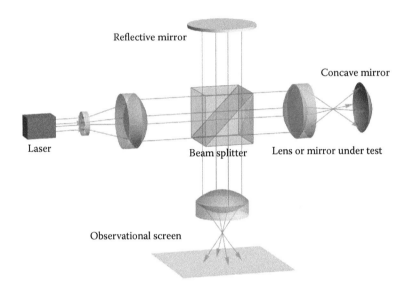

Figure 4.4 A schematic of a Twyman–Green interferometer, which uses a single beam splitter and interferes wavefronts returning from the interferometer's two arms.

pattern. The Twyman–Green interferometer in Figure 4.4 is similar, except that the single beam splitter is used twice. The Twyman–Green interferometer is common in optical shops for testing optical components.[3] The collimated transmitted beam is shown focusing after a lens, retro reflecting from a concave mirror, and double passing the lens before reflecting from the beam splitter. This configuration can be used to test either the lens, the shape of the mirror, or the combination of the two elements.

Figure 4.5 shows a *Mach–Zehnder interferometer* configured such that the polarization states and the relative amplitude of the two exiting beams can be adjusted arbitrarily. Linearly polarized light from a laser is collimated and passes through a half waveplate. A polarizing beam splitter transmits p-polarized light to mirror 1 and s-polarized light to mirror 2. Both paths have a *two-retarder polarization controller*. First, the orientation of the quarter wave

4.4 Interference of Nearly Parallel Monochromatic Plane Waves

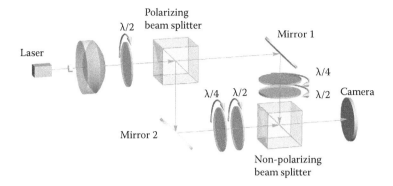

Figure 4.5 Mach–Zehnder interferometer configured to study the interference of polarized light. The rotatable quarter wave (λ/4) and half wave (λ/2) linear retarders in each path allow any pair of polarization ellipses to be generated. The initial rotatable half wave retarder adjusts the ratio of the flux in each arm. When this polarization beam splitter (PBS) outputs *p*-polarized light, the beam splitter transmits all the light to mirror 1. When the PBS outputs *s*-polarized light, the polarizing beam splitter transmits all the flux to mirror 2.

linear retarder adjusts the ellipticity of the light to any desired value from left circular through linear to right circular. Then, the orientation of the half wave linear retarder adjusts the orientation of the major axis of each polarization state's ellipse. The two beams are combined with a non-polarizing beam splitter and their interference pattern is measured by a camera, film, or some other detector. The angle between the beams is adjusted by tilting the mirrors; this controls the *spatial frequency* of the fringes. The *phase* is adjusted by translating one of the mirrors; this moves the fringes. The relative amplitude of the two beams is adjusted by rotating the initial half wave linear retarder to send more light into one channel and less into the other; this changes the *fringe visibility*. With this interferometer, *interference between two arbitrary polarization states* and *arbitrary amplitudes* can be created. Section 4.4 examines some of the resulting interference patterns.

4.4 Interference of Nearly Parallel Monochromatic Plane Waves

Consider the interference of two plane waves propagating in nearly parallel directions as shown in Figure 4.6. The first beam has polarization vector $\mathbf{E}_1$ and the second beam has polarization vector $\mathbf{E}_2$. Where the beams overlap, they interfere and the *electric fields add*,

$$\mathbf{E}(\mathbf{r},t) = \mathcal{R}e\left(\mathbf{E}_1 e^{i\left(\frac{2\pi}{\lambda}\mathbf{k}_1\cdot\mathbf{r} - \omega t - \phi_1\right)} + \mathbf{E}_2 e^{i\left(\frac{2\pi}{\lambda}\mathbf{k}_2\cdot\mathbf{r} - \omega t - \phi_2\right)}\right). \quad (4.3)$$

Let the *normalized propagation vectors* $\mathbf{k}_1$ and $\mathbf{k}_2$ lie very near the *z*-axis in the *x*–*z* plane, so the *z*-components of the **E**-field are very close to zero. The propagation angles measured from the *z*-axis are ζ_1 and ζ_2. The propagators in the phase become

$$\mathbf{k}_1 \cdot \mathbf{r} = \begin{pmatrix} \sin\zeta_1 \\ 0 \\ \cos\zeta_1 \end{pmatrix} \cdot \begin{pmatrix} x \\ y \\ z \end{pmatrix} = x\sin\zeta_1 + z\cos\zeta_1, \quad \mathbf{k}_2 \cdot \mathbf{r} = \begin{pmatrix} \sin\zeta_2 \\ 0 \\ \cos\zeta_2 \end{pmatrix} \cdot \begin{pmatrix} x \\ y \\ z \end{pmatrix} = x\sin\zeta_2 + z\cos\zeta_2. \quad (4.4)$$

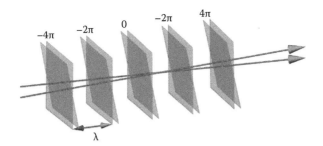

Figure 4.6 Interference of two plane waves, blue and pink, propagating at a small angle to each other and the z-axis.

Thus, the **E**-field is

$$\mathbf{E}(z,t) = \mathcal{R}e\left[e^{-i\omega t} \left(\begin{pmatrix} E_{1,x} \\ E_{1,y} \\ E_{1,z} \end{pmatrix} e^{i\frac{2\pi}{\lambda}(x\sin\zeta_1 + z\cos\zeta_1) - i\phi_1} + \begin{pmatrix} E_{2,x} \\ E_{2,y} \\ E_{2,z} \end{pmatrix} e^{i\frac{2\pi}{\lambda}(x\sin\zeta_2 + z\cos\zeta_2) - i\phi_2} \right) \right]. \quad (4.5)$$

The interference pattern can be viewed in any plane. For simplicity, let $z = 0$. As the propagation vectors approach the z-axis, the z-components approach zero. Then, the z-components can be dropped and the **E**-vectors can be replaced with 2×1 Jones vectors. The **E**-field at the observation plane becomes

$$\mathbf{E}(0,t) = \mathcal{R}e\left[e^{-i\omega t} \left(\begin{pmatrix} E_{1,x} \\ E_{1,y} \end{pmatrix} e^{i\left(\frac{2\pi}{\lambda}x\sin\zeta_1 - \phi_1\right)} + \begin{pmatrix} E_{2,x} \\ E_{2,y} \end{pmatrix} e^{i\left(\frac{2\pi}{\lambda}x\sin\zeta_2 - \phi_2\right)} \right) \right]. \quad (4.6)$$

First, consider the case where the two interfering beams are in the *same polarization state* but may have different amplitudes A_1 and A_2. The phase of the x-component can be used as the *reference phase* ($\phi_1 = 0$) and the y-phase becomes the *relative phase* ($\phi_2 = \phi$). The normalized Jones vector, now labeled as $(F_{1,x}, F_{2,x})$, is being factored out, so Equation 4.6 becomes

$$\mathbf{E}(0,t) = \mathcal{R}e\left[e^{-i\omega t} \begin{pmatrix} F_{1,x} \\ F_{1,y} \end{pmatrix} \left(A_1 e^{i\left(\frac{2\pi}{\lambda}x\sin\zeta_1\right)} + A_2 e^{i\left(\frac{2\pi}{\lambda}x\sin\zeta_2 - \phi\right)} \right) \right]. \quad (4.7)$$

The *intensity distribution* $P(x)$ in the interference pattern is

$$\begin{aligned} P(x) &= \frac{\varepsilon_0 c}{2} \mathbf{E}^\dagger \mathbf{E} \\ &= \frac{\varepsilon_0 c}{2} \left(A_1 e^{-i\left(\frac{2\pi}{\lambda}x\sin\zeta_1\right)} + A_2 e^{-i\left(\frac{2\pi}{\lambda}x\sin\zeta_2 - \phi\right)} \right) \left(A_1 e^{i\left(\frac{2\pi}{\lambda}x\sin\zeta_1\right)} + A_2 e^{i\left(\frac{2\pi}{\lambda}x\sin\zeta_2 - \phi\right)} \right) \\ &= \frac{\varepsilon_0 c}{2} \left(A_1^2 + A_2^2 + A_1 A_2 e^{i\frac{2\pi x}{\lambda}(\sin\zeta_1 - \sin\zeta_2) + i\phi} + A_1 A_2 e^{-i\frac{2\pi x}{\lambda}(\sin\zeta_1 - \sin\zeta_2) - i\phi} \right) \\ &= \frac{\varepsilon_0 c}{2} \left\{ A_1^2 + A_2^2 + 2A_1 A_2 \cos\left[\frac{2\pi x}{\lambda}(\sin\zeta_1 - \sin\zeta_2) + \phi\right] \right\}. \end{aligned} \quad (4.8)$$

4.4 Interference of Nearly Parallel Monochromatic Plane Waves

This yields a cosinusoidal flux variation, the interference pattern, with a period of $\Lambda = \lambda / (\sin \zeta_1 - \sin \zeta_2)$. The minimum flux ($P_{min}$) and the maximum flux ($P_{max}$) are

$$P_{min} = \frac{\varepsilon_0 c}{2}\left(A_1^2 + A_2^2 - 2A_1 A_2\right) \text{ and } P_{max} = \frac{\varepsilon_0 c}{2}\left(A_1^2 + A_2^2 + 2A_1 A_2\right). \tag{4.9}$$

The quality of the fringes is described by the metric *fringe visibility V*,

$$V = \frac{P_{max} - P_{min}}{P_{max} + P_{min}} = \frac{2A_1 A_2}{A_1^2 + A_2^2}, \tag{4.10}$$

which depends on the relative flux between the two beams. The fringe visibility is a maximum, $V = 1$, when the amplitudes are equal, $A_1 = A_2$, and thus the minimum flux is 0. These are the easiest fringes to detect. As the ratio of the amplitudes A_1/A_2 increases or decreases, P_{min} increases, the fringe visibility decreases, and the fringes become harder to detect. As the ratio A_1/A_2 approaches zero, P_{min} approaches P_{max}, and the fringes become difficult to measure due to the small intensity modulation. Figure 4.7 shows interference fringes of different visibility resulting from different ratios of the two fluxes. The higher the fringe visibility, the easier the fringes are to measure and the better the signal to noise of the interferometric measurement will be. For the remainder of the chapter, the term $\frac{\varepsilon_0 c}{2}$ will be dropped from flux equations and the equations are presented in normalized flux units.

Next, interference between beams with different polarization states is examined in Examples 4.1 and 4.2.

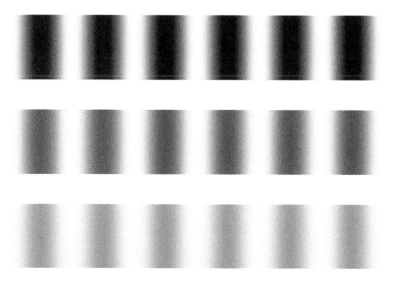

Figure 4.7 Fringe patterns between two beams with identical polarization states obtained by interfering two beams of equal flux, $P_1/P_2 = 1$, (top) yielding a fringe visibility of one, (middle) $P_1/P_2 = 1/3$ with a visibility of 0.5, and (bottom) $P_1/P_2 = 9/11$ with a visibility of 0.1.

Example 4.1 Interference of Horizontally and Vertically Polarized Light

Consider the interference between a horizontally polarized, 0°, and a vertically polarized, 90°, plane wave propagating in the x–z plane close to and symmetrically about the z-axis, $\zeta_1 = -\zeta_2 = \zeta$. Assume the beams have equal amplitudes; hence, the Jones vectors are $\mathbf{E}_1 = (1, 0)$ and $\mathbf{E}_2 = (0, 1)$. The **E**-field in the $z = 0$ plane, Equation 4.7, becomes

$$\mathbf{E}(0,t) = \mathcal{R}e\left[e^{-i\omega t} \left(\begin{pmatrix} 1 \\ 0 \end{pmatrix} e^{i\left(\frac{2\pi}{\lambda} x \sin\zeta\right)} + \begin{pmatrix} 0 \\ 1 \end{pmatrix} e^{-i\left(\frac{2\pi}{\lambda} x \sin\zeta - \phi\right)} \right) \right]. \quad (4.11)$$

This produces a Jones vector varying in the x-direction. The propagators (exponents) describe a linearly varying relative phase between the two beams. The resulting interference pattern is plotted in Figure 4.8. When the x- and y-polarized beams are in phase, the light is polarized at 45°. As the relative phase varies, polarization state is modulating along a great circle on the Poincaré sphere in the S_2 and S_3 plane from 45°, to right circular, to 135°, to left circular, and so on. These are *polarization fringes*. The intensity distribution in the polarization fringe pattern is constant,

$$P(x) = \mathbf{E}^\dagger \mathbf{E} = 1, \quad (4.12)$$

so the fringe visibility is zero. A camera will register no intensity fluctuations and thus no information beyond a constant flux. Thus, many sources stated that *orthogonal polarization states don't interfere*. These orthogonal states do interfere, the polarization state is modulating, but there are no intensity fringes. Thus, it is more precise to state that *orthogonal polarization states do not produce interference (intensity) fringes, only polarization fringes*. When a linear polarizer oriented at 45° is inserted into the polarization fringes, as in Figure 4.9,

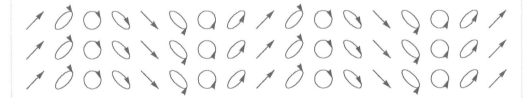

Figure 4.8 The interference of a horizontally polarized and vertically polarized beam of equal amplitude produces a periodic polarization state. The flux in these polarization fringes is constant. The maximum flux ($P_{\max}$) and the minimum flux ($P_{\min}$) are equal and the fringe visibility is zero.

Figure 4.9 Insert a linear polarizer into the polarization fringes of Figure 4.8 and intensity fringes (bright and dark bands) are recovered and can be measured.

4.4 Interference of Nearly Parallel Monochromatic Plane Waves

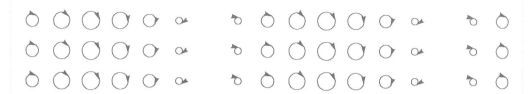

Figure 4.10 Insert a circular polarizer into the fringes of Figure 4.8 and the intensity fringes are recovered at a different phase.

intensity fringes are recovered. A circular polarizer recovers intensity fringes at a different phase, as shown in Figure 4.10. Many interferometers interfere orthogonally polarized beams and then use a polarizer to generate the fringes.

A similar pattern of polarization fringes is formed from the interference of two plane waves with 45° and 135° polarization and equal amplitudes as shown in Figure 4.11. Now, the polarization state evolves on the Poincaré sphere along the great circle through the S_1- and S_3-axes.

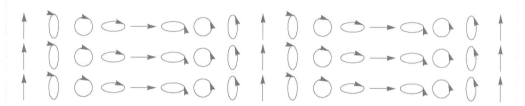

Figure 4.11 The interference of 45° and 135° tilted plane waves yields a pattern similar to the fringes of Figure 4.8. The polarization evolves along a different great circle around the Poincaré sphere in the S_1 and S_3 plane.

Example 4.2 Interference of Right and Left Circularly Polarized Light

A particularly useful set of polarization fringes is formed from the interference of right and left circularly polarized light. The **E**-field, from Equation 4.7, becomes

$$\mathbf{E}(0,t) = \frac{1}{\sqrt{2}} \mathcal{R}e \left[e^{-i\omega t} \left(\begin{pmatrix} 1 \\ i \end{pmatrix} e^{i\left(\frac{2\pi}{\lambda} x \sin\zeta + \phi\right)} + \begin{pmatrix} 1 \\ -i \end{pmatrix} e^{-i\left(\frac{2\pi}{\lambda} x \sin\zeta + \phi\right)} \right) \right]$$

$$= \sqrt{2} \mathcal{R}e \left[e^{-i\omega t} \begin{pmatrix} \cos\left(\frac{2\pi}{\lambda} x \sin\zeta + \phi\right) \\ \sin\left(\frac{2\pi}{\lambda} x \sin\zeta + \phi\right) \end{pmatrix} \right], \quad (4.13)$$

which is a rotating linearly polarized state with period $\Lambda = \lambda/(2 \sin\zeta)$, as seen in Figure 4.12. When a linear polarizer is inserted, intensity fringes are recovered. When the polarizer is rotated, as in Figure 4.13, the fringes move with the polarizer. This provides a simple method of moving an interference pattern, such as the moving or stepped fringes used in phase-stepping interferometry.

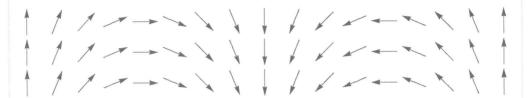

Figure 4.12 Addition of equal amplitude right and left circularly polarized beams with a 5% frequency difference yields a rotating linearly polarized state.

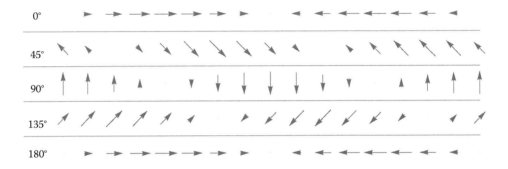

Figure 4.13 Inserting a polarizer at 0° (top) into the polarization fringes of Figure 4.12 yields intensity fringes. Rotating the polarizer (remaining rows) causes the fringe pattern to move, providing a convenient method to shift an interference pattern.

4.5 Interference of Plane Waves at Large Angles

In interferometers, the two beams are typically propagating very close to each other, within milliradians. For hologram writing, the angles are often much larger. The principles of Section 4.4 apply for arbitrary propagation directions as well, but the fringe visibility depends on the orientation of the polarization relative to the plane containing the propagation vectors $\mathbf{k}_1$ and $\mathbf{k}_2$. Figure 4.14 shows two cases of monochromatic plane waves propagating at 90° to each other. In the case on the left, the linear polarization of both beams is along the z-direction, $\mathbf{k}_1 \times \mathbf{k}_2$, and the fields $\mathbf{E}_1$ and $\mathbf{E}_2$ are parallel. In this case, the beams can constructively and destructively interfere, producing interference fringes with good visibility. In the case on the right, $\mathbf{E}_1$ and $\mathbf{E}_2$ are in the plane defined by $\mathbf{k}_1$ and $\mathbf{k}_2$ but are perpendicular to each other, $\mathbf{E}_1 \cdot \mathbf{E}_2 = 0$. These fields cannot produce intensity fringes.

Similarly, consider the case of Figure 4.15 where two circularly polarized beams, $\mathbf{E}_1$ and $\mathbf{E}_2$,

$$\mathbf{E}_1 = \frac{1}{\sqrt{2}} \begin{pmatrix} 1 \\ i \\ 0 \end{pmatrix}, \quad \mathbf{E}_2 = \frac{1}{\sqrt{2}} \begin{pmatrix} 1 \\ 0 \\ i \end{pmatrix} \tag{4.14}$$

4.5 Interference of Plane Waves at Large Angles

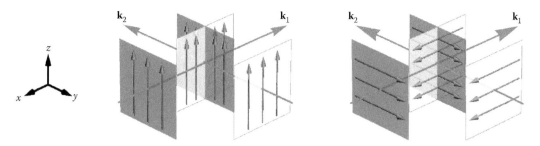

Figure 4.14 Two plane waves propagating at 90° from each other with aligned polarizations (left) and orthogonal polarizations (right).

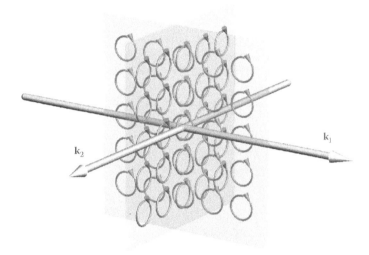

Figure 4.15 In the interference of two circularly polarized beams propagating at 90° to each other, the components along $\mathbf{k}_1 \times \mathbf{k}_2$ (vertical) can produce interference fringes while the components in the $\mathbf{k}_1$ and $\mathbf{k}_2$ plane cannot.

are propagating at a relative angle of 90°. In this case, the z-components of each beam will interfere, but the orthogonal components will not. Assuming both beams have a flux of 1, the average flux over a period of the interference pattern is 2. The flux of the x- and y-components would add separately while the amplitudes of the two z-components would add or subtract. The maximum and minimum fluxes, P_{max} and P_{min}, and the fringe visibility are

$$P_{max} = \left(\frac{1}{\sqrt{2}}\begin{pmatrix}1\\i\\0\end{pmatrix} + \frac{1}{\sqrt{2}}\begin{pmatrix}1\\0\\i\end{pmatrix}\right)^{\dagger}\left(\frac{1}{\sqrt{2}}\begin{pmatrix}1\\i\\0\end{pmatrix} + \frac{1}{\sqrt{2}}\begin{pmatrix}1\\0\\i\end{pmatrix}\right) = 3,$$

$$P_{min} = \left(\frac{1}{\sqrt{2}}\begin{pmatrix}1\\i\\0\end{pmatrix} + \frac{1}{\sqrt{2}}\begin{pmatrix}-1\\0\\-i\end{pmatrix}\right)^{\dagger}\left(\frac{1}{\sqrt{2}}\begin{pmatrix}1\\i\\0\end{pmatrix} + \frac{1}{\sqrt{2}}\begin{pmatrix}-1\\0\\-i\end{pmatrix}\right) = 1, \quad (4.15)$$

$$V = \frac{P_{max} - P_{min}}{P_{max} + P_{min}} = \frac{3-1}{3+1} = \frac{1}{2}.$$

In this case, fringes of visibility $V = \frac{1}{2}$ would result.

4.6 Polarization Considerations in Holography

The previous section described how, when beams are propagating at angles near 90° to each other, the polarization components in the plane of the propagation vectors have *reduced fringe forming capability*, because they are not parallel. The fringe visibility can be high for one polarization state and low for the other. This has important implications for setting the polarization in holographic setups.

Holography is a method for recording optical waves and recreating the waves at a later time. A *hologram* is an optical element, such as a transparency, containing a coded record of the optical wave. This coded record is often an *intensity interference pattern*. Since the recorded wave can be a complex wavefront, holograms are capable of creating *three-dimensional views* of complex objects. Large numbers of holographic setups have been described for a wide variety of tasks. One of the most common configurations will be considered in this section to highlight the polarization issues that are common to most holographic recording setups.

Figure 4.16 shows a common configuration for recording holograms. A laser is split at a *variable beam splitter*. Each beam is *spatially filtered*. The *test beam* illuminates the sample and the *scattered light* illuminates the hologram. The *reference beam* is a clean spherical wave that illuminates the hologram. The interference between the test and sample beams is recorded in photographic film or a similar holographic media and developed to form an amplitude or phase pattern. Later, when the hologram is illuminated with a copy of the reference beam, several diffraction orders are generated at and propagate beyond the hologram, including one order that will be a continuation of the test beam. This holographic wavefront contains a three-dimensional view of the object.

To produce holographic fringes with good visibility, the polarization states of the reference beam and the test beam should be nearly the same. In Figure 4.16, the best hologram is obtained when the polarization state is perpendicular to the plane of the page, as in Figure 4.14 (left). If the polarization is in the plane of the page, similar to Figure 4.14 (right), the fringe visibility will be poor. At some parts of the hologram, the reference and sample beam polarizations might be orthogonal, and no fringes would occur in that area.

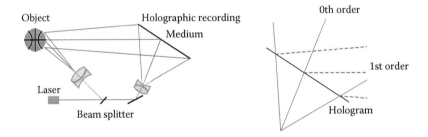

Figure 4.16 A typical optical setup for recording holograms. A laser beam with a good coherence length is split at beam splitters. Part of the light transmitted through the beam splitters is diverging, and its spherical wave illuminates the holographic recording medium. The larger fraction of the light is spatially filtered and illuminates a scattering object. The scattered light, which can be considered as a large ensemble of spherical waves emanating from the object, is incident on the holographic recording medium. The interference pattern between the object's wave and reference wave is recorded as the hologram. Later when the hologram is illuminated with a spherical wave (right), several diffracted orders result from interaction with the fine fringes in the hologram. The 0th-order beam is the continuation of the incident spherical wave. The more interesting beams are the ±1st-order diffracted beams. The first order beam is the continuation of the object wave, which, when viewed, projects a three-dimensional image of the object. The other order contains a distorted view of the object.

4.7 The Addition of Polarized Beams

4.7.1 Addition of Polarized Light of Two Different Frequencies

The addition of monochromatic beams of *two different frequencies* does not produce a *polarization ellipse*. The tip of the electric field vector traces a more general shape that can be considered as a *time-varying polarization state* that appears as an evolving ellipse. For example, consider two plane waves propagating along the z-axis. Let $\mathbf{E}_1(\mathbf{r}, t)$ be an x-polarized beam of angular frequency ω_1 combined with $\mathbf{E}_2(\mathbf{r}, t)$, which is a y-polarized beam of equal amplitude but different angular frequency ω_2. The resulting field is the sum of the individual electric fields,

$$\mathbf{E}(\mathbf{r},t) = \mathcal{R}e\left(\hat{\mathbf{x}}\mathbf{E}_1\, e^{i(k_1 z - \omega_1 t - \phi_x)} + \hat{\mathbf{y}}\mathbf{E}_2\, e^{i(k_2 z - \omega_2 t - \phi_y)}\right). \tag{4.16}$$

For a 10% frequency difference, $\omega_1 = 1.1\omega_2$, the electric field traces the pattern shown in Figure 4.17. For each short instant, the electric field is tracing a figure that is nearly an ellipse, but the ellipticity is steadily changing. This particular shape is an example of a *Lissajous figure*, the type of curve formed by the parametric equation

$$\begin{pmatrix} x(t) \\ y(t) \end{pmatrix} = \begin{pmatrix} a\,\cos(2\pi c t) \\ b\,\cos(2\pi d t + \phi) \end{pmatrix}, \tag{4.17}$$

where c and d are two integers, and a and b are arbitrary amplitudes. This wave has a constantly varying polarization ellipse because the phase difference between the $\mathbf{E}_1$ and $\mathbf{E}_2$ components is changing, $\omega_1 t - \omega_2 t = 1.1\omega_2 t - \omega_2 t = 0.1\omega_2 t$. For these two different frequencies, the phase difference increases linearly in time. This beam starts as 45° linearly polarized with the two beams in phase. The light becomes elliptically polarized as the x- and y-components become out of phase and steadily increases in its ellipticity until it is right circularly polarized. The polarization ellipse evolves into 135° polarized light when the two beams are 180° out of phase and continues changing, becoming left circularly polarized. After 10 periods of ω_1 and 11 periods of ω_2, the beams are in

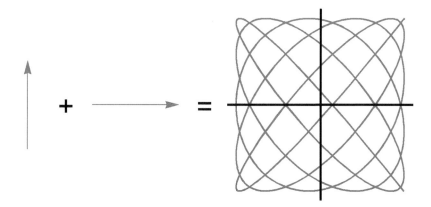

Figure 4.17 Addition of equal horizontally and vertically polarized beams with a 10% frequency difference yields a time-varying polarization state that, when averaged, behaves as unpolarized light.

phase again and the ellipse is 45° linearly polarized. The *time-varying Stokes parameters* for this evolving shape are

$$\mathbf{S}(t) = \begin{pmatrix} 2 \\ 0 \\ 2\cos(0.1\omega_2 t) \\ 2\sin(0.1\omega_2 t) \end{pmatrix}, \quad (4.18)$$

which is rapidly fluctuating, completing a cycle every 10 periods of ω_1, about 2×10^{-14} s for visible light. S_1 is zero because the wave always has equal *x*- and *y*-components.

To measure the *Stokes parameters* of this beam, a series of polarized flux measurements are performed. Each measurement requires much more than 10 optical periods; hence, the time-averaged Stokes parameters $\langle \mathbf{S}(t) \rangle$,

$$\mathbf{S} = \langle \mathbf{S}(t) \rangle = \begin{pmatrix} 2 \\ 0 \\ \langle 2\cos(0.1\omega_2 t) \rangle \\ \langle 2\sin(0.1\omega_2 t) \rangle \end{pmatrix} = \begin{pmatrix} 2 \\ 0 \\ 0 \\ 0 \end{pmatrix}, \quad (4.19)$$

are measured, yielding the Stokes parameters for *unpolarized light*. Although the measured Stokes parameters indicate unpolarized light, the beam has horizontal and vertical linearly polarized components at different frequencies. To a *Stokes polarimeter*, this beam is indistinguishable from unpolarized light.

In conclusion, *the polarization state measured by a Stokes polarimeter when combining two laser beams of different frequencies is the sum of the Stokes parameters of the two individual laser beams*. For the previous example, the Stokes parameter equation becomes

$$\mathbf{S} = \mathbf{H} + \mathbf{V} = \begin{pmatrix} 1 \\ 1 \\ 0 \\ 0 \end{pmatrix} + \begin{pmatrix} 1 \\ -1 \\ 0 \\ 0 \end{pmatrix} = \begin{pmatrix} 2 \\ 0 \\ 0 \\ 0 \end{pmatrix}. \quad (4.20)$$

Figure 4.18 shows the addition of equal beams of right and left *circularly polarized* light with a 5% frequency difference. This flower petal-like pattern describes a state that is nearly linear but steadily rotating in orientation, moving through 360° in about 20 optical periods. Optical detectors are too slow to follow this rapid evolution of the polarization state. For this wave, the component fluxes P_H and P_V are equal. Similarly, P_{45} and P_{135} are present in equal amounts, so $S_2 = 0$. Finally, careful analysis shows that the electric field vector spends equal time moving in the clockwise and counterclockwise directions, so $S_3 = P_R - P_L = 0$. These example waves are coherent, not incoherent, because the beam is periodic. The interference that occurs is too fast to observe, but the interference can be inferred and understood from the equations. Similarly, incoherent waves also interfere; however, the waves are random; the variations that occur in any particular circumstance cannot be known through measurement.

Figure 4.19 shows the results of adding several other pairs of monochromatic beams with a 10% frequency difference.

4.7 The Addition of Polarized Beams

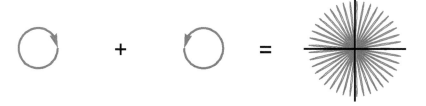

Figure 4.18 Addition of equal right and left circularly polarized beams with a 5% frequency difference yields a rotating nearly linearly polarized state.

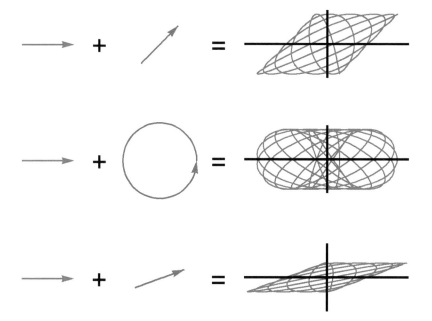

Figure 4.19 Three examples of the sum of non-orthogonally polarized beams with different frequencies. (Top) Interference of horizontal and 45° linearly polarized beams with 10% frequency difference yields a partially polarized beam at 22.5°. (Middle) Interference of horizontal and right circularly polarized beams with 10% frequency difference yields a partially elliptically polarized beam with a horizontal fast axis. (Bottom) Summation of horizontal and 20° linearly polarized beams with 10% frequency difference yields a nearly polarized beam oriented at 10°.

4.7.2 Addition of Polychromatic Beams

In this section, the measurement of a *polychromatic* unpolarized beam will be examined in detail and numerically modeled in time. A horizontally polarized polychromatic wave is added to a vertically polarized polychromatic wave, and the measurement of the resulting unpolarized beam by a Stokes polarimeter is simulated. Section 4.7.1 showed how two plane wave components at different frequencies generate rapidly evolving polarization states, reducing the *degree of polarization*. Similarly, in the next example, since most of the pairs of frequency components in two white light beams are different, the degree of polarization is greatly reduced when polychromatic beams are added.

The wavelengths in white light span the visual response of the eye, roughly from 400 to 700 nm, and the frequencies span the range from 750 to 430 terahertz (THz, 10^{12} Hz). The light's electric field is a superposition of the electric fields at each of the constituent frequencies. Thus, a polychromatic beam can be written as an integral over frequency of monochromatic beams.

A collimated white light beam propagating in the z-direction that has transmitted through a horizontal polarizer has an electric field

$$\mathbf{E}_x(\mathbf{r},t) = \mathcal{R}e\left(\hat{\mathbf{x}} \int_0^\infty A_x(\omega) e^{i(kz-\omega t-\phi_x(\omega))} d\omega\right), \quad (4.21)$$

where $A_x(\omega)$ is the real amplitude of this white light beam as a function of angular frequency and $\phi_x(\omega)$ is the phase of each frequency component at $t = 0$. Any component in the y-direction has been removed by the polarizer. The resulting polychromatic electric field is determined by integrating the electric field spectrum, $E_x(\omega)$, with respect to frequency. Similarly, another white light beam transmitted through a y-polarizer has the electric field

$$\mathbf{E}_y(\mathbf{r},t) = \mathcal{R}e\left(\hat{\mathbf{y}} \int_0^\infty A_y(\omega) e^{i(kz-\omega t-\phi_y(\omega))} d\omega\right). \quad (4.22)$$

When these two beams are combined, the polarization state and its Stokes parameters are

$$\mathbf{E}(\mathbf{r},t) = \mathbf{E}_x(\mathbf{r},t) + \mathbf{E}_y(\mathbf{r},t). \quad (4.23)$$

When performing simulations, the integral is usually replaced by the sum of discrete frequency components of the form

$$\mathbf{E}_1(z,t) = \hat{\mathbf{x}}\,\mathcal{R}e\left[\sum_{q=1}^Q \left(A_{x,q} e^{i(k_q z-\omega_q t-\phi_{x,q})}\right)\right], \quad (4.24)$$

$$\mathbf{E}_2(z,t) = \hat{\mathbf{y}}\,\mathcal{R}e\left[\sum_{r=1}^R \left(A_{y,r} e^{i(k_r z-\omega_r t-\phi_{y,r})}\right)\right]. \quad (4.25)$$

For this simulation, $Q = R = 8$ randomly generated frequencies, amplitudes, and phases are used as represented in Figure 4.20. Figure 4.21 shows the x-polarized (left first panel) and y-polarized (left second panel) electric fields for the first 3×10^{-14} s. Note that these waves are non-periodic because of their polychromatic nature. The widths and heights of individual oscillations vary between pairs of peaks. These two waves combine to form the *unpolarized* state as seen in Figure 4.21 (right). The instantaneous fluxes in the electric field components, $P(\mathbf{r}, t)$, in watts per meter squared, conveyed by this wave are proportional to the square of the component electric fields (the Poynting vector),

$$P(\mathbf{r},t) = \frac{\varepsilon_0 c}{2} |\mathbf{E}(\mathbf{r},t)|^2. \quad (4.26)$$

Figure 4.22 shows the x and y instantaneous fluxes (in units of $\varepsilon_0 c/2$), which are always positive. For flux measurements through x-polarizers (green) and y-polarizers (orange), these signals are integrated by an optical detector producing the photocurrent. The S_0 Stokes parameter is equal to the

4.7 The Addition of Polarized Beams

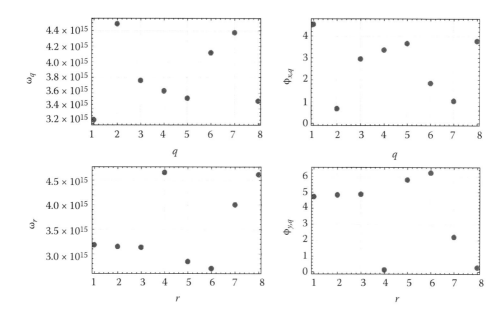

Figure 4.20 Randomly generated frequencies and phases used in Equation 4.25 for the example shown in Figure 4.21.

Figure 4.21 A short duration (10^{-13} s) showing the oscillations of an example polychromatic $\mathbf{E}_1(0, t)$, x-polarized and $\mathbf{E}_2(0, t)$, y-polarized beams (left). The resulting electric field vector traces a random figure (middle), an evolving polarization ellipse (right).

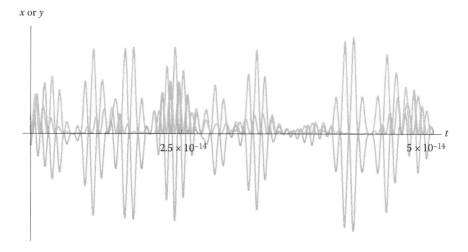

Figure 4.22 The instantaneous flux through a horizontal (green) and a vertical (orange) polarizer. The purple curve is the difference of the horizontal flux minus the vertical flux. Integrating this purple function yields S_1.

integral of the sum of the green and orange curves. Similarly, S_1 is equal to their difference, shown in Figure 4.22 in purple, a signal that goes positive and negative and has an integral that approaches zero for long integration times.

The components of the instantaneous flux through 45° and 135° polarizers is

$$\left|\hat{\mathbf{E}}_{45}^{\dagger} \cdot \mathbf{E}\right|^2 = \frac{(A_x + A_y)^2}{2} \quad \text{and} \quad \left|\hat{\mathbf{E}}_{135}^{\dagger} \cdot \mathbf{E}\right|^2 = \frac{(A_x - A_y)^2}{2}. \tag{4.27}$$

The circularly polarized fluxes can be simulated by delaying the y-polarized component by a quarter of a wavelength, $\mathbf{E}_y(t, \pi/2)$, before calculating the flux through 45° and 135° linear polarizers. The quarter wavelength delay is achromatic; it is not a delay in time; every frequency is delayed by one quarter of a period, as shown in Figure 4.23. The polarized fluxes in the four basis states, 45°, 135°, right and left circularly polarized, are plotted in Figure 4.24, and the instantaneous differences are plotted in blue. The integrals of the blue curves in Figure 4.22 and Figure 4.24 are the Stokes parameters S_1, S_2, and S_3.

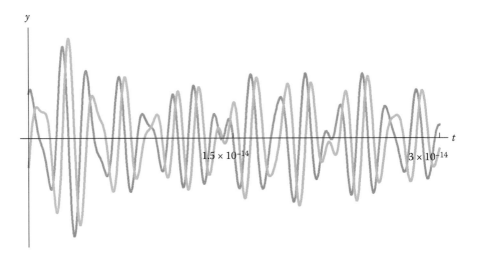

Figure 4.23 $\mathbf{E}_y(t)$, orange, and $\mathbf{E}_y(t, \pi/2)$, green, where every frequency component has been delayed by a quarter of a wavelength, which involves a longer time delay for longer wavelengths and a shorter delay for shorter wavelengths.

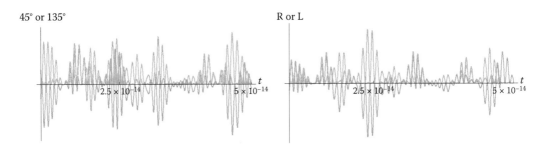

Figure 4.24 The instantaneous flux through (left) 45° (green) and 135° (orange) polarizers, and (right) (green) and left (orange) circular polarizers over 3×10^{-14} s. The blue curves show the difference as a function of time. Integrals of the blue curves yield S_2 (left) and S_3 (right). These integrals tend to become zero over times for this wave.

4.7 The Addition of Polarized Beams

$$\mathbf{E}_y(t, \pi/2) = \mathcal{R}e\left[\hat{\mathbf{y}} \sum_{r=1}^{R} \left(A_{y,r} e^{i(k_r z - \omega_r t + \pi/2)}\right)\right]. \tag{4.28}$$

This simulation demonstrated how the addition of horizontal and vertical polychromatic polarized beams of equal irradiance yield an approximately unpolarized beam. The answer is only "approximately unpolarized" because the result of this calculation is random, depending on the initial parameters and integration time. When the simulation is rerun with many different initial conditions, the answers are distributed with unpolarized light as the mean, or the most likely result. By this method, the addition of beams of arbitrary polarization states can be simulated and the resulting polarization states are well approximated by the sum of the Stokes parameters of the two separate beams. Thus, it is seen how the addition of Stokes parameters describes the addition of polychromatic waves.

4.7.3 A Gaussian Wave Packet Example

Another example of the addition of *polychromatic* waves will be given, which produces mode locked pulses with rapidly varying polarization states. Pulses with time-varying polarization states are used in *quantum optics* and *spectroscopy* to get fine control over quantum states and atomic transitions. *Complex pulses* with rapidly changing frequencies and polarizations can put atoms or molecules into particular *quantum states* with unique *density matrices*.

An approximately Gaussian wave packet is formed from the combination of $Q = 21$ parallel monochromatic plane waves with different frequencies, each linearly polarized at a different orientation. The center frequency of $\omega_0 = 3 \times 10^{15}$ radians/s corresponds to $\lambda_0 = 627.9$ nm. The 21 frequencies ω_q are separated by $\Delta\omega = 3 \times 10^{13}$ radians/s with $q = -10, -9, \ldots, 10$, where

$$\omega_q = 3 \times 10^{15} + 3 \times 10^{13} \, q. \tag{4.29}$$

The real amplitudes A_q vary as a Gaussian function (Figure 4.25),

$$A_q = \frac{\exp\left(\dfrac{-q^2}{18}\right)}{\sqrt{2\pi}}. \tag{4.30}$$

These choices are arbitrary, but the conclusions to be drawn are quite general in nature.

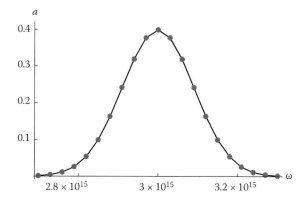

Figure 4.25 The 21 real amplitudes for the wave packet form a Gaussian envelope.

110   Chapter 4: Interference of Polarized Light

The phases are all set equal to zero at $t = 0$. Each frequency component is linearly polarized. The center frequency ω_0 is polarized at 0°. Then, there is an 18° ($\pi/10$ radians) rotation of the linear polarization angle χ between each frequency,

$$\chi_q = \frac{\pi q}{10}. \tag{4.31}$$

Hence, the corresponding time-dependent expression for the wave's electric field is

$$\mathbf{E}(t) = \sum_{q=-10}^{10} \mathcal{R}e\left[e^{-i\omega_q t} A_q \begin{pmatrix} \cos(\pi q/10) \\ \sin(\pi q/10) \end{pmatrix} \right]. \tag{4.32}$$

Figure 4.26 shows the x- and y-electric fields as a function of time. Since the amplitude distribution is approximately Gaussian, the waveform is also nearly Gaussian. Since the frequencies are discrete and periodic, the waveform is also periodic (a Fourier series) with a period of 21×10^{-12} s as seen in Figure 4.27. A Stokes polarimeter would measure the polarization state $\mathbf{S}$ as the sum of the Stokes parameters for each frequency.

$$\mathbf{S} = \sum_{q=-10}^{10} \mathbf{S}_q = \sum_{q=-10}^{10} A_q^2 \begin{pmatrix} 1 \\ \cos(\pi q/5) \\ \sin(\pi q/5) \\ 0 \end{pmatrix} = \begin{pmatrix} 1 \\ \cos(\pi q/5) \\ \sin(\pi q/5) \\ 0 \end{pmatrix} = \begin{pmatrix} 3 \\ 0.5 \\ 0 \\ 0 \end{pmatrix}. \tag{4.33}$$

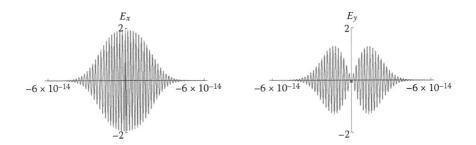

Figure 4.26 The x- and y-electric fields within one wave packet.

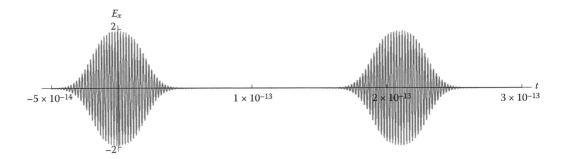

Figure 4.27 The x-component of the electric field for two pulses from the sequence of pulses.

4.7 The Addition of Polarized Beams

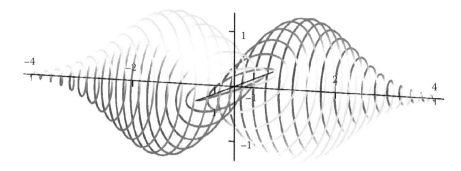

Figure 4.28 The *x*- and *y*-components versus time spiral around the axis in a steadily evolving polarization ellipse shown in a 3D view. The color cycles from red, green, blue, to red every 2.1×10^{-15} s, the approximate period. The first half of the pulse has left helicity (counterclockwise). After passing through horizontal polarization, the second half of the pulse has right helicity. Time is marked along the *z*-axis in units of 10^{-14} s.

Figure 4.28 provides a 3D view of the time-varying polarization ellipse for one Gaussian pulse width. The field is linearly polarized at 0° at the center of the pulse and then begins spiraling in ellipses of increasing ellipticity, before the flux of the pulse fades out. Figure 4.29 shows the instantaneous fluxes that constitute the Stokes parameters. Each of the frequency components has a set of Stokes parameters $\mathbf{S}_q$ listed in Table 4.1. The Stokes parameters for this beam is the sum of the component Stokes parameters, $\mathbf{S} = (3, 0.5, 0, 0)$, with *DoP* = 1/6.

Mode locked lasers produce *pulse trains* containing a *comb of frequencies*, such as Figure 4.25. This pulse begins by rotating clockwise, which will tend to cause a transition from state *A* into state *B*, changing the *angular momentum* of a molecule. A horizontally oscillating field can now drive the molecule to state *C*, and finally the counterclockwise field may drive it into another state *D*.

Mode locked lasers typically produce linearly polarized light. One method to produce this complex pulse is to utilize a thick optically active plate, such as C-cut *quartz*, with its *dispersion of optical activity*. This will rotate the plane of polarization of different frequencies by different amounts, to produce pulses with polarizations like this example. Placing the optically active plate inside the laser cavity can increase the polarization rotation by the *cavity Q-factor*. Other combinations of retarders can produce a wide array of pulse polarization properties.

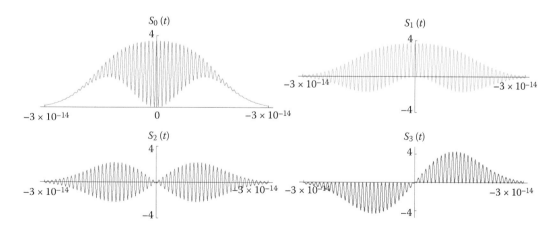

Figure 4.29 The time-dependent contributions to the Stokes parameters: (upper left) S_0, (upper right) S_1, (lower left) S_2, and (lower right) S_3. The Stokes parameters for this beam are proportional to the integrals of these functions.

Table 4.1 Stokes Parameters for Gaussian Wave Packet Spectral Components

Index	S_0	S_1	S_2	S_3
−10	0.00000238	0.00000238	0	0
−9	0.0000196	0.0000159	0.0000115	0
−8	0.000129868	0.0000401316	0.000123512	0
−7	0.000687587	−0.000212476	0.000653935	0
−6	0.00291502	−0.0023583	0.00171341	0
−5	0.0098957	−0.0098957	0	0
−4	0.0268993	−0.021762	−0.015811	0
−3	0.0585498	−0.0180929	−0.0556842	0
−2	0.102047	0.0315343	−0.0970525	0
−1	0.142418	0.115219	−0.0837113	0
0	0.159155	0.159155	0	0
1	0.142418	0.115219	0.0837113	0
2	0.102047	0.0315343	0.0970525	0
3	0.0585498	−0.0180929	0.0556842	0
4	0.0268993	−0.021762	0.015811	0
5	0.0098957	−0.0098957	0	0
6	0.00291502	−0.0023583	−0.00171341	0
7	0.000687587	−0.000212476	−0.000653935	0
8	0.000129868	0.0000401316	−0.000123512	0
9	0.0000196413	0.0000158901	−0.0000115449	0
10	0.00000238	0.00000238	0	0

The *relative phases* of the comb of frequencies are very important for shaping the pulse. If the 21 phases ϕ_q are chosen randomly over the interval $(-\pi, \pi)$, as in Figure 4.30 (left), the pulse is no longer Gaussian but has an irregular shape as in Figure 4.30 (right).

$$\mathbf{E}(t) = \sum_{q=-10}^{10} \mathcal{R}e\left[e^{-i(\omega_q t + \phi_q)} A_q \begin{pmatrix} \cos(\pi q/10) \\ \sin(\pi q/10) \end{pmatrix} \right] \tag{4.34}$$

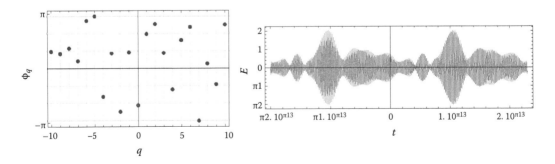

Figure 4.30 When the 21 Gaussian pulse amplitudes of Figure 4.25 are given random phases (left), the pulse shape is corrupted and irregular (right). Two periods are shown.

4.8 Conclusion

In conclusion, when adding two monochromatic beams with different wavelengths and different polarization states, the resulting polarization state evolves rapidly in time. The smaller the frequency difference, the longer the pattern takes to repeat. A polarimeter averages over many of these cycles in making its measurements, and the degree of polarization measured is reduced by averaging over polarization states. The Stokes parameters of the combined beam are equal to the sum of the Stokes parameters of the two individual laser beams.

A monochromatic beam resembles a single note from a piano, while an incoherent beam is similar to the sound played by leaning on many piano keys with your arm. Adding Jones vectors is similar to playing the exact same note on several different pianos. These sound waves will constructively and destructively interfere, and interference fringes of sound can be set up in the room. Adding Stokes parameters resembles having several small children banging on pianos in the same room.

4.9 Problem Sets

4.1 Find the degree of polarization of the three beams in Figure 4.19.

4.2 What polarization states result from adding the following Jones vectors for right $\mathbf{R} = (1,-i)/\sqrt{2}$ and left $\mathbf{L} = (1,i)/\sqrt{2}$ circularly polarized light with the following relative phases ϕ:
   a. In phase, $\phi = 0$.
   b. Out of phase, $\phi = \pi$.
   c. In quadrature, $\phi = \pi/2$.
   Assume the phase of $\mathbf{R}$ is fixed and the phase of $\mathbf{L}$ changes.

4.3 Write a program to plot the polarization ellipse for $\mathbf{E} = \begin{pmatrix} a_x + i b_x \\ a_y + i b_y \end{pmatrix}$ by drawing a series of lines from (x_j, y_j) to (x_{j+1}, y_{j+1}) where $\begin{pmatrix} x_j \\ y_j \end{pmatrix} = \mathcal{R}e\left[e^{-i 2\pi t_j} \begin{pmatrix} a_x + i b_x \\ a_y + i b_y \end{pmatrix} \right]$ for $t = (0, \Delta t, 2\Delta t, \ldots, 1)$.

4.4 When monochromatic 0° linearly polarized light is interfered with 90° linearly polarized light, the Jones vector function for the polarization fringes is of the form $\mathbf{E}(x) = e^{-i 2\pi x/\Lambda} \begin{pmatrix} 1 \\ 0 \end{pmatrix} + e^{i 2\pi x/\Lambda} \begin{pmatrix} 0 \\ 1 \end{pmatrix}$, where Λ is the period of the fringes.
   a. Plot the fringes over two periods.
   b. What is the resulting flux $P(x)$?
   c. Plot the path of the fringes on the Poincaré sphere.

4.5
   a. Plot the polarization fringes for the interference of monochromatic right and left circularly polarized beams of equal amplitude A_0 over two periods.
   b. Plot the path of the fringes on the Poincaré sphere.
   c. Plot the path of the fringes on the Poincaré sphere when the amplitudes change to $A_R = 2A_0/3$ and $A_L = 4A_0/3$. What is the resulting flux $P(x)$?
   d. Repeat (c.) for $A_R = A_0/3$ and $A_L = 5A_0/3$.

4.6 Consider the polarization fringe pattern $\mathbf{E}(x) = \begin{pmatrix} \cos(2\pi x/\Lambda) \\ i\sin(2\pi x/\Lambda) \end{pmatrix}$. This pattern can only result from the interference $\mathbf{E}(x) = e^{-i2\pi x/\Lambda}\mathbf{E}_1 + e^{i2\pi x/\Lambda}\mathbf{E}_2$ of two particular polarization states $\mathbf{E}_1$ and $\mathbf{E}_2$ in a particular amplitude ratio $|\mathbf{E}_1|/|\mathbf{E}_2|$.
   a. Plot the polarization fringes $\mathbf{E}(x)$.
   b. Find the unique states $\mathbf{E}_1$ and $\mathbf{E}_2$.

4.7 Estimate the two unique Jones vectors that interfere to yield the following interference patterns. Estimate the ratios of their amplitudes and fluxes.

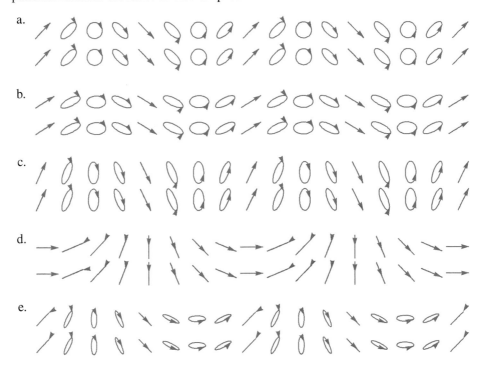

4.8 Consider the interference between the monochromatic linear $\mathbf{E}_1 = \begin{pmatrix} 1 \\ 0 \end{pmatrix}$ and elliptical $\mathbf{E}_2 = \begin{pmatrix} \cos\zeta \\ i\sin\zeta \end{pmatrix}$ beams.
   a. Find the ellipticity of $\mathbf{E}_2$ as a function of ζ.
   b. Find the fringe visibility V of the polarization fringes as a function of ζ for $0 \le \zeta \le \pi$.
   c. Plot the polarization fringes for $\zeta = 0, \pi/6, \pi/3, \pi/2, 3\pi/4$, and π.

4.9 Three lasers with frequencies $\upsilon_1 = 480$ THz, $\upsilon_2 = 500$ THz, and $\upsilon_3 = 520$ THz are combined with dichroic filters and are propagating with the same propagation vector. The associated Jones vectors are $\mathbf{E}_1 = (1, 0)$, $\mathbf{E}_2 = (\cos(\pi/3), \sin(\pi/3))$, and $\mathbf{E}_3 = (\cos(2\pi/3), \sin(2\pi/3))$.
   a. What is the resulting polarization state $\mathbf{E}_\alpha$? Use Jones vectors or Stokes parameters as appropriate.
   b. What is the degree of polarization of $\mathbf{E}_\alpha$?
   A fourth beam $\mathbf{E}_4$ is added at a frequency $\upsilon_4 = 480$ THz with the Jones vector $\mathbf{E}_4 = (0, i)$.
   c. What is the resulting polarization state $\mathbf{E}_\beta$?

4.9 Problem Sets

    d. What is the degree of polarization of $\mathbf{E}_\beta$?
    e. Now, beam $\mathbf{E}_1$ is removed. What is the degree of polarization of the combination of beams $\mathbf{E}_2$, $\mathbf{E}_3$, and $\mathbf{E}_4$?

4.10 Two polarization states, $\mathbf{E}_1 = (2, 0)$ and $\mathbf{E}_2 = (-1, i)$, are interfered. What relative phase yields the brightest beam? What relative phase yields the faintest beam? If two plane waves propagate at an angle, what will the fringe visibility be?

4.11 Young's double slit with polarizers. Young's double slit is configured with a right circular polarizer over one slit and a left circular polarizer over the other slit. The slits are very narrow and have equal width. The slits are separated by 10 wavelengths. The slits are equally illuminated with nearly monochromatic linearly polarized light at normal incidence.
    a. Describe the interference pattern formed on a screen in the far field as a function of the angle θ from the centerline between the slits.
    b. Draw the polarization fringes.

4.12 This problem simulates the addition of two incoherent polarized beams, one horizontally polarized, one vertically polarized. This example provides an example of why the superposition of polychromatic beams should be treated by the sum of Stokes parameters. Both beams are described by the polychromatic plane wave equations:

$$\mathbf{E}_1(z,t) = \mathcal{R}e\left[\hat{\mathbf{x}} \sum_{q=1}^{Q} A_{x,q} e^{i(k_q z - 2\pi v_q t + \phi_{x,q})}\right]$$

$$\mathbf{E}_2(z,t) = \mathcal{R}e\left[\hat{\mathbf{y}} \sum_{q=1}^{Q} A_{y,q} e^{i(k_q z - 2\pi v_q t + \phi_{y,q})}\right]$$

Frequencies for visible light span the range from 430 to 750 THz. For simplicity, we can drop the THz and consider eight frequencies distributed between 430 and 750 Hz. For both x- and y-components separately, generate a set of eight frequencies in our spectral band. Choose eight phases in the range from 0 to 2π. Let the eight amplitudes all equal 1.
    a. Tabulate all your values in two tables. First, present Table 1 containing your parameters for $\mathbf{E}_1(\mathbf{r}, t)$ with $1 \le q \le 8$ and Table 2 containing parameters for $\mathbf{E}_2(\mathbf{r}, t)$ with $1 \le r \le 8$.
    b. Plot the instantaneous values of $E_x(0, t, \phi_x = 0)$, $E_y(0, t, \phi_y = 0)$, and $E_y(0, t, \phi_y = \pi/2)$, which is the E_y signal where each component is shifted by a quarter of a wavelength, as a function of time for the x-polarized beam and separately for the y-polarized beam. Use enough points to resolve the oscillations and plot a list of at least 400 values.
    c. Plot the flux as a function of time transmitted through the six basis polarizers, P_H, P_V, P_{45}, P_{135}, P_R, and P_L.
    d. Perform numerical integrals to calculate the Stokes parameters. As the integration time is increased, how rapidly do the integrals converge? Your Stokes parameters will likely converge to unpolarized light, but this is not guaranteed.
    e. To understand this better, set your first y-frequency equal to your first x-frequency, and leave all the rest of the frequencies as before. Recalculate the Stokes parameters.
    f. Finally, set all eight y-frequencies equal to the eight x-frequencies. Then, set all the y-phases equal to the corresponding x-phases plus $\pi/2$. Recalculate the Stokes parameters.
    g. Draw some conclusions from parts a to part e.

4.13 a. Consider the phase shifting Twyman–Green interferometer (as shown below) which uses a PBS in conjunction with two linear quarter wave retarders to get the light through the system with minimal loss. Once the beams are recombined, an analyzer is used to get the fringes. What are the Jones matrices for the two paths before 45° linear polarizer? What is the Jones matrix for the entire system? Be sure to account for unmatched phase between the arms.

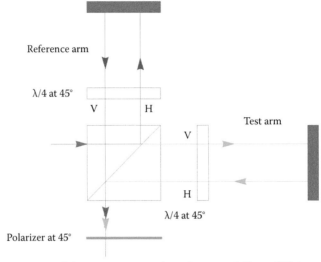

b. Suppose one of the quarter waveplates is rotated (from 45°) by some small angle δ. What is the new Jones matrix for that arm? For the system? How will this affect the measured phase?

c. Suppose the retardance of one of the waveplates is off by a small amount δ. What is the new Jones matrix for that arm? For the overall system? How will this affect the measured phase?

d. When $\mathbf{L}_{45}$ is incident, plot the polarization (before the $\mathbf{LP}_{45}$) as a function of the phase shift.

e. Plot fringe contrast as a function of input polarization.

f. For what polarization is fringe contrast maximized?

g. For what polarization state is fringe contrast minimized?

References

1. O. S. Heavens and R. W. Ditchburn, *Insight into Optics*, New York: John Wiley & Sons (1991).
2. E. Hecht, *Optics*, 4th edition, Addison Wesley (2002), pp. 386–387.
3. D. Malacara, (ed.), *Optical Shop Testing*, Vol. 59, New York: John Wiley & Sons (2007).
4. E. P. Goodwin and J. C. Wyant, Field guide to interferometric optical testing, *SPIE* (2006).
5. C. M. Vest, *Holographic Interferometry*, Vol. 476, New York: John Wiley & Sons (1979), p. 1.
6. P. K. Rastogi, *Holographic Interferometry*, Vol. 1, Heidelberg/Berlin: Springer (1994).
7. S. Tolansky, *An Introduction to Interferometry*, New York: John Wiley & Sons (1973).
8. P. L. Polavarapu, (ed.), *Principles and Applications of Polarization-Division Interferometry*, New York: John Wiley & Sons (1998).
9. A. Ya Karasik and V. A. Zubov, *Laser Interferometry Principles*, Boca Raton, FL: CRC Press (1995).
10. M. P. Rimmer and J. C. Wyant, Evaluation of large aberrations using a lateral-shear interferometer having variable shear, *Applied Optics* 14.1 (1975): 142–150.

5

Jones Matrices and Polarization Properties

5.1 Introduction

A *polarization element* is any optical element used to alter or control the polarization state of light and to transform light between polarization states. The most common polarization elements are *polarizers* and *retarders*. Mirrors, lenses, thin films, and nearly all optical elements alter polarization to some extent, but usually are not considered as polarization elements because this is not their primary role, but a side effect.

The polarization properties of polarization elements and optical elements are classified into three broad categories:

- *Diattenuation*, polarization-dependent amplitude change
- *Retardance*, polarization-dependent phase change
- *Depolarization*, random reduction in the degree of polarization

This chapter addresses the following class of problem; given a set of polarization properties, calculate the corresponding *Jones matrix*; this is the *forward problem*. The Jones matrix provides a powerful method to describe sequences of polarization elements and the intrinsic polarization properties of ray paths through optical systems. Chapter 14 continues the discussion of Jones matrices for the *inverse problem*; given an arbitrary Jones matrix, determine its polarization properties.

This chapter develops descriptions of polarizers, retarders, diattenuation, and retardance using Jones matrices. Sequences of polarization elements are modeled by matrix multiplication. Depolarization cannot be described with Jones matrices; hence, this topic is deferred until Chapter 6 where it is treated using the Mueller calculus. Chapter 9 extends the Jones matrix method into a three-dimensional representation, the *polarization ray tracing calculus.*

First, in Sections 5.2 and 5.3, the basic polarization effects and material properties are described. This is followed by the presentation of the various Jones matrices that model these properties. Then, the diattenuation and retardance components of more complex matrices are analyzed in terms of *Hermitian matrices*, *unitary matrices*, and the *singular value decomposition*.

5.2 Dichroic and Birefringent Materials

Many *polarization elements* are constructed from materials whose optical properties are described by *refractive indices* and *absorption coefficients* that depend on the polarization state. There are many examples of such polarizing materials including most crystals, such as *calcite* and *quartz*, oriented materials such as *sheet polarizers*, and *nanostructured materials* such as *diffraction gratings*, *holograms*, and *wire grid polarizers*.

When the absorption coefficient of a material depends on the polarization state, the material is *dichroic* and displays *dichroism*. Dichroism was first discovered by Biot in 1815 in the semiprecious gemstone tourmaline.[1] Figure 5.1 demonstrates the dichroism in four pieces of tourmaline illuminated from below with 0° and 90° polarized light. In the visible spectrum, light in the ordinary mode is absorbed much more than in the extraordinary mode.[2,3] A common sheet polarizer is another example of a *dichroic polarizer*.[4,5]

When the refractive index of a material depends on the polarization state, the material is *birefringent* and the *birefringence* δ is the difference between the two principal refractive indices, $\Delta n = n_1 - n_2$. A major use for birefringent materials is constructing *retarders*.

5.3 Diattenuation and Retardance

This section provides basic definitions of polarization properties of polarization elements, which leads into a discussion of their matrix representations in the next section.

Figure 5.1 Dichroism was first discovered in tourmaline. The dichroic gemstone tourmaline has a different absorption spectrum for orthogonal polarizations and partially polarizes the transmitted light. (Left) Through a 0° polarizer, these four cut and polished pieces of tourmaline are brighter and transmit more light. (Right) Through a 90° polarizer, the color shifts because the absorption bands are polarization dependent, and the stones have a deeper color.

5.3 Diattenuation and Retardance

5.3.1 Diattenuation

Diattenuation—The intensity transmittance of an element is a function of the incident polarization state, as shown in Figure 5.2.[6] The *diattenuation D* is a metric for the strength of a polarizer or partial polarizer. The diattenuation is defined in terms of the maximum T_{max} (over all polarization states) and the minimum T_{min} intensity transmittances as

$$D = \frac{T_{max} - T_{min}}{T_{max} + T_{min}}, \quad 0 \leq D \leq 1. \tag{5.1}$$

The diattenuation has the useful property that D varies from 1 for a polarizer to 0 for an element that transmits all polarization states equally, such as a retarder or a non-polarizing interaction. When unpolarized light is incident on a polarization element, the exiting degree of polarization is equal to the diattenuation of the element.

Polarizer—An optical element designed to transmit light in a specified polarization state independent of the incident polarization state. The transmission of the orthogonal state is near zero. Unpolarized incident light will exit polarized.

Ideal polarizer—The transmitted polarization state is independent of the incident polarization state; only one polarization state can exit. The orthogonal polarization state, the extinguished state, has a transmittance $T_{min} = 0$; thus, the diattenuation is

$$D = \frac{T_{max} - 0}{T_{max} + 0} = 1. \tag{5.2}$$

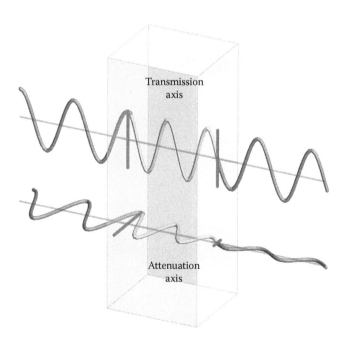

Figure 5.2 A diattenuator transmits one polarization state, aligned with the diattenuator's transmission axis, with the maximum transmittance, and the orthogonal polarization, aligned with the attenuation axis, with the minimum transmission. The "di" in diattenuator refers to two, here the two eigenpolarizations with a different transmission. In the sheet polarizer example, the attenuated state is exponentially decaying in amplitude due to a greater absorption coefficient for light polarized in the direction of the attenuation axis.

So for ideal polarizers, $D = 1$. Polarizers are never ideal and their diattenuation is close to but less than one.

Linear polarizer—A device that, when placed in an incident unpolarized beam, produces a beam of light whose electric field vector is oscillating primarily in one plane, with only a small component in the perpendicular plane.[7]

Polarization independent transmission—When all incident polarization states are transmitted with equal attenuation, then $T_{max} = T_{min}$, so the diattenuation is

$$D = \frac{T_{max} - T_{max}}{T_{max} + T_{max}} = 0. \tag{5.3}$$

An example is the ideal retarder; the polarization states change upon transmission but T_{max} and T_{min} are equal and $D = 0$.

Extinction ratio—E, the ratio of the maximum to minimum transmission, another common metric for the quality of a polarizer,

$$E = \frac{T_{max}}{T_{min}} = \frac{1+D}{1-D}. \tag{5.4}$$

The ideal polarizer has $E = \infty$.

Diattenuator—An optical element or polarization element that displays diattenuation. Polarizers have a diattenuation very close to one, but nearly all optical interfaces have some diattenuation. Examples of diattenuators include the following: polarizers and dichroic materials (sheet polarizer, tourmaline), as well as metal and dielectric interfaces with reflection and transmission differences described by Fresnel equations; thin films (homogeneous and isotropic); and diffraction gratings.

Polarization-dependent loss—PDL, the extinction ratio specified in decibels,

$$PDL = 10 \, Log_{10} \frac{T_{max}}{T_{min}}. \tag{5.5}$$

Polarization-dependent loss is a common way of specifying diattenuation in fiber optics.

Dichroism—A material property that induces diattenuation during propagation. Dichroism is associated with a difference in absorption coefficients for orthogonal polarization states.

5.3.2 Retardance

Eigenpolarization—A polarization state that exits an element or system in the same polarization state as the incident state. The exiting state is unaltered except for a possible change of amplitude and absolute phase. Every non-depolarizing polarization element has two eigenpolarizations. Any incident light not in an eigenpolarization state is transmitted in a polarization state different from the incident state. Eigenpolarizations are eigenvectors of the corresponding Jones or Mueller matrix.

Ideal retarder—A lossless optical element with two eigenpolarization states with different *optical path lengths* (*OPL*) or *phase changes*. Figure 5.3 shows the different delays for two modes propagating through a half wave retarder.

Retarder—An optical element with polarization properties close to an ideal retarder. A retarder will have some transmission loss, and typically a minor amount of diattenuation. The transmission of the two eigenpolarizations should be equal (or the element would also be a diattenuator). As an example, birefringent retarders, such as waveplates, divide incident light into two modes

5.3 Diattenuation and Retardance

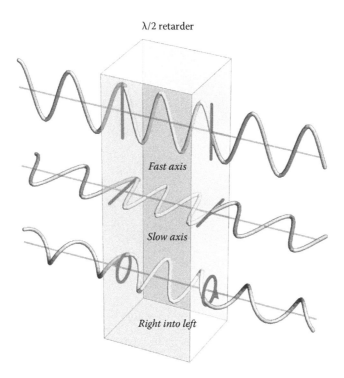

Figure 5.3 In a retarder, two modes, the top and middle waves, have different optical path lengths. The state entering polarized along the fast axis emerges first, here after 2¼ periods. The state along the slow axis emerges later, here after 2¾ periods. Light in any other polarization state emerges in a different polarization state. For this half wave linear retarder, right circular polarized light emerges left circularly polarized.

with orthogonal polarizations and delay one mode with respect to the other due to birefringence, the refractive index difference between the modes. Other retarding interactions include the following: reflections from metals, reflection and transmission at multilayer thin films, stress birefringence, and interactions with diffraction gratings. These interactions also are often both retarding and diattenuating.

Retardance—δ, the optical path difference between the eigenpolarizations. Retardance is commonly specified in four different units (see Table 5.1). In this book, retardance is generally specified in radians, so $\delta = 2\pi$ indicates one wavelength of optical path difference. π radians of retardance is 180°, $\lambda/2$, or ½ wave. The usage is generally obvious from the context.

The most common retarders in practice are *quarter wave linear retarders* and *half wave linear retarders*. Quarter wave linear retarders have a retardance of $\pi/2$ radians and are most commonly used to convert between linear and circularly polarized light. Half wave linear retarders have a retardance of π and are commonly used to rotate the plane of linear polarization.

Table 5.1 Common Units for Retardance

Radians	For one wave of retardance $\delta = 2\pi$
Degrees	For one wave of retardance $\delta = 360°$
Nanometer	For one wave of retardance $\delta = \lambda$; for 550 nm light, $\delta = 550$ nm
Waves	For one wave of retardance $\delta = 1$

Polarized Light and Optical Systems

Figure 5.4 An axis (brown) through the center of a Poincaré sphere.

Linear retarder—A retarder with fast and slow eigenpolarizations that are linearly polarized and orthogonal. Linear retarders are usually made from plates of *birefringent crystals*, particularly *calcite, magnesium fluoride, quartz, rutile,* and *yttrium vanadate*.

Circular retarder—A retarder with fast and slow eigenpolarizations that are right and left circularly polarized. *Optically active liquids* such as *glucose* and *dextrose solutions* are circular retarders.

Elliptical retarder—A retarder with fast and slow eigenpolarizations that are orthogonal elliptical polarization states. *Crystalline quartz* is an elliptical retarder, linear retarder, or circular retarder depending on how the plate is cut from the crystal.

Waveplate—Also called *retardation plate*—is a retarder constructed from a plane parallel plate or plates of linearly birefringent material. Waveplates are almost always linear retarders.

Fast axis—The eigenpolarization associated with the smaller *OPL* and the state of polarization that emerges first. For a birefringent retarder, it is the mode associated with the lower refractive index. The fast axis can be linear, elliptical, or circular. For a linear retarder, the axis is a line at a particular angle, such as 0° and 180°. For an elliptical or circular retarder, it is the corresponding elliptical polarization.

Slow axis—The eigenpolarization associated with the larger *OPL*, and for a birefringent retarder, the mode associated with the higher refractive index.

Note the term *axis* may sound as if it only refers to *linear polarization states*, but as used here, the *fast* and *slow* eigenpolarizations may also be *elliptical* or *circular*, and the term "axis" is still applied. An axis through the center of the Poincaré sphere,* shown in Figure 5.4, can pass through two orthogonal linear states, two orthogonal elliptical states, or the two orthogonal circular states.

Birefringence—A material property that induces retardance associated with propagation.

The definitions above make reference only to elements in *transmission* for simplicity. Where the term *transmission* has been used, the definitions apply to *any light beam exiting* an optical element, transmitted, reflected, diffracted by a diffraction grating, scattered from a surface or volume, and so on.

---

* Poincaré sphere is described in Chapter 3.

Math Tip 5.1   Matrix Multiplication

The matrix product of matrices **A** and **B**, **A B** or **A·B**, is calculated element by element as the inner product of the corresponding row of **A** and column of **B**.[8–10] The *i,j*th element of **A·B** is

$$(\mathbf{A} \cdot \mathbf{B})_{i,j} = \sum_{k} A_{i,k} B_{k,j}. \tag{5.6}$$

The dot · is inserted to indicate matrix multiplication when it clarifies the meaning of an equation; elsewhere, where the operation of matrix multiplication is obvious, the dot is dropped.

Matrix multiplication is order dependent since, in general, matrices do not commute,

$$\sum_{k} A_{i,k} B_{k,j} \neq \sum_{k} B_{i,k} A_{k,j}, \text{ so } \mathbf{A} \cdot \mathbf{B} \neq \mathbf{B} \cdot \mathbf{A}. \tag{5.7}$$

When **A** · **B** = **B** · **A**, **A** and **B** are referred to as *commuting matrices*.

Matrix multiplication is *associative*; adjacent matrices can be combined with parenthesis in any order and multiplied,

$$(\mathbf{A} \cdot \mathbf{B}) \cdot \mathbf{C} = \mathbf{A} \cdot (\mathbf{B} \cdot \mathbf{C}). \tag{5.8}$$

5.4   Jones Matrices

The mathematics of Jones matrices provides a systematic method to describe these polarization effects and is particularly useful to calculate the interactions of sequences of optical elements. Jones matrices provide an easy method to characterize the polarization properties of polarization elements and optical elements. Together, the Jones matrix and the Jones vector comprise the *Jones calculus*; calculus refers to a system for calculation, in this case polarization calculation.[11–17] The incident light is described by a two-element Jones vector, **E** (see Chapter 2). The polarization element is described by a Jones matrix **J**, a 2 × 2 matrix of complex elements.

The *Jones matrix* relates an arbitrary incident polarization state $\mathbf{E}_0$ to the corresponding exiting polarization state **E**′ by matrix–vector multiplication,

$$\begin{aligned}\mathbf{E}' &= \mathbf{J} \cdot \mathbf{E}_0 \\ &= \begin{pmatrix} j_{xx} & j_{xy} \\ j_{yx} & j_{yy} \end{pmatrix} \begin{pmatrix} E_x \\ E_y \end{pmatrix} = \begin{pmatrix} j_{x \leftarrow x} & j_{x \leftarrow y} \\ j_{y \leftarrow x} & j_{y \leftarrow y} \end{pmatrix} \begin{pmatrix} E_x \\ E_y \end{pmatrix} \\ &= \begin{pmatrix} j_{xx} E_x + j_{xy} E_y \\ j_{yx} E_x + j_{yy} E_y \end{pmatrix} = \begin{pmatrix} E'_x \\ E'_y \end{pmatrix}.\end{aligned} \tag{5.9}$$

The *dot* in Equation 5.9 represents matrix–vector multiplication. The Jones matrix can characterize the polarization transformations of a single polarization element or a sequence of polarization

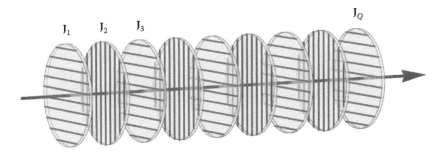

Figure 5.5 Light propagating through a series of Q polarization elements with Jones matrices $\mathbf{J}_q$.

elements. The units of the *Jones vector elements* E_x and E_y are volts per meter, the units of the electric field. The Jones matrix elements j_{xx}, j_{xy}, j_{yx}, and j_{yy}, are dimensionless.

A sequence of polarization elements as shown in Figure 5.5 has the Jones matrix $\mathbf{J}$,

$$\mathbf{J} = \mathbf{J}_Q \cdot \mathbf{J}_{Q-1} \cdots \mathbf{J}_2 \cdot \mathbf{J}_1 = \prod_{q=1}^{Q} \mathbf{J}_{Q-q+1} \tag{5.10}$$

determined by matrix multiplication. In this *product notation*, $\prod$ indicates the multiplication of a list of terms, in this case, matrix multiplication. The first element that interacts with the light is on the right side of the matrix multiplication, and the last element that interacts with the light is on the left side of the matrix multiplication. A benefit of using matrices is that the system is characterized for all incident polarization states.

5.4.1 Eigenpolarizations

Eigenvalues and *eigenvectors* assist in understanding and classifying the polarization properties of Jones matrices. The eigenvector of a matrix is a vector that maintains its direction after multiplication by the matrix; the output eigenvector is a constant times the input eigenvector. Every Jones matrix has two eigenvectors $\mathbf{E}_q$ and $\mathbf{E}_r$, and two associated eigenvalues ξ_q and ξ_r, which satisfy

$$\begin{aligned}\mathbf{J} \cdot \mathbf{E}_q = \xi_q \mathbf{E}_q = \rho_q e^{-i\phi_q} \mathbf{E}_q, \\ \mathbf{J} \cdot \mathbf{E}_r = \xi_r \mathbf{E}_r = \rho_r e^{-i\phi_r} \mathbf{E}_r.\end{aligned} \tag{5.11}$$

The amplitudes, ρ_q and ρ_r, and phases, ϕ_q and ϕ_r, of the eigenvectors may have changed during the interaction, but its state, the orientations and ellipticities, are unchanged. The eigenvalues are the corresponding change in amplitude and phase. The eigenvectors of a Jones matrix are also called *eigenpolarizations* because they represent the two and only two polarization states that are unchanged upon transmission.[18] Matrices proportional to the identity matrix are exceptions; all states are eigenpolarizations.

5.4 Jones Matrices

Math Tip 5.2 Eigenvalues and Eigenvectors of a 2 × 2 Matrix

To solve Equation 5.11 for the two eigenvalues, ξ_q, ξ_r, the *characteristic equation* is set up,

$$\det(\mathbf{J} - \xi \mathbf{I}) = \det\begin{pmatrix} j_{xx} - \xi & j_{xy} \\ j_{yx} & j_{yy} - \xi \end{pmatrix}$$
$$= (j_{xx} - \xi)(j_{yy} - \xi) - j_{xy} j_{yx} \qquad (5.12)$$
$$= 0.$$

$\mathbf{I}$ is the 2 × 2 identity matrix. The *determinant* of a matrix, det($\mathbf{J}$), is non-zero if the matrix has an inverse and its magnitude provides the scale factor by which the matrix transforms areas. The determinant of a 2 × 2 matrix is

$$\det\left[\begin{pmatrix} m_{11} & m_{12} \\ m_{21} & m_{22} \end{pmatrix}\right] = m_{11} m_{22} - m_{12} m_{21}. \qquad (5.13)$$

The right side of Equation 5.12 is a quadratic equation in ξ

$$\xi^2 - (j_{xx} + j_{yy})\xi + j_{xx} j_{yy} - j_{xy} j_{yx} = 0. \qquad (5.14)$$

The two solutions, *eigenvalues* ξ_q and ξ_r, are found in terms of the Jones matrix elements as

$$\xi_q, \xi_r = \frac{j_{xx} + j_{yy}}{2} \pm \frac{1}{2}\sqrt{(j_{xx} - j_{yy})^2 + 4 j_{xy} j_{yx}}. \qquad (5.15)$$

The eigenvalues are substituted back into Equation 5.11 to solve for the *eigenvectors*. Since eigenvectors are not required to be normalized, the eigenvectors can be set to *trial vectors*

$$\mathbf{E}_q = \begin{pmatrix} 1 \\ q \end{pmatrix} \text{ and } \mathbf{E}_r - \begin{pmatrix} 1 \\ r \end{pmatrix}, \qquad (5.16)$$

and q and r are found to be

$$q = \frac{-j_{xx} + j_{yy} + \sqrt{(j_{xx} - j_{yy})^2 + 4 j_{xy} j_{yx}}}{2 j_{xy}} = \frac{\xi_q - j_{xx}}{j_{xy}} = \frac{j_{yx}}{\xi_q - j_{yy}}, \qquad (5.17)$$

and

$$r = \frac{-j_{xx} + j_{yy} + \sqrt{(j_{xx} - j_{yy})^2 + 4 j_{xy} j_{yx}}}{2 j_{xy}} = \frac{\xi_r - j_{xx}}{j_{xy}} = \frac{j_{yx}}{\xi_r - j_{yy}}. \qquad (5.18)$$

The two normalized eigenpolarizations are

$$\hat{\mathbf{E}}_q = \frac{e^{-i\zeta_1}}{\sqrt{1+|q|^2}} \begin{pmatrix} 1 \\ q \end{pmatrix} \text{ and } \hat{\mathbf{E}}_r = \frac{e^{-i\zeta_2}}{\sqrt{1+|r|^2}} \begin{pmatrix} 1 \\ r \end{pmatrix}. \quad (5.19)$$

Phases ζ_1 and ζ_2 are arbitrary. The overall phase ζ of an eigenvector is undetermined; it is included here as a reminder that it is a free parameter. For complex-valued vectors, the condition for normalization is $\mathbf{E}_q^\dagger \cdot \mathbf{E}_q = 1$.

In the special case of Equation 5.17 where

$$j_{yx} = 0 \text{ and } \xi_q - j_{yy} = 0, \text{ then } \mathbf{E}_q = \begin{pmatrix} 0 \\ 1 \end{pmatrix}. \quad (5.20)$$

Similarly, when

$$j_{yx} = 0 \text{ and } \xi_r - j_{y,y} = 0, \text{ then } \mathbf{E}_r = \begin{pmatrix} 0 \\ 1 \end{pmatrix}. \quad (5.21)$$

An $n \times n$ square matrix will have n eigenvalues and n eigenvectors.

Jones matrices are classified into two classes, *homogeneous* and *inhomogeneous*, based on the orthogonality of the eigenpolarizations.[19] Two complex-valued vectors are orthogonal when $\mathbf{E}_q^\dagger \cdot \mathbf{E}_r = 0$.

Homogeneous polarization element—An element whose eigenpolarizations are orthogonal. Its eigenpolarizations are the states of maximum and minimum transmittance and also of maximum and minimum *OPL*. Homogeneous Jones matrices have comparatively simple properties. A homogeneous element is classified as linear, circular, or elliptical depending on the form of the eigenpolarizations. Hence, when $\mathbf{E}_q$ and $\mathbf{E}_r$ are linearly polarized states, **J** is a linear element: a linear diattenuator, a linear retarder, or a combination linear diattenuator and retarder. When $\mathbf{E}_q$ and $\mathbf{E}_r$ are circularly polarized states, **J** is a circular element. Similarly, when $\mathbf{E}_q$ and $\mathbf{E}_r$ are elliptically polarized states, **J** is an elliptical element.

Inhomogeneous polarization element—An element whose eigenpolarizations are not orthogonal, $\mathbf{E}_q^\dagger \cdot \mathbf{E}_r \neq 0$. This occurs when the diattenuation axis and retardance axis are not aligned. Inhomogeneous Jones matrices have more complex properties and *cannot be simply classified as linear, circular, or elliptical elements*. The eigenpolarizations of inhomogeneous matrices are generally not the states of maximum and minimum transmittance. Such inhomogeneous elements will also display different polarization characteristics for forward and backward propagating beams. A detailed treatment of inhomogeneous matrices is contained in Chapter 14.

Example 5.1 Inhomogeneous Matrix

The matrix $\begin{pmatrix} 1 & -1/2 \\ 0 & 1/2 \end{pmatrix}$ has eigenvalues $\xi_q = 1$ and $\xi_r = \frac{1}{2}$ associated with eigenvectors $\mathbf{E}_q = \begin{pmatrix} 1 \\ 0 \end{pmatrix}$, linearly polarized light at 0°, and $\mathbf{E}_r = \begin{pmatrix} 1 \\ 1 \end{pmatrix}$, linearly polarized light at 45°. Thus, it is inhomogeneous since $\mathbf{E}_q^\dagger \cdot \mathbf{E}_r \neq 0$.

5.4 Jones Matrices

Table 5.2 Notation for Ideal Polarization Element Matrices

CR(δ)	Circular retarder, retardance δ	Equation 5.48
ED(d_H, d_{45}, d_R)	Elliptical diattenuator, diattenuation components d_H, d_{45}, d_L	Equation 5.56
EP(ε, ψ)	Elliptical polarizer, ellipticity ε, orientation ψ	Equation 5.30
ER($\delta_H, \delta_{45}, \delta_R$)	Elliptical retarder, retardance components $\delta_H, \delta_{45}, \delta_L$	Equation 5.60
LD(t_1, t_2, θ)	Linear diattenuator (partial polarizer), amplitude transmittances t_1, t_2 ($t_1 \geq t_2$), t_1 axis at θ	Equation 5.33
LP(θ)	Linear polarizer, transmission axis at θ	Equation 5.29
LR(δ, θ)	Linear retarder, retardance δ, oriented with fast axis at θ	Equation 5.41
R(θ)	Rotation matrix for change of basis by an angle θ	Equation 5.23

5.4.2 Jones Matrix Notation

The same symbols are used for Jones matrices and Mueller matrices, since they always occur separately, and their multiplications produce parallel outputs. Retarder Jones matrices are assumed to be in the symmetric phase convention unless otherwise specified (Table 5.2).

5.4.3 Rotation of Jones Matrices

Many optical systems rotate polarizers or retarders as shown in Figure 5.6 to change and modulate polarization states. Given a polarization element's initial Jones matrix **J**, the new Jones matrix **J**(θ) when the element is rotated about the light propagation direction by angle θ is

$$\mathbf{J}(\theta) = \mathbf{R}(\theta) \cdot \mathbf{J} \cdot \mathbf{R}(-\theta), \tag{5.22}$$

where the *Jones rotation matrix* **R** is the same as the two-dimensional Cartesian rotation matrix,

$$R(\theta) = \begin{pmatrix} \cos\theta & -\sin\theta \\ \sin\theta & \cos\theta \end{pmatrix}. \tag{5.23}$$

This is an example of a *unitary change of basis*. Similarly, a Jones vector **E** rotated by θ becomes $\mathbf{E}(\theta) = \mathbf{R}(\theta) \cdot \mathbf{E}$.

Figure 5.6 A polarization element being rotated about the incident light beam and the element's surface normal.

A way to remember the location of the minus sign on the j_{xy} and j_{yx} elements in $\mathbf{R}(\theta)$ is to rotate $\mathbf{H} = \begin{pmatrix} 1 \\ 0 \end{pmatrix}$ into $\widehat{\mathbf{45}} = \frac{1}{\sqrt{2}} \begin{pmatrix} 1 \\ 1 \end{pmatrix}$ using

$$\mathbf{R}(45°) \cdot \mathbf{H} = \frac{1}{\sqrt{2}} \begin{pmatrix} 1 & -1 \\ 1 & 1 \end{pmatrix} \cdot \begin{pmatrix} 1 \\ 0 \end{pmatrix} = \frac{1}{\sqrt{2}} \begin{pmatrix} 1 \\ 1 \end{pmatrix}. \tag{5.24}$$

The Jones rotation matrices $\mathbf{R}(\theta)$ have the following properties:

1. For a series of rotations, the angles are additive,

$$\mathbf{R}(\alpha) \cdot \mathbf{R}(\beta) = \mathbf{R}(\alpha + \beta). \tag{5.25}$$

2. $\mathbf{R}(-\theta)$ is the inverse of $\mathbf{R}(\theta)$,

$$\mathbf{R}(\alpha) \cdot \mathbf{R}(-\alpha) = \mathbf{R}(\alpha) \cdot \mathbf{R}^{-1}(\alpha) = \mathbf{R}(\alpha - \alpha) = \mathbf{R}(0) = \begin{pmatrix} 1 & 0 \\ 0 & 1 \end{pmatrix}. \tag{5.26}$$

3. The $\mathbf{R}$'s commute, $\mathbf{R}(\alpha) \cdot \mathbf{R}(\beta) = \mathbf{R}(\beta) \cdot \mathbf{R}(\alpha)$.

When both the incident state and the polarization element rotate together by θ,

$$[\mathbf{R}(\theta) \cdot \mathbf{J} \cdot \mathbf{R}(-\theta)] \cdot [\mathbf{R}(\theta) \cdot \mathbf{E}] = \mathbf{R}(\theta) \cdot \mathbf{J} \cdot \mathbf{E}, \tag{5.27}$$

the two rotations $\mathbf{R}(-\theta) \cdot \mathbf{R}(\theta)$ cancel, which provides a way to remember that rotating a matrix is $[\mathbf{R}(\theta) \cdot \mathbf{J} \cdot \mathbf{R}(-\theta)]$, not the reverse.

5.5 Polarizer and Diattenuator Jones Matrices

5.5.1 Polarizer Jones Matrices

Ideal polarizers completely transmit one polarization state and completely attenuate the orthogonal polarization state. The Jones matrix for a *horizontal linear polarizer*, $\mathbf{LP}(0)$, transmits all of the *x*-component of the incident light and blocks the *y*-component,

$$\mathbf{LP}(0°) \cdot \begin{pmatrix} E_x \\ E_y \end{pmatrix} = \begin{pmatrix} 1 & 0 \\ 0 & 0 \end{pmatrix} \cdot \begin{pmatrix} E_x \\ E_y \end{pmatrix} = \begin{pmatrix} E_x \\ 0 \end{pmatrix}. \tag{5.28}$$

The Jones matrices $\mathbf{LP}(0)$ for linear polarizers oriented to pass light polarized at an angle θ can be calculated from Equation 5.22,[20]

$$\mathbf{LP}(\theta) = \mathbf{R}(\theta) \cdot \mathbf{LP}(0) \cdot \mathbf{R}(-\theta) = \begin{pmatrix} \cos^2 \theta & \cos \theta \sin \theta \\ \cos \theta \sin \theta & \sin^2 \theta \end{pmatrix}. \tag{5.29}$$

5.5 Polarizer and Diattenuator Jones Matrices

Table 5.3 Polarizer Jones Matrices

Transmission Axis	Polarizer Jones Matrices
Horizontal linear polarizer **HLP**, **L**(0°)	$\begin{pmatrix} 1 & 0 \\ 0 & 0 \end{pmatrix}$
Vertical linear polarizer **VLP**, **L**(90°)	$\begin{pmatrix} 0 & 0 \\ 0 & 1 \end{pmatrix}$
45° Linear polarizer **L**(45°)	$\dfrac{1}{2}\begin{pmatrix} 1 & 1 \\ 1 & 1 \end{pmatrix}$
135° Linear polarizer **L**(135°)	$\dfrac{1}{2}\begin{pmatrix} 1 & -1 \\ -1 & 1 \end{pmatrix}$
Right circular polarizer **RCP**	$\dfrac{1}{2}\begin{pmatrix} 1 & i \\ -i & 1 \end{pmatrix}$
Left circular polarizer **LCP**	$\dfrac{1}{2}\begin{pmatrix} 1 & -i \\ i & 1 \end{pmatrix}$

Table 5.3 lists the Jones matrices for the six basis polarization states. Any table of Jones matrices must adopt a convention for the absolute phase. The *absolute phase* of a Jones matrix is a phase that multiplies the whole Jones matrix. Here, the overall absolute phase of the Jones matrix is chosen so that the j_{xx} is real and positive. Any of these matrices can be multiplied by -1 or by $e^{i\phi}$ and the action of the polarizer matrix remains the same, although the exiting phase will be different.

To test if a Jones matrix represents a polarizer, calculate the eigenvalues. If one of the eigenvalues is zero, the element is a polarizer.

Example 5.2 Operation of a Linear Polarizer on the Eigenpolarization States

The linear polarizer **LP**(θ) transmits all of the linearly polarized state $\hat{\mathbf{L}}(\theta)$, which is an eigenpolarization with eigenvalue $\xi_1 = 1$,

$$\mathbf{LP}(\theta) \cdot \hat{\mathbf{L}}(\theta) = \begin{pmatrix} \cos^2\theta & \cos\theta\sin\theta \\ \cos\theta\sin\theta & \sin^2\theta \end{pmatrix} \cdot \begin{pmatrix} \cos\theta \\ \sin\theta \end{pmatrix} = \begin{pmatrix} \cos\theta \\ \sin\theta \end{pmatrix} = 1 \cdot \begin{pmatrix} \cos\theta \\ \sin\theta \end{pmatrix}.$$

The linear polarizer **LP**(θ) completely attenuates the state $\hat{\mathbf{L}}(\theta+90°)$, which is an eigenpolarization with eigenvalue $\xi_2 = 0$,

$$\mathbf{LP}(\theta) \cdot \hat{\mathbf{L}}(\theta+90°) = \begin{pmatrix} \cos^2\theta & \cos\theta\sin\theta \\ \sin\theta\cos\theta & \sin^2\theta \end{pmatrix} \cdot \begin{pmatrix} \sin\theta \\ -\cos\theta \end{pmatrix} = \begin{pmatrix} 0 \\ 0 \end{pmatrix} = 0 \cdot \begin{pmatrix} \sin\theta \\ -\cos\theta \end{pmatrix}.$$

The Jones matrix for the ideal *elliptical polarizer* that passes all of the state $\mathbf{EP}(\varepsilon, \psi)$, Equation 2.41, with *ellipticity* ε oriented at angle ψ, and blocks the orthogonal state $\mathbf{EP}(-\varepsilon, \psi + \pi/2)$ is

$$\mathbf{EP}(\varepsilon, \psi) = \frac{1}{1+\varepsilon^2} \begin{pmatrix} \cos^2 \psi + \varepsilon^2 \sin^2 \psi & i\varepsilon + (\varepsilon^2 - 1)\cos\psi\sin\psi \\ -i\varepsilon - (\varepsilon^2 - 1)\cos\psi\sin\psi & \varepsilon^2 \cos^2 \psi + \sin^2 \psi \end{pmatrix}. \quad (5.30)$$

5.5.2 Linear Diattenuator Jones Matrices

Diattenuators are *partial polarizers*. One eigenpolarization has amplitude transmittance t_1 and the orthogonal eigenpolarization has amplitude transmittance t_2. Because of the two different transmittances, or attenuations, the term *diattenuation* is used to refer to the process of polarization-dependent transmittance. Thus, *diattenuator* refers to an optical element with diattenuation. The incident states of maximum and minimum transmittance of a *homogeneous diattenuator* are orthogonal polarization states. A *pure diattenuator* has only diattenuation and thus zero retardance.

To generate the Jones matrices for diattenuators, first consider a horizontal linear diattenuator, $\mathbf{LD}(t_{max}, t_{min}, 0°)$, with amplitude transmission t_{max} for the x-component, and t_{min} for the y-component ($t_{max} > t_{min}$, t_{max} and t_{min} are *positive definite*, i.e., real and non-negative), with the transmission axis for t_{max} oriented at 0°,

$$\mathbf{LD}(t_1, t_2, 0°) = \mathbf{LD}(t_{max}, t_{min}, 0) = \begin{pmatrix} t_{max} & 0 \\ 0 & t_{min} \end{pmatrix} = \begin{pmatrix} \sqrt{T_{max}} & 0 \\ 0 & \sqrt{T_{min}} \end{pmatrix}, \quad (5.31)$$

where $T_{max} = t_{max}^2$ and $T_{min} = t_{min}^2$ are the corresponding intensity transmittances. The effect on incident light is to attenuate the x- and y-components by different amounts,

$$\mathbf{LD}(t_1, t_2, 0) \cdot \mathbf{E} = \begin{pmatrix} t_{max} & 0 \\ 0 & t_{min} \end{pmatrix} \cdot \begin{pmatrix} E_x \\ E_y \end{pmatrix} = \begin{pmatrix} t_{max} E_x \\ t_{min} E_y \end{pmatrix} = \begin{pmatrix} E'_x \\ E'_y \end{pmatrix} = \mathbf{E}', \quad (5.32)$$

where $\mathbf{E}'$ is the exiting state. The linear diattenuator with the transmitting axis at an angle θ is

$$\begin{aligned}\mathbf{LD}(t_1, t_2, \theta) &= \mathbf{R}(\theta) \cdot \mathbf{LD}(t_1, t_2, 0) \cdot \mathbf{R}(-\theta) \\ &= \begin{pmatrix} t_{max} \cos^2\theta + t_{min} \sin^2\theta & (t_{max} - t_{min})\cos\theta\sin\theta \\ (t_{max} - t_{min})\cos\theta\sin\theta & t_{max}\sin^2\theta + t_{min}\cos^2\theta \end{pmatrix}.\end{aligned} \quad (5.33)$$

The diattenuation of $\mathbf{LD}(t_1, t_2, \theta)$ is found by converting the amplitude transmittances to intensity transmittances by squaring,

$$D\left[\mathbf{LD}(t_{max}, t_{min}, 0°)\right] = \frac{t_{max}^2 - t_{min}^2}{t_{max}^2 + t_{min}^2}. \quad (5.34)$$

5.5 Polarizer and Diattenuator Jones Matrices

The relationship between t_{min}, t_{max} and the diattenuation is

$$t_{min} = t_{max}\sqrt{\frac{1-D}{1+D}}. \tag{5.35}$$

Figure 5.7 plots the four elements of the Jones matrix for linear diattenuators with t_{max} equal to 1.

Example 5.3 Derive the Jones Matrix for a Linear Diattenuator

A Jones matrix can be derived from a specification linking two input Jones vectors to two output Jones vectors. The following example derives a particular linear diattenuator with its transmission axis oriented at 45°. The Jones matrix is defined by the following two conditions:

1. All of the incident 45° linearly polarized light is transmitted as 45° linearly polarized light.
2. ¼ of the incident 135° linearly polarized light amplitude (or 1/16 of the flux) is transmitted as 135° linearly polarized light.
   a. Set up the two Jones matrix equations for these conditions and find the corresponding linear equations for the Jones matrix elements. Then solve for the individual elements.
      Since

$$\begin{pmatrix} a & b \\ c & d \end{pmatrix} \cdot \frac{1}{\sqrt{2}}\begin{pmatrix} 1 \\ 1 \end{pmatrix} = \frac{1}{\sqrt{2}}\begin{pmatrix} 1 \\ 1 \end{pmatrix},$$

$$\begin{pmatrix} a & b \\ c & d \end{pmatrix} \cdot \frac{1}{\sqrt{2}}\begin{pmatrix} 1 \\ -1 \end{pmatrix} = \frac{1}{4}\frac{1}{\sqrt{2}}\begin{pmatrix} 1 \\ -1 \end{pmatrix},$$

$a + b = 1$, $c + d = 1$, $a - b = 1/4$, and $c - d = -1/4$. Hence,

$$\mathbf{J} = \begin{pmatrix} a & b \\ c & d \end{pmatrix} = \frac{1}{8}\begin{pmatrix} 5 & 3 \\ 3 & 5 \end{pmatrix}.$$

   b. Find the determinant, eigenvalues, eigenvectors, and matrix inverse.

$$\text{Determinant} = \det(\mathbf{J}) = ad - bc = \frac{5}{8} \times \frac{5}{8} - \frac{3}{8} \times \frac{3}{8} = \frac{1}{4}.$$

Eigenvalues: $\det(\mathbf{J} - \xi\mathbf{I}) = 0 = (a-\xi)(d-\xi) - bc$, $\xi_q = 1$, $\xi_r = 1/4$.

Eigenvectors: $\mathbf{J} \cdot \mathbf{E_q} = \xi_q \mathbf{E_q}$, $\mathbf{E_q} = 1\begin{pmatrix} E_x \\ E_y \end{pmatrix} = \frac{1}{8}\begin{pmatrix} 5E_x + 3E_y \\ 3E_x + 5E_y \end{pmatrix}$, $\mathbf{E_q} = \begin{pmatrix} 1 \\ 1 \end{pmatrix}$, $\mathbf{E_r} = \begin{pmatrix} 1 \\ -1 \end{pmatrix}$.

Matrix inverse: $\mathbf{J}^{-1} = \dfrac{1}{2}\begin{pmatrix} 5 & -3 \\ -3 & 5 \end{pmatrix}$.

c. Show that when linearly polarized light is incident, the transmitted light is always linearly polarized and find the corresponding angle of polarization ψ.

Multiply the Jones vector for linearly polarized light oriented at angle θ by **J**:

$$\frac{1}{8}\begin{pmatrix} 5 & 3 \\ 3 & 5 \end{pmatrix}\begin{pmatrix} \cos\theta \\ \sin\theta \end{pmatrix} = \frac{1}{8}\begin{pmatrix} 5\cos\theta + 3\sin\theta \\ 3\cos\theta + 5\sin\theta \end{pmatrix}, \quad \psi = \arctan\left(\frac{3\cos\theta + 5\sin\theta}{5\cos\theta + 3\sin\theta}\right).$$

The final polarization state only has real elements; therefore, it must be linear. There are no complex terms and therefore no phase differences between the *x*- and *y*-components to generate ellipticity; both the *x*- and *y*-components go to zero simultaneously, twice per period.

Example 5.4 Polarizer Tests

The matrix

$$\mathbf{J}_1 = \begin{pmatrix} 0 & 1/2 \\ 0 & 1/2 \end{pmatrix} \tag{5.36}$$

has eigenvalues $\xi_q = \tfrac{1}{2}$ and $\xi_r = 0$. Since one of the eigenvalues is 0, the corresponding eigenvector (1, 0) is completely blocked and only one state exits (1, 1). Since the eigenvectors (1, 0) and (1, 1) are not orthogonal, the polarizer is inhomogeneous.

The matrix

$$\mathbf{J}_1 = \frac{1}{200}\begin{pmatrix} 99 & 101 \\ 101 & 99 \end{pmatrix} \tag{5.37}$$

has eigenvalues $\xi_q = 1$ and $\xi_r = 1/100$. Since neither eigenvalue is 0, the element is not a polarizer but a diattenuator. The eigenvectors (1, 1) and (1, −1) are orthogonal, so the diattenuator is homogeneous. The diattenuation is

$$D = \frac{1^2 - \left(\dfrac{1}{100}\right)^2}{1^2 + \left(\dfrac{1}{100}\right)^2} = \frac{999}{1001}. \tag{5.38}$$

5.6 Retarder Jones Matrices

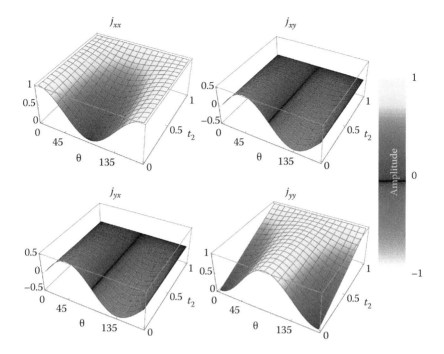

Figure 5.7 The Jones matrix elements are plotted for linear diattenuators with $t_1 = 1$, **LD**$(1, t_2, \theta)$. All elements are real. Ideal polarizers are along the front edge. The identity matrix ($t_1 = t_2 = 1$, non-diattenuating Jones matrices) are across the back. Cyan indicates 1, black indicates 0, and brown and yellow indicate values less than 1.

5.6 Retarder Jones Matrices

Ideal retarders introduce a different phase change into two orthogonal eigenpolarizations. An ideal retarder is lossless. Real retarders have some loss and usually a very small amount of diattenuation as well.

5.6.1 Linear Retarder Jones Matrices

The linear retarder **LR**(δ, θ) has two orthogonal eigenpolarizations, $\hat{\mathbf{L}}(\theta)$ and $\hat{\mathbf{L}}(\theta + 90°)$. The linearly polarized state $\hat{\mathbf{L}}(\theta)$ remains unchanged upon transmission through the retarder. Only the phase of this eigenpolarization changes. It changes by ϕ_1, yielding an eigenvalue $\xi_1 = e^{-i\phi_1}$, the minus sign arising from the decreasing phase convention as described in Section 2.17. Since this is a retarder, the phase change ϕ_2 for the orthogonal state $\hat{\mathbf{L}}(\theta + 90°)$ is different from ϕ_1, yielding the eigenvalue $\xi_2 = e^{-i\phi_2}$. The retardance is $\delta = |\phi_1 - \phi_2|$, the absolute value of the phase difference between the two eigenpolarizations. The associated Jones matrix for $\theta = 0$ is

$$\mathbf{LR}(\delta, 0) = \begin{pmatrix} e^{-i\phi_1} & 0 \\ 0 & e^{-i\phi_2} \end{pmatrix}. \quad (5.39)$$

A retarder has two degrees of freedom, the *absolute phase change* and the *retardance*, but often only one, the retardance, is specified. Thus, when constructing the Jones matrix for a retarder, there is a *choice of phase* associated with a *phase convention*:

A. *Slow axis unchanged convention*, advances the phase of the fast eigenpolarization, leaving the slow eigenpolarization's phase unchanged
B. *Fast axis unchanged convention*, delays the phase of the slow eigenpolarization, leaving the fast eigenpolarization's phase unchanged
C. The *symmetric phase convention*, which advances the phase of the fast eigenpolarization and retards the phase of the slow eigenpolarization symmetrically
D. Selecting some other division of the phase between the eigenpolarizations

For a linear retarder with a horizontally polarized fast axis, these four choices appear as follows:

$$\text{A. } \begin{pmatrix} e^{-i\delta} & 0 \\ 0 & 1 \end{pmatrix}, \text{ B. } \begin{pmatrix} 1 & 0 \\ 0 & e^{i\delta} \end{pmatrix}, \text{ C. } \begin{pmatrix} e^{-i\delta/2} & 0 \\ 0 & e^{i\delta/2} \end{pmatrix}, \text{ D. } \begin{pmatrix} e^{-i(\delta/2+\phi)} & 0 \\ 0 & e^{i(\delta/2-\phi)} \end{pmatrix}. \tag{5.40}$$

The symmetric phase form integrates best with the description of polarization properties using Pauli matrices presented in Chapter 14; hence, this is the form preferred here, but hand calculations are frequently easier with forms A and B. The Jones matrices for *linear retarders* **LR**(δ, θ) with a fast axis orientation θ in the symmetric phase convention is

$$\begin{aligned}\mathbf{LR}(\delta, \theta) &= \mathbf{R}(\theta) \begin{pmatrix} e^{-i\delta/2} & 0 \\ 0 & e^{i\delta/2} \end{pmatrix} \mathbf{R}(-\theta) \\ &= \begin{pmatrix} e^{-i\delta/2} \cos^2 \theta + e^{i\delta/2} \sin^2 \theta & -i\sin\left(\dfrac{\delta}{2}\right)\sin(2\theta) \\ -i\sin\left(\dfrac{\delta}{2}\right)\sin(2\theta) & e^{i\delta/2} \cos^2 \theta + e^{-i\delta/2} \sin^2 \theta \end{pmatrix}.\end{aligned} \tag{5.41}$$

Table 5.4 tabulates the common Jones matrices for linear and circular quarter and half wave retarders. Figure 5.8 plots four elements of the linear retarder Jones matrix for all retardances and orientations. Because the Jones matrix elements are complex, both amplitude and phase are represented. Amplitude is represented by the height of the 2D surface, and color encodes the phase of the element. Across the front edge of the plot of the four elements, the retardance $\delta = 0$, and the Jones matrix is the identity matrix. Retarders along the left and right sides (i.e., $\theta = 0°$ or $180°$) are horizontal retarders; vertical retarders are on the line up the middle (i.e., $\theta = 90°$). Across the back side (i.e., $\delta = 180°$) are all the half wave linear retarders.

The Jones matrices for *quarter wave linear retarders* in the symmetric phase form **LR**($\pi/2$, θ) are

$$\mathbf{LR}(\pi/2, \theta) = \frac{1}{\sqrt{2}} \begin{pmatrix} 1 - i\cos 2\theta & -2i\cos\theta\sin\theta \\ -2i\cos\theta\sin\theta & 1 + i\cos 2\theta \end{pmatrix}. \tag{5.42}$$

Quarter wave retarders transform between linearly polarized light oriented at ±45° from the fast axis and circularly polarized light. *Half linear wave retarders* **LR**(π, θ) in the symmetric phase convention have the Jones matrix

$$\mathbf{LR}(\pi, \theta) = \begin{pmatrix} -i\cos 2\theta & -i\sin 2\theta \\ -i\sin 2\theta & i\cos 2\theta \end{pmatrix}. \tag{5.43}$$

5.6 Retarder Jones Matrices

Table 5.4 Retarder Jones Matrices in the Decreasing Phase Convention for Three Absolute Phase Conventions: Fast Axis Unchanged, Symmetric Phase Convention, and Slow Axis Unchanged

	Retarders: Quarter Wave		
Fast Axis	Fast Axis Unchanged	Symmetric Phase Convention	Slow Axis Unchanged
Horizontal, 0° **LR** ($\pi/2$, 0°)	$\begin{pmatrix} 1 & 0 \\ 0 & i \end{pmatrix}$	$\frac{1}{\sqrt{2}}\begin{pmatrix} 1-i & 0 \\ 0 & 1+i \end{pmatrix}$	$\begin{pmatrix} -i & 0 \\ 0 & 1 \end{pmatrix}$
Vertical, 90° **LR** ($\pi/2$, 90°)	$\begin{pmatrix} i & 0 \\ 0 & 1 \end{pmatrix}$	$\frac{1}{\sqrt{2}}\begin{pmatrix} 1+i & 0 \\ 0 & 1-i \end{pmatrix}$	$\begin{pmatrix} 1 & 0 \\ 0 & -i \end{pmatrix}$
45° **LR** ($\pi/2$, 45°)	$\frac{1}{2}\begin{pmatrix} 1+i & 1-i \\ 1-i & 1+i \end{pmatrix}$	$\frac{1}{\sqrt{2}}\begin{pmatrix} 1 & -i \\ -i & 1 \end{pmatrix}$	$\frac{1}{2}\begin{pmatrix} 1-i & -1-i \\ -1-i & 1-i \end{pmatrix}$
135° **LR** ($\pi/2$, 135°)	$\frac{1}{2}\begin{pmatrix} 1+i & -1+i \\ -1+i & 1+i \end{pmatrix}$	$\frac{1}{\sqrt{2}}\begin{pmatrix} 1 & i \\ i & 1 \end{pmatrix}$	$\frac{1}{2}\begin{pmatrix} 1-i & 1+i \\ 1+i & 1-i \end{pmatrix}$
Right circular QWRCR	$\frac{1}{2}\begin{pmatrix} 1+i & 1+i \\ -1-i & 1+i \end{pmatrix}$	$\frac{1}{\sqrt{2}}\begin{pmatrix} i & i \\ -i & i \end{pmatrix}$	$\frac{1}{2}\begin{pmatrix} 1-i & -1+i \\ 1-i & 1-i \end{pmatrix}$
Left circular QWLCR	$\frac{1}{2}\begin{pmatrix} 1+i & -1-i \\ 1+i & 1+i \end{pmatrix}$	$\frac{1}{\sqrt{2}}\begin{pmatrix} i & -i \\ i & i \end{pmatrix}$	$\frac{1}{2}\begin{pmatrix} 1-i & 1-i \\ -1+i & 1-i \end{pmatrix}$

	Retarders: Half Wave		
Fast Axis	Fast Axis Unchanged	Symmetric Phase Convention	Slow Axis Unchanged
Horizontal, 0° **LR** (π, 0°)	$\begin{pmatrix} 1 & 0 \\ 0 & -1 \end{pmatrix}$	$\begin{pmatrix} -i & 0 \\ 0 & i \end{pmatrix}$	$\begin{pmatrix} -1 & 0 \\ 0 & 1 \end{pmatrix}$
Vertical, 90° **LR** (π, 90°)	$\begin{pmatrix} -1 & 0 \\ 0 & 1 \end{pmatrix}$	$\begin{pmatrix} i & 0 \\ 0 & -i \end{pmatrix}$	$\begin{pmatrix} 1 & 0 \\ 0 & -1 \end{pmatrix}$
45° **LR** (π, 45°)	$\begin{pmatrix} 0 & 1 \\ 1 & 0 \end{pmatrix}$	$\begin{pmatrix} 0 & -i \\ -i & 0 \end{pmatrix}$	$\begin{pmatrix} 0 & -1 \\ -1 & 0 \end{pmatrix}$
135° **LR** (π, 135°)	$\begin{pmatrix} 0 & -1 \\ -1 & 0 \end{pmatrix}$	$\begin{pmatrix} 0 & i \\ i & 0 \end{pmatrix}$	$\begin{pmatrix} 0 & 1 \\ 1 & 0 \end{pmatrix}$
Right circular **CR** ($-\pi$)	$\begin{pmatrix} 0 & i \\ -i & 0 \end{pmatrix}$	$\begin{pmatrix} 0 & 1 \\ -1 & 0 \end{pmatrix}$	$\begin{pmatrix} 0 & -i \\ i & 0 \end{pmatrix}$
Left circular **LR** (π)	$\begin{pmatrix} 0 & -i \\ i & 0 \end{pmatrix}$	$\begin{pmatrix} 0 & -1 \\ 1 & 0 \end{pmatrix}$	$\begin{pmatrix} 0 & i \\ -i & 0 \end{pmatrix}$

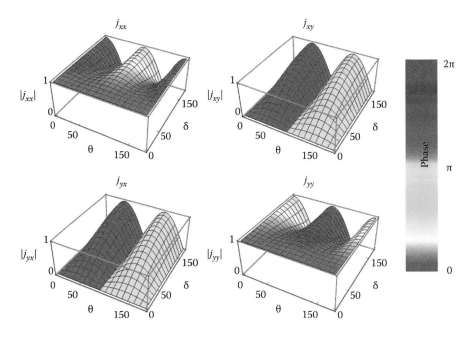

Figure 5.8 The four Jones matrix elements for all linear retarders plotted for the symmetric phase form. The orientation varies across the front (degrees) and the retardance varies from front to back (degrees). Height is the magnitude of the element. Color encodes the phase of the element cyclically from red for 0 phase, through green to cyan for ½ a wave of phase through purple and pink and back to red for 1 wave of phase.

To transform $\mathbf{LR}(\pi, \theta)$ into the *fast axis unchanged convention*, the matrix is multiplied by i,

$$i\mathbf{LR}(\pi, \theta) = \begin{pmatrix} \cos 2\theta & \sin 2\theta \\ \sin 2\theta & -\cos 2\theta \end{pmatrix}. \tag{5.44}$$

Both matrices, $i\,\mathbf{LR}(\pi, \theta)$ or $\mathbf{LR}(\pi, \theta)$, represent half wave linear retarders. One common use of half wave linear retarders is rotating a linearly polarized state into a new orientation. Consider a half wave linear retarder with its fast axis oriented at θ and an incident linear state oriented at $(\theta - \alpha)$, that is, α radians on one side of the fast axis. The output state becomes

$$i\mathbf{LR}(\pi, \theta) \cdot \begin{pmatrix} \cos(\theta - \alpha) \\ \sin(\theta - \alpha) \end{pmatrix} = \begin{pmatrix} \cos(\theta + \alpha) \\ \sin(\theta + \alpha) \end{pmatrix}, \tag{5.45}$$

which is rotated to α radians on the other side of the fast axis; the state is rotated by 2α. Thus, half wave retarders can convert between any two linearly polarized states by orienting the fast axis midway between the input and desired output states. For incident linearly polarized light, a rotating half wave linear retarder will rotate the output linear polarization at twice the retarder's angular velocity. For incident elliptically polarized light, the major axis of the exiting ellipse rotates at twice the angular velocity. Half wave linear retarders convert right circular into left circularly polarized light, and vice versa,

$$i\mathbf{LR}(\pi, \theta) \cdot \begin{pmatrix} 1 \\ \pm i \end{pmatrix} = e^{\pm 2i\theta} \begin{pmatrix} 1 \\ \mp i \end{pmatrix}. \tag{5.46}$$

If the half wave retarder is rotating at an angular velocity of ω radians per second,

5.6 Retarder Jones Matrices

$$i\mathbf{LR}(\pi, \omega t) \cdot \begin{pmatrix} 1 \\ \pm i \end{pmatrix} = e^{\pm 2i\omega t} \begin{pmatrix} 1 \\ \mp i \end{pmatrix}, \quad (5.47)$$

the light acquires a linear phase shift in time, signifying that the light is shifted in angular frequency by twice the retarder's angular velocity, either up or down in frequency. Thus, *rotating retarders* can make imperceptibly small changes in light's color by *imparting angular momentum* to or *subtracting angular momentum* from the light.

To express the Jones matrix of a linear retarder in the form that advances the phase of the fast eigenpolarization, leaving the slow eigenpolarization unchanged, multiply the symmetric phase form by $e^{-i\delta/2}$, yielding $e^{-i\delta/2}\mathbf{LR}(\delta, \theta)$. Similarly, to express the Jones matrix of a linear retarder in the form that delays the phase of the slow eigenpolarization, leaving the fast eigenpolarization unchanged, use $e^{i\delta/2}\mathbf{LR}(\delta, \theta)$.

5.6.2 Circular Retarder Jones Matrices

Circular retarders have Jones matrices $\mathbf{CR}(\delta)$, which are *Cartesian rotation matrices*, except that the trigonometric argument is $\delta/2$,

$$\mathbf{CR}(\delta) = \begin{pmatrix} \cos\dfrac{\delta}{2} & \sin\dfrac{\delta}{2} \\ -\sin\dfrac{\delta}{2} & \cos\dfrac{\delta}{2} \end{pmatrix}. \quad (5.48)$$

Note that the one wave circular retarder has the Jones matrix

$$\mathbf{CR}(2\pi) = \begin{pmatrix} -1 & 0 \\ 0 & -1 \end{pmatrix}, \quad (5.49)$$

where the −1, an overall phase of π, results from the choice of the symmetric phase convention. Jones matrices for retarders with integer numbers of waves of retardance, one wave ($\delta = 2\pi$), two waves ($\delta = 4\pi$), and so on, are all proportional to the identity matrix and do not cause polarization state change. One common cause of circular retardance is *optical activity*.

5.6.3 Vortex Retarders

Next, vortex retarders will be explored as an example of the application of Jones matrices. An optical vortex is a phase singularity about a zero of an electromagnetic field.[21] Figure 5.9 (left) shows a vortex in circularly polarized light. In this example, the phase varies by 2π about a zero at the center of the pupil phase map. The phase is continuous around the zero, but discontinuous crossing the zero. This is a screw dislocation; in this map, the phase has the form $\arctan(y/x)$. Other vortices can take the form $m \arctan(y/x)$, where m is the order of the vortex. Figure 5.9 (right) shows an ordinary zero of the field for comparison.

A *vortex retarder* is a spatially varying half wave retarder where the fast axis of the retarder rotates as a function of the angle $\phi = \arctan(y/x)$ from the x axis.[22–24] The vortex retarder with spatially varying Jones matrix in polar coordinates is

$$\mathbf{J}(m\phi) = \begin{pmatrix} \cos(m\phi) & \sin(m\phi) \\ \sin(m\phi) & -\cos(m\phi) \end{pmatrix}, \quad (5.50)$$

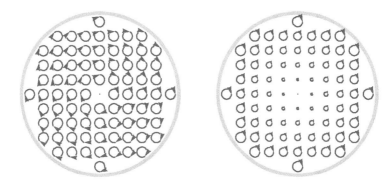

Figure 5.9 Pupil map of an optical vortex (left) where the phase of the light, indicated by the position of the arrow around the circle, varies by 2π around a zero of the field. The field is discontinuous at this vortex. (Right) An ordinary zero of the field, where the polarization state is constant, there is no discontinuity, and no optical vortex.

where m is the order of the vortex retarder. Figure 5.10 shows pupil maps of the retardance for orders $m = 1, 2, 3$, and 4. The value of Equation 5.50 at the origin depends on the direction ϕ from which the origin is approached. Thus, Equation 5.50 is not continuous at the origin. Vortex retarders can be fabricated from liquid crystal polymers. The retardance varies faster and faster as the origin is approached; hence, fabricated vortex retarders will use an opaque black spot over the center to conceal a small region of disorder at the vertex.

Figure 5.11 shows how the plane of exiting polarization varies for collimated incident light linearly polarized at 0°. These polarization patterns are examples of *polarization vortices* at the origin. At a polarization vortex, the polarization state rotates by a multiple of π when moving in a small circle around a zero in the field.[25] When these vortex retarders are positioned between 0° and 90° linear polarizers, the transmitted intensity is as shown in Figure 5.12. When this light is brought to a focus, interesting

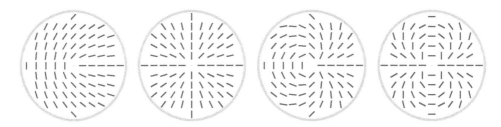

Figure 5.10 Pupil maps of the retardance (shown in pink with the orientation of the lines indicating the fast axis orientation) for vortex retarders with (from left) $m = 1, 2, 3$, and 4. These vortex retarders are half wave linear retarders everywhere whose fast axes rotate $m/2$ times around the pupil; the fast axis orientations are constant along the radial lines. The discontinuity in retardance at the origin is marked with a small dot.

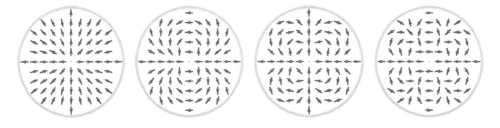

Figure 5.11 Pupil maps of the exiting polarization state for incident 0° polarized light for vortex retarders with (from left) $m = 1, 2, 3$, and 4. Exiting light is linearly polarized everywhere with constant amplitude except at the origin where a discontinuity, the vortex, is present.

Polarized Light and Optical Systems

5.7 General Diattenuators and Retarders

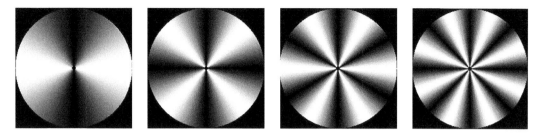

Figure 5.12 Pupil intensity for the vortex retarder between horizontal and vertical linear polarizers for $m = 1, 2, 3$, and 4 varies cosinusoidally around the pupil.

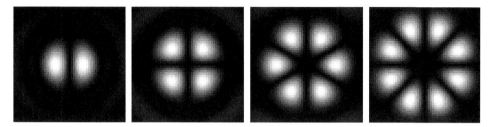

Figure 5.13 Point spread functions of the vortex retarder between horizontal and vertical linear polarizers for $m = 1, 2, 3$, and 4. They are dark in the center and are surrounded by a set of bright islands that have been compared to propellers. If the second polarizer is rotated, the propellers rotate with the polarizer. The calculation of point spread functions is treated in Chapter 16.

images are formed as shown in Figure 5.13. These intensity images are calculated as Fourier transforms of the exiting light distribution, which are then squared, to calculate the flux distributions, following the methods of Chapter 16. Vortex retarders have applications for creating *optical tweezers* and in creating special point spread functions in microlithography.[26,27] The pattern of light in Figure 5.11 (left), radially polarized light, is particularly interesting when imaging very high numerical aperture beams since it can be brought into a tighter (smaller) focus than a uniformly polarized beam.[28,29]

5.7 General Diattenuators and Retarders

The following sections extend the equations for homogeneous Jones matrices developed in the earlier sections for linear and circular eigenpolarizations to elliptical eigenpolarizations. First, an alternative form for linear diattenuators is examined. This section is more algebraically challenging but the resulting equations can be quite useful.

5.7.1 Linear Diattenuators

The diattenuator Jones matrix (Equation 5.31) takes a simple and interesting form when the average amplitude transmission, the average of its eigenvalues, is set to $\cosh(d)$, and where $2d$ is the hyperbolic arctangent of the diattenuation D, that is, $d = \tanh^{-1}(D)/2$ and

$$\mathbf{LD}(D,0) = \begin{pmatrix} \cosh(d) + \sinh(d) & 0 \\ 0 & \cosh(d) - \sinh(d) \end{pmatrix}. \tag{5.51}$$

Equation 5.51 is simple and instructive due to the high degree of symmetry. The eigenvalues are plotted in Figure 5.14 where one of the eigenvalues, the j_{xx} matrix element, is greater than 1, implying *gain* in this form.

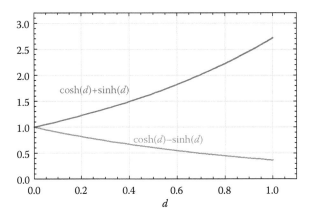

Figure 5.14 The two eigenvalues for the simplest form of the linear diattenuator Jones matrix, Equation 5.51, as a function of the parameter $d = \tanh^{-1}(D)/2$. Usually, $\mathbf{LD}(D, 0)$ is then multiplied by a constant to scale the two eigenvalues into the range from 0 to 1.

The Jones matrix for a *linear diattenuator* at an arbitrary orientation is found from the rotation operation (Equation 5.22) to be

$$\mathbf{LD}(D_H, D_{45}) = \frac{1}{D}\begin{pmatrix} D\cosh(d) + D_H \sinh(d) & D_{45}\sinh(d) \\ D_{45}\sinh(d) & D\cosh(d) - D_H \sinh(d) \end{pmatrix}, \quad (5.52)$$

where D_H and D_{45} are horizontal and 45° diattenuation-related parameters.

5.7.2 Elliptical Diattenuators

An *elliptical diattenuator* transmits two elliptical eigenpolarizations with the polarization and phase unchanged but with different amplitude transmittances ξ_q and ξ_r, the two eigenvalues. The elliptical diattenuator Jones matrix actually has the simplest form when expressed in terms of the *Stokes parameters*, not the Jones vectors, for these two elliptical eigenpolarizations. Let the diattenuation D be expressed in terms of three *diattenuation components*, a horizontal component D_H, a 45° component D_{45}, and a left circular component D_L,

$$D = \sqrt{D_H^2 + D_{45}^2 + D_L^2}, \quad 0 \leq D \leq 1, \quad (5.53)$$

where the un-normalized eigenpolarizations as Stokes parameters are

$$\mathbf{S}_q = \begin{pmatrix} D \\ D_H \\ D_{45} \\ -D_L \end{pmatrix} \text{ and } \mathbf{S}_r = \begin{pmatrix} D \\ -D_H \\ -D_{45} \\ D_L \end{pmatrix}. \quad (5.54)$$

The corresponding eigenpolarizations as normalized Jones vectors are $\mathbf{E}_q$ and $\mathbf{E}_r$,

$$\mathbf{E}_q = \frac{1}{\sqrt{2D(D_H+D)}}\begin{pmatrix} D_H - D \\ D_{45} + iD_L \end{pmatrix} \text{ and } \mathbf{E}_r = \frac{1}{\sqrt{2D(D_H+D)}}\begin{pmatrix} D_H + D \\ D_{45} + iD_L \end{pmatrix}. \quad (5.55)$$

5.7 General Diattenuators and Retarders

The Jones matrix for this *elliptical diattenuator*, $\mathbf{ED}(D_H, D_{45}, D_L)$, is most easily expressed in terms of the matrix exponential and hyperbolic arctangent as

$$\mathbf{ED}(D_H, D_{45}, D_L) = \exp\left[\operatorname{arctanh}(D) \frac{D_H \sigma_1 + D_{45} \sigma_2 + D_L \sigma_3}{2D}\right]. \tag{5.56}$$

This form is discussed further in Section 14.4.5. The two *amplitude transmittances*, the eigenvalues, are

$$\begin{aligned}\xi_q &= \cosh\left[\frac{\operatorname{arctanh}(D)}{2}\right] + \sinh\left[\frac{\operatorname{arctanh}(D)}{2}\right] = \cosh(d) + \sinh(d), \\ \xi_r &= \cosh\left[\frac{\operatorname{arctanh}(D)}{2}\right] - \sinh\left[\frac{\operatorname{arctanh}(D)}{2}\right] = \cosh(d) - \sinh(d).\end{aligned} \tag{5.57}$$

As shown in Figure 5.14, one of the amplitude transmittances in Equation 5.57 will be greater than 1, indicating a Jones matrix with gain. Hence, to describe a conventional optical element, one without gain, the matrix $\mathbf{ED}(D_H, D_{45}, D_L)$ needs to be multiplied by a constant less than $1/\xi_q$. For example, to set the elliptical diattenuator's maximum amplitude transmission to t_{max}, scale the matrix as follows:

$$\frac{t_{max}}{\xi_q} \mathbf{ED}(D_H, D_{45}, D_L). \tag{5.58}$$

Example 5.5 Elliptical Diattenuator Example

The elliptical diattenuator Jones matrix with the maximum transmitted state Stokes parameters $(\sqrt{3}, 1, 1, -1)$ and diattenuation components $D_H = D_{45} = D_L = 1/2$, for a diattenuation magnitude of $D = \sqrt{3}/2$, is

$$\mathbf{ED}\left(\frac{1}{2}, \frac{1}{2}, \frac{1}{2}\right) = \begin{pmatrix} 2\sqrt{\frac{2}{3}} & \frac{1-i}{\sqrt{6}} \\ \frac{1+i}{\sqrt{6}} & \sqrt{\frac{2}{3}} \end{pmatrix}, \tag{5.59}$$

with eigenvalues (amplitude transmittances) $\sqrt{2+\sqrt{3}}$ and $\sqrt{2-\sqrt{3}}$. If the maximum amplitude transmission should be $t_{max} = \frac{1}{2}$, then the matrix should be scaled to $\frac{1}{2\sqrt{2+\sqrt{3}}} \mathbf{ED}\left(\frac{1}{2}, \frac{1}{2}, \frac{1}{2}\right)$.

5.7.3 Elliptical Retarders

An *elliptical retarder* transmits two elliptically polarized states unchanged with different phase changes. Choosing those phase changes symmetrically as $\phi_q = -\delta/2$ and $\phi_r = \delta/2$, as was done for linear retarders, elliptical retarder Jones matrix $\mathbf{ER}(\delta_H, \delta_{45}, \delta_L)$ has the simplest form when expressed in terms of the Stokes parameters for these two eigenpolarizations using the Pauli matrices,*

$$\mathbf{ER}(\delta_H, \delta_{45}, \delta_L) = \exp\left[-i(\delta_H \sigma_1 + \delta_{45} \sigma_2 + \delta_L \sigma_3)/2\right]$$

$$= \begin{pmatrix} \cos\left(\dfrac{\delta}{2}\right) - i\sin\left(\dfrac{\delta}{2}\right)\dfrac{\delta_H}{\delta} & -i\sin\left(\dfrac{\delta}{2}\right)\dfrac{(\delta_{45} - i\delta_L)}{\delta} \\ -i\sin\left(\dfrac{\delta}{2}\right)\dfrac{(\delta_{45} + i\delta_L)}{\delta} & \cos\left(\dfrac{\delta}{2}\right) + i\sin\left(\dfrac{\delta}{2}\right)\dfrac{\delta_H}{\delta} \end{pmatrix}, \quad (5.60)$$

where the retardance magnitude is given in terms of the three retardance components: a horizontal component δ_H, a 45° component δ_{45}, and a left circular component δ_L,

$$\delta = \sqrt{\delta_H^2 + \delta_{45}^2 + \delta_L^2}. \quad (5.61)$$

The eigenpolarizations are given by Equations 5.54 and 5.55 with retardance components δ_H, δ_{45}, and δ_L substituted for the corresponding diattenuation components: d_H, d_{45}, and d_L. Elliptical retarders are covered in greater detail in Chapter 14.

Example 5.6 Elliptical Diattenuator and Elliptical Retarder

The elliptical diattenuator with diattenuation components $D_H = 0.1$, $D_{45} = 0.2$, and $D_L = -0.3$ with un-normalized Stokes parameter eigenpolarizations $\mathbf{S}_1 = (0.374, 0.1, 0.2, 0.3)$ and $\mathbf{S}_2 = (0.374, -0.1, -0.2, -0.3)$ is

$$\mathbf{ED}(0.1, 0.2, -0.3) = \begin{pmatrix} 1.07 & 0.11 + 0.16i \\ 0.11 - 0.16i & 0.97 \end{pmatrix}. \quad (5.62)$$

All numbers are rounded to three decimal places in this section. The eigenvalues are $\xi_1 = 1.22$ and $\xi_2 = 0.82$; one of the eigenvalues is greater than 1 as expected. The eigenpolarization states, including the effect of the diattenuation, are plotted in Figure 5.15. For a maximum amplitude transmittance of one, the Jones matrix becomes

$$\frac{\mathbf{ED}(0.1, 0.2, -0.3)}{1.22} = \begin{pmatrix} 0.88 & 0.09 + 0.13i \\ 0.09 - 0.13i & 0.79 \end{pmatrix}. \quad (5.63)$$

The corresponding elliptical retarder $\mathbf{ER}(0.1, 0.2, -0.3)$ with retardance components, $\delta_H = 0.1$, $\delta_{45} = 0.2$, and $\delta_L = -0.3$ is

$$\mathbf{ER}(0.1, 0.2, -0.3) = \begin{pmatrix} 0.98 - 0.05i & 0.15 - 0.10i \\ -0.15 - 0.10i & 0.98 + 0.05i \end{pmatrix}. \quad (5.64)$$

---

* The derivation of the elliptical retarder equation and the Pauli matrices are explained further in Chapter 14.

5.8 Non-Polarizing Jones Matrices for Amplitude and Phase Change

The eigenvalues in Cartesian in polar form are

$$\xi_1 = 0.98 - 0.19i = e^{-0.19i} \text{ and } \xi_2 = 0.98 + 0.19i = e^{0.19i}. \tag{5.65}$$

The difference between the phases of the eigenvalues is the retardance δ,

$$\delta = 0.19 - (-0.19) = \sqrt{0.1^2 + 0.2^2 + 0.3^2} = 0.37. \tag{5.66}$$

A combination elliptical diattenuator and retarder with common eigenpolarizations can be generated by multiplying **ER**(0.1, 0.2, −0.3) times **ED**(0.1, 0.2, −0.3).

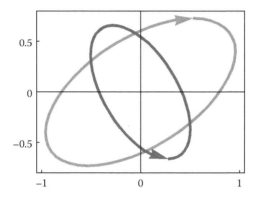

Figure 5.15 The elliptical diattenuator eigenpolarizations, scaled to the eigenvalues (amplitude transmittances), have opposite helicities and orthogonal major axes. The phases are the same as the incident phases.

5.8 Non-Polarizing Jones Matrices for Amplitude and Phase Change

The *diagonal Jones matrix* **A**(ρ) with two equal real elements describes polarization-independent *change of amplitude*, such as *absorption*, or polarization-independent reflection,

$$\mathbf{A}(\rho) \cdot \mathbf{E} = \begin{pmatrix} \rho & 0 \\ 0 & \rho \end{pmatrix} \cdot \begin{pmatrix} E_x \\ E_y \end{pmatrix} = \begin{pmatrix} \rho E_x \\ \rho E_y \end{pmatrix} = \begin{pmatrix} E'_x \\ E'_y \end{pmatrix} = \rho \mathbf{E} = \mathbf{E}', \tag{5.67}$$

where $0 \le \rho \le 1$ is the amplitude transmittance. The Jones matrix $\mathbf{\Phi}(\phi)$ for an overall *change of phase* is

$$\mathbf{\Phi}(\phi) = e^{-i\phi} \begin{pmatrix} 1 & 0 \\ 0 & 1 \end{pmatrix}. \tag{5.68}$$

This phase change is independent of the polarization state, such as propagation through the atmosphere or glass. A Jones matrix function can be constructed using this phase Jones matrix for spherical aberration, coma, or any other wavefront aberrations. Given an *OPL*, the corresponding Jones matrix is

$$\mathbf{\Phi}(\phi) = e^{i\frac{2\pi}{\lambda}OPL} \begin{pmatrix} 1 & 0 \\ 0 & 1 \end{pmatrix}. \tag{5.69}$$

Polarized Light and Optical Systems

Φ expresses propagation through an isotropic material of thickness t and refractive index n. For example, the phase Jones matrix of glass is

$$e^{i\frac{2\pi}{\lambda}nt}\begin{pmatrix} 1 & 0 \\ 0 & 1 \end{pmatrix}. \tag{5.70}$$

Example 5.7 Nilpotent Matrix

A matrix $\mathbf{N}$ is nilpotent if

$$\mathbf{N}^2 = \begin{pmatrix} 0 & 0 \\ 0 & 0 \end{pmatrix}.$$

Two identical polarization elements with nilpotent matrices in series would pass no light. Consider the nilpotent matrix generated from three polarizer Jones matrices,

$$\mathbf{N} = \mathbf{LP}(90°) \cdot \mathbf{LP}(45°) \cdot \mathbf{LP}(0°) = \begin{pmatrix} 0 & 0 \\ 1/2 & 0 \end{pmatrix}.$$

The two eigenvalues of $\mathbf{N}$ are 0 and 0. Its eigenvectors are also degenerate (the same), $(0, 1)$, the only incident state with zero transmission. It is easily shown that the state with maximum transmission is $\begin{pmatrix} 1 \\ 0 \end{pmatrix}$, where

$$\mathbf{N} \cdot \begin{pmatrix} 1 \\ 0 \end{pmatrix} = \begin{pmatrix} 0 \\ 1/2 \end{pmatrix}.$$

5.9 Matrix Properties of Jones Matrices

The properties of Jones matrices mirror the properties of two basic classes of matrices: Hermitian matrices and unitary matrices.[30]

5.9.1 Hermitian Matrices: Diattenuation

The Jones matrices $\mathbf{H}$ representing diattenuation are *Hermitian matrices*; the adjoint or conjugate transpose of a Hermitian matrix equals the matrix itself,

$$\mathbf{H} = \mathbf{H}^\dagger. \tag{5.71}$$

Thus, the elements of a Hermitian matrix satisfy

$$\begin{pmatrix} j_{xx} & j_{xy} \\ j_{yx} & j_{yy} \end{pmatrix} = \begin{pmatrix} j_{xx}^* & j_{yx}^* \\ j_{xy}^* & j_{yy}^* \end{pmatrix}. \tag{5.72}$$

5.9 Matrix Properties of Jones Matrices

Thus, the diagonal elements of a Hermitian matrix must be real since

$$j_{xx} = j_{xx}^*, \quad j_{yy} = j_{yy}^*, \tag{5.73}$$

and the off-diagonal elements must be complex conjugates of each other,

$$j_{xy} = j_{yx}^* = u + iv \text{ and } j_{yx} = j_{xy}^* = u - iv. \tag{5.74}$$

If the Hermitian matrix is expressed in terms of four real coefficients, h_0, h_1, h_2, and h_3, related to the Pauli matrices,

$$\mathbf{H} = \begin{pmatrix} h_0 + h_1 & h_2 - ih_3 \\ h_2 + ih_3 & h_0 - h_1 \end{pmatrix}, \tag{5.75}$$

then **H**'s eigenvalues are

$$\xi_q, \xi_r = h_0 \pm \sqrt{h_1^2 + h_2^2 + h_3^2}, \tag{5.76}$$

with corresponding eigenvectors

$$\mathbf{E}_q = \begin{pmatrix} h_1 + \sqrt{h_1^2 + h_2^2 + h_3^2} \\ h_2 + ih_3 \end{pmatrix}, \quad \mathbf{E}_r = \begin{pmatrix} h_1 - \sqrt{h_1^2 + h_2^2 + h_3^2} \\ h_2 + ih_3 \end{pmatrix}. \tag{5.77}$$

Hence, the diattenuation D of **H** is

$$D = \frac{\xi_q^2 - \xi_r^2}{\xi_q^2 + \xi_r^2} = \frac{2h_0\sqrt{h_1^2 + h_2^2 + h_3^2}}{h_0^2 + h_1^2 + h_2^2 + h_3^2}. \tag{5.78}$$

The eigenvalues of a Hermitian matrix are real. Thus, Hermitian matrices represent measurements, just as a polarizer can be inserted in a beam with a radiometer to detect the power in a particular polarization state. The eigenvectors of Hermitian matrices are orthogonal. Hermitian matrices are stretching and compressing matrices and will deform a circle or sphere of unit vectors into an ellipsoid of vectors (which may have complex elements). The maximum and minimum length vectors generated are the eigenvectors and have lengths scaled by the eigenvalues.

Hermitian matrices are diagonalizable by a unitary transformation. After the unitary transformation, the eigenvalues of the Hermitian matrix are the diagonal elements.

5.9.2 Unitary Matrices and Unitary Transformations: Retarder

Retarder Jones matrices are unitary matrices, denoted **U**. For unitary Jones matrices, the matrix adjoint equals the matrix inverse,

$$\mathbf{U}^{-1} = \begin{pmatrix} j_{xx} & j_{xy} \\ j_{yx} & j_{yy} \end{pmatrix}^{-1} = \begin{pmatrix} j_{xx} & j_{xy} \\ j_{yx} & j_{yy} \end{pmatrix}^\dagger = \mathbf{U}^\dagger = \begin{pmatrix} j_{xx}^* & j_{yx}^* \\ j_{xy}^* & j_{yy}^* \end{pmatrix}. \tag{5.79}$$

The eigenvalues of a retarder Jones matrix, ξ_q and ξ_r, like all unitary matrices, have a magnitude of 1, so they can be expressed as

$$\xi_q = e^{-i\phi_q}, \quad \xi_r = e^{-i\phi_r}. \tag{5.80}$$

The retardance of a retarder Jones matrix **U** is the magnitude of the difference between the phases of the eigenvalues,

$$\delta = \left|\arg(\xi_q) - \arg(\xi_r)\right| = \left|\phi_q - \phi_r\right|. \tag{5.81}$$

The rows of **U** are normalized orthogonal vectors,

$$\mathbf{U}_i^\dagger \cdot \mathbf{U}_j = \delta_{ij} \tag{5.82}$$

and thus form an orthonormal basis. Similarly, the columns of **U** are also orthonormal. Unitary matrices have a unit determinant, $\det(\mathbf{U}) = 1$. Unitary matrices represent rotations. Thus, a unitary matrix will rotate a polarization state, $\mathbf{U} \cdot \mathbf{E}$ without changing the magnitude of the vector

$$|\mathbf{U} \cdot \mathbf{E}| = |\mathbf{E}|. \tag{5.83}$$

The Jones rotation matrix

$$\mathbf{R}(\theta) = \begin{pmatrix} \cos\theta & -\sin\theta \\ \sin\theta & \cos\theta \end{pmatrix} \tag{5.84}$$

is a unitary matrix. $\mathbf{R}(\theta)$ has real elements and rotates a real vector into another real vector. Real unitary matrices such as $\mathbf{R}(\theta)$ are labeled *orthogonal matrices*.[31] The first column of **R** contains the state into which the vector (1, 0) will rotate. Similarly, the second column of **R** is the state into which (0, 1) rotates. Other unitary matrices with complex elements perform complex rotations; a real vector can rotate into a complex vector.

A *unitary transformation* rotates a matrix **J** into a new basis by left multiplying a unitary matrix and right multiplying by the matrix inverse,

$$\mathbf{U} \cdot \mathbf{J} \cdot \mathbf{U}^{-1}. \tag{5.85}$$

To transform **J** from an *x–y* basis into an arbitrary basis, **U** contains rows of vectors conjugate to the new orthonormal basis vectors.

A unitary transformation of **J** preserves the matrix's eigenvalues and determinant. It rotates **J**'s eigenvectors as a group, preserving their dot products and thus the angles between the eigenvectors are unchanged.

A Hermitian Jones matrix **H** can always be diagonalized by a unitary transformation,

$$\mathbf{U} \cdot \mathbf{H} \cdot \mathbf{U}^{-1} = \begin{pmatrix} \rho_q & 0 \\ 0 & \rho_r \end{pmatrix}. \tag{5.86}$$

After diagonalization, the two real eigenvalues ρ_q and ρ_r of **H** are on the diagonal. The rows of $\mathbf{U}^{-1}$ and the columns of **U** contain the normalized eigenvectors of **H** that are always orthogonal.

5.9 Matrix Properties of Jones Matrices

Example 5.8 Unitary Transformation

An important *unitary transformation* between linear and circular bases is performed by the matrix **B**,

$$\mathbf{B} = \frac{1}{\sqrt{2}} \begin{pmatrix} 1 & i \\ 1 & -i \end{pmatrix}, \tag{5.87}$$

where $\mathbf{B}^{-1} = \mathbf{B}^{\dagger} = \frac{1}{\sqrt{2}} \begin{pmatrix} 1 & 1 \\ -i & i \end{pmatrix}$.

B transforms a Jones vector (E_x, E_y) expressed in the x–y linear basis into the R–L circular basis;

$$\begin{pmatrix} E_R \\ E_L \end{pmatrix} = \mathbf{B} \cdot \begin{pmatrix} E_x \\ E_y \end{pmatrix} = \frac{1}{\sqrt{2}} \begin{pmatrix} E_x + iE_y \\ E_x - iE_y \end{pmatrix}. \tag{5.88}$$

In the R–L circular basis, right circularly polarized light has the unit Jones vector (1, 0) and the left circularly polarized Jones vector is (0, 1).

As an example of a unitary transformation, the left circular polarizer Jones matrix in the x–y linear basis is $\mathbf{LCP} = \frac{1}{2} \begin{pmatrix} 1 & -i \\ i & 1 \end{pmatrix}$. The corresponding Jones matrix in the R–L circular basis is

$$\mathbf{B} \cdot \frac{1}{2} \begin{pmatrix} 1 & -i \\ i & 1 \end{pmatrix} \cdot \mathbf{B}^{-1} = \frac{1}{\sqrt{2}} \begin{pmatrix} 1 & i \\ 1 & -i \end{pmatrix} \cdot \frac{1}{2} \begin{pmatrix} 1 & -i \\ i & 1 \end{pmatrix}$$

$$\cdot \frac{1}{\sqrt{2}} \begin{pmatrix} 1 & 1 \\ -i & i \end{pmatrix} = \begin{pmatrix} 0 & 0 \\ 0 & 1 \end{pmatrix}, \tag{5.89}$$

where, for example, a beam of arbitrary polarization is analyzed as follows:

$$\begin{pmatrix} 0 & 0 \\ 0 & 1 \end{pmatrix} \begin{pmatrix} E_R \\ E_L \end{pmatrix} = \begin{pmatrix} 0 \\ E_L \end{pmatrix}. \tag{5.90}$$

A circular retarder with retardance δ in the x–y linear basis has the Jones matrix

$\mathbf{CR}(\delta) = \begin{pmatrix} \cos\frac{\delta}{2} & \sin\frac{\delta}{2} \\ -\sin\frac{\delta}{2} & \cos\frac{\delta}{2} \end{pmatrix}$. The corresponding Jones matrix in the R–L circular basis is

$$\mathbf{B} \cdot \begin{pmatrix} \cos\frac{\delta}{2} & \sin\frac{\delta}{2} \\ -\sin\frac{\delta}{2} & \cos\frac{\delta}{2} \end{pmatrix} \cdot \mathbf{B}^{-1} = \begin{pmatrix} \cos\frac{\delta}{2} - i\sin\frac{\delta}{2} & 0 \\ 0 & \cos\frac{\delta}{2} + i\sin\frac{\delta}{2} \end{pmatrix} = \begin{pmatrix} e^{-i\delta/2} & 0 \\ 0 & e^{+i\delta/2} \end{pmatrix}. \tag{5.91}$$

Thus, it is seen how unitary transforms can be used to perform Jones matrix operations in bases other than the x–y basis, for example, in the R–L basis or in any other orthogonal basis that might be convenient.

Example 5.9 Diagonalizing a Hermitian Matrix

The elliptical diattenuator $\mathbf{H}_1$,

$$\mathbf{H}_1 = \frac{\mathbf{ED}\left(0, \frac{-1+e^2}{\sqrt{2}(1+e^2)}, -\frac{-1+e^2}{\sqrt{2}(1+e^2)}\right)}{\sqrt{e}} = \begin{pmatrix} \dfrac{1+e}{2e} & \dfrac{(1+i)\sinh\dfrac{1}{2}}{\sqrt{2e}} \\ \dfrac{(1-i)\sinh\dfrac{1}{2}}{\sqrt{2e}} & \dfrac{1+e}{2e} \end{pmatrix}, \quad (5.92)$$

with Stokes eigenpolarizations

$$\mathbf{S}_q = \begin{pmatrix} 1 \\ 0 \\ 1/\sqrt{2} \\ 1/\sqrt{2} \end{pmatrix} \text{ and } \mathbf{S}_r = \begin{pmatrix} 1 \\ 0 \\ -1/\sqrt{2} \\ -1/\sqrt{2} \end{pmatrix} \quad (5.93)$$

will be diagonalized by a unitary transformation. One eigenpolarization is elliptically polarized halfway between 45° and right circularly polarized; the other eigenpolarization is orthogonal, halfway between 135° and left circularly polarized. The factor $1/\sqrt{e}$ adjusts the maximum and minimum amplitude transmittances without changing the diattenuation; it is an *absorption factor* described in Equation 5.58. These Stokes eigenpolarizations transform from Equation 5.55 into normalized Jones eigenpolarizations $\mathbf{E}_q$ and $\mathbf{E}_r$,

$$\mathbf{E}_q = \begin{pmatrix} \dfrac{1+i}{2} \\ \dfrac{1}{\sqrt{2}} \end{pmatrix}, \mathbf{E}_r = \begin{pmatrix} -\dfrac{1+i}{2} \\ \dfrac{1}{\sqrt{2}} \end{pmatrix}. \quad (5.94)$$

Hence, the unitary matrix $\mathbf{U}_1$, which diagonalizes $\mathbf{H}_1$, is constructed using the conjugates of $\mathbf{E}_q$ and $\mathbf{E}_r$ as its rows,

$$\mathbf{U}_1 = \frac{1}{2}\begin{pmatrix} 1-i & \sqrt{2} \\ -1+i & \sqrt{2} \end{pmatrix}. \quad (5.95)$$

Performing the unitary transformation $\mathbf{U}_1 \, \mathbf{H}_1 \, \mathbf{U}_1^{-1}$ on $\mathbf{H}_1$ yields a diagonal matrix,

$$\mathbf{U}_1 \mathbf{H}_1 \mathbf{U}_1^{-1} = \frac{1}{2}\begin{pmatrix} 1-i & \sqrt{2} \\ -1+i & \sqrt{2} \end{pmatrix} \cdot \begin{pmatrix} \dfrac{1+e}{2e} & \dfrac{(1+i)\sinh\dfrac{1}{2}}{\sqrt{2e}} \\ \dfrac{(1-i)\sinh\dfrac{1}{2}}{\sqrt{2e}} & \dfrac{1+e}{2e} \end{pmatrix} \quad (5.96)$$

$$\cdot \frac{1}{2}\begin{pmatrix} 1+i & -1-i \\ \sqrt{2} & \sqrt{2} \end{pmatrix} = \begin{pmatrix} 1 & 0 \\ 0 & 1/e \end{pmatrix}.$$

The two amplitude transmittances of the elliptical diattenuator $\mathbf{H}_1$, its eigenvalues 1 and $1/e$, are now located on the diagonal.

5.9.3 Polar Decomposition: Separating Retardance from Diattenuation

An arbitrary Jones matrix can have a mixture of diattenuation and retardance. If the diattenuator and retarder have the same eigenvectors, the Jones matrix $\mathbf{J}$ is homogeneous and the unitary and Hermitian components commute; the *commutator* $[\mathbf{H}, \mathbf{U}]$ is zero,

$$[\mathbf{H}, \mathbf{U}] = \mathbf{H} \cdot \mathbf{U} - \mathbf{U} \cdot \mathbf{H} = 0. \quad (5.97)$$

Such a homogeneous Jones matrix can be diagonalized by a *unitary transformation*,

$$\mathbf{U} \cdot \mathbf{J} \cdot \mathbf{U}^{-1} = \begin{pmatrix} \xi_q & 0 \\ 0 & \xi_r \end{pmatrix} \quad (5.98)$$

yielding the eigenvalues on the diagonal of the diagonal matrix and the eigenvectors in the columns of $\mathbf{U}$.

In general, a matrix $\mathbf{J}$ with an arbitrary mix of diattenuation and retardance will have a non-zero commutator, $[\mathbf{H}, \mathbf{U}] \neq 0$. Then, $\mathbf{J}$ cannot be diagonalized by a unitary transformation and $\mathbf{J}$ has *non-orthogonal eigenpolarizations*; this defines an *inhomogeneous matrix*. Also, the eigenvalues do not necessarily represent the maximum and minimum amplitude transmissions, and the homogeneous Jones matrix equation for diattenuation is not generally true,

$$D \neq \frac{\xi_q^2 - \xi_r^2}{\xi_q^2 + \xi_r^2}. \quad (5.99)$$

An inhomogeneous matrix or homogeneous matrix can be expressed using the *polar decomposition* as a product of a unitary matrix and a Hermitian matrix or, alternatively, as a Hermitian matrix times a unitary matrix:[32–35]

$$\mathbf{J} = \mathbf{U} \cdot \mathbf{H} = \mathbf{H}' \cdot \mathbf{U}. \tag{5.100}$$

In the two forms of the polar decomposition, the unitary matrix remains the same, but the Hermitian matrices $\mathbf{H}$ and $\mathbf{H}'$ differ depending on the order of the decomposition,[34] as shown in Figure 5.16. The polar decomposition components can be calculated by two different methods.

By the first method, $\mathbf{J}$ can be decomposed by the *singular value decomposition* into the product of a unitary matrix $\mathbf{W}$, a diagonal matrix of singular values $\mathbf{D}$, and the adjoint of a unitary matrix $\mathbf{V}^\dagger$,[36]

$$\mathbf{J} = \mathbf{W} \cdot \mathbf{D} \cdot \mathbf{V}^\dagger = \begin{pmatrix} w_{x,1} & w_{x,2} \\ w_{y,1} & w_{y,2} \end{pmatrix} \begin{pmatrix} \Lambda_1 & 0 \\ 0 & \Lambda_2 \end{pmatrix} \begin{pmatrix} v^*_{x,1} & v^*_{y,1} \\ v^*_{x,2} & v^*_{y,2} \end{pmatrix}, \tag{5.101}$$

where the singular values are sorted as $\Lambda_1 \geq \Lambda_2$. The maximum amplitude transmittance for $\mathbf{J}$ is Λ_1, corresponding to input state $(v_{x,1}\ v_{y,1})$ and output state $(w_{x,1}\ w_{y,1})$. The minimum amplitude transmittance corresponds to the second singular value, conjugate of the second row of $\mathbf{V}^\dagger$, and the second column of $\mathbf{W}$. Hence, the diattenuation of $\mathbf{J}$ is found from the singular values as

$$D = \frac{\Lambda_1^2 - \Lambda_2^2}{\Lambda_1^2 + \Lambda_2^2}. \tag{5.102}$$

More on the singular value decomposition can be found in Math Tip 9.2. The Hermitian components of the polar decomposition of $\mathbf{J}$ are calculated as follows:

$$\mathbf{H} = \mathbf{V}\mathbf{D}\mathbf{V}^\dagger, \quad \mathbf{H}' = \mathbf{W}\mathbf{D}\mathbf{W}^\dagger. \tag{5.103}$$

Then the unitary part, the pure retarder, is

$$\mathbf{U} = \mathbf{W}\mathbf{V}. \tag{5.104}$$

The *second method* calculates the squares of the Hermitian parts as

$$\mathbf{J}^\dagger \mathbf{J} = \mathbf{H}^2, \text{ and } \mathbf{J}\mathbf{J}^\dagger = \mathbf{H}'^2. \tag{5.105}$$

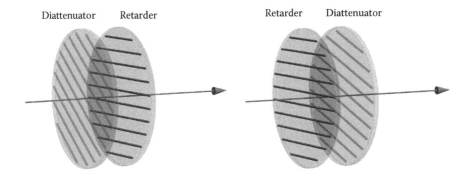

Figure 5.16 An inhomogeneous Jones matrix can be expressed using the polar decomposition as (left) a diattenuator followed by a retarder, or (right) an identical retarder followed by a different diattenuator.

5.9 Matrix Properties of Jones Matrices

This leads to another pair of expressions for the Hermitian or diattenuating parts of the polar decomposition involving *matrix square roots* (Equation 14.36),

$$\mathbf{H} = \sqrt{\mathbf{J}^\dagger \mathbf{J}}, \quad \mathbf{H}' = \sqrt{\mathbf{J}\mathbf{J}^\dagger}. \quad (5.106)$$

Equation 5.106 requires that $\mathbf{H}$ and $\mathbf{J}$ be nonsingular, with a non-zero determinant. Then, $\mathbf{U} = \mathbf{J}\mathbf{H}^{-1} = \mathbf{H}'^{-1}\mathbf{J}$.

For an arbitrary Jones matrix $\mathbf{J}$, the pure retarder Jones matrix $\mathbf{U}$ can be regarded as the retarding part of $\mathbf{J}$ and either $\mathbf{H}$ or $\mathbf{H}'$ can be regarded as the diattenuating part. Note that the polar decomposition is order dependent; either the retarder comes first followed by the diattenuator, or vice versa. Inhomogeneous matrices are further discussed in Section 14.6. An order-independent representation of polarization components, the *exponential form of the polarization components*, is presented in Section 14.4.5.

Example 5.10 Polar Decomposition Example

Consider the inhomogeneous matrix

$$\mathbf{J} = \frac{1}{4}\begin{pmatrix} 2-2i & 2-2i \\ -1-i & 1+i \end{pmatrix} \quad (5.107)$$

with eigenvalues $\approx (0.868e^{-1.025i}, 0.576e^{1.025i})$. The polarization ellipses for the incident normalized eigenvectors and the exiting eigenvectors are plotted in Figure 5.17. The singular value decomposition components of $\mathbf{J}$ are

$$\mathbf{W} = \begin{pmatrix} e^{-i\pi/4} & 0 \\ 0 & e^{i\pi/4} \end{pmatrix} = \mathbf{LR}\left(\pi/2, 0\right), \quad (5.108)$$

$$\mathbf{D} = \begin{pmatrix} 1 & 0 \\ 0 & 1/2 \end{pmatrix} = \mathbf{LD}\left(1, 1/2, 0\right), \quad (5.109)$$

and

$$\mathbf{V}^\dagger = \frac{1}{\sqrt{2}}\begin{pmatrix} 1 & 1 \\ -1 & 1 \end{pmatrix} = \mathbf{CR}\left(\pi/2, 0\right). \quad (5.110)$$

The Hermitian parts are

$$\mathbf{H} = \sqrt{\mathbf{J}^\dagger \mathbf{J}} = \mathbf{VDV}^\dagger = \frac{1}{4}\begin{pmatrix} 3 & 1 \\ 1 & 3 \end{pmatrix}, \text{ and}$$

$$\mathbf{H}' = \sqrt{\mathbf{J}\mathbf{J}^\dagger} = \mathbf{WDW}^\dagger = \begin{pmatrix} 1 & 0 \\ 0 & 1/2 \end{pmatrix}. \quad (5.111)$$

Both have eigenvalues of 1 and ½, for a diattenuation of 3/5; thus, this is the diattenuation of **J** according to the polar decomposition. The unitary part is

$$\mathbf{U} = \mathbf{W}\,\mathbf{V}^\dagger = \frac{1}{2}\begin{pmatrix} 1-i & 1-i \\ -1-i & 1+i \end{pmatrix}, \qquad (5.112)$$

with eigenvalues $\pi/3$ and $-\pi/3$, for a retardance of $2\pi/3$. This is the retardance of **J** according to the polar decomposition. The eigenvectors of **H**, **H′**, and **U** are plotted in Figure 5.18.

J is expressed by the polar decomposition in two ways as **U H** and **H′U**,

$$\begin{aligned}\mathbf{J} = \mathbf{U}\,\mathbf{H} &= \frac{1}{2}\begin{pmatrix} 1-i & 1-i \\ -1-i & 1+i \end{pmatrix}\cdot\frac{1}{4}\begin{pmatrix} 3 & 1 \\ 1 & 3 \end{pmatrix} \\ &= \mathbf{H'}\,\mathbf{U} = \begin{pmatrix} 1 & 0 \\ 0 & \tfrac{1}{2} \end{pmatrix}\cdot\frac{1}{2}\begin{pmatrix} 1-i & 1-i \\ -1-i & 1+i \end{pmatrix} \\ &= \frac{1}{4}\begin{pmatrix} 2-2i & 2-2i \\ -1-i & 1+i \end{pmatrix}.\end{aligned} \qquad (5.113)$$

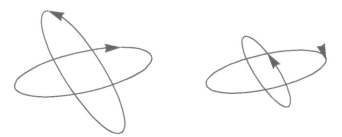

Figure 5.17 Normalized incident eigenpolarizations (left) and exiting eigenpolarizations (right). Note that major axes are not orthogonal; hence, eigenpolarizations are not orthogonal. Both diattenuation and retardance are present.

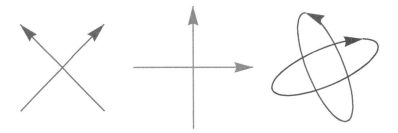

Figure 5.18 Normalized eigenpolarizations of the Hermitian (diattenuating) part **H** (left, red), of the Hermitian part **H′** (center, green) and the unitary (retarding) part **U** (right, blue) of the inhomogeneous Jones matrix **J**.

5.10 Increasing Phase Sign Convention

Sections 2.17 and 2.18 introduced and discussed the sign conventions for phase. This book uses the *decreasing phase sign convention*. The other sign convention, the *increasing phase sign convention*, is not used in this book and will be described in this section. Subscript *M* will be used in this section to indicate quantities in the increasing phase sign convention where the sign of ϕ changes. The monochromatic plane wave then takes the form

$$\mathbf{E}_M(z,t) = \mathbf{Re}\left[\begin{pmatrix} E_x \\ E_y \end{pmatrix} e^{i(\omega t - kz + \phi)}\right] = \begin{pmatrix} E_x \\ E_y \end{pmatrix} \cos(\omega t - kz + \phi). \quad (5.114)$$

The only change in the plane wave equation with the increasing phase sign convention is the sign of $\omega t - kz + \phi$. Note that since cosine is an *even function*, the sign of the argument to cosine does not matter, $\cos(\theta) = \cos(-\theta)$.

In the increasing phase sign convention, left and right circularly polarized light Jones vectors are

$$\mathbf{L}_M = \frac{ae^{i\phi}}{\sqrt{2}}\begin{pmatrix} 1 \\ -i \end{pmatrix}, \quad \mathbf{R}_M = \frac{ae^{i\phi}}{\sqrt{2}}\begin{pmatrix} 1 \\ i \end{pmatrix}. \quad (5.115)$$

Note that the Jones vector for left circularly polarized light in the increasing phase convention is the Jones vector for right circularly polarized light in the decreasing phase convention and vice versa. *This can become very confusing, if the convention is not stated clearly*!

The choice of sign convention affects the signs of Jones matrix elements. Table 5.5 compares several common Jones matrices in both decreasing phase sign convention and the increasing phase sign convention. They are the conjugate of each other.

5.11 Conclusion

The Jones calculus provides a succinct and flexible method for performing *diattenuation* and *retardance* calculations. The Jones calculus is set up for treating problems with *coherent light*; in particular *adding Jones vectors* simulates the addition of coherent beams, and adding Jones matrices simulates combining laser beams taking multiple paths in an interferometer, as described in Chapter 4.

The analysis of Jones matrices with orthogonal eigenvectors, *homogeneous matrices*, is simpler than the more general case, and the eigenvalues of homogeneous Jones matrices directly relate to retardance and diattenuation. *Diattenuators* have *Hermitian Jones matrices* and *retarders* have *unitary Jones matrices*.

The definition of Jones vectors and Jones matrices assumes the light is *monochromatic* and thus *coherent*. For this reason, Jones matrices work well for ray tracing calculations for polarization. But more generally, Jones matrices work well to describe the diattenuation and retardance of optical systems or sequences of polarization elements, and in this context, their application is not restricted to monochromatic light. Because Jones matrices contain an *absolute phase*, they integrate easily with optical design calculations, making them an effective tool for handling polarization aberration problems in optical design.

For *ideal polarization elements*, the polarization properties are readily defined. For *real polarization elements*, the precise definition of the polarization properties is more subtle, since polarization elements, such as polarizers and retarders, may display some diattenuation, some retardance, and maybe even some depolarization. These properties will vary with wavelength and angle of incidence, and may also vary with position on the element.

The Jones matrices for retarders are most elegant in the *symmetrical phase convention*, where the phase of the fast state is advanced by half the retardance and the phase of the slow state is delayed

Table 5.5 Jones Matrices in the Decreasing and Increasing Phase Conventions

Fast Axis	Retarders: Quarter Wave	
	Decreasing Phase Convention Fast Axis Unchanged	Increasing Phase Convention Fast Axis Unchanged
Horizontal, 0° **QWHLR** or LR ($\pi/2$, 0°)	$\begin{pmatrix} 1 & 0 \\ 0 & i \end{pmatrix}$	$\begin{pmatrix} 1 & 0 \\ 0 & -i \end{pmatrix}$
Vertical, 90° **QWVLR** LR ($\pi/2$, 90°)	$\begin{pmatrix} i & 0 \\ 0 & 1 \end{pmatrix}$	$\begin{pmatrix} -i & 0 \\ 0 & 1 \end{pmatrix}$
45° **QW45LR** LR ($\pi/2$, 45°)	$\frac{1}{2}\begin{pmatrix} 1+i & 1-i \\ 1-i & 1+i \end{pmatrix}$	$\frac{1}{2}\begin{pmatrix} 1-i & 1+i \\ 1+i & 1-i \end{pmatrix}$
135° **QW135LR** LR ($\pi/2$, 135°)	$\frac{1}{2}\begin{pmatrix} 1+i & -1+i \\ -1+i & 1+i \end{pmatrix}$	$\frac{1}{2}\begin{pmatrix} 1-i & -1-i \\ -1-i & 1-i \end{pmatrix}$
Right circular **QWRCR**	$\frac{1}{2}\begin{pmatrix} 1+i & 1+i \\ -1-i & 1+i \end{pmatrix}$	$\frac{1}{2}\begin{pmatrix} 1-i & 1-i \\ -1+i & 1-i \end{pmatrix}$
Left circular **QWLCR**	$\frac{1}{2}\begin{pmatrix} 1+i & -1-i \\ 1+i & 1+i \end{pmatrix}$	$\frac{1}{2}\begin{pmatrix} 1-i & -1+i \\ 1-i & 1-i \end{pmatrix}$

Fast Axis	Retarders: Half Wave	
	Decreasing Phase Convention	Increasing Phase Convention
Half wave horizontal Linear retarder **HWHLR**	$\begin{pmatrix} 1 & 0 \\ 0 & -1 \end{pmatrix}$	$\begin{pmatrix} 1 & 0 \\ 0 & -1 \end{pmatrix}$
Half wave vertical Linear retarder **HWVLR**	$\begin{pmatrix} -1 & 0 \\ 0 & 1 \end{pmatrix}$	$\begin{pmatrix} -1 & 0 \\ 0 & 1 \end{pmatrix}$
Half wave 45° Linear retarder **HW45LR**	$\begin{pmatrix} 0 & 1 \\ 1 & 0 \end{pmatrix}$	$\begin{pmatrix} 0 & 1 \\ 1 & 0 \end{pmatrix}$
Half wave 135° Linear retarder **HW135LR**	$\begin{pmatrix} 0 & -1 \\ -1 & 0 \end{pmatrix}$	$\begin{pmatrix} 0 & -1 \\ -1 & 0 \end{pmatrix}$
Half wave right Circular retarder **HWRCR**	$\begin{pmatrix} 0 & i \\ -i & 0 \end{pmatrix}$	$\begin{pmatrix} 0 & -i \\ i & 0 \end{pmatrix}$
Half wave left Circular retarder **HWLCR**	$\begin{pmatrix} 0 & -i \\ i & 0 \end{pmatrix}$	$\begin{pmatrix} 0 & i \\ -i & 0 \end{pmatrix}$

(Continued)

5.12 Problem Sets

Table 5.5 (Continued) Jones Matrices in the Decreasing and Increasing Phase Conventions

	Polarizers	
Fast Axis	Decreasing Phase Convention	Increasing Phase Convention
Horizontal linear Polarizer **HLP**	$\begin{pmatrix} 1 & 0 \\ 0 & 0 \end{pmatrix}$	$\begin{pmatrix} 1 & 0 \\ 0 & 0 \end{pmatrix}$
Vertical linear Polarizer **VLP**	$\begin{pmatrix} 0 & 0 \\ 0 & 1 \end{pmatrix}$	$\begin{pmatrix} 0 & 0 \\ 0 & 1 \end{pmatrix}$
45° Linear Polarizer **L45P**	$\dfrac{1}{2}\begin{pmatrix} 1 & 1 \\ 1 & 1 \end{pmatrix}$	$\dfrac{1}{2}\begin{pmatrix} 1 & 1 \\ 1 & 1 \end{pmatrix}$
135° Linear Polarizer **L135P**	$\dfrac{1}{2}\begin{pmatrix} 1 & -1 \\ -1 & 1 \end{pmatrix}$	$\dfrac{1}{2}\begin{pmatrix} 1 & -1 \\ -1 & 1 \end{pmatrix}$
Right circular Polarizer **RCP**	$\dfrac{1}{2}\begin{pmatrix} 1 & i \\ -i & 1 \end{pmatrix}$	$\dfrac{1}{2}\begin{pmatrix} 1 & -i \\ i & 1 \end{pmatrix}$
Left circular Polarizer **LCP**	$\dfrac{1}{2}\begin{pmatrix} 1 & -i \\ i & 1 \end{pmatrix}$	$\dfrac{1}{2}\begin{pmatrix} 1 & i \\ -i & 1 \end{pmatrix}$

The absolute phase is chosen; thus, fast axis undergoes zero phase change and the slow axis is delayed.

by half the retardance. Generally, the overall phase of the polarization element is not known, such as for purchased elements; then, the symmetric phase form for retarders is an elegant but arbitrary choice for the Jones matrix. Many references provide the Jones matrix in the other retarder conventions: (1) slow eigenpolarization unchanged or (2) fast eigenpolarization unchanged. These retarder Jones matrices are plotted in amplitude and phase (color coded) in Figure 5.19 for comparison: (left) *slow eigenpolarization unchanged*, which advances the phase of the fast eigenpolarization, or (right) *fast eigenpolarization unchanged*, which delays the phase of the slow eigenpolarization. The only difference from Figure 5.8 is the phase.

5.12 Problem Sets

Math Tip 5.3 Phase in Problem Sets

Normally, when specifying polarization elements for problem sets, the *absolute phase* of the element is not specified (the problem text would get tedious, long winded, and confusing if it was). Thus, if a problem calls for a *quarter linear wave retarder* with a horizontal fast axis, the Jones matrix might be any one of

$$\begin{pmatrix} 1 & 0 \\ 0 & i \end{pmatrix}, \quad \frac{1}{\sqrt{2}}\begin{pmatrix} 1-i & 0 \\ 0 & 1+i \end{pmatrix}, \quad \begin{pmatrix} -i & 0 \\ 0 & 1 \end{pmatrix}, \quad e^{-i\phi}\begin{pmatrix} 1 & 0 \\ 0 & i \end{pmatrix}. \quad (5.116)$$

Each of these matrices is correct if the absolute phase is not specified. Further, the text suggests that the second matrix, in the *symmetric retarder convention*, is the most canonical, but clearly it is not the easiest form to use for hand calculation. Hence, when grading problems, the answer should be allowed with any phase. For example, if you think the answer is the identity matrix and you get any of the following forms,

$$\begin{pmatrix} 1 & 0 \\ 0 & 1 \end{pmatrix}, \begin{pmatrix} i & 0 \\ 0 & i \end{pmatrix}, \frac{1}{\sqrt{2}}\begin{pmatrix} 1-i & 0 \\ 0 & 1-i \end{pmatrix}, \begin{pmatrix} -1 & 0 \\ 0 & -1 \end{pmatrix}, \begin{pmatrix} e^{-i\phi} & 0 \\ 0 & e^{-i\phi} \end{pmatrix}, \quad (5.117)$$

stop, you are done!

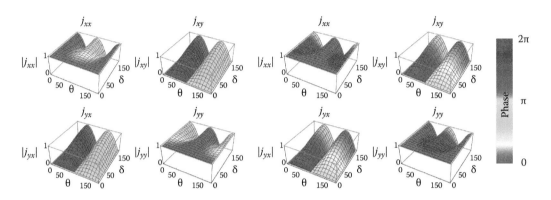

Figure 5.19 The Jones matrix elements for all linear retarders in the form that advances the phase of the fast eigenpolarization leaving the slow eigenpolarization unchanged (left) are constant (red) along the left and right edges of the j_{xx} element and up the middle of the j_{yy} element. Conversely, the Jones matrix elements for all linear retarders in the form that delays the phase of the slow eigenpolarization leaving the fast eigenpolarization unchanged (right) are constant (red) along the left and right edges of the j_{yy} element and up the middle of the j_{xx} element. This figure should be compared to Figure 5.8.

5.1 Matrix multiply the following sequences of Jones matrices. Then, identify the resulting Jones matrix in the tables of polarizer and retarder Jones matrices. The matrix may differ by a constant or in phase (i.e., −1 and i), particularly depending on the phase chosen for retarders.
   a. Two linear retarders **LR**[$\pi/2$,0]
   b. **LR**[$\pi/2$,90°] followed by **LR**[$\pi/2$,0]
   c. **LR**[$\pi/4$,90°], **LP**[0°], and **LR**[$\pi/4$,0]
   d. Three linear polarizers **LP**[0°], **LP**[45°], and **LP**[0°]
   e. A linear polarizer between two circular retarders **CR**[$-\pi/2$], **LP**[0°], and **CR**[$\pi/2$]
   f. Two left circular diattenuators **CD**[1,0] with diattenuation 1
   g. Circular retarders and circular diattenuator **CR**[$-\pi/2$], **CD**[1,0], and **CR**[$\pi/2$]
   h. **CR**[π] followed by **LR**[π,0]
   i. **CR**[π], **LR**[π, 45°], and **LR**[π, 0]
   j. **LR**[$\pi/2$, −45°], **CR**[π], and $-i$**LR**[$\pi/2$, 45°]
   k. **LR**[$\pi/2$, −45°], **CR**[$\pi/2$], and $-i$**LR**[$\pi/2$, 45°]

5.2 Find the Jones matrices for the following retarders in the three phase conventions, (1) fast axis unchanged, (2) symmetric phase convention, (3) slow axis unchanged.
   a. **LR**[π, 45°]

5.12 Problem Sets

   b. **LR[π/2, 135°]**
   c. **LR[π/8, 0]**
   d. **LR[π/2, 30°]**
   e. **LR[π/4, 45°]**
   f. **LR[π/3, 45°]**

5.3 Given the Jones matrix $\mathbf{J}_1 = \begin{pmatrix} -\dfrac{1}{2} & \dfrac{1}{2} - \dfrac{i}{\sqrt{2}} \\ -\dfrac{1}{2} - \dfrac{i}{\sqrt{2}} & \dfrac{3}{2} \end{pmatrix}$.

   a. Show that $\mathbf{J}_1$ represents a polarizer by calculating its eigenvalues.
   b. Why do these eigenvalues demonstrate this matrix as a polarizer?
   c. Find the transmitted and extinguished states as un-normalized Jones vectors.
   d. Is the element homogeneous?

5.4 Derive Malus' law for transmission of linearly polarized light through a linear polarizer by the operation of the Jones matrix for a linear polarizer **LP(θ)** light polarized at an angle θ–α.

5.5 Analyze the polarization properties of the Jones matrix $\mathbf{J} = \begin{pmatrix} \dfrac{1}{2} + \dfrac{i}{4} & \dfrac{1}{2} - \dfrac{i}{4} \\ \dfrac{1}{2} - \dfrac{i}{4} & \dfrac{1}{2} + \dfrac{i}{4} \end{pmatrix}$ from its eigenvalues and eigenvectors.

5.6 Find the eigenvalues and eigenpolarizations of the following Jones matrices. Classify the matrices as retarders, polarizers, diattenuators, or combinations. Classify the matrices as linear, circular, or elliptical polarization elements:

   a. $\dfrac{1}{2}\begin{pmatrix} 1 & i \\ -i & 1 \end{pmatrix}$

   b. $\dfrac{1}{2}\begin{pmatrix} 1+i & 1-i \\ 1-i & 1+i \end{pmatrix}$

   c. $\begin{pmatrix} 0 & -1 \\ 1 & 0 \end{pmatrix}$

   d. $\dfrac{1}{8}\begin{pmatrix} 5 & 3 \\ 3 & 5 \end{pmatrix}$

   e. $\begin{pmatrix} 1 & 0 \\ 0 & 1/2 \end{pmatrix}$

   f. $\dfrac{1}{2}\begin{pmatrix} 1-i & -1+i \\ \sqrt{2} & \sqrt{2} \end{pmatrix}$

   g. $\dfrac{1}{4}\begin{pmatrix} 3 & 1 \\ 1 & 3 \end{pmatrix}$

   h. $\dfrac{1}{8}\begin{pmatrix} 5 & -\sqrt{3} \\ -\sqrt{3} & 7 \end{pmatrix}$

   i. $\dfrac{1}{2\sqrt{2}}\begin{pmatrix} 1+\sqrt{3} & -1+\sqrt{3} \\ -1+\sqrt{3} & 1+\sqrt{3} \end{pmatrix}$

j. $\dfrac{1}{\sqrt{2}}\begin{pmatrix} i & -1 \\ 1 & -i \end{pmatrix}$

k. $\dfrac{1}{\sqrt{2}}\begin{pmatrix} 0 & -1+i \\ 1+i & 0 \end{pmatrix}$

l. $\dfrac{1}{4}\begin{pmatrix} 3 & i \\ -i & 3 \end{pmatrix}$

m. $\dfrac{1}{3}\begin{pmatrix} 1 & 2 \\ 2 & 1 \end{pmatrix}$

n. $\dfrac{1}{3}\begin{pmatrix} 1 & -2 \\ 2 & 1 \end{pmatrix}$

5.7 Verify the following equation for the diattenuation of a Jones matrix using the matrix for **LD**(1, t_{min}, π/4).

$$D = \sqrt{1 - \dfrac{4|\det \mathbf{J}|^2}{Tr(\mathbf{J}^\dagger \cdot \mathbf{J})^2}} \qquad (5.118)$$

5.8 Find the transmittance of an arbitrary Jones matrix, $\mathbf{J} = \begin{pmatrix} j_{11} & j_{12} \\ j_{21} & j_{22} \end{pmatrix}$, for unpolarized incident light.
   a. Calculate the transmitted flux for 0° and 90° polarization states. The transmittance is the ratio of the transmitted flux to the incident flux.
   b. Unpolarized light can be treated as an incoherent combination of any two orthogonal polarization states. From part (a), find the transmittance for unpolarized light.
   c. Calculate the transmitted flux for 45° and 135° polarization states, and from these values, find the transmittance for unpolarized light.

5.9 Show that the diattenuation of the elliptical diattenuator Jones matrix of Equation 5.56 is equal to D by considering the eigenvalues (Equation 5.57).

5.10 Show that the product of the two polarizers **LP**(0) **LP**(π/4) is not a linear polarizer of the form in Equation 5.29.

5.11 a. Find the Jones matrix for a 180° rotation of a polarization state.
   b. What happens to a coherent beam of light when the plane of polarization is rotated by 180°°? Does the Jones vector change?
   c. How does the Jones vector and the state change for linearly polarized light?
   d. How does the Jones vector and the state change for circularly polarized light?

5.12 From the definition of matrix multiplication,

$$(\mathbf{A} \cdot \mathbf{B})_{i,j} = \sum_k A_{i,k} B_{k,j}$$

   a. Show that matrix multiplication is associative $(\mathbf{A} \cdot \mathbf{B}) \cdot \mathbf{C} = \mathbf{A} \cdot (\mathbf{B} \cdot \mathbf{C})$.
   b. Show that if $\mathbf{A} \cdot \mathbf{B} = \mathbf{C}$, then for the adjoints, that $\mathbf{A}^\dagger \cdot \mathbf{B}^\dagger = \mathbf{C}^\dagger$.

5.13 Examine the periodicity of the retarder equations.

a. Show that the linear retarder equation $\mathbf{LR}_1(\delta,0°) = \begin{pmatrix} e^{-i\delta} & 0 \\ 0 & 1 \end{pmatrix}$ is periodic with a period 2π, $\mathbf{LR}_1(\delta, 0°) = \mathbf{LR}_1(\delta + 2\pi, 0°)$.

b. Show that the symmetric form of the retarder equation, given here for arbitrary orientation,

$$\mathbf{LR}(\delta,\theta) = \begin{pmatrix} e^{-\frac{i\delta}{2}}\cos^2\theta + e^{\frac{i\delta}{2}}\sin^2\theta & -i\sin\left(\frac{\delta}{2}\right)\sin(2\theta) \\ -i\sin\left(\frac{\delta}{2}\right)\sin(2\theta) & e^{\frac{i\delta}{2}}\cos^2\theta + e^{-\frac{i\delta}{2}}\sin^2\theta \end{pmatrix}$$

is not periodic with period 2π, $\mathbf{LR}(\delta, \theta) \neq \mathbf{LR}(\delta + 2\pi, \theta)$. Explain the difference. Why is this acceptable?

5.14 a. Show that the half wave linear retarder Jones matrix $\mathbf{LR}(\pi,\theta) = \begin{pmatrix} \cos 2\theta & \sin 2\theta \\ \sin 2\theta & -\cos 2\theta \end{pmatrix}$ converts left circularly polarized light into right circularly polarized light for all angles θ.

b. How does the exiting state vary with angle θ?

5.15 A general linear diattenuator oriented at 45° can be written with Pauli spin matrices as $\rho_0\sigma_0 + \rho_2\sigma_2$ (see Chapter 14). Find the Jones matrix. Calculate the amplitude transmittances for 45° and 135° linearly polarized light as a function of ρ_0 and ρ_2. Calculate the two intensity transmittances and the diattenuation.

5.16 Show that the diattenuation of the linear retarder Jones matrix

$$\mathbf{LR}(\delta,\theta) = \begin{pmatrix} e^{-\frac{i\delta}{2}}\cos^2\theta + e^{\frac{i\delta}{2}}\sin^2\theta & -i\sin\left(\frac{\delta}{2}\right)\sin(2\theta) \\ -i\sin\left(\frac{\delta}{2}\right)\sin(2\theta) & e^{\frac{i\delta}{2}}\cos^2\theta + e^{-\frac{i\delta}{2}}\sin^2\theta \end{pmatrix}$$

is always zero.

5.17 Find the Jones matrix for the linear diattenuator that transmits 4/9 of the flux polarized at 30° as 30° linearly polarized light and transmits 1/9 of the flux polarized at 120° as 120° linearly polarized light.

5.18 Give an example of a nilpotent 2×2 matrix $\mathbf{N}$ with two eigenvalues equal to zero. For one particular state $\mathbf{E}_1$, $\mathbf{N}$ transmits all the incident flux, but the light exits in the orthogonal polarization state.

5.19 a. What is the Jones matrix for a partial polarizer that transmits all of the x-polarized flux (as x-polarized light) and half the y-polarized flux as y-polarized light?

b. What is the Jones matrix $\mathbf{D}(\theta)$ when the transmission axis is rotated counterclockwise to an angle θ?

c. Consider left circularly polarized light incident on a series of two of these polarizers, $\mathbf{D}(\pi/2)\,\mathbf{D}(0)$. What is the Jones matrix for the sequence of two diattenuators?
d. Find the exiting polarization state. What is the flux relative to the incident state? What is the ellipticity and angle of the polarization ellipse?
e. Find the Jones matrix for the sequence of three diattenuators $\mathbf{D}(\pi/2)\,\mathbf{D}(\pi/4)\,\mathbf{D}(0)$.
f. Which polarization state has the maximum transmission?
g. Find the diattenuation D.

5.20 a. Find a polarization state $\mathbf{E}_2$ orthogonal to $\mathbf{E}_1 = \begin{pmatrix} a \\ b+ic \end{pmatrix}$.

b. Show that there is no single matrix $\begin{pmatrix} j_{xx} & j_{xy} \\ j_{yx} & j_{yy} \end{pmatrix}$ that will transform $\mathbf{E}_1$ into $\mathbf{E}_2$ for all values a, b, and c. Thus, there is no polarization element, *the orthogonalizer*, that will always transform an arbitrary input state into the orthogonal state.

5.21 A particular optical instrument requires a polarization element that transforms all of the incident horizontal linearly polarized light into 45° linearly polarized light, and transforms all of the incident 45° linearly polarized light into right circularly polarized light.
a. Find a Jones Matrix $\mathbf{J}$ with these properties. Note that the matrix will change based on the phases assumed.
b. What are the eigenvalues of $\mathbf{J}$?
c. Is $\mathbf{J}$ homogeneous? Are the eigenpolarizations orthogonal?
d. Does $\mathbf{J}$ have gain? Does more power ever exit than enters?

5.22 a. Derive a method to find the arbitrary Jones matrix that makes the following two arbitrary transformations,

$$\begin{pmatrix} j_{xx} & j_{xy} \\ j_{yx} & j_{yy} \end{pmatrix}\begin{pmatrix} a \\ b \end{pmatrix} = \begin{pmatrix} s \\ t \end{pmatrix},\ \text{and}\ \begin{pmatrix} j_{xx} & j_{xy} \\ j_{yx} & j_{yy} \end{pmatrix}\begin{pmatrix} c \\ d \end{pmatrix} = \begin{pmatrix} u \\ v \end{pmatrix}.$$

Create a four-by-four matrix that is a function of a, b, c, and d to operate on the vector $\begin{pmatrix} j_{xx} \\ j_{xy} \\ j_{yx} \\ j_{yy} \end{pmatrix}$ yielding $\begin{pmatrix} s \\ t \\ u \\ v \end{pmatrix}$. Then solve this equation for $j_{xx}, j_{xy}, j_{yx},$ and j_{yy}.

b. Find the Jones matrix to transform $(1, 0)$ into $(0, 1)$ and to transform $(0, 1)$ into $(1, -1/\sqrt{2})$

5.23 Figure 5.11 shows a radially polarized beam of light created with a vortex retarder and an input polarized beam. How could a tangentially polarized beam be created?

5.24 Show that to first order in the diattenuation, the horizontal linear diattenuator has the Jones matrix,

$$\mathbf{LD}\left(1+\frac{d}{2}, 1-\frac{d}{2}, 0°\right) = \begin{pmatrix} 1+\dfrac{d}{2} & 0 \\ 0 & 1-\dfrac{d}{2} \end{pmatrix}.$$

5.25 Show that Equation 5.19 fails for matrices of the form $\mathbf{J} = \begin{pmatrix} 1 & 0 \\ a & b \end{pmatrix}$. Find an alternative equation for the eigenpolarizations for this special case.

5.26 Derive the elliptical polarizer Jones matrix in Equation 5.30, by finding the unitary matrix $\mathbf{U}$ that transforms (1, 0) into $\mathbf{E}(\varepsilon, \psi)$ and transforms (0, 1) into the orthogonal state. Then, use $\mathbf{U}$ to perform a unitary change of basis on $\mathbf{LP}(0)$ to generate $\mathbf{EP}(\varepsilon, \psi)$. Verify the operation of $\mathbf{EP}(\varepsilon, \psi)$ on $\mathbf{E}(\varepsilon, \psi)$ and on the orthogonal state.

References

1. F. Pezzotta and B. M. Laurs, Tourmaline: The kaleidoscopic gemstone, *Elements* 7.5 (2011): 333–338.
2. M. Grabau, Polarized light enters the world of everyday life, *J. Appl. Phys.* 9.4 (1938): 215–225.
3. W. R. Phillips, *Mineral Optics: Principles and Techniques*, WH Freeman (1971).
4. E. Land, Some aspects of the development of sheet polarizers, *J. Opt. Soc. Am.* 41 (1951): 957–962.
5. W. A. Shurcliff, *Polarized Light. Production and Use*, Cambridge, MA: Harvard University Press (1966).
6. R. A. Chipman, Polarization analysis of optical systems, *Opt. Eng.* 28.2 (1989): 280–290.
7. J. M. Bennett, Polarizers, in *Handbook of Optics*, Vol. I, Chapter 13, 3rd edition, ed. M. Bass, New York: McGraw-Hill (2010).
8. G. E. Shilov, *Linear Algebra* (translated and edited by Richard A. Silverman), Dover (1977).
9. D. Poole, *Linear Algebra: A Modern Introduction*, Cengage Learning (2014).
10. G. Strang, *Linear Algebra and Its Applications*, Belmont, CA: Thomson, Brooks/Cole (2006).
11. R. Jones, A new calculus for the treatment of optical systems, *J. Opt. Soc. Am.* 31 (1941): 500–503.
12. H. Hurwitz Jr. and R. Jones, A new calculus for the treatment of optical systems, *J. Opt. Soc. Am.* 31 (1941): 493–495.
13. R. Jones, A new calculus for the treatment of optical systems, *J. Opt. Soc. Am.* 31 (1941): 488–493.
14. R. Jones, A new calculus for the treatment of optical systems. IV, *J. Opt. Soc. Am.* 32 (1942): 486–493.
15. R. Jones, A new calculus for the treatment of optical systems VI. Experimental determination of the matrix, *J. Opt. Soc. Am.* 37 (1947): 110.
16. R. Jones, A new calculus for the treatment of optical systems V. A more general formulation, and description of another calculus, *J. Opt. Soc. Am.* 37 (1947): 107.
17. R. Jones, A new calculus for the treatment of optical systems. VII. Properties of the N-matrices, *J. Opt. Soc. Am.* 38 (1948): 671–683.
18. R. M. A. Azzam and N. M. Bashara, *Ellipsometry and Polarization Light*, New York: North-Holland (1977), p. 97.
19. S.-Y. Lu and R. A. Chipman, Homogeneous and inhomogeneous Jones matrices, *JOSA A* 11.2 (1994): 766–773.
20. D. Goldstein, *Polarized Light, Revised and Expanded*, Vol. 83, CRC Press (2011).
21. J. F. Nye and M. V. Berry, Dislocations in wave trains, *Proceedings of the Royal Society of London A: Mathematical, Physical and Engineering Sciences*, Vol. 336. No. 1605. The Royal Society (1974).
22. S. McEldowney, D. Shemo, R. Chipman, and P. Smith, Creating vortex retarders using photoaligned liquid crystal polymers, *Opt. Lett.* 33 (2008): 134–136.
23. S. McEldowney, D. Shemo, and R. Chipman, Vortex retarders produced from photo-aligned liquid crystal polymers, *Opt. Express* 16 (2008): 7295–7308.
24. P. Piron, P. Blain, S. Habraken, and D. Mawet, Polarization holography for vortex retarders recording, *Appl. Opt.* 52 (2013): 7040–7048.
25. A. Boivin, J. Dow, and E. Wolf, Energy flow in the neighborhood of the focus of a coherent beam, *J. Opt. Soc. Am.* 57 (1967): 1171–1175.
26. K. T. Gahagan and G. A. Swartzlander, Optical vortex trapping of particles, *Opt. Lett.* 21.11 (1996): 827–829.
27. J. Curtis and D. Grier, Modulated optical vortices, *Opt. Lett.* 28 (2003): 872–874.
28. S. Quabis, R. Dorn, M. Eberler, O. Glockl, and G. Leuchs, Focusing light to a tighter spot, *Opt. Commun.* 179 (2000): 1–7.

29. R. Dorn, S. Quabis, and G. Leuchs, Sharper focus for radially polarized light beam, *Phys. Rev. Lett.* 91 (2003): 23.
30. G. Arfken, *Mathematical Methods for Physicists*, Chapter 4.4, New York: Academic (1970).
31. G. Arfken, *Mathematical Methods for Physicists*, Chapter 4.3, New York: Academic (1970).
32. P. Lancaster and M. T. Tismenetsky, *The Theory of Matrices*, 2nd edition, New York: Academic (1985).
33. R. A. Horn and C. R. Johnson, *Topics in Matrix Analysis*, Cambridge: Cambridge University Press (1991).
34. S. Lu and R. Chipman, Homogeneous and inhomogeneous Jones matrices, *J. Opt. Soc. Am. A* 11 (1994): 766–773.
35. S. Lu and R. Chipman, Interpretation of Mueller matrices based on polar decomposition, *J. Opt. Soc. Am. A* 13 (1996): 1106–1113.
36. H. H. Barrett and K. J. Myers, *Foundations of Image Science*, Section 1.5, New York: Wiley (2004).

6

Mueller Matrices

6.1 Introduction

The *Mueller matrix* provides a systematic way of representing all of the polarization properties of a sample: the *diattenuation*, *retardance*, and *depolarization*. The Mueller matrix has a simple definition; it is the 4 × 4 matrix that relates a set of incident Stokes parameters to the exiting Stokes parameters, for any type of sample. Despite this simple definition, the properties of Muller matrices are quite complex.

With the introduction of the Mueller matrix, the range of polarization phenomena under description is now greatly expanded. With the *Jones calculus*, calculations were limited to *non-depolarizing interactions*; all fully polarized incident states emerge as fully polarized. Indeed, for the majority of optical systems, lenses, telescopes, microscopes, and fiber optic systems, these non-depolarizing interactions are the desired condition. Most optical systems operate extremely close to the non-depolarizing condition.

In a *depolarizing interaction*, polarized light becomes partially polarized; the degree of polarization is reduced. In a microscope or telescope, such depolarization is usually associated with defects such as dirt, fingerprints, scratches, or a bad optical coating. For such defects, the Mueller matrix, not the Jones matrix, is appropriate to characterize the polarization effects. But more generally when polarized light scatters from everything, paint, paper, dirt, rocks, plastics, the degree of polarization decreases and some depolarization is present. Thus, the Mueller matrix expands the variety of light–matter interactions that can be described.

6.2 The Mueller Matrix

The Mueller matrix, **M**, is a 4 × 4 matrix that transforms the incident Stokes parameters, **S**, into the exiting Stokes parameters, **S'**, by matrix vector multiplication,[1,2]

$$\mathbf{M} \cdot \mathbf{S} = \mathbf{S'} = \begin{pmatrix} M_{00} & M_{01} & M_{02} & M_{03} \\ M_{10} & M_{11} & M_{12} & M_{13} \\ M_{20} & M_{21} & M_{22} & M_{23} \\ M_{30} & M_{31} & M_{32} & M_{33} \end{pmatrix} \begin{pmatrix} S_0 \\ S_1 \\ S_2 \\ S_3 \end{pmatrix} = \begin{pmatrix} S'_0 \\ S'_1 \\ S'_2 \\ S'_3 \end{pmatrix}. \tag{6.1}$$

From the definition of matrix–vector multiplication, the exiting Stokes parameters **S'** are

$$\mathbf{S'} = \begin{pmatrix} S'_0 \\ S'_1 \\ S'_2 \\ S'_3 \end{pmatrix} = \begin{pmatrix} S_0 M_{00} + S_1 M_{01} + S_2 M_{02} + S_3 M_{03} \\ S_0 M_{10} + S_1 M_{11} + S_2 M_{12} + S_3 M_{13} \\ S_0 M_{20} + S_1 M_{21} + S_2 M_{22} + S_3 M_{23} \\ S_0 M_{30} + S_1 M_{31} + S_2 M_{32} + S_3 M_{33} \end{pmatrix}. \tag{6.2}$$

Each element of the incident **S** is related to the four elements of **S'** by the elements of **M**. Since the elements of **S** and **S'** are *irradiances*, the elements of **M** are dimensionless ratios of irradiances. Since irradiances are real, the elements of **M** are *real valued*, not complex numbers. Our convention numbers the subscripts from 0 to 3 to match the corresponding Stokes parameter subscripts.

Example 6.1 Meaning of Mueller Matrix Columns

When *unpolarized light* is incident, the 0th column of the Mueller matrix (counting from 0 to 3) describes the *exiting polarization state*,

$$\mathbf{M} \cdot \mathbf{S} = \mathbf{S'} = \begin{pmatrix} M_{00} & M_{01} & M_{02} & M_{03} \\ M_{10} & M_{11} & M_{12} & M_{13} \\ M_{20} & M_{21} & M_{22} & M_{23} \\ M_{30} & M_{31} & M_{32} & M_{33} \end{pmatrix} \begin{pmatrix} 1 \\ 0 \\ 0 \\ 0 \end{pmatrix} = \begin{pmatrix} M_{00} \\ M_{10} \\ M_{20} \\ M_{30} \end{pmatrix}. \tag{6.3}$$

This is also the average *exiting polarization state*, since when 0° or 90° polarized light is incident, the exiting state is the 0th column plus or minus the 1st column,

$$\mathbf{M} \cdot \mathbf{S} = \mathbf{S'} = \begin{pmatrix} M_{00} & M_{01} & M_{02} & M_{03} \\ M_{10} & M_{11} & M_{12} & M_{13} \\ M_{20} & M_{21} & M_{22} & M_{23} \\ M_{30} & M_{31} & M_{32} & M_{33} \end{pmatrix} \begin{pmatrix} 1 \\ 1 \\ 0 \\ 0 \end{pmatrix} = \begin{pmatrix} M_{00} \\ M_{10} \\ M_{20} \\ M_{30} \end{pmatrix} + \begin{pmatrix} M_{01} \\ M_{11} \\ M_{21} \\ M_{31} \end{pmatrix}, \tag{6.4}$$

6.4 Non-Polarizing Mueller Matrices

$$\mathbf{M} \cdot \mathbf{S} = \mathbf{S}' = \begin{pmatrix} M_{00} & M_{01} & M_{02} & M_{03} \\ M_{10} & M_{11} & M_{12} & M_{13} \\ M_{20} & M_{21} & M_{22} & M_{23} \\ M_{30} & M_{31} & M_{32} & M_{33} \end{pmatrix} \begin{pmatrix} 1 \\ -1 \\ 0 \\ 0 \end{pmatrix} = \begin{pmatrix} M_{00} \\ M_{10} \\ M_{20} \\ M_{30} \end{pmatrix} - \begin{pmatrix} M_{01} \\ M_{11} \\ M_{21} \\ M_{31} \end{pmatrix}. \tag{6.5}$$

Thus, the 0th column is the average of these two output states. When 45° or 135° is incident, the exiting state is the 0th column plus or minus the 2nd column. Similarly, when right and left circularly polarized light are incident, the exiting state is the 0th column plus or minus the 3rd column. Thus, the 0th column is seen as the average output.

6.3 Sequences of Polarization Elements

The effect of a *series of polarization elements* or interactions is described by *matrix multiplication* of their individual Mueller matrices $\mathbf{M}_q$, where q is the index describing the order the elements are encountered. The first element encountered is on the right side of the sequence of matrices being multiplied. The final element $\mathbf{M}_Q$ is the leftmost matrix in the matrix product,

$$\mathbf{M} = \mathbf{M}_Q \cdot \mathbf{M}_{Q-1} \ldots \mathbf{M}_2 \cdot \mathbf{M}_1 = \prod_{q=1}^{Q} \mathbf{M}_{Q-q+1}. \tag{6.6}$$

Thus, the properties of sequences of polarization effects are readily calculated using the simple operation of matrix multiplication. In evaluating cascades of Mueller matrices, the *associative rule for matrix multiplication* can be used,

$$(\mathbf{M}_3 \cdot \mathbf{M}_2) \cdot \mathbf{M}_1 = \mathbf{M}_3 \cdot (\mathbf{M}_2 \cdot \mathbf{M}_1), \tag{6.7}$$

and adjacent matrices can be grouped in any order for multiplication. Example 6.2 contains a sequence of polarization elements example.

6.4 Non-Polarizing Mueller Matrices

A *non-polarizing optical element* does not change the polarization state of any incident polarization state, only the *amplitude* and/or *phase* change. The Mueller matrix for a *non-absorbing, non-polarizing sample* is the 4 × 4 *identity matrix*, $\mathbf{I}$,

$$\mathbf{I} = \begin{pmatrix} 1 & 0 & 0 & 0 \\ 0 & 1 & 0 & 0 \\ 0 & 0 & 1 & 0 \\ 0 & 0 & 0 & 1 \end{pmatrix}. \tag{6.8}$$

I is the Mueller matrix for *vacuum* and the approximate Mueller matrix for *air*. For a neutral density filter or an element with polarization-independent absorption or loss, the Mueller matrix has $T_{max} = T_{min} = T$, and the resulting Mueller matrix is proportional to the identity matrix and can be expressed in terms of our notation for linear diattenuators, **LD**$(T_{max}, T_{min}, \theta)$ (Equation 6.53), as

$$\mathbf{LD}(T,T,0) = T \begin{pmatrix} 1 & 0 & 0 & 0 \\ 0 & 1 & 0 & 0 \\ 0 & 0 & 1 & 0 \\ 0 & 0 & 0 & 1 \end{pmatrix}. \tag{6.9}$$

The Stokes parameters do not contain an *absolute phase* term, unlike Jones vectors. The Jones matrix for absolute phase change ϕ and *optical path length* is

$$e^{i\phi} \begin{pmatrix} 1 & 0 \\ 0 & 1 \end{pmatrix}. \tag{6.10}$$

There is *no corresponding Mueller matrix for phase change*. If phase change or optical path length needs to be calculated, such calculations need to be performed in addition to the Mueller calculations. Changes in relative phase between the polarization components of the light, the *retardance*, is calculated within the Mueller calculus.

6.5 Rotating Polarization Elements about the Light Direction

When a polarization element with Mueller matrix **M** is rotated about the incident beam of light by an angle θ, the angle of incidence is unchanged. For example, consider a normal-incidence beam passing through an element rotating about its normal, the resulting Mueller matrix $\mathbf{M}(\theta)$ is

$$\mathbf{M}(\theta) = \mathbf{R}_M(\theta) \cdot \mathbf{M} \cdot \mathbf{R}_M(-\theta)$$

$$= \begin{pmatrix} 1 & 0 & 0 & 0 \\ 0 & \cos 2\theta & -\sin 2\theta & 0 \\ 0 & \sin 2\theta & \cos 2\theta & 0 \\ 0 & 0 & 0 & 1 \end{pmatrix} \cdot \begin{pmatrix} M_{00} & M_{01} & M_{02} & M_{03} \\ M_{10} & M_{11} & M_{12} & M_{13} \\ M_{20} & M_{21} & M_{22} & M_{23} \\ M_{30} & M_{31} & M_{32} & M_{33} \end{pmatrix} \cdot \begin{pmatrix} 1 & 0 & 0 & 0 \\ 0 & \cos 2\theta & \sin 2\theta & 0 \\ 0 & -\sin 2\theta & \cos 2\theta & 0 \\ 0 & 0 & 0 & 1 \end{pmatrix}. \tag{6.11}$$

$\mathbf{R}_M(\theta)$, the *Mueller rotation matrix*,

$$\mathbf{R}_M(\theta) = \begin{pmatrix} 1 & 0 & 0 & 0 \\ 0 & \cos 2\theta & -\sin 2\theta & 0 \\ 0 & \sin 2\theta & \cos 2\theta & 0 \\ 0 & 0 & 0 & 1 \end{pmatrix}, \tag{6.12}$$

is the matrix for rotations about the optical axis. $\mathbf{R}_M(\theta)$ was introduced in Chapter 3 along with a discussion of the *non-orthogonal coordinates for the Stokes parameters*.

For vectors, a rotational change of basis is accomplished by left multiplication with a rotation matrix. Hence, Stokes parameters are rotated with respect to their coordinate system by the transformation

6.5 Rotating Polarization Elements about the Light Direction

$$\mathbf{S}(\theta) = \mathbf{R}_M(\theta) \cdot \mathbf{S} = \begin{pmatrix} 1 & 0 & 0 & 0 \\ 0 & \cos 2\theta & -\sin 2\theta & 0 \\ 0 & \sin 2\theta & \cos 2\theta & 0 \\ 0 & 0 & 0 & 1 \end{pmatrix} \cdot \begin{pmatrix} S_0 \\ S_1 \\ S_2 \\ S_3 \end{pmatrix} = \begin{pmatrix} S_0 \\ S_1 \cos 2\theta - S_2 \sin 2\theta \\ S_1 \sin 2\theta + S_2 \cos 2\theta \\ S_3 \end{pmatrix}. \quad (6.13)$$

The 2θ occurs because the S_1 and S_2 axes are only 45° apart. A rotation of 180° returns the Stokes parameters to their original state so $\mathbf{R}_M(180°)$ must equal the identity matrix. By definition, θ is *positive* for rotations that initially move an x-component toward the $+y$ axis, rotations that are counterclockwise looking into the beam.

For matrices, a rotational change of basis requires *two rotation matrices of opposite sign*, one multiplied from the left and one from the right. This is an example of a *unitary transformation* of a matrix. Note the $+\theta$ in the first rotation matrix and $-\theta$ in the second rotation matrix in Equation 6.11. These signs are understood by comparing a matrix–vector multiplication in un-rotated coordinates

$$\mathbf{M} \cdot \mathbf{S} = \mathbf{S}', \quad (6.14)$$

with the multiplication of the rotated matrix and rotated Stokes parameters, each shown initially in brackets,

$$\begin{aligned} \left[\mathbf{R}_M(\theta) \cdot \mathbf{M} \cdot \mathbf{R}_M(-\theta) \right] \cdot \left[\mathbf{R}_M(\theta) \cdot \mathbf{S} \right] \\ = \mathbf{R}_M(\theta) \cdot \mathbf{M} \cdot \left[\mathbf{R}_M(-\theta) \cdot \mathbf{R}_M(\theta) \right] \cdot \mathbf{S} \\ = \mathbf{R}_M(\theta) \cdot \mathbf{M} \cdot \mathbf{S} \\ = \mathbf{R}_M(\theta) \cdot \mathbf{S}'. \end{aligned} \quad (6.15)$$

Since matrix multiplication is *associative*, $(\mathbf{A} \cdot \mathbf{B}) \cdot \mathbf{C} = \mathbf{A} \cdot (\mathbf{B} \cdot \mathbf{C})$, the parenthesis can be rearranged into the second expression. $\mathbf{R}_M(\theta)$ is a unitary rotation matrix and an opposite rotation must undo the effect of a rotation,

$$\begin{aligned} & \mathbf{R}_M(\theta) \cdot \mathbf{R}_M(-\theta) \\ & = \begin{pmatrix} 1 & 0 & 0 & 0 \\ 0 & \cos 2\theta & -\sin 2\theta & 0 \\ 0 & \sin 2\theta & \cos 2\theta & 0 \\ 0 & 0 & 0 & 1 \end{pmatrix} \cdot \begin{pmatrix} 1 & 0 & 0 & 0 \\ 0 & \cos 2\theta & \sin 2\theta & 0 \\ 0 & -\sin 2\theta & \cos 2\theta & 0 \\ 0 & 0 & 0 & 1 \end{pmatrix} \\ & = \begin{pmatrix} 1 & 0 & 0 & 0 \\ 0 & 1 & 0 & 0 \\ 0 & 0 & 1 & 0 \\ 0 & 0 & 0 & 1 \end{pmatrix}, \end{aligned} \quad (6.16)$$

yielding the identity matrix. Hence,

$$\mathbf{R}_M(-\theta) = \mathbf{R}_M^{-1}(\theta). \quad (6.17)$$

6.6 Retarder Mueller Matrices

Retarders are polarization elements that have two polarization states that are transmitted in the incident polarization state (*eigenpolarizations*) but with different optical path lengths (*phases*). This section presents Mueller matrix formulas for ideal retarders, where it is assumed that the two eigenpolarizations are transmitted without loss and the element has *no diattenuation*.

Birefringent retarders operate by dividing the incident light into two modes with orthogonal polarization states and different refractive indices. Propagation delays one mode with respect to the other, resulting in two optical path lengths, and an optical path difference, the retardance, as was seen in Figure 5.3. Other retardance mechanisms include reflections from metals, reflection and transmission through multilayer thin films, stress birefringence, and interactions with diffraction gratings. These interactions also are often *diattenuating*.

A retarder is specified by the optical path difference between the eigenpolarizations (the *retardance* δ) and the eigenpolarization states, the state with the smaller optical path length (the *fast axis*) and the larger optical path length (the *slow axis*). Retardance is generally expressed in radians, so $\delta = 2\pi$ indicates one wavelength of optical path difference. Note that *axis* implies a *linear polarization state*, but the fast eigenpolarization may also be *elliptical* or *circular* and the term "*axis*" can still be applied.

The action of retarders on Stokes parameters is visualized as a rotation of the *Poincaré sphere*. The line passing through the two retarder eigenpolarizations forms the *rotation axis* on the Poincaré sphere. The magnitude of the *rotation* is the *retardance*. This description of retarders is elaborated at length in Section 6.8.

In the Mueller calculus, retarders are represented by *real unitary matrices* of the form

$$\mathbf{M}_{\text{Retarder}} = \begin{pmatrix} 1 & 0 & 0 & 0 \\ 0 & & & \\ 0 & & 3\times 3 \text{ Rotation Matrix} & \\ 0 & & & \end{pmatrix}, \quad (6.18)$$

where, except for the M_{00} element, the first row and column are zeros. Real unitary matrices are called *orthogonal matrices*; the rows of an orthogonal matrix form a set of *orthogonal unit vectors*, as do the set of columns. The definition of a *unitary matrix* $\mathbf{U}$ is a matrix whose *adjoint* equals its *matrix inverse*,

$$\mathbf{U}^{\dagger} = (\mathbf{U}^{\text{T}})^* = \mathbf{U}^{-1}. \quad (6.19)$$

For a real matrix, the complex conjugate of a matrix equals the matrix, so the transpose of an orthogonal matrix $\mathbf{O}$ equals its inverse,

$$\mathbf{O}^{\text{T}} = \mathbf{O}^{-1}. \quad (6.20)$$

Thus, $\mathbf{M}^{\text{T}} = \mathbf{M}^{-1}$ tests if a Mueller matrix is a *pure retarder*. Orthogonal matrices such as retarder Mueller matrices are *rotation matrices*. The lower right 3×3 elements form a rotation matrix in (S_1, S_2, S_3) space, where (S_1, S_2, S_3) are the three Stokes parameters. This is how retarders mathematically rotate the Poincaré sphere. The *retardance* δ of a pure retarder Mueller matrix is

6.6 Retarder Mueller Matrices

$$\delta = \cos^{-1}\left(\frac{M_{00} + M_{11} + M_{22} + M_{33}}{2} - 1\right) = \cos^{-1}\left(\frac{\text{Tr}(\mathbf{M})}{2} - 1\right). \quad (6.21)$$

Tr indicates the *trace* of a matrix, the sum of the diagonal elements. A horizontal linear retarder with retardance, δ, has the Mueller matrix,

$$\mathbf{LR}(\delta, 0) = \begin{pmatrix} 1 & 0 & 0 & 0 \\ 0 & 1 & 0 & 0 \\ 0 & 0 & \cos\delta & \sin\delta \\ 0 & 0 & -\sin\delta & \cos\delta \end{pmatrix}. \quad (6.22)$$

When $\mathbf{LR}(\delta, 0)$ operates on a set of Stokes parameters, the first two elements, S_0 and S_1, are left unchanged; they are the non-zero elements in the two eigenpolarizations, horizontal and vertical linearly polarized light. A *linear retarder* $\mathbf{LR}(\delta, 0)$ with its *fast axis* oriented at an angle θ can be derived using the *Mueller rotation operation*, as shown in Equation 6.11, and has the Mueller matrix

$$\mathbf{LR}(\delta, \theta) = \begin{pmatrix} 1 & 0 & 0 & 0 \\ 0 & \cos^2 2\theta + \cos\delta \sin^2 2\theta & (1 - \cos\delta)\cos 2\theta \sin 2\theta & -\sin\delta \sin 2\theta \\ 0 & (1 - \cos\delta)\cos 2\theta \sin 2\theta & \cos\delta \cos^2 2\theta + \sin^2 2\theta & \cos 2\theta \sin\delta \\ 0 & \sin\delta \sin 2\theta & -\cos 2\theta \sin\delta & \cos\delta \end{pmatrix}. \quad (6.23)$$

A *circular retarder*, $\mathbf{CR}(\delta)$, with retardance, δ, has the Mueller matrix

$$\mathbf{CR}(\delta) = \begin{pmatrix} 1 & 0 & 0 & 0 \\ 0 & \cos\delta & \sin\delta & 0 \\ 0 & -\sin\delta & \cos\delta & 0 \\ 0 & 0 & 0 & 1 \end{pmatrix}. \quad (6.24)$$

This Mueller matrix can describe *optically active materials* and *Faraday rotation*.

The Mueller matrices for *quarter wave retarders*, $\delta = \pi/2$, and *half wave retarders*, $\delta = \pi$, with fast axes corresponding to the basis polarization states are given in Table 6.1.

A *half wave linear retarder* with *fast axis* at angle θ, $\mathbf{HWLR}(\theta)$, has the Mueller matrix

$$\mathbf{HWLR}(\theta) = \begin{pmatrix} 1 & 0 & 0 & 0 \\ 0 & \cos 4\theta & \sin 4\theta & 0 \\ 0 & \sin 4\theta & -\cos 4\theta & 0 \\ 0 & 0 & 0 & -1 \end{pmatrix}. \quad (6.25)$$

The Mueller matrix elements for half wave linear retarders are plotted in Figure 6.1 (right). The Mueller matrix is the *same* for a horizontal half wave linear retarder and a vertical half wave linear retarder because both half wave retarders transform all incident Stokes parameters equally. The Mueller matrix for a *quarter wave linear retarder* with fast axis at angle θ, $\mathbf{QWLR}(\theta)$, is

Table 6.1 Quarter Wave and Half Wave Retarder Mueller Matrices for the Basis Polarization States

Type of Retarder	Symbol	Mueller Matrix
Horizontal quarter wave linear retarder	HQWLR	$\begin{pmatrix} 1 & 0 & 0 & 0 \\ 0 & 1 & 0 & 0 \\ 0 & 0 & 0 & 1 \\ 0 & 0 & -1 & 0 \end{pmatrix}$
Vertical quarter wave linear retarder	VQWLR	$\begin{pmatrix} 1 & 0 & 0 & 0 \\ 0 & 1 & 0 & 0 \\ 0 & 0 & 0 & -1 \\ 0 & 0 & 1 & 0 \end{pmatrix}$
45° Quarter wave linear retarder	QWLR(45°)	$\begin{pmatrix} 1 & 0 & 0 & 0 \\ 0 & 0 & 0 & -1 \\ 0 & 0 & 1 & 0 \\ 0 & 1 & 0 & 0 \end{pmatrix}$
135° Quarter wave linear retarder	QWLR(135°)	$\begin{pmatrix} 1 & 0 & 0 & 0 \\ 0 & 0 & 0 & 1 \\ 0 & 0 & 1 & 0 \\ 0 & -1 & 0 & 0 \end{pmatrix}$
Quarter wave right circular retarder	QWRCR	$\begin{pmatrix} 1 & 0 & 0 & 0 \\ 0 & 0 & 1 & 0 \\ 0 & -1 & 0 & 0 \\ 0 & 0 & 0 & 1 \end{pmatrix}$
Quarter wave left circular retarder	QWLCR	$\begin{pmatrix} 1 & 0 & 0 & 0 \\ 0 & 0 & -1 & 0 \\ 0 & 1 & 0 & 0 \\ 0 & 0 & 0 & 1 \end{pmatrix}$
Horizontal or vertical half wave linear retarder (same matrix)	HHWLR	$\begin{pmatrix} 1 & 0 & 0 & 0 \\ 0 & 1 & 0 & 0 \\ 0 & 0 & -1 & 0 \\ 0 & 0 & 0 & -1 \end{pmatrix}$
45° or 135° Half wave linear retarder	HWLR(45°)	$\begin{pmatrix} 1 & 0 & 0 & 0 \\ 0 & -1 & 0 & 0 \\ 0 & 0 & 1 & 0 \\ 0 & 0 & 0 & -1 \end{pmatrix}$
Right or left half wave circular retarder	RHWCR	$\begin{pmatrix} 1 & 0 & 0 & 0 \\ 0 & -1 & 0 & 0 \\ 0 & 0 & -1 & 0 \\ 0 & 0 & 0 & 1 \end{pmatrix}$

6.6 Retarder Mueller Matrices

$$\mathbf{QWLR}(\theta) = \begin{pmatrix} 1 & 0 & 0 & 0 \\ 0 & \cos^2 2\theta & \cos 2\theta \sin 2\theta & -\sin 2\theta \\ 0 & \cos 2\theta \sin 2\theta & \sin^2 2\theta & \cos 2\theta \\ 0 & \sin 2\theta & -\cos 2\theta & 0 \end{pmatrix}$$

$$= \begin{pmatrix} 1 & 0 & 0 & 0 \\ 0 & \frac{1}{2}(1+\cos 4\theta) & \frac{1}{2}\sin 4\theta & -\sin 2\theta \\ 0 & \frac{1}{2}\sin 4\theta & \frac{1}{2}(1-\cos 4\theta) & \cos 2\theta \\ 0 & \sin 2\theta & -\cos 2\theta & 0 \end{pmatrix}. \tag{6.26}$$

The Mueller matrix elements are plotted in Figure 6.1 (left). The Mueller matrix elements for all linear retarders $\mathbf{LR}(\delta, \theta)$ are plotted in two dimensions in Figure 6.2.

The general *elliptical retarder Mueller matrix* can be expressed in terms of *retardance components*, $(\delta_H, \delta_{45}, \delta_R)$. The magnitude of the retardance δ and the associated Stokes eigenpolarization are

$$\delta = \sqrt{\delta_H^2 + \delta_{45}^2 + \delta_R^2}, \quad \mathbf{S}_{\text{fast}} = \begin{pmatrix} 1 \\ \delta_H/\delta \\ \delta_{45}/\delta \\ \delta_R/\delta \end{pmatrix}. \tag{6.27}$$

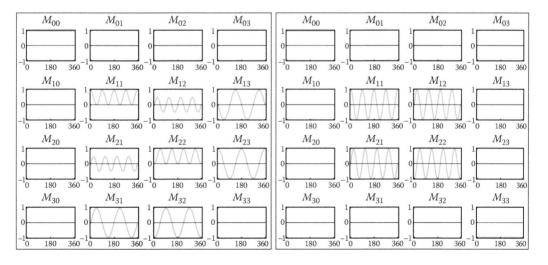

Figure 6.1 (Left) The Mueller matrix elements of the family of quarter wave linear retarders. The abscissa is the retarder fast axis orientation, from 0° to 360°. (Right) The Mueller matrix elements of a half wave linear retarder as a function of the fast axis orientation.

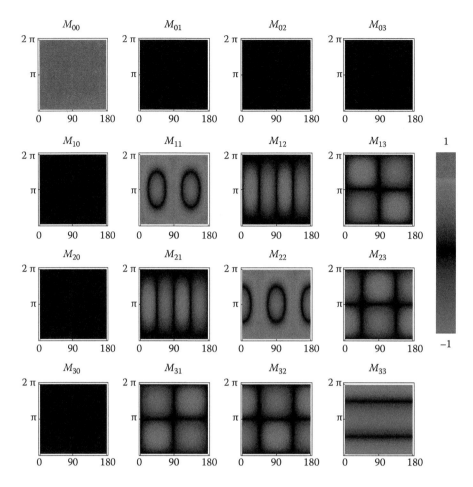

Figure 6.2 Density plots of the Mueller matrix elements of the linear retarders. Retardance in radians is along the y-axis. Fast axis orientation in degrees is along the x-axis. Coloring represents the value of the Mueller matrix element. The bottom (0 waves) and top (1 wave) sides of each figure represents the identity matrix, $\delta = (0, 2\pi)$. Similarly, a horizontal line across the center contains all the half wave linear retarder matrix elements, $\delta = \pi$.

The *elliptical retarder's Mueller matrix* is

$$\mathbf{ER}(\delta_H, \delta_{45}, \delta_R) = \begin{pmatrix} 1 & 0 & 0 & 0 \\ 0 & \dfrac{\delta_H^2 + (\delta_{45}^2 + \delta_R^2)C}{\delta^2} & \dfrac{\delta_{45}\delta_H T}{\delta^2} + \dfrac{\delta_R S}{\delta} & \dfrac{\delta_H \delta_R T}{\delta^2} - \dfrac{\delta_{45} S}{\delta} \\ 0 & \dfrac{\delta_{45}\delta_H T}{\delta^2} - \dfrac{\delta_R S}{\delta} & \dfrac{\delta_{45}^2 + (\delta_R^2 + \delta_H^2)C}{\delta^2} & \dfrac{\delta_R \delta_{45} T}{\delta^2} + \dfrac{\delta_H S}{\delta} \\ 0 & \dfrac{\delta_H \delta_R T}{\delta^2} + \dfrac{\delta_{45} S}{\delta} & \dfrac{\delta_R \delta_{45} T}{\delta^2} - \dfrac{\delta_H S}{\delta} & \dfrac{\delta_R^2 + (\delta_{45}^2 + \delta_H^2)C}{\delta^2} \end{pmatrix}, \quad (6.28)$$

where $C = \cos\delta$, $S = \sin\delta$, $T = 1 - \cos\delta$. The Mueller matrix for *half wave elliptical retarders*, **HWR**, simplifies to the following form,

6.6 Retarder Mueller Matrices

$$\mathbf{HWR}(r_1,r_2,r_3) = \begin{pmatrix} 1 & 0 & 0 & 0 \\ 0 & -1+2r_1^2 & 2r_2r_1 & 2r_1r_3 \\ 0 & 2r_2r_1 & -1+2r_2^2 & 2r_2r_3 \\ 0 & 2r_3r_1 & 2r_2r_3 & -1+2r_3^2 \end{pmatrix}, \quad (6.29)$$

where $\sqrt{r_1^2+r_2^2+r_3^2}=1$, $r_1=\frac{\delta_H}{\pi}$, $r_2=\frac{\delta_{45}}{\pi}$, $r_3=\frac{\delta_R}{\pi}$. Figure 6.3 shows density plots of the Mueller matrix elements for all quarter wave and half wave elliptical retarders.

The *retardance parameters* (δ_H, δ_{45}, δ_R) of a retarder Mueller matrix can be calculated from Equations 6.27 and 6.28 by adding the off-diagonal elements,

$$(\delta_H, \delta_{45}, \delta_R) = \frac{\delta}{2\sin\delta}(M_{23}-M_{32}, M_{31}-M_{13}, M_{12}-M_{21}). \quad (6.30)$$

For retardances very near zero, Equation 6.30 can be replaced by

$$(\delta_H, \delta_{45}, \delta_R) = \frac{1}{2}(M_{23}-M_{32}, M_{31}-M_{13}, M_{12}-M_{21}). \quad (6.31)$$

Equation 6.30 does not work for half wave retarders due to a zero denominator; for this special case, the following equation, which is derived from the diagonal elements, works well,

$$(\delta_H, \delta_{45}, \delta_R) = \pi\left(\sqrt{\frac{M_{11}+1}{2}}, \text{sign}(M_{12})\sqrt{\frac{M_{22}+1}{2}}, \text{sign}(M_{13})\sqrt{\frac{M_{33}+1}{2}}\right). \quad (6.32)$$

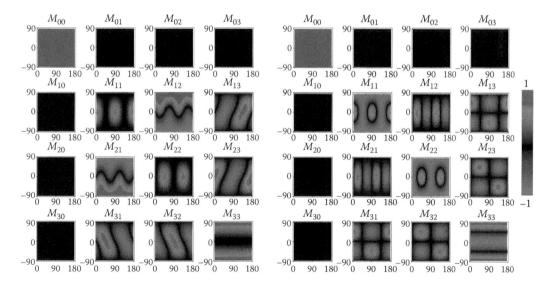

Figure 6.3 Density plots of quarter wave (left) and half wave (right) elliptical retarder Mueller matrix elements. The x-axis is orientation. The y-axis is Poincaré sphere latitude for the fast axis, varying from left circular across the bottom, to linear across the center, to right circular at the top. Coloring represents the value of the Mueller matrix element. Each square represents a flattened Poincaré sphere. Linear retarders fall along the line across the center.

Since the square root always returns a positive value, the sign of the off-diagonal elements is used to determine the octant of the retarder axis. Equation 6.32 always returns an axis with a positive δ_H. For half wave retarders, the retarder with an orthogonal axis also has the same Mueller matrix.

> **Example 6.2** Mueller Matrix Multiplication Example
>
> A half wave linear retarder with a 0° fast axis followed by a half wave linear retarder with a 45° fast axis becomes a half wave circular retarder,
>
> $$\mathbf{HWLR}(45°) \cdot \mathbf{HWLR}(0) = \begin{pmatrix} 1 & 0 & 0 & 0 \\ 0 & 0 & 1 & 0 \\ 0 & 1 & 0 & 0 \\ 0 & 0 & 0 & -1 \end{pmatrix} \cdot \begin{pmatrix} 1 & 0 & 0 & 0 \\ 0 & 1 & 0 & 0 \\ 0 & 0 & -1 & 0 \\ 0 & 0 & 0 & -1 \end{pmatrix}$$
>
> $$= \begin{pmatrix} 1 & 0 & 0 & 0 \\ 0 & -1 & 0 & 0 \\ 0 & 0 & -1 & 0 \\ 0 & 0 & 0 & 1 \end{pmatrix} = \mathbf{LHWCR} = \mathbf{RHWCR}.$$
>
> (6.33)
>
> The Mueller matrices of both left and right half wave circular retarders are equal, so the helicity is not determined.

6.7 Polarizer and Diattenuator Mueller Matrices

6.7.1 Basic Polarizers

A *polarizer* is specified by the unique exiting polarization state; the orthogonal polarization state is blocked. An *ideal polarizer* transmits 100% of the transmitted polarization state and completely blocks the orthogonal polarization state. These two states are the two *eigenpolarizations* of the polarizer.

> **Example 6.3** Mueller Matrix for a Horizontal Linear Polarizer
>
> The Mueller matrix for an ideal horizontal linear polarizer **HLP** operates on an arbitrary **S** yielding Stokes parameters where $S_0' = S_1'$, which must be horizontal linearly polarized light,
>
> $$\mathbf{HLP} \cdot \mathbf{S} = \frac{1}{2}\begin{pmatrix} 1 & 1 & 0 & 0 \\ 1 & 1 & 0 & 0 \\ 0 & 0 & 0 & 0 \\ 0 & 0 & 0 & 0 \end{pmatrix} \cdot \begin{pmatrix} S_0 \\ S_1 \\ S_2 \\ S_3 \end{pmatrix} = \begin{pmatrix} S_0' \\ S_1' \\ S_2' \\ S_3' \end{pmatrix} = \frac{1}{2}\begin{pmatrix} S_0 + S_1 \\ S_0 + S_1 \\ 0 \\ 0 \end{pmatrix} = \frac{S_0 + S_1}{2}\begin{pmatrix} 1 \\ 1 \\ 0 \\ 0 \end{pmatrix}. \quad (6.34)$$
>
> Since the first two rows of **HLP** are equal, the first two elements of **S′** are equal, the S_2 and S_3 characteristics of the incident light are lost, and the exiting light is always horizontal linearly polarized. Horizontal and vertical linearly polarized light are the two eigenpolarizations of this matrix,

6.7 Polarizer and Diattenuator Mueller Matrices

$$\mathbf{HLP} \cdot \mathbf{H} = \frac{1}{2} \begin{pmatrix} 1 & 1 & 0 & 0 \\ 1 & 1 & 0 & 0 \\ 0 & 0 & 0 & 0 \\ 0 & 0 & 0 & 0 \end{pmatrix} \cdot \begin{pmatrix} 1 \\ 1 \\ 0 \\ 0 \end{pmatrix} = 1 \begin{pmatrix} 1 \\ 1 \\ 0 \\ 0 \end{pmatrix},$$

$$\mathbf{HLP} \cdot \mathbf{V} = \frac{1}{2} \begin{pmatrix} 1 & 1 & 0 & 0 \\ 1 & 1 & 0 & 0 \\ 0 & 0 & 0 & 0 \\ 0 & 0 & 0 & 0 \end{pmatrix} \cdot \begin{pmatrix} 1 \\ -1 \\ 0 \\ 0 \end{pmatrix} = \begin{pmatrix} 0 \\ 0 \\ 0 \\ 0 \end{pmatrix} = 0 \begin{pmatrix} 1 \\ -1 \\ 0 \\ 0 \end{pmatrix},$$

(6.35)

with eigenvalues 1 and 0, the intensity transmittances for these two states. Any 4 × 4 matrix has 4 eigenvectors and eigenvalues. The other two eigenvalues and eigenvectors of **HLP** are as follows:

$$\frac{1}{2} \begin{pmatrix} 1 & 1 & 0 & 0 \\ 1 & 1 & 0 & 0 \\ 0 & 0 & 0 & 0 \\ 0 & 0 & 0 & 0 \end{pmatrix} \cdot \begin{pmatrix} 0 \\ 0 \\ 1 \\ 0 \end{pmatrix} = 0 \begin{pmatrix} 0 \\ 0 \\ 1 \\ 0 \end{pmatrix}, \text{ and}$$

$$\frac{1}{2} \begin{pmatrix} 1 & 1 & 0 & 0 \\ 1 & 1 & 0 & 0 \\ 0 & 0 & 0 & 0 \\ 0 & 0 & 0 & 0 \end{pmatrix} \cdot \begin{pmatrix} 0 \\ 0 \\ 0 \\ 1 \end{pmatrix} = 0 \begin{pmatrix} 0 \\ 0 \\ 0 \\ 1 \end{pmatrix}.$$

(6.36)

These two eigenvectors are not valid Stokes parameters since they violate the condition on Stokes parameters that $S_0 < \sqrt{S_1^2 + S_2^2 + S_3^2}$; hence, these two eigenvectors do not count as eigenpolarizations.

Using $\mathbf{M}_R(\theta)$, the *Mueller matrix for a linear polarizer* with transmission axis at θ is readily calculated from **HLP**, as

$$\mathbf{LP}(\theta) = \mathbf{R}_M(\theta) \cdot \mathbf{HLP} \cdot \mathbf{R}_M(-\theta)$$

$$= \begin{pmatrix} 1 & 0 & 0 & 0 \\ 0 & \cos 2\theta & -\sin 2\theta & 0 \\ 0 & \sin 2\theta & \cos 2\theta & 0 \\ 0 & 0 & 0 & 1 \end{pmatrix} \cdot \frac{1}{2} \begin{pmatrix} 1 & 1 & 0 & 0 \\ 1 & 1 & 0 & 0 \\ 0 & 0 & 0 & 0 \\ 0 & 0 & 0 & 0 \end{pmatrix} \cdot \begin{pmatrix} 1 & 0 & 0 & 0 \\ 0 & \cos 2\theta & \sin 2\theta & 0 \\ 0 & -\sin 2\theta & \cos 2\theta & 0 \\ 0 & 0 & 0 & 1 \end{pmatrix}$$

$$= \begin{pmatrix} 1 & \cos 2\theta & \sin 2\theta & 0 \\ \cos 2\theta & \cos^2 2\theta & \sin 2\theta \cos 2\theta & 0 \\ \sin 2\theta & \sin 2\theta \cos 2\theta & \sin^2 2\theta & 0 \\ 0 & 0 & 0 & 0 \end{pmatrix}.$$

(6.37)

LP(θ) transmits the linear polarization state oriented at θ and blocks the state at $\theta + 90°$.

Table 6.2 Polarizer Mueller Matrices for the Basis Polarization States

Type of Polarizer	Symbol	Mueller Matrix
Horizontal linear polarizer	HLP	$\dfrac{1}{2}\begin{pmatrix} 1 & 1 & 0 & 0 \\ 1 & 1 & 0 & 0 \\ 0 & 0 & 0 & 0 \\ 0 & 0 & 0 & 0 \end{pmatrix}$
Vertical linear polarizer	VLP	$\dfrac{1}{2}\begin{pmatrix} 1 & -1 & 0 & 0 \\ -1 & 1 & 0 & 0 \\ 0 & 0 & 0 & 0 \\ 0 & 0 & 0 & 0 \end{pmatrix}$
45° Linear polarizer	LP(45°)	$\dfrac{1}{2}\begin{pmatrix} 1 & 0 & 1 & 0 \\ 0 & 0 & 0 & 0 \\ 1 & 0 & 1 & 0 \\ 0 & 0 & 0 & 0 \end{pmatrix}$
135° Linear polarizer	LP(135°)	$\dfrac{1}{2}\begin{pmatrix} 1 & 0 & -1 & 0 \\ 0 & 0 & 0 & 0 \\ -1 & 0 & 1 & 0 \\ 0 & 0 & 0 & 0 \end{pmatrix}$
Right circular polarizer	RCP	$\dfrac{1}{2}\begin{pmatrix} 1 & 0 & 0 & 1 \\ 0 & 0 & 0 & 0 \\ 0 & 0 & 0 & 0 \\ 1 & 0 & 0 & 1 \end{pmatrix}$
Left circular polarizer	LCP	$\dfrac{1}{2}\begin{pmatrix} 1 & 0 & 0 & -1 \\ 0 & 0 & 0 & 0 \\ 0 & 0 & 0 & 0 \\ -1 & 0 & 0 & 1 \end{pmatrix}$

Table 6.2 contains the Mueller matrices for the six ideal polarizers that transmit the six basis sets of Stokes parameters. Figure 6.4 graphs the 16 Mueller matrix elements for a linear polarizer as a function of transmission axis angle θ.

Consider an elliptical polarizer that transmits the polarization state with the major axis of the ellipse oriented at θ and with the transmitting eigenpolarization located at latitude η on the Poincaré sphere, where $-\pi/2 \leq \eta \leq \pi/2$. The *elliptical polarizer Mueller matrix* $\mathbf{EP}(\theta, \eta)$ is

$$\mathbf{EP}(\theta,\eta) = \frac{1}{2}\begin{pmatrix} 1 & \cos 2\theta \cos\eta & \sin 2\theta \cos\eta & \sin\eta \\ \cos 2\theta \cos\eta & \cos^2 2\theta \cos^2\eta & \dfrac{1}{2}\sin 4\theta \cos^2\eta & \cos 2\theta \cos\eta \sin\eta \\ \sin 2\theta \cos\eta & \dfrac{1}{2}\sin 4\theta \cos^2\eta & \sin^2 2\theta \cos^2\eta & 2\cos\theta \sin\theta \cos\eta \sin\eta \\ \sin\eta & \cos 2\theta \sin\eta \cos\eta & \sin 2\theta \cos\eta \sin\eta & \sin^2\eta \end{pmatrix}. \qquad (6.38)$$

6.7 Polarizer and Diattenuator Mueller Matrices

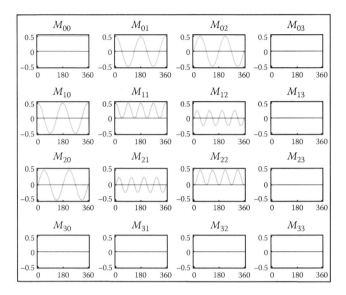

Figure 6.4 The 16 Mueller matrix elements for the set of linear polarizers as a function of transmission axis angle θ.

Problem 6.7 in Section 6.15 describes the derivation. Alternatively, consider an elliptical polarizer that transmits the state with Poincaré sphere coordinates (d_H, d_{45}, d_R) and blocks the orthogonal state. The transmitted eigenpolarization is

$$\mathbf{S}_1 = (1, d_H, d_{45}, d_R), \quad d_H^2 + d_{45}^2 + d_R^2 = 1. \tag{6.39}$$

This *elliptical polarizer Mueller matrix* is

$$\mathbf{EP}(d_H, d_{45}, d_R) = \frac{1}{2}\begin{pmatrix} 1 & d_H & d_{45} & d_R \\ d_H & d_H^2 & d_H d_{45} & d_H d_R \\ d_{45} & d_H d_{45} & d_{45}^2 & d_{45} d_R \\ d_R & d_H d_R & d_R d_{45} & d_R^2 \end{pmatrix}. \tag{6.40}$$

6.7.2 Transmittance and Diattenuation

Polarizers and *partial polarizers* are characterized by the property *diattenuation*, which describes the magnitude of the variation of the transmitted irradiance as a function of the incident polarization state. The diattenuation magnitude, D, usually referred to as the *diattenuation*, is a function of the maximum, T_{max}, and minimum, T_{min}, transmittances of a polarization element or optical interaction,

$$D = \frac{T_{max} - T_{min}}{T_{max} + T_{min}} = \frac{\sqrt{M_{01}^2 + M_{02}^2 + M_{03}^2}}{M_{00}}, \quad 0 \le D \le 1. \tag{6.41}$$

The diattenuation has the useful property that D varies from 1 for a *polarizer* to 0 for an element that *transmits all polarization states equally*, such as a retarder or a *non-polarizing interaction*.

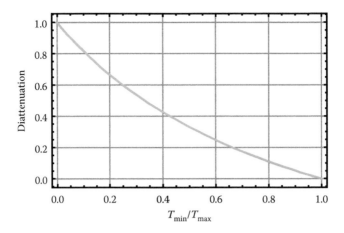

Figure 6.5 Relationship between the diattenuation and the extinction ratio, T_{min}/T_{max}.

The transmitted irradiance of a Mueller matrix and its diattenuation depends only on the first row, $M_0 = (M_{00}, M_{01}, M_{02}, M_{03})$, because these are the only elements that affect S'_0. The diattenuation is not linear in the extinction ratio T_{min}/T_{max} as shown in Figure 6.5.

To find T_{max} and T_{min}, first the incident Stokes parameters are normalized, so $s_0 = 1$. The *normalized Stokes parameters* are defined by a three-element vector **s** that represents the coordinate of the state on the unit Poincaré sphere,

$$\hat{\mathbf{S}} = \frac{\mathbf{S}}{S_0} = \begin{pmatrix} s_0 \\ s_1 \\ s_2 \\ s_3 \end{pmatrix} = \begin{pmatrix} 1 \\ \mathbf{s} \end{pmatrix}, \quad \mathbf{s} = (s_1, s_2, s_3). \tag{6.42}$$

For *unpolarized light*, the magnitude of **s** is zero, $|\mathbf{s}| = 0$. For a completely polarized state, $|\mathbf{s}| = 1$. The introduction of **s** allows the polarization state to be specified irrespective of the flux.

The *transmittance*, $T(\mathbf{s})$, of a device with Mueller matrix **M** is the ratio of the exiting flux to the incident flux,

$$T(\mathbf{M}, \mathbf{s}) = \frac{S'_0}{S_0} = \mathbf{M} \cdot \hat{\mathbf{S}} = M_{00} + M_{01} s_1 + M_{02} s_2 + M_{03} s_3, \tag{6.43}$$

which depends on the *dot product* of the first row of the Mueller matrix with the incident Stokes parameters. The dependence of the transmission on incident polarization state is characterized by a three-element *diattenuation parameter*, **d**, defined as

$$\mathbf{d} = \frac{(M_{01}, M_{02}, M_{03})}{M_{00}} = (d_H, d_{45}, d_R). \tag{6.44}$$

The diattenuation parameters have three components corresponding to the *three components of the Stokes parameters, x/y, 45°/135°, right/left*, each of which characterize how the transmission varies with each of the Stokes parameter component. The diattenuation parameter set, **d**, is often called the *diattenuation vector* or *analyzer vector*, but similar to the Stokes parameters, **d** is not a true vector. Diattenuation parameters do not add. For a Stokes three-element vector, **s**, the transmission function, *T*, is

6.7 Polarizer and Diattenuator Mueller Matrices

$$T(\mathbf{M},\mathbf{s}) = \frac{S_0'}{S_0} = \mathbf{M} \cdot \hat{\mathbf{S}} = M_{00} + M_{01}s_1 + M_{02}s_2 + M_{03}s_3 = M_{00}(1+\mathbf{d}\cdot\mathbf{s}). \tag{6.45}$$

The *average transmission*, calculated by averaging over all sets of polarized Stokes parameters, is M_{00}. The average transmission is also the transmission for unpolarized incident light, $\mathbf{s}_U = (0, 0, 0)$. The polarization-dependent variation of the transmission is contained in the dot product term between the incident Stokes three-element vector and the diattenuation vector, $\mathbf{s}\cdot\mathbf{d}$. The maximum transmission, $T_{\max}$, occurs when the dot product is maximized, which occurs when $\mathbf{s}$ and $\mathbf{d}$ are parallel, and the magnitude of S_0' is also maximized. The incident Stokes parameters with maximum transmittance, $\mathbf{S}_{\max}$, and minimum transmittance, $\mathbf{S}_{\min}$, are

$$\mathbf{S}_{\max} = \frac{\mathbf{d}}{|\mathbf{d}|} = \frac{1}{D}\begin{pmatrix} D \\ d_H \\ d_{45} \\ d_R \end{pmatrix} \quad \text{and} \quad \mathbf{S}_{\min} = \frac{-\mathbf{d}}{|\mathbf{d}|} = \frac{1}{D}\begin{pmatrix} D \\ -d_H \\ -d_{45} \\ -d_R \end{pmatrix}, \tag{6.46}$$

yielding

$$T_{\max} = M_{00}(1+D) \quad \text{and} \quad T_{\min} = M_{00}(1-D). \tag{6.47}$$

Therefore, the diattenuation of any Mueller matrix is

$$D(\mathbf{M}) = \frac{T_{\max} - T_{\min}}{T_{\max} + T_{\min}} = \frac{\sqrt{M_{01}^2 + M_{02}^2 + M_{03}^2}}{M_{00}}. \tag{6.48}$$

For an ideal polarizer, the minimum transmission is zero, $D = 1$, $T_{\min} = M_{00}(1 - D) = 0$. *Linear polarization sensitivity* or *linear diattenuation* $LD(\mathbf{M})$ characterizes the variation of intensity transmittance with incident linear polarization states:

$$LD(\mathbf{M}) = \frac{\sqrt{M_{01}^2 + M_{02}^2}}{M_{00}}. \tag{6.49}$$

Linear polarization sensitivity is frequently specified as a performance parameter in remote sensing systems designed to measure incident power independently of any linearly polarized component present in scattered earth-light.[3] $LD(\mathbf{M}) = 1$ identifies $\mathbf{M}$ as a linear analyzer; $\mathbf{M}$ is not necessarily a linear polarizer, but may represent a linear polarizer followed by some other polarization element.

Diattenuation in fiber optic components and systems is often characterized by the *polarization dependent loss*, *PDL*, specified in decibels:

$$PDL(\mathbf{M}) = 10\text{Log}_{10}\frac{T_{\max}}{T_{\min}}. \tag{6.50}$$

Glan Thompson polarizers routinely achieve a *PDL* of 60 dB. Many different formulations of sheet polarizer are also fabricated and sold at a lower cost. *Sheet polarizers* are *dichroic polarizers* that depend on the difference in absorption for light polarized along and perpendicular to a molecular absorption axis, and thus have a strong wavelength dependence for their *PDL*. The *PDL* varies from over 50 dB for the best sheet polarizers to 20 dB for some low-cost sheet polarizers. *Polarcor** dichroic polarizers consist of nanometer-size silver crystals aligned in glass, and can achieve a *PDL* of 60 dB in the near infrared.

When sequences of polarizers have parallel transmission axes, the net polarization-dependent loss is the sum of the individual polarization-dependent losses. The polarization-dependent loss of two diattenuators with orthogonal transmission axes is calculated by subtracting the polarization-dependent loss of one from the other.

6.7.3 Polarizance

The *polarizance*, $P(\mathbf{M})$, is the degree of polarization *DoP* of the exiting light for unpolarized incident light, $\mathbf{U}$,[4]

$$P(\mathbf{M}) = DoP(\mathbf{M} \cdot \mathbf{U}) = \frac{\sqrt{M_{10}^2 + M_{20}^2 + M_{30}^2}}{M_{00}}. \tag{6.51}$$

The exiting polarization state, $\mathbf{S}_p(\mathbf{M})$, is the first column of $\mathbf{M}$,

$$\mathbf{S}_p(\mathbf{M}) = \mathbf{M} \cdot \mathbf{U} = \begin{pmatrix} M_{00} & M_{01} & M_{02} & M_{03} \\ M_{10} & M_{11} & M_{12} & M_{13} \\ M_{20} & M_{21} & M_{22} & M_{23} \\ M_{30} & M_{31} & M_{32} & M_{33} \end{pmatrix} \cdot \begin{pmatrix} 1 \\ 0 \\ 0 \\ 0 \end{pmatrix} = \begin{pmatrix} M_{00} \\ M_{10} \\ M_{20} \\ M_{30} \end{pmatrix}. \tag{6.52}$$

The *polarizance* does not necessarily equal the *diattenuation*. Nor does $\mathbf{S}_p$ necessarily equal $\mathbf{S}_{\max}$, the incident state of maximum transmittance. *Polarizance parameters* or the *polarizance vector* is defined as $(p_H, p_{45}, p_R) = (M_{10}, M_{20}, M_{30})/M_{00}$, the normalized Stokes parameters when unpolarized light is incident.

6.7.4 Diattenuators

The Mueller matrix for a *partial polarizer* or homogeneous diattenuator with intensity transmittances T_x and T_y along the x and y axes is $\mathbf{LD}(T_x, T_y, 0)$, where

$$\mathbf{LD}(T_x, T_y, 0) \cdot \begin{pmatrix} S_0 \\ S_1 \\ S_2 \\ S_3 \end{pmatrix} = \frac{1}{2} \begin{pmatrix} T_x + T_y & T_x - T_y & 0 & 0 \\ T_x - T_y & T_x + T_y & 0 & 0 \\ 0 & 0 & 2\sqrt{T_x T_y} & 0 \\ 0 & 0 & 0 & 2\sqrt{T_x T_y} \end{pmatrix} \cdot \begin{pmatrix} S_0 \\ S_1 \\ S_2 \\ S_3 \end{pmatrix} = \begin{pmatrix} S_0' \\ S_1' \\ S_2' \\ S_3' \end{pmatrix}. \tag{6.53}$$

Ideal diattenuators have two different intensity transmittances, $T_{\max}$ and $T_{\min}$, for two orthogonal linear eigenpolarizations, thus the name *"di" "attenuator."* A *linear diattenuator* $\mathbf{LD}(T_x, T_y, \theta)$ oriented at angle θ has the Mueller matrix

---

* Polarcor is a trademark of Corning Inc.

6.7 Polarizer and Diattenuator Mueller Matrices

$$\begin{aligned}
&\mathbf{LD}(T_{max},T_{min},\theta)\\
&=\mathbf{R}_M(\theta)\cdot\mathbf{LD}(T_{max},T_{min},0)\cdot\mathbf{R}_M(-\theta)\\
&=\frac{1}{2}\begin{pmatrix} A & B\cos 2\theta & B\sin 2\theta & 0 \\ B\cos 2\theta & A\cos^2 2\theta + C\sin^2 2\theta & (A-C)\cos 2\theta\sin 2\theta & 0 \\ B\sin 2\theta & (A-C)\cos 2\theta\sin 2\theta & C\cos^2 2\theta + A\sin^2 2\theta & 0 \\ 0 & 0 & 0 & C \end{pmatrix},
\end{aligned} \quad (6.54)$$

where

$$A = T_{max} + T_{min}, \quad B = T_{max} - T_{min}, \quad C = 2\sqrt{T_{max}\,T_{min}}. \quad (6.55)$$

This is an example of a *unitary transformation* by the unitary matrix $\mathbf{R}_M(\theta)$.

Figure 6.6 is a density plot of all linear diattenuator Mueller matrices $\mathbf{LD}(1, T_{min}, \theta)$. Ideal diattenuators have *no retardance*, although, in practice, most diattenuators have some retardance. An example of a pure linear diattenuator without retardance is transmission into a transparent dielectric;

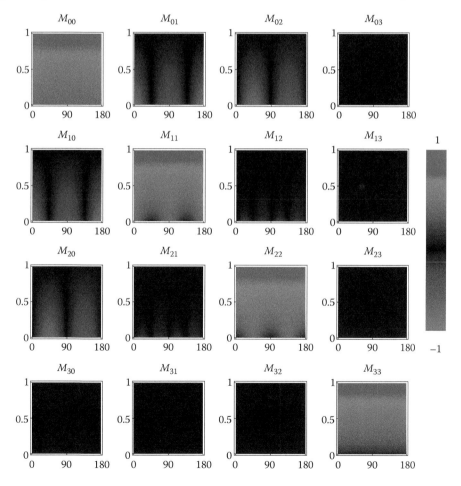

Figure 6.6 Mueller matrices for all linear diattenuators with $T_{max} = 1$. The *x*-axis is the transmission axis orientation θ, which varies from 0° to 180°, and the *y*-axis is T_{min}, which varies from 0 (polarizers) along the bottom to 1 (non-polarizers with identity matrix) along the top.

T_{max} and T_{min} are then given by intensity Fresnel coefficients. Reflection at metal surfaces acts as a *diattenuator with retardance*.

Ideal diattenuator Mueller matrices are *Hermitian matrices*; they have real eigenvalues. A Hermitian matrix equals the complex conjugate of its matrix transpose, its *Hermitian adjoint*, $\mathbf{H} = \mathbf{H}^\dagger = (\mathbf{H}^T)^*$. Since Mueller matrices are real, $\mathbf{H}^* = \mathbf{H}$, ideal diattenuator Mueller matrices equal their *transpose*,

$$\mathbf{H} = \mathbf{H}^T \tag{6.56}$$

and are symmetric about the diagonal. An *ideal horizontal linear polarizer* has zero transmission along one axis; $T_x = 1$, $T_y = 0$, and has the following Mueller matrix,

$$\mathbf{LD}(1,0,0) = \frac{1}{2}\begin{pmatrix} 1 & 1 & 0 & 0 \\ 1 & 1 & 0 & 0 \\ 0 & 0 & 0 & 0 \\ 0 & 0 & 0 & 0 \end{pmatrix}. \tag{6.57}$$

A real polarizer will have $T_x < 1$; thus,

$$\mathbf{LD}(T_x,0,0) = \frac{T_x}{2}\begin{pmatrix} 1 & 1 & 0 & 0 \\ 1 & 1 & 0 & 0 \\ 0 & 0 & 0 & 0 \\ 0 & 0 & 0 & 0 \end{pmatrix}. \tag{6.58}$$

The general equation for a *diattenuator Mueller matrix* $\mathbf{D}$, either linear, elliptical, or circular, expressed in terms of the first row of the Mueller matrix is

$$\begin{aligned}\mathbf{D}(d_H, d_{45}, d_R, T_{Avg}) \\ = T_{Avg}\begin{pmatrix} 1 & d_H & d_{45} & d_R \\ d_H & A & 0 & 0 \\ d_{45} & 0 & A & 0 \\ d_R & 0 & 0 & A \end{pmatrix} + \frac{T_{Avg}(1-A)}{D^2}\begin{pmatrix} 0 & 0 & 0 & 0 \\ 0 & d_H^2 & d_{45}d_H & d_H d_R \\ 0 & d_{45}d_H & d_{45}^2 & d_{45}d_R \\ 0 & d_H d_R & d_{45}d_R & d_R^2 \end{pmatrix},\end{aligned} \tag{6.59}$$

where $D = \sqrt{d_H^2 + d_{45}^2 + d_R^2}$, $A = \sqrt{1 - d_H^2 - d_{45}^2 - d_R^2}$, $T_{Avg} = \frac{T_{max} + T_{min}}{2}$.

6.8 Poincaré Sphere Operations

6.8.1 The Operation of Retarders on the Poincaré Sphere

The principal reason for the widespread use of the *Poincaré sphere*, introduced in Section 3.13, is the simple geometric description it provides for the operation of *retarders*. The basic model of a retarder is an element that splits the light into two orthogonal polarization states (modes) and applies a different phase shift to each mode (see Figure 5.3). This *optical path difference* is the *retardance*, δ.

6.8 Poincaré Sphere Operations

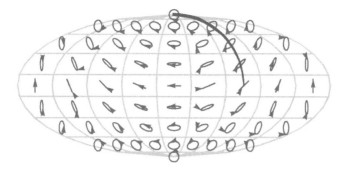

Figure 6.7 Vertical quarter wave linear retarder operating on 45° linearly polarized light, which brings the polarization to right circularly polarized. The trajectory is drawn thicker toward the exiting state.

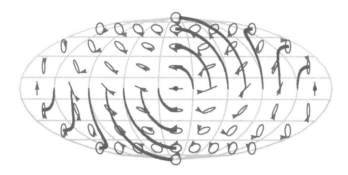

Figure 6.8 A vertical quarter wave linear retarder operating on a set of linearly polarized states rotates the polarization 90° about the S_1 axis.

The Mueller matrix for a retarder contains a 3 × 3 rotation matrix as was shown in Equation 6.18. The Mueller matrix for an elliptical retarder **ER**, shown in Equation 6.28, is expressed in terms of three retardance components, $(\delta_H, \delta_{45}, \delta_R)$, which define the Stokes parameters of the retarder fast axis. This Mueller matrix operates on the incident Stokes parameters by rotating their initial location on the Poincaré sphere about an axis through the retarder fast and slow states by an angle equal to δ. For an incident polarization state propagating through a birefringent retarder, as the retarder's retardance steadily increases from 0 to δ, the state evolves along a circular arc about the retarder axis.

Figure 6.7 shows the trajectory when 45° linearly polarized light is incident on a *vertical quarter wave linear retarder*. The polarization evolves along an arc from (0, 1, 0), through $(0, -1, 1)/\sqrt{2}$, ending at (0, 0, 1). In this case, the rotation is about the S_1-axis; the direction the polarization evolves is given by applying the *left-hand rule* about the fast axis. This assumes that the Poincaré sphere is drawn with right-handed coordinates such that*

$$S_1 \times S_2 = S_3. \tag{6.60}$$

By placing your left hand on the sphere above the fast axis with your thumb pointing out from the fast axis, make a fist and the direction your fingers curl indicates the direction the polarization state evolves around the axis of rotation for $\delta > 0$.

Figure 6.8 shows the action of a *quarter wave linear retarder* with a vertical fast axis operating on different linearly polarized incident states. Each incident state moves 90° from the equator along an arc about the *retarder axis* around the sphere.

---

* The right-handed cross product can be easily verified. By placing your right-hand fingers along $+S_1$ and sweeping them to $+S_2$, your thumb will point toward S_3; this test checks that the coordinate system is right handed.

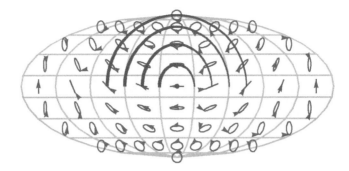

Figure 6.9 A vertical half wave linear retarder operating on a set of linearly polarized states. It rotates the polarization 180° about the S_1 axis, mapping linearly polarized input states into linearly polarized output states.

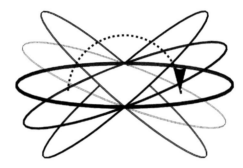

Figure 6.10 A 180° rotation of a state on the equator about an axis passing through the equator brings the state back to another state on the equator.

Figure 6.9 shows the action of a half wave linear retarder with a 0° fast axis operating on several linear polarized states. Each linear state moves along a 180° trajectory and ends back on the equator, as another linear state. Reasoning with the Poincaré sphere, it is easily seen that all half wave linear retarders bring a state on the equator back onto another state on the equator, leaving the fast and slow axes fixed, as shown in Figure 6.10. Thus, half wave linear retarders are used to rotate the orientation of linearly polarized light.

Figure 6.11 shows the action of a one-wave *circular retarder* ($\delta = 2\pi$). The rotation axis now passes through the poles. For circular retarders, the Poincaré sphere rotates like the earth with an axis through the poles.

Sequences of retarders are analyzed as a series of rotations of the Poincaré sphere. Several important examples will be presented. Figure 6.12 shows the action of a half wave linear retarder with a horizontal fast axis followed by another half wave linear retarder with a 45° fast axis on several polarization states. When right circularly polarized light is incident, the first retarder moves it to the left circular pole, and then the second retarder returns it back to the right circular pole. Thus, right circularly polarized light is an eigenpolarization of this retarder combination. 45° polarized light is rotated to 135° and then is unchanged by the half wave retarder with a 45° fast axis, as shown by the black line in Figure 6.12. The overall rotation is 180°; hence, this retarder combination acts as a *half wave circular retarder*.

For any *sequence of rotations of a sphere*, two points on the surface must end up back where they started. Thus, a *sequence of rotations* has the overall result of a single rotation about a rotation axis. For propagation through a sequence of retarders, the overall polarization transformations are equivalent to the operation of a single retarder, usually with an elliptical fast axis. The two states

6.8 Poincaré Sphere Operations

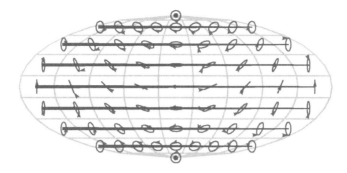

Figure 6.11 Action of a one-wave circular retarder rotates the Poincaré sphere through 360° about the poles, so all states exit as the incident polarization state.

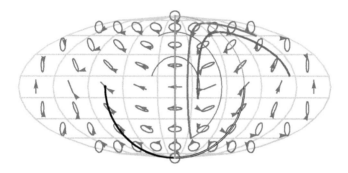

Figure 6.12 A sequence of two ½ wave linear retarders at 0° and 45° is a half wave circular retarder. This sequence corresponds to the Pauli matrices multiplication $\sigma_1 \times \sigma_2 = i\sigma_3$ (see Equation 6.108).

that end up where they started are the *eigenpolarizations* of the *compound retarder*. In fact, *elliptical retarders* are often most easily constructed from a set of two or more linear retarders.

Circular retarders can also be constructed from *linear retarders*. Consider a linear retarder with retardance δ oriented at 45°. When this retarder is placed between a horizontal quarter wave retarder and a vertical quarter wave linear retarder, the fast and slow axes of the δ retarder rotate to the circularly polarized poles; the three-element compound retarder forms a circular retarder with retardance δ. If the retardance of the horizontal and vertical linear retarders is adjusted to an angle μ, the fast and slow axes move to a latitude of ±μ on the Poincaré sphere.

Example 6.4 Sequence of Three Quarter Wave Retarders

Consider a sequence of three quarter wave retarders with fast axes of 0°, 45°, and then left circular state. The trajectories on the Poincaré sphere are shown in Figure 6.13 for two incident elliptical states, one for an eigenstate, $(1, 1/\sqrt{2}, 0, 1/\sqrt{2})$, and another state (1, 0.383, 0, 0.924) with a horizontally oriented major axis. The eigenstate follows a spherical triangle returning to the initial location; the first arc of the trajectory is the thinnest; the final arc is the thickest. For the other state, light becomes linearly polarized after the horizontal quarter wave retarder and then moves to a slightly elliptical state with a 45° orientation. Finally, the state moves along a circle of latitude due to the circular retarder and ends up elliptically polarized with a horizontally oriented major axis. This output state is the input state *rotated 180°* about the eigenpolarization; hence, the sequence is a *half wave elliptical retarder*.

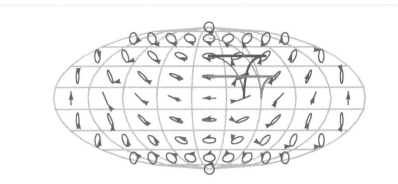

Figure 6.13 Trajectories are shown on the Poincaré sphere for a sequence of three quarter wave retarders, (1) horizontal, (2) 45°, and (3) left circular retarders. The two sequences of arcs correspond to two different incident polarization states. The horizontal retarder's trajectory is the thinnest and the circular retarder's trajectory is the thickest. One of the trajectories' sequences returns to its starting point; this is an eigenvector or eigenpolarization state for the compound retarder. Because the other trajectory begins and ends symmetrically with respect to the eigenstate, the assembly of three retarders acts as a half wave retarder with elliptical eigenstates.

6.8.2 The Operation of a Rotating Linear Retarder

One important polarization modulator is the *rotating linear retarder*. Consider horizontal linearly polarized light incident on a *spinning quarter wave linear retarder*. Figure 6.14 (center) shows the evolution of polarization state as the retarder's fast axis rotates. When the fast axis is oriented horizontally, the incident polarization state is an eigenpolarization and thus the state is unchanged. When the fast axis moves to 10° from horizontal, the polarization evolves through the retarder along a small 90° arc, moving away from the equator along a 45° angle. As the retarder angle increases, the trajectory approaches the 45° longitude circle. As the angle reaches 45°, the state moves through the right circularly polarized pole. When the fast axis is vertical, the light has returned to horizontal state, and the state has moved around the top half of a figure eight in the northern hemisphere. During the next 90° rotation of the fast axis, the polarization trajectory moves through the southern hemisphere through the left circularly polarized light, completing the figure eight trajectory as shown in Figure 6.14 (center). Figure 6.14 (left) shows the smaller, narrower figure eight trajectory for a *rotating 1/8 wave linear retarder*. Figure 6.14 (right) shows the trajectory for a *rotating 3/8 wave linear retarder* that starts in front and then circles around the back side of the sphere. The trajectory for a *rotating 1/2 wave linear retarder* circles the equator twice in 180° of retarder rotation.

Figure 6.14 The action of a rotating linear retarder on horizontal linearly polarized light is shown for three values of retardance, 1/8 wave (left), 1/4 wave (center), and 3/8 wave (right). The resulting polarization states trace a vertical figure eight on the surface of the Poincaré sphere. The 3/8 wave retarder has a smaller average distance from all the sphere states than the 1/4 wave retarder; the 3/8 wave is close to the optimum retardance value for a rotating retarder polarimeter.

6.8.3 The Operation of Polarizers and Diattenuators

The *transmittance*, or the *fraction of flux transmitted* through dichroic *polarizers* and *diattenuators*, can be depicted on and inside the Poincaré sphere as a series of planes perpendicular to the diattenuator's transmission axis. For example, the action of an ideal horizontal linear polarizer on an arbitrary set of normalized Stokes parameters has the Mueller matrix equation,

$$\mathbf{HLP} \cdot \hat{\mathbf{S}} = \frac{1}{2} \begin{pmatrix} 1 & 1 & 0 & 0 \\ 1 & 1 & 0 & 0 \\ 0 & 0 & 0 & 0 \\ 0 & 0 & 0 & 0 \end{pmatrix} \cdot \begin{pmatrix} 1 \\ s_1 \\ s_2 \\ s_3 \end{pmatrix} = \frac{1}{2} \begin{pmatrix} 1+s_1 \\ 1+s_1 \\ 0 \\ 0 \end{pmatrix}. \tag{6.61}$$

The transmittance, $(1 + s_1)/2$, corresponds to a series of planes through the Poincaré sphere perpendicular to the S_1 axis as shown in Figure 6.15. Each plane indicates the set of polarization states with the same transmittance. The plane through the middle of the sphere indicates Stokes parameters with a transmittance of ½, including 45° polarized, 135° polarized, left and right circularly polarized, and unpolarized. The plane tangent to the sphere at (1, 0, 0) represents the transmittance of one at the polarizer's transmission axis. The plane tangent to the sphere at (1, 0, 0) represents the transmittance of zero at its extinction axis. The same construction can be performed for any polarizer. For a diattenuator, the transmittances are scaled to vary linearly between T_{max} and T_{min}.

Next, consider the evolution of the Stokes parameters during propagation through a dichroic material, such as a sheet polarizer. The incident polarization state will move toward the transmission axis on the Poincaré sphere. Figure 6.16 shows some of the corresponding trajectories through a dichroic diattenuator for several different incident polarization states. The diattenuation is zero at the entrance and increases with the propagation distance. For a polarized input beam, the polarization state evolves along the great circles from the *absorption axis* toward the *transmission axis*. The arc lengths are shorter when the initial state is closer to the transmitted state or the attenuated state. The trajectories for the two incident states on the axis, the transmitted state and the attenuated state, have zero length; thus, those states do not move.

Figure 6.15 For diattenuators, lines perpendicular to the transmission axis also indicate a set of states with equal transmission. The transmission varies linearly from T_{min} to T_{max} for planes perpendicular to the axis.

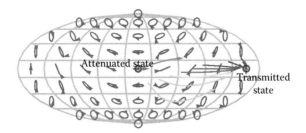

Figure 6.16 Evolution of polarization states through a dichroic medium with a diattenuation of 0.6 where the transmitted state (transmission axis) is vertically polarized and the attenuated state is horizontally polarized is plotted on the Poincaré sphere for several initial polarization states. At the beginning of each trajectory, the diattenuation is zero. The diattenuation steadily increases along the arrow. At the midpoint of each arrow, the diattenuation is 0.3; at the end or tip of the arrow, the diattenuation is 0.6. The trajectories would continue to the transmitted state for a diattenuation of one.

6.8.4 Indicating Polarization Properties

Many commercial polarimeters use the Poincaré sphere to represent the eigenstates associated with diattenuation properties, (d_H, d_{45}, d_R), polarizance properties (p_H, p_{45}, p_R), and retardance properties $(\delta_H, \delta_{45}, \delta_R)$, where indices H, 45, and R indicate the corresponding Stokes parameter components. Two examples are shown in this section.

Example 6.5 Sequence of Polarizers

A horizontal linear polarizer followed by a 45° linear polarizer has a Mueller matrix

$$\mathbf{LP}(45°) \cdot \mathbf{LP}(0°) = \frac{1}{2}\begin{pmatrix} 1 & 0 & 1 & 0 \\ 0 & 0 & 0 & 0 \\ 1 & 0 & 1 & 0 \\ 0 & 0 & 0 & 0 \end{pmatrix} \cdot \frac{1}{2}\begin{pmatrix} 1 & 1 & 0 & 0 \\ 1 & 1 & 0 & 0 \\ 0 & 0 & 0 & 0 \\ 0 & 0 & 0 & 0 \end{pmatrix} = \frac{1}{4}\begin{pmatrix} 1 & 1 & 0 & 0 \\ 0 & 0 & 0 & 0 \\ 1 & 1 & 0 & 0 \\ 0 & 0 & 0 & 0 \end{pmatrix}. \quad (6.62)$$

The eigenpolarizations are vertically polarized light and 135° polarized light; hence, this is an *inhomogeneous polarizer* Mueller matrix. The diattenuation parameters from the top row are

$$(M_{01}, M_{02}, M_{03}) = (1, 0, 0). \quad (6.63)$$

The polarizance parameters from the left column are

$$(M_{10}, M_{20}, M_{30}) = (0, 1, 0). \quad (6.64)$$

These properties are represented with the Poincaré sphere in Figure 6.17. The retardance is undefined for a polarizer and so is not shown.

6.8 Poincaré Sphere Operations

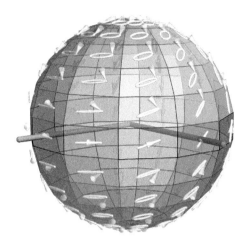

Figure 6.17 Poincaré sphere representation of the polarization properties of a particular inhomogeneous polarizer formed from a horizontal polarizer followed by a 90° polarizer. The brown line indicates the diattenuation axis and the blue line represents the polarizance axis.

Example 6.6 A Linear Diattenuator Followed by a Retarder

A horizontal linear diattenuator, $T_{max} = 1$, $T_{min} = 0.5$, with diattenuation = 1/3, followed by a 45° quarter wave linear retarder has the Mueller matrix

$$\mathbf{LR}(\pi/2, 45°) \cdot \mathbf{LD}(1, 1/2, 0°)$$

$$= \begin{pmatrix} 1 & 0 & 0 & 0 \\ 0 & 0 & 0 & -1 \\ 0 & 0 & 1 & 0 \\ 0 & 1 & 0 & 0 \end{pmatrix} \cdot \frac{1}{4} \begin{pmatrix} 3 & 1 & 0 & 0 \\ 1 & 3 & 0 & 0 \\ 0 & 0 & \dfrac{4}{\sqrt{2}} & 0 \\ 0 & 0 & 0 & \dfrac{4}{\sqrt{2}} \end{pmatrix}$$

$$= \frac{1}{4} \begin{pmatrix} 3 & 1 & 0 & 0 \\ 0 & 0 & 0 & \dfrac{-4}{\sqrt{2}} \\ 0 & 0 & \dfrac{4}{\sqrt{2}} & 0 \\ 1 & 3 & 0 & 0 \end{pmatrix} \qquad (6.65)$$

with diattenuation parameters (1/3, 0, 0), polarizance parameters (0, 0, 1/3), and retardance parameters (0, π/2, 0). These properties are represented on the Poincaré sphere as in Figure 6.18.

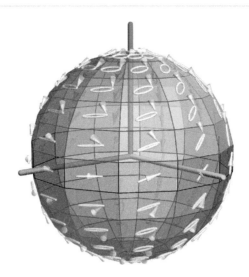

Figure 6.18 Poincaré sphere representation of the polarization properties of the Mueller matrix of Equation 6.65. The brown line indicates the diattenuation axis, the blue line represents the polarizance axis, and the green line indicates the retardance fast axis.

6.9 Weak Polarization Elements

Weak polarization elements cause only small changes to the polarization state. Weak polarization elements have Mueller matrices close to the *identity matrix times a constant* (to account for absorption or transmission losses). The properties of weak Mueller matrices are much simpler than those of general Mueller matrices because the retardance, diattenuation, and depolarization are close to zero. Some important examples of such weakly polarizing elements would be the lens surfaces and mirror surfaces in lenses, microscopes, and telescopes, where the polarization properties are not zero due to Fresnel equations, antireflection coatings, or mirrored surfaces, but the diattenuation and retardance effects are usually well below 0.05.

The structure of the Mueller calculus and the properties of these weak elements can be explored by performing Taylor series on the polarization property Mueller matrix expressions with respect to diattenuation or retardance. Weak retarders have a retardance near zero. Performing a Taylor series expansion on the general equation for an elliptical retarder (Equation 6.28) and keeping the first-order terms yield the following simple *weak retarder Mueller matrix*,

$$\lim_{\delta_H,\delta_{45},\delta_R \to 0} \mathbf{ER}(\delta_H,\delta_{45},\delta_R) = \begin{pmatrix} 1 & 0 & 0 & 0 \\ 0 & 1 & \delta_R & -\delta_{45} \\ 0 & -\delta_R & 1 & \delta_H \\ 0 & \delta_{45} & -\delta_H & 1 \end{pmatrix}. \quad (6.66)$$

Similarly, a first-order Taylor series expansion on the general diattenuator expression yields the *weak diattenuator Mueller matrix*,

6.9 Weak Polarization Elements

$$\lim_{d_H, d_{45}, d_R \to 0} \mathbf{D}(d_H, d_{45}, d_R, T_{Avg}) = T_{Avg} \begin{pmatrix} 1 & d_H & d_{45} & d_R \\ d_H & 1 & 0 & 0 \\ d_{45} & 0 & 1 & 0 \\ d_R & 0 & 0 & 1 \end{pmatrix}. \quad (6.67)$$

Combining these two expressions yields the *weak diattenuator and retarder Mueller matrix*

$$\mathbf{WDR}(d_H, d_{45}, d_R, \delta_H, \delta_{45}, \delta_R, T_{Avg}) = T_{Avg} \begin{pmatrix} 1 & d_H & d_{45} & d_R \\ d_H & 1 & \delta_R & -\delta_{45} \\ d_{45} & -\delta_R & 1 & \delta_H \\ d_R & \delta_{45} & -\delta_H & 1 \end{pmatrix}. \quad (6.68)$$

These three equations are only correct to first order. Higher-order terms, which are present when these parameters are not infinitesimal, are calculated from the exact equations presented earlier.

Weak diattenuators are symmetric in the top row and first column. Weak retarders are anti-symmetric in the off-diagonal lower right 3 × 3 elements. The presence of anti-symmetric components in the top row and column and symmetric components in the lower right 3 × 3 elements of weak polarization element Mueller matrices indicates the presence of *depolarization*, as discussed in Section 6.11.4.

Example 6.7 *Weakly Polarized Mueller Matrix*

Consider a weakly polarized Mueller matrix

$$\mathbf{M} = \begin{pmatrix} 1 & 0.02 & 0 & 0.01 \\ 0.02 & 1 & -0.005 & 0 \\ 0 & 0.005 & 1 & -0.01 \\ 0.01 & 0 & 0.01 & 1 \end{pmatrix}. \quad (6.69)$$

The diattenuation has a horizontal component $d_H = 0.02$ and a right circular component $d_R = 0.01$. The diattenuation magnitude is

$$d = \sqrt{0.02^2 + 0.01^2} \approx 0.022. \quad (6.70)$$

The Stokes parameters for the state with maximum transmission are any vector proportional to

$$\mathbf{S}_{max} \propto \begin{pmatrix} 0.022 \\ 0.02 \\ 0 \\ 0.01 \end{pmatrix}. \quad (6.71)$$

The retardance has a horizontal component $\delta_H = 0.01$ and a right circular component $\delta_R = 0.005$. Since these parameters are proportional to the diattenuation parameters, the matrix is *homogeneous* and has *orthogonal eigenpolarizations* shared by the diattenuation part and the retardance part. The retardance magnitude is

$$\delta = \sqrt{0.01^2 + 0.005^2} \approx 0.011. \tag{6.72}$$

6.10 Non-Depolarizing Mueller Matrices

Non-depolarizing Mueller matrices are the set of Mueller matrices for which completely polarized incident light with $DoP(\mathbf{S}) = 1$ transmits as completely polarized light for all incident polarization states. Non-depolarizing Mueller matrices have a *depolarization index* of one, which will be described in Equation 6.79. Non-depolarizing Mueller matrices are a subset of the Mueller matrices. *Jones matrices* can only represent non-depolarizing interactions. The non-depolarizing Mueller matrices are those Mueller matrices with corresponding Jones matrices; thus, non-depolarizing Mueller matrices are also called *Mueller–Jones matrices*.[5]

An *ideal polarizer* is non-depolarizing; when the incident beam is polarized, the exiting beam is polarized. Similarly, an ideal retarder is non-depolarizing. The non-depolarizing Mueller matrices comprise the Mueller matrices for the matrix product of all arbitrary sequences of diattenuation and retardance. A *Mueller–Jones* matrix must satisfy the following condition for all θ and η,

$$DoP(\mathbf{M} \cdot \mathbf{S}) = DoP\left(\begin{pmatrix} M_{00} & M_{01} & M_{02} & M_{03} \\ M_{10} & M_{11} & M_{12} & M_{13} \\ M_{20} & M_{21} & M_{22} & M_{23} \\ M_{30} & M_{31} & M_{32} & M_{33} \end{pmatrix} \cdot \begin{pmatrix} 1 \\ \cos 2\theta \cos \eta \\ \sin 2\theta \cos \eta \\ \sin \eta \end{pmatrix}\right) = 1. \tag{6.73}$$

One necessary, but not sufficient, condition for non-depolarizing Mueller matrices is[6]

$$\text{Tr}(\mathbf{M} \cdot \mathbf{M}^T) = 4 M_{00}^2, \tag{6.74}$$

where Tr is the trace of a matrix, the sum of the diagonal elements. Tr $(\mathbf{M} \cdot \mathbf{M}^T)$ equals the sum of the squares of all the matrix elements,

$$\begin{aligned}\text{Tr}(\mathbf{M} \cdot \mathbf{M}^T) = & M_{00}^2 + M_{01}^2 + M_{02}^2 + M_{03}^2 + M_{10}^2 + M_{11}^2 + M_{12}^2 + M_{13}^2 \\ & + M_{20}^2 + M_{21}^2 + M_{22}^2 + M_{23}^2 + M_{30}^2 + M_{31}^2 + M_{32}^2 + M_{33}^2.\end{aligned} \tag{6.75}$$

In a typical imaging optical system, *depolarization* is an undesirable characteristic for lens and mirror surfaces, filters, and polarization elements. Depolarization is associated with scattering, and optical surfaces are carefully fabricated and coated to minimize scattering. Depolarization is generally very small in high-quality optical surfaces. Thus, the majority of optical surfaces are well described by non-depolarizing Mueller matrices.

6.11 Depolarization

Depolarization is the reduction of the degree of polarization (*DoP*) of light. Depolarization was first described by David Brewster in 1815.[7] In the Mueller calculus, depolarization can be pictured as a *coupling of polarized into unpolarized light*. For polarized incident light, the exiting Stokes parameters for a depolarized beam have a *DoP* < 1 and can be mathematically separated into a fully polarized and an unpolarized set of Stokes parameters.

Optical elements with considerable scattering depolarize light to some extent. Similarly rough metal surfaces, painted surfaces, and natural surfaces such as rock, grass, and sand partially depolarize light. *Integrating spheres* (Figure 6.19) are frequently used in the laboratory to depolarize light into nearly unpolarized light as are thin plates of opal. *Milk* and other turbid fluids do a fine job depolarizing. When an integrating sphere is illuminated with laser light, the exiting light is a *speckle pattern* and the exiting polarization state is scrambled and resembles a *depolarized beam*. Figure 6.20 shows an ellipse map of a simulated polarized speckle pattern. The *x*-polarized light

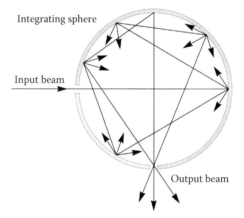

Figure 6.19 An integrating sphere is a good depolarizer. The interior is coated with a highly reflective diffuse reflector. Most light rays scatter multiple times before reaching the output port, ensuring that each exiting ray contains light from a large variety of paths and scattering angles.

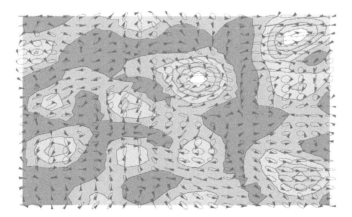

Figure 6.20 A simulation of a speckle pattern from a laser-illuminated integrating sphere or another depolarizer. Laser speckle patterns are polarized at each point because the beam is monochromatic. However, the amplitude and polarization state vary from speckle to speckle. Here, the polarization ellipses are mapped over a contour plot of the intensity distribution where brown is darkest and white is brightest. Speckle patterns have large regions of low flux and much smaller bright regions, the speckles.[8,9]

Table 6.3 Degrees of Freedom in Polarization Matrices

	Jones Matrix	Mueller Matrix
Transmission	1	1
Absolute phase	1	0
Diattenuation	3	3
Retardance	3	3
Depolarization	0	9

forms one speckle pattern and the *y*-polarized light forms a second pattern. These combine and are described by a Jones vector at each point of a randomly polarized field. For monochromatic illumination, the pattern is polarized at each point, but averaging the Stokes parameters over an area would yield to a depolarized measurement.

The Mueller matrix has 16 independent *degrees of freedom* (*DoF*), as shown in Table 6.3. Of the 16 degrees of freedom, one corresponds to loss, three to diattenuation, and three to retardance. The remaining *nine degrees of freedom* describe *depolarization*.

The depolarization associated with the three diagonal elements of the Mueller matrix, M_{11}, M_{22}, and M_{33}, tend to be the most significant of the nine degrees of freedom. The *ideal depolarizer Mueller matrix*, **ID**, transforms all incident beams into unpolarized light,

$$\mathbf{ID} \cdot \mathbf{S} = \begin{pmatrix} 1 & 0 & 0 & 0 \\ 0 & 0 & 0 & 0 \\ 0 & 0 & 0 & 0 \\ 0 & 0 & 0 & 0 \end{pmatrix} \cdot \begin{pmatrix} S_0 \\ S_1 \\ S_2 \\ S_3 \end{pmatrix} = \begin{pmatrix} S_0 \\ 0 \\ 0 \\ 0 \end{pmatrix}. \tag{6.76}$$

Only *unpolarized light* exits such a device. Although this matrix is an idealization, some devices such as integrating spheres approach this limit of nearly complete depolarization.

The *partial depolarizer Mueller matrix* depolarizes all incident states equally,

$$\mathbf{PD} \cdot \mathbf{S} = \begin{pmatrix} 1 & 0 & 0 & 0 \\ 0 & d & 0 & 0 \\ 0 & 0 & d & 0 \\ 0 & 0 & 0 & d \end{pmatrix} \cdot \begin{pmatrix} S_0 \\ S_1 \\ S_2 \\ S_3 \end{pmatrix} = (1-d) \begin{pmatrix} S_0 \\ 0 \\ 0 \\ 0 \end{pmatrix} + d \begin{pmatrix} S_0 \\ S_1 \\ S_2 \\ S_3 \end{pmatrix}. \tag{6.77}$$

All fully polarized incident states exit partially polarized with $DoP(\mathbf{PD} \cdot \mathbf{S}) = d$. The *diagonal depolarizer Mueller matrix* **DD** represents a variable partial depolarizer; the degree of polarization of the exiting light is a function of the incident state, with an exiting *DoP* of a for S_1, b for S_2, and c for S_3,

$$\mathbf{DD} = \begin{pmatrix} 1 & 0 & 0 & 0 \\ 0 & a & 0 & 0 \\ 0 & 0 & b & 0 \\ 0 & 0 & 0 & c \end{pmatrix}. \tag{6.78}$$

6.11 Depolarization

Physically, depolarization is closely related to scattering and usually has its origin in retardance or diattenuation, which are rapidly varying in time, space, or wavelength.

6.11.1 The Depolarization Index and the Average Degree of Polarization

Two depolarization metrics, the *depolarization index* and the *average degree of polarization*, have been introduced to describe the degree to which a Mueller matrix depolarizes incident states.[10-12]

The *depolarization index* $DI(\mathbf{M})$ is the Euclidian distance, indicated by $\| \; \|$, of the normalized Mueller matrix $\mathbf{M}/M_{00}$ from the *ideal depolarizer*:

$$DI(\mathbf{M}) = \left\| \frac{\mathbf{M}}{M_{00}} - \mathbf{ID} \right\| = \frac{\sqrt{\left(\sum_{i,j} M_{ij}^2\right) - M_{00}^2}}{\sqrt{3}\, M_{00}}. \tag{6.79}$$

$DI(\mathbf{M})$ varies from 0 for the ideal depolarizer to 1 for all non-depolarizing Mueller matrices, including all pure diattenuators, pure retarders, and any sequences composed from them. The form of the depolarization index equation is similar to the equation for degree of polarization for Stokes parameters.

The *average degree of polarization*, or *AverageDoP*, is the arithmetic mean of the degree of polarization of the exiting light for polarized incident light averaged over the Poincaré sphere,

$$AverageDoP(\mathbf{M}) = \frac{\int_0^\pi \int_{-\pi/2}^{\pi/2} DoP(\mathbf{M} \cdot \mathbf{S}(\theta, \eta)) \cos(\eta)\, d\eta\, d\theta}{2\pi}, \tag{6.80}$$

where, for Stokes parameters $\mathbf{S}(\theta, \eta)$, θ is the orientation of the major axis, and η is the latitude in radians on the Poincare sphere,

$$\mathbf{S}(\theta, \eta) = \begin{pmatrix} 1 \\ \cos(2\theta)\cos(\eta) \\ \sin(2\theta)\cos(\eta) \\ \sin(\eta) \end{pmatrix}. \tag{6.81}$$

The *AverageDoP* varies from 0 to 1, summarizing the depolarizing properties in a single number. When *AverageDoP* is equal to 1, the exiting light is always completely polarized, indicating a *non-depolarizing Mueller matrix*. Values near 1 indicate a small amount of depolarization. When *AverageDoP* equals 0, the exiting light is completely depolarized; only unpolarized light exits the interaction. The *DI* and *AverageDoP* of an aperture with different polarizers are shown in Figure 6.21; a detailed calculation is shown in Example 6.9.

The *DI* and the *AverageDoP* are usually close in value. The *AverageDoP* is the easier metric to understand; it provides the mean *DoP* of the exiting light averaged over the Poincaré sphere, the expected value. The *DI* has a clear geometric meaning in the Mueller matrix configuration space, being the fractional distance of a Mueller matrix along a line segment from the ideal depolarizer to the hypersphere of non-depolarizing Mueller matrices.

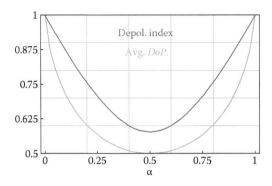

Figure 6.21 Depolarization index (red) and average depolarization (green) of a family of Mueller matrices formed from an aperture covered with a fraction α of horizontal polarizer and a fraction $1 - \alpha$ of vertical polarizer.

6.11.2 Degree of Polarization Surfaces and Maps

Insight into the nine degrees of freedom of depolarization can be found by examining the variations of *DoP* with incident state using *degree of polarization surfaces* and *degree of polarization maps*.[13]

The *DoP* surface for a Mueller matrix, **M**, is formed by moving each set of normalized Stokes parameters, **S**, formed from the three Stokes parameters (S_1, S_2, S_3) on the surface of the Poincaré sphere radially inward to a distance $DoP(\mathbf{S}' = \mathbf{M} \cdot \mathbf{S})$ from the origin. It is plotted for all incident **S** on the surface of the Poincaré sphere,

$$DoP\ \text{Surface}(\mathbf{M},\mathbf{S}) = \frac{\sqrt{S_1'(\mathbf{M},\mathbf{S})^2 + S_2'(\mathbf{M},\mathbf{S})^2 + S_3'(\mathbf{M},\mathbf{S})^2}}{S_0'(\mathbf{M},\mathbf{S})}(S_1,S_2,S_3). \qquad (6.82)$$

The *DoP* surface results from the product of a *scalar*, the *DoP*, and a *vector*, (S_1, S_2, S_3), forming a three-dimensional surface. For a non-depolarizing Mueller matrix, the exiting *DoP* is 1 for all incident states; hence, the *DoP* surface is the *unit sphere*; the Poincaré sphere does not shrink in this case.

The *DoP* map represents *DoP* as a contour plot on a flat map of the Poincaré sphere. To create the *DoP map*, the surface of the Poincaré sphere is parameterized in terms of the orientation of the major axis θ and the degree of circular polarization *DoCP* as

$$\mathbf{S}(\theta, DoCP) = \begin{pmatrix} 1 \\ \cos(2\theta)\cos(\sin^{-1} DoCP) \\ \sin(2\theta)\cos(\sin^{-1} DoCP) \\ DoCP \end{pmatrix}. \qquad (6.83)$$

The *DoP* of the light exiting the Mueller matrix is plotted as a contour plot as a function of θ and *DoCP*. The *DoP surface* and the *DoP map* represent the same information, the exiting *DoP* as a function of incident polarization state.

A depolarizing example is the Mueller matrix $\mathbf{M}_1$ formed from an aperture covered one-half with a horizontal polarizer and the other half with a vertical polarizer,

6.11 Depolarization

$$\mathbf{M}_1 = \frac{1}{2}\begin{pmatrix} 1 & 1 & 0 & 0 \\ 1 & 1 & 0 & 0 \\ 0 & 0 & 0 & 0 \\ 0 & 0 & 0 & 0 \end{pmatrix} + \frac{1}{2}\begin{pmatrix} 1 & -1 & 0 & 0 \\ -1 & 1 & 0 & 0 \\ 0 & 0 & 0 & 0 \\ 0 & 0 & 0 & 0 \end{pmatrix} = \begin{pmatrix} 1 & 0 & 0 & 0 \\ 0 & 1 & 0 & 0 \\ 0 & 0 & 0 & 0 \\ 0 & 0 & 0 & 0 \end{pmatrix}. \quad (6.84)$$

The *DoP* surface and *DoP* map for $\mathbf{M}_1$ are plotted in Figure 6.22. The two maxima of *DoP* occur when (1) vertically polarized light is incident and only vertically polarized light exits and (2) when horizontally polarized light is incident and only horizontally polarized light exits. For incident states residing on a circle of the Poincaré sphere midway between horizontal and vertical, the exiting light is completely depolarized and is the incoherent sum of half horizontally polarized and half vertically polarized light.

As another example, consider an element with a Mueller matrix $\mathbf{M}_2$ where one-half of the aperture is covered with a horizontal linear polarizer and the other half is covered by a 45° linear polarizer,

$$\mathbf{M}_2 = \frac{1}{2}\begin{pmatrix} 1 & 1 & 0 & 0 \\ 1 & 1 & 0 & 0 \\ 0 & 0 & 0 & 0 \\ 0 & 0 & 0 & 0 \end{pmatrix} + \frac{1}{2}\begin{pmatrix} 1 & 0 & 1 & 0 \\ 0 & 0 & 0 & 0 \\ 1 & 0 & 1 & 0 \\ 0 & 0 & 0 & 0 \end{pmatrix} = \begin{pmatrix} 1 & 0.5 & 0.5 & 0 \\ 0.5 & 0.5 & 0 & 0 \\ 0.5 & 0 & 0.5 & 0 \\ 0 & 0 & 0 & 0 \end{pmatrix}. \quad (6.85)$$

The *DoP* surface and *DoP* map for $\mathbf{M}_2$ are plotted in Figure 6.23. The two states of maximum *DoP*, (1,−1,0,0) and (1,0,−1,0), are the states that are blocked by one of the two polarizers. It is seen that the maxima of the *DoP* map do not need to be orthogonal.

DoP maps and surfaces may exhibit one or two maxima and one or two minima. These maxima or minima may also be degenerate for entire circles of incident states around the Poincaré sphere.

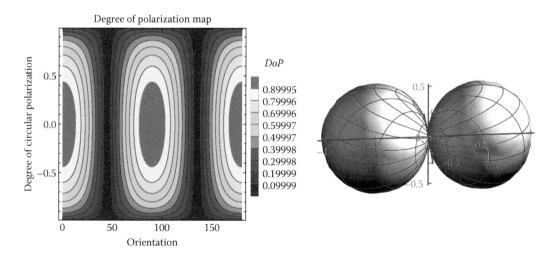

Figure 6.22 (Left) Degree of polarization surface and (right) degree of polarization map for $\mathbf{M}_1$, a Mueller matrix formed from the sum of horizontal and 90° linear polarizers.

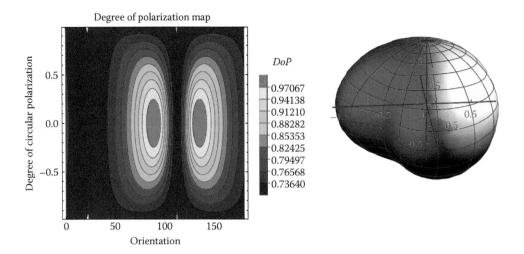

Figure 6.23 (Left) Degree of polarization map and (right) degree of polarization surface for $\mathbf{M}_2$, a Mueller matrix formed from the sum of horizontal and 45° linear polarizers.

6.11.3 Testing for Physically Realizable Mueller Matrices

To be physically realizable,[14,15] a Mueller matrix needs to operate on all possible sets of Stokes parameters producing valid exiting Stokes parameters; otherwise, the Mueller matrix is not physically realizable. Only a subset of 4 × 4 matrices are *physically realizable Mueller matrices*. Mueller matrices that are not physically realizable can be an issue in measured Mueller matrices. Polarimeters always have noise. Retarders, diattenuators, and their combinations lie right on the boundary between physically realizable and not physically realizable Mueller matrices. Thus, a small amount of noise can shift a Mueller matrix with a depolarization index of one into the non-physically realizable region. Goldstein provides algorithms and examples of moving non-physically realizable Mueller matrices to the closest physically realizable matrices.[16]

For a Mueller matrix to be physically realizable, several conditions must apply. First, the output flux must be non-negative, $S_0' \geq 0$. Second, for all input polarization states, the output degree of polarization must lie between zero and one,

$$0 \leq DoP(\mathbf{M} \cdot \mathbf{S}) \leq 1. \tag{6.86}$$

If the degree of polarization is greater than one, the output state is non-physical. One test for physicality is to test that the depolarization index lies between zero and one,

$$0 \leq DI(\mathbf{M}) \leq 1. \tag{6.87}$$

Several comprehensive and systematic tests have been developed. Givens and Kostinski[17] evaluate the physicality of Mueller matrices as follows. Defining the Lorentz metric matrix $\mathbf{G}$ as

$$\mathbf{G} = \begin{pmatrix} 1 & 0 & 0 & 0 \\ 0 & -1 & 0 & 0 \\ 0 & 0 & -1 & 0 \\ 0 & 0 & 0 & -1 \end{pmatrix}, \tag{6.88}$$

then the *Mueller coherence matrix*,

6.11 Depolarization

$$\mathbf{D} = \mathbf{G}\mathbf{M}^T\mathbf{G}\mathbf{M} = \begin{pmatrix} a_{00} & a_{01} & a_{02} & a_{03} \\ a_{10} & a_{11} & a_{12} & a_{13} \\ a_{20} & a_{21} & a_{22} & a_{23} \\ a_{30} & a_{31} & a_{32} & a_{33} \end{pmatrix}, \quad (6.89)$$

where

$$a_{00} = M_{00}^2 - M_{10}^2 - M_{20}^2 - M_{30}^2$$
$$a_{01} = M_{00}M_{01} - M_{10}M_{11} - M_{20}M_{21} - M_{30}M_{31}$$
$$a_{02} = M_{00}M_{02} - M_{10}M_{12} - M_{20}M_{22} - M_{30}M_{32}$$
$$a_{03} = M_{00}M_{03} - M_{10}M_{13} - M_{20}M_{23} - M_{30}M_{33}$$

$$a_{10} = -M_{00}M_{01} + M_{10}M_{11} + M_{20}M_{21} + M_{30}M_{31}$$
$$a_{11} = -M_{01}^2 + M_{11}^2 + M_{21}^2 + M_{31}^2$$
$$a_{12} = -M_{01}M_{02} + M_{11}M_{12} + M_{21}M_{22} + M_{31}M_{32}$$
$$a_{13} = -M_{01}M_{03} + M_{11}M_{13} + M_{21}M_{23} + M_{31}M_{33}$$

$$a_{20} = -M_{00}M_{02} + M_{10}M_{12} + M_{20}M_{22} + M_{30}M_{32}$$
$$a_{21} = -M_{01}M_{02} + M_{11}M_{12} + M_{21}M_{22} + M_{31}M_{32}$$
$$a_{22} = -M_{02}^2 + M_{12}^2 + M_{22}^2 + M_{32}^2$$
$$a_{23} = -M_{02}M_{03} + M_{12}M_{13} + M_{22}M_{23} + M_{32}M_{33}$$

$$a_{30} = -M_{00}M_{03} + M_{10}M_{13} + M_{20}M_{23} + M_{30}M_{33}$$
$$a_{31} = -M_{01}M_{03} + M_{11}M_{13} + M_{21}M_{23} + M_{31}M_{33}$$
$$a_{32} = -M_{02}M_{03} + M_{12}M_{13} + M_{22}M_{23} + M_{32}M_{33}$$
$$a_{33} = -M_{03}^2 + M_{13}^2 + M_{23}^2 + M_{33}^2$$

is calculated. In order to be a physically realizable Mueller matrix, (1) all the eigenvalues of **D** must be real, and (2) the eigenvalue associated with the largest eigenvector must be a valid set of Stokes parameters.

Example 6.8 Testing Mueller Matrices for Physicality

Consider an example of a family of depolarizing matrices $\mathbf{M}_3$ with a single degree of freedom β,

$$\mathbf{M}_3 = \begin{pmatrix} 1 & 0.1 & 0 & 0 \\ \beta & 0.9 & 0 & 0 \\ 0 & 0 & 0.9 & 0 \\ 0 & 0 & 0 & 0.8 \end{pmatrix}. \quad (6.90)$$

Over what range of β, the M_{10} element, is $\mathbf{M}_3$ a valid physically realizable Mueller matrix? The associated Mueller coherence matrix

$$\mathbf{D}_3 = \begin{pmatrix} 1-\beta^2 & 0.1-0.9\beta & 0 & 0 \\ -0.1+0.9\beta & 0.8 & 0 & 0 \\ 0 & 0 & 0.81 & 0 \\ 0 & 0 & 0 & 0.64 \end{pmatrix} \quad (6.91)$$

has eigenvalues

$$\lambda = \begin{cases} 0.81, \\ 0.64, \\ 0.9 - 0.5\beta^2 - 0.5\sqrt{0.72\beta - 3.64\beta^2 + \beta^4}, \\ 0.9 - 0.5\beta^2 + 0.5\sqrt{0.72\beta - 3.64\beta^2 + \beta^4}. \end{cases} \quad (6.92)$$

The discriminant $0.72\beta - 3.64\beta^2 + \beta^4 \geq 0$ for the range $0 \leq \beta \leq 0.2$; hence, all four eigenvalues are only real over this range of β. It is readily verified that the associated eigenvector for the fourth eigenvalues has physical Stokes parameters over this range. Thus, $\mathbf{M}_3$ is a valid Mueller matrix only when $0 \leq \beta \leq 0.2$.

6.11.4 Weak Depolarizing Elements

Section 6.9 described the Mueller matrices for weak non-depolarizing elements in the vicinity of the identity matrix in terms of three diattenuation and three retarding degrees of freedom. This formalism readily extends to the *nine remaining degrees of freedom* and thus the *nine forms of depolarization*.[18] One simple way to view the depolarizing degrees of freedom is as follows. For weak Mueller matrices, the depolarizing degrees of freedom are associated with the following matrix element combinations

$$\mathbf{WDepol} = \begin{pmatrix} 1 & e_1 & e_2 & e_3 \\ -e_1 & 1-g_1 & f_3 & -f_2 \\ -e_2 & f_3 & 1-g_2 & f_1 \\ -e_3 & -f_2 & f_1 & 1-g_3 \end{pmatrix}, \quad (6.93)$$

where the nine forms have been grouped into three families. g_1, g_2, and g_3 are associated with the *diagonal* and labeled *diagonal depolarization*. e_1, e_2, and e_3 are associated with *antisymmetric values* in the diattenuation elements and are labeled *amplitude depolarization*. f_1, f_2, and f_3 are associated with symmetric values in the retardance elements and are labeled *phase depolarization*. Each of the three forms, *amplitude depolarization*, *phase depolarization*, and *diagonal depolarization*,

6.11 Depolarization

has a term associated with S_1, a term associated with S_2, and a term associated with S_3. When the e, f, and g components are close to 0, the matrix **WDepol** is close to the *identity matrix* and the depolarization can be regarded as *weak*. Because of the symmetry of these elements, **WDepol** has no diattenuation and no retardance. As the strength of the depolarization increases, the terms do not remain purely linear. Depolarizing matrices outside the weak limit with particular depolarizing terms can be generated by raising **WDepol** to arbitrary powers,

$$\mathbf{WDepol}^N = \begin{pmatrix} 1 & e_1 & e_2 & e_3 \\ -e_1 & 1-g_1 & f_3 & f_2 \\ -e_2 & f_3 & 1-g_2 & f_1 \\ -e_3 & f_2 & f_1 & 1-g_3 \end{pmatrix}^N. \quad (6.94)$$

Care must be taken, because the matrices of Equation 6.93 are only physically realizable in the limit as the coefficients equal to zeros; thus, Equation 6.93 is slightly non-physical, and the resulting strong matrices need small adjustments for physicality.

6.11.5 The Addition of Mueller Matrices

The *matrix product of Mueller matrices* shown in Equation 6.6 represents sequences of polarization elements. On the other hand, the *addition of Mueller matrices* represents polarization elements side by side sharing an aperture. Whenever two different non-depolarizing Mueller matrices are added, *depolarization* must be introduced. In general, across an aperture, different Stokes parameters exit and are combined, reducing the degree of polarization.

Mueller matrix functions can be integrated over time or space to simulate time- or space-varying polarization processes, as in the following two examples.

Example 6.9 Two Polarizers over an Aperture

Consider an aperture where one fraction, α, of it is covered by a horizontal linear polarizer and the remainder $(1 - \alpha)$ of it is filled with a vertical linear polarizer. The resulting Mueller matrix is

$$\alpha \mathbf{HLP} + (1-\alpha) \mathbf{VLP} = \frac{\alpha}{2} \begin{pmatrix} 1 & 1 & 0 & 0 \\ 1 & 1 & 0 & 0 \\ 0 & 0 & 0 & 0 \\ 0 & 0 & 0 & 0 \end{pmatrix} + \frac{(1-\alpha)}{2} \begin{pmatrix} 1 & -1 & 0 & 0 \\ -1 & 1 & 0 & 0 \\ 0 & 0 & 0 & 0 \\ 0 & 0 & 0 & 0 \end{pmatrix} = \frac{1}{2} \begin{pmatrix} 1 & 2\alpha-1 & 0 & 0 \\ 2\alpha-1 & 1 & 0 & 0 \\ 0 & 0 & 0 & 0 \\ 0 & 0 & 0 & 0 \end{pmatrix}.$$

(6.95)

When $\alpha = \frac{1}{2}$, the aperture is half horizontally polarized and half vertically polarized; the combination acts as a partial depolarizer with a depolarization index of $1/\sqrt{3}$. S_2 and S_3 are depolarized but S_1 exits unchanged.

Example 6.10 A Spinning Quarter Wave Linear Retarder

A rapidly spinning quarter wave linear retarder, similar to the case described in Figure 6.14, has a time-averaged Mueller matrix

$$\int_0^{2\pi} \mathbf{QWLR}(\theta)\, d\theta = \int_0^{2\pi} \begin{pmatrix} 1 & 0 & 0 & 0 \\ 0 & \cos^2 2\theta & \cos 2\theta \sin 2\theta & -\sin 2\theta \\ 0 & \cos 2\theta \sin 2\theta & \sin^2 2\theta & \cos 2\theta \\ 0 & \sin 2\theta & -\cos 2\theta & 0 \end{pmatrix} d\theta = \begin{pmatrix} 1 & 0 & 0 & 0 \\ 0 & \frac{1}{2} & 0 & 0 \\ 0 & 0 & \frac{1}{2} & 0 \\ 0 & 0 & 0 & 0 \end{pmatrix}$$

(6.96)

with a depolarization index of $1/\sqrt{6}$ and an average DoP of $\pi/8$, which differ by ≈ 0.016. This Mueller matrix completely depolarizes circularly polarized light and reduces the degree of linear polarization by half.

6.12 Relating Jones and Mueller Matrices

6.12.1 Transforming Jones Matrices into Mueller Matrices Using Tensor Product

Along with Mueller matrices, Jones matrices (Chapter 5) form a very useful representation of sample polarization, particularly because Jones matrices have simpler properties and are more easily manipulated and interpreted. The complication in mapping Mueller matrices onto Jones matrices and vice versa is that Mueller matrices cannot represent absolute phase and Jones matrices cannot represent depolarization. Only *non-depolarizing Mueller matrices* or the *Mueller–Jones matrices* have corresponding Jones matrices. All Jones matrices have a corresponding Mueller matrix. However, since the absolute phase is not represented in Mueller matrices, many Jones matrices with different absolute phase can be mapped to the same Mueller matrix.

Both Jones matrices and Mueller matrices can calculate the polarization properties of sequences of non-depolarizing interactions, the effect of cascading a series of diattenuators and retarders. When this same polarization element sequence is calculated by Jones matrices and alternatively by Mueller matrices, the answer contains the same diattenuating and retarding properties. Either method is suitable.

6.12 Relating Jones and Mueller Matrices

Math Tip 6.1 The Tensor Product

The tensor product of two 2×2 matrices, $\mathbf{A} \otimes \mathbf{B}$, is the matrix[5,19–21]

$$\mathbf{A} \otimes \mathbf{B} = \begin{pmatrix} a_{11} & a_{12} \\ a_{21} & a_{22} \end{pmatrix} \otimes \begin{pmatrix} b_{11} & b_{12} \\ b_{21} & b_{22} \end{pmatrix}$$

$$= \begin{pmatrix} a_{11}\begin{pmatrix} b_{11} & b_{12} \\ b_{21} & b_{22} \end{pmatrix} & a_{12}\begin{pmatrix} b_{11} & b_{12} \\ b_{21} & b_{22} \end{pmatrix} \\ a_{21}\begin{pmatrix} b_{11} & b_{12} \\ b_{21} & b_{22} \end{pmatrix} & a_{22}\begin{pmatrix} b_{11} & b_{12} \\ b_{21} & b_{22} \end{pmatrix} \end{pmatrix} \quad (6.97)$$

$$= \begin{pmatrix} a_{11}b_{11} & a_{12}b_{12} & a_{12}b_{11} & a_{12}b_{12} \\ a_{11}b_{21} & a_{11}b_{22} & a_{12}b_{21} & a_{12}b_{22} \\ a_{21}b_{11} & a_{21}b_{12} & a_{22}b_{11} & a_{22}b_{12} \\ a_{21}b_{21} & a_{21}b_{22} & a_{22}b_{21} & a_{22}b_{22} \end{pmatrix}.$$

A Jones matrix $\mathbf{J}$ is transformed into the equivalent Mueller matrix $\mathbf{M}$ by a unitary transformation of the *tensor product* $\mathbf{J}^T \otimes \mathbf{J}^\dagger$ and the unitary matrix $\mathbf{U}$,

$$\mathbf{U} = \frac{1}{\sqrt{2}}\begin{pmatrix} 1 & 0 & 0 & 1 \\ 1 & 0 & 0 & -1 \\ 0 & 1 & 1 & 0 \\ 0 & i & -i & 0 \end{pmatrix} = (\mathbf{U}^{-1})^\dagger. \quad (6.98)$$

Note that each of the four rows are each flattened versions of the Pauli matrices, but the fourth row is the negative of σ_3.* The Mueller matrix corresponding to Jones matrix $\mathbf{J}$ is[†]

$$\mathbf{M} = \mathbf{U} \cdot (\mathbf{J}^* \otimes \mathbf{J}) \cdot \mathbf{U}^{-1}. \quad (6.99)$$

All Jones matrices of the form $\mathbf{J}' = e^{-i\phi} \mathbf{J}$ transform to the same Mueller matrix. Consider the Jones matrix with its complex elements expressed in polar coordinate form,

$$\mathbf{J} = \begin{pmatrix} j_{xx} & j_{xy} \\ j_{yx} & j_{yy} \end{pmatrix} = \begin{pmatrix} \rho_{xx}e^{-i\phi_{xx}} & \rho_{xy}e^{-i\phi_{xy}} \\ \rho_{yx}e^{-i\phi_{yx}} & \rho_{yy}e^{-i\phi_{yy}} \end{pmatrix}. \quad (6.100)$$

---

* In this book for Jones matrices, a positive σ_3 is associated with left circular polarization (see Sections 14.6.1 and 14.6.2). In the Stokes parameters, a positive S_3 is associated with right circular and elliptical polarization. Thus, a minus sign is needed for conversion between these two components.
† In some works on Stokes parameters and Mueller matrices, a positive value of the last Stokes parameter S_3 indicates left circularly polarized light, not right circularly polarized light, as it does in this book.

The tensor product $\mathbf{J}^* \otimes \mathbf{J}$ is calculated as Equation 6.97,

$$(\mathbf{J}^* \otimes \mathbf{J}) = \begin{pmatrix} \rho_{xx}e^{i\phi_{xx}}\begin{pmatrix} \rho_{xx}e^{-i\phi_{xx}} & \rho_{xy}e^{-i\phi_{xy}} \\ \rho_{yx}e^{-i\phi_{yx}} & \rho_{yy}e^{-i\phi_{yy}} \end{pmatrix} & \rho_{xy}e^{i\phi_{xy}}\begin{pmatrix} \rho_{xx}e^{-i\phi_{xx}} & \rho_{xy}e^{-i\phi_{xy}} \\ \rho_{yx}e^{-i\phi_{yx}} & \rho_{yy}e^{-i\phi_{yy}} \end{pmatrix} \\ \rho_{yx}e^{i\phi_{yx}}\begin{pmatrix} \rho_{xx}e^{-i\phi_{xx}} & \rho_{xy}e^{-i\phi_{xy}} \\ \rho_{yx}e^{-i\phi_{yx}} & \rho_{yy}e^{-i\phi_{yy}} \end{pmatrix} & \rho_{yy}e^{i\phi_{yy}}\begin{pmatrix} \rho_{xx}e^{-i\phi_{xx}} & \rho_{xy}e^{-i\phi_{xy}} \\ \rho_{yx}e^{-i\phi_{yx}} & \rho_{yy}e^{-i\phi_{yy}} \end{pmatrix} \end{pmatrix}$$

$$= \begin{pmatrix} \rho_{xx}^2 & \rho_{xx}\rho_{xy}e^{i(\phi_{xx}-\phi_{xy})} & \rho_{xy}\rho_{xx}e^{i(\phi_{xy}-\phi_{xx})} & \rho_{xy}^2 \\ \rho_{xx}\rho_{yx}e^{i(\phi_{xx}-\phi_{yx})} & \rho_{xx}\rho_{yy}e^{i(\phi_{xx}-\phi_{yy})} & \rho_{xy}\rho_{yx}e^{i(\phi_{xy}-\phi_{yx})} & \rho_{xy}\rho_{yy}e^{i(\phi_{xy}-\phi_{yy})} \\ \rho_{yx}\rho_{xx}e^{i(\phi_{yx}-\phi_{xx})} & \rho_{yx}\rho_{xy}e^{i(\phi_{yx}-\phi_{xy})} & \rho_{yy}\rho_{xx}e^{i(\phi_{yy}-\phi_{xx})} & \rho_{yy}\rho_{xy}e^{i(\phi_{yy}-\phi_{xy})} \\ \rho_{yx}^2 & \rho_{yx}\rho_{yy}e^{i(\phi_{yx}-\phi_{yy})} & \rho_{yy}\rho_{yx}e^{i(\phi_{yy}-\phi_{yx})} & \rho_{yy}^2 \end{pmatrix}. \qquad (6.101)$$

When $\mathbf{J}^* \otimes \mathbf{J}$ is transformed by $\mathbf{U}$ and $\mathbf{U}^{-1}$, it gives the Mueller matrix elements as in Equation 6.102.

$$\mathbf{U} \cdot (\mathbf{J}^* \otimes \mathbf{J}) \cdot \mathbf{U}^{-1} = \begin{pmatrix} M_{00} \\ M_{01} \\ M_{02} \\ M_{03} \\ M_{10} \\ M_{11} \\ M_{12} \\ M_{13} \\ M_{20} \\ M_{21} \\ M_{22} \\ M_{23} \\ M_{30} \\ M_{31} \\ M_{32} \\ M_{33} \end{pmatrix} = \begin{pmatrix} \frac{1}{2}\left(\rho_{xx}^2 + \rho_{xy}^2 + \rho_{yx}^2 + \rho_{yy}^2\right) \\ \frac{1}{2}\left(\rho_{xx}^2 - \rho_{xy}^2 + \rho_{yx}^2 - \rho_{yy}^2\right) \\ \rho_{xx}\rho_{xy}\cos(\phi_{xx}-\phi_{xy}) + \rho_{yx}\rho_{yy}\cos(\phi_{yx}-\phi_{yy}) \\ \rho_{xx}\rho_{xy}\sin(\phi_{xx}-\phi_{xy}) + \rho_{yx}\rho_{yy}\sin(\phi_{yx}-\phi_{yy}) \\ \frac{1}{2}\left(\rho_{xx}^2 + \rho_{xy}^2 - \rho_{yx}^2 - \rho_{yy}^2\right) \\ \frac{1}{2}\left(\rho_{xx}^2 - \rho_{xy}^2 - \rho_{yx}^2 + \rho_{yy}^2\right) \\ \rho_{xx}\rho_{xy}\cos(\phi_{xx}-\phi_{xy}) - \rho_{yx}\rho_{yy}\cos(\phi_{yx}-\phi_{yy}) \\ \rho_{xx}\rho_{xy}\sin(\phi_{xx}-\phi_{xy}) - \rho_{yx}\rho_{yy}\sin(\phi_{yx}-\phi_{yy}) \\ \rho_{xx}\rho_{yx}\cos(\phi_{xx}-\phi_{yx}) + \rho_{xy}\rho_{yy}\cos(\phi_{xy}-\phi_{yy}) \\ \rho_{xx}\rho_{yx}\cos(\phi_{xx}-\phi_{yx}) - \rho_{xy}\rho_{yy}\cos(\phi_{xy}-\phi_{yy}) \\ \rho_{xy}\rho_{yx}\cos(\phi_{xy}-\phi_{yx}) + \rho_{xx}\rho_{yy}\cos(\phi_{xx}-\phi_{yy}) \\ -\rho_{xy}\rho_{yx}\sin(\phi_{xy}-\phi_{yx}) + \rho_{xx}\rho_{yy}\sin(\phi_{xx}-\phi_{yy}) \\ -\rho_{xx}\rho_{yx}\sin(\phi_{xx}-\phi_{yx}) - \rho_{xy}\rho_{yy}\sin(\phi_{xy}-\phi_{yy}) \\ -\rho_{xx}\rho_{yx}\sin(\phi_{xx}-\phi_{yx}) + \rho_{xy}\rho_{yy}\sin(\phi_{xy}-\phi_{yy}) \\ -\rho_{xy}\rho_{yx}\sin(\phi_{xy}-\phi_{yx}) - \rho_{xx}\rho_{yy}\sin(\phi_{xx}-\phi_{yy}) \\ -\rho_{xy}\rho_{yx}\cos(\phi_{xy}-\phi_{yx}) + \rho_{xx}\rho_{yy}\cos(\phi_{xx}-\phi_{yy}) \end{pmatrix}. \qquad (6.102)$$

6.12 Relating Jones and Mueller Matrices

Example 6.11 Numerical Example of Converting a Jones Matrix to a Mueller Matrix

Consider a Jones matrix

$$\mathbf{J} = \begin{pmatrix} \dfrac{1}{4} & \dfrac{1}{4} \\ \dfrac{e^{i\frac{\pi}{2}}}{\sqrt{2}} & \dfrac{e^{-i\frac{\pi}{2}}}{\sqrt{2}} \end{pmatrix} = \begin{pmatrix} \dfrac{1}{4} & \dfrac{1}{4} \\ \dfrac{i}{\sqrt{2}} & \dfrac{-i}{\sqrt{2}} \end{pmatrix}. \quad (6.103)$$

This Jones matrix represents a linear diattenuator oriented at 45° with a diattenuation of 0.778, followed by a 120° retarder oriented at 22.5° and ellipticity = −0.391 ≈ π/8. In the Pauli matrix decomposition, this retarder has a linear retardance of 88° oriented at 22.5° and a circular retardance of 81°. The retarder is then followed by another diattenuator with diattenuation 0.778 oriented at 0°. This particular **J** provides simple transformation without involving trivial Jones or Mueller matrices.

The tensor product

$$(\mathbf{J}^* \otimes \mathbf{J}) = \begin{pmatrix} \dfrac{1}{4}\begin{pmatrix} \dfrac{1}{4} & \dfrac{1}{4} \\ \dfrac{e^{i\frac{\pi}{2}}}{\sqrt{2}} & \dfrac{e^{-i\frac{\pi}{2}}}{\sqrt{2}} \end{pmatrix} & \dfrac{1}{4}\begin{pmatrix} \dfrac{1}{4} & \dfrac{1}{4} \\ \dfrac{e^{i\frac{\pi}{2}}}{\sqrt{2}} & \dfrac{e^{-i\frac{\pi}{2}}}{\sqrt{2}} \end{pmatrix} \\ \dfrac{e^{-i\frac{\pi}{2}}}{\sqrt{2}}\begin{pmatrix} \dfrac{1}{4} & \dfrac{1}{4} \\ \dfrac{e^{i\frac{\pi}{2}}}{\sqrt{2}} & \dfrac{e^{-i\frac{\pi}{2}}}{\sqrt{2}} \end{pmatrix} & \dfrac{e^{i\frac{\pi}{2}}}{\sqrt{2}}\begin{pmatrix} \dfrac{1}{4} & \dfrac{1}{4} \\ \dfrac{e^{i\frac{\pi}{2}}}{\sqrt{2}} & \dfrac{e^{-i\frac{\pi}{2}}}{\sqrt{2}} \end{pmatrix} \end{pmatrix}, \quad (6.104)$$

which contracts to

$$(\mathbf{J}^* \otimes \mathbf{J}) = \begin{pmatrix} \left(\dfrac{1}{4}\right)^2 & \left(\dfrac{1}{4}\right)^2 & \left(\dfrac{1}{4}\right)^2 & \left(\dfrac{1}{4}\right)^2 \\ \dfrac{e^{i\frac{\pi}{2}}}{4\sqrt{2}} & \dfrac{e^{-i\frac{\pi}{2}}}{4\sqrt{2}} & \dfrac{e^{i\frac{\pi}{2}}}{4\sqrt{2}} & \dfrac{e^{-i\frac{\pi}{2}}}{4\sqrt{2}} \\ \dfrac{e^{-i\frac{\pi}{2}}}{4\sqrt{2}} & \dfrac{e^{-i\frac{\pi}{2}}}{4\sqrt{2}} & \dfrac{e^{i\frac{\pi}{2}}}{4\sqrt{2}} & \dfrac{e^{i\frac{\pi}{2}}}{4\sqrt{2}} \\ \dfrac{e^{-i\frac{\pi}{2}}}{\sqrt{2}}\dfrac{e^{i\frac{\pi}{2}}}{\sqrt{2}} & \dfrac{e^{-i\frac{\pi}{2}}}{\sqrt{2}}\dfrac{e^{-i\frac{\pi}{2}}}{\sqrt{2}} & \dfrac{e^{i\frac{\pi}{2}}}{\sqrt{2}}\dfrac{e^{i\frac{\pi}{2}}}{\sqrt{2}} & \dfrac{e^{i\frac{\pi}{2}}}{\sqrt{2}}\dfrac{e^{-i\frac{\pi}{2}}}{\sqrt{2}} \end{pmatrix}$$

$$= \begin{pmatrix} \dfrac{1}{16} & \dfrac{1}{16} & \dfrac{1}{16} & \dfrac{1}{16} \\ \dfrac{i}{4\sqrt{2}} & -\dfrac{i}{4\sqrt{2}} & \dfrac{i}{4\sqrt{2}} & -\dfrac{i}{4\sqrt{2}} \\ -\dfrac{i}{4\sqrt{2}} & -\dfrac{i}{4\sqrt{2}} & \dfrac{i}{4\sqrt{2}} & \dfrac{i}{4\sqrt{2}} \\ \dfrac{1}{2} & \dfrac{-1}{2} & \dfrac{-1}{2} & \dfrac{1}{2} \end{pmatrix}. \qquad (6.105)$$

The corresponding Mueller matrix is

$$\mathbf{M} = \mathbf{U} \cdot (\mathbf{J}^* \otimes \mathbf{J}) \cdot \mathbf{U}^{-1} = \begin{pmatrix} \dfrac{9}{16} & 0 & -\dfrac{7}{16} & 0 \\ -\dfrac{7}{16} & 0 & \dfrac{9}{16} & 0 \\ 0 & 0 & 0 & -\dfrac{1}{2\sqrt{2}} \\ 0 & -\dfrac{1}{2\sqrt{2}} & 0 & 0 \end{pmatrix}. \qquad (6.106)$$

6.12.2 Conversion of Jones Matrices to Mueller Matrices Using Pauli Matrices

An equivalent method to convert Jones matrices to Mueller matrices utilizes dot products with two Pauli matrices to determine each Mueller matrix element, $M_{i,j}$,

$$M_{ij} = \frac{1}{2}\text{Tr}(\boldsymbol{\sigma}_i \cdot \mathbf{J}^* \cdot \boldsymbol{\sigma}_j \cdot \mathbf{J}^T), \tag{6.107}$$

where Tr is the trace of the matrix and

$$\boldsymbol{\sigma}_0 = \begin{pmatrix} 1 & 0 \\ 0 & 1 \end{pmatrix}, \quad \boldsymbol{\sigma}_1 = \begin{pmatrix} 1 & 0 \\ 0 & -1 \end{pmatrix}, \quad \boldsymbol{\sigma}_2 = \begin{pmatrix} 0 & 1 \\ 1 & 0 \end{pmatrix}, \quad \text{and} \quad \boldsymbol{\sigma}_3 = \begin{pmatrix} 0 & -i \\ i & 0 \end{pmatrix} \tag{6.108}$$

are the identity matrix and Pauli matrices.

Example 6.12 Calculation of a Component in the Jones Matrix

Calculation of the M_{13} component from the Jones matrix in Equation 6.100.

$$\begin{aligned} M_{13} &= \frac{1}{2}\text{Tr}(\boldsymbol{\sigma}_1 \cdot \mathbf{J}^* \cdot \boldsymbol{\sigma}_3 \cdot \mathbf{J}^T) \\ &= \frac{1}{2}\text{Tr}\left(\begin{pmatrix} 1 & 0 \\ 0 & -1 \end{pmatrix} \cdot \begin{pmatrix} \rho_{xx}e^{i\phi_{xx}} & \rho_{xy}e^{i\phi_{xy}} \\ \rho_{yx}e^{i\phi_{yx}} & \rho_{yy}e^{i\phi_{yy}} \end{pmatrix} \cdot \begin{pmatrix} 0 & -i \\ i & 0 \end{pmatrix} \cdot \begin{pmatrix} \rho_{xx}e^{-i\phi_{xx}} & \rho_{yx}e^{-i\phi_{yx}} \\ \rho_{xy}e^{-i\phi_{xy}} & \rho_{yy}e^{-i\phi_{yy}} \end{pmatrix} \right) \\ &= \rho_{xx}\rho_{xy}\sin(\phi_{xx} - \phi_{xy}) + \rho_{yx}\rho_{yy}\sin(\phi_{yy} - \phi_{yx}), \end{aligned} \tag{6.109}$$

which matches the result from the first method in Equation 6.102.

6.12.3 Transforming Mueller Matrices into Jones Matrices

Non-depolarizing Mueller matrices are transformed into the equivalent Jones matrices using the following relations:

$$\mathbf{J} = \begin{pmatrix} j_{xx} & j_{xy} \\ j_{yx} & j_{yy} \end{pmatrix} = \begin{pmatrix} \rho_{xx}e^{-i\phi_{xx}} & \rho_{xy}e^{-i\phi_{xy}} \\ \rho_{yx}e^{-i\phi_{yx}} & \rho_{yy}e^{-i\phi_{yy}} \end{pmatrix}, \tag{6.110}$$

where the amplitudes are

$$\rho_{xx} = \sqrt{\frac{M_{00} + M_{01} + M_{10} + M_{11}}{2}}, \quad \rho_{xy} = \sqrt{\frac{M_{00} - M_{01} + M_{10} - M_{11}}{2}},$$
$$\rho_{yx} = \sqrt{\frac{M_{00} + M_{01} - M_{10} - M_{11}}{2}}, \quad \rho_{yy} = \sqrt{\frac{M_{00} - M_{01} - M_{10} + M_{11}}{2}},$$
(6.111)

and the relative phases are

$$\begin{cases} \phi_{xx} - \phi_{xy} = \tan^{-1}\left(\dfrac{M_{03} + M_{13}}{M_{02} + M_{12}}\right), \\ \phi_{yx} - \phi_{xx} = \tan^{-1}\left(\dfrac{M_{30} + M_{31}}{M_{20} + M_{21}}\right), \\ \phi_{yy} - \phi_{xx} = \tan^{-1}\left(\dfrac{M_{32} - M_{23}}{M_{22} + M_{33}}\right). \end{cases}$$
(6.112)

The phase ϕ_{xx} is not determined and is the "reference phase" for the other ϕ. Because of the large number of Mueller matrix elements, and the constraints between non-depolarizing elements, these equations are not unique.

A special case occurs when $j_{xx} = 0$; both the numerator and denominator of the $\tan^{-1}$ are 0 and the phase equations in Equation 6.112 fail. The transformation equations can be recast in closely related forms and use the phase of another Jones matrix element as the "reference phase."

Example 6.13 Cascading Two Linear Diattenuators with a Rotation

Two identical linear diattenuators (partial linear polarizers) have $T_{max} = 1.0$ and $T_{min} = 0.707$. Find the Mueller matrix $\mathbf{M}(\theta)$ as the transmission axis of the second diattenuator is rotated.

a. What is the maximum and minimum diattenuation as a function of θ?
b. Do any of the combinations form elliptical diattenuators?
c. What is the polarizance as a function of θ?

The mathematical expressions are relatively complex, so it only makes sense to work this problem using a calculator or computer software.

6.12 Relating Jones and Mueller Matrices

a. By multiplying the two linear diattenuator expressions, it yields the complex expression

$$\mathbf{M}(\theta) = \mathbf{LD}(T_{max}, T_{min}, \theta) \cdot \mathbf{LD}(T_{max}, T_{min}, 0)$$

$$= \frac{1}{2}\begin{pmatrix} A & B\cos 2\theta & B\sin 2\theta & 0 \\ B\cos 2\theta & A\cos^2 2\theta + C\sin^2 2\theta & A - C\cos 2\theta \sin 2\theta & 0 \\ B\sin 2\theta & (A-C)\cos 2\theta \sin 2\theta & C\cos^2 2\theta + A\sin^2 2\theta & 0 \\ 0 & 0 & 0 & C \end{pmatrix} \frac{1}{2}\begin{pmatrix} A & B & 0 & 0 \\ B & A & 0 & 0 \\ 0 & 0 & C & 0 \\ 0 & 0 & 0 & C \end{pmatrix}$$

$$= \frac{1}{4}\begin{pmatrix} A^2 + B^2 \cos 2\theta & 2AB\cos^2 \theta & BC\sin 2\theta & 0 \\ 2B\cos^2 \theta(C + (A-C)\cos 2\theta) & B^2\cos 2\theta + A^2\cos^2 2\theta + AC\sin^2 2\theta & \frac{1}{2}(A - 2C)\sin 4\theta & 0 \\ B(A + (A-C)\cos 2\theta)\sin 2\theta & (B^2 + A(A-C)\cos 2\theta)\sin 2\theta & C(C\cos^2 2\theta + A\sin^2 2\theta) & 0 \\ 0 & 0 & 0 & C^2 \end{pmatrix}$$

(6.113)

where

$$A = T_{max} + T_{min}, \quad B = T_{max} - T_{min}, \quad C = 2\sqrt{T_{max} T_{min}}. \tag{6.114}$$

Substituting $T_{max} = 1.0$ and $T_{min} = 0.707 = 1/\sqrt{2}$ yields expressions for the 16 elements:

$$\mathbf{M} = \begin{pmatrix} M_{00} \\ M_{01} \\ M_{02} \\ M_{03} \\ M_{10} \\ M_{11} \\ M_{12} \\ M_{13} \\ M_{20} \\ M_{21} \\ M_{22} \\ M_{23} \\ M_{30} \\ M_{31} \\ M_{32} \\ M_{33} \end{pmatrix} = \frac{1}{8}\begin{pmatrix} 3 + 2\sqrt{2} + (3 - 2\sqrt{2})\cos 2\theta \\ 2\cos^2 \theta \\ -2^{3/4}\left(-2 + \sqrt{2}\right)\sin 2\theta \\ 0 \\ \cos 2\theta + \cos^2 2\theta + 2 \cdot 2^{1/4}\left(-1 + \sqrt{2}\right)\sin^2 2\theta \\ (3 - 2\sqrt{2})\cos 2\theta + (3 + 2\sqrt{2})\cos^2 2\theta + 2 \cdot 2^{1/4}\left(1 + \sqrt{2}\right)\sin^2 2\theta \\ -\dfrac{\left(-2\sqrt{2} + 2 \cdot 2^{3/4}\right)\sin 4\theta}{2^{1/4}} \\ 0 \\ (1 + (1 + 2 \cdot 2^{1/4} - 2 \cdot 2^{3/4})\cos 2\theta)\sin 2\theta \\ -\left(-3 + 2\sqrt{2} + \left(-3 + 2 \cdot 2^{1/4} - 2\sqrt{2} + 2 \cdot 2^{3/4}\right)\cos 2\theta\right)\sin 2\theta \\ 2 \cdot 2^{3/4}\left(2^{3/4}\cos^2 2\theta + \left(1 + \dfrac{1}{\sqrt{2}}\right)\sin^2 2\theta\right) \\ 0 \\ 0 \\ 0 \\ 0 \\ 4\dfrac{1}{\sqrt{2}} \end{pmatrix} \tag{6.115}$$

b. The diattenuation of an individual plate is

$$D = \frac{1-\frac{1}{\sqrt{2}}}{1+\frac{1}{\sqrt{2}}} = \frac{\sqrt{2}-1}{\sqrt{2}+1} = 3-2\sqrt{2} \approx 0.1716, \qquad (6.116)$$

a rather weak partial polarizer. The maximum diattenuation occurs when the transmission axes of the two plates are parallel.

$$\begin{aligned} T_{max} &= 1 \times 1 = 1 \text{ and } T_{min} = 0.7 \times 0.7 = 1/2, \\ D_{max} &= (1-1/2)/(1+1/2) = 1/3. \end{aligned} \qquad (6.117)$$

Likewise, the minimum diattenuation occurs when the transmission axes are perpendicular and the diattenuations cancel,

$$\begin{aligned} T_{max} &= 1 \times 0.7 = 0.7 \text{ and } T_{min} = 0.7 \times 1 = 0.7, \\ D_{min} &= (0.7-0.7)/(0.7+0.7) = 0. \end{aligned} \qquad (6.118)$$

c. Since $M_{03} = 0$, the diattenuation is always linear.
d. The polarizance $P(\theta)$ is purely linear since $M_{30} = 0$, and is graphed in Figure 6.24.

$$\begin{aligned} P(\theta) &= \frac{\sqrt{M_{10}^2 + M_{20}^2}}{M_{00}} = \frac{\sqrt{\left(\cos 2\theta + \cos^2 2\theta + 2^{1/4}\left(-1+\sqrt{2}\right)\sin^2 2\theta\right)^2 + \left(\left(1+\left(1+2^{1/4}-2^{3/4}\right)\cos 2\theta\right)\sin 2\theta\right)^2}}{32\left(\frac{1}{4}\left(1+\frac{1}{\sqrt{2}}\right)^2 + \frac{1}{4}\left(1-\frac{1}{\sqrt{2}}\right)^2 \cos 2\theta\right)} \\ &= \frac{\sqrt{-13+12\sqrt{2}+4\cos 2\theta + (17-12\sqrt{2})\cos 4\theta}}{8\sqrt{2}\left(\frac{1}{4}\left(1+\frac{1}{\sqrt{2}}\right)^2 + \frac{1}{4}\left(1-\frac{1}{\sqrt{2}}\right)^2 \cos 2\theta\right)}. \end{aligned} \qquad (6.119)$$

When unpolarized light is incident, the exiting light is unpolarized for θ = 90° + n 180°.

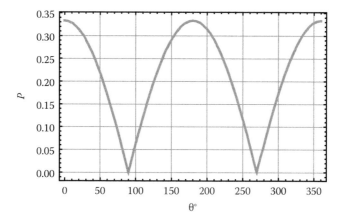

Figure 6.24 Polarizance of the diattenuator combination as the second element is rotated.

6.13 Ray Tracing with Mueller Matrices

Mueller matrices are frequently used for ray tracing, particularly *incoherent ray tracing*. Jones matrices are preferred for ray tracing imaging systems because they contain an absolute phase. Many systems need an incoherent ray trace, such as illumination optics and simulations of scattering systems. In illumination systems, for example, automobile headlamps that are faceted into many lenslets, light rays reach each part of the illuminated surface from many different paths with very different optical path lengths. The interference does not need to be calculated between these different rays, since the quality of a spherical wavefront is not being evaluated. Thus, summing Mueller matrices or Stokes parameters is appropriate. Similarly, in simulating light propagating through aerosols in the atmosphere, animal tissue, turbid media, or scattered light calculations in optical systems, the flux, direction, and polarization state are needed, but the optical path length is not used for determining interference. For these purposes, Mueller matrices are commonly used to describe these interactions.

The *s*-component leaving one interaction is usually not aligned with the *s*-component of the next interaction, which often happens with skew rays; Mueller rotation operations (Equation 6.11) need to be applied at each of these interactions. Each ray will generally have a different geometrical transformation, which is explored in Chapter 17, and this will need to be addressed to combine the Mueller matrices from different ray paths, which, for example, combine at a detector pixel.

The Mueller matrices for scattering can also be used in a Mueller matrix ray trace. There are many different models for the Mueller matrices for scattering from various surfaces or volume scattering.[22,23] These Mueller matrix calculations will not be described here. A library of Mueller matrices for scattering is available from the National Institute of Standards and Technology.[24]

6.13.1 Mueller Matrices for Refraction

Reflections and refractions at homogenous and isotropic interfaces (e.g., typical glass or metal interfaces) have s- and p-eigenpolarizations. The associated polarization properties are a combination of diattenuation and retardance, whose axes are aligned with the s- and p-planes. Since the light changes direction, a different coordinate system is needed to represent the incident and exiting Stokes parameters in global coordinates.

In what follows, s is aligned with $+S_1$ and p is aligned with $-S_1$. T_s is the s-intensity transmittance and T_p is the p-intensity transmittance. The retardance between the s- and p-states is δ. T_s, T_p, and δ are determined from Fresnel equations (Chapter 8) or from a thin film coating calculation such as Sections 13.2 and 13.3.1.[4] The Mueller matrix representing refraction is the product of the diattenuator and retarder Mueller matrices,

$$\mathbf{LDR}(D,\delta,0) = \frac{1}{2}\begin{pmatrix} T_s+T_p & T_s-T_p & 0 & 0 \\ T_s-T_p & T_s+T_p & 0 & 0 \\ 0 & 0 & 2\sqrt{T_sT_p}\cos\delta & 2\sqrt{T_sT_p}\sin\delta \\ 0 & 0 & -2\sqrt{T_sT_p}\sin\delta & 2\sqrt{T_sT_p}\cos\delta \end{pmatrix}$$

$$= \frac{T_s+T_p}{2}\begin{pmatrix} 1 & D & 0 & 0 \\ D & 1 & 0 & 0 \\ 0 & 0 & \sqrt{1-D^2}\cos\delta & \sqrt{1-D^2}\sin\delta \\ 0 & 0 & -\sqrt{1-D^2}\sin\delta & \sqrt{1-D^2}\cos\delta \end{pmatrix},$$

(6.120)

where D is the diattenuation. For transmission at an uncoated interface, the retardance δ is 0. For thin film-coated interfaces, such as anti-reflection coatings or beam splitter coatings, the retardance is non-zero. Refraction Mueller matrices are homogeneous; the eigenpolarizations, the s- and p-polarizations, are orthogonal.

6.13.2 Mueller Matrices for Reflection

In most optics notation, including this chapter, a sign change occurs in the coordinate system after reflection to maintain a right-handed coordinate system for changing direction of the propagation vector. Imagine a Stokes polarimeter measuring in transmission. Now, to measure in reflection, the polarimeter rotates around about the vertical y-axis, perpendicular to the z-axis along which the incident light propagates. By moving the polarimeter around the y-axis to measure the beam in reflection, it is seen the 45° component has changed sign. In addition, the helicity (i.e., handedness) of all circular and elliptical states also changes sign upon the reflection, since right circular polarization reflects as left circular polarization, and vice versa.

After reflection, the S_2 component of Stokes parameters (linearly polarized light at 45°/135°) and the S_3 component (circularly polarized light) change sign. The S_2 component changes sign during reflection (diffuse or specular) because the z-component of the light propagation vector (the component parallel to the sample surface normal) changes sign. To maintain a right-handed coordinate system, one of the transverse coordinates must change sign as well. Choosing x, then spatial coordinates (x, y, z) switch to $(-x, y, -z)$ after reflection or backscatter from a sample; z is the direction of propagation before reflection that changes to $-z$ after reflection. The change of coordinates dictates that a beam polarized at an angle of 45° that reflects polarized in the same global plane is described as having a 135° orientation in the coordinates after reflection.

R_s is the s-intensity reflectance, R_p is the p-intensity reflectance, and δ is the retardance as calculated from Fresnel equations or thin film equations. The Mueller matrix for reflection is

$$\mathbf{M}_{\text{refl}}(R_s, R_p, \delta) = \frac{1}{2} \begin{pmatrix} R_s + R_p & R_s - R_p & 0 & 0 \\ R_s - R_p & R_s + R_p & 0 & 0 \\ 0 & 0 & -2\sqrt{R_s R_p}\cos\delta & -2\sqrt{R_s R_p}\sin\delta \\ 0 & 0 & 2\sqrt{R_s R_p}\sin\delta & -2\sqrt{R_s R_p}\cos\delta \end{pmatrix}. \quad (6.121)$$

With this convention for reflection, the equation for rotating the Mueller matrix, **M**, of a sample measured by a polarimeter in a reflection configuration about its normal changes to

$$\mathbf{M}_{\text{refl}}(\theta) = \mathbf{R}(\theta) \cdot \mathbf{M}_{\text{refl}} \cdot \mathbf{R}(\theta)$$

$$\begin{pmatrix} 1 & 0 & 0 & 0 \\ 0 & \cos\theta & -\sin\theta & 0 \\ 0 & \sin\theta & \cos\theta & 0 \\ 0 & 0 & 0 & 1 \end{pmatrix} \cdot \begin{pmatrix} M_{00} & M_{01} & M_{02} & M_{03} \\ M_{10} & M_{11} & M_{12} & M_{13} \\ M_{20} & M_{21} & M_{22} & M_{23} \\ M_{30} & M_{31} & M_{32} & M_{33} \end{pmatrix} \cdot \begin{pmatrix} 1 & 0 & 0 & 0 \\ 0 & \cos\theta & -\sin\theta & 0 \\ 0 & \sin\theta & \cos\theta & 0 \\ 0 & 0 & 0 & 1 \end{pmatrix}, \quad (6.122)$$

compared to Equation 6.11 for Mueller matrices in transmission. For example, the Mueller matrix of a transmission polarizer with its transmission axis oriented at 20° and the Mueller matrix of a reflection polarizer oriented at 20° (for the incident light) are different since polarized light exits the reflection polarizer oriented at −20° in the reflection coordinates (20° in the incident coordinates). In essence, the reflection polarizer Mueller matrix is analyzing at 20° but polarizing at −20°. For the special cases of linear polarizer matrices oriented at 0° or 90° and linear retarders oriented at 0° or 90°, this transformation results in the same Mueller matrices for transmission and reflection.

The normalized reflection Mueller matrices for weakly polarizing reflecting samples, those with diattenuation, retardance, and depolarization close to zero, are close to the Mueller matrix for an ideal reflector,

$$\mathbf{M}_{\text{refl}} = \begin{pmatrix} 1 & 0 & 0 & 0 \\ 0 & 1 & 0 & 0 \\ 0 & 0 & -1 & 0 \\ 0 & 0 & 0 & -1 \end{pmatrix}. \quad (6.123)$$

$\mathbf{M}_{\text{refl}}$ is also the Mueller matrix for reflection or scatter in the absence of polarization effects.

6.14 The Origins of the Mueller Matrix

Hans Müller, a Swiss-born professor of physics, developed the Mueller matrix concept in the early 1940s[4] and has the Mueller matrix named in his honor. He described it in a classified report,[25] later declassified, and in detail in notes for his physics course in 1943.[26] His only publication related to the 4 × 4 matrix that bears his name was a short meeting abstract for the Optical Society of America.[27] His graduate student Nathan Parke developed the Mueller matrix concept further in his dissertation[28] and a related publication.[1] R. Clark Jones then compared the Mueller and Jones matrices in one of his series of papers on the Jones calculus[29] and made reference to the "recently declassified" report that Müller had authored.

Before all of this work in 1929, Paul Soleillet had developed a set of four linear equations to relate incident and exiting Stokes parameters, equivalent to the Mueller matrix except that a matrix formalism was not employed.[30] These linear equations were also placed into a matrix formalism by Francis Perrin in 1942.[31]

6.15 Problem Sets

6.1 Show that the following Mueller matrices are equal:
   a. **LR**(δ, 45°) and **LR**($2\pi - \delta$, 135°)
   b. **LP**(θ) and **LP**($\theta + \pi$)
   c. **LD**(1, t, $\pi/8$) and **LD**(1, t, $9\pi/8$)

6.2 a. Find the Mueller matrices for linear polarizers at 0°, 45°, and 90°: **LP**(0), **LP**($\pi/4$), and **LP**($\pi/2$).
   b. Calculate **LP**(0)·**LP**($\pi/2$) and **LP**(0)·**LP**($\pi/4$)·**LP**($\pi/2$).
   c. Find the Mueller matrices for quarter wave linear retarders with fast axes at 0°, 45°, and 90°: **LR**($\pi/2$, 0), **LR**($\pi/2$, $\pi/4$), and **LR**($\pi/2$, $\pi/2$).
   d. Do **LP**(0) and **LR**($\pi/2$, 0) commute, that is, is **LP**(0)·**LR**($\pi/2$, 0) = **LR**($\pi/2$, 0)·**LP**(0) true?
   e. Do **LP**(0) and **LR**($\pi/2$, $\pi/4$) commute?

6.3 Find **LR**($\pi/2$, 0)·**LP**($\pi/4$)·**LR**($\pi/2$, $\pi/2$). Show this is a circular polarizer.

6.4 Show that the Mueller matrix for a quarter wave linear retarder with fast axis orientation θ equals the Mueller matrix for a three-quarter wave linear retarder with fast axis orientation $\theta \pm \pi/2$.

6.5 Show that the diattenuator Mueller matrix equation (Equation 6.59) reduces to the linear polarizer equation (Equation 6.37) when $T_{max} = 1$ and $T_{min} = 0$.

6.6 Analyze the properties of the Mueller matrix $\mathbf{M} = \begin{pmatrix} 1 & 0 & 0 & 0 \\ 0 & 0 & 1 & 0 \\ 0 & 1 & 0 & 0 \\ 0 & 0 & 0 & -1 \end{pmatrix}$.
   a. Identify it as a polarizer or retarder Mueller matrix.
   b. How does the transmitted flux vary with the incident Stokes parameters?
   c. What is the action on circularly polarized light?
   d. Plot the action on linearly polarized light as the axis changes.
   Use the eigenvalues and associated eigenvectors in your description.

6.7 Show that the elliptically polarized Stokes parameters $\mathbf{S}(\theta, \eta) = (1, \cos 2\theta \cos \eta, \sin 2\theta \cos \eta, \sin \eta)$ is an eigenpolarization of **EP**(θ, η).

6.8 Calculate the Mueller matrix **EP**(θ, η) for an elliptical polarizer in Equation 6.38 that transmits the Stokes parameters (1, $\cos 2\theta \cos \eta$, $\sin 2\theta \cos \eta$, $\sin \eta$).
   a. Find the linear retarder Mueller matrix **U** that transforms the linearly polarized state (1, $\cos 2\theta$, $\sin 2\theta$, 0) into the target state (1, $\cos 2\theta \cos \eta$, $\sin 2\theta \cos \eta$, $\sin \eta$).
   b. Starting with the linear polarizer Mueller matrix **LP**(θ), apply a unitary transform using **U** to transform **LP**(θ) into **EP**(θ, η).
   c. Verify the eigenvalues and eigenvectors of **EP**(θ, η).

6.15 Problem Sets

6.9 a. How are the eigenvalues of an ideal circular retarder Mueller matrix equation (Equation 6.24) related to the retardance δ?

b. How are the eigenvalues of a linear retarder Mueller matrix related to its retardance δ?

6.10 Multiply the Mueller matrices for an $\mathbf{LR}(\pi/2, 0)$, followed by an $\mathbf{LR}(\pi/2, \pi/4)$, and finally a $\mathbf{CR}(\pi/2)$. Determine the retardance and the eigenpolarizations.

6.11 a. Describe the set of all Mueller matrices that transform arbitrary incident Stokes parameters into unpolarized light. Which elements must be zero? Which elements can be non-zero?

b. How many degrees of freedom does this set have?

c. Which of these Mueller matrices completely block certain incident states of polarization?

6.12 Consider the partially polarized beam with the Stokes parameters $\mathbf{S} = (1, s_1, s_2, s_3)$.

a. Which diattenuator Mueller matrix $\mathbf{M}_D$ transforms $\mathbf{S}$ into unpolarized light?

b. Explain why this diattenuator can reduce, not increase, the degree of polarization.

c. What is the relationship between the degree of polarization of $\mathbf{S}$ and the diattenuation of $\mathbf{M}_D$?

d. What is the effect of $\mathbf{M}_D$ on the orthogonal polarization state with the same *DoP*?

6.13 What is the periodicity in orientation θ of the equation for the Mueller matrix of a half wave linear retarder in Equation 6.25?

6.14 Show that the Mueller matrix for a half wave elliptical retarder with retardance components $(\delta_H, \delta_{45}, \delta_R)$ in Equation 6.28,

$$\mathbf{HWR}(r_1, r_2, r_3) = \begin{pmatrix} 1 & 0 & 0 & 0 \\ 0 & -1+2r_1^2 & 2r_2 r_1 & 2r_1 r_3 \\ 0 & 2r_2 r_1 & -1+2r_2^2 & 2r_2 r_3 \\ 0 & 2r_3 r_1 & 2r_2 r_3 & -1+2r_3^2 \end{pmatrix},$$

where $\sqrt{r_1^2 + r_2^2 + r_3^2} = 1$, $r_1 = \dfrac{\delta_H}{\pi}$, $r_2 = \dfrac{\delta_{45}}{\pi}$, and $r_3 = \dfrac{\delta_R}{\pi}$, is the same Mueller matrix as for an elliptical retarder with orthogonal retardance components $(-\delta_H, -\delta_{45}, -\delta_R)$.

6.15 Develop a set of equations for transforming Mueller matrices into Jones matrices for the special case where Jones matrix element $j_{xx} = 0$.

6.16 Using the equation for the general elliptical retarder in Equation 6.28, show that the following two Mueller matrices are equal, $\mathbf{ER}(\delta_H, \delta_{45}, \delta_R) = \mathbf{ER}\left(\dfrac{2\pi(\delta_H, \delta_{45}, \delta_R)}{\sqrt{\delta_H^2 + \delta_{45}^2 + \delta_R^2}} - (\delta_H, \delta_{45}, \delta_R) \right)$.

6.17 Given a polarized beam with major axis orientation θ and latitude on the Poincaré sphere of η, find the orientations for a half wave linear retarder that will convert the state into the orthogonal state.

6.18 Create an example of a retarder Mueller matrix multiplying a set of Stokes parameters that produces an arc of states propagating through a retarder on the Poincaré sphere following the left-hand rule. With your left thumb aligned along the fast axis of a retarder emerging from the Poincaré sphere, the motion of states around the rotation axis follows the direction of the left-hand fingers.

6.19 Using the Poincaré sphere, deduce all of the retarders that will convert horizontal linearly polarized light into 45° linearly polarized light. What is the minimum retardance that will perform the transformation? How is the fast axis oriented?

6.20 A linear retarder with retardance 90° with fast axis oriented at 45° is placed between a horizontal quarter wave retarder and a vertical quarter wave linear retarder. The assembly forms a circular retarder. What is the retardance of the assembly? How do the properties of the assembly vary as the fast axis of the central quarter wave retarder is rotated from 0° to 180°?

6.21 a. Show by reasoning with the Poincaré sphere that, when a linear retarder with retardance δ oriented at 45° is placed between a horizontal quarter wave linear retarder and a vertical quarter wave linear retarder, the assembly forms a circular retarder.
   b. What is the retardance of the retarder? Explain the magnitude of the retardance using the Poincaré sphere.
   c. Work the problem with Jones matrices, and show that this sequence performs a unitary transformation on the middle retarder.

6.22 Consider a linear diattenuator $\mathbf{LD}(T_{max}, T_{min}, \pi/4)$. What is the condition on the incident and exiting Stokes parameters such that the entering and exiting degrees of polarization are equal? Explain how a non-polarizing element, the diattenuator, can decrease the degree of polarization of certain input states.

6.23 For a partially linear polarized beam with Stokes parameters, $(S_0, S_1, S_2, 0)$ where $S_0 > \sqrt{S_1^2 + S_2^2}$, show that the maximum amount of light that can be transmitted through an ideal polarizer is more than the polarized flux $\sqrt{S_1^2 + S_2^2}$. How much more flux can be transmitted?

6.24 Determine the linear diattenuator that converts the state $(4, 1, 1, 0)$ into unpolarized light. Let $T_{max} = 1$ for simplicity. Do diattenuators increase the degree of polarization? Explain why the degree of polarization is reduced here.

6.25 Derive the Mueller matrix $\mathbf{LR}(\delta,0)$ for a linear retarder with a retardance δ radians and a fast axis orientation at 0. Set up and solve a set of at least 16 linear equations for the matrix elements M_{ij}. Each set of four equations is of the form

$$\mathbf{LR}(\delta,0°) \cdot \mathbf{S} = \mathbf{S'} = \begin{pmatrix} M_{00} & M_{01} & M_{02} & M_{03} \\ M_{10} & M_{11} & M_{12} & M_{13} \\ M_{20} & M_{21} & M_{22} & M_{23} \\ M_{30} & M_{31} & M_{32} & M_{33} \end{pmatrix} \begin{pmatrix} S_0 \\ S_1 \\ S_2 \\ S_3 \end{pmatrix} = \begin{pmatrix} S'_0 \\ S'_1 \\ S'_2 \\ S'_3 \end{pmatrix},$$

and should relate appropriate incident and exiting Stokes parameters. Jones calculus is needed to generate incident and exiting pairs of states with the needed S_2 and S_3 components.

6.26 Calculate and compare the depolarization index and the average degree of polarization for the family of diatteunator Mueller matrices $\mathbf{Ms}$ as a function of s, t, and u, where

$$\mathbf{Ms} = \begin{pmatrix} 1 & s & t & u \\ 0 & 0 & 0 & 0 \\ 0 & 0 & 0 & 0 \\ 0 & 0 & 0 & 0 \end{pmatrix}.$$

6.15 Problem Sets

6.27 a. Calculate the depolarization index for the family of Mueller matrices $\mathbf{M}\alpha$ connecting the ideal depolarizer and the identity matrix as a function of the mixing ratio α

$$\mathbf{M}\alpha = \alpha \mathbf{I} + (1-\alpha)\mathbf{ID} = \begin{pmatrix} 1 & 0 & 0 & 0 \\ 0 & \alpha & 0 & 0 \\ 0 & 0 & \alpha & 0 \\ 0 & 0 & 0 & \alpha \end{pmatrix}.$$

b. Calculate the average degree of polarization for the family of Mueller matrices $\mathbf{M}\alpha$ connecting the ideal depolarizer and the identity matrix as a function of the mixing ratio α.

c. Calculate the depolarization index for the family of partial depolarizers $\mathbf{M}\alpha\gamma$ that depolarize linearly polarized light to a DoP of α and circularly polarized light to γ, as a function of α and γ, where

$$\mathbf{M}\alpha\gamma = \begin{pmatrix} 1 & 0 & 0 & 0 \\ 0 & \alpha & 0 & 0 \\ 0 & 0 & \alpha & 0 \\ 0 & 0 & 0 & \gamma \end{pmatrix}.$$

d. Calculate the average degree of polarization for $\mathbf{M}\alpha\gamma$ as a function of the mixing ratio α.

e. For which values of α and γ are the depolarization index and the average degree of polarization equal?

f. Which is greater, the depolarization index or the average degree of polarization?

g. In the neighborhood of the ideal depolarizer, which varies linearly and which varies quadratically?

6.28 Which forms of depolarization are expected due to the following polarization element defects? Add or integrate Mueller matrices to evaluate the following properties:

a. A crystalline quarter wave linear retarder with a 45° fast axis has a wedge, a thickness variation. For example, add 89° and 91° retarders, or integrate from 89° to 91°.

b. A polymer linear retarder with a 0° fast axis has a fast axis variation about 0° due to the stretching process.

c. A sheet polarizer with a 45° transmission axis has transmission axis variation about 45° due to the stretching process.

d. A sheet polarizer with a 90° transmission axis has diattenuation variations due to thickness variations.

e. A sheet polarizer with a 45° transmission axis has transmission axis variation about 45° due to the stretching process.

6.29 A system in production needs a quarter wave linear retarder to convert right circularly polarized light into horizontal linearly polarized light. The final polarization state should lie within a 0.03 radian circle of horizontal linearly polarized light. Use the Poincaré sphere to tolerate the allowable error in retarder magnitude and orientation.

References

1. N. G. Parke, Optical algebra, *J. Math. Phys.* 28.1 (1949): 131–139.
2. R. A. Chipman, Mueller matrices, in *Handbook of Optics*, Vol. I, Chapter 14, ed. M. Bass, McGraw-Hill (2009).
3. P. W. Maymon and R. A. Chipman, *Linear Polarization Sensitivity Specifications for Spaceborne Instruments*, San Diego International Society for Optics and Photonics (1992).
4. W. A. Shurcliff, *Polarized Light, Production and Use*, Cambridge, MA: Harvard University Press (1962).
5. C. Brosseau, *Fundamentals of Polarized Light: A Statistical Optics Approach*, Wiley-Interscience (1998), 228 pp.
6. J. J. Gil and E. Bernabeu, Obtainment of the polarizing and retardation parameters of a non-depolarizing optical system from the polar decomposition of its Mueller matrix, *Optik* 76.2 (1987): 67–71.
7. D. Brewster, Experiments on the depolarisation of light as exhibited by various mineral, animal, and vegetable bodies, with a reference of the phenomena to the general principles of polarization, *Philos. Trans. R. Soc. London* 105 (1815): 29–53.
8. J. Christopher Dainty, ed. *Laser Speckle and Related Phenomena*, Vol. 9, Springer Science & Business Media (2013).
9. R. Frieden, *Probability, Statistical Optics, and Data Testing: A Problem Solving Approach*, Vol. 10, Springer Science & Business Media (2012).
10. J. J. Gil, and E. Bernabeu, A depolarization criterion in Mueller matrices, *J. Mod. Opt.* 32.3 (1985): 259–261.
11. J. J. Gil and E. Bernabeu, Depolarization and polarization indices of an optical system, *J. Mod. Opt.* 33.2 (1986): 185–189.
12. R. A. Chipman, Depolarization index and the average degree of polarization, *Appl. Opt.* 44.13 (2005): 2490–2495.
13. B. DeBoo, J. Sasian, and R. Chipman, Degree of polarization surfaces and maps for analysis of depolarization, *Opt. Express*, 12(20) (2004): 4941–4958.
14. S. R. Cloude, Conditions for the physical realisability of matrix operators in polarimetry, *Proc. Soc. Photo - Opt. Instrum. Eng.* 1166 (1989): 177–185.
15. S. Cloude, *Polarisation: Applications in Remote Sensing*, Oxford University Press (2010).
16. D. Goldstein, *Polarized Light*, 3rd edition, Section 8.4.2. New York, NY: Marcel Dekker (2011).
17. C. R. Givens and A. B. Kostinski, A simple necessary and sufficient condition on physically realizable Mueller matrices, *J. Mod. Opt.* 40.3 (1993): 471–481.
18. H. D. Noble, S. C. McClain, and R. A. Chipman, Mueller matrix roots depolarization parameters, *Appl. Opt.* 51 (2012): 735–744.
19. R. Simon, The connection between Mueller and Jones matrices of polarization optics, *Opt. Commun.* 42.5 (1982): 293–297.
20. K. Kim, L. Mandel et al., Relationship between Jones and Mueller matrices for random media. *J. Opt. Soc. Am. A* 4 (1987): 433–437.
21. D. Goldstein, *Polarized Light*, 2nd edition New York, NY: Marcel Dekker (2003), 166 pp.
22. J. C. Stover, *Optical Scattering: Measurement and Analysis*, SPIE Press (2012).
23. T. Germer, Polarized light diffusely scattered under smooth and rough interfaces, *Optical Science and Technology, SPIE's 48th Annual Meeting*, International Society for Optics and Photonics (2003).
24. T. Germer (http://www.nist.gov/pml/div685/grp06/scattering_scatmech.cfm, accessed February 29, 2016).
25. H. Mueller, Memorandum on the polarization optics of the photoelastic shutter, Report No. 2 of the OSRD project OEMsr-576, Nov. 15, 1943. A declassified report.
26. H. Mueller, Informal notes about 1943 on Course 8.26 at Massachusetts Institute of Technology.
27. H. Mueller, The foundations of optics, Proceedings of the Winter Meeting of the Optical Society of America, *J. Opt. Soc. Am.* 38 (1948): 661.
28. N. G. Parke III, Matrix Optics, PhD thesis, Department of Physics, Massachusetts Institute of Technology (1948), 181 pp.
29. R. Clark Jones, A new calculus for the treatment of optical system, V. A more general formulation, and description of another calculus, *J. Opt. Soc. Am* 37(2) (1947): 107–110.
30. P. Soleillet, Parameters characterising partially polarized light in fluorescence phenomena, *Ann. Phys.* 12(10) (1929): 23–97.
31. F. Perrin, Polarization of light scattered by isotropic opalescent media, *J. Chem. Phys.* 10.7 (1942): 415–427.

7

Polarimetry

7.1 Introduction

Humans naturally have little appreciation for the richness of polarization in our surroundings. The human eye can barely sense polarization; the response to different polarization states varies by less than 3%, too little to notice. To measure, visualize, and utilize this polarization information, *polarimeters* are needed. Many different types of polarimeters have been developed for different wavelengths, speeds, and applications. Polarization images frequently contain unexpected and fascinating information. *Polarized sunglasses* begin to give us a glimpse of the polarization in our surroundings, but only a small glimpse because the sunglasses' polarizers are at a fixed angle. For a more striking view of the polarization in the environment, place a horizontal polarizer over one eye and a vertical polarizer over the other. With orthogonal polarization states entering our two eyes, many objects take on an unnatural appearance, as the brain tries to interpret "unnatural differences" in the signals coming from the two eyes. Moving water is particularly interesting to view through orthogonal polarizers. The polarization of water fountains, ocean waves, and other turbulent flows is large enough and rapidly fluctuating so that the view through orthogonal polarizers into our two eyes is particularly difficult for the visual system to reconcile. Viewing the world through orthogonal polarizers indicates the large polarization content surrounding us. To quantify such polarization signatures, polarimeters are used to measure the degree of polarization, the angle of polarization, and the wavelength dependence of polarization. Many types of polarimeters and their applications are surveyed in this chapter.

This chapter develops methods and algorithms for measuring the *Stokes parameters* and *Mueller matrix* elements. A general formulation for the measurement and data reduction from a series of radiometric measurements with polarization elements is presented. This data reduction is a linear estimation problem and lends itself to efficient solution using linear algebra, usually with a least-squares estimator to find the best match to the data. An important consideration is to keep the *null space*, a set of signal patterns that will not be generated by actual Stokes parameters, from entering into the data analysis. Similar developments of the polarimetric data reduction process are found in other references.[1-4]

Several types of *Stokes polarimeters* are discussed. Rotating element polarimetry, oscillating element polarimetry, and phase modulation polarimetry are methods that acquire a series of measurements over time to obtain the Stokes parameters.[5] Other techniques, such as division-of-amplitude polarimetry and division-of-wavefront polarimetry, can measure all four elements of the Stokes parameters simultaneously. This chapter includes a discussion on optimizing Stokes polarimeters.

The principles of polarization measurements are surveyed. One of the primary difficulties in performing accurate polarization measurements is the systematic errors due to non-ideal polarization elements. Therefore, the polarimetric measurement and data reduction process is formulated to incorporate arbitrary polarization elements calibrated by measurement of their transmitted and analyzed Stokes parameters. Polarimeter optimization is addressed through the minimization of the *condition number*.

Throughout this chapter, quantities are formulated in terms of the Stokes parameters and Mueller matrix, as these usually comprise the most appropriate representation of polarization for radiometric measurements.

One of the first polarimeters was described by David Brewster, the inventor of the kaleidoscope.[6]

7.2 What Does the Polarimeter See?

One reason polarization images are interesting is because of the stark differences from color images, such as the red, green, and blue images from a camera or the way our eyes render scenes. In summary, color conveys an *energy* while polarization conveys a *direction*.

Color arises from the energy levels in atoms and molecules. Incident light fields cause the charges to oscillate, and these motions interact with the rotational, vibrational, and electronic levels of the molecules, with some of the light being absorbed as heat, while the oscillating charges radiate the remaining energy away as transmitted, reflected, or scattered. Thus, the color we see, the color measured with spectrometers, is a spectral fingerprint of the energy levels in a scattering medium. In conventional images, all of the contrast in brightness and color arises from these spectral properties.

The oscillating charges primarily radiate as small molecular sized dipoles, charge oscillating back and forth sinusoidally about its equilibrium like a charge on a spring. Such dipoles have a specific radiation pattern. Dipoles cannot radiate along the dipole axis, the line along which the charges are oscillating. The strongest radiation occurs in the plane perpendicular to the dipole. Thus, when viewing a plane of dipoles all oscillating in a plane perpendicular to the line of sight, the light will be polarized along the dipole axis. When unpolarized light is observed, the charges are oscillating equally in all directions projected onto the plane transverse to our line of sight. When the light is highly polarized, the molecular charges are oscillating primarily along one axis, when projected onto our view. Hence, the polarization information occurs in two steps. First, the incident light causes the charges to oscillate. Polarized incident light typically makes the charges oscillate primarily along the direction of the incident light's electric field. Some materials can rapidly spread the direction of charge oscillation over a range of angles, but most materials respond by oscillating in a direction close to the incident field. Once the charges are in oscillation, a polarimeter can observe how the charge oscillation is distributed from its perspective. If the oscillation directions are uneven, the light is partially polarized.

Thus, the polarization signature is more of a geometric signature, responding to the polarization of the incident light and our view of the charge oscillations. Color is dependent on quantum mechanical energy levels in the material. Thus, color and intensity are often poorly correlated with the polarization. The polarization image contains information on the orientation of a scattering surface, its refractive index and texture, as well as the polarization state of the light incident on the surface.

7.3 Polarimeters

Polarimeters are optical instruments for measuring the polarization properties of light beams and samples. *Polarimetry*, the science of polarization measurement, is most simply characterized as *radiometry with polarization elements*. Accurate polarimetry requires careful attention to all the issues necessary for accurate radiometry, together with many additional polarization issues that must be mastered to accurately determine polarization properties from polarimetric measurements.

Typical applications of polarimeters include the following: remote sensing of the Earth and astronomical bodies, calibration of polarization elements, measurements of the thickness and refractive indices of thin films (ellipsometry), spectropolarimetric studies of materials, and alignment of polarization-critical optical systems such as liquid crystal displays and projectors.

7.3.1 Light-Measuring Polarimeters

Light-measuring polarimeters measure the polarization state of a beam of light and its polarization characteristics: the Stokes parameters, the direction of oscillation of the electric field vector for a linearly polarized beam, the helicity of a circularly polarized beam, the elliptical parameters of an elliptically polarized beam, and the degree of polarization.

A light-measuring polarimeter utilizes a set of polarization elements placed in a beam of light in front of a radiometer. The light beam is analyzed by this set of polarization state analyzers, and a set of flux measurements is acquired. The polarization characteristics of the light beam are determined from these measurements using data reduction algorithms in Section 7.4.1.

7.3.2 Sample-Measuring Polarimeters

Sample-measuring polarimeters determine the relationship between the polarization states incident upon and exiting a sample, and infer the polarization characteristics of the sample, its diattenuation, retardance, and depolarization, from the measurements. The term *exiting beam* is general and in different measurements might describe beams that are transmitted, reflected, diffracted, scattered, or otherwise modified. The term *sample* is also an inclusive term used in a broad sense to describe a general light–matter interaction or sequence of such interactions and applies to practically anything.

Measurements are acquired using a set of polarization elements located between a source and sample, and the exiting beams are analyzed with a separate set of polarization elements between the sample and radiometer. Some of the samples of great interest include surfaces, thin films on surfaces, polarization elements, optical elements, optical systems, natural scenes, biological samples, and industrial samples.

Accurate polarimetric measurements can be made only if the *polarization generator* and *polarization analyzer* are well calibrated; the polarization states exiting the generator must be well known, as are the states analyzed by the analyzer. To perform accurate polarimetry, the polarization elements need not be ideal or of the highest quality. If the *Mueller matrices of the polarization components* are known from careful calibration, the systematic errors due to *non-ideal polarization elements* are removed during the data reduction in Section 7.4.3.

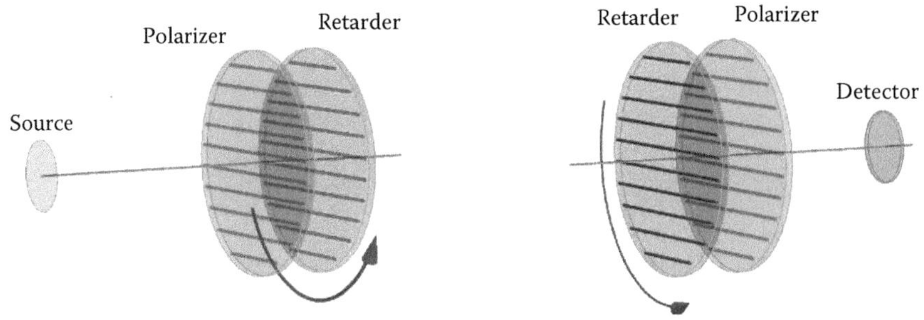

Figure 7.1 (Left) A *polarization generator*, which in this example consists of a light source, a polarizer, and a rotating retarder, can generate a set of calibrated polarization states. (Right) A *polarization analyzer*, which in this example consists of a rotating retarder, a polarizer, and a detector, measures a set of calibrated polarized flux components in an incident beam.

7.3.3 Complete and Incomplete Polarimeters

A light-measuring polarimeter is *complete* if the four Stokes parameters can be determined from its measurements. An *incomplete* light-measuring polarimeter cannot be used to determine the four Stokes parameters, but measures a subset. For example, a polarimeter that employs a rotating polarizer in front of a detector does not determine the circular polarization content, S_3, of a beam, and is incomplete. Similarly, a sample-measuring polarimeter is *complete* if it is capable of measuring the full Mueller matrix, and *incomplete* otherwise. Complete polarimeters are referred to as *Stokes polarimeters* or *Mueller polarimeters*.

7.3.4 Polarization Generators and Analyzers

A *polarization generator* consists of a light source, optical elements, and polarization elements to produce a beam of *known polarization state*. An example is shown in Figure 7.1 (left). A polarization generator is specified by the Stokes parameters **S** of the exiting beam. A *polarization analyzer* is a configuration of polarization elements, optical elements, and a detector that performs a *flux measurement* of a particular polarization component in an incident beam. Figure 7.1 (right) shows an example, a rotating retarder polarization analyzer. A polarization analyzer is characterized by a Stokes-like *analyzer vector* **A** that specifies the incident polarization state being analyzed, the state that produces the maximal response at the detector. *Sample-measuring polarimeters* require polarization generators and polarization analyzers, while *light-measuring polarimeters* only require polarization analyzers. Frequently, the terms *polarization generator* and *polarization analyzer* refer just to the polarization elements in the generator and analyzer. It is important to distinguish between elliptical (and circular) generators and elliptical analyzers for a given state because they generally have different polarization characteristics and different Mueller matrices.

7.4 Mathematics of Polarimetric Measurement and Data Reduction

7.4.1 Stokes Polarimetry

The *polarization analyzer* consists of the polarization elements used for analyzing the polarization state, any other optical elements (lenses, mirrors, etc.) following the polarization elements, and the polarimeter's detector. The polarization effects from all elements are included in the measurement and data reduction procedures. A polarization analyzer is characterized by an *analyzer*

7.4 Mathematics of Polarimetric Measurement and Data Reduction

vector containing four elements, defined analogously to the Stokes parameters. Let P_H be the flux measured by the detector (the current or voltage generated) when one unit of horizontally polarized light is incident. Similarly, P_V, P_{45}, P_{135}, P_R, and P_L are the detector's flux measurements for the corresponding incident polarized beams with unit flux. Then, the polarization analyzer's analyzer vector **A** is

$$\mathbf{A} = \begin{pmatrix} a_0 \\ a_1 \\ a_2 \\ a_3 \end{pmatrix} = \begin{pmatrix} P_H + P_V \\ P_H - P_V \\ P_{45} - P_{135} \\ P_R - P_L \end{pmatrix}. \tag{7.1}$$

Note that in the absence of noise, $P_H + P_V = P_{45} + P_{135} = P_R + P_L$. The response P of the polarization analyzer to an arbitrary polarization state **S** is a dot product of the analyzer vector with the incident Stokes parameters,

$$P = \mathbf{A} \cdot \mathbf{S} = a_0 S_0 + a_1 S_1 + a_2 S_2 + a_3 S_3. \tag{7.2}$$

The analyzer vector will be the top row of the Mueller matrix for the polarization analyzer.

In what follows, it will be assumed that the Stokes parameters being measured are constant in time.

A Stokes parameters measurement is a set of measurements acquired with a set of polarization analyzers placed into the beam of light. Let the total number of analyzers be Q, with each analyzer $\mathbf{A}_q$ specified by index $q = 0, 1,..., Q - 1$. Assume the incident Stokes parameters are the same for all measurements. The qth measurement generates an output, a flux measurement $P_q = \mathbf{A}_q \cdot \mathbf{S}$. A *polarimetric measurement matrix* **W** is defined as a $Q \times 4$ matrix with the qth row containing the analyzer vector $\mathbf{A}_q$,

$$\mathbf{W} = \begin{pmatrix} a_{00} & a_{01} & a_{02} & a_{03} \\ a_{10} & a_{11} & a_{12} & a_{13} \\ \vdots & & & \\ a_{Q-1,0} & a_{Q-1,1} & a_{Q-1,2} & a_{Q-1,3} \end{pmatrix}. \tag{7.3}$$

The Q measured fluxes are arranged in a measurement *flux vector*, $\mathbf{P} = (P_0\ P_1\ \ldots\ P_{Q-1})^T$. **P** is related to **S** by the *polarimetric measurement equation*,

$$\mathbf{P} = \begin{pmatrix} P_0 \\ P_1 \\ \vdots \\ P_{Q-1} \end{pmatrix} = \mathbf{W} \cdot \mathbf{S} = \begin{pmatrix} a_{00} & a_{01} & a_{02} & a_{03} \\ a_{10} & a_{11} & a_{12} & a_{13} \\ \vdots & & & \\ a_{Q-1,0} & a_{Q-1,1} & a_{Q-1,2} & a_{Q-1,3} \end{pmatrix} \cdot \begin{pmatrix} S_0 \\ S_1 \\ S_2 \\ S_3 \end{pmatrix}. \tag{7.4}$$

To calculate the Stokes parameters from the data, the inverse of **W** is determined and applied to the measured data. The measured value for the incident Stokes parameters $\mathbf{S}_m$ is related to the data by the *polarimetric data reduction matrix* $\mathbf{W}^{-1}$,

$$\mathbf{W}^{-1} \cdot \mathbf{P} = \mathbf{S}_m. \tag{7.5}$$

This is the *polarimetric data reduction equation*. During the setup of a Stokes polarimeter, the analyzer vectors are established. This calibration determines the rows of **W** and thus **W**$^{-1}$. **W**$^{-1}$ is then applied routinely to the measurements to calculate the Stokes parameters.

Three considerations in the solution of the polarimetric measurement equation are the *existence*, *rank*, and *uniqueness* of the matrix inverse **W**$^{-1}$. The rank of **W** is the number of linearly independent rows. The most straightforward case occurs when four measurements are performed. If $Q = 4$ and if four linearly independent analyzer vectors are used, then **W** is of rank four, and then **W**$^{-1}$ exists and is unique and nonsingular. The polarimeter's data reduction is performed by Equation 7.5; the polarimeter measures all four elements of the incident Stokes parameters. To be linearly independent, the determinant of **W** must be zero, so the matrix inverse exists.

During calibration of the polarimeter, the objective is the accurate determination of **W**.

Example 7.1 Polarimeter with Four Polarizers

Consider a Stokes polarimeter taking four measurements with ideal polarizers placed in front of a radiometer. In sequence, the four analyzers are (1) a horizontal polarizer, (2) a vertical polarizer, (3) a 45° polarizer, and (4) a right circular polarizer. Arranging the four analyzer vectors as the rows of the *polarimetric measurement matrix*, **W** is

$$\mathbf{P} = \begin{pmatrix} P_0 \\ P_1 \\ P_2 \\ P_3 \end{pmatrix} = \mathbf{W} \cdot \mathbf{S} = \frac{1}{2} \begin{pmatrix} 1 & 1 & 0 & 0 \\ 1 & -1 & 0 & 0 \\ 1 & 0 & 1 & 0 \\ 1 & 0 & 0 & 1 \end{pmatrix} \cdot \begin{pmatrix} S_0 \\ S_1 \\ S_2 \\ S_3 \end{pmatrix}. \quad (7.6)$$

The one-half arises from the one-half that occurs in the Mueller matrix of ideal polarizers, since polarizers transmit one-half of unpolarized light. Because the matrix **W** is square, the inverse of **W** is unique,

$$\mathbf{W}^{-1} = \begin{pmatrix} 1 & 1 & 0 & 0 \\ 1 & -1 & 0 & 0 \\ -1 & -1 & 2 & 0 \\ -1 & -1 & 0 & 2 \end{pmatrix}. \quad (7.7)$$

The four flux measurements are labeled as P_H, P_V, P_{45}, and P_R. The set of the polarimeter's flux measurements are placed in a vector, the *flux vector* **P**. The Stokes parameters **S** are calculated by the matrix multiplication of the *polarimetric data reduction matrix* **W**$^{-1}$ on the flux vector,

$$\mathbf{S} = \begin{pmatrix} S_0 \\ S_1 \\ S_2 \\ S_3 \end{pmatrix} = \mathbf{W}^{-1} \cdot \mathbf{P} = \begin{pmatrix} 1 & 1 & 0 & 0 \\ 1 & -1 & 0 & 0 \\ -1 & -1 & 2 & 0 \\ -1 & -1 & 0 & 2 \end{pmatrix} \cdot \begin{pmatrix} P_H \\ P_V \\ P_{45} \\ P_R \end{pmatrix} = \begin{pmatrix} P_H + P_V \\ P_H - P_V \\ -P_H - P_V + 2P_{45} \\ -P_H - P_V + 2P_R \end{pmatrix}. \quad (7.8)$$

7.4 Mathematics of Polarimetric Measurement and Data Reduction

If the polarimeter measured a flux vector **P** = (2, 1, 2, 2), for example, then the measured polarization state is found as

$$\mathbf{W}^{-1} \cdot \mathbf{P} = \begin{pmatrix} 1 & 1 & 0 & 0 \\ 1 & -1 & 0 & 0 \\ -1 & -1 & 2 & 0 \\ -1 & -1 & 0 & 2 \end{pmatrix} \begin{pmatrix} 2 \\ 1 \\ 2 \\ 2 \end{pmatrix} = \begin{pmatrix} 3 \\ 1 \\ 1 \\ 1 \end{pmatrix}. \tag{7.9}$$

When $Q > 4$, more than four measurements are taken, **W** is not square and the inverse $\mathbf{W}^{-1}$ is not unique; multiple $\mathbf{W}^{-1}$ exist. The four Stokes parameters **S** are overdetermined; there are more equations than unknowns. In the absence of noise, the different $\mathbf{W}^{-1}$ yield the same **S**. However, not every set of flux measurements corresponds to a meaningful set of Stokes parameters; some potential flux vectors lie in the *null space*; they are flux vectors that the polarimeter cannot generate (see Math Tip 7.1), but which noise can generate. Since noise is always present, the optimum $\mathbf{W}^{-1}$ is desired. The least-squares estimate for $\mathbf{S}_m$ utilizes a particular matrix inverse, the *pseudoinverse* $\mathbf{W}_P^{-1}$ of **W**,

$$\mathbf{W}_P^{-1} = (\mathbf{W}^T \cdot \mathbf{W})^{-1} \cdot \mathbf{W}^T. \tag{7.10}$$

The optimal estimate of **S** given by a set of flux measurements is

$$\mathbf{S} = \mathbf{W}_P^{-1} \cdot \mathbf{P} = (\mathbf{W}^T \cdot \mathbf{W})^{-1} \cdot \mathbf{W}^T \cdot \mathbf{P}. \tag{7.11}$$

When **W** is of rank three or less, the polarimeter is *incomplete*. The optimal matrix inverse is the pseudoinverse, but only three or fewer *properties* of the Stokes parameters elements are determined; the projection of the Stokes parameters onto three or fewer directions is measured. If these directions align with the Stokes basis vectors, then these Stokes parameters elements are measured, but in general, linear combinations of the parameters are measured.

Many polarimeters utilize rotating or oscillating polarization elements. For these polarimeters, the analyzer vectors contain trigonometric functions of the rotation angles or oscillating parameters. The flux at the detector is a periodic function. The Stokes parameters are then related to the Fourier series coefficients of the flux.

Four measurements are necessary to measure the four Stokes parameters. Often more than four measurements are used in a Stokes polarimeter to increase accuracy and reduce the effects of noise in the measurement. When four measurements are taken, any noise changes the measurement in proportion to the noise. When more than four measurements are used, the data can be estimated, and the effect of noise is reduced.

7.4.2 Measuring Mueller Matrix Elements

Measuring individual Mueller matrix elements is examined first; elements are measured in groups of four. In Section 7.4.3, an algorithm is developed for measuring the entire Mueller matrix using the pseudoinverse.

The four Mueller matrix elements M_{00}, M_{01}, M_{10}, and M_{11} can be measured using four measurements with ideal horizontal (**H**) and vertical (**V**) linear polarizers. Four measurements P_0, P_1, P_2, and P_3 are taken with (generator/analyzer) settings of (**H/H**), (**V/H**), (**H/V**), and (**V/V**), determining the following combinations of Mueller matrix elements,

$$\begin{aligned} P_0 &= \left(M_{00} + M_{01} + M_{10} + M_{11}\right)/4, \\ P_1 &= \left(M_{00} - M_{01} + M_{10} - M_{11}\right)/4, \\ P_2 &= \left(M_{00} + M_{01} - M_{10} - M_{11}\right)/4, \\ P_3 &= \left(M_{00} - M_{01} - M_{10} + M_{11}\right)/4. \end{aligned} \quad (7.12)$$

These four equations are solved for the Mueller matrix elements yielding

$$\begin{pmatrix} M_{00} \\ M_{01} \\ M_{10} \\ M_{11} \end{pmatrix} = \begin{pmatrix} P_0 + P_1 + P_2 + P_3 \\ P_0 - P_1 + P_2 - P_3 \\ P_0 + P_1 - P_2 - P_3 \\ P_0 - P_1 - P_2 + P_3 \end{pmatrix}. \quad (7.13)$$

Other Mueller matrix elements are determined using different combinations of generator and analyzer states. The four matrix elements at the corners of a rectangle in the Mueller matrix (M_{00}, M_{0i}, M_{j0}, M_{ji}) can be determined from four measurements using a ± i-generator and a ± j-analyzer. For example, a pair of right and left circularly polarizing generators and a pair of 45° and 135° oriented analyzers determine elements (M_{00}, M_{03}, M_{20}, M_{23}).

If an unpolarized generator is used, then two elements, M_{00} and M_{0i}, can be measured in two measurements with a pair of analyzers for Stokes parameter $\pm S_i$. The only element that can be measured individually in a single measurement is M_{00} using an unpolarized generators and analyzer, which is essentially a pure radiometric non-polarizing measurement.

7.4.3 Mueller Data Reduction Matrix

This section develops data reduction equations to calculate Mueller matrices from arbitrary sequences of measurements.[2,7,8] The algorithm uses either ideal or calibrated values for the polarization generator and analyzer vectors. The data reduction equations are straightforward matrix–vector multiplication on a data vector. This method is an extension of the data reduction methods presented in Section 7.4.1 on Stokes polarimetry.

A Mueller matrix polarimeter takes Q measurements identified by index $q = 0, 1, \ldots Q - 1$. For the qth measurement, the generator produces a beam with Stokes parameters $\mathbf{S}_q$ and the beam exiting

7.4 Mathematics of Polarimetric Measurement and Data Reduction

the sample is analyzed by analyzer vector $\mathbf{A}_q$. The measured flux P_q is related to the sample Mueller matrix by

$$\begin{aligned}
P_q &= \mathbf{A}_q^T \mathbf{M} \mathbf{S}_q \\
&= \begin{pmatrix} a_{q0} & a_{q1} & a_{q2} & a_{q3} \end{pmatrix} \cdot \begin{pmatrix} M_{00} & M_{01} & M_{02} & M_{03} \\ M_{10} & M_{11} & M_{12} & M_{13} \\ M_{20} & M_{21} & M_{22} & M_{23} \\ M_{30} & M_{31} & M_{32} & M_{33} \end{pmatrix} \begin{pmatrix} S_{q0} \\ S_{q1} \\ S_{q2} \\ S_{q3} \end{pmatrix} \\
&= \sum_{j=0}^{3} \sum_{k=0}^{3} a_{qj} m_{jk} S_{qk} \\
&= a_{q0} S_{q0} M_{00} + a_{q0} S_{q1} M_{01} + a_{q0} S_{q2} M_{02} + a_{q0} S_{q3} M_{03} \\
&\quad + a_{q1} S_{q1} M_{10} + a_{q1} S_{q1} M_{11} + a_{q1} S_{q2} M_{12} + a_{q1} S_{q3} M_{13} \\
&\quad + a_{q2} S_{q2} M_{20} + a_{q2} S_{q1} M_{21} + a_{q2} S_{q2} M_{22} + a_{q2} S_{q3} M_{23} \\
&\quad + a_{q3} S_{q3} M_{30} + a_{q3} S_{q1} M_{31} + a_{q3} S_{q2} M_{32} + a_{q3} S_{q3} M_{33}.
\end{aligned} \tag{7.14}$$

To apply the methods used previously for Stokes polarimeters, this equation is rewritten as a vector–vector dot product by flattening the Mueller matrix into a 16 × 1 *Mueller vector* $\vec{\mathbf{M}} = \begin{pmatrix} M_{00} & M_{01} & M_{02} & M_{03} & M_{10} & \cdots & M_{33} \end{pmatrix}^T$. Then, a 16 × 1 *polarimetric measurement vector* $\mathbf{W}_q$, equivalent to the Stokes analyzer vector, for the qth measurement is defined as follows:

$$\begin{aligned}
\mathbf{W}_q &= \begin{pmatrix} w_{q00} & w_{q01} & w_{q02} & w_{q03} & w_{q10} & \cdots & w_{q33} \end{pmatrix}^T \\
&= \begin{pmatrix} a_{q0} S_{q0} & a_{q0} S_{q1} & a_{q0} S_{q2} & a_{q0} S_{q3} & a_{q1} S_{q0} & \cdots & a_{q3} S_{q3} \end{pmatrix}^T
\end{aligned} \tag{7.15}$$

where $w_{qjk} = a_{qj} s_{qk}$. The qth measured flux is the dot product

$$P_q = \mathbf{W}_q \cdot \vec{\mathbf{M}} = \begin{pmatrix} a_{q0} S_{q0} \\ a_{q0} S_{q1} \\ a_{q0} S_{q2} \\ a_{q0} S_{q3} \\ a_{q1} S_{q0} \\ a_{q1} S_{q1} \\ \vdots \\ a_{q3} S_{q3} \end{pmatrix} \cdot \begin{pmatrix} M_{00} \\ M_{01} \\ M_{02} \\ M_{03} \\ M_{10} \\ M_{11} \\ \vdots \\ M_{33} \end{pmatrix} = \begin{pmatrix} w_{q,00} \\ w_{q,01} \\ w_{q,02} \\ w_{q,03} \\ w_{q,10} \\ w_{q,11} \\ \vdots \\ w_{q,33} \end{pmatrix} \cdot \begin{pmatrix} M_{00} \\ M_{01} \\ M_{02} \\ M_{03} \\ M_{10} \\ M_{11} \\ \vdots \\ M_{33} \end{pmatrix}. \tag{7.16}$$

The full sequence of Q Mueller polarimeter measurements is described by the $Q \times 16$ *polarimetric measurement matrix* **W**, where the qth row is $\mathbf{W}_q$. This *polarimetric measurement equation* for the Mueller matrix polarimeter relates the flux vector **P** to the sample *Mueller vector* as

$$\mathbf{P} = \mathbf{W} \cdot \vec{\mathbf{M}} = \begin{pmatrix} P_1 \\ P_2 \\ \vdots \\ P_Q \end{pmatrix} = \begin{pmatrix} w_{1,00} & w_{1,01} & \cdots & w_{1,33} \\ w_{2,00} & w_{2,01} & \cdots & w_{2,33} \\ \vdots & & & \\ w_{Q,00} & w_{Q,01} & \cdots & w_{Q,33} \end{pmatrix} \cdot \begin{pmatrix} M_{00} \\ M_{01} \\ \vdots \\ M_{33} \end{pmatrix}. \tag{7.17}$$

If **W** contains 16 linearly independent rows, all 16 elements of the Mueller matrix can be determined. When $Q = 16$, the matrix inverse is unique and the Mueller matrix elements are determined from the *polarimetric data reduction equation*,

$$\vec{\mathbf{M}} = \mathbf{W}^{-1} \cdot \mathbf{P}. \tag{7.18}$$

When $Q > 16$, $\vec{\mathbf{M}}$ is overdetermined and the optimal (least-squares) *polarimetric data reduction equation* for $\vec{\mathbf{M}}$ uses the pseudoinverse $\mathbf{W}_P^{-1}$ of **W**,

$$\vec{\mathbf{M}} = (\mathbf{W}^T \cdot \mathbf{W})^{-1} \cdot \mathbf{W}^T \cdot \mathbf{P} = \mathbf{W}_P^{-1} \cdot \mathbf{P}. \tag{7.19}$$

The advantages of this procedure are as follows. First, this procedure does not assume that the set of polarization state generator and analyzer have any particular form. For example, the polarization elements in the generator and analyzer do not need to be rotated in uniform angular increments but can comprise an arbitrary sequence. Second, the polarization elements are not assumed to be ideal polarization elements. If the polarization generator and analyzer vectors are determined through a calibration procedure, the effects of non-ideal polarization elements are corrected in the data reduction. Third, the procedure readily treats overdetermined measurement sequences (more than 16 measurements for the full Mueller matrix), providing a least-squares solution. Finally, a matrix–vector form of data reduction is readily implemented and easily understood.

Example 7.2 Data Reduction for an Example Mueller Matrix Polarimeter

Consider a polarimeter with filter wheels over the generator and analyzer, each with an open setting, a horizontal linear polarizer, a 45° polarizer, and a right circular polarizer. The generator filter wheel moves to its first position, open with an unpolarized generated state, while the analyzer wheel rotates through its four positions and four measurements are taken. The generator filter wheel then moves to its second, third, and fourth positions and

7.4 Mathematics of Polarimetric Measurement and Data Reduction

the measurement sequence repeats for 16 measurements. The *polarimetric measurement matrix* **W** is

$$\mathbf{W} = \frac{1}{4} \begin{pmatrix} 4 & 0 & 0 & 0 & 0 & 0 & 0 & 0 & 0 & 0 & 0 & 0 & 0 & 0 & 0 & 0 \\ 2 & 0 & 0 & 0 & 2 & 0 & 0 & 0 & 0 & 0 & 0 & 0 & 0 & 0 & 0 & 0 \\ 2 & 0 & 0 & 0 & 0 & 0 & 0 & 0 & 2 & 0 & 0 & 0 & 0 & 0 & 0 & 0 \\ 2 & 0 & 0 & 0 & 0 & 0 & 0 & 0 & 0 & 0 & 0 & 0 & 2 & 0 & 0 & 0 \\ 2 & -2 & 0 & 0 & 0 & 0 & 0 & 0 & 0 & 0 & 0 & 0 & 0 & 0 & 0 & 0 \\ 1 & -1 & 0 & 0 & 1 & -1 & 0 & 0 & 0 & 0 & 0 & 0 & 0 & 0 & 0 & 0 \\ 1 & -1 & 0 & 0 & 0 & 0 & 0 & 0 & 1 & -1 & 0 & 0 & 0 & 0 & 0 & 0 \\ 1 & -1 & 0 & 0 & 0 & 0 & 0 & 0 & 0 & 0 & 0 & 1 & -1 & 0 & 0 & 0 \\ 2 & 0 & -2 & 0 & 0 & 0 & 0 & 0 & 0 & 0 & 0 & 0 & 0 & 0 & 0 & 0 \\ 1 & 0 & -1 & 0 & 1 & 0 & -1 & 0 & 0 & 0 & 0 & 0 & 0 & 0 & 0 & 0 \\ 1 & 0 & -1 & 0 & 0 & 0 & 0 & 0 & 1 & 0 & -1 & 0 & 0 & 0 & 0 & 0 \\ 1 & 0 & -1 & 0 & 0 & 0 & 0 & 0 & 0 & 0 & 0 & 1 & 0 & -1 & 0 \\ 2 & 0 & 0 & -2 & 0 & 0 & 0 & 0 & 0 & 0 & 0 & 0 & 0 & 0 & 0 & 0 \\ 1 & 0 & 0 & -1 & 1 & 0 & 0 & -1 & 0 & 0 & 0 & 0 & 0 & 0 & 0 & 0 \\ 1 & 0 & 0 & -1 & 0 & 0 & 0 & 0 & 1 & 0 & 0 & -1 & 0 & 0 & 0 & 0 \\ 1 & 0 & 0 & -1 & 0 & 0 & 0 & 0 & 0 & 0 & 0 & 0 & 1 & 0 & 0 & -1 \end{pmatrix}. \quad (7.20)$$

The *polarimetric data reduction matrix* is

$$\mathbf{W}^{-1} = \begin{pmatrix} 1 & 0 & 0 & 0 & 0 & 0 & 0 & 0 & 0 & 0 & 0 & 0 & 0 & 0 & 0 & 0 \\ -1 & 0 & 0 & 0 & 2 & 0 & 0 & 0 & 0 & 0 & 0 & 0 & 0 & 0 & 0 & 0 \\ -1 & 0 & 0 & 0 & 0 & 0 & 0 & 0 & 2 & 0 & 0 & 0 & 0 & 0 & 0 & 0 \\ -1 & 0 & 0 & 0 & 0 & 0 & 0 & 0 & 0 & 0 & 0 & 0 & 2 & 0 & 0 & 0 \\ -1 & 2 & 0 & 0 & 0 & 0 & 0 & 0 & 0 & 0 & 0 & 0 & 0 & 0 & 0 & 0 \\ 1 & -2 & 0 & 0 & -2 & 4 & 0 & 0 & 0 & 0 & 0 & 0 & 0 & 0 & 0 & 0 \\ 1 & -2 & 0 & 0 & 0 & 0 & 0 & 0 & -2 & 4 & 0 & 0 & 0 & 0 & 0 & 0 \\ 1 & -2 & 0 & 0 & 0 & 0 & 0 & 0 & 0 & 0 & 0 & 0 & -2 & 4 & 0 & 0 \\ -1 & 0 & 2 & 0 & 0 & 0 & 0 & 0 & 0 & 0 & 0 & 0 & 0 & 0 & 0 & 0 \\ 1 & 0 & -2 & 0 & -2 & 0 & 4 & 0 & 0 & 0 & 0 & 0 & 0 & 0 & 0 & 0 \\ 1 & 0 & -2 & 0 & 0 & 0 & 0 & 0 & -2 & 0 & 4 & 0 & 0 & 0 & 0 & 0 \\ 1 & 0 & -2 & 0 & 0 & 0 & 0 & 0 & 0 & 0 & 0 & 0 & -2 & 0 & 4 & 0 \\ -1 & 0 & 0 & 2 & 0 & 0 & 0 & 0 & 0 & 0 & 0 & 0 & 0 & 0 & 0 & 0 \\ 1 & 0 & 0 & -2 & -2 & 0 & 0 & 4 & 0 & 0 & 0 & 0 & 0 & 0 & 0 & 0 \\ 1 & 0 & 0 & -2 & 0 & 0 & 0 & 0 & -2 & 0 & 0 & 4 & 0 & 0 & 0 & 0 \\ 1 & 0 & 0 & -2 & 0 & 0 & 0 & 0 & 0 & 0 & 0 & 0 & -2 & 0 & 0 & 4 \end{pmatrix}. \quad (7.21)$$

The singular values of $\mathbf{W}^{-1}$ are

$$\begin{pmatrix} 7.464 & 5.464 & 5.464 & 5.464 & 5.464 & 4 & 4 & 4 & 4 & 2 & 2 & 1.464 & 1.464 & 1.464 & 1.464 & 0.536 \end{pmatrix}^T. \quad (7.22)$$

Hence, the *condition number** is about $7.46/0.54 \approx 13.9$, which is far from optimum.

---

* Condition number is defined in Section 7.11.

The tools of this section have provided a method to measure the Mueller matrices of samples and are applicable to transmission, reflection, and scattering measurements. Often, the measurement of the Mueller matrix is an intermediate step and the desired information is diattenuation or retardance. In this case, the methods of Sections 6.6, 6.11.3, 6.12, and 14.6 can be applied.

7.4.4 Null Space and the Pseudoinverse

The *null space* and *pseudoinverse* of a matrix are important tools to understand the algorithms for measuring accurate Mueller matrices and Stokes parameters. The null space and pseudoinverse identify how noise propagates through the polarimetric data analysis and are applied to minimize noise in polarimeters. The section provides an example of the data reduction of a Stokes polarimeter in Example 7.3. This result is then applied to explain null space in Math Tip 7.1. The pseudoinverse is explained in Math Tip 7.2 and then applied in Example 7.4 to describe how to reduce noise in this polarimeter measurement.

Example 7.3 Stokes Polarimeter with Six Measurements

Consider a Stokes polarimeter that acquires six measurements: P_H, P_V, P_{45}, P_{135}, P_R, and P_L. The six analyzer vectors are arranged in the rows of $\mathbf{W}$,

$$\mathbf{W} = \frac{1}{2}\begin{pmatrix} 1 & 1 & 0 & 0 \\ 1 & -1 & 0 & 0 \\ 1 & 0 & 1 & 0 \\ 1 & 0 & -1 & 0 \\ 1 & 0 & 0 & 1 \\ 1 & 0 & 0 & -1 \end{pmatrix}. \quad (7.23)$$

Since $\mathbf{W}$ is not a square matrix, it does not have a unique matrix inverse. The pseudoinverse $\mathbf{W}_P^{-1}$ is

$$\mathbf{W}_P^{-1} = \begin{pmatrix} \frac{1}{3} & \frac{1}{3} & \frac{1}{3} & \frac{1}{3} & \frac{1}{3} & \frac{1}{3} \\ 1 & -1 & 0 & 0 & 0 & 0 \\ 0 & 0 & 1 & -1 & 0 & 0 \\ 0 & 0 & 0 & 0 & 1 & -1 \end{pmatrix}. \quad (7.24)$$

7.4 Mathematics of Polarimetric Measurement and Data Reduction

The polarimetric data reduction process on a series of six measurements becomes

$$\mathbf{W}_P^{-1} \cdot \mathbf{P} = \begin{pmatrix} \frac{1}{3} & \frac{1}{3} & \frac{1}{3} & \frac{1}{3} & \frac{1}{3} & \frac{1}{3} \\ 1 & -1 & 0 & 0 & 0 & 0 \\ 0 & 0 & 1 & -1 & 0 & 0 \\ 0 & 0 & 0 & 0 & 1 & -1 \end{pmatrix} \cdot \begin{pmatrix} P_H \\ P_V \\ P_{45} \\ P_{135} \\ P_R \\ P_L \end{pmatrix} = \begin{pmatrix} \frac{1}{3}(P_H + P_V + P_{45} + P_{135} + P_R + P_L) \\ P_H - P_V \\ P_{45} - P_{135} \\ P_R - P_L \end{pmatrix}.$$

(7.25)

Note that S_0, the total flux, can be calculated in multiple ways from the measurements, including from the sums of the orthogonal polarized fluxes,

$$S_0 = P_H + P_V = P_{45} + P_{135} = P_R + P_L.$$

(7.26)

The pseudoinverse averages these three different estimates of S_0 in the first row of $\mathbf{W}_P^{-1}$. Using Equation 7.26, other matrix inverses can be generated. Three obvious alternatives follow from Equation 7.26:

$$\mathbf{W}_1^{-1} = \begin{pmatrix} 1 & 1 & 0 & 0 & 0 & 0 \\ 1 & -1 & 0 & 0 & 0 & 0 \\ 0 & 0 & 1 & -1 & 0 & 0 \\ 0 & 0 & 0 & 0 & 1 & -1 \end{pmatrix},$$

$$\mathbf{W}_2^{-1} = \begin{pmatrix} 0 & 0 & 1 & 1 & 0 & 0 \\ 1 & -1 & 0 & 0 & 0 & 0 \\ 0 & 0 & 1 & -1 & 0 & 0 \\ 0 & 0 & 0 & 0 & 1 & -1 \end{pmatrix},$$

(7.27)

$$\mathbf{W}_3^{-1} = \begin{pmatrix} 0 & 0 & 0 & 0 & 1 & 1 \\ 1 & -1 & 0 & 0 & 0 & 0 \\ 0 & 0 & 1 & -1 & 0 & 0 \\ 0 & 0 & 0 & 0 & 1 & -1 \end{pmatrix}.$$

In the absence of noise, all three matrix inverses would calculate the same Stokes parameters. For the matrix inverse, all possible top rows are generated by the equation

$$(a \quad a \quad b \quad b \quad 1-a-b \quad 1-a-b),$$

(7.28)

with arbitrary real a and b.

Math Tip 7.1 Null Space

The polarimeter of Example 7.3 can only generate four patterns of fluxes with the six flux measurements: S_0 generates the first column of **W**, S_1 generates the second column of **W**, S_2 generates the third column, and S_3 generates the fourth column. In the absence of noise, all flux vectors the polarimeter measures should be linear combinations of these four columns. Since the flux vector has six elements, two more basis vectors are needed to fully span a six-dimensional space, for example,

$$\mathbf{N}_1 = \begin{pmatrix} a & a & 0 & 0 & -a & -a \end{pmatrix}, \text{ and } \mathbf{N}_2 = \begin{pmatrix} 0 & 0 & b & b & -b & -b \end{pmatrix} \quad (7.29)$$

or any pair of linear combination of ($\alpha\, \mathbf{N}_1 + \beta\, \mathbf{N}_2$). For a matrix **M**, of dimensions ($m \times n$), the null space is the set of vectors $\mathbf{N}_i$ such that

$$\mathbf{M} \cdot \mathbf{N}_i = \begin{pmatrix} 0 \\ 0 \\ \vdots \\ 0 \end{pmatrix}. \quad (7.30)$$

The matrix–vector product is a zero vector of length equal to the number of columns of **M**.

In the case of a polarimeter, the null space of interest is associated with $\mathbf{W}_P^{-1}$. When **W** multiplies a Stokes parameters vector, it cannot generate a vector in the null space of $\mathbf{W}_P^{-1}$. In Example 7.3, $\mathbf{N}_1$ and $\mathbf{N}_2$ in Equation 7.29 span the null space.

Note that when a null space vector multiplies the inverse of **M**, the result is also a zero vector,

$$\mathbf{N}_1 \cdot \mathbf{W} = \begin{pmatrix} a & a & 0 & 0 & -a & -a \end{pmatrix} \cdot \frac{1}{2} \begin{pmatrix} 1 & 1 & 0 & 0 \\ 1 & -1 & 0 & 0 \\ 1 & 0 & 1 & 0 \\ 1 & 0 & -1 & 0 \\ 1 & 0 & 0 & 1 \\ 1 & 0 & 0 & -1 \end{pmatrix} = \begin{pmatrix} 0 \\ 0 \\ 0 \\ 0 \end{pmatrix}. \quad (7.31)$$

Similarly,

$$\mathbf{N}_2 \cdot \mathbf{W} = \begin{pmatrix} 0 \\ 0 \\ 0 \\ 0 \end{pmatrix}. \quad (7.32)$$

This means that the null space vectors can be added in any linear combination, ($\alpha\, \mathbf{N}_1 + \beta\, \mathbf{N}_2$), to any of the four rows of $\mathbf{W}_P^{-1}$ and the resulting matrix remains a matrix inverse. Thus, two vectors can be added in any amount to each of the four rows of $\mathbf{W}_P^{-1}$ and the resulting matrix

7.4 Mathematics of Polarimetric Measurement and Data Reduction

is an inverse of the $\mathbf{W}$ in Equation 7.24. Hence, $\mathbf{W}^{-1}$ has eight degrees of freedom. Among this large space of matrix inverses, which is the best inverse for a polarimeter? As explained further in Example 7.4, the pseudoinverse provides the most robust data reduction in the presence of noise.

For Example 7.3, some easily understood top rows of other matrix inverses include the following vectors:

$$(1\ 1\ 0\ 0\ 0\ 0), (0\ 0\ 1\ 1\ 0\ 0), \text{ and } (0\ 0\ 0\ 0\ 1\ 1). \tag{7.33}$$

These expressions correspond to calculating the total flux S_0 as $P_H + P_V = P_{45} + P_{135} = P_R + P_L$.

Math Tip 7.2 Pseudoinverse

The matrix inverses of rectangular matrices are not unique. Among the set of matrix inverses, the pseudoinverse $\mathbf{W}_P^{-1}$ is unique because none of its rows contain any component of the null space; each row is orthogonal to the null space,

$$\mathbf{W}_P^{-1} \cdot \mathbf{N}_i = \begin{pmatrix} 0 \\ 0 \\ \vdots \\ 0 \end{pmatrix}, \tag{7.34}$$

where $\mathbf{N}_i$ is any null space vector.[9,10] Thus, using the pseudoinverse for the polarimetric data reduction operation, Equations 7.5 and 7.25 provide an optimal least-squares solution that eliminates as much noise from the data reduction as possible. The pseudoinverse can be calculated by several algorithms. First, the pseudoinverse is available in all standard mathematics packages such as Mathematica and MATLAB®. A simple and straightforward method to calculate the pseudoinverse of a Stokes polarimetric measurement matrix $\mathbf{W}$ with m × 4 elements operates on the components of $\mathbf{W}$'s singular value decomposition,

$$\mathbf{W} = \mathbf{U}\mathbf{D}\mathbf{V}^\dagger$$

$$= \begin{pmatrix} u_{00} & u_{01} & u_{02} & & u_{0,m-1} \\ u_{10} & u_{11} & u_{12} & \cdots & u_{1,m-1} \\ u_{20} & u_{21} & u_{22} & & u_{2,m-1} \\ \vdots & & \ddots & & \\ u_{m-1,0} & u_{m-1,1} & u_{m-1,2} & & u_{m-1,m-1} \end{pmatrix} \begin{pmatrix} \Lambda_0 & 0 & 0 & 0 \\ 0 & \Lambda_1 & 0 & 0 \\ 0 & 0 & \Lambda_2 & 0 \\ 0 & 0 & 0 & \Lambda_3 \\ & & \vdots & \\ 0 & 0 & 0 & 0 \end{pmatrix} \begin{pmatrix} v_{00}^* & v_{01}^* & v_{02}^* & v_{03}^* \\ v_{10}^* & v_{11}^* & v_{12}^* & v_{13}^* \\ v_{20}^* & v_{21}^* & v_{22}^* & v_{23}^* \\ v_{30}^* & v_{31}^* & v_{32}^* & v_{33}^* \end{pmatrix},$$

$$\tag{7.35}$$

with **U** an $m \times m$ unitary matrix, **D** an $m \times 4$ diagonal matrix, and **V** an 4×4 unitary matrix. **D** has four singular values, Λ_i, on the diagonal followed by rows of zeros to fill out the $m \times 4$ elements. The pseudoinverse $\mathbf{W}_P^{-1}$ is

$$\mathbf{W}_P^{-1} = \mathbf{V}\mathbf{D}^{-1}\mathbf{U}^\dagger$$

$$= \begin{pmatrix} v_{00} & v_{10} & v_{20} & v_{30} \\ v_{01} & v_{11} & v_{21} & v_{31} \\ v_{02} & v_{12} & v_{22} & v_{32} \\ v_{03} & v_{13} & v_{23} & v_{33} \end{pmatrix} \begin{pmatrix} \frac{1}{\Lambda_0} & 0 & 0 & 0 & 0 & & 0 \\ 0 & \frac{1}{\Lambda_1} & 0 & 0 & 0 & \cdots & 0 \\ 0 & 0 & \frac{1}{\Lambda_2} & 0 & 0 & & 0 \\ 0 & 0 & 0 & \frac{1}{\Lambda_3} & 0 & & 0 \end{pmatrix}$$

$$\begin{pmatrix} u_{00}^* & u_{10}^* & u_{20}^* & & u_{m-1,0}^* \\ u_{01}^* & u_{11}^* & u_{21}^* & \cdots & u_{m-1,1}^* \\ u_{02}^* & u_{12}^* & u_{22}^* & & u_{m-1,2}^* \\ \vdots & & & \ddots & \\ u_{0,m-1}^* & u_{1,m-1}^* & u_{2,m-1}^* & & u_{m-1,m-1}^* \end{pmatrix}.$$

(7.36)

Example 7.4 A Modulated Polarimeter Example

To explore the concepts of the multiplicity of matrix inverses, the pseudoinverse, the null space, and the effects of noise, consider the following overdetermined polarimeter. A rotating retarder polarimeter has a quarter wave retarder rotating in front of a horizontal linear polarizer, acquiring 16 measurements every 45°, starting from $\theta_0 = 0$ and ending at 315°. The analyzer's Mueller matrix equation is

$$\mathbf{HLP} \cdot \mathbf{QWLR}(\theta) = \frac{1}{2} \begin{pmatrix} 1 & 1 & 0 & 0 \\ 1 & 1 & 0 & 0 \\ 0 & 0 & 0 & 0 \\ 0 & 0 & 0 & 0 \end{pmatrix} \cdot \begin{pmatrix} 1 & 0 & 0 & 0 \\ 0 & \frac{1}{2}(1+\cos 4\theta) & \frac{1}{2}\sin 4\theta & -\sin 2\theta \\ 0 & \frac{1}{2}\sin 4\theta & \frac{1}{2}(1-\cos 4\theta) & \cos 2\theta \\ 0 & \sin 2\theta & -\cos 2\theta & 0 \end{pmatrix}$$

$$= \frac{1}{2} \begin{pmatrix} 1 & \frac{1}{2}(1+\cos 4\theta) & \frac{1}{2}\sin 4\theta & -\sin 2\theta \\ 1 & \frac{1}{2}(1+\cos 4\theta) & \frac{1}{2}\sin 4\theta & -\sin 2\theta \\ 0 & 0 & 0 & 0 \\ 0 & 0 & 0 & 0 \end{pmatrix}. \quad (7.37)$$

The analyzer vector $\mathbf{A}(\theta)$ is the top row of the Mueller matrix,

$$\mathbf{A}(\theta) = \frac{1}{2} \begin{pmatrix} 1 \\ \frac{1}{2}(1+\cos 4\theta) \\ \frac{1}{2}\sin 4\theta \\ -\sin 2\theta \end{pmatrix}. \quad (7.38)$$

The four elements of the analyzer vector for a continuously rotating retarder are plotted in Figure 7.2. The detected flux from S_0 is unmodulated by the rotating retarder. S_1 (green) generates four periods of cosine modulation within 360° with a constant (DC) offset of one-half. S_2 (cyan) has four periods of sinusoidal modulation. S_3 has two periods of minus sinusoidal modulation. The modulation of S_3 is twice as great as the modulation of S_1 and S_2; hence, for a retardance of $\pi/2$, the quarter wave rotating retarder polarimeter measures S_3 twice as accurately as S_1 and S_2 because of its larger modulation.

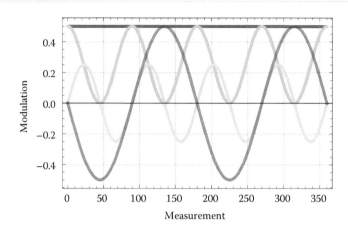

Figure 7.2 Elements of the analyzer vector as the quarter wave retarder rotates: (red) $\mathbf{A}_0$, (green) $\mathbf{A}_1$, (cyan) $\mathbf{A}_2$, (purple) $\mathbf{A}_3$.

Sampling the analyzer vector Equation 7.38 and Figure 7.2 every 15° yields Equation 7.39, the *polarimetric measurement matrix* **W**, shown with an accompanying bar chart graph (Figure 7.3),

$$\mathbf{W} = \frac{1}{2}\begin{pmatrix} 1 & 1 & 0 & 0 \\ 1 & 1/2 & 1/2 & -1/\sqrt{2} \\ 1 & 0 & 0 & -1 \\ 1 & 1/2 & -1/2 & -1/\sqrt{2} \\ 1 & 1 & 0 & 0 \\ 1 & 1/2 & 1/2 & 1/\sqrt{2} \\ 1 & 0 & 0 & 1 \\ 1 & 1/2 & -1/2 & 1/\sqrt{2} \\ 1 & 1 & 0 & 0 \\ 1 & 1/2 & 1/2 & -1/\sqrt{2} \\ 1 & 0 & 0 & -1 \\ 1 & 1/2 & -1/2 & -1/\sqrt{2} \\ 1 & 1 & 0 & 0 \\ 1 & 1/2 & 1/2 & 1/\sqrt{2} \\ 1 & 0 & 0 & 1 \\ 1 & 1/2 & -1/2 & 1/\sqrt{2} \end{pmatrix}. \quad (7.39)$$

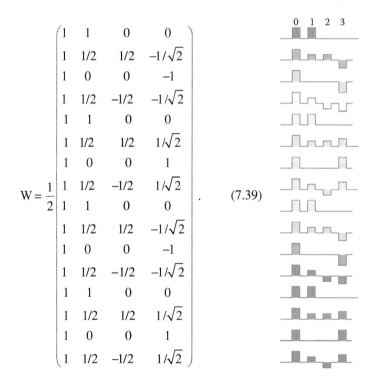

Figure 7.3 A graphic view of the elements of the polarimetric measurement matrix for the rotating retarder example. The last eight rows are the same as the first eight rows, which can be beneficial for noise reduction. Note that the second half of the measurements are a repeat of the first half. There is nothing wrong with this, and in fact such duplication can reduce the effect of slow drifts and low-frequency noise on the polarization measurement and, if done properly, can completely remove the effect of an overall linear drift in brightness or responsivity.

7.4 Mathematics of Polarimetric Measurement and Data Reduction

The polarimetric data reduction matrix $\mathbf{W}_P^{-1}$ is the pseudoinverse of $\mathbf{W}$ and is plotted in Figure 7.4,

$$\mathbf{W}_P^{-1} = (\mathbf{W}^T \cdot \mathbf{W})^{-1} \cdot \mathbf{W}^T$$

$$= \frac{1}{8} \begin{pmatrix} -1 & 1 & 3 & 1 & -1 & 1 & 3 & 1 \\ 4 & 0 & -4 & 0 & 4 & 0 & -4 & 0 \\ 0 & 4 & 0 & -4 & 0 & 4 & 0 & -4 \\ 0 & -\sqrt{2} & -2 & -\sqrt{2} & 0 & \sqrt{2} & 2 & \sqrt{2} \\ -1 & 1 & 3 & 1 & -1 & 1 & 3 & 1 \\ 4 & 0 & -4 & 0 & 4 & 0 & -4 & 0 \\ 0 & 4 & 0 & -4 & 0 & 4 & 0 & -4 \\ 0 & -\sqrt{2} & -2 & -\sqrt{2} & 0 & \sqrt{2} & 2 & \sqrt{2} \end{pmatrix}. \quad (7.40)$$

As an example of a simulated measurement, if the polarimeter measures the flux vector $\mathbf{P} = (0, 1, 2, 1, 0, 1, 2, 1, 0, 1, 2, 1, 0, 1, 2, 1)$, the measured Stokes parameters $\mathbf{S}$ are

$$\mathbf{S} = \mathbf{W}_P^{-1} \cdot \mathbf{P} = \begin{pmatrix} 4 \\ -4 \\ 0 \\ 0 \end{pmatrix}, \quad (7.41)$$

or vertically polarized light with a flux of four.

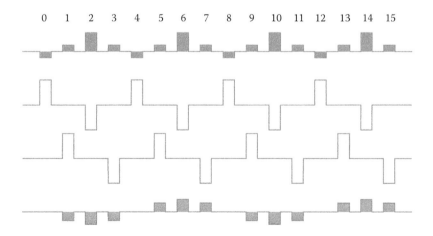

Figure 7.4 Graphic representation of the pseudoinverse for the rotating retarder example.

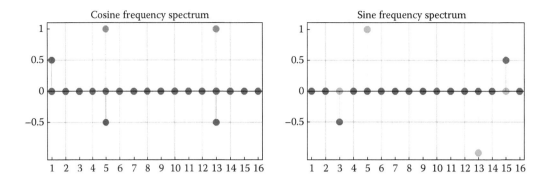

Figure 7.5 Discrete Fourier transforms of the first row of $\mathbf{W}_P^{-1}$ (red, cosine, for S_0), the second row (orange, cosine, for S_1), the third row (green, sine, for S_2), and the fourth row (blue, sine, for S_3). The elements of the DFT count from one (the DC or constant term), to two (the fundamental frequency, one sine or cosine per period), to N (here 16, the negative fundamental frequency). Each sine or cosine consists of two delta functions.

A *frequency domain analysis* can explain the effect of noise on the polarimetric measurement and assist in visualizing the *null space*. The frequency content is calculated by taking the *discrete Fourier transform* (DFT, Math Tip 16.1) of the rows of $\mathbf{W}_P^{-1}$, plotted in Figure 7.5. These DFTs are related to the frequencies in the columns of $\mathbf{W}$ since each of the Stokes parameters generates specific frequencies in the flux vector (Figure 7.2). The data reduction needs to pick those frequencies out of $\mathbf{P}$. The third row of $\mathbf{W}_P^{-1}$, as seen by its green DFT (right side, sine frequency spectrum), is easily understood; to calculate S_2, the third row of Equation 7.40 picks out, DFT-like, the amplitude of the fourth harmonic component ($n = 5, 13$ in the DFT) of the flux vector. Similarly, the fourth row of $\mathbf{W}_P^{-1}$, as seen by its blue DFT, calculates S_3 from the amplitude of the second harmonic component of the flux vector. Since that component is modulated twice as much by a quarter wave retarder as S_1 and S_2, its DFT components in Equation 7.40 are half as large as S_3. The DFT of the top row of $\mathbf{W}_P^{-1}$, the red DFT, is initially puzzling. The flux from S_0 is not modulated; it is constant. Why does the red DFT contain fourth harmonic components ($n = 5, 13$)? S_1 modulates the fourth harmonic cosinusoidally but also has a DC component, the constant offset in the green curve of Figure 7.2. Thus, the data reduction needs to determine how much of the DC part of the flux vector is due to S_1 and remove that from the total DC to calculate the part of the DC component of the flux due to S_0.

The noise present in the flux vector and its effect on the measurement are understood through the DFT of the noise. Sampling the noise into 16 values, there will be a DC component, eight cosinusoidal components, and seven sinusoidal components (the eighth sinusoidal component, at the Nyquist frequency, is seen to sample to zero at all elements). These are plotted in Figure 7.6. Each of these 16 functions, considered as a vector, is orthogonal to all the other functions, forming a useful basis for the polarimeter signals.

Consider when the third harmonic, either sinusoidal or cosinusoidal for example, is present in the noise. The third harmonic is orthogonal to all the rows of the data reduction matrix, $\mathbf{W}_P^{-1}$. Thus, this component of the noise doesn't affect the polarimeter's accuracy! The third harmonic does not affect the measurement. The pseudoinverse automatically removes any third harmonic in $\mathbf{P}$ from consideration. Similarly, noise in the first harmonic, fifth, sixth, seventh, and the cosinusoidal part of the second harmonic are orthogonal to all the rows of $\mathbf{W}_P^{-1}$ and don't affect the measured Stokes parameters. The null space of $\mathbf{W}_P^{-1}$ is spanned by these 12 functions or "vectors," cosinusoidal frequencies: 1, 2, 3, 5, 6, 7, and 8, and sinusoidal frequencies: 1, 3, 5, 6, and 7. Only noise that lies in the modulation frequencies of the polarimeter, DC, fourth (cosinusoidal and sinusoidal), and the sinusoidal part of the second harmonic affects the measured Stokes parameters. Hence, three quarters of randomly distributed noise, $(16 - 4)/16$, should be filtered out by $\mathbf{W}_P^{-1}$ since it lies in the null space of $\mathbf{W}_P^{-1}$.

7.4 Mathematics of Polarimetric Measurement and Data Reduction

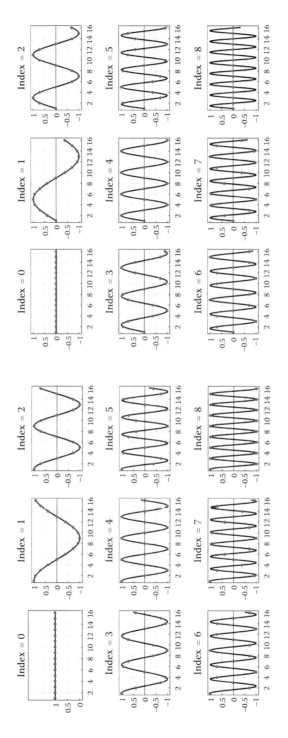

Figure 7.6 Sampled cosines (left) and sinusoids (right) for a real sampled function with 16 points form the 16 function basis set in the discrete Fourier transform. The sinusoids with index zero and eight are not part of the basis since they uniformly sample to zero.

What about all the matrix inverses other than $\mathbf{W}_P^{-1}$? Vectors corresponding to frequencies at the first harmonic, fifth, sixth, seventh, and the cosinusoidal part of the second harmonic can be added to any and all rows of $\mathbf{W}_P^{-1}$ and the result is another matrix inverse, $\mathbf{W}^{-1}$, since these components are orthogonal to all the columns of $\mathbf{W}$. Consider a matrix inverse $\mathbf{W}_A^{-1}$ where some proportion of these components are added to some of the rows. The resulting matrix inverse will now respond to the corresponding noise frequency if it is present in the flux vector. All that has been accomplished is to make the data reduction respond to signals that should always be noise. This *reduces the accuracy* of the polarimeter. Thus, the *pseudoinverse makes the data reduction immune to as much of the noise as possible*, and using any other matrix inverse is generally a *bad idea*. Another way to state this point, the pseudoinverse provides a *least-squares fit* of the data to the signals generated by the Stokes parameters. The polarimeter has no means to distinguish between noise and signal at the four frequencies that it is utilizing, but using the pseudoinverse, it can eliminate noise at frequencies different from the signal modulation. In fact, it can eliminate as much of the noise as possible.

7.5 Classes of Polarimeters

Polarimeters operate by acquiring measurements with a set of polarization analyzers. The following sections classify polarimeters into four broad methods in which multiple measurements are most often acquired. A complete Stokes polarimeter requires a minimum of four flux measurements with a set of linearly independent polarization generators in order to set up four equations with four unknowns, the four Stokes parameters. Many Stokes polarimeters use more than four flux measurements to improve the signal-to-noise ratio and/or reduce systematic errors. Polarimeters can be combined with *monochromators* or *spectrometers* to measure *Stokes parameter spectra* or *Mueller matrix spectra*.

7.5.1 Time-Sequential Polarimeters

In a time-sequential polarimeter, the series of flux measurements are taken sequentially in time. Between measurements, the polarization generator and analyzer are changed. Time-sequential polarimeters frequently employ rotating polarization elements or filter wheels containing a set of analyzers.

7.5.2 Modulated Polarimeters

Modulated polarimeters contain a polarization analyzer that varies periodically in time, space, or wavelength. Modulated polarimeters encode polarization information in channels at high frequency in the measurement.[11] *Spatially modulated polarimeters* such as micro-grid polarimeters or birefringent wedge prism polarimeters[12] can take "snapshot" polarization images of transient or rapidly varying phenomena. *Time-modulated polarimeters* employ retarders with periodically varying elements, such as retarders uniformly rotating in motors,[13] photoelastic modulators,[14] and sinusoidally varying liquid crystals.[15] Such polarimeters can be employed to capture high spatial resolution images of temporally static scenes.

Spectrally modulated polarimeters use the retardance dispersion of thick birefringent crystals in front of a polarizer to vary the analyzer vector with wavelength and encode the polarization into spectral information, which is then passed to a spectrometer where a spectrum is measured. Unpolarized light has its spectrum unmodified. As the degree of polarization increases, roughly cosinusoidal modulations of the spectrum occur, and the amplitude of these "channel spectra" indicates the degree of polarization.

Typically, spatially modulated polarimeters are considered to have error artifacts when measuring scenes that vary spatially in polarization or intensity,[16] while temporally modulated polarimeters are considered to have errors measuring temporally varying scenes.[17] Spectrally modulated polarimeters[18] have similar artifacts when the flux or polarization of measured object varies with wavelength. Nonetheless, Tyo has shown that if the corresponding polarization or intensity variations satisfy bandlimited constraints, artifact-free polarization measurements can be performed.[19,20]

7.5.3 Division of Amplitude

Division-of-amplitude polarimeters utilize beam splitters to divide the measured beam and direct component beams to multiple analyzers and detectors. A division-of-amplitude polarimeter can acquire its multiple measurements simultaneously, providing advantages for measurements of *rapidly changing scenes* or measurements from *moving platforms*. Many division-of-amplitude polarimeters use polarizing beam splitters to simultaneously divide and analyze the beam.

7.5.4 Division of Aperture

Division-of-aperture polarimeters use multiple polarization analyzers operating side by side. The aperture of the polarimeter beam is subdivided. Each beam propagates through a separate polarization analyzer to a separate detector. The detectors are usually synchronized to acquire measurements simultaneously. This is similar in principle to the polarizing glasses used in 3D movie systems, where different analyzers are placed over each eye, usually a right and left circular analyzer, presenting two different perspective views simultaneously to each eye.

7.5.5 Imaging Polarimeters

When the polarimeter's detector is a focal plane array, a series of images acquired with different analyzers (the raw images) can be reduced to measure a Stokes vector image or a Mueller matrix image.

Imaging polarimeters are particularly susceptible to misalignment of the raw images, since polarization properties are determined from the difference between flux measurements. Such misalignment causes polarization artifacts in the image on account of spurious polarization mixed with the actual polarization. Raw image misalignments occur due to source motion, polarimeter motion, vibration, and beam wander from a slight wedge in rotating components. Polarization artifacts are largest in areas where the image intensity is changing the fastest, around object edges, and near point sources. The edges of objects are usually where the angles of incidence and angles of scatter are larger. The largest polarization is typically expected around these areas. Because of vibration, image motion, and image misalignment, these are also the areas where the data are most suspect. Other errors result from imperfect polarization elements and detector noise.

When the source flux fluctuates between raw images, a uniform polarization error occurs across the entire image. Source fluctuations are a serious problem in outdoor Stokes imagery because sunlight fluctuates due to cloud motion.

Many polarization images and spectra, even those published, are inaccurate. In our polarization laboratory where rigorous polarimeter operating procedures are in place, many data acquisition runs are evaluated and then discarded as dubious and remeasured. There are problems with stray light, monochromator calibration and drift, ambient lights coming on in the room during measurements, misaligned samples, new software problems, and issues related to the building ventilation systems and power supply. It is recommended that all polarization measurements be approached with a degree of skepticism until the measurement system and measurement circumstances are clearly understood and appropriate tests and calibrations are provided.

7.6 Stokes Polarimeter Configurations

Polarimeters are constructed with a large variety of polarization generators depending on cost, speed, accuracy, size, and other considerations. All polarimeters incorporate one or more polarizers to analyze the polarization. Most polarimeters also use retarders to provide diversity in the analyzed polarization states. These retarders may be waveplates that are rotated or retarders with adjustable retardance such as liquid crystal devices, electro-optical modulators, and other retardance modulators. The next set of sections surveys some of the more common polarimeter configurations and comments on their key characteristics.

7.6.1 Simultaneous Polarimetric Measurement

Two classes of polarimeters can perform simultaneous polarimetric measurement: *division-of-aperture polarimeters* and *division-of-amplitude polarimeters*. The polarimeter configurations described earlier in Section 7.4 are sequential-in-time polarimeters. A single detector is used to make a sequence of measurements as the polarization analyzer is changed. Sequential-in-time polarimeters have significant polarization artifacts when the light is rapidly fluctuating during the measurement sequence.

Some polarimeters are developed to make all measurements simultaneously. Light sources, such as explosions, or rocket plumes, change their brightness or move too fast to be accurately measured by a sequential-in-time polarimeter. In division-of-wavefront polarimetry, the incident wavefront is divided spatially and simultaneous measurements are made at different points in space. Alternatively, in division-of-amplitude polarimetry, the amplitude of the incident wavefront is divided with beam splitters and multiple analyzers and detectors are used in parallel. These two classes of polarimeters generally have no moving parts.

7.6.1.1 Division-of-Aperture Polarimetry

Division-of-aperture polarimetry analyzes different parts of the wavefront with separate polarization elements. For example, multiple cameras can be pointed at the same scene with different polarizers integrated into each camera or radiometer. The first division-of-aperture polarimeter operated in space used a pair of bore-sighted cameras (cameras pointing in exactly the same direction) that were flown on the space shuttle.[21,22] A linear polarizer was placed in front of each camera with the polarizer's axes oriented orthogonal to each other. The detectors must be spatially registered so that the field of view of the detector elements on each focal plane is well aligned.

7.6.1.2 Division-of-Focal-Plane Polarimetry

Division-of-focal-plane polarimetry uses arrays of polarization analyzers incorporated above a focal plane such that neighboring pixels analyze different polarization states. These are also known as *micro-polarizer array polarimeters*, since they incorporate arrays of pixel-sized micro-polarizers integrated with a focal plane array, such as a CCD or CMOS image sensor.[23–25] An example *micro-polarizer array* configuration is shown in Figure 7.7 with a repeating array of four linear polarizers oriented at 0°, 45°, 90°, and 135° positioned above a focal plane. The earliest configurations used infrared *wire grid polarizers*[26] that are easier to fabricate than visible wire grid polarizers because of the larger wire sizes and wire pitch. As wire grid polarizer technology advanced, visible wire grid polarizers and micro-polarizer arrays were developed.[27] Many polarizer configurations can be used and the analyzers can be linear, elliptical, or circular.[15] Flux measurements from several pixels are needed to calculate the Stokes parameters at a small group of pixels, and then the

7.6 Stokes Polarimeter Configurations

Figure 7.7 A micro-polarizer array has a repeating set of polarizers and is suitable to be placed over a focal plane array such that adjacent pixels analyze different polarization states.

data reduction is repeated across the image. An example imaging Mueller matrix measurement of a micro-polarizer array is shown in Figure 7.8 where the M_{01} and M_{10} elements show the 0° and 90° polarizer elements and the M_{02} and M_{20} elements show the 45° and 135° polarizer elements. For this micro-polarizer array, the polarizer elements are all linear; no circular components were measured, and thus this is an incomplete polarimeter.

The advantage of the micro-polarizer array polarimeter method is the small size and simultaneous measurement of the Stokes parameters available from the polarization element array. The reduction in resolution of the detector by the number of different polarization elements and the spatial displacement of information within the polarization element pattern are disadvantages. The first micro-polarizer arrays were fabricated as separate components and then carefully aligned above the focal plane. The separation between the array and the detector array causes crosstalk for larger cones of light as more light passes through a neighboring polarizer before striking the intended pixel; that is, light through one polarizer element can reach neighboring pixels. Light reflected from the detector surface can also reflect again from the micro-polarizer array back toward the detector, causing problems. Micro-polarizer arrays directly fabricated on the CMOS or CCD surface minimize this shadowing between neighboring pixels and also minimize multiple reflections.

Micro-polarizer arrays of linear elements are common. Arrays of elliptically polarizing elements to measure full Stokes images using *dichroic dyes* have also been developed as shown in Figure 7.9. A thin photopolymerizable polymer layer is spun onto the substrate, and then a pattern of polarized UV light produces a pixelated alignment layer. A liquid crystal polymer acting as a retarder is spun on as thin film, the fast axis oriented with an alignment layer, which is then polymerized into a hard

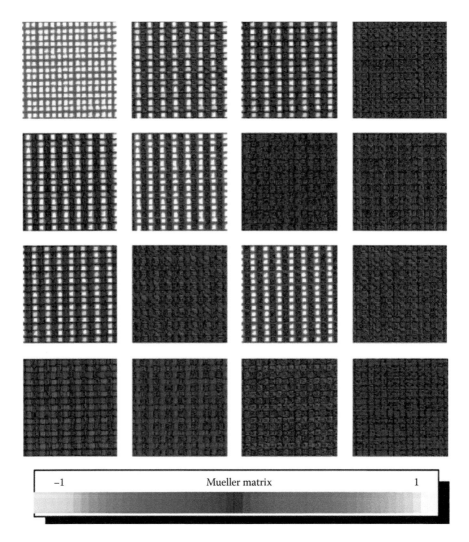

Figure 7.8 A Mueller matrix image of a micro-polarizer array. The M_{00} image (upper left) shows the apertures of the polarizers as yellow and opaque regions as dark red. In the M_{01} and M_{10} images, the yellow squares are 0° polarizers and the blue squares are 90° polarizers. Similarly in the M_{02} and M_{20} images, the yellow squares are 45° polarizers and the blue squares are 135° polarizers. Ideally, the last column and the last row would be identically zero. Quantitative analysis of this image provides precise measurements of polarizer diattenuation and orientation errors.

film with UV light. An isolation layer with a smooth upper surface is added to form a substrate for the polarizer. The process is repeated to fabricate the polarizer layer, consisting of a liquid polymer doped with dyes. Examples of polarized images including degree of linear polarization (*DoLP*) measured by a full Stokes division-of-focal-plane polarimeter incorporating this technology are shown in Figures 7.10 and 7.11.

7.6 Stokes Polarimeter Configurations

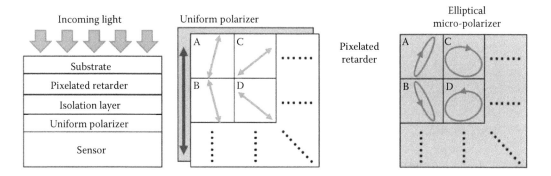

Figure 7.9 (Left) A full Stokes micro-polarizer array is fabricated on a substrate with a pixelated retarder layer, an isolation layer, and then a uniform polarizer layer, and placed in proximity with the sensor. (Middle) The pixelated linear retarder array is fabricated with four orientations: A, B, C, and D, as shown. Light propagates through the pixelated retarder and then through a uniform linear polarizer. (Right) The corresponding analyzed states are represented by ellipses: A, B, C, and D, which are optimally located on the Poincaré sphere. (Courtesy of OSA *Optics Express*; Figure 2 in W.-L. Hsu, G. Myhre, K. Balakrishnan, N. Brock, M. Ibn-Elhaj, and S. Pau, Full-Stokes imaging polarimeter using an array of elliptical polarizer, *Opt. Express* 22(3), 3063–3074 (2014).[28])

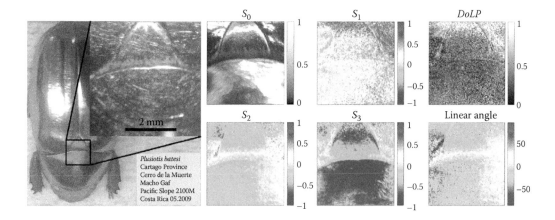

Figure 7.10 Full Stokes image of a beetle, *Plusiotis batesi*, reflects circularly polarized light from its shell taken with the full Stokes division-of-focal-plane polarimeter. (Left) Image of the whole beetle, and a small square indicating the measured region near the head, enlarged in the upper right of the frame. (Right) Measured polarization parameters: S_0, S_1, $DoLP$, S_2, S_3, and the linear angle. The S_3 image shows the remarkable characteristic of this beetle's exoskeleton, which reflects unpolarized incident light into light that is nearly left circularly polarized. (Courtesy of OSA Optic Express; Figure 9 in W.-L. Hsu, J. Davis, K. Balakrishnan, M. Ibn-Elhai, S. Kroto, N. Brock, and S. Pau, Polarization microscope using a near infrared full-Stokes imaging polarimeter, *Opt. Express* 23(4), 4357–4368 (2015).[29])

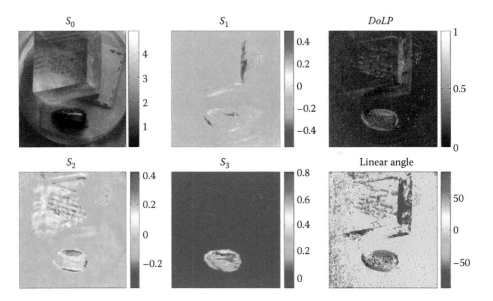

Figure 7.11 (Top left) Intensity image, S_0, of beetle, dark object at bottom, and calcite crystal, upper center. The beetle shows some linear polarization in S_1, *DoLP*, and S_2. The striking effect is the beetle's substantial circular polarization seen in S_3 (lower middle). Note the absence of any other circular polarization in the image. (Courtesy of OSA *Optics Express*; Figure 8 in W.-L. Hsu, G. Myhre, K. Balakrishnan, N. Brock, M. Ibn-Elhaj, and S. Pau, Full-Stokes imaging polarimeter using an array of elliptical polarizer, *Opt. Express* 22(3), 3063–3074 (2014).[28])

7.6.1.3 Division-of-Amplitude Polarimetry

In *division-of-amplitude polarimetry* (DOAP), the amplitude of the incident beam is divided into several components with beam splitters, then analyzed, and finally detected at multiple detectors. For example, a *polarizing beam splitter* or *Wollaston prism* can divide the beam into two orthogonal components that can be measured separately. Several variations of division-of-amplitude polarimeters will be described.

DOAPs may employ as few as two detectors to analyze two orthogonally polarized components of light using a polarizing beam splitter, or it can measure the complete Stokes parameters using four detectors. To analyze more than two Stokes parameters, the first beam splitter of a division-of-amplitude polarimeter *must be partially polarizing*, not completely polarizing. If the initial beam splitters are polarizing (transmit a degree of polarization of one), the light is completely analyzed into two states at the first beam splitter, and subsequent beam splitters cannot measure any additional linearly independent information.

A four-channel DOAP based on Gamiz,[30] shown in Figure 7.12, uses a partially polarizing beam splitter, two retarders, and two polarizing beam splitters. Measurements are made at four detectors. The polarization elements can be chosen to analyze at four points on the Poincaré sphere located at the vertices of a regular tetrahedron. This optimizes the condition number of the polarimetric data reduction matrix (see Section 7.11). The ideal values for the partially polarizing beam splitter are found to be close to 80% and 20% for the parallel and perpendicular components. A quarter wave retarder before detectors 1 and 2 is oriented at 45° and the half wave retarder before detectors 3 and 4 is oriented at 22.5°. For an imaging polarimeter, care must be taken to ensure spatial registration of the detectors and equalization of detector response. Ideally, the four focal planes are aligned to 1/100 of the pixel pitch, and magnification and distortion are controlled at a similar level. Imaging Ronchi rulings and similar periodic patterns and observing Moiré fringes on the four detectors can facilitate such fine alignment between channels. A drawback to the use of different detectors to

7.6 Stokes Polarimeter Configurations

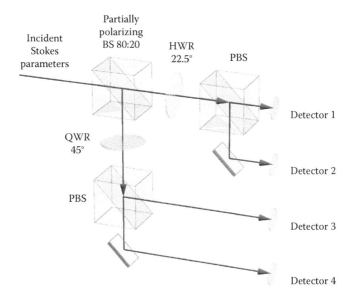

Figure 7.12 A four-channel division-of-amplitude polarimeter configured so four beams exit in parallel. The first element is a partially polarizing beam splitter that transmits 0.8 of the p-polarized flux and 0.2 of the s-polarized flux; the remaining flux reflects. PBS is a polarizing beam splitter, QWR is a quarter wave retarder, and HWR is a half wave retarder.

reconstruct the Stokes parameters is that small errors in calibration and drift between the detectors lead to *DoLP* measurement uncertainty. Polarimeters of this class have proven extremely difficult to operate at 1% or better polarization accuracy.

Another DOAP configuration, shown in Figure 7.13, outputs four beams analyzed into I_0, I_{90}, I_{45}, and I_{135}, all exiting parallel and in the same polarization state, polarized in the plane of the page. Let's consider the four analyzed beams one at a time. The incident light is split by a non-polarizing beam splitter. The reflected beam internally reflects and is split by a polarizing beam splitter. (1) The p-polarized transmitted beam, now analyzed and polarized in the plane of the

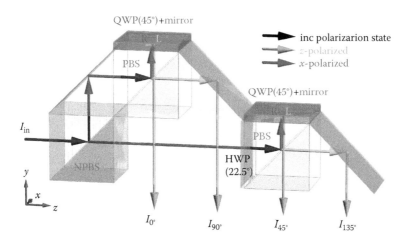

Figure 7.13 A division-of-amplitude polarimeter for linear polarization with four beams with equal path lengths exiting in the same direction.

page (90°) internally reflects again and exits out the bottom. (2) The *s*-polarized reflected beam, now analyzed and polarized along *z*, transmits through a quarter waveplate at 45°, becoming right circularly polarized, and normally reflects from the mirror at the top into a left circularly polarized beam. After passing through the quarter waveplate a second time, the beam is now *p*-polarized so that it transmits through the polarizing beam splitter and exits out the bottom polarized along *z* but analyzing the incident beam at 0°. Returning to the beam transmitted by the non-polarizing beam splitter (horizontal black arrow), this passes through a two-cube-width spacer block and passes through a half wave linear retarder oriented at 22.5°; this rotates the linear polarizations by 45°. The beam is incident on a polarizing beam splitter. (3) The light component incident on the DOAP at 45° is now oriented at 0°, and this *s*-polarized light reflects upward. Another quarter waveplate at 45° followed by a mirror transforms the beam into a downward propagating 90° polarized beam, which, since it is *p*-polarized at the polarizing beam splitter, is transmitted and exits out the bottom of the DOAP, and contains half the 45° component of the incident light. (4) Returning to the second PBS, the light component incident on the DOAP at 135° is now oriented at 90°, and this *p*-polarized light is transmitted. This beam internally reflects at the far-right side of the DOAP and exits oriented at 90° but analyzing the incident 135° light. Thus, at the exit port, the leftmost beam analyzes I_0, then I_{90}, then I_{45}, and to the right I_{135}, but all are polarized along *z*. Note that all four beams have an equal path length; the path length is equal to five times the width of the component cubes.

A different type of DOAP is the four-detector photopolarimeter, a complete Stokes polarimeter, described by Azzam,[31] where a light beam strikes four detectors in sequence, as shown in Figure 7.14. Part of the light striking the first detector is reflected to the second detector, with a larger fraction of the *p*-polarized light transmitted into and measured at the first detector, and more of the *s*-polarized light specularly reflected to the second detector. Similarly, the second detector is rotated to preferentially measure the incident 45° component. Likewise, the third detector partially analyzes the light, reflecting still a different state to the fourth detector. The last detector, at normal incidence, absorbs substantially all the remaining light. Any reflected light headed back toward the polarimeter's entrance is partially detected by the first three detectors boosting the overall sensitivity. The detectors are coated with thin films designed for specific retardances on reflection, providing the polarimeter with sensitivity to the incident circular polarization as well as linear polarizations. The signal measured by each detector is proportional to the fraction of the light that it absorbs, and that fraction is a linear combination of the Stokes parameters. The polarimeter requires out-of-plane reflections in order to measure S_2 and S_3. This type of polarimeter is very angle of incidence–sensitive; the incident angle should be maintained to better than 0.1° for 1% accuracy.

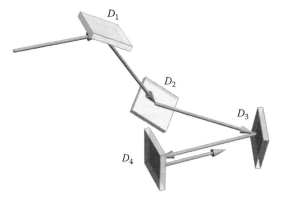

Figure 7.14 In Azzam's four-detector photopolarimeter, light is partially detected at each detector, and the reflected light from each detector has its polarization changed.

7.6.2 Rotating Element Polarimetry

Many Stokes polarimeters use rotating polarization elements, either rotating linear retarders or rotating polarizers, to modulate the detected flux in time. Depending on the configuration, different combinations of Stokes parameters can be measured.

7.6.2.1 Rotating Analyzer Polarimeters

The rotating analyzer polarimeter (Figure 7.15) measures the linear Stokes parameters by rotating a polarizer in front of a detector. The polarimeter's analyzer vector as a function of the analyzer's transmission axis θ, is

$$\mathbf{A}(\theta) = \frac{1}{2}\begin{pmatrix} 1 & \cos 2\theta & \sin 2\theta & 0 \end{pmatrix}^T. \tag{7.42}$$

The superscript T for transpose indicates a column vector. The trajectory of the analyzer is shown in green on the Poincaré sphere in Figure 7.15 (right) as a circle around the equator. Because the last element of $\mathbf{A}$ is zero, this is an incomplete polarimeter, which cannot measure S_3. When the measurements are taken in equal angular increments over a multiple of 180°, a data reduction algorithm based on Fourier series is easily developed. If Q equally spaced flux measurements in angle spanning a multiple of 180° are acquired,

$$\theta_q = q\Delta\theta, \quad \Delta\theta = n\frac{180°}{Q}, \text{ where } q = 0, 1, 2, \ldots, Q-1, \tag{7.43}$$

the signal is periodic. The analyzer vector operating on the first three Stokes parameters is

$$\begin{aligned} P(\theta) &= \mathbf{A}(\theta) \cdot \mathbf{S} \\ &= \frac{1}{2}\begin{pmatrix} 1 & \cos n\theta & \sin n\theta \end{pmatrix}^T \cdot \begin{pmatrix} S_0 & S_1 & S_2 \end{pmatrix}^T \\ &= \frac{1}{2}(S_0 + S_1 \cos n\theta + S_2 \sin n\theta). \end{aligned} \tag{7.44}$$

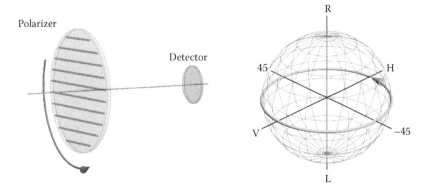

Figure 7.15 (Left) Rotating analyzer polarimeter rotates a polarizer in front of a detector or camera and measures S_0, S_1, and S_2. (Right) The analyzer vector for a rotating polarizer polarimeter moves around the equator of the Poincaré sphere.

Thus, if a polarizer is rotated through 360° ($n = 2$) and fluxes are acquired at equally spaced angles, the data can be expressed as the Fourier series

$$P(\theta) = a_0 + a_2 \cos 2\theta + b_2 \sin 2\theta, \tag{7.45}$$

where the signal is modulated on the second harmonic, and the first harmonic coefficients are zero ($a_1 = 0 = b_1$). The measured Stokes parameters are

$$S_0 = 2a_0, \ S_1 = 2a_2, \ S_2 = 2b_2. \tag{7.46}$$

The orientation of the polarization, ψ, and the degree of linear polarization, $DoLP$, are

$$\psi = \frac{1}{2} \arctan\left(\frac{S_2}{S_1}\right) = \frac{1}{2} \arctan\left(\frac{b_2}{a_2}\right), \ DoLP = \frac{\sqrt{S_1^2 + S_2^2}}{S_0} = \frac{\sqrt{b_2^2 + a_2^2}}{a_0}. \tag{7.47}$$

One issue with the rotating polarizer polarimeter is that most detectors have some polarization sensitivity; they respond slightly differently to different linear polarization states of equal flux. For a photodiode, for example, the responsivity may vary by 0.25% to 2% as the plane of linear polarization is rotated. To perform polarimetry at the 1% accuracy level, such detector sensitivity and the sensitivity (diattenuation) of any other optical components need to be incorporated into the analyzer vector through calibration and thus improve the accuracy of the polarimetric data reduction matrix. This detector sensitivity is addressed with the next polarimeter configuration.

7.6.2.2 Rotating Analyzer Plus Fixed Analyzer Polarimeter

One method to address the residual polarization sensitivity of detectors is to combine a rotating analyzer with a second fixed analyzer. Thus, the system transmits only a single polarization state through any optics after the second analyzer (fold mirrors, grating, etc.) to the detector, as shown in Figure 7.16. A disadvantage is that now the average transmission through the pair of analyzers decreases from one-half to one-quarter. Another disadvantage is that the flux will always be low for a source polarized nearly orthogonal to the second analyzer, making measurements less accurate near that state.

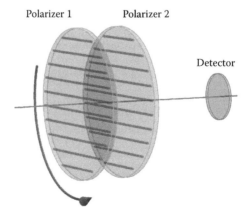

Figure 7.16 Polarimeter with rotating analyzer plus a fixed analyzer before the detector removes issues with diattenuation in the detector or optics following the second polarizer since it transmits a fixed polarization state following polarizer 2.

7.6 Stokes Polarimeter Configurations

7.6.2.3 Rotating Retarder and Fixed Analyzer Polarimeters

The most common rotating element polarimeter is the rotating retarder polarimeter, because of the ease of assembly and calibration, and because all four *Stokes parameters* are measured, the polarimeter is complete. The incident light transmits through a rotating linear retarder, followed by a fixed linear polarizer; then, the light is incident on a detector and a series of flux measurements are acquired while the retarder is rotated to a series of angles θ_q. This configuration is illustrated in Figure 7.17.

The modulated signal is composed of two frequencies, which can be expressed as the following Fourier series:

$$P = \frac{a_0}{2} + \frac{1}{2}\sum_{n=1}^{2}\left(a_{2n}\cos(2n\theta) + b_{2n}\sin(2n\theta)\right), \tag{7.48}$$

where θ is the azimuthal angle of the retarder. The analyzer vector is calculated by multiplying linearly polarized light by the Mueller matrix of a linear retarder. Assuming an ideal polarizer with its transmission axis oriented at 0°, the analyzer vector as a function of retarder orientation and retardance δ is the first row ($()_0$ indicates the top (0th) row of a Mueller matrix) of the matrix

$$\mathbf{A}(\theta) = \left(\mathbf{LP}(0)\cdot\mathbf{LR}(\delta,\theta)\right)_0$$

$$= \frac{1}{2}\begin{pmatrix}1\\1\\0\\0\end{pmatrix}^T \cdot \begin{pmatrix} 1 & 0 & 0 & 0 \\ 0 & \cos^2 2\theta + \cos\delta\sin^2 2\theta & (1-\cos\delta)\cos 2\theta\sin 2\theta & -\sin\delta\sin 2\theta \\ 0 & (1-\cos\delta)\cos 2\theta\sin 2\theta & \cos\delta\cos^2 2\theta + \sin^2 2\theta & \cos 2\theta\sin\delta \\ 0 & \sin\delta\sin 2\theta & -\cos 2\theta\sin\delta & \cos\delta \end{pmatrix} \tag{7.49}$$

$$= \frac{1}{2}\begin{pmatrix} 1 \\ \cos^2 2\theta + \cos\delta\sin^2 2\theta \\ (1-\cos\delta)\cos 2\theta\sin 2\theta \\ -\sin\delta\sin 2\theta \end{pmatrix}.$$

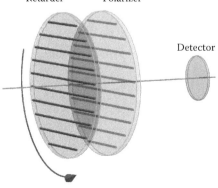

Figure 7.17 Rotating retarder, fixed polarizer polarimeter.

For example, if the retarder's retardance is $\delta = \pi/2$, the analyzer vector becomes

$$\mathbf{A}(\theta) = \frac{1}{2}\begin{pmatrix} 1 \\ \cos^2 2\theta \\ \cos 2\theta \sin 2\theta \\ -\sin 2\theta \end{pmatrix} = \frac{1}{2}\begin{pmatrix} 1 \\ \dfrac{1+\cos 2\theta}{2} \\ \dfrac{\sin 4\theta}{2} \\ -\sin 2\theta \end{pmatrix}. \tag{7.50}$$

The expression on the right is given strictly as the sum of terms $\sin(n\theta)$ and $\cos(n\theta)$, so the flux $P(\theta)$ can be manipulated into a Fourier series with the Stokes parameters related to the Fourier coefficients as

$$\begin{aligned} P(\theta) &= \mathbf{A}(\theta) \cdot \mathbf{S} \\ &= \frac{1}{2}S_0 + \frac{1}{4}S_1 + \frac{\cos 4\theta}{4}S_1 + \frac{\sin 4\theta}{4}S_2 - \frac{\sin 2\theta}{2}S_3 \\ &= a_0 + b_2 \sin 2\theta + a_4 \cos 4\theta + b_4 \sin 4\theta. \end{aligned} \tag{7.51}$$

Therefore, a *Fourier series* of a sequence of flux measurements taken over 360° has the *Fourier coefficients*

$$P(\theta) = a_0 + b_2 \sin 2\theta + a_4 \cos 4\theta + b_4 \sin 4\theta. \tag{7.52}$$

Other frequency components should be zero or contain noise. By manipulating the Fourier coefficients, the Stokes parameters are found to be

$$S_0 = 2(a_0 - a_4),\ S_1 = 4a_4,\ S_2 = 4b_4,\ S_3 = -2b_2. \tag{7.53}$$

If the sequence of flux measurements are taken over 180° of rotation, the signal is modulated as $\cos 2\theta$, $\sin 2\theta$, and $\sin \theta$.

One advantage of rotating retarder polarimeter configurations is that the detector observes only a single polarization state; hence, any polarization sensitivity of the detector will not cause signal modulation or compromise the polarimeter's accuracy. Additional optical elements may be located anywhere in the optical system, but it is preferable when possible to locate them after the polarizer. When strongly polarizing elements like *fold mirrors* and *diffraction gratings* are placed after the polarizer, their polarization properties do not affect the signal. Another advantage is that since the linear polarizer is not rotated, only a single rotary stage is used.

Section 6.8.2 describes the *Poincaré sphere* trajectories for rotating retarders of different retardances. If the rotating retarder is a half wave linear retarder, the trajectory circles the equator and the polarimeter is incomplete. The last element of the analyzer vector is zero and the polarimeter will not measure S_3. The analyzer analyzes a rotating linear polarization state; the polarimeter acts like a rotating polarizer polarimeter. In general, the rotating retarder polarimeter becomes inaccurate for values of the retardance close to zero and near a half wave as the *polarimetric measurement matrix* becomes nearly singular.

7.6 Stokes Polarimeter Configurations

Example 7.5 Rotating Quarter Wave Retarder Polarimeter

The previous section analyzed the data in terms of the Fourier series of the flux values; here, an example of the polarimetric data reduction matrix method is provided for a rotating quarter waveplate. Let the polarimeter take eight measurements in 180° of rotation, so flux measurements are acquired at angles of 0°, 22.5°, 45°, 67.5°, 90°, 112.5°, 135°, and 157.5°. The polarimetric measurement matrix is

$$\mathbf{W} = \frac{1}{4}\begin{pmatrix} 2 & 2 & 0 & 0 \\ 2 & 1 & 1 & -\sqrt{2} \\ 2 & 0 & 0 & -2 \\ 2 & 1 & -1 & -\sqrt{2} \\ 2 & 2 & 0 & 0 \\ 2 & 1 & 1 & \sqrt{2} \\ 2 & 0 & 0 & 2 \\ 2 & 1 & -1 & \sqrt{2} \end{pmatrix}. \qquad (7.54)$$

Note that the first row equals the fifth row since the analyzer vector is the same when the retarder's fast axis is parallel or perpendicular to the polarizer's transmission axis. This redundancy does not matter to the **W** matrix analysis method; it utilizes whatever arbitrary sequence of measurements are provided. The Stokes parameters are determined from the *polarimetric data reduction matrix*, $\mathbf{W}_P^{-1}$, operating on the flux vector of measurements **P**,

$$\mathbf{S} = \mathbf{W}_P^{-1} \cdot \mathbf{P}$$

$$= \frac{1}{4}\begin{pmatrix} -1 & 1 & 3 & 1 & -1 & 1 & 3 & 1 \\ 4 & 0 & -4 & 0 & 4 & 0 & -4 & 0 \\ 0 & 4 & 0 & -4 & 0 & 4 & 0 & -4 \\ 0 & -\sqrt{2} & -2 & -\sqrt{2} & 0 & \sqrt{2} & 2 & \sqrt{2} \end{pmatrix} \begin{pmatrix} P_0 \\ P_{22.5} \\ P_{45} \\ P_{67.5} \\ P_{90} \\ P_{112.5} \\ P_{135} \\ P_{157.5} \end{pmatrix}. \qquad (7.55)$$

The Fourier components are seen in the columns of **W** and the rows of $\mathbf{W}_P^{-1}$. The second and third rows of $\mathbf{W}_P^{-1}$ have two periods of cosine and sine respectively. The bottom row has one period of sine.

7.6.3 Variable Retarder and Fixed Polarizer Polarimeter

Variable retarders allow the magnitude of retardance to be controlled while the fast and slow states remain fixed as shown in Figure 7.18 (left). A polarimeter with a single variable linear retarder as seen in Figure 7.18 (right), such as a *liquid crystal cell*, an *electro-optic modulator*, or a *photoelastic modulator*, in front of a linear polarizer can analyze states along a single circle on the *Poincaré sphere*.[8] Usually, the variable retarder's fast axis is located 45° from the polarizer's transmission axis to get the longest arc on the Poincaré sphere. As an example, for a retarder fast axis at 0° and a polarizer transmission axis at 45°, the analyzer vector $\mathbf{A}(\delta)$ is the top row of the Mueller matrix for the polarization elements,

$$\mathbf{A}(\delta) = \left(\mathbf{LP}(\pi/4) \cdot \mathbf{LR}(\delta,0)\right)_0 = \frac{1}{2}\begin{pmatrix}1\\0\\1\\0\end{pmatrix}^T \cdot \mathbf{LR}(\delta,0)$$

(7.56)

$$= \frac{1}{2}\begin{pmatrix}1\\1\\0\\0\end{pmatrix}^T \cdot \begin{pmatrix}1 & 0 & 0 & 0\\0 & 1 & 0 & 0\\0 & 0 & \cos\delta & \sin\delta\\0 & 0 & -\sin\delta & \cos\delta\end{pmatrix} = \frac{1}{2}\begin{pmatrix}1\\0\\\cos\delta\\\sin\delta\end{pmatrix}.$$

Thus, for a variable retarder with its fast axis located at 0° followed by a 45° polarizer, S_0, S_2, and S_3 can be measured but S_1 cannot be measured.

With two variable retarders, a *complete analyzer* can be constructed, capable of measuring all states (Figure 7.19). The optimal orientation places the second fast axis at 45° from the first retarder's transmission axis. For example, choosing orientations 45° for the first variable retarder's fast

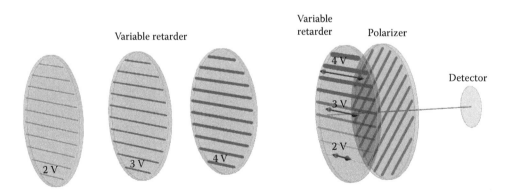

Figure 7.18 (Left) Liquid crystal variable linear retarders have electrically adjustable retardance indicated by line thickness. (Right) Polarimeter with a variable retarder in front of a linear polarizer. Optimally, the retarder's fast axis is located at 45° from the polarizer's transmission axis. The variation of voltage in time to the liquid crystal cell is indicated by the different voltages and line thicknesses on the variable retarder.

7.6 Stokes Polarimeter Configurations

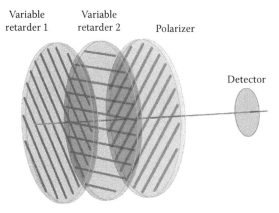

Figure 7.19 A polarization analyzer formed from two variable linear retarders can form a complete Stokes polarimeter, one capable of analyzing all polarization states. Optimally, the fast axes of retarders 1 and 2 are 45° apart, and the polarizer's transmission axis is 45° from fast axis 2.

axis, 0° for the second variable retarder's fast axis, and polarizer transmission axis 135° yields the analyzer vector

$$\mathbf{A}(\delta_1,\delta_2) = \left(\mathbf{LP}(-\pi/4) \cdot \mathbf{LR}(\delta_2,0) \cdot \mathbf{LR}(\delta_1,\pi/4)\right)_0$$

$$= \frac{1}{2}\begin{pmatrix} 1 \\ 0 \\ -1 \\ 0 \end{pmatrix}^T \begin{pmatrix} 1 & 0 & 0 & 0 \\ 0 & 1 & 0 & 0 \\ 0 & 0 & \cos\delta_2 & \sin\delta_2 \\ 0 & 0 & -\sin\delta_2 & \cos\delta_2 \end{pmatrix} \cdot \begin{pmatrix} 1 & 0 & 0 & 0 \\ 0 & \cos\delta_1 & 0 & -\sin\delta_1 \\ 0 & 0 & 1 & 0 \\ 0 & \sin\delta_1 & 0 & \cos\delta_1 \end{pmatrix} \quad (7.57)$$

$$= \frac{1}{2}\begin{pmatrix} 1 \\ -\sin\delta_1 \sin\delta_2 \\ -\cos\delta_2 \\ -\cos\delta_1 \sin\delta_2 \end{pmatrix}.$$

7.6.4 Photoelastic Modulator Polarimeters

One common form of variable linear retarder is the *photoelastic modulator* or *PEM*. Applying force to transparent materials induces retardance through *stress-induced birefringence*, as described in Chapter 25. Variable retarders can be fabricated by applying force to transparent elements, but because of the small values of the *stress-optic coefficient* in glasses and other suitable materials, large forces are necessary. A clever way to overcome this limitation, the *PEM*, was invented by Badoz in the 1960s[32] and further developed by Kemp.[33] The PEM was commercialized in the 1970s by Hinds Instruments, which developed an array of PEM components and PEM-based ellipsometric and polarimetric instruments.[34]

To form a PEM, a highly resonant element shape is fabricated. Elements with *mechanical quality factors* (*Q* factors) in excess of 10^4 can easily be fabricated. When excited with a sound wave from a *piezoelectric transducer* at the element's fundamental frequency, the amplitude of the standing acoustic wave within the crystal rapidly builds up to 10^4 of the amplitude of the driving signal and significant birefringence and retardance are achieved with reasonable driving signals of less than 0.5 W. The resulting retardance is cosinusoidally modulated and maintains a very steady frequency of retardance oscillation. For retardance modulation along the x–y axes, the PEM's time-dependent Mueller matrix $\mathbf{LR}(t, \delta_0, \theta = 0)$ is

$$\mathbf{LR}(t,\delta_0,\theta=0) = \begin{pmatrix} 1 & 0 & 0 & 0 \\ 0 & 1 & 0 & 0 \\ 0 & 0 & \cos(\delta_0 \cos(\omega t)) & \sin(\delta_0 \cos(\omega t)) \\ 0 & 0 & -\sin(\delta_0 \cos(\omega t)) & \cos(\delta_0 \cos(\omega t)) \end{pmatrix}, \quad (7.58)$$

where ω is 2π times the frequency of the sound wave and δ_0 is the magnitude of the cosinusoidal retardance. By varying the amplitude of the driving sound wave, the amplitude of retardance modulation δ_0 can be adjusted to, for example, a quarter wave, a half wave, or far larger amplitudes.

PEMs are combined with quarter wave retarders to create *circular retardance modulators* that will oscillate the orientation of incident linearly polarized light. The circular retardance modulators consist of a PEM variable linear retarder between crossed quarter wave linear retarders, described by the Mueller matrix equation

$$\mathbf{CR}(t) = \mathbf{LR}(\pi/2, \pi/4) \cdot \mathbf{LR}(\delta_0 \cos(\omega t), 0) \cdot \mathbf{LR}(\pi/2, -\pi/4)$$

$$= \begin{pmatrix} 1 & 0 & 0 & 0 \\ 0 & 0 & 0 & -1 \\ 0 & 0 & 1 & 0 \\ 0 & 1 & 0 & 0 \end{pmatrix} \begin{pmatrix} 1 & 0 & 0 & 0 \\ 0 & 1 & 0 & 0 \\ 0 & 0 & \cos(\delta_0 \cos(\omega t)) & \sin(\delta_0 \cos(\omega t)) \\ 0 & 0 & -\sin(\delta_0 \cos(\omega t)) & \cos(\delta_0 \cos(\omega t)) \end{pmatrix} \begin{pmatrix} 1 & 0 & 0 & 0 \\ 0 & 0 & 0 & 1 \\ 0 & 0 & 1 & 0 \\ 0 & -1 & 0 & 0 \end{pmatrix}$$

$$= \begin{pmatrix} 1 & 0 & 0 & 0 \\ 0 & \cos(\delta_0 \cos(\omega t)) & \sin(\delta_0 \cos(\omega t)) & 0 \\ 0 & -\sin(\delta_0 \cos(\omega t)) & \cos(\delta_0 \cos(\omega t)) & 0 \\ 0 & 0 & 0 & 1 \end{pmatrix}.$$

(7.59)

When linearly polarized light is incident, the output is a linearly polarized beam with an oscillating orientation, sinusoidally rocking back and forth. For example, if the retardance amplitude δ_0 is a quarter of a wave with 0° light incident, aligned along the 3:00 to 9:00 axis on a clock, the output orientation oscillates sinusoidally from 12:00 to 6:00. Equation 7.59 is an example of a unitary transformation rotating the eigenstates of a polarization element on the Poincaré sphere.

7.6 Stokes Polarimeter Configurations

PEMs are combined with linear polarizers to analyze linear polarization. The PEM circular retardance modulator followed by a 0° linear polarizer has the analyzer vector

$$\mathbf{A}(t) = \left(\mathbf{LP}(0) \cdot \mathbf{CR}(\delta_0)\right)_0$$

$$= \frac{1}{2}\begin{pmatrix} 1 & 1 & 0 & 0 \\ 1 & 1 & 0 & 0 \\ 0 & 0 & 0 & 0 \\ 0 & 0 & 0 & 0 \end{pmatrix} \cdot \begin{pmatrix} 1 & 0 & 0 & 0 \\ 0 & \cos(\delta_0 \cos(\omega t)) & \sin(\delta_0 \cos(\omega t)) & 0 \\ 0 & -\sin(\delta_0 \cos(\omega t)) & \cos(\delta_0 \cos(\omega t)) & 0 \\ 0 & 0 & 0 & 1 \end{pmatrix}_0$$

$$= \frac{1}{2}\begin{pmatrix} 1 \\ 1 + \cos(\delta_0 \cos(\omega t)) \\ \sin(\delta_0 \cos(\omega t)) \\ 0 \end{pmatrix}. \tag{7.60}$$

The *Multi-Angle Imager for Aerosols* polarimeter of Section 7.6.5 has two channels, one with a 0° polarizer following a PEM, and the other with a 45° polarizer following the same PEM in parallel. For the 45° channel, the analyzer vector is

$$\mathbf{A}(t) = \left(\mathbf{LP}(\pi/4) \cdot \mathbf{CR}(\delta_0)\right)_0$$

$$= \frac{1}{2}\begin{pmatrix} 1 & 0 & 1 & 0 \\ 0 & 0 & 0 & 0 \\ 1 & 0 & 1 & 0 \\ 0 & 0 & 0 & 0 \end{pmatrix} \cdot \begin{pmatrix} 1 & 0 & 0 & 0 \\ 0 & \cos(\delta_0 \cos(\omega t)) & \sin(\delta_0 \cos(\omega t)) & 0 \\ 0 & -\sin(\delta_0 \cos(\omega t)) & \cos(\delta_0 \cos(\omega t)) & 0 \\ 0 & 0 & 0 & 1 \end{pmatrix}_0$$

$$= \frac{1}{2}\begin{pmatrix} 1 \\ 1 - \sin(\delta_0 \cos(\omega t)) \\ \cos(\delta_0 \cos(\omega t)) \\ 0 \end{pmatrix}. \tag{7.61}$$

Some PEM benefits include large apertures, low operating voltages, high optical power acceptance (for use with *high-power lasers*), and wide angular acceptance.[35] Because the retardance arises from a standing wave, there is spatial variation of retardance, approximately quadratic about the center of the element. PEMs are commonly constructed from glass, fused silica, ZnSe, and CaF_2 and have transmittance over a wide spectral range. Because of the high frequency stability, polarimetric sensitivities (i.e., precision) of better than 10^{-5} have been obtained.[36–38]

Typical PEM element sizes of tens of centimeters yield fundamental frequencies in the range of 20 to 80 kHz. Hence, PEM polarimeters utilize high-speed detectors measuring at megahertz rates to resolve the modulation. Slow frequency operation of PEM modulators is not practical as the element size would be unreasonably large. Two PEMs can be operated as a pair with slightly different frequencies. Then, the retardance signal comprises a high-frequency signal at the average frequency and a low-frequency signal at the beat frequency, providing low-frequency modulation that can be measured with cameras or other slower detectors. PEMs are often operated as electro-optic circular retardance modulators between pairs of quarter wave retarders.

7.6.5 The MSPI and MAIA Imaging Polarimeters

Two examples of PEM-based imaging polarimeters are the *AirMSPI* and *MAIA* polarimeters developed at the Jet Propulsion Laboratory for remote sensing of aerosols in the atmosphere. These polarimeters measure the first three Stokes parameters using circular retardance modulators in front of a linear polarizer as described in Section 7.6.4 to modulate the incident polarization.

AirMSPI, shown in Figure 7.20 (left), is short for the *Airborne Multi-Angle Spectro Polarimetric Imager*. AirMSPI was developed as a prototype polarimeter for a future NASA Earth Sciences mission to study aerosols, clouds, and land surfaces from space. A second-generation instrument, AirMSPI-2, measures linear polarization in five spectral bands centered at 445, 645, 865, 1620, and 2185 nm. The optical layout with the dual PEM modulators and side-by-side SWIR and UV-VIS-NIR focal planes is shown in Figure 7.20 (right).

AirMSPI developed the technologies for highly accurate *DoLP* measurements achieving accuracies of better than 0.5%. Then, AirMSPI flew over 100 flights on NASA's high-altitude *ER-2* aircraft acquiring polarimetric images of the atmosphere and Earth's surface from up to an altitude of 20 km. These high-altitude images provided space-like data to assist development of the imaging polarimeter and the data reduction algorithms, allowing the polarimeter to demonstrate the value of the data and to prove its suitability for space operation.[39]

The *Multi-Angle Imager for Aerosols*, MAIA, is a follow-on imaging polarimeter to AirMSPI. MAIA was selected by NASA as part of its *Earth Venture Instrument* program in 2016 to be built and placed in Earth orbit as a remote sensing facility.[40] MAIA takes radiometric and polarimetric measurements needed to characterize the quantities, sizes, and compositions of particulate matter in *air pollution*. The MAIA research combines space-based aerosol measurements with health and hospital records to develop the relationships between health problems like cardiovascular and respiratory diseases and premature deaths with the multiyear record of aerosols and air pollution to be acquired by MAIA.

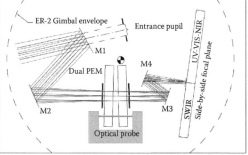

Figure 7.20 (Left) AirMSPI-2 (black bump) mounted under the nose of the NASA ER-2 plane. (Right) AirMSPI-2 optical layout. (Courtesy of NASA/JPL-Caltech.)

MAIA is a pushbroom imager that builds up images from rows of pixels as the polarimeter moves along the ground track. MAIA has a set of long and narrow bandpass filters arranged over rows of pixels on its focal plane and acquires images in 14 spectral bands from 365 to 2125 nm. Three of the bands, 445, 665, and 1035 nm, are polarimetric with a 0° wire grid polarizer over one set of pixel rows and a 45° wire grid polarizer over an adjacent set. Two PEMS near the aperture stop modulate the linear polarization with a period of 36 ms. The focal planes are read out about 1000 times per second. Unpolarized light is unmodulated while completely polarized light is about 70% modulated. S_0 and S_1 (*I* and *Q*) are determined from the modulation amplitude through the 0° polarizers while S_0 and S_2 (*I* and *U*) are determined with the 45° polarizers.

One serious issue with space-based radiometric and polarimetric measurements is the *variation in pixel gain* over time as detectors age and are exposed to space radiation.[41,42] Reductions of pixel gain of over 5% are not uncommon during the lifetime of satellite imaging systems. Adjacent pixels will degrade by different amounts. Thus, polarization measurement concepts that use different pixels to measure Stokes parameters are challenged to achieve 1% accuracy long term in space. MSPI and MAIA address this pixel degradation issue by using PEMs to measure both S_0 and S_1 at a single pixel, which provides an accurate normalized $s_1 = S_1/S_0$ even if the gain changes. On another row, S_0 and S_2 are measured, providing an accurate $s_2 = S_2/S_0$ long term. These are combined for the degree of polarization measurement $DoLP = \sqrt{s_1^2 + s_2^2}$ and angle of linear polarization $AoLP = $ arctan (s_2/s_1)/2, and these polarization metrics remain accurate even with large differential pixel gain changes.

AirMSPI and MAIA use a *CMOS focal plane* for wavelengths below 1000 nm and a *HgCdTe focal plane* above 1000 nm. These focal planes are not fast enough to measure the PEM modulation at 42 kHz. To generate a slower polarization modulation, two PEMs with a small frequency difference of about 25 kHz with aligned axes are placed in the beam. Now, the retardance of PEM1 and PEM2 constructively and destructively interfere. When the two PEMs are modulated with the same retardance amplitude δ_0, the dual-PEM retardance is rewritten as the product of a high- and low-frequency modulation,

$$\delta(t) = \delta_0 \sin(\omega_1 t - \varphi_1) + \delta_0 \sin(\omega_2 t - \varphi_2) = 2\delta_0 \cos(\omega_b t - \eta) \sin(\bar{\omega} t - \bar{\varphi}), \qquad (7.62)$$

where ω_1 and ω_2 are the two PEM's resonant frequencies with phases φ_1 and φ_2. The average of the two PEM frequencies, $\bar{\omega}$, is a high-frequency carrier wave, and $\bar{\varphi}$ is the average of φ_1 and φ_2. This high-frequency oscillation will be averaged over by the detector. The low-frequency retardance modulation occurs at the beat frequency ω_b, which is half the difference frequency, with phase η, which is resolved by the focal planes. The detector acquires about 20 measurements per frame time, π/ω_b. The dual PEMs are placed between crossed quarter waveplates at ±45° and operated as a circular retardance modulator.

7.6.6 Example Atmospheric Polarization Images

Atmospheric polarization is one of the most interesting and important targets for *remote sensing polarimeters*. Figure 7.21 shows two *DoLP* images taken with a full-sky polarimeter at *Montana State University* by Joseph Shaw. The circumference of the images is the horizon and the zenith is in the center. This *full sky polarimeter* uses *liquid crystal variable retarders* and a lens with a wide field of view to measure the linear polarization of the sky.[43,44] An obscuration blocks a small region around the sun to avoid too much light on the focal plane. The polarization of the clear blue sky is dominated by Rayleigh scattering.[45,46] (Left) The degree of linear polarization on a clear day peaks along the vertical yellow band about 90° from the sun. For this particular sky, the *DoLP* reaches a maximum of about 65%. The *DoLP* is low around the sun in the center of the blue area on the left. Beyond the band of peak polarization, the *DoLP* decreases and becomes nearly unpolarized at the anti-solar point, which is only sunlight illuminated just at sunrise and sunset. This nominal

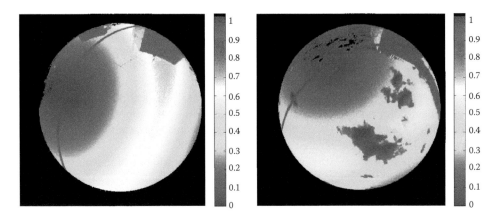

Figure 7.21 (Left) Full sky degree of linear polarization images in false color for a clear sky measured at 530 ± 10 nm shows the typical cloudless sky *DoLP* pattern in the sky, with low *DoLP* around the sun, left side, with the *DoLP* increasing radially from the sun and reaching a peak *DoLP* around 90° from the sun. (Right) In a sky with a few clouds around 90° from the sun, the clouds are seen to be nearly unpolarized surrounded by a region of reduced *DoLP*. (Courtesy of Professor Joseph Shaw, Montana State University, Optical Technology Center.)

polarization distribution of the clear sky is modeled by the *Rayleigh scattering* equations from atmospheric gases, nitrogen, oxygen, argon, and the minor constituents. In a theoretical clear, dry sky, the peak *DoLP* can reach as high as 85% (red in legend). This peak *DoLP* decreases (blue) due to *multiple scattering*. Clouds are dominated by multiple scattering that substantially reduces the *DoLP*. (Right) In another full sky image, a few clouds are visible near the high *DoLP* band as blue irregular patches; these are areas of greatly reduced *DoLP*. On this day, the overall *DoLP* is reduced due to aerosols and the peak *DoLP* only reaches about 60%.

Aerosols are the small particles held aloft in the Earth's atmosphere, including *water droplets*, *sulfate particles* from volcanoes, *dust*, and particulates from industrial processes (*air pollution*). Aerosols typically range from a few tenths of a micrometer to a few micrometers in size, but sizes outside of this range are also observed. Larger particles are harder to keep aloft and fall out in a short time scale. Scattering from aerosols is a significant factor in the *Earth's energy balance* relating the incident flux of *sunlight* to the energy radiated back into space. The scattering from spherical droplets is described by the *Mie theory*,[47–49] which has considerable polarization dependence at scattering angles from near 140° (the angle of rainbows) to 180° (retroreflection). Considerable development has gone into exploiting the polarization signatures of aerosols to determine their concentration, size distribution, and real and imaginary refractive indices.[50] Multi-angle multi-spectral polarization measurements can even estimate the distribution of particulate matter, dust, smoke, and so on, in the atmosphere.[51,52]

The polarization signals from clouds are predominantly from the first scatter, and while the light is nearly unpolarized at many scattering angles, it is quite polarized in the cloudbow and supernumerary bows.[53,54] Scattering from thin aerosols is partially polarized. *Hyperspectral imaging* combined with Mie scattering theory can determine the mean particle size and the imaginary part of the refractive index of an aerosol, providing an indication of water purity. Adding multi-angle polarimetric data at visible and shortwave infrared wavelengths can determine the aerosol's real part of refractive index, n_r, with much greater sensitivity than intensity measurements alone. This sensitivity to aerosol properties has been demonstrated with the airborne *Research Scanning Polarimeter* (*RSP*)[55,56] and analyzed through theoretical sensitivity studies.[57] Aerosols were measured from space with the *Polarization and Directionality of Earth's Reflectances* (*POLDER*), a single-channel polarimeter with eight spectral bands, three of which were polarimetric, and a *DoLP*

7.6 Stokes Polarimeter Configurations

uncertainty of ~2%.[58] The *Aerosol Polarimeter Sensor* (*APS*) instrument for NASA's Glory mission, using similar design concepts to the airborne RSP, was designed to provide very accurate multi-angle polarimetric measurements (linear polarization uncertainty ~0.2%), but in a coarse resolution (6–20 km) due to non-imaging operation.[59] Unfortunately, Glory failed to reach orbit after launch in 2011.

Several examples of the large and complex aerosol polarization signatures available to downward-looking polarimeters will now be shown from AirMSPI (Section 7.6.5). Figure 7.22 is a typical cloud polarization image where the AirMSPI polarimeter has performed a backward sweep (i.e., the sweep starts from ahead of the airplane to nadir and ends looking backward). The two images are narrowest at the nadir. The rendered intensity image (left) and *DoLP* image (right) have been georegistered so that the width of the swath corresponds to the distance spanned by the image on the ground. The RGB intensity image renders the density and brightness of clouds with conspicuous dark breaks between clouds where the Earth's surface is visible. There is no significant color to the clouds; they are uniformly white because the cloud droplets are large relative to the wavelength of visible light, so each wavelength is scattered by about the same amount. The *DoLP* image is colored based on the degree of linear polarization of each waveband. Black areas have unpolarized light in the three wavebands, 470, 660, and 865 nm. White areas, such as the *primary cloudbow* at a scattering angle of about 140°, have a high *DoLP* in all three wavebands. The colored rings, the *supernumerary bows*, are polarized by scattering from the cloud's water droplets, located on the *DoLP* image to the upper left of the primary cloudbow. The *supernumerary bows* are centered on the antisolar point, the *glory*, and the *DoLP* changes rapidly with wavelength and angle. The blue ring is more polarized at 470 nm than the other wavelengths, the red ring is more polarized at 865 nm, and so on. The sequence of colors in this region from 138° to 180° depends on the droplet sizes, which, in this scene, closely match Mie scattering from a distribution of spherical water droplets with a median diameter of 12 μm, providing an example of how imaging polarimetry measures aerosol properties.

In practice, when analyzing downward-looking polarimetric data, the multi-angle intensity data determine the first approximation to the three-dimensional distribution of aerosols. The shorter wavelengths are more sensitive to the density of smaller aerosol particles and the larger wavelengths

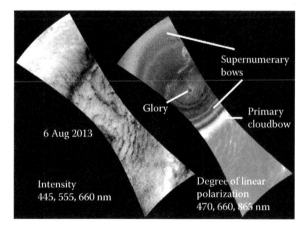

Figure 7.22 AirMSPI intensity and *DoLP* images of clouds over the Pacific Ocean show the high *DoLP* of the primary cloudbow at about 140° from the sun, highly polarized in all three wavebands, thus appearing white. Inside the primary cloudbow, the degree of polarization oscillates with viewing angle, creating several colored rings with the degree of polarization peaking at different angles for 470 nm (colored blue), 660 (colored green), and 865 (colored red) nm. At the center of the rings is the glory, light retroreflected from the droplets. This angular distribution of *DoLP* matches closely to Mie scattering from a distribution of spherical water droplets with a median diameter of 12 μm. These data were obtained from the NASA Langley Research Center Atmospheric Science Data Center. (Courtesy of NASA/JPL-Caltech.)

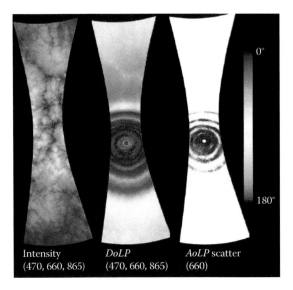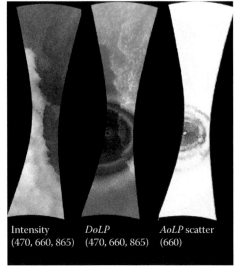

Figure 7.23 A pair of AirMSPI cloud images demonstrating Mie scattering from (left) uniform and (right) non-uniform droplet distributions. (Left set, left) Intensity image of clouds colored as 470 nm (blue), 660 nm (green), and 865 nm (red). (Left set, middle) *DoLP* image using the same coloring scheme. The cloudbow is the narrow white band above and below the center where all three wavebands are highly polarized. Moving toward the center are the highly colored (spectrally polarized) supernumerary bands where the *DoLP* varies with wavelength in angular bands. (Left set, right) The angle of linear polarization (*AoLP*) varies in circular bands about the glory; its legend is on the right side. (Right set, left) Intensity image showing a brighter cloud grouping on the lower left and a darker cloud grouping on the upper right. (Right set, middle) The boundary between the two cloud groupings is conspicuous in the *DoLP*. The colored supernumerary rings have a discontinuity across the cloud boundary. The red, green, and blue rings have a larger radius on the left side, showing clearly a difference in droplet size in the two cloud groups. (Right set, right) The corresponding *AoLP* image. These data were obtained from the NASA Langley Research Center Atmospheric Science Data Center. (Courtesy of NASA/JPL-Caltech.)

are more sensitive to the density of the larger aerosol particles. Then, forward light scattering models combine the polarization and intensity measurements to estimate particle size and refractive index, and can even distinguish liquid water from ice and irregularly shaped solid particles of dust, sulfate, and pollution, using Mie scattering fits and other scattering models.

Figure 7.23 compares AirMSPI data sets for clouds with (left) uniform droplet sizes and (right) two adjacent cloud groups with significantly different droplet sizes. On the right, the size of the supernumary bows changes passing from one cloud group to the other.

7.7 Sample-Measuring Polarimeters

Figure 7.24 shows a schematic layout of a sample-measuring polarimeter to determine the polarization characteristics of a sample, including diattenuation, retardance, and depolarization. The sample's Mueller matrix **M** elements can be obtained through a sequence $q = 0, 1,..., Q - 1$ of polarimetric measurements. The polarization state generator (PSG) prepares a set of polarization states with a sequence of Stokes parameters $\mathbf{S}_q$. The Stokes parameters exiting the sample are $\mathbf{M} \cdot \mathbf{S}_q$. These exiting states are analyzed by the qth polarization state analyzer (PSA) $\mathbf{A}_q$, yielding the qth measured flux $P_q = \mathbf{A}_q^T \mathbf{M} \mathbf{S}_q$. Each measured flux is assumed to be a linear function of the sample's Mueller matrix elements (nonlinear optical interactions such as frequency doubling are not addressed by this method). A set of linear equations is developed from the set of polarimetric measurements to solve for the Mueller matrix elements.

7.7 Sample-Measuring Polarimeters

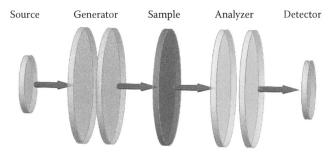

Figure 7.24 A sample-measuring polarimeter consists of a source, a polarization state generator (PSG), the sample, a polarization state analyzer (PSA), and the detector.

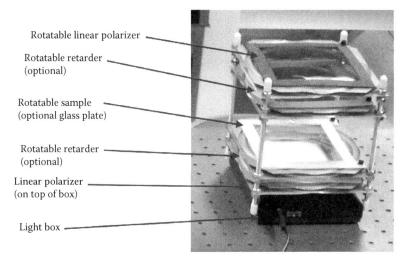

Figure 7.25 Picture of a polariscope fabricated from typical hardware store components, sheet polarizers, and retarders. With the optional retarders, a circular polariscope and other elliptical polariscopes can be configured.

7.7.1 Polariscopes

Polariscopes are relatively simple polarimeters that examine samples placed between pairs of polarizers, as shown in Figure 7.25. The principal application is screening samples for retardance or stress birefringence and polariscopes excel at this task. Commonly, two orthogonal linear polarizers create a dark field (dark background). Then, any polarization alteration caused by the sample placed between the polarizers causes light leakage, which is readily observed in the dark field. With a polariscope, a visual observer can detect very small levels of retardance, or easily estimate the distribution of birefringence in a sample. Any configuration of two linear, elliptical, or circular polarizers can be used. Next, the most important polariscope configurations are described to understand the polarization properties each configuration can reveal. Section 25.6.1 (in Chapter 25) contains example polariscope images and a comparison of polariscope configurations (see Figures 25.18 through 25.25). Detailed treatments of polariscopes can be found in several references[60–63] with comprehensive treatments by Theocaris and Gdoutos[64] and Aben and Guillemet.[65]

7.7.1.1 Linear Polariscope

The *linear polariscope* places the sample between two linear polarizers, usually sheet polarizers.* The polarizers can be oriented orthogonally for a dark field, parallel for a bright field measurement,

---

* The linear polariscope is also sensitive to one component of linear diattenuation and to the circular diattenuation, corresponding to S_2 and S_3, but is much less commonly used for this application.

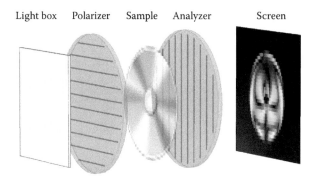

Figure 7.26 The crossed linear polarizer polariscope places a sample between orthogonal linear polarizers that are typically large sheet polarizers. A large diffuse light source is used, so the sample is easily viewed as one's head moves around. An optional lens (not shown) can be used to project the polariscope's image onto a screen or camera. This polariscope has a dark background, so any polarization change induced by the sample stands out with high visibility. Here, a transparent compact disk (CD) blank, before finishing with metallization and stamping, is shown as the sample with an actual CD polariscope image on the screen.

or anywhere in between. The classic *crossed linear polariscope* has crossed polarizers and a dark field as seen in Figure 7.26. Consider a horizontal polarizer illuminating the sample and a vertical polarizer analyzing the exiting light. The transmitted flux P depends on the sample's Mueller matrix elements as follows:

$$P = \mathbf{A}^T \cdot \mathbf{M} \cdot \mathbf{S} = \frac{1}{2}\begin{pmatrix} 1 & -1 & 0 & 0 \end{pmatrix} \cdot \begin{pmatrix} m_{00} & m_{01} & m_{02} & m_{03} \\ m_{10} & m_{11} & m_{12} & m_{13} \\ m_{20} & m_{21} & m_{22} & m_{23} \\ m_{30} & m_{31} & m_{32} & m_{33} \end{pmatrix} \cdot \begin{pmatrix} 1 \\ 1 \\ 0 \\ 0 \end{pmatrix} \quad (7.63)$$

$$= \frac{m_{00} + m_{01} - m_{10} - m_{11}}{2}.$$

Hence, when $m_{00} + m_{01} - m_{10} - m_{11} = 0$, the field remains dark. If the sample is a linear retarder, $\mathbf{LR}(\delta,\theta)$, the flux equation becomes

$$P = \mathbf{A}^T \cdot \mathbf{LR}(\delta,\theta) \cdot \mathbf{S}$$

$$= \frac{1}{2}\begin{pmatrix} 1 & -1 & 0 & 0 \end{pmatrix} \cdot \begin{pmatrix} 1 & 0 & 0 & 0 \\ 0 & \cos^2 2\theta + \cos\delta \sin^2 2\theta & (1-\cos\delta)\cos 2\theta \sin 2\theta & -\sin\delta \sin 2\theta \\ 0 & (1-\cos\delta)\cos 2\theta \sin 2\theta & \cos\delta \cos^2 2\theta + \sin^2 2\theta & \cos 2\theta \sin\delta \\ 0 & \sin\delta \sin 2\theta & -\cos 2\theta \sin\delta & \cos\delta \end{pmatrix} \cdot \begin{pmatrix} 1 \\ 1 \\ 0 \\ 0 \end{pmatrix} \quad (7.64)$$

$$= \frac{1 - (\cos^2 2\theta + \cos\delta \sin^2 2\theta)}{2} = \frac{\sin^2(\delta/2)\sin^2 2\theta}{4}.$$

7.7 Sample-Measuring Polarimeters

The crossed linear polariscope has a $\sin^2 2\theta$ response to the orientation of linear retardance, with no response (leaked flux) when the retardance fast axis is at 0° or 90°, and a maximum response to orientations at 45° or 135°. Thus, the crossed linear polariscope finds half the linear retardance but misses the other half of the linear retardance. This is easily addressed if the sample can be measured, then rotated 45° and measured again, detecting retardance contributions from both the 0° and 45° components. Thus, with two measurements, a good view of the linear retardance due to stress birefringence can be obtained, showing how the magnitude and orientation of retardance vary across the sample. Taking a second-order Taylor series expansion of Equation 7.64 in δ about zero yields $\delta^2 \sin^2 2\theta/4$, showing the transmitted flux is quadratic in the retardance for small retardances.

The crossed polarizer linear polariscope is also sensitive to *circular retardance* as shown in the following equation,

$$P = \mathbf{A}^T \cdot \mathbf{CR}(\delta_R) \cdot \mathbf{S}$$

$$= \frac{1}{2}\begin{pmatrix} 1 & 1 & 0 & 0 \end{pmatrix} \cdot \begin{pmatrix} 1 & 0 & 0 & 0 \\ 0 & \cos\delta_R & \sin\delta_R & 0 \\ 0 & -\sin\delta_R & \cos\delta_R & 0 \\ 0 & 0 & 0 & 1 \end{pmatrix} \begin{pmatrix} 1 \\ 1 \\ 0 \\ 0 \end{pmatrix} \quad (7.65)$$

$$= \frac{1-\cos\delta_R}{2} \approx \frac{\cos^2\delta_R}{4}.$$

Thus, any single polariscope measurement measures two, δ_{45} and δ_R, of the three retardance components (δ_0, δ_{45}, δ_R) but cannot measure both components of linear retardance at once. δ_0 is not measured since 0° light is incident on the sample, and this is an eigenpolarization of a 0° or 90° oriented linear retarder.

The *parallel polarizer linear polariscope* shown in Figure 7.27 has a bright field, not a dark field in the background; thus, small retardances are more difficult to see visually with this configuration. This parallel polarizer linear polariscope is the basis for a useful *retardance measurement technique*. The transmittance for a linear retarder rotating between parallel polarizers is plotted in Figure 7.28 for a set of retardances. The maximum transmittance T_{max} and minimum transmittance T_{min} are measured and the retardance is readily determined as

$$\delta = \arccos\left(\frac{2T_{min} - T_{max}}{T_{max}}\right). \quad (7.66)$$

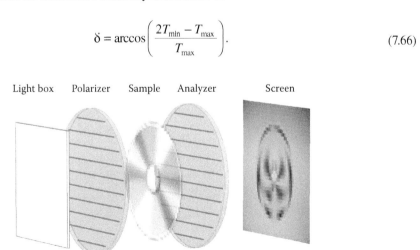

Figure 7.27 Parallel polarizer linear polariscope schematic shown with a compact disk (CD) as the sample.

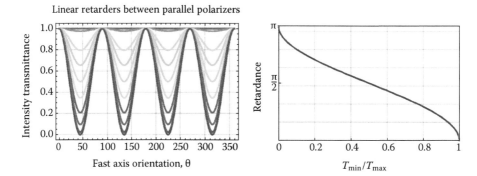

Figure 7.28 (Left) Intensity transmittance as a linear retarder rotates between parallel polarizers. Measuring the maximum and minimum transmittances, T_{max} and T_{min}, provides a useful retardance and axis orientation measurement technique. Curves show the intensity modulation for 0° retardance (red, top), 18° (orange, second from top), 36°, ... 90° (cyan, middle), ... 180° (magenta, bottom) retarders. (Right) The retardance in radians as a function of the ratio T_{min}/T_{max}.

Figure 7.28 (right) plots the retardance as a function of T_{min}/T_{max}; since the function is vertical at $T_{min}/T_{max} = 0$ and 1, the test is not very sensitive for retardances near zero or one-half wave. The retardance axis is the orientation of T_{max}, but the fast and slow axes are not distinguished by this test; thus, the axis might be the fast axis or might be the slow axis.* The transmittance functions in Figure 7.28 (left) are the same for δ or $2\pi - \delta$, $2\pi + \delta$, and so on, so the *order of the retardance* is not determined either. Determining the order of retardance is addressed in Chapter 26 (Multi-Order Retarders and the Mystery of Discontinuities).

7.7.1.2 Circular Polariscope

The circular polariscope places the sample between two orthogonal circular polarizers, and any polarization change is observed as light leakage against a dark background. Figure 7.29 shows a schematic with a right circular polarizer constructed from a horizontal linear polarizer followed by a quarter wave linear retarder at 135°, the sample, and a left circular polarizer formed from a quarter wave linear retarder at 45° followed by a vertical linear polarizer. This polariscope will measure the total linear retardance, $\sqrt{\delta_0^2 + \delta_{45}^2}$. Any circular retardance, δ_R, or circular diattenuation has circularly polarized eigenpolarizations, and thus, it cannot be observed. The flux equation, P, for the circular polariscope and a linearly retarding sample with linear retardance δ at an orientation θ is

$$P = \mathbf{A}^T \cdot \mathbf{LR}(\delta, \theta) \cdot \mathbf{S}$$

$$= \frac{1}{2} \begin{pmatrix} 1 & 0 & 0 & -1 \end{pmatrix} \cdot \begin{pmatrix} 1 & 0 & 0 & 0 \\ 0 & \cos^2 2\theta + \cos\delta \sin^2 2\theta & (1-\cos\delta)\cos 2\theta \sin 2\theta & -\sin\delta \sin 2\theta \\ 0 & (1-\cos\delta)\cos 2\theta \sin 2\theta & \cos\delta\cos^2 2\theta + \sin^2 2\theta & \cos 2\theta \sin\delta \\ 0 & \sin\delta \sin 2\theta & -\cos 2\theta \sin\delta & \cos\delta \end{pmatrix} \cdot \begin{pmatrix} 1 \\ 0 \\ 0 \\ 1 \end{pmatrix} \quad (7.67)$$

$$= \frac{1 - \cos\delta}{2} \approx \frac{\delta^2}{4}.$$

---

* The fast and slow axes can be distinguished by tilting the retarder and analyzing the retardance variation.

7.7 Sample-Measuring Polarimeters

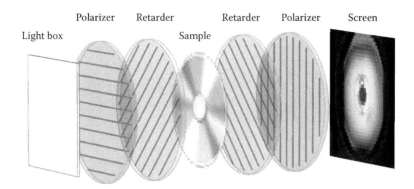

Figure 7.29 Circular polariscope schematic with a CD as the sample. The sample is placed between orthogonal circular polarizers. The first retarder is oriented at 135° from the first linear polarizer, which together comprise the generator's circular polarizer. The second linear polarizer is oriented at 45° from the second retarder, which together comprise the analyzer's circular polarizer.

There is no θ dependence in the transmitted flux $(1 - \cos \delta)/2$. The linear retardance is measured equally independent of its orientation. If a sample with stress birefringence, which is almost entirely linear retardance, is rotated, the polariscope pattern rotates, but the leakage associated with retardance doesn't change. No additional polarization information is obtained by rotating the sample. In practice, the sheet circular polarizers commonly used for this polariscope have retardance dispersion, and the crossed circular polarizers leak more red and violet light, so the circular polariscope has a purple rather than black background.

7.7.1.3 Interference Colors

Retarders have dispersion with a larger retardance in the blue and smaller retardance in the red, as is seen in Figures 19.5, 19.6, and 19.7.* This dispersion causes color variation in birefringent samples viewed in polariscopes; such color variation is known as *interference colors*. The Michael–Levy interference color chart (Figure 7.30) shows the colors versus wavelength in nanometers, labeled across the bottom of the chart.[66,67] These interference colors are seen in Figure 7.31 for a crossed polarizer linear polariscope with a birefringent wedge with linearly varying retardance. The darkest band near the top is where zero retardance occurs. For small retardances, the leaked light is faint white on a black background, which appears as gray. The next white band is at a half wave or about 275 nm of retardance. When the light steps through one wave of retardance, it is completely extinguished, since the incident polarization is returned to its incident polarization state. The blue light with its shorter wavelength will be extinguished first, around 450 nm, creating a brownish color. Then, between 500 and 700 nm, green is extinguished, followed by yellow and red, creating the strong color pattern seen in this region in Figure 7.31. The next nearly white band (yellow) occurs at three half waves or around 825 nm. As the retardance increases, some very saturated colors are created, yellow, magenta, purple, cyan, and yellow again. As the order of retardance (number of waves) further increases, the colors become more pastel and slowly fade in strength.

Interference colors can be used to estimate the retardance of birefringent materials. If the material is known, the thickness of a birefringent sample can be estimated using the Michael–Levy interference color chart (Figure 7.30), where straight lines from the origin show the colors for different

---

* This refers to single element or simple retarders. Compound retarders formed from multiple plates can be designed to have more retardance in the red than in the blue.

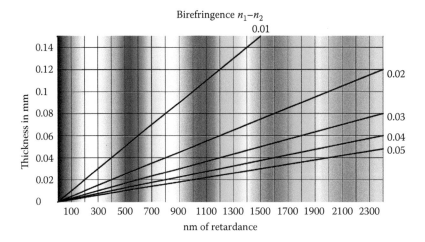

Figure 7.30 The Michael–Levy interference color chart shows the interference colors as a function of retardance in nanometers, crystal thickness, and birefringence (as a slope of black lines). (Courtesy of OSA Optics & Photonics News; a figure in R. Chipman and A. Peinado, The mystery of the birefringent butterfly, *Opt. Photon. News* 24.10, 52–57 (2013).[69])

Figure 7.31 (Left) Interference colors are seen with a linearly varying retarder, a quartz wedge, placed in a crossed polarizer polariscope. The thickness increases from the top left corner toward the lower right corner. The black band along the narrow top edge corresponds to zero waves of retardance. (Right) When the quartz wedge is placed in the parallel polarizer polariscope, the fringes shift. The dark fringe is now centered at one half wave of retardance; this band is dark since the half wave of quartz is rotating 0° into 90° where it is extinguished.

birefringences labeled along the right side of the figure and thicknesses along the left side of the figure.[68] Thus, a yellow color for a material with birefringence of 0.01 indicates a thickness of about 0.085 mm.

Interference colors is used in Figure 7.32 to create art, a colorful butterfly, by polishing gypsum to different thicknesses.[69] Figure 7.33 maps the butterfly's thickness function, where the estimated spectrum for the different thicknesses is shown on the right.

7.7 Sample-Measuring Polarimeters

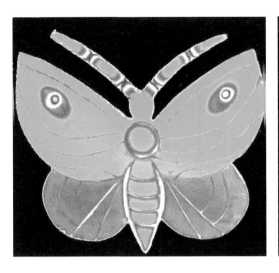

Figure 7.32 (Left) A colorful image of a butterfly is observed in the crossed linear polarizer polariscope. The butterfly was formed by polishing a birefringent material, gypsum, to varying thicknesses. This piece of historic birefringent art is named *Gips-bild* (*Gypsum figure* in English). (Right) Viewed outside the polariscope, the subject is transparent. The butterfly outline is faintly visible but no colors are present. (Courtesy of OSA Optics & Photonics News; a figure in R. Chipman and A. Peinado, The mystery of the birefringent butterfly, *Opt. Photon. News* 24.10, 52–57 (2013).[69])

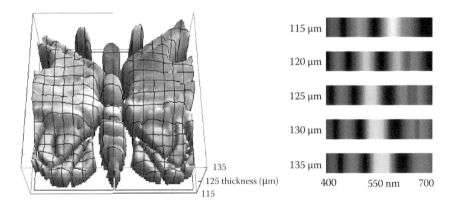

Figure 7.33 (Left) Thickness function for the butterfly with thicknesses shown in height as well as color. (Right) Spectra for various thicknesses of gypsum used in the butterfly showing the interference fringes as a function of wavelength. (Courtesy of OSA Optics & Photonics News; a figure in R. Chipman and A. Peinado, The mystery of the birefringent butterfly, *Opt. Photon. News* 24.10, 52–57 (2013).[69])

7.7.1.4 Polariscope with Tint Plate

The linear polariscope and circular polariscope do not distinguish between equal retardances with fast axis orientations that are 90° apart. In the Mueller calculus, these are positive and negative retardances with equal Mueller matrices; that is,

$$\mathbf{LR}(\delta, \theta + 90°) = \mathbf{LR}(-\delta, \theta). \tag{7.68}$$

Viewing retarding samples in these polariscopes, the orientation of the fast and slow axes is ambiguous to 90°. This ambiguity can be resolved by adding a retarder called a *tint plate* inside

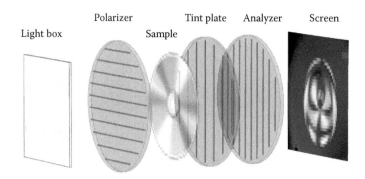

Figure 7.34 The sensitive tint plate polariscope adds a retarder before the analyzer. The retarder's fast axis is parallel to the analyzer's transmission axis; then, sample areas with zero stress remain black between crossed polarizers, but the colors are not symmetric about zero, becoming yellowish to one side and pinkish to the other side of zero (45° vs. 135° orientations).

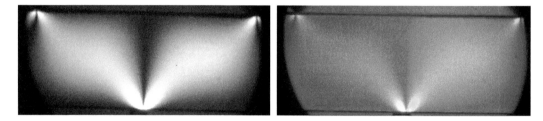

Figure 7.35 Stress is created in a rectangle of glass by pushing upward from the center with an adjustable screw, while the top of the glass is held against two pins on the left and right. (Left) Between crossed 0° and 90° crossed polarizers, the transmitted flux is zero where the retardance is either vertically or horizontally oriented. When the retardance axis rotates from vertical, either clockwise or counterclockwise, light leaks through the analyzer, but the sense of rotation is not observed, the leakage is an even function about vertical. (Right) Adding a quarter wave tint plate, now clockwise rotation becomes bluish and counterclockwise rotation becomes yellowish, thus providing a view of the orientation of retardance changing by 90° when the retardance passes through zero.

the polariscope, as seen in Figure 7.34.[70] To resolve "positive" and "negative" retardances (45° versus 135°), a retarder is added after the sample and before the polarizer in a configuration called the *sensitive tint plate polariscope* as seen in Figure 7.34. The dispersion of the tint plate combined with the retardance of the sample colors retardances near zero differently. To keep zero retardance of the sample in a black transmitting state, the tint plate retarder's fast axis is aligned along the polarizer's axis. The tint plate's retardance and dispersion then add to the retardance of the sample, shifting the location of the black zero order fringe and coloring positive retardances and negative retardances bluish and yellowish.* The tint plate coloration is compared to the crossed polarizer polarimeter in Figure 7.35. Other tint plate polariscope patterns can be seen in Figures 25.18–25.21, 25.24, and 25.25.

7.7.1.5 Conoscope

The retardance of birefringent samples varies with direction of propagation through the sample. This variation can be visualized with a *conoscope*, a combination of a polariscope with optics that

---

* Which sign or retardance is bluish and which is yellowish depends on the orientation of the tint plates's fast axis.

7.7 Sample-Measuring Polarimeters

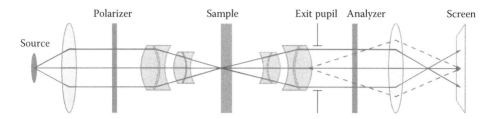

Figure 7.36 A conoscope focuses polarized light through a sample to reveal how the polarization changes as a function of angle. Thus, the exit pupil of the second microscope objective is viewed directly or imaged onto a screen.

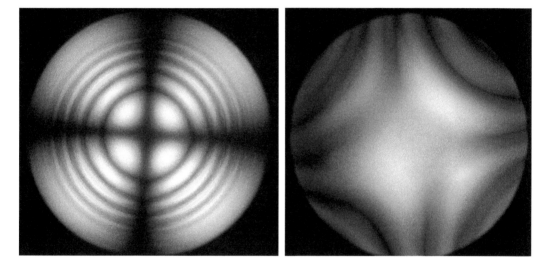

Figure 7.37 (Left) Conoscopic image through a thick calcite C-plate with zero retardance at the center and quadratically increasing retardance with angle. Note how the change of colors in the radial direction follows the Michael–Levy interference color chart (Figure 7.30) since the retardance is increasing in the radial direction. A C-plate has a radially oriented retardance orientation, so the vertical and horizontal dark fringes occur because the retardance is parallel or perpendicular to the two polarizers. (Right) Conoscopic image through a quartz A-plate with a tint plate added to show the sign of the retardance. The retardance is decreasing along the 45° diagonal, colored yellowish, and is increasing along the 135° diagonal, colored bluish.

focuses strongly through a sample, as shown in Figure 7.36. Thick birefringent elements with many waves of retardance produce complex patterns, such as the conoscopic image of a *calcite A-plate* in Figure 7.37. Dark fringes occur in two conditions: (1) where the retardance fast or slow axis is aligned with the crossed polarizers, and/or (b) when the retardance passes through zero or an integer number of waves. The coloration is an ideal example of the *Michael–Levy interference color chart* (Figure 7.30), which can be used to estimate the retardance. Figure 7.37 shows a conoscope image using a sensitive tint plate of a quartz retarder to make the directions of increasing and decreasing retardance visible. Figure 7.38 shows how a conoscope can easily (1) identify a *biaxial crystal* (Section 19.5 and Figure 19.20) from the presence of two white circles and (2) locate the two optic axes through the middle of them.

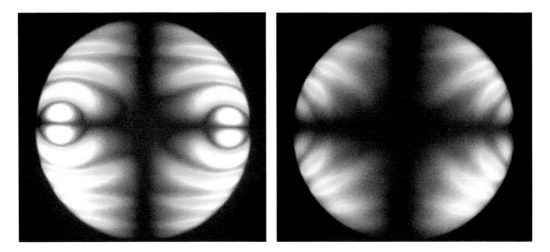

Figure 7.38 (Left) A conoscopic image of a biaxial crystal *muskovite* has two white circles with black lines through them indicating the directions of the two optic axes, the directions of zero birefringence. (Right) For *topaz* with stronger birefringence, the two optic axes are located outside the 30° field of view to the left and right of this conoscopic image.

7.7.2 Mueller Polarimetry Configurations

A *Mueller matrix polarimeter* measures all of the polarization properties of a sample, diattenuation, retardance, and depolarization. All of the polarization analyzers described for Stokes polarimeters can be incorporated into the polarization analyzer section of a *Mueller matrix polarimeter* including *rotating retarders*, *micro-polarizer arrays*, *division of amplitude*, and *division of aperture* analyzers. Most of these analyzers can be reversed and used as polarization generators as well.

The Mueller matrix polarimeter measures the Mueller matrix of everything between the polarization state generator and the polarization state analyzer. This is the *polarization critical region*, where any significant polarization from beam splitters, mirrors, lenses, and so on, needs to be characterized and accounted for in data reduction. When possible, lenses, mirrors, and other elements are located outside of this region.

Depending on the application, Mueller matrices may be measured in transmission, reflection, scattering, or retroreflection configurations as shown in Figure 7.39, by reconfiguring the polarization generator and polarization analyzer.

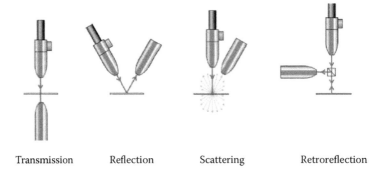

Figure 7.39 A general-purpose Mueller matrix polarimeter with a polarization generator (top component) and a portable polarization analyzer (left, bottom component) can be flexibly configured to measure a sample (blue) in transmission, reflection, scattering, and retroreflection (Section 7.7.2.2). (Courtesy of Axometrics, Huntsville, Alabama.)

7.7 Sample-Measuring Polarimeters

For characterizing optical elements, coatings, and ellipsometric measurements, the Mueller matrix is often measured as a function of angle of incidence. A range of angles can be measured simultaneously with a Mueller matrix imaging polarimeter by focusing through or on the sample, and imaging the camera at the second lens's exit pupil, as shown in Figure 7.40 for transmission. Figure 7.41 shows the reflection case. Figure 7.42 (left) shows a Mueller matrix polarimeter that can measure a wide angle of incidence by using microscope objectives. Figure 7.42 (right) shows a viewing angle widening film measured from 18° angle of incidence on the right side of the pupil to a 72° angle of incidence on the left side of the pupil. These films are used in liquid crystal (LC) cells for displays to increase the field of view of the cell and reduce color variations with angle. The variation of this Mueller matrix as a function of angle of incidence is designed to compensate the natural variation of retardance of a nematic liquid crystal cell with angle.

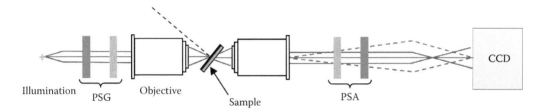

Figure 7.40 An example Mueller matrix imaging polarimeter configured for transmission angle of incidence measurements, such as ellipsometric measurements, uses two microscope objectives to obtain polarization change as a function of angle of incidence. In this example, the sample is tilted to acquire larger angles of incidence. To map angles of incidence, the CCD is focused on the microscope objective's exit pupil.

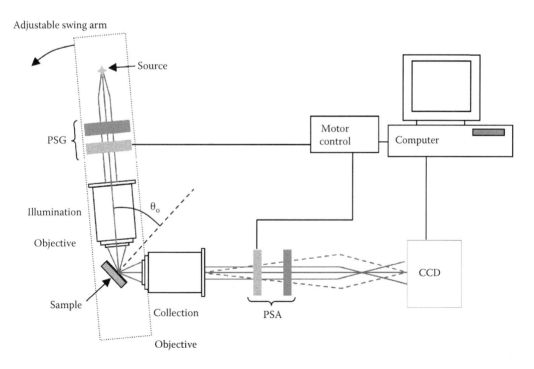

Figure 7.41 A Mueller matrix imaging polarimeter configured to measure reflection properties as a function of angle of incidence using two microscope objectives in the sample compartment. The microscope objective's exit pupil is imaged onto the CCD so that each pixel receives light that reflected at a different angle of incidence and azimuth.

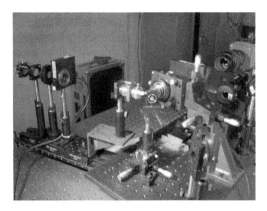 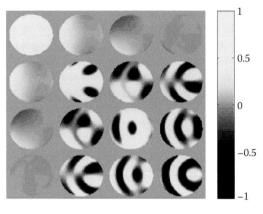

Figure 7.42 (Left) A picture of an angle of incidence Mueller matrix polarimeter with two microscope objectives in the center illuminating and collecting reflected light from a sample. The polarization generator with a rotating retarder is on the left side. The analyzer and camera are on the right side. (Right) An example angle of incidence Mueller matrix image of a field widening film for liquid crystal projector systems. The film is engineered to vary its retardance as a function of angle in opposition to the LC cell, greatly increasing the field of view of the combined device.

7.7.2.1 Dual Rotating Retarder Polarimeter

The *dual rotating retarder Mueller matrix polarimeter* shown in Figure 7.43 is one of the most common Mueller matrix polarimeters. Light from the source first passes through (1) a fixed linear polarizer, then (2) through a rotating linear retarder, (3) the sample, (4) a rotating linear retarder, and finally (5) through a fixed linear polarizer. In the most common configuration, the polarizers are parallel, and the retarders are rotated in angular increments of five-to-one. This five-to-one ratio encodes all 16 Mueller matrix elements onto the amplitudes and phases of 12 distinct frequencies in the detected signal. This configuration was first described by Azzam[1] and provided an explanation of how the ratios one-to-one, two-to-one, three-to-one, and four-to-one all yield incomplete polarimeters. Thus, five-to-one is the first integer ratio yielding a complete Mueller matrix polarimeter. Many other rotation ratios are also used. The data reduction can be performed using the polarimetric data reduction matrix method of the previous section, or alternatively, the detected signal can be Fourier analyzed and the Mueller matrix elements can be calculated from the Fourier coefficients.[71]

This dual rotating retarder Mueller matrix polarimeter configuration has several design advantages. Since the polarizers do not move, the polarizer in the generator accepts only one polarization

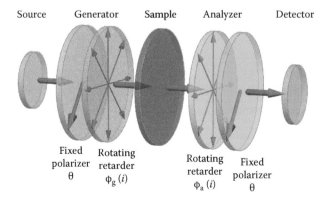

Figure 7.43 The dual rotating retarder polarimeter consists of a source, a fixed linear polarizer, a retarder that rotates in steps, the sample, a second retarder that rotates in steps, a fixed linear polarizer, and the detector.

Polarized Light and Optical Systems

state from the source optics, making the measurement immune to source polarization and polarization aberrations from optics before the polarizer. If the polarizer did rotate while the incident beam is elliptically polarized, a systematic modulation of intensity would be introduced, which would require compensation in the data reduction. Similarly, the polarizer in the analyzer does not rotate; only one polarization state is transmitted through the analyzing optics and onto the detector. Any diattenuation in the analyzing optics and any polarization sensitivity in the detector will not affect the measurements.

Optimal values for the retardances are near $2\pi/3$ rad ($\lambda/3$ waveplates). If $\delta_1 = \delta_2 = \pi$ rad (half wave linear retarders), only linear states are generated and analyzed, and the last row and column of the sample Mueller matrix are not measured.

The dual rotating retarder Mueller matrix polarimeter is frequently calibrated by taking a measurement of air as the two retarders rotate, a generalization of Equation 7.66. Since the two retarders rotate at different rates, the individual retardances can be determined, along with the starting orientations of the retarder axes, and the relative orientation of the final polarizer in a small angle approximation, which is sufficient for small errors[72] or, in a more general method, suitable for larger errors.[73,74]

7.7.2.2 Polarimetry Near Retroreflection

Some reflective optical components need to be tested near normal incidence, such as *corner cube retroreflectors*, *liquid crystal on silicon* (LCoS) panels (Section 24.4.6), and other reflective spatial light modulators. *Retroreflection* testing requires the insertion of a low-polarization, ideally non-polarizing, beam splitter in front of the sample as shown in Figure 7.44. Ideally, the non-polarizing beam splitter has equal *s*- and *p*-amplitudes and equal *s*- and *p*-phases.[75]

In Figure 7.44, a portion of the beam from the polarization state generator reflects from a non-polarizing beam splitter and is normally incident on the sample; the remainder is removed in a beam dump. The light reflected from the sample is divided at the beam splitter and the transmitted

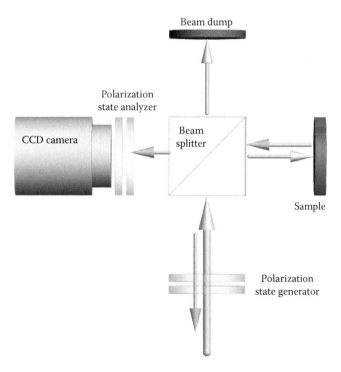

Figure 7.44 Imaging polarimeter configured for retroreflection testing using a non-polarizing beam splitter and beam dump.

portion continues through the polarization analyzer to the focal plane. The focal plane acquires a series of raw images of the sample, and from the set of raw images, the Mueller matrix image of all the optics in the polarization critical region is calculated pixel by pixel.

To obtain the Mueller matrix image of the sample, contributions from the reflection off the non-polarizing beam splitter and transmission through the non-polarizing beam splitter must be calibrated and removed. The ideal non-polarizing beam splitter should have no polarization, and its retardance and diattenuation should be zero; the Mueller matrix would be the identity matrix for both reflection and transmission. In practice, commercially available non-polarizing beam splitters always have some diattenuation and retardance.

The sample Mueller matrix, $\mathbf{M}_S$, is determined from the measured Mueller matrix, $\mathbf{M}_{measured}$, where $\mathbf{M}_T$ represents the beam splitter in transmission, and $\mathbf{M}_R$ represents the beam splitter in reflection,

$$\mathbf{M}_{measured} = \mathbf{M}_T \cdot \mathbf{M}_S \cdot \mathbf{M}_R. \tag{7.69}$$

$\mathbf{M}_T$ and $\mathbf{M}_R$ are measured during sample compartment calibration at each wavelength. $\mathbf{M}_S$ is determined as

$$\mathbf{M}_S = (\mathbf{M}_T)^{-1} \cdot \mathbf{M}_{measured} \cdot (\mathbf{M}_R)^{-1}. \tag{7.70}$$

The compensation must be cautiously applied, with all instrumental variables, such as collimation, vignetting, stray light, and angle of incidence, carefully considered.

The same method is applicable to lenses, mirrors, and other supplemental optics used to manipulate the beams through the sample compartment. Once the Mueller matrices for the optics before the sample, $\mathbf{M}_1$, and after the sample, $\mathbf{M}_2$, are calibrated, their matrix inverses can be applied during data reduction,

$$\mathbf{M}_S = (\mathbf{M}_2)^{-1} \cdot \mathbf{M}_{measured} \cdot (\mathbf{M}_1)^{-1}. \tag{7.71}$$

7.8 Interpreting Mueller Matrix Images

As examples of Mueller matrix measurements, several samples are examined. Figure 7.45 shows the Mueller matrix image of a sheet of a low-quality *dichroic polarizer*. If this was an ideal polarizer, the M_{00}, M_{01}, M_{10}, and M_{11} element images would be solid cyan and all the other elements would be

Figure 7.45 The normalized Mueller matrix image of a 12 × 12 mm area of a very low cost polarizer. It reveals many defects, including variations of the polarizer orientation and polarization leakage.

7.8 Interpreting Mueller Matrix Images 277

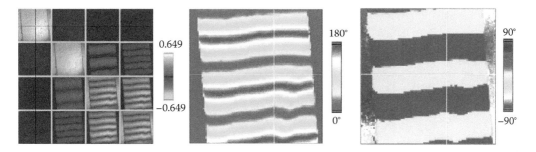

Figure 7.46 (Left) Mueller matrix image of a quartz wedge producing a linearly varying retardance. (Middle) Blue bands indicate an integer number of waves of retardance; red bands are half wave positions. The bands are not straight because the thickness of the wedge is not uniform. (Right) The orientation of the retardance changes by 90° turning from red to blue every time the retardance passes through an integer or a half integer value of retardance.

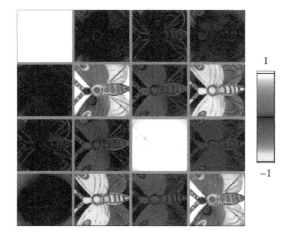

Figure 7.47 Mueller matrix image of the gypsum butterfly demonstrates linear retardance at 45° in elements M_{11}, M_{13}, M_{31}, and M_{33}. (Courtesy of OSA Optics & Photonics News; a figure in R. Chipman and A. Peinado, The mystery of the birefringent butterfly, *Opt. Photon. News* 24.10, 52–57 (2013).[69])

black. This polarizer was fabricated by spraying dichroic liquid crystal molecules onto a moving plastic substrate where the molecules self-aligned. Here, the alignment is far from perfect. In the M_{02} and M_{20} elements, red indicates a counterclockwise rotation of the transmission axis and cyan indicates a clockwise rotation. The values in the M_{22} and M_{33} elements indicate significant polarizer leakage, indicating that the diattenuation is less than one. There is a large market for low-cost polarizers for inexpensive LC displays, such as for children's toys, and such Mueller matrix data can simulate how such a component will perform relative to the specifications and requirements.

Figure 7.46 shows a Mueller matrix image of a quartz wedge. The retardance is varying linearly from top to bottom, producing periodic horizontal stripes in the M_{22}, M_{23}, M_{32}, and M_{33} elements.

Figure 7.47 shows the Mueller matrix image of the gypsum butterfly. The thickness data of Figure 7.33 were derived from this image. The butterfly image is seen in elements M_{11}, M_{13}, M_{31}, and M_{33}, the elements associated with linear retardance at 45° with elements M_{13} and M_{31} having the opposite sign (Section 6.6). The weak signals in elements M_{12}, M_{21}, M_{23}, and M_{32} indicate that the sample's retardance axis was slightly rotated from 45° in the polarimeter.

Next, the conoscopic Mueller matrix images of two uniaxial C-plates* are compared, a calcite plate without optical activity and a quartz plate with additional optical activity. A C-plate has the

---

* Definitions of C-plate is described in Figure 19.12.

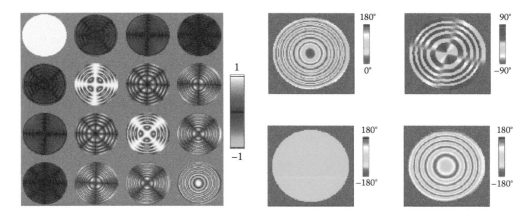

Figure 7.48 (Left) Mueller matrix conoscopic image of a calcite C-plate showing the rapid variation of retardance with angle of incidence over a 30° cone of incident light. (Right, top left) Linear retardance. (Right, top right) Retardance orientation. (Right, bottom left) Circular retardance. (Right, bottom right) Retardance magnitude.

optic axis oriented perpendicular to the two faces of a birefringent plate, such that light normally incident experiences no linear birefringence. Figure 7.48 shows the Mueller matrix image of a calcite C-plate with a rapidly varying retardance for a 30° cone of incident beam. At normal incidence, there is no birefringence, so the retardance is zero, and the Mueller matrix is an identity matrix, indicated by yellow at the center pixels on the diagonal elements, and black on the off-diagonal elements. As the angle of incidence increases from zero, the birefringence increases symmetrically in all directions; the linear retardance (Figure 7.48, right, upper left frame) increases quadratically. The retardance orientation (Figure 7.48, right, upper right frame) rotates uniformly through 360° around the center. When the retardance reaches 180°, the innermost red ring, the retardance orientation changes by 90°, first annular band, and the retardance increases across this annular band until it reaches 360° (one wave) and the second annular band begins. For this cone at one wave of retardance, the Mueller matrix is the identity matrix again, the first yellow circle in the Mueller matrix image. At the second and third circles, the retardance is two waves and three waves. At the edge, the retardance has reached about four and a quarter waves. The circular retardance image (Figure 7.48, right, lower left frame) is zero, indicating that the retardance is purely linear. The polarization aberrations of a C-plate are discussed in Section 21.6.2, which are similar to this example with zero optical activity.

For comparison, Figure 7.49 shows (left) a measured Mueller matrix as a function of angle of incidence through a quartz C-plate and (right) the associated retardance parameters. Quartz is a uniaxial material, and this measurement is centered right down the optic axis; hence, the linear retardance is zero in the center. But unlike a normal uniaxial material, quartz also has optical activity (circular retardance) for propagation near the optic axis. At normal incidence, the measured Mueller matrix is approximately

$$\begin{pmatrix} 1 & 0 & 0 & 0 \\ 0 & -1 & 0 & 0 \\ 0 & 0 & -1 & 0 \\ 0 & 0 & 0 & 1 \end{pmatrix}, \qquad (7.72)$$

the Mueller matrix of a half wave circular retarder. From the center to the edge, at 30° angle of incidence, the retardance increases by seven waves as seen in the retardance magnitude image

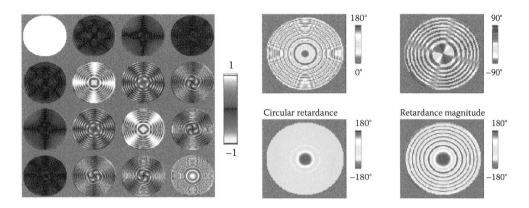

Figure 7.49 (Left) Mueller matrix conoscopic image of a quartz C-plate showing the rapid variation of retardance with angle of incidence over a 30° cone of incident light. (Right, top left) Linear retardance. (Right, top right) Retardance orientation. Optical activity is seen near the optic axis in the center of the circular retardance figure (right, bottom left). (Right, bottom right) Retardance magnitude.

(Figure 7.49, right, lower right). The circular retardance (Figure 7.49, right, lower left) decreases away from normal incidence approaching zero. The linear retardance magnitude, zero on the axis of a C-plate, increases quadratically and symmetrically from the center (Figure 7.49, right, upper left).* The orientation of linear retardance varies by 360° for each full rotation. Every time the retardance magnitude passes through 0 or π, the retardance orientation changes by 90°, explaining the annular bands. Whenever the retardance magnitude passes $2n\pi$, the center of each green band in Figure 7.49 (right, lower right), the corresponding Mueller matrix must be the identity matrix, the diagonal elements of the Mueller matrix have yellow rings, and the off-diagonal elements have black rings. Some weak linear and circular diattenuation are also visible in the top row of the Mueller matrix image.

7.9 Calibrating Polarimeters

The goal of polarimeter calibration is to obtain an accurate data reduction matrix; thus, the polarimeter can be operated reliably and accurately. When the polarization elements used in the generator and analyzer and their orientations are known to sufficient accuracy, then the **W** matrix is easily created and an accurate data reduction is straightforward.

Calibration is rarely so easy. It is impractical to base a polarimeter's calibration strictly on polarization component manufacturer's specifications; the tolerances are too large. The retardance of commercial retarders is specified to only a percent or so, and their retardance varies with temperature and wavelength. But retarders can be accurately calibrated by rotating them between parallel polarizers (Section 7.7.1.1). Well-specified polarizers are easier to obtain, such as Glan–Taylor polarizers with an extinction ratio of greater than 10^5; the orientation of the transmission axis still needs to be determined relative to the polarimeter's intended coordinate system for x, y, 45°, and 135°.

---

* Some moiré fringing is evident toward the edge. The retardance is varying rapidly and each pixel is averaging over an increasing range of angles toward the edge.

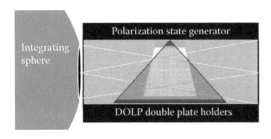

Figure 7.50 (Left) Schematic for a polarization state generator for calibrating with low *DoLP* states. It uses an integrating sphere to generate unpolarized light that passes through a pair of oppositely tilted glass plates to create beams with small *DoLP*. Two pairs of plates are shown: (cyan) smaller tilt for smaller *DoLP*, and (blue) larger tilt for greater *DoLP*. The beams from the integrating sphere have a range of angles of incidence. Beams traveling upward will strike the first plate at a larger angle and then strike the second plate at a smaller angle, compensating first-order Fresnel coefficient variations. (Right) Picture of four sets of paired glass plates for generating low *DoLP* light.

Many polarimeters are calibrated by rotating high-quality "calibration" linear polarizers, such as Glan–Thompson prism polarizers, which can produce a good calibration data set from light with a degree of polarization of one. When it is desired to calibrate near or at unpolarized light, then suitable calibration sources must be constructed. Integrating spheres are generally used as unpolarized sources, although it is difficult to demonstrate that they are unpolarized to a few tenths of a percent. For generating beams with a small degree of linear polarization, tilted glass plates have been used in unpolarized beams, and the *DoLP* is either calculated from the Fresnel equations or measured with rotating polarizers.[76] One issue is that as the angle of incidence of a beam changes across a tilted glass plate, the *DoLP* will change linearly, which can be a source of error. Two glass plates tilted in opposite directions provide a higher-order solution, canceling the linear *DoLP* variation as shown in Figure 7.50.

Another systematic error is the drift of the generator and analyzer vectors between calibration and operation. The temperature of the polarimeter may change between calibration and operation, which changes retardances, as well as slightly changing light paths through the optics. The angles of stepper motors and other rotary stages are known to drift. Liquid crystal cells often drift significantly over periods of days. Whatever the cause, if the polarimetric measurement matrix at the time of measurement is $\mathbf{W}_{drift}$, then the error in the Stokes parameters or Mueller matrix elements takes the form

$$\mathbf{M}_{measured} = \mathbf{W}_P^{-1} \cdot \mathbf{W}_{drift} \cdot \mathbf{M}_S, \tag{7.73}$$

where $\mathbf{M}_S$ is the Mueller matrix or Stokes parameters of the sample and $\mathbf{M}_{measured}$ is the measurement.

7.10 Artifacts in Polarimetric Images

Artifacts in polarimeters are apparent polarization features that are not real but a result of the *systematic errors* in the polarimeter. A simple way to understand artifacts is to picture the polarimeter measuring a completely unpolarized scene. Any polarization present in the polarimetric image is thus due to some error in the polarimeter operation or data reduction. Polarization artifacts are frequently associated with motion in the scene, *intensity gradients* near edges of the scene, and drift of the polarimeter's analyzer vectors between calibration and operation. Artifacts are ubiquitous

7.10 Artifacts in Polarimetric Images

in polarimetric data. A thorough understanding can avoid erroneous interpretation of polarimetric data in cases where artifacts are expected.

7.10.1 Pixel Misalignment

Since imaging polarimeters need to acquire several images of a scene through different analyzers, there is an opportunity for error when the fields of view for a particular pixel do not coincide in position, time, or wavelength.

Consider a *rotating retarder imaging polarimeter* taking sequential measurements in time. If the retarder has some wedge, non-parallel surfaces, it acts as a weak prism deflecting the light. As the retarder rotates between measurements, the image undergoes a small circle on the focal plane, and as a result, at each pixel, the field of view wobbles slightly. If the object is of uniform intensity and polarization, this circling motion makes little difference. But when there is an intensity gradient, particularly around point sources and edges, the measured light flux will fluctuate and the data reduction will miscalculate and generate a polarization artifact.[77]

The polarization artifacts are much larger with division-of-focal-plane polarimeters since adjacent pixels have different analyzers, and several pixels are needed to perform each polarization measurement.[78] As an example, Figure 7.51 is a 660 nm image of a parking lot taken with the MSPI polarimeter. The corresponding *DoLP* is shown in Figure 7.52 (left). In Figure 7.52 (right), the *DoLP* data have been reduced to simulate a micro-polarizer array imaging polarimeter with a polarizer mask of the form of Figure 7.7. All the differences between the right and left sides are polarization artifacts from the micro-polarizer array. In the right *DoLP* image, artifacts occur around the edges of all the objects and shadows. Figure 7.53 compares the measured MSPI S_2 image to the corresponding S_2 image with the micro-polarizer array polarimeter.

Figure 7.51 An irradiance image of cars in a parking lot.

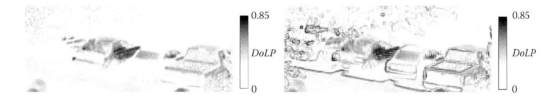

Figure 7.52 (Left) The *DoLP* image of Figure 7.51. White areas have low *DoLP*. (Right) The *DoLP* image as measured by a division-of-focal-plane polarimeter. Any difference between the left and right images is a polarization artifact. The artifacts are significant around all bright edges, such as around the edges of the cars and the edges of their shadows, as well as artifacts in the tree. The division-of-focal-plane polarization artifacts have a signature checkerboard pattern.

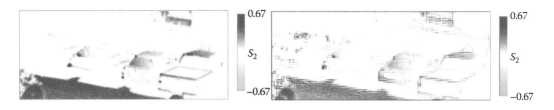

Figure 7.53 (Left) The measured MSPI S_2 image and (right) the S_2 image measured with a division-of-focal-plane polarimeter. The difference between the two images characterizes the polarization artifacts for this image. Again, the checkerboard structure is seen in the image on the right.

7.11 Optimizing Polarimeters

When designing a new polarimeter, guidance is needed in selecting a good set of polarization generators and analyzers. It seems obvious that all the polarization analyzers should not be located together on the *Poincaré sphere*, but should be spread out. Standard methods from linear algebra, the *singular valued decomposition* and *condition number*, provide guidance on generating numerically stable data reduction matrices.[71,79–82] For example, a Stokes parameter measurement using four analyzer states whose vertices form a regular *tetrahedron* on the Poincaré sphere gives an optimum choice for a four-measurement Stokes polarimeter. This configuration provides nearly equal coverage over the entire sphere. The tetrahedron can have different orientations on the sphere; there is not a single optimum regular tetrahedron orientation. Figure 7.54 shows how the vertices of the tetrahedron can be reached by a polarizer and a rotating linear retarder with a retardance near 133° (green trajectory). Another way of generating a regular tetrahedron is described in Figure 7.12.

The *rank* and *null space* of the polarimetric measurement matrix **W** identifies a polarimeter as complete or incomplete. The rank of **W** should be 4 for a complete Stokes polarimeter and 16 for a complete Mueller matrix polarimeter. Any polarization state that lies partially or wholly in the null space of **W** cannot be measured. The **W** for a complete polarimeter has no null space.* When the measured **M** has components in the null space, the data reduction returns a nearby reconstruction in the range of **W**.

Each row of **W** forms one basis vector in the reconstruction of **M**; that is, the measured intensity at each polarimeter state is the projection of **M** onto the corresponding basis vector. For an effective reconstruction, there should be minimum correlation between basis vectors; they should be linearly independent, widely distributed, and well balanced in magnitude. For an overspecified system with $Q > 16$, the basis vectors provide redundant coverage of the polarization space, improving performance in the presence of noise. Basis states may be chosen to lie more densely in directions where most information about **M** is desired. For example, polarimeters to measure stress birefringence are most interested in linear retardance; hence, the basis states can be selected to improve the signal-to-noise ratio on those parameters at the expense of diattenuation and depolarization accuracy.

---

* In contrast, the inverse $\mathbf{W}^{-1}$ will have a null space when the number of measurements exceeds the number of parameters measured, when **W** is over constrained.

7.11 Optimizing Polarimeters

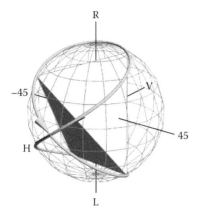

Figure 7.54 The Poincaré sphere trajectory for a rotating retarder analyzer with the optimum retardance will pass through the four vertices of a regular tetrahedron. The trajectory crosses itself at the polarizer position.

For a general-purpose polarimeter that measures a wide variety of arbitrary **M**, the polarimetric measurement matrix should be as far from singular as possible; it should be *well conditioned*.

Insight into the conditioning of **W** is obtained from its singular value decomposition (SVD), which was introduced to polarimeter design by Tyo[80] and Sabatke et al.[79] The SVD factors any $N \times K$ matrix **W** as

$$\mathbf{W} = \mathbf{U} \cdot \mathbf{D} \cdot \mathbf{V}^T = \mathbf{U} \cdot \begin{pmatrix} \mu_1 & & & & \\ & \mu_2 & & & \\ & & \ddots & & \\ & & & \mu_{K-1} & \\ & & & & \mu_K \\ 0 & 0 & \cdots & 0 & 0 \\ & & \vdots & & \end{pmatrix} \cdot \mathbf{V}^T, \qquad (7.74)$$

where **U** and **V** are $N \times N$ and $K \times K$ unitary matrices and **D** is an $N \times K$ diagonal matrix. The diagonal elements μ_k are the singular values. The rank of **W** is the number of non-zero singular values. Those columns of **U** associated with non-zero singular values form an orthonormal basis for the range of **W**; those columns of **V** associated with zero-valued singular values form an orthonormal basis for the null space of **W**. The columns of **V** associated with non-zero singular values form an orthonormal basis that spans the full vector space of **W** and thus reconstructs **M**. Each singular value gives the relative strength of the corresponding vector in this basis set, and the columns of **U** form a mapping from the **V** basis set back to the original basis set of **W**. Since

$$\mathbf{P} = \mathbf{W} \cdot \mathbf{M} = \mathbf{U} \cdot \begin{pmatrix} u_1 & 0 & 0 & 0 \\ 0 & u_2 & 0 & 0 \\ & & \ddots & \\ 0 & 0 & 0 & u_{16} \\ & & \vdots & \end{pmatrix} \cdot \mathbf{V}^T \cdot \mathbf{M}, \qquad (7.75)$$

the rows of **U** corresponding to zero-valued singular values describe sets of flux measurements that are not generated by any Mueller matrix; thus, their presence in a polarimetric measurement can only be due to noise (see Example 7.4). Based on this interpretation, any basis vector in **V** that is associated with a relatively small singular value is near the null space and likely has little information content; such small singular values predominantly amplify noise into the reconstruction of **M**. Error sources that produce projections (flux vectors) that are similar to the flux vectors generated by the basis vectors in **V** (particularly those which correspond to large singular values) will couple strongly into the reconstruction of **M**. The L_2 *condition number* is equal to the ratio of the largest to smallest singular values,[83] and thus minimizing the condition number is equivalent to equalizing, to the extent possible, the range of singular values so that the basis vectors have a wide distribution and similar weight.

For a four-measurement Stokes polarimeter, the Stokes vectors representing each of the four analyzer states, when plotted on the Poincaré sphere, define a tetrahedron that is generally irregular. The volume of the tetrahedron is proportional to the determinant of **W** and is maximized when the vertices form a regular tetrahedron (Figure 7.54). In this case, the maximum distance from a vertex to any point on the sphere is minimized, and the condition number is also at a minimum.

Two examples of applying the condition number to Mueller matrix polarimeters follow; the first is an example of an optimum polarimeter, the second is nearly singular.

Example 7.6 Dual Tetrahedron Mueller Polarimeter

Consider a Mueller matrix polarimeter that uses four generator states $\mathbf{V}_1$, $\mathbf{V}_2$, $\mathbf{V}_3$, and $\mathbf{V}_4$, located at the vertices of a regular tetrahedron on the Poincaré sphere, with associated Stokes vectors

$$\mathbf{V}_1 = (1,\ 1,\ 0,\ 0),$$
$$\mathbf{V}_2 = \left(1, -\frac{1}{3}, \frac{2\sqrt{2}}{3}, 0\right),$$
$$\mathbf{V}_3 = \left(1,\ -\frac{1}{3}, \frac{-\sqrt{2}}{3}, \sqrt{\frac{2}{3}}\right), \quad (7.76)$$
$$\mathbf{V}_4 = \left(1,\ -\frac{1}{3}, -\frac{\sqrt{2}}{3}, -\sqrt{\frac{2}{3}}\right).$$

The analyzer states are also chosen as $\mathbf{V}_1$, $\mathbf{V}_2$, $\mathbf{V}_3$, $\mathbf{V}_4$. Sixteen measurements are acquired at each of the combinations of generator and analyzer. This polarimeter is one member of the set of 16-measurement Mueller matrix polarimeters with minimum condition number, so it can be considered an optimum configuration. The corresponding 16 singular values are

$$\left(4,\ \frac{4\sqrt{3}}{3},\ \frac{4\sqrt{3}}{3},\ \frac{4\sqrt{3}}{3},\ \frac{4\sqrt{3}}{3},\ \frac{4\sqrt{3}}{3},\ \frac{4\sqrt{3}}{3},\ \frac{4}{3},\ \frac{4}{3},\ \frac{4}{3},\ \frac{4}{3},\ \frac{4}{3},\ \frac{4}{3},\ \frac{4}{3},\ \frac{4}{3},\ \frac{4}{3}\right), \quad (7.77)$$

and the condition number, equal to the quotient of the first and last singular values, is 3. This is likely the optimum condition number for a Mueller matrix polarimeter. Each of the 16 columns of **U** represents a different orthogonal component used to reconstruct a measured Mueller matrix. In the presence of white noise, the Mueller matrix component corresponding to the first column will be measured with the highest signal-to-noise ratio, about $\sqrt{3}$ times better than the next six components (columns) from **U**, and about 3 times better than the last nine Mueller matrix components. The component measured most accurately is M_{00}. The relative accuracy of the various matrix element measurements by the optimum polarimeter is shown in the matrix

$$\begin{pmatrix} 1 & \frac{1}{\sqrt{3}} & \frac{1}{\sqrt{3}} & \frac{1}{\sqrt{3}} \\ \frac{1}{\sqrt{3}} & \frac{1}{3} & \frac{1}{3} & \frac{1}{3} \\ \frac{1}{\sqrt{3}} & \frac{1}{3} & \frac{1}{3} & \frac{1}{3} \\ \frac{1}{\sqrt{3}} & \frac{1}{3} & \frac{1}{3} & \frac{1}{3} \end{pmatrix}. \qquad (7.78)$$

Thus, the retardance elements are measured less accurately than the diattenuation and polarizance elements.

Example 7.7 Nearly Singular Mueller Polarimeter

As an example of a polarimeter with a nearly singular polarimetric measurement matrix, the second row of **W** (generate $\mathbf{V}_1$, analyze $\mathbf{V}_2$),

$$(1, \ -0.333, \ 0.943, \ 0, \ 1, \ -0.333, \ 0.943, \ 0, \ 0, \ 0, \ 0, \ 0, \ 0, \ 0, \ 0, \ 0), \qquad (7.79)$$

will be replaced with a vector

$$(1, \ 1, \ 0.005, \ 0, \ 1, \ 1, \ 0.0005, \ 0, \ 0, \ 0, \ 0, \ 0, \ 0, \ 0, \ 0, \ 0), \qquad (7.80)$$

nearly equal to the first row of **W**,

$$(1, \ 1, \ 0, \ 0, \ 1, \ 1, \ 0, \ 0, \ 0, \ 0, \ 0, \ 0, \ 0, \ 0, \ 0, \ 0), \qquad (7.81)$$

so that these two rows are nearly linearly dependent. Examining the resulting singular values,

$$\begin{aligned}(4.056, \ 2.727, \ 2.309, \ 2.309, \ 2.309, \ 2.309, \ 2.022, \ 1.499, \\ 1.333, \ 1.333, \ 1.333, \ 1.333, \ 1.333, \ 1.333, \ 1.333, \ 0.0004),\end{aligned} \qquad (7.82)$$

the last singular value is close to zero and the condition number is about 100. Whenever the measured flux contains the pattern corresponding to the last row of $\mathbf{V}^T$, this component will be amplified by about 100 during the data reduction relative to the other 15 components of the Mueller matrix and will usually dominate the measurement. In the presence of random noise, the measured Mueller matrix will often be close to the 16th column of **U** (partitioned into a 4 × 4 "Mueller matrix"), and so the measurement will be inaccurate.

In summary, components corresponding to very small singular values are greatly amplified in the matrix inverse and can overwhelm the remainder of the Mueller matrix in the polarimetric data reduction.

7.12 Problem Sets

7.1 A rotating retarder polarimeter rotates a quarter wave linear retarder in front of a linear polarizer oriented at 0°. It acquires six flux measurements P_q at retarder fast axis angles $\theta_q = 0°, 30°, 60°, 90°, 120°,$ and $150°$.
   a. Calculate the polarimetric measurement matrix $\mathbf{W}$.
   b. Are any rows equal? Is this a problem? Is the polarimeter complete?
   c. How does the measured flux of S_0, an unpolarized component, modulate?
   d. Find the polarimetric data reduction matrix $\mathbf{W}_P^{-1}$ from the pseudoinverse of $\mathbf{W}$.
   e. Which frequencies are present in each row as the retarder rotates through 180 degrees?
   f. Why isn't the row that calculates S_0 constant?
   g. Calculate the Stokes parameters corresponding to the set of flux measurements

   $$\mathbf{P} = \left(5,\ 5 - \frac{\sqrt{3}}{2},\ 5 - \frac{\sqrt{3}}{2},\ 5,\ 5 + \frac{\sqrt{3}}{2},\ 5 + \frac{\sqrt{3}}{2}\right).$$

   h. Calculate the Stokes parameters corresponding to the set of flux measurements $\mathbf{P}_N = (0, 1, -1, 0, -1, 1)$. Of course, a negative flux is not physically realizable. This is an example of a *null space vector*.
   i. Show that $\mathbf{P}_N$ can be added to any row of $\mathbf{W}_P^{-1}$ and the resulting matrix is also a matrix inverse of $\mathbf{W}$.
   j. What is the relationship of $\mathbf{P}_N$ to the columns of $\mathbf{W}$?

7.2 Consider a polarimeter with a rotating quarter wave linear retarder in front of a linear polarizer oriented at 0°. We will use this polarimeter to view the effects of noise in our signal. The $\mathbf{W}$ and $\mathbf{W}^{-1}$ matrices are as follows:

$$\mathbf{W} = \frac{1}{8}\begin{pmatrix} 4 & 4 & 0 & 0 \\ 4 & 1 & \sqrt{3} & -2\sqrt{3} \\ 4 & 1 & -\sqrt{3} & -2\sqrt{3} \\ 4 & 4 & 0 & 0 \\ 4 & 1 & \sqrt{3} & 2\sqrt{3} \\ 4 & 1 & -\sqrt{3} & 2\sqrt{3} \end{pmatrix} \qquad \mathbf{W}^{-1} = \frac{1}{3}\begin{pmatrix} -1 & 2 & 2 & -1 & 2 & 2 \\ 4 & -2 & -2 & 4 & -2 & -2 \\ 0 & 2\sqrt{3} & -2\sqrt{3} & 0 & 2\sqrt{3} & -2\sqrt{3} \\ 0 & -\sqrt{3} & -\sqrt{3} & 0 & \sqrt{3} & \sqrt{3} \end{pmatrix}.$$

   a. Consider measuring a Stokes vector that changes in time. If the measurements are made instantaneously at times t_1 through t_6, how might you mathematically represent the flux measurements made by this polarimeter? Note that it is not correct to write $\mathbf{W} \cdot \mathbf{S}(t)$ to represent the six measurements. Find a valid representation.

   b. Make $\mathbf{S}(t) = \mathbf{S} + \mathbf{n}(t) = \begin{pmatrix} \sqrt{3} \\ 1 \\ 1 \\ 1 \end{pmatrix} + \cos\left(\frac{\pi}{6}t\right)\begin{pmatrix} 0.1 \\ 0.03 \\ 0.03 \\ 0.03 \end{pmatrix}$, where $\mathbf{S}$ is composed of a constant vector and a "noise" vector. Let six measurement be taken at times $t_n =$

n seconds for $n = 1,2,3...,6$. Find the set of flux measurements and calculate the final Stokes parameters computed by this polarimeter. By what percentage does your measured Stokes vector differ from the constant part of the signal?

c. Next, take 12 measurements instead of 6, using the retarder orientations $\theta_m = (30 \times m)°$ for $m = 0, 1, 2, ..., 11$. Find the measured Stokes parameters assuming the signal continues to follow the form from (b). Use measurement times $t_n = n$ seconds for $n = 1,2,3...,12$. By what percentage do your measured Stokes parameters differ from the constant part of the signal? What is it about the noise that gives this result?

d. Using the matrices from (c), add the vector $\begin{pmatrix} \sqrt{3} & 1 & 0 & -1 & -\sqrt{3} & -2 & -\sqrt{3} & -1 & 0 & 1 & \sqrt{3} & 2 \end{pmatrix}$ to each of the rows of $\mathbf{W}^{-1}$. Is $\mathbf{W}^{-1}$ still an inverse of $\mathbf{W}$? What is the measured Stokes parameters using this new $\mathbf{W}^{-1}$? Again, use measurement times $t_n = n$ seconds for $n = 1,2,3...,12$. By what percentage do your measured Stokes parameters differ from the constant part of the signal?

7.3 A rotating polarizer polarimeter is used with a side-on photomultiplier that has a 10% greater responsivity R for horizontal polarized light for vertically polarized light, varying as $R(\theta) = R_0 (1.05 + 0.05 \cos2\theta)$. Describe how to correct the error from this sensitivity during the data reduction.

7.4 In a polarimeter, light passes first through a rotating quarter wave linear retarder and then through a rotating polarizer before being detected. The polarizer is rotated at three times the retarder fast axis angle θ.
a. Determine the analyzer vector as a function of θ.
b. Express each of the analyzer vector elements as a Fourier cosine and sine series over 2π of rotation of the retarder.
c. During operation, the quarter wave retarder is rotated through 2π and the polarizer is rotated through 6π, and Fourier cosine coefficients $c_0, c_1, c_2, c_3, ...$ and Fourier sine coefficients $s_1, s_2, s_3, ...$ are calculated from an equally spaced list of flux measurements. (Note that sine does not have a DC Fourier coefficient, s_0; it would be zero everywhere). Find the relationship between the Stokes parameters S_0, S_1, S_2, S_3, and the Fourier coefficients of the signal.

7.5 Sequential-in-time polarimeters have reduced accuracy when measuring fluctuating sources. Consider an unpolarized source that has a ±10% variation of flux, sinusoidal in time with a period of 1 s,

$$P(t) = S_0 \left[1 + 0.1 \sin(2\pi t)\right].$$

Let a rotating polarizer polarimeter that rotates through 360° in $T = 2$ s acquiring 16 measurements measure the beam.
a. What will the error in the Stokes parameters be?
b. If instead the polarimeter acquires its measurements in $T = 1.5$ s, what will the error in the Stokes parameters be?
c. How fast would the polarimeter need to acquire its measurements to ensure a Stokes parameter accuracy of 1% (one-tenth the amplitude of the source fluctuations)?
d. Generalize your result. If a source is fluctuating with an amplitude ψ, with a period of T or less, how fast do the measurements need to be acquired for an accuracy of β%?

7.6 Compare the rotating polarizer polarimeter analyzer vector to the rotating analyzer plus fixed 0° analyzer polarimeter analyzer vector. When horizontal and vertical polarized light are incident, what will be the ratio of total flux measured? Which will measure vertical polarized light more accurately? How much more accurately?

7.7 A polarimeter measures several of the Mueller matrix elements using the following combinations of ideal generators and analyzers. The following eight flux measurements are taken: $P_{H,H}$, $P_{H,V}$, $P_{V,H}$, $P_{V,V}$, $P_{45,45}$, $P_{45,135}$, $P_{135,45}$, and $P_{135,135}$.
  a. Which Mueller matrix elements can be measured? Which cannot?
  b. Determine the polarimetric measurement matrix, **W**, for the polarimeter, **W·M** = **P**, where **P** is a vector of eight measured fluxes and **M** is a vector of the Mueller matrix elements that can be measured. Discard the unmeasurable Mueller matrix elements from the **M** vector.
  c. Provide any valid polarimetric data reduction matrix, $\mathbf{W}^{-1}$, for this polarimeter. A matrix inverse operation is not necessary.
  d. What are the dimensions (columns, rows) of $\mathbf{W}^{-1}$?
  e. Describe the concept of the *null space* of a matrix inverse.
  f. What is the number of dimensions of the null space of $\mathbf{W}^{-1}$?
  g. Determine one null space vector of $\mathbf{W}^{-1}$.

7.8 A rotating partial polarizer with $T_{max} = 1$ and $T_{min} = 0.5$ is used in a polarimeter.
  a. What is the analyzer vector as a function of angle θ?
  b. The polarimeter takes measurements at 0°, 45°, 90°, and 135°. Then, the polarizer is placed at 0°, and a quarter wave retarder with its fast axis is inserted in front at ±45° for the right and left partial analyzer measurements. What is the polarimetric measurement matrix **W**, and the polarimetric data reduction matrix $\mathbf{W}^{-1}$?

7.9 A Stokes polarimeter takes *M* flux measurements of an unpolarized beam of light that has intensity fluctuations $P(m) = (100 + 100 \sin(2\pi\, m/M)/2$. The symbol *m* is the index number for the measurement being made by the polarimeter. This makes S_0 different for each measurement taken. Let the polarimeter use a rotating analyzer that makes the following four measurements in sequence: H, 45, V, 135 and let *m* = 1, 2, 3, 4 for the four measurements.
  a. Write out the analyzer vector for this polarimeter in terms of *m*. Also, find the **W** and $\mathbf{W}^{-1}$ matrices for this polarimeter.
  b. Generate a list of magnitudes for the S_1 calculated by this polarimeter for *M* = 1, 2, 3, …, 32.
  c. Which value of *M* produces the largest S_1? Does this value correspond to any frequencies on the analyzer vector? If so, how?
  d. We know that the beam is unpolarized; thus, the values for S_1 are polarimeter artifacts. Using the plot from (b) explain the impact of different frequencies from S_0 on the S_1 output?

7.10 A circular polariscope has a right circular polarizer, an arbitrary Mueller matrix sample, and a left circular analyzer.
  a. Set up the Mueller matrix equation for this polariscope and determine the fraction of the flux incident on the sample that is transmitted through the analyzer as a function of the sample's Mueller matrix elements.
  b. Which Mueller matrix elements do not cause any change in the dark background?
  c. Which Mueller matrix elements do cause light leakage?
  d. Determine the amount of light leakage (transmission) for a linear retarder of retardance δ oriented at 0°.
  e. Show that the leakage for a linear retarder is independent of fast axis orientation.
  f. What is the leakage for a circular retarder of retardance δ_R?

7.12 Problem Sets

7.11 Consider the polarization artifacts in an unpolarized image. A Stokes imaging polarimeter takes six images of a scene to measure the Stokes image using the following six analyzers in sequence: H, V, 45, 135, R, L, and uses the pseudoinverse for the data reduction. Each image exposure is very short, but the time between images is 0.1 s. The image is 32 × 32 pixels. Assume the scene is unpolarized, but that the scene's image is moving across the CCD at a rate of 10 pixels per second in the −x-direction. Calculate the measured Stokes image for the following test scenes where $1 \le m, n \le 32$.

a. $I(m,n) = 100 + 100 \sin\left(\dfrac{2\pi m}{16}\right)$

b. $I(m,n) = 32\sqrt{m/32}$

c. $I(m,n) = 100 e^{-((m-16)^2 + (n-16)^2)/64}$

7.12 A micro-polarizer array Stokes imaging polarimeter is measuring an unpolarized star with an Airy disk one 1 pixel in diameter. The center of a star can lie anywhere with respect to the pixels. What are the largest and smallest polarization errors that can occur and where are the respective centers of the stellar images?

7.13 A polarimeter makes three measurements with (1) an open radiometer (no polarization elements), (2) an ideal linear polarizer with transmission axis at 22.5°, and (3) an ideal linear polarizer with transmission axis at 112.5°.

a. Set up the polarimetric measurement matrix **W**. How many linearly independent rows does it have (what is the matrix rank)?
b. Calculate the flux measurement vectors PH and P45 for two unit flux incident states, 0° linear and 45° linearly polarized light.
c. Calculate the polarimetric data reduction matrix W^{-1} using the pseudoinverse.
d. Calculate the incident Stokes vectors from PH and P45.
e. This is an incomplete polarimeter. Show on the 3D Poincaré sphere the sets of polarization states that yield the same measured Stokes parameters.
f. Compare the number of rows of W^{-1} to the rank and comment on the total amount of information.
g. Are any of the three measurements redundant?

7.14 The nine forms of depolarization (*e* amplitude, *f* phase, *g* diagonal), when small, can be expressed as the weak depolarizing Mueller matrix

$$\mathbf{M} = \begin{pmatrix} 1 & e_1 & e_2 & e_3 \\ -e_1 & 1-g_1 & f_3 & f_2 \\ -e_2 & f_3 & 1-g_2 & f_1 \\ -e_3 & f_2 & f_1 & 1-g_3 \end{pmatrix}.$$

a. How sensitive is the crossed linear polarizer polarimeter to the nine forms of weak depolarization?
b. How does the sensitivity vary when the sample is rotated by $\pi/4$?
c. How sensitive is the crossed circular polarizer polarimeter to the nine forms of weak depolarization?

7.15 Consider measuring linear diattenuation in polariscopes.
   a. Show the equation for a linear diattenuator with $T_{max} = 1$, transmission axis $\theta = 0$, and diattenuation D by operating on $(1,1,0,0)$ and $(1,-1,0,0)$,

$$\mathbf{LD}(T_{max} = 1, D, 0) = \frac{1}{2}\begin{pmatrix} \frac{1}{1+D} & \frac{D}{1+D} & 0 & 0 \\ \frac{D}{1+D} & \frac{1}{1+D} & 0 & 0 \\ 0 & 0 & \sqrt{\frac{1-D}{1+D}} & 0 \\ 0 & 0 & 0 & \sqrt{\frac{1-D}{1+D}} \end{pmatrix}.$$

   b. How sensitive is the crossed linear polarizer polarimeter to weak linear diattenuation as a function of diattenuation D and angle of transmission θ? Find an exact expression.
   c. Perform a Taylor series to second order in D.
   d. How sensitive is the crossed circular polarizer polarimeter to weak linear diattenuation as a function of diattenuation D and angle of transmission θ? Find an exact expression.
   e. Perform a Taylor series to second order in D.

7.16 Consider measuring six Mueller matrix elements.
   a. What is the minimum number of measurements needed to measure the six Mueller matrix elements M00, M01, M20, M21, M30, and M31?
   b. Identify the pairs of generators and analyzers necessary for this measurement.
   c. Label your flux component measurements $\mathbf{P} = (P_0, P_1, P_2, ...)$ and provide a polarimetric data reduction matrix $\mathbf{W}^{-1}$ to calculate these Mueller matrix elements from the flux vector $\mathbf{P}$.

7.17 Consider a Stokes polarimeter with two analyzers located on the equator of the Poincaré sphere and two analyzers in the S_1 and S_3 plane parameterized by an angle α:

$$\mathbf{A}_1 = (1, \cos\alpha, \sin\alpha, 0), \mathbf{A}_2 = (1, \cos\alpha, -\sin\alpha, 0),$$
$$\mathbf{A}_3 = (1, \cos(\pi-\alpha)\alpha, 0, \sin\alpha), \mathbf{A}_4 = (1, \cos(\pi-\alpha)\alpha, 0, -\sin\alpha).$$

   a. Find the polarimetric measurement matrix W.
   b. Show that this is a complete polarimeter as long as $\sin\alpha \neq 0$ and $\cos\alpha \neq 0$.
   c. What are the singular values and the polarimetric data reduction matrix $\mathbf{W}^{-1}$ for $\alpha = \pi/4$?
   d. What are the singular values and the polarimetric data reduction matrix $\mathbf{W}^{-1}$ for $\alpha = \pi/90$?
   e. For (d), when measured from the center of the Poincaré sphere, what is the largest possible largest angle between a state on the surface of the sphere and the nearest analyzer?
   f. Find the value for α that optimizes the condition number (approximate solution is fine).

g. Plot the analyzer vectors for this solution on the Poincaré sphere. What geometric figure do they form?
h. For (f) when measured from the center of the Poincaré sphere, what is the largest angle between two analyzers on the surface of the sphere?

Note that the condition number does not have a quadratic minimum. It has a discontinuity in the slope at its minimum, because the values of two of the singular values are crossing. First, one singular value is in the denominator, and then the other, switching at the minimum.

7.18 A polarimeter measures the linear Stokes parameters (S_0, S_1, S_2) by rotating a polarizer **LP**(θ) front of a detector to angles $\theta = 0°, 30°, 60°, 90°, 120°, 150°$.
   a. Find the 6 × 3 polarimetric measurement matrix **W** to act on (S_0, S_1, S_2) calculating the flux vector **P** = $(P_1, P_2, P_3, P_4, P_5, P_6)$.
   b. Calculate the data reduction matrix as the pseudoinverse of **W**P^{-1}.

The detector is replaced with another detector, which is also a weak diattenuator, it is only 90% as sensitive to 135° polarized light as 45° polarized light.

   c. Calculate the Mueller matrix **LD**[1,9/10,π/4] for this weak diattenuator, and the Mueller matrix for the combination **LD**[1,9/10, π/4].**LP**[0].
   d. Calculate the new polarimetric measurement matrix, call it **W1**, for the polarimeter with diattenuating detector.
   e. Plot the flux vectors when {1,1,0}, {1,–1,0}, {1,0,1}, and {1,0,–1} are incident. If the detector is changed, but the data reduction is not changed, then polarization measurement errors occur.
   f. To simulate this error, calculate the 3 × 3 matrix **W**P^{-1} . **W1**.

This matrix relates the incident state to the measurement. Ideally **W**P^{-1}. **W1** would be the identity matrix.

   g. Calculate measured state when {1,1,0}, {1,–1,0}, {1,0,1}, and {1,0,–1} are incident.
   h. Calculate the error when {1,1,0}, {1,–1,0}, {1,0,1}, and {1,0,–1} are incident.

Interpret the results.
   i. Which states are measured with the wrong flux?
   j. Which states are measured with the wrong degree of polarization?
   k. Which states are measured with the wrong angle of polarization?

Acknowledgments

The authors would like to thank the following for their collaborations that have contributed to this chapter: Neil Beaudry, Christine Bradley, Brian de Boo, David Chenault, Dave Diner, Ann Elsner, Michael Garay, Anna-Britt Mahler, Alba Peinado, Larry Pezzaniti, Matt Smith, Paula Smith, Karen Twietmeyer, Justin Wolfe, and Feng Xu.

References

1. R. M. A. Azzam, Photopolarimetric measurement of the Mueller matrix by Fourier analysis of a single detected signal, *Opt. Lett.* 2.6 (1978): 148–150.
2. M. A. F. Thiel, Error calculation of polarization measurements, *JOSA* 66.1 (1976): 65–67.
3. R. M. A. Azzam and N. M. Bashara, *Ellipsometry and Polarized Light*, North-Holland (1987).
4. J. O. Stenflo, Optimization of the LEST polarization modulation system, *LEST Foundation, Technical Report* 44 (1991).
5. P. S. Hauge, Recent developments in instrumentation in ellipsometry, *Surf. Sci.* 96.1-3 (1980): 108–140.

6. D. Brewster, On the compensations of polarized light, with the description of a polarimeter, for measuring degrees of polarization, *Trans. R. Irish Acad.* (1843): 377–392.
7. R. M. A. Azzam, Mueller-matrix ellipsometry: A review, *Optical Science, Engineering and Instrumentation 97, International Society for Optics and Photonics* (1997).
8. J. M. Bueno, Polarimetry using liquid-crystal variable retarders: Theory and calibration, *J. Opt. A: Pure Appl. Opt.* 2.3 (2000): 216.
9. E. H. Moore, On the reciprocal of the general algebraic matrix, *Bull. Am. Math. Soc.* 26 (1920): 294–295.
10. R. Penrose, A generalized inverse for matrices, *Proc. Cambridge Philos. Soc.* 51(3) (1955): 406–413.
11. A. S. Alenin and J. S. Tyo, Generalized channeled polarimetry, *J. Opt. Soc. Am. A* 31 (2014): 1013–1022.
12. K. Oka and T. Kaneko, Compact complete imaging polarimeter using birefringent wedge prisms, *Opt. Express* 11 (2003): 1510–1519.
13. D. H. Goldstein, Mueller matrix dual-rotating retarder polarimeter, *Appl. Opt.* 31 (1992): 6676–6683.
14. D. J. Diner, A. Davis, B. Hancock, G. Gutt, R. A. Chipman, and B. Cairns, Dual-photoelastic–modulator-based polarimetric imaging concept for aerosol remote sensing, *Appl. Opt.* 46 (2007): 8428–8445.
15. G. Myhre, W.-L. Hsu, A. Peinado, C. LaCasse, N. Brock, R. A. Chipman, and S. Pau, Liquid crystal polymer full-stokes division of focal plane polarimeter, *Opt. Express* 20 (2012): 27393–27409.
16. B. M. Ratliff, J. K. Boger, M. P. Fetrow, J. S. Tyo, and. T. Black, Image processing methods to compensate for IFOV errors in microgrid imaging polarimeters, in *Proc. SPIE vol. 6240: Polarization: Measurement, Analysis, and Remote Sensing VII*, eds. D. H. Goldstein and D. B. Chenault, Bellingham, WA: SPIE (2006), p. 62400E.
17. L. Gendre, A. Foulonneau, and L. Bigué, High-speed imaging acquisition of stokes linearly polarized components using a single ferroelectric liquid crystal modulator, in *Proc. SPIE vol. 7461: Polarization Science and Remote Sensing IV*, eds. J. A. Shaw and J. S. Tyo, Bellingham, WA: SPIE (2009), p. 74610G.
18. M. W. Kudenov, M. E. L. Jungwirth, E. L. Dereniak, and G. R. Gerhart, White light Sagnac interferometer for snapshot linear polarimetric imaging, *Opt. Express* 17 (2009): 22520–22534.
19. J. S. Tyo, C. F. LaCasse, and B. M. Ratliff, Total elimination of sampling errors in polarization imagery obtained with integrated microgrid polarimeters, *Opt. Lett.* 34 (2009): 3187–3189.
20. C. F. LaCasse, J. S. Tyo, and R. A. Chipman, Spatio-temporally modulated polarimetry, in *Proc. SPIE vol. 8160: Polarization Science and Remote Sensing V*, eds. J. A. Shaw and J. S. Tyo, Bellingham, WA: SPIE (2011), p. 816020.
21. S. A. Israel and M. J. Duggin, Characterization of terrestrial features using Space-Shuttle-based polarimetry, *Geoscience and Remote Sensing Symposium, 1992. IGARSS'92. International*. Vol. 2. IEEE (1992).
22. Walter G. Egan, W. R. Johnson, and V. S. Whitehead, Terrestrial polarization imagery obtained from the Space Shuttle: Characterization and interpretation. *Appl. Opt.* 30.4 (1991): 435–442.
23. C. S. Chun, Fleming, D. L., Harvey, W. A., and Torok, E. J., Polarization-sensitive thermal imaging sensor, In *SPIE's 1995 International Symposium on Optical Science, Engineering, and Instrumentation*, International Society for Optics and Photonics (1995), pp. 438–444.
24. G. P. Nordin, J. T. Meier, P. C. Deguzman, and M. W. Jones, Micropolarizer array for infrared imaging polarimetry, *J. Opt. Soc. Am. A* 16(5) (1999): 1168–1174.
25. G. Myhre, A. Sayyad, and S. Pau, Patterned color liquid crystal polymer polarizers, *Opt. Express* 18.26 (2010): 27777–27786.
26. P. Yeh, A new optical model for wire grid polarizers, *Opt. Commun.* 26.3 (1978): 289–292.
27. S. Arnold, et al. 52.3: An improved polarizing beamsplitter LCOS projection display based on wire-grid polarizers, *SID Symposium Digest of Technical Papers*, Vol. 32, No. 1, Blackwell Publishing Ltd (2001).
28. W.-L. Hsu, G. Myhre, K. Balakrishnan, N. Brock, M. Ibn-Elhaj, and S. Pau, Full-Stokes imaging polarimeter using an array of elliptical polarizer, *Opt. Express* 22(3) (2014): 3063–3074.
29. W.-L. Hsu, J. Davis, K. Balakrishnan, M. Ibn-Elhai, S. Kroto, N. Brock, and S. Pau, Polarization microscope using a near infrared full-Stokes imaging polarimeter, *Opt. Express* 23(4) (2015): 4357–4368.
30. V. L. Gamiz, Performance of a four channel polarimeter with low light level detection, *Proc. SPIE* 3121 (1997): 35–46.
31. R. Azzam, Arrangement of four photodetectors for measuring the state of polarization of light, *Opt. Lett.* 10 (1985): 309–311.
32. M. Billardon and J. Badoz, Birefringence modulator, *CR Acad. Sci. Ser. B* 262 (1966): 1672–1675.
33. J. C. Kemp, Piezo-optical birefringence modulators: New use for a long-known effect, *JOSA* 59.8 (1969): 950–954.
34. http://www.hindsinstruments.com/

35. J. O. Stenflo and H. Povel, Astronomical polarimeter with 2-D detector arrays, *Appl. Opt.* 24 (1985): 3893–3898.
36. H. Povel, H. Aebersold, and J. O. Stenflo, Charge-coupled device image sensor as a demodulator in a 2-D polarimeter with a piezoelastic modulator, *Appl. Opt.* 29 (1990): 1186–1190.
37. H. Povel, C. U. Keller, and I.-A. Yadigaroglu, Two-dimensional polarimeter with a charge-coupled-device image sensor and a piezoelastic modulator, *Appl. Opt.* 33 (1994): 4254–4260.
38. A. M. Gandorfer and H. P. Povel, First observations with a new imaging polarimeter, *Astron. Astrophys.* 328 (1997): 381–389.
39. F. Xu, G. Harten, D. J. Diner, O. V. Kalashnikova, F. C. Seidel, C. J. Bruegge, and O. Dubovik, Coupled retrieval of aerosol properties and land surface reflection using the Airborne Multiangle SpectroPolarimetric Imager (AirMSPI), *J. Geophys. Res. Atmos.* (2017).
40. Y. Liu and D. J. Diner, Multi-angle imager for aerosols: A satellite investigation to benefit public health, *Public Health Rep.* 132.1 (2017): 14–17.
41. P. M. Teillet et al., A generalized approach to the vicarious calibration of multiple Earth observation sensors using hyperspectral data, *Remote Sens. Environ.* 77.3 (2001): 304–327.
42. K. J. Thome et al., Vicarious calibration of ASTER via the reflectance-based approach, *IEEE Trans. Geosci. Remote Sens.* 46.10 (2008): 3285–3295.
43. N. J. Pust and J. A. Shaw, Dual-field imaging polarimeter using liquid crystal variable retarders, *Appl. Opt.* 45 (2006): 5470–5478.
44. N. J. Pust and J. A. Shaw, Digital all-sky polarization imaging of partly cloudy skies, *Appl. Opt.* 47 (2008): H190–H198.
45. S. Chandrasekhar and D. D. Elbert, The illumination and polarization of the sunlit sky on Rayleigh scattering, *Trans. Am. Philos. Soc.* 44(6) (1954) 643–728.
46. A. T. Young, Rayleigh scattering, *Appl. Opt.* 20.4 (1981): 533–535.
47. G. Mie, Beiträge zur Optik trüber Medien, speziell kolloidaler Metallösungen, *Ann. Phys.* 330(3) (1908): 377–445.
48. H. C. van de Hulst, *Light Scattering by Small Particles*, New York: John Wiley & Sons (1957).
49. C. F. Bohren and D. R. Huffmann, *Absorption and Scattering of Light by Small Particles*, New York: Wiley-Interscience (2010).
50. A. A. Kokhanovsky, *Satellite Aerosol Remote Sensing over Land*, ed. G. Leeuw, Berlin: Springer (2009).
51. M. I. Mishchenko et al., Past, present, and future of global aerosol climatologies derived from satellite observations: A perspective, *J. Quant. Spectr. Radiative Transfer* 106.1 (2007): 325–347.
52. M. I. Mishchenko, L. D. Travis, and A. A. Lacis, *Multiple Scattering of Light by Particles: Radiative Transfer and Coherent Backscattering*, Cambridge University Press (2006).
53. G. P. Konnen, *Polarized Light in Nature*, Cambridge: Cambridge University Press (1985).
54. K. L. Coulson, *Polarization and Intensity of Light in the Atmosphere*, Hampton: A. Deepak (1988).
55. B. Cairns, L. D. Travis, and E. E. Russell, The research scanning polarimeter: Calibration and ground-based measurements, *Proc. SPIE.* 3754 (1999).
56. J. Chowdhary, B. Cairns, and L. D. Travis, Case studies of aerosol retrievals over the ocean from multi-angle, multispectral photopolarimetric remote sensing data, *J. Atmos. Sci.* 59.3 (2002): 383–397.
57. D. A. Chu, Y. J. Kaufman, G. Zibordi, J. D. Chern, J. Mao, C. Li, and B. N. Holben, Global monitoring of air pollution over land from the Earth Observing System-Terra Moderate Resolution Imaging Spectroradiometer (MODIS), *J. Geophys. Res. Atmos.* 108(D21) (2003).
58. J. L. Deuzé et al., Analysis of the POLDER (POLarization and Directionality of Earth's Reflectances) airborne instrument observations over land surfaces, *Remote Sens. Environ.* 45.2 (1993): 137–154.
59. M. I. Mishchenko, B. Cairns, J. E. Hansen, L. D. Travis, G. Kopp, C. F. Schueler, B. A. Fafaul, R. J. Hooker, H. B. Maring, and T. Itchkawich, Accurate monitoring of terrestrial aerosols and total solar irradiance: Introducing the Glory Mission, *Bull. Am. Meteorol. Soc.* 88(5) (2007): 677–691.
60. R. M. A. Azzam and N. M. Bashara, *Ellipsometry and Polarized Light*, Amsterdam: Elsevier (1987).
61. H. Aben, J. Anton, and A. Errapart, Modern photoelasticity for residual stress measurement in glass, *Strain* 44.1 (2008): 40–48.
62. H. Aben, *Integrated Photoelasticity*, McGraw-Hill International Book Company (1979).
63. T. S. Narasimhamurty, *Photoelastic and Electro-optic Properties of Crystals*, Springer Science & Business Media (2012).
64. P. S. Theocaris and E. E. Gdoutos, *Matrix Theory of Photoelasticity*, Springer-Verlag (1979).
65. H. Aben and C. Guillemet, *Photoelasticity of Glass*, Springer Science & Business Media (2012).
66. A. Michel-Lévy and A. Lacroix, *Les Minéraux des Roches*, Paris: Librairie Polytechnique (1888).

67. https://www.mccrone.com/mm/the-michel-levy-interference-color-chart-microscopys-magical-color-key/#sthash.rx7qUDgR.dpuf (accessed April 23, 2017).
68. W. D. Nesse, *Introduction of Optical Mineralogy*, 2nd edition, University of Cambridge (1991).
69. R. Chipman and A. Peinado, The mystery of the birefringent butterfly, *Opt. Photon. News* 24.10 (2013): 52–57.
70. H. Aben, and I. C. Guillemet, Basic photoelasticity, in *Photoelasticity of Glass*, Berlin Heidelberg: Springer (1993), pp. 51–68.
71. M. H. Smith, Optimization of a dual-rotating-retarder Mueller matrix polarimeter, *Appl. Opt.* 41.13 (2002): 2488–2493.
72. D. H. Goldstein and R. A. Chipman, Error analysis of a Mueller matrix polarimeter, *JOSA A* 7.4 (1990): 693–700.
73. D. B. Chenault, J. L. Pezzaniti, and R. A. Chipman, Mueller matrix algorithms, in *Proc. SPIE 1746, Polarization Analysis and Measurement*, 231 (1992).
74. L. Broch, and L. Johann, Optimizing precision of rotating compensator ellipsometry, *Phys. Stat. Sol. C* 5(5) (2008): 1036–1040.
75. M. Tilsch and K. Hendrix, Optical interference coatings design contest 2007: Triple bandpass filter and nonpolarizing beam splitter, *Appl. Opt.* 47 (2008): C55–C69.
76. A. Mahler and R. Chipman, Polarization state generator: a polarimeter calibration standard, *Appl. Opt.* 50 (2011): 1726–1734.
77. R. A. Chipman, Polarimeter calibration error gets far out of control, in *Proc. SPIE 9583, An Optical Believe It or Not: Key Lessons Learned IV*, 95830H (2015).
78. M. Novak et al., Analysis of a micropolarizer array-based simultaneous phase-shifting interferometer, *Appl. Opt.* 44.32 (2005): 6861–6868.
79. D. S. Sabatke, A. M. Locke, M. R. Descour, W. C. Sweatt, J. P. Garcia, E. L. Dereniak, S. A. Kemme, and G. S. Phipps, Figures of merit for complete Stokes polarimeter optimization, in *Proc. SPIE 4133, Polarization Analysis, Measurement, and Remote Sensing III*, 75 (2000).
80. J. S. Tyo, Design of optimal polarimeters: Maximization of signal-to-noise ratio and minimization of systematic error, *Appl. Opt.* 41(4) (2002): 619–630.
81. K. Twietmeyer, GDx-MM: An imaging Mueller matrix retinal polarimeter, College of Optical Sciences, Tucson, The University of Arizona, PhD thesis, 347 (2007).
82. K. Twietmeyer and R. A. Chipman, Optimization of Mueller matrix polarimeters in the presence of error sources, *Opt. Express* 16(15) (2008): 11589–11603.
83. R. A. Horn and C. R. Johnson, *Matrix Analysis*, Cambridge: Cambridge University Press (1985).

8

Fresnel Equations

8.1 Introduction

The *Fresnel equations* describe the behavior of light at an interface between two media. The amplitude of the light divides between the reflected and refracted beam based on the refractive indices of the two media. The resulting *amplitude coefficients*, described by the Fresnel coefficients, are used within our various polarization matrices to include the effects of reflection and refraction in polarization ray tracing calculations. When wavefronts propagate through optical systems, the resulting polarization effects generate diattenuation and retardance aberrations.

One of the most straightforward polarization changes that occur in an optical system are the changes at dielectric to dielectric and dielectric to metal interfaces. When light strikes an interface, it divides into reflected and transmitted beams. The amount of light reflected and transmitted depends on its polarization state and orientation with respect to the interface. The Fresnel equations relate the complex amplitudes of the reflected and transmitted beams to the optical properties of the surface and the angle of incidence. They are used during polarization ray tracing to calculate the changes in polarization as light propagates through optical systems to determine the diattenuation and retardance associated with ray intercepts and ray paths through optical systems.

Optical surfaces with coatings, such as antireflection coatings and beam splitter coatings, have more complex amplitude reflection and transmission coefficients that are calculated by the application of the Fresnel coefficients to multilayer stacks of optical materials as discussed in Chapter 13 (Thin Films). The calculation of the amplitude transmission coefficients for anisotropic materials, such as calcite and sapphire, is derived in Chapter 19.

Our interest is on the effect of interfaces on light propagation. The Fresnel equations can be derived rigorously from *Maxwell's equations* applied to a homogeneous and isotropic interface. This derivation is well covered in many other sources. Here, we will focus on the polarization effects, the consequences of Fresnel equations, their magnitudes and forms, and how to integrate these polarization effects into optical design.

This chapter presents the Fresnel equations and then applies the Fresnel equations to describe polarization changes that occur at dielectric interfaces, total internal reflection, and reflection from metals, as in mirrors. The changes in polarization that occur on reflection and refraction are characterized. Then, the resulting polarization aberrations are described for several example optical systems, such as a Cassegrain telescope in Chapter 12 and an astronomical telescope in Chapter 27.

8.2 Propagation of Light

8.2.1 Plane Waves and Rays

Given a *plane wave* incident on a plane interface, what are the amplitudes and phases of the reflected and refracted plane waves? Optical design is concerned with plane, spherical, and other wavefronts. In optical ray tracing, light is treated as a *light ray*. A small patch of a wavefront can be assumed to locally have a plane wavefront. The restriction to plane waves is for the purpose of the derivation. Similarly, an optical interface is plane, spherical, or aspheric surface. But in the vicinity of the light ray, the interface can be considered as a plane surface for the purpose of ray tracing. Thus, the change of the light ray's polarization upon reflection and refraction can be calculated using the local properties of the interface; this works remarkably well as an approximation.

8.2.2 Plane of Incidence

The *plane of incidence* contains the *wavevector* $\mathbf{k}_i$ of an incident plane wave and the *surface normal* $\boldsymbol{\eta}$, as shown in Figure 8.1. The incident medium has refractive index n_i. The transmitted medium has refractive index n_t. The plane wave is incident with an *angle of incidence* θ_i measured from the surface normal. A reflected beam propagates back into the incident medium in the direction $\mathbf{k}_r$. A transmitted beam propagates into the transmitted medium in the direction $\mathbf{k}_t$ at an *angle of refraction* θ_t.

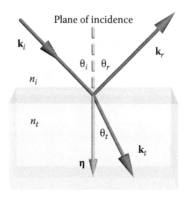

Figure 8.1 The configuration for reflection and refraction of a plane wave at an interface. The plane of incidence is the plane containing the incident wavevector $\mathbf{k}$ and the surface normal $\boldsymbol{\eta}$.

8.2.3 Homogeneous and Isotropic Interfaces

An interface divides part of an optical system into two volumes with different properties, such as a lens surface or the surface of a mirror. In this chapter, it is assumed that the material is isotropic, homogeneous, non-absorbing, non-magnetic, and has no free charges on the interface.

The following provides the terminology of different types of material and interface:

Homogeneous interface—Interface with uniform optical properties such as constant refractive index $n + i\kappa$ over the clear aperture, which may be real (transparent medium) or complex (absorbing or metal)

Inhomogeneous interface—Non-uniform interface with varying composition or varying coating thickness

Isotropic material—The same refractive index in all directions and polarizations within the material

Anisotropic material—Birefringent or optically active materials; light experiences varying refractive index as a function of the light's polarization direction

8.2.4 Light Propagation in Media

Light propagates in a vacuum at the speed of light c,

$$c = 299\ 792\ 458 \text{ m/s}. \tag{8.1}$$

When light propagates through a material, the light's electric and magnetic field drives the material's charges, the electrons and protons, which then oscillate at the light's frequency. In an absorbing material, the atoms and molecules readily make transitions into other electronic and vibrational quantum states of the material, and the light is absorbed into heat or scattered into other directions. Transparent materials, like air, water, and glass, do not have available transitions near the light's frequency.

Consider light refracting into glass. Monochromatic light drives the charges at the surface into simple harmonic oscillation. The charges are accelerated by the light, and accelerating charges radiate. For a monochromatic plane wave or spherical wave, a whole volume of charge is set oscillating, and the coherent radiation from the volume radiates a beam of light into the forward direction, along the direction of the original beam of light. The charges in the transparent material do not oscillate quite in phase with the incident light; thus, the atoms radiate light that is slightly delayed. The net result of the superposition of the incident light and light reradiated by the atoms is that the light slows down. This is the source of the refractive index of the light.

The *refractive index n* (also called *index of refraction*) of transparent material is the ratio of the velocity of light in vacuum to the velocity V of light in a medium,

$$n = c/V. \tag{8.2}$$

The angle of refraction θ_t is related to the angle of incidence θ_i by *Snell's law*

$$n_i \sin\theta_i = n_t \sin\theta_t. \tag{8.3}$$

Snell's law embodies the *conservation of momentum*. It also describes the *phase matching* that must occur at an interface between the two media, as shown in Figure 8.2.

There is an equilibrium between the light field and the oscillation of the transparent materials charges. The electric field of the light is reduced, while some of the energy is tied up in the charge oscillation. The light's field is reduced, but its flux is not. When light is incident on an interface,

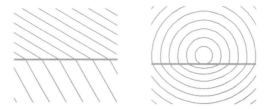

Figure 8.2 Intersection of an incident set of wavefronts on the interface for a plane wave (left) and a spherical wave (right) matches the intersections of the transmitted set of wavefronts. The incident and exiting wavefronts have different wavelengths and propagation velocities. Since they have the same frequency, they match at the interface.

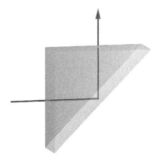

Figure 8.3 Internal reflection occurs when light reflects from a glass–air interface.

the charges of the atoms and molecules are set into motion. The reflected and transmitted *amplitude coefficients* of the light's electric field are calculated by solving Maxwell's equations at the interface.

Reflection associated with transparent material is divided into two cases, external and internal. External reflection occurs when $n_i < n_t$, such as when light reflects from air to glass interface. Internal reflection occurs when $n_t < n_i$, such as light reflecting inside glass from an air interface, as shown in Figure 8.3.

When light reflects from metal surface, the electric field of the reflected light is affected by the complex refractive index of the metal. Metals have free charges that easily move between atoms when an electric field is present or a voltage is applied. Optically, this is the cause of the metal's complex refractive index, $n + i\kappa$. The imaginary part of the refractive index governs how rapidly a wave is absorbed propagating through a material. The incident light's energy is substantially absorbed within a few tens of nanometers when reflecting from metals with κ greater than 2.

8.3 Fresnel Equations

8.3.1 *s*- and *p*-Polarization Components

The incident light's electric field is oscillating in a plane *transverse* to **k**. It can be divided into a *p*-polarized component oscillating in the plane of incidence and an *s*-polarized component oscillating perpendicular to the plane of incidence as shown in Figure 8.4. A basis vector for the *s*-component $\hat{\mathbf{s}}$ is constructed as

$$\hat{\mathbf{s}} = \frac{\mathbf{k} \times \boldsymbol{\eta}}{\|\mathbf{k} \times \boldsymbol{\eta}\|}, \tag{8.4}$$

8.3 Fresnel Equations

Figure 8.4 The electric fields for *s*- and *p*-polarized components of a beam of light incident on an interface. (Left) *s*-component perpendicular to the plane of incidence. (Right) *p*-component in the plane of incidence. The *s*-component lies on the surface while the *p*-component has a component into the surface.

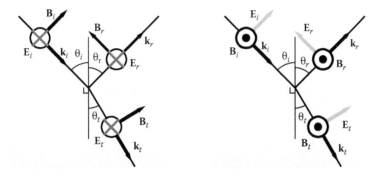

Figure 8.5 (Left) Transverse electric field orthogonal to the plane of incidence. (Right) Transverse magnetic field orthogonal to the plane of incidence.

which lies on the interface. Since the *s*-component's electric field is perpendicular to the plane of incidence, it is also referred to as *transverse electric*, or the *TE mode*, as shown in Figure 8.5. The basis vector for the *p*-component $\hat{\mathbf{p}}$ is

$$\hat{\mathbf{p}} = \hat{\mathbf{k}} \times \hat{\mathbf{s}}, \tag{8.5}$$

which lies in the plane of incidence. The *p*-component is also known as the *transverse magnetic*, or the *TM mode*, because the corresponding magnetic field is transverse to the plane of incidence. The set $(\hat{\mathbf{s}}, \hat{\mathbf{p}}, \hat{\mathbf{k}})$ forms a right-handed orthonormal basis. The incident light's polarization vector **E** is expressed in terms of the *s*- and *p*-components as

$$\mathbf{E} = E_s \hat{\mathbf{s}} + E_p \hat{\mathbf{p}}. \tag{8.6}$$

An arbitrary incident plane wave can be decomposed into *s*- and *p*-components that are then reflected and refracted separately. At normal incidence, the definitions of *s* and *p* break down, the cross product in Equation 8.4 goes to zero, and $\hat{\mathbf{s}}$ becomes undefined. Any pair of orthogonal unit vectors on the surface can be chosen since $\hat{\mathbf{s}}$ and $\hat{\mathbf{p}}$ have become *degenerate*.

8.3.2 Amplitude Coefficients

Consider a plane wave incident at a homogeneous and isotropic interface. The incident medium is assumed to have no absorption; thus, n_i is real. n_t can be real or complex. The incident plane wave interacts with the interface, giving rise to reflected and transmitted plane waves. An *s*-polarized incident plane wave reflects and transmits into an *s*-polarized reflected wave and an *s*-polarized transmitted wave. The same is true for an incident *p*-polarized plane wave. These two incident states, *s* and *p*, are the eigenpolarizations for reflection and refraction. Other incident polarization states, combinations of *s* and *p* states, are not in general reflected and refracted in the incident state. Therefore, the description of reflection and refraction is simplest in terms of the *s*- and *p*-components.

The amplitudes of the plane waves before and after the interface are related by *amplitude coefficients* r_s, t_s, r_p, and t_p. The incident and reflected electric field amplitudes are related as*

$$\begin{cases} \mathbf{E}_{s,r} = r_s \mathbf{E}_{s,i} = \rho_{s,r} e^{-i\phi_{s,r}} \mathbf{E}_{s,i} \\ \mathbf{E}_{s,t} = t_s \mathbf{E}_{s,i} = \rho_{s,t} e^{-i\phi_{s,t}} \mathbf{E}_{s,i} \end{cases} \text{ and } \begin{cases} \mathbf{E}_{p,r} = r_p \mathbf{E}_{p,i} = \rho_{p,r} e^{-i\phi_{p,r}} \mathbf{E}_{p,i} \\ \mathbf{E}_{p,r} = t_p \mathbf{E}_{p,i} = \rho_{p,t} e^{-i\phi_{p,t}} \mathbf{E}_{p,i} \end{cases}. \tag{8.7}$$

8.3.3 The Fresnel Equations

The amplitude coefficients for an interface associated with two dielectric or an incident dielectric and metal substrate are the *Fresnel coefficients*. There are four coefficients, an *s*- and *p*-coefficient for reflection and an *s*- and *p*-coefficient for transmission, given as follows[1-4]:

$$r_s(\theta_i) = \frac{n_i \cos\theta_i - n_t \cos\theta_t}{n_i \cos\theta_i + n_t \cos\theta_t} = \frac{-\sin(\theta_i - \theta_t)}{\sin(\theta_i + \theta_t)}, \tag{8.8}$$

$$t_s(\theta_i) = \frac{2 n_i \cos\theta_i}{n_i \cos\theta_i + n_t \cos\theta_t} = \frac{2 \sin\theta_t \cos\theta_i}{\sin(\theta_i + \theta_t)}, \tag{8.9}$$

$$r_p(\theta_i) = \frac{n_t \cos\theta_i - n_i \cos\theta_t}{n_t \cos\theta_i + n_i \cos\theta_t} = \frac{\tan(\theta_i - \theta_t)}{\tan(\theta_i + \theta_t)}, \text{ and} \tag{8.10}$$

$$t_p(\theta_i) = \frac{2 n_i \cos\theta_i}{n_t \cos\theta_i + n_i \cos\theta_t} = \frac{2 \sin\theta_t \cos\theta_i}{\sin(\theta_i + \theta_t)\cos(\theta_i - \theta_t)}. \tag{8.11}$$

Using Snell's law, the angle of refraction is eliminated, yielding an alternative form for these four coefficients dependent only on angle of incidence,

$$r_s(\theta_i) = \frac{\cos\theta_i - \sqrt{n^2 - \sin^2\theta_i}}{\cos\theta_i + \sqrt{n^2 - \sin^2\theta_i}}, \tag{8.12}$$

$$t_s(\theta_i) = \frac{2\cos\theta_i}{\cos\theta_i + \sqrt{n^2 - \sin^2\theta_i}}, \tag{8.13}$$

---

* These equations are presented in the decreasing phase convention used throughout the book, as described in Sections 2.17 and 2.18. In the increasing phase convention, the complex amplitude coefficient is the complex conjugate or $\rho e^{i\phi}$. Of course, these two conventions produce the equivalent electric field **E**.

8.3 Fresnel Equations

$$r_p(\theta_i) = \frac{n^2 \cos\theta_i - \sqrt{n^2 - \sin^2\theta_i}}{n^2 \cos\theta_i + \sqrt{n^2 - \sin^2\theta_i}}, \text{ and} \quad (8.14)$$

$$t_p(\theta_i) = \frac{2n \cos\theta_i}{n^2 \cos\theta_i + \sqrt{n^2 - \sin^2\theta_i}}, \quad (8.15)$$

where $n = n_t/n_i$.

Fresnel coefficients are a function of refractive index. Thus, they can appear in two forms: the form for decreasing phase convention calculated by $n + i\kappa$, and the form for increasing phase convention calculated with $n - i\kappa$.

8.3.4 Intensity Coefficients

The Fresnel amplitude coefficients relate the electric field amplitudes of the various wavefronts. The Fresnel intensity coefficients relate the fraction of flux in the reflected and refracted beams to the incident flux. For the reflected beam, the intensity coefficients are just the magnitude squared of the amplitude coefficients,

$$R_s(\theta_i) = |r_s(\theta_i)|^2, \text{ and} \quad (8.16)$$

$$R_p(\theta_i) = |r_p(\theta_i)|^2. \quad (8.17)$$

For the transmission coefficients, two additional effects must be considered. First, for waves propagating inside materials other than vacuum, the light waves induce charge motions that contribute to the flux. The flux in a material of refractive index n is proportional to n times the flux in vacuum. Thus, a plane wave with amplitude of 1 V/m in silicon, $n = 4$, conveys four times the energy per unit area as a plane wave in vacuum with the same 1 V/m amplitude. Second, the area of a beam changes on refraction as shown in Figure 8.6. The ratio of the refracted area to incident area is

$$\frac{A_t}{A_i} = \frac{\cos\theta_t}{\cos\theta_i}. \quad (8.18)$$

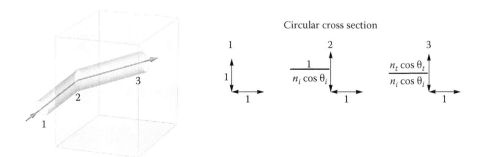

Figure 8.6 (Left) A beam with area A_i incident on an interface (2) refracts into a beam with area A_t. (1) A circular beam with a unit radius in the incident medium, (2) has a cross section on the interface that is stretched along the plane of incidence. (3) The beam cross section after refraction.

Note that the cross-sectional area of a beam does not change on reflection because $\theta_i = \theta_r$.

Taking these two factors into account, the Fresnel intensity coefficients become

$$T_s(\theta_i) = \frac{n_t \cos\theta_t}{n_i \cos\theta_i} |t_s(\theta_i)|^2 \text{ and} \tag{8.19}$$

$$T_p(\theta_i) = \frac{n_t \cos\theta_t}{n_i \cos\theta_i} |t_p(\theta_i)|^2, \tag{8.20}$$

where n_t/n_i accounts for different power conveyed in the different media, and $\cos\theta_t/\cos\theta_i$ accounts for the change in beam cross section.

Diattenuation of the transmitted and reflected light at isotropic surface is defined by the s- and p-Fresnel intensity transmission and reflection coefficients,

$$D_t = \frac{T_s - T_p}{T_s + T_p} \text{ and } D_r = \frac{R_s - R_p}{R_s + R_p}. \tag{8.21}$$

Example 8.1 Reflection and Transmission at Glass

Consider a beam incident in air $n_i = 1$ reflecting and transmitting from a glass interface with refractive index $n_t = 1.5$. Figure 8.7 (left) shows the Fresnel amplitude transmission coefficients as a function of θ_i. Near normal incidence, the refracted **E** into the glass is 0.8 of the incident **E**. This amplitude monotonically decreases as θ_i increases and reaches zero at grazing incidence. Figure 8.7 (right) shows the Fresnel amplitude reflection coefficients. These have magnitude 0.2 at normal incidence; $r_s(0)$ and $r_p(0)$ are equal in magnitude, but because of the sign convention for Fresnel coefficients, discussed below, they are opposite in sign, with $r_s(0) < 0$. $r_s(\theta_i)$ monotonically decreases to -1 at $\theta_i = 90°$; all the light reflects at grazing incidence. $r_p(\theta_i)$ decreases, crossing zero at about $\theta_i = 57°$, which is Brewster's angle discussed in Section 8.3.6.

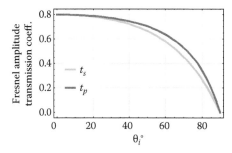 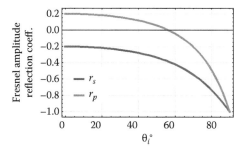

Figure 8.7 (Left) Amplitude transmission coefficients t_s (green) and t_p (blue) as a function of incident angle at an interface with incident refractive index $n_i = 1$ and transmitted refractive index $n_t = 1.5$. (Right) Amplitude reflection coefficients r_s (red) and r_p (orange) as a function of incident angle.

After accounting for the additional factors in the computation of the intensity transmittance, Equations 8.19 and 8.20, the Fresnel intensity transmission coefficients have a different

8.3 Fresnel Equations

form compared to the amplitude coefficients. Figure 8.8 shows the intensity coefficients. The s intensity reflectance R_s monotonically increases from 0.04 to 1. The p intensity reflectance R_p decreases slowly from 0.04, reaching 0 at Brewster's angle. At normal incidence, a fraction of the flux ($T = 0.96$) of both the s- and p-components are refracted into the glass. The flux transmittance of the p-component T_p increases, initially quadratically to 1 at Brewster's angle, before decreasing rapidly to 0 at grazing incidence. The intensity transmittance of the s-component T_S decreases monotonically from 0.96 to 0.

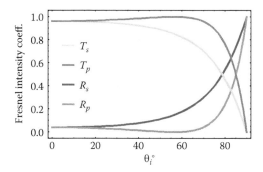

Figure 8.8 Intensity transmission coefficients T_s (green) and T_p (blue) and intensity reflection coefficients R_s (red) and R_p (orange) as a function of incident angle at an interface with incident refractive index $n_i = 1$ and transmitted refractive index $n_t = 1.5$.

Figure 8.9 shows the corresponding diattenuation for refraction and reflection. Several features are worth repeating. First, at Brewster's angle, the reflected light is fully polarized; the diattenuation is 1, only a small fraction of the flux, ~0.17 is reflected. Also at Brewster's angle, 100% transmission is attained for the p-component; this is very useful in situations such as inside laser cavities, where loss must be minimized. Near normal incidence, the diattenuation, and thus the polarization change, is small. Further, since both refractive indices are real, the Fresnel coefficients are real and there is no retardance, only diattenuation. Note that as the reflected p-component passes through Brewster's angle, it undergoes a 180° phase change.

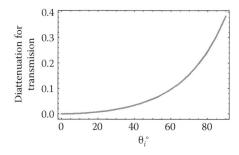
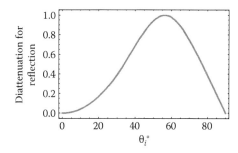

Figure 8.9 At an interface with incident refractive index $n_i = 1$ and transmitted refractive index $n_t = 1.5$, (left) diattenuation upon transmission as a function of incident angle and (right) diattenuation upon reflection as a function of incident angle.

8.3.5 Normal Incidence

At normal incidence, the incident light lies along the normal of the interface and the difference between s and p disappears. The electric field for any incident polarization state lies along the interface, the definition of s-polarization. At normal incidence, the plane of incidence is undefined, so s and p become degenerate. Taking the limit of the Fresnel equations as θ_i approaches zero, the s and p amplitude transmission coefficients become equal,

$$t(0) = t_p(0) = t_s(0) = \frac{2n_i}{n_t + n_i}. \tag{8.22}$$

Similarly, the two amplitude reflection coefficients become

$$r(0) = r_p(0) = -r_s(0) = \frac{n_t - n_i}{n_t + n_i}. \tag{8.23}$$

Thus, the normal incidence intensity transmission and reflection coefficients are

$$T(0) = T_p(0) = T_s(0) = \frac{n_t}{n_i} \left| \frac{2n_i}{n_t + n_i} \right|^2, \tag{8.24}$$

$$R(0) = R_p(0) = R_s(0) = \left| \frac{n_t - n_i}{n_t + n_i} \right|^2. \tag{8.25}$$

The derivations of Fresnel amplitude coefficients use *right-handed coordinates* for both the incident and reflected beams and $(\hat{\mathbf{s}}, \hat{\mathbf{p}}, \hat{\mathbf{k}})$ in almost all references. Since the $\hat{\mathbf{k}}$ basis vector reverses direction, this condition requires a change of sign in one of the transverse basis vectors (s or p) at the interface after reflection, as shown in Figure 8.10. This change of direction requires an extra minus sign in one of the reflection amplitude coefficients to conform to this convention. Since our polarization ray tracing algorithms use global coordinates, this minus sign is not needed. One must be very careful in choosing the sign of amplitude coefficient and stay consistent throughout the ray tracing calculation.

Consider linearly polarized light at or near normal incidence with its polarization oriented at 45° from the incident right-handed coordinates as shown in Figure 8.11. The light causes electrons to oscillate along this 45° and the reflected light is polarized in this same plane. Globally, the plane

Figure 8.10 Examples of right-handed coordinate systems for light at normal incidence. (Left) Incident light basis vectors and (right) reflected light basis vectors. Because the propagation vector **k** changes direction, one of the coordinates, **x** or **y**, must also change sign to retain the right-handed convention; **y** is shown changing here.

8.3 Fresnel Equations

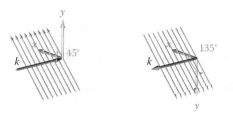

Figure 8.11 In a right-handed coordinate system containing **k**, 45° linear polarization externally reflects into 135° linear polarization.

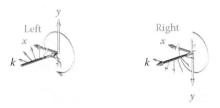

Figure 8.12 In a right-handed coordinate system containing **k**, left circular polarization always reflects into right circular polarization at normal incidence from an isotropic interface.

of polarization of the incident and reflected light are in the same plane. Expressed in right-handed coordinates, this incident light is 45° polarized and the reflected light is 135° polarized! The Fresnel *s*-coefficient in Equations 8.8 and 8.12 incorporate this minus sign.*

Figure 8.12 shows left circularly polarized light incident on an interface and the state following reflection. The light's field moves the reflector's charges in a clockwise direction rotating from $(+x \rightarrow +y)$ to $(-x \rightarrow +y)$. The field of the reflected light continues to rotate in the same direction as a function of time (small orange arrows), but now the direction of propagation is reversed. The helicity of the wave has changed from left to right circularly polarized.

8.3.6 Brewster's Angle

A special angle related to the Fresnel equations is the *Brewster's angle*. Brewster's angle θ_B is defined by the angle of incidence θ_i where the *p*-reflectance is zero, $r_p = 0$, which occurs at

$$\theta_B = \tan^{-1} \frac{n_t}{n_i}. \tag{8.26}$$

Figure 8.13 (left) shows Brewster's angle at varying n_t/n_i. Note that at $n_t/n_i = 1$, Brewster's angle is 45°.

Since no *p*-polarized light is reflected at Brewster's angle and only *s*-polarized light is reflected, the reflected light is polarized. When unpolarized light is incident on a surface at Brewster's angle, the reflected light is linearly polarized transverse to the plane of incidence; the reflected light is *s*-polarized. Therefore, the degree of polarization becomes one at Brewster's angle, as shown in Figure 8.14.

This characteristic provides a convenient method to produce a polarizer; reflecting a nearly collimated beam from a dielectric medium at Brewster's angle can work as a polarizer. Although the reflected flux is not so great, about 0.17 for $n_t/n_i = 1.5$, the polarizer is extremely easy to construct.

---

* This is the most common Fresnel equation sign convention.

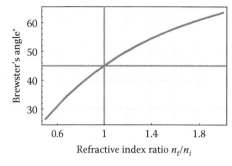
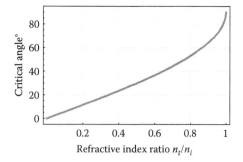

Figure 8.13 (Left) Brewster's angle θ_B and (right) the critical angle θ_C as a function of the ratio of refractive indices n_t/n_i.

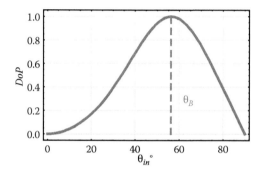

Figure 8.14 Degree of polarization of unpolarized incident light reflected at an air–glass interface for $n_t/n_i = 1.5$.

The Brewster's angle phenomena occurs at the special incidence angle where the angle between the reflected and refracted beams is 90°, $\theta_r + \theta_t = 90°$. The direction of motion of the charges in the refracting medium is 90° from the angle of refraction. Dipole oscillators do not radiate along that axis. Thus, at Brewster's angle, the charges in the refracting material cannot radiate (i.e., reflect) light back into the incident medium, and 100% of the *p*-polarized light is refracted into the refracting medium. This can be extremely useful. For example, many elements are placed into laser cavities at Brewster's angle and operated with *p*-polarized light to minimize losses and maximize gain.

8.3.7 Critical Angle

Another special angle related to the Fresnel equations is the *critical angle*. A distinction is made between *external reflection* where $n_i < n_t$ and *internal reflection* where $n_i > n_t$. Internal reflection occurs inside glass for example. During internal reflection, only light incident within a range of θ_i less than the critical angle can escape into the lower index medium. The critical angle occurs when the angle of refraction from Snell's law is equal to 90°; the refracted light exits the interface at grazing angle. Solving Snell's law for $\theta_t = 90°$ yields an incident angle, the critical angle θ_C of

$$\sin \theta_C = \frac{n_t}{n_i}. \tag{8.27}$$

For θ_i greater than the θ_C, all incident energy is internally reflected, a condition called *total internal reflection* or TIR. TIR is extremely useful for designing low loss optics. Many optical systems

8.3 Fresnel Equations

such as binoculars use TIR prisms to minimize loss. Fold mirrors are often replaced with prisms. The entering and exiting faces are antireflection coated to minimize reflection loss, and the TIR at the hypotenuse is 100%, yielding far less loss than reflection from a metal mirror. A comprehensive list of TIR prism designs is found in Wolfe.[5] Figure 8.13 (right) plots the critical angle at varying n_t/n_i. For glasses with $1.5 < n_i < 2$, the θ_C falls in the range of 40° to 30°; thus, higher index glasses undergo TIR over a larger range of angles, and such higher indices are often necessary for TIR devices. Much of the sparkle of diamond and cubic zirconia gemstones comes from TIR associated with their high refractive indices.

Below the critical angle, dielectric materials with real refractive index have real Fresnel coefficients and no phase change occurs on internal reflection. The phase changes beyond the critical angle are

$$\tan\frac{\phi_s}{2} = \frac{\sqrt{\sin^2\theta_r - \sin^2\theta_C}}{\cos\theta_r} \quad \text{and} \quad \tan\frac{\phi_p}{2} = \frac{\sqrt{\sin^2\theta_r - \sin^2\theta_C}}{\cos\theta_r \sin^2\theta_C}. \tag{8.28}$$

Retardance is defined in Section 5.3.2 as the optical path difference between eigenpolarizations, here the *s*- and *p*-polarized light. Calculating the phase difference using the phases of the Fresnel coefficients

$$\Delta = Arg[r_s] - Arg[r_p] = \phi_s - \phi_p \tag{8.29}$$

produces a number, Δ, which contains an extra π phase due to geometrical rotation upon reflection, as described in Figures 8.10 through 8.12. Here, in this normal incidence example, $\Delta = \pi$, due to the change of the sign of the *s*-basis vector. *s*- and *p*-light should have the same phase change. This extra π phase must be removed to obtain the *optical path difference*, the physical phase difference,

$$\delta = Arg[-r_s] - Arg[r_p] = (Arg[r_s] + \pi) - Arg[r_p] = \phi_s - \phi_p + \pi. \tag{8.30}$$

The issues of this additional π for polarization ray tracing are discussed in Section 17.6.1.

8.3.8 Intensity and Phase Change with Incident Angle

Reflections as a function of incident angle are treated in this section. Figure 8.15 plots the Fresnel *s* and *p* reflection coefficients for external and internal reflections for an air–glass ($n = 1.5$) interface.

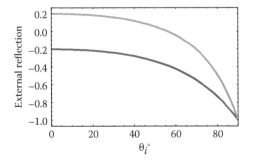
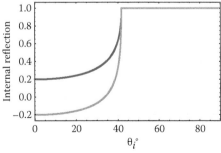

Figure 8.15 Fresnel external (left) and internal (right) amplitude reflection coefficients for *s*-polarization (red) and *p*-polarization (orange) versus the angle of incidence.

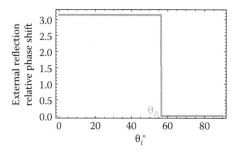 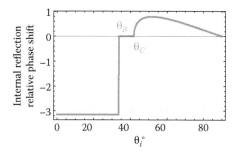

Figure 8.16 The relative phase shift for external (left) and internal (right) reflection calculated from the Fresnel coefficients as a function of the angle of incidence.

The Fresnel coefficients[6] are defined in the incident and exiting local coordinates using the sign convention in broadest use,[7] which are both right handed. At normal incidence, s and p reflection coefficients have the opposite sign due to this choice of local coordinates.

Figure 8.16 shows relative phase shifts between s- and p-polarizations for external and internal reflections. A residual Fresnel π phase difference exists as the consequence of coordinate change during reflection. At Brewster's angle θ_B, the p Fresnel reflection coefficient passes through 0 and changes sign, causing an additional π phase shift upon reflection for the incident angles greater than θ_B. The transmitted light is partially polarized while the reflected beam is completely polarized. At critical angle θ_C, the s- and p-polarized light has equal amplitude but different phase change and results in non-zero retardance.

8.3.9 Jones Matrices with Fresnel Coefficients

The Jones matrices for rays reflecting from or refracting into isotropic media incorporate the Fresnel amplitude coefficients. First, consider a Jones matrix defined in s–p coordinates. In this basis, the reflection and refraction Jones matrices are diagonal matrices,

$$J_{reflect} = \begin{pmatrix} r_s & 0 \\ 0 & r_p \end{pmatrix} \quad \text{and} \quad J_{refract} = \begin{pmatrix} t_s & 0 \\ 0 & t_p \end{pmatrix} \qquad (8.31)$$

for a single isotropic surface, because s-polarized light couples entirely to s-polarized light and p-polarized light couples entirely to p-polarized light. Since Jones matrices are expressed in right-handed coordinate systems, the phase shifts are the Fresnel coefficients phase shifts. For angles of incidence smaller than θ_B, the Jones matrix for reflection shows π phase shift due to the right-handed local coordinate choice before and after the reflection, as shown in Figure 8.10. This value is not the proper retardation because it includes the π geometric transformation of the local coordinates as described in Section 17.6.1. The construction of the polarization ray tracing **P** matrices for reflection and refraction incorporating the Fresnel coefficients is covered in Section 10.6.

8.4 Fresnel Refraction and Reflection

8.4.1 Dielectric Refraction

In the Fresnel equations, only the ratio of the refractive indices matters; the coefficients only depend on the ratio. Figure 8.17 shows the effect of the refractive index ratio, $n = n_t/n_i$ on the

8.4 Fresnel Refraction and Reflection

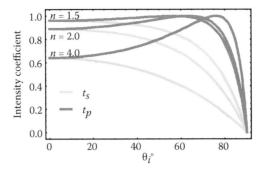

Figure 8.17 Intensity transmission coefficients t_s (green) and t_p (blue) as a function of incident angle for an interface as the ratio of refractive indices n_2/n_1 changes: 1.5, 2, and 4.

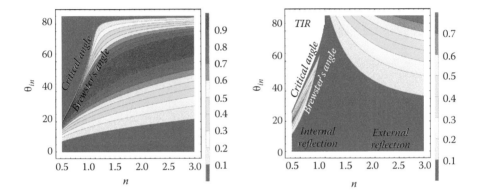

Figure 8.18 (Left) Transmission diattenuation for dielectric interface with varying refractive index ratio n at varying incident angles θ_{in}. Brewster's angle occurs down the middle of the violet band. (Right) Reflection diattenuation for dielectric interface with varying index n at varying incident angle θ_{in}.

Fresnel intensity transmission coefficients. As the refractive index ratio increases, the normal incidence transmission decreases and Brewster's angle increases. The separation between the s- and p-coefficients occurs more rapidly at higher index. Again, no retardance is present with these purely real refractive indices.

The diattenuation of a dielectric interface is plotted versus refractive index ratio and angle of incidence in the colored contour plot of Figure 8.18. Diattenuation is zero at normal incidence, along the bottom of the graph, at grazing incidence, along the top of the graph, and for total internal reflection, the red region along the upper left side of the contour plot. Along vertical cross sections, diattenuation is seen to rise monotonically to a peak of one, at Brewster's angle, and then falls monotonically back to zero, at the critical angle for $n_1/n_2 < 1$, or monotonically back to zero at grazing incidence for $n_1/n_2 > 1$.

8.4.2 External Reflection

External reflection occurs when light reflects from a less dense medium to a denser medium, such as from an air–glass interface. Figure 8.19 depicts external reflection of the electric field for three different polarizations: s-polarization states, an equal combination of s- and p-polarization states, and p-polarization states. Since the Fresnel sign conventions can be confusing, it is helpful to visualize the fields just before and just after the interface in three dimensions.

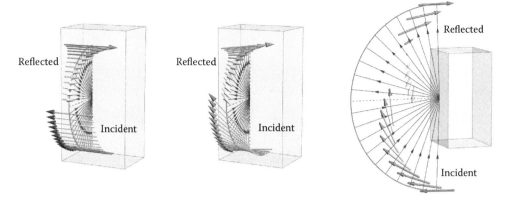

Figure 8.19 External reflection for the incident s-polarized states (left), 45° polarization states (center), and p-polarized states (right) as angles of incidence change from 7° to 88° near grazing incidence. Since the p-polarization figures would overlap for different angles of incidence, each incident and reflected pair has been spatially translated for better visualization. The blue arrow indicates the reflected electric field at Brewster's angle. Each incident electric field vector (black arrowhead) and corresponding reflected electric field vector (colored arrowhead) pair is plotted in the same color. The arrows represent the oscillations of the electric field vector in time. The length of the arrow represents the amplitude of the electric field vector.

Care is needed to properly interpret the signs of the Fresnel coefficients. When calculated properly, polarization ray tracing matrix will yield zero retardance upon reflection at normal incidence and π phase shift for external reflection in the range, $\theta_B < \theta_i < \theta_C$. The retardance for internal reflection changes rapidly for the angle of incidence greater than θ_C as shown in Figure 8.16. The reflected electric field vector has smaller amplitude than the corresponding incident electric field. The s-component of the electric field steadily increases from 0.2 to 1.0 as the angle of incidence increases. The p-component of the electric field changes its sign after Brewster's angle and has zero reflectance at the Brewster's angle. This is the origin of π retardance for reflections in the range $\theta_B < \theta_i < \theta_C$.

Figure 8.20 shows the external reflection of circularly polarized light. Since both s- and p-components have a π phase shift upon reflection when $\theta_i < \theta_B$, left circularly polarized light reflects as right circularly polarized light at normal incidence. Note that "left" and "right" are defined locally relative to the propagation direction. The incident and reflected light circulate in the same direction when viewed in 3D global coordinate for $\theta_i < \theta_B$. As θ_i approaches θ_B, the amplitude of the reflected p-component is lower than the reflected s-component. Thus, it reflects as right elliptically polarized light. When $\theta_i = \theta_B$, the reflected light becomes purely s-polarized. For $\theta_i > \theta_B$, both reflected s- and p-components have increasing amplitude toward grazing incidence, while the p-component has an extra π shift before Brewster's angle. Thus, the left circularly polarized light reflects as left elliptically polarized light. When $\theta_i \approx 90°$, left circularly polarized light reflects as left circularly polarized light.

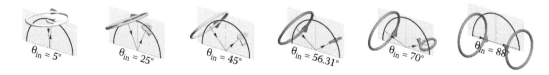

Figure 8.20 Left circularly incident light at various incident angles externally reflect from an air–glass interface. The reflected light around normal incident is right elliptically polarized. It becomes linearly polarized at Brewster's angle ($\theta_B = 56.31°$) and switches to left elliptically polarized above θ_B.

8.4.3 Internal Reflection

Internal reflection is reflection at an interface from a larger to a smaller refractive index, such as reflection inside glass at a glass–air interface. When light is incident above the critical angle, the intensity reflectance for both *s*- and *p*-polarizations are 1; the reflection is lossless. This highly desirable property is responsible for the widespread use of right angle prisms to reflect beams through 90°, the popularity of prisms in binoculars, and many other applications. Since the *s*- and *p*-intensity reflectances are equal above the critical angle, the diattenuation is zero, but large retardances are present, as shown in the following internal reflection example.

Figure 8.21 shows the intensity reflection coefficients R_s and R_p and intensity transmission coefficients T_s and T_p for a ray propagating from glass with index $n_1 = 1.5$ to air with index $n_2 = 1$. At normal incidence, 4% of the flux reflects. R_s increases monotonically to 1 at the critical angle = 41.81°. Above θ_C, the reflectance remains 1 until grazing incidence, $\theta_i = 90°$. R_p decreases to 0 at Brewster's angle, $\theta_B = 33.69°$. Beyond Brewster's angle, the intensity reflectance rises rapidly to 1.

The amplitude reflection coefficients, r_s and r_p, become complex when $\theta_i > \theta_C$. Figure 8.22 shows that the magnitude of r_s and r_p are both 1 above critical angle with changing phase. The Fresnel phase of r_s and r_p are shown in Figure 8.23. The π Fresnel phase in r_p at normal incidence is due to the right-handed Fresnel coordinate. Hence, both *s*- and *p*-components have zero phase shift upon internal reflection. The retardance is the phase difference between *s*- and *p*-components. Notice that the retardance has infinite slope at θ_C.

Figure 8.24 depicts the internal reflected electric field for three different polarizations: *s*-polarization states, the combination of *s*- and *p*-polarization states, and *p*-polarization states. The reflected *p*-polarized light starts to have non-zero phase shift at $\theta_i = \theta_B$, while the *s*-polarized light starts to

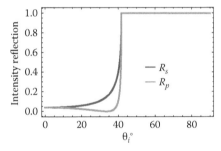 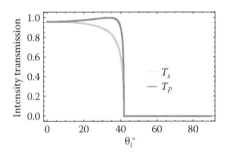

Figure 8.21 (Left) Intensity reflection coefficients R_s (red) and R_p (orange) and (right) intensity transmission coefficients T_s (green) and T_p (blue) as a function of incident angle.

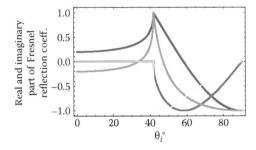 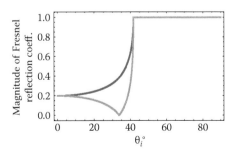

Figure 8.22 (Left) Fresnel amplitude reflection coefficient $\mathcal{R}e(r_s)$ (red), $\mathcal{R}e(r_p)$ (orange), $\mathcal{I}m(r_s)$ (purple), and $\mathcal{I}m(r_p)$ (cyan) as a function of incident angle. (Right) Absolute magnitude of Fresnel amplitude reflection coefficient of r_s (red) and r_p (orange) as a function of incident angle.

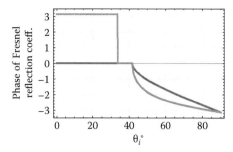 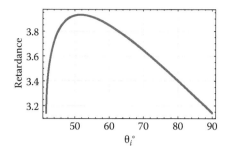

Figure 8.23 (Left) Fresnel phase shift of r_s (red) and r_p (orange) as a function of incident angle. (Right) Retardance upon internal reflection as a function of incident angle.

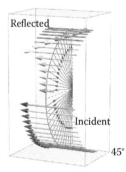 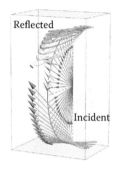 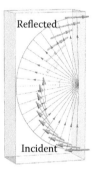

Figure 8.24 (Left) The incident and reflected states for internal reflection for s-polarized incident light from 7° to 88°. (Center) Internal reflection for 45° polarized incident light from 0.6° to 86°. The phase shift between s and p beyond Brewster's angle produces elliptically polarized reflected light. (Right) Internal reflection for p-polarized incident light. Reflected s- and p-polarized components are in phase from normal incidence to Brewster's angle, and then the phase changes (the arrowheads shift from the end of the line segments).

have phase shift at $\theta_i = \theta_C$. The different phase shift between these two components produces retardance. Therefore, the electric field with both components reflects to elliptically polarized light, as shown in Figure 8.24 (middle).

Figure 8.25 shows the internal reflection of circularly polarized light. At normal incidence, both s- and p-components have zero phase shift upon reflection with a reversed propagation direction; hence, left circularly polarized light reflects to right circularly polarized light. As θ_i approaches θ_B, the amplitude of the reflected p-component is lower than the reflected s-component. Thus, it reflects as right elliptically polarized light. When $\theta_i = \theta_B$, the reflected light becomes purely s-polarized. For $\theta_i > \theta_B$, both reflected s- and p-components have increasing amplitude toward grazing incidence, while the p-component has an extra π shift. Thus, the left circularly polarized light reflects as left elliptically polarized light. When $\theta_i \approx 90°$, left circularly polarized light reflects as left circularly polarized light.

Figure 8.25 Internal reflection for left circularly polarized light. The reflected light around normal incident is right elliptically polarized. It becomes linearly polarized at Brewster's angle ($\theta_B = 33.69°$) and switches to left elliptically polarized above θ_B. Its ellipticity increases to left circularly polarized light at critical angle ($\theta_C = 41.81°$).

8.4 Fresnel Refraction and Reflection

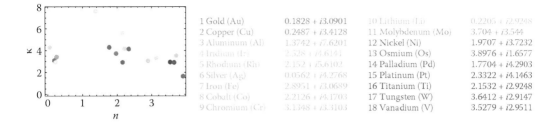

Figure 8.26 Complex refractive indices of common metals at 633 nm contain a real part n and an imaginary part κ.

8.4.4 Metal Reflection

Light reflecting from a smooth metal surface is affected by its complex refractive index, $n + i\kappa$, as opposed to the real refractive index of dielectric surface. Whereas transparent dielectrics like glasses and optical plastics have refractive indices greater than one with very small imaginary parts, the refractive indices of metals have a substantial imaginary component. Figure 8.26 lists the refractive indices of some common metals at 633 nm and locates these refractive indices on the (n, κ) complex plane.

The change of amplitude and phase of reflection from a smooth metal surface is calculated using the same Fresnel equations that apply to dielectrics, except that the complex value of the metal's refractive index is used. The resulting Fresnel coefficients are complex valued. When r_s and r_p are expressed in polar form, $r_s = \rho_s e^{-i\phi_s}$, $r_p = \rho_p e^{-i\phi_p}$, the magnitudes ρ_s and ρ_p describe a change in amplitude of the electric field components and ϕ_s and ϕ_p describe changes in phase. Since the transmitted part of the wave is rapidly absorbed near the surface of the metal, the transmission coefficients do not apply, but the reflection coefficients work fine.

Example 8.2 Fresnel Coefficients for Aluminum

The most common metal reflector is aluminum because of its high reflectivity in the visible and near infrared, low cost, and ease of machining. Aluminum doesn't corrode, is resistant to humidity, and doesn't tarnish due to trace atmospheric sulfur like silver does. The spectral variation of aluminum's refractive index is shown in Figure 8.27.

Consider reflection from a smooth aluminum surface at 633 nm where the complex refractive index is $1.374 + i7.620$. The amplitude reflection coefficients, shown in Figure 8.28 for s and p, are about 0.955 at normal incidence. The s-component has the greater amplitude reflectance

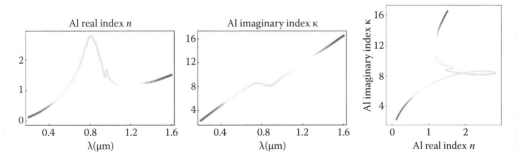

Figure 8.27 (Left) The real part of the refractive index of aluminum versus wavelength, (center) the imaginary part, and (right) a parametric plot of real versus imaginary.

and increases monotonically to one at grazing incidence, $\theta_i = 90°$. The p-coefficient decreases, quadratically at first, and then reaches a minimum around $\theta_i = 82°$ of about 0.83, before rapidly increasing to one at grazing incidence (Figure 8.29). This difference in amplitude reflectance causes the diattenuation for reflection (Figure 8.28, right). Figure 8.29 (left) shows the phase shift from the amplitude reflection coefficients for s-polarized (red) and p-polarized (orange) light. The difference in phase shift results in the retardance for reflection (right). Measurement techniques for the phase change at metallic reflection are discussed by Medicus et al.[8]

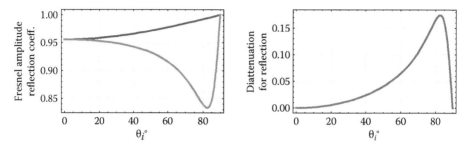

Figure 8.28 (Left) Amplitude reflection coefficients r_s (red), r_p (orange) and (right) diattenuation for reflection from an Al surface as a function of the angle of incidence.

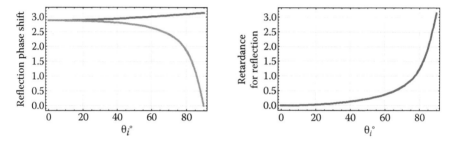

Figure 8.29 (Left) Phase shift, ϕ_s (red) and ϕ_p (orange), upon reflection from Al, and (right) retardance as a function of incident angle.

Light transmitted across the front surface of a metal propagates only a small distance into the metal being absorbed within the first few nanometers. The fraction of this absorbed light is calculated as $1 - R$, where R is the intensity reflection coefficient. Values for the absorption of aluminum as a function of angle of incidence for the s- and p-components are plotted in Figure 8.30.

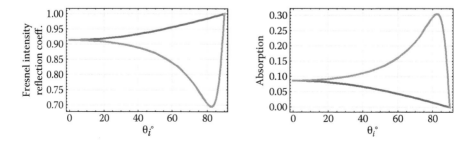

Figure 8.30 Reflection (left) and absorption (right) of s (red) and p (orange) polarizations during reflection from Al as a function of incident angle.

8.4 Fresnel Refraction and Reflection

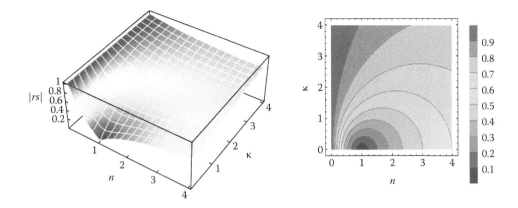

Figure 8.31 Reflectance of metals at normal incidence as a function of complex refractive index.

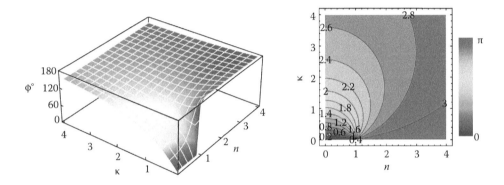

Figure 8.32 Phase change upon reflection for metals for normal incidence.

8.4.4.1 Normal Incidence Reflectance

The normal incidence reflectance of metals from air with $n = 1$ is shown in Figure 8.31 as a function of complex refractive index. The reflectance always increases with increasing κ. It also increases with increasing n when n is larger than 1, or decreasing n when n is less than 1.

The corresponding phase shift is shown in Figure 8.32. These phase shifts are mostly just below π, except when n is less than the air index 1 and κ is less than 2. When $\kappa = 0$, this phase change becomes the same as external reflection at dielectric surface, which is always π for $n > 1$.

8.4.4.2 Retardance and Diattenuation of Metal at Non-Normal Incidence

The retardance of metals at 30° incident angle is shown in Figure 8.33. The reflection retardance decreases with increasing κ. For $n > 1$ and $\kappa > 1$, retardance decreases with both n and κ.

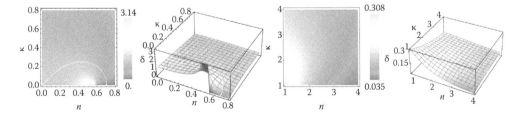

Figure 8.33 Retardance (in radians) for reflection from Al at a 30° incident angle. (Left) Shown in false color and 3D plots (left two) for $0 < n < 0.8$ and $0 < \kappa < 0.8$. The right two figures are for $1 < n < 4$ and $1 < \kappa < 4$.

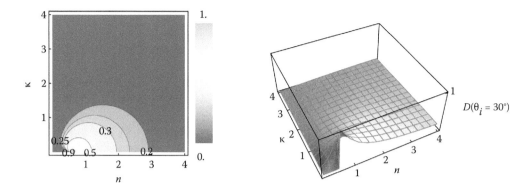

Figure 8.34 Diattenuation for reflection from metals at 30° incident angle with varying refractive index.

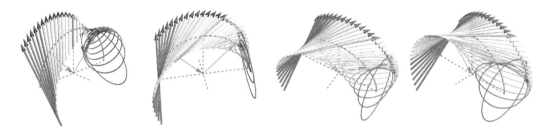

Figure 8.35 45° linearly polarized incident light reflects from an air–metal surface at various angles. Black arrowheads indicate incident light, and colored arrowheads indicate reflected light. The arrow pair at each incident angle has the same color.

The diattenuation reflecting off metals at 30° incident angle is shown in Figure 8.34. The diattenuation decreases to 0 as κ increases. When $\kappa = 0$, diattenuation decreases with increasing n.

Figure 8.35 shows the polarization change for light incident on a metal polarized at 45° to the plane of incidence, reflecting from air to Al at various angles of incidence. An unambiguous discussion of the handedness of the reflectance from a metal surface is discussed by Swindell.[9]

8.5 Approximate Representations of Fresnel Coefficients

The Fresnel coefficients are the first and simplest example of amplitude coefficients; later amplitude coefficients for multilayer films, anisotropic interfaces, and diffractive optical structures will be introduced. The functional forms of Equations 8.8 through 8.11 are complex and the coefficients are difficult to manipulate. Their behavior is not obvious from inspection characteristics of the Fresnel coefficients.

When applying the Fresnel equations, the goal is frequently to *understand* optical path length, amplitude, phase, diattenuation, and retardance and apply the equations to find suitable configurations of optical elements and coatings to achieve optical system specifications. This *understanding* is enhanced when the optical system properties can be expressed in terms of simply defined functions. It is frequently helpful to take complex equations, like the Fresnel equations, and replace them with approximate but simpler functions, usually polynomials, which, although approximate, maintain a high degree of accuracy. Then, by reasoning with these approximate functions, the source of aberration can be more clearly described, and methods for aberration compensation can be more easily constructed. Often, approximate functions can improve communication. Within optical design and

8.5 Approximate Representations of Fresnel Coefficients

analysis software, the exact equations will still be used, but for gaining an understanding of sources of polarization aberration, approximate functions can be enabling.

Two types of approximate functions will be used depending on the circumstances, (1) Taylor series and (2) function fits. A *Taylor series* represents a function in the neighborhood of a point as a series of polynomial terms calculated from the derivatives to the function. In contrast, a *fit* describes a function over a range as a sum of basis functions with weights adjusted to minimize the difference between the fit and the function.

8.5.1 Taylor Series for the Fresnel Coefficients

The *Taylor series* of a function $f(x)$ in the neighborhood of a point x_0 is defined as

$$f(x) = \sum_{n=0}^{\infty} \frac{\partial^n f(x_0)}{\partial x^n} \frac{(x-x_i)^n}{n!}$$

$$= f(x_0) + f'(x_0)(x-x_0) + f''(x_0)\frac{(x-x_0)^2}{2!} + f'''(x_i)\frac{(x-x_i)^3}{3!} + \ldots$$

(8.32)

The first two terms describe the tangent line through the point. The third term adds a quadratic component, and so on. For example, Figure 8.36 shows the convergence of linear, quadratic, and cubic fits to an r_s Fresnel coefficient, each extending the range of accurate representation.

The Fresnel equations are even functions, symmetric about zero angle of incidence. Thus, the Fresnel coefficient Taylor series about normal incidence have only even terms. Relatively simple and useful closed form expressions for the quadratic coefficients are calculated as follows:

$$r_s(i) = \frac{n_1 - n_2}{n_1 + n_2} + \frac{i^2 n_1 (n_1 - n_2)}{n_2 (n_1 + n_2)}, \quad r_p(i) = \frac{n_2 - n_1}{n_1 + n_2} + \frac{i^2 n_1 \left(n_1^2 - n_1 n_2\right)}{n_2 (n_1 + n_2)},$$

(8.33)

$$t_s(i) = \frac{2n_1}{n_1 + n_2} + \frac{i^2 n_1 (n_1 - n_2)}{n_2 (n_1 + n_2)}, \quad t_p(i) = \frac{2n_1}{n_1 + n_2} + \frac{i^2 n_1^2 (n_1 - n_2)}{n_2^2 (n_1 + n_2)}.$$

(8.34)

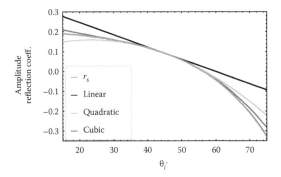

Figure 8.36 Linear, quadratic, and cubic Taylor series fits about an angle of incidence of 45° to the r_s Fresnel coefficient from air into $n = 1.5$, showing the increasing range of accurate representation with increasing order.

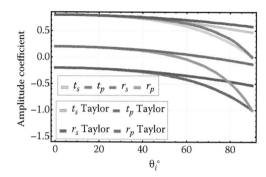

Figure 8.37 Comparison of Fresnel amplitude coefficients for reflection and refraction with quadratic Taylor series fits. The fits are excellent to beyond $\theta = 30°$.

Figure 8.37 compares the second-order Taylor series fits about normal incidence to the Fresnel amplitude coefficients. The deviations are of fourth order and the fits are quite accurate to 30°.

8.6 Conclusion

The Fresnel equations describe the polarization state changes on reflection and refraction. In optical systems, each beam of light is typically an aberrated spherical wave, and when it is incident on a surface, the angle of incidence and plane of incidence varies across the beam. Thus, because of the Fresnel equations, the magnitude and orientation of the diattenuation and retardance also vary across the beam. These variations are referred to as *polarization aberrations*, just as wavefront aberrations are variations of the optical path length and phase across beams and images. The polarization aberrations are divided into diattenuation aberrations and retardance aberrations. These polarization aberrations are generally not large and frequently have a minor or negligible effect on the optical system performance. But it is important to understand their genesis and effect.

In Chapter 12, several optical systems and their polarization aberrations will be analyzed. In each case, the optical system is described and a beam of light is chosen for analysis. The angles of incidence are calculated and used to produce maps of the diattenuation aberration and retardance aberration. The point spread functions are shown as Jones matrices for coherent light or Mueller matrices for incoherent light and the distribution of polarization is shown for sample input polarizations.

8.7 Problem Sets

8.1 Find the relationship between refractive indices for which the intensity reflection coefficients at normal incidence equal the intensity transmission coefficients.

8.2 Show that if a beam is incident on a plane parallel plate from air at Brewster's angle, it also strikes the rear surface at Brewster's angle.

8.3 Show that for reflection in air from a dielectric medium, both $R_s(\theta)$ and $R_p(\theta)$ approach the same value as $\theta \to 90°$. Find the slopes of $R_s(\theta)$ and $R_p(\theta)$ as $\theta \to 90°$.

8.4 Verify the following relation between the Fresnel s and p amplitude coefficients for the angle of incidence θ: $r_p = \dfrac{r_s(r_s - \cos 2\theta)}{1 - r_s \cos 2\theta}$.

8.5 Find the real refractive index n where at normal incidence from air $t_p = r_p$. What is the ratio of reflection and transmitted fluxes, R_p/T_p?

8.6 Find the real refractive index n where at normal incidence from air $t_p = \kappa\, r_p$, where κ is a real constant. What is the ratio of reflection and transmitted fluxes, R_p/T_p, when $\kappa = 3$?

8.7 Light refracts from air into a glass interface with refractive index $n = \sqrt{3}$.
   a. Find Brewster's angle.
   b. At Brewster's angle, what fraction of the p-polarized flux refracts?
   c. At Brewster's angle, is the reflected beam or the refracted beam 100% polarized?
   d. What is the angle between the reflected and refracted beam at Brewster's angle?
   e. For light exiting this interface, glass into air, what is the value of the critical angle?

8.8 Design a variable diattenuator using a tilted parallel plate piece of glass with $n = 2$. A collimated beam is incident on a tilted glass plate and refracts into and then out of the plate. By tilting the glass plate, a variable diattenuation D is introduced. If unpolarized light is incident, the degree of polarization of the exiting light is equal to the plate's diattenuation. Provide a lookup table for the inverse function $\theta(D)$, the angle of incidence as a function of the desired diattenuation. Ignore multiple reflections. Comment on possible implementation issues when this device is constructed.

8.9 Consider uncoated dielectric interface without absorption, that is, with real refractive index. Define weakly diattenuating as a diattenuation less than 0.05 and strongly diattenuating as a diattenuation greater than 0.5.
   a. Over what range of refractive index ratios $n = n_2/n_1$ with $1/3 < n < 3$, and what range of angles of incidence $0° < \theta < 90°$ are dielectric interfaces weakly diattenuating in transmission?
   b. Strongly diattenuating in reflection?
   c. Weakly diattenuating in reflection?
   d. Strongly reflecting in reflection?

8.10 a. Does an uncoated dielectric interface described by Fresnel equations have diattenuation?
   b. Does an uncoated dielectric interface described by Fresnel equations have retardance?
   c. What is the form of the diattenuation and retardance: linear, elliptical, or circular?

8.11 The complex refractive index of silver at three wavelengths is as follows:

400 nm	0.138 + 2.005i
710 nm	0.168 + 4.286i
1460 nm	0.374 + 9.449i

   a. At 710 nm, plot the s and p intensity reflection coefficients for angle of incidence between 0° and 90°. At about what angle of incidence is the diattenuation greatest? Does the diattenuation increase toward longer wavelength or shorter wavelength?
   b. At 710 nm, plot the s and p absolute phase changes on reflection for angle of incidence between 0° and 90°. Plot the retardance of angle of incidence between 0° and 90°. For small angles of incidence, the retardance is approximately $\delta(\theta) = \delta_0 + \delta_2(\theta)$. Determine δ_0 and δ_2.

c. Does the retardance at small angles increase toward longer wavelength or shorter wavelength?
d. How would one change the refractive index to maximize reflectance and minimize retardance (close to π). What types of materials have complex indices like this?

8.12 For a set of n points, an order $n - 1$ polynomial can be found that passes exactly through all the points. For example, consider five points $(x_0, f(x_0))$, $(x_1, f(x_1))$, $(x_2, f(x_2))$, $(x_3, f(x_3))$, and $(x_4, f(x_4))$, to be fit to the fourth-order polynomial $f(x) = c_0 + x\, c_1 + x^2\, c_2 + x^3\, c_3 + x^4\, c_4$. The points and coefficients can be related by the matrix equation:

$$\begin{pmatrix} m_{00} & m_{01} & m_{02} & m_{03} & m_{04} \\ m_{10} & m_{11} & m_{12} & m_{13} & m_{14} \\ m_{20} & m_{21} & m_{22} & m_{23} & m_{24} \\ m_{30} & m_{31} & m_{32} & m_{33} & m_{34} \\ m_{40} & m_{41} & m_{42} & m_{43} & m_{44} \end{pmatrix} \begin{pmatrix} c_0 \\ c_1 \\ c_2 \\ c_3 \\ c_4 \end{pmatrix} = \begin{pmatrix} f(x_0) \\ f(x_1) \\ f(x_2) \\ f(x_3) \\ f(x_4) \end{pmatrix}.$$

a. Find the matrix coefficients m_{ij} in this equation to calculate the c's from the set of $f(x)$.
b. Show the matrix equation that calculates the polynomial coefficients $c_0, \ldots c_4$.
c. Provide the equation for fitting three points to a quadratic equation yielding c_0, c_1, and c_2. Provide all nine matrix elements as a function of $(x_1, f(x_1))$, $(x_2, f(x_2))$, and $(x_3, f(x_3))$.
d. Provide the equation for fitting three points to an even order fourth-order equation yielding c_0, c_2, and c_4. Provide all nine matrix elements as functions of $(x_0, f(x_0))$, $(x_2, f(x_2))$, and $(x_4, f(x_4))$.
e. Fit the amplitude transmission coefficients t_s and t_p of an air/silicon ($n = 4$) interface for the angles 0°, 30°, and 45°.
f. For which component, t_s or t_p, is the fourth-order term more significant?

8.13 Perform the least squares polynomial fit with $f(\theta) = a_0 + a_2\theta^2 + a_4\theta^4$ to the amplitude coefficients for $n_1 = 1.0$ and $n_2 = 1.5$, as shown in Figure 8.7.

8.14 Find the Jones matrix for a real half wave retarder, fast axis at 45°, of sapphire at $\lambda = 589$ nm where $n_O = 1.76817$ and $n_E = 1.76009$. Chapter 5 presented the Jones matrices for ideal retarders; now consider a real retarder. Find the thickness t, then the absolute phases for the ordinary and extraordinary rays. Because of these optical path lengths, the retarder's Jones matrix will not be in *symmetric, fast axis unchanged*, or *slow axis unchanged* form (Table 5.4). Evaluate the Fresnel equations at normal incidence for the ordinary and extraordinary modes and include the small resulting diattenuation in the Jones matrix.

References

1. R. M. A. Azzam and N. M. Bashara, *Ellipsometry and Polarized Light*, North-Holland, Elsevier Science (1987).
2. M. Born and E. Wolf, *Principles of Optics*, 7th (expanded) edition (1999), pp. 790–852.
3. D. H. Goldstein, *Polarized Light*, 3rd edition, Boca Raton, FL: CRC Press (2010).
4. J. A. Stratton, *Electromagnetic Theory*, New York: John Wiley & Sons (2007).

5. W. B. Wolfe, Nondispersive prisms, *The Handbook of Optics 2, Handbook of Optics*, eds. M. Bass, E. Van Stryland, D. Williams, and W. Wolfe, New York: McGraw-Hill, 1996, 4–1 (1994).
6. E. Hecht, *Optics*, Addison-Wesley (2002), pp. 113–122.
7. M. Born and E. Wolf, *Principles of Optics*, Cambridge University Press (2003).
8. K. M. Medicus, A. Fricke, J. E. Brodziak Jr., and A. D. Davies, The effect of phase change on reflection on optical measurements, in *Proc. SPIE 5879, Recent Developments in Traceable Dimensional Measurements III*, 587906 (2005).
9. W. Swindell, Handedness of polarization after metallic reflection of linearly polarized light, *J. Opt. Soc. Am.* 61 (1971): 212–215.

9

Polarization Ray Tracing Calculus

The objective of *polarization ray tracing* is to calculate the evolution of the polarization state of rays through an optical system and to determine the polarization properties, such as diattenuation and retardance, associated with these ray paths. By tracing many rays, the *polarization aberrations* associated with an optical system can be assessed, and the behavior of a particular optical system, associated coating designs, polarization elements, and other components can be compared with the optical system's polarization specifications. Moreover, the knowledge of polarization aberrations due to coatings, individual surfaces, polarization elements, and other components can also aid in informing the design and defining polarization specifications. Polarization ray tracing enables detailed insights into how individual surfaces or components affect system polarization performance.

Among different polarization ray tracing matrix methods, *Jones matrices* have been used in optical design for at least 20 years.[1,2] Polarization ray tracing, which calculates the Jones matrix associated with an arbitrary ray path through an optical system, was introduced to calculate the *polarization aberration function*.[3] The Jones matrix deals with *Jones vectors*, which specifically refer to a monochromatic plane wave, describing the electric field and the polarization ellipse with respect to an x–y coordinate system in the *transverse plane*. If the plane wave is not propagating along the z-axis, then the x–y coordinates are referred to as "*local coordinates*" associated with a particular transverse plane. However, to use Jones vectors and matrices in optical design for the ray tracing of highly curved beams, local coordinate systems are required for each ray, and each of its ray segments, to define the direction of the Jones vector's x- and y-components in space, and these local coordinate systems lead to complications due to the intrinsic *singularities* of local coordinates. In the experience of the authors, working in Jones vector local coordinates leads to a cascade of minor complications, both in handling rays near the singularities and in describing *high numerical aperture beams*. Such issues are intrinsic to any choice of local coordinates. According to the *Winding Number Theorem*,[4] it is impossible to define a continuous and differentiable vector field constrained to lie on the surface of a sphere over the entire sphere without at least two zeros in the field; a set of latitude vectors or conversely a set of longitude vectors provide two examples, where the zeros occur at the poles. All local coordinate choices have such singularities.

In this chapter, a different *polarization ray tracing calculus* is presented to solve the problem of singularities in local coordinates. Polarization effects at each ray intercept are described by a three-by-three *polarization ray tracing matrix*, **P**, which is a generalized version of a two-by-two Jones matrix. With a three-by-three matrix, arbitrary propagation directions are easily accommodated and the problem of singularities in local coordinates is avoided (see Chapter 11, The Jones Pupil and Local Coordinate Systems). Polarization effects are propagated along ray paths through optical systems by matrix multiplication of the **P** matrices for each ray intercept. In image space, arrays of **P** matrices can then be used to determine the resultant polarization state on curved surfaces such as spherical and aberrated wavefronts as a three-dimensional (3D) electric field. 3D polarization ray tracing methods have been mentioned in several manuscripts, but not developed into full mathematical methods.[5–7] 3D polarization ray tracing algorithms in Refs. 8 and 9 are the basis of this chapter and the polarization ray tracing code, "Polaris-M,"[10,11] which was developed at the Polarization Laboratory at the University of Arizona and has now been commercialized by Airy Optics, Inc.

In Section 9.1, the 3D polarization ray tracing matrix is developed. Sections 9.2 and 9.3 further develop the **P** matrix with examples. Section 9.4 derives an algorithm to calculate the diattenuation of a given **P** matrix. An interferometer is used as an example in Section 9.5, and the issue of adding **P** matrices is addressed in Section 9.6. A hollow corner cube retroreflector is analyzed in Section 9.7.

9.1 Definition of Polarization Ray Tracing Matrix, P

The *polarization ray tracing matrix* **P** characterizes the change in a three-element electric field vector due to interaction with an optical surface, element, a sequence of optical elements, or an entire optical system. Consider the evolution of the polarization state of a ray through an optical system with N interfaces labeled by index q. Assume for the moment that all the materials are isotropic so that polarization changes will occur only at interfaces and not along ray segments. (This restriction will be removed later.) The ray will reflect or refract at interface $q = 1, 2, 3$, and so on. The ray exits interface $q - 1$ propagating in direction $\hat{\mathbf{k}}_{q-1}$ with electric field vector $\mathbf{E}_{q-1}$ and is incident upon interface q. At interface q, the polarization state will be modified, perhaps by a polarization element, a reflection, or a refraction, and the ray will exit interface q with propagation vector $\hat{\mathbf{k}}_q$ and electric field vector $\mathbf{E}_q$. The incident electric field vector $\mathbf{E}_{q-1}$ and exiting electric field vector $\mathbf{E}_q$ are linearly related by the polarization ray tracing matrix for the qth ray intercept $\mathbf{P}_q$, as shown in Figure 9.1,

$$\mathbf{E}_q = \begin{pmatrix} E_{x,q} \\ E_{y,q} \\ E_{z,q} \end{pmatrix} = \mathbf{P}_q \mathbf{E}_{q-1} = \begin{pmatrix} p_{xx,q} & p_{xy,q} & p_{xz,q} \\ p_{yx,q} & p_{yy,q} & p_{yz,q} \\ p_{zx,q} & p_{zy,q} & p_{zz,q} \end{pmatrix} \begin{pmatrix} E_{x,q-1} \\ E_{y,q-1} \\ E_{z,q-1} \end{pmatrix}. \tag{9.1}$$

Similarly, the incident *propagation vector* $\hat{\mathbf{k}}_{q-1}$ and exiting propagation vector $\hat{\mathbf{k}}_q$ are also linearly related by the polarization ray tracing matrix $\mathbf{P}_q$.*

---
* Propagation vectors are normalized throughout this book.

9.2 Formalism of Polarization Ray Tracing Matrix Using Orthogonal Transformation 325

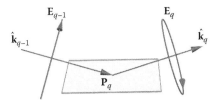

Figure 9.1 For the ray or plane wave incident at the qth optical interface, the matrix $\mathbf{P}_q$ relates the incident polarization state, $\mathbf{E}_{q-1}$, and propagation vector, $\mathbf{k}_{q-1}$, to the exiting polarization state, $\mathbf{E}_q$, and propagation vector, $\mathbf{k}_q$.

The net polarization effect of a series of isotropic optical elements is represented by cascading the $\mathbf{P}_q$ matrices for each ray intercept to yield polarization ray tracing matrix $\mathbf{P}_{Total}$,

$$\mathbf{P}_{Total} = \mathbf{P}_N \mathbf{P}_{N-1} \cdots \mathbf{P}_q \cdots \mathbf{P}_2 \mathbf{P}_1 = \prod_{q=N,-1}^{1} \mathbf{P}_q. \tag{9.2}$$

The polarization ray tracing matrix is a 3D generalization of the Jones matrix and Equation 9.2 has the same form as the Jones matrix product equation, Equation 5.10, from Chapter 5 (Jones Matrices and Polarization Properties).

In a conventional ray trace, to calculate a ray's contribution to the wavefront aberration, the optical path lengths along all ray segments between the optical system's entrance pupil and exit pupil are summed (Chapter 10, Optical Ray Tracing). Similarly, the polarization ray tracing matrix describes the polarization-dependent transmission and phase contributions to the optical path length. Polarization-dependent transmission and phase contributions may be due to coated and uncoated interfaces, diffraction gratings, holographic elements, and other polarization effects.

$\mathbf{P}_q$ as defined in Equation 9.1 is under-constrained; it doesn't uniquely define $\mathbf{P}_q$. In Equation 9.1, the transformation of all polarization states can be described as linear combinations of the transformations of any two linearly independent basis vectors, $\mathbf{E}_a$ and $\mathbf{E}_b$,

$$\mathbf{E}'_a = \mathbf{P}_q \mathbf{E}_a, \quad \mathbf{E}'_b = \mathbf{P}_q \mathbf{E}_b. \tag{9.3}$$

The relationship in Equation 9.3 yields six equations, one for each row, but $\mathbf{P}_q$ has nine elements. Thus, Equation 9.1 does not fully constrain $\mathbf{P}_q$. To uniquely define $\mathbf{P}_q$, Equation 9.3 is augmented with an additional condition relating the incident and the exiting propagation vectors,

$$\hat{\mathbf{k}}_q = \mathbf{P}_q \hat{\mathbf{k}}_{q-1}. \tag{9.4}$$

9.2 Formalism of Polarization Ray Tracing Matrix Using Orthogonal Transformation

Ray tracing calculations using the polarization ray tracing calculus involve frequent transformations between the *global coordinates* where the optical system and each $\mathbf{P}$ are defined and the *local coordinates* where the physics of polarization elements, anisotropic materials, thin film interfaces, diffraction gratings, reflection, refraction, and other phenomena are formulated. *Orthogonal transformations* between different coordinate systems, such as s–p coordinates, are straightforward and ubiquitous. This section explains the coordinate transformation notation.

Orthogonal matrices, also known as real *unitary matrices*, describe rotations of *orthogonal coordinate systems*. In our case, orthogonal matrices transform between a local coordinate basis selected for a calculation at an interface and the global coordinate basis and vice versa. In general, a separate pair of basis vectors is needed before and after the interface due to the change of ray direction.

For reflection and refraction from surfaces, the *s*- and *p*-polarization states along with the propagation vector form a natural basis, the $(\hat{\mathbf{s}}, \hat{\mathbf{p}}, \hat{\mathbf{k}})$ basis, as shown in Figure 9.2. For smooth interfaces of isotropic media, $\hat{\mathbf{s}}$ and $\hat{\mathbf{p}}$ are defined to be perpendicular and parallel to the plane of incidence, respectively, and thus are the eigenpolarizations of the Fresnel equations. The surface local coordinates incident on $(\hat{\mathbf{s}}_q, \hat{\mathbf{p}}_q, \hat{\mathbf{k}}_{q-1})$ and exiting from $(\hat{\mathbf{s}}'_q, \hat{\mathbf{p}}'_q, \hat{\mathbf{k}}_q)$ the qth intercept are

$$\hat{\mathbf{s}}_q = \frac{\hat{\mathbf{k}}_{q-1} \times \hat{\boldsymbol{\eta}}_q}{|\hat{\mathbf{k}}_{q-1} \times \hat{\boldsymbol{\eta}}_q|}, \quad \hat{\mathbf{p}}_q = \hat{\mathbf{k}}_{q-1} \times \hat{\mathbf{s}}_q, \quad \hat{\mathbf{s}}'_q = \hat{\mathbf{s}}_q, \quad \hat{\mathbf{p}}'_q = \hat{\mathbf{k}}_q \times \hat{\mathbf{s}}_q, \tag{9.5}$$

where $\hat{\boldsymbol{\eta}}$ is the surface normal of the qth surface. Our convention will direct $\hat{\boldsymbol{\eta}}$ away from the incident medium into the next medium. There is a special case for normally incident rays where $\hat{\boldsymbol{\eta}}_q$ is parallel to $\hat{\mathbf{k}}_{q-1}$ such that $(\hat{\boldsymbol{\eta}}_q \times \hat{\mathbf{k}}_{q-1} = 0)$ and $\hat{\mathbf{s}}_q$ of Equation 9.5 is undefined. In this case, any local coordinates that form an orthonormal right-handed coordinate system can be used for normal incidence. One simple choice is to reuse $\hat{\mathbf{s}}'_{q-1}$ from the previous ray intercept and calculate $\hat{\mathbf{p}}_q$,

$$\hat{\mathbf{s}}_q = \hat{\mathbf{s}}'_q = \hat{\mathbf{s}}'_{q-1}, \quad \hat{\mathbf{p}}_q = \hat{\mathbf{k}}_{q-1} \times \hat{\mathbf{s}}_q, \quad \hat{\mathbf{p}}'_q = \hat{\mathbf{k}}_q \times \hat{\mathbf{s}}_q. \tag{9.6}$$

Figure 9.2 shows the local coordinate bases for the incident ray along $\hat{\mathbf{k}}_{In}$ at a planar interface with a surface normal $\hat{\boldsymbol{\eta}}$. $(\hat{\mathbf{s}}'_T, \hat{\mathbf{p}}'_T, \hat{\mathbf{k}}_T)$ are the local coordinate bases for the refracted ray and $(\hat{\mathbf{s}}'_R, \hat{\mathbf{p}}'_R, \hat{\mathbf{k}}_R)$ are the local coordinate bases for the reflected ray.

The incident $(\hat{\mathbf{s}}_q)$ and exiting $(\hat{\mathbf{s}}'_q)$ vectors for the qth interface are always the same; thus, $\hat{\mathbf{s}}_q$ is used for both vectors; only the $\hat{\mathbf{p}}_q$ vector changes. The orthogonal matrices are

$$\mathbf{O}_{in,q}^{-1} = \begin{pmatrix} \hat{s}_{x,q} & \hat{s}_{y,q} & \hat{s}_{z,q} \\ \hat{p}_{x,q} & \hat{p}_{y,q} & \hat{p}_{z,q} \\ \hat{k}_{x,q-1} & \hat{k}_{y,q-1} & \hat{k}_{z,q-1} \end{pmatrix}, \quad \mathbf{O}_{out,q} = \begin{pmatrix} \hat{s}_{x,q} & \hat{p}'_{x,q} & \hat{k}_{x,q} \\ \hat{s}_{y,q} & \hat{p}'_{y,q} & \hat{k}_{y,q} \\ \hat{s}_{z,q} & \hat{p}'_{z,q} & \hat{k}_{z,q} \end{pmatrix}. \tag{9.7}$$

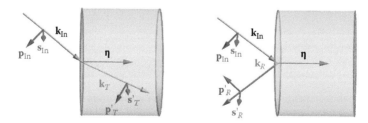

Figure 9.2 Local coordinate bases for the incident and refracted rays (left) and incident and reflected rays (right).

9.2 Formalism of Polarization Ray Tracing Matrix Using Orthogonal Transformation

$\mathbf{O}_{in,q}^{-1}$ operates on the incident electric field $\mathbf{E}_{q-1}$, defined in the global coordinate system, and calculates a projection of $\mathbf{E}_{q-1}$ onto the incident sp local coordinate system $(\hat{\mathbf{s}}_q, \hat{\mathbf{p}}_q, \hat{\mathbf{k}}_{q-1})$; that is,

$$\mathbf{E}_{sp,q-1} = \begin{pmatrix} E_{s,q-1} \\ E_{p,q-1} \\ 0 \end{pmatrix} = \mathbf{O}_{in,q}^{-1} \begin{pmatrix} E_{x,q-1} \\ E_{y,q-1} \\ E_{z,q-1} \end{pmatrix}. \tag{9.8}$$

$\mathbf{O}_{out,q}$ rotates the global coordinate system $(\hat{\mathbf{x}}, \hat{\mathbf{y}}, \hat{\mathbf{z}})$ to the exiting sp' local coordinate system $(\hat{\mathbf{s}}_q, \hat{\mathbf{p}}'_q, \hat{\mathbf{k}}_q)$ and operates on the electric field $\mathbf{E}_{sp',q} = \begin{pmatrix} E_{s,q} \\ E_{p',q} \\ 0 \end{pmatrix}$, defined in the exiting local coordinate system $(\hat{\mathbf{s}}_q, \hat{\mathbf{p}}'_q, \hat{\mathbf{k}}_q)$, and calculates the exiting electric $\mathbf{E}_q$ in the global coordinate system,

$$\mathbf{E}_q = \mathbf{O}_{out,q} \cdot \begin{pmatrix} E_{s,q} \\ E_{p',q} \\ 0 \end{pmatrix} = \begin{pmatrix} \hat{s}_{x,q} & \hat{p}'_{x,q} & \hat{k}_{x,q} \\ \hat{s}_{y,q} & \hat{p}'_{y,q} & \hat{k}_{y,q} \\ \hat{s}_{z,q} & \hat{p}'_{z,q} & \hat{k}_{z,q} \end{pmatrix} \cdot \begin{pmatrix} E_{s,q} \\ E_{p',q} \\ 0 \end{pmatrix} = E_{s,q} \begin{pmatrix} \hat{s}_{x,q} \\ \hat{s}_{y,q} \\ \hat{s}_{z,q} \end{pmatrix} + E_{p',q} \begin{pmatrix} \hat{p}'_{x,q} \\ \hat{p}'_{y,q} \\ \hat{p}'_{z,q} \end{pmatrix}$$

$$= E_{s,q} \hat{\mathbf{s}}_q + E_{p',q} \hat{\mathbf{p}}'_q. \tag{9.9}$$

Math Tip 9.1 Orthogonal Transformation

An *orthogonal transformation* is a rotation from one coordinate system to another. This is performed with a rotation matrix, a *unitary matrix* that maps one set of orthonormal coordinates $(\hat{\mathbf{a}}, \hat{\mathbf{b}}, \hat{\mathbf{c}})$ into another $(\hat{\mathbf{d}}, \hat{\mathbf{e}}, \hat{\mathbf{f}})$. A *rotation matrix* $\mathbf{R}_1$ that has $(\hat{\mathbf{a}}, \hat{\mathbf{b}}, \hat{\mathbf{c}})$ as its columns,

$$\mathbf{R}_1 = \begin{pmatrix} a_x & b_x & c_x \\ a_y & b_y & c_y \\ a_z & b_z & c_z \end{pmatrix}, \tag{9.10}$$

rotates global coordinates $(\hat{\mathbf{x}}, \hat{\mathbf{y}}, \hat{\mathbf{z}})$ to $(\hat{\mathbf{a}}, \hat{\mathbf{b}}, \hat{\mathbf{c}})$,

$$\begin{pmatrix} a_x & b_x & c_x \\ a_y & b_y & c_y \\ a_z & b_z & c_z \end{pmatrix} \begin{pmatrix} 1 \\ 0 \\ 0 \end{pmatrix} = \begin{pmatrix} a_x \\ a_y \\ a_z \end{pmatrix}, \begin{pmatrix} a_x & b_x & c_x \\ a_y & b_y & c_y \\ a_z & b_z & c_z \end{pmatrix} \begin{pmatrix} 0 \\ 1 \\ 0 \end{pmatrix} = \begin{pmatrix} b_x \\ b_y \\ b_z \end{pmatrix}, \begin{pmatrix} a_x & b_x & c_x \\ a_y & b_y & c_y \\ a_z & b_z & c_z \end{pmatrix} \begin{pmatrix} 0 \\ 0 \\ 1 \end{pmatrix} = \begin{pmatrix} c_x \\ c_y \\ c_z \end{pmatrix}. \tag{9.11}$$

Similarly, the inverse of $\mathbf{R}_1$ rotates $(\hat{\mathbf{a}}, \hat{\mathbf{b}}, \hat{\mathbf{c}})$ back to $(\hat{\mathbf{x}}, \hat{\mathbf{y}}, \hat{\mathbf{z}})$. When $(\hat{\mathbf{a}}, \hat{\mathbf{b}}, \hat{\mathbf{c}})$ are real-valued vectors, the inverse of $\mathbf{R}_1$ is also the transpose of $\mathbf{R}_1$.

$$\mathbf{R}_1^{-1} = \begin{pmatrix} a_x & a_y & a_z \\ b_x & b_y & b_z \\ c_x & c_y & c_z \end{pmatrix},$$

$$\begin{pmatrix} a_x & a_y & a_z \\ b_x & b_y & b_z \\ c_x & c_y & c_z \end{pmatrix} \begin{pmatrix} a_x \\ a_y \\ a_z \end{pmatrix} = \begin{pmatrix} 1 \\ 0 \\ 0 \end{pmatrix}, \begin{pmatrix} a_x & a_y & a_z \\ b_x & b_y & b_z \\ c_x & c_y & c_z \end{pmatrix} \begin{pmatrix} b_x \\ b_y \\ b_z \end{pmatrix} = \begin{pmatrix} 0 \\ 1 \\ 0 \end{pmatrix}, \begin{pmatrix} a_x & a_y & a_z \\ b_x & b_y & b_z \\ c_x & c_y & c_z \end{pmatrix} \begin{pmatrix} c_x \\ c_y \\ c_z \end{pmatrix} = \begin{pmatrix} 0 \\ 0 \\ 1 \end{pmatrix}. \quad (9.12)$$

Using Equation 9.10, an orthogonal transformation matrix that rotates $(\hat{\mathbf{x}}, \hat{\mathbf{y}}, \hat{\mathbf{z}})$ to $(\hat{\mathbf{d}}, \hat{\mathbf{e}}, \hat{\mathbf{f}})$ can be calculated,

$$\mathbf{R}_2 = \begin{pmatrix} d_x & e_x & f_x \\ d_y & e_y & f_y \\ d_z & e_z & f_z \end{pmatrix}. \quad (9.13)$$

Example 9.1 Find the Orthogonal Transformation Matrix That Rotates $(\hat{\mathbf{a}}, \hat{\mathbf{b}}, \hat{\mathbf{c}})$ to $(\hat{\mathbf{d}}, \hat{\mathbf{e}}, \hat{\mathbf{f}})$

Using equations in Math Tip 0.1, $\mathbf{R}_2 \mathbf{R}_1^{-1}$ rotates $(\hat{\mathbf{a}}, \hat{\mathbf{b}}, \hat{\mathbf{c}})$ to $(\hat{\mathbf{d}}, \hat{\mathbf{e}}, \hat{\mathbf{f}})$ since $\mathbf{R}_1^{-1}$ transforms $(\hat{\mathbf{a}}, \hat{\mathbf{b}}, \hat{\mathbf{c}})$ to $(\hat{\mathbf{x}}, \hat{\mathbf{y}}, \hat{\mathbf{z}})$ and $\mathbf{R}_2$ transforms $(\hat{\mathbf{x}}, \hat{\mathbf{y}}, \hat{\mathbf{z}})$ to $(\hat{\mathbf{d}}, \hat{\mathbf{e}}, \hat{\mathbf{f}})$. Therefore, $\mathbf{R}_2 \mathbf{R}_1^{-1} \hat{\mathbf{a}} = \hat{\mathbf{d}}, \mathbf{R}_2 \mathbf{R}_1^{-1} \hat{\mathbf{b}} = \hat{\mathbf{e}}, \mathbf{R}_2 \mathbf{R}_1^{-1} \hat{\mathbf{c}} = \hat{\mathbf{f}}$.

Example 9.2 Orthogonal Transformation Matrix for a Flat Glass Plate

A ray propagating along $\hat{\mathbf{k}}_{In} = (0, \sin(\pi/6), \cos(\pi/6))$ refracts into a flat glass plate with $\hat{\boldsymbol{\eta}} = (0, 0, 1)$ and refracts along $\hat{\mathbf{k}}_T = \left(0, \frac{1}{3}, \frac{2\sqrt{2}}{3}\right)$, as shown in Figure 9.2 (left). Calculate the orthogonal transformation matrices $\mathbf{O}_{in}^{-1}$ and $\mathbf{O}_{out}$.

9.2 Formalism of Polarization Ray Tracing Matrix Using Orthogonal Transformation

First, calculate local coordinate vectors,

$$\hat{\mathbf{s}} = \frac{\hat{\mathbf{k}}_{In} \times \hat{\boldsymbol{\eta}}}{|\hat{\mathbf{k}}_{In} \times \hat{\boldsymbol{\eta}}|} = (1,0,0),$$

$$\hat{\mathbf{p}} = \hat{\mathbf{k}}_{In} \times \hat{\mathbf{s}} = \left(0, \frac{\sqrt{3}}{2}, -\frac{1}{2}\right),$$

$$\hat{\mathbf{s}}' = \hat{\mathbf{s}} = (1,0,0),$$

$$\hat{\mathbf{p}}' = \hat{\mathbf{k}}_T \times \hat{\mathbf{s}}' = \left(0, \frac{2\sqrt{2}}{3}, -\frac{1}{3}\right).$$

Then, the orthogonal transformation matrices are

$$\mathbf{O}_{in}^{-1} = \begin{pmatrix} \hat{s}_x & \hat{s}_y & \hat{s}_z \\ \hat{p}_x & \hat{p}_y & \hat{p}_z \\ \hat{k}_{x,In} & \hat{k}_{y,In} & \hat{k}_{z,In} \end{pmatrix} = \begin{pmatrix} 1 & 0 & 0 \\ 0 & \frac{\sqrt{3}}{2} & \frac{-1}{2} \\ 0 & \frac{1}{2} & \frac{\sqrt{3}}{2} \end{pmatrix},$$

$$\mathbf{O}_{out} = \begin{pmatrix} \hat{s}_x & \hat{p}'_x & \hat{k}_{x,T} \\ \hat{s}_y & \hat{p}'_y & \hat{k}_{y,T} \\ \hat{s}_z & \hat{p}'_z & \hat{k}_{z,T} \end{pmatrix} = \begin{pmatrix} 1 & 0 & 0 \\ 0 & \frac{2\sqrt{2}}{3} & \frac{1}{3} \\ 0 & \frac{-1}{2} & \frac{2\sqrt{2}}{3} \end{pmatrix}.$$

The physics of *reflection* and *refraction* at dielectric, metal, and multilayer coated interfaces is described in terms of the incident $(\hat{\mathbf{s}}, \hat{\mathbf{p}})$ components. $\mathbf{P}_q$ for a refraction or reflection can be derived by using $\mathbf{J}_{t,q}$ and $\mathbf{J}_{r,q}$, which are defined in a local $\hat{\mathbf{s}}$ and $\hat{\mathbf{p}}$ basis, and Equation 9.7,

$$\mathbf{J}_{t,q} = \begin{pmatrix} \alpha_{s,t,q} & 0 & 0 \\ 0 & \alpha_{p,t,q} & 0 \\ 0 & 0 & 1 \end{pmatrix}, \mathbf{J}_{r,q} = \begin{pmatrix} \alpha_{s,r,q} & 0 & 0 \\ 0 & \alpha_{p,r,q} & 0 \\ 0 & 0 & 1 \end{pmatrix}. \tag{9.14}$$

Note that the Jones matrices for reflection and refraction are diagonal matrices when expressed in *sp* coordinates. The subscript *t* indicates refraction, *r* indicates reflection, *s* indicates *s*-polarization, and *p* indicates *p*-polarization. $\alpha_{s,t,q}$ and $\alpha_{p,t,q}$ are *s*- and *p*-amplitude transmission coefficients and $\alpha_{s,r,q}$ and $\alpha_{p,r,q}$ are *s*- and *p*-amplitude reflection coefficients. For an uncoated interface between two isotropic media, the coefficients are calculated from the Fresnel equations. For coated interfaces, the coefficients are calculated from multilayer coating calculations (Chapter 13, Thin Films).[12–14] The polarization ray tracing matrices for refraction and reflection, respectively, are

$$\mathbf{P}_q = \mathbf{O}_{out,q} \mathbf{J}_{t,q} \mathbf{O}_{in,q}^{-1} \text{ and } \mathbf{P}_q = \mathbf{O}_{out,q} \mathbf{J}_{r,q} \mathbf{O}_{in,q}^{-1}. \qquad (9.15)$$

Example 9.3 $\mathbf{P}_q$ for a Refraction

Using Equations 9.8, 9.9, and 9.15, the exiting electric field $\mathbf{E}_q$ in the global coordinate system after a refraction is related to the incident field $\mathbf{E}_{q-1}$ through a series of coordinate transformations:

$$\mathbf{E}_q = \mathbf{P}_q \mathbf{E}_{q-1}$$

$$= \mathbf{O}_{out,q} \begin{pmatrix} \alpha_{s,t,q} & 0 & 0 \\ 0 & \alpha_{p,t,q} & 0 \\ 0 & 0 & 1 \end{pmatrix} \mathbf{O}_{in,q}^{-1} \begin{pmatrix} E_{x,q} \\ E_{y,q} \\ E_{z,q} \end{pmatrix}$$

$$= \mathbf{O}_{out,q} \begin{pmatrix} \alpha_{s,t,q} & 0 & 0 \\ 0 & \alpha_{p,t,q} & 0 \\ 0 & 0 & 1 \end{pmatrix} \begin{pmatrix} E_{s,q-1} \\ E_{p,q-1} \\ 0 \end{pmatrix} = \mathbf{O}_{out,q} \cdot \begin{pmatrix} \alpha_{s,t,q} E_{s,q-1} \\ \alpha_{p,t,q} E_{p,q-1} \\ 0 \end{pmatrix} = \mathbf{O}_{out,q} \cdot \begin{pmatrix} E_{s,q} \\ E_{p',q} \\ 0 \end{pmatrix}$$

$$= \begin{pmatrix} \hat{s}_{x,q} & \hat{p}'_{x,q} & \hat{k}_{x,q} \\ \hat{s}_{y,q} & \hat{p}'_{y,q} & \hat{k}_{y,q} \\ \hat{s}_{z,q} & \hat{p}'_{z,q} & \hat{k}_{z,q} \end{pmatrix} \begin{pmatrix} E_{s,q} \\ E_{p',q} \\ 0 \end{pmatrix} = E_{s,q} \hat{\mathbf{s}}_q + E_{p',q} \hat{\mathbf{p}}'_q.$$

1. $\mathbf{O}_{in,q}^{-1}$ projects the incident electric field $\mathbf{E}_{q-1}$ defined in the global coordinate system onto the incident local coordinate bases $(\hat{\mathbf{s}}_q, \hat{\mathbf{p}}_q, \hat{\mathbf{k}}_{q-1})$, resulting in $\mathbf{E}_{sp,q-1}$. Note that $\mathbf{E}_{sp,q-1}$ is a Jones vector defined in $(\hat{\mathbf{s}}_q, \hat{\mathbf{p}}_q, \hat{\mathbf{k}}_{q-1})$ with an additional zero.
2. $\mathbf{J}_{t,q}$ (or $\mathbf{J}_{r,q}$) calculates the exiting Jones vector with an additional zero, $\mathbf{E}_{sp',q}$, defined in $(\hat{\mathbf{s}}_q, \hat{\mathbf{p}}'_q, \hat{\mathbf{k}}_q)$.
3. $\mathbf{O}_{out,q}$ converts $\mathbf{E}_{sp',q}$ to the exiting electric field $\mathbf{E}_q$ in the global coordinate system.

Polarization transformations during reflection and refraction have *s*- and *p*-polarizations as eigenpolarizations. Transformations due to diffraction gratings, holograms, sub-wavelength gratings, and other non-isotropic interfaces often couple some *s*- into *p-polarization* and vice versa. Thus, the Jones matrices for gratings, holograms, wire grid polarizers, sub-wavelength gratings, and other non-isotropic interfaces can have off-diagonal elements. Therefore, in general, the $\mathbf{P}_q$ matrix for a given ray intercept is

$$\mathbf{P}_q = \mathbf{O}_{out,q} \mathbf{J}_q \mathbf{O}_{in,q}^{-1}, \qquad (9.16)$$

where $\mathbf{J}_q = \begin{pmatrix} j_{11} & j_{12} & 0 \\ j_{21} & j_{22} & 0 \\ 0 & 0 & 1 \end{pmatrix}$ represents the Jones matrix for the interaction in its arbitrarily chosen local coordinate system.*

---

* For gratings and holographic elements, different diffraction orders would naturally have different $\mathbf{J}_q$.

9.3 Retarder Polarization Ray Tracing Matrix Examples

For interactions that do not change the ray direction, for example, sheet polarizers and retarders, the surface local coordinates are arbitrarily chosen to be perpendicular to the propagation vector and Equation 9.16 becomes

$$\mathbf{P}_q = \mathbf{O}_{in,q} \mathbf{J}_q \mathbf{O}_{in,q}^{-1}. \tag{9.17}$$

9.3 Retarder Polarization Ray Tracing Matrix Examples

The matrix $\mathbf{P}$ is a polarization ray tracing matrix for a **single** ray since orthogonal transformation matrices (Equation 9.7) are different for each ray, unless the beam of light is collimated and all surfaces are planar surfaces. The $\mathbf{P}$ matrix in Equation 9.16 is dependent not only on the corresponding Jones matrix (optical element) but also on the propagation vector.

Figure 9.3 shows quarter wave *linear retarders* acting upon beams with different propagation vectors (blue arrows). Each of the three rays is normally incident onto a retarder element. The Jones matrix for a horizontal fast axis quarter wave retarder with $\hat{\boldsymbol{\eta}} = (0,0,1)$ is $\begin{pmatrix} e^{-i\pi/4} & 0 \\ 0 & e^{i\pi/4} \end{pmatrix}$. The Jones matrix is specified in a symmetric phase convention where the fast axis polarization state is advanced by an eighth of a wave and the slow axis is delayed by an eighth of a wave. The corresponding $\mathbf{P}$ matrix for the same quarter wave retarder with $\hat{\boldsymbol{\eta}} = (0,1,0)$ or $\hat{\boldsymbol{\eta}} = (1,0,0)$ is different. Table 9.1 shows the $\mathbf{P}$ matrices for all three cases.

Note that $\mathbf{P}$ for a ray propagating along the z-axis is the same as the Jones matrix padded with zeros and a one in the lower right element, but in general, $\mathbf{P}$ matrices are different from Jones matrices. The $\mathbf{P}$'s in Table 9.1 relate the phase to the corresponding component of the electric field in global coordinates.

Example 9.4 Calculate the $\mathbf{P}$ Matrix for a Ray Reflected from a Flat Glass Plate Using Fresnel Reflection Coefficients

A ray propagating along $\hat{\mathbf{k}}_{In} = (0, \sin(\pi/6), \cos(\pi/6))$ reflects from the flat glass plate ($n = 1.5$) with $\hat{\boldsymbol{\eta}} = (0,0,1)$ and refracts along $\hat{\mathbf{k}}_R$, as shown in Figure 9.2 (right). Calculate the $\mathbf{P}$ matrix for this reflection.

First, calculate $\hat{\mathbf{k}}_R$ using the law of reflection,

$$\hat{\mathbf{k}}_R = (0, \sin(\pi/6), -\cos(\pi/6)).$$

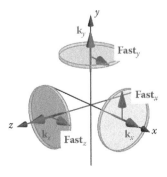

Figure 9.3 Quarter wave linear retarders for rays propagating along x-, y-, and z-axes are shown. Blue arrows indicate the propagation vectors and red arrows indicate the fast axis of each retarder.

Table 9.1 Polarization Ray Tracing Matrices for Identical Horizontal Fast Axis Linear Quarter Wave Retarders with Normally Incident Rays But Having Three Different Surface Normal Vectors

J	$\hat{\boldsymbol{\eta}}$	$\hat{\mathbf{k}}_{In}$	$\hat{\mathbf{k}}_{Out}$	P
$\begin{pmatrix} e^{-i\pi/4} & 0 \\ 0 & e^{i\pi/4} \end{pmatrix}$	$\begin{pmatrix} 0 \\ 0 \\ 1 \end{pmatrix}$	$\begin{pmatrix} 0 \\ 0 \\ 1 \end{pmatrix}$	$\begin{pmatrix} 0 \\ 0 \\ 1 \end{pmatrix}$	$\begin{pmatrix} e^{-i\pi/4} & 0 & 0 \\ 0 & e^{i\pi/4} & 0 \\ 0 & 0 & 1 \end{pmatrix}$
$\begin{pmatrix} e^{-i\pi/4} & 0 \\ 0 & e^{i\pi/4} \end{pmatrix}$	$\begin{pmatrix} 0 \\ 1 \\ 0 \end{pmatrix}$	$\begin{pmatrix} 0 \\ 1 \\ 0 \end{pmatrix}$	$\begin{pmatrix} 0 \\ 1 \\ 0 \end{pmatrix}$	$\begin{pmatrix} e^{-i\pi/4} & 0 & 0 \\ 0 & 1 & 0 \\ 0 & 0 & e^{i\pi/4} \end{pmatrix}$
$\begin{pmatrix} e^{-i\pi/4} & 0 \\ 0 & e^{i\pi/4} \end{pmatrix}$	$\begin{pmatrix} 1 \\ 0 \\ 0 \end{pmatrix}$	$\begin{pmatrix} 1 \\ 0 \\ 0 \end{pmatrix}$	$\begin{pmatrix} 1 \\ 0 \\ 0 \end{pmatrix}$	$\begin{pmatrix} 1 & 0 & 0 \\ 0 & e^{-i\pi/4} & 0 \\ 0 & 0 & e^{i\pi/4} \end{pmatrix}$

Then, calculate local coordinate vectors,

$$\hat{\mathbf{s}} = \frac{\hat{\mathbf{k}}_{In} \times \hat{\boldsymbol{\eta}}}{|\hat{\mathbf{k}}_{In} \times \hat{\boldsymbol{\eta}}|} = (1,0,0),$$

$$\hat{\mathbf{p}} = \hat{\mathbf{k}}_{In} \times \hat{\mathbf{s}} = \left(0, \frac{\sqrt{3}}{2}, -\frac{1}{2}\right),$$

$$\hat{\mathbf{s}}' = \hat{\mathbf{s}} = (1,0,0),$$

$$\hat{\mathbf{p}}' = \hat{\mathbf{k}}_R \times \hat{\mathbf{s}}' = \left(0, -\frac{\sqrt{3}}{2}, -\frac{1}{2}\right).$$

Thus, the orthogonal transformation matrices are

$$\mathbf{O}_{in}^{-1} = \begin{pmatrix} \hat{s}_x & \hat{s}_y & \hat{s}_z \\ \hat{p}_x & \hat{p}_y & \hat{p}_z \\ \hat{k}_{x,In} & \hat{k}_{y,In} & \hat{k}_{z,In} \end{pmatrix} = \begin{pmatrix} 1 & 0 & 0 \\ 0 & \frac{\sqrt{3}}{2} & \frac{-1}{2} \\ 0 & \frac{1}{2} & \frac{\sqrt{3}}{2} \end{pmatrix},$$

$$\mathbf{O}_{out} = \begin{pmatrix} \hat{s}_x & \hat{p}'_x & \hat{k}_{x,R} \\ \hat{s}_y & \hat{p}'_y & \hat{k}_{y,R} \\ \hat{s}_z & \hat{p}'_z & \hat{k}_{z,R} \end{pmatrix} = \begin{pmatrix} 1 & 0 & 0 \\ 0 & \frac{-\sqrt{3}}{2} & \frac{1}{2} \\ 0 & \frac{-1}{2} & \frac{-\sqrt{3}}{2} \end{pmatrix}.$$

9.3 Retarder Polarization Ray Tracing Matrix Examples

The angle of incidence is $\pi/6$; hence, the Fresnel reflection coefficients are $r_s = -0.2404$ and $r_p = 0.1589$. Thus, the Jones matrix for this reflection is

$$\mathbf{J} = \begin{pmatrix} r_s & 0 \\ 0 & r_p \end{pmatrix} = \begin{pmatrix} -0.2404 & 0 \\ 0 & 0.1589 \end{pmatrix}.$$

Using Equation 9.15, the **P** matrix is

$$\mathbf{P} = \mathbf{O}_{out}\mathbf{J}_r\mathbf{O}_{in}^{-1} = \begin{pmatrix} \hat{s}_x & \hat{p}'_x & \hat{k}_{x,R} \\ \hat{s}_y & \hat{p}'_y & \hat{k}_{y,R} \\ \hat{s}_z & \hat{p}'_z & \hat{k}_{z,R} \end{pmatrix} \begin{pmatrix} r_s & 0 & 0 \\ 0 & r_p & 0 \\ 0 & 0 & 1 \end{pmatrix} \begin{pmatrix} \hat{s}_x & \hat{s}_y & \hat{s}_z \\ \hat{p}_x & \hat{p}_y & \hat{p}_z \\ \hat{k}_{x,In} & \hat{k}_{y,In} & \hat{k}_{z,In} \end{pmatrix}$$

$$= \begin{pmatrix} 1 & 0 & 0 \\ 0 & \frac{-\sqrt{3}}{2} & \frac{1}{2} \\ 0 & \frac{-1}{2} & \frac{-\sqrt{3}}{2} \end{pmatrix} \begin{pmatrix} -0.240408 & 0 & 0 \\ 0 & 0.1589 & 0 \\ 0 & 0 & 1 \end{pmatrix} \begin{pmatrix} 1 & 0 & 0 \\ 0 & \frac{\sqrt{3}}{2} & \frac{-1}{2} \\ 0 & \frac{1}{2} & \frac{\sqrt{3}}{2} \end{pmatrix}$$

$$= \begin{pmatrix} -0.240408 & 0 & 0 \\ 0 & 0.130825 & 0.501818 \\ 0 & -0.501818 & -0.710275 \end{pmatrix}.$$

Example 9.5 The **P** Matrix for a Half Wave Retarder

Consider a ray with $\hat{\mathbf{k}} = (0, \sin 10°, \cos 10°)$ that passes through a half waveplate that has its fast axis oriented α from the local x-axis, surface normal $\hat{\boldsymbol{\eta}} = (0, \sin 10°, \cos 10°)$, and Jones matrix $\begin{pmatrix} \cos 2\alpha & \sin 2\alpha \\ \sin 2\alpha & -\cos 2\alpha \end{pmatrix}$, as shown in Figure 9.4. The finite thickness of the half waveplate is ignored in this example.

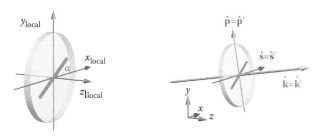

Figure 9.4 (Left) A half waveplate has its fast axis (pink line) oriented at α from the x_{local} axis. (Right) The plate is tilted 10° from the z-axis. A normal incident ray (brown arrow) passes through the tilted waveplate in the global xyz coordinate systems.

For normal incidence, *s*- and *p*-polarizations are degenerate. $\hat{\mathbf{s}}$ is chosen to be (1, 0, 0). Then, $\hat{\mathbf{p}} = (0, \cos 10°, -\sin 10°)$, calculated by Equation 9.6, since the ray is undeviated. For this waveplate example, $\hat{\mathbf{s}} = \hat{\mathbf{s}}'$, $\hat{\mathbf{p}} = \hat{\mathbf{p}}'$, and $\hat{\mathbf{k}} = \hat{\mathbf{k}}'$. By Equation 9.15, the **P** matrix for a ray normally incident on the tilted waveplate is

$$\mathbf{P} = \begin{pmatrix} 1 & 0 & 0 \\ 0 & \cos 10° & \sin 10° \\ 0 & -\sin 10° & \cos 10° \end{pmatrix} \begin{pmatrix} \cos 2\alpha & \sin 2\alpha & 0 \\ \sin 2\alpha & -\cos 2\alpha & 0 \\ 0 & 0 & 1 \end{pmatrix} \begin{pmatrix} 1 & 0 & 0 \\ 0 & \cos 10° & \sin 10° \\ 0 & -\sin 10° & \cos 10° \end{pmatrix}^{-1}$$

$$= \begin{pmatrix} \cos 2\alpha & \cos 10° \sin 2\alpha & -\sin 10° \sin 2\alpha \\ \cos 10° \sin 2\alpha & -\cos^2 10° \cos 2\alpha + \sin^2 10° & \cos^2 \alpha \sin 20° \\ -\sin 10° \sin 2\alpha & \cos^2 \alpha \sin 20° & \cos^2 10° - \cos 2\alpha \sin^2 10° \end{pmatrix}.$$

The incident *p*-polarization $\begin{pmatrix} 0 \\ \cos 10° \\ -\sin 10° \end{pmatrix}$, originally oriented at $\pi/2 - \alpha$ from the fast axis, is rotated to $\mathbf{P} \cdot \begin{pmatrix} 0 \\ \cos 10° \\ -\sin 10° \end{pmatrix} = \begin{pmatrix} \sin 2\alpha \\ -\cos 10° \cos 2\alpha \\ \sin 10° \cos 2\alpha \end{pmatrix}$, $\cos^{-1}(\sin 2\alpha)$ from the fast axis. The incident *s*-polarization, originally oriented at α from the fast axis, is rotated to

$\mathbf{P} \cdot \begin{pmatrix} 1 \\ 0 \\ 0 \end{pmatrix} = \begin{pmatrix} \cos 2\alpha \\ \cos 10° \sin 2\alpha \\ -\sin 10° \sin 2\alpha \end{pmatrix}$, 2α from the fast axis as shown in Figure 9.5.

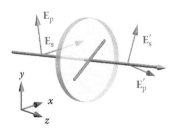

Figure 9.5 The incident electric field $\mathbf{E}_s$ (pink) and $\mathbf{E}_p$ (blue) passes through a half waveplate shown in Figure 9.4 and rotates to $\mathbf{E}'_s$ and $\mathbf{E}'_p$.

9.4 Diattenuation Calculation Using Singular Value Decomposition

One objective of a polarization ray trace is to understand how an optical system changes the polarization properties of incident light. Methods to analyze the polarization properties of Jones matrices are discussed in Chapters 5 and 14 and many texts.[15–17] This section derives a related algorithm to define diattenuation of an arbitrary polarization ray tracing matrix, **P**. The algorithm presented

9.4 Diattenuation Calculation Using Singular Value Decomposition

assumes that the ray begins and ends in a medium with a refractive index of one, such as air or vacuum, but these results are readily generalized to other object and image space refractive indices.

Diattenuation, $\mathcal{D}$, characterizes the difference of the maximum, T_{max}, and minimum, T_{min}, intensity transmittances possible when considering all incident polarization states,

$$\mathcal{D} = \frac{T_{max} - T_{min}}{T_{max} + T_{min}} \text{ and } 0 \leq \mathcal{D} \leq 1. \quad (9.18)$$

An ideal polarizer has diattenuation equal to one; one incident polarization state is fully transmitted, but another is fully attenuated. An element with a diattenuation of zero attenuates all polarization states equally.

The diattenuation calculation for general **P** matrices is complicated by the fact that the eigenvectors of **P** generally do not represent physical polarization states because, in most cases, rays enter and exit optical elements in different directions.[18] Therefore, the *singular value decomposition* (SVD), not the eigenvectors, is appropriate to calculate the diattenuation of **P**.

Math Tip 9.2 Singular Value Decomposition (SVD)

The SVD[19,20] of a 3 × 3 square matrix **B** is the matrix decomposition

$$\mathbf{B} = \mathbf{U}\mathbf{D}\mathbf{V}^\dagger = \begin{pmatrix} u_{x,0} & u_{x,1} & u_{x,2} \\ u_{y,0} & u_{y,1} & u_{y,2} \\ u_{z,0} & u_{z,1} & u_{z,2} \end{pmatrix} \begin{pmatrix} \Lambda_0 & 0 & 0 \\ 0 & \Lambda_1 & 0 \\ 0 & 0 & \Lambda_2 \end{pmatrix} \begin{pmatrix} v^*_{x,0} & v^*_{y,0} & v^*_{z,0} \\ v^*_{x,1} & v^*_{y,1} & v^*_{z,1} \\ v^*_{x,2} & v^*_{y,2} & v^*_{z,2} \end{pmatrix}, \quad (9.19)$$

where **U** and **V** are *unitary matrices* and **D** is a *diagonal matrix* with non-negative real elements. † indicates *Hermitian adjoint*,

$$\mathbf{V}^\dagger = (\mathbf{V}^*)^T. \quad (9.20)$$

The columns of **U** are orthonormal eigenvectors of $\mathbf{BB}^\dagger$ and the columns of **V** are orthonormal eigenvectors of $\mathbf{B}^\dagger\mathbf{B}$.

The matrices $\mathbf{BB}^\dagger$ and $\mathbf{B}^\dagger\mathbf{B}$ are Hermitian and share the same eigenvalues. The diagonal elements of **D** are named the *singular values* Λ_i[21] and are the non-negative real square roots of the eigenvalues from $\mathbf{BB}^\dagger$ or $\mathbf{B}^\dagger\mathbf{B}$. Convention orders the singular values in descending order,

$$\Lambda_0 \geq \Lambda_1 \geq \Lambda_2 \geq 0. \quad (9.21)$$

The order of singular values could be rearranged if the columns of **U** and **V** are also rearranged. Our preferred convention is placing the incident and exiting propagation vectors in the first columns of **U** and **V**. This can lead to a reordering of the singular values.

Since **P** was constructed such that $\mathbf{P}\hat{\mathbf{k}}_{q-1} = \hat{\mathbf{k}}_q$, (1) one of **P**'s singular values is always one, (2) the associated column of **V** is $\hat{\mathbf{k}}_0$, and (3) the associated column of **U** is $\hat{\mathbf{k}}_Q$,

$$\mathbf{P} = \mathbf{U}\mathbf{D}\mathbf{V}^\dagger = \begin{pmatrix} k_{x,Q} & u_{x,1} & u_{x,2} \\ k_{y,Q} & u_{y,1} & u_{y,2} \\ k_{z,Q} & u_{z,1} & u_{z,2} \end{pmatrix} \begin{pmatrix} 1 & 0 & 0 \\ 0 & \Lambda_1 & 0 \\ 0 & 0 & \Lambda_2 \end{pmatrix} \begin{pmatrix} k_{x,0}^* & k_{y,0}^* & k_{z,0}^* \\ v_{x,1}^* & v_{y,1}^* & v_{z,1}^* \\ v_{x,2}^* & v_{y,2}^* & v_{z,2}^* \end{pmatrix}. \quad (9.22)$$

The other two columns of **V**, $\mathbf{v}_1$ and $\mathbf{v}_2$, are the two special polarization states in the incident transverse plane that generate the maximum and minimum transmitted flux. $\mathbf{v}_1$ and $\mathbf{v}_2$ are always orthogonal. Similarly, the last two columns of **U** are the two corresponding *orthogonal polarization states* $\mathbf{u}_1$ and $\mathbf{u}_2$ in the exiting transverse plane.

Relationships between **P**, its singular values, and these special polarization states, are

$$\mathbf{P}\mathbf{v}_1 = \Lambda_1 \mathbf{u}_1, \mathbf{P}\mathbf{v}_2 = \Lambda_2 \mathbf{u}_2, \text{ and } \mathbf{P}\hat{\mathbf{k}}_0 = \hat{\mathbf{k}}_Q. \quad (9.23)$$

When $\Lambda_1 \neq \Lambda_2$, $\mathbf{v}_1$ and $\mathbf{v}_2$ are the only two orthogonal incident polarization states that remain orthogonal when they emerge from **P** as $\mathbf{u}_1$ and $\mathbf{u}_2$, respectively. Therefore, these two orthogonal sets of states, $(\mathbf{v}_1, \mathbf{v}_2)$ and $(\mathbf{u}_1, \mathbf{u}_2)$, form a *canonical basis* for incident and exiting polarization states. The SVD leads directly to these special polarization states, which are similar to eigenpolarizations.

An arbitrary normalized incident polarization state **E** can be expressed as a linear combination of $\mathbf{v}_1$ and $\mathbf{v}_2$ as

$$\mathbf{E} = \alpha \mathbf{v}_1 + \beta \mathbf{v}_2, \quad (9.24)$$

where α and β are in general complex, $\sqrt{|\alpha|^2 + |\beta|^2} = 1$. The transmitted electric field vector after **P** is **P E**. Therefore, the *flux* of the transmitted electric field is

$$I_{Trans} = |\mathbf{P}\mathbf{E}|^2 = \mathbf{E}^\dagger \mathbf{P}^\dagger \mathbf{P} \mathbf{E}. \quad (9.25)$$

From Equations 9.23 and 9.25, the flux of the transmitted electric field is

$$I_{Trans} = |\alpha|^2 \Lambda_1^2 + |\beta|^2 \Lambda_2^2 = |\alpha|^2 \left(\Lambda_1^2 - \Lambda_2^2\right) + \Lambda_2^2. \quad (9.26)$$

Since both $|\alpha|^2 \left(\Lambda_1^2 - \Lambda_2^2\right)$ and Λ_2^2 are positive and $\Lambda_1 \geq \Lambda_2$ by construction, the maximum intensity transmittance occurs when the incident state is $\mathbf{v}_1$, and the minimum intensity occurs when the incident state is $\mathbf{v}_2$; that is,

$$I_{Trans} = \begin{cases} I_{max} = \Lambda_1^2 \text{ if } |\alpha|^2 = 1 \\ I_{min} = \Lambda_2^2 \text{ if } |\alpha|^2 = 0 \end{cases}, \quad (9.27)$$

for any polarization ray tracing matrix **P**. Thus, the *diattenuation* of **P** is

$$\mathcal{D} = \frac{\Lambda_1^2 - \Lambda_2^2}{\Lambda_1^2 + \Lambda_2^2}, \quad (9.28)$$

and $\mathbf{v}_1 = \mathbf{v}_{max}$ and $\mathbf{v}_2 = \mathbf{v}_{min}$ are the incident polarization states for which **P** gives the maximum and minimum transmittance. $\mathbf{u}_1 = \mathbf{P}\mathbf{v}_1$ and $\mathbf{u}_2 = \mathbf{P}\mathbf{v}_2$ are the corresponding exiting polarization states. For an ideal polarizer, one incident polarization state is fully attenuated; thus, the singular value $\Lambda_2 = 0$.

The derivation of an algorithm to define the *retardance* of **P** matrices is introduced in a separate chapter because of its complexity (see Chapter 17, Parallel Transport and the Calculation of Retardance).

9.5 Example—Interferometer with a Polarizing Beam Splitter

In this section, an *interferometer* with a *polarizing beam splitter* (PBS) is used as an example for the polarization ray tracing calculus. Figure 9.6 shows the schematic of the interferometer. In the arms of the interferometer, combinations of quarter wave retarders and mirrors are used to route light through the PBSs such that light does not reflect back toward the laser, which is what would happen without the quarter wave retarders. This happens because circularly polarized light changes its handedness when it reflects from mirrors; that is, left reflects as right, and vice versa. Performing this calculation in three dimensions helps model these changes in a straightforward way.

In this example, only the polarizing elements' **P** matrices are calculated. For simplicity, all rays are propagating along the axes and normally incident at each mirror and retarder.

Consider a laser generating vertically polarized light. Before the PBS, all rays pass through a half wave linear retarder (HWLR) with fast axis at 22.5° from the horizontal axis. This produces a beam polarized along 45° such that an equal amount of *s*- and *p*-polarization is incident on the PBS. The Jones matrix for the HWLR at 22.5° is

$$\mathbf{J}_{HWLR} = R\left[\frac{\pi}{8}\right] \cdot \begin{pmatrix} 1 & 0 \\ 0 & -1 \end{pmatrix} \cdot R\left[\frac{\pi}{8}\right]^{-1} = \frac{1}{\sqrt{2}} \begin{pmatrix} 1 & 1 \\ 1 & -1 \end{pmatrix}. \tag{9.29}$$

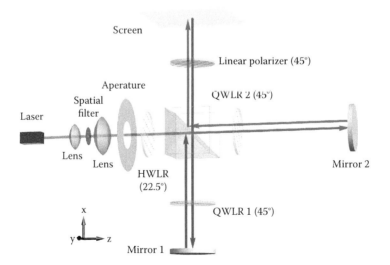

Figure 9.6 Interferometer with a polarizing beam splitter and quarter wave retarders in each arm. Circularly polarized light is incident on each mirror. Incident left circularly polarized light reflects from mirrors as right circularly polarized and vice versa. Thus, the *s*-polarized light that reflects downward from the polarizing beam splitter is converted to *p*-polarized light on its return path and transmits toward the screen. The two beams exiting upward from the PBS are in orthogonal polarization states and would not produce intensity interference fringes without the linear polarizer projecting them into a common 45° polarization state.

Since $\hat{\boldsymbol{\eta}}_1 = \hat{\mathbf{k}}_0 = \hat{\mathbf{z}}$, the ray does not bend at the interface; hence, $\mathbf{O}_{in,1} = \mathbf{O}_{out,1}$. Therefore, the $\mathbf{P}$ matrix for the first ray intercept (HWLR) is

$$\mathbf{P}_{HWLR} = \mathbf{O}_{out,1} \mathbf{J}_{HWLR} \mathbf{O}_{in,1}^{-1}$$
$$= \begin{pmatrix} 1 & 0 & 0 \\ 0 & 1 & 0 \\ 0 & 0 & 1 \end{pmatrix} \cdot \frac{1}{\sqrt{2}} \begin{pmatrix} 1 & 1 & 0 \\ 1 & -1 & 0 \\ 0 & 0 & \sqrt{2} \end{pmatrix} \cdot \begin{pmatrix} 1 & 0 & 0 \\ 0 & 1 & 0 \\ 0 & 0 & 1 \end{pmatrix} = \frac{1}{\sqrt{2}} \begin{pmatrix} 1 & 1 & 0 \\ 1 & -1 & 0 \\ 0 & 0 & \sqrt{2} \end{pmatrix}. \quad (9.30)$$

The next polarizing element is the PBS, which divides the beam into two paths: the *reference* path (which reflects and transmits) and the *test* path (which transmits and reflects). Since $\mathbf{P}$ is dependent on propagation direction, each path has unique $\mathbf{P}$ matrices.

9.5.1 Ray Tracing the Reference Path

First consider the reference path. The light is split into *s*- and *p*-components at the hypotenuse of the PBS with a surface normal $\hat{\boldsymbol{\eta}} = (1,0,1)/\sqrt{2}$. Thus,

$$\hat{\mathbf{k}}_{reference,2} = -\hat{\mathbf{x}}, \; \hat{\mathbf{s}}_{reference,2} = \hat{\mathbf{s}}'_{reference,2} = \hat{\mathbf{y}}, \; \hat{\mathbf{p}}_{reference,2} = -\hat{\mathbf{x}}, \; \hat{\mathbf{p}}'_{reference,2} = -\hat{\mathbf{z}}. \quad (9.31)$$

The subscript *reference* indicates rays along the reference path. The $\mathbf{P}$ matrix for reflection from the PBS is

$$\mathbf{P}_{PBS(R)} = \begin{pmatrix} | & | & | \\ \hat{\mathbf{s}}'_{reference,2} & \hat{\mathbf{p}}'_{reference,2} & \hat{\mathbf{k}}_{reference,2} \\ | & | & | \end{pmatrix} \cdot \begin{pmatrix} r_s & 0 & 0 \\ 0 & r_p & 0 \\ 0 & 0 & 1 \end{pmatrix} \cdot \begin{pmatrix} - & \hat{\mathbf{s}}_{reference,2} & - \\ - & \hat{\mathbf{p}}_{reference,2} & - \\ - & \hat{\mathbf{k}}_1 & - \end{pmatrix}$$
$$= \begin{pmatrix} 0 & 0 & -1 \\ 1 & 0 & 0 \\ 0 & -1 & 0 \end{pmatrix} \begin{pmatrix} 1 & 0 & 0 \\ 0 & 0 & 0 \\ 0 & 0 & 1 \end{pmatrix} \begin{pmatrix} 0 & 1 & 0 \\ -1 & 0 & 0 \\ 0 & 0 & 1 \end{pmatrix} = \begin{pmatrix} 0 & 0 & -1 \\ 0 & 1 & 0 \\ 0 & 0 & 0 \end{pmatrix}. \quad (9.32)$$

Here, lines indicate that vectors occupy the corresponding matrix rows or columns.

Note that the middle matrix, a 3D Jones matrix, is in its local $(\hat{\mathbf{s}}, \hat{\mathbf{p}}, \hat{\mathbf{k}})$ coordinates. $\mathbf{P}_{PBS(R)}$ is in global coordinates and shows explicitly that the PBS reflects the propagation vector from $\hat{\mathbf{z}}$ to $-\hat{\mathbf{x}}$, reflects $\hat{\mathbf{y}}$ polarization into $\hat{\mathbf{y}}$ polarization, and removes the $\hat{\mathbf{x}}$ polarized component of the incident ray.

The next element is a quarter wave linear retarder (QWLR) with a fast axis at 45° from the *y* toward the *z* axis, $(0, 1/\sqrt{2}, 1/\sqrt{2})$, with Jones matrix

$$\mathbf{J}_{QWLR(45)}$$
$$= \mathbf{R}[-45°] \cdot \frac{1}{\sqrt{2}} \begin{pmatrix} 1-i & 0 \\ 0 & 1+i \end{pmatrix} \cdot \mathbf{R}[-45°]^{-1} = \frac{1}{\sqrt{2}} \begin{pmatrix} 1-i & 0 \\ 0 & 1+i \end{pmatrix} = \frac{1}{\sqrt{2}} \begin{pmatrix} 1 & -i \\ -i & 1 \end{pmatrix}. \quad (9.33)$$

9.5 Example—Interferometer with a Polarizing Beam Splitter

The ray is normally incident on the QWLR; thus, we choose local coordinates following Equation 9.6; $\hat{s}'_{reference,2}$ is used as $\hat{s}_{reference,3}$ and $\hat{p}_{reference,3}$ and $\hat{p}'_{reference,3}$ are calculated;

$$\hat{k}_{reference,3} = -\hat{x}, \hat{s}_{reference,3} = \hat{s}'_{reference,3} = \hat{s}'_{reference,2} = \hat{y}, \hat{p}_{reference,3} = \hat{p}'_{reference,3} = -\hat{z}. \tag{9.34}$$

Thus, the **P** matrix for the **QWLR** is

$$\mathbf{P}_{QWLR} = \mathbf{O}_{out} \cdot \mathbf{J}_{QWLR(45)} \cdot \mathbf{O}_{in}^{-1}$$

$$= \begin{pmatrix} 0 & 0 & -1 \\ 1 & 0 & 0 \\ 0 & -1 & 0 \end{pmatrix} \cdot \frac{1}{\sqrt{2}} \begin{pmatrix} 1 & -i & 0 \\ -i & 1 & 0 \\ 0 & 0 & \sqrt{2} \end{pmatrix} \begin{pmatrix} 0 & 1 & 0 \\ 0 & 0 & -1 \\ -1 & 0 & 0 \end{pmatrix} = \frac{1}{\sqrt{2}} \begin{pmatrix} \sqrt{2} & 0 & 0 \\ 0 & 1 & i \\ 0 & i & 1 \end{pmatrix}. \tag{9.35}$$

The next element is an ideal mirror that retroreflects the ray. Since the ray is normally incident on the mirror,

$$\hat{k}_{reference,4} = \hat{x}, \hat{s}_{reference,4} = \hat{s}'_{reference,4} = \hat{s}'_{reference,3} = \hat{y},$$
$$\hat{p}_{reference,4} = -\hat{z}, \hat{p}'_{reference,4} = \hat{z}. \tag{9.36}$$

Using $\mathbf{J}_{mirror} = \begin{pmatrix} 1 & 0 \\ 0 & -1 \end{pmatrix}$, the **P** matrix for the ideal mirror is

$$\mathbf{P}_{M,1} = \mathbf{O}_{out} \cdot \mathbf{J}_{mirror} \cdot \mathbf{O}_{in}^{-1}$$

$$= \begin{pmatrix} 0 & 0 & 1 \\ 1 & 0 & 0 \\ 0 & 1 & 0 \end{pmatrix} \cdot \begin{pmatrix} 1 & 0 & 0 \\ 0 & -1 & 0 \\ 0 & 0 & 1 \end{pmatrix} \cdot \begin{pmatrix} 0 & 1 & 0 \\ 0 & 0 & -1 \\ -1 & 0 & 0 \end{pmatrix} = \begin{pmatrix} -1 & 0 & 0 \\ 0 & 1 & 0 \\ 0 & 0 & 1 \end{pmatrix}. \tag{9.37}$$

Then, the ray propagates back to **QWLR** in the opposite direction (along +x). The local coordinates are

$$\hat{k}_{reference,5} = \hat{x}, \hat{s}_{reference,5} = \hat{s}'_{reference,5} = \hat{s}'_{reference,4} = \hat{y}, \hat{p}_{reference,5} = \hat{p}'_{reference,5} = \hat{z}. \tag{9.38}$$

The fast axis of the **QWLR** is still at $(0, 1/\sqrt{2}, 1/\sqrt{2})$, but since the propagation direction is reversed, the fast axis orientation in Jones matrix local coordinates is at 135°. Hence,

$$\mathbf{P}_{QWLR,2} = \mathbf{O}_{out} \cdot \mathbf{J}_{QWLR(135)} \cdot \mathbf{O}_{in}^{-1}$$

$$= \begin{pmatrix} 0 & 0 & 1 \\ 1 & 0 & 0 \\ 0 & 1 & 0 \end{pmatrix} \cdot \frac{1}{\sqrt{2}} \begin{pmatrix} 1 & i & 0 \\ i & 1 & 0 \\ 0 & 0 & \sqrt{2} \end{pmatrix} \begin{pmatrix} 0 & 1 & 0 \\ 0 & 0 & 1 \\ 1 & 0 & 0 \end{pmatrix} = \frac{1}{\sqrt{2}} \begin{pmatrix} \sqrt{2} & 0 & 0 \\ 0 & 1 & i \\ 0 & i & 1 \end{pmatrix}. \tag{9.39}$$

Next, the ray transmits through the PBS with surface normal $\hat{\boldsymbol{\eta}} = (1,0,1)/\sqrt{2}$,

$$\hat{\mathbf{k}}_{reference,6} = \hat{\mathbf{x}}, \hat{\mathbf{s}}_{reference,6} = \hat{\mathbf{s}}'_{reference,6} = -\hat{\mathbf{y}}, \hat{\mathbf{p}}_{reference,5} = \hat{\mathbf{p}}'_{reference,5} = -\hat{\mathbf{z}}. \quad (9.40)$$

The PBS reflects $\hat{\mathbf{s}}_{reference,6}$ state out of the beam, transmits $\hat{\mathbf{p}}_{reference,6}$, and transmits $\hat{\mathbf{k}}_{reference,5}$ undeviated,

$$\mathbf{P}_{PBS(T)} = \begin{pmatrix} 0 & 0 & 1 \\ -1 & 0 & 0 \\ 0 & -1 & 0 \end{pmatrix} \cdot \begin{pmatrix} 0 & 0 & 0 \\ 0 & 1 & 0 \\ 0 & 0 & 1 \end{pmatrix} \cdot \begin{pmatrix} 0 & -1 & 0 \\ 0 & 0 & -1 \\ 1 & 0 & 0 \end{pmatrix} = \begin{pmatrix} 1 & 0 & 0 \\ 0 & 0 & 0 \\ 0 & 0 & 1 \end{pmatrix}. \quad (9.41)$$

$\mathbf{P}_{PBS(T)}$ transmits the $\hat{\mathbf{z}}$ polarization, removes $\hat{\mathbf{y}}$ polarization, and does not deviate the propagation vector of the incident ray.

9.5.2 Ray Tracing through the Test Path

For the test path, the PBS transmits the *p*-polarization and the propagation vector does not change. The subscript *test* indicates rays along the test path. Using the surface normal of the PBS,

$$\hat{\mathbf{k}}_{test,2} = \hat{\mathbf{z}}, \hat{\mathbf{s}}_{test,2} = \hat{\mathbf{s}}'_{test,2} = \hat{\mathbf{y}}, \hat{\mathbf{p}}_{test,2} = \hat{\mathbf{p}}'_{test,2} = -\hat{\mathbf{x}}, \quad (9.42)$$

$$\mathbf{P}_{PBS(T)} = \begin{pmatrix} 0 & -1 & 0 \\ 1 & 0 & 0 \\ 0 & 0 & 1 \end{pmatrix} \cdot \begin{pmatrix} 0 & 0 & 0 \\ 0 & 1 & 0 \\ 0 & 0 & 1 \end{pmatrix} \cdot \begin{pmatrix} 0 & 1 & 0 \\ -1 & 0 & 0 \\ 0 & 0 & 1 \end{pmatrix} = \begin{pmatrix} 1 & 0 & 0 \\ 0 & 0 & 0 \\ 0 & 0 & 1 \end{pmatrix}. \quad (9.43)$$

$\mathbf{P}_{PBS(T)}$ transmits the $\hat{\mathbf{x}}$ polarization and blocks the $\hat{\mathbf{y}}$ polarization. Then, the light propagates through a **QWLR** oriented at 45° between the *x*- and *y*-axis. The local coordinates and the corresponding **P** matrix are

$$\hat{\mathbf{k}}_{test,3} = \hat{\mathbf{z}}, \hat{\mathbf{s}}_{test,3} = \hat{\mathbf{s}}'_{test,3} = \hat{\mathbf{s}}'_{test,2} = \hat{\mathbf{y}}, \hat{\mathbf{p}}_{test,2} = \hat{\mathbf{p}}'_{test,2} = -\hat{\mathbf{x}}, \quad (9.44)$$

$$\mathbf{P}_{QWLR} = \mathbf{O}_{out} \cdot \mathbf{J}_{QWLR(45)} \cdot \mathbf{O}_{in}^{-1}$$

$$= \begin{pmatrix} 0 & -1 & 0 \\ 1 & 0 & 0 \\ 0 & 0 & 1 \end{pmatrix} \cdot \frac{1}{\sqrt{2}} \begin{pmatrix} 1 & -i & 0 \\ -i & 1 & 0 \\ 0 & 0 & \sqrt{2} \end{pmatrix} \cdot \begin{pmatrix} 0 & 1 & 0 \\ -1 & 0 & 0 \\ 0 & 0 & 1 \end{pmatrix} = \frac{1}{\sqrt{2}} \begin{pmatrix} 1 & i & 0 \\ i & 1 & 0 \\ 0 & 0 & \sqrt{2} \end{pmatrix}. \quad (9.45)$$

Assuming the test mirror is also an ideal mirror, the light reflects, changing its propagation direction from *z* to *−z*.

$$\hat{\mathbf{k}}_{test,4} = \hat{\mathbf{z}}, \hat{\mathbf{s}}_{test,4} = \hat{\mathbf{s}}'_{test,4} = \hat{\mathbf{s}}'_{test,3} = \hat{\mathbf{y}}, \hat{\mathbf{p}}_{test,4} = -\hat{\mathbf{x}}, \hat{\mathbf{p}}'_{test,4} = \hat{\mathbf{x}}. \quad (9.46)$$

9.5 Example—Interferometer with a Polarizing Beam Splitter

$$\mathbf{P}_{M,2} = \mathbf{O}_{out} \cdot \mathbf{J}_{mirror} \cdot \mathbf{O}_{in}^{-1}$$

$$= \begin{pmatrix} 0 & 1 & 0 \\ 1 & 0 & 0 \\ 0 & 0 & -1 \end{pmatrix} \cdot \begin{pmatrix} 1 & 0 & 0 \\ 0 & -1 & 0 \\ 0 & 0 & 1 \end{pmatrix} \cdot \begin{pmatrix} 0 & 1 & 0 \\ -1 & 0 & 0 \\ 0 & 0 & 1 \end{pmatrix} = \begin{pmatrix} 1 & 0 & 0 \\ 0 & 1 & 0 \\ 0 & 0 & -1 \end{pmatrix}. \quad (9.47)$$

The light propagates back to **QWLR** in the opposite direction. Similar to the reference path case, the fast axis of the **QWLR** in local coordinates is now at 135°.

$$\hat{\mathbf{k}}_{test,5} = -\hat{\mathbf{z}}, \hat{\mathbf{s}}_{test,5} = \hat{\mathbf{s}}'_{test,5} = \hat{\mathbf{s}}'_{test,4} = \hat{\mathbf{y}}, \hat{\mathbf{p}}_{test,5} = \hat{\mathbf{p}}'_{test,5} = \hat{\mathbf{x}}, \quad (9.48)$$

$$\mathbf{P}_{QWLR,2} = \mathbf{O}_{out} \cdot \mathbf{J}_{QWLR(135)} \cdot \mathbf{O}_{in}^{-1}$$

$$= \begin{pmatrix} 0 & 1 & 0 \\ 1 & 0 & 0 \\ 0 & 0 & -1 \end{pmatrix} \cdot \frac{1}{\sqrt{2}} \begin{pmatrix} 1 & i & 0 \\ i & 1 & 0 \\ 0 & 0 & \sqrt{2} \end{pmatrix} \cdot \begin{pmatrix} 0 & 1 & 0 \\ 1 & 0 & 0 \\ 0 & 0 & -1 \end{pmatrix} = \frac{1}{\sqrt{2}} \begin{pmatrix} 1 & i & 0 \\ i & 1 & 0 \\ 0 & 0 & \sqrt{2} \end{pmatrix}. \quad (9.49)$$

Finally, the ray reaches the PBS the second time. This time, the ray gets reflected and changes its propagation direction from −z into x-direction.

$$\hat{\mathbf{k}}_{test,6} = \hat{\mathbf{x}}, \hat{\mathbf{s}}_{test,6} = \hat{\mathbf{s}}'_{test,6} = \hat{\mathbf{y}}, \hat{\mathbf{p}}_{test,6} = \hat{\mathbf{x}}, \hat{\mathbf{p}}'_{test,6} = \hat{\mathbf{z}}, \quad (9.50)$$

$$\mathbf{P}_{PBS(R)} = \begin{pmatrix} 0 & 0 & 1 \\ 1 & 0 & 0 \\ 0 & 1 & 0 \end{pmatrix} \cdot \begin{pmatrix} 1 & 0 & 0 \\ 0 & 0 & 0 \\ 0 & 0 & 1 \end{pmatrix} \cdot \begin{pmatrix} 0 & 1 & 0 \\ 1 & 0 & 0 \\ 0 & 0 & -1 \end{pmatrix} = \begin{pmatrix} 0 & 0 & -1 \\ 0 & 1 & 0 \\ 0 & 0 & 0 \end{pmatrix}. \quad (9.51)$$

$\mathbf{P}_{PBS(R)}$ reflects the $\hat{\mathbf{y}}$ polarization, while blocking the $\hat{\mathbf{z}}$ polarization.

9.5.3 Ray Tracing through the Analyzer

Exiting the PBS, both paths have rays propagating along the +x-axis and pass through a linear polarizer at 45° in the y–z plane. Using $\mathbf{J}_{LP[45]} = \begin{pmatrix} 0.5 & 0.5 \\ 0.5 & 0.5 \end{pmatrix}$ and local coordinates, $\hat{\mathbf{k}}_7 = \hat{\mathbf{x}}, \hat{\mathbf{s}}_7 = \hat{\mathbf{s}}'_7 = \hat{\mathbf{y}}, \hat{\mathbf{p}}_7 = \hat{\mathbf{p}}'_7 = \hat{\mathbf{z}}$, the polarizer $\mathbf{P}_{LP}$ is

$$\mathbf{P}_{LP} = \mathbf{O}_{out} \mathbf{J}_{LP[45]} \mathbf{O}_{in}^{-1}$$

$$= \begin{pmatrix} 0 & 0 & 1 \\ 1 & 0 & 0 \\ 0 & 1 & 0 \end{pmatrix} \cdot \begin{pmatrix} 0.5 & 0.5 & 0 \\ 0.5 & 0.5 & 0 \\ 0 & 0 & 1 \end{pmatrix} \cdot \begin{pmatrix} 0 & 1 & 0 \\ 0 & 0 & 1 \\ 1 & 0 & 0 \end{pmatrix} = \begin{pmatrix} 1 & 0 & 0 \\ 0 & 0.5 & 0.5 \\ 0 & 0.5 & 0.5 \end{pmatrix}. \quad (9.52)$$

9.5.4 Cumulative P Matrix for Both Paths

A matrix sequence for the reference path (red path) is listed from right to left since the sequence of matrix multiplications is written from right to left:

$$\text{LP[45]} \leftarrow \text{PBS(T)} \leftarrow \text{QWLR} \leftarrow \text{Mirror} \leftarrow \text{QWLR} \leftarrow \text{PBS(R)} \leftarrow \text{HWLR}$$

$$\begin{pmatrix} 1 & 0 & 0 \\ 0 & \frac{1}{2} & \frac{1}{2} \\ 0 & \frac{1}{2} & \frac{1}{2} \end{pmatrix} \begin{pmatrix} 1 & 0 & 0 \\ 0 & 0 & 0 \\ 0 & 0 & 1 \end{pmatrix} \begin{pmatrix} 1 & 0 & 0 \\ 0 & \frac{1}{\sqrt{2}} & \frac{i}{\sqrt{2}} \\ 0 & \frac{i}{\sqrt{2}} & \frac{1}{\sqrt{2}} \end{pmatrix} \begin{pmatrix} -1 & 0 & 0 \\ 0 & 1 & 0 \\ 0 & 0 & 1 \end{pmatrix} \begin{pmatrix} 1 & 0 & 0 \\ 0 & \frac{1}{\sqrt{2}} & \frac{i}{\sqrt{2}} \\ 0 & \frac{i}{\sqrt{2}} & \frac{1}{\sqrt{2}} \end{pmatrix} \begin{pmatrix} 0 & 0 & -1 \\ 0 & 1 & 0 \\ 0 & 0 & 0 \end{pmatrix} \begin{pmatrix} \frac{1}{\sqrt{2}} & \frac{1}{\sqrt{2}} & 0 \\ \frac{1}{\sqrt{2}} & \frac{-1}{\sqrt{2}} & 0 \\ 0 & 0 & 1 \end{pmatrix}. \quad (9.53)$$

The cumulative **P** for the reference path is

$$\mathbf{P}_{reference} = \begin{pmatrix} 0 & 0 & 1 \\ \frac{i}{2\sqrt{2}} & \frac{-i}{2\sqrt{2}} & 0 \\ \frac{i}{2\sqrt{2}} & \frac{-i}{2\sqrt{2}} & 0 \end{pmatrix}, \quad (9.54)$$

and the electric field from the reference path at the observation screen is

$$\mathbf{E}_{reference} = \mathbf{P}_{reference} \cdot \begin{pmatrix} E_x \\ E_y \\ 0 \end{pmatrix} = \begin{pmatrix} 0 \\ \frac{i}{2\sqrt{2}}(E_x - E_y) \\ \frac{i}{2\sqrt{2}}(E_x - E_y) \end{pmatrix}. \quad (9.55)$$

A matrix sequence for the test path (blue path) is

$$\text{LP[45]} \leftarrow \text{PBS(R)} \leftarrow \text{QWLR} \leftarrow \text{Mirror} \leftarrow \text{QWLR} \leftarrow \text{PBS(T)} \leftarrow \text{HWLR}$$

$$\begin{pmatrix} 1 & 0 & 0 \\ 0 & \frac{1}{2} & \frac{1}{2} \\ 0 & \frac{1}{2} & \frac{1}{2} \end{pmatrix} \begin{pmatrix} 0 & 0 & -1 \\ 0 & 1 & 0 \\ 0 & 0 & 0 \end{pmatrix} \begin{pmatrix} 1 & 0 & 0 \\ 0 & \frac{1}{\sqrt{2}} & \frac{i}{\sqrt{2}} \\ 0 & \frac{i}{\sqrt{2}} & \frac{1}{\sqrt{2}} \end{pmatrix} \begin{pmatrix} 1 & 0 & 0 \\ 0 & 1 & 0 \\ 0 & 0 & -1 \end{pmatrix} \begin{pmatrix} 1 & 0 & 0 \\ 0 & \frac{1}{\sqrt{2}} & \frac{i}{\sqrt{2}} \\ 0 & \frac{i}{\sqrt{2}} & \frac{1}{\sqrt{2}} \end{pmatrix} \begin{pmatrix} 1 & 0 & 0 \\ 0 & 0 & 0 \\ 0 & 0 & 1 \end{pmatrix} \begin{pmatrix} \frac{1}{\sqrt{2}} & \frac{1}{\sqrt{2}} & 0 \\ \frac{1}{\sqrt{2}} & \frac{-1}{\sqrt{2}} & 0 \\ 0 & 0 & 1 \end{pmatrix}. \quad (9.56)$$

The cumulative **P** for the test path is

9.5 Example—Interferometer with a Polarizing Beam Splitter

$$\mathbf{P}_{test} = \begin{pmatrix} 0 & 0 & 1 \\ \dfrac{i}{2\sqrt{2}} & \dfrac{i}{2\sqrt{2}} & 0 \\ \dfrac{i}{2\sqrt{2}} & \dfrac{i}{2\sqrt{2}} & 0 \end{pmatrix}, \qquad (9.57)$$

and the electric field from the test path is

$$\mathbf{E}_{test} = \mathbf{P}_{test} \cdot \begin{pmatrix} E_x \\ E_y \\ 0 \end{pmatrix} = \begin{pmatrix} 0 \\ \dfrac{i}{2\sqrt{2}}(E_x + E_y) \\ \dfrac{i}{2\sqrt{2}}(E_x + E_y) \end{pmatrix}. \qquad (9.58)$$

For horizontally polarized incident light from the laser, a sequence of electric field vectors can be calculated for both paths by multiplying the laser light's polarization vector $\mathbf{E}_0 = (1,0,0)$ by each $\mathbf{P}$ matrix. Following the polarization state through the interferometer's reference arm, the sequence of global polarization states is

LP[45] ← PBS(T) ← QWLR ← Mirror ← QWLR ← PBS(R) ← HWLR ← $\mathbf{E}_{in}$

$$\begin{pmatrix} 0 \\ \dfrac{i}{2\sqrt{2}} \\ \dfrac{i}{2\sqrt{2}} \end{pmatrix} \begin{pmatrix} 0 \\ 0 \\ \dfrac{i}{\sqrt{2}} \end{pmatrix} \begin{pmatrix} 0 \\ 0 \\ \dfrac{i}{\sqrt{2}} \end{pmatrix} \begin{pmatrix} 0 \\ \dfrac{1}{2} \\ \dfrac{i}{2} \end{pmatrix} \begin{pmatrix} 0 \\ \dfrac{1}{2} \\ \dfrac{i}{2} \end{pmatrix} \begin{pmatrix} 0 \\ \dfrac{1}{\sqrt{2}} \\ 0 \end{pmatrix} \begin{pmatrix} \dfrac{1}{\sqrt{2}} \\ \dfrac{1}{\sqrt{2}} \\ 0 \end{pmatrix} \begin{pmatrix} 1 \\ 0 \\ 0 \end{pmatrix}. \qquad (9.59)$$

The corresponding sequence for the test path is

LP[45] ← PBS(R) ← QWLR ← Mirror ← QWLR ← PBS(T) ← HWLR ← $\mathbf{E}_{in}$

$$\begin{pmatrix} 0 \\ \dfrac{i}{2\sqrt{2}} \\ \dfrac{i}{2\sqrt{2}} \end{pmatrix} \begin{pmatrix} 0 \\ \dfrac{i}{\sqrt{2}} \\ 0 \end{pmatrix} \begin{pmatrix} 0 \\ \dfrac{i}{\sqrt{2}} \\ 0 \end{pmatrix} \begin{pmatrix} \dfrac{1}{2} \\ \dfrac{i}{2} \\ 0 \end{pmatrix} \begin{pmatrix} \dfrac{1}{2} \\ \dfrac{i}{2} \\ 0 \end{pmatrix} \begin{pmatrix} \dfrac{1}{\sqrt{2}} \\ 0 \\ 0 \end{pmatrix} \begin{pmatrix} \dfrac{1}{\sqrt{2}} \\ \dfrac{1}{\sqrt{2}} \\ 0 \end{pmatrix} \begin{pmatrix} 1 \\ 0 \\ 0 \end{pmatrix}. \qquad (9.60)$$

Since the $\mathbf{P}$ matrix is constructed so that the exiting *propagation vector* is returned when the incident propagation vector is multiplied to the $\mathbf{P}$ matrix, a sequence of propagation vectors for both paths can be calculated in a similar way. The sequence of propagation vectors for the reference path is

$$\mathbf{LP[45]} \leftarrow \mathbf{PBS(T)} \leftarrow \mathbf{QWLR} \leftarrow \mathbf{Mirror} \leftarrow \mathbf{QWLR} \leftarrow \mathbf{PBS(R)} \leftarrow \mathbf{HWLR} \leftarrow \hat{\mathbf{k}}_0$$

$$\begin{pmatrix}1\\0\\0\end{pmatrix} \begin{pmatrix}1\\0\\0\end{pmatrix} \begin{pmatrix}1\\0\\0\end{pmatrix} \begin{pmatrix}1\\0\\0\end{pmatrix} \begin{pmatrix}-1\\0\\0\end{pmatrix} \begin{pmatrix}-1\\0\\0\end{pmatrix} \begin{pmatrix}0\\0\\1\end{pmatrix} \begin{pmatrix}0\\0\\1\end{pmatrix}, \quad (9.61)$$

and the sequence of propagation vectors of the test path is

$$\mathbf{LP[45]} \leftarrow \mathbf{PBS(R)} \leftarrow \mathbf{QWLR} \leftarrow \mathbf{Mirror} \leftarrow \mathbf{QWLR} \leftarrow \mathbf{PBS(T)} \leftarrow \mathbf{HWLR} \leftarrow \hat{\mathbf{k}}_0$$

$$\begin{pmatrix}1\\0\\0\end{pmatrix} \begin{pmatrix}1\\0\\0\end{pmatrix} \begin{pmatrix}0\\0\\-1\end{pmatrix} \begin{pmatrix}0\\0\\-1\end{pmatrix} \begin{pmatrix}0\\0\\1\end{pmatrix} \begin{pmatrix}0\\0\\1\end{pmatrix} \begin{pmatrix}0\\0\\1\end{pmatrix} \begin{pmatrix}0\\0\\1\end{pmatrix}. \quad (9.62)$$

If the test path has an unknown sample described by a Jones matrix instead of an ideal mirror, the matrix sequence becomes

$$\mathbf{LP[45]} \leftarrow \mathbf{PBS(R)} \leftarrow \mathbf{QWLR} \leftarrow \mathbf{Sample} \leftarrow \mathbf{QWLR} \leftarrow \mathbf{PBS(T)} \leftarrow \mathbf{HWLR}$$

$$\begin{pmatrix}1 & 0 & 0\\0 & \tfrac{1}{2} & \tfrac{1}{2}\\0 & \tfrac{1}{2} & \tfrac{1}{2}\end{pmatrix} \begin{pmatrix}0 & 0 & -1\\0 & 1 & 0\\0 & 0 & 0\end{pmatrix} \begin{pmatrix}1 & 0 & 0\\0 & \tfrac{1}{\sqrt{2}} & \tfrac{i}{\sqrt{2}}\\0 & \tfrac{i}{\sqrt{2}} & \tfrac{1}{\sqrt{2}}\end{pmatrix} \begin{pmatrix}-1 & 0 & 0\\0 & j_{yy} & j_{yz}\\0 & j_{zy} & j_{zz}\end{pmatrix} \begin{pmatrix}1 & 0 & 0\\0 & \tfrac{1}{\sqrt{2}} & \tfrac{i}{\sqrt{2}}\\0 & \tfrac{i}{\sqrt{2}} & \tfrac{1}{\sqrt{2}}\end{pmatrix} \begin{pmatrix}1 & 0 & 0\\0 & 0 & 0\\0 & 0 & 1\end{pmatrix} \begin{pmatrix}\tfrac{1}{\sqrt{2}} & \tfrac{1}{\sqrt{2}} & 0\\\tfrac{1}{\sqrt{2}} & -\tfrac{1}{\sqrt{2}} & 0\\0 & 0 & 1\end{pmatrix}. \quad (9.63)$$

With a horizontally polarized incident light, the test path electric field is

$$\mathbf{E}_{out} = \frac{(j_{yx} - j_{xy}) + i(j_{xx} + j_{yy})}{4\sqrt{2}}\begin{pmatrix}0\\1\\1\end{pmatrix}. \quad (9.64)$$

9.6 The Addition Form of Polarization Ray Tracing Matrices

This section introduces a method for *coherent* combination (i.e., addition) of *polarization ray tracing matrices*. In Section 9.1, Equation 9.4 was chosen to uniquely define the **P** matrix for a given ray. However, Equation 9.4 is not the only choice. The transformation of all polarization states on a transverse plane perpendicular to the propagation vector $\hat{\mathbf{k}}_{q-1}$ can be described as

9.6 The Addition Form of Polarization Ray Tracing Matrices

linear combinations of the transformations of *any* two linearly independent basis vectors, $\mathbf{E}_a$ and $\mathbf{E}_b$,

$$\mathbf{E}'_a = \mathbf{P}_q \mathbf{E}_a, \quad \mathbf{E}'_b = \mathbf{P}_q \mathbf{E}_b. \tag{9.65}$$

The relationship in Equation 9.65 yields six equations, one for each row, but $\mathbf{P}_q$ has nine elements. Thus, Equation 9.1 does not fully constrain $\mathbf{P}_q$. To uniquely define $\mathbf{P}_q$, an additional set of three constraints is applied by relating the incident and exiting propagation vectors,

$$\mathbf{P}_q \hat{\mathbf{k}}_{q-1} = \gamma \hat{\mathbf{k}}_q. \tag{9.66}$$

The choice of γ is arbitrary, but only two values, either 0 or 1, allow $\mathbf{P}_q$ to be repeatedly cascaded and maintain the value of γ. Both choices of γ describe the same polarization effects. With $\gamma = 1$, which was the choice made in Section 9.1, only ideal polarizers have singular matrices,

$$\mathbf{P}_q \hat{\mathbf{k}}_{q-1} = \hat{\mathbf{k}}_q. \tag{9.67}$$

$\gamma = 1$ is the principal definition for $\mathbf{P}$ used for the majority of this book. With the inclusion of Equation 9.67, $\mathbf{P}_q$ is now uniquely defined for each ray.

With $\gamma = 0$, $\mathbf{P}_q$ is always singular and thus $\mathbf{P}_q^{-1}$ never exists. One of the singular values of $\mathbf{P}_q$ will always be zero, as will one of the eigenvalues. When the $\mathbf{P}$ matrix is defined with this convention, $\gamma = 0$, the *addition form of the polarization ray tracing matrix* is indicated with the overscript cup $\breve{\mathbf{P}}$,

$$\breve{\mathbf{P}}_q \hat{\mathbf{k}}_{q-1} = 0. \tag{9.68}$$

The definition Equation 9.68 simplifies the addition of beams such as with interferometers. Consider a ray entering a Mach–Zehnder interferometer generating two exiting rays with $\mathbf{P}_A$ and $\mathbf{P}_B$. For an incident polarization state $\mathbf{E}_{in}$, the exiting state is the sum of the exiting states for the individual beams

$$\mathbf{E}_{out} = (\mathbf{P}_A + \mathbf{P}_B) \mathbf{E}_{in} = \mathbf{E}'_A + \mathbf{E}'_B. \tag{9.69}$$

$\mathbf{E}_{out}$ can be written as

$$\mathbf{E}_{out} = \mathbf{P} \mathbf{E}_{in}, \tag{9.70}$$

where $\mathbf{P}$ is the polarization ray tracing matrix of the interferometer. But applying Equation 9.4 to the propagation vector $\hat{\mathbf{k}}_{in}$ yields

$$\mathbf{P} \hat{\mathbf{k}}_{in} = (\mathbf{P}_A + \mathbf{P}_B) \hat{\mathbf{k}}_{in} = 2\hat{\mathbf{k}}_{out}, \tag{9.71}$$

which is not what is desired, since $\mathbf{P} \hat{\mathbf{k}}_{in}$ should be $\hat{\mathbf{k}}_{out}$. For combining parallel beams during ray tracing through interferometers or *birefringent filters*, a change can be made in the polarization ray tracing matrix definition, to avoid the issue of doubling the $\mathbf{k}$ vector, by using the alternative definition, $\breve{\mathbf{P}}$, from Equation 9.68. The difference between $\mathbf{P}$ and $\breve{\mathbf{P}}$ is the dyad $\mathbf{D}$ formed by the outer product between $\hat{\mathbf{k}}_{in}$ and $\hat{\mathbf{k}}_{out}$,

$$\mathbf{D} = \begin{pmatrix} \hat{k}_{x,in}\hat{k}_{x,out} & \hat{k}_{y,in}\hat{k}_{x,out} & \hat{k}_{z,in}\hat{k}_{x,out} \\ \hat{k}_{x,in}\hat{k}_{y,out} & \hat{k}_{y,in}\hat{k}_{y,out} & \hat{k}_{z,in}\hat{k}_{y,out} \\ \hat{k}_{x,in}\hat{k}_{z,out} & \hat{k}_{y,in}\hat{k}_{z,out} & \hat{k}_{z,in}\hat{k}_{z,out} \end{pmatrix},$$

$$\mathbf{D}\hat{\mathbf{k}}_{in} = \begin{pmatrix} \hat{k}_{x,in}\hat{k}_{x,out} & \hat{k}_{y,in}\hat{k}_{x,out} & \hat{k}_{z,in}\hat{k}_{x,out} \\ \hat{k}_{x,in}\hat{k}_{y,out} & \hat{k}_{y,in}\hat{k}_{y,out} & \hat{k}_{z,in}\hat{k}_{y,out} \\ \hat{k}_{x,in}\hat{k}_{z,out} & \hat{k}_{y,in}\hat{k}_{z,out} & \hat{k}_{z,in}\hat{k}_{z,out} \end{pmatrix} \begin{pmatrix} \hat{k}_{x,in} \\ \hat{k}_{y,in} \\ \hat{k}_{z,in} \end{pmatrix} = \begin{pmatrix} \hat{k}_{x,out} \\ \hat{k}_{y,out} \\ \hat{k}_{z,out} \end{pmatrix} = \hat{\mathbf{k}}_{out}.$$

(9.72)

Thus, the equation

$$\check{\mathbf{P}} = \mathbf{P} - \mathbf{D} \tag{9.73}$$

allows for easy transformation between the polarization ray tracing matrix, which is convenient for multiplying $\mathbf{P}$, and the form $\check{\mathbf{P}}$, which is useful for addition.

9.6.1 Combining P Matrices for the Interferometer Example

The $\mathbf{P}$ matrix of the interferometer system in Section 9.5 can be calculated using the $\check{\mathbf{P}}$ for each path. $\mathbf{P}_{test}$ and $\mathbf{P}_{ref}$ are constructed using Equation 9.4. One beam enters the interferometer and splits into two beams, and both exit the system along the same direction. This system is a good candidate to add $\mathbf{P}_{test}$ and $\mathbf{P}_{ref}$ to calculate the combined $\mathbf{P}$ matrix using $\check{\mathbf{P}}$. The dyad matrix for $\hat{\mathbf{k}}_{in} = \hat{\mathbf{z}}$ and $\hat{\mathbf{k}}_{out} = \hat{\mathbf{x}}$ is

$$\mathbf{D} = \begin{pmatrix} 0 & 0 & 1 \\ 0 & 0 & 0 \\ 0 & 0 & 0 \end{pmatrix}. \tag{9.74}$$

Then, $\check{\mathbf{P}}$ can be calculated for both paths,

$$\check{\mathbf{P}}_{test} = \mathbf{P}_{test} - \mathbf{D} = \begin{pmatrix} 0 & 0 & 0 \\ \dfrac{i}{2\sqrt{2}} & \dfrac{i}{2\sqrt{2}} & 0 \\ \dfrac{i}{2\sqrt{2}} & \dfrac{i}{2\sqrt{2}} & 0 \end{pmatrix}, \check{\mathbf{P}}_{ref} = \mathbf{P}_{ref} - \mathbf{D} = \begin{pmatrix} 0 & 0 & 0 \\ \dfrac{i}{2\sqrt{2}} & \dfrac{-i}{2\sqrt{2}} & 0 \\ \dfrac{i}{2\sqrt{2}} & \dfrac{-i}{2\sqrt{2}} & 0 \end{pmatrix}. \tag{9.75}$$

The combined $\mathbf{P}$ matrix for the system $\mathbf{P}_{combined}$ is

9.7 Example—A Hollow Corner Cube

$$\mathbf{P}_{combined} = \breve{\mathbf{P}}_{test} + \breve{\mathbf{P}}_{ref} + \mathbf{D} = \begin{pmatrix} 0 & 0 & 1 \\ \dfrac{i}{\sqrt{2}} & 0 & 0 \\ \dfrac{i}{\sqrt{2}} & 0 & 0 \end{pmatrix}. \tag{9.76}$$

Using $\mathbf{P}_{combined}$, $\mathbf{E}_{out}$ can be calculated directly and provides the same result as adding $\mathbf{E}_{test}$ and $\mathbf{E}_{ref}$, which are calculated from $\mathbf{P}_{test}$ and $\mathbf{P}_{ref}$, respectively,

$$\mathbf{E}_{out} = \mathbf{P}_{combined} \cdot \mathbf{E}_{in} = \mathbf{P}_{combined} \cdot \begin{pmatrix} E_x \\ E_y \\ 0 \end{pmatrix} = \dfrac{iE_x}{\sqrt{2}} \begin{pmatrix} 0 \\ 1 \\ 1 \end{pmatrix} \tag{9.77}$$

$$= \mathbf{P}_{test} \cdot \mathbf{E}_{test} + \mathbf{P}_{ref} \cdot \mathbf{E}_{ref}.$$

9.7 Example—A Hollow Corner Cube

Corner cubes are commonly used as *retroreflectors*, and their polarization properties are well studied.[22–24] A hollow aluminum-coated corner cube provides an example of an inhomogeneous polarization component, an element in which the diattenuation and retardance are not aligned. This example focuses on calculating the **P** matrix of the corner cube and diattenuation of the system.

Figure 9.7 shows the hollow corner cube consisting of three mutually perpendicular aluminum mirrors. The entrance face is intercept number 1 and the three mutually perpendicular reflecting surfaces are intercepts 2, 3, and 4. A refractive index of $0.77 + 6.06i$ is assumed for aluminum at 500 nm.

A collimated beam of incident light can take six different ray paths depending on its entry location.[25] Figure 9.8 shows one of the ray paths with propagation vectors marked with black arrows. The incident and the exiting propagation vectors are anti-parallel and aligned with the z-axis.

Figure 9.9 shows the corner cube with the propagation vectors in black, *s*-local coordinate vectors in solid red, and *p*-local coordinate vectors in dashed blue in three different views. The left figure shows how the local coordinate bases (*s*, *p*) change as the ray propagates from the same viewpoint as in Figure 9.8. The middle and right figures are rotation about the *y*-axis.

Each reflecting surface of the corner cube is specified by its normal $\hat{\mathbf{\eta}}_q$. $\hat{\mathbf{k}}_{q-1}$ determines the local coordinates for each intercept q and $\mathbf{P}_q$ is uniquely defined. Table 9.2 summarizes this ray's propagation vectors, local coordinates, and **P** matrix.

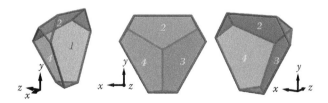

Figure 9.7 A corner cube retroreflector (CCR) with surfaces labeled from 1 for the front surface to 4.

Figure 9.8 Top view of a single ray path through a hollow aluminum-coated CCR. The three reflecting surfaces of the corner cube are mutually perpendicular.

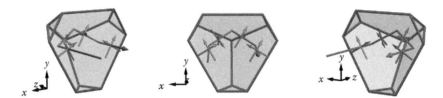

Figure 9.9 Three different views of the corner cube showing how local coordinate vectors (*s*-vectors in solid red, *p*-vectors in blue, and propagation vectors in black) change as the ray propagates through the corner cube.

The net polarization ray tracing matrix $\mathbf{P}_{cc}$ (*cc* for corner cube) for this ray path is calculated by cascading the three $\mathbf{P}$ matrices in Table 9.2,

$$\mathbf{P}_{cc} = \mathbf{P}_3\,\mathbf{P}_2\,\mathbf{P}_1 = \begin{pmatrix} 0.39+0.78i & 0.01+0.02i & 0 \\ -0.02i & 0.40+0.78i & 0 \\ 0 & 0 & -1 \end{pmatrix}. \tag{9.78}$$

The SVD of $\mathbf{P}_{cc}$ gives

$$\mathbf{U}_{cc} = \begin{pmatrix} 0 & 0.63+0.15i & 0.74+0.17i \\ 0 & 0.37-0.66i & -0.32+0.57i \\ 1 & 0 & 0 \end{pmatrix},$$

$$\mathbf{D}_{cc} = \begin{pmatrix} 1 & 0 & 0 \\ 0 & 0.88 & 0 \\ 0 & 0 & 0.87 \end{pmatrix}, \tag{9.79}$$

$$\mathbf{V}_{cc} = \begin{pmatrix} 0 & 0.43-0.49i & 0.47-0.6i \\ 0 & -0.41-0.64i & 0.38+0.52i \\ -1 & 0 & 0 \end{pmatrix}.$$

As shown in Equation 9.19, $\mathbf{V}_{cc}$ and $\mathbf{U}_{cc}$ have the incident and exiting propagation vectors as their first columns. Table 9.3 lists the maximum and minimum intensity transmittances assuming the incident electric field's intensity is one and the diattenuation of the corner cube is calculated from the singular values of the $\mathbf{P}_{cc}$ matrix.

The last two columns of $\mathbf{V}_{cc}$ and $\mathbf{U}_{cc}$ represent two incident polarization states ($\mathbf{v}_1$, $\mathbf{v}_2$) and exiting states ($\mathbf{u}_1$, $\mathbf{u}_2$) with the maximum and the minimum intensity transmittances. $\mathbf{v}_1$, $\mathbf{v}_2$ and $\mathbf{u}_1$, $\mathbf{u}_2$ are

9.7 Example—A Hollow Corner Cube

Table 9.2 Propagation Vectors, Local Coordinate Basis Vectors, Surface Normal Vectors, and Polarization Ray Tracing Matrices Associated with a Ray Path through an Aluminum-Coated Hollow Corner Cube

q	$\hat{\mathbf{k}}_{q-1}$	$\hat{\mathbf{k}}_q$	$\hat{\mathbf{p}}_q$	$\hat{\mathbf{p}}'_q$	$\hat{\mathbf{s}}_q$	$\hat{\boldsymbol{\eta}}_q$	$\mathbf{P}_q$
1	$\begin{pmatrix} 0 \\ 0 \\ -1 \end{pmatrix}$	$\begin{pmatrix} -\frac{2\sqrt{2}}{3} \\ 0 \\ -\frac{1}{3} \end{pmatrix}$	$\begin{pmatrix} -1 \\ 0 \\ 0 \end{pmatrix}$	$\begin{pmatrix} -\frac{1}{3} \\ 0 \\ \frac{2\sqrt{2}}{3} \end{pmatrix}$	$\begin{pmatrix} 0 \\ -1 \\ 0 \end{pmatrix}$	$\begin{pmatrix} \sqrt{\frac{2}{3}} \\ 0 \\ -\frac{1}{\sqrt{3}} \end{pmatrix}$	$\begin{pmatrix} 0.26+0.16i & 0 & 0.94 \\ 0 & -0.96-0.18i & 0 \\ -0.75-0.46i & 0 & 0.33 \end{pmatrix}$
2	$\begin{pmatrix} -\frac{2\sqrt{2}}{3} \\ 0 \\ -\frac{1}{3} \end{pmatrix}$	$\begin{pmatrix} -\frac{\sqrt{2}}{3} \\ \sqrt{\frac{2}{3}} \\ \frac{1}{3} \end{pmatrix}$	$\begin{pmatrix} -\frac{1}{6} \\ \frac{\sqrt{3}}{2} \\ \frac{\sqrt{2}}{3} \end{pmatrix}$	$\begin{pmatrix} \frac{5}{6} \\ \frac{1}{2\sqrt{3}} \\ \frac{\sqrt{2}}{3} \end{pmatrix}$	$\begin{pmatrix} -\frac{1}{2\sqrt{3}} \\ -\frac{1}{2} \\ \sqrt{\frac{2}{3}} \end{pmatrix}$	$\begin{pmatrix} -\frac{1}{\sqrt{6}} \\ -\frac{1}{\sqrt{2}} \\ -\frac{1}{\sqrt{3}} \end{pmatrix}$	$\begin{pmatrix} 0.25-0.08i & 0.44+0.33i & 0.7+0.24i \\ -0.95-0.05i & -0.04+0.08i & 0.23+0.14i \\ -0.15 & 0.72+0.27i & -0.57-0.01i \end{pmatrix}$
3	$\begin{pmatrix} -\frac{\sqrt{2}}{3} \\ \sqrt{\frac{2}{3}} \\ \frac{1}{3} \end{pmatrix}$	$\begin{pmatrix} 0 \\ 0 \\ 1 \end{pmatrix}$	$\begin{pmatrix} \frac{1}{6} \\ -\frac{1}{2\sqrt{3}} \\ \frac{2\sqrt{2}}{3} \end{pmatrix}$	$\begin{pmatrix} \frac{1}{2} \\ -\frac{\sqrt{3}}{2} \\ 0 \end{pmatrix}$	$\begin{pmatrix} -\frac{\sqrt{3}}{2} \\ -\frac{1}{2} \\ 0 \end{pmatrix}$	$\begin{pmatrix} -\frac{1}{\sqrt{6}} \\ \frac{1}{\sqrt{2}} \\ -\frac{1}{\sqrt{3}} \end{pmatrix}$	$\begin{pmatrix} -0.65-0.09i & -0.53-0.15i & 0.37+0.23i \\ -0.53-0.15i & -0.04+0.08i & -0.65-0.4i \\ -0.47 & 0.82 & 0.33 \end{pmatrix}$

Table 9.3 The Maximum Intensity of Output and Associated Incident Electric Field, the Minimum Intensity of Transmitted Electric Field and Associated Incident Electric Field, and the Diattenuation for the Ray through the Corner Cube System

I_{max}		$\mathbf{v}_1$		I_{min}		$\mathbf{v}_2$		$\mathcal{D}$
0.774	$\begin{pmatrix} 0.65e^{-0.85i} \\ 0.76e^{-2.15i} \\ 0 \end{pmatrix} = e^{-2.15i}$	$\begin{pmatrix} 0.65e^{1.30i} \\ 0.76 \\ 0 \end{pmatrix}$		0.757	$\begin{pmatrix} 0.76e^{-0.91i} \\ 0.65e^{0.94i} \\ 0 \end{pmatrix} = e^{-0.91i}$	$\begin{pmatrix} 0.76 \\ 0.65e^{1.85i} \\ 0 \end{pmatrix}$		0.014

elliptically polarized; hence, this path through the corner cube acts as a weak elliptical diattenuator with a diattenuation of 0.014. $\mathbf{v}_1$ and $\mathbf{v}_2$ are the only pair of orthogonal incident polarization states that remain orthogonal upon exit. Figure 9.10 shows the polarization states associated with the maximum and the minimum intensity transmittances represented in local coordinate systems, $(\hat{\mathbf{s}}_0, \hat{\mathbf{p}}_0, \hat{\mathbf{k}}_0)$ and $(\hat{\mathbf{s}}_3, \hat{\mathbf{p}}_3, \hat{\mathbf{k}}_3)$, so that each ellipse's propagation vector is coming out of the page. The red ellipses in Figure 9.10 (left) are the exiting (i.e., $\mathbf{u}_1$) polarization state (shown on the left), which has maximum transmittance, and the corresponding incident (i.e., $\mathbf{v}_1$) polarization state (shown on the right). Note that the handedness of the $\mathbf{v}_1$ and $\mathbf{u}_1$ are opposite to each other; after an odd number of reflections, the left-handed incident polarization state exits right handed since the handedness of the polarization state is defined in local coordinate systems. Similarly, the blue ellipses in Figure 9.10 (right) are the exiting (i.e., $\mathbf{u}_2$) polarization state, which has minimum transmittance, and the corresponding incident (i.e., $\mathbf{v}_2$) polarization state. The handedness of $\mathbf{v}_2$ and $\mathbf{u}_2$ is also opposite to each other.

The use of local coordinate systems to describe Jones vectors with opposite propagation directions would complicate the discussion of these polarization states and transformations. In global coordinates, as shown in Figure 9.11, the direction of rotation of the electric field is in the same

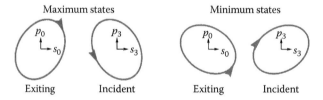

Figure 9.10 The polarization ellipses represented in local coordinate. (Left) The exiting state from the corner cube with the maximum intensity and corresponding incident state are shown in red. (Right) The exiting state with the minimum intensity and corresponding incident state are shown in blue.

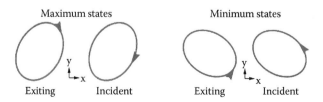

Figure 9.11 The polarization ellipses looking into the corner cube. (Left) The exiting state from the corner cube with the maximum intensity and corresponding incident state are shown in red. (Right) The exiting state with the minimum intensity and corresponding incident state are shown in blue.

direction for the corresponding incident and exiting states; $\mathbf{v}_1$ and $\mathbf{u}_1$ have the same handedness and $\mathbf{v}_2$ and $\mathbf{u}_2$ have the same handedness due to the anti-parallel propagation vectors, $\hat{\mathbf{k}}_0$ and $\hat{\mathbf{k}}_3$. All states in Figure 9.11 are represented in global coordinates (x–y plane) looking into the corner cube; since $\hat{\mathbf{k}}_3 = \hat{\mathbf{z}}$, the exiting electric fields are coming out of the page while the incident electric fields are going into the page.

All states in Figure 9.11 are represented in an x–y plane looking into the corner cube; since $\hat{\mathbf{k}}_3 = \hat{\mathbf{z}}$ and $\hat{\mathbf{k}}_0$ and $\hat{\mathbf{k}}_3$ are anti-parallel, the exiting polarization ellipses are coming out of the page while the incident polarization ellipses are going into the page.

9.8 Conclusion

The three-by-three polarization ray tracing matrices enable ray tracing in global coordinates, which provide an easy basis to interpret polarization properties for optical systems where rays are constantly changing directions. The interferometer in Section 9.5 and the corner cube in Section 9.7 highlight how global coordinates provide a straightforward basis. Using local coordinates, different analysts may make different choices, complicating the interpretation of complex geometries. It remains straightforward to convert results from global coordinates into other interesting local coordinate bases.

A formalism for polarization ray tracing using three-by-three matrices has been developed and the relationship to the Jones calculus has been shown. Algorithms for reflection, refraction, and polarization elements are summarized with specific examples.

If the optical system includes anisotropic or birefringent media, the propagation portions take the form of retarder matrices for birefringent media and/or diattenuation matrices for dichroic media. Denoting the propagation effect from ray interface q to $q+1$ as $\mathbf{A}_{q+1,q}$, the polarization ray tracing matrix for a ray through an optical system with anisotropic media (i.e., Equation 9.2) becomes

$$\mathbf{P}_{Total} = \mathbf{P}_N \mathbf{A}_{N,N-1} \mathbf{P}_{N-1} \cdots \mathbf{A}_{3,2} \mathbf{P}_2 \mathbf{A}_{2,1} \mathbf{P}_1 = \mathbf{P}_N \prod_{q=N-1,-1}^{1} \mathbf{A}_{q+1,q} \mathbf{P}_q. \tag{9.80}$$

This formulation works well for stress birefringent and weakly anisotropic materials. In strongly birefringent materials like calcite and rutile, birefraction between the two modes (ordinary and extraordinary) causes ray doubling. In this case, each of the separated rays refracting into a birefringent material needs a separate polarization ray tracing matrix. Since the modes in these birefringent media have a single polarization, the ray intercept $\mathbf{P}$s have the form of a polarizer. This polarizer matrix selects the incident state that couples into the specified mode. Further comments on ray tracing in anisotropic materials are beyond the scope of the present chapter and are covered in Chapter 19 (Birefringent Ray Trace).

The calculation of the diattenuation of a $\mathbf{P}$ matrix is achieved via the SVD. The maximum and minimum transmittances are related to singular values of the $\mathbf{P}$ matrix. The two unitary matrices of the SVD provide two canonical entering and exiting polarization states that are orthogonal to each other and related by the singular values. The incident propagation vector and the exiting propagation vector are related by $\mathbf{P}$ due to the constraint $\mathbf{P}\hat{\mathbf{k}}_{q-1} = \hat{\mathbf{k}}_q$ applied to the definition of $\mathbf{P}$. $\mathbf{P}$ and a propagation vector dyad are introduced to simplify the addition of $\mathbf{P}$ matrices for parallel rays.

A step-by-step example of performing ray tracing with the $\mathbf{P}$ matrix is illustrated using an interferometer system with a PBS. Finally, a numeric example of ray tracing through a hollow aluminum corner cube was presented.

9.9 Problem Sets

9.1 Consider a unit vector $\mathbf{V} = (v_x, v_y, v_z)$.
   a. Find a unit vector orthogonal to $\mathbf{V}$. Is it unique?
   b. Find the matrix $\mathbf{Dy}$ that maps $\mathbf{V}$ into $\mathbf{V}$, and maps all vectors orthogonal to $\mathbf{V}$ to zero vectors. Show that your matrix works properly. This type of projection matrix is called a dyad.

9.2 Given two orthonormal bases of three vectors, $(\mathbf{A}, \mathbf{B}, \mathbf{C})$ and $(\mathbf{E}, \mathbf{F}, \mathbf{G})$, define a 3 × 3 unit dyad, $\mathbf{H}$, which maps unit vector $\mathbf{A}$ into $\mathbf{E}$ while mapping the two orthogonal components $\mathbf{B}$ and $\mathbf{C}$ to zero,

$$\mathbf{H} \cdot \begin{pmatrix} A_x & B_x & C_x \\ A_y & B_y & C_y \\ A_z & B_z & C_z \end{pmatrix} = \begin{pmatrix} E_x & 0 & 0 \\ E_y & 0 & 0 \\ E_z & 0 & 0 \end{pmatrix}.$$

Find the following dyads:
   a. Maps (0, 1, 0) into (0, 1, 0).
   b. Maps (0, 1, 0) into (1, 0, 0).
   c. Maps (0, 0, 1) into (1, 0, 0).
   d. Maps (0, 0, 1) into (0, 0, −1).
   e. Maps (0, 0, 1) into (α, β, γ).
   f. Maps (α, β, γ) into (α, β, γ).
   g. Maps (α, β, γ) into $(\delta, \varepsilon, \zeta)$.
   h. What are the eigenvalues of a unit dyad?
   i. What is the determinant of a dyad?
   j. Demonstrate that the rows of (g) are linearly dependent.
   k. Are dyad matrices unitary or Hermitian?

9.3 Light is propagating with the propagation vector $\mathbf{k}_1 = (-2, -3, 3)/\sqrt{2^2 + 3^2 + 3^2}$.
   a. Create the dyad matrix $\mathbf{K}_1$, for $\mathbf{k}_1$. Find two basis vectors for the transverse plane by the following procedure. Pick two simple but arbitrary vectors, $\mathbf{v}_1$ and $\mathbf{v}_2$, in different directions.
   b. Form the x-basis vector $\mathbf{x}_1$ by subtracting $\mathbf{k}_1 \cdot \mathbf{v}_1$ from $\mathbf{v}_1$, to get an orthogonal vector, and normalize. Verify whether $\mathbf{x}_1$ is orthogonal to $\mathbf{k}_1$ and normalized.
   c. Form $\mathbf{X}_1$, the dyad matrix for $\mathbf{x}_1$.
   d. Draw pictures of the various vectors in the $(\mathbf{k}_1, \mathbf{v}_1)$ plane and explain the mathematical operation.
   e. Form the y-basis vector $\mathbf{y}_1$ by subtracting $\mathbf{K}_1 \cdot \mathbf{v}_2$ and $\mathbf{X}_1 \cdot \mathbf{v}_2$ from $\mathbf{v}_2$, to get an orthogonal vector, and normalizing. Verify whether $\mathbf{y}_1$ is orthogonal to $\mathbf{k}_1$. This is an example of the Gram–Schmidt orthonormalization algorithm in three dimensions.
   f. Is the orthonormal basis $(\mathbf{k}_1, \mathbf{x}_1, \mathbf{y}_1)$ left handed or right handed?
   g. Write a polarization vector $\mathbf{E}_1$ for right circularly polarized light propagating along $\mathbf{K}_1$.

9.4 Orthogonal matrices
   a. Find the orthogonal matrix $\mathbf{O}_1$ to rotate $\mathbf{x}$ into $\mathbf{u}$, $\mathbf{y}$ into $\mathbf{v}$, $\mathbf{z}$ into $\mathbf{w}$, where $\mathbf{x} = (1, 0, 0)$, $\mathbf{y} = (0, 1, 0)$, $\mathbf{z} = (0, 0, 1)$, $\mathbf{u} = \left(\frac{3}{5}, \frac{4}{5}, 0\right)$, $\mathbf{v} = (0, 0, 1)$, $\mathbf{w} = \left(\frac{4}{5}, -\frac{3}{5}, 0\right)$.

9.9 Problem Sets

b. Find the orthogonal matrix $\mathbf{O}_2$ to rotate $\mathbf{r}$ into $\mathbf{x}$, $\mathbf{s}$ into $\mathbf{y}$, $\mathbf{t}$ into $\mathbf{z}$, where $\mathbf{r} = (0, 12, 5)/13$, $\mathbf{s} = (1, 0, 0)$, $\mathbf{t} = (0, 5, -12)/13$.

c. Find the orthogonal matrix $\mathbf{O}_3$ to rotate $\mathbf{r}$ into $\mathbf{u}$, $\mathbf{s}$ into $\mathbf{v}$, $\mathbf{t}$ into $\mathbf{w}$.

9.5 A flat glass plate with normal $\boldsymbol{\eta} = (\sin(\pi/6), 0, \cos(\pi/6))$ has an incident beam propagating with $\mathbf{k}_{in} = (0, \sin(\pi/6), \cos(\pi/6))$. The refractive light has propagation vector

$$\mathbf{k}_1 = \frac{1}{\sqrt{2}}\left(\frac{1}{4}, \frac{1}{2}, \frac{3\sqrt{3}}{4}\right).$$

a. Find the **s**-basis vector.
b. Find the **p**-basis vector.
c. Find the orthogonal matrix to rotate the global $(\mathbf{x}, \mathbf{y}, \mathbf{z})$ coordinates into the local $(\mathbf{k}, \mathbf{s}, \mathbf{p})$.
d. Find the $\mathbf{s}'$ and $\mathbf{p}'$ basis vectors after refraction.
e. Find the orthogonal matrix to rotate the global $(\mathbf{x}, \mathbf{y}, \mathbf{z})$ to local $(\mathbf{k}, \mathbf{s}', \mathbf{p}')$ coordinates.
f. Find the angle of incidence and angle of refraction.
g. If the incident refractive index is 1, find the index of refraction of the medium.

Refraction at this interface has the Jones matrix $\mathbf{J} = \begin{pmatrix} 3/4 & 0 \\ 0 & 6\sqrt{2}/11 \end{pmatrix}$.

h. Find the associated polarization ray tracing matrix $\mathbf{P}$.
i. What is the associated singular value decomposition?

9.6
a. What are the three singular values for the polarization ray tracing matrix of an ideal polarizer?
b. What are the three singular values for the polarization ray tracing matrix of an ideal retarder?

9.7 Find the matrix inverse $\mathbf{P}^{-1}$ of the singular value decomposition in terms of $\mathbf{U}$, $\mathbf{D}$, and $\mathbf{V}$ where $\mathbf{P} = \mathbf{U}\mathbf{D}\mathbf{V}^\dagger$.

9.8 Given the following singular value decomposition,

$$\begin{pmatrix} 0 & 0 & 1 \\ \frac{1}{\sqrt{2}} & \frac{1}{\sqrt{2}} & 0 \\ \frac{i}{\sqrt{2}} & \frac{-i}{\sqrt{2}} & 0 \end{pmatrix} \begin{pmatrix} \frac{3}{8} & 0 & 0 \\ 0 & \frac{1}{8} & 0 \\ 0 & 0 & 1 \end{pmatrix} \begin{pmatrix} 0 & \frac{1}{\sqrt{2}} & \frac{-i}{\sqrt{2}} \\ 0 & \frac{1}{\sqrt{2}} & \frac{i}{\sqrt{2}} \\ 1 & 0 & 0 \end{pmatrix}.$$

a. Find the diattenuation.
b. Along which axes does the light enter and exit?
c. Is this a linear, elliptical, or circular element?

9.9 In order to calculate reflected and/or refracted polarization state at an intercept, we often decompose the polarization state into s- and p-polarizations. When a ray enters the intercept at 0 angle of incidence, that is, normal incident to the surface, there is no difference between s- and p-polarizations. In this problem, we calculate Jones matrices and polarization ray tracing ($\mathbf{P}$) matrices to describe reflection at normal incidence and compare the results.

a. Consider a ray propagating in air ($n_1 = 1$) along $\mathbf{k}_1 = (0, 0, 1)$ incident on a glass surface with index $n_1 = 1.52$ and a surface normal $\boldsymbol{\eta} = (0, 0, 1)$. Write a Jones matrix for reflection using Fresnel reflection coefficients, $\begin{pmatrix} r_s & 0 \\ 0 & r_p \end{pmatrix}$.

b. Using the Jones matrix calculated in part (a), calculate the reflected Jones vector $\mathbf{E}_{out}$ for a right circularly polarized incident Jones vector $\mathbf{E}_{in} = (1, -i)/\sqrt{2}$.

c. What polarization state is $\mathbf{E}_{out}$ in?

d. Now, calculate the P matrix for the same ray.

e. Using the P matrix in part (d), calculate the polarization vector $\mathbf{E}$ for right circularly polarized incident light, vector $(1, -i, 0)/\sqrt{2}$, reflecting at normal incidence.

f. Compare the results from parts (c) and (e). Which polarization state is the reflected electric field vector in? Are (c) and (e) in conflict? Explain.

9.10 For light propagating along the z-axis, a polarization element has the Jones matrix $\mathbf{J} = \begin{pmatrix} j_{11} & j_{12} \\ j_{21} & j_{22} \end{pmatrix}$. Now, the element is placed in a system where light is propagating along $\mathbf{k} = (1, 1, 1)/\sqrt{3}$ and the x-axis has been moved to $\mathbf{x}_1 = (0, -1, 1)/\sqrt{2}$.

a. Find the polarization ray tracing matrix **P** as a function of the Jones matrix elements.

b. If **J** is a quarter wave linear retarder with the fast axis at 45°, find **P**.

c. Which incident polarization vector $\mathbf{E}_1$ will yield left circularly polarized output with flux 16?

d. Find the singular value decomposition of $\mathbf{P} = \mathbf{VDW}^\dagger$.

9.11 Light propagating in the direction $\mathbf{k}_0 = (1, 0, 2)/\sqrt{5}$ is incident on a wire grid polarizer with normal $\boldsymbol{\eta}_0 = (1, \sqrt{4}, 1)/\sqrt{6}$. All of the p-polarized light is transmitted through the polarizer and exits in the same direction, $\mathbf{k}_0$.

a. Find the polarization ray tracing matrix $\mathbf{P}_t$ for the polarizer in transmission.

b. Find the reflected propagation vector $\mathbf{k}_r$.

c. If all of the s-polarized light is reflected, find the matrix for reflection $\mathbf{P}_r$.

d. The wire grid polarizer is replaced with another wire grid polarizer, which reflects 0.8 of the s-polarized light amplitude and delays the phase by $\pi/3$. It reflects 0.1 of the p-polarized light and delays the phase by $\pi/2$. Find the polarization ray tracing matrix for reflection.

9.12 A reflection linear quarter wave retarder has a surface normal $\boldsymbol{\eta}_1 = (0, 0, 1)$. It advances the phase of the horizontal light (fast axis is **x**) by $\pi/4$ and retards the orthogonal state by $\pi/4$. The incident ray has a propagation vector $\mathbf{k}_0 = (0, \sin 30°, \cos 30°)$. Calculate the 3D polarization ray tracing matrix.

a. Calculate $(\mathbf{s}_1, \mathbf{p}_1, \mathbf{k}_0)$ and $(\mathbf{s}'_1, \mathbf{p}'_1, \mathbf{k}'_0)$.

b. Calculate **P** matrix using a Jones matrix for the horizontal fast axis linear quarter wave retarder, $\mathbf{J} = \begin{pmatrix} e^{-i\pi/4} & 0 \\ 0 & e^{i\pi/4} \end{pmatrix}$.

c. When right circularly polarized light is incident, what is the orientation of the major axis in three dimensions of the reflected polarization state?

9.13 Light with a propagation vector $\mathbf{k} = (1, 1, 0)/\sqrt{2}$ is normally incident on the front face of a Wollaston prism. At the hypotenuse with normal $\boldsymbol{\eta} = (\cos 55°, \sin 55°, 0)$, the s-component of the light, which is polarized in the z-direction, is refracted and exits the

9.9 Problem Sets

cube propagating in the direction $\mathbf{k}_\alpha = (\cos 48°, \sin 48°, 0)$. The orthogonal component of the incident light, the **p**-component at the interface, is refracted in the opposite direction and exits the cube in the direction $\mathbf{k}_\beta = (\cos 42°, \sin 42°, 0)$. Assume both beams lose 9% of their incident flux, 4% entering the cube, 1% reflection loss at the hypotenuse, and 4% exiting into air. Our equations will analyze just the polarization ray tracing matrices between the incident and exiting beams in air. We will not write **P** matrices at the hypotenuse.

a. Find the **s**-, **p**-, and **k**-components of the incident light, in air.
b. Find the matrix for the amplitude transmissions of the Wollaston polarizer for beam α, including the polarizing effect of the interface and the end-to-end losses. The Fresnel coefficients for the beams at the interfaces are as $(as_\alpha, ap_\alpha, as_\beta, ap_\beta)$.
c. Find the orthogonal transformation for beam α from global to local coordinates

$$\mathbf{O}_\alpha^{-1} = \begin{pmatrix} s_x & s_y & s_z \\ p_x & p_y & p_z \\ k_x & k_y & k_z \end{pmatrix}.$$

d. Find the orthogonal transformation for beam α from local to global coordinates

$$\mathbf{O}_\alpha = \begin{pmatrix} s_x & p'_x & k\alpha_x \\ s_y & p'_y & k\alpha_y \\ s_z & p'_z & k\alpha_z \end{pmatrix}.$$

e. Find the polarization ray tracing matrix for the α polarizer path.
f. When left circularly polarized light is incident, with $\mathbf{EL} = \left(\frac{1}{2} - \frac{i}{2}, \frac{-1}{2} + \frac{i}{2}, \frac{1+i}{\sqrt{2}} \right)$, find the **E** vector for the exiting light. What is the exiting flux and polarization state?
g. Find the polarization ray tracing matrix for the β polarizer path.

9.14 Light propagating along the y-axis encounters a roof mirror with mirror normals $\boldsymbol{\eta}_1$ and $\boldsymbol{\eta}_2$. $\mathbf{k}_0 = (0,1,0)$, $\boldsymbol{\eta}_1 = (1,1,0)/\sqrt{2}$, $\boldsymbol{\eta}_2 = (-1,1,0)/\sqrt{2}$. One half of the beam reflects from mirror 1 and mirror 2 and exits along $\mathbf{k}_2 = (0, -1, 0)$.
a. Assuming Fresnel reflection coefficients *rs* and *rp* for each mirror, find the polarization ray tracing matrices for the individual reflections, $\mathbf{P}_1$ and $\mathbf{P}_2$, and for the entire path **P**. Keep *rs* and *rp* as variables.
b. The other half of the beam strikes mirror 2 before mirror 1 and exits in the same direction. Find the polarization ray tracing matrices for the individual reflections and for the entire path.
c. Find the Jones vector, which will emerge as right circularly polarized light.

9.15 A linear polarizer with a transmission axis at 45° has a surface normal $\boldsymbol{\eta}_1 = (0,0,1)$. The incident ray has a propagation vector $\mathbf{k}_0 = (0, \sin 30°, \cos 30°)$. The polarizer does not deviate the propagation vector for the transmitted ray.
a. Using $\boldsymbol{\eta}_1$ and $\mathbf{k}_0$, calculate $(\mathbf{s}_1, \mathbf{p}_1, \mathbf{k}_0)$ and $(\mathbf{s}'_1, \mathbf{p}'_1, \mathbf{k}'_0)$.
b. Generate the orthogonal matrices $\mathbf{O}_{in}$ and $\mathbf{O}_{out}$.
c. Calculate the **P** matrix using the Jones matrix for a linear polarizer at 45°.

9.16 An adjustable three-mirror reflector takes incident light along the z-axis, reflects it three times such that the light also exits along the z-axis. The mirrors are ideal reflectors, $rs = 1$, $rp = 1$. As the adjustment θ is performed, the propagation vectors after each mirror vary as follows:

$$\mathbf{k}_0 = (0, 1, 0),$$
$$\mathbf{k}_1 = (\sin\theta, 0, \cos\theta),$$
$$\mathbf{k}_2 = (0, \sin\theta, \cos\theta),$$
$$\mathbf{k}_3 = \mathbf{k}_0.$$

a. For the case of $\theta = 45°$, find the first mirror's reflection basis vectors $\mathbf{s}_1, \mathbf{p}_1, \mathbf{p}'_1$. You can use the alternative expression for the **s**-basis vector $\mathbf{s}_q = \dfrac{\mathbf{k}_q \times \mathbf{k}_{q-1}}{|\mathbf{k}_q \times \mathbf{k}_{q-1}|}$.

b. Create the orthogonal transformation matrices $\mathbf{O}_{in,1}$ and $\mathbf{O}_{out,1}$ for $\theta = 45°$.

c. Calculate the polarization ray tracing matrix $\mathbf{P}_1$.

d. Let the incident light be *x*-polarized. Find the orientation of the exiting polarized light for $\theta = 45°$.

e. Create a program that calculates $\mathbf{P}_1, \mathbf{P}_2, \mathbf{P}_3$ and a total ray tracing matrix $\mathbf{P}$ when a value for θ is input.

f. For *x*-polarized incident light, plot the orientation of the exiting polarized light as a function of θ.

g. For what θ is the light approximately 45° polarized?

h. For what angle is the light *y*-polarized?

9.17 The polarization ray tracing matrix for a ray path is defined by the three matrix equations

$$\mathbf{E}_{a,1} = \mathbf{P} \cdot \mathbf{E}_{a,0}$$
$$\mathbf{E}_{b,1} = \mathbf{P} \cdot \mathbf{E}_{b,0}$$
$$\mathbf{k}_1 = \mathbf{P} \cdot \mathbf{k}_0,$$

where $\mathbf{E}_{a,0}$ and $\mathbf{E}_{b,0}$ are two of the columns of $\mathbf{V}$, which are attenuated by Λ_1 and Λ_2, respectively, when propagating along the ray path. Now, light is sent the reverse direction along the ray path, and the two states defined by the columns of $\mathbf{U}$ are attenuated by Λ_1 and Λ_2, respectively. Find the **PRT** matrix for the opposite direction $\breve{\mathbf{P}}$. Because the attenuation is the same in both directions, $\breve{\mathbf{P}}$ is not quite the matrix inverse of $\mathbf{P}$.

9.18 Find $\breve{\mathbf{P}}$ for the following $\mathbf{P}$. For each, find the corresponding incident and exiting propagation vectors, $\mathbf{k}_0$ and $\mathbf{k}_1$, and the diattenuation.

a. $\mathbf{P}_1 = \begin{pmatrix} 1 & 0 & 0 \\ 0 & 0.9 & 0 \\ 0 & 0 & 0.5 \end{pmatrix}$ b. $\mathbf{P}_2 = \begin{pmatrix} 0.5 & 0 & 0 \\ 0 & 0.9 & 0 \\ 0 & 0 & 1 \end{pmatrix}$ c. $\mathbf{P}_3 = \begin{pmatrix} 0 & 0 & 1 \\ 0 & 0.9 & 0 \\ 0.5 & 0 & 0 \end{pmatrix}$

d. $\mathbf{P}_4 = \begin{pmatrix} 0 & 0.9 & 0 \\ 0 & 0 & 1 \\ 0.5 & 0 & 0 \end{pmatrix}$ e. $\mathbf{P}_5 = \dfrac{1}{8}\begin{pmatrix} 7 & -1 & 0 \\ -1 & 7 & 0 \\ 0 & 0 & 4 \end{pmatrix}$ f. $\mathbf{P}_6 = \dfrac{1}{8}\begin{pmatrix} 1 & 7 & 0 \\ -7 & -1 & 0 \\ 0 & 0 & 4 \end{pmatrix}$

g. $\mathbf{P}_7 = \begin{pmatrix} 0.714 & 0.01 & -0.52 \\ -0.102 & 0.827 & 0.296 \\ 0.289 & -0.106 & 0.394 \end{pmatrix}$ h. $\mathbf{P}_8 = \begin{pmatrix} -0.67 & 0.183 & 0.052 \\ 0.01 & 0.036 & -0.822 \\ 0.19 & -0.719 & -0.311 \end{pmatrix}$

9.19 Consider the phase-shifting Twyman–Green interferometer, as shown below, using a PBS in conjunction with two linear quarter wave retarders to get the light through the system with minimal loss.
  a. Once the beams are recombined, an analyzer is used to get the fringes. What are the **P** matrices for the two paths before the 45° linear polarizer? What is the **P** matrix for the entire system? Be sure to account for unmatched phase between the arms.

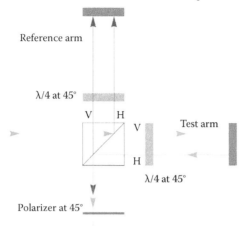

  b. Suppose one of the quarter waveplates is rotated (from 45°) by some small angle δ. What is the new **P** matrix for that arm? For the system? How will this affect the measured phase?
  c. Suppose the retardance of one of the waveplates is off by a small amount δ. What is the new **P** matrix for that arm? For the overall system? How will this affect the measured phase?

References

1. E. Waluschka, A polarization ray trace, *Opt. Eng.* 28 (1989): 86–89.
2. R. A. Chipman, Mechanics of polarization ray tracing, *Opt. Eng.* 34 (1995): 1636–1645.
3. R. A. Chipman, Polarization analysis of optical systems, *Opt. Eng.* 28 (1989): 90–99.
4. S. G. Krantz, The index or winding number of a curve about a point, in *Handbook of Complex Variables*, Boston, MA: Birkhäuser (1999), pp. 49–50.
5. P. Torok, P. Varga, Z. Laczik, and G. R. Booker, Electromagnetic diffraction of light focused through a planar interface between materials of mismatched refractive indices: An integral representation, *J. Opt. Soc. Am. A* 12 (1995): 325–332.
6. P. Torok and T. Wilson, Rigorous theory for axial resolution in confocal microscopes, *Opt. Commun.* 137 (1997): 127–135.
7. P. Torok, P. D. Higdon, and T. Wilson, Theory for confocal and conventional microscopes imaging small dielectric scatterers, *J. Mod. Opt.* 45 (1997): 1681–1698.
8. G. Yun, K. Crabtree, and R. A. Chipman, Three-dimensional polarization ray-tracing calculus I, definition and diattenuation, *Appl. Opt.* 50 (2011): 2855–2865.
9. G. Yun, S. McClain, and R. A. Chipman, Three-dimensional polarization ray-tracing calculus II, retardance, *Appl. Opt.* 50 (2011): 2866–2874.
10. R. A. Chipman, Challenges for polarization ray tracing, in *International Optical Design Conference*, OSA Technical Digest (CD), paper IWA4, OSA (2010).
11. W. S. T. Lam, S. McClain, G. Smith, and R. Chipman, Ray tracing in biaxial materials, in *International Optical Design Conference*, OSA Technical Digest (CD), paper IWA1, OSA (2010).
12. H. A. Macleod, *Thin-Film Optical Filters*, McGraw-Hill (1986), pp. 179–209.
13. P. H. Berning, Theory and calculations of optical thin films, *Phys. Thin Films* 1 (1963): 69–121.
14. M. Born and E. Wolf, *Principles of Optics*, Cambridge University Press (1999), pp. 63–74.

15. S. Lu and R. A. Chipman, Homogeneous and inhomogeneous Jones matrices, *J. Opt. Soc. Am. A* 11 (1994): 766–773.
16. R. M. A. Azzam and N. M. Bashara, *Ellipsometry and Polarized Light*, North Holland, (1977), pp. 67–84.
17. R. A. Chipman, Polarization Aberrations, PhD Dissertation, University of Arizona (1987).
18. S. Lu and R. A. Chipman, Interpretation of Mueller matrices based on polar decomposition, *J. Opt. Soc. Am. A* 13 (1996): 1106–1113.
19. R. Barakat, Conditions for the physical realizability of polarization matrices characterizing passive systems, *J. Mod. Opt.* 34 (1987): 1535–1544.
20. P. Lancaster and M. Tismenetsky, *The Theory of Matrices*, Academic (1985), pp. 1535–1544.
21. G. H. Gloub and C. F. Van Loan, *Matrix Computations*, Johns Hopkins University Press (1996), pp. 70–73.
22. E. R. Peck, Polarization properties of corner reflectors and cavities, *J. Opt. Soc. Am.* 52 (1962): 253–257.
23. M. A. Acharekar, Derivation of internal incidence angles and coordinate transformations between internal reflections for corner reflectors at normal incidence, *Opt. Eng.* 23 (1984): 669–674.
24. R. Kalibjian, Stokes polarization vector and Mueller matrix for a corner-cube reflector, *Opt. Commun.* 240 (2004): 39–68.
25. J. Liu and R. M. A. Azzam, Polarization properties of corner-cube retroreflectors: Theory and experiment, *Appl. Opt.* 36(7) (1997).

10

Optical Ray Tracing

10.1 Introduction

Ray tracing is the primary technique for calculating the path of light through optical systems. Conventional *geometrical ray tracing* calculates the *optical path length* of sets of rays representing *wavefronts* and simulates the image quality of imaging systems, for analyzing *illumination systems*, and for modeling the propagation of light in *scattering systems* such as the atmosphere or biological tissue. To include the effects of thin films, polarization elements, diffraction gratings, and liquid crystals in the simulation, the methods of geometrical ray tracing are generalized to *polarization ray tracing*. The optical path lengths of rays are still calculated, but polarization matrices are incorporated into the calculation to track changes to the amplitude and polarization.

The goals of this chapter are as follows:

- Introduce conventional geometrical ray tracing
- Provide algorithms for polarization ray tracing to track the change of polarization state as light propagates through optical systems
- Calculate polarization properties of ray paths by applying the *polarization ray tracing matrix*
- Determine and interpret the polarization properties of wavefronts with *the polarization aberration function* and the *Jones pupil*

Together with Chapter 16 (Image Formation with Polarization Aberration), Chapter 10 provides tools to understand "how polarized light interacts with optical systems."

Polarization elements and the polarization properties of optical elements like mirrors, beam splitters, and diffraction gratings can be described individually. But these elements work in concert in all types of optical systems such as cameras, spectrometers, polarimeters, telecommunications systems, and microlithography systems.

As polarization has increased in importance, the need to simulate more than reflection and refraction has placed a greater burden on the optical designer. For example, the coating engineer may understand the polarization properties of a coating, but doesn't have the analytic tools to calculate how light rays will interact with these coatings on dozens of lens and mirror surfaces. Similarly, a liquid crystal cell designer has specialized simulation tools for liquid crystals, but these tools don't provide an end-to-end optical system simulation. In contrast, optical designers have always had the tools and taken responsibility for the end-to-end simulation of optics with their optical design software.

As a result, optical design programs need to incorporate an increasing number of science modules (simulation routines) to calculate the physical effects of coatings, diffraction gratings, liquid crystals, stress birefringence, and more during the optical design process. This chapter starts with conventional optical systems, reflections and refractions, lenses, and mirrors. In later chapters, other classes of optical elements are introduced: polarizers, retarders, crystal elements, diffractive optical elements, liquid crystal cells, and elements with stress birefringence. The *polarization ray tracing calculus* becomes the tool for tying everything together for end-to-end simulation and becomes a systematic tool to understand how diattenuation and retardance effects interact with each other when these optical elements are cascaded.

10.2 Goals for Ray Tracing

The principal objective of ray tracing imaging systems is to characterize the exiting wavefronts and their properties. The shape of a wavefront exiting an imaging system relative to a spherical wavefront is the *wavefront aberration function*. Similarly, the relationship of the amplitude and polarization of the exiting wavefront to the entering wavefront is the *polarization aberration function*. These functions can be calculated by ray tracing a grid of rays through an optical system. Usually, a two-dimensional grid of optical path length values forms the wavefront aberration function. Similarly, a grid of polarization ray tracing matrices will comprise the polarization aberration function.

In most imaging systems, spherical waves enter and nearly spherical waves exit the system, forming images on an image surface. In geometrical ray tracing, the spherical wavefront emitted from a light source is divided into small patches; each patch is assigned a light ray. The light ray's propagation vector is aligned with the normal to the wavefront. Each of these rays propagates until the intersection of the ray with the first optical surface, called the *ray intercept*, is found. There, the ray *reflects* or *refracts*, using the law of reflection or *Snell's law*, and a new propagation vector is calculated. The intercept of this second ray segment is found at the second surface and the reflection or refraction process is repeated. At each surface, the ray intercept must be within the surface's aperture; otherwise, the ray is terminated if it is outside of the aperture or if it reaches the image plane. The image quality is determined from the ray trace. In this chapter, the ray tracing process is considered in six steps:

1. Properties of light rays
2. Specification of the optical system
3. Description of the light beams
4. Geometrical ray tracing algorithms
5. Polarization ray tracing
6. Analysis of aberrations and image quality

10.2 Goals for Ray Tracing

An example of a polarization ray trace analysis will highlight the analysis steps to prepare us for the path through the remainder of this chapter. A cell phone lens (U.S. Patent 7,535,658) will be used to describe each step of the optical system specification, geometrical and polarization ray tracing, and analysis.

Example 10.1 Geometrical and Polarization Ray Tracing of the Example Cell Phone Lens

Before discussing the details of ray tracing, the main polarization ray trace concepts are presented for the example cell phone lens system. Figure 10.1 shows a cross section through the center of the cell phone lens and the shapes of its aspheric lenses. The lens prescription is tabulated in Section 10.14. By ray tracing grids of rays through optical systems, the image quality and polarization characteristics of the lens are accurately assessed.

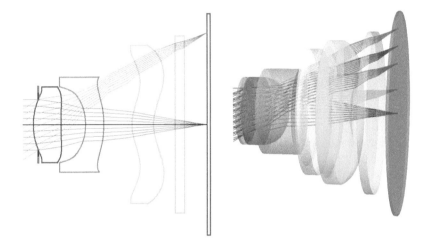

Figure 10.1 (Left) The y–z cross section of a cell phone lens system with aspheric lenses. The aperture stop and entrance pupil are just before the first lens surface. Rays from an on-axis field are red and those from an off-axis field are brown. (Right) A three-dimensional view of the example cell phone lens with five object field points. Rays from each object point are shown in a different color. Figures from Polaris-M.

Light enters the system from the left, refracts through four lenses and an infra-red (IR) blocking filter, and focuses on a CMOS image sensor on the right. In Figure 10.1 (left), light rays are shown from two distant object points: the on-axis field and an off-axis field. At each surface q, a surface equation $z_q(x, y)$, an aperture function $p_q(x, y, z)$, and the refractive index of the material following the surface, $n_q(\lambda)$ are defined. This provides the description of the optical system necessary for geometric ray tracing. For polarization ray tracing, the coatings for each surface must also be specified. In this case, the lens will be analyzed with uncoated interfaces. Often, when the coatings are not known, quarter wave coatings of MgF_2, a very common antireflection coating, are used as the default.

The *variation of the optical path length* (*OPL*) of rays from a single object point through the system describes how the wavefront exiting the optical system deviates from the desired spherical wavefront. Figure 10.2 (left) shows a bundle of rays from three field points propagating through the cell phone lens and focusing on the image surface. A perfect system would induce zero OPL variation relative to the spherical reference wavefront and have a perfect spherical exiting wavefront and a flat *OPL* variation across the pupil. In most imaging systems, the *OPL* variation is not flat. For example, Figure 10.2 (right) shows the variation of the *OPL* for the ray grid, which has an increasing optical path length along one axis while decreasing along the orthogonal axis, in an aberration pattern called *astigmatism*.

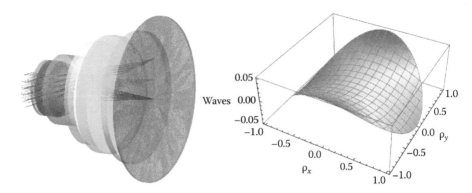

Figure 10.2 (Left) Three grids of rays (red for on-axis field, green for 10° off-axis field, and blue for 20° off-axis field) are traced through the example cell phone lens. (Right) An example deviation of the wavefront from a spherical wavefront is plotted across the pupil in a wavefront aberration function plot for an off-axis field in a pattern known as astigmatism.

The exiting wavefronts also have polarization-dependent aberration. For uncoated and thin film-coated interfaces, the aberration arises from the *variation of angle of incidence* (*AoI*). Figure 10.3 shows a series of *angle of incidence maps* at each surface of the cell phone lens, where the location of each line represents the ray location within the bundle of rays. The length of the line represents the magnitude of the incident angle of that ray at each surface. The orientation of the line represents the orientation of the plane of incidence. For our

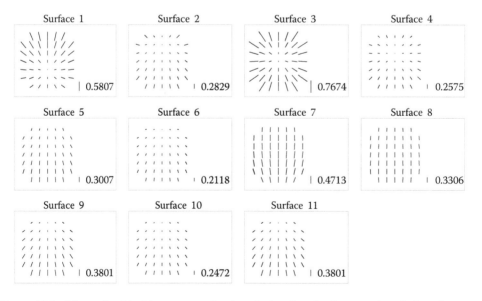

Figure 10.3 The angle of incidence maps of each optical surface for the example cell phone lens system for a 10° field. The lines in the map represent the plane of incidence, the plane of *p*-polarized light. Labels at the right bottom in each map indicate the maximum incident angle in radians.

example cell phone lens, the incident angles are largest at surface 3, 1, and then 7. Rays at normal incidence are seen at surfaces 1, 2, 3, 4, 6, 9, and 10, where the ray at normal incidence appears as a line with zero length (dot) in the *AoI* map.

The cell phone lens system might be fabricated *uncoated* (without coatings) or with many different choices of *antireflection coatings*. The diattenuation of an uncoated cell phone lens is plotted surface by surface in the *diattenuation maps* shown in Figure 10.4. The diattenuation of each ray is calculated from its angle of incidence using the Fresnel equations of Chapter 8. Similar to the *AoI* map, the length of each line in the diattenuation map represents the magnitude of diattenuation. The orientation of the line represents the orientation of the diattenuation axis, the linear polarization state with the maximum transmittance. For our example system, only linear diattenuation is

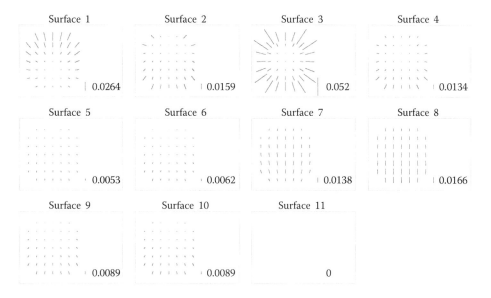

Figure 10.4 Diattenuation maps for each surface of the example cell phone lens system for a 10° field. Labels at the bottom right indicate the maximum diattenuation.

present, and surface 3 contributes the most diattenuation. If the lens surfaces were antireflection coated, the diattenuation would change and a small amount of retardance would be introduced.

The cumulative diattenuation at the image plane can be calculated ray by ray from the cumulative polarization ray tracing matrix described in Chapter 9. Figure 10.5 (left) overlays the diattenuation maps from Figure 10.4 of each surface adjusted to the same scale;

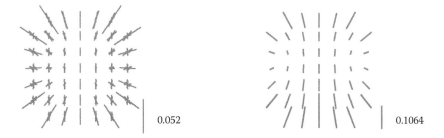

Figure 10.5 (Left) The diattenuation maps for each surface from Figure 10.4 are aligned and overlaid. (Right) The cumulative diattenuation is displayed at the exit pupil of the cell phone lens.

Figure 10.5 (right) shows the corresponding cumulative diattenuation map for the entire lens, from entrance pupil to exit pupil (see Section 10.5.3 for definitions of entrance and exit pupil).

Another important description of the polarization aberration is the *Jones pupil*, the Jones matrix for each ray, depicting its variation over the pupil. A polarization aberration-free system has an identity matrix as its Jones pupil. Detailed descriptions of the Jones pupil are included in Chapter 16 (Image Formation with Polarization Aberration). Figure 10.6 shows the Jones pupil for our example cell phone lens at the exit pupil for the 10° field. The diagonal elements indicate an amplitude transmittance of about 0.8 for x coupled into x-polarized and for y coupled into y-polarized light. The off-diagonal elements are small, below 0.02; thus, this Jones matrix is close to the identity matrix times a constant of 0.8. An uncoated lens has

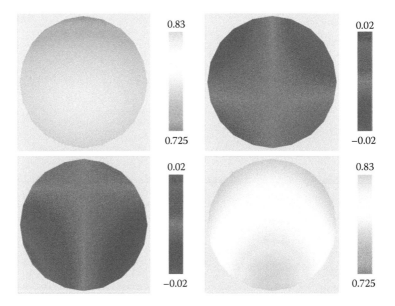

Figure 10.6 Jones pupil of the example cell phone lens system for a 10° field beam.

no retardance and thus no imaginary part to the Jones matrices; hence, this particular Jones pupil is purely real; in general, Jones pupils are complex.

The *AoI* map, diattenuation map, and Jones pupil are useful metrics to evaluate polarization aberrations contributed from each optical surface as well as accumulated through the optical system. Being able to generate these maps can assist the optical designer in understanding which surfaces produce the most polarization aberration, and if possible, how these polarization aberrations might compensate each other. Polarization aberrations are treated further in Chapter 12 (Fresnel Aberrations) and Chapter 15 (Paraxial Polarization Aberrations). This chapter focuses on algorithms for the calculation of the Jones pupil.

10.3 Specification of Optical Systems

Optical systems are collections of optical elements for manipulating light. The systems consist of a series of optical elements: lenses, mirrors, gratings, and so on. The optical elements divide the space into a series of volumes, the insides of the optical elements or the spaces between them. Figure 10.7 shows the example system. These volumes are separated by optical interfaces or surfaces that can be specified by optical surface parameters listed in Table 10.1.

10.3 Specification of Optical Systems

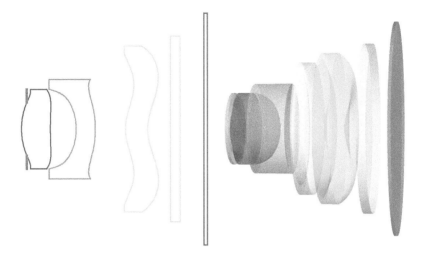

Figure 10.7 (Left) An optical system comprising an aperture stop, four aspheric lenses, a filter shown in blue, and an image sensor shown in magenta. (Right) 3D view of the same system.

Table 10.1 Surface Parameters at Each Optical Interface

Location of surface vertex	**v**
Surface normal of surface vertex	**a**
Surface equation	$z(x,y)$ or $f(x,y,z)$
Aperture equation	$p(x,y,z)$
Refractive index of incident material	n_i
Refractive index of transmitting medium	n_x
Thickness (distance to next surface)	t

To keep our discussion simple, the optical system description that follows applies to radially symmetric imaging systems. The optical system will be described in a global (x, y, z) coordinate system. Light will initially propagate with increasing z-coordinate. The z-axis will be the *optical axis*, the axis of symmetry. The centers of curvature of all spherical surfaces lie on the z-axis. All the vertices and axes of radially symmetric surfaces, conic and aspheric surfaces, are also coincident with the optical axis.

In a sequential ray trace, the order the optical wave encounters the surfaces will be specified by surface index q,

$$q = 0, 1, 2, \ldots Q. \tag{10.1}$$

The object surface corresponds to $q = 0$, $q = 1$ could be the entrance pupil, $q = Q - 1$ could be the exit pupil, and $q = Q$ is the image surface. Non-sequential ray racing is discussed in Section 10.8.

The optical system is designed for a particular wavelength range $\lambda_{min} \leq \lambda \leq \lambda_{max}$ or set of wavelengths. In general, a reference wavelength λ_{ref} is specified and used for determination of focal lengths, pupil locations and diameters, and similar calculations. The space between surface q and $q + 1$ has a real *refractive index* function $n_q(\lambda)$ if it is an isotropic transparent material, such as glass. If a material has significant *absorption*, it also has an imaginary *absorption coefficient* $\kappa_q(\lambda)$. For anisotropic materials such as calcite, $n_q(\lambda)$ is replaced with the dielectric tensor $\varepsilon_q(\lambda)$. Isotropic materials and homogeneous interfaces are assumed until Chapter 19 on anisotropic materials.

10.3.1 Surface Equations

The shape of each surface is described by a *surface equation* $z_q(x, y)$, or in implicit form $f_q(x, y, z) = 0$. For example, a spherical surface with radius of curvature R (concave if $R > 0$) and vertex on the z-axis at z_0 is represented by one of the equations

$$z(x, y) = z_0 + R - \sqrt{R^2 - x^2 - y^2} \quad \text{or} \quad f(x, y, z) = -z + z_0 + R - \sqrt{R^2 - x^2 - y^2} = 0. \quad (10.2)$$

The first form $z(x,y)$ describes the *sag* or *saggita*, the height from a reference plane.

Radially symmetric aspheric (non-spherical) surfaces, such as paraboloids (Figure 10.8) and ellipsoids, have a symmetry axis through their vertex $\mathbf{v}_q$ with an axis vector $\mathbf{a}_q$. A system may have tilts such that $\mathbf{a}_q$ is not parallel to the optical axis, or decenters, which move $\mathbf{v}_q$ off the axis.

Non-rotationally symmetric surfaces are common; for example, plastic molding provides an inexpensive and quick way to mass produce complex aspheric surfaces, such as progressive eyeglass lenses or the lenses used in cell phone lenses and laser printers. Such surfaces require different surface equations such as aspheric coefficients or Q-type polynomial coefficients.[1]

10.3.2 Apertures

Each optical surface has a finite extent. The light transmitting or reflecting region on the optical surface is defined by its *aperture* (Figure 10.9). Rays inside the aperture reflect or refract, while rays

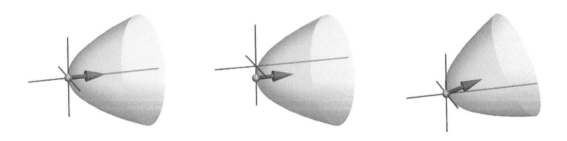

Figure 10.8 A parabolic surface specified by its vertex (yellow sphere) and its surface normal direction at the vertex, axis vector (red arrow). (Middle) Decentered parabolic surface. (Right) Tilted parabolic surface.

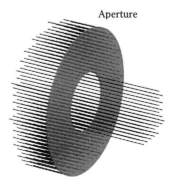

Figure 10.9 An on-axis beam passes through an aperture.

10.3 Specification of Optical Systems

outside the aperture are blocked or *vignetted*. The aperture for each surface is specified by a one-zero aperture function $p_q(x, y, z)$,

$$p_q(x, y, z) = \begin{cases} 1 & \text{inside the aperture} \\ 0 & \text{outside the aperture} \end{cases}. \quad (10.3)$$

10.3.3 Optical Interfaces

When light interacts with an optical surface, the physical light–surface interaction should be applied. Interfaces in optical systems may be uncoated or coated with one of a myriad of types of *coatings*, such as antireflection coatings, reflection enhancing coatings, and beamsplitting coatings. In general, coatings do not influence the geometrical ray trace; the path of rays and their optical path length are not altered much by the coatings. Thus, most geometrical ray tracing is performed without specifying coatings. But the results of a polarization ray trace are dependent on the coatings, and these coatings must be part of the optical system specification in order to obtain an accurate polarization simulation.

Coatings may consist of a single layer, but typical coatings comprise a *multilayer thin film stack*, such as described in Section 13.3 (Multilayer Thin Films). Thin films are usually modeled as homogeneous and isotropic structures. A homogeneous coating has a uniform thickness over the aperture. An isotropic coating has the same refractive index for all directions of the light's electric field; the thin film material is not birefringent. Homogeneous and isotropic interfaces are by far the most common, but many interesting devices can be made from anisotropic coatings or inhomogeneous (spatially varying) coatings. In addition to coated interfaces, other optical surfaces may have periodic structures, such as diffraction gratings and diffractive optical elements.

10.3.4 Dummy Surfaces

Dummy surfaces are surfaces that can be added to the optical system description; they are interfaces that do not correspond to the optical element surfaces. The *entrance pupil* and *exit pupil* are examples of surfaces of interest that are not lens or mirror surfaces. Dummy surfaces have the same refractive index before and after the surface and thus do not deviate light rays. Frequently, it is desirable to find the locations of rays on particular surfaces, such as the entrance pupil, so a dummy surface is used. Figure 10.10 shows a dummy surface in front of a lens, which is useful for calculating the beam sizes. A dummy surface might be virtual, such that the ray in the space between two lenses may be extended to some distant plane to examine, for example, where

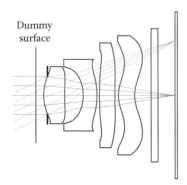

Figure 10.10 Example of a dummy surface, which does not deviate the rays, in front of an optical system.

a virtual image may form. Then, the ray may back up to refract at the next sequential surface. Another common use for dummy surfaces is to simplify tilts and decenters in optical systems by breaking the coordinate system changes into steps. Dummy surfaces also expedite performing calculations in multiple coordinate systems by changing the coordinate system on a dummy surface.

10.4 Specifications of Light Beams

Imaging systems are designed to accept and image a certain spatial extent of the incident light, described by area and solid angle, referred to as the *etendué*, defined below. Only a limited solid angle of the light from an object will pass through the optical system without being blocked. The beams through the simplest optical systems are specified by the extent of the object (often a nominal size) and the entrance pupil. Then, all rays from the object that lie within the solid angle Ω subtended by the entrance pupil are transmitted through the optical system to the image, while rays outside the object or entrance pupil are blocked.

Light rays are geometric objects that obey the laws of reflection and refraction. Light rays are purely *geometric objects*. Propagating through an optical system, the *light ray* is divided into *ray segments*, each segment with a starting and ending position and a direction, as shown in Figure 10.11.

As geometric constructs, light rays approximate the behavior of light. They indicate *normals to the wavefronts* and the *Poynting vector*, the direction of energy flow. A narrow laser beam, although not as infinitesimally thin as a light ray, would propagate through the lens of Figure 10.11 along a path centered on a light ray. Light rays are invaluable. But as with many abstractions, the light ray concept is not fully rigorous from the perspective of diffraction and quantum optics. Light rays work well in some cases, not so well in others, such as when the size of structures is similar to or smaller than the wavelength of light. In optical design, light rays are indispensable.

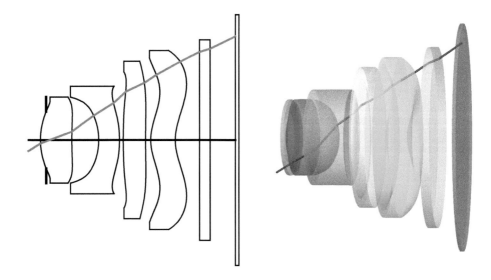

Figure 10.11 (Left) Light ray (gray) propagating through the example lens with 5 optical elements and 11 ray segments, refracting at each surface, terminating at an image sensor. (Right) Same system in 3D.

10.5 System Descriptions

10.5.1 Object Plane

In ray tracing, light rays start at the object, surface $q = 0$. The object is typically modeled as a circular or square flat area perpendicular to the optical axis. This may represent a scene in front of a camera, a light source, or some other light emitting or scattering region. It is not always flat and may be curved. Similarly, the limits of the image may be specified, for example, the boundary of a CCD, a screen, or some other element.

10.5.2 Aperture Stop

Of all the apertures in the system (including edges of optical elements), the aperture that limits the on-axis beam through an optical system is called the *aperture stop*. Many camera lenses have an aperture stop with an adjustable diameter, adjusting the *f number* setting of the camera. *Vignetting* occurs when rays are blocked by apertures other than the aperture stop. Typically in a lens, the on-axis rays are only blocked by the aperture stop. As the object moves off-axis, vignetting occurs; at some point, rays begin to be blocked from the edges of the beam by other apertures, and this vignetting steadily increases as the beam moves off-axis. Some optical systems are intentionally designed with optical elements large enough to avoid any vignetting over their field of view.

As an example, Figure 10.12 shows a lens form known as a "double Gauss" lens. The on-axis beam passes within all the lens surfaces and are limited at the aperture in the middle of the lens, making it the aperture stop. The rays from the bottom of the 10° and 14° fields are limited or vignetted at the bottom of the second lens, which is drawn slightly oversized. Similarly, the rays from the top of the 10° and 14° fields are vignetted at the top of the last lens element. Thus, it is seen that the 10° and 14° beams do not fill the aperture stop since additional truncation of the beams occurred away from the aperture stop.

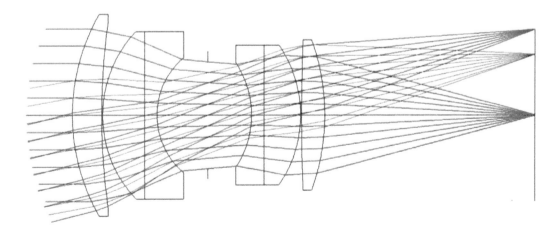

Figure 10.12 Lens with an aperture stop located in the space between the two lens groups, from U.S. Patent 2,532,751. On-axis beam is shown in red, 10° off-axis beam is shown in green, and 14° off-axis beam is shown in blue.

10.5.3 Entrance and Exit Pupils

The *entrance pupil* is the image of the aperture stop in object space. The entrance pupil identifies which rays in the object space pass through the aperture stop. Its diameter is the Entrance Pupil Diameter (*EPD*). Because of the importance of quantities associated with the entrance pupil, they are indicated by the subscript E. Table 10.2 lists the four surfaces of particular interest and their respective subscripts. In ray tracing, it is only necessary to trace the rays that pass through the aperture stop; thus, rays can be deleted in object space if they fall outside the entrance pupil.

Similarly, the image of the aperture stop in image space is the *exit pupil*. In an imaging system, nearly spherical waves pass through the exit pupil and converge on the corresponding image points. The *reference sphere* for a particular image point is the sphere that passes through the center of the exit pupil with its center of curvature on the image point. The reference sphere is used as the basis of a coordinate system for the wavefront aberration of the exiting wavefronts, since it represents the ideal exiting wavefront. Figure 10.13 shows on-axis and off-axis wavefronts entering the example cell phone lens system through the entrance pupil. Entrance and exit pupils may be outside a lens (real) or inside the lens (virtual). Virtual entrance and exit pupils are shown in Figure 10.14. Reflection and refraction are not occurring at those locations.

10.5.4 Importance of the Exit Pupil

The exit pupil is a surface of fundamental importance in ray tracing because of its role in calculating the *point spread functions* (PSF) of the optical system. Diffraction calculation requires the amplitude and phase of the wavefront to be accurately described on a spherical surface centered on the image point. With ray tracing, the optical path length, and thus the wavefront, can be calculated on a spherical surface.

Table 10.2 Notation for Images and Pupils

	Identifying Subscript	Surface Number, q
Object surface	O	0
Entrance pupil	E	1
Exit pupil	X	$Q-1$
Image plane	I	Q

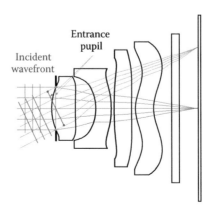

Figure 10.13 An entrance pupil and two collimated incident wavefronts entering the cell phone lens. The entrance pupil defines in object space the rays that will and will not make it through the aperture stop.

10.5 System Descriptions

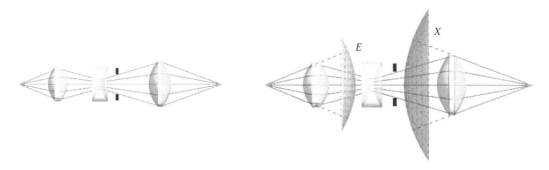

Figure 10.14 (Left) A point object is imaged through an optical system to a point image. The aperture stop is placed between the second and third lenses. (Right) Virtual entrance pupil (E) is the gray surface on the left. The virtual exit pupil (X) is the brown surface on the right. Since the rays to these pupils (dotted lines) are extensions of the real ray segments (solid lines), these pupils are *virtual*. Real pupils are physically accessible, that is, E before the first lens or X after the last lens.

Figure 10.15 Cross sections of the amplitude of a monochromatic wavefront propagating in the vicinity of the exit pupil. Before (left three) and after (right three) the exit pupil (center figure, rectangle), the wavefront's amplitude has many wiggles and oscillations due to Fresnel diffraction, increasing away from the exit pupil. At the exit pupil, the amplitude is most constant.

What is needed is a surface with known amplitude where the amplitude has a simple description. At a typical exit pupil, a sharply focused image of the aperture stop is formed, which has uniform brightness within the pupil and is dark outside the exit pupil boundary. In the vicinity of the exit pupil, the light's amplitude evolves by the equations of Fresnel diffraction (near-field diffraction) and develops many oscillations, particularly around the boundary. Figure 10.15 shows a typical example of Fresnel diffraction before (left three graphs) and after (right three graphs) an exit pupil (middle graph). The exit pupil is nearly uniform with a sharp transition between the bright and dark areas. Fresnel diffraction ringing is seen before and after the exit pupil, with oscillations increasing further from the exit pupil. Therefore, the amplitude is easier to describe at the exit pupil than elsewhere.

Optical design programs default, in their basic modes of PSF calculation, to assume a one–zero exit pupil amplitude function, one inside the pupil and zero outside the pupil. Hence, the most common form of PSF calculation by ray tracing programs can be summarized as follows: (1) ray tracing from the entrance pupil to the exit pupil to calculate the phase and the shape of the wavefront, (2) assume a constant amplitude within the exit pupil, and (3) perform diffraction calculations (Fourier transforms) from the exit pupil to the image plane to obtain the PSF. This algorithm is simple and performs remarkably well. The electric field at other planes can also be calculated from the exit pupil by Fresnel diffraction.

If the diffraction calculations need to be performed from a different surface from the exit pupil, then all the wiggles and oscillations of the amplitude, as seen in Figure 10.15, need to be determined. These amplitude variations cannot be calculated by ray tracing alone, but require diffraction calculations, far more complex calculations than ray tracing. Thus, it is nearly miraculous how simply and efficiently ray tracing from the entrance pupil to the exit pupil couples into diffraction and PSF calculations.

What about variations of amplitude in the exit pupil? Rays lose different amounts of flux passing through optical systems due to reflections and absorptions. Such amplitude aberrations are referred to as *apodization*. Also, equally spaced rays in the aperture can become non-uniformly spaced in the exit pupil (*pupil aberration*), redistributing flux and causing apodization. Small and smooth variations of amplitude in the exit pupil, 20% or less, have very little effect on the size of the PSF. Typically, apodization has less effect on the PSF than a tenth of a wave of wavefront aberration. Thus, the wavefront aberration is very important, and the apodization is much less important. In fact, apodization is still not routinely included in the calculation of the PSF in most geometrical ray tracing simulations.

The optical system's apodization is calculated by polarization ray tracing and is typically included in the PSF calculation using polarization ray tracing. Conventional geometric ray tracing ignores the apodization, assuming a uniformly illuminated pupil, and produce PSFs that are accurate for a majority of the traditional optical systems. Polarization ray tracing will produce a more accurate PSF.

10.5.5 Marginal and Chief Rays

A *marginal ray* starts from the on-axis object point and passes through the edge of the entrance pupil and aperture stop. The heights of the marginal ray at each surface are identified as y_q. The marginal ray forms an angle u with the optical axis. A *chief ray* is a ray from the edge of the object that passes through the center of the entrance pupil and through the center of the aperture stop, where it crosses the optical axis. Quantities associated with the chief ray are indicated by a *horizontal line* over the quantity. The height of the chief ray at the object is $\bar{y}_0$ and the angle of the chief ray with respect to the optical axis is $\bar{u}_0$. The height of the chief ray at the entrance pupil is $\bar{y}_E = 0$ since it is defined to pass through the center of the pupil. Figure 10.16 shows the marginal rays from an on-axis object point, an off-axis chief ray passing through the center of the pupil, and an *axial ray* propagating along the optical axis.

Figure 10.17 (left) shows an off-axis collimated ray fan entering the entrance pupil in object space where the chief ray passes through the center of the entrance pupil. Figure 10.17 (right) shows another off-axis ray fan entering the lens from object space, propagating through the first surface, second surface, and so on, until the image space. The rays in image space have been continued backward (dotted lines) to the virtual exit pupil. The reference sphere (blue) is centered on the image point and passes through the center of the exit pupil.

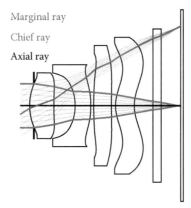

Figure 10.16 Fan of rays for an on-axis beam and an off-axis beam traced through the cell phone lens. The entrance pupil is the black aperture located against the first surface. The axial ray (black) propagates down the axis; it is the central ray in the on-axis beam. The marginal rays (red) go through the edge of the entrance pupil in the on-axis beam. The chief ray (magenta) is the central ray in the off-axis beam and is aimed at the center of the entrance pupil.

10.5 System Descriptions

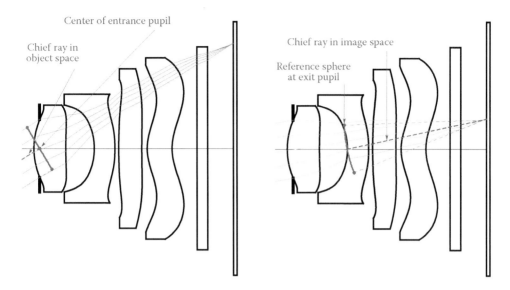

Figure 10.17 (Left) The chief ray and entrance pupil in object space for a larger field angle. For a collimated input beam, the reference sphere at the entrance pupil is flat. (Right) Off-axis ray fan with a reference sphere centered on the image point and passing through the center of the exit pupil. The blue dotted lines continue from the image space rays un-deviated backward to reach the exit pupil.

10.5.6 Numerical Aperture and Lagrange Invariant

Several metrics describe the angle and volume of light that can pass through an optical system. The *numerical aperture* (*NA*) for a radially symmetric optical system, shown in Figure 10.18, describes the cone of an on-axis beam in terms of the real marginal ray angle U,

$$NA = n|\sin U|, \qquad (10.4)$$

where the object space refractive index is n at the reference wavelength. Equation 10.4 is approximated to $n|u|$ in paraxial optics when the real marginal ray angle is small.

The *linearity of paraxial optics* provides a relationship between the paraxial ray heights and angles of the marginal and chief rays propagating through the optical system in terms of the *Lagrange invariant* (*H*),

$$H = n\,\bar{u}\,y - n\,u\,\bar{y}. \qquad (10.5)$$

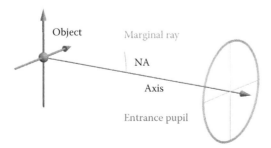

Figure 10.18 The numerical aperture is the sine of the angle from the axis to the marginal ray.

H is invariant as paraxial beams, propagate, refract, and reflect; thus, the Lagrange invariant is a conserved quantity in a paraxial system.[2] The Lagrange invariant has units of length times angle in radians, evaluated in the meridional plane. The Lagrange invariant can be evaluated anywhere along the beam, and in any space (object space, inside the first lens, etc.). At pupils and at the stop, the chief ray height $(\bar{y})$ is zero, then $H = n\,\bar{u}\,y$. At images or objects, the marginal ray height (y) is zero, and therefore $H = -n\,u\,\bar{y}$. Hence, as the beam's y and $\bar{y}$ become smaller, the angles u and $\bar{u}_0$ must become larger and vice versa.

10.5.7 Etendué Ξ

The total light processing capacity of an optical system in three dimensions is described by its *etendué* Ξ, also known as *throughput*. *Etendué* is defined in object space as

$$\Xi = n^2 A_O \Omega_E = n^2 A_E \Omega_O \approx n^2 (A_O A_E)/(4\pi L^2), \tag{10.6}$$

where A_O is the area of the object, A_E is the area of the entrance pupil, Ω_E is the solid angle of the entrance pupil viewed from the object, Ω_O is the solid angle of the object viewed from the entrance pupil, and L is the separation of the object and entrance pupil. Etendué is the generalization of the Lagrange invariant into the third dimension. Etendué, like focal length, pupil sizes, and positions, is typically defined in terms of paraxial quantities. The etendué provides a definition for the concept of how many light rays can be transmitted through an optical system. Each ray is considered as a differential area and differential solid angle; this can be integrated yielding total "*number*" of light rays, *number* being an area times a solid angle in a *four-dimensional phase space* of light rays. The etendué is related to the square of the Lagrange invariant.

Figure 10.19 (left) shows a volume between an object and entrance pupil as completely filled with rays. In Figure 10.19 (right), the area of the object times the solid angle subtended by the pupil (the set of yellow rays) is equal to the area of the pupil times the solid angle subtended by the object.* This quantity, the etendué, is preserved by paraxial reflection and refraction. After the first surface, a "first space" of light rays can be defined. In this first space, a first image (image of the object by the first surface) and a first pupil are located. The image and pupil may be real, located between the first and second surface, or virtual, located by extending the light rays to infinity beyond the first and second surface. Within the first space, the etendué is unchanged, as it is in the second space, and so on, until reaching the image space.

The etendué is an invariant of the optical system and cannot be reduced. Figure 10.20 shows three beams with equal etendué. The circles on the left are objects, images, or pupils and the cones represent the outer limit of the rays passing through each point on the objects, images, or pupils. When the area of an image or pupil is reduced, the solid angle of the light cones must correspondingly increase, and vice versa. These example beams are all *telecentric* because the central rays for each cone are all parallel to the axis and the corresponding pupil is at infinity. Typically, all the central rays (chief rays) converge or diverge from the center of a pupil.

The cone of light in an optical system fills a certain volume of a *four-dimensional phase space* that is conserved through reflection and refraction. This phase space can be defined in different ways. For example, the phase space can be defined by object coordinates (two dimensions) and entrance pupil coordinates (two dimensions). It can be defined by image coordinates (two dimensions) and the angular coordinates of the ray at the image (two dimensions), and so on, throughout the system. It can also be defined by object coordinates (two dimensions), pupil coordinates (two dimensions), and the distance between the object and pupil.

---

* The object and pupil can be reversed, pupil on left, image on right, depending on the specifics of the beams in a particular space.

10.5 System Descriptions

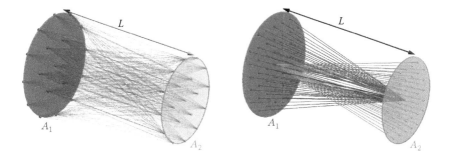

Figure 10.19 (Left) The space between an object or image (purple) and an entrance or exit pupil (light blue) is typically filled with rays, which define the system's etendué. (Right) The area of the image times the solid angle subtended by the pupil is equal to the area of the pupil times the solid angle subtended by the object. A_1 and A_2 are A_O and A_E, or A_E and A_O.

Figure 10.20 Three beams of light with the same etendué $\Xi = A\,\Omega$. (Center) An area A where the rays from each point fill a solid angle Ω. (Left) An area $A/4$ where the rays from each point fill $4\,\Omega$. (Right) An area $4A$ where the rays from each point fill $\Omega/4$.

If the image is smaller than the object, then the solid angle of the exit pupil seen from the image is correspondingly larger than the entrance pupil solid angle, conserving the etendué. The etendué of an optical system cannot be reduced without eliminating rays. The "phase space of rays" is *incompressible*. Thermodynamically, it is not possible to reduce the etendué of a light beam. Consider the etendué as analogous to *entropy*, which also cannot be reduced.

Scattering will increase the etendué of a beam. Consider a small area of a scatterer, such as diffuser or sandpaper, illuminated by a small solid angle of light. Following the scatterer, the light fills a much larger solid angle, exiting the same area, thus increasing the etendué, defining the volume needed to contain the light rays. Optical systems cannot shape this scattered beam back into the original volume, which requires reducing its etendué.

Large etendué is always desirable. The cost of camera lenses, microscope objectives, telescopes, and so forth, scales rapidly with etendué. Thus, etendué is one of the most important optical specifications.

10.5.8 Polarized Light

Ray tracing of imaging systems uses a model where a *spherical wave* from an object point enters an optical system. The *wavefront* is divided into an array of light rays, each representing a patch of wavefront as in Figure 10.21; each spherical wavefront patch is shown with a propagation vector. The incident light's polarization may have a simple distribution, like a uniform plane wave or the simple field of a dipole, or may have an arbitrarily complex distribution, like Figure 10.22 (left). For an arbitrarily polarized wavefront, each wavefront patch's (each ray's) polarization state is easily described by a polarization vector **E**.

Historically, it is common to describe the polarization states in ray tracing with *Jones vectors* where a different *local coordinate* system is needed for the E_x and E_y components of each ray

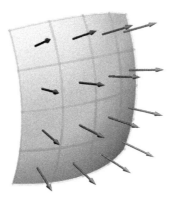

Figure 10.21 A spherical wavefront with propagation vectors is shown with a set of wavefront patches. Light rays propagate along the propagation vectors.

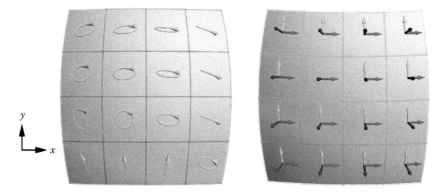

Figure 10.22 (Left) An arbitrary polarized wavefront with polarization ellipses (cyan) representing a grid of wavefront patches. Each of the polarization states can be described as a three-element polarization vector in the global coordinates. (Right) A set of local x (blue), y (green), and k (black) basis vectors are introduced to define the Jones vector x- and y-components for a set of rays associated with a spherical wavefront. These coordinates are local because they are different for each ray.

as shown in Figure 10.22 (right). This set of local coordinates, shown in blue and green, is *local* because it is different for every ray. Using Jones vectors with local coordinates is frequently done in polarization ray tracing and other analyses. As a ray propagates through an optical system changing directions, a different set of Jones vector local coordinates are needed for each ray and every ray intercept; the local coordinates move through the system along with the ray. For optical design, a better method is to describe the polarization states across the wavefront in a single global polarization state with three element polarization vectors **E**.

In many polarization analyses, the details of the source polarization are not needed. In polarization ray tracing, the polarization properties of the ray paths are calculated as polarization ray tracing matrices or Jones matrices. These matrices and their properties apply to any incident polarization state. Hence, the incident polarization does not need to be specified to perform the ray trace analysis.

10.6 Ray Tracing

The goal of a ray trace is to follow the path of the light through the optical system, applying the laws of reflection and refraction, and then to assemble information on the rays in image space, suitable to assess the image quality and performance of the optical system. The wavefront and polarization

10.6 Ray Tracing

Table 10.3 Ray Quantities at Each Ray Intercept for a Geometrical Ray Trace

Ray intercept	$\mathbf{r}_q$
Surface normal	$\boldsymbol{\eta}_q$
Propagation vector after surface	$\mathbf{k}_q$
Length of ray segment	$t_{q-1,q}$
Optical path length of ray segment	$OPL_{q-1,q}$

for each object point and wavelength need to be traced separately since the rays have different paths through the system. The variation of wavefront as a function of object coordinate and wavelength, the system's aberrations, is the most significant calculation in ray tracing. Table 10.3 lists the quantities to be calculated at each ray intercept for a conventional geometrical ray. Figure 10.32 shows an example ray propagating through the cell phone lens system; the ray intercepts, propagation vectors, and normals at each ray intercepts are tabulated in Table 10.5. Section 10.8 addresses non-sequential ray tracing where the order the light passes the surfaces is not specified at the beginning of the ray trace.

10.6.1 Ray Intercept

The algorithm to trace a single ray and calculate the optical path due to reflections and refractions from the object to exit pupil is as follows. For each ray, the object coordinate $\mathbf{H}_0$ and wavelength are specified. To ray trace the wavefront through an optical system, an area enclosing the entrance pupil is divided into segments, for example, a set of square or rectangular patches as shown in Figure 10.21. The center of each patch is located at entrance pupil coordinates $\boldsymbol{\rho}_E$ with normalized propagation vector $\mathbf{k}_0$ along the line from $\mathbf{H}_0$ to $\boldsymbol{\rho}_E$,

$$\mathbf{k}_0 = \frac{\boldsymbol{\rho}_E - \mathbf{H}_0}{|\boldsymbol{\rho}_E - \mathbf{H}_0|} = \begin{pmatrix} L \\ M \\ N \end{pmatrix}. \tag{10.7}$$

The components of the propagation vector, (L, M, N), are the *direction cosines* of the propagation direction, the cosines of the angles the ray forms with the x-, y-, and z-axes. Now that the ray has been aimed at the optical system, it is traced to its ray intercept with the first surface, $\mathbf{r}_1$, then its ray intercept with the second surface, $\mathbf{r}_2$, and so on.

Figure 10.23 shows the geometry for calculating the ray intercept at surface $q + 1$ and the ray's optical path length from q to $q + 1$, OPL_q. The ray exits the previous surface, q, at ray intercept $\mathbf{r}_q$ along $\mathbf{k}_q$. The equation for the ray path in parametric form as a function of the distance t_q along the ray is

$$\mathbf{r}_{q+1}(t) = \mathbf{r}_q + t_q \mathbf{k}_q. \tag{10.8}$$

The ray propagates toward surface $q + 1$. The refractive index for the current ray segment is n_q and the ray segment after refraction or reflection is n_{q+1}. The surface is located with a vertex $\mathbf{v}_{q+1}$, which may be on the optical axis ($z = 0$) or decentered off the optical axis. A rotationally symmetric surface has an axis $\mathbf{a}_{q+1}$, which will not be parallel to the optical axis if the surface is tilted. The surface is defined by surface equation $f(x, y, z) = 0$.

The process for finding intercepts is understood by substituting the x-, y-, and z-components of Equation 10.8 into the surface equation,

$$f\left(r_{q,x} + t_q k_{q,x},\ r_{q,y} + t_q k_{q,y},\ r_{q,z} + t_q k_{q,z}\right), \tag{10.9}$$

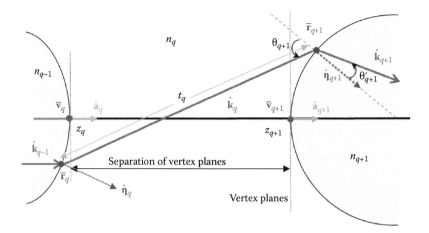

Figure 10.23 To find the ray intercept $\mathbf{r}_{q+1}$, first the length of the ray from $\mathbf{r}_q$ is determined by finding the intersection of the ray with the surface $q + 1$. Then, the law of reflection or refraction is applied to determine the propagation vector $\mathbf{k}_{q+1}$ in the next medium.

and adjusting t_q, until the first zero of f, right side of Equation 10.2, is found in the forward direction. The ray segment length t_q is the distance between $\mathbf{r}_q$ and $\mathbf{r}_{q+1}$,

$$t_q = |\mathbf{r}_{q+1} - \mathbf{r}_q| = \sqrt{(x_{q+1} - x_q)^2 + (y_{q+1} - y_q)^2 + (z_{q+1} - z_q)^2}. \tag{10.10}$$

The numerical methods for finding ray intercepts are outside the scope of this book. The algorithm for finding the ray intercept depends on the type of surface. A widely used algorithm for ray intercepts with spherical surfaces was published by Spencer; his form is preferred because it avoids divide by zero problems and programs efficiently.[3–5] Other algorithms calculate ray intercepts for conic surfaces,[3,6] aspheric surfaces,[7] and general freeform optical surfaces.[8,9]

10.6.2 Multiplicity of Ray Intercepts with a Surface

The ray tracing algorithm needs to select the proper ray intercept with a surface. Many rays (lines) have two or more intersections with a given surface equation, as shown in Figure 10.24 (left) for a circle. Other shapes, such as aspherics, Figure 10.24 (right), can have more than two ray intercepts per line. The ray intercept algorithm must include logic to evaluate and select the correct ray intercept, the intercept encountered first and which is inside the surface's aperture.

10.6.3 Optical Path Length

The *optical path length* for the ray segment from $\mathbf{r}_q$ to $\mathbf{r}_{q+1}$ is the length t_q times the refractive index n_q,

$$OPL_q = n_q t_q = n_q \sqrt{(x_{q+1} - x_q)^2 + (y_{q+1} - y_q)^2 + (z_{q+1} - z_q)^2}. \tag{10.11}$$

The number of wavelengths between the two ray intercepts is OPL_q/λ. The change in phase along the ray segment is

10.6 Ray Tracing

Figure 10.24 (Left) A ray intercepting a spherical surface can have two ray intercepts (green), one (blue), or none. (Right) More complex surfaces, such as aspheric surfaces, can have many ray intercepts.

$$\Delta\phi_q = \frac{2\pi n_q t_q}{\lambda}, \quad (10.12)$$

which is positive in the *decreasing phase convention* because phase decreases in time and increases in space.

The optical path length, as shown in Figure 10.25, for the qth ray through the system from the entrance pupil, E, to the exit pupil, X, is

$$OPL = \sum_{q=E}^{X} n_q t_q. \quad (10.13)$$

Figure 10.25 depicts the individual OPL contributions for a marginal ray through the example cell phone lens.

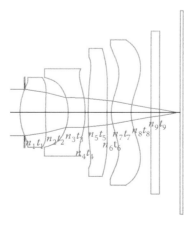

Figure 10.25 The optical path length of a ray through an optical system is the sum of the optical path lengths of the individual ray segments.

10.6.4 Reflection and Refraction

For both reflection and refraction, the normal vector $\hat{\boldsymbol{\eta}}_{q+1}$ to the surface at the ray intercept $\mathbf{r}_{q+1}$ is calculated. For a surface specified in implicit form, $f(x, y, z) = 0$, the normal is

$$\hat{\boldsymbol{\eta}}_{q+1} = \text{Normalize}\left[\left(\frac{\partial f(x,y,z)}{\partial x}, \frac{\partial f(x,y,z)}{\partial y}, \frac{\partial f(x,y,z)}{\partial z}\right)\right] \tag{10.14}$$

evaluated at a ray intercept $(x_{q+1}, y_{q+1}, z_{q+1})$.

In parametric form, the surface is described by a function of coordinates in a plane (x, y) as $(X(x, y), Y(x, y), Z(x, y))$. This is often the plane tangent to the surface vertex centered on the optical axis. The surface normal for the parametric form is

$$\hat{\boldsymbol{\eta}}_{q+1} = \text{Normalize}\left[\frac{\partial(X(x,y), Y(x,y), Z(x,y))}{\partial x} \times \frac{\partial(X(x,y), Y(x,y), Z(x,y))}{\partial y}\right]. \tag{10.15}$$

See Problem 10.2 for example.

The plane of incidence contains the ray propagation vector $\hat{\mathbf{k}}_q$ and the surface normal $\hat{\boldsymbol{\eta}}_{q+1}$ as shown in Figure 10.26 (left). The convention for the surface normal chosen here, shown in Figure 10.26 (left), points the normal into the transmitted medium into which the ray would refract.

The *angle of incidence* θ_{q+1} at a ray intercept $\mathbf{r}_{q+1}$ is the angle between the ray propagation vector from the previous surface and the surface normal vector,

$$\theta_{q+1} = \cos^{-1}(\hat{\mathbf{k}}_q \cdot \hat{\boldsymbol{\eta}}_{q+1}). \tag{10.16}$$

From the law of reflection, the propagation vector for the reflected ray is

$$\mathbf{k}_{q+1} = \mathbf{k}_q - 2(\mathbf{k}_q \cdot \hat{\boldsymbol{\eta}}_{q+1})\hat{\boldsymbol{\eta}}_{q+1}. \tag{10.17}$$

Snell's law of refraction is defined in terms of the angle of incidence θ_{q+1} and angle of refraction θ'_{q+1} as

$$n_{q+1} \sin\theta'_{q+1} = n_q \sin\theta_{q+1}. \tag{10.18}$$

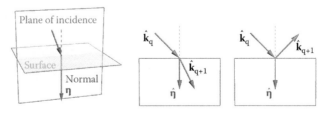

Figure 10.26 (Left) At a ray intercept, the plane of incidence is defined by the incident ray $\hat{\mathbf{k}}_q$ and the normal $\hat{\boldsymbol{\eta}}$. The normal is perpendicular to the surface and chosen to point toward the transmitted medium for both refracted (middle) and reflected rays (right).

10.6 Ray Tracing

Using Equations 10.16 and 10.18,

$$\cos\theta'_{q+1} = \sqrt{1-\left(\frac{n_q}{n_{q+1}}\right)^2(1-\cos^2\theta_{q+1})}. \quad (10.19)$$

The propagation vector for the refracted ray is

$$\mathbf{k}_{q+1} = \frac{n_q}{n_{q+1}}\mathbf{k}_q - \left(\frac{n_q}{n_{q+1}}\cos\theta_{q+1} - \cos\theta'_{q+1}\right)\hat{\boldsymbol{\eta}}_{q+1}. \quad (10.20)$$

Snell's law can be expressed in vector form without any reference to specific coordinates as

$$n_q\left(\hat{\mathbf{k}}_q \times \hat{\boldsymbol{\eta}}\right) = n_{q+1}\left(\hat{\mathbf{k}}_{q+1} \times \hat{\boldsymbol{\eta}}\right), \quad (10.21)$$

where $\hat{\boldsymbol{\eta}}$ is the unit vector normal to the interface and $\hat{\mathbf{k}}_q$ and $\hat{\mathbf{k}}_{q+1}$ are unit vectors parallel to the incident and transmitted propagation vectors, respectively.

10.6.5 Polarization Ray Tracing

Conventional ray tracing calculates the shape of wavefronts, but not their amplitudes and polarization states. This is not because the pioneers of optical design in the 1800s and early 1900s were not interested in amplitude and polarization, they were, but because the additional complications of including amplitudes and polarization in the ray trace calculation were substantial, complex, and difficult to justify, especially before the computer age in the time of rooms full of hand calculators.

For polarization optical design, optical path lengths are supplemented with calculations of the polarization ray tracing matrix, **P**. This allows the contributions of thin films, uncoated interfaces, gratings, anisotropic materials, and other optical elements to be included in the ray trace. **P** contributes the information on the amplitude, the diattenuation, and the retardance of the ray path. With **P**, the polarization ellipse in the entrance pupil is related to the polarization ellipse in the exit pupil for all polarization states. Each **P** can be converted into a Jones matrix, describing the eight degrees of freedom for the polarization of the ray path.

Table 10.4 lists the additional information obtained from polarization ray tracing at each ray intercept, beyond the quantities calculated in a conventional geometrical ray trace shown in

Table 10.4 Additional Information for Each Ray Intercept at Surface q from a Polarization Ray Trace

Ray intercept basis	$s_q = s'_q, p_q, p'_q$
Amplitude coefficients	$(t_{s,q}, t_{p,q}, r_{s,q}, r_{p,q}) = (\alpha_{t,s,q}, \alpha_{t,p,q}, \alpha_{r,s,q}, \alpha_{t,p,q})$
Jones matrix for ray intercept in s–p basis	$\mathbf{J}_{t,q} = \begin{pmatrix} \alpha_{t,s,q} & 0 \\ 0 & \alpha_{t,p,q} \end{pmatrix}$ and $\mathbf{J}_{r,q} = \begin{pmatrix} \alpha_{r,s,q} & 0 \\ 0 & \alpha_{r,p,q} \end{pmatrix}$
Polarization ray tracing matrix for ray intercept in global coordinates	$\mathbf{P}_q = \mathbf{O}_{out,q} \cdot \mathbf{J}_q \cdot \mathbf{O}_{in,q}^{-1}$ $= \begin{pmatrix} s'_{x,q} & p'_{x,q} & k'_{x,q} \\ s'_{y,q} & p'_{y,q} & k'_{y,q} \\ s'_{z,q} & p'_{z,q} & k'_{z,q} \end{pmatrix} \cdot \begin{pmatrix} \alpha_{s,q} & 0 & 0 \\ 0 & \alpha_{p,q} & 0 \\ 0 & 0 & 1 \end{pmatrix} \cdot \begin{pmatrix} s_{x,q} & s_{y,q} & s_{z,q} \\ p_{x,q} & p_{y,q} & p_{z,q} \\ k_{x,q} & k_{y,q} & k_{z,q} \end{pmatrix}$

Table 10.3. The amplitude transmission along the ray path is calculated, which was unavailable from the geometrical ray trace. The variation of amplitude across a wavefront is called *apodization* or *amplitude aberration*. The polarization ray trace calculates the dependence of the phase on the incident polarization state, the retardance aberration; different incident states have different optical path lengths, and thus different aberrations and interferograms. Finally, the apodization varies with the incident polarization state; this is the diattenuation aberration. Thus, the polarization ray trace provides eight times as much information about the optical systems aberrations as the wavefront aberration: the transmitted amplitude (one degree of freedom), the retardance aberration (three degrees of freedom), and the diattenuation aberration (three degrees of freedom).

The steps for the polarization ray trace are as follows. The first three steps consist of the geometrical ray trace, described in previous sections. To polarization ray trace a ray through surface $q + 1$, the ray parameters exiting surface q are needed. The steps to polarization ray tracing are outlined below:

1. Calculate the ray's intersection at surface $q + 1$, and the optical path length between surface q to surface $q + 1$ (Section 10.6.1).
2. Calculate surface normal $\mathbf{\eta}_{q+1}$ at ray intercept (Section 10.6.2).
3. Calculate reflected and refracted propagation vectors (Section 10.6.2).
4. Calculate the incident and exiting basis vectors, s and p, at the ray intercept (Section 10.6.6).
5. Calculate orthogonal matrices to rotate the incident basis into the exiting basis (Table 10.4).
6. Calculate the amplitude coefficients and the corresponding Jones matrix at the ray intercept, embodying the physics of the interaction (Section 10.6.7).
7. Calculate the polarization ray tracing matrix $\mathbf{P}$ and parallel transport matrix $\mathbf{Q}$ at the ray intercept (Table 10.4).
8. This process repeats until the last surface.

10.6.6 s- and p-Components

At each ray intercept, a local basis is needed to describe the polarization in the ray's local coordinate. The standard basis is the *s*- and *p*-basis, which, along with the propagation vector ($\hat{\mathbf{k}}_q$), forms a natural basis triplet, the $(\hat{\mathbf{s}}, \hat{\mathbf{p}}, \hat{\mathbf{k}})$ basis. The *s*-component is the component of the light field perpendicular to the plane of incidence and the *p*-component is the component in the plane of incidence, as shown in Figure 10.27. Figure 10.28 shows this orthonormal basis triplet before $(\hat{\mathbf{s}}_{q+1}, \hat{\mathbf{p}}_{q+1}, \hat{\mathbf{k}}_q)$ and after $(\hat{\mathbf{s}}'_{q+1}, \hat{\mathbf{p}}'_{q+1}, \hat{\mathbf{k}}_{q+1})$ the $q + (q + 1)^{th}$ ray intercept,

$$\hat{\mathbf{s}}_{q+1} = \frac{\hat{\mathbf{k}}_q \times \hat{\mathbf{\eta}}_{q+1}}{|\hat{\mathbf{k}}_q \times \hat{\mathbf{\eta}}_{q+1}|}, \quad \hat{\mathbf{p}}_{q+1} = \hat{\mathbf{k}}_q \times \hat{\mathbf{s}}_{q+1}, \quad \text{and} \quad \hat{\mathbf{s}}'_{q+1} = \hat{\mathbf{s}}_{q+1}, \quad \hat{\mathbf{p}}'_{q+1} = \hat{\mathbf{k}}_{q+1} \times \hat{\mathbf{s}}_{q+1}. \tag{10.22}$$

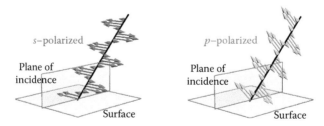

Figure 10.27 *s*- and *p*-polarization components at the ray intercept. Local tangent to the surface is shown as a yellow plane. The plane of incidence is the blue rectangle.

10.6 Ray Tracing

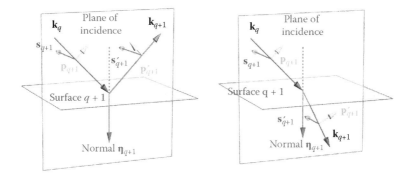

Figure 10.28 The propagation vectors, surface normal, and *s*- and *p*-basis vectors are shown before and after an interface for (left) reflection and (right) refraction.

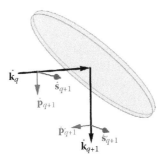

Figure 10.29 The *s*- and *p*-basis vectors for a reflection ray intercept.

For refraction, the $\hat{\mathbf{s}}_q$ basis vector is the same before and after the surface; only the $\hat{\mathbf{p}}_q$ vector changes. For reflection as is shown in Figure 10.29, the Fresnel coefficient for *s*-polarization changes sign to maintain a right-handed triplet of basis vectors. At isotropic interfaces, the *s*- and *p*-polarization components are the eigenpolarizations for the Fresnel refraction and reflection equations. The *p*-basis vector can also be obtained by applying Gram–Schmidt orthogonalization to $\hat{\mathbf{k}}_{inc}$ and $\hat{\boldsymbol{\eta}}$,

$$\hat{\mathbf{p}} = \hat{\mathbf{k}}_{inc} \times \hat{\mathbf{s}} = \frac{\hat{\mathbf{k}}_{inc} - \left(\hat{\mathbf{k}}_{inc} \cdot \hat{\boldsymbol{\eta}}\right)\hat{\boldsymbol{\eta}}}{\sqrt{1 - \left(\hat{\mathbf{k}}_{inc} \cdot \hat{\boldsymbol{\eta}}\right)^2}}. \tag{10.23}$$

10.6.7 Amplitude Coefficients and Interface Jones Matrix

The light's electric field before and after an interface is described by amplitude coefficients. Before and after the interface refers to both reflection and refraction. The equations for the amplitude coefficients are different for different classes of interfaces. The amplitude coefficients for uncoated interfaces are described in Sections 8.3.2 and 8.3.3. The amplitude coefficients for coated interfaces and multilayer thin films are described in Section 13.3.1. The amplitude coefficients for anisotropic materials, such as uniaxial and biaxial crystals and optically active materials, are described in Section 19.7. The amplitude coefficients for diffractive optical elements, including diffraction gratings, wire grid polarizers, and holograms, are calculated by RCWA as described in Section 23.4. With the amplitude relations, the polarization ellipse after an interface can be calculated from the polarization ellipse before the interface, as shown in Figure 10.30.

One pair of amplitude coefficients relate the incident light's *s*-polarized component to the *s*- and *p*-components of the exiting light. The second pair of amplitude coefficients relates the incident light's *p*-polarized component to the *s*- and *p*-components of the exiting light. This set of amplitude coefficients can be expressed as a Jones matrix for refraction, $\mathbf{J}_{t,q}$, or reflection, $\mathbf{J}_{r,q}$,

$$\mathbf{J}_{t,q} = \begin{pmatrix} \alpha_{t,s \leftarrow s,q} & \alpha_{t,s \leftarrow p,q} \\ \alpha_{t,p \leftarrow s,q} & \alpha_{t,p \leftarrow p,q} \end{pmatrix} \text{ and } \mathbf{J}_{r,q} = \begin{pmatrix} \alpha_{r,s \leftarrow s,q} & \alpha_{r,s \leftarrow p,q} \\ \alpha_{r,p \leftarrow s,q} & \alpha_{r,p \leftarrow p,q} \end{pmatrix}, \quad (10.24)$$

where the subscript *t* is for refraction (transmission), *r* is for reflection, *s* is for *s*-polarization, and *p* is for *p*-polarization. A Jones matrix describes the relationship between the incident polarization and the exiting polarization. The Jones matrix is defined in a local coordinate system where the local *z*-axis is along the ray propagation and local *x* and *y* are contained in the transverse plane, where polarization ellipses lie. The most common local coordinate system is the $(\hat{\mathbf{s}}, \hat{\mathbf{p}}, \hat{\mathbf{k}})$ basis; reflection and refraction at dielectric, metal, and multilayer coated interfaces are described in terms of $(\hat{\mathbf{s}}, \hat{\mathbf{p}})$ components. Jones matrix relates the incident polarization state $\mathbf{E}_{q-1}$ to the corresponding exiting polarization state $\mathbf{E}_q$ by matrix–vector multiplication,

$$\mathbf{E}_{q,t} = \begin{pmatrix} E'_s \\ E'_p \end{pmatrix} = \mathbf{J}_t \cdot \mathbf{E}_{q-1} = \begin{pmatrix} j_{s \leftarrow s} & j_{s \leftarrow p} \\ j_{p \leftarrow s} & j_{p \leftarrow p} \end{pmatrix} \begin{pmatrix} E_s \\ E_p \end{pmatrix}. \quad (10.25)$$

The Fresnel coefficients provide polarization-dependent relationships between the incident amplitude and reflected/transmitted amplitude. The Fresnel coefficients are often represented using *s*- and *p*-eigenpolarizations of the ray interface. For an incident polarization state at a ray intercept, the incident electric field is decomposed into *s*- and *p*-components, and the corresponding

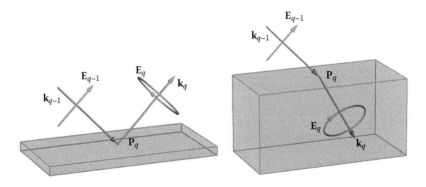

Figure 10.30 For the *q*th optical interface, the matrix $\mathbf{P}_q$ relates the incident polarization state $\mathbf{E}_{q-1}$ and propagation vector $\mathbf{k}_{q-1}$ to the exiting polarization state $\mathbf{E}_q$ and propagation vector $\mathbf{k}_q$.

Fresnel transmission/reflection coefficients are applied to each component.* See Chapter 8 for a more detailed discussion on Fresnel equations.

For simple reflections and refractions, the Jones matrices are diagonal matrices as shown below,

$$\mathbf{J}_{t,q} = \begin{pmatrix} \alpha_{t,s,q} & 0 \\ 0 & \alpha_{t,p,q} \end{pmatrix} \text{ and } \mathbf{J}_{r,q} = \begin{pmatrix} \alpha_{r,s,q} & 0 \\ 0 & \alpha_{r,p,q} \end{pmatrix}, \quad (10.26)$$

where $\alpha_{t,s,q}$ and $\alpha_{t,p,q}$ are s- and p-amplitude transmission coefficients and $\alpha_{r,s,q}$ and $\alpha_{r,p,q}$ are reflection coefficients. For an uncoated interface, these amplitude coefficients are calculated from the Fresnel equations. For coated interfaces, the coefficients are calculated from multilayer coating calculations (Section 13.3.1).[1,10,11]

Jones matrices for gratings, holograms, sub-wavelength gratings, and other non-isotropic interfaces in general have off-diagonal elements,

$$\mathbf{J}_q = \begin{pmatrix} j_{11} & j_{12} \\ j_{21} & j_{22} \end{pmatrix}. \quad (10.27)$$

$\mathbf{J}_q$ represents the Jones matrix for the interaction in a local coordinate system.

10.6.8 Polarization Ray Tracing Matrix

The three-by-three *polarization ray tracing matrix* $\mathbf{P}$ characterizes the polarization changes and propagation direction changes at a ray intercept. The $\mathbf{P}$ matrix is a generalized Jones matrix in three dimensions and keeps track of all three components of the electric field of the light,

$$\mathbf{E}_q = \begin{pmatrix} E_{x,q} \\ E_{y,q} \\ E_{z,q} \end{pmatrix} = \mathbf{P}_q \mathbf{E}_{q-1} = \begin{pmatrix} P_{xx,q} & P_{xy,q} & P_{xz,q} \\ P_{yx,q} & P_{yy,q} & P_{yz,q} \\ P_{zx,q} & P_{zy,q} & P_{zz,q} \end{pmatrix} \begin{pmatrix} E_{x,q-1} \\ E_{y,q-1} \\ E_{z,q-1} \end{pmatrix}. \quad (10.28)$$

More detailed discussion on the formalism of the $\mathbf{P}$ matrix and its advantages over the conventional Jones matrix is discussed in Chapter 9.

The extended Jones matrices for refraction $\mathbf{J}_{t,q}$ and reflection $\mathbf{J}_{r,q}$ in s- and p-coordinates are straightforward generalizations of the s- and p-Jones matrices in Equation 10.26,

$$\mathbf{J}_{t,q} = \begin{pmatrix} \alpha_{t,s,q} & 0 & 0 \\ 0 & \alpha_{t,p,q} & 0 \\ 0 & 0 & 1 \end{pmatrix} \text{ and } \mathbf{J}_{r,q} = \begin{pmatrix} \alpha_{r,s,q} & 0 & 0 \\ 0 & -\alpha_{r,p,q} & 0 \\ 0 & 0 & -1 \end{pmatrix}. \quad (10.29)$$

---

* Some ray tracing software calculates the average of s- and p-components to estimate the transmitted/reflected amplitudes. Since the s- and p-components are different for each ray propagation direction and the surface normal of the ray intercept, the simple average of two components cannot correctly describe the polarization ray tracing of an optical system.

See Example 10.2. Similarly, a more general Jones matrix (Equation 10.27) for an element in transmission is

$$\mathbf{P}_q = \begin{pmatrix} j_{11} & j_{12} & 0 \\ j_{21} & j_{22} & 0 \\ 0 & 0 & 1 \end{pmatrix}. \tag{10.30}$$

For reflecting devices, the j_{21} and j_{22} elements undergo an additional sign change as described in Section 8.3.7. Then, the matrices of Equations 10.29 and 10.30 are rotated into global coordinates using the algorithms for orthogonal transformations of Section 9.2,

$$\mathbf{P}_q = \mathbf{O}_{\text{out},q} \cdot \mathbf{J}_q \cdot \mathbf{O}_{\text{in},q}^{-1} = \begin{pmatrix} s'_{x,q} & p'_{x,q} & k_{x,q} \\ s'_{y,q} & p'_{y,q} & k_{y,q} \\ s'_{z,q} & p'_{z,q} & k_{x,q} \end{pmatrix} \cdot \begin{pmatrix} r_{s,q} & 0 & 0 \\ 0 & r_{p,q} & 0 \\ 0 & 0 & 1 \end{pmatrix} \cdot \begin{pmatrix} s_{x,q} & s_{y,q} & s_{z,q} \\ p_{x,q} & p_{y,q} & p_{z,q} \\ k_{x,q-1} & k_{y,q-1} & k_{z,q-1} \end{pmatrix}. \tag{10.31}$$

The polarization ray tracing algorithm with the **P** matrix is demonstrated (1) for normal incidence in Example 10.2 and (2) using two fold mirrors as shown in Figure 10.31 in Example 10.3.

Example 10.2 Polarization Ray Tracing Matrices at Normal Incidence

For normal incident refraction, set $\mathbf{k}_{\text{in}} = (0, 0, 1) = \mathbf{k}_t$ for simplicity. The incident and transmission orthogonal matrices are

$$\mathbf{O}_{\text{in}} = \begin{pmatrix} s_x & p_x & k_{x,0} \\ s_y & p_y & k_{y,0} \\ s_z & p_z & k_{z,0} \end{pmatrix} = \begin{pmatrix} 1 & 0 & 0 \\ 0 & 1 & 0 \\ 0 & 0 & 1 \end{pmatrix} \text{ and}$$

$$\mathbf{O}_{\text{out},t} = \begin{pmatrix} s'_x & p'_x & k_{x,1} \\ s'_y & p'_y & k_{y,1} \\ s'_z & p'_z & k_{z,1} \end{pmatrix} = \begin{pmatrix} 1 & 0 & 0 \\ 0 & 1 & 0 \\ 0 & 0 & 1 \end{pmatrix}, \tag{10.32}$$

respectively. Given the refraction Jones matrix $\begin{pmatrix} t_s & 0 \\ 0 & t_p \end{pmatrix}$, the refraction **P** matrix is simply

$$\mathbf{P}_{\text{refract},0°} = \begin{pmatrix} t_s & 0 & 0 \\ 0 & t_p & 0 \\ 0 & 0 & 1 \end{pmatrix}. \tag{10.33}$$

10.6 Ray Tracing

For normal incident reflection, $\mathbf{k}_{in} = (0, 0, 1) = -\mathbf{k}_t$ and the reflection Jones matrix $\begin{pmatrix} r_s & 0 \\ 0 & r_p \end{pmatrix}$, the reflection orthogonal matrix is

$$\mathbf{O}_{out,r} = \begin{pmatrix} s'_x & p'_x & k_{x,1} \\ s'_y & p'_y & k_{y,1} \\ s'_z & p'_z & k_{z,1} \end{pmatrix} = \begin{pmatrix} 1 & 0 & 0 \\ 0 & -1 & 0 \\ 0 & 0 & -1 \end{pmatrix}, \qquad (10.34)$$

and the reflection **P** matrix is

$$\mathbf{P}_{reflect,0°} = \begin{pmatrix} r_s & 0 & 0 \\ 0 & -r_p & 0 \\ 0 & 0 & -1 \end{pmatrix}. \qquad (10.35)$$

Example 10.3 Two Gold Fold Mirrors

Consider two gold-coated mirrors with a ray at 765 nm that reflects from *M1* then *M2*. The diattenuation and retardance are calculated individually for each reflection as well as for the combined effect of the two reflections.

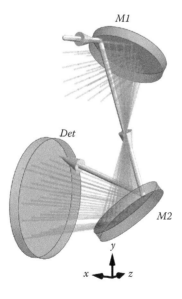

Figure 10.31 A grid of rays in a converging beam (gray) are shown reflecting from mirror, *M1*, then mirror, *M2*, before striking a detector, *Det*. Polarization ray trace details are provided for the top ray, highlighted in thick orange arrows

The first mirror is oriented with normal $\hat{\boldsymbol{\eta}}_1 = (0, 1, 1)/\sqrt{2}$, and the second mirror is oriented with $\hat{\boldsymbol{\eta}}_2 = (-1, -1, 0)/\sqrt{2}$. At the first reflection, the incident propagation vector $\hat{\mathbf{k}}_1 = (-0.195, -0.195, 0.961)$ with angle of incidence $\theta_1 = 57.184°$ reflects into $\hat{\mathbf{k}}_2 = (-0.195, -0.961, 0.195)$. For the incident ray, $\hat{\mathbf{s}}_1 = (-0.973, 0.164, -0.164)$ and $\hat{\mathbf{p}}_1 = (-0.126, -0.967, -0.221)$ from Equation 10.22. The exiting s-basis vector is unchanged, $\hat{\mathbf{s}}'_1 = (-0.973, 0.164, -0.164)$, but the exiting p-component rotates to $\hat{\mathbf{p}}'_1 = (0.126, -0.221, -0.967)$. The Fresnel coefficients r_{s1} and r_{p1} for the first reflection are $0.992 e^{i2.918}$ and $0.975 e^{-i0.751}$. The polarization ray tracing matrix $\mathbf{P}_1$ for the first reflection is

$$\mathbf{P}_1 = \mathbf{O}_{\text{out},1} \cdot \mathbf{J}_1 \cdot \mathbf{O}_{\text{in},1}^{-1}$$

$$= \begin{pmatrix} s'_{1x} & p'_{1x} & k_{2x} \\ s'_{1y} & p'_{1y} & k_{2y} \\ s'_{1z} & p'_{1z} & k_{2z} \end{pmatrix} \cdot \begin{pmatrix} r_{s1} & 0 & 0 \\ 0 & r_{p1} & 0 \\ 0 & 0 & 1 \end{pmatrix} \cdot \begin{pmatrix} s_{1x} & s_{1y} & s_{1z} \\ p_{1x} & p_{1y} & p_{1z} \\ k_{1x} & k_{1y} & k_{1z} \end{pmatrix} \quad (10.36)$$

$$= \begin{pmatrix} -0.889 + 0.219i & 0.106 + 0.046i & -0.361 + 0.054i \\ 0.361 - 0.054i & 0.314 - 0.137i & -0.863 - 0.039i \\ -0.106 - 0.046i & 0.655 - 0.628i & 0.314 - 0.137i \end{pmatrix}.$$

In this equation, the matrix $\mathbf{O}_{\text{in},1}^{-1}$ operates on the incident electric field $\mathbf{E}$ in global coordinates, rotating it into the local $(\hat{\mathbf{s}}_1, \hat{\mathbf{p}}_1, \hat{\mathbf{k}}_1)$ basis yielding $(E_s, E_p, 0)$, the projection of $\mathbf{E}$ onto the s–p incident local coordinates. $\mathbf{J}_1$ is an extended Jones matrix defined in Equation 10.29. Then, matrix $\mathbf{O}_{\text{out},1}$ rotates the resultant field from the local $(\hat{\mathbf{s}}'_1, \hat{\mathbf{p}}'_1, \hat{\mathbf{k}}'_2)$ coordinate back to the global Cartesian coordinate. In this process, the electric field $\mathbf{E}$ in $(\hat{\mathbf{s}}_1, \hat{\mathbf{p}}_1, \hat{\mathbf{k}}_1)$ basis is mapped to $(r_{s1}\hat{\mathbf{s}}'_1, r_{p1}\hat{\mathbf{p}}'_1, \hat{\mathbf{k}}'_2)$. Since both orthogonal matrices, $\mathbf{O}_{\text{in},1}$ and $\mathbf{O}_{\text{out},1}$, are defined by right-handed orthonormal basis, two of the three exiting basis, $\hat{\mathbf{k}}_2$ and $\hat{\mathbf{p}}'_1$, have a π phase change due to the changing local coordinate in reflection, where $\hat{\mathbf{k}}_2$ is understood as reflection and the π phase in $\hat{\mathbf{p}}'_1$ is added to $\text{Arg}[r_{p1}]$.

The diattenuation and retardance associated with the $\mathbf{P}$ matrix are calculated by the algorithms described in Chapters 9 and 17. Diattenuation is calculated from the singular value decomposition of $\mathbf{P}_1$,

$$\mathbf{P}_1 = \mathbf{U}_1 \boldsymbol{\Sigma}_1 \mathbf{V}_1^\dagger$$

$$= \begin{pmatrix} -0.195 & -0.973 e^{i2.918} & 0.126 e^{-i0.750} \\ -0.961 & 0.164 e^{i2.918} & -0.221 e^{-i0.750} \\ 0.195 & -0.164 e^{i2.918} & -0.967 e^{-i0.750} \end{pmatrix} \cdot \begin{pmatrix} 1 & 0 & 0 \\ 0 & 0.992 & 0 \\ 0 & 0 & 0.975 \end{pmatrix} \quad (10.37)$$

$$\cdot \begin{pmatrix} -0.195 & -0.973 & -0.126 \\ -0.195 & 0.164 & -0.967 \\ 0.961 & -0.164 & -0.221 \end{pmatrix}^\dagger,$$

10.6 Ray Tracing

where the singular values 0.992 and 0.975 correspond to $|r_{s1}|$ and $|r_{p1}|$. Thus, the diattenuation is $D = (|r_{s1}|^2 - |r_{p1}|^2)/(|r_{s1}|^2 + |r_{p1}|^2) = 0.018$. The maximum transmission axis is (−0.973, 0.164, −0.164), which is the second column of $\mathbf{V}_1$ and $\hat{\mathbf{s}}_1$ in the entrance space, or $e^{i2.918}$ (−0.973, 0.164, −0.164), which is the second column of $\mathbf{U}_1$ and $e^{i2.918}\hat{\mathbf{s}}'_1$ in the space following the M1.

To calculate the retardance induced by M1, the geometrical transformation of the ray path is first removed from $\mathbf{P}_1$; this in essence straightens out the path.* The geometrical transformation matrix $\mathbf{Q}_1$ (introduced in Chapter 17) for M1 is

$$\mathbf{Q}_1 = \mathbf{O}_{\text{out},1} \cdot \mathbf{I}_{\text{reflect1}} \cdot \mathbf{O}_{\text{in},1}^{-1} = \begin{pmatrix} s'_{1x} & p'_{1x} & k_{2x} \\ s'_{1y} & p'_{1y} & k_{2y} \\ s'_{1z} & p'_{1z} & k_{2z} \end{pmatrix} \cdot \begin{pmatrix} 1 & 0 & 0 \\ 0 & -1 & 0 \\ 0 & 0 & 1 \end{pmatrix} \cdot \begin{pmatrix} s_{1x} & s_{1y} & s_{1z} \\ p_{1x} & p_{1y} & p_{1z} \\ k_{1x} & k_{1y} & k_{1z} \end{pmatrix} = \begin{pmatrix} 1 & 0 & 0 \\ 0 & 0 & -1 \\ 0 & -1 & 0 \end{pmatrix}.$$

(10.38)

$\mathbf{Q}_1$ describes a non-polarizing reflection. The M1 retardance is calculated from the eigenvalues of $\mathbf{U}_{1Q} \cdot \mathbf{U}_{1Q}^{-1}$, where $\mathbf{Q}_1^{-1}\mathbf{P}_1 = \mathbf{U}_{1Q}\mathbf{\Sigma}_{1Q}\mathbf{V}_{1Q}^{-1}$ by singular value decomposition. The phase difference between the eigenvalues is 0.527 radian, the M1 retardance magnitude. The eigenvector corresponding to the smaller eigen-phase is (0.126, 0.967, 0.221), which is the fast axis and p-polarization in the entrance space.

For the second reflection, the $\mathbf{P}_2$ matrix and the corresponding diattenuation and retardance are calculated in the same manner. The reflected $\hat{\mathbf{k}}_3$ is (0.961, 0.195, 0.195) and the reflection angle is 35.16°. The incident $\hat{\mathbf{s}}_2$ and $\hat{\mathbf{p}}_2$ are (0.239, −0.239, −0.941) and (0.951, −0.137, 0.277). The exiting $\hat{\mathbf{s}}'_2$ and $\hat{\mathbf{p}}'_2$ are (0.239, −0.239, −0.941) and (−0.137, 0.951, −0.277). The Fresnel reflection coefficients r_{s2} and r_{p2} for the second reflections are $0.988e^{i2.80}$ and $0.982e^{-i0.507}$. Then,

$$\mathbf{P}_2 = \begin{pmatrix} -0.352 + 0.081i & -0.855 - 0.028i & 0.365 - 0.056i \\ 0.792 - 0.450i & -0.352 + 0.081i & 0.054 - 0.052i \\ -0.054 + 0.052i & -0.365 + 0.056i & -0.853 + 0.327i \end{pmatrix} \text{ and } \mathbf{Q}_2 = \begin{pmatrix} 0 & -1 & 0 \\ -1 & 0 & 0 \\ 0 & 0 & 1 \end{pmatrix}.$$

(10.39)

The corresponding diattenuation is 0.006 with maximum reflection axis (0.239, −0.239, −0.941) along s-polarization, and the retardance is 0.169 radian with the fast axis (0.951, −0.137, 0.277) along p-polarization in the entrance space of M2.

---

* To demonstrate the calculation of retardance, the geometric transformation matrix $\mathbf{Q}$ from Chapter 17 is used here.

Combining the effects from both mirrors, the cumulative retardance and diattenuation are obtained from the cumulative **P** matrix, which is calculated by cascading $\mathbf{P}_1$ and $\mathbf{P}_2$ through matrix multiplication:

$$\mathbf{P} = \mathbf{P}_2 \cdot \mathbf{P}_1 = \begin{pmatrix} -0.056 - 0.124i & -0.109 - 0.165i & 0.967 - 0.059i \\ -0.737 + 0.625i & 0.008 - 0.006i & 0.055 + 0.125i \\ 0.013 - 0.013i & -0.468 + 0.820i & 0.110 + 0.163i \end{pmatrix} \text{ and}$$

$$\mathbf{Q} = \mathbf{Q}_2 \cdot \mathbf{Q}_1 = \begin{pmatrix} 0 & 0 & 1 \\ -1 & 0 & 0 \\ 0 & -1 & 0 \end{pmatrix}.$$

(10.40)

Then, applying the same method described before, the overall diattenuation is 0.012 with maximum transmission axis $(0.977e^{-i2.447}, 0.118e^{i0.445}, 0.175e^{-i2.414})$, and retardance is 0.365 radian with fast axis $(0.076e^{i0.186}, 0.974, 0.213e^{i0.013})$ in the entrance space.

Example 10.4 Ray Tracing Parameters of a Ray Refracting through a Cell Phone Lens

An incident ray with incident propagation vector (0.0858, 0.1730, 0.98117) enters the example cell phone lens at (0, 0, 0) on the first lens element, as shown in Figure 10.32. Following the ray tracing procedure outlined in Section 10.6.5, the polarization ray tracing parameters are calculated from ray parameters shown in Table 10.10. The polarization ray tracing parameters are shown in Table 10.5.

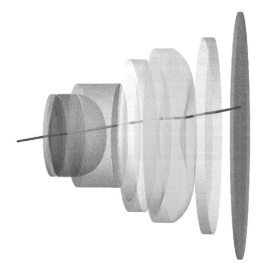

Figure 10.32 A ray propagates through the example cell phone lens.

In Table 10.5, the Jones matrix with Fresnel s and p transmission coefficients, the polarization ray tracing matrix **P**, and the parallel transport matrix **Q** for each ray intercept are given. This example system has uncoated lens surfaces.

10.6 Ray Tracing

Table 10.5 Jones Matrix with Fresnel *s* and *p* Transmission Coefficients, **P** Matrix, and **Q** Matrix for the Cell Phone Lens System

q	Jones Matrix $(\mathbf{J}_q)$	**P** Matrix $(\mathbf{P}_q)$	**Q** Matrix $(\mathbf{Q}_q)$
0 (STOP)	$\begin{pmatrix} 1 & 0 \\ 0 & 1 \end{pmatrix}$	$\begin{pmatrix} 1 & 0 & 0 \\ 0 & 1 & 0 \\ 0 & 0 & 1 \end{pmatrix}$	$\begin{pmatrix} 1 & 0 & 0 \\ 0 & 1 & 0 \\ 0 & 0 & 1 \end{pmatrix}$
1	$\begin{pmatrix} 0.77 & 0 \\ 0 & 0.77 \end{pmatrix}$	$\begin{pmatrix} 0.767 & 0.002 & -0.013 \\ 0.002 & 0.771 & -0.025 \\ 0.045 & 0.090 & 0.992 \end{pmatrix}$	$\begin{pmatrix} 1.000 & -0.001 & -0.032 \\ -0.001 & 0.998 & -0.065 \\ 0.032 & 0.065 & 0.997 \end{pmatrix}$
2	$\begin{pmatrix} 1.24 & 0 \\ 0 & 1.24 \end{pmatrix}$	$\begin{pmatrix} 1.234 & -0.002 & 0.023 \\ -0.002 & 1.230 & 0.046 \\ -0.056 & -0.113 & 1.003 \end{pmatrix}$	$\begin{pmatrix} 0.999 & -0.001 & 0.035 \\ -0.001 & 0.998 & 0.071 \\ -0.035 & -0.071 & 0.997 \end{pmatrix}$
3	$\begin{pmatrix} 0.77 & 0 \\ 0 & 0.77 \end{pmatrix}$	$\begin{pmatrix} 0.769 & 0.003 & 0.013 \\ 0.003 & 0.775 & 0.026 \\ 0.026 & 0.052 & 0.991 \end{pmatrix}$	$\begin{pmatrix} 1.000 & 0.000 & -0.008 \\ 0.000 & 1.000 & -0.015 \\ 0.008 & 0.0151 & 1.000 \end{pmatrix}$
4	$\begin{pmatrix} 1.23 & 0 \\ 0 & 1.24 \end{pmatrix}$	$\begin{pmatrix} 1.232 & -0.004 & -0.002 \\ -0.004 & 1.226 & -0.004 \\ -0.039 & -0.079 & 1.009 \end{pmatrix}$	$\begin{pmatrix} 1.000 & 0.000 & 0.017 \\ 0.000 & 0.999 & 0.034 \\ -0.017 & -0.034 & 0.999 \end{pmatrix}$
5	$\begin{pmatrix} 0.79 & 0 \\ 0 & 0.79 \end{pmatrix}$	$\begin{pmatrix} 0.790 & 0.003 & -0.018 \\ 0.003 & 0.793 & -0.035 \\ 0.050 & 0.101 & 0.990 \end{pmatrix}$	$\begin{pmatrix} 0.999 & -0.001 & -0.038 \\ -0.001 & 0.997 & -0.076 \\ 0.038 & 0.076 & 0.996 \end{pmatrix}$

(*Continued*)

Table 10.5 (Continued) Jones Matrix with Fresnel s and p Transmission Coefficients, **P** Matrix, and **Q** Matrix for the Cell Phone Lens System

q	Jones Matrix ($\mathbf{J}_q$)	P Matrix ($\mathbf{P}_q$)	Q Matrix ($\mathbf{Q}_q$)
6	$\begin{pmatrix} 1.21 & 0 \\ 0 & 1.21 \end{pmatrix}$	$\begin{pmatrix} 1.209 & -0.003 & 0.022 \\ -0.003 & 1.205 & 0.044 \\ -0.055 & -0.110 & 1.003 \end{pmatrix}$	$\begin{pmatrix} 0.999 & -0.001 & 0.035 \\ -0.001 & 0.998 & 0.070 \\ -0.035 & -0.070 & 0.997 \end{pmatrix}$
7	$\begin{pmatrix} 0.76 & 0 \\ 0 & 0.77 \end{pmatrix}$	$\begin{pmatrix} 0.760 & 0.001 & -0.057 \\ 0.001 & 0.762 & -0.114 \\ 0.082 & 0.165 & 0.983 \end{pmatrix}$	$\begin{pmatrix} 0.997 & -0.006 & -0.078 \\ -0.006 & 0.988 & -0.158 \\ 0.078 & 0.158 & 0.984 \end{pmatrix}$
8	$\begin{pmatrix} 1.24 & 0 \\ 0 & 1.26 \end{pmatrix}$	$\begin{pmatrix} 1.243 & -0.001 & 0.077 \\ -0.001 & 1.241 & 0.155 \\ -0.107 & -0.216 & 0.985 \end{pmatrix}$	$\begin{pmatrix} 0.997 & -0.007 & 0.081 \\ -0.007 & 0.987 & 0.164 \\ -0.081 & -0.164 & 0.983 \end{pmatrix}$
9	$\begin{pmatrix} 0.79 & 0 \\ 0 & 0.79 \end{pmatrix}$	$\begin{pmatrix} 0.790 & 0.003 & -0.014 \\ 0.003 & 0.794 & -0.027 \\ 0.047 & 0.095 & 0.990 \end{pmatrix}$	$\begin{pmatrix} 0.999 & -0.001 & -0.034 \\ -0.001 & 0.998 & -0.068 \\ 0.034 & 0.068 & 0.997 \end{pmatrix}$
10	$\begin{pmatrix} 1.21 & 0 \\ 0 & 1.22 \end{pmatrix}$	$\begin{pmatrix} 1.211 & -0.003 & 0.020 \\ -0.003 & 1.206 & 0.041 \\ -0.055 & -0.110 & 1.004 \end{pmatrix}$	$\begin{pmatrix} 0.999 & -0.001 & 0.034 \\ -0.001 & 0.998 & 0.068 \\ -0.034 & -0.068 & 0.997 \end{pmatrix}$
11 (DET)	$\begin{pmatrix} 1 & 0 \\ 0 & 1 \end{pmatrix}$	$\begin{pmatrix} 1 & 0 & 0 \\ 0 & 1 & 0 \\ 0 & 0 & 1 \end{pmatrix}$	$\begin{pmatrix} 1 & 0 & 0 \\ 0 & 1 & 0 \\ 0 & 0 & 1 \end{pmatrix}$

10.7 Wavefront Analysis

Optical designers have developed many metrics to evaluate the performance of optical systems, principal among them is the wavefront aberration function that plays a fundamental role in the evaluation of imaging systems. The expansion of the wavefront aberration about the optical axis into the Seidel aberrations provides a language for discussing the shape and form of aberration (see Section 10.7.5). A different expansion of a single wavefront into a set of orthonormal functions over a circular aperture, the Zernike polynomials, is also common (see Section 10.7.6). Later chapters extend these methods for the description of polarization aberrations.

10.7.1 Normalized Coordinates

Section 10.6 describes a single ray at a series of interfaces. Now, the interest shifts to characterizing large numbers of rays at the object, image, and pupils. Each ray starts on the object surface at object vector $\mathbf{h_O}$ and intersects the entrance pupil at aperture vector $\boldsymbol{\rho}_E$, as shown in Figure 10.33. For the description of aberrations, the magnitude of $\boldsymbol{\rho}$ is commonly normalized to one at the edge of the pupil and $\mathbf{h}$ normalized to one at the edge of the object.

Each of these rays is paired to an exiting ray in image space at the exit pupil. The description of wavefront aberrations is simplified when normalized coordinates are used for the object and the pupil. In Figure 10.33, a circular object is shown with a circular entrance pupil, but normalized coordinates are useful with arbitrarily shaped objects and pupils, provided the unit vectors are defined at both locations.

10.7.2 Wavefront Aberration Function

An individual ray trace provides information on the path, phase, and polarization of a single ray. To evaluate an optical system, one needs to understand the properties of rays that fill the pupils, originate from the entire object, and span the spectral range. The most important tool for evaluating the imaging optical system's performance has been the *wavefront aberration function*, which describes the deviation of wavefronts in the exit pupil from an ideal spherical surface. In conventional optical design, the wavefront aberration function is evaluated by tracing sets of rays through the optical system.

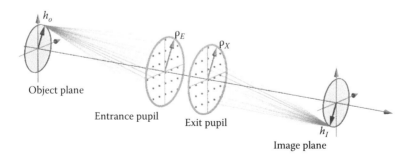

Figure 10.33 Normalized coordinates for aberration theory. Rays in object space start at a normalized object coordinate $\mathbf{h_O}$ and intersect the entrance pupil at a normalized coordinate $\boldsymbol{\rho}_E$. An exit aperture vector $\boldsymbol{\rho}_X$ on the exit pupil and a field vector $\mathbf{h_I}$ on the image plane are most common.

For each object **h** and wavelength λ, the deviation of the wavefront from spherical reference surface across the exit pupil as a function of entrance pupil coordinates ρ_E is a scalar valued function $W(\mathbf{h}, \boldsymbol{\rho}, \lambda)$. W is commonly expressed in waves and described on a plane or spherical surface that passes through the center of the exit pupil. The electric field of a scalar wavefront on this reference sphere surface parameterized by (ρ, ϕ) or (x, y) is

$$\mathbf{E}(\rho,\phi) = E_0 \exp\left(-i2\pi W(\rho,\phi)/\lambda\right) = \mathbf{E}(x,y) = E_0 \exp\left(-i2\pi W(x,y)/\lambda\right). \tag{10.41}$$

The wavefront aberration function is the phase expressed in waves.

In conventional optical design, the wavefront aberration function is the principal quantity calculated to characterize the quality of image formation. The wavefront aberration function is the optical path length along each ray from a spherical surface in the entrance pupil to a reference sphere in the exit pupil. Usually, the optical path length of the chief ray between the entrance pupil and exit pupil is subtracted from all of the optical path lengths to normalize the wavefront aberration function to zero in the center of the pupil, making it easier to interpret and compare. The wavefront aberration describes the chromatic aberration, spherical aberration, coma, and all the other geometrical aberrations of the optical system.

In order to design, optimize, and analyze an optical system by ray tracing, a large number of rays are traced to sample the image quality; rays are traced from a set of object points that sample many locations in the pupil over the range of wavelengths. In order to form an image of good resolution, all the rays from each object point need to converge to a small area in the image, and the optical path length of those rays needs to be equal to within a fraction of a wavelength. Additionally, for a given object point, the images of light for all the wavelengths should form at the same image point to avoid *chromatic aberration*, blurring as a function of wavelength. Thus, a major task of an optical design program is to take the algorithm for tracing a single ray and systematically apply the algorithm to grids of rays from a set of object points and wavelengths.

10.7.3 Polarization Aberration Function

While the wavefront aberration function describes the geometrical contribution to the shape and phase of the exiting wavefront, the system's interfaces, thin films, filters, diffraction gratings, crystals, liquid crystals, and other components make contributions to the phase that are not directly related to the optical path lengths. These changes to the phase, amplitude, and polarization state are calculated using our polarization ray tracing algorithms. In the polarization ray tracing algorithms, the total polarization effects of the optical system is stored in a polarization ray tracing matrix $\mathbf{P}_{Total}$, which is calculated by cascading polarization ray tracing matrices $\mathbf{P}_q$ along the ray paths for each ray intercept from the entrance pupil to the exit pupil,

$$\mathbf{P}_{Total} = \mathbf{P}_N \mathbf{P}_{N-1} \cdots \mathbf{P}_q \cdots \mathbf{P}_2 \mathbf{P}_1 = \prod_{q=N,-1}^{1} \mathbf{P}_q. \tag{10.42}$$

The polarization aberration function is **P** as a function of ray coordinates, $\mathbf{P}(\mathbf{h}, \boldsymbol{\rho}, \lambda)$. Optical coatings adjust the wavefront by adding small amounts of defocus, spherical aberration, coma, and other aberrations. Since coatings also have retardance, its contribution to the polarization aberrations are different for different incident polarization states. Therefore, they cannot be described solely by the wavefront aberration function, which is only a scalar function. Figure 10.34 (left) shows the polarization aberration function for the example cell phone lens at its 10° field relating the polarization vectors in the incident collimated beam to the polarization vectors in the exit pupil. Since the rays

10.7 Wavefront Analysis

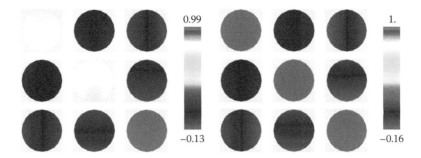

Figure 10.34 (Left) Polarization aberration function **P**(**h**, **ρ**, λ) for the cell phone lens on-axis at **h** = (0, 0) at 10° field. (Right) Corresponding geometric transformation function **Q**(**h**, **ρ**, λ) that relates several quantities between the entrance and exit pupil, (1) propagation vectors, and (2) polarization vectors for a non-polarizing optical system, where the light polarization ellipse only changes direction at each ray intercept, rotating about the *s*-basis vector.

change direction between the entrance and exit pupils, part of the polarization aberration function is a rotation matrix that changes the ray's direction. Another part incorporates the loss of amplitude at each uncoated ray intercept. Finally, this function contains the accumulated diattenuation of the uncoated interfaces; Figure 10.5 (right) was derived from this **P**(**h**, **ρ**, λ). The diattenuation function or diattenuation map is calculated from **P**(**h**, **ρ**, λ) using the algorithms of Section 9.4. If this lens also contains sources of retardance, the retardance would also be contained in **P**(**h**, **ρ**, λ) and calculated by the algorithms of Chapter 17.

The Jones pupil, **J**(**h**, **ρ**, λ), relates Jones vectors defined in the entrance to exit pupils. Since the associated wavefront at these pupil surfaces is usually spherical, local coordinates must be defined over each sphere. The polarization associated to each ray is well described in the associated local coordinate. The most common local coordinate system for spherical wavefront is the double pole coordinate system (see Figures 11.8 through 11.10). The choice of local coordinates is discussed at length in Chapter 11. In the double pole coordinate system, as seen in the center of Figure 11.10 (left), usually the anti-pole is the most uniform part of the coordinates and is located along the chief ray. The singularity, seen in the center of Figure 11.10 (right), is placed in the opposite direction, backward along the chief ray.

Figure 10.6 shows the Jones pupil for the example cell phone lens at 10° field, which describes the diattenuation, loss of amplitude, phase changes from interfaces, and the retardance. In the case of an uncoated lens, the phase changes and retardance are zero; hence, this lens' Jones pupils contain only real values.

10.7.4 Evaluation of the Aberration Function

The wavefront aberration function of Section 10.7.2 is a continuous function defined for all **h** and **ρ**. In ray tracing, the wavefront aberration function must be sampled by tracing many rays. Thus, a set of pupil arrays for quantities of interest are calculated to assess wavefront and image quality.

For each chosen **h** and wavelength λ, a grid of ray intercepts, $\mathbf{r}_{i,j}$, that overfill the entrance pupil are defined (Figure 10.35); indices i and j run from $-M$ to M, yielding a $2M + 1$ by $2M + 1$ grid of rays.* The ray with indices $(i, j) = (0, 0)$ is typically the chief ray.[†]

Each ray is traced through the system to a reference surface in image space. The ray intercepts $\mathbf{r}_{X,i,j}$ and propagation vectors $\mathbf{k}_{X,i,j}$ at the reference surface are gathered into exit pupil arrays,

---

* A square grid is shown, but polar grids and other grids appropriate to particular geometries are also common.
† Odd numbers of rays aren't required, but putting a ray down the very center of the pupil and center of array is convenient.

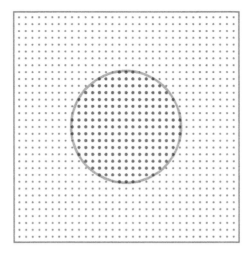

Figure 10.35 Ray grid on the entrance pupil surface shown overfilling the pupil. The blue circle is the boundary of the entrance pupil with rays inside the pupil shown in red.

$$\mathbf{r}_X = \begin{pmatrix} \mathbf{r}_{-M,-M} & \mathbf{r}_{-M,-M+1} & \cdots & \mathbf{r}_{-M,M} \\ \vdots & & \ddots & \vdots \\ \mathbf{r}_{M,-M} & \mathbf{r}_{M,-M+1} & \cdots & \mathbf{r}_{M,M} \end{pmatrix}, \quad \mathbf{k}_X = \begin{pmatrix} \mathbf{k}_{-M,-M} & \mathbf{k}_{-M,-M+1} & \cdots & \mathbf{k}_{-M,M} \\ \vdots & & \ddots & \vdots \\ \mathbf{k}_{M,-M} & \mathbf{k}_{M,-M+1} & \cdots & \mathbf{k}_{M,M} \end{pmatrix}. \quad (10.43)$$

A spot diagram, a representation of the transverse ray aberrations of the optic, can be calculated from the set of ray intercepts $\mathbf{r}_{X,i,j}$ and propagation vectors $\mathbf{k}_{X,i,j}$ by extending the ray paths to the image plane and recording the set of "spots." A smaller spot is desirable; the rays all converging to a point is ideal.

The first objective is determining an optical path difference (*OPD*) map in the exit pupil, which describes the shape of the wavefront exiting the optical system and how far the wavefront departs from spherical. The exit pupil array is calculated by summing the optical path lengths of each ray segment from the object to the exit pupil.

The sampled *OPL* at the exit pupil, $OPL_{X,i,j}$, is the basis for wavefront aberration function,

$$\mathbf{OPL}_X = \begin{pmatrix} OPL_{X,-M,-M} & OPL_{X,-M,-M+1} & \cdots & OPL_{X,-M,M} \\ \vdots & & \ddots & \vdots \\ OPL_{X,M,-M} & OPL_{X,M,-M+1} & \cdots & OPL_{X,M,M} \end{pmatrix}. \quad (10.44)$$

The sampled optical path difference, *OPD*, at the exit pupil subtracts the chief ray from all the *OPL*s:

$$OPD_{X,i,j} = OPL_{X,i,j} - OPL_{X,0,0}. \quad (10.45)$$

The wavefront aberration function in waves, $W(\mathbf{r}_{i,j}) = \mathbf{OPD}/\lambda$, is now represented as a sampled function at a set of exit pupil coordinates.

10.7 Wavefront Analysis

The reference sphere is constructed by defining an image location and constructing a sphere passing through the center of the exit pupil centered on this image point. Then, the ray intercepts of a grid of rays in image space are found on the reference sphere, and the optical lengths from the object or, equivalently, the entrance pupil (which differs by a constant for all rays) is tabulated. The *OPL* for the chief ray can be subtracted from all the *OPL*s, leaving the deviation from the reference sphere, *OPD*, for all the rays traced in the ray grid.

$$\begin{pmatrix} 0 & 0 & 0 & 0 & 0 & 0 & 0 & 0 & 0 \\ 0 & 0 & OPD_{-3,-2} & OPD_{-3,-1} & OPD_{-3,0} & OPD_{-3,1} & OPD_{-3,2} & 0 & 0 \\ 0 & OPD_{-2,-3} & OPD_{-2,-2} & OPD_{-2,-1} & OPD_{-2,0} & OPD_{-2,1} & OPD_{-2,2} & OPD_{-2,3} & 0 \\ 0 & OPD_{-1,-3} & OPD_{-1,-2} & OPD_{-1,-1} & OPD_{-1,0} & OPD_{-1,1} & OPD_{-1,2} & OPD_{-1,3} & 0 \\ 0 & OPD_{0,-3} & OPD_{0,-2} & OPD_{0,-1} & OPD_{0,0} & OPD_{0,1} & OPD_{0,2} & OPD_{0,3} & 0 \\ 0 & OPD_{1,-3} & OPD_{1,-2} & OPD_{1,-1} & OPD_{1,0} & OPD_{1,1} & OPD_{1,2} & OPD_{1,3} & 0 \\ 0 & OPD_{2,-3} & OPD_{2,-2} & OPD_{2,-1} & OPD_{2,0} & OPD_{2,1} & OPD_{2,2} & OPD_{2,3} & 0 \\ 0 & 0 & OPD_{3,-2} & OPD_{3,-1} & OPD_{3,0} & OPD_{3,1} & OPD_{3,2} & 0 & 0 \\ 0 & 0 & 0 & 0 & 0 & 0 & 0 & 0 & 0 \end{pmatrix} \quad (10.46)$$

Usually the ray grid overfills the entrance pupil and aperture stop; hence, some of the rays around the outside of the grid are vignetted and do not reach the exit pupil.

To construct the reference sphere, the image location must be defined. Often, the image location is defined as the paraxial image location. Other criteria can be used such as locating the image in the plane where the root mean square (RMS) spot size is a minimum, the center of gravity of the spot. For a well-formed image, all of these criteria yield reference spheres that are close to each other, and all are appropriate for aberration definition.

Similarly, the *polarization aberration function* for each element of the ray grid is organized into an array representing the sampled polarization aberration function,

$$\mathbf{P}_X = \begin{pmatrix} 0 & 0 & 0 & 0 & 0 & 0 & 0 & 0 & 0 \\ 0 & 0 & \mathbf{P}_{-3,-2} & \mathbf{P}_{-3,-1} & \mathbf{P}_{-3,0} & \mathbf{P}_{-3,1} & \mathbf{P}_{-3,2} & 0 & 0 \\ 0 & \mathbf{P}_{-2,-3} & \mathbf{P}_{-2,-2} & \mathbf{P}_{-2,-1} & \mathbf{P}_{-2,0} & \mathbf{P}_{-2,1} & \mathbf{P}_{-2,2} & \mathbf{P}_{-2,3} & 0 \\ 0 & \mathbf{P}_{-1,-3} & \mathbf{P}_{-1,-2} & \mathbf{P}_{-1,-1} & \mathbf{P}_{-1,0} & \mathbf{P}_{-1,1} & \mathbf{P}_{-1,2} & \mathbf{P}_{-1,3} & 0 \\ 0 & \mathbf{P}_{0,-3} & \mathbf{P}_{0,-2} & \mathbf{P}_{0,-1} & \mathbf{P}_{0,0} & \mathbf{P}_{0,1} & \mathbf{P}_{0,2} & \mathbf{P}_{0,3} & 0 \\ 0 & \mathbf{P}_{1,-3} & \mathbf{P}_{1,-2} & \mathbf{P}_{1,-1} & \mathbf{P}_{1,0} & \mathbf{P}_{1,1} & \mathbf{P}_{1,2} & \mathbf{P}_{1,3} & 0 \\ 0 & \mathbf{P}_{2,-3} & \mathbf{P}_{2,-2} & \mathbf{P}_{2,-1} & \mathbf{P}_{2,0} & \mathbf{P}_{2,1} & \mathbf{P}_{2,2} & \mathbf{P}_{2,3} & 0 \\ 0 & 0 & \mathbf{P}_{3,-2} & \mathbf{P}_{3,-1} & \mathbf{P}_{3,0} & \mathbf{P}_{3,1} & \mathbf{P}_{3,2} & 0 & 0 \\ 0 & 0 & 0 & 0 & 0 & 0 & 0 & 0 & 0 \end{pmatrix}. \quad (10.47)$$

Equation 10.47 shows an array with seven rays across the pupil diameter, but of course the size can be arbitrarily selected.

The polarization aberration function relates polarization states in the entrance and exit pupil in three dimensions in an unambiguous way. The optical designer and colleagues also need the polarization represented in flattened forms to print, display on computer monitors, and project onto screens. Polarization aberration functions defined over pairs of spherical surfaces, such as reference spheres in the entrance and exit pupils, are converted into Jones matrix functions by defining local coordinate systems over the pair of spherical surfaces and rotating the polarization aberration functions into each sphere's local coordinate systems, as described at length in Chapter 11. The transformation from **P** into **J** is performed with two rotation matrices,

$$\mathbf{J} = \mathbf{O}_X^{-1} \cdot \mathbf{P} \cdot \mathbf{O}_E = \begin{pmatrix} \hat{x}_{x,X} & \hat{x}_{y,X} & \hat{x}_{z,X} \\ \hat{y}_{z,X} & \hat{y}_{z,X} & \hat{y}_{z,X} \\ \hat{k}_{x,X} & \hat{k}_{y,X} & \hat{k}_{z,X} \end{pmatrix} \begin{pmatrix} p_{xx} & p_{xy} & p_{xz} \\ p_{yx} & p_{yy} & p_{yz} \\ p_{zx} & p_{zy} & p_{zz} \end{pmatrix} \begin{pmatrix} \hat{x}_{x,E} & \hat{y}_{x,E} & \hat{k}_{x,E} \\ \hat{x}_{y,E} & \hat{y}_{y,E} & \hat{k}_{y,E} \\ \hat{x}_{z,E} & \hat{y}_{z,E} & \hat{k}_{z,E} \end{pmatrix} = \begin{pmatrix} J_{11} & J_{12} & 0 \\ J_{21} & J_{22} & 0 \\ 0 & 0 & 1 \end{pmatrix}.$$

(10.48)

Matrix $\mathbf{O}_E$ rotates the entrance pupil local coordinates for the specified ray into global coordinate $(\hat{x}_E, \hat{y}_E, \hat{k}_E)$. The first column contains the x-, y-, and z-components of the x-local coordinate; the second column, the y-local coordinate; and the third column, the propagation vector at the entrance pupil. This orthogonal matrix indicates rotation from local to global coordinate is represented as an orthogonal matrix and rotation from global to local coordinate is represented as the inverses of orthogonal matrix. Matrix $\mathbf{O}_X^{-1}$ rotates the global coordinates of **P** into the local coordinate of the exit pupil. The first row of $\mathbf{O}_X^{-1}$ contains the ray's exit pupil x-local coordinate; the second row, the y-local coordinate; and the third row, the propagation vector.

By applying rotations associated with the local coordinate for each ray intercept in object and image space, each $\mathbf{P}_{i,j}$ is transformed into a Jones matrix $\mathbf{J}_{i,j}$, and the array of Jones matrices is the *Jones pupil*,

$$\mathbf{J}_X = \begin{pmatrix} 0 & 0 & 0 & 0 & 0 & 0 & 0 & 0 & 0 \\ 0 & 0 & \mathbf{J}_{-3,-2} & \mathbf{J}_{-3,-1} & \mathbf{J}_{-3,0} & \mathbf{J}_{-3,1} & \mathbf{J}_{-3,2} & 0 & 0 \\ 0 & \mathbf{J}_{-2,-3} & \mathbf{J}_{-2,-2} & \mathbf{J}_{-2,-1} & \mathbf{J}_{-2,0} & \mathbf{J}_{-2,1} & \mathbf{J}_{-2,2} & \mathbf{J}_{-2,3} & 0 \\ 0 & \mathbf{J}_{-1,-3} & \mathbf{J}_{-1,-2} & \mathbf{J}_{-1,-1} & \mathbf{J}_{-1,0} & \mathbf{J}_{-1,1} & \mathbf{J}_{-1,2} & \mathbf{J}_{-1,3} & 0 \\ 0 & \mathbf{J}_{0,-3} & \mathbf{J}_{0,-2} & \mathbf{J}_{0,-1} & \mathbf{J}_{0,0} & \mathbf{J}_{0,1} & \mathbf{J}_{0,2} & \mathbf{J}_{0,3} & 0 \\ 0 & \mathbf{J}_{1,-3} & \mathbf{J}_{1,-2} & \mathbf{J}_{1,-1} & \mathbf{J}_{1,0} & \mathbf{J}_{1,1} & \mathbf{J}_{1,2} & \mathbf{J}_{1,3} & 0 \\ 0 & \mathbf{J}_{2,-3} & \mathbf{J}_{2,-2} & \mathbf{J}_{2,-1} & \mathbf{J}_{2,0} & \mathbf{J}_{2,1} & \mathbf{J}_{2,2} & \mathbf{J}_{2,3} & 0 \\ 0 & 0 & \mathbf{J}_{3,-2} & \mathbf{J}_{3,-1} & \mathbf{J}_{3,0} & \mathbf{J}_{3,1} & \mathbf{J}_{3,2} & 0 & 0 \\ 0 & 0 & 0 & 0 & 0 & 0 & 0 & 0 & 0 \end{pmatrix}.$$

(10.49)

Data reduction is applied to each Jones matrix to calculate a diattenuation pupil map, retardance pupil map, an amplitude map, and related polarization property distributions. Similarly, if the incident polarization state is specified, the exiting polarization state can be calculated for a grid of rays. The transmittance of the optical system can be evaluated by calculating the transmission from the entrance to exit pupil.

The complexity of the wavefront determines the minimum number of rays needed in the ray grid to properly sample the wavefront and represent all its topographic features as shown in Figure 10.36.

10.7.5 Seidel Wavefront Aberration Expansion

A language is needed for describing wavefront aberration functions to provide an efficient way to communicate the shapes and properties of wavefronts, such as how many peaks, holes, ridges, and

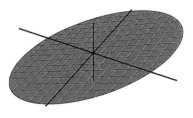 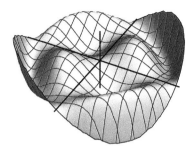

Figure 10.36 (Left) A simple wavefront needs smaller numbers of rays. A complex wavefront (right) needs more rays to accurately describe the shape.

valleys they have. The shape of a wavefront is often described as low-order polynomials, which arise naturally from the laws of reflection and refraction of spherical and aspheric surfaces.

The Seidel wavefront aberration expansion is obtained by expanding $W(\mathbf{h}, \boldsymbol{\rho}, \lambda)$ in a polynomial in the object and pupil coordinates, as shown in Equation 10.50. The scalar wavefront aberration function is the scalar products of the aperture vector and field vector, that is, $\mathbf{h} \cdot \mathbf{h}$, $\mathbf{h} \cdot \boldsymbol{\rho}$, and $\boldsymbol{\rho} \cdot \boldsymbol{\rho}$. The reference spheres are defined as the paraxial image locations for the reference wavelength,

$$W(\mathbf{h},\boldsymbol{\rho}) = \sum_{\alpha,\beta,\gamma=0}^{\infty} W_{\alpha,\beta,\gamma}(\mathbf{h}\cdot\mathbf{h})^{\alpha}(\mathbf{h}\cdot\boldsymbol{\rho})^{\beta}(\boldsymbol{\rho}\cdot\boldsymbol{\rho})^{\gamma}. \qquad (10.50)$$

$W_{\alpha,\beta,\gamma}$ are the aberration coefficients describing the magnitude of each aberration term. For a radially symmetric lens and mirror system, symmetry limits the terms present to the following reduced set, indexed by j, m, and n,

$$W(\mathbf{h},\boldsymbol{\rho}) = \sum_{j,m,n=0}^{\infty} W_{k,l,m}(\mathbf{h}\cdot\mathbf{h})^{j}(\mathbf{h}\cdot\boldsymbol{\rho})^{m}(\boldsymbol{\rho}\cdot\boldsymbol{\rho})^{n}, \qquad (10.51)$$

where j, m, and n run over the integers and indices k and l are defined as

$$k = 2j + m, \text{ and } l = 2n + m. \qquad (10.52)$$

The order of the aberration is $2(j + m + n)$, which is always even. Table 10.6 shows the first 10 terms of Seidel aberrations, where ϕ is the angle from x to y measured counterclockwise. The first six terms are plotted in Figure 10.37.

The Seidel aberrations describe the shape of the wavefront in simple and easy-to-understand functions with well-understood meanings. The fourth-order Seidel aberration coefficients of radially symmetric systems can be easily calculated from a paraxial ray trace, in an algorithm dating from the 1800s.[12] Each Seidel term is calculated as the sum of marginal and chief ray parameters over the surfaces. As the aberrations propagate from surface to surface, the aberrations generate additional higher-order aberration components. Thus, the calculation of sixth- and higher-order aberrations is complicated by the evaluation of these induced aberration terms and does not proceed

Table 10.6 Zeroth-, Second-, and Fourth-Order Seidel Aberrations

Name	Vector Form	Polar Form	j	m	n
Zeroth Order					
Piston	$W_{0,0,0}$	$W_{0,0,0}$	0	0	0
Second Order					
Quadratic piston	$W_{2,0,0}(\mathbf{h}\cdot\mathbf{h})$	$W_{2,0,0}\,H^2$	2	0	0
Tilt	$W_{1,1,1}(\mathbf{h}\cdot\boldsymbol{\rho})$	$W_{1,1,1}\,H\rho\cos\phi$	0	2	0
Defocus	$W_{0,2,0}(\boldsymbol{\rho}\cdot\boldsymbol{\rho})$	$W_{0,2,0}\,\rho^2$	0	0	2
Fourth Order					
Spherical aberration	$W_{0,4,0}(\boldsymbol{\rho}\cdot\boldsymbol{\rho})^2$	$W_{4,0,0}\,\rho^4$	4	0	0
Coma	$W_{1,3,1}(\mathbf{h}\cdot\boldsymbol{\rho})(\boldsymbol{\rho}\cdot\boldsymbol{\rho})$	$W_{1,3,1}\,H\rho^3\cos\phi$	1	3	1
Astigmatism	$W_{2,2,2}(\mathbf{h}\cdot\boldsymbol{\rho})^2$	$W_{2,2,2}\,H^2\rho^2\cos^2\phi$	2	2	2
Field curvature	$W_{2,2,0}(\mathbf{h}\cdot\mathbf{h})(\boldsymbol{\rho}\cdot\boldsymbol{\rho})$	$W_{2,2,0}\,H^2\rho^2$	2	2	0
Distortion	$W_{3,1,1}(\mathbf{h}\cdot\mathbf{h})(\mathbf{h}\cdot\boldsymbol{\rho})$	$W_{3,1,1}\,H^3\rho\cos\phi$	3	1	1
Quartic piston	$W_{4,0,0}(\mathbf{h}\cdot\mathbf{h})^2$	$W_{4,0,0}\,H^4$	4	0	0

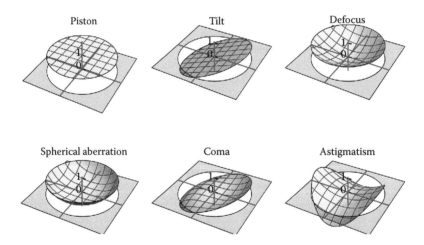

Figure 10.37 The Seidel zeroth-order aberration (piston), second-order aberrations (tilt and defocus), and fourth-order aberrations (spherical aberration, coma, and astigmatism).

as a simple sum over surfaces. Sasián[13,14] provides equations to calculate the Seidel aberrations through sixth order from the paraxial ray trace. The Seidel aberrations for off-axis systems with tilts and decenters cannot be described by the reduced number of terms in Equation 10.51 and Table 10.6, but require the fuller set of terms in Equation 10.50, for example, as shown in the *nodal* (or *vector*) *aberration theory* of Thompson.[15–20]

The Seidel aberrations provide a description of the wavefront aberration over the field of view of a radially symmetric optical system. For many purposes, it is only necessary to describe a single wavefront at a time. For example, interferometers measure a single wavefront, or a sequence of wavefronts, but not continuous wavefronts as functions of object coordinate.

The Seidel aberration coefficients can also be determined from a set of wavefronts by curve fitting. The wavefronts might be determined by ray tracing or from a set of interferograms.

10.7 Wavefront Analysis

One disadvantage of using polynomials as a basis set for surfaces is that they are not orthogonal. Consider a set of Seidel coefficients for the second- and fourth-order terms determined by curve fitting. Next start over and determine the Seidel coefficients through sixth order. The two sets of second- and fourth-order coefficients will be different because the Seidel aberration terms are not orthogonal. Also, the wavefront in the exit pupil of lenses, telescopes, and other imaging systems occurs in an infinite variety of shapes. The wavefront can be discontinuous as occurs with Fresnel lenses and other optical elements with steps in their surface heights. The wavefront can fold over on itself and become multivalued; then, there are caustics within the wavefront.

10.7.6 Zernike Polynomials

The *Zernike polynomials* are a separate set of basis functions, different from the Seidel aberrations, for describing wavefronts over a circular aperture.[21–23] The Zernike polynomials $Z_n(\rho, \phi)$, or in Cartesian coordinates $Z_n(x, y)$, are constructed as an *orthonormal set* of scalar functions over the unit circle,

$$\int_0^{2\pi}\int_0^1 Z_m(\rho,\phi)Z_n(\rho,\phi)\rho\,d\rho\,d\phi = \delta_{m,n}. \tag{10.53}$$

A circular wavefront with aberration in waves, $W(\rho, \phi)$, which has been scaled to a diameter of two, can be expressed in terms of a set of Zernike coefficients, z_n, as

$$z_n = \int_0^{2\pi}\int_0^1 W(\rho,\phi)Z_n(\rho,\phi)\rho\,d\rho\,d\phi. \tag{10.54}$$

Properties of Zernike polynomials are compiled in many references.[24,25] Table 10.7 lists the Zernike polynomials in polar coordinates. These are plotted in Figure 10.38. Several orderings of the Zernike polynomials are in use; hence, this numbering is not universal. The remainder of this section assumes that wavefront aberration functions with circular apertures have been scaled to a radius of one (diameter of two).

The Zernike polynomials are *complete*, meaning any smooth and differentiable function can be described by including enough orders. Therefore, with an infinite number of terms, any *delta function* within the unit circle can be constructed.

The Zernike polynomials are used in representing *aberrations* and *interferogram data*.[26] Because of their orthogonality, the values of the coefficients do not depend on the number of terms evaluated. The Zernike coefficients, $(z_1, z_2, \ldots)$ for a wavefront are found by projecting the Zernike polynomials, one at a time, onto the wavefront aberration function,

$$z_i = \iint_{\text{unit circle}} W(x,y)Z_i(x,y)\,dx\,dy. \tag{10.55}$$

Since the Zernike polynomials are orthogonal, they provide a simple way to remove *tilt* and *defocus* from a wavefront. If a wavefront is expressed in Zernike polynomials,

$$W(\rho,\phi) = \sum_1^N z_n Z_n(\rho,\phi), \tag{10.56}$$

Table 10.7 Table of the First 36 Zernike Polynomials in Polar Coordinate Form

1	2	3	4	5	6
1	$2\rho\cos\phi$	$2\rho\sin\phi$	$\sqrt{3}(2\rho^2-1)$	$\sqrt{6}\,\rho^2\sin(2\phi)$	$\sqrt{6}\,\rho^2\cos(2\phi)$
7	8	9	10	11	12
$2\sqrt{2}\,\rho(3\rho^2-2)\sin\phi$	$2\sqrt{2}\,\rho(3\rho^2-2)\cos\phi$	$2\sqrt{2}\,\rho^3\sin(3\phi)$	$2\sqrt{2}\,\rho^3\cos(3\phi)$	$\sqrt{5}(6\rho^4-6\rho^2+1)$	$\sqrt{10}\,\rho^2(4\rho^2-3)\cos(2\phi)$
13	14	15	16	17	18
$\sqrt{10}\,\rho^2(4\rho^2-3)\sin(2\phi)$	$\sqrt{10}\,\rho^4\cos(4\phi)$	$\sqrt{10}\,\rho^4\sin(4\phi)$	$2\sqrt{3}\,\rho(10\rho^4-12\rho^2+3)\cos\phi$	$2\sqrt{3}\,\rho(10\rho^4-12\rho^2+3)\sin\phi$	$2\sqrt{3}\,\rho^3(5\rho^2-4)\cos(3\phi)$
19	20	21	22	23	24
$2\sqrt{3}\,\rho^3(5\rho^2-4)\sin(3\phi)$	$2\sqrt{3}\,\rho^5\cos(5\phi)$	$2\sqrt{3}\,\rho^5\sin(5\phi)$	$\sqrt{7}(20\rho^6-30\rho^4+12\rho^2-1)$	$\sqrt{14}\,\rho^2(15\rho^4-20\rho^2+6)\sin(2\phi)$	$\sqrt{14}\,\rho^2(15\rho^4-20\rho^2+6)\cos(2\phi)$
25	26	27	28	29	30
$\sqrt{14}\,\rho^4(6\rho^2-5)\sin(4\phi)$	$\sqrt{14}\,\rho^4(6\rho^2-5)\cos(4\phi)$	$\sqrt{14}\,\rho^6\sin(6\phi)$	$\sqrt{14}\,\rho^6\cos(6\phi)$	$4\rho(35\rho^6-60\rho^4+30\rho^2-4)\sin\phi$	$4\rho(35\rho^6-60\rho^4+30\rho^2-4)\cos\phi$
31	32	33	34	35	36
$4\rho^3(21\rho^4-30\rho^2+10)\sin(3\phi)$	$4\rho^3(21\rho^4-30\rho^2+10)\cos(3\phi)$	$4\rho^5(7\rho^2-6)\sin(5\phi)$	$4\rho^5(7\rho^2-6)\cos(5\phi)$	$4\rho^7\sin(7\phi)$	$4\rho^7\cos(7\phi)$

10.7 Wavefront Analysis 403

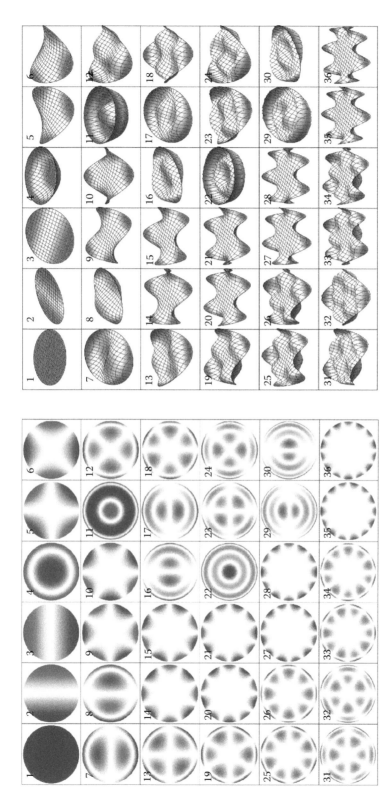

Figure 10.38 The Zernike polynomials are a set of orthonormal functions over the unit circle. Red indicates positive values, blue negative, and white is zero.

then the wavefront aberration after removing tilt and defocus, W_{fit}, that is, with respect to the best fit sphere, thus minimizing the RMS wavefront aberration, is obtained by dropping the first four terms, z_1, z_2, z_3, and z_4,

$$W_{fit}(\rho,\phi) = \sum_{5}^{N} z_n Z_n(\rho,\phi). \tag{10.57}$$

Although the standard use of the Zernike polynomials is the description of wavefronts, the Zernike polynomials can also describe amplitude, diattenuation, retardance, or any general scalar parameter. Thus, a set of eight Zernike coefficients can be used to represent a *Jones pupil*.

The Zernike polynomials evaluated at a grid of points within a circle, such as a set of equally spaced exit pupil locations, will be almost but not quite orthogonal, since their *orthogonality* is defined by integrals over the unit circle (Equation 10.53) and the grid sampling only approximates the integral. If precise orthonormality is desired, sets of sampled Zernike polynomials can be re-orthogonalized by applying Gram–Schmidt orthogonalization to the sampled function set. A similar Gram–Schmidt orthonormalization procedure is used to generate Zernike-like polynomials over aperture shapes other than the unit circle, generating new basis sets that have orthonormality.[27,28]

Zernike polynomials can be generalized to represent vector functions over a unit circle.[29,30] Linear diattenuation and linear retardance are not vector functions since they repeat every 180° of rotation, similar to line segments repeat after rotation. For the representation of linear diattenuation and linear retardance distributions over circular pupils, Ruoff and Totzeck[31] introduced a useful set of basis functions they named *orientation Zernike polynomials*, discussed further in Section 15.5.2.

Example 10.5 Example System: Wavefront Aberration Function

The wavefront aberration of the example cell phone lens for the on-axis wavefront is shown in Figure 10.39 as (left) a surface plot and (right) a colored contour plot; this wavefront shows defocus and spherical aberration. As the object and image move off-axis, the wavefront develops some coma and astigmatism and has the form of Figure 10.40 at the 10° field.

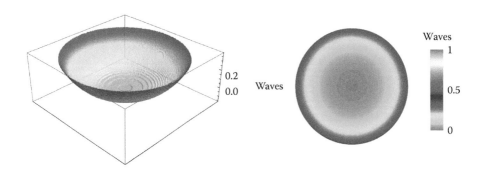

Figure 10.39 On-axis wavefront aberration for the cell phone lens shows spherical aberration.

10.7 Wavefront Analysis

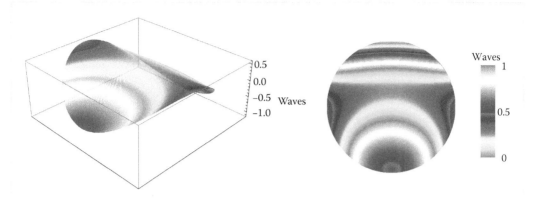

Figure 10.40 Off-axis wavefront aberration for the 10° field shows coma, astigmatism, and other aberrations.

10.7.7 Wavefront Quality

The ideal wavefront exiting an imaging system is a spherical wavefront with a uniform amplitude and a uniform polarization state. Wavefronts that are nearly spherical to within a quarter of a wavelength of optical path length are usually considered as *diffraction limited*; such images are very close to the ideal Airy disk, perhaps a little broader with a peak intensity reduced by a few percent.[32] One metric for wavefront quality is the RMS wavefront aberration, the square root of the wavefront aberration integrated over the aperture and normalized by the area

$$\Delta W_{RMS} = \frac{\sqrt{\iint_{pupil} W^2(x,y)\,dx\,dy}}{\iint_{pupil} dx\,dy}. \tag{10.58}$$

The RMS wavefront aberration has units of waves. A constant one wave of piston wavefront aberration (Zernike Z_1) has an RMS of one wave. As the RMS decreases, the point spread function becomes more compact and the peak brightness at the center of the aperture increases and the PSF approaches the ideal diffraction-limited PSF.

Another image metric for near diffraction-limited wavefronts and images is the Strehl ratio, S, the ratio of the intensity at the center of the PSF divided by the diffraction-limited PSF. The center of the Fourier transform is just the average of a function; thus, the Strehl ratio becomes

$$S = \frac{\left|\iint_{pupil} e^{i2\pi W(x,y)/\lambda}\,dx\,dy\right|}{\left|\iint_{pupil} dx\,dy\right|} \approx e^{-\Psi^2}, \quad \Psi = \left(\frac{2\pi}{\lambda}\right)^2 \left\langle \left(W(x,y) - \bar{W}\right)^2 \right\rangle, \tag{10.59}$$

where $\bar{W}$ is the average value of the wavefront aberration, and $\langle\,\rangle$ indicates the average value.

10.7.8 Polarization Quality

Many optical components or systems are desired to be non-polarizing; ideally, the diattenuation and retardance would be zero everywhere. For example, optics located between pairs of polarizers are ideally non-polarizing. Jones matrices with small amounts of diattenuation and retardance can be analyzed by placing them in the form of a non-polarizing part, and a sum of Pauli matrices.

$$\begin{aligned}\mathbf{J} &= \begin{pmatrix} j_{xx} & j_{xy} \\ j_{yx} & j_{yy} \end{pmatrix} = c_0 \begin{pmatrix} 1 & 0 \\ 0 & 1 \end{pmatrix} + c_1 \begin{pmatrix} 1 & 0 \\ 0 & -1 \end{pmatrix} + c_2 \begin{pmatrix} 0 & 1 \\ 1 & 0 \end{pmatrix} + c_3 \begin{pmatrix} 0 & -i \\ i & 0 \end{pmatrix} \\ &= \begin{pmatrix} c_0 + c_1 & c_2 - ic_3 \\ c_2 + ic_3 & c_0 - c_1 \end{pmatrix} = c_0 \left(\sigma_0 + \frac{c_1 \sigma_1 + c_2 \sigma_2 + c_3 \sigma_3}{c_0} \right) \\ &= c_0 \left(\sigma_0 + \frac{(a_1 + ib_1)\sigma_1 + (a_2 + ib_2)\sigma_2 + (a_3 + ib_3)\sigma_3}{c_0} \right). \end{aligned}$$

(10.60)

In the form on the bottom line of Equation 10.60, $c_0 = \rho_0 e^{-i\phi_0}$ represents the non-polarizing part of the Jones matrix. For small a and b, a_1, a_2, and a_3 indicate the 0°, 45°, and left circular components of diattenuation, and b_1, b_2, and b_3 indicate the 0°, 45°, and left circular components of retardance, respectively. Hence, for small a and b the RMS diattenuation is the integral

$$D_{RMS} = \frac{\sqrt{\iint\limits_{pupil} \left(a_1^2(x,y) + a_2^2(x,y) + a_3^2(x,y)\right) dx\, dy}}{\iint\limits_{pupil} dx\, dy},$$

(10.61)

which, for a non-polarizing system, would ideally be zero. For small $D_{RMS} < 0.5$, D_{RMS} and diattenuation magnitude are nearly equal and then diverge for larger D_{RMS} values. For small values, D_{RMS} can be interpreted as an average diattenuation, expressing the average of the diattenuation magnitude over the pupil. The RMS retardance in radians is

$$\delta_{RMS} = \frac{\sqrt{\iint\limits_{pupil} \left(b_1^2(x,y) + b_2^2(x,y) + b_3^2(x,y)\right) dx\, dy}}{\iint\limits_{pupil} dx\, dy}.$$

(10.62)

Thus, the RMS polarization aberration, Pol_{RMS}, indicates how far the pupil is from non-polarizing,

$$Pol_{RMS} = \frac{\sqrt{D_{RMS}^2 + \delta_{RMS}^2}}{\iint\limits_{pupil} dx\, dy}.$$

(10.63)

For a system specified to be non-polarizing, the coatings and optical prescription should be optimized to drive Pol_{RMS} close to zero. Then, minimal polarization change will occur between the entrance and exit pupils.

10.8 Non-Sequential Ray Trace

The sequential ray tracing procedure described above should be modified for systems where the light does not always intercept the surfaces in a single prescribed order, first surface 1, then 2, then 3, and so on. For example, in a lenslet array, each lenslet surface is a separate optical surface. During a non-sequential ray trace, when a ray exits a surface, the ray tracing algorithm must find the next ray intercept. The order that the ray encounters the surfaces must be determined by the ray tracing algorithm during the ray trace; it is not specified at the beginning because different rays will intersect the surfaces in different orders. Most illumination systems are non-sequential, as are corner cubes and many complex prisms.

Consider a ray exiting surface q from ray intercept $\mathbf{r}_q$ with $\mathbf{k}_q$. In non-sequential ray tracing, the next ray intercept $\mathbf{r}_{q+1}$ is found as follows:

1. Find intersection of the ray with each surface in the optical system $q = 1, \ldots Q$. For many surfaces, multiple ray intersections may occur; that is, the intersection equation has multiple roots.
2. Sort each ray intercept by distance $\mathbf{r}_q$ and choose the first ray intercept encountered.

If the next ray intercept ($\mathbf{r}_{q+1}$) is inside the aperture, the ray reflects or refracts based on the surface specification. If $\mathbf{r}_{q+1}$ is outside of the aperture, the ray gets terminated.

In systems traced by sequential ray tracing, the goal of the calculation is often the wavefront aberration function. For non-sequential systems, the optical system usually divides the incident wavefront into many pieces that exit the optic in different locations and directions; the wavefront gets cut up and rearranged. Thus, for many of these systems, such as illumination systems, the objective is to calculate a flux distribution, rather than wavefront aberration.

10.9 Coherent and Incoherent Ray Tracing

Ray tracing is performed in *coherent* and *incoherent ray tracing modes*. Coherent or incoherent here refers to the algorithms for calculating phase of the light, not the coherency of the incident light. The calculation of the phase and optical path length of wavefronts through an optical system is an example of *coherent ray tracing*, ray tracing that determines the phase of the resulting light. The coherent mode uses the optical path lengths of all the rays when combining the light; this is needed in all image formation calculations. If the calculation is capable of determining the phase of coherent light, then it is also suitable for determining the propagation and image formation of incoherent light through the optical system as well. The incoherent mode combines the light beams incoherently. Thus, *Jones matrices* are the *coherent polarization formalism* and *Mueller matrices* are the *incoherent formalism*.

The ray tracing calculations are usually different for illumination systems compared to imaging systems. In illumination systems and scattered light analysis, treating the light as nearly spherical wavefronts is often not useful or applicable, because light can take so many paths through an illumination system such as automobile headlight or molded plastic waveguide. Ray tracing methods for illumination emphasize the calculation of the flux and the light propagation direction. The rays reach the illuminated surface with a large range of optical path lengths, and the combination of polarized beams tend to behave more like the addition of Stokes parameters than the interference and image formation of a nearly spherical wavefront. Thus, for these illumination and scattered light ray tracing applications, Stokes parameters are a more suitable representation of the light's polarization state than Jones or polarization vectors. Similarly, Mueller matrices are a more appropriate representation than Jones matrices or polarization ray tracing matrices.

In some optical system simulations, the phase of the light is not needed or may not even be well defined. In the majority of illumination systems, the light is polychromatic, which washes out interference effects. The light may take many different paths to a particular point on the illuminated surface. The effects of interference between the different polychromatic beams are not observable when the optical path length differences are greater than a few waves. The illuminated spot is well approximated by the sum of intensities rather than the sum of amplitudes, or when polarization is considered, by the sum of Stokes parameters rather than the sum of Jones vectors.

A similar situation occurs in the simulation of stray light. In scattered light problems, the optical path lengths of rays reach a surface range over differences of thousands of wavelengths, even for rays that may scatter from adjacent locations on the next to the last surface. Such ray paths and polarization properties are generally evaluated by Monte Carlo methods, and the light's full electric field cannot practically be calculated. In scattered light simulation, it is not appropriate to combine the rays coherently; hence, incoherent algorithms such as Mueller matrices are used for the final combination of all the traced rays.

Consider a telescope in orbit looking at the surface of the Earth. The sun is outside the telescope's field of view, but some sunlight may enter the telescope tube. By simulating the light scattering from the telescope baffles and other mechanical structures, as well as the reflections and refractions of the optical elements, the stray light due to the sun can be estimated at the focal plane, typically by Monte Carlo methods. The optical path lengths of the stray light rays incident on a particular pixel may vary by centimeters, even meters. Thus, even though the optical path lengths and phases of the rays may be calculated, the individual phases of a randomly generated set of rays are not particularly useful for simulating stray light measurements of the telescopes. By replacing the sun with a laser shining into the telescope barrel, a speckle pattern will result at the focal plane instead. Although this speckle pattern depends on the phases of all the light converging on a point, the speckle pattern cannot be simulated in detail since the speckle distribution depends on the roughness of all the baffles' surfaces. The detailed topography of the mechanical surfaces will never be known to an accuracy of a tenth of a wave. Hence, even for this coherent laser light, the incoherent ray trace is appropriate for determining the flux levels at the focal plane.

Thus, although it sounds counterintuitive at first, coherent light is incoherently combined for scattering and speckle simulations. For example, the light exiting a laser-illuminated integrating sphere forms a speckle pattern; the light combines coherently but in a random way, forming an irregular speckle pattern. This pattern cannot be exactly calculated because it requires sub-wavelength knowledge of all the bumps and fine topography of the scattering surfaces, as shown in Figure 10.41. To exactly calculate such a speckle pattern, all these bumps would need to be well sampled by the

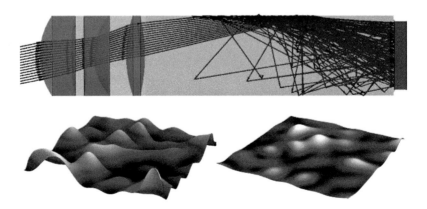

Figure 10.41 (Top) Off-axis laser light scatters from inside a lens barrel creating glare on the focal plane. (Bottom) The co- and crossed-polarized speckle patterns.

scattering polarization ray trace, which would take trillions of rays. But the statistics of the laser's speckle pattern can be calculated, and the speckle statistics, the intensity statistics, polarization statistics, and speckle size statistics can be calculated from a comparatively small number of rays. Hence, the probability distribution of speckle is calculated, but not the specific pattern.

In conclusion, coherent ray tracing and incoherent ray tracing refer to the *coherence and incoherence of the calculation*, respectively, and not the coherence and incoherence of the light. A coherent ray trace calculates phases and wavefronts and can simulate image formation and point spread functions. An incoherent calculation may not necessarily calculate the phase of the ray path; sometimes, only a bidirectional reflectance distribution function or a Mueller matrix may be known for a scattering surface, and the phase change at ray intercepts may not be known or need to be calculated.

10.9.1 Polarization Ray Tracing with Mueller Matrices

The Mueller matrices for reflection and refraction, shown in Section 6.13, are frequently used for incoherent ray tracing. Since Mueller matrices are defined in the ray's local coordinate system, the ray tracing coordinate system for Mueller matrices needs to be defined.

Consider a Mueller polarimeter consisting of a polarization generator that illuminates a sample, and a polarization analyzer that collects the light exiting the sample in a particular direction. We wish to characterize the polarization modification properties of the sample for a particular incident and exiting beam through the Mueller matrix. The incident polarization states are specified by Stokes vectors defined relative to an **xy**-coordinate system orthogonal to the propagation direction of the incident light. Similarly, the exiting light's Stokes vector is defined relative to an **x′y′**-coordinate system orthogonal to its propagation direction. For transmission measurements where the beam exits undeviated, the orientations of **xy** and **x′y′** will naturally be chosen to be aligned, **x** = **x′** and **y** = **y′**. The global orientation of **xy** is arbitrary, and the measured Mueller matrix varies systematically if **xy** and **x′y′** are rotated together.

When the exiting beam emerges in a different direction from the incident beam, orientations must be specified for both sets of coordinates. For measurements of reflection from a surface, a logical choice sets **xy** and **x′y′** to the *sp*-orientations for the two beams. Other Mueller matrix measurement configurations may have other obvious arrangements for the coordinates. All choices, however, are arbitrary and lead to different Mueller matrices. Let a Mueller matrix **M** be defined relative to a particular **xy** and **x′y′**. Let another Mueller matrix $\mathbf{M}(\theta_1, \theta_2)$ for the same measurement conditions have its **x** axis rotated by θ_1 and **x′** axis rotated by θ_2, where $\theta > 0$ indicates a counterclockwise rotation looking into the beam (**x** into **y**). These Mueller matrices are related by the equation

$$\mathbf{M}(\theta_1,\theta_2) = \begin{pmatrix} 1 & 0 & 0 & 0 \\ 0 & \cos 2\theta_2 & -\sin 2\theta_2 & 0 \\ 0 & \sin 2\theta_2 & \cos 2\theta_2 & 0 \\ 0 & 0 & 0 & 1 \end{pmatrix}$$

$$\cdot \begin{pmatrix} M_{00} & M_{01} & M_{02} & M_{03} \\ M_{10} & M_{11} & M_{12} & M_{13} \\ M_{20} & M_{21} & M_{22} & M_{23} \\ M_{30} & M_{31} & M_{32} & M_{33} \end{pmatrix} \begin{pmatrix} 1 & 0 & 0 & 0 \\ 0 & \cos 2\theta_1 & \sin 2\theta_1 & 0 \\ 0 & -\sin 2\theta_1 & \cos 2\theta_1 & 0 \\ 0 & 0 & 0 & 1 \end{pmatrix}.$$

(10.64)

When $\theta_1 = \theta_2$, the coordinates rotate together, the eigenvalues are preserved, the circular polarization properties are preserved, and the linear properties are shifted in orientation. When $\theta_1 \neq \theta_2$, the matrix properties are qualitatively different; the eigenvalues of the matrix change. If the eigenpolarizations of **M** were orthogonal, they may not remain orthogonal. After we perform data reduction on the matrix, the basic polarization properties couple in a complex fashion. For example, linear diattenuation in **M** yields a circular retardance component in $\mathbf{M}(\theta_1, \theta_2)$. The selection of the coordinate systems for the incident and exiting beams is not important for describing exiting polarization states but is crucial for properly identifying polarization characteristics of the sample.

10.10 The Use of Polarization Ray Tracing

In our presentation, polarization ray tracing has been formulated using polarization ray tracing matrices. The polarization ray tracing calculations can also be performed with Jones matrices and a series of local coordinates. Polarization optical design has several principal objectives:

1. To find good configurations of optical and polarization elements that meet optical specifications
2. To calculate the evolution of polarization states through optical systems
3. To calculate the polarization properties associated with ray paths through optical systems, their diattenuation, and retardance
4. To compare the effects of various coating designs on optical system performance
5. To simulate measurements that can be performed upon the optical system, such as to predict interferograms
6. To predict polarization aberration measurements, such as Mueller matrix images, taken through optical systems
7. To assess the effects of polarization aberrations on image formation, such as predicting the point spread function and its polarization distribution for different incident polarization states
8. To predict the dependence of the optical transfer function on the incident polarization state
9. To tolerance optical systems by determining the sensitivity to design perturbations

The most important method for these analyses is polarization ray tracing, which calculates the polarization properties along ray paths through optical systems and propagates the polarization state along these ray paths. Polarization ray tracing is performed one wavelength at a time, calculating the polarization properties of ray paths and the amplitude, phase, and polarization state of the exiting light. The most important of these quantities is the phase of the light, calculated from the optical path length along the ray.

The alternative method for polarization ray tracing is to perform a geometrical ray trace (conventional optical design) while, in addition, calculating Jones matrices for each ray intercept and ray segment. The conventional ray trace calculation determines the optical path length from the entrance to exit pupil. The ray intercept Jones matrices are multiplied together to determine the polarization properties associated with the ray path, its diattenuation, and retardance. The Jones matrix for each ray intercept describes the effect of the given interface at the ray's wavelength and angle of incidence. The interface might contain thin film coatings, a polarization element, diffraction grating, anisotropic materials, or other effects. These ray-intercept Jones matrices describe the polarization-dependent amplitude transmittance through the interface, the diattenuation of the ray intercept. Conventional ray tracing calculates the phase of the light via the optical path length, the number of wavelengths along the ray path from an incident reference surface. The ray-intercept Jones matrices contain polarization-dependent adjustments to the phase of the ray, in addition to the optical path length. Metallic reflections, thin film coatings, and other optical interfaces contribute to the phase of the light, and ray-intercept Jones matrices provide a straightforward method for including these phases in the ray trace.

The use of Jones matrices for polarization ray tracing requires that local coordinate systems be defined for each ray segment. Because of the many, many local coordinate issues, some of a fundamental nature, the next chapter is devoted later to local coordinate issues. Our recommendation is that polarization ray tracing be performed in global coordinates, and the next chapter develops a more general polarization ray tracing calculus for this purpose based on a three-dimensional generalization of the Jones matrix.

There is great value in having both methods, a two-dimensional Jones calculus and a three-dimensional polarization ray tracing calculus. Jones matrices are simple and widely understood. Jones matrices provide a straightforward method for characterizing the polarization properties of optical systems. Because Jones matrices describe phase, they integrate well with conventional optical design for describing polarization effects in optical systems. Jones matrices will be used in this book for a wide variety of problems. In particular, the wavefront aberration function of conventional optical design will be supplemented with a Jones matrix as a function of ray coordinates providing a *polarization aberration function* or *Jones pupil* for exiting wavefronts. Many ray tracing quantities are more easily represented and explained with Jones matrices. They just require more care and attention to detail. The serious student of polarization optical design methods needs to fully understand the issues associated with the local coordinate systems.

10.11 Brief History of Polarization Ray Tracing

Optical design and ray tracing algorithms arose to simulate the physics of reflection and refraction in optical systems. The associated polarization effects were, for the most part, safely ignored for decades. This brief history is not meant to be comprehensive, but to bring some context to the present work.

The fundamental physics behind polarization ray tracing is old. Drude,[33–35] Stratton,[36] Azzam and Bashara,[37] Born and Wolf,[38] and many others derive relations for the changes to the polarization of light at each non-normal refraction and reflection. This introduces diattenuation and retardance, which apodize and change the wavefront. The magnitude of the degraded performance depends on the particular opto-mechanical layout selected for the optical system architecture and the mirror coatings.

One of the first detailed analyses of optical system polarization aberrations was performed by Inoué in 1957.[39] Inoué was studying the birefringent structures inside cells with polarizing microscopes. By viewing organisms in a microscope with the sample between crossed polarizers, otherwise transparent cellular structures could be made visible. The polarization changes introduced by the microscope objective lenses let some light through the crossed polarizers, limiting the contrast, thus reducing the ability to see the fainter features.

Breckinridge[40] developed methods to analyze the polarization of grating spectrographs. Some polarization is introduced by the metal mirror coatings, but the dominant contribution arises due to the different diffraction efficiencies for light polarized parallel and perpendicular to the diffraction grating rulings. Title[41,42] developed methods for the design of Lyot and other birefringent filters and performed a comprehensive study into the properties and polarization aberrations of compound (multi-element) retarders.

Chipman,[43,44] while performing polarization ray tracing of lens and mirror systems, noticed the similarity of the resulting diattenuation and retardance patterns to the Seidel wavefront aberration functions and introduced a series of polarization aberrations comparable to defocus, tilt, and piston, shown in Figure 10.42. McGuire and Chipman[45,46] extended this work to higher-order polarization aberrations and developed methods for analyzing the polarization-dependent point spread functions and optical transfer functions of optical systems.

Waluschka[47] introduced a polarization ray tracing algorithm in 1988 that systematized the calculation of polarization aberrations due to coatings. Optical Research Associates added a polarization ray tracing capability to the CODE V optical analysis program in 1993, by allowing the program to perform a coating calculation at each intercept, and cascading the results to calculate the output polarization state on a ray-by-ray basis. Shortly thereafter, a coating polarization ray tracing

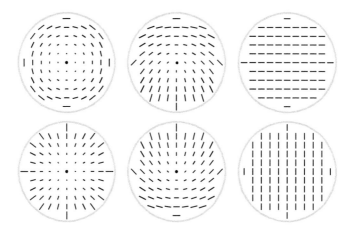

Figure 10.42 The second-order diattenuation and retardance polarization aberration (left) defocus, (center) tilt, and (right) piston. The top row shows the negative form and the bottom row shows the positive form.

capability was included in the ZEMAX program from Focus Software (now named Zemax LLC). Other commercial codes continued the trend of adding polarization ray tracing capabilities.

Ruoff and Totzeck introduced an expansion for polarization aberrations called vector Zernike polynomials, whose functional forms were closely tied to the Zernike polynomials, and applied these aberrations to the analysis of polarization aberrations in microlithography systems.[31,48]

Young and Chipman, while working on polarization ray tracing algorithms, systematized a 3 × 3 polarization matrix system and named it the polarization ray tracing (PRT) calculus (Chapter 9).[49,50] The PRT calculus has the benefit of eliminating the local coordinate systems that were needed when ray tracing with Jones matrices. In the process, they discovered skew aberrations, a polarization rotation that varies across the exit pupil for skew rays.[51]

Considering astronomical instrumentation, several important examples can be provided. Witzel et al.[52] characterized the polarization transmissivity of the Very Large Telescope conceptual design. Hines et al.[53] analyzed the Hubble Space Telescope NICMOS instrument. Ovelar et al.[54] modeled the Extremely Large Telescope conceptual design for instrumental polarization. Breckinridge and Oppenheimer[55] and Breckinridge[56] established that the shape of the PSF image for the astronomical telescope depends on polarization aberrations. McGuire and Chipman[57,58] and Young et al.[59,60] developed analytic tools and models to analyze polarization aberrations and their effect on image formation.

This book focuses on the PRT calculus for polarization ray tracing, but uses the Jones calculus for representing the polarization of spherical wavefronts on flat computer screens and on paper. The PRT calculus is the basis for a polarization ray tracing program, Polaris-M,[61] developed at the University of Arizona by Chipman and colleagues and licensed to Airy Optics, Inc. Polaris-M is used for most of the figures throughout this book.

10.12 Summary and Conclusion

Polarization ray tracing is performed by augmenting the conventional ray tracing algorithms with polarization matrices for each ray intercept and for anisotropic elements, for each anisotropic ray segment. The optical path length is calculated conventionally as the sum of the optical path lengths for each segment. The polarization properties of ray paths are obtained from the matrix product of the polarization matrices. Polarization ray tracing a grid of rays across the pupil provides the polarization aberration that augments the description of the wavefront aberration. In particular, the retardance aberration describes how the wavefront aberration changes with incident polarization state. This polarization aberration function can be used to determine the polarization properties of the point spread function.

10.13 Problem Sets

10.1 Describe the uses of dummy surfaces.

10.2 Calculate the surface normal to the following surfaces at arbitrary x and y:
   a. $f(x, y) = x^3$
   b. $g(x, y) = xy$
   c. $h(x, y) = (x^2 + y^2) y^3$

10.3 A ray propagating along $\mathbf{k} = (1, 1, 0)/\sqrt{2}$ is incident on paraboloid $f(x, y) = (x^2 + y^2)/2$ at $r = (1, 1, z)$.
   a. Find the s-basis vector.
   b. Find the p-basis vector.
   c. Find the reflected propagation vector $\mathbf{k}_1$.

10.4 Finding ray intercepts with complex shapes is an essential step in ray tracing. Here, we find an equation for the simplest surface in closed form. A plane can be specified by any point $\mathbf{p}_0 = (p_{0x}, p_{0y}, p_{0z})$ on a plane and a normal to the plane $\boldsymbol{\eta} = (\eta_x, \eta_y, \eta_z)$. When a dot product is taken between a vector from $\mathbf{p}_0$ to any point $\mathbf{p} = (p_x, p_y, p_z)$ on the surface and the normal vector, the dot product will equal zero.
   a. Find an explicit equation for the plane in the form of $p_z (p_x, p_y)$ as a function of the components of $\mathbf{p}$ and $\boldsymbol{\eta}$. This works as long as $\boldsymbol{\eta}$ is not perpendicular to $(0, 0, 1)$.

   A ray can be specified by a point on the ray $\mathbf{r}_0$, the propagation vector formed from direction cosines $\mathbf{k}$, and a distance t along the ray from $\mathbf{r}_0$, as $\mathbf{r}(t) = (r_{0x}, r_{0y}, r_{0z}) + (k_x, k_y, k_z) t$.
   b. Find an expression for the ray intercept at the plane by first solving for t in terms of the components of $\mathbf{p}_0$, $\boldsymbol{\eta}$, $\mathbf{r}_0$, and $\mathbf{k}$. Hint: First find the distance along the ray and then calculate the ray intercept.
   c. Find the s-basis vector for the ray.

10.5 Calculate the refracted propagation vector for an incident ray with $\mathbf{k}_{q-1} = (0, 0, -1)$ incident at a ray intercept with normal $\boldsymbol{\eta}_q = -(\sin\theta, 0, \cos\theta)$ pointing into the second medium where $n_{q-1} = 1$ and $n_q = 2$.
   a. Show that the angle of incidence and angle of refraction obey Snell's law.
   b. Show that $\mathbf{k}_{q-1}$, $\mathbf{k}_q$, and $\boldsymbol{\eta}_q$ lie in the same plane.

10.6 a. List all quantities that need to be calculated at each ray intercept during a polarization ray trace.
   b. Describe the sequence of operations performed.

10.7 A marginal ray propagating through an uncoated lens has a p-intensity transmission coefficient $T_{1,\max}$ and s-intensity coefficient $T_{1,\min}$. Similarly, at the second surface, the intensity coefficients are $T_{2,\max}$ and $T_{2,\min}$.
   a. What are the diattenuations D_1 and D_2 for the ray intercepts?
   b. Since this is a marginal ray, the p-component at the first interface is the p-component at the second interface. What is $T_{\max}$ and $T_{\min}$ for the combination of these two ray intercepts?
   c. Find the diattenuation as a function of $T_{1,\max}$, $T_{1,\min}$, $T_{2,\max}$, and $T_{2,\min}$.
   d. Find the diattenuation D as a function of D_1 and D_2.
   e. Find a first-order Taylor series expansion for D as a function of D_1 and D_2.

10.8 Calculate the three-dimension polarization ray tracing matrices for the three fold mirror aluminum coated system:

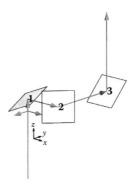

a. The mirrors are aligned so that the surface normal at each mirror are defined as $(-1, 0, 1)/\sqrt{2}$, $(1, -1, 0)/\sqrt{2}$, and $(0, 1, -1)/\sqrt{2}$. For a given incident propagation vector $(0, 0, 1)$, calculate the propagation vectors after each mirror.
b. Using surface normal and propagation vectors, calculate the *spk* basis vectors at all three interfaces.
c. Calculate orthogonal transformation matrices at all three interfaces.
d. Calculate **P** matrices at all three interfaces using $r_s = -0.947 - 0.2191i$ and $r_p = 0.8491 + 0.4150i$.
e. Calculate the overall **P** matrix for the system.
f. Calculate the diattenuation using singular values of the overall **P** matrix and compare with the diattenuation of a single mirror.
g. Find the orientation of the exiting light's **E**-field for an incident linearly polarized light beam oriented at θ.

10.9 A single element uncoated lens has a refractive index $n = 1.517$. A ray trace is performed from an on-axis object through the two lens surfaces to the exit pupil. The ray trace output is shown below, listing the ray intercepts, direction cosines of propagation vectors, and the surface normal at each ray intercepts.

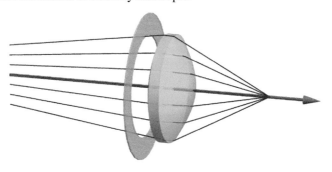

	X	Y	Z	L	M	N	Inc°	Ref.°	Length	Surf L	Surf M	Surf N
OBJ	0	0	−50	0	0.084	0.996				0	0	−1
STO	0	4.2	0	0	0.084	0.996	4.802	4.802	0	0	0	−1
2	0	4.337	1.628	0	−0.058	1.516	20.025	13.048	1.634	0	0.263	−0.965
3	0	4.175	5.894	0	−0.331	0.944	27.692	44.818	4.268	0	−0.431	−0.902
4	0	6.558	−0.911	0	−0.331	0.944	19.301	19.301	−7.210	0	0	−1
IMG	0	0.000	17.814	0	−0.331	0.944	19.301		19.840	0	0	−1

a. Calculate the polarization ray tracing matrix $\mathbf{P}_1$ for the first ray intercept at surface 2.
b. Calculate the polarization ray tracing matrix $\mathbf{P}_2$ for the second ray intercept at surface 3.
c. Find the overall polarization ray tracing matrix $\mathbf{P}$ for the two ray intercepts.
d. Perform a singular value decomposition on $\mathbf{P}$.
e. Relate the component matrices of the singular values of $\mathbf{P}$ to the components of $\mathbf{P}_1$ and $\mathbf{P}_2$.
f. Calculate the diattenuation of this ray path.
g. Find the Jones matrix for the ray path. Let the local x coordinate be $(1, 0, 0)$ for the incident and exiting Jones vectors.

10.10 Light propagating through a Fresnel rhomb retarder has four ray intercepts with angles of incidence $0°$, $51.79°$, $51.79°$, and $0°$. The glass has a refractive index $n = 1.497$.
  a. Using Fresnel equations, plot r_s and r_p as a function of angle around $51.79°$.
  b. Let the light enter the rhomb at $(0, 0, 1)$. All surface normals are in the y–z plane. Find the polarization ray tracing matrices for the four ray intercepts.
  c. Find the cumulative polarization ray tracing matrix $\mathbf{P}$ through the rhomb.
  d. Find the phase changes for x- and y-polarized light from entrance to exit.
  e. When the Fresnel rhomb is rotated about the z-axis by angle ξ, find a unitary transformation to apply to $\mathbf{P}$ to calculate $\mathbf{P}(\xi)$.

10.11 a. What is so special about the exit pupil of an optical system?
  b. What makes the exit pupil so useful in optical analysis?
  c. Microscopes and binoculars have external (real) exit pupils for coupling efficiently to the eye. If you are holding one of these systems, how can the exit pupil be located?

10.12 a. What is the definition of etendué, and what does this quantity mean physically?
  b. What is the significance of etendué for an optical system?

10.13 Create and evaluate the following model of the polarization of a rainbow:
A ray with propagation vector $\mathbf{k}_0$ refracts into a sphere of water with index 1.333. The light internally reflects once and then refracts back out of the sphere, as depicted below. For a small range of incident angles, the transmission along this light path is quite bright, creating rainbows. This situation is simply analyzed by realizing that the light path is symmetric about the back internal reflection. Thus, the angles in air at 1 and 3 are equal, as are the angles of incidence and reflection in water. Set up a unit sphere centered at the origin. Analyze ray paths in the x–z plane. Let the internal reflection occur at the bottom of the sphere, $\mathbf{r}_2 = (0, 0, 1)$ in the figure below, where the angles of incidence and reflection are parameterized as ξ. Start from the back reflection, polarization ray trace backward and forward. After solving the problem in terms of the internal angle ξ, transform the solution into a description in terms of the angle θ between the incident sunlight and the light returned from the raindrop. The rainbow corresponds to only a small range of ξ and θ where the return from the droplet is bright. The dispersion of the rainbow is due to the variation of the refractive index of water with wavelength, not analyzed here.

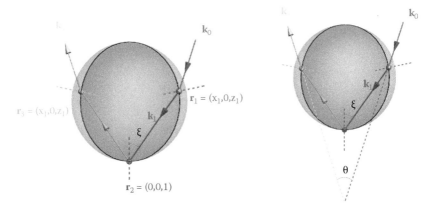

a. Find the ray intercept $\mathbf{r}_1$, where the first refraction occurs, in terms of ξ.
b. Find the propagation vector $\mathbf{k}_1$ after the first refraction, and the propagation vector $\mathbf{k}_2$, after the internal reflection in terms of ξ.
c. Find the surface normal η_1 at the first ray intercept, in terms of ξ.
d. Find polarization ray tracing matrices $\mathbf{P}_1$, $\mathbf{P}_2$, and $\mathbf{P}_3$ as a function of ξ at all three ray intercepts.
e. Find the total $\mathbf{P} = \mathbf{P}_3\,\mathbf{P}_2\,\mathbf{P}_1$ as a function of ξ.
f. Find the angle θ between $\mathbf{k}_3$ and $-\mathbf{k}_0$ in terms of ξ.
g. Plot the total diattenuation D as a function of ξ.
h. Plot the transmission of the entire ray path as a function of θ.

10.14 Calculate the RMS wavefront aberration of the following Seidel aberrations over a circular aperture or radius one. For example, tilt $W(\rho,\phi) = \rho\cos\phi$. The pupil has an area

$$A = \int_0^{2\pi}\!\!\int_0^1 \rho\,d\rho\,d\phi = \pi.$$ The RMS wavefront error for tilt is $\dfrac{1}{A}\displaystyle\int_0^{2\pi}\!\!\int_0^1 \rho^2 \cos^2\phi\,d\rho\,d\phi = 1/4$.

a. Defocus ρ^2
b. Spherical aberration ρ^4
c. Coma $\rho^3 \cos\phi$
d. Astigmatism $\rho^2 \cos^2\phi$

10.14 Appendix: Cell Phone Lens Prescription

The prescription of the example cell phone lens is taken from U.S. Patent 7,535,658, a typical four-element injection-molded plastic lens. The ray trace of an example ray is included below to support the calculation of its Jones matrix, $\mathbf{J}$.

The radii of curvature, thicknesses, and refractive indices are shown in Figure 10.43 and tabulated in Table 10.8. The lenses' focal length is 5.57 mm, with a back focal length of 1.53 mm operating at F/2.8. At the left side of the lens, pressed against the first surface, is the aperture stop. Since the aperture stop is in object space, it is also the entrance pupil. The flat surfaces 9 and 10 are an IR blocking filter, necessary since CMOS sensors are very sensitive to IR light. The filter is needed since the camera needs to render scenes in the correct balance of red, green, and blue, without distorting color influence from the infrared.

The lens surfaces are aspheric on both surfaces, biaspheric elements; hence, less surfaces are necessary to reduce the aberrations. The surfaces are specified by coefficients of the form

10.14 Appendix: Cell Phone Lens Prescription

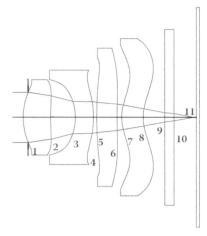

Figure 10.43 The example lens in cross-section view with surfaces numbered, and the example ray path shown.

$$z = \frac{cr^2}{1+\sqrt{1-\kappa c^2 r^2}} + A_3 r^3 + A_4 r^4 + A_5 r^5 + \ldots A_{10} r^{10} \qquad (10.65)$$

tabulated in Table 10.9. This prescription differs from the majority of aspheric surface prescriptions, which use only even powers of r, in the inclusion of odd powers of r: r^3, r^5, and so on. A common opinion is that the odd powers add almost nothing to the even powers in ability to correct aberrations; nonetheless, expressing the surface figure with odd and even terms was the choice of this lens designer.

Table 10.10 shows a ray trace table for the ray of Figure 10.43 with a ray position, a propagation vector, and a surface normal at each ray intercept q. These data are used with the algorithms of Section 10.6.8 to calculate polarization ray tracing matrices and a Jones matrix for the ray path in Example 10.4.

Table 10.8 Lens Prescription of U.S. Patent 7,535,658

Surface #	R	t	n
STOP		−0.20	
1	1.962	1.19	1.471
2	33.398	0.93	
3	−2.182	0.75	1.603
4	−6.367	0.10	
5	5.694	0.89	1.510
6	9.192	0.16	
7	1.674	0.85	1.510
8	1.509	0.70	
9	0	0.40	1.516
10	0	0.64	

Note: R is radius of curvature, t is surface spacing, and n is refractive index.

Table 10.9 Aspheric Surface Prescriptions for the Cell Phone Lens

Surface #	κ	A_3 (10^{-2})	A_4 (10^{-2})	A_5 (10^{-2})	A_6 (10^{-2})	A_7 (10^{-2})	A_8 (10^{-2})	A_9 (10^{-2})	A_{10} (10^{-2})
1	2.153	−1.895	2.426	−5.123	0.08371	0.7850	0.4091	−0.7732	−0.4265
2	40.18	−0.4966	−1.434	−0.6139	−0.009284	0.6438	−0.5720	−2.385	1.108
3	2.105	−4.388	−2.555	5.160	−4.307	−2.831	3.162	4.630	−4.877
4	3.382	−11.31	−7.863	10.94	0.6228	−2.216	−0.589	−0.4123	0.1041
5	−221.1	−7.876	7.020	0.1575	−0.9958	−0.7322	0.06914	0.2540	−0.0765
6	0.9331	0.9694	−0.2516	−0.3606	−0.02497	−0.0684	−0.01414	0.02932	−0.007284
7	−7.617	7.429	−6.933	−0.5811	0.2396	−0.2100	−0.03119	−0.005552	0.0007969
8	−2.707	0.1767	−4.652	1.625	−0.3522	−0.07106	0.03825	0.006271	−0.00263

10.14 Appendix: Cell Phone Lens Prescription

Table 10.10 Ray Position, Propagation Vector, and Surface Normal at Each Ray Intercept q

q	Ray Intercept ($\mathbf{r}_q$)	Propagation Vector ($\mathbf{k}_q$)	Surface Normal ($\boldsymbol{\eta}_q$)
0 (STOP)	$\begin{pmatrix} -0.086 \\ -0.173 \\ -0.981 \end{pmatrix}$	$\begin{pmatrix} 0.086 \\ 0.173 \\ 0.981 \end{pmatrix}$	$\begin{pmatrix} 0 \\ 0 \\ 1 \end{pmatrix}$
1	$\begin{pmatrix} 0 \\ 0 \\ 0 \end{pmatrix}$	$\begin{pmatrix} 0.054 \\ 0.109 \\ 0.993 \end{pmatrix}$	$\begin{pmatrix} 0 \\ 0 \\ 1 \end{pmatrix}$
2	$\begin{pmatrix} 0.049 \\ 0.099 \\ 0.901 \end{pmatrix}$	$\begin{pmatrix} 0.089 \\ 0.179 \\ 0.980 \end{pmatrix}$	$\begin{pmatrix} -0.005 \\ -0.010 \\ 1.000 \end{pmatrix}$
3	$\begin{pmatrix} 0.119 \\ 0.241 \\ 1.680 \end{pmatrix}$	$\begin{pmatrix} 0.081 \\ 0.164 \\ 0.983 \end{pmatrix}$	$\begin{pmatrix} 0.069 \\ 0.139 \\ 0.988 \end{pmatrix}$
4	$\begin{pmatrix} 0.168 \\ 0.339 \\ 2.271 \end{pmatrix}$	$\begin{pmatrix} 0.098 \\ 0.197 \\ 0.976 \end{pmatrix}$	$\begin{pmatrix} 0.054 \\ 0.109 \\ 0.993 \end{pmatrix}$
5	$\begin{pmatrix} 0.182 \\ 0.367 \\ 2.408 \end{pmatrix}$	$\begin{pmatrix} 0.061 \\ 0.122 \\ 0.991 \end{pmatrix}$	$\begin{pmatrix} -0.012 \\ -0.024 \\ 1.000 \end{pmatrix}$
6	$\begin{pmatrix} 0.222 \\ 0.448 \\ 3.064 \end{pmatrix}$	$\begin{pmatrix} 0.095 \\ 0.191 \\ 0.977 \end{pmatrix}$	$\begin{pmatrix} -0.006 \\ -0.012 \\ 1.000 \end{pmatrix}$
7	$\begin{pmatrix} 0.246 \\ 0.495 \\ 3.306 \end{pmatrix}$	$\begin{pmatrix} 0.017 \\ 0.034 \\ 0.999 \end{pmatrix}$	$\begin{pmatrix} -0.125 \\ -0.252 \\ 0.960 \end{pmatrix}$
8	$\begin{pmatrix} 0.257 \\ 0.519 \\ 4.012 \end{pmatrix}$	$\begin{pmatrix} 0.098 \\ 0.197 \\ 0.976 \end{pmatrix}$	$\begin{pmatrix} -0.130 \\ -0.261 \\ 0.957 \end{pmatrix}$
9	$\begin{pmatrix} 0.318 \\ 0.640 \\ 4.610 \end{pmatrix}$	$\begin{pmatrix} 0.065 \\ 0.130 \\ 0.989 \end{pmatrix}$	$\begin{pmatrix} 0 \\ 0 \\ 1 \end{pmatrix}$
10 (DET)	$\begin{pmatrix} 0.337 \\ 0.679 \\ 4.910 \end{pmatrix}$	$\begin{pmatrix} 0.098 \\ 0.197 \\ 0.976 \end{pmatrix}$	$\begin{pmatrix} 0 \\ 0 \\ 1 \end{pmatrix}$

References

1. G. Forbes, Better ways to specify aspheric shapes can facilitate design, fabrication and testing alike, JMA1, OSA/IODC/OF&T (2010).
2. J. E. Greivenkamp, *Field Guide to Geometrical Optics*, Vol. 1, Bellingham, Washington: SPIE Press (2004).
3. G. H. Spencer and M. V. R. K. Murty, General ray-tracing procedure, *JOSA* 52.6 (1962): 672–676.
4. D. Malacara-Hernández and Z. Malacara-Hernández, *Handbook of Optical Design*, Boca Raton, FL: CRC Press (2013).
5. W. T. Welford, *Aberrations of Optical Systems*, CRC Press (1986).
6. T. Y. Baker, Ray tracing through non-spherical surfaces, *Proc. Phys. Soc.* 55.5 (1943): 361.
7. W. A. Allen and J. R. Snyder, Ray tracing through uncentered and aspheric surfaces, *JOSA* 42.4 (1952): 243–249.
8. M. A. J. Sweeney and R. H. Bartels, Ray tracing free-form B-spline surfaces, *IEEE Comput. Graphics Appl.* 6.2 (1986): 41–49.
9. S. Ortiz et al., Three-dimensional ray tracing on Delaunay-based reconstructed surfaces, *Appl. Opt.* 48.20 (2009): 3886–3893.
10. H. A. Macleod, *Thin-Film Optical Filters*, McGraw-Hill (1986), pp. 179–209.
11. P. H. Berning, Theory and calculations of optical thin films, *Phys. Thin Films*, 1 (1963): 69–121.
12. D. C. O'Shea, *Elements of Modern Optical Design*, CH6, New York: John Wiley & Sons (1985).
13. J. Sasián, Theory of sixth-order wave aberrations, *Appl. Opt.* 49.16 (2010): D69–D95.
14. J. Sasián, *Introduction to Aberrations in Optical Imaging Systems*, Cambridge University Press (2013).
15. K. P. Thompson, Multinodal fifth-order optical aberrations of optical systems without rotational symmetry: The comatic aberrations, *JOSA A* 27.6 (2010): 1490–1504.
16. K. P. Thompson, Aberration fields in tilted and decentered optical systems, Dissertation, Optical Sciences, University of Arizona (1980).
17. K. Thompson, Description of the third-order optical aberrations of near-circular pupil optical systems without symmetry, *JOSA A* 22.7 (2005): 1389–1401.
18. K. P. Thompson, Multinodal fifth-order optical aberrations of optical systems without rotational symmetry: Spherical aberration, *JOSA A* 26.5 (2009): 1090–1100.
19. K. P. Thompson, Multinodal fifth-order optical aberrations of optical systems without rotational symmetry: The astigmatic aberrations, *JOSA A* 28.5 (2011): 821–836.
20. K. Fuerschbach, J. P. Rolland, and K. P. Thompson, Theory of aberration fields for general optical systems with freeform surfaces, *Opt. Express* 22.22 (2014): 26585–26606.
21. F. Zernike, Diffraction theory of the knife-edge test and its improved form, the phase-contrast method, *Monthly Notices R. Astron. Soc.* 94, (1934): 377–384.
22. A. B. Bhatia and E. Wolf, On the circle polynomials of Zernike and related orthogonal sets, in *Mathematical Proceedings of the Cambridge Philosophical Society*, Vol. 50, No. 01, Cambridge University Press (1954).
23. V. N. Mahajan, Zernike circle polynomials and optical aberrations of systems with circular pupils, *Appl. Opt.* 33.34 (1994): 8121–8124.
24. C. J. Kim and R. R. Shannon, Catalog of Zernike polynomials, *Appl. Opt. Opt. Eng.* 10 (1987): 193–221.
25. J. C. Wyant and K. Creath, Basic wavefront aberration theory for optical metrology, *Appl. Opt. Opt. Eng.* 11.s 29 (1992): 2.
26. V. N. Mahajan, Zernike polynomial and wavefront fitting, *Optical Shop Testing*, 3rd edition (2017), pp. 498–546.
27. W. Swantner and W. Chow, Gram–Schmidt orthonormalization of Zernike polynomials for general aperture shapes, *Appl. Opt.* 33.10 (1994): 1832–1837.
28. R. Upton and B. Ellerbroek, Gram–Schmidt orthogonalization of the Zernike polynomials on apertures of arbitrary shape, *Opt. Lett.* 29.24 (2004): 2840–2842.
29. C. Zhao and J. H. Burge, Orthonormal vector polynomials in a unit circle, Part I: Basis set derived from gradients of Zernike polynomials, *Opt. Express* 15.26 (2007): 18014–18024.
30. C. Zhao and J. H. Burge, Orthonormal vector polynomials in a unit circle, Part II: Completing the basis set. *Opt. Express* 16.9 (2008): 6586–6591.
31. J. Ruoff and M. Totzeck, Orientation Zernike polynomials: A useful way to describe the polarization effects of optical imaging systems, *J. Micro/Nanolithogr. MEMS MOEMS* 8.3 (2009): 031404–031404.
32. J. B. Develis, G. B. Parrent, and B. Thompson, *The New Physical Optics Notebook: Tutorials in Fourier Optics*, Vol. 61. New York: SPIE Optical Engineering Press (1989).

33. P. Drude, Zur elektronentheorie der metalle, *Ann. Phys.* 06.3 (1900): 566–613.
34. P. Drude, Zur elektronentheorie der metalle; II. Teil. galvanomagnetische und thermomagnetische effecte, *Ann. Phys.* 308.11 (1900): 369–402.
35. P. Drude, C. Riborg, and R. A. Millikan, *The Theory of Optics*. Translated from German by C. R. Mann and R. A. Millikan, London, New York (1902).
36. J. A. Stratton, *Electromagnetic Theory*, McGraw-Hill (1941).
37. R. M. A. Azzam and N. M. Bashara, *Ellipsometry and Polarized Light*, North-Holland: Elsevier Science Publishing Co., Inc. (1987).
38. M. Born and E. Wolf, *Principles of Optics: Electromagnetic Theory of Propagation, Interference and Diffraction of Light*, CUP Archive (2000).
39. S. Inoué and W. Lewis Hyde, Studies on depolarization of light at microscope lens surfaces II. The simultaneous realization of high resolution and high sensitivity with the polarizing microscope, *J. Biophys. Biochem. Cytol.* 3.6 (1957): 831–838.
40. J. B. Breckinridge, Polarization properties of a grating spectrograph, *Appl. Opt.* 10 (1971): 286–294.
41. A. M. Title, Improvement of birefringent filters. 2: Achromatic waveplates, *Appl. Opt.* 14 (1975): 229–237.
42. A. M. Title and H. E. Ramsey, Improvements in birefringent filters. 6: Analog birefringent elements, *Appl. Opt.* 19 (1980): 2046–2058.
43. R. A. Chipman, Polarization aberrations of lenses, in *Proc. Int. Lens Design Conf.*, ed. W. H. Taylor, SPIE, Vol. 554 (1985), pp. 82–87.
44. R. A. Chipman, Polarization Aberrations, Dissertation, Optical Sciences, University of Arizona, Tucson (1987).
45. J. P. McGuire and R. A. Chipman, Polarization aberrations I: Rotationally symmetric optical systems, *Appl. Opt.*, 33(2) (1994): 5080–5100.
46. J. P. McGuire and R. A. Chipman, Polarization aberrations II: Tilted and decentered optical systems, *Appl. Opt.*, 33(2) (1994): 5101–5107.
47. E. Waluschka, A polarization ray trace, *Opt. Eng.* 28 (1989): 86–89.
48. J. Ruoff and M. Totzeck, Using orientation Zernike polynomials to predict the imaging performance of optical systems with birefringent and partly polarizing components, in *International Optical Design Conference, International Society for Optics and Photonics* (2010).
49. G. Yun, K. Crabtree, and R. Chipman, Three-dimensional polarization ray-tracing calculus I: Definition and diattenuation, *Appl. Opt.* 50 (2011): 2855–2865.
50. G. Yun, S. McClain, and R. Chipman, Three-dimensional polarization ray-tracing calculus II: Retardance, *Appl. Opt.* 50 (2011): 2866–2874.
51. G. Yun, K. Crabtree, and R. Chipman, Skew aberration: A form of polarization aberration, *Opt. Lett.* 36 (2011): 4062–4064.
52. G. Witzel et al., The instrumental polarization of the Nasmyth focus polarimetric differential imager NAOS/CONICA (NACO) at the VLT. Implications for time-resolved polarimetric measurements of Sagittarius A*, *Astron. Astrophys.* 525 (2011): A130.
53. D. C. Hines, G. D. Schmidt, and G. Schneider, Analysis of polarized light with NICMOS, *Publ. Astron. Soc. Pacific* 112.773 (2000): 983.
54. M. de Juan Ovelar et al., Modeling the instrumental polarization of the VLT and E-ELT telescopes with the M&m's code, in *SPIE Astronomical Telescopes + Instrumentation, International Society for Optics and Photonics* (2012).
55. J. B. Breckinridge and B. R. Oppenheimer, Polarization effects in reflecting coronagraphs for white-light applications in astronomy, *Astrophys. J.* 600.2 (2004): 1091.
56. J. B. Breckinridge, Self-induced polarization anisoplanatism, in *SPIE Optical Engineering + Applications*, Proc. SPIE, 8860, 886012 (2013).
57. J. P. McGuire Jr. and R. A. Chipman, Diffraction image formation in optical systems with polarization aberrations I: Formulation and example. *J. Opt. Soc. Am. A* 7(9) (1990): 1614–1626.
58. J. P. McGuire Jr. and R. A. Chipman, Diffraction image formation in optical systems with polarization aberrations II: Amplitude response matrices for rotationally symmetric systems, *J. Opt. Soc. Am. A*. 8 (1991): 833–840.
59. G. Yun, K. Crabtree, and R. A. Chipman, Three-dimensional polarization ray-tracing calculus I: Definition and diattenuation, *Appl. Opt.* 50.18 (2011): 2855–2865.
60. G. Yun, S. C. McClain, and R. A. Chipman, Three-dimensional polarization ray-tracing calculus II: Retardance, *Appl. Opt.* 50.18 (2011): 2866–2874.
61. R. A. Chipman and W. S. T. Lam, The Polaris-M ray tracing program, in *SPIE Optical Engineering+ Applications*, International Society for Optics and Photonics (2015).

11

The Jones Pupil and Local Coordinate Systems

11.1 Introduction: Local Coordinates for Entrance and Exit Pupils

Given the distribution of polarization states on a spherical or nearly spherical wavefront, transferring this information to a computer screen involves choices on how to *flatten* the information, the same issues involved in printing maps of the Earth. In addition, if a pair of coordinates are defined over the spherical surface, then Jones vectors can be used for specification of polarization, and algorithms can translate back and forth between the two-dimensional Jones vector **E** representation and the three-dimensional (3D) polarization ray tracing vector description. A *latitude and longitude* system for the surface basis vectors on a sphere seems an obvious and straightforward choice; the fields radiated by *dipoles* align with lines of longitude. But another coordinate system for spherical surfaces, the double pole coordinates, also occurs naturally, for example, when lenses bring collimated linearly polarized wavefronts to focus. A related issue is defining how a linearly polarized spherical wavefront curves around the sphere. The radiation from a dipole is linearly polarized everywhere. But should every wavefront that is linearly polarized everywhere be considered as a *linearly polarized wavefront*? A collimated linearly polarized wavefront has all the electric fields aligned in a single direction; a wavefront with a wavy or irregular orientation, such as the vortex fields of Figure 5.11, should not be considered as *linearly polarized* but as polarized linearly everywhere. This chapter explores two different forms of linearly polarized wavefronts, the *dipole* and the *double pole*, and uses these forms to create two useful coordinate systems for spherical surfaces.

The description of the polarization with 3D global coordinates is robust, is straightforward to calculate, and provides an excellent method for a computer ray trace. The results, however, can be difficult to visualize in three dimensions because optical designers view ray tracing results on

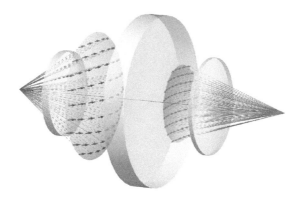

Figure 11.1 Linearly polarized incident (red) and exiting (orange) spherical wavefronts of a grid of rays. The optical system between the two wavefronts is shown.

computer screens and print the information onto paper. Visualizing the spherical waves, as shown in Figure 11.1, on a plane surface requires transforming a 3D vector field onto two dimensions. For example, the most common method to represent the polarization aberrations is as a Jones matrix function in two-dimensional pupil coordinates, the *Jones pupil*. The Jones pupil[1] is commonly used in industry and is generated as output from commercial optical design software. To fully understand the Jones pupil, it is necessary to master the subtleties of *local coordinate systems* used to represent 3D data on flat surfaces.

This chapter presents two methods for converting the polarization representation from 3D to 2D, and vice versa, and analyzes how the description of the aberrations differs whether using the one coordinate system over the others. The common latitude and longitude description of coordinates on a sphere is used to generate basis vectors for the surface of a sphere, named *dipole coordinates* here. This and another coordinate system, *double pole coordinates*, which arises naturally in optical systems, are developed and applied to optical design problems. These systems are derived from two important models for the radiation of electromagnetic waves: the *dipole radiator model* that describes the polarization of light emerging from ideal linear polarizer, and the *double pole model* that describes the polarization of light exiting ideal lenses.

As illustrated in Chapter 9 (Polarization Ray Tracing Calculus) and Chapter 10 (Optical Ray Tracing), the image quality of an optical system is often determined by tracing grids of rays. When the optical path length for each ray in the grid is calculated at the exit pupil, the wavefront information is obtained as the wavefront aberration function. In order to obtain the polarization and wavefront information, grids of *polarization ray tracing matrices* **P** or *Jones matrices*[2-7] are traced along each ray. To use Jones matrices, x- and y-coordinates, named *local coordinates*, must be defined for the incident Jones vector and separately for the exiting Jones vector. A cell phone lens example shown in Section 11.7 illustrates the importance of choosing appropriate local coordinates when converting the polarization properties from the entrance pupil to the exit pupil, and the **P** pupil, into the Jones pupil.

11.2 Local Coordinates

The objective of a local coordinate system is to provide basis vectors that reflect local circumstances. For example, in optical design programs, separate x–y–z local coordinates for each reflecting and refracting surface are often set up with the origin of each local coordinate set located at the vertex of the surface; thus, the surface can be most easily defined about the axis of the surface.

11.2 Local Coordinates

Chapter 2 (Polarized Light) and Chapter 5 (Jones Matrices and Polarization Properties) developed the Jones calculus, where the Jones vector is the description of polarization state,

$$\mathbf{J} = \begin{pmatrix} E_x \\ E_y \end{pmatrix}. \tag{11.1}$$

where E_x and E_y are complex amplitudes[2–5,8] along two directions, x and y, that form an orthonormal x–y–k basis set with the propagation direction $\mathbf{k}$. If the plane wave is not propagating along the z-axis, then the "x–y" coordinates are referred to as "local coordinates" associated with a particular transverse plane. By moving a local coordinate system from one ray segment to the next ray segment, the polarization state can be described through an optical system with a series of local coordinates. To use Jones vectors to describe polarization on a spherical wavefront, a system for defining local coordinates is needed to define E_x and E_y at each point. Similarly, Jones matrices relating polarization between an entering and exiting wavefront require two sets of basis vectors, one on each sphere.

It is common in optical design software to use 2 × 2 Jones matrices to describe the optical elements, polarization effects, and the overall polarization properties of optical systems. Jones matrices contain and describe polarization effects such as the polarization elements used to transform between polarization states.[2–5,8] For beams with small numerical aperture (NA), such as paraxial beams or laser beams, the wavefront is not very curved and the z-component of the field is a small fraction of the total field. In this case, the use of a single set of x–y coordinates for the wavefront's Jones vectors[9] is satisfactory. To use Jones vectors and matrices for ray tracing highly curved beams, local coordinates are required for each ray and each of its ray segments to define the Jones vector's x- and y-components in three dimensions. Thus, the goal is to provide the equations to transform between local coordinates on a sphere and the global coordinates and understand the consequences of different choices of local coordinates.

The *unit propagation sphere*, shown in Figure 11.2, is a sphere of normalized propagation vectors where each propagation vector $\hat{\mathbf{k}}$ is indicated by a point on this unit sphere. For a given $\hat{\mathbf{k}}$, the local coordinates provide a rule for local basis vectors in the *transverse plane*, a local x and a local y, which lie on the surface of the sphere. These local basis vectors are used to specify Jones vectors, Stokes parameters, or other quantities in the transverse plane of $\hat{\mathbf{k}}$.

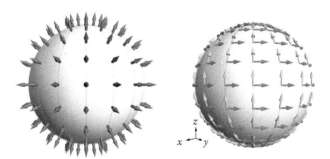

Figure 11.2 (Left) The propagation sphere is a unit sphere with propagation vectors, shown in black, emerging normal to the sphere. Local coordinates on the sphere are generated by various algorithms that assign local x- and y-basis vectors for each propagation vector. (Right) An example of local x- and y-basis vectors are shown on the unit propagation sphere as red and green arrows. The x–y–z axes shown at the corner identify the global x–y–z coordinate. These basis vectors were chosen using a latitude and longitude system with the poles located along the z-axis.

11.3 Dipole Coordinates

The most familiar coordinate system for the surface of a sphere is the *latitude/longitude coordinate system*, here called the *dipole coordinates*. Dipole coordinates are defined relative to a polar axis, specified by an axis vector, $\hat{\mathbf{a}}_{Loc}$, which defines the location of the two singularities or poles. Figure 11.3 shows the dipole coordinates on a unit sphere with $\hat{\mathbf{a}}_{Loc} = (0, 0, 1)$; the two poles are at $(0, 0, 1)$ and $(0, 0, -1)$. Each unit propagation vector $\hat{\mathbf{k}}$ is expressed in terms of latitude, θ, measured from the x–y plane, and longitude, ϕ, measured from the x–z plane, as

$$\hat{\mathbf{k}} = (k_x, k_y, k_z) = (\cos\phi \sin\theta, \sin\phi \sin\theta, \cos\theta). \tag{11.2}$$

The local x-basis vector, $\hat{\mathbf{x}}_{Loc}$, taken along a line of constant latitude, is calculated as the normalized derivative of Equation 11.2 with respect to ϕ. The local y-basis vector, $\hat{\mathbf{y}}_{Loc}$, taken along a line of constant longitude, is the cross product of $\hat{\mathbf{k}}$ and the latitude vector, $\hat{\mathbf{x}}_{Loc}$. Hence, the dipole local coordinates when the axis is along z, $\hat{\mathbf{a}}_{Loc} = (0, 0, 1)$, is

$$\begin{aligned}\hat{\mathbf{x}}_{Loc} &= \frac{(-k_y, k_x, 0)}{\sqrt{k_x^2 + k_y^2}} = (-\sin\phi, \cos\phi, 0) \text{ and} \\ \hat{\mathbf{y}}_{Loc} &= \hat{\mathbf{k}} \times \hat{\mathbf{x}}_{Loc} = \frac{\left(k_x k_z, k_y k_z, -k_x^2 - k_y^2\right)}{\sqrt{k_x^2 + k_y^2}} = (\cos\theta \cos\phi, \cos\theta \sin\phi, -\sin\theta).\end{aligned} \tag{11.3}$$

$(\hat{\mathbf{x}}_{Loc}, \hat{\mathbf{y}}_{Loc}, \hat{\mathbf{k}})$ forms a right-handed local coordinate for the transverse plane defined by $\hat{\mathbf{k}}$.

Through this formalism, a Jones vector can be defined for an arbitrary $\hat{\mathbf{k}}$ using $(\hat{\mathbf{x}}_{Loc}, \hat{\mathbf{y}}_{Loc}, \mathbf{k})$. When $\hat{\mathbf{k}} = \pm \hat{\mathbf{a}}_{Loc}$, the light is propagating along the axis and the coordinates described in Equation 11.3 become singular. Figure 11.4 shows the dipole coordinates viewed along the axis; the local x- and y-coordinates vary rapidly as $\hat{\mathbf{k}}$ approaches the pole.

The dipole coordinate for the axis along the x-axis or y-axis is easily obtained by transposing the individual elements in $\hat{\mathbf{x}}_{Loc}$ and $\hat{\mathbf{y}}_{Loc}$ in Equation 11.3. For example, when $\hat{\mathbf{a}}_{Loc} = (1, 0, 0)$,

$$\begin{aligned}\hat{\mathbf{x}}_{Loc} &= \frac{(0, -k_z, k_y)}{\sqrt{k_y^2 + k_z^2}} \text{ and} \\ \hat{\mathbf{y}}_{Loc} &= \hat{\mathbf{k}} \times \hat{\mathbf{x}}_{Loc} = \frac{\left(k_y^2 + k_z^2, -k_x k_y, -k_x k_z\right)}{\sqrt{k_y^2 + k_z^2}}.\end{aligned} \tag{11.4}$$

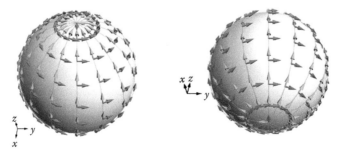

Figure 11.3 Dipole coordinates on a unit sphere for a dipole along the z-axis are shown in two different views. The red arrows are the local x-basis vectors and the green arrows are the local y-basis vectors.

11.3 Dipole Coordinates

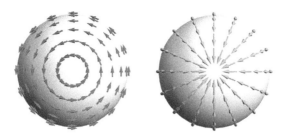

Figure 11.4 (Left) Dipole local *x*-coordinates and (right) dipole local *y*-coordinates change direction rapidly near the poles for small changes in direction.

For some problems, it may be desirable to locate the axis in an arbitrary direction with axis vector $\hat{\mathbf{a}}_{Loc} = (a_x, a_y, a_z)$. The local *x*-basis vector, $\hat{\mathbf{x}}_{Loc}$, along the new latitude lines, and new local *y*-basis vector, $\hat{\mathbf{y}}_{Loc}$, for arbitrary **k** are as follows:

$$\hat{\mathbf{x}}_{Loc} = \frac{\hat{\mathbf{a}} \times \hat{\mathbf{k}}}{|\hat{\mathbf{a}} \times \hat{\mathbf{k}}|} = \frac{(a_y k_z - a_z k_y, a_z k_x - a_x k_z, a_x k_y - a_y k_x)}{\sqrt{(a_y k_x - a_x k_y)^2 + (a_z k_x - a_x k_z)^2 + (a_z k_y - a_y k_z)^2}},$$

$$\hat{\mathbf{y}}_{Loc} = \frac{\hat{\mathbf{k}} \times \hat{\mathbf{a}} \times \hat{\mathbf{k}}}{|\hat{\mathbf{a}} \times \hat{\mathbf{k}}|} = \frac{\begin{pmatrix} -a_y k_x k_y + a_x k_y^2 - a_z k_x k_z + a_x k_z^2 \\ a_y k_x^2 - a_x k_x k_y - a_z k_y k_z + a_y k_z^2 \\ a_z k_x^2 + a_z k_y^2 - a_x k_x k_z - a_y k_y k_z \end{pmatrix}}{\sqrt{(a_y k_x - a_x k_y)^2 + (a_z k_x - a_x k_z)^2 + (a_z k_y - a_y k_z)^2}}. \quad (11.5)$$

Example 11.1 Local to Global Polarization State

Consider left circularly polarized light propagating in the direction $\hat{\mathbf{k}} = (6, 3, 2)/7$. Choosing the local coordinate axis, $\hat{\mathbf{a}}_{Loc} = (0, 0, 1)$, the state can be specified by a Jones vector, for example,

$$\mathbf{E} = \frac{1}{\sqrt{2}} \begin{pmatrix} 1 \\ i \end{pmatrix}. \quad (11.6)$$

By Equation 11.2, the latitude is $\theta = \tan^{-1}\left(\frac{3\sqrt{5}}{2}\right)$ and the longitude is $\phi = \tan^{-1}\left(\frac{1}{2}\right)$. The dipole local coordinates for this $\hat{\mathbf{k}}$ from Equation 11.3 are as follows:

$$\hat{\mathbf{x}}_{Loc} = \frac{1}{\sqrt{5}} \begin{pmatrix} -1 \\ 2 \\ 0 \end{pmatrix}, \hat{\mathbf{y}}_{Loc} = \frac{1}{7\sqrt{5}} \begin{pmatrix} -4 \\ -2 \\ 15 \end{pmatrix}. \quad (11.7)$$

Thus, the polarization state in global coordinates, the polarization vector **E**, is

$$\mathbf{E} = \frac{\hat{\mathbf{x}}_{Loc} + i\hat{\mathbf{y}}_{Loc}}{\sqrt{2}} = \frac{1}{7\sqrt{10}} \begin{pmatrix} -7 - 4i \\ 14 - 2i \\ 15i \end{pmatrix}. \qquad (11.8)$$

Figure 11.5 shows these vectors relative to the unit propagation axis.

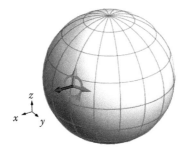

Figure 11.5 The polarization ellipse for left circularly polarized light is shown in 3D with the (black) propagation vector, (red) the local x dipole basis vector, and (green) the local y dipole basis vector.

Since ray tracing programs need to trace rays in all directions, a coordinate system should work for all $\hat{\mathbf{k}}$ or all choices of axis vector. But the definition of dipole coordinates has a pair of singularities along the axis. The usual way to handle this singularity is to define a special case for a small region about the poles. For example, within 10^{-6} radians, the basis vectors could be chosen as $\hat{\mathbf{x}}_{Loc} = (1,0,0)$ and $\hat{\mathbf{y}}_{Loc} = (0,1,0)$. This is not pleasing or elegant, but something must be done. Figure 11.6 shows this choice; the region around a pole viewed along the axis vector now has an arbitrary pair of basis vectors defined at the pole.

The singularities at the poles are a big bother in local coordinate systems. There is no local coordinate system without singularities, because singularities are a consequence of the winding number theorem[10] (Math Tip 11.1). Fortunately, this *singularity-at-the-pole problem* can be avoided during ray tracing by using 3D polarization ray tracing calculus. However, the issue of poles still occurs in

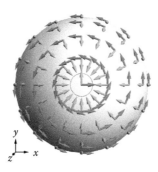

Figure 11.6 The dipole coordinate system viewed from $\hat{\mathbf{a}}_{Loc}$ (i.e., singular point). To avoid singularity and divide by zero problems, the basis vectors can be set to constants within a very small patch, shown in yellow, about the origin.

11.3 Dipole Coordinates

transferring results onto flat surfaces, just as poles create issues in making maps of the entire globe. The 3D polarization ray tracing calculus is described as robust because it avoids singularities.

Math Tip 11.1 Winding Number Theorem

According to the winding number theorem,[10] it is impossible to define a continuous and differentiable vector field constrained to lie on the surface of a sphere over the entire sphere without at least two zeros in the field. A set of latitude vectors or conversely a set of longitude vectors provide two examples, where the zeros occur at the poles. All local coordinate choices for Jones vectors covering the entire sphere have such singularities.

Figure 11.7 shows $\hat{\mathbf{x}}_{Loc}$ from the dipole coordinates on a unit sphere with $\hat{\mathbf{a}}_{Loc} = (0,0,1)$, in different viewpoints.

The first dipole local coordinate example above placed the poles along the z-axis. Depending on the problem, $\hat{\mathbf{a}}_{Loc}$ can be chosen in an arbitrary direction to simplify a particular problem. Typically, $\hat{\mathbf{a}}_{Loc}$ will be placed 90° from the optical axis of the optical system to move the singularities out of the way. Previously, the dipole local coordinate example was defined for the $\hat{\mathbf{a}}_{Loc} = (0,0,1)$. Now, let's find a method to generate dipole local coordinates for an arbitrary axis vector.

Example 11.2 Dipole Coordinates When $\hat{\mathbf{a}}_{Loc} = (0, 1, 0)$

In this example, the dipole coordinates for $\hat{\mathbf{a}}_{Loc} = (0, 1, 0)$ are calculated using an alternative algorithm to Equation 11.5.

1. Calculate a vector angle, α, between $\hat{\mathbf{z}} = (0,0,1)$, the default axis of the dipole coordinates, and the $\hat{\mathbf{a}}_{Loc}$ vector: $\alpha = \cos^{-1}(\hat{\mathbf{z}} \cdot \hat{\mathbf{a}}_{Loc}) = \cos^{-1}(0) = \pi/2$, where $\hat{\mathbf{z}} = \hat{\mathbf{k}}$.
2. Define a rotation axis vector $\vec{r} = \hat{\mathbf{z}} \times \hat{\mathbf{a}}_{Loc} = (0,0,1) \times \hat{\mathbf{a}}_{Loc} = (-1,0,0)$.
3. Calculate a rotation matrix, $\mathbf{R}$, with a rotation axis $\vec{r}$ and rotation angle α (counterclockwise about $\vec{r}$): $\mathbf{R} = \begin{pmatrix} 1 & 0 & 0 \\ 0 & 0 & 1 \\ 0 & -1 & 0 \end{pmatrix}$.
4. Calculate $\hat{\mathbf{k}}_{new} = \mathbf{R} \cdot \hat{\mathbf{k}}$, where $\hat{\mathbf{k}} = (\cos\phi \sin\theta, \sin\phi \sin\theta, \cos\theta)$: $\hat{\mathbf{k}}_{new} = \mathbf{R} \cdot \hat{\mathbf{k}} = (\cos\phi \sin\theta, \cos\theta, -\sin\phi \sin\theta)$.
5. $\hat{\mathbf{x}}_{Loc}$ is found by differentiating $\hat{\mathbf{k}}_{new}$ with respect to ϕ, $\hat{\mathbf{x}}_{Loc} = (-\sin\phi, 0, -\cos\phi,)$, and $\hat{\mathbf{y}}_{Loc} = \hat{\mathbf{k}}_{new} \times \hat{\mathbf{x}}_{Loc} = (-\cos\phi \cos\theta, \sin\theta, \sin\phi \cos\theta)$.

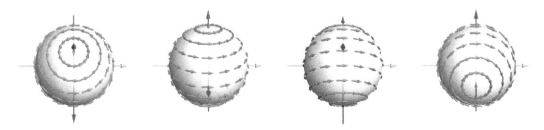

Figure 11.7 Dipole x-coordinates (red arrows) on a unit sphere for a dipole along the z-axis in different viewpoints. Global x, y, and z are shown in red, green, and blue arrows, respectively. (Left) Viewed from near $+z$ and (right) viewed from near $-z$.

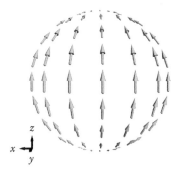

Figure 11.8 The field from a dipole oriented along z is aligned with the latitude basis vectors creating a vertically polarized spherical wavefront. The vectors are not unit vectors; the amplitude varies as the cosine of the angle from the equator.

To revisit the question posted in the beginning of this chapter, $\hat{\mathbf{x}}_{Loc}$ and $\hat{\mathbf{y}}_{Loc}$ can conveniently be used to describe polarization of a linearly polarized spherical wave radiated by an *oscillating electric dipole*. By apodizing $\hat{\mathbf{y}}_{Loc,i}$ by $\sin \beta_i$, where β_i is the angle between $\hat{\mathbf{k}}_i$ and the $\hat{\mathbf{a}}_{Loc}$ vector, the **E** field radiated by a linearly polarized dipole on a spherical wavefront shown in Figure 11.8 only has a component along $\hat{\mathbf{y}}_{Loc}$ and has no component along $\hat{\mathbf{x}}_{Loc}$. Thus, dipole local coordinates can simplify problems where the sources are radiating dipoles.

$$\mathbf{E}_i = \hat{\mathbf{y}}_{Loc,i} \sin \beta_i. \tag{11.9}$$

11.4 Double Pole Coordinates

Another useful local coordinate system on a sphere is the *double pole coordinates*. Dipole coordinates may be the most familiar coordinate system for a sphere, but the double pole coordinates are the most useful spherical local coordinates in ray tracing for two reasons. (1) Double pole coordinates match the form of polarization that naturally occurs with lens and mirror systems. (2) Double pole coordinates scale (stretch) in such a way as to match magnification changes and changes of NA.

It would be nice to eliminate the two troublesome *poles* of the dipole coordinates. One method to try to eliminate the poles is shown in Figure 11.9. A point on the sphere, named the *anti-pole*, is chosen. A basis vector direction is chosen at the anti-pole, illustrated (left) by the red vector. This basis vector is translated along a great circle (Figure 11.9, middle) to define the basis vectors along an arc. This parallel transport operation is repeated along every arc (right); the basis vector maintains its angle with respect to every great circle through the anti-pole.* Observe that the poles at the top and bottom of the sphere have been eliminated. The basis vectors have been defined over an entire hemisphere without a singularity, which cannot be done with the dipole coordinates!

Have the singularities been eliminated? Figure 11.10 shows the basis vector set continued onto the back side of the sphere. The sphere now has one singularity diametrically opposite the anti-pole. The basis vector rotates by 4π around this singularity; hence, it counts as a *double pole*. Remember, the basis vectors rotated by 2π around the dipole's poles. Thus, unfortunately, as the *winding number theorem*[10] stated, local coordinates on a sphere must have two or more *zeros*. The double pole coordinates place both zeros in the same location. One advantage is that, now, this double pole coordinate system can cover almost the entire sphere, almost 4π steradians, without a singularity!

---

* Parallel transport is discussed in detail in Chapter 17.

11.4 Double Pole Coordinates

Figure 11.9 (Left) To construct the double pole basis vector set, a point named the anti-pole is selected and a basis vector direction is chosen, red vector. (Middle) The vector from the anti-pole is translated along a great circle, vertical red line, to define the basis vectors along an arc. (Right) Repeating this parallel transport operation over all great circles defines the basis vector, a local *x*-coordinate, for example, over the sphere. The second basis vector is orthogonal to the first basis vector and the normal to the sphere.

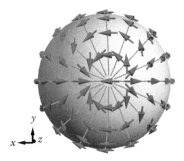

Figure 11.10 Rear side of the sphere for the double pole coordinates shows an interesting double singularity at the pole where the basis vector rotates by 4π (720°) for a single circuit around the double singularity.

One way to think about the double pole vector function is to consider moving the two poles of the dipole coordinates toward one another, stretching the dipole vector functions, until the poles coincide in one location. The two poles would finally merge, generating a *double pole*, which yields effectively one singular point. Figure 11.11 shows local *x* and local *y* for the double pole coordinates with a singular point, the double pole, at (0, 0, −1), and the opposite point, the anti-pole at (0, 0, 1). Does this contradict the winding number theorem? No, the double pole function has *double singularity* in each local coordinate. Look carefully at Figure 11.11 (right). The double pole basis vectors rotate by 4π during a 2π rotation around the singular point; hence, this singularity counts as two

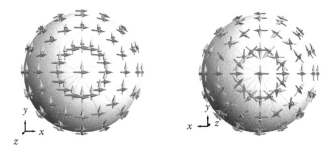

Figure 11.11 The pair of double pole basis vectors (red and green) on a unit propagation sphere with the double pole at (0, 0, −1). (Left) On the *front side* of the double pole basis (viewed from +*z*), no singularity is visible. (Right) The *back side* of the double pole basis, viewed from −*z*, has its doubly degenerate singular point at (0, 0, −1), where the basis vectors rotate by 4π around the pole.

zeros.* The left figure shows double pole coordinates on the *front side* of the sphere where there is no singular point, while the right figure shows the *back side* where a double pole (i.e., a doubly degenerate singular point) exists.

To generate the equations for the double pole basis vectors $(\hat{\mathbf{x}}_{Loc}, \hat{\mathbf{y}}_{Loc})$ over the unit propagation sphere, two vectors are needed to define the coordinate system. First, the unit vector $\hat{\mathbf{a}}_{Loc}$ defines the direction to the anti-pole. Second, the vector $\hat{\mathbf{x}}_o$ defines the direction of the first basis vector at the anti-pole. $\hat{\mathbf{x}}_o$ is the red vector in Figure 11.9 (left). The local y vector at the anti-pole is $\hat{\mathbf{y}}_o = \hat{\mathbf{a}}_{Loc} \times \hat{\mathbf{x}}_o$, forming a right-handed basis set $(\hat{\mathbf{x}}_o, \hat{\mathbf{y}}_o, \hat{\mathbf{a}}_{Loc})$. By applying the algorithm of Figure 11.9, double pole coordinates for an arbitrary propagation vector, **k**, are generated by a rotation of the two basis vectors defined at the anti-pole along a rotation axis $\vec{\mathbf{r}}$ by angle θ,

$$(\hat{\mathbf{x}}_{Loc}, \hat{\mathbf{y}}_{Loc}) = \left(\mathbf{R} \cdot \hat{\mathbf{x}}_o, \mathbf{R} \cdot (\hat{\mathbf{a}}_{Loc} \times \hat{\mathbf{x}}_o) \right), \tag{11.10}$$

where $\theta = -\cos^{-1}(\hat{\mathbf{k}} \cdot \hat{\mathbf{a}}_{Loc})$, $\vec{\mathbf{r}} = \hat{\mathbf{k}} \times \hat{\mathbf{a}}_{Loc}$, and **R** is a rotation matrix with a rotation axis $\vec{\mathbf{r}}$ by θ. This $(\hat{\mathbf{x}}_{Loc}, \hat{\mathbf{y}}_{Loc}, \hat{\mathbf{k}})$ forms a right-handed coordinate system. For example, the local coordinate set for $\hat{\mathbf{a}}_{Loc} = (0, 0, 1)$ with *x*-basis vector $(1, 0, 0)$ is

$$\hat{\mathbf{x}}_{Loc} = \left(1 - \frac{k_x^2}{1+k_z}, \frac{-k_x k_y}{1+k_z}, -k_x \right), \tag{11.11}$$

$$\hat{\mathbf{y}}_{Loc} = \left(\frac{-k_x k_y}{1+k_z}, 1 - \frac{k_y^2}{1+k_z}, -k_y \right). \tag{11.12}$$

A geometric construction to visualize the basis vector orientations is shown in Figure 11.12. The white and black circles are orthogonal at all their intersection points.

When a linearly polarized beam focuses through an ideal lens, the polarization of the beam exiting the lens is naturally described by the *double pole coordinates*, as shown in Figure 11.13. A non-polarizing lens changes the direction of polarization ellipse at each ray intercept by folding the polarization states around the corresponding *s*-basis vector.[11] Thus, a non-polarizing element rotates the transverse plane of polarization about the *s* polarization state but does not change the polarization ellipse in any other way. When this operation is applied to an aberration-free lens, an array on incident linearly polarized vectors will be transformed into an array of vectors in the exit pupil in the form of the double pole basis vectors.

Example 11.3 Parabolic Reflector Creating a Double Pole Polarization Pattern

The double pole coordinate system occurs naturally in some optical systems. Figure 11.14 (left and middle) shows a parabolic reflector system that focuses incident collimated rays into a converging spherical wavefront with a large solid angle. As the paraboloid is extended to the left toward infinity, the solid angle grows until the size of the wavefront approaches a complete sphere, a 4π steradian wavefront converging to the focal point as shown in Figure 11.14 (right). For a linearly polarized collimated beam incident on this paraboloid with a non-polarizing mirror ($r_s = r_p$), the polarization distribution on this spherical wavefront has the double pole form.

---

* Fun fact: This is also known as the "hairy ball theorem," which is appropriately named if you consider a tennis ball with small fibers that can be brushed in certain directions. If you try to comb the hairs to form the minimum number of singularities, you will come up with the double pole coordinates (https://en.wikipedia.org/wiki/Hairy_ball_theorem).

11.4 Double Pole Coordinates

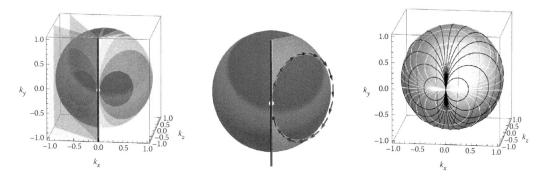

Figure 11.12 A set of circles depicting the orientation of the double pole basis vectors around the sphere is constructed by first placing a line (black) through the double pole (center of white spot) tangent to the unit propagation sphere. (Left) A series of planes are constructed through the line intersecting the unit propagation sphere (k_x, k_y, k_z) generating a set of circles of varying sizes that all intersect at the double pole. (Middle) The orientation of the $\hat{\mathbf{y}}_{Loc}$ basis vector is shown aligned with one of the circles. (Right) The orientation of the basis vectors over the sphere is shown for $\hat{\mathbf{x}}_{Loc}$ in white and $\hat{\mathbf{y}}_{Loc}$ in black. These are the equivalent of the lines of constant latitude and lines of constant longitude in the dipole coordinate system.

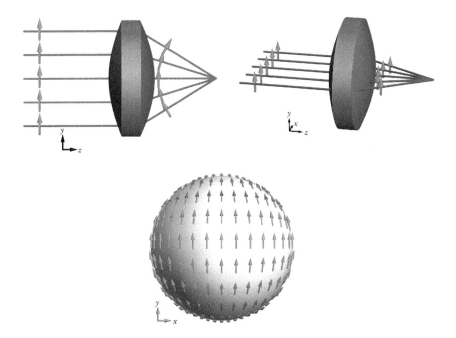

Figure 11.13 (Top left) A collimated line of linearly y-polarized rays in the yz-plane enters and refracts through an ideal non-polarizing lens, and the exiting polarization vectors bend around the spherical wavefront after rotating at the first and second refraction about the local s-vectors, which for these rays all lie in the x-direction. (Top right) The polarization vectors incident in the xz-plane also rotate about the s-basis vectors at each ray intercept. For these rays, the s-basis vector is along y for all the ray intercepts, and the polarization is along y, so the overall polarization upon exit does not rotate. (Bottom) Polarization of the spherical wave exiting the non-polarizing lens has the double pole pattern, with the yz- and xz-cross sections shown in the top row.

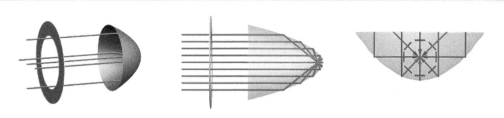

Figure 11.14 (Left) 3D view of a parabolic reflector with an aperture. (Middle) Cross-sectional view of a longer parabolic reflector where light reaches the focal plane with a solid angle greater than 2π steradians, and (right) how rays approach focal point from all possible angles with a polarization state drawn in blue.

Surrounding the focal point, each ray's propagation vector maps to a point on a unit sphere and each ray's polarization state can be plotted on the sphere at that point. Figure 11.15 (left) shows a 3D view of the reflected polarization state over the 2π steradians near the vertex [right side of Figure 11.14 (middle) or bottom half of Figure 11.14 (right)]. Figure 11.15 (right) shows the reflected polarization state over the 2π steradians facing the incident light direction. The left figure shows the smooth side of the sphere that corresponds to the wavefront on the mirror-vertex side of the focal point. The right figure shows the back side of the wavefront where the polarization is rapidly changing around the focal point.

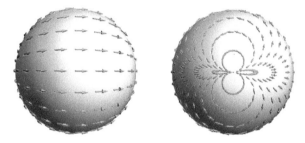

Figure 11.15 The polarization of a 4π steradian spherical wavefront from a non-polarizing reflecting paraboloid illuminated with horizontal linearly polarized converging to the focal point. (Left) The side of the wavefront near the parabolic reflector's vertex and (right) the back side of the wavefront facing the opening of the paraboloid are shown. These patterns correspond to the double pole local coordinates.

One very important characteristic of the double pole coordinate system is its *scale invariance* to magnification changes about the anti-pole. Referring to Figure 11.9, the coordinates can be expanded or contracted about the anti-pole, but since the orientation is constant along the great circles, the coordinate system remains the same. Consider a beam entering an optical system with a numerical aperture na_1 and exiting with na_2 with the anti-pole located along the chief ray for each wavefront. The beam's solid angle is magnified or minified between entering and exiting. If the entering beam's polarization has the double pole pattern, then the exiting beam also has the double pole pattern. The double pole coordinates are scale invariant to magnification about the anti-pole, making it preferred for the description of spherical wavefronts in imaging systems; in this case, the double pole axis is located opposite to the chief ray and the double pole singularity will not be encountered. Because of this scale invariance, the local coordinate system described as a wavefront in double pole coordinates remains the same as a spherical wave moves from surface to surface with changing NA.

Figure 11.16 shows double pole coordinates with $\hat{\mathbf{a}}_{Loc} = (0, 0, 1)$ and $\hat{\mathbf{x}}_{Loc} = (1, 0, 0)$ on a unit sphere, viewed in four different angles with red vectors for $\hat{\mathbf{x}}_{Loc}$. The $\hat{\mathbf{y}}_{Loc}$ are rotated

11.4 Double Pole Coordinates

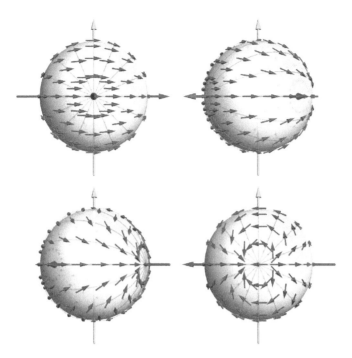

Figure 11.16 Various views of one of the double pole basis vectors are shown starting from the front side of the sphere to the back side with its doubly degenerate singular point. Note the 720° rotation of $\hat{\mathbf{x}}_{Loc}$ about the double pole. Global x, y, and z are shown in red, green, and blue arrows, respectively.

90° counterclockwise from the $\hat{\mathbf{x}}_{Loc}$. Gray lines show great circles that pass through axis $\hat{\mathbf{a}}_{Loc}$. The first figure is viewed from the front side of the sphere (i.e., from $\hat{\mathbf{a}}_{Loc}$); thus, there is no singularity on this side of the sphere. The last figure views along the double pole $-\hat{\mathbf{a}}_{Loc}$ placing the singular point at the center of the view.

Since the front side of the sphere has a smooth and continuous vector field, the double pole coordinates are the local coordinates of choice when converting the 3D polarization ray tracing matrix (defined in Chapter 9) to the Jones pupil or vice versa with the anti-pole along the chief ray. The double pole coordinates, $(\hat{\mathbf{x}}_{Loc}, \hat{\mathbf{y}}_{Loc})$, are ideal for describing the orientation of the polarization on a spherical wavefront; when a collimated linearly polarized wave converges or diverges into a spherical wave through an ideal non-polarizing lens or parabolic mirror, the wavefront polarization has the form of double pole coordinates. Figure 11.17 shows a vertically polarized spherical wavefront described using double pole $\hat{\mathbf{y}}_{Loc}$ coordinates.

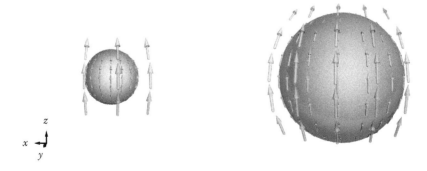

Figure 11.17 The double pole y-coordinates are used to create a linearly polarized spherical wavefront for two different numerical apertures.

Example 11.4 Double Pole Coordinates When $\hat{\mathbf{a}}_{Loc} = (0, 1, 0)$

Calculate the double pole coordinates when $\hat{\mathbf{a}}_{Loc} = (0, 1, 0)$ and $\hat{\mathbf{x}}_o = (1,0,0)$ for arbitrary $\hat{\mathbf{k}} = (k_x, k_y, k_z)$ using Equation 11.10.

First, the rotation matrix, $\mathbf{R}$, for rotation axis $\vec{\mathbf{r}} = \hat{\mathbf{k}} \times \hat{\mathbf{a}}_{Loc} = (-k_z, 0, k_x)$ through an angle of $\theta = -\cos^{-1}(\hat{\mathbf{k}} \cdot \hat{\mathbf{a}}_{Loc}) = -\cos^{-1}(k_y)$ is

$$\mathbf{R} = \begin{pmatrix} k_y - \dfrac{k_x^2}{1+k_y} & k_x & \dfrac{-k_x k_y}{1+k_y} \\ -k_x & k_y & -k_z \\ \dfrac{-k_x k_z}{1+k_y} & k_z & 1 - \dfrac{k_z^2}{1+k_y} \end{pmatrix}. \qquad (11.13)$$

Then, the double pole local coordinates can be calculated as

$$\hat{\mathbf{x}}_{Loc} = \left(1 - \dfrac{k_x^2}{1+k_y}, -k_x, \dfrac{-k_x k_z}{1+k_y}\right) \text{ and } \hat{\mathbf{y}}_{Loc} = \left(\dfrac{k_x k_z}{1+k_y}, k_z, \dfrac{k_z^2}{1+k_y} - 1\right). \qquad (11.14)$$

11.5 High Numerical Aperture Wavefronts

Polarization variations across curved wavefronts are intrinsic. When focusing light beams with very *high NA beams*, approaching or exceeding a hemisphere or 2π steradians, these polarization variations are frequently detrimental[12,13] For example, consider an *x*-linearly polarized spherical beam of the double pole coordinate form as shown in Figure 11.18. In the *y–z* plane (vertical in the left figure, horizontal cut through the middle figure), the polarization is uniformly polarized in the *x*-direction, despite the wavefront curvature. However, along the *x–z* plane, the light field must tip upward and downward to remain tangential to the surface of the sphere, as shown around the circumference of the center figure, and the middle cross section through the right figure. The *x* polarized light continues rotating from the center until at the edge of the pupil (near the *x*-axis) the light becomes polarized along the optical axis, since the light is now propagating along $(0, \pm 1, 0)$. Similarly, along $(1, 0, 0)$, the light remains *x* polarized to the edge of the pupil as shown in the right figure, but tips around the edge of the pupil. Thus, around the edge of the pupil, the polarization varies as shown in

Figure 11.18 A high NA spherical wave linearly polarized in the *x*-direction viewed (left) along the optical axis, $(0, 0, 1)$, (center) viewed along $(0, 1, 0)$, and (right) viewed along $(1, 0, 0)$. This polarization is the double pole *x*-basis vector. The double pole is located on the back side of the sphere in the middle at $(0,0,-1)$. Global *x*, *y*, and *z* are shown in red, green, and blue arrows, respectively.

11.6 Converting P Pupils to Jones Pupils

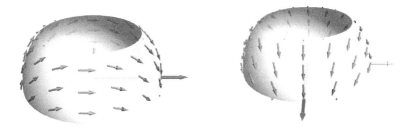

Figure 11.19 Distribution of *x*-linearly polarized light around the edge of a hemispherical wavefront. Global *x*, *y*, and *z* are shown in red, green, and blue arrows, respectively.

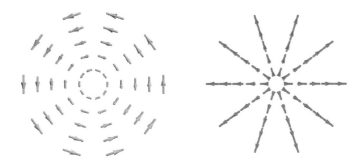

Figure 11.20 The polarization distributions in (left) a tangentially polarized wavefront, and (right) a radially polarized wavefront.

Figure 11.19. The result of this polarization variation is a modification to the point spread function where it becomes elongated in one direction, much like astigmatism.

Many polarization distributions are of interest in high NA imaging such as the radial[14–16] and tangential polarization distributions shown in Figure 11.20. Note that these electromagnetic waves cannot be extended to the center (i.e., a singular point) without discontinuity.

11.6 Converting P Pupils to Jones Pupils

This section presents an algorithm for converting grids of *polarization ray tracing matrices* (**P**) into *Jones pupils*. The **P** matrix pupil map, introduced in Section 10.7.3, provides a concise and complete 3D description of how an optical system transforms an incident polarized wavefront into the corresponding exiting polarized wavefront. However, it can be challenging to visualize the polarization changes in three dimensions; it is helpful to project the wavefront and polarization states onto flat surfaces such as computer screens or printed pages like this book's. Figure 11.21 shows incident and exiting spherical wavefronts with some polarization aberration. To aid visualization, the polarization ellipses can be transformed to a transverse plane that is perpendicular to the chief ray's

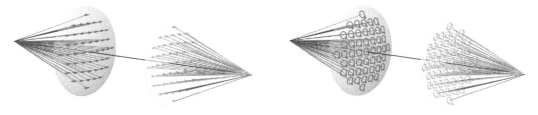

Figure 11.21 Linearly and circularly polarized incident (blue) and exiting (green) spherical wavefronts for a grid of rays.

propagation vector, by transforming a three-element polarization vector into a Jones vector. Optical designers often want to evaluate the polarization aberration or Jones pupil at the exit pupil of the system. Therefore, the focus of this section is to calculate a Jones pupil, which will be represented in local coordinates, at the exit pupil of the system.

Typically, a *polarization ray trace* calculates the properties of a grid of rays through an optical system, yielding a grid of **P** matrices. The 3 × 3 **P** matrix is a generalized version of the 2 × 2 Jones matrix. Jones matrices are convenient for data reduction into diattenuation and retardance aberrations. Further, Jones pupil output is widely understood since many commercial optical design software packages use Jones matrices to perform polarization ray tracing. Converting the 3D **P** into the 2 × 2 Jones matrices requires an x and y basis for the incident state and *another* x' and y' basis for the exiting state, one local coordinate system for the incident wavefront with its grid of incident rays and one for the exiting wavefront with the corresponding grid of exiting rays. This conversion, transforming a matrix from global coordinates to local coordinates, can be performed in many different ways. A pair of incident and exit local coordinate systems must be selected; dipole, double pole, and so on, and the $\hat{\mathbf{a}}_{Loc}$ must be specified. These selections are usually arbitrary, often default specifications. But careless choice of local coordinates can lead to misleading polarization behaviors, *Jones pupil artifacts*. An appropriate choice of local coordinates may simplify the representation of the wavefronts. In particular, it is desirable to move the singularities of the local coordinates out of the solid angle subtended by the wavefront. The best local coordinates tend to closely match the polarization of the beams; if the polarization distribution is close to a dipole pattern, dipole coordinates may be clearest. Many linearly polarized wavefronts resemble the double pole pattern; hence, a double pole local coordinate system with an anti-pole along the chief ray may be simplest. Other basis vector patterns can be created and applied to simplify other problems.

In Chapter 9, orthogonal transformation matrices, $\mathbf{O}_{in}$ and $\mathbf{O}_{out}$, were introduced to convert the Jones matrix into the **P** matrix. $\mathbf{O}_{in}$ transforms the incident local coordinates to the global coordinates and $\mathbf{O}_{out}$ transforms the exiting local coordinates to the global coordinates at a given ray intercept. The **P** is constructed by two orthogonal transformation matrices and a Jones matrix. The Jones pupil for a grid of rays with ray indices (i, j) at the exit pupil, $\mathbf{J}_{ij,X}$, can be calculated from the **P** pupil at the exit pupil, $\mathbf{P}_{ij,X}$, using the orthogonal matrices $\mathbf{O}_{ij,E}$ and $\mathbf{O}_{ij,X}$, where E indicates entrance pupil and X indicates exit pupil,

$$\mathbf{J}_{ij,X} = \mathbf{O}_{ij,X}^{-1} \mathbf{P}_{ij,X} \mathbf{O}_{ij,E}. \tag{11.15}$$

$\mathbf{O}_{ij,E}$ transforms the entrance pupil local coordinates to the global coordinates, and $\mathbf{O}_{ij,X}^{-1}$ transforms the global coordinates to the exit pupil local coordinates. Thus, $\mathbf{J}_{ij,X}$ has basis vector sets in both the entrance and the exit pupils.

The entrance and exit pupil local coordinates can be chosen arbitrarily, but should be consistent throughout the ray grid. For example, the dipole coordinate, double pole coordinate, or the $(\hat{\mathbf{s}}, \hat{\mathbf{p}}, \hat{\mathbf{k}})$ coordinates[17] can be used as local coordinates. For the given ray propagation vectors at the entrance pupil $(\hat{\mathbf{k}}_E)$ and the exit pupil $(\hat{\mathbf{k}}_X)$, the orthogonal transformation matrices have $(\mathbf{v}_{ij}, \mathbf{w}_{ij})$, x–y local coordinates of the entrance and exit pupils, as their column vectors

$$\mathbf{O}_{ij,E} = \begin{pmatrix} v_{ij,x,E} & w_{ij,x,E} & k_{ij,x,E} \\ v_{ij,y,E} & w_{ij,y,E} & k_{ij,y,E} \\ v_{ij,z,E} & w_{ij,z,E} & k_{ij,z,E} \end{pmatrix},$$

$$\mathbf{O}_{ij,X} = \begin{pmatrix} v_{ij,x,X} & w_{ij,x,X} & k_{ij,x,X} \\ v_{ij,y,X} & w_{ij,y,X} & k_{ij,y,X} \\ v_{ij,z,X} & w_{ij,z,X} & k_{ij,z,X} \end{pmatrix}, \tag{11.16}$$

where i and j are the ray indices at the pupils, $(\mathbf{v}_{ij,q}, \mathbf{w}_{ij,q}, \mathbf{k}_{ij,q})$ forms the right-handed local coordinates, $(\mathbf{v}_{ij,q} \times \mathbf{w}_{ij,q} = \mathbf{k}_{ij,q}$, and $q = E$ (for entrance pupil) or X (for exit pupil). Since the orthogonal transformation matrices depend on the local coordinates, the choice of local coordinates determines the resulting Jones pupil.

In most cases, double pole coordinates is the most appropriate coordinate system to use since it describes almost the entire 4π steradian sphere without a singular point.

11.7 Example: Cell Phone Lens Aberrations

In this section, the example cell phone lens system[18] of Chapter 10, shown in Figure 11.22, is used as an example to demonstrate the conversion between an array of **P** matrices and the Jones pupil, and explores the artifacts in the Jones pupil due to the choice of local coordinates. This example chooses an uncoated lens for analysis.

A set of **P** matrices for a grid of rays across the wavefront forms a **P** matrix pupil at the exit pupil as shown in Figure 11.23 for 589 nm. This **P** matrix pupil can be converted into a Jones matrix pupil by applying Equation 11.15 on a ray-by-ray basis. The choice of the form of basis for $(\mathbf{v}_i, \mathbf{w}_i)$ yields different Jones pupils. Figure 11.24 shows the Jones pupil rendered using a pair of double pole coordinates (left), dipole coordinates (middle), and s–p coordinates (right) on the entrance and exit pupils. The local x-coordinates for three local coordinate systems are shown in the second row of Figure 11.24. For double pole and dipole coordinates, local x-coordinates are smooth and well-defined without a singular point on the exit pupil. Since the NA of the lens is not too large, the

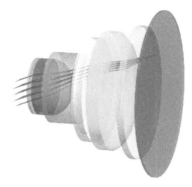

Figure 11.22 A cell phone lens system with 20° off-axis field.

Figure 11.23 A **P** matrix pupil for the cell phone lens at 20° off-axis field.

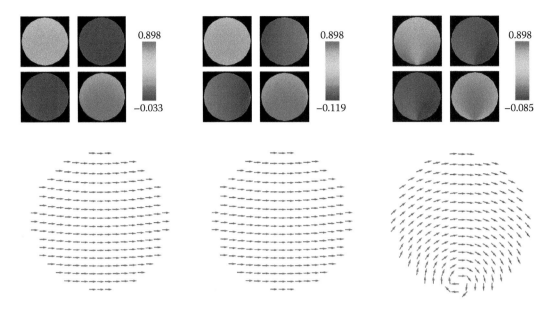

Figure 11.24 (Top) Jones pupil of the cell phone lens system using (left) double pole coordinates, (middle) dipole coordinates, and (right) s–p coordinates to convert from the **P** matrix pupil. (Bottom) Local x-coordinates across the exit pupil for the (left) double pole coordinates, (middle) dipole coordinates, and (right) s–p coordinates.

differences between double pole and dipole coordinates are small across the pupil. The **P** matrix pupil converts to smooth Jones matrix pupils in both cases. For the s–p coordinate example, the entrance pupil local coordinates use s and p for the first lens surface and the exit pupil local coordinates use s and p for the final lens surface. These s–p local x-coordinates have a singular point on the bottom center about the ray at normal incidence at the last surface, where local x-coordinates vary rapidly, yielding rapidly varying local x-coordinates on that singular point. As a result, the s–p Jones pupil contains rapid change around the same singular point. Such singular behavior in the Jones pupil might be misinterpreted as an interesting polarization effect such as a vortex, when no vortex is present, illustrating the importance of a good coordinate system choice.

The top left figure shows the amplitude of the Jones pupil using double pole coordinates with $\hat{\mathbf{a}}_{Loc} = \hat{\mathbf{k}}_{chief}$ so that the front side of the exiting wavefront, thus the Jones pupil, has no singularity. The phase of the Jones pupil is zero for this uncoated example system. The top middle figure shows the amplitude of the Jones pupil using dipole coordinates with $\hat{\mathbf{a}}_{Loc}$ along $\pm\hat{\mathbf{k}}_{chief}$. The top right figure shows the amplitude of the Jones pupil using s–p coordinates at the first and last surfaces, where the bottom right figure shows the x-coordinates (i.e., s-coordinates) on the last surface. The s–p coordinates, described in Chapter 9, are the most common local coordinate systems for describing the polarization state for single surface refraction and/or reflection, since the s and p polarization components are the eigenpolarizations of those light–matter interactions. Here, the s–p coordinates provide a misleading view of the polarization transformations.

11.8 Wavefront Aberration Function Difference between Dipole and Double Pole Coordinates

The choice of local coordinate basis does not just affect the polarization description; it also affects the wavefront aberrations calculated! This is most obvious when the light is circularly polarized. The electric field vector of circularly polarized light rotates once per period; hence, the phase of circularly polarized light is in the direction of the electric field vector at a reference time. When the orientation of the basis vector from which the reference phase (i.e., zero phase) is measured rotates, then the circularly polarized

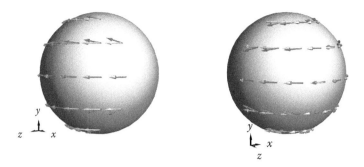

Figure 11.25 Two different views showing the comparison of $\hat{\mathbf{X}}_{Loc}$ in the dipole coordinates (orange, pole on y-axis) and the double pole basis (red, pole at (0, 0, −1)). The rotation between the two coordinate systems is clearer around the edges of the fields. The red vectors are shifted slightly downward for comparison.

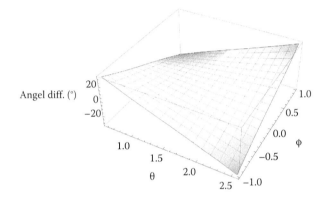

Figure 11.26 The angle between the two coordinate systems about the z-axis is toroidal.

light's phase is different in this second coordinate system. This effect causes a circular polarized wavefront in dipole coordinates to appear astigmatic when represented in double pole coordinates, and vice versa.

Figure 11.25 shows $\hat{\mathbf{x}}_{Loc}$ in the dipole coordinates ($\hat{\mathbf{a}}_{Loc}$ = (0, 1, 0)) in orange and the double pole basis ($\hat{\mathbf{a}}_{Loc}$ = (0, 0, 1)) in red in two different views. A systematic rotation is observed between the two coordinate systems. Figure 11.26 plots this rotation angle between the two coordinate systems and shows its toroidal shape, the shape of astigmatism. Changing the Jones pupil between these two different local coordinates will alter the description of astigmatism for circularly polarized light. For example, if dipole coordinates are used for the entrance pupil and the double pole coordinates are used for the exit pupil for a system without astigmatism, the Jones pupil will display astigmatism, which varies as Figure 11.26 across the pupil. θ is latitude and ϕ is longitude, where a propagation vector can be written as $\hat{k} = (k_x, k_y, k_z) = (\cos\phi\sin\theta, \sin\phi\sin\theta, \cos\theta)$.

This example illustrates how the choice of local coordinates affects the wavefront aberration function for circularly polarized light and the description of polarization for other states. It provides guidance on picking local coordinates that are appropriate for one's problem. Changing local coordinates changes the way you interpret the system aberrations, even though nothing has changed regarding the light's propagation.

11.9 Conclusion

This chapter analyzes two choices for local coordinates on a sphere, the double pole and the dipole coordinates, which are suitable for describing the polarization of spherical waves. One important task of local coordinates is to allow the translation of results on a spherical wavefront onto a computer screen or a flat piece of paper.

The definition of these local coordinates was motivated by forms of polarized waves, radiation from an oscillating dipole in the case of the dipole coordinates, and a polarized beam focused through a lens or paraboloid in the case of the double pole coordinates. Polarized wavefronts can be described by any local coordinate system. The choice of coordinates should be used to simplify the representation where possible.

Local coordinates defined over the full sphere must have singularities. These singularities are a bother, but they are not a true problem. Where possible, the singularities are chosen to lie outside the physical wavefronts. When a coordinate system singularity lies within a wavefront, then within that coordinate system representation, it appears that the polarization state is varying rapidly, when actually it is the coordinates that are varying. The situation resembles the situation with map projections such as Mercator and Mollweide around the north and south poles. Any map projection stretches the Earth's surface unevenly.

11.10 Problem Sets

11.1 Explain the difference between local and global coordinates.

11.2 What are the advantages of the double pole coordinate system over the dipole coordinate system?

11.3 To define the double pole coordinate system an axis vector, $\hat{\mathbf{a}}_{Loc}$, a reference local x vector $\hat{\mathbf{x}}_o$ is needed. Why does the definition of the dipole coordinates use an $\hat{\mathbf{a}}_{Loc}$ vector but not a local x vector?

11.4 a. Show for the dipole basis with axis $\mathbf{a} = (0, 0, 1)$ that the local coordinate basis vectors x_{Loc} and y_{Loc} are orthogonal to the propagation vector $\mathbf{k} = (k_x, k_y, k_z)$.
  b. Show that the local coordinate basis vectors are normalized.
  c. Show that x_{Loc} and y_{Loc} are orthogonal to each other.
  d. Show that the cross product of x_{Loc} and y_{Loc} is the propagation vector $\mathbf{k}$.

11.5 a. Show for the double pole basis with axis $\mathbf{a} = (0, 0, 1)$ that the local coordinate basis vectors x_{Loc} and y_{Loc} are orthogonal to the propagation vector $\mathbf{k} = (k_x, k_y, k_z)$.
  b. Show that the local coordinate basis vectors are normalized.
  c. Show that x_{Loc} and y_{Loc} are orthogonal to each other.
  d. Show that the cross product of x_{Loc} and y_{Loc} is the propagation vector $\mathbf{k}$.

11.6 Light is propagating in the direction $\hat{\mathbf{k}} = (3, 4, 5)/5\sqrt{2}$.
  a. Create a local coordinate system $(\hat{\mathbf{x}}_{Loc}, \hat{\mathbf{y}}_{Loc})$ for the transverse plane where the $\hat{\mathbf{x}}_{Loc}$ is in the x–z plane.
  b. Create a local coordinate system $(\hat{\mathbf{x}}_{Loc}, \hat{\mathbf{y}}_{Loc})$ for the transverse plane where the $\hat{\mathbf{y}}_{Loc}$ is in the y–z plane.

11.7 Is the dipole coordinate system rotationally symmetric about its $\hat{\mathbf{a}}_{Loc}$? Is the double pole coordinate system rotationally symmetric about the $\hat{\mathbf{a}}_{Loc}$?

11.8 Consider a great circle through the pole in a double pole local coordinate system and the local coordinates along this circle. How do these local coordinate basis vectors rotate with respect to this plane?

11.9 a. Draw how one of the double pole local coordinates varies about a great circle in the plane perpendicular to the axis for $\hat{\mathbf{a}}_{Loc} = (0, 0, 1)$ on the x–y plane.

11.10 Problem Sets   443

b. How does this change as the plane moves closer to the double pole remaining perpendicular to the $\hat{\mathbf{a}}_{Loc}$ vector? Further from the double pole?

11.10 Linearly polarized light is oriented at 60° as shown in the figure.

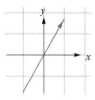

a. Write the normalized Jones vector in this x–y basis.
b. Write the Jones vector in local x'–y' coordinates, rotated 45° counterclockwise from the global x–y basis.

11.11 A region of the dipole local y-coordinates and a region of the double pole local y-coordinate basis sets are plotted over a spherical cap in the following two figures. Identify which figure contains a dipole coordinate field and which contains a double pole coordinate field. Where are the poles located for each set?

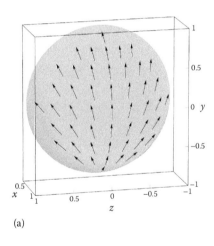
(a)

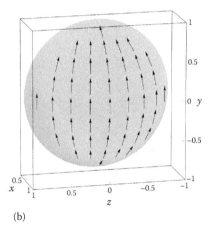
(b)

11.12 The winding number theorem states that continuous and differentiable vector fields on a sphere must have at least two zeros. Find the zeros for the following vector fields:
a. The sum of a vector field described by $\hat{\mathbf{x}}_{Loc}$ of the dipole coordinates with an $\hat{\mathbf{a}}_{Loc} = (0, 1, 0)$ and a vector field described by $\hat{\mathbf{x}}_{Loc}$ of the dipole coordinates with an $\hat{\mathbf{a}}_{Loc} = (1, 0, 0)$.
b. The sum of the double pole $\hat{\mathbf{x}}_{Loc}$ coordinates with an $\hat{\mathbf{a}}_{Loc} = (0, 0, 1)$ and $\hat{\mathbf{x}}_o = (1, 0, 0)$, and the double pole $\hat{\mathbf{x}}_{Loc}$ coordinates with an $\hat{\mathbf{a}}_{Loc} = (0, 0, -1)$ and $\hat{\mathbf{x}}_o = (1, 0, 0)$.

11.13 Numerical calculation for a dipole coordinate system
a. Calculate the dipole coordinate vectors for $\hat{\mathbf{a}}_{Loc} = (0, 0, 1)$ at latitude = 90° (equator) and 0° < longitude < 360° in every 20°.
b. Repeat for latitude = 45°.
c. Repeat for latitude = 0.1°.

11.14 A spherical wave exits a 0.8 NA microscope objective, and its polarization distribution is identical to the double pole $\hat{\mathbf{y}}_{Loc}$ coordinates centered on the z-axis and $\hat{\mathbf{x}}_o = (1, 0, 0)$. If this polarization pattern is represented in dipole coordinates with an $\hat{\mathbf{a}}_{Loc} = (0, 1, 0)$, what will be the maximum angle between the polarization state and the dipole coordinate system?

11.15 A complete 4π steradian spherical wave has polarization that matches $\hat{\mathbf{x}}_{Loc}$ of the double pole coordinates with $\hat{\mathbf{a}}_{Loc} = (0, 0, 1)$ and $\hat{\mathbf{x}}_o = (1, 0, 0)$. Looking from the origin toward $(0, 0, 1)$ through a rotating linear polarizer, qualitatively describe how the transmitted intensity changes as a function of the polarizer's transmission axis θ measured from $(1, 0, 0)$.

11.16 A fold mirror with a metal coating ($n = 1.015192 + i6.6273$) has a surface normal $\boldsymbol{\eta} = (0, -1/\sqrt{2}, 1/\sqrt{2})$. Consider a cone of rays incident on the metal mirror with a chief ray along $\hat{\mathbf{k}}_{chief,1} = (0, 0, 1)$ and the edge ray along $\hat{\mathbf{k}}_1 = (0, \sin 30°, \cos 30°)$. The metal mirror reflects the chief ray; $\hat{\mathbf{k}}_{chief,2} = (0, 1, 0)$.

a. For the edge ray with $\hat{\mathbf{k}}_1 = (0, \sin 30°, \cos 30°)$, calculate the reflected ray's propagation vector, $\hat{\mathbf{k}}_2$.

b. Using Equation 11.17, calculate the orthogonal transformation matrices for the incident ray and the reflected ray using the double pole coordinate system.

c. A **P** matrix for this ray is given:

$$\mathbf{P}_{Out} = \begin{pmatrix} -1.00851 + 0.0763605i & 0 & 0 \\ 0 & 0.680102 - 0.429658i & 0.607343 + 0.248063i \\ 0 & -0.177972 + 0.74419i & 0.680102 - 0.429658i \end{pmatrix}.$$

Calculate a Jones matrix using (a) and (b).

References

1. J. Ruoff and M. Totzeck, Orientation Zernike polynomials: A useful way to describe the polarization effects of optical imaging systems, *J. Micro/Nanolithogr. MEMS MOEMS* 8 (2009): 031404.
2. R. C. Jones, A new calculus for the treatment of optical systems I, *J. Opt. Soc. Am.* 31 (1941): 488–493, 493–499, 500–503.
3. R. C. Jones, A new calculus for the treatment of optical systems. IV, *J. Opt. Soc. Am.* 32 (1942): 486–493.
4. R. C. Jones, A new calculus for the treatment of optical systems. V. A more general formulation, and description of another calculus, *J. Opt. Soc. Am.* 37 (1947): 107–110, 110–112.
5. R. C. Jones, A new calculus for the treatment of optical systems. VII. Properties of the N-matrices, *J. Opt. Soc. Am.* 38 (1948): 671–683.
6. G. Yun, K. Crabtree, and R. Chipman, Three-dimensional polarization ray tracing calculus I: Definition and diattenuation, *Appl. Opt.* 50 (2011): 2855–2865.
7. G. Yun, S. McClain, and R. A. Chipman, Three-dimensional polarization ray tracing, retardance, *Appl. Opt.* 50, 2855–2865.
8. R. C. Jones, A new calculus for the treatment of optical systems, *J. Opt. Soc. Am.* 46 (1956): 126–131.
9. W. Singer, M. Totzeck, and H. Gross, Physical image formation, in *Handbook of Optical Systems*, New York: Wiley (2005), pp. 613–620.
10. S. G. Krantz, The index or winding number of a curve about a point, in *Handbook of Complex Variables*, Boston, MA (1999), pp. 49–50.
11. M. Mansuripur, *Classical Optics and Its Applications*, figures 3.1 and 3.2, Cambridge, UK: Cambridge University Press (2009).

12. T. Chen and T. D. Milster, Properties of induced polarization evanescent reflection with a solid immersion lens (SIL), *Opt. Exp.* 15(3) (2007): 1191–1204.
13. T. D. Milster, J. S. Jo, K. Hirota, K. Shimura, and Y. Zhang, The nature of the coupling field in optical data storage using solid immersion lenses, *Jap. J. Appl. Phys.* 38 (1999): 1793–1794.
14. M. O. Scully and M. S. Zubairy, Simple laser accelerator: Optics and particle dynamics, *Phys. Rev. A* 44 (1991): 2656.
15. R. Dorn, S. Quabis, and G. Leuchs, Sharper focus for a radially polarized light beam, *Phys. Rev. Lett.* 91 (2003): 233901.
16. T. Grosjean, D. Courjon, and C. Bainier, Smallest lithographic marks generated by optical focusing systems, *Opt. Lett.* 32 (2007): 976–978.
17. E. Waluschka, A polarization ray trace, *Opt. Eng.* 28 (1989): 86–89.
18. M. Taniyama, U.S. Patent 7,535,658 B2 (May 19, 2009).

12

Fresnel Aberrations

12.1 Introduction

This chapter explores the *polarization aberrations* of uncoated lenses and metal mirrors. The Fresnel equations describe the change of polarization at uncoated interfaces. In optical systems, because of the Fresnel equations, polarization variations known as polarization aberrations occur across the exit pupil, which can be called *Fresnel aberrations*. Such polarization aberrations are usually visualized in terms of the diattenuation and retardance variations associated with ray paths.

The aberration of an optical system is its deviation from ideal performance. In an imaging system with ideal spherical or plane wave illumination, the desired output is spherical wavefronts centered on the correct image point with constant amplitude and constant polarization state. Deviations from spherical wavefronts arise from variations of optical path length of rays through the optic due to the geometry of the optical surfaces and the laws of reflection and refraction. The deviations from spherical wavefronts are known as the wavefront aberration function. Deviations from constant amplitude arise from differences in reflection or refraction efficiency between rays. Such amplitude variations are amplitude aberration or apodization. Polarization change also occurs at each reflecting and refracting surface due to differences between the s- and p-components of the light's reflectance and transmission coefficients. Across a set of rays, the angles of incidence and, thus, the polarization change vary, so that a uniformly polarized input beam has polarization variations when exiting.[1,2] For many optical systems, the desired polarization output would be a constant polarization state with no polarization change transiting the system; such ray paths through an optical system can be described by identity Jones matrices. A useful definition of polarization aberrations is deviations from this identity matrix caused by the interaction of the light with the optical system.

In this hierarchy, wavefront aberrations are by far the most important aberration, as variations of optical path length of small fractions of a wavelength can greatly reduce the image quality. The relative priority of wavefront aberrations is so great that for the first 40 years of computer-aided optical design, amplitude and polarization aberration were not calculated by the leading commercial optical design software packages. The variations of amplitude and polarization found in high-performance astronomical systems cause much less change to the image quality than the wavefront aberrations, but as the community prepares to image and measure the spectrum and polarization of exoplanets and similar demanding tasks, these amplitude and polarization effects can no longer be ignored. For example, Stenflo[3] has discussed limitations on the accuracy of solar magnetic field measurements due to polarization aberration.

In a system of reflecting and refracting elements, amplitude and polarization aberration contributions arise from the Fresnel coefficients for uncoated or reflecting metal surfaces and by related amplitude coefficients for thin film-coated interfaces. Polarization aberration, also called *instrumental polarization*, refers to all polarization changes of the optical system and the variations with pupil coordinate, object location, and wavelength. The term *Fresnel aberrations* refers to those polarization aberrations that arise strictly from the Fresnel equations, that is, systems of metal-coated mirrors and uncoated lenses.[1,4–6] Multilayer coated surfaces produce polarization aberrations with similar functional forms and may have larger or smaller magnitudes.

This chapter applies the Fresnel equations to describe polarization changes that occur in several example optical systems, an uncoated lens, a fold mirror, a combination of fold mirrors, a Cassegrain telescope, and a Fresnel rhomb retarder. Further analysis of a Cassegrain telescope with a fold mirror is included in Chapter 27, where a list of design rules are derived.

12.2 Uncoated Single-Element Lens

This section will show the many different ways of representing polarization aberration of a fast singlet lens and demonstrate its impact on image formation. Figure 12.1 shows a set of meridional rays refracting through a fast single-element lens coming to focus on the optical axis. The aperture stop that is also the entrance pupil (gray annulus) is located in object space a small distance in front of the lens. On-axis images formed with fast (high numerical aperture) spherical lenses have large amounts of spherical aberration and highly aberrated point spread functions. To correct this, the surfaces of this lens have been aspherized, they are not spherical surfaces, in such a way that the wavefront aberration of the lens for the on-axis image is negligible (a few milliwaves, λ/1000). The prescription of the lens element shown in Figure 12.1 is listed in Table 12.1.

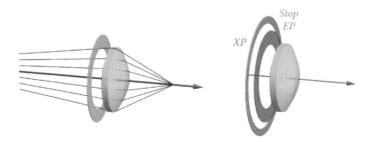

Figure 12.1 A fast single-element lens is our first Fresnel aberration example. The entrance pupil (*EP*, gray) limits the ray bundle from the on-axis object point. On the right, the image of the entrance pupil is the exit pupil (*XP*, brown), which is associated with rays in image space, the space of the rays after they refract from the last surface. In this case, the exit pupil is virtual, located to the left of the final surface, and larger than the entrance pupil, since here the pupil magnification is greater than one.

12.2 Uncoated Single-Element Lens

Table 12.1 Lens Prescription for a Fast Singlet in Figure 12.1

Lens refractive index	1.51674
Object location	−50 mm from stop
Stop location	−1 mm from first lens surface
Image location	10.7725 mm from seconds lens surface
Effective focal length	10 mm
Entrance pupil diameter	14 mm

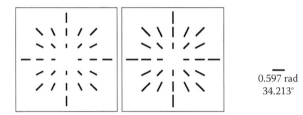

Figure 12.2 Angle of incidence map for front (left) and back (middle) surface is radially oriented and increases approximately linearly. (Right) The scale shows the maximum angle of incidence in radians and degrees.

From a ray trace of an on-axis object point with a grid of rays, the angle of incidence (*AoI*) and orientation of the plane of incidence (*PoI*) are calculated for each surface and plotted as a set of *angle of incidence maps* (*AoI* map) in Figure 12.2. In the *AoI* map, the length of each line segments represents the magnitude of the *AoI*, and the line's orientation indicates the orientation of the *plane of incidence* (*PoI*), which is also the p-polarization orientation for the ray. On the left side of Figure 12.2, for the first lens surface, the *AoI* is zero for the ray down the center of the axis through the center of the entrance pupil, represented by a line with zero length. Moving out from the center of the pupil, the angles of incidence increase with distance from the center of the pupil (the radial pupil coordinate). Four marginal rays, rays from the on-axis field through the edge of the pupil, are represented, top, bottom, left, and right, and from the legend shown on the right, their length corresponds to an angle of incidence of ~0.45 radians ≈ 34°. These lines are all radial, indicating that the *PoI* for all of these rays is radially oriented and contains the optical axis. The middle *AoI* map is plotted for the second lens surface where the angles of incidence are a little larger; the marginal ray angle of incidence is about 0.6 radians. Again, the plane of incidence is radially oriented. For both surfaces, the angle of incidence increases approximately linearly with the radial pupil coordinate.

Math Tip 12.1 Approximately Linear and Quadratic Functions

It is standard in aberration theory that when a function is described as linear, it means the function is approximately linear, in the sense that the linear term in a Taylor series dominates. This avoids repeatedly using the term *approximately linear* throughout the remainder of the chapter. Similar interpretation applies to the terms *quadratic* and *cubic*.

From the Fresnel equations for an air ($n = 1$)-to-glass (such as, $n = 1.51674$) interface, the corresponding diattenuation for the ray's intercepts at the first surface is calculated, and the corresponding diattenuation magnitude and orientation are plotted as diattenuation maps in Figure 12.3 (left). In the *diattenuation map*, the length of each line represents the diattenuation magnitude and the

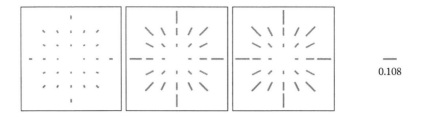

Figure 12.3 Diattenuation maps for the front surface, back surface, and the combination of front and back surfaces of the singlet. The scale on the right shows the maximum diattenuation of the three maps.

line's orientation is the orientation of the maximum transmission state. Since the *p*-Fresnel coefficient is larger than the *s*-Fresnel coefficient for refraction at air-to-glass interfaces, the diattenuation is all oriented radially in the plane of incidence. The magnitude of the diattenuation increases quadratically from zero in the middle of the pupil. At the second surface in the middle of Figure 12.3, the light is undergoing internal refraction and refraction from glass into air, and the diattenuation effects are much larger, increasing quadratically from the center of the pupil to a diattenuation of about 0.1 for the marginal rays. The right side graphs the diattenuation for the combination of the two surfaces; the magnitude is dominated by the second surface. This radially oriented linear diattenuaion increasing quadratically from the center of the pupil is the form of an uncoated lens' polarization aberration on-axis.

Figure 12.4 shows the exiting state from the singlet system shown in Figure 12.1 for 0° linearly polarized incident light, horizontally oriented electric field. Polarization aberration is visible as the amplitude and polarization state of the light vary across the pupil. Moving from the center to the right or left of the pupil, the amplitude of the light increases quadratically, since the *p*-transmission Fresnel coefficient increases quadratically with angle of incidence from the transmission down the axis. Moving from the center to the top or bottom, the amplitude decreases quadratically, since the light is *s*-polarized. Along the *x*-axis, the 0° polarized light is along the diattenuation axis and being preferentially transmitted. Along the *y*-axis, the light is polarized perpendicular to the diattenuation axis and is preferentially reflected. In the corners of the pupil, the incident 0° polarized light is at an angle to the 45° diattenuation axis, and so the exiting polarization rotates toward the diattenuation axis. The light incident parallel to or perpendicular to the diattenuation axis is in an eigenpolarization state, and the polarization state of the light exiting the lens is unchanged by the polarization aberration. Other incident states will exit the lens with the state rotated toward the transmission axis.

When the lens is illuminated with 45° polarized light, the light incident along the ±45° directions in the pupil is in an eigenpolarization. Along the 45° directions, the amplitude increases quadratically from the center. Along the 135° directions, the amplitude decreases quadratically from the center as shown in Figure 12.5 (left). When left circularly polarized light is incident, the polarization

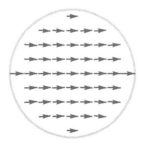

Figure 12.4 Polarization ellipse map of the singlet's exit pupil for a 0° polarized incident state.

12.2 Uncoated Single-Element Lens

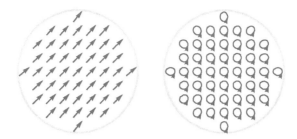

Figure 12.5 Exiting states from the singlet for the 45° (left) and left circularly polarized (right) incident states.

distribution exiting the lens is shown in Figure 12.5 (right). The light transitions from circular in the middle of the pupil to increasingly elliptical at the edge of the pupil. The major axis of each ellipse is aligned with the diattenuation axes shown in Figure 12.3.

If the lens is placed in between crossed linear polarizers (oriented at 0° and 90°), as shown in Figure 12.6 (left), the 0° component of the light is removed and only the 90° component is transmitted, as shown in Figure 12.7 (left). The pupil is now dark along the *x*- and *y*-axes and the transmitted light is the brightest at the edge of the pupil along the ±45° directions. The corresponding with one "n" flux distribution (amplitude squared) is referred to as the *Maltese cross*, seen in Figure 12.7 (middle). The Maltese cross is an important pattern, commonly seen with most radially symmetric optical systems placed between crossed polarizers. Along the diagonal, the amplitude of the leaked light varies as ρ squared (ρ is pupil coordinate) while the transmitted flux varies as ρ to the fourth power. If the lens is stopped down to a smaller aperture, the leaked flux rapidly diminishes. For a small numerical aperture, the Maltese cross

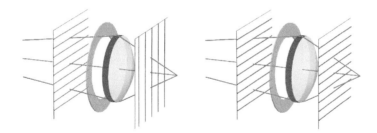

Figure 12.6 Singlet lens illuminated through a 0° polarizer and analyzed through a 90° analyzer (left) and analyzed through a 0° analyzer (right); the brown annulus is the entrance pupil.

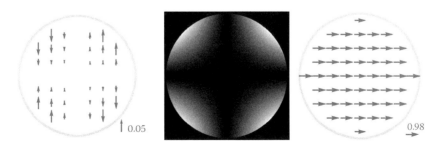

Figure 12.7 Polarization exiting the single-element lens between crossed polarizers (left) is dark along the *x*- and *y*-axes (left). This is easily seen placing most lenses in a polariscope with crossed linear polarizers. The pattern observed, called the *Maltese cross*, leaks light through the four diagonals of the pupil (middle). Between parallel linear polarizers (right), only a small amplitude variation is observed.

may be difficult to observe. If the two polarizers rotate together, the Maltese cross will rotate with the polarizers. When the polarizers are parallel, as in Figure 12.6 (right), the polarization pupil map has a uniform polarization state and nearly uniform amplitude, as shown in Figure 12.7 (right).

When the fast singlet lens is analyzed by polarization ray tracing a grid of rays, as described in Chapter 9, the polarization aberration function is calculated, shown in Figure 12.8 (left). This displays the elements of the polarization ray tracing matrices for the on-axis beam through the singlet. It maps the curved input spherical wavefront onto the curved exiting spherical wavefront and transforms the polarization states in three dimensions. The *xy*, *xz*, *yx*, and *zx* elements are principally concerned with these curvatures of the wavefronts. The diattenuation is most clearly visible in the *xy* and *yx* elements. The polarization aberration function can be flattened into a Jones pupil choosing the local coordinates for the incident and exiting spherical waves, that is, specifying the local *x*- and *y*-coordinates to be used for describing Jones vectors and Jones matrices over the two wavefronts as explained in Chapter 10. Using double pole coordinates with the poles at $-z$ yields the Jones pupil of Figure 12.8 (right). In the off-diagonal elements, *xy* and *yx*, the first and third quadrants are 180° out of phase with the second and fourth quadrants. The pupil average of the *xy* and *yx* pupils is zero. This has an implication in image formation.

The images of an on-axis point source when the lens is placed between different pairs of linear polarizers are shown in the four elements of Figure 12.9. Each of these is an amplitude response function, the electric field amplitude distribution as described in Section 16.4. Taken together, the four functions form the *amplitude response matrix*, **ARM**, which can operate on a Jones vector for a point source to determine the Jones vector distribution in its image. The elements of the **ARM** are calculated by Fourier transformation of the four elements of the Jones pupil as described in Chapter 16. The J_{11} function of Figure 12.9 corresponds to the amplitude distribution for the lens between a 0° polarizer and 0° analyzer. This function is very close to the Airy function for the diffraction-limited image of a circular aperture. It differs because the *xx* element of the Jones pupil is not quite uniformly illuminated. The J_{21} function corresponds to Figure 12.6 (left) with the lens illuminated through a 0° polarizer and the output analyzed through a 90° analyzer (polarizer). This image consists of four main lobes surrounding a dark center, and looks nothing like an Airy disk. The center is dark because the central value of the Fourier transform is the average value of the input function; in this case, the averages of both the *xy* and *yx* Jones pupil elements are zero.

This **ARM** describes how the images of point sources have spatial variations of polarization state within their point spread functions. The image of a point source polarized at 0° is shown in Figure 12.10 as a Jones vector distribution. The image in the center and along the *x*- and *y*-axes is

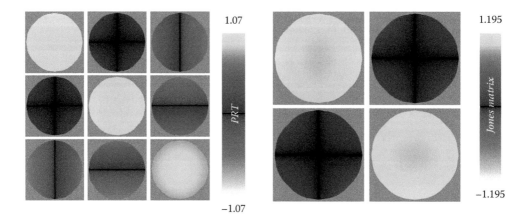

Figure 12.8 (Left) The polarization aberration function and (right) the Jones Pupil for the fast single-element lens.

12.2 Uncoated Single-Element Lens

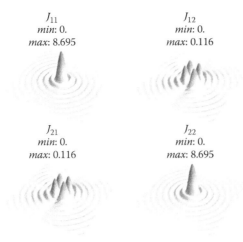

Figure 12.9 Amplitude response function of the fast singlet lens contains the amplitude of the **E** field distributions for combinations of 0° and 90° illumination analyzed through combinations of 0° and 90° analyzers and takes the form of a Jones matrix function.

Figure 12.10 (Left) Amplitude response Jones vector (J_{11} and J_{21} of Figure 12.9) and (right) its corresponding states plotted for horizontally polarized incident light.

polarized at 0°; elsewhere, it is a mixture of 0° and 90° components. Figure 12.11 (right) zooms in on the center of the image's polarization state map where the polarization orientation varies between ±1.46°.

The image of a point source is the *point spread function*, PSF. In the presence of polarization aberration, the PSF is generalized to the *point spread matrix*, **PSM**, a Mueller matrix function that operates on the Stokes parameters of the object point yielding a Stokes function for the image point as described in Chapter 16. The **PSM** for the uncoated singlet lens is shown in two views in Figures

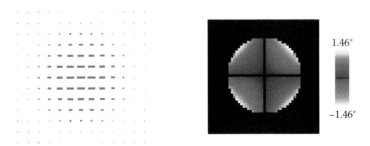

Figure 12.11 (Left) Zoom in on the polarization states in the center of the Figure 12.10 (right). (Right) The orientation of the polarizations state varies by ±1.5° within the first zero of the Airy disk.

12.12 and 12.13 (left). Figure 12.12 is a 3D plot with the maximum and minimum values indicated; for example, the M_{00} element peaks at 76 while the M_{01}, M_{02}, M_{10}, and M_{20} elements peak at 0.64, about 100 times lower. Figure 12.13 (left) is a colored raster plot showing blue for positive and red for negative values. Down the diagonal are four nearly equal functions that are close to the Airy-squared functions, approximating an identity matrix, indicating that the overall polarization of a

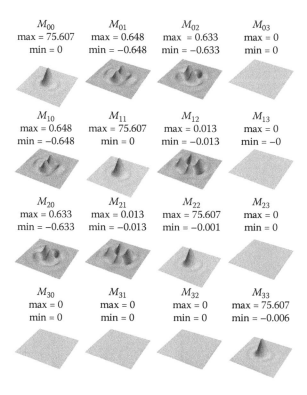

Figure 12.12 The Mueller matrix point spread function for the uncoated lens as a surface plot. Elements are scaled differently to show the details. The maximum and minimum values are shown individually in each element.

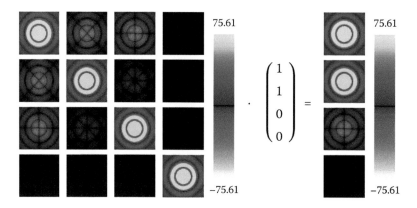

Figure 12.13 The Mueller matrix PSF is shown as a raster plot (left) operating on the Stokes parameters for horizontally polarized light (vector in middle) yielding the polarization distribution in the image as Stokes parameters. The S_1 distribution is nearly equal to the S_0 distribution, with S_2 showing a small component coupled into a 45° component (blue) and a 135° component (red), describing the spatial variation of the polarization in the image. The raster plot is saturated in order to show the details in the dimmer component.

point source is not very different from the incident polarization state. When this **PSM** operates on the Stokes parameters for 0° light, the Stokes parameters of the image formed are shown in Figure 12.13 (right). The S_2 element, third row, shows that the red areas would be brighter when analyzed by a 45° than by a 135° polarizer, while the blue areas would have the opposite occurring.

12.3 Fold Mirror

Fold mirrors coated with metals such as aluminum or silver are a significant source of polarization aberration in many systems.[7] Since the angles of incidence at fold mirrors are generally large, the polarization aberrations tend to be larger than polarization aberrations in on-axis systems. The refractive indices of metals are complex; hence, the reflection causes both diattenuation and retardance. The Fresnel coefficients for gold are plotted in Figure 12.14. The s- and p-phases in Figure 12.14 are plotted in the global coordinate phase convention, phases are equal at normal incidence, not separated by π. Subtracting the phases yields the retardance graph in Figure 12.14 (bottom right). The retardance has a larger effect on the polarization state than the diattenuation in Figure 12.14 (bottom left).

Figure 12.15 (left) shows an example system with a grid of rays at 500 nm reflecting from a gold fold mirror ($n = 0.855 + 1.895\,i$) over a range of angles from $5° < \theta < 80°$. The corresponding pupil map of the magnitude and orientation of the angle and plane of incidence are shown in Figure 12.15 (right). The polarization aberration function in Figure 12.16 maps the polarization vector of the incident into the exiting polarization vector, containing both changes of direction, phase, amplitude, and polarization. After a choice of local coordinates for incident and exiting spherical waves, the Jones pupil (Figure 12.17) is calculated, which represents the phase, amplitude, and polarization

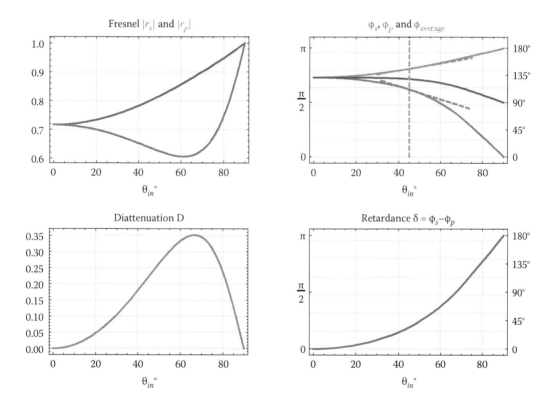

Figure 12.14 The magnitudes (top left) and phases (top right) for the Fresnel coefficients for gold. The difference between the phases is the linear retardance (bottom right), which increases quadratically from normal incidence.

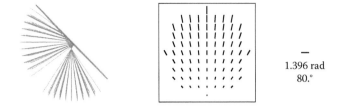

Figure 12.15 A converging grid of rays reflecting from a gold mirror (left) has the angle of incidence pupil map on the right.

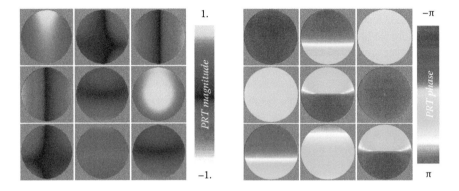

Figure 12.16 The amplitude (left) and phase (right) of the **P** matrix pupil.

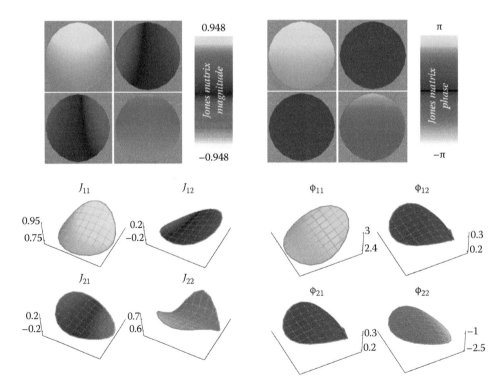

Figure 12.17 (Top row) The Jones pupil for the fold mirror is brighter for the x-component (s-polarized along the y-axis) as seen in the xx magnitude than the y-component in the yy magnitude. (Bottom row) 3D plots of the Jones pupil.

12.3 Fold Mirror

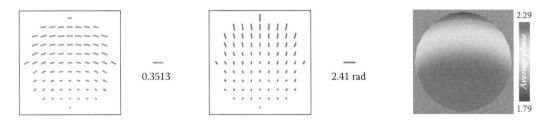

Figure 12.18 The diattenuation (left) is small at the bottom of the pupil near normal incidence reaching a maximum near 65° angle of incidence. The transmission axis rotates from the left to the right side of the pupil. The retardance (middle) increases to nearly a half wave at the top of the pupil. The fast axis is orthogonal with the diattenuation's transmission axis. The average phase between s and p polarizations is shown on the right.

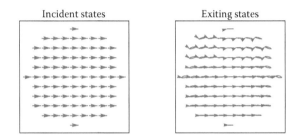

Figure 12.19 An ellipse map for incident 0° polarized light (left) and the reflected state (right). Arrowheads indicate the phases.

changes. Figure 12.17 (bottom row) contains a 3D surface representation of the Jones pupil using double pole coordinates* with the pole opposite to the chief ray direction.

From the Jones pupil, the diattenuation pupil map (Figure 12.18, left) and retardance pupil map (Figure 12.18, right) are calculated. In the *retardance map*, the length of the line segment represents the retardance magnitude and the line's orientation is the fast axis orientation. Along the middle of the pupil, the eigenpolarizations are 0° and 90° rotating counterclockwise to the left side and clockwise to the right side. This is why the xy and yx elements of the Jones pupil change sign crossing the y-axis. In this example, the diattenuation and retardance are both smallest at the bottom of the pupil where the beam is approaching normal incidence and increasing toward the top. The diattenuation reaches a maximum around $\theta_{in} = 65°$ and then decreases. The retardance increases all the way to the top of the pupil where it reaches ~2.41 radians, or about 0.43 waves of retardance.

The polarization state exiting for 0° and 90° incident polarizations is seen in Figures 12.19 and 12.20. The changes are the least for 0° and 90° incident polarizations since these are eigenpolarizations along the y-axis, the meridional plane, and these incident states are closest to eigenpolarizations overall across the remainder of the pupil. The reflected light becomes increasingly elliptical, moving away from the meridional plane. The helicity changes sign crossing the meridional axis.

Consider incident states that decomposed into equal amounts of 0° and 90°, such as 45° and circular polarized light. For these states, the change of polarization upon reflection is the largest. Figure 12.21 shows pupil maps for 45° light incident and the corresponding reflected states. The polarization changes are very large in the upper half of the pupil. Moving up the meridional axis, the light becomes increasingly elliptically polarized and the major axis of the ellipse rotates from 45° to 135°. Swindell discusses the handedness of this reflected state.[8]

---

* Section 11.4, our basic coordinate system for spherical waves.

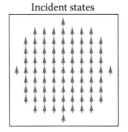
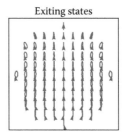

Figure 12.20 An ellipse map for incident 90° polarized light (left) and the reflected state (right). The reflected phase has a large change from the bottom to the top of the pupil of more than $3\pi/4$.

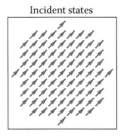
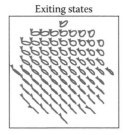

Figure 12.21 An ellipse map for incident 45° polarized light (left) and the reflected state (right). The ellipses are drawn looking into the beams; thus, near the bottom, 45° polarized light reflects as 135°.

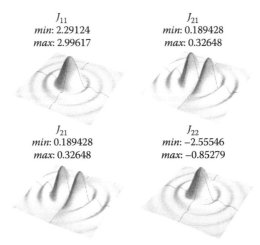

Figure 12.22 The modulus of the **ARM** for the fold mirror. The centers of the 11 and 22 responses are both shifted in opposite directions with respect to the geometric center by small amounts since their phases, as seen in Figures 12.14 and 12.17, change in opposite directions.

The modulus of the **ARM*** is plotted in Figure 12.22. When viewed between two 0° polarizers, the 11 element, a close comparison would show that the image is not that close to an Airy disk since it is slightly elliptical and the diffraction rings are not complete along the x-axis. Between two 90° polarizers, the amplitude in the 22 element is observed, which resembles the 11 image rotated by 90°. The peaks of the 11 and 22 images are shifted slightly in opposite directions owing to the

---

* Described in Section 16.4.

12.3 Fold Mirror

opposite linear components of the *s*- and *p*-phase shifts seen in Figure 12.14 (top right). In Figure 12.22, the peak of J_{22} is slightly behind the black *x*-axis cross section while the peak of J_{11} is slightly in front of this cross section. The image between crossed polarizers is very interesting. The pupil averaged value of the *xy* and *yx* Jones pupil elements of Figure 12.17 is zero; thus, the center of the 12 and 21 **ARM** elements is zero. Both image components have two main lobes straddling the center along the meridional plane. The energy missing from the diffraction rings of the 11 and 22 elements is found in the rings of the 21 and the 12 elements. The 22 element is more aberrated than the 11 element due to the larger phase variation of the *p*-component seen in Figure 12.14.

The polarization structure of the image formed with 0° incident light is shown (Figure 12.23) as a 3D plot of the *x*- and *y*-components (left) and pupil polarization ellipse map (right). The central lobe is wider than it is tall, similar to elliptically shaped images formed with astigmatism. The polarization state is seen to be mostly slightly elliptical with a major axis rotating steadily counterclockwise from the left to right sides of the pupil.

The polarization changes to the image structure are much larger with 45° light incident as shown in Figure 12.24. The light at the center of the central lobe is elliptically polarized with the major axis at 135° due to reflection.

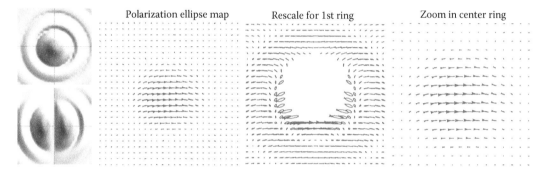

Figure 12.23 The amplitude response Jones vector images formed from 0° polarized incident light are shown on the left, with the *x*-amplitude on top and the *y*-amplitude at the bottom. The peak of the *y*-amplitude is 1/8 the peak of the *x*-amplitude. The corresponding ellipse maps are shown in three different scalings. The polarization ellipse map on the left shows a polarization variation across the Airy disk. By excluding the higher amplitude inside the first ring, the second ellipse map rescales to show the ellipticity at the edge of the first ring. The third ellipse map zooms into the center region, showing the steady polarization rotation from left to right.

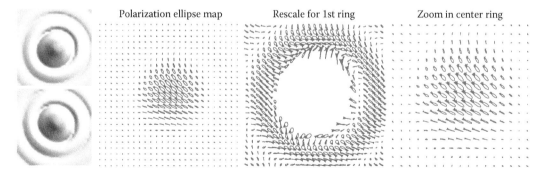

Figure 12.24 The amplitude response Jones vector images formed from 45° polarized incident light are shown on the left, with the modulus of the *x*-amplitude on top and the *y*-amplitude at the bottom. The peak of the *y*-amplitude is 1/1.2 the peak of the *x*-amplitude. The corresponding ellipse maps are shown in three different scalings. The polarization ellipse map on the left shows a polarization variation across the Airy disk. The second ellipse map is rescaled to show the ellipticity at the edge of the first ring. The third ellipse map zooms into the center region showing the steady polarization rotation from left to right.

Polarized Light and Optical Systems

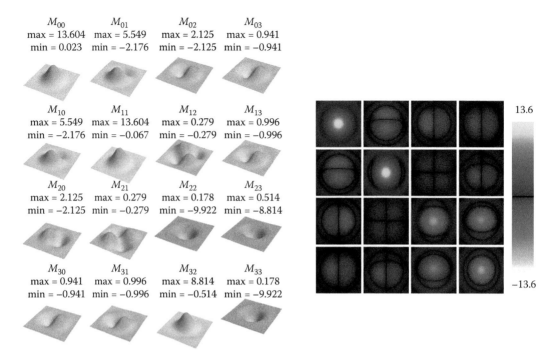

Figure 12.25 Mueller point spread matrix plotted in two different forms.

Transforming the Jones pupil of Figure 12.17 into a Mueller matrix yields the point spread matrix in Figure 12.25 shown in 3D (left) and raster plot (right) views. The first column shows the Stokes parameters when the object point is unpolarized. The center of the image is nearly unpolarized, because the centers of the (1,0), (2,0), and (3,0) functions are nearly zero. The linearly polarized component rotates around the image center from the 10 and 20 functions, while the top has left ellipticity and the bottom has right ellipticity from the (3,0) element.

The response for 0° illumination is found by adding the zeroth and first columns. Its flux is calculated by adding the (0,0) and (0,1) elements, and it is seen that the peak of the 0° image has moved toward $+x$. Subtracting the (0,0) and (0,1) elements, it is seen that the peak of the 90° image has moved toward $-y$ but with 90° illumination, and the brightest part of the 90° image has moved toward $-x$. The (2,2) element is negative in the center because 45° reflects as 135° as seen in Figure 12.24. From the 00 and (0,2) elements, it is seen that for 45° and 135° illumination, the image peaks shift in opposite directions in y; similarly, image peak shifts occur for right and left circularly polarized light.

The integral of the (0,0) element provides the net flux (unnormalized) in the image for unpolarized illumination; about 8% of the light has been lost to absorption. Other integrals provide the net flux through various polarizer combinations. For example, when the mirror is preceded by a 45° polarizer (transmission axis at $(1, 1, 0)/\sqrt{2}$ in global coordinates) and followed by a 135° polarizer $((0, 1, 1)/\sqrt{2})$ after reflection, the net flux is the integral of $M_{00} + M_{02} - M_{20} - M_{22}$, which is twice the integral of M_{02}.

This section has shown that the polarization aberrations due to a gold mirror are substantial. Similar results are obtained for aluminum and other mirrors. A large numerical aperture was used to make the effects large. For smaller numerical apertures, the magnitude of the effects approximately linearly. This example demonstrates why fold mirrors cause many of our polarization problems in optical systems.

12.4 Combination of Fold Mirror Systems

Section 12.3 discussed the polarization changes of fold mirrors caused by diattenuation and retardance associated with the Fresnel coefficients. The resultant polarization aberration can be compensated by using individual components with aberrations of the opposite sign, minimizing the resultant aberration. Here, sequences of one, two, and four gold-coated fold mirrors are considered, to understand configurations that reduce the polarization aberrations.

Figure 12.26 shows a flat mirror $M1$ reflecting a 0.2 NA converging beam. This beam is scanned by rotating the mirror about the z-axis; this corresponds to the incident ray at the center of the beam keeping an angle of incidence of 45°. A grid of rays evenly sample the converging wavefront, and this incident beam propagates from the xy-plane toward the z-direction. In Figure 12.26, the mirror is rotated about the z-axis by 180° to five orientations, and the mirror normals are tabulated in Table 12.2.

The plane of incidence of each ray rotates as $M1$ scans. An AoI map of the mirror viewed from the entrance space is shown in Figure 12.27. Shorter lines in Figure 12.27 correspond to a ray moving closer to normal incidence. The ray at the center has a 45° AoI. The AoI varies linearly from

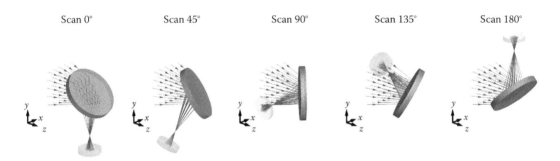

Figure 12.26 Ray trace figures showing a converging beam reflecting from a fold mirror (blue) toward a detector (gray). The image location scans as the orientation of the fold mirror rotates around the incident beam, centered on the z-axis, by 0°, 45°, 90°, 135°, and 180° from (0, 1, 1).

Table 12.2 Surface Normals for the Mirror of Figure 12.6

Scan Angle	0°	45°	90°	135°	180°
Surface normal	$\dfrac{1}{\sqrt{2}}\begin{pmatrix}0\\1\\1\end{pmatrix}$	$\dfrac{1}{2}\begin{pmatrix}1\\1\\\sqrt{2}\end{pmatrix}$	$\dfrac{1}{\sqrt{2}}\begin{pmatrix}1\\0\\1\end{pmatrix}$	$\dfrac{1}{2}\begin{pmatrix}1\\-1\\\sqrt{2}\end{pmatrix}$	$\dfrac{1}{\sqrt{2}}\begin{pmatrix}0\\-1\\1\end{pmatrix}$

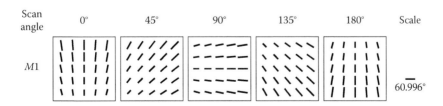

Figure 12.27 The one-fold mirror system's angle of incidence map shown in Figure 12.26. The legend on the right shows the largest incident angle in degrees for all frames.

across the pupil in one direction, while in the perpendicular direction, the *PoI* rotates from one side to the other linearly. Therefore, as *M*1 scans around, this *AoI* pattern rotates.

Figure 12.28 shows the view from the entrance space of the diattenuation for reflection from gold for each ray. The diattenuation map is oriented orthogonal to the *AoI* map since the state with maximum reflectance is along the *s*-polarization. As shown in Figure 12.14 (bottom left), the magnitude of diattenuation increases quadratically with *AoI* as the rays move from normal incidence.

Figure 12.29 shows the variations of retardance across the pupil, the retardance maps, resulting from scanning *M*1. Since $\phi_{rp} < \phi_{rs}$ for the gold mirror, the fast axis for reflection is *p*-polarized. Thus, the retardance map has the same orientation as the *AoI* map, which is orthogonal to the diattenuation axes shown in Figure 12.29. As shown in Figure 12.14, the retardance of gold varies quadratically with incident angle.

To reduce the polarization aberrations of the fold mirror, another fold mirror can be added in the reflection path. The second mirror is oriented to have the *s*-polarized component of the axial ray on the first mirror become *p*-polarized on the second mirror, and vice versa. This is the *crossed mirror configuration*. The polarization aberration for one incident ray in a converging beam can be completely compensated. About this ray, such a mirror pair has small and linearly varying polarization aberration about the center.

In this two-fold mirror system, the first mirror *M*1 is fixed with surface normal at (0, 1, 1), while the second mirror *M*2 rotates about the incident beam along the *y*-axis. *M*2's surface normal rotates about the *M*1–*M*2 axis by 180° from (1, 1, 0), as shown across Figure 12.30. The *M*2's surface normals are listed in Table 12.3.

As shown in Figure 12.31, when *M*2 scans, its plane of incidence changes with respect to *M*1. The *PoI* axes shown on *M*2 are propagated back to the entrance space of *M*1; therefore, all maps

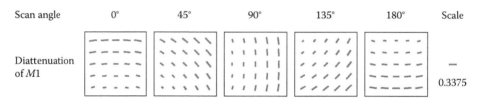

Figure 12.28 Diattenuation map for the scanning fold mirror. The maximum diattenuation is shown in the scale on the right.

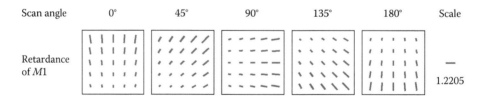

Figure 12.29 Retardance map for the scanning fold mirror. The maximum retardance in radians is shown in the scale on the right.

Figure 12.30 Ray trace figures showing a converging beam reflecting from fold mirror *M*1 (blue), then a scanning fold mirror *M*2 (magenta), and then incident on a detector (gray).

12.4 Combination of Fold Mirror Systems

Table 12.3 Surface Normals of $M2$ in Figure 12.30

Scan Angle	0°	45°	90°	135°	180°
Surface normal	$\dfrac{1}{\sqrt{2}}\begin{pmatrix}1\\1\\0\end{pmatrix}$	$\dfrac{1}{2}\begin{pmatrix}1\\\sqrt{2}\\-1\end{pmatrix}$	$\dfrac{1}{\sqrt{2}}\begin{pmatrix}0\\1\\-1\end{pmatrix}$	$\dfrac{1}{2}\begin{pmatrix}-1\\\sqrt{2}\\-1\end{pmatrix}$	$\dfrac{1}{\sqrt{2}}\begin{pmatrix}-1\\1\\0\end{pmatrix}$

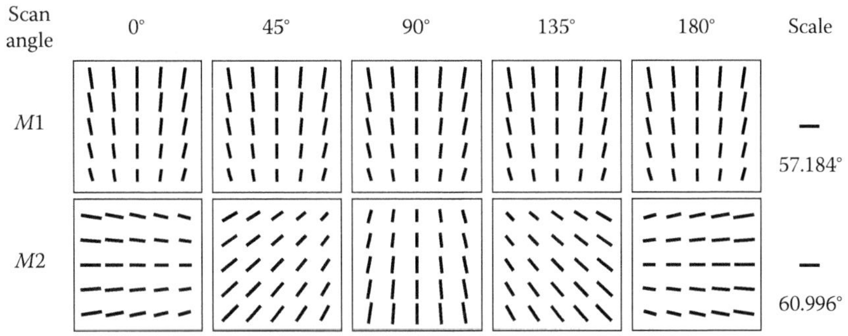

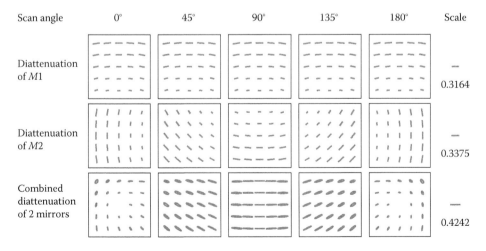

Figure 12.31 The two fold mirrors' angle of incidence maps shown in Figure 12.30. $M2$ scans about the y-axis while $M1$ is fixed. The crossed mirror configuration occurs at 0° and 180°. The legend on the right shows the largest incident angle in degrees for all frames. The $M2$ maps have been propagated back to the entrance space of $M1$.

are compared directly in object space. The center ray's incident angles at both mirrors are kept at 45°. As $M2$ scans about $M1$, the p-polarization at the center ray of the two mirrors changes from orthogonal, to parallel, and back to orthogonal. In the crossed mirror configuration (scan angle 0° and 180°), the *Pol*s between $M1$ and $M2$ are close to orthogonal across the beams and then become less parallel as $M2$ scans.

The overall **P** matrix is calculated for each ray path within the field. The diattenuation and retardance are then obtained through data reduction of these **P** matrices, as explained in Chapters 9 and 17. The diattenuation of $M1$ and $M2$ and the cumulative diattenuation from both mirrors as $M2$ scans are shown in Figure 12.32. All maps directly compare because all the diattenuation axes

Figure 12.32 Diattenuation map for $M1$ and $M2$, and the total system. The maximum diattenuation is shown in the scale on the right.

Polarized Light and Optical Systems

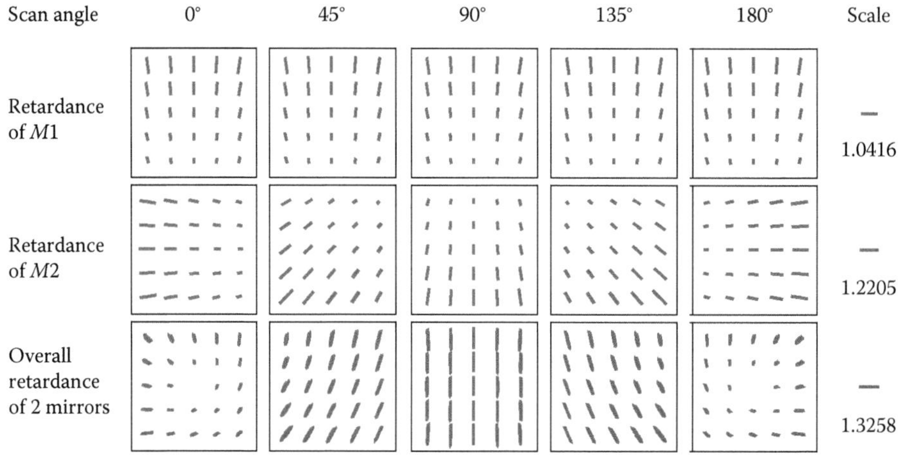

Figure 12.33 Retardance map for $M1$ and $M2$, and the total system. The maximum retardance is shown in the scale on the right in radians.

are viewed from the entrance space. At a 90° scan angle, the diattenuation orientations of $M1$ and $M2$ are close to parallel and the overall diattenuation magnitude is maximized. At 0° and 180° scan angles, the diattenuation orientations are closest to orthogonal over the pupil; hence, the cumulative diattenuation reaches a minimum. The corresponding retardance maps shown in Figure 12.33 have similar variations, except that they are approximately orthogonal to the diattenuation maps.

At 0° and 180° scan angles, the polarizations of the two mirrors are nearly orthogonal with similar magnitudes; thus, the maximum diattenuation and retardance cancel each other, producing complete cancellation at the center with a linear variation of diattenuation and retardance through the center. Their axes change by 90° crossing the center. Such polarization aberration forms are termed *linear diattenuation tilt* and *linear retardance tilt*[9] where the axis orientation varies through 180° moving around the node; these will be discussed in Chapter 15.

The null of the cumulative polarization moves away from the center when $M2$ scans away from 0°. This null moves outside the field to the upper left when $M2$ scans to 45°. It returns to the center of the field when $M2$ scans through 180°. When $M2$ scans to 90°, the overall polarization magnitude of the two mirrors adds, and doubles the overall aberration when compared to the single mirror aberration.

The diattenuation of the individual mirrors is linear, not elliptical, everywhere. However, a small amount of elliptical diattenuation and retardance are present in the cumulative aberration and are observed in Figures 12.32 and 12.33 toward the edge of the pupil as slight ellipticities. Such elliptical polarization occurs when two linear diattenuators or retarders have axes neither parallel nor perpendicular and thus generate some circular polarization.[10] When the polarization axes are 45° or 135° apart, the maximum circular component is generated. This coupling of linear into circular polarization is discussed in Section 14.3.

Polarization aberrations of a single fold mirror are reduced by the *crossed mirror configuration*. However, a linear variation remains even though their plane of incidence is orthogonal at the center. Further polarization reduction can be achieved by four mirrors arranged as a *double-crossed mirror system*. In the system of Figure 12.34, $M1$ and $M2$ are fixed, and their normals are at (0, 1, 1) and (−1, −1, 0), respectively. The third mirror $M3$ and fourth mirror $M4$ are fixed with respect to each other, but the pair scan around the x-axis, which is the axis of the light exiting the $M1$–$M2$ assembly. The normals of $M3$ and $M4$ start at (1, 1, 0) and (0, 1, 1). As shown across Figure 12.34, the $M3$–$M4$ detector assembly rotates about the x-axis through 180°. The normals of $M3$ are listed in Table 12.4. Figure 12.35 shows the angle of incidence maps for all four mirrors. $M1$ and $M2$ are

12.4 Combination of Fold Mirror Systems

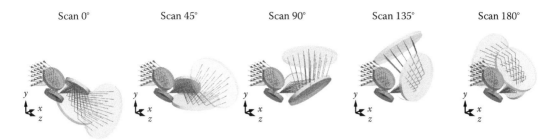

Figure 12.34 Ray trace figures showing a converging beam reflecting from M1 (blue), M2 (magenta), M3 (yellow), then M4 (green), and finally reaching a detector (gray). M1–M2 assembly is fixed, while the M3–M4 assembly scans about the x-axis.

Table 12.4 Surface Normals of M3 in Figure 12.34

Scan Angle	0°	45°	90°	135°	180°
Surface normal	$\frac{1}{\sqrt{2}}\begin{pmatrix}1\\1\\0\end{pmatrix}$	$\frac{1}{2}\begin{pmatrix}\sqrt{2}\\1\\1\end{pmatrix}$	$\frac{1}{\sqrt{2}}\begin{pmatrix}1\\0\\-1\end{pmatrix}$	$\frac{1}{2}\begin{pmatrix}\sqrt{2}\\-1\\-1\end{pmatrix}$	$\frac{1}{\sqrt{2}}\begin{pmatrix}1\\-1\\0\end{pmatrix}$

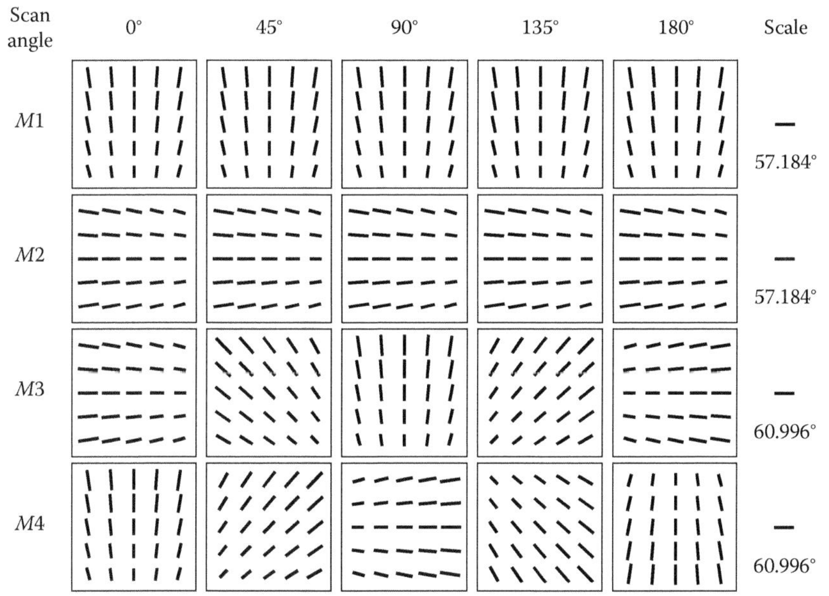

Figure 12.35 The angle of incidence maps of the four mirrors in Figure 12.34. M1 and M2 are fixed while M3 and M4 scan as a pair. The legend on the right shows the largest incident angle for all frames in degrees.

fixed in the crossed mirror configuration while M3 and M4 rotate as a pair in the *crossed mirror configuration*.

The retardance and diattenuation maps for the double-crossed mirror system are shown in Figures 12.36 and 12.37. Since M1 is orthogonal to M2, and M3 is orthogonal to M4 in all scan configurations, there is always a polarization null at the center of the field while the polarizations increase linearly toward the edge of the field. At 0° scan angle, M3–M4 and M1–M2 have the same

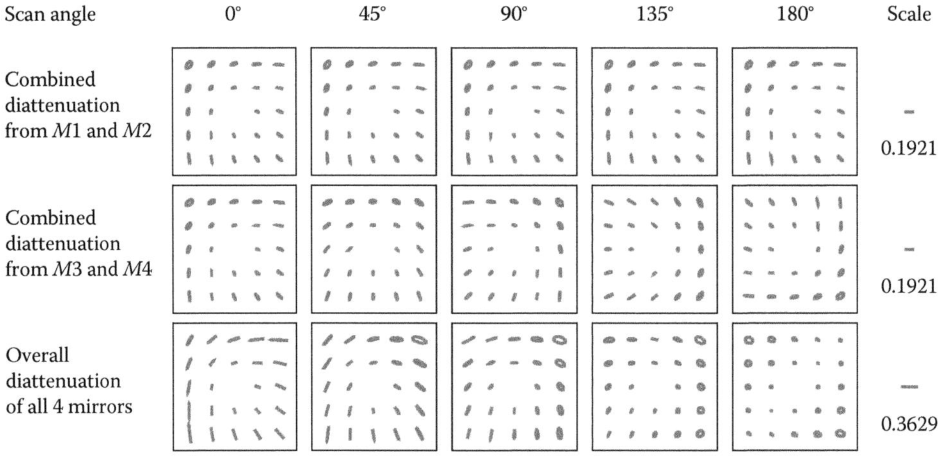

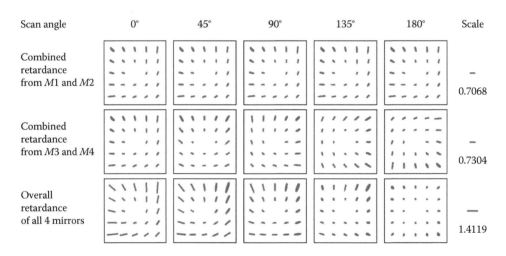

Figure 12.36 Diattenuation maps for the $M1$–$M2$ assembly (top row) and $M3$–$M4$ assembly (middle row), and the total diattenuation from all four mirrors (bottom row). The maximum diattenuation is shown in the scale on the right.

Figure 12.37 Retardance maps for the $M1$–$M2$ assembly (top row) and $M3$–$M4$ assembly (middle row), and the total system (bottom row). The maximum retardance is shown in the scale on the right in radians.

magnitude and orientation in both the diattenuation and retardance; hence, the cumulative polarizations are twice that of the individual two-mirror assembly and are dominated by linear tilt. As $M3$–$M4$ scans from $0°$ to $180°$, the overall aberration steadily decreases as $\cos^2(\phi/2)$, where ϕ is the scan angle. At a $180°$ scan angle, the aberration maps of $M1$–$M2$ and $M3$–$M4$ are nearly orthogonal to each other across the field and the overall polarization aberration is minimized; the linear variations from the two-mirror assemblies across the field are opposite and compensate. The magnitude of the residual aberration is quadratic across the field for this four-mirror combination at $180°$ scan angle and well corrected for a large field of view.

A summary of the cumulative polarizations of the single, the two crossed, and the double-crossed gold mirror systems is tabulated in Tables 12.5 and 12.6.

These results for gold mirrors generalize to most reflective coatings. For most of the common reflective coatings, the diattenuation and retardance maps are similar in form to Figures 12.28, 12.29, 12.32, 12.33, 12.36, and 12.37 of the gold mirrors shown here, except for a change of scale;

12.4 Combination of Fold Mirror Systems

Table 12.5 For a 0.2 NA Illumination, the Maximum, Average, and Chief Ray Diattenuation of the Single, the Two-Crossed, and the Double-Crossed Gold Mirror Systems

	\multicolumn{5}{c}{Maximum Diattenuation}
	0°	45°	90°	135°	180°
4 Mirrors	0.3629	0.3198	0.2812	0.2376	0.2055
2 Mirrors	0.1921	0.3591	0.4242	0.3591	0.1921
1 Mirror	0.3164	0.3375	0.3164	0.3375	0.3164
	\multicolumn{5}{c}{Average Diattenuation}
	0°	45°	90°	135°	180°
4 Mirrors	0.2147	0.2015	0.1684	0.1256	0.0927
2 Mirrors	0.1221	0.3086	0.4169	0.3086	0.1221
1 Mirror	0.2255	0.2255	0.2255	0.2255	0.2255
	\multicolumn{5}{c}{Chief Ray Diattenuation}
	0°	45°	90°	135°	180°
4 Mirrors	0	0	0	0	0
2 Mirrors	0	0.3090	0.4218	0.3090	0
1 Mirror	0.2212	0.2212	0.2212	0.2212	0.2212

Table 12.6 For a 0.2 NA Illumination, the Maximum, Average, and Chief Ray Retardance in Radians of the Single, the Two-Crossed, and the Double-Crossed Gold Mirror Systems

	Maximum Retardance				
	0°	45°	90°	135°	180°
Double-crossed mirrors	1.4119	1.2654	0.9919	0.8800	0.3587
Crossed mirrors	0.7068	1.1315	1.3258	1.1315	0.7068
Single mirror	1.0416	1.2205	1.0416	1.2205	1.0416
	Average Retardance				
	0°	45°	90°	135°	180°
Double-crossed mirrors	0.7699	0.7104	0.5613	0.3569	0.1836
Crossed mirrors	0.3926	0.9097	1.2470	0.9097	0.3926
Single mirror	0.6465	0.6465	0.6465	0.6465	0.6465
	Chief Ray Retardance				
	0°	45°	90°	135°	180°
Double-crossed mirrors	0	0	0	0	0
Crossed mirrors	0	0.8435	1.1939	0.8435	0
Single mirror	0.5969	0.5969	0.5969	0.5969	0.5969

the polarization maps are primarily linear and quadratic with the incident angle. Figure 12.38 shows the diattenuation and retardance of uncoated metal mirrors with real and imaginary refractive indices spanning most dielectric and metal surfaces. Over this range of refractive indices, the retardance has similar functional forms with a predominantly increasing linear component at 45°. On the other hand, the diattenuation also has similar functional forms but varying magnitude. At 45°, the behavior is predominantly linear and increasing except for the upper left case in Figure 12.38.

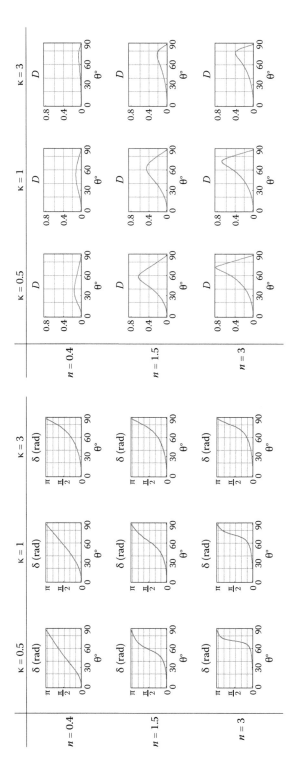

Figure 12.38 Retardance maps in radians (left) and diattenuation maps (right) as a function of incident angle for metal reflections for a 3 × 3 grid of refractive indices $n + i\kappa$.

12.5 Cassegrain Telescope

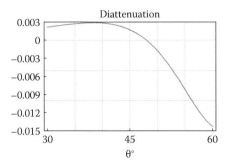 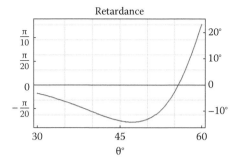

Figure 12.39 Diattenuation and retardance as a function of angle reflecting from 0.430 μm Al_2O_3 coating ($n = 1.6033$) on a silver substrate ($n + i\kappa = 0.0314 + i\, 5.308$) at 500 nm.

To generate the polarization patterns with the form shown in Figures 12.28, 12.29, 12.32, 12.33, 12.36, and 12.37 but with arbitrary magnitudes from a spherical wavefront that reflects from a flat mirror, the following conditions apply. (1) The variation of the diattenuation or retardance is predominantly linear over the range of incident angles, and (2) the diattenuation or retardance magnitude is much less than one. If the variations are not linear, the two mirror aberrations will not be linear, and then the compensated four-mirror aberrations (Figures 12.36 and 12.37, bottom right) will not be quadratic. The aberrations in the present example already start to challenge these assumptions.

An example coating that does not satisfy condition (1) is shown in Figure 12.39, where the diattenuation and retardance variation are not linear about 45°. Regarding (1), the nonlinear higher-order Taylor series expansion terms of the diattenuation and retardance create nonlinearities in diattenuation and retardance maps. Regarding (2), as the magnitudes increase, the interactions of non-aligned diattenuation and retardance between surfaces generate larger amounts of elliptical retardance and diattenuation, which are observed as the small ellipticities in the bottom rows of Figures 12.32 and 12.33. Operating fold mirrors at higher incident angles, toward grazing incidence, generates larger retardance.

In conclusion, the magnitude of diattenuation or retardance of two mirrors is reduced when cascading components with orthogonal diattenuation axes (maximum transmission axis) or orthogonal retardance axes (fast axis). The polarization aberration of one mirror is well compensated by another mirror crossed such that the *p*-polarized light exiting the first mirror is *s*-polarized on the second mirror. This *crossed mirror configuration* corrects the polarization aberration at one point, leaving a residual linear variation across the field.

Additional mirrors, a four-mirror system constructed from two pairs of crossed mirrors, can further reduce the polarization aberration of a *crossed mirror assembly*. The minimum polarization is obtained when the linear variation of *M*1–*M*2 is opposite from the linear variation of *M*3–*M*4. Such a configuration produces zero polarization at the center of the field with a small residual quadratic variation. This *double-crossed mirror system* has the lowest polarization aberrations across a large field of view among the one-, two-, and four-mirror systems shown in this section.

The polarization aberrations of the metal mirrors are a function of Fresnel coefficients calculated from the metal refractive index. Increasing refractive index decreases the magnitude of retardance. Decreasing the imaginary part of the refractive index and increasing the real part of the refractive index increases the magnitude of diattenuation. The polarization magnitudes are scaled by the metal coating, while the patterns of the polarization orientations remain the same with the polarization nulls at the center of the four-mirror configurations.

12.5 Cassegrain Telescope

The polarization aberrations of radially symmetric mirror systems, like Cassegrain telescopes, have radially symmetric patterns of diattenuation and retardance. Figure 12.40 shows the Cassegrain

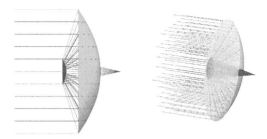

Figure 12.40 Cassegrain telescope with collimated light entering from the left, reflecting from a paraboloid-shaped primary mirror (cyan), then secondary mirror (purple), and focusing behind the primary mirror. This system has zero wavefront aberration, but the metal coatings induce polarization aberration.

Table 12.7 Cassegrain Telescope Specifications

Primary Mirror	
Radius R_1	5688 mm
Conic constant κ_1	−1
F/#	2.3862
Secondary mirror	
Radius R_2	1228.28 mm
Conic constant κ_2	−1.780
Mirror separation	2317.80
Image plane	3674.77 from secondary mirror
Entrance pupil diameter	8323.3 mm
Exit pupil diameter	1743.44, −485.50 from secondary mirror
Aluminum index	0.958 + 6.690 i
Wavelength λ	0.55 μm

telescope, which will be analyzed as an example. Table 12.7 lists the telescope's specifications. Collimated light enters from the right along the optical z-axis and reaches the primary mirror, shown as a cyan paraboloid in Figure 12.40 (right), which serves as the aperture stop and entrance pupil. The reflected light converges onto the secondary mirror (shown as a purple hyperboloid) and reflects and passes through a hole in the primary mirror, coming to focus at an image plane behind the primary mirror. The radii and conic constants are chosen so that an on-axis collimated beam has zero geometrical wavefront aberration; all of the on-axis rays have the same optical path length from the entrance pupil to the exit pupil and image point. The angle of incidence over the pupil is plotted for the primary and secondary mirrors in Figure 12.41 (top); the angles increase linearly from the center of the pupil and are radially oriented. Applying the Fresnel coefficients for aluminum (Figure 12.41, bottom) to the angles of Figure 12.41 yields the diattenuation pupil maps of Figure 12.42 for the primary and secondary. The combined diattenuation effect from the entrance to exit pupil (12.42 right) is tangentially oriented and increases quadratically. Similarly, applying the Fresnel phase coefficients yields the retardance pupil maps of Figure 12.43. The combination of Figures 12.42 (right) and 12.43 (right) yields the Jones pupil of Figure 12.44.

The phase changes due to the coatings (Figure 12.44, right) on the diagonal (x into x and y into y) are toroidal, curving upward quadratically along one axis and downward along the other. This toroidal form indicates that astigmatism is introduced by the aluminum reflection's retardance!! For

12.5 Cassegrain Telescope

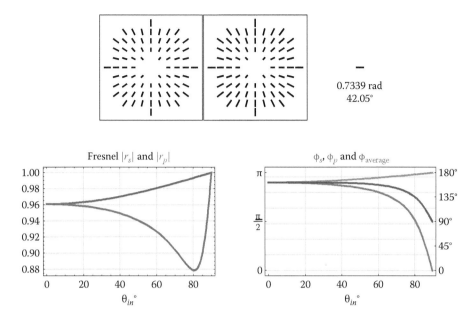

Figure 12.41 (Top) The angle of incidence functions at the primary mirror (top left) and secondary mirror (top right). (Bottom) The magnitudes and phases for the Fresnel coefficients for aluminum.

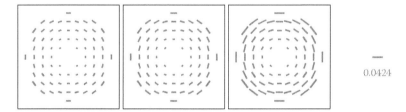

Figure 12.42 Pupil maps of the diattenuation on the primary mirror (left) and secondary mirror (middle) are tangentially oriented because the *s*-component has the greater reflectance. The map on the right is the combined diattenuation from the entrance to the exit pupil. The diattenuation reaches more than 0.04 for the marginal rays at the edge of the pupil.

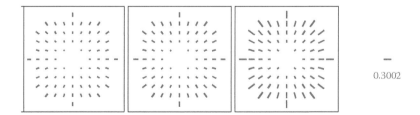

Figure 12.43 Retardance pupil maps for the primary (left), secondary (middle), and combination (right) are radially oriented; the *p*-polarized light exits before the *s*-polarized component. The retardance increases quadratically from the center of the pupil to about λ/20 for the marginal rays.

x into *x*, the peak-to-valley astigmatism is about 0.3 radians. For *y* into *y*, the magnitude is the same, but the astigmatism orientation is rotated by 90°. Hence, when the Cassegrain telescope is placed between parallel polarizers, astigmatism is measured in the exiting wavefront, and this astigmatism rotates with the polarizer pair. Such telescope-associated astigmatism is discussed by Reiley[11] in his dissertation.[12]

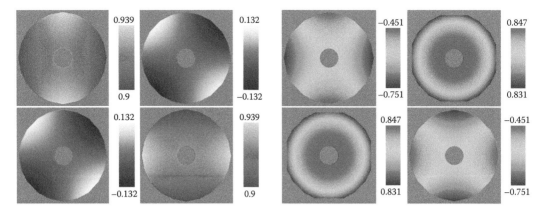

Figure 12.44 The magnitude of the Jones pupil (left) and the phase (right) for the Cassegrain telescope on-axis shows how the *x*- and *y*-components of the incident light will be apodized and coupled by the aluminum reflections. The reflectance of *x*-polarized light into *x*-polarized light increases toward the top and bottom of the pupil where it is *s*-polarized and decreases toward the left and right sides. The pattern for *y*-polarized light into *y*-polarized is rotated by 90°. The reflectance of *x* into *y* and *y* into *x* is zero along the *x*- and *y*-axes and increases along the ±45° directions. These results differ from the single-element lens due to the metal reflections.

This mirror-induced astigmatism is surprising to many familiar with wavefront aberration theory. For radially symmetric systems, there is no astigmatism on-axis. Instead, for lenses, Cassegrain telescopes, and other systems, the astigmatism increases quadratically from a zero (node) in the center of the field of view. Thus, conventionally, when this astigmatism was observed in interferograms, one would suspect that the astigmatism might have been polished into one of the mirrors, that the mirror was not radially symmetric. In such a case, the astigmatism's orientation should rotate with the offending mirror. But because this is polarization aberration, if the mirrors are rotated, the astigmatism doesn't rotate; it rotates when the polarization is rotated. Further, if the coating is changed from aluminum to gold or some other metal, the magnitude of the astigmatism changes, which would not happen with wavefront aberrations.

In addition to introducing astigmatism, the mirror's retardance and diattenuation couple light into the orthogonal polarization state, as is indicated in the off-diagonal elements of the Jones pupil (Figure 12.44). Like the uncoated lens, this coupled polarization is seen to be apodized into the Maltese cross form, with an amplitude at the edge of the pupil of ~0.13 of the incident light for this F/#, wavelength, and metal. With crossed polarizers, when this Maltese cross light is brought to focus in the image plane, it forms the interesting point spread function components shown in the off-diagonal elements of Figure 12.45 containing four islands of light with dark bands oriented horizontally and vertically through the center of the PSF. The magnitude is plotted here; the pairs of islands on opposite diagonals have opposite amplitudes, positive and negative, and the absolute value is easier to visualize. Thus, when 0° linearly polarized light is incident, a 0° linear polarizer analyzes the J_{11} amplitude, which is astigmatic (at a scale not obvious in this figure). When the analyzing polarizer is rotated to 90°, a much weaker pattern of four islands is observed, the J_{21} amplitude. For intermediate orientations, the image passes through a continuum of intermediate patterns.

The PSFs of the Cassegrain telescope can be measured by placing the telescope in a Mueller matrix imaging polarimeter, measuring its response to incoherent plane waves. This Mueller matrix point spread matrix is simulated in Figure 12.46. The first column contains the Stokes image of the unpolarized PSF. The total flux in the M_{00} element is rotationally symmetric because the astigmatism of all the polarization orientations averages to a radially symmetric distribution, which is larger than the diffraction-limited pattern. The M_{10} and M_{20} elements show that away from the center, which is unpolarized, the PSF becomes partially polarized with a spatially varying polarization.

12.5 Cassegrain Telescope

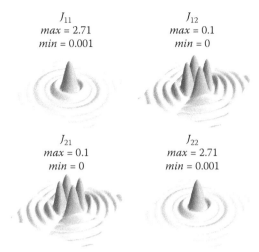

Figure 12.45 The amplitude response functions for the Cassegrain telescope show the spatial variations of polarization state in the image of point objects for several polarization combinations. For x into x and y into y, the amplitude distributions are close to the Airy disk, but slightly elliptical due to the astigmatism induced by the differences between the metals' s- and p-phases. The x into y and y into x responses show four islands around the first Airy disk.

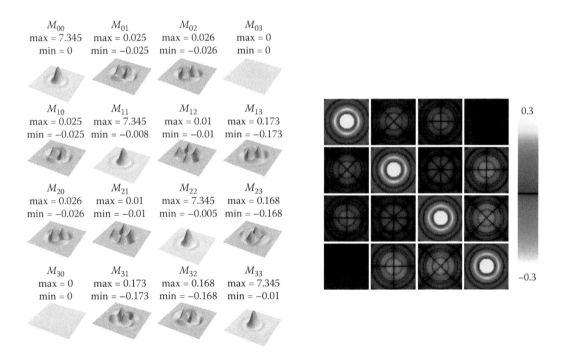

Figure 12.46 The point spread matrix for the Cassegrain telescope's on-axis image shows the polarization distributions within the PSF for arbitrary incoherent incident polarization states. The surface plots (left) are individually scaled to show fine details in the polarization distributions. The polarization coupling due to retardance, M_{13}, M_{23}, M_{31}, and M_{32}, is much larger than the coupling due to diattenuation, M_{01}, M_{02}, M_{10}, and M_{20}. The raster plots (right) are overexposed in the center to reveal the faint details in the polarization structure.

12.6 Fresnel Rhomb

One clever application of the Fresnel equations is the Fresnel rhomb, a total internal reflection (TIR)-based quarter wave retarder.[13] Figure 12.47 (left) shows a Fresnel rhomb, a prism with a parallelogram cross section and two TIRs. For a refractive index around $n = 1.5$, the retardance associated with TIR reaches a maximum near an eighth of a wave at an angle of incidence close to 52°. Thus, two internal reflections can provide approximately a quarter of a wave of retardance. The Fresnel rhomb is easy and inexpensive to fabricate since it doesn't use birefringent materials, just a block of glass chosen with minimal stress birefringence. The beam enters and exits in parallel directions. Because of the lateral displacement between entering and exiting rays, it is inconvenient to rotate a Fresnel rhomb.

Figure 12.47 examines some design choices for the Fresnel rhomb. Figure 12.47 (middle) plots the TIR retardance versus angle of incidence for refractive indices in the range of 1.495 to 1.5. A horizontal black line indicates one eighth wave of retardance (~0.785), the target retardance. Figure 12.47 (right) shows that for a refractive index of 1.4965 and an angle of 51.79°, the retardance curve is just tangent to the eighth wave retardance line; thus, retardance is compensated in angle at this condition. One glass with a refractive index close to 1.4965 is Schott N-PK52A. Slightly larger refractive indices yield two eighth wave crossings a fraction of a degree apart.

Since the retardance depends on the refractive index, the wavelength dependence of the Fresnel rhomb depends only on the variation of refractive index with wavelength. This is plotted for Schott N-PK52A in Figure 12.48 showing that the retardance varies by less than 2.5° (0.04 radians) over

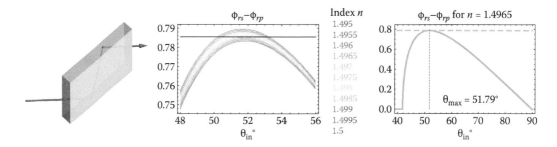

Figure 12.47 (Left) A Fresnel rhomb (green) is a prism with a parallelogram cross section and two internal reflections. A ray path (red) shows a ray at normal incidence (for minimum diattenuation) at the entering and exiting faces. (Middle) A refractive index $n = 1.4965$ gives a $\pi/4$ retardance at wavelength 589 nm. (Right) For index $n = 1.4965$, a maximum retardance is attained at incident angle $\theta_{in} = 51.79°$.

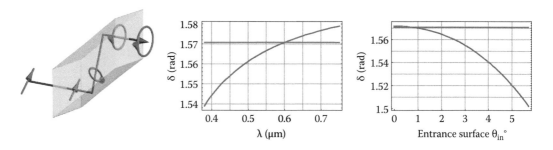

Figure 12.48 (Left) A polarization ray trace through an N-PK52A Fresnel rhomb shows the evolution of a 45° polarized incident state exiting as right circularly polarized, depicted as blue polarization ellipses. (Middle) The ray trace yields the retardance spectrum (in blue). The red line represents the target quarter wave retardance. (Right) The retardance versus incident angle is shown from 0° to 5° in air at the entrance surface.

the visible. This is much better than birefringent waveplates where the retardance varies as the variation of the birefringence times the thickness divided by the wavelength $\delta = 2\pi\,\Delta n\,t/\lambda$.[14] For waveplates, the $1/\lambda$ term yields a rapid variation of retardance with wavelength. Therefore, the retardance of Fresnel rhombs is far more achromatic than waveplates. Using the polarization ray trace, the variation of retardance with wavelength (left) and angle of incidence (right) for the Schott N-PK52A rhomb is plotted in Figure 12.48.

Many clever variations of Fresnel rhombs have been developed, often to modify the achromatization or control the field variations of retardance. A summary of these Fresnel rhomb-related designs is found in the work of Bennett.[14]

12.7 Conclusion

Systems of uncoated lenses and/or metal mirrors naturally have polarization aberrations that follow from the Fresnel equations. These polarization aberrations have a form similar to the angle of incidence maps for the surfaces. These examples, shown in this chapter, were evaluated by the polarization ray tracing method presented in Chapter 9 using the Polaris-M program. A grid of rays are traced and their polarization ray tracing matrices are calculated. The diattenuation and retardance magnitude and orientation are calculated for the rays to represent the polarization aberrations. These parameters are then flattened using the algorithms of Chapter 11 and represented in the double pole coordinate system so that the diattenuation and retardance maps can be shown on a flat page or computer screen. For the singlet lens and Cassegrain telescope examples shown, the polarization aberrations are not particularly large because their effect on the point spread function is small, and the wavefront aberrations are a much higher priority.

To understand the magnitude of the polarization aberrations and determine when it is a problem, it is necessary to perform a polarization ray trace. This will show if the polarization aberrations are small enough to not be a concern, or to determine if the polarization aberrations may cause problems, such as loss of resolution or leakage at polarizers.

This chapter presents the polarization aberrations of uncoated systems as a first step toward understanding more complex interfaces presented in later chapters. In high-quality optical systems, such as lenses for movie production, LC projectors, microlithography, and other applications, uncoated surfaces such as those analyzed in this chapter are the exception, not the rule. Most lens surfaces are antireflection coated. Many mirrors have multilayer reflection-enhancing coatings. Other coatings are incorporated as spectral filters, beam splitters, and other functions. In that case, the examples here, the single-element lens, multiple fold mirrors, and Cassegrain telescope, would have different magnitudes of the diattenuation and retardance, although the form would be similar, because the diattenuation and retardance of thin films are still aligned with the s- and p-planes. Thus, they need to be polarization ray traced using their specified coatings to determine the new magnitudes of the diattenuation and retardance maps. To gain some understanding of such cases, Chapter 13 analyzes some typical optical thin films to provide examples and guidance for including the coatings in the polarization ray trace and polarization aberration specification.

12.8 Problem Sets

12.1 The flux distribution (amplitude squared), seen in Figure 12.7 (left) is referred to the *Maltese cross*. The Maltese cross is an important pattern, commonly seen with most radially symmetric optical systems placed between crossed polarizers. Along the diagonal, the amplitude of the leaked light varies as ρ^2 while the transmitted flux varies as ρ^4, where ρ is pupil coordinate. If the lens is stopped down to a smaller aperture, the leaked flux rapidly diminishes. For a small numerical aperture, the Maltese cross may be difficult to observe. If the two polarizers rotate together, the Maltese cross will rotate with the polarizers.

a. Consider the *Maltese cross*, seen in Figure 12.7 (center). Why does the pattern rotate when the two polarizer are rotated together? What happens if just one polarizer is rotated by a small amount? If the lens alone is rotated between crossed polarizers, what happens to the pattern?

b. Describe the pattern of the transmitted flux when a cylindrical lens is rotated between crossed polarizers.

c. Consider the total transmitted flux in the Maltese cross. If the diameter of the pupil is reduced to one half, by what fraction is the transmitted flux reduced?

12.2 Find the refractive indices for which the intensity reflection coefficients at normal incidence equal the intensity transmission coefficients.

12.3 Show that if a beam is incident on a plane parallel plate from air at Brewster's angle, that it is incident on the rear surface at Brewster's angle.

12.4 Find the Jones matrix for a real half wave retarder, fast axis at 45°, of sapphire at $\lambda = 589$ nm where $n_O = 1.76817$ and $n_E = 1.76009$. Chapter 5 (Jones Matrices and Polarization Properties) presented the Jones matrices for ideal retarders; now, consider a real retarder. Find the thickness t and calculate the absolute phases for the ordinary and extraordinary rays. Because of these optical path lengths, the retarder's Jones matrix will not be in *symmetric*, *fast axis unchanged*, or *slow axis unchanged* form (Table 5.4). Evaluate the Fresnel equations at normal incidence for the ordinary and extraordinary modes and include the small resulting diattenuation in the Jones matrix.

12.5 Show that, for reflection in air from a dielectric, both $R_s(\theta)$ and $R_p(\theta)$ approach the same value as θ approaches 90°. Find the slopes of $R_s(\theta)$ and $R_p(\theta)$ as θ approaches 90°.

12.6 An excellent fit to the Fresnel amplitude coefficients r_s, r_p, t_s, and t_p, for $n_0 = 1$ and $n_1 = 1.5$ over the entire angle of incidence range $0 < \theta < 90°$ can be obtained with an even fourth-order fit as shown in Figure 12.49. Perform the least squares polynomial fit with $f(\theta) = a_0 + a_2 \theta^2 + a_4 \theta^4$ using the method of Section 8.5. For each amplitude coefficient, make a table of its values at $0°, 2°, 4°, \ldots 90°$. Then, fit each tabulated function $f(\theta)$ to find the coefficients a_0, a_2, and a_4. For example, the first fit is $r_s = -8 \times 10^{-9} \theta^4 - 0.0000306 \theta^2 - 0.203699$.

12.7 Verify $r_p = \dfrac{r_s(r_s - \cos 2\theta)}{1 - r_s \cos 2\theta}$ and $r_s = \dfrac{1}{2}\cos 2\theta(1 - r_p) + \sqrt{r_p + \dfrac{1}{4}\cos^2 2\theta(1 - r_p)^2}$. Hint: Start with the r_s as a function of θ, and r_p as a function of θ.[15]

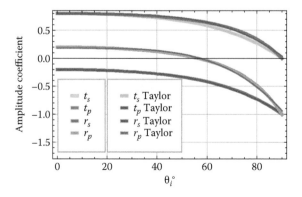

Figure 12.49 Comparison of Fresnel amplitude coefficients to a set of fourth-order fits yields very close fits, almost on top of each other.

References

1. H. K. and S. Inoué, Diffraction images in the polarizing microscope, *J. Opt. Soc. Am.* 49 (1959): 191–198.
2. R. A. Chipman, Polarization analysis of optical systems, *Opt. Eng.* 28(2) (1989): 90–99.
3. J. O. Stenflo, The measurement of solar magnetic fields, *Rep. Prog. Phys.* 41.6 (1978): 865.
4. R. A. Chipman, Polarization aberrations, PhD dissertation, College of Optical Sciences, University of Arizona (1987).
5. R. A. Chipman, Polarization analysis of optical systems II, *Proc. SPIE* 1166 (1989): 79–99.
6. J. P. McGuire and R. A. Chipman, Polarization aberrations. 1. Rotationally symmetric optical systems, *Appl. Opt.* 33 (1994): 5080–5100 and Polarization aberrations. 2. Tilted and decentered optical systems. *Appl. Opt.* 33 (1994): 5101–5107.
7. D. J. Reiley and R. A. Chipman, Coating-induced wave-front aberrations: On-axis astigmatism and chromatic aberration in all-reflecting systems, *Appl. Opt.* 33 (1994): 2002–2012.
8. W. Swindell, Handedness of polarization after metallic reflection of linearly polarized light, *J. Opt. Soc. Am.* 61 (1971): 212–215.
9. R. A. Chipman, Polarization aberrations, PhD dissertation, The University of Arizona (1987).
10. D. B. Chenault and R. A. Chipman, Measurements of linear diattenuation and linear retardance spectra with a rotating sample spectropolarimeter, *Appl. Opt.* 32 (1993): 3513–3519.
11. D. J. Reiley and R. A. Chipman, Coating-induced wave-front aberrations: On-axis astigmatism and chromatic aberration in all-reflecting systems, *Appl. Opt.* 33(10) (1994): 2002–2012.
12. D. J. Reiley, Polarization in Optical Design, dissertation, Physics, University of Alabama in Huntsville (1993).
13. R. J. King, Quarter-wave retardation systems based on the Fresnel rhomb principle, *J. Sci. Instrum.* 43.9 (1966): 617.
14. J. M. Bennett, A critical evaluation of rhomb-type quarter wave retarders, *Appl. Opt.* 9.9 (1970): 2123–2129.
15. R. M. A. Azzam, Direct relation between Fresnel's interface reflection coefficients for the parallel and perpendicular polarizations, *J. Opt. Soc. Am.* 69(7) (1979).

13

Thin Films

13.1 Introduction

Optical *thin films* are ubiquitous in optical systems. The majority of refracting surfaces have *antireflection coatings* to reduce loss. Films are placed over metal mirrors to protect the metal surface and boost their reflectance. Many thin films are used to spectrally filter light, to change its spectral content, such as *bandpass filters* that block light outside a given spectral range. *Beam splitter coatings* are used to divide or combine wavefronts; beam splitter coatings are necessary in interferometers and many other applications. Many beam splitter coatings are designed to be non-polarizing, so as to not change the polarization state. Other beam splitter coatings are designed as *polarizing beam splitters*, to reflect one polarization state and transmit the orthogonal state.

All coatings affect the polarization state of incident light. *Amplitude coefficients* characterize the polarization properties of the coating; these amplitude coefficients for thin films relate electric fields across interfaces in the same way as the Fresnel coefficients of Chapter 12; they characterize the amplitude and phase changes of the s- and p-polarized components of the light.

This chapter reviews the optics of *homogeneous and isotropic* thin films. First, the reflectance and transmission of single-layer thin films are derived and studied. Then, the general algorithms for the optical properties of an arbitrary multilayer thin film are presented, and the polarization performance of several important families of coatings is studied from the perspective of the optical engineer and lens designer.

Surfaces with thin film coatings, such as antireflection coatings and beam splitter coatings, have amplitude reflection and transmission coefficients that are calculated by the application of the Fresnel coefficients to multilayer stacks of optical materials. Four types of interfaces are listed below:

- *Homogeneous interface*—Properties of the interface are constant over the clear aperture.
- *Inhomogeneous interface*—The composition changes in space or the coating thickness changes over the clear aperture.
- *Isotropic interface*—Refractive indices of all materials are the same in all directions and for all polarizations.
- *Anisotropic interface*—Birefringent or optically active materials are used and the refractive index depends on the polarization state. Anisotropy may arise from the use of birefringent materials, from strain, and also from coating microstructure, since many deposited coatings grow as arrays of microscopic pillars, which induces form birefringence.[1,2] Also, normally isotropic materials become anisotropic when subjected to strong electric or magnetic fields.

13.2 Single-Layer Thin Films

Consider a plane wave of light incident on a thin layer of refractive index n_1 and thickness t on a substrate of refractive index n_2, as shown in Figure 13.1. The optical film is characterized by the amplitude reflection and transmission coefficients of the s and p *eigenpolarizations*. The plane wave is represented by a light ray in the figure. As the ray propagates through the film, part of the flux reflects and refracts at each interface, with some of the light undergoing multiple reflections. Each path through the film is referred to as a *partial wave*. Each partial wave has its own amplitude and phase, and the partial waves will constructively or destructively interfere. Thus, the overall amplitude reflection and transmission are determined by the summation of all the partial reflected or transmitted waves. These transmissions and reflections are wavelength and incident angle dependent.

The relative amplitude and phase carried by each partial *wave* is calculated by *Fresnel equation* and the phase thickness of the film. The individual complex amplitude coefficients of the ray and top/bottom interface interaction are given by Equations 13.4 through 13.11. The ray angles in each medium are obtained from *Snell's law* (Equation 13.1). These angles can be rewritten as in Equations 13.2 and 13.3. Figure 13.2 shows the Fresnel coefficients' subscripts at each interface: *air→film* (01), *film→substrate* (12), and *film→air* (10).

$$n_0 \sin\theta_0 = n_1 \sin\theta_1 = n_2 \sin\theta_2, \qquad (13.1)$$

$$\cos\theta_1 = \sqrt{1 - \frac{n_0^2 \sin^2\theta_0}{n_1^2}}, \qquad (13.2)$$

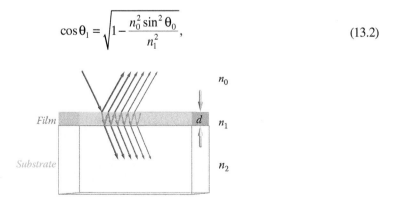

Figure 13.1 A single layer of thin film coating with index n_1 on a substrate with index n_2. Multiple reflections within the film generate a set of reflected and transmitted partial waves.

13.2 Single-Layer Thin Films

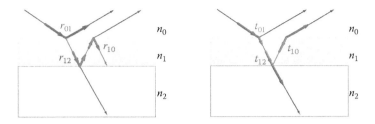

Figure 13.2 Fresnel amplitude coefficients of each interface for reflections (left) and transmissions (right).

$$\cos\theta_2 = \sqrt{1 - \frac{n_0^2 \sin^2\theta_0}{n_2^2}}, \tag{13.3}$$

$$r_{01s} = \frac{n_0 \cos\theta_0 - n_1 \cos\theta_1}{n_0 \cos\theta_0 + n_1 \cos\theta_1} = -r_{10s}, \tag{13.4}$$

$$r_{01p} = \frac{n_1 \cos\theta_0 - n_0 \cos\theta_1}{n_1 \cos\theta_0 + n_0 \cos\theta_1} = -r_{10p}, \tag{13.5}$$

$$r_{12s} = \frac{n_1 \cos\theta_1 - n_2 \cos\theta_2}{n_1 \cos\theta_1 + n_2 \cos\theta_2}, \tag{13.6}$$

$$r_{12p} = \frac{n_2 \cos\theta_1 - n_1 \cos\theta_2}{n_2 \cos\theta_1 + n_1 \cos\theta_2}, \tag{13.7}$$

$$t_{01s} = \frac{2n_0 \cos\theta_0}{n_0 \cos\theta_0 + n_1 \cos\theta_1}, \tag{13.8}$$

$$t_{01p} = \frac{2n_0 \cos\theta_0}{n_1 \cos\theta_0 + n_0 \cos\theta_1}, \tag{13.9}$$

$$t_{12s} = \frac{2n_1 \cos\theta_1}{n_1 \cos\theta_1 + n_2 \cos\theta_2}, \tag{13.10}$$

$$t_{12p} = \frac{2n_1 \cos\theta_1}{n_2 \cos\theta_1 + n_1 \cos\theta_2}, \tag{13.11}$$

The *phase thickness* 2β is the phase difference between adjacent rays for both transmission and reflection (derived in the Appendix), where

$$\beta(n_0, n_1, \theta_0, d) = \frac{2\pi}{\lambda} n_1 d \cos\theta_1. \tag{13.12}$$

The s and p amplitude reflectance and transmittance coefficients for a *single-layer coating* calculated by the summation of all partial wave complex amplitudes (derived in the Appendix) are shown in Equations 13.13 through 13.16.

$$r_s(\theta_i, n_0, n_1, n_2, d) = \frac{r_{01s} + r_{12s}e^{i2\beta}}{1 - r_{12s}r_{10s}e^{i2\beta}}$$
$$= |r_s|e^{-i\phi_{rs}} \tag{13.13}$$

$$r_p(\theta_i, n_0, n_1, n_2, d) = \frac{r_{01p} + r_{12p}e^{i2\beta}}{1 - r_{12p}r_{10p}e^{i2\beta}}$$
$$= |r_p|e^{-i\phi_{rp}} \tag{13.14}$$

$$t_s(\theta_i, n_0, n_1, n_2, d) = \frac{t_{01s}t_{12s}}{1 - r_{10s}r_{12s}e^{i2\beta}}$$
$$= |t_s|e^{-i\phi_{ts}} \tag{13.15}$$

$$t_p(\theta_i, n_0, n_1, n_2, d) = \frac{t_{01p}t_{12p}}{1 - r_{10p}r_{12p}e^{i2\beta}}$$
$$= |t_p|e^{-i\phi_{tp}} \tag{13.16}$$

The absolute values indicate the fraction of the incident amplitude that takes each path. The arguments of the amplitude coefficients, ϕ, indicate the phase change between the incident and exiting beam. The phase is measured relative to the incident surface for reflection and the emerging surface for transmission.

These thin film coefficients (Equations 13.13 through 13.16) are the Fresnel-like amplitude coefficients that are used in the polarization ray tracing **P** matrix (Chapter 9) for each ray intercept during a polarization ray trace as in Equation 9.15,

$$\mathbf{P}_t = \mathbf{O}_{transmit} \begin{pmatrix} t_s & 0 & 0 \\ 0 & t_p & 0 \\ 0 & 0 & 1 \end{pmatrix} \mathbf{O}_{in}^{-1} \text{ and } \mathbf{P}_r = \mathbf{O}_{reflect} \begin{pmatrix} r_s & 0 & 0 \\ 0 & r_p & 0 \\ 0 & 0 & 1 \end{pmatrix} \mathbf{O}_{in}^{-1}, \tag{13.17}$$

for refraction and reflection.

13.2.1 Antireflection Coatings

Antireflection coatings are designed to improve the transmission of optical elements. A significant side benefit is that antireflection coatings almost always reduce polarization aberration as well. The most common antireflection coating is a quarter wave thick (optical path length *n t*) layer of a

13.2 Single-Layer Thin Films

low-index material. For a quarter wave single-layer coating $n_1 d = \lambda/4$, at normal incidence $\theta_1 = \theta_2 = \theta_3 = 0°$, the reflectance is

$$R = \left(\frac{n_0 n_2 - n_1^2}{n_0 n_2 + n_1^2} \right)^2 \qquad (13.18)$$

and

$$R = 0 \text{ when } n_1 = \sqrt{n_0 n_2}. \qquad (13.19)$$

With a bare glass of index 1.5 in air, zero reflectance is obtained with a quarter wave coating of index $\sqrt{1.5} = 1.225$. Thin film coating materials with refractive indices as low as 1.225 are not readily available. The most common low-index coating material is magnesium fluoride, MgF_2. A single-layer antireflection coating is commonly made of a quarter wave of MgF_2; reflectance $R = 1.4\%$ on glass. The s and p intensity coefficients, diattenuation, and retardance are compared with the uncoated surface in the following.

The intensity transmission of MgF_2 film on glass as a function of the film thickness, shown in Figure 13.3. Has maxima at quarter and three-quarter wave thicknesses.

With a quarter wave MgF_2 film, the coated glass has over 98% transmission and less than 2% reflection at normal incidence, as shown in Figure 13.4. The transmission *diattenuation* decreases

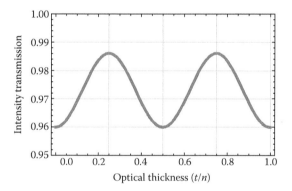

Figure 13.3 Normal incidence transmission through MgF_2 coating on an $n = 1.5$ substrate as a function of the coating's optical thickness (t/n).

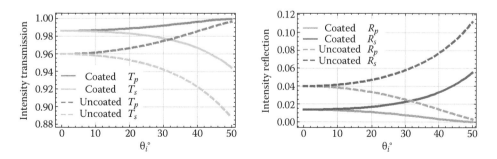

Figure 13.4 s and p (left) transmission and (right) reflection intensity coefficients of the $\lambda/4$ MgF_2 coating on glass at wavelength 0.55 μm.

and the reflection diattenuation increases slightly, as shown in Figure 13.5, because of this quarter wave coating. The phase change on transmission for the *s*- and *p*-components are almost identical and are quadratic for small angles, as shown in Figure 13.6, indicating a very small contribution to the overall power of the lens incorporating this coating (Section 13.4). Both transmission and reflection *retardance* increases quadratically with angle, as shown in Figure 13.7.

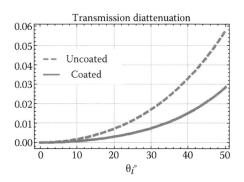 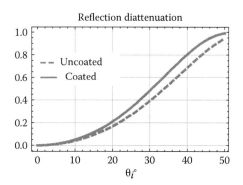

Figure 13.5 (Left) Transmission and (right) reflection diattenuation of the $\lambda/4$ MgF$_2$ coating on glass at wavelength 0.55 µm.

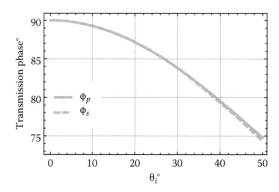

Figure 13.6 Phase change on transmission for the *s*- and *p*-components of the $\lambda/4$ MgF$_2$ coating at wavelength 0.55 µm.

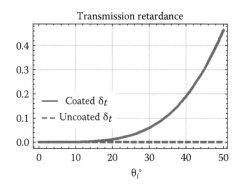 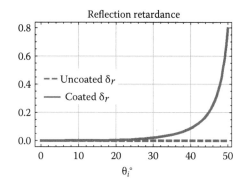

Figure 13.7 (Left) Transmission and (right) reflection retardance in degrees of the $\lambda/4$ MgF$_2$ coating on glass at wavelength 0.55 µm are both negligible.

13.2 Single-Layer Thin Films

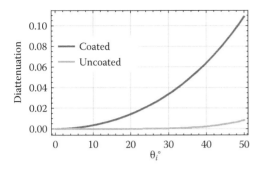
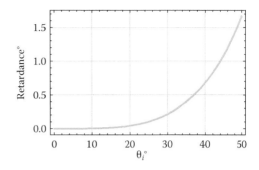

Figure 13.8 (Left) Diattenuation and (right) retardance of the quarter wave MgF_2 coating ($t = \lambda/(4n_1)$) on a substrate with refractive index $n_{sub} = n_{MgF_2}^2 = 1.9044$, $n_0 = 1$ and $n_1 = 1.38$.

13.2.2 Ideal Single-Layer Antireflection Coating

The ideal *single-layer antireflection coating*, where the layer index is the geometric mean of the incident and substrate indices (Equations 13.19), has remarkably low diattenuation and retardance. Figure 13.8 shows the performance of a quarter wave of MgF_2 on a substrate with a refractive index equal to the MgF_2 index squared. On the left, the diattenuation is compared to the uncoated diattenuation as a function of angle of index. The diattenuation remains almost on the *x*-axis from zero to beyond $\theta = 30°$. As shown by Azzam,[3] the variation of diattenuation with angle of incidence is sixth order; thus, there is no quadratic or fourth-order variation in the diattenuation with angle. Similarly on the right, the retardance remains remarkably low with angle, being of fourth order; the retardance has no quadratic variation for this index and thickness.

13.2.3 Metal Beam Splitters

Beam splitters divide the amplitude of incident light into transmitted and reflected partial waves. Here, the single-layer equations analyze a single film of aluminum on a glass substrate at normal incidence and at a 45° incident angle. The performance of *aluminum beam splitters* operating at an angle of incidence of 45° will be explored to determine how close the performance can come to ideal non-polarizing behavior.

At normal incidence, reflectance and transmission are approximately equal for a ~4.2-nm-thick aluminum film as seen in Figure 13.9.

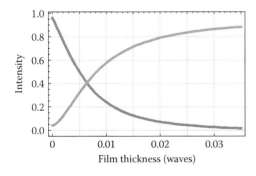
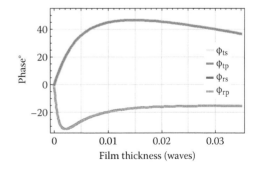

Figure 13.9 (Left) Intensity transmission (orange) and reflection (blue), and (right) phase transmission and reflection for normal incidence as a function of thickness for an aluminum film ($n_1 = 1.2 + 7.26i$) on glass ($n_2 = 1.5$) at 0.6 μm wavelength.

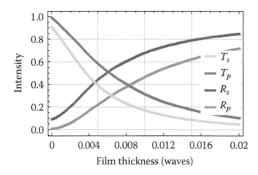 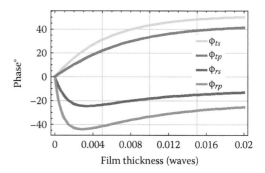

Figure 13.10 (Left) Intensity transmission and reflection for aluminum film ($n_1 = 1.2 + 7.26i$) on glass substrate ($n_2 = 1.5$) for 45° incident angle at 0.6 μm. (Right) Phase changes for transmission and reflection for the aluminum film on glass substrate.

The intensity and phase for transmission and reflection of an aluminum film at a 45° incident angle are shown in Figure 13.10. The s- and p-transmission coefficients and the s- and p-reflection coefficients never cross, demonstrating that a *non-polarizing beam splitter* is not possible with a single aluminum layer. However, the four coefficients are very close to each other at a ~3.6-nm-thick aluminum film.

13.3 Multilayer Thin Films

We now consider the calculation of the reflectance and transmission of a *multilayer thin film* coating. The objective is the determination of the amplitude coefficients, r_s, r_p, t_s, and t_p as functions of wavelength λ and angle of incidence θ. These amplitude coefficients are needed to construct the *P matrix* for *polarization ray tracing*.

Consider a homogeneous and isotropic thin film consisting of a set of Q parallel layers. The incident medium is assumed transparent. The coating thicknesses can be formed into a vector of thicknesses **T**,

$$\mathbf{T} = \left(t_1, t_2, \ldots, t_q, \ldots, t_Q\right). \tag{13.20}$$

These thicknesses are in unit of length, such as millimeters, and are thus called *metric thicknesses* to distinguish them from the *optical thicknesses* specified in waves. The calculations will be performed one wavelength at a time; thus, at each wavelength, the refractive indices of each layer can also be arranged into a vector Λ,

$$\mathbf{\Lambda} = (n_1, n_2, \ldots, n_q, \ldots, n_Q). \tag{13.21}$$

Specification of the coating includes the thickness of each layer and their complex refractive indices.

Figure 13.11 shows an eight-layer film stack, and each layer has a thickness and index associated to it. For the coating on curved substrates, the film will be treated as having locally flat layers parallel to the tangent plane to the substrate.

13.3 Multilayer Thin Films

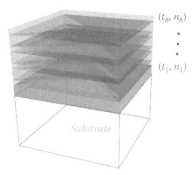

Figure 13.11 Schematic of an eight-layer multilayer thin film stack on a substrate. Each layer is specified by its thickness t and refractive index n.

13.3.1 Algorithms

The *amplitude coefficients* for homogeneous and isotropic *multilayer coatings* can be calculated by a straightforward matrix method first introduced by Abelès.[4,5] This algorithm is systematized by the *characterized matrix* for transparent substrates. This algorithm is in widespread use and is at the core of most commercial thin film simulation software.[6]

The algorithm uses characteristic matrices specified for each film layer to calculate the transmission and reflections of the multilayer film assembly. For Q layers of films,

$$\begin{pmatrix} B \\ C \end{pmatrix} = \prod_{q=1}^{Q} \begin{pmatrix} \cos\beta_q & i\sin\beta_q/\eta_q \\ i\eta_q \sin\beta_q & \cos\beta_q \end{pmatrix} \begin{pmatrix} 1 \\ \eta_m \end{pmatrix}, \quad (13.22)$$

where the phase thickness of layer q is

$$\beta_q = \frac{2\pi n_q d_q \cos\theta_q}{\lambda}, \quad (13.23)$$

with complex refractive index η_q for layer q. η_m is the characteristic admittance of the substrate. The s- and p-polarizations for the TE and TM components have different characteristic matrices and are calculated separately with different characteristic admittances:

$$\eta_{s,q} = \sqrt{\frac{\varepsilon_o}{\mu_o}} n_q \cos\theta_q \text{ and } \eta_{p,q} = \sqrt{\frac{\varepsilon_o}{\mu_o}} \frac{n_q}{\cos\theta_q}. \quad (13.24)$$

The overall reflection and transmission is calculated by the matrix product Equations 13.22 for each layer, which operates on the substrate vector. The amplitude reflection coefficient is

$$r = \frac{\eta_0 - Y}{\eta_0 + Y} = \frac{\eta_0 B - C}{\eta_0 B + C} \quad (13.25)$$

where surface admittance $Y = \dfrac{C}{B}$, and the phase change upon reflection is

$$\phi_r = -\arctan\left[\dfrac{\mathrm{Im}\left[\eta_m(BC^*-CB^*)\right]}{\eta_m^2 BB^* - CC^*}\right]. \tag{13.26}$$

The complex transmission coefficient is

$$t = \left(\dfrac{2\eta_0}{\eta_0 B + C}\right)^* \tag{13.27}$$

and the corresponding phase change of transmission measured relative to the emerging surface is

$$\phi_t = -\arctan\left[\dfrac{-\mathrm{Im}\left[\eta_o B + C\right]}{\mathrm{Re}\left[\eta_o B + C\right]}\right]. \tag{13.28}$$

The intensity coefficient is the ratio of the exiting to the incident flux. The reflection and transmission intensity are

$$R = \left(\dfrac{\eta_0 B - C}{\eta_0 B + C}\right)\left(\dfrac{\eta_0 B - C}{\eta_0 B + C}\right)^* \text{ and } T = \dfrac{4\eta_0 \,\mathrm{Re}[\eta_m]}{(\eta_0 B + C)(\eta_0 B + C)^*}, \tag{13.29}$$

respectively. Absorption is zero if all of the film's refractive indices are real; otherwise, the absorption is

$$A = \dfrac{4\eta_0 \,\mathrm{Re}[BC^* - \eta_m]}{(\eta_0 B + C)(\eta_0 B + C)^*}. \tag{13.30}$$

13.3.2 Quarter and Half Wave Films

Some simple relationships can be used in thin film design when the film thickness is a multiple of a quarter wave; these thicknesses have the maximum and minimum effect on reflection and transmission. When the dielectric lossless film is half wave thick with $\beta = m\pi/4$, where $m = 0, 2, 4, \ldots$, its characteristic matrix becomes

$$\pm \begin{pmatrix} 1 & 0 \\ 0 & 1 \end{pmatrix}. \tag{13.31}$$

This *absentee layer* has no effect on the reflectance and transmittance for light at the wavelength for which the film is half wave. A film is quarter wave thick when $m = 1, 3, 5, \ldots$ and the corresponding characteristic matrix is

$$\pm \begin{pmatrix} 0 & i/\eta \\ i\eta & 0 \end{pmatrix}. \tag{13.32}$$

13.3 Multilayer Thin Films

The *surface admittance* Y becomes

$$\frac{\eta_1^2 \eta_3^2 \eta_5^2 \cdots}{\eta_2^2 \eta_4^2 \cdots \eta_m^2} \quad \text{or} \quad \frac{\eta_2^2 \eta_4^2 \cdots \eta_m^2}{\eta_1^2 \eta_3^2 \eta_5^2 \cdots} \tag{13.33}$$

for an odd or even number of quarter wave films. Many common film designs use *quarter wave layers*, which are labeled as

- L   low-index quarter wave layer
- M   medium-index quarter wave layer
- H   high-index quarter wave layer

For a low-index quarter wave layer, the partial wave that reflects from the first surface of the layer and the partial wave that reflects from the second surface are 180° out of phase, minimizing reflection. Some example quarter wave designs are the following:

- Air L Glass   quarter wave antireflection coating
- Air HLHLHLHL Glass   alternating eight-layer coating

13.3.3 Reflection-Enhancing Coatings

Coatings with high reflectance can be readily fabricated from alternating high- and low-index layers of quarter wave thickness. These *reflective coatings* provide an alternative to metal coatings and can provide much higher reflectance; however, the spectral bandwidth and polarization properties can be quite different. Reflection-enhancing coatings provide a good example of how to interpret the results of thin film calculations to interpret the effects of coatings on wavefront and polarization aberrations.

Figure 13.12 (left) shows a schematic of a two-layer dielectric multilayer coating with an alternating high- and low-index layer on an n-BK7 substrate. With our coating shorthand notation, the coating is Air HL Glass, a quarter wave of high index on a quarter wave of low-index material. For this example, *hafnium oxide*, HfO_2 is chosen as the high index material ($n_H = 1.94$) and MgF_2 is the low-index material ($n_L = 1.39$).

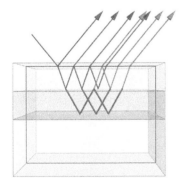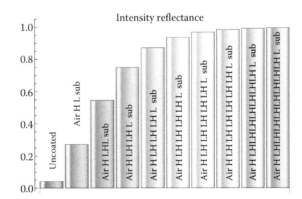

Figure 13.12 (Left) Depiction of the family of partial waves reflecting from an Air HL Glass coating. For the Air HL Glass coating at normal incidence, all the partial waves exit the top surface in phase, enhancing reflectivity. (Right) Reflectance from $(HL)^N$ coatings increases rapidly with the number of layer pairs exceeding 99% with nine layer pairs.

In the quarter wave antireflection coating (Air L Glass), the partial waves reflecting from the air-to-MgF$_2$ interface and the partial wave reflecting from the MgF$_2$-to-glass interface are π out of phase after exiting the coating, thus suppressing reflection. For the Air HL Glass coating at normal incidence, the partial waves from the Air–H, H–L, and L–H interfaces, shown in Figure 13.12 (left), are all in phase after reflecting and exiting the top of the coating, thus enhancing reflection, but because the amplitude coefficients are small, only ~25% of the light is reflected from Air HL Glass. This structure is now used as a building block to build high reflection coatings, because as HL layer pairs are added, all the reflected partial waves remain in phase.

Figure 13.12 (right) shows the increasing normal incidence reflection at $\lambda_0 = 0.55$ μm as the number of HL layers increases. With a four-layer coating (Air HLHL Glass), reflectance increases to 50%. By the time the number of layers increases to 10 alternating layers of HfO$_2$ and MgF$_2$ (Air HLHLHLHLHL Glass), the normal incidence reflection is about 95%, and at 18 alternating layers, reflectivity is over 99%.

This sounds excellent! Why would *metal reflectors* be used if such high reflectance is available? This high reflectivity is only obtained over a limited range of wavelengths and angles, and outside this range, the reflectivity can be much worse than a metal mirror. Designs with large numbers of layers and high reflectivity have rapid variation of *amplitude coefficients* with angle and wavelength; hence, this coating family provides a good example of how to interpret the thin film coating calculation output to understand the wavefront aberrations and polarization aberrations that coatings can cause.

For the 10-layer HfO$_2$ and MgF$_2$ reflective coating (Air HLHLHLHLHL Glass) at λ_0, the reflectivity is shown in Figure 13.13.

For 18 alternating layers of HfO$_2$ and MgF$_2$ (Air HLHLHLHLHLHLHLHLHL Glass), the resultant reflection is almost 100% from 0° to ~30° incidence, as shown in Figure 13.14 (left). If the mirror is used beyond 40°, the reflectivity rapidly drops and the coating becomes highly diattenuating. The corresponding phase of the *s*- and *p*-light is shown in Figure 13.14 (right). The phases change much more rapidly than a metal film with angle. The difference, $\phi_p - \phi_s$, is the retardance, which increases rapidly as the reflectivities change, with the *p*-component undergoing 180° of phase change in about 50°. The average phase change $(\phi_p + \phi_s)/2$ is plotted in magenta, and should be considered as a contribution to the wavefront aberration of any system using this coating. At the edge of the high reflectivity zone, 40° angle of incidence, the average phase has changed by 0.5 radians or ~ $\lambda/12$. As is discussed later, since the average phase is changing nearly quadratically, this coating's quadratic average phase change is a source of defocus and chromatic aberration when the angle of

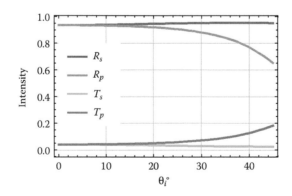

Figure 13.13 The reflectivity of a 10-layer reflection-enhancing coating (Air HLHLHLHLHL Glass) at 0.55 μm has a normal incidence reflectivity of 94%, a little bit higher than aluminum (~92%). Some light transmits into the glass substrate (bottom, T_s, T_p).

13.3 Multilayer Thin Films

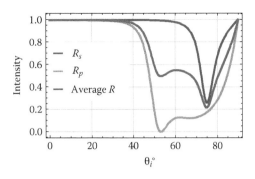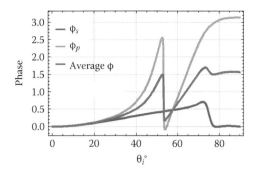

Figure 13.14 (Left) Intensity reflectance and (right) phase change on reflection of an 18-layer reflection-enhancing coating versus angle of incidence at 0.55 μm. The s and p light are reflecting from different effective depths.

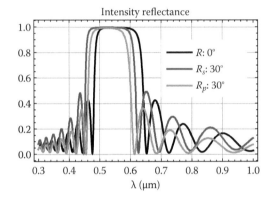

Figure 13.15 Intensity reflectance of the 18-layer reflection-enhancing coating versus wavelength for 0° and 30°. Note the "*blue shift*" of the spectral bandpass to shorter wavelengths as the angle of incidence changes.

incidence changes across the coating, such as (1) a collimated beam at a spherical mirror or (2) a converging or diverging beam at a flat mirror.

Figure 13.15 shows the behavior of this 18-layer reflection-enhancing coating design versus wavelength. Note that this film behaves like a *bandpass filters* with 0.1 μm bandwidth (0.5–0.6 μm) from 0° to 30°. Thus, the excellent reflectance at the design wavelength comes at a cost of reduced spectral bandwidth. As the angle of incidence increases, the width of the *p*-bandpass reduces faster than the *s*-bandpass; thus, there is a large diattenuation at the edges of the bandpass. The phases, the coating-induced defocus, and chromatic aberration introduced by this coating are explored further in Section 13.4.

For these many-layer reflection coatings, the reflectance is small at each interface and the light penetrates deep into the coating. A large number of *partial waves* have comparable amplitude, for example, all the singly reflected beams, or all the triply reflected beams, and so on. There is an *effective depth* where the average light reflects. For the highest reflecting coatings, such as the 18-layer coating, almost no light reaches the bottom of the coating. The result of this effective depth for off-axis rays can be a noticeable offset d_{eff} between an incident and reflected ray, as shown in Figure 13.16. The phase difference between *s*- and *p*-phases is an indication that the *s*-polarized component with its higher amplitude reflectance reflects closer to the surface on the average while the *p*-component reflects from deeper in the coating on the average.

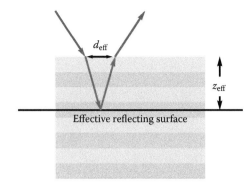

Figure 13.16 Light reflecting from a multilayer film can be given an effective depth of z_{eff} based on the mean depth the partial rays reflect from the structure. The exiting ray has a mean offset d_{eff} from the incident ray.

13.3.4 Polarizing Beam Splitters

An ideal *polarizing beam splitter* (PBS) divides incident light into its *s*- and *p*-components, transmitting the *p*-component and reflecting the *s*-component. Real beam splitters approach this ideal but are limited by the laws of multilayer coatings as applied to real materials. Thus, polarizing beam splitters have a limited range of angle of incidence and wavelength over which they are effective, that is, where they exceed an extinction ratio specification such as 1000:1. When illuminated with spherical waves, these behaviors show up as polarization aberrations due to the variation of angle of incidence. The perfect polarizing beam splitter coating could be considered as one that is an ideal antireflection coating for the *p*-component and an ideal reflector for the *s*-component. It is difficult to find polarizing beam-splitting coating designs that simultaneously have a high extinction ratio, a broad wavelength range, and a large range of angles of incidence.

The light flux through a polarizing beam splitter can be considered in three classes:[7]

1. Light that takes the correct path in the correct polarization state
2. Light that takes the incorrect path
3. Light that takes the correct path but ends up in the incorrect polarization state

One approach for PBS coating design is to make a thin film version of the pile-of-plates polarizer, operating near Brewster's angle. At Brewster's angle, all of the *p*-polarized component is transmitted and some of the *s*-polarized component is reflected. By stacking a group of plates, that is, thin film interfaces near Brewster's angle, the overall reflectance of the *s*-component is increased. For the pile-of-plates polarizer, the reflected beam is purely *s*-polarized and thus has a high degree of polarization. Some of the *s*-polarized light leaks into the transmitted beam; hence, it has a lower degree of polarization and extinction ratio. Thus, it is generally the case for a PBS that the reflected beam has the higher extinction ratio and diattenuation while the transmitted beam is of lower quality. Thus, when using a PBS as a polarizer, the reflected beam is preferred.

Many optical systems prefer PBS with angles of incidence around 45°, so the transmitted and reflected beams exit orthogonal to each other. The 45° PBS are called *polarization beam splitting cubes*, shown in Figure 13.17. The polarizing beam-splitting coating is fabricated on the hypotenuse of a right prism of high-index glass, which is then glued to another right prism. The faces are usually antireflection coated to increase overall throughput.

To best exploit Brewster's angle, many PBS coating designs operate at angles of incidence around 55°–60°, which provides easier designs with larger angular and wavelength bandwidths. The literature contains many designs for PBS coatings for various wavelengths and angles. An early theory for PBS coatings was developed by MacNeille.[8] An example of the MacNeille beam splitter is

13.3 Multilayer Thin Films

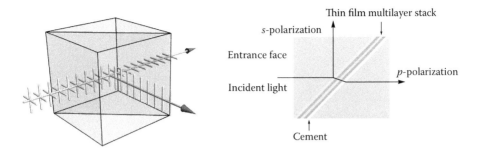

Figure 13.17 Schematic of a polarization beam splitting cube (left) showing *s*-polarized light reflecting and *p*-polarized light transmitting. (Right) One of the two prisms is coated and cemented to the other prism.

analyzed on SF5 glass at 550 nm to understand the type of behaviors expected and show how the performance transforms into polarization aberrations. This coating alternates high- and low-index layers of ZnS and MgF_2 films with thicknesses:

$$t_{ZnS} = \frac{\lambda}{4n_{ZnS}\sqrt{n_{ZnS}^2 + n_{MgF_2}^2}} \text{ and } t_{MgF_2} = \frac{\lambda}{4n_{MgF_2}\sqrt{n_{ZnS}^2 + n_{MgF_2}^2}}. \tag{13.34}$$

Figure 13.18 (left) shows the performance as a function of the number of layers. T_p is weakly dependent on the number of layers since it is the transmitted state at Brewster's angle. The *s*-reflectance steadily increases from 1 pair to 10 layer pairs and then oscillates as additional layer pairs are added. With this design, high *p*-transmission and high *s*-reflection are obtained between 0.55 and 0.7 μm.

The performance of a 37-layer MacNeille coating optimized for 550 nm is shown in Figure 13.19 for a short wavelength, the optimized wavelength, and a long wavelength. Only the desired *s*-reflected and *p*-transmitted beams are shown. At the design wavelength of 550 nm, both *p*-transmission and *s*-reflection are high (~100%) and steady between 41° and 47°, providing good polarizing beam-splitting performance over a 6° angular range. At 440 nm, the *s*-reflection is below 0.2; hence, the majority of both beams are transmitted and there is no polarizing beam-splitting function. At 730 nm, there is a beam-splitting function below 46° and the *s*-reflectance is high, but the *p*-transmittance oscillates with wavelength. The remaining *p*-light is reflected; thus, the quality and the *degree of polarization* of the reflected beam are poor.

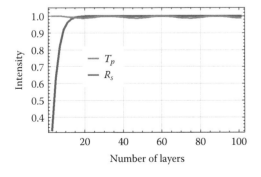
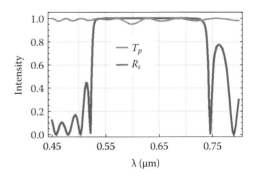

Figure 13.18 (Left) Reflected and transmitted intensity of the MacNeille coating as the number of layers increases, optimized at 0.55 μm. (Right) Reflected and transmitted intensity of the 23-layer MacNeille coating. The coating thickness is calculated for wavelength 0.55 μm.

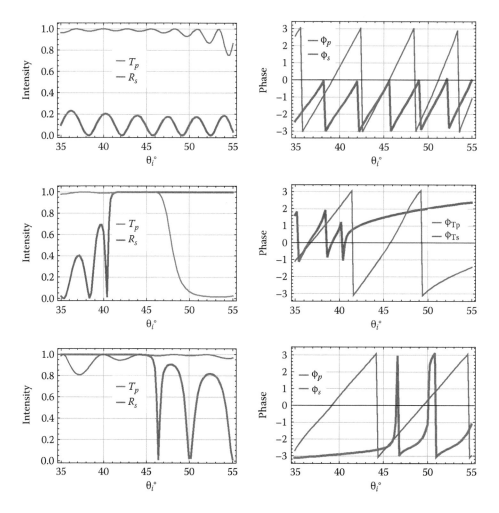

Figure 13.19 Intensity and phase of a 37-layer MacNeille coating as a function of incident angle in glass at wavelength 440 nm (top row), 550 nm (middle row), and 730 nm (bottom row). Coating thickness is optimized for 550 nm. (Right) Vertical lines indicate 2π phase discontinuities that occur due to the arctangents in Equations 13.26 and 13.28.

Math Tip 13.1 Polynomial Curve Fitting

The equations for multilayer film amplitude coefficients are complicated enough to make *analytical manipulation* nearly impossible. For the purposes of understanding aberrations and estimating the effects on image formation, treating the coefficients as linear, quadratic, or cubic *polynomials* provides tremendous insight, facilitating interfacing coating performance with the description of aberrations in a series of orders, second-order wavefront aberrations, fourth order, and so on. Thus, fitting the amplitude coefficients to polynomials provides forms that are easily manipulated and dovetail with aberration theory. The *Taylor series approximations* (Section 8.5.1) provide an algorithm to calculate polynomials with a close fit in the neighborhood of a point by matching the value, first derivative, second derivative, and so on, through a given order. These Taylor series fits diverge in value, moving away from that point, typically as the next-order polynomial in the series. Thus, a quadratic fit will typically diverge cubically from the function it is fitting, and so forth.

13.3 Multilayer Thin Films

For many purposes, it is better to fit a function over the entire range of interest by using a *least square fit* instead of matching derivatives with a Taylor series. The square root of the square of the difference (*RMS* root mean square) between the approximate and exact functions is a common metric for the fit, and the coefficients are calculated to minimize the *RMS* difference. Let $g(x)$ be an approximation to the exact function $f(x)$. The *RMS* is calculated on a continuous basis as

$$RMS = \frac{1}{x_2 - x_1} \sqrt{\int_{x_1}^{x_2} \left(f(x) - g(x)\right)^2 dx}. \tag{13.35}$$

Similarly, the *RMS* can be calculated for a discrete set of N point as

$$RMS = \frac{1}{x_N - x_1} \sqrt{\sum_{n=1}^{N} \left(f(x_n) - g(x_n)\right)^2}. \tag{13.36}$$

The advantage of including the square root in the metric is that the units and scale of the *RMS* correspond to the function being fit.

Many different functions are used for curve fitting. Sines and cosines are the basis for Fourier series fits. Many forms of polynomials such as Legendre and Chebyshev polynomials provide excellent basis functions for certain problems. Here, our concern is to perform simple polynomial fitting to provide a convenient representation for Fresnel and other thin film functions. It is desired to find the polynomial equation of order N

$$f(x) = a_0 + a_1 x + a_2 x^2 + a_3 x^3 + \ldots + a_N x^N = \sum_{n=0}^{N} a_n x^n, \tag{13.37}$$

which passes closest to a set of $M + 1$ data points $dx_1, dx_2, \ldots, dx_m, \ldots, dx_M$. In general, if $M + 1 = N$, then a polynomial can be found that passes exactly through the set of data points. If $M + 1 > N$, the equation that passes closest in the least square sense of minimizing the square of the errors is sought.

$$\Delta(a_0, a_1, a_2, \ldots, a_N) = \sum_{n=0}^{N} \left[f(x_m) - dx_m\right]^2. \tag{13.38}$$

If $M + 1 = N$, the calculation of polynomial coefficients can be formulated as the matrix equation

$$\begin{pmatrix} 1 & x_1 & x_1^2 & \cdots & x_1^N \\ 1 & x_2 & x_2^2 & & x_2^N \\ 1 & x_3 & x_3^2 & & x_3^N \\ \vdots & & & \ddots & \\ 1 & x_N & x_N^2 & & x_N^N \end{pmatrix} \begin{pmatrix} a_0 \\ a_1 \\ a_2 \\ \vdots \\ a_N \end{pmatrix} = \begin{pmatrix} d_0 \\ d_1 \\ d_2 \\ \vdots \\ d_N \end{pmatrix} = \mathbf{X}\vec{a} = d, \tag{13.39}$$

where each row vector product evaluates $f(x)$ at one point. For example, for a cubic equation fit to four data points,

$$\begin{pmatrix} 1 & x_1 & x_1^2 & x_1^3 \\ 1 & x_2 & x_2^2 & x_2^3 \\ 1 & x_3 & x_3^2 & x_3^3 \\ 1 & x_4 & x_4^2 & x_4^3 \end{pmatrix} \begin{pmatrix} a_0 \\ a_1 \\ a_2 \\ a_3 \end{pmatrix} = \begin{pmatrix} d_0 \\ d_1 \\ d_2 \\ d_3 \end{pmatrix} = \mathbf{X}\vec{a}. \quad (13.40)$$

The solution is found by operating on the data with the matrix inverse of $\mathbf{X}$,

$$\mathbf{X}^{-1}\vec{d} = \vec{a} = \begin{pmatrix} a_0 & a_1 & a_2 & a_3 \end{pmatrix}^T, \quad (13.41)$$

yielding the polynomial coefficients. Provided all the x_n are unique, the solution is exact and unique and passes through all the data. If $M + 1 > N$, the *pseudoinverse* $\mathbf{X}_p^{-1}$:

$$\mathbf{X}_p^{-1} = (\mathbf{X}^T \cdot \mathbf{X})^{-1} \cdot \mathbf{X}^T \quad (13.42)$$

provides the least squares best-fit solution. Figure 13.20 shows an example of fits of the form

$$f(\theta) = a_0 + a_2\theta^2 + a_4\theta^4 \quad (13.43)$$

to the four *Fresnel amplitude coefficients*. The coefficients a_0, a_2, and a_4 have been adjusted to minimize the area between the amplitude coefficients and the fit curves. All of these fits cross the exact functions three times. Unlike the Taylor series, the values do not match at the origin. Polynomial fits often provide superior approximations over Taylor series, but the method should be chosen to fit the problem.

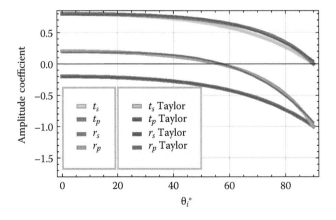

Figure 13.20 Comparison between Fresnel amplitude coefficients and the fourth-order polynomial fits over the range $0° \leq \theta \leq 90°$.

13.4 Contributions to Wavefront Aberrations

Thin films contribute to the *wavefront aberration*, *apodization*, and *polarization aberrations* of optical systems. Evaluating these contributions has been one of the main rationales for adding polarization ray tracing capabilities to *optical design programs*. Consider Figure 13.21, which shows a lens with films drawn in purple on the entering and exiting surfaces, with thicknesses exaggerated for clarity. Note how the films can be considered as meniscus lenses attached to the main lens. Equal radii meniscus lenses have small but non-zero power from the lensmaker's equation. The equations of *paraxial optics* could be used to calculate the effect of such thin lenses on the focal length and the longitudinal *chromatic aberration* of the lens. Would such an algorithm yield the proper correction to the focal length, particularly considering the multiple reflections that occur within the films? A better approach is to consider the lens, the two films, and their paraxial behavior as quadratic phase changes, which can be expressed as

$$\Phi(\theta) \approx \Phi_2 \theta^2. \tag{13.44}$$

Φ_2 characterizes the quadratic phase change Φ in radians per radian of angle of incidence θ.

Next, consider the *defocus* and chromatic aberration from a reflecting film, the 18-layer *reflection-enhancing film* described in Section 13.3.3 and graphed in Figure 13.14. To see the defocus contribution at the reference wavelength of 550 nm, Figure 13.22 shows the *s*- and *p*-phases with the average phase shifts (purple) with a quadratic fit to the average phase for small angles (green) where the average phase is seen to be well fit to a quadratic for angles less than 30°. This quadratic contribution indicates the defocus wavefront aberration the coating will contribute when coated on an on-axis mirror and illuminated on-axis. The magnitude of the defocus will depend on the quadratic evaluated for the marginal ray angle of incidence. This defocus varies with wavelength and thus introduces a small longitudinal chromatic aberration. In the regions of the spectrum (Figure 13.15, black) where the reflectance is low or changing rapidly, such as from 475 to 490 nm, the constructive and destructive interference between the partial waves changes rapidly with angle, and the phase changes and thus coating-induced aberrations tend to be larger. Figure 13.23 shows the *s*, *p*, and average phase shifts for small angles (0° to 30°) over this spectral range. At 475 nm, the reflectance is very low and the average quadratic phase variation (magenta) is negative. By 480 nm, the average quadratic phase is positive and large, and decreases from 480 to 500 nm, which is in the high reflectance bandpass.

Figure 13.21 The thin films (purple) on a glass lens, shown here with exaggerated thickness, act as very weak lenses in combination with the glass lens.

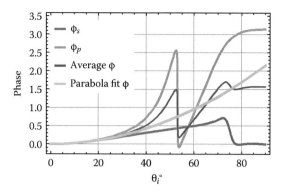

Figure 13.22 For the 18-layer enhanced reflection coating at its reference wavelength of 550 nm, the *s*-phase (orange) and *p*-phase change (red) are approximately quadratic and nearly equal for small angles. The average phase (purple) is well fit by a quadratic (green) for angles less than 30°, indicating that the coating will contribute defocus when illuminated on-axis on a radially symmetric mirror.

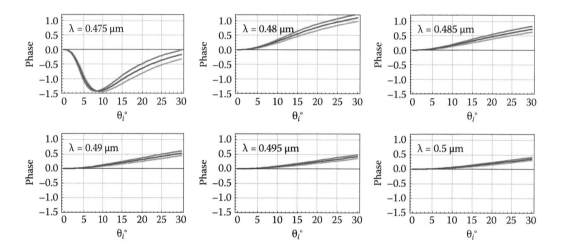

Figure 13.23 The *s* (orange), *p* (red), and average phase change (purple) for the spectral region from 475 to 500 nm, where the reflectivity is changing rapidly. From 480 to 500 nm, the nearly quadratic variation is seen to steadily decrease. This change causes chromatic aberration. At 475 nm where the reflectivity is very low, the phase variation has significant higher-order terms.

The higher-order residual at 475 nm is a contribution to the *spherical aberration* and *higher-order aberrations* from the coating. Figure 13.24 plots the average phase for 0, 0.2, and 0.4 radians as a function of wavelength; the closer the curves, the less defocus will be introduced.

Simple algorithms can evaluate the quadratic phase magnitude coefficient. For example, using *finite differences* to evaluate the *s*- and *p*-phases at 0° and θ_0, a small angle of incidence, the quadratic phase change coefficient, Φ_2 (Equation 13.44), is approximated as

$$\Phi_2 = \left[\frac{\phi_p(\theta_0) + \phi_s(\theta_0)}{2} - \phi(0)\right] \bigg/ \theta_0^2. \tag{13.45}$$

13.5 Phase Discontinuities

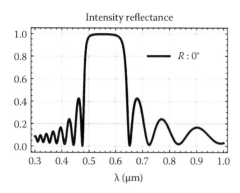
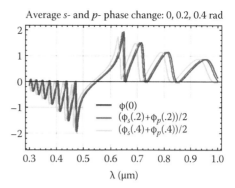

Figure 13.24 (Left) Intensity reflectance of the 18-layer enhanced reflection coating. (Right) The average phase change on reflection versus wavelength at 0 (purple), 0.2 (blue), and 0.4 (green) radians. The 0 and 0.2 curves are nearly equal. Where the three curves intersect, such as near 520 and 540 nm, the coating's defocus contribution is zero.

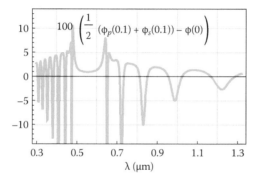

Figure 13.25 Variation of the quadratic phase coefficient, Φ_2, for the 18-layer reflection-enhancing coating of Figure 13.15 indicates the relative amount of defocus at different wavelengths, that is, chromatic aberration. Units are radians of defocus wavefront aberration per radian of marginal ray angle at the coating.

The variation of this defocus contribution is plotted in Figure 13.25 using $\theta_0 = 0.1$ radians. The corresponding longitudinal *chromatic aberration* is seen to be highly wavelength dependent. Thus, Figure 13.24 (right) contains information for the calculation of both the *retardance aberration* contribution from the thin film and the *wavefront aberration* contribution.

Thus, it is seen how contributions of thin films to lens power and wavefront aberration can be calculated from their amplitude coefficients. Usually, the contribution of the thin films to the lens power is very small. There are circumstances where these contributions should be checked, particularly where large numbers of partial waves are significant, or if coatings are used in angular or spectral regions where large oscillations in amplitude occur.

13.5 Phase Discontinuities

Abrupt jumps in the phase of *amplitude coefficients* are another issue with thin film calculations and their interpretation. The phases of the complex amplitude coefficients are generally returned within the range of $-\pi$ to π by the *arctan* function of Equation 13.26. Physically, as the angle of incidence or wavelength varies, the phases can vary over many waves. Thus, in phase plots,

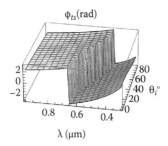 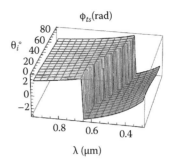

Figure 13.26 The phase changes on transmission through the antireflection coating example show phase discontinuities along a band from 700 nm at grazing incidence to 550 nm at normal incidence. In this case, the *s*-(left) and *p*-(right) phase discontinuities are very close together.

particularly of thick films, one or more 2π *phase discontinuities* may occur in the thin film program output but the discontinuities are not real. The phase discontinuities for the quarter wave MgF_2 as antireflection coating are plotted in Figure 13.26 where phase discontinuities occur, *retardance discontinuities* must also occur. Figure 13.24 shows phase discontinuities around 53°.

These phase discontinuities can prove troublesome for an analysis, such as calculating *absolute phase* or *wavefront aberration*; a *phase unwrapping* operation can be performed separately on the *s*-phase and *p*-phase. The image formation calculations for the *point spread function* and *optical transfer function* described in Chapter 16 require complex valued amplitude and wavefront maps from the exit pupil to be Fourier transformed. In this Fourier transformation operation, the input is an array of complex values in *Cartesian form*, $z = x + i\, y$. Fortunately, this *Fourier transform* operation is not affected by 2π phase jumps in the *exit pupil* function; thus, the unwrapping of phase values is not needed for this important operation.

Interesting phase unwrapping issues regarding thick retarders, similar to the issues of this section, are also a frequent issue, and are treated in Chapter 26.

Example 13.1 Phase Discontinuities in Overcoated Gold

Phase discontinuities in *amplitude coefficient* calculations are a larger problem in thicker coatings, since the *optical path length* of the light varies more with wavelength and angle of incidence. As an example, consider a thick protective dielectric overcoating on *gold*. Gold is an excellent reflector in the infrared. However, since gold is so soft, it cannot be easily cleaned. Thus, gold is frequently overcoated with a hard transparent dielectric to provide a protective and cleanable overcoating. Aluminum oxide, as in *sapphire*, is a preferred choice. Figure 13.27 (top row) plots the *s*-phase (left) and *p*-phase (right) as a function of angle of incidence ($0° < \theta < 90°$) and thickness ($0 < t < 0.250$ μm) and periodic phase changes are seen as the overcoat thickness increases. A phase unwrapping algorithm can unwrap the phase beyond $\pm\pi$, as shown in Chapter 26.

13.6 Conclusion

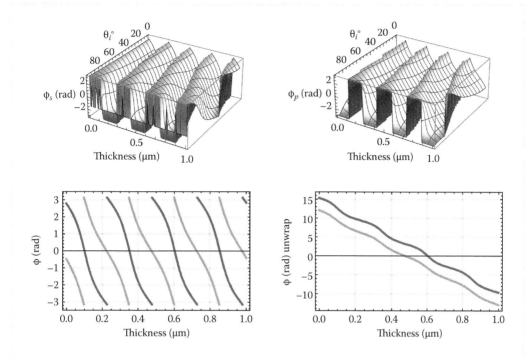

Figure 13.27 (Top row) The phase changes for aluminum oxide Al_2O_3 overcoated on gold shows periodic phase jump artifacts, phase discontinuities of 2π, as a function of thickness. (Bottom row) s-phase (red) and p-phase (orange) at 30° incident angle. (Bottom left) Phase wrapped between $+\pi$ and $-\pi$. (Bottom right) Phase unwrapped about 0.5 μm thickness.

13.6 Conclusion

The equations for single-layer and multilayer thin films have been reviewed and applied to a series of example coatings of particular interest in the polarization analysis of optical systems. The polarization properties are described by *s*- and *p*-amplitude reflection and transmission coefficients, in parallel with the Fresnel coefficients.

Single-layer anti-reflection coatings, such as quarter wave of magnesium fluoride on glass, are seen to both boost transmission and greatly reduce diattenuation, solving two problems at once. Beam splitters that divide the flux equally can be made from thin films of metal, a few nanometers thick, but such single-layer coatings have large differences in their *s*- and *p*-properties, yielding significant diattenuation and retardance. Making non-polarizing beam splitter coatings requires far more complex coatings. *Further, for many vendors and in many catalogs, non-polarizing only means that the s- and p-reflectances and transmittances are equal. It does not always mean that the s- and p-phase changes are equal.* Thus, right circularly polarized light will generally reflect and transmit into beams that are not circular but elliptically polarized, unless more complex, phase matching designs are used.

Efficient reflection coatings can be produced from multilayer quarter wave coatings with alternating high- and low-index layers. Such coatings can have far higher reflectivities than bare metal coatings. Because the reflection from each layer is small, a large number of layers are required. Thus, the light goes deep into these coatings and a very large number of partial waves carry significant flux. Since the light goes so deep into the coatings, these high reflectivity designs are much more angle and polarization sensitive than metal coatings, and can have substantial phase variations that contribute to the wavefront aberration. Thus, thick coatings in general can contribute defocus, chromatic aberration, coma, astigmatism, and other aberrations in measurable amounts, and should be analyzed carefully in systems with tight specifications.

Similar coating designs can be used for polarizing beam splitters. Polarizing beam splitters tend to have far more *p*-polarized light contaminating the transmitted beam than *s*-polarized light contaminating the reflected beam; thus, when polarizing beam splitters are used as polarizers, the reflected beam is generally preferred. Polarizing beam splitters generally have significant phase variations with angle and can also introduce aberration into converging and diverging beams.

These examples are intended to demonstrate methods for analyzing the plots of the magnitudes and phases of the amplitude coefficients to estimate the wavefront aberrations, apodization, diattenuation, and retardance due to coatings. Then, different coatings can be compared, and the coating aberration information can be integrated with other optical design information. The wavefront aberrations of coatings in general may be small, but they are ignored at the optical designer's potential peril, especially now that optical analysis software makes it easier to polarization ray trace.

13.7 Appendix: Derivation of Single-Layer Equations

Consider an incident beam of light with amplitude E_{inc}. The amplitude reflectance at the air/film, film/substrate, and film/air interfaces are the interface Fresnel coefficients r_{01}, r_{12}, and r_{10}. Similarly, the amplitude transmission Fresnel coefficients through the air/film and film/substrate interfaces are t_{01}, t_{12}, and t_{10}. Figure 13.28 shows the first three reflected partial waves. The amplitudes of the reflected partial waves are as follows:

$$\begin{aligned}
E_I &= r_{01} E_{inc} \\
E_{II} &= t_{01} r_{12} t_{10} E_{inc} e^{i\alpha} = t_{01} t_{10} r_{12} e^{i\alpha} E_{inc} \\
E_{III} &= t_{01} r_{12} r_{10} r_{12} t_{10} E_{inc} e^{i2\alpha} = t_{01} t_{10} r_{12}^2 r_{10} e^{i2\alpha} E_{inc} \\
&\vdots \\
E_N &= t_{01} t_{10} r_{12} (r_{12} r_{10} e^{i\alpha})^N E_{inc}.
\end{aligned} \qquad (13.46)$$

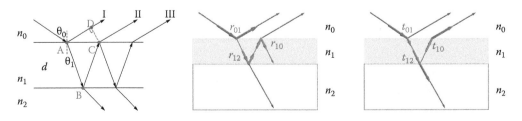

Figure 13.28 (Left) Schematic of a ray refracting and reflecting at a single-layer thin film yielding multiple partial rays. The Fresnel amplitude coefficients of each interface interaction for reflections (middle) and transmissions (right) are shown.

13.7 Appendix: Derivation of Single-Layer Equations

Thus, the amplitudes of the partial waves form a geometrical series (a, $a\,x$, $a\,x^2$, $a\,x^3$, ...) where each successive term is multiplied by $r_{12}\,r_{10}$. Let $\alpha = 2\beta$ be the phase:

$$\alpha = 2\beta = \frac{2\pi}{\lambda}\Delta OPL_{I,II}, \text{ where}$$

$$\Delta OPL_{I,II} = OPL_{ABC} - OPL_{AD}, \text{ where } OPL_{ABC} = 2n_1 \frac{d}{\cos\theta_1} \text{ and } OPL_{AD} = 2n_0 d \tan\theta_1 \sin\theta_0$$

$$\Delta OPL_{I,II} = \frac{2n_1 d - 2n_0 d \sin\theta_1 \sin\theta_0}{\cos\theta_1}, \text{ where } n_0 \sin\theta_0 = n_1 \sin\theta_1$$

$$\Delta OPL_{I,II} = \frac{2n_1 d(1-\sin^2\theta_1)}{\cos\theta_1} = 2n_1 d \cos\theta_1, \tag{13.47}$$

where constructive interference happens when $\Delta OPL_{I,II} = 2n_1 d \cos\theta_1 = m\lambda$. The total reflectance is the sum of all E as $N \to \infty$, which is the sum of the geometrical series

$$a + ax + ax^2 + ax^3 + \ldots = a\sum_{n=0}^{\infty} x^n = \frac{a}{1-x}, \tag{13.48}$$

where $-1 < x < 1$. This must be repeated once for the s- and once for the p-amplitude reflection coefficients,

$$r = r_{01} + t_{01}t_{10}r_{12}e^{i\alpha} \lim_{N\to\infty}\left(1 + r_{12}r_{10}e^{i\alpha} + r_{12}^2 r_{10}^2 e^{i2\alpha} \ldots + r_{12}^{N-2} r_{10}^{N-2} e^{i(N-2)\alpha}\right)$$

$$= r_{01} + t_{01}t_{10}r_{12}e^{i\alpha}\sum_{n=0}^{\infty}\left(r_{12}r_{10}e^{i\alpha}\right)^n = r_{01} + t_{01}t_{10}r_{12}e^{i\alpha}\sum_{n=0}^{\infty}\left(-r_{12}r_{01}e^{i\alpha}\right)^n$$

$$= r_{01} + \frac{t_{01}t_{10}r_{12}e^{i\alpha}}{1-r_{12}r_{10}e^{i\alpha}} = \frac{r_{01} - r_{01}r_{12}r_{10}e^{i\alpha} + t_{01}t_{10}r_{12}e^{i\alpha}}{1-r_{12}r_{10}e^{i\alpha}} \tag{13.49}$$

$$= \frac{r_{01} - r_{12}e^{i\alpha}(r_{01}r_{10} - t_{01}t_{10})}{1-r_{12}r_{10}e^{i\alpha}} = \frac{r_{01} + r_{12}e^{i\alpha}(-r_{01}r_{10} + t_{01}t_{10})}{1-r_{12}r_{10}e^{i\alpha}}$$

$$= \frac{r_{01} + r_{12}e^{i\alpha}(r_{01}r_{01} + t_{01}t_{10})}{1-r_{12}r_{10}e^{i\alpha}} = \frac{r_{01} + r_{12}e^{i\alpha}}{1-r_{12}r_{10}e^{i\alpha}} = \frac{r_{01} + r_{12}e^{i\alpha}}{1+r_{12}r_{01}e^{i\alpha}}.$$

For non-absorbing film, the overall reflected and transmitted flux shown in Figure 13.29 must sum to the incident flux, and the following relations hold:

$$\begin{cases} 1 = r_{01}r_{01} + t_{01}t_{10} \\ 0 = r_{01}t_{01} + t_{01}r_{10} \end{cases}, \text{ then } \begin{cases} 1 = r_{01}^2 + t_{01}t_{10} \\ r_{01} = -r_{10} \end{cases}. \tag{13.50}$$

For a film surrounded by two identical media, $n_0 = n_2$, so $r_{12} = r_{10}$ and the reflection coefficient becomes

$$r = \frac{r_{01} + r_{10}e^{i\alpha}}{1-r_{10}r_{10}e^{i\alpha}} = \frac{r_{01} - r_{01}e^{i\alpha}}{1-r_{01}^2 e^{i\alpha}} = \frac{r_{01}(1-e^{i\beta})}{1-r_{01}^2 e^{i\beta}}. \tag{13.51}$$

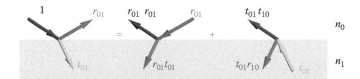

Figure 13.29 (Left) An incident lightray (black arrow) with amplitude 1 reflects (blue arrow) and refracts (green arrow) at the interface with amplitude coefficients r_{01} and t_{01}. (Middle) Light incident into the interface at the opposite direction of the reflected light, with the same amplitude coefficient r_{01}. It refracts (purple) and reflects (black) into the opposite direction as the incident light of the left figure. The two exiting beam amplitude coefficients are altered as the two exiting rays in the left figure. (Right) Similar to the middle figure, A ray is incident in the opposite direction of the refracted light.

Similarly, for the amplitude transmission coefficients, t_s and t_p, the following equation is evaluated twice for s and p:

$$t = t_{01}t_{12}\left(1 + r_{12}r_{10}e^{i\alpha} + r_{12}^2 r_{10}^2 e^{i2\alpha} + \ldots\right) = t_{01}t_{12}\sum_{n=0}^{\infty}(r_{12}r_{10}e^{i\alpha})^n \qquad (13.52)$$

$$= \frac{t_{01}t_{12}}{1 - r_{10}r_{12}e^{i\alpha}} = \frac{t_{01}t_{12}}{1 + r_{01}r_{12}e^{i\alpha}}.$$

13.8 Problem Sets

13.1 Write the characteristic matrix for a quarter wave layer of MgF_2 at 60° and calculate the s- and p-amplitude coefficients.

13.2 Figure 13.6 shows the s- and p-phase change on transmission for a MgF_2 antireflection coating. The optical path lengths and phase changes for the partial waves for each of the two polarizations are equal. Thus, where does the phase change and retardance arise?

13.3 Consider a high-index ($n_1 = 2.895$) film with quarter wave optical thickness (i.e., half the film thickness period) on a low-index ($n_2 = 1.386$) substrate at wavelength $\lambda = 488$ nm. When using these two materials to construct a beam-splitting coating (produce equal power in transmission and reflection), what is the operating angle?

13.4 Find the condition for zero reflectance from a single-layer thin film at normal incidence. Consider Equations 13.13 and 13.14. Set the amplitude reflectivity coefficients equal to 0, $a_s(\theta = 0, n_0, n_1, n_2, t_1) = a_p(\theta = 0, n_0, n_1, n_2, t_1) = 0$ and solve for the thickness t_1 and refractive index n_1 that yield zero reflection at normal incidence.

13.5 For a set of n points, an order $n - 1$ polynomial that passes exactly through all the points can be found. For example, consider five points $(x_0, f(x_0)), (x_1, f(x_1)), \ldots (x_4, f(x_4))$, to be fit to the fourth-order polynomial $f(x) = c_0 + x\,c_1 + x^2\,c_2 + x^3\,c_3 + x^4\,c_4$. The points and coefficients can be related by the matrix equation

$$\begin{pmatrix} m_{00} & m_{01} & m_{02} & m_{03} & m_{04} \\ m_{10} & m_{11} & m_{12} & m_{13} & m_{14} \\ m_{20} & m_{21} & m_{22} & m_{23} & m_{24} \\ m_{30} & m_{31} & m_{32} & m_{33} & m_{34} \\ m_{40} & m_{41} & m_{42} & m_{43} & m_{44} \end{pmatrix} \begin{pmatrix} c_0 \\ c_1 \\ c_2 \\ c_3 \\ c_4 \end{pmatrix} = \begin{pmatrix} f(x_0) \\ f(x_1) \\ f(x_2) \\ f(x_3) \\ f(x_4) \end{pmatrix}.$$

a. Find the matrix coefficients m_{ij} in this equation to calculate the c's from the set of $f(x)$.
b. Show the matrix equation that calculates the polynomial coefficients, $c_0,\ldots$.
c. Provide the equation for fitting three points to a quadratic equation yielding c_0, c_1, and c_2. Provide all nine matrix elements as functions of $(x_1, f(x_1))$, $(x_2, f(x_2))$, and $(x_3, f(x_3))$.

13.6 Continue on the previous problem.
a. Provide the equation for fitting three points to an even-order fourth-order equation yielding c_0, c_2, and c_4. Provide all nine matrix elements as functions of $(x_0, f(x_0))$, $(x_2, f(x_2))$, and $(x_4, f(x_4))$.
b. Fit the amplitude transmission coefficients t_s and t_p for an air–silicon interface for the angles.
c. For which component, t_s and t_p, is the fourth-order term more significant?

References

1. I. J. Hodgkinson, F. Horowitz, H. A. Macleod, M. Sikkens, and J. J. Wharton, Measurement of the principal refractive indices of thin films deposited at oblique incidence, *JOSA A* 2(10) (1985): 1693–1697.
2. I. J. Hodgkinson and Q. H. Wu, *Birefringent Thin Films and Polarizing Elements*, Singapore: World Scientific (1997).
3. R. M. A. Azzam and M. M. K. Howlader, Fourth- and sixth-order polarization aberrations of antireflection-coated optical surfaces, *Opt. Lett.* 26 (2001): 1607–1608.
4. F. Abelès, Researches sur la propagation des ondes électromagnétiques sinusoïdales dans les milieus stratifies. Applications aux couches minces, *Ann. Phys. Paris*, 12ième Series 5 (1950): 596–640.
5. F. Abelès, Researches sur la propagation des ondes électromagnétiques sinusoïdales dans les milieus stratifies. Applications aux couches minces, *Ann. Phys. Paris*, 12ième Series 5 (1950): 706–784.
6. H. A. Macleod, *Thin-Film Optical Filters*, 3rd edition, Taylor & Francis (2001).
7. J. L. Pezzaniti and R. A. Chipman, Angular dependence of polarizing beam-splitter cubes, *Appl. Opt.* 33.10 (1994): 1916–1929.
8. S. M. MacNeille, Beam splitter, U.S. Patent 2,403,731 (1946).

14

Jones Matrix Data Reduction with Pauli Matrices

14.1 Introduction

This chapter develops the algorithm for calculating the *diattenuation* and *retardance* components of *Jones matrices*. These components are analogous to the Stokes parameters, except that they characterize the strength and eigenpolarizations of the diattenuating part and retarding part of the Jones matrix. Expressing Jones matrices as combinations of simple diattenuators and retarders provides a helpful way to increase understanding and ease communication.

The derivation of Jones matrices for various types of retarders, polarizers, and diattenuators is straightforward, and covered in Chapter 5. This chapter addresses the inverse problem; given a Jones matrix, calculate its polarization properties: the *retardance, diattenuation, amplitude*, and *phase*. Jones matrices have four complex matrix elements, each with a real and imaginary part, for a total of eight *degrees of freedom*, listed in Table 14.1. Two degrees of freedom, ρ_0 and ϕ_0, are *non-polarizing*; the incident and exiting polarization states are unchanged by pure *amplitude* or *phase* interactions. Three degrees of freedom describe diattenuation, D_H, D_{45}, and D_L, its strength and eigenpolarizations. The last three degrees of freedom describe retardance, δ_H, δ_{45}, and δ_L, its strength and eigenpolarizations.

In this chapter, the components D_H, D_{45}, D_L, δ_H, δ_{45}, and δ_L for homogeneous matrices are derived in an *order-independent form* through the use of *matrix exponentials* and *matrix logarithms*. The three retardance components can be represented in a three-dimensional *retarder space*. Similarly, the three diattenuation components can be represented in a *diattenuation space*. Overlaying these spaces leads to a simple understanding of *homogeneous* and *inhomogeneous polarization elements* (Section 14.7).

Table 14.1 Classification of the Eight Degrees of Freedom in the Jones Matrix

ρ_0	Amplitude, non-polarizing
ϕ_0	Phase, non-polarizing
D_H	Linear diattenuation, horizontal or vertical component
D_{45}	Linear diattenuation, 45° or 135° component
D_L	Circular diattenuation, left or right component
δ_H	Linear retardance, horizontal or vertical component
δ_{45}	Linear retardance, 45° or 135° component
δ_L	Circular retardance, left or right component

Weak polarization elements with small diattenuation and retardance are also analyzed. The properties of Jones matrices are particularly simple in the neighborhood of the identity matrix. Weak elements deserve particular attention since most light interactions at antireflection-coated lens surfaces and at metal mirror surfaces are relatively weakly polarizing. Optical systems are full of weakly polarizing interactions.

The *Jones vector* and the *Jones matrix* were introduced in Chapters 2 and 4 as a system for polarization calculations for light propagating along the *z*-axis. The incident light is described by a two-element Jones vector, **E** (see Chapter 2). A *polarization element* is described by a Jones matrix **J**, a 2 × 2 matrix of complex elements. The basic Jones matrix equation takes the following form[1]:

$$\mathbf{J} \cdot \mathbf{E} = \mathbf{E}' = \begin{pmatrix} j_{xx} & j_{xy} \\ j_{yx} & j_{yy} \end{pmatrix} \cdot \begin{pmatrix} E_x \\ E_y \end{pmatrix} = \begin{pmatrix} E'_x \\ E'_y \end{pmatrix}. \quad (14.1)$$

Frequently, in polarization research, a sample's Jones matrix is calculated, which leads directly to the question *what polarization properties are associated with the Jones matrix? Is this similar to another sample? Does it have more or less polarization?* Comparison of the matrices on an element-by-element basis can be helpful, but far more meaningful is a comparison on the basis of established polarization properties. *What kinds of diattenuation or retardance does the sample display and in what quantities?* Sections 14.4.5 and 14.6 provide the algorithm for calculating polarization properties in an order-independent way, the diattenuation and retardance are *shuffled together*, as opposed to the retardance occurring *before* the diattenuation due to matrix multiplication or vice versa.

This chapter applies two representations of Jones matrices. The first representation expresses the Jones matrix as the sum of *Pauli matrices*, σ_1, σ_2, and σ_3, and the identity matrix σ_0,

$$\mathbf{J} = \begin{pmatrix} j_{xx} & j_{xy} \\ j_{yx} & j_{yy} \end{pmatrix} = c_0 \begin{pmatrix} 1 & 0 \\ 0 & 1 \end{pmatrix} + c_1 \begin{pmatrix} 1 & 0 \\ 0 & -1 \end{pmatrix} + c_2 \begin{pmatrix} 0 & 1 \\ 1 & 0 \end{pmatrix} + c_3 \begin{pmatrix} 0 & -i \\ i & 0 \end{pmatrix}$$
$$= c_0 \sigma_0 + c_1 \sigma_1 + c_2 \sigma_2 + c_3 \sigma_3, \quad (14.2)$$

where c_0, c_1, c_2, and c_3 are the *complex Pauli coefficients*. The second representation treats the Jones matrix as the exponential of the sum of Pauli matrices,

$$\mathbf{J} = \begin{pmatrix} j_{xx} & j_{xy} \\ j_{yx} & j_{yy} \end{pmatrix} = e^{b_0 \sigma_0 + b_1 \sigma_1 + b_2 \sigma_2 + b_3 \sigma_3}. \quad (14.3)$$

Here b_0, b_1, b_2, and b_3 are the *exponential Pauli coefficients*, which can be calculated by taking the matrix logarithm of **J**

$$\ln(\mathbf{J}) = b_0 \boldsymbol{\sigma}_0 + b_1 \boldsymbol{\sigma}_1 + b_2 \boldsymbol{\sigma}_2 + b_3 \boldsymbol{\sigma}_3. \tag{14.4}$$

Equations 14.3 and 14.4 are special because they help reveal the *structure* of Jones matrices and the relationships of the polarization properties. Throughout this book, diattenuation and retardance have been discussed and analyzed. Equation 14.3 is a natural and unique representation of Jones matrices because it leads to simple expressions for diattenuation and retardance. In physics, such natural and unique representations are termed *canonical forms*.

14.2 Pauli Matrices and Jones Matrices

The Pauli matrices were introduced into quantum mechanics by Wolfgang Pauli to describe the interaction of the angular momentum of electrons and nuclei with external magnetic fields.[2–4] Light is a quantum phenomenon, and even though a quantum formulation is not used here, it is natural that the underlying mathematics of quantum mechanics should appear in the polarization calculus.[5]

14.2.1 Pauli Matrix Identities

The Pauli matrices, $\boldsymbol{\sigma}_1$, $\boldsymbol{\sigma}_2$, and $\boldsymbol{\sigma}_3$, are defined as*,[6]

$$\boldsymbol{\sigma}_1 = \begin{pmatrix} 1 & 0 \\ 0 & -1 \end{pmatrix}, \boldsymbol{\sigma}_2 = \begin{pmatrix} 0 & 1 \\ 1 & 0 \end{pmatrix}, \boldsymbol{\sigma}_3 = \begin{pmatrix} 0 & -i \\ i & 0 \end{pmatrix}. \tag{14.5}$$

These are supplemented with the addition of the 2 × 2 identity matrix, indicated by subscript 0,

$$\boldsymbol{\sigma}_0 = \begin{pmatrix} 1 & 0 \\ 0 & 1 \end{pmatrix}, \tag{14.6}$$

to form a basis for 2 × 2 complex matrices. The general rules for Pauli matrix multiplication are as follows. Let $\alpha = 1$, 2, or 3. The square of each Pauli matrix is the identity matrix,

$$\boldsymbol{\sigma}_\alpha \cdot \boldsymbol{\sigma}_\alpha = \boldsymbol{\sigma}_\alpha^2 = \boldsymbol{\sigma}_0, \tag{14.7}$$

i.e., the Pauli matrices are matrix square roots of the identity matrix. As square roots of the identity matrix, each Pauli matrix is a half wave retarder Jones matrix as listed in Table 14.2.

The matrix multiplication of two Pauli matrices leads to $\pm i$ times the third Pauli matrix. For example, multiplying the first two Pauli matrices yields

$$\boldsymbol{\sigma}_1 \cdot \boldsymbol{\sigma}_2 = \begin{pmatrix} 1 & 0 \\ 0 & -1 \end{pmatrix} \begin{pmatrix} 0 & 1 \\ 1 & 0 \end{pmatrix} = \begin{pmatrix} 0 & 1 \\ -1 & 0 \end{pmatrix} = i \begin{pmatrix} 0 & -i \\ i & 0 \end{pmatrix} = i\boldsymbol{\sigma}_3, \tag{14.8}$$

---

* In quantum mechanics, the following subscript notation for the Pauli matrices is more common:

$$\boldsymbol{\sigma}_x = \begin{pmatrix} 0 & 1 \\ 1 & 0 \end{pmatrix}, \boldsymbol{\sigma}_y = \begin{pmatrix} 0 & -i \\ i & 0 \end{pmatrix}, \boldsymbol{\sigma}_z = \begin{pmatrix} 1 & 0 \\ 0 & -1 \end{pmatrix}.$$

Our subscript notation is chosen to coordinate with the numbering of the Stokes parameters and the labeling of diattenuation and retardance components.

Table 14.2 Pauli Matrices and the Identity Matrix as Polarization Element

σ_0	Identity matrix, non-polarizing, non-absorbing
σ_1	Half wave of linear retardance between 0° and 90° light
σ_2	Half wave of linear retardance between 45° and 135° light
σ_3	Half wave of circular retardance between left and right light

while reversing the order of matrix multiplication yields

$$\sigma_2 \cdot \sigma_1 = \begin{pmatrix} 0 & 1 \\ 1 & 0 \end{pmatrix} \begin{pmatrix} 1 & 0 \\ 0 & -1 \end{pmatrix} = \begin{pmatrix} 0 & -1 \\ 1 & 0 \end{pmatrix} = -i \begin{pmatrix} 0 & -i \\ i & 0 \end{pmatrix} = -i\sigma_3. \quad (14.9)$$

Let (α, β, γ) be an *even permutation* of $(1, 2, 3)$, either $(1, 2, 3)$, $(2, 3, 1)$, or $(3, 1, 2)$. Then, for the even permutations,

$$\sigma_\alpha \cdot \sigma_\beta = i\sigma_\gamma. \quad (14.10)$$

For the *odd permutations*: $(1, 3, 2)$, $(2, 1, 3)$, or $(3, 2, 1)$,

$$\sigma_\alpha \cdot \sigma_\beta = -i\sigma_\gamma. \quad (14.11)$$

Thus, matrix multiplication of Pauli matrices is *anti-commutative*,

$$\sigma_\alpha \cdot \sigma_\beta = -\sigma_\beta \cdot \sigma_\alpha. \quad (14.12)$$

14.2.2 Expansion in a Sum of Pauli Matrices

The set of Jones matrices have a remarkable structure when represented as a sum of the Pauli matrices. The Pauli matrices and identity matrix form a complete basis for the set of 2 × 2 complex matrices, such that any Jones matrix can be expressed as a sum,

$$\mathbf{J} = \begin{pmatrix} j_{xx} & j_{xy} \\ j_{yx} & j_{yy} \end{pmatrix} = c_0 \begin{pmatrix} 1 & 0 \\ 0 & 1 \end{pmatrix} + c_1 \begin{pmatrix} 1 & 0 \\ 0 & -1 \end{pmatrix} + c_2 \begin{pmatrix} 0 & 1 \\ 1 & 0 \end{pmatrix} + c_3 \begin{pmatrix} 0 & -i \\ i & 0 \end{pmatrix},$$

$$= \begin{pmatrix} c_0 + c_1 & c_2 - ic_3 \\ c_2 + ic_3 & c_0 - c_1 \end{pmatrix} = c_0\sigma_0 + c_1\sigma_1 + c_2\sigma_2 + c_3\sigma_3, \quad (14.13)$$

where c_0, c_1, c_2, and c_3 are the *complex Pauli coefficients*,

$$c_0 = \frac{j_{xx} + j_{yy}}{2}, \quad c_1 = \frac{j_{xx} - j_{yy}}{2}, \quad c_2 = \frac{j_{xy} + j_{yx}}{2}, \quad c_3 = \frac{i(j_{xy} - j_{yx})}{2}. \quad (14.14)$$

The complex Pauli coefficients can also be expressed using the trace operator Tr as

$$c_i = \frac{1}{2}\text{Tr}[\mathbf{J} \cdot \sigma_i], \quad (14.15)$$

Polarized Light and Optical Systems

14.2 Pauli Matrices and Jones Matrices

where the trace operator is the sum of the diagonal elements, $\text{Tr}[\mathbf{J}] = j_{xx} + j_{yy}$. c_0 is associated with the identity matrix and thus describes the *polarization independent amplitude change and phase change*.

14.2.3 Pauli Sign Convention

The Pauli matrices are defined so that

$$\boldsymbol{\sigma}_1\boldsymbol{\sigma}_2 = i\boldsymbol{\sigma}_3, \quad \boldsymbol{\sigma}_2\boldsymbol{\sigma}_3 = i\boldsymbol{\sigma}_1, \text{ and } \boldsymbol{\sigma}_3\boldsymbol{\sigma}_1 = i\boldsymbol{\sigma}_2. \tag{14.16}$$

In the decreasing phase convention, positive $\boldsymbol{\sigma}_3$ indicates a left circularly polarized component. Hence, a positive real f indicates left circular (or, in the presence of real $\boldsymbol{\sigma}_1$ and $\boldsymbol{\sigma}_2$ terms, elliptical) diattenuation,

$$\mathbf{J} = \boldsymbol{\sigma}_0 + f\boldsymbol{\sigma}_3 = \begin{pmatrix} 1 & 0 \\ 0 & 1 \end{pmatrix} + f\begin{pmatrix} 0 & -i \\ i & 0 \end{pmatrix}. \tag{14.17}$$

Similarly, a positive imaginary g indicates left circular or elliptical retardance,

$$\mathbf{J} = \boldsymbol{\sigma}_0 + ig\boldsymbol{\sigma}_3 = \begin{pmatrix} 1 & 0 \\ 0 & 1 \end{pmatrix} + ig\begin{pmatrix} 0 & -i \\ i & 0 \end{pmatrix}. \tag{14.18}$$

This is different from our convention for Stokes parameters where a positive S_3 indicates a right circular component. Thus, care and extra minus signs are needed when converting from these Jones matrices to Mueller matrices, as described in Section 6.12.1.

14.2.4 Pauli Coefficients of a Polarization Element Rotated about the Optical Axis

The transformation for Jones matrices rotated about the light propagation direction by angle θ is

$$\mathbf{J}(\theta) = \mathbf{R}(\theta) \cdot \mathbf{J} \cdot \mathbf{R}(-\theta), \tag{14.19}$$

where $\mathbf{R}$ is the two-dimensional *Cartesian rotation matrix*,

$$\mathbf{R}(\theta) = \begin{pmatrix} \cos\theta & -\sin\theta \\ \sin\theta & \cos\theta \end{pmatrix}. \tag{14.20}$$

Consider a Jones matrix expressed as a sum of Pauli matrices, Equation 14.13. $\boldsymbol{\sigma}_0$ and $\boldsymbol{\sigma}_3$ are *invariant under rotation*[*]:

$$\mathbf{R}(\theta) \cdot \boldsymbol{\sigma}_0 \cdot \mathbf{R}(-\theta) = \boldsymbol{\sigma}_0, \tag{14.21}$$

$$\mathbf{R}(\theta) \cdot \boldsymbol{\sigma}_3 \cdot \mathbf{R}(-\theta) = \boldsymbol{\sigma}_3. \tag{14.22}$$

---

[*] See Problem 14.4.

Rotation couples σ_1 and σ_2 into each other,

$$\mathbf{R}(\theta) \cdot \sigma_1 \cdot \mathbf{R}(-\theta) = \sigma_1 \cos 2\theta + \sigma_2 \sin 2\theta, \qquad (14.23)$$

$$\mathbf{R}(\theta) \cdot \sigma_2 \cdot \mathbf{R}(-\theta) = -\sigma_1 \sin 2\theta + \sigma_2 \cos 2\theta. \qquad (14.24)$$

After rotation by the angle θ, the Jones matrix $\mathbf{J}(\theta)$ becomes

$$\begin{aligned}\mathbf{J}' &= \mathbf{R}(\theta) \cdot \mathbf{J} \cdot \mathbf{R}(-\theta) \\ &= c_0 \sigma_0 + (c_1 \cos 2\theta - c_2 \sin 2\theta)\sigma_1 + (c_1 \sin 2\theta + c_2 \cos 2\theta)\sigma_2 + c_3 \sigma_3.\end{aligned} \qquad (14.25)$$

Example 14.1 Rotation of Pauli Coefficients

Consider the Jones matrix $\mathbf{J}_1$ formed from the sequence of a quarter wave right circular retarder, a linear diattenuator with transmission axis at 22.5° and transmissions of 1 and 0.5, and a quarter wave linear retarder with a fast axis at 60°,

$$\begin{aligned}\mathbf{J}_1 &= \mathbf{LR}\left(\frac{\pi}{2}, \frac{\pi}{3}\right) \mathbf{LD}\left(1, \frac{1}{2}, \frac{\pi}{8}\right) \mathbf{CR}\left(\frac{\pi}{4}\right) \\ &= \begin{pmatrix} 0.707 + 0.354\,i & -0.612\,i \\ -0.612\,i & 0.707 - 0.354\,i \end{pmatrix} \begin{pmatrix} 0.927 & 0.177 \\ 0.177 & 0.573 \end{pmatrix} \begin{pmatrix} 0.924 & 0.383 \\ -0.383 & 0.924 \end{pmatrix} \\ &= \begin{pmatrix} 0.558 + 0.313\,i & 0.366 - 0.183\,i \\ -0.04 - 0.463\,i & 0.422 - 0.528\,i \end{pmatrix}.\end{aligned}$$

(14.26)

Then, the Pauli *c-coefficients* are $c_0 = 0.49 - 0.108\,i$, $c_1 = 0.068 + 0.421\,i$, $c_2 = 0.163 - 0.323\,i$, and $c_3 = -0.14 + 0.203\,i$, which are plotted on the Argand (complex) plane in Figure 14.1 (left). Figure 14.1 (right) shows the transformation of the coefficients for a rotating polarization element.

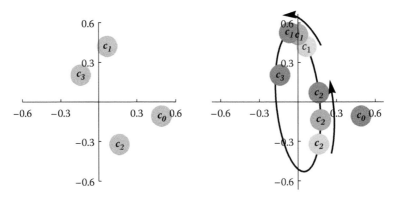

Figure 14.1 (Left) The Pauli *c*-coefficients for matrix $\mathbf{J}_1$ are plotted on the Argand plane. (Right) As the polarization element is rotated about the incident axis, c_1 and c_2 move around an ellipse once per 180° of rotation, while c_0 and c_3 remain fixed. After 45°, c_2 has moved to c_1's initial location, and c_1 has moved to $-c_2$'s initial location. The lighter c_1 and c_2 disks indicate the coefficients without rotation. The medium darkness disks are after 11.25°, and the darkest disks are the coefficients after 22.5° of rotation.

14.2.5 Eigenvalues and Eigenvectors and Matrix Functions for the Pauli Sum Form

Jones matrices, when expressed as a Pauli matrix sum, have a particularly elegant expression for the eigenvalues ξ_q and ξ_r. The *characteristic equation* for the *eigenvalues* is

$$\det(\mathbf{J} - \xi \boldsymbol{\sigma}_0) = \det \begin{pmatrix} c_0 + c_1 - \xi & c_2 - ic_3 \\ c_2 + ic_3 & c_0 - c_1 - \xi \end{pmatrix} = 0 \qquad (14.27)$$
$$= c_0^2 - c_1^2 - c_2^2 - c_3^2 - 2c_0 \xi + \xi^2,$$

a quadratic equation with roots

$$\xi_q, \xi_r = c_0 \pm \sqrt{c_1^2 + c_2^2 + c_3^2}. \qquad (14.28)$$

Note that this squaring operation is c_i^2, not the more common $|c_i|^2$. This simplicity is an indication that this complex Pauli sum representation is fundamental. The two *eigenvectors* (eigenpolarizations), $\mathbf{E}_q$ and $\mathbf{E}_r$, are

$$\mathbf{E}_q = \begin{pmatrix} c_1 + \sqrt{c_1^2 + c_2^2 + c_3^2} \\ c_2 + ic_3 \end{pmatrix}, \quad \mathbf{E}_r = \begin{pmatrix} c_1 - \sqrt{c_1^2 + c_2^2 + c_3^2} \\ c_2 + ic_3 \end{pmatrix}. \qquad (14.29)$$

The *determinant* of a Jones matrix has a similarly compact expression in Pauli sum coefficients,

$$\det(\mathbf{J}) = j_{xx} j_{yy} - j_{xy} j_{yx} = c_0^2 - c_1^2 - c_2^2 - c_3^2, \qquad (14.30)$$

as does the *matrix inverse*

$$\mathbf{J}^{-1} = \frac{1}{\det \mathbf{J}} \begin{pmatrix} j_{yy} & -j_{xy} \\ -j_{yx} & j_{xx} \end{pmatrix} = \frac{c_0 \boldsymbol{\sigma}_0 - c_1 \boldsymbol{\sigma}_1 - c_2 \boldsymbol{\sigma}_2 - c_3 \boldsymbol{\sigma}_3}{\det \mathbf{J}}, \qquad (14.31)$$

and *matrix transpose*

$$\mathbf{J}^T = \begin{pmatrix} j_{xx} & j_{yx} \\ j_{xy} & j_{yy} \end{pmatrix} = c_0 \boldsymbol{\sigma}_0 + c_1 \boldsymbol{\sigma}_1 + c_2 \boldsymbol{\sigma}_2 - c_3 \boldsymbol{\sigma}_3. \qquad (14.32)$$

The *Hermitian adjoint* is

$$\mathbf{J}^\dagger = \begin{pmatrix} j_{xx}^* & j_{yx}^* \\ j_{xy}^* & j_{yy}^* \end{pmatrix} = c_0^* \boldsymbol{\sigma}_0 + c_1^* \boldsymbol{\sigma}_1 + c_2^* \boldsymbol{\sigma}_2 + c_3^* \boldsymbol{\sigma}_3. \qquad (14.33)$$

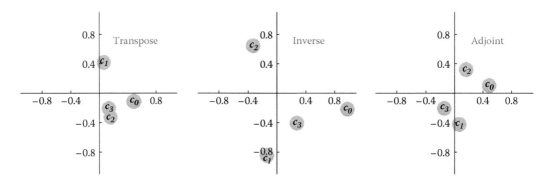

Figure 14.2 Geometrical pictures of operations on the complex Pauli coefficients are shown. (Left) For the transpose, c_3 moves symmetrically across the origin. (Center) The determinant of $\mathbf{J}_1$ is ½, so for the matrix inverse of $\mathbf{J}_1$, first c_1, c_2, and c_3 move symmetrically across the origin, then all four coefficients double their distance from the origin. (Right) For the adjoint of $\mathbf{J}_1$, all of the coefficients move symmetrically across the x-axis.

Figure 14.2 shows how the c-coefficients transform for the example matrix $\mathbf{J}_1$ of Equation 14.26 as the transpose, inverse, and adjoint operations are performed.

$\mathbf{J}$ is a pure diattenuator (hermitian) if $\mathbf{J} = \mathbf{J}^\dagger$. Thus, c_0, c_1, c_2, and c_3 are real for a diattenuator (with no added phase). $\mathbf{J}$ is a pure retarder (unitary) if $\mathbf{J}^{-1} = \mathbf{J}^\dagger$. Thus, c_0 is real and c_1, c_2, and c_3 are imaginary.

The Pauli summation form for matrix powers of Jones matrices includes, for the *matrix square*,

$$\mathbf{J}^2 = \left(c_0^2 + c_1^2 + c_2^2 + c_3^2\right)\boldsymbol{\sigma}_0 + 2c_0c_1\boldsymbol{\sigma}_1 + 2c_0c_2\boldsymbol{\sigma}_2 + 2c_0c_3\boldsymbol{\sigma}_3, \tag{14.34}$$

for the *matrix cube*,

$$\mathbf{J}^3 = c_0\left[c_0^2 + 3\left(c_1^2 + c_2^2 + c_3^2\right)\right]\boldsymbol{\sigma}_0 + \left(3c_0^2 + c_1^2 + c_2^2 + c_3^2\right)(c_1\boldsymbol{\sigma}_1 + c_2\boldsymbol{\sigma}_2 + c_3\boldsymbol{\sigma}_3), \tag{14.35}$$

and for the *matrix square root*,

$$\mathbf{J}^{1/2} = \frac{\Gamma}{2}\boldsymbol{\sigma}_0 + \frac{(c_1\boldsymbol{\sigma}_1 + c_2\boldsymbol{\sigma}_2 + c_3\boldsymbol{\sigma}_3)}{\Gamma}, \tag{14.36}$$

where

$$\Gamma = \sqrt{c_0 + \sqrt{c_1^2 + c_2^2 + c_3^2}} + \sqrt{c_0 - \sqrt{c_1^2 + c_2^2 + c_3^2}}. \tag{14.37}$$

Note that *matrix root* are not unique; Equation 14.36 is the equation for the principal square root, the matrix root that lies between $\mathbf{J}$ and the identity matrix.

14.2.6 Canonical Summation Form

The *Pauli coefficients* are related to the linear and circular forms of diattenuation and retardance associated with $\mathbf{J}$ (Table 14.1). The coefficient of the identity matrix, c_0, is related to non-polarizing behavior. The magnitude $|c_0|$ is related to change in amplitude, the argument $\text{Arg}[c_0]$ to a change in

14.3 Sequences of Polarization Elements

Table 14.3 Relationship of Real, $\mathfrak{Re}$, and Imaginary, $\mathfrak{Im}$, Parts of the Pauli d Coefficients to the Polarization Properties

Abs (c_0)	Amplitude
Arg (c_0)	Phase
$\mathfrak{Re}(f_1)$	Linear diattenuation along the x- and y-axes
$\mathfrak{Im}(f_1)$	Linear retardance along the x- and y-axes
$\mathfrak{Re}(f_2)$	Linear diattenuation along the ±45° axes
$\mathfrak{Im}(f_2)$	Linear retardance along the ±45° axes
$\mathfrak{Re}(f_3)$	Circular diattenuation
$\mathfrak{Im}(f_3)$	Circular retardance

overall phase. The real and imaginary parts of the other coefficients are similarly related to *diattenuation* and *retardance* of the other basis pairs as listed in Table 14.3, when the expansion is written in a form where c_0, the absolute amplitude and phase change, i.e., the polarization-independent parts of the Jones matrix, have been factored out,

$$\mathbf{J} = c_0 \left(\boldsymbol{\sigma}_0 + \frac{c_1\boldsymbol{\sigma}_1 + c_2\boldsymbol{\sigma}_2 + c_3\boldsymbol{\sigma}_3}{c_0} \right) = c_0(\boldsymbol{\sigma}_0 + f_1\boldsymbol{\sigma}_1 + f_2\boldsymbol{\sigma}_2 + f_3\boldsymbol{\sigma}_3). \tag{14.38}$$

The relationship between these f-coefficients and the diattenuation and retardance components is developed in Section 14.4.

14.3 Sequences of Polarization Elements

When polarization elements are cascaded, new forms of polarization, not present in the polarization elements, are generated. For example, a 0° linear retarder followed by a 45° linear retarder has elliptical eigenpolarizations; thus, some circular retardance has been generated by the interaction. This coupling follows from the Pauli matrix identity (Equation 14.8). Conversely, a sequence of a diattenuator followed by a retarder generates a diattenuation component, following from the relations such as $\boldsymbol{\sigma}_1 \cdot i\boldsymbol{\sigma}_2 = -\boldsymbol{\sigma}_3$.

In this section, these polarization couplings are organized and presented with examples. This classification is later applied in polarization ray tracing and polarization aberration theory in Chapter 15 to help simplify the understanding of the polarization properties of optical systems. These couplings are tabulated in Table 14.4. A modified Pauli matrix multiplication table provides a description of the polarization couplings that occur.

Table 14.4 Interactions of Diattenuating and Retarding Properties

	D_H	D_{45}	D_L	δ_H	δ_{45}	δ_L
D_H		δ_L	$-\delta_{45}$		D_L	$-D_{45}$
D_{45}	$-\delta_L$		δ_H	$-D_L$		D_H
D_L	δ_{45}	$-\delta_H$		D_{45}	$-D_H$	
δ_H		D_L	$-D_{45}$		δ_L	$-\delta_{45}$
δ_{45}	$-D_L$		D_H	$-\delta_L$		δ_H
δ_L	D_{45}	$-D_H$		δ_{45}	$-\delta_H$	

Left side: first property; top: second property; cell: additional form of polarization generated.

Example 14.2 Sequences of Two Quarter Wave Linear Retarders Generating a Circular Retardance Component

The properties of a sequence of polarization elements depend on the order the components are encountered. Consider two quarter wave retarders, $\mathbf{LR}(\pi/2, 0)$ and $\mathbf{LR}(\pi/2, \pi/4)$, expressed in Pauli summation form. When $\mathbf{LR}(\pi/2, 0)$ is encountered first, the Jones matrix product is

$$\begin{aligned}\mathbf{LR}(\pi/2,\pi/4)\cdot\mathbf{LR}(\pi/2,0) &= \frac{1}{\sqrt{2}}\begin{pmatrix} 1 & -i \\ -i & 1 \end{pmatrix}\cdot\frac{1}{\sqrt{2}}\begin{pmatrix} 1-i & 0 \\ 0 & 1+i \end{pmatrix} \\ &= \frac{(\boldsymbol{\sigma}_0 - i\boldsymbol{\sigma}_2)}{\sqrt{2}}\cdot\frac{(\boldsymbol{\sigma}_0 - i\boldsymbol{\sigma}_1)}{\sqrt{2}} = \frac{\boldsymbol{\sigma}_0 - i\boldsymbol{\sigma}_1 - i\boldsymbol{\sigma}_2 - i\boldsymbol{\sigma}_3}{2} \\ &= \boldsymbol{\sigma}_0\cos\frac{2\pi}{3} - i\sin\frac{2\pi}{3}\frac{(i\boldsymbol{\sigma}_1 + i\boldsymbol{\sigma}_2 + i\boldsymbol{\sigma}_3)}{2} \\ &= \frac{1}{2}\begin{pmatrix} 1-i & 1-i \\ -1-i & 1+i \end{pmatrix},\end{aligned} \qquad (14.39)$$

which is an elliptical retarder with retardance $\delta = 2\pi/3$. The fast Jones eigenpolarization and corresponding Stokes parameters for the fast axis (eigenpolarization), both unnormalized for simplicity, are

$$\mathbf{E}_{fast} = \begin{pmatrix} (1+i)(1+\sqrt{3}) \\ 2 \end{pmatrix}, \mathbf{S}_{fast} = \left(\sqrt{3}, 1, 1, 1\right). \qquad (14.40)$$

Hence, the combination of two linear retarders in Equation 14.39 can generate a circular retardance component. Reversing the order, when $\mathbf{LR}(\pi/2, \pi/4)$ is encountered first,

$$\begin{aligned}\mathbf{LR}(\pi/2,0)\cdot\mathbf{LR}(\pi/2,\pi/4) &= \frac{(\boldsymbol{\sigma}_0 - i\boldsymbol{\sigma}_1)}{\sqrt{2}}\cdot\frac{(\boldsymbol{\sigma}_0 - i\boldsymbol{\sigma}_2)}{\sqrt{2}} = \frac{\boldsymbol{\sigma}_0 - i\boldsymbol{\sigma}_1 - i\boldsymbol{\sigma}_2 + i\boldsymbol{\sigma}_3}{2} \\ &= \boldsymbol{\sigma}_0\cos\frac{2\pi}{3} - i\sin\frac{2\pi}{3}\frac{(i\boldsymbol{\sigma}_1 + i\boldsymbol{\sigma}_2 - i\boldsymbol{\sigma}_3)}{2}.\end{aligned} \qquad (14.41)$$

The result is also an elliptical retarder with retardance $\delta = 2\pi/3$, but the unnormalized Stokes vector fast axis (eigenpolarization) now has the opposite circular component,

$$\mathbf{S}_{fast} = \left(\sqrt{3}, 1, 1, -1\right). \qquad (14.42)$$

This time, the opposite circular component was generated. In general, when two different forms of polarization are encountered in series, a third form of polarization is introduced. This is most easily seen using the expansion into sums of Pauli matrices. In the first retarder sequence example, Equation 14.39, the term $i\boldsymbol{\sigma}_2\cdot i\boldsymbol{\sigma}_1 = i\boldsymbol{\sigma}_3$ has coupled the combination of horizontal and 45° retardance into circular retardance. When the retarders are reversed, the opposite circular retardance is introduced because $i\boldsymbol{\sigma}_1\cdot i\boldsymbol{\sigma}_2 = -i\boldsymbol{\sigma}_3$.

14.3 Sequences of Polarization Elements

Example 14.3 Sequences of Two Diattenuators Coupling into a Circular Retardance Component

Similarly, a sequence of horizontal diattenuation followed by 45° diattenuation must yield a circular retardance component because $\sigma_1 \cdot \sigma_2 = i\sigma_3$. It is not immediately obvious why non-parallel diattenuation sequences should generate retardance, so consider the example of two linear diattenuators with diattenuation $D = 3/5$. The first has a 0° transmission axis, and the second has a 45° transmission axis. The Jones matrix products in Cartesian form and Pauli coefficient form are

$$\mathbf{J} = \mathbf{LD}(1,1/2,\pi/4) \cdot \mathbf{LD}(1,1/2,0)$$

$$= \frac{1}{4}\begin{pmatrix} 3 & 1 \\ 1 & 3 \end{pmatrix} \cdot \begin{pmatrix} 1 & 0 \\ 0 & 1/2 \end{pmatrix} = \frac{1}{8}\begin{pmatrix} 6 & 2 \\ 1 & 3 \end{pmatrix} \qquad (14.43)$$

$$= \frac{3\sigma_0 + 3\sigma_2}{4} \cdot \frac{3\sigma_0 + 3\sigma_1}{4} = \frac{9\sigma_0 + 3\sigma_1 + 3\sigma_2 + i\sigma_3}{16}.$$

The circular component is clear in the Pauli matrix sum of $\mathbf{J}$ as the term $i\sigma_3/16$. The eigenpolarizations are linearly polarized light oriented at 11.7° and 119.3°; hence, the Jones matrix is inhomogeneous since the eigenpolarizations are not orthogonal (not 90° apart). $\mathbf{J}$ rotates the plane of polarization of linearly polarized light as shown in Figure 14.3 (left). Figure 14.3 (right) shows the angle the light is rotated as a function of the orientation of the incident polarization. This rotation is asymmetric with more rotation toward smaller angles (counterclockwise looking into the beam) than clockwise rotation. The average rotation is −5.3°. Since the counterclockwise rotation dominates, a circular retardance component occurs in the Pauli matrix coefficients.

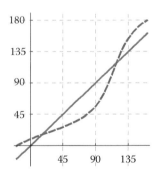 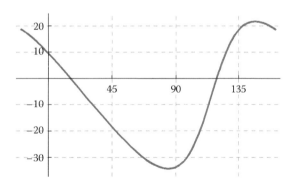

Figure 14.3 (Left) Plane of polarization of light incident on (solid) and exiting the linear diattenuator sequence $\mathbf{J}$. Deviation from the green diagonal line indicates rotation of the plane of polarization. (Right) Rotation of the plane of polarization by $\mathbf{J}$ is asymmetric; the area is larger below the axis than above.

The cascading of Jones matrices in the complex Pauli sum form can also be calculated by matrix multiplication of the coefficients as follows. For example, the product of the three matrices

$$(a_0\boldsymbol{\sigma}_0 + a_1\boldsymbol{\sigma}_1 + a_2\boldsymbol{\sigma}_2 + a_3\boldsymbol{\sigma}_3) = (b_0\boldsymbol{\sigma}_0 + b_1\boldsymbol{\sigma}_1 + b_2\boldsymbol{\sigma}_2 + b_3\boldsymbol{\sigma}_3) \\ \cdot (c_0\boldsymbol{\sigma}_0 + c_1\boldsymbol{\sigma}_1 + c_2\boldsymbol{\sigma}_2 + c_3\boldsymbol{\sigma}_3) \cdot (d_0\boldsymbol{\sigma}_0 + d_1\boldsymbol{\sigma}_1 + d_2\boldsymbol{\sigma}_2 + d_3\boldsymbol{\sigma}_3)$$

(14.44)

has coefficients (a_0, a_1, a_2, a_3), which can be calculated by arranging the first Jones matrix's Pauli sum coefficients in a vector, and the remaining matrices' coefficients are arranged into matrices that embody the rules for Pauli matrix multiplication as follows:

$$\begin{pmatrix} a_0 \\ a_1 \\ a_2 \\ a_3 \end{pmatrix} = \begin{pmatrix} b_0 & b_1 & b_2 & b_3 \\ b_1 & b_0 & -ib_3 & ib_2 \\ b_2 & ib_3 & b_0 & -ib_1 \\ b_3 & -ib_2 & ib_1 & b_0 \end{pmatrix} \cdot \begin{pmatrix} c_0 & c_1 & c_2 & c_3 \\ c_1 & c_0 & -ic_3 & ic_2 \\ c_2 & ic_3 & c_0 & -ic_1 \\ c_3 & -ic_2 & ic_1 & c_0 \end{pmatrix} \cdot \begin{pmatrix} d_0 \\ d_1 \\ d_2 \\ d_3 \end{pmatrix}.$$

(14.45)

14.4 Exponentiation and Logarithms of Matrices

14.4.1 Exponentiation of Matrices

The principal task here is to define and calculate properties to the Jones matrix elements in a simple, unique way. It has been shown how representing Jones matrices as the sum of Pauli matrices provides simple forms for the common matrix operations, such as the inverse and adjoint. The *matrix exponential* is an elegant representation of the polarization properties that provides an *order-independent algorithm*. Similarly, the matrix exponential and matrix logarithm provide algorithms for the representation of diattenuation Jones matrices and retarder matrices in terms of their components.

The exponential function e^x is defined in terms of the series expansion

$$e^x = \exp(x) = 1 + x + \frac{x^2}{2!} + \frac{x^3}{3!} + \ldots = \sum_{n=0}^{\infty} \frac{x^n}{n!}.$$

(14.46)

The exponential of a matrix is defined by the same series expansion but applied to a matrix, with the identity matrix substituted for one,

$$e^{\mathbf{M}} = \exp(\mathbf{M}) = \boldsymbol{\sigma}_0 + \mathbf{M} + \frac{\mathbf{M}^2}{2!} + \frac{\mathbf{M}^3}{3!} + \ldots = \sum_{n=0}^{\infty} \frac{\mathbf{M}^n}{n!}.$$

(14.47)

For example, the matrix exponential of the identity matrix is easily calculated because any power of the identity matrix is the identity matrix

$$\boldsymbol{\sigma}_0 = \boldsymbol{\sigma}_0^2 = \boldsymbol{\sigma}_0^3 = \boldsymbol{\sigma}_0^n.$$

(14.48)

Thus, the matrix exponential of the identity matrix is e times the identity matrix,

$$e^{\boldsymbol{\sigma}_0} = \boldsymbol{\sigma}_0 \left(1 + 1 + \frac{1}{2!} + \frac{1}{3!} + \ldots \right) = \boldsymbol{\sigma}_0 \sum_{n=0}^{\infty} \frac{1}{n!} = \begin{pmatrix} e & 0 \\ 0 & e \end{pmatrix}.$$

(14.49)

14.4 Exponentiation and Logarithms of Matrices

Consider Jones matrices expressed as the matrix exponential of the sum of Pauli matrices, as this provides relationships corresponding to the eight Jones matrix degrees of freedom (Table 14.1),

$$\mathbf{J} = \begin{pmatrix} j_{xx} & j_{xy} \\ j_{yx} & j_{yy} \end{pmatrix} = e^{b_0 \sigma_0 + b_1 \sigma_1 + b_2 \sigma_2 + b_3 \sigma_3}. \tag{14.50}$$

The coefficients b_0, b_1, b_2, and b_3 are the exponential Pauli coefficients of Equation 14.3.

Although Equation 14.47 defines the matrix exponential, in general, this is not the best algorithm to calculate the matrix exponential, due to the rate of convergence for some matrices as well as issues of the accumulation of numerical roundoff errors. The calculation of matrix exponentials has many potential pitfalls beyond the scope of this discussion. Anyone programming their own matrix exponential function should consult the extensive literature.[7–11]

14.4.2 Logarithms of Matrices

To apply the matrix exponential formalism to a Jones matrix $\mathbf{J}$, it is necessary to find the *matrix logarithm*, $\mathbf{M}$, which, when exponentiated (Equation 14.47), equals $\mathbf{J}$. The logarithm function is defined as the inverse of the exponential function. Hence, when a matrix $\mathbf{K}$ is equal to the matrix exponential of the matrix $\mathbf{M}$, $\mathbf{K} = e^{\mathbf{M}}$, then $\mathbf{M}$ is the matrix logarithm of $\mathbf{K}$, $\mathbf{M} = \ln(\mathbf{K})$.

The scalar logarithm function, plotted in Figure 14.4, has a singularity at $x = 0$, so the logarithm cannot be usefully expanded about the origin. The most common series expansion for the logarithm is the Taylor series expansion about $x = 1$,

$$\ln(x) = (x-1) - \frac{(x-1)^2}{2} + \frac{(x-1)^3}{3} + \ldots = \sum_{n=1}^{\infty} \frac{(x-1)^n (-1)^{n-1}}{n}. \tag{14.51}$$

The matrix version of Equation 14.51 is an expansion about the identity matrix,

$$\ln(\mathbf{J}) = -\sum_{n=1}^{\infty} \frac{(\sigma_0 - \mathbf{J})^n (-1)^{n-1}}{n}$$

$$= \sigma_0 - \mathbf{J} - \frac{(\sigma_0 - \mathbf{J})^2}{2} + \frac{(\sigma_0 - \mathbf{J})^3}{3} - \frac{(\sigma_0 - \mathbf{J})^4}{4} + \ldots \tag{14.52}$$

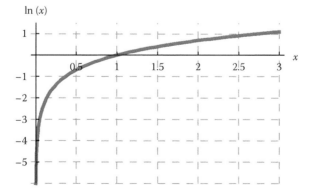

Figure 14.4 $\ln(x)$ has singularity at $x = 0$. The standard Taylor series for $\ln(x)$ is performed about $x = 1$.

Equation 14.52 can be used to evaluate the matrix logarithm and calculate b_0, b_1, b_2, and b_3. In practice, Equation 14.52 converges very slowly.[12] The denominator increases slowly (compare to Equation 14.46), and the terms alternate in sign. Also, Equation 14.52 does not converge for all Jones matrices. For example, for pure retarders, Equation 14.52 only converges for $|\delta| < 2\pi/3$. Equation 14.52 may define the matrix logarithm, but computationally, it is not the best algorithm for matrix logarithm computation.

Details on algorithms for the general calculation of matrix logarithms are beyond the scope of this book and involve many complications.[13,14] Many clever matrix logarithm algorithms have been developed, and one of the best is

$$\ln(\mathbf{M}, K, n) \cong 2^{n+1} \sum_{k=0}^{K} \frac{\left[\left(\mathbf{M}^{\frac{1}{n}} - \boldsymbol{\sigma}_0\right)\left(\mathbf{M}^{\frac{1}{n}} + \boldsymbol{\sigma}_0\right)^{-1}\right]^{2k+1}}{2k+1}. \tag{14.53}$$

Excellent convergence is usually obtained with $K > 4$ and $n > 4$.

14.4.3 Retarder Matrices

The exponentiation of the Pauli matrices generates the matrix forms for all the diattenuators and retarders. First, imaginary valued coefficients $i\Delta$ are considered and then real coefficients α are considered in the next section.

The exponentiation of $\boldsymbol{\sigma}_1$ with an imaginary coefficient $i\Delta$ yields a wonderful form,

$$\begin{aligned}
e^{i\Delta\boldsymbol{\sigma}_1} &= \sum_{n=0}^{\infty} \frac{(i\Delta\boldsymbol{\sigma}_1)^n}{n!} = \boldsymbol{\sigma}_0 + i\Delta\boldsymbol{\sigma}_1 + \frac{(i\Delta\boldsymbol{\sigma}_1)^2}{2!} + \frac{(i\Delta\boldsymbol{\sigma}_1)^3}{3!} + \ldots \\
&= \boldsymbol{\sigma}_0\left(1 - \frac{\Delta^2}{2!} + \frac{\Delta^4}{4!} + \ldots\right) + i\boldsymbol{\sigma}_1\left(\Delta - \frac{\Delta^3}{3!} + \frac{\Delta^5}{5!} + \ldots\right) \\
&= \begin{pmatrix} \cos\Delta + i\sin\Delta & 0 \\ 0 & \cos\Delta - i\sin\Delta \end{pmatrix} = \begin{pmatrix} e^{i\Delta} & 0 \\ 0 & e^{-i\Delta} \end{pmatrix} \\
&= \boldsymbol{\sigma}_0 \cos\Delta + i\boldsymbol{\sigma}_1 \sin\Delta = \mathbf{LR}(\Delta/2, 0).
\end{aligned} \tag{14.54}$$

This is the Jones matrix for a retarder $\mathbf{LR}(\Delta/2, 0)$ with a vertical fast axis and retardance $\delta = \Delta/2$. Thus, a retarder with a 0° ($\delta > 0$) or a 90° ($\delta < 0$) fast axis has the exponential form

$$\mathbf{LR}(\delta_H, 0) = \exp\left(\frac{-i\delta_H}{2}\boldsymbol{\sigma}_1\right) = \boldsymbol{\sigma}_0 \cos\frac{\delta_H}{2} - i\boldsymbol{\sigma}_1 \sin\frac{\delta_H}{2}. \tag{14.55}$$

Similarly, the exponential Jones matrices for 45° retarders $\mathbf{LR}(\delta_{45}, 0)$ and left circular retarders $\mathbf{LR}(\delta_L, \pi)$ are generated by the matrix exponentials of the other two Pauli matrices:

$$\mathbf{LR}(\delta_{45}, 0) = \exp\left(\frac{-i\delta_{45}}{2}\boldsymbol{\sigma}_2\right) = \boldsymbol{\sigma}_0 \cos\frac{\delta_{45}}{2} - i\boldsymbol{\sigma}_2 \sin\frac{\delta_{45}}{2}, \tag{14.56}$$

$$\mathbf{CR}(\delta_L, \pi) = \exp\left(\frac{-i\delta_L}{2}\boldsymbol{\sigma}_3\right) = \boldsymbol{\sigma}_0 \cos\frac{\delta_L}{2} - i\boldsymbol{\sigma}_3 \sin\frac{\delta_L}{2}. \tag{14.57}$$

14.4 Exponentiation and Logarithms of Matrices

Combining Equations 14.55, 14.56, and 14.57, all retarder Jones matrices $\mathbf{ER}(\delta_H, \delta_{45}, \delta_L)$ can be expressed as the matrix exponential of a sum of Pauli spin matrices with purely imaginary Pauli exponential coefficients. This matrix exponential can be manipulated into the trigonometric form shown on the second line,*

$$\mathbf{ER}(\delta_H, \delta_{45}, \delta_L) = e^{-i(\delta_H \sigma_1 + \delta_{45}\sigma_2 + \delta_L \sigma_3)/2}$$
$$= \sigma_0 \cos\left(\frac{\delta}{2}\right) - i\sin\left(\frac{\delta}{2}\right)\left(\frac{\delta_H \sigma_1 + \delta_{45}\sigma_2 + \delta_L \sigma_3}{\delta}\right), \quad (14.58)$$

where the retardance magnitude δ is

$$\delta = \sqrt{\delta_H^2 + \delta_{45}^2 + \delta_L^2}. \quad (14.59)$$

This retarder Jones matrix is in the symmetric phase convention (Section 5.6.1). The minus sign is due to the decreasing phase convention. Thus, for any unitary matrix $\mathbf{U}$, the retardance components (Table 14.1) are calculated from the matrix logarithm of $\mathbf{U}$,

$$2i \ln(\mathbf{U}) = \phi\, \sigma_0 + \delta_H \sigma_1 + \delta_{45}\sigma_2 + \delta_L \sigma_3, \quad (14.60)$$

where ϕ describes absolute phase, which may be present. Since the exponential of the sum of Pauli matrices and retardance coefficients yields the elliptical retarder Jones matrix, then taking the matrix logarithm of an ideal elliptical retarder Jones matrix,

$$\ln\left[\mathbf{ER}(\delta_H, \delta_{45}, \delta_L)\right] = \ln\left[e^{-i(\delta_H \sigma_1 + \delta_{45}\sigma_2 + \delta_L \sigma_3)/2}\right] = -i\frac{(\delta_H \sigma_1 + \delta_{45}\sigma_2 + \delta_L \sigma_3)}{2}, \quad (14.61)$$

returns an expression with the retardance components.

14.4.4 Diattenuator Matrices

Next, the exponential form for diattenuators is analyzed. Consider the exponentiation of Pauli matrices with real coefficients. The exponentiation of $\alpha_1 \sigma_1$ yields

$$\exp(\alpha_1 \sigma_1) = \begin{pmatrix} \exp(\alpha_1) & 0 \\ 0 & \exp(-\alpha_1) \end{pmatrix}$$
$$= \begin{pmatrix} \cosh\alpha_1 + \sinh\alpha_1 & 0 \\ 0 & \cosh\alpha_1 - \sinh\alpha_1 \end{pmatrix} = \sigma_0 \cosh\alpha_1 + \sigma_1 \sinh\alpha_1. \quad (14.62)$$

This is the general form for horizontal diattenuators. The amplitude transmittances, the eigenvalues, associated with the eigenpolarizations of $\exp(\alpha_1 \sigma_1)$, t_{max} and t_{min}, are

$$t_{max} = \cosh\alpha_1 + |\sinh\alpha_1|, \quad t_{min} = \cosh\alpha_1 - |\sinh\alpha_1|. \quad (14.63)$$

---
* See Problem 14.11.

The intensity transmittances, T_{max} and T_{min}, are the squares of the eigenvalues, the diagonal elements,

$$T_{max} = (\cosh\alpha_1 + \sinh\alpha_1)^2, \quad T_{min} = (\cosh\alpha_1 - \sinh\alpha_1)^2. \tag{14.64}$$

This leads to an expression for the diattenuation D,

$$D = \frac{T_{max} - T_{min}}{T_{max} + T_{min}} = \frac{2\cosh\alpha_1 \sinh\alpha_1}{2(\cosh^2\alpha_1 + \sinh^2\alpha_1)} = \frac{1}{2}\tanh(2\alpha_1). \tag{14.65}$$

Thus, the exponential form for a horizontal diattenuator, an "elliptical" diattenuator with a horizontal component, $\mathbf{ED}(D_H, 0, 0)$, expressed in terms of diattenuation component D_H is

$$\begin{aligned}\mathbf{ED}(D_H, 0, 0) &= \exp\left(\frac{\text{arctanh}\, D_H}{2}\boldsymbol{\sigma}_1\right) \\ &= \boldsymbol{\sigma}_0 \cosh\left(\frac{\text{arctanh}\, D_H}{2}\right) + \boldsymbol{\sigma}_1 \sinh\left(\frac{\text{arctanh}\, D_H}{2}\right).\end{aligned} \tag{14.66}$$

The average intensity transmittance, T_U, for $\mathbf{ED}(D_H, 0, 0)$, the transmittance for unpolarized light, is

$$T_U = \frac{(\cosh\alpha_1)^2 + (\sinh\alpha_1)^2}{2}, \tag{14.67}$$

which is greater than or equal to one; hence, the matrix $\mathbf{ED}$ will need to be multiplied by a constant to yield the desired transmittance for unpolarized light, which will be between zero and one. General elliptical diattenuators, $\mathbf{ED}(D_H, D_{45}, D_L)$, defined in terms of three real diattenuation components D_H, D_{45}, D_L, as is described in Section 5.7.2, with diattenuation magnitude

$$D = \sqrt{D_H^2 + D_{45}^2 + D_L^2}, \quad 0 \leq D \leq 1 \tag{14.68}$$

are obtained by the matrix exponential equation exponentiating $D_H\boldsymbol{\sigma}_1 + D_{45}\boldsymbol{\sigma}_2 + D_L\boldsymbol{\sigma}_3$

$$\mathbf{ED}(D_H, D_{45}, D_L) = \exp\left[\text{arctanh}(D)\frac{D_H\boldsymbol{\sigma}_1 + D_{45}\boldsymbol{\sigma}_2 + D_L\boldsymbol{\sigma}_3}{2D}\right], \tag{14.69}$$

with the normalizing factor $\text{arctanh}(D)/(2D)$. $\mathbf{ED}(D_H, D_{45}, D_L)$ is then multiplied by any desired amplitude ρ and phase $e^{-i\phi}$, to set the two non-polarizing degrees of freedom, amplitude and phase. For example, if $\mathbf{ED}(D_H, D_{45}, D_L)$ is divided by its larger eigenvalue, then the maximum amplitude transmission is one, and thus $T_{max} = 1$.

As diattenuators approach polarizers, $D \to 1$, the arctanh approaches infinity and Equation 14.69 diverges. Thus, the logarithm of a polarizer Jones matrix is undefined, just as 1/0 is undefined. The square root of a diattenuator Jones matrix is a weaker diattenuator, and as higher-order matrix roots are taken, the root of a diattenuator Jones matrix approaches the identity matrix as the root order approaches infinity. The larger the diattenuation, the more slowly the identity matrix is approached. But the matrix square root of a polarizer, a singular matrix, is the same polarizer matrix. For example, the matrix square root for a 0° linear polarizer is itself,

$$\begin{pmatrix} 1 & 0 \\ 0 & 0 \end{pmatrix} \cdot \begin{pmatrix} 1 & 0 \\ 0 & 0 \end{pmatrix} = \begin{pmatrix} 1 & 0 \\ 0 & 0 \end{pmatrix}; \text{ therefore, } \sqrt{\begin{pmatrix} 1 & 0 \\ 0 & 0 \end{pmatrix}} = \begin{pmatrix} 1 & 0 \\ 0 & 0 \end{pmatrix}. \tag{14.70}$$

14.4 Exponentiation and Logarithms of Matrices

Thus, the matrix roots of polarizers do not approach the identity matrix. As a result, the matrix logarithm of a polarizer matrix is undefined, much as the logarithm of zero is negative infinity.

Hence, the set of diattenuator Jones matrices that can be analyzed by the matrix exponential/matrix logarithm method for component analysis is an *open set* that can only approach the set of polarizer matrices. To analyze polarizer matrices, which are *singular matrices*, these matrices need to be *bumped* into nearby diattenuators (such as with a diattenuation of 0.9999). This small translation can be accomplished by evaluating eigenvalues and eigenvectors, changing one of the eigenvalues from 0 to a small number such as 0.0001, and regenerating the matrix as a very strong diattenuator.

14.4.5 Polarization Properties of Homogeneous Jones Matrices

Section 14.4.3 demonstrated how the retardance and eigenstates of unitary matrices can be simply calculated from the matrix logarithm of a unitary matrix. Similarly, Section 14.4.4 showed a similar calculation for a Hermitian matrix's diattenuation and eigenstates. This matrix logarithm algorithm can be extended to analyze the combined diattenuation and retardance of homogeneous Jones matrices, where the eigenpolarizations $\mathbf{E}_q$ and $\mathbf{E}_r$ are orthogonal, $\mathbf{E}_q^\dagger \cdot \mathbf{E}_r = 0$, and will work with minor error for Jones matrices that are nearly homogeneous, $\mathbf{E}_q^\dagger \cdot \mathbf{E}_r \approx 0$ (the diattenuation and retardance eigenstates are nearly identical). Two matrices, $\mathbf{A}$ and $\mathbf{B}$, that share the same eigenvectors will commute; the result of their matrix multiplication is order independent,

$$\mathbf{A} \cdot \mathbf{B} = \mathbf{B} \cdot \mathbf{A}. \tag{14.71}$$

For such commuting matrices, the commutation operation indicated by square braces $[\mathbf{A}, \mathbf{B}]$ is zero,

$$[\mathbf{A}, \mathbf{B}] = \mathbf{A} \cdot \mathbf{B} - \mathbf{B} \cdot \mathbf{A} = 0. \tag{14.72}$$

If $\mathbf{A}$ is a unitary matrix and $\mathbf{B}$ is a Hermitian matrix, then $\mathbf{J}_H = \mathbf{A} \cdot \mathbf{B} = \mathbf{B} \cdot \mathbf{A}$ is a matrix with both diattenuation and retardance and has orthogonal eigenpolarizations; $\mathbf{J}_H$ is a homogeneous Jones matrix.

Now, the algorithm for calculating the components of homogeneous Jones matrices can be assembled from the separate retardance and diattenuation expressions. Combining the results of Sections 14.4.3 and 14.4.4, a Jones matrix $\mathbf{J}$ can be expressed in terms of the eight degrees of freedom (Table 14.1). For the first step in the Jones matrix data reduction, the *identity matrix* is factored out of the exponential and the exponential Pauli coefficients are written in real and imaginary parts,

$$\mathbf{J} = \exp(b_0 \boldsymbol{\sigma}_0) \exp\left[\frac{(d_H - i\delta_H)}{2} \boldsymbol{\sigma}_1 + \frac{(d_{45} - i\delta_{45})}{2} \boldsymbol{\sigma}_2 + \frac{(d_L - i\delta_L)}{2} \boldsymbol{\sigma}_3\right]. \tag{14.73}$$

These coefficients are obtained from the *matrix logarithm* of $\mathbf{J}$ when expressed as a sum of Pauli matrices,

$$\ln(\mathbf{J}) = b_0 \boldsymbol{\sigma}_0 + \frac{(d_H - i\delta_H)}{2} \boldsymbol{\sigma}_1 + \frac{(d_{45} - i\delta_{45})}{2} \boldsymbol{\sigma}_2 + \frac{(d_L - i\delta_L)}{2} \boldsymbol{\sigma}_3. \tag{14.74}$$

To obtain the exponential form of the diattenuation and retardance components, the *Pauli logarithm coefficients* are separated into real and imaginary parts. The imaginary parts are the *retardance components.* The real components need to be scaled by a factor of arctanh(*D*)/2*D* from Equation 14.69 to obtain the *diattenuation components*, so the hermitian diattenuator Jones matrix is

$$\mathbf{J} = \exp(b_0 \boldsymbol{\sigma}_0) \exp\left[\operatorname{arctanh}(D) \frac{D_H \boldsymbol{\sigma}_1 + D_{45} \boldsymbol{\sigma}_2 + D_L \boldsymbol{\sigma}_3}{2D}\right], \tag{14.75}$$

where the *diattenuation D* is

$$D = \sqrt{D_H^2 + D_{45}^2 + D_L^2}. \tag{14.76}$$

The minus signs on the retardance components follow from the *decreasing phase sign convention*. Equation 14.75 generates an arbitrary homogeneous Jones matrix from its polarization properties. Jones matrices in the increasing phase convention would use a plus sign. The second half of the exponential is a *unitary matrix* **U**,

$$\mathbf{U} = \exp\left(\frac{-i\delta_H}{2} \boldsymbol{\sigma}_1 + \frac{-i\delta_{45}}{2} \boldsymbol{\sigma}_2 + \frac{-i\delta_L}{2} \boldsymbol{\sigma}_3\right), \tag{14.77}$$

representing a pure retarder with retardance $\delta = \sqrt{\delta_H^2 + \delta_{45}^2 + \delta_L^2}$. $\mathbf{Sr}_1 = (\delta, \delta_H, \delta_{45}, \delta_L)$ and $\mathbf{Sr}_2 = (\delta, -\delta_H, -\delta_{45}, -\delta_L)$ are the retardance eigenpolarizations expressed as Stokes parameters. This is derived in Section 14.4.3.

All homogeneous Jones matrices can be generated in terms of amplitude, ρ, phase, φ, diattenuation, *D*, and retardance, δ, components from the exponential form

$$\mathbf{J} = \rho e^{-i\phi} \exp\left[\operatorname{arctanh}(D) \frac{(D_H \boldsymbol{\sigma}_1 + D_{45} \boldsymbol{\sigma}_2 + D_L \boldsymbol{\sigma}_3)}{2D} - \frac{i(\delta_H \boldsymbol{\sigma}_1 + \delta_{45} \boldsymbol{\sigma}_2 + \delta_L \boldsymbol{\sigma}_3)}{2}\right]. \tag{14.78}$$

The diattenuation components are restricted to the range

$$D = \sqrt{D_H^2 + D_{45}^2 + D_L^2}, \quad 0 \le D < 1, \tag{14.79}$$

but there is no restriction on the range of the other five components.

Conversely, the polarization properties of a nonsingular but otherwise arbitrary homogeneous Jones matrix expressed in terms of the components of Table 14.1 is found through the matrix logarithm of **J**,

$$\ln(\mathbf{J}) = (\ln(\rho) - i\phi)\boldsymbol{\sigma}_0 + \frac{(d_H - i\delta_H)}{2}\boldsymbol{\sigma}_1 + \frac{(d_{45} - i\delta_{45})}{2}\boldsymbol{\sigma}_2 + \frac{(d_L - i\delta_L)}{2}\boldsymbol{\sigma}_3. \tag{14.80}$$

The real parts of the Pauli logarithm coefficients in the exponential in Equation 14.75 is a *Hermitian matrix* **H**,

$$\mathbf{H} = \exp\left(\operatorname{arctanh}(D) \frac{D_H \boldsymbol{\sigma}_1 + D_{45} \boldsymbol{\sigma}_2 + D_L \boldsymbol{\sigma}_3}{2D}\right), \tag{14.81}$$

14.4 Exponentiation and Logarithms of Matrices

which represents a pure diattenuator; **H** represents the diattenuating part of **J**. The diattenuation of **H** is

$$D = \sqrt{D_H^2 + D_{45}^2 + D_L^2} = \tanh\left(2\sqrt{d_H^2 + d_{45}^2 + d_L^2}\right) = \tanh(2d); \quad (14.82)$$

hence,

$$d = \operatorname{arctanh}\left(\frac{D}{2}\right). \quad (14.83)$$

Sd$_1$ = (D, D_H, D_{45}, D_L) and **Sd**$_2$ = (D, $-D_H$, $-D_{45}$, $-D_L$) are the diattenuation eigenpolarizations expressed as Stokes parameters. This is derived in Section 14.4.4. Finally, the *identity matrix term*

$$c_0 = \exp(b_0) = \rho_0 \exp(-i\phi_0) \quad (14.84)$$

contains the *polarization-independent amplitude change* ρ_0 and the *polarization-independent phase change* ϕ_0. Thus, using the *exponential form of the polarization components*, the amplitude, phase, diattenuation, and retardance of a Jones matrix are characterized by the eight coefficients ϕ_0, ρ_0, D_H, D_{45}, D_L, δ_H, δ_{45}, and δ_L, defined in Equations 14.74 and 14.75. Since the coefficients in these two equations are added, the polarization properties occur in an order-dependent form, unlike the polar decomposition of Section 5.9.3 where either (1) the diattenuation occurs before the retardance or (2) the retardance occurs before the diattenuation. The *exponential form of the polarization components* and Equations 14.74 and 14.75 are explored in greater detail in Sections 14.4.5 and 14.6.

The polarization components of a homogeneous Jones matrix as defined by Equations 14.78 and 14.80 are now in an order-independent form; the polarization properties are mixed together. None of the properties occur before or after the other properties, unlike the polar decomposition of a Jones matrix (**J** = **U** · **H** = **H**′ · **U**) or the singular value decomposition **J** = **W** · **D** · **V**†; both decompositions are discussed in Section 5.9. This is shown in Figure 14.5 where a series of infinitesimal Jones matrices representing a small contribution for each of the eight components of Table 14.1 are arranged in a repeating series whose product is the specified Jones matrix **J**.

Singular matrices, polarizers with a determinant of zero, can be handled with care as described at the end of Section 14.4.4.

Figure 14.5 In the *exponential form of the polarization components*, the amplitude, phase, diattenuation, and retardance of a Jones matrix are characterized by the eight coefficients with eight corresponding Jones matrices. Since none of the polarization effects occur before any of the others, this representation can be pictured as taking each form of polarization, slicing its contribution into many infinitesimal pieces, **J**$_1$, **J**$_2$, ..., **J**$_8$, and shuffling the pieces together into a repeating pattern as shown. Now, none of the effects occur before or after the other effects, since if a few of the components, such as the first **J**$_1$ and **J**$_2$, are moved from the beginning of the sequence to the end, the Jones matrix for the overall sequence barely changes, since these are differential matrices only infinitesimally different from the identity matrix.

Example 14.4 Components of a Homogeneous Jones Matrix

The eight polarization components of the homogeneous Jones matrix $\mathbf{K}$ will be calculated,

$$\mathbf{K} = \begin{pmatrix} 0.4655 - 0.3041i & -0.4826 - 0.4212\,i \\ 0.5744 - 0.2834i & 0.4196 + 0.0482\,i \end{pmatrix}. \tag{14.85}$$

The matrix logarithm of $\mathbf{K}$ is

$$\ln(\mathbf{K}) = \begin{pmatrix} -0.1975 - 0.2618\,i & -0.6803 - 0.6813\,i \\ 0.8905 - 0.3659\,i & -0.3026 + 0.2618\,i \end{pmatrix}. \tag{14.86}$$

The real parts are associated with diattenuation and the imaginary parts are associated with retardance. Expressing $2 \ln(\mathbf{K})$ as a sum of Pauli coefficients, and separating the real and imaginary parts, yields

$$\begin{aligned}\ln(\mathbf{K}) &= -0.25\boldsymbol{\sigma}_0 + (0.105103\boldsymbol{\sigma}_1 + 0.210205\boldsymbol{\sigma}_2 + 0.315308\boldsymbol{\sigma}_3)/2 \\ &\quad + i(-0.5236\boldsymbol{\sigma}_1 - 1.0472\boldsymbol{\sigma}_2 - 1.5708\boldsymbol{\sigma}_3)/2.\end{aligned} \tag{14.87}$$

The imaginary Pauli coefficients are the three retardance coefficients times minus two:

$$(\delta_H, \delta_{45}, \delta_R) = (0.5236, 1.0472, 1.5708) = \left(\frac{\pi}{6}, \frac{\pi}{3}, \frac{\pi}{2}\right). \tag{14.88}$$

Their ratios describe the form of the retardance eigenpolarizations as Stokes parameters. The diattenuation is obtained from twice the real Pauli coefficients as

$$D = \tanh\left(2\sqrt{0.105103^2 + 0.210205^2 + 0.315308^2}\right) = 0.374 \tag{14.89}$$

with the three diattenuation components (Table 14.1)

$$(D_H, D_{45}, D_L) = \frac{1}{10}(1, 2, 3). \tag{14.90}$$

These describe the form of the diattenuation eigenpolarizations as Stokes parameters, yielding a net diattenuation of

$$D = \sqrt{D_H^2 + D_{45}^2 + D_L^2} = \frac{\sqrt{14}}{10}. \tag{14.91}$$

14.4 Exponentiation and Logarithms of Matrices

This Hermitian matrix, without the amplitude term (−0.05),

$$\mathbf{H} = \exp(0.105103\boldsymbol{\sigma}_1 + 0.210205\boldsymbol{\sigma}_2 + 0.315308\boldsymbol{\sigma}_3), \tag{14.92}$$

has amplitude transmittances of 1.2173 and 0.8215. The first real coefficient, −0.5, is associated with $\boldsymbol{\sigma}_0$ and describes the amplitude transmittance of

$$\rho = e^{-0.25} = 0.7788. \tag{14.93}$$

Example 14.5 Retarder Matrix

Consider the unitary matrix that results from the matrix product of a 45° linear retarder with retardance $\delta_1 = \pi/2$, followed by a right circular retarder with retardance $\delta_2 = \pi/3$,

$$\mathbf{CR}(\pi/3) \cdot \mathbf{LR}(\pi/2, 45°) = \frac{1}{2\sqrt{2}} \begin{pmatrix} \sqrt{3} - i & -\sqrt{3}i + 1 \\ -\sqrt{3}i - 1 & \sqrt{3} - i \end{pmatrix}. \tag{14.94}$$

Taking the matrix logarithm yields

$$\ln\left[\mathbf{CR}(\pi/3) \cdot \mathbf{CR}(\pi/2, 45°)\right] = \frac{1}{\sqrt{5}} \begin{pmatrix} -i\tau & (1 - i\sqrt{3})\tau \\ -2(-1)^{1/3}\tau & i\tau \end{pmatrix}, \quad \tau = \arctan\sqrt{\frac{5}{3}}. \tag{14.95}$$

The exponential Pauli coefficients are

$$\frac{i\tau}{\sqrt{5}}\left(0, -1, -\sqrt{3}, 1\right). \tag{14.96}$$

Hence, the three retardance components (Table 14.1) are

$$(\delta_H, \delta_{45}, \delta_L) = \frac{2\tau}{\sqrt{5}}\left(1, \sqrt{3}, -1\right). \tag{14.97}$$

Note how the combination of 45° and circular retardance has generated 0° retardance. Taking the matrix exponential of these terms,

$$\exp\left[\frac{-i}{2}(\delta_H \boldsymbol{\sigma}_1 + \delta_{45}\boldsymbol{\sigma}_2 + \delta_L \boldsymbol{\sigma}_3)\right] = \frac{1}{2\sqrt{2}}\begin{pmatrix} \sqrt{3} - i & -\sqrt{3}i + 1 \\ -\sqrt{3}i - 1 & \sqrt{3} - i \end{pmatrix} \tag{14.98}$$

$$= \mathbf{CR}(\pi/3) \cdot \mathbf{LR}(\pi/2, \pi/4),$$

returns to the original unitary matrix (Equation 14.94).

Example 14.6   Diattenuator Matrix

Consider the Hermitian matrix for the linear diattenuator that transmits ¾ of the amplitude for 22½° linearly polarized light and ¼ of the amplitude for 112½° linearly polarized light,

$$\mathbf{LD}(3/4,1/4,\pi/8) = \frac{1}{8}\begin{pmatrix} 4+\sqrt{2} & \sqrt{2} \\ \sqrt{2} & 4-\sqrt{2} \end{pmatrix}. \tag{14.99}$$

The diattenuation is

$$D = \frac{\left(\frac{3}{4}\right)^2 - \left(\frac{1}{4}\right)^2}{\left(\frac{3}{4}\right)^2 + \left(\frac{1}{4}\right)^2} = \frac{4}{5}. \tag{14.100}$$

Taking the matrix logarithm yields

$$\ln\left[\mathbf{LD}(3/4,1/4,\pi/8)\right] = \begin{pmatrix} \frac{1}{4}\left(-8\ln 2 + \sqrt{2}\ln 3 + \ln 9\right) & \frac{\ln 3}{2\sqrt{2}} \\ \frac{\ln 3}{2\sqrt{2}} & \frac{1}{4}\left(-8\ln 2 - \sqrt{2}\ln 3 + \ln 9\right) \end{pmatrix}$$

$$\approx \begin{pmatrix} -0.449 & 0.388 \\ 0.388 & -1.225 \end{pmatrix}. $$

$$\tag{14.101}$$

The exponential Pauli coefficients are

$$\left(\ln\left(\frac{\sqrt{3}}{4}\right), \frac{\ln 3}{2\sqrt{2}}, \frac{\ln 3}{2\sqrt{2}}, 0\right). \tag{14.102}$$

The first coefficient is associated with $\boldsymbol{\sigma}_0$ and describes the amplitude transmittance of

$$\exp\left[\ln\left(\frac{\sqrt{3}}{4}\right)\right] = \frac{\sqrt{3}}{4}. \tag{14.103}$$

14.5 Elliptical Retarders and the Retarder Space

The three diattenuation components (Table 14.1) are

$$(D_H, D_{45}, D_L) = \left(\frac{2\sqrt{2}}{5}, \frac{2\sqrt{2}}{5}, 0\right), \qquad (14.104)$$

for a net diattenuation of

$$D = \sqrt{D_H^2 + D_{45}^2 + D_L^2} = \frac{4}{5}. \qquad (14.105)$$

Observe that diattenuation at 22.5° generates equal D_H and D_{45} components.

14.5 Elliptical Retarders and the Retarder Space

The description of retarders with three complex exponential Pauli coefficients (Equation 14.58) leads directly to a geometric picture of retarders. Retarders can be represented as points (δ_H, δ_{45}, δ_L) in a three-dimensional *retarder space*, as in Figure 14.6.

Points in the δ_H–δ_{45} plane represent linear retarders while points along the δ_L-axis represent pure circular retarders. In this space, all quarter wave elliptical retarders lie on a sphere of radius $\pi/2$, all half wave retarders on a sphere of radius π, and so on. All Jones matrices on spheres of radius $2\pi n$, where n is an integer, the retarder order, have the identity matrix times a phase as their Jones matrix, as does the point at the origin. The retarder space is similar to the Poincaré sphere except the retardance components are plotted instead of the Stokes parameters. In retarder space, there is no limit on the range of retardance; the value of the retardance components can assume any magnitude.

Retarder Jones matrices have an ambiguity with regard to retarder order n, the integer part or half integer part of the number of waves of retardance. There is a family of retarders that will perform an equivalent transformation on all polarization states. For example, a retarder with retardance $\delta = 0$ leaves all polarization states unchanged, as do retarders with $\delta = 2\pi$, or $n2\pi$;

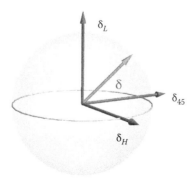

Figure 14.6 A retarder space with δ_H, δ_{45}, and δ_L axes, showing a retarder as a pink arrow from the origin. The distance from the origin is the retardance magnitude δ. Linear retarders lie in the equatorial plane, the plane of the gray circle.

all polarization states are returned to the incident polarization state and their Jones matrices are identical.

As another example, a quarter wave retarder rotates the Poincaré sphere $\pi/2$ radians clockwise about a fast axis. A three-quarter wave retarder with an orthogonal fast axis that rotates the Poincaré sphere $3\pi/2$ radians counterclockwise has the same Jones or Mueller matrix. Physically, the device is different. Compared to a quartz quarter wave retarder, it might also be quartz, rotated 90°, and three times as thick, but as black boxes, both retarders perform identical transformations. As a third example, half wave retarders with orthogonal axes rotate the Poincaré sphere by half a rotation in opposite directions and have the same Jones matrices (Table 5.4). Thus, in general, all Jones matrices with retardance $2\pi(n + \delta)$ and a particular normalized fast axis $(\delta_H, \delta_{45}, \delta_L)$ and all Jones matrices with retardance $2\pi(m - \delta)$ and the orthogonal normalized fast axis $(-\delta_H, -\delta_{45}, -\delta_L)$ have the same Jones matrix (m and n integers).

Figure 14.7 shows two groups of identical retarder Mueller matrices with different absolute phases in the retarder space. Each sphere represents retarders with retardance $n\pi$. The spheres of radius (retardance) $2n\pi$ correspond to multiple wave retarders with identity Mueller matrices. All red points (**A** and **A'**) are identical retarder Mueller matrices and are equally spaced by 2π. Similarly, the green points (**B** and **B'**) show another set of identical Mueller matrices in the retarder space.

For each Mueller matrix with retardance δ, there are a series of Mueller matrices with the retardance $2n\pi + \delta$ with the same fast axes $(\delta_H, \delta_{45}, \delta_L)$ and with the retardance $2n\pi - \delta$ with the orthogonal fast axes $(-\delta_H, -\delta_{45}, -\delta_L)$. The eigenpolarizations $\mathbf{V}_1$ and $\mathbf{V}_2$ of the corresponding Jones matrix, $\mathbf{ER}(\delta_H, \delta_{45}, \delta_L)$, are

$$\mathbf{v}_1 = \frac{1}{\sqrt{2\delta(\delta_H + \delta)}} \begin{pmatrix} \delta_H + \delta \\ \delta_{45} - i\delta_L \end{pmatrix}, \quad \mathbf{v}_2 = \frac{1}{\sqrt{2\delta(\delta_H - \delta)}} \begin{pmatrix} \delta_H - \delta \\ \delta_{45} - i\delta_L \end{pmatrix}, \quad (14.106)$$

when $\delta \neq \delta_H$. These are the fast and slow axes. For the special case, $\mathbf{ED}(\delta_H, 0, 0)$, which has a divide by zero issue in Equation 14.106, the eigenpolarizations become

$$\mathbf{v}_1 = \begin{pmatrix} 1 \\ 0 \end{pmatrix}, \quad \mathbf{v}_2 = \begin{pmatrix} 0 \\ 1 \end{pmatrix}. \quad (14.107)$$

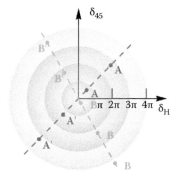

Figure 14.7 Retarder space is a Poincaré sphere-like space that represents retarder fast axes $(\delta_H, \delta_{45}, \delta_L)$ in a three-dimensional space. Two groups of Mueller matrices (**A** and **B**) with the same retardance modulo 2π are shown in the retarder space. The points in each group are separated by 2π and share the same fast and slow axes. Retarders on the one-wave and two-wave retardance spheres have identity matrix Mueller matrices.

14.6 Polarization Properties of Inhomogeneous Jones Matrices

The eigenpolarizations of the corresponding retarder Mueller matrix are more straightforward, with the Stokes parameters, S_1 and S_2,

$$S_1 = \frac{1}{\delta} \begin{pmatrix} \delta \\ \delta_H \\ \delta_{45} \\ \delta_L \end{pmatrix}, \quad S_2 = \frac{1}{\delta} \begin{pmatrix} \delta \\ -\delta_H \\ -\delta_{45} \\ -\delta_L \end{pmatrix}. \tag{14.108}$$

In Equations 14.106 and 14.108, 1 and 2 stand for the fast and slow modes, but which is fast and which is slow is often undetermined from the Jones and Mueller matrices alone.*

Considering the representation of retarders in 3-space, it is tempting to think of a *retarder vector*, but $(\delta_H, \delta_{45}, \delta_L)$ does not behave as a vector. Sequences of retarders do not obey vector addition but instead behave as unit quaternions, which cascade following Equation 14.45.[15,16] In the limit of *very small retardances*, however, when $\delta \ll 1$, the retardance of a sequence of weak retarders is the sum of these "*vectors*." This can be useful in understanding retardance accumulation in lenses and other weakly polarizing optical systems, as is explored in Chapter 15 (Paraxial Polarization Aberrations).

14.6 Polarization Properties of Inhomogeneous Jones Matrices

Unfortunately, the simple matrix exponential and matrix logarithm relationships for homogeneous Jones matrices do not extend to inhomogeneous Jones matrices, due to the noncommutation of their unitary and hermitian parts. Consider the matrix product of the exponential of two matrices, U and H, in terms of the multiplication of the leading terms of their matrix exponential expansions (Equation 14.47),

$$\exp(H) \cdot \exp(U) = \left(\sum_{n=0}^{\infty} \frac{H^n}{n!} \right) \cdot \left(\sum_{n=0}^{\infty} \frac{U^n}{n!} \right)$$

$$\sigma_0 + (H + U) + \left(\frac{H^2}{2!} + H \cdot U + \frac{U^2}{2!} \right) + \left(\frac{H^3}{3!} + \frac{H^2 \cdot U}{2!} + \frac{H \cdot U^2}{2!} + \frac{U^3}{3!} \right) + \dots, \tag{14.109}$$

which contains a number of noncommmuting terms $H \cdot U$, $\dfrac{H^2 \cdot U}{2!}$, $\dfrac{H \cdot U^2}{2!}$, and so on. The form of $\exp(H) \cdot \exp(U)$ was an important result in the development of Lie algebras, a result known as the Baker–Hausdorff–Campbell formula.[17,18] A convenient form of the Baker–Hausdorff–Campbell formula was given by Dynkin,[19,20]

$$\log\bigl(\exp(A) \cdot \exp(B)\bigr) = (A + B) + \frac{1}{2}[A, B] + \frac{1}{12}\bigl([A, [A, B]] + [B, [A, B]]\bigr)$$

$$+ \frac{1}{24}\bigl[B, [A, [A, B]]\bigr] + \dots. \tag{14.110}$$

---

* The orientation and ellipticity of these eigenpolarizations are determined in Problem 14.19.

When **A** and **B** commute, then the product of their matrix exponentials is simple,

$$\exp(\mathbf{A}) \cdot \exp(\mathbf{B}) = \exp(\mathbf{A} + \mathbf{B}). \tag{14.111}$$

This is the form of our matrix exponential expression for homogeneous Jones matrices (Equation 14.78) yielding a simple result for diattenuation and retardance components in terms of the matrix logarithm. The Baker–Hausdorff–Campbell formula indicates that such a simple result for diattenuation and retardance cannot be obtained for the diattenuation and retardance of an inhomogeneous Jones matrix.

Example 14.7 Homogeneous and Inhomogeneous Matrix Examples

Consider applying Equation 14.78 to two examples. First, consider a homogeneous example where $D_H = D = 1/5$ and $\delta_H = \delta = \pi/4$. The Jones matrix $\mathbf{J}_{Homo}$ becomes

$$\mathbf{J}_{Homo} = \exp\left[\operatorname{arctanh}(1/5)\frac{(1/5)\boldsymbol{\sigma}_1/5}{2(1/5)} - \frac{i\pi\boldsymbol{\sigma}_1/4}{2}\right] \approx \exp\left[(0.98698 - 0.39270i)\boldsymbol{\sigma}_1\right]$$

$$= \begin{pmatrix} 0.782542 - 0.782542\,i & 0 \\ 0 & 0.638943 + 0.638943\,i \end{pmatrix}. \tag{14.112}$$

The eigenvalues ξ_1 and ξ_2 lie on the diagonal and have phases $-\pi/8$ and $\pi/8$ for a net retardance of $\pi/4$. The diattenuation

$$D = \frac{T_{max} - T_{min}}{T_{max} + T_{min}} = \frac{|\xi_1|^2 - |\xi_2|^2}{|\xi_1|^2 + |\xi_2|^2} = 0.2 \tag{14.113}$$

is 1/5 as intended.

Next, construct an inhomogeneous example by rotating the retardance from 0° to 45°. Now, the diattenuation is still $D_H = D = 1/5$, but the retardance becomes $\delta_{45} = \delta = \pi/4$. The Jones matrix becomes

$$\mathbf{J}_{Inh} = \exp\left[\operatorname{arctanh}(1/5)\frac{(1/5)\boldsymbol{\sigma}_1/5}{2(1/5)} - \frac{i\pi\boldsymbol{\sigma}_2/4}{2}\right] \approx \exp\left[\begin{pmatrix} 0.101366 & -0.785398\,i \\ -0.785398i & -0.101366 \end{pmatrix}\right]$$

$$= \begin{pmatrix} 0.803161 & -0.708371\,i \\ -0.708371\,i & 0.620311 \end{pmatrix}. \tag{14.114}$$

The maximum and minimum amplitude transmittances are found from the singular valued decomposition's singular values as $\Lambda_1 \approx 1.0956$, $\Lambda_2 \approx 0.912746$. The diattenuation of $\mathbf{J}_{Inh}$ is

$$D = \frac{T_{max} - T_{min}}{T_{max} + T_{min}} = \frac{|\Lambda_1|^2 - |\Lambda_2|^2}{|\Lambda_1|^2 + |\Lambda_2|^2} = 0.1806, \tag{14.115}$$

which does not equal 1/5. Thus, it is seen that Equation 14.78 fails for inhomogeneous examples. The diattenuation magnitude is not correct.

14.7 Diattenuation Space and Inhomogeneous Polarization Elements

Taking the matrix logarithm function does not produce the diattenuation and retardance components for an inhomogeneous Jones matrix because of the order-dependent terms in the Baker–Hausdorff–Campbell relation (Equation 14.114). Hence, there is not a clear algebraic path from the matrix logarithm to the diattenuation and retardance components, but the matrix logarithm can be used as a seed for an optimization algorithm.

An alternative path to diattenuation and retardance components, the polar decomposition (Section 5.9.3), does provide an algorithm to define retardance and diattenuation components,

$$\mathbf{J} = \mathbf{U} \cdot \mathbf{H} = \mathbf{H}' \cdot \mathbf{U}. \tag{14.116}$$

The Jones matrix is separated into hermitian and unitary parts, $\mathbf{U}$ and $\mathbf{H}$ or equivalently $\mathbf{H}'$ and $\mathbf{U}$. These can be analyzed using matrix logarithms to yield retardance and diattenuation components, see Sections 14.4.3 and 14.4.4.

14.7 Diattenuation Space and Inhomogeneous Polarization Elements

Inhomogeneous Jones matrices will be explored and the relationship of the diattenuation components and retardance components will be examined for homogeneous and inhomogeneous Jones matrices. Jones matrices can be classified into two classes, *homogeneous Jones matrices*, with orthogonal eigenpolarizations, and *inhomogeneous Jones matrices*, with non-orthogonal eigenpolarizations, as was introduced in Section 5.4.1.[21,22] Two Jones vectors are orthogonal when $\vec{\mathbf{E}}_q^\dagger \cdot \vec{\mathbf{E}}_r = 0$. When the eigenpolarizations are orthogonal, the corresponding Jones matrix has comparatively simple properties. Orthogonal eigenpolarization states on the surface of the Poincaré sphere are indicated by arrows in Figure 14.8 (left), where the arrows point in opposite directions. For a homogeneous Jones matrix, the eigenpolarizations are also the states of maximum and minimum transmission. Homogeneous polarizers and homogeneous retarders are *named* after the form of the eigenpolarizations. Hence, when $\mathbf{E}_q$ and $\mathbf{E}_r$ are linear polarized states, the $\mathbf{J}$ is a linear element: a linear diattenuator, a linear retarder, or a combination linear diattenuator and retarder. When $\mathbf{E}_q$ and $\mathbf{E}_r$ are circularly polarized states, $\mathbf{J}$ is a circular element. Similarly, when $\mathbf{E}_q$ and $\mathbf{E}_r$ are elliptical polarized states, then $\mathbf{J}$ is an elliptical element.

When the eigenpolarizations are not orthogonal, that is, when

$$\hat{\mathbf{E}}_q^\dagger \cdot \hat{\mathbf{E}}_r = \left(E_{x,q}^* E_{y,q}^* \right) \left(E_{x,r} E_{y,r} \right) = E_{x,q}^* E_{x,r} + E_{y,q}^* E_{y,r} \neq 0, \tag{14.117}$$

then $\mathbf{J}$ is *inhomogeneous*. Figure 14.8 (right) shows the eigenstates and states of maximum and minimum transmission for the example inhomogeneous Jones matrix of Example 14.8. Inhomogeneous

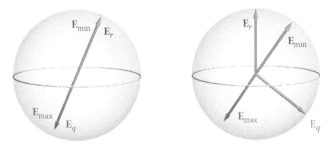

Figure 14.8 (Left) For a homogeneous Jones matrix, the orthogonal eigenpolarization states $\vec{\mathbf{E}}_q$ and $\vec{\mathbf{E}}_r$ point in opposite directions on the Poincaré sphere, and these states are the same as the incident states of maximum and minimum transmission, $\mathbf{E}_{max}$ and $\mathbf{E}_{min}$. (Right) For inhomogeneous Jones matrices, $\vec{\mathbf{E}}_q$ and $\vec{\mathbf{E}}_r$ are not orthogonal and are different from incident states of maximum and minimum transmittance $\mathbf{E}_{max}$ and $\mathbf{E}_{min}$, which are always orthogonal.

Jones matrices have more complex properties than homogenous matrices and cannot be simply classified as linear, circular, or elliptical elements.

Example 14.8 Inhomogeneous Matrix Example

The Jones matrix $\mathbf{J}_1$, which transforms (1, 0) into (1, 0) with eigenvalue 1, and transforms $(1,1)/\sqrt{2}$ into $(1,1)/\sqrt{2}$ with an eigenvalue of 1/3, is

$$\mathbf{J}_I = \begin{pmatrix} 1 & -2/3 \\ 0 & 1/3 \end{pmatrix}. \tag{14.118}$$

These eigenvectors are different from the states of maximum transmission, $\mathbf{E}_{Max} \approx (-0.811, 0.584)$, $\mathbf{E}_{Min} \approx (0.584, 0.811)$, which are orthogonal.

14.7.1 Superposing the Diattenuation and Retardance Spaces

Like the retardance components, the three diattenuation components, (D_H, D_{45}, D_L), can also be plotted in a three-dimensional space, the *diattenuation space*, as in Figure 14.9. Diattenuation space is limited to a radius of one, like the Poincaré sphere. The surface represents polarizers, the interior diattenuators. The center represents the identity matrix, a matrix with zero diattenuation. Linear polarizers lie along the outside in the D_H, D_{45} plane, shown by the gray circle. Circular polarizers are located at the two poles.

When the diattenuation space and retardance space are superposed, as in Figure 14.10, and the retardance and diattenuation components plotted, then for homogeneous Jones matrices, the components point along the same axis, either in the same or in opposite directions. Thus, if the cross products of the diattenuation components and the retardance components is zero, the Jones matrix is homogeneous.

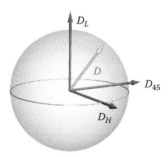

Figure 14.9 The diattenuation space with D_H, D_{45}, and D_L axes lies within a unit sphere. The diattenuation D is the distance from the origin to a set of diattenuation components (D_H, D_{45}, D_L). Linear polarizers lie along the gray circle.

14.8 Weak Polarization Elements

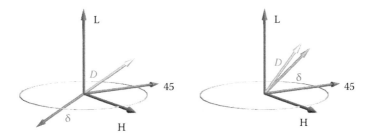

Figure 14.10 After overlapping the retardance space and the diattenuation space, (left) for a homogeneous Jones matrix, the diattenuation components and the retardance components point along the same axis, either in the same or opposite directions. (Right) For an inhomogeneous matrix, the diattenuation components and retardance components are not aligned along the same axis; $(D_H, D_{45}, D_L) \neq k\,(\delta_H, \delta_{45}, \delta_L)$, for any constant k.

14.8 Weak Polarization Elements

Weak polarization elements cause only small changes to the polarization state. For example, the polarization associated with many ray paths through optical systems is weak. Antireflection coatings and reflections at small angles are typically weakly polarizing. In camera lenses and other similar optics, the diattenuation is generally small, $D \ll 1$, as is the retardance $\delta \ll 1$.

Consider the matrix exponential equation (Equation 14.47) for a matrix $\mathbf{M}$, which is close to zero, because the diattenuation components (D_H, D_{45}, D_L) and retardance components $(\delta_H, \delta_{45}, \delta_L)$ are close to zero. If the elements of $\mathbf{M}$ are of the order $\approx 10^{-4}$, the elements $\mathbf{M}^2$ are of the order $\approx 10^{-8}$ (Equation 14.47), and so on, for the higher orders; hence, the higher-order terms are not significant. It is not necessary to assume that the amplitude transmittance ρ_0 and phase change $-\phi_0$ are small parameters, they can assume any value. Thus, for weak polarization elements, the Jones matrix can be approximated by only the first-order term of the matrix exponential series expansion (Equation 14.47),

$$\rho_0 e^{-i\phi_0} \exp(\mathbf{M}) \approx \rho_0 e^{-i\phi_0}(\sigma_0 + \mathbf{M}) \approx \rho_0 e^{-i\phi_0} \begin{pmatrix} 1 + M_{x,x} & M_{x,y} \\ M_{y,x} & 1 + M_{y,y} \end{pmatrix},$$

(14.119)

$$M_{x,x}, M_{x,y}, M_{y,x}, M_{y,y} \approx 0.$$

The corresponding Jones matrix is close to the identity matrix times a complex constant, $\rho_0 e^{-i\phi_0}$. For weak polarization elements, polarization properties are greatly simplified, particularly in the Pauli representation, and can be directly determined from the complex Pauli sum coefficients by the relation, correct to first order,

$$\mathbf{J} \approx \rho_0 e^{-i\phi_0}\left(\sigma_0 + \frac{D_H - i\delta_H}{2}\sigma_1 + \frac{D_{45} - i\delta_{45}}{2}\sigma_2 + \frac{D_L - i\delta_L}{2}\sigma_3\right).$$

(14.120)

14.9 Summary and Conclusion

The elements of the Jones matrix relate the incident x- and y-components of the light to the exiting x- and y-components. Transformations into Pauli matrix sums separate the Jones matrix into components associated with (1) x- and y-oriented diattenuation and retardance, (2) 45°- and 135°-oriented diattenuation and retardance, and (3) the left and right circular diattenuation and retardance. With the Pauli basis, the interactions of polarization effects becomes clear, for example, how x-diattenuation and 45° retardance create a circular diattenuation component and how the circular diattenuation component changes sign when the two elements are reversed. Taking the exponential of a linear combination of Pauli matrices leads directly to an expression for the retardance components and, with a bit of further manipulation, to the diattenuation components. Unlike the polar decomposition, this Pauli matrix exponential provides an order-independent decomposition into polarization properties. Thus, taking the matrix logarithm of a Jones matrix leads to simple algorithms for the retardance and diattenuation components of Jones matrices. The first-order series expansion of the matrix exponential provides equations for weak polarization elements, which will be useful in Chapter 15 (Paraxial Polarization Aberrations).

The purpose of using linear and circular diattenuation and retardance components is to understand the types of retardance and diattenuation associated with polarization matrices and optical systems; in particular, it can facilitate communication between workers, between optical design and metrology departments for example, or between companies, optical system integrators, and vendors as another example. It is not necessary to use this chapter's (D_H, D_{45}, D_L, δ_H, δ_{45}, δ_L, ρ, ϕ) basis to work polarization problems.[23,24] Perhaps certain applications might prefer some other basis, even one where diattenuation and retardance, or hermitian and unitary parts, were mixed rather than separated, if it simplified a problem. Such approaches date back to the dawn of the Jones calculus.[25–27]

The main result of this chapter, the definition of the polarization components of Table 14.1 by matrix exponentials and logarithms (Equations 14.78 and 14.80) presented here, is not unique among all researchers. Similar values for the polarization components can be obtained from the polar decomposition or the singular value decomposition (Section 5.9.3).

The calculation of polarization components of Jones matrices should also be considered in light of the extensive literature on polarization properties and components of Mueller matrices, where the problem is considerably more complicated due to the inclusion of depolarization.[28–50] For example, the retardance components in the Axometrics Mueller matrix polarimeter are obtained from the polar decomposition of Mueller matrices.[51] The particular retardance values obtained are close to the exponential Pauli coefficients, and the spirit of the two definitions is similar. In the case of the polar decomposition, the diattenuation comes before the retardance, which comes before any depolarization of a Mueller matrix. With the exponential Pauli coefficients used here, the properties are mixed such that neither property comes first; they are continuously blended. This exponential Pauli definition is analogous to what Noble et al. have done with Mueller matrix roots to define degrees of freedom in the Mueller matrix.[52,53]

The authors find value in many of the approaches to polarization component definition. The Pauli exponential definition is developed in detail here because it is theoretically elegant and thus canonical as an order-independent definition. The important issue for good scientific communication is to reference parameters derived from polarization matrices back to the definitions used. Then, multiple definitions can exist, they always will, but when necessary, properties can be translated between different sets of definitions.

14.10 Problem Sets

14.1 What are the eigenvalues and eigenvectors of σ_1, σ_2, and σ_3?

14.2 Are σ_1, σ_2, and σ_3 unitary? Are σ_1, σ_2, and σ_3 hermitian?

14.3 Calculate the following functions of the Pauli spin matrices in terms of sums of Pauli spin matrices,
   a. $e^{\alpha \sigma_2}$
   b. $\cos(\beta \sigma_1)$

14.4 A Jones matrix $\mathbf{J}$ is expressed in the form $\exp(a\,\sigma_1 + b\,\sigma_2)$, where a and b can be complex and σ_1 and σ_2 are Pauli matrices.
   a. Convert this matrix into a non-exponential form.
   b. What polarization element does this represent if a and b are real?
   c. What polarization element does this represent if a and b are imaginary?
   d. What is the orientation of the polarization element?
   e. If the element is homogeneous?, what are the relations between the real parts of a and b and the imaginary parts of a and b?
   f. Expand $\mathbf{J}$ to the third order in a Taylor series and show that no order-dependent terms (such as terms of the form $\sigma_1 \sigma_2 = i \sigma_3$) are present.

14.5 Find the complex Pauli coefficients (Equation 14.14) for the following Jones matrices. For the retarders, use the symmetric phase form of the retarder.
   a. LP(0)
   b. LP($\pi/4$)
   c. LP($\pi/2$)
   d. RCP
   e. LCP
   f. LR (π,0)
   g. LR (π,$\pi/4$)
   h. LR (π,$\pi/2$)
   i. LR (π,$\pi/8$)
   j. CR(π)
   k. LR ($\pi/2$,$\pi/4$)
   l. LR ($\pi/2$,$\pi/2$)
   m. LR ($\pi/3$,$\pi/2$)

14.6 Perform the unitary transformation $\mathbf{R}(\pi/2) \cdot \mathbf{J} \cdot \mathbf{R}(\pi/2)$ on each of the Jones matrices of Problem 14.5. For the retarders, use the symmetric phase form of the retarder. Then, find the complex Pauli coefficients.

14.7 Prove the relationship $\mathbf{E}_q^\dagger \cdot \mathbf{E}_r = \left(\mathbf{E}_r^\dagger \cdot \mathbf{E}_q\right)^*$ for the adjoints and complex conjugates of complex two element vectors in the Cartesian form $\mathbf{E}_q = (E_{x,q} - iF_{x,q}, E_{y,q} - iF_{y,q})$, and $\mathbf{E}_r = (E_{x,r} + iF_{x,r}, E_{y,r} + iF_{y,r})$. Repeat the proof with the polar forms $\mathbf{E}_q = \left(\rho_{x,q} e^{-i\phi_{x,q}}, \rho_{y,q} e^{-i\phi_{y,q}}\right)$ and $\mathbf{E}_r = \left(\rho_{x,r} e^{-i\phi_{x,r}}, \rho_{y,r} e^{-i\phi_{y,r}}\right)$.

14.8 Verify the equations for rotation of the Pauli matrices:

$\mathbf{R}(\theta) \cdot \sigma_0 \cdot \mathbf{R}(-\theta) = \sigma_0$, $\qquad \mathbf{R}(\theta) \cdot \sigma_1 \cdot \mathbf{R}(-\theta) = \sigma_1 \cos 2\theta + \sigma_2 \sin 2\theta$,

$\mathbf{R}(\theta) \cdot \sigma_2 \cdot \mathbf{R}(-\theta) = -\sigma_1 \sin 2\theta + \sigma_2 \cos 2\theta$, $\qquad \mathbf{R}(\theta) \cdot \sigma_3 \cdot \mathbf{R}(-\theta) = \sigma_3$.

14.9 Show by matrix multiplication that the matrix inverse of $\mathbf{J} = c_0\boldsymbol{\sigma}_0 + c_1\boldsymbol{\sigma}_1 + c_2\boldsymbol{\sigma}_2 + c_3\boldsymbol{\sigma}_3$ is $\mathbf{J}^{-1} = \dfrac{c_0\boldsymbol{\sigma}_0 - c_1\boldsymbol{\sigma}_1 - c_2\boldsymbol{\sigma}_2 - c_3\boldsymbol{\sigma}_3}{c_0^2 - c_1^2 - c_2^2 - c_3^2}$ for both orders of matrix multiplication: $\mathbf{J}\,\mathbf{J}^{-1}$ and $\mathbf{J}^{-1}\mathbf{J}$.

14.10 Consider the diattenuator Jones matrix $\mathbf{J}_{d1}$ and retarder Jones matrix $\mathbf{J}_{r1}$ where each of the Pauli coefficients of the diattenuator is proportional to the corresponding Pauli coefficient of $\mathbf{J}_{r1}$.

$$\mathbf{J}_{d1} = \begin{pmatrix} 1 & 0 \\ 0 & 1 \end{pmatrix} + 0.1\begin{pmatrix} 1 & 0 \\ 0 & -1 \end{pmatrix} + 0.2\begin{pmatrix} 0 & 1 \\ 1 & 0 \end{pmatrix} + 0.3\begin{pmatrix} 0 & -i \\ i & 0 \end{pmatrix}$$

$$\mathbf{J}_{r1} = \begin{pmatrix} 1 & 0 \\ 0 & 1 \end{pmatrix} + 0.1\,i\begin{pmatrix} 1 & 0 \\ 0 & -1 \end{pmatrix} + 0.2\,i\begin{pmatrix} 0 & 1 \\ 1 & 0 \end{pmatrix} + 0.3\,i\begin{pmatrix} 0 & -i \\ i & 0 \end{pmatrix}$$

a. Are $\mathbf{J}_{d1}$ and $\mathbf{J}_{r1}$ homogeneous?
b. Show that $\mathbf{J}_{d1}$ and $\mathbf{J}_{r1}$ have the same eigenvectors.
c. Show that Jd1 and Jr1 commute, [Jd1, Jr1] = Jd1·Jr1 − Jr1·Jd1 = 0.
d. Now, let

$$\mathbf{J}_{d2} = \begin{pmatrix} 1 & 0 \\ 0 & 1 \end{pmatrix} - 0.2\begin{pmatrix} 1 & 0 \\ 0 & -1 \end{pmatrix} - 0.4\begin{pmatrix} 0 & 1 \\ 1 & 0 \end{pmatrix} - 0.6\begin{pmatrix} 0 & -i \\ i & 0 \end{pmatrix}.$$

Show $[\mathbf{J}_{d1}, \mathbf{J}_{d2}] = 0$ and $[\mathbf{J}_{r1}, \mathbf{J}_{d2}] = 0$.
e. Explain why these matrices all commute.

14.11 A sequence of a horizontal linear retarder followed by a 45° linear retarder generates a component of circular retardance.
a. Calculate the following matrix product. Then, determine the amount of circular retardance from the product,

$$\mathbf{LR}(\delta_{45}, \pi/4) \cdot \mathbf{LR}(\delta_H, 0) = \left(\cos\frac{\delta_{45}}{2}\boldsymbol{\sigma}_0 - i\sin\frac{\delta_{45}}{2}\boldsymbol{\sigma}_2\right)\left(\cos\frac{\delta_H}{2}\boldsymbol{\sigma}_0 - i\sin\frac{\delta_H}{2}\boldsymbol{\sigma}_1\right).$$

b. Plot the amount of circular retardance as a function of δ_H when $\delta_H = \delta_{45}$ and $0 \le \delta_H \le 2\pi$.
c. When $\delta_H = \delta_{45} = \pi$, what are the retardance eigenstates and where do they lie on the Poincaré sphere? What is the total retardance?
d. When $\delta_H = \delta_{45} = \pi/2$, what are the retardance eigenstates and where do they lie on the Poincaré sphere? What is the total retardance?

14.12 Show the following relation for elliptical retarders from Equation 14.58:

$$\sin\left(\frac{\delta_H\boldsymbol{\sigma}_1 + \delta_{45}\boldsymbol{\sigma}_2 + \delta_L\boldsymbol{\sigma}_3}{2}\right) = \sin\left(\frac{\delta}{2}\right)\frac{(\delta_H\boldsymbol{\sigma}_1 + \delta_{45}\boldsymbol{\sigma}_2 + \delta_R\boldsymbol{\sigma}_3)}{\delta}.$$

14.13 Find the Jones matrix for the following weak diattenuators and retarders, and express as a sum of Pauli matrices:
a. J1, a linear retarder with $\delta_1 = 0.02$ radians of retardance and a fast axis $\theta_1 = 0$.
b. J2, a linear retarder with $\delta_2 = 0.04$ radians of retardance and $\theta_2 = 45°$.
c. J_3, a circular diattenuator with amplitude transmissions of $t_R = 1.01$ and $t_L = 0.99$.

14.10 Problem Sets

14.14 Find the commutator $[\mathbf{C}, \mathbf{D}]$ of $\mathbf{C} = c_0\sigma_0 + c_1\sigma_1 + c_2\sigma_2 + c_3\sigma_3$, $\mathbf{D} = d_0\sigma_0 + d_1\sigma_1 + d_2\sigma_2 + d_3\sigma_3$ and express the result as the sum of Pauli matrices, $e_0\sigma_0 + e_1\sigma_1 + e_2\sigma_2 + e_3\sigma_3$.

14.15 Expand $\exp[i(\delta_H\sigma_1 + \delta_{45}\sigma_2 + \delta_L\sigma_3)/2]$ into its first four orders using Equation 14.47. Collect the odd and even orders and show that it is the beginning of the series for

$$\sigma_0 \cos\left(\frac{\delta}{2}\right) + i\sin\left(\frac{\delta}{2}\right)\left(\frac{\delta_H\sigma_1 + \delta_{45}\sigma_2 + \delta_L\sigma_3}{\delta}\right).$$

14.16 Show that for $\mathbf{J} = c_0\sigma_0 + c_1\sigma_1 + c_2\sigma_2 + c_3\sigma_3$,

$$e^{\mathbf{J}} = \begin{pmatrix} e^{c_0}\left(\cosh\Psi + \dfrac{c_1 \sinh\Psi}{\Psi}\right) & \dfrac{e^{c_0}(c_2 - ic_3)\sinh\Psi}{\Psi} \\ \dfrac{e^{c_0}(c_2 + ic_3)\sinh\Psi}{\Psi} & e^{c_0}\left(\cosh\Psi - \dfrac{c_1 \sinh\Psi}{\Psi}\right) \end{pmatrix},$$

where $\Psi = \sqrt{c_1^2 + c_2^2 + c_3^2}$.

14.17 Using Problem 14.15, show the identity $\det(e^A) = e^{\text{Tr}(A)}$.

14.18 A dichroic sheet polarizer has the Jones matrix $\mathbf{DSP} = \dfrac{1}{2500}\begin{pmatrix} 421 & 420 \\ 420 & 421 \end{pmatrix}$.
   a. What is the diattenuation and the orientation of the transmission axis?
   b. If the polarizer could be sliced exactly in half, into two equal sheets, what would the Jones matrix be for each individual sheet?

14.19 Find the ellipticity of the eigenpolarizations of Equation 14.106 as a function of (δ_H, δ_{45}, δ_R). Find the orientation of the major axis of the eigenpolarizations.

14.20 What are the constraints on complex Pauli coefficients c_0, c_1, c_2, and c_3 such that $\mathbf{J}$ has orthogonal eigenvectors?

14.21 Show that the parameter α as a function of the diattenuation in Equation 14.65 is

$$\alpha \approx \frac{D}{2} + \frac{D^3}{6} + O[D^5]. \tag{14.121}$$

14.22 A horizontal linear diattenuator with amplitude transmittances 1 and t_y acts on linearly polarized light $\mathbf{B}(\theta)$. By what angle is the plane of polarization rotated?

14.23 If an homogeneous Jones matrix has eigenpolarizations

$$\hat{\mathbf{E}}_q = \begin{pmatrix} 1 \\ \alpha \end{pmatrix}, \quad \hat{\mathbf{E}}_r = \begin{pmatrix} 1 \\ \beta \end{pmatrix}, \tag{14.122}$$

what is the relationship between α and β?

14.24 $\mathbf{J} = (i\sigma_0 + i\sigma_2)$ has an imaginary σ_2 component. Is $\mathbf{J}$ a diattenuator or a retarder? Calculate the eigenvalues to confirm the classification. Explain why c_0 is factored out of Equation 14.38.

14.25 Show that $\sigma_\alpha \sigma_\beta = \delta_{\alpha\beta}\sigma_0 + i\varepsilon_{\alpha\beta\gamma}\sigma_\gamma$ is true for all nine pairs of α and β; that is, $(\alpha, \beta) = (1, 1), (1, 2), (1, 3), (2, 1), \ldots (3, 3)$. The Kroniker delta symbol $\delta_{\alpha\beta}$ is defined as

$$\delta_{\alpha\beta} = \begin{cases} 0 \text{ if } \alpha \neq \beta, \\ 1 \text{ if } \alpha = \beta. \end{cases} \quad (14.123)$$

The Levi–Civita symbol $\varepsilon_{\alpha\beta\gamma}$ is defined for even and odd permutations of (1, 2, 3) as

$$\varepsilon_{\alpha\beta\gamma} = \begin{cases} 1 & \text{if } (\alpha,\beta,\gamma) = (1,2,3),(2,3,1), \text{ or } (3,1,2) \\ -1 & \text{if } (\alpha,\beta,\gamma) = (1,3,2),(2,1,3), \text{ or } (3,2,1). \\ 0, & \text{otherwise} \end{cases} \quad (14.124)$$

References

1. R. Clark Jones, A new calculus for the treatment of optical systems, *JOSA* 31.7 (1941): 488–493.
2. W. Pauli, *General Principles of Quantum Mechanics*, Springer Science & Business Media (2012).
3. P. A. M. Dirac, *The Principles of Quantum Mechanics*, No. 27, Oxford University Press (1981).
4. C. Cohen-Tannoudji, B. Diu, F. Laloe, *Quantum Mechanics*, 2nd edition, Wiley (1992).
5. J. R. Oppenheimer, Note on light quanta and the electromagnetic field, *Phys. Rev.* 38.4 (1931): 725.
6. R. M. A. Azzam and N. M. Bashara, *Ellipsometry and Polarized Light*, North-Holland, Elsevier Science (1987).
7. C. Moler and C. Van Loan, Nineteen dubious ways to compute the exponential of a matrix, *SIAM Rev.* 20.4 (1978): 801–836.
8. C. Moler and C. Van Loan, Nineteen dubious ways to compute the exponential of a matrix, twenty-five years later, *SIAM Rev.* 45.1 (2003): 3–49.
9. N. J. Higham, The scaling and squaring method for the matrix exponential revisited, *SIAM J. Matrix Anal. Appl.* 26.4 (2005): 1179–1193.
10. N. J. Higham, *Functions of Matrices: Theory and Computation*, Siam (2008).
11. N. J. Higham and A. H. Al-Mohy, Computing matrix functions, *Acta Numerica* 19 (2010): 159–208.
12. N. J. Higham, Evaluating Padé approximants of the matrix logarithm, *SIAM J. Matrix Anal. Appl.* 22.4 (2001): 1126–1135.
13. S. H. Cheng, N. J. Higham, C. S. Kenney, and A. J. Laub, Approximating the logarithm of a matrix to specified accuracy, *SIAM J. Matrix Anal. Appl.* 22(4) (2001): 1112–1125.
14. A. H. Al-Mohy and N. J. Higham, Improved inverse scaling and squaring algorithms for the matrix logarithm, *SIAM J. Sci. Comput.* 34.4 (2012): C153–C169.
15. M. Martinelli and R. A. Chipman, Endless polarization control algorithm using adjustable linear retarders with fixed axes, *J. Lightwave Technol.* 21.9 (2003): 2089.
16. J. B. Kuipers, *Quaternions and Rotation Sequences*, Vol. 66, Princeton: Princeton University Press (1999).
17. H. Poincaré, *Compt. Rend. Acad. Sci. Paris* 128 (1899): 1065–1069; *Camb. Philos. Trans.* 18 (1899): 220–255.
18. H. Baker, *Proc. Lond. Math. Soc.* (1) 34 (1902): 347–360; H. Baker, *Proc. Lond. Math. Soc.* (1) 35 (1903): 333–374; H. Baker, *Proc. Lond. Math. Soc.* (Ser 2) 3 (1905): 24–47.
19. E. Borisovich Dynkin, Вычисление коэффициентов в формуле Campbell–Hausdorff [Calculation of the coefficients in the Campbell–Hausdorff formula], *Doklady Akademii Nauk SSSR* (in Russian) 57 (1947): 323–326.
20. N. Jacobson, *Lie Algebras*, John Wiley & Sons (1966).
21. J. J. Gil and E. Bernabeu, Obtainment of the polarizing and retardation parameters of a non-depolarizing optical system from the polar decomposition of its Mueller matrix, *Optik* 76 (1987): 67.
22. S.-Y. Lu and R. A. Chipman, Homogeneous and inhomogeneous Jones matrices, *JOSA A* 11.2 (1994): 766–773.
23. O. Arteaga and A. Canillas, Pseudopolar decomposition of the Jones and Mueller–Jones exponential polarization matrices, *JOSA A* 26.4 (2009): 783–793.

24. O. Arteaga and A. Canillas, Analytic inversion of the Mueller–Jones polarization matrices for homogeneous media, *Opt. Lett.* 35.4 (2010): 559–561.
25. R. Clark Jones, A new calculus for the treatment of optical systems. VII. Properties of the N-matrices, *JOSA* 38.8 (1948): 671–683.
26. R. Clark Jones, New calculus for the treatment of optical systems. VIII. Electromagnetic theory, *JOSA* 46.2 (1956): 126–131.
27. D. G. M. Anderson and R. Barakat, Necessary and sufficient conditions for a Mueller matrix to be derivable from a Jones matrix, *JOSA A* 11.8 (1994): 2305–2319.
28. S. R. Cloude, Group theory and polarisation algebra, *Optik* 75.1 (1986): 26–36.
29. J. J. Gil and E. Bernabeu, Obtainment of the polarizing and retardation parameters of a non-depolarizing optical system from the polar decomposition of its Mueller matrix, *Optik* 76 (1987): 67.
30. S. R. Cloude, Uniqueness of target decomposition theorems in radar polarimetry, in *Direct and Inverse Methods in Radar Polarimetry*, Springer Netherlands (1992), pp. 267–296.
31. S.-Y. Lu and R. A. Chipman, Interpretation of Mueller matrices based on polar decomposition, *JOSA A* 13.5 (1996): 1106–1113.
32. S. R. Cloude, Lie groups in electromagnetic wave propagation and scattering, *J. Electromag. Waves Appl.* 6.7 (1992): 947–974.
33. R. Ossikovski, A. De Martino, and S. Guyot, Forward and reverse product decompositions of depolarizing Mueller matrices, *Opt. Lett.* 32.6 (2007): 689–691.
34. J. J. Gil, Polarimetric characterization of light and media, *Eur. Phys. J. Appl. Phys.* 40.01 (2007): 1–47.
35. R. Ossikovski et al., Depolarizing Mueller matrices: How to decompose them? *Phys. Stat. Solidi A* 205.4 (2008): 720–727.
36. R. Ossikovski, Analysis of depolarizing Mueller matrices through a symmetric decomposition, *JOSA A* 26.5 (2009): 1109–1118.
37. S. N. Savenkov, Jones and Mueller matrices: Structure, symmetry relations and information content, in *Light Scattering Reviews* 4 (2009), pp. 71–119.
38. S. Cloude, *Polarisation: Applications in Remote Sensing*, Oxford University Press (2009).
39. F. Boulvert et al., Decomposition algorithm of an experimental Mueller matrix, *Opt. Commun.* 282.5 (2009): 692–704.
40. B. N. Simon et al., A complete characterization of pre-Mueller and Mueller matrices in polarization optics, *JOSA A* 27.2 (2010): 188–199.
41. N. Ghosh, M. F. G. Wood, and I. A. Vitkin, Influence of the order of the constituent basis matrices on the Mueller matrix decomposition-derived polarization parameters in complex turbid media such as biological tissues, *Opt. Commun.* 283.6 (2010): 1200–1208.
42. V. Devlaminck and P. Terrier, Non-singular Mueller matrices characterizing passive systems, *Optik* 121.21 (2010): 1994–1997.
43. R. Ossikovski, Differential matrix formalism for depolarizing anisotropic media, *Opt. Lett.* 36.12 (2011): 2330–2332.
44. O. Arteaga, E. Garcia-Caurel, and R. Ossikovski, Anisotropy coefficients of a Mueller matrix, *JOSA A* 28.4 (2011): 548–553.
45. T. A. Germer, Realizable differential matrices for depolarizing media, *Opt. Lett.* 37.5 (2012): 921–923.
46. R. Ossikovski, Differential and product Mueller matrix decompositions: A formal comparison. *Opt. Lett.* 37.2 (2012): 220–222.
47. J. J. Gil, I. San José, and R. Ossikovski, Serial–parallel decompositions of Mueller matrices, *JOSA A* 30.1 (2013): 32–50.
48. J. J. Gil, Transmittance constraints in serial decompositions of depolarizing Mueller matrices: The arrow form of a Mueller matrix, *JOSA A* 30.4 (2013): 701–707.
49. S. R. Cloude, Depolarization synthesis: Understanding the optics of Mueller matrix depolarization, *JOSA A* 30.4 (2013): 691–700.
50. J. J. Gil, Review on Mueller matrix algebra for the analysis of polarimetric measurements, *J. Appl. Remote Sens.* 8.1 (2014): 081599.
51. S.-Y. Lu and R. A. Chipman, Interpretation of Mueller matrices based on polar decomposition, *JOSA A* 13.5 (1996): 1106–1113.
52. H. D. Noble and R. A. Chipman, Mueller matrix roots algorithm and computational considerations, *Opt. Express* 20.1 (2012): 17–31.
53. H. D. Noble, Mueller Matrix Roots, dissertation, University of Arizona (2011).

15

Paraxial Polarization Aberrations

15.1 Introduction

This chapter develops methods for the description of the polarization aberrations of optical systems with an emphasis on radially symmetric optical systems such as camera lenses, microscope objectives, and telescopes.

Aberrations are deviations from ideal and desired behavior. In conventional optical design, spherical wavefronts are desired. A uniform spherical wavefront with circular profile focuses to an *Airy disk*; any deviation of the wavefront from spherical, any aberration, increases the size of this image and thus reduces the resolution and information content of an image. To better understand and communicate aberration information, *wavefront aberration* terms are defined as a basis set of functions that can be added to provide a close fit to the target wavefront, as in Section 10.7.

Similarly, *polarization aberrations* are deviations from a uniform amplitude and uniform polarization state. To transmit wavefronts with a uniform polarization state for arbitrary inputs, the ray paths need to be free from diattenuation and retardance. Thus, polarization aberrations can be described as the deviations from an identity Jones matrix for all rays.

This chapter provides algorithms to analyze the polarization of *weakly polarizing optical elements* such as lenses and mirrors and to investigate how these polarizations interact. For very weakly polarizing elements, the *Pauli matrix* components can be added, providing great simplification and insight. For *paraxial optics*, three polarization aberrations occur, aberrations that are the diattenuation or retardance equivalents of *defocus* (quadratic), *tilt* (linear), and *piston* (constant), as shown in Figure 15.1. For lens and mirror interfaces, the variation of diattenuation

543

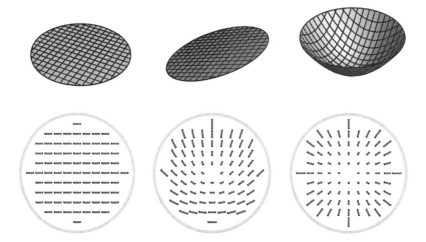

Figure 15.1 Wavefront aberration (top row) and polarization aberration (bottom row) for piston (left), tilt (middle), and defocus (right).

and retardance with angle of incidence about normal incidence is mainly quadratic. Thus, after describing each interface by a quadratic variation of diattenuation and retardance, the polarization defocus, tilt, and piston coefficients for a radially symmetric system are readily calculated by a paraxial calculation.

In geometrical optics, the *paraxial region* comprises those rays for which the *Seidel aberrations*, spherical aberration, coma, and astigmatism, are not significant, perhaps less than a tenth of a wave. For the paraxial polarization aberrations, the paraxial region spans the set of rays near the axis where second-order fits to the amplitude coefficients are accurate; see Chapter 8 for a discussion of the region of accuracy for second-order fits to Fresnel equations. *The polarization paraxial region turns out to be much larger than the paraxial region for rays and wavefront.* Thus, the paraxial polarization algorithms developed here are applicable to a large number of lens and mirror systems. Paraxial optical calculations provide a great simplification over ray tracing equations. An additional simplification for the polarization calculations is that the rays can be treated as propagating along the z-axis, neglecting the z-component of the electric field; hence, Jones matrices will be used for the paraxial analysis instead of 3×3 polarization ray tracing matrices.

The development of paraxial polarization aberrations follows these steps. A summary of the *paraxial ray trace equations* are presented in the Appendix to this chapter. The paraxial ray trace is used to calculate the *angle of incidence* across the wavefront at each interface. The *amplitude coefficients*, such as *Fresnel equations* for uncoated interfaces, are fit to quadratic equations. The combination of the paraxial angle of incidence with these quadratic amplitude relations leads to surface contributions of the polarization aberration forms, three for linear diattenuation and three for linear retardance, at each interface. These surface contributions can be summed over the interfaces to obtain a paraxial polarization aberration for the optical system. Several examples will demonstrate the utility of the paraxial polarization aberrations. Finally, *Zernike polynomials* are generalized for the discussion of higher-order polarization aberrations.

One interesting result is a binodal form for the polarization aberration, a form where the magnitudes of the off-axis retardance and diattenuation aberrations have two zeros in the pupil; a similar result occurs with astigmatism in tilted or decentered systems where it is called *binodal astigmatism*.[1]

As is common in aberration theory, when the terms *linear* and *quadratic* and so on, are used, *approximately linear* and *approximately quadratic* are implied.

15.2 Polarization Aberrations

The *aberration* of an optical system is its deviation from ideal performance. In an imaging system with ideal spherical or plane wave illumination, the desired output is spherical wavefronts with constant amplitude and constant polarization state centered on the correct image point. Deviations from spherical wavefronts arise from variations of *optical path length* of rays through the optic due to the geometry of the optical surfaces and the laws of reflection and refraction; this is described by the *wavefront aberration function*. Deviations from constant amplitude arise from differences in reflection or refraction efficiency between rays. Amplitude variations are *amplitude aberration* or *apodization*. Polarization change also occurs at each reflecting and refracting surface due to differences of reflectance and transmission coefficients between the *s*- and *p*-components of the light. Across a set of rays, the angles of incidence and thus the polarization change varies, so that a uniformly polarized input beam has polarization variations when exiting.[2,3] For many optical systems, the desired polarization output would be a constant polarization state with no polarization change transiting the system; such ray paths through an optical system can be described by identity Jones matrices. Deviations from this identity matrix are referred to as polarization aberrations.

Polarization aberration, also called *instrumental polarization*, refers to all polarization changes of the optical system and the variations with pupil coordinate ρ, object location **H**, and wavelength λ, **J**(**H**, ρ, λ). The term *Fresnel aberrations* (Chapter 12) refers to those polarization aberrations that arise strictly from the Fresnel equations, that is, systems of metal-coated mirrors and uncoated lenses.[2,3,4,5] Multilayer coated surfaces produce polarization aberrations with similar functional forms to uncoated surfaces with larger or smaller magnitudes, depending on the coating and wavelength. For systems with homogeneous and isotropic interfaces, polarization aberrations are predominantly related to linear diattenuation and retardance; their magnitudes are described by even functions of angle of incidence.

Since the *Jones matrix* has eight degrees of freedom, variations from the *Jones pupil* identity matrix can be described with an expansion into a set of eight *Zernike polynomials* for either the real and imaginary parts, or the amplitudes and phases of each element. In reflecting and refracting optical systems, most of the deviation from the identity matrix Jones pupil due to polarization aberration occurs as linear diattenuation and linear retardance. *Linear retardance* and *linear diattenuation* do not behave as vectors, and expansion of retardance aberrations into vector Zernike polynomials is not appropriate. Instead, a new mathematical object, *orientors*, is introduced, which provides a better description for linear retardance aberrations. Similarly, orientors provide a useful basis for the expansion of linear diattenuation aberrations. Further explanation of the orientors is provided in Section 15.5.2.

Lens and mirror surfaces are typically *weak polarization elements*. Weak polarization elements, as described in Section 14.8, have a small amount of diattenuation and retardance, such that the diattenuation D and retardance δ are much less than one,

$$|D| \ll 1 \text{ and } |\delta| \ll 1. \tag{15.1}$$

The Jones matrices for weak polarization elements are usefully expressed as follows. First, the Jones matrix **J** is expressed as the sum of Pauli matrices as

$$\mathbf{J} = \begin{pmatrix} j_{xx} & j_{xy} \\ j_{yx} & j_{yy} \end{pmatrix} = c_0 \begin{pmatrix} 1 & 0 \\ 0 & 1 \end{pmatrix} + c_1 \begin{pmatrix} 1 & 0 \\ 0 & -1 \end{pmatrix} + c_2 \begin{pmatrix} 0 & 1 \\ 1 & 0 \end{pmatrix} + c_3 \begin{pmatrix} 0 & -i \\ i & 0 \end{pmatrix}$$

$$= \begin{pmatrix} c_0 + c_1 & c_2 - ic_3 \\ c_2 + ic_3 & c_0 - c_1 \end{pmatrix} = c_0 \boldsymbol{\sigma}_0 + c_1 \boldsymbol{\sigma}_1 + c_2 \boldsymbol{\sigma}_2 + c_3 \boldsymbol{\sigma}_3. \tag{15.2}$$

Then, the coefficient c_0 of the identity matrix $\boldsymbol{\sigma}_0$ is factored out

$$\mathbf{J} = c_0 \left(\boldsymbol{\sigma}_0 + \frac{c_1 \boldsymbol{\sigma}_1 + c_2 \boldsymbol{\sigma}_2 + c_3 \boldsymbol{\sigma}_3}{c_0} \right). \tag{15.3}$$

c_0 characterizes the average change in the amplitude and phase of the light, which are readily identified in polar coordinates,

$$c_0 = \rho_0 e^{-i\phi_0}. \tag{15.4}$$

Next, the remaining coefficients are separated into real and imaginary parts,

$$\mathbf{J} \approx \rho_0 e^{-i\phi_0} \left(\boldsymbol{\sigma}_0 + \frac{D_H - i\delta_H}{2} \boldsymbol{\sigma}_1 + \frac{D_{45} - i\delta_{45}}{2} \boldsymbol{\sigma}_2 + \frac{D_L - i\delta_L}{2} \boldsymbol{\sigma}_3 \right). \tag{15.5}$$

With the addition of the factor of one-half, the D's and δ's are now in units of diattenuation and retardance. This equation is the *canonical form* for the Jones matrix of a weak polarization element, as discussed in Chapter 14. The minus signs are chosen to remain consistent with the decreasing phase sign convention used throughout. Equation 15.5 shows the first-order series expansion of the general exponential form for the Jones matrix for diattenuators and retarders written in terms of three diattenuation components and three retardance components see Equation 14.78.

Equation 15.5, the *canonical form*, is the preferred and most useful form, for a Jones matrix for a *ray intercept*, providing a simple expression in terms of three diattenuation and three retardance components. The utility of this form for the Jones matrix is that when the diattenuation and retardance components are small, they can simply be added, as the *order-dependent terms* are very small. For ray intercepts, the interface Jones matrices are also linear, not elliptical or circular diattenuation or retardance, and the D_L and δ_L are generally zero.

15.2.1 Interaction of Weakly Polarizing Jones Matrices

The Jones matrix for a sequence of polarization interactions is the matrix product of the Jones matrices for the individual interactions. For *weak polarization interactions*, the resultant Jones matrix can be simplified to the sum of the Pauli coefficients for the individual interactions. The Jones matrix for a sequence of two weakly polarizing interactions, which can be two ray intercepts in a weakly polarizing optical system, is studied to observe how the polarization properties interact. This leads to a simple equation for sequences of weak polarization elements. Since most reflecting and refracting interfaces are *linear diattenuators* and *linear retarders*, not circular or elliptical, we start with an example of the Jones matrices of two linear interfaces.

Consider a light ray entering and exiting a lens with *antireflection coatings* (Section 13.2.1). Antireflection coatings are weakly polarizing with linear eigenpolarizations; thus, the diattenuation, D, and retardance, δ, coefficients are small,

$$D_{H,1}, D_{45,1}, D_{H,2}, D_{45,2} \ll 1, \quad \delta_{H,1}, \delta_{45,1}, \delta_{H,2}, \delta_{45,2} \ll 1. \tag{15.6}$$

15.2 Polarization Aberrations

The second subscript refers to the first or second ray intercept. There are no σ_3 components. The H and 45 components depend on the *planes of incidence* for the ray intercepts. The Jones matrices for the two ray intercepts in canonical form are

$$\mathbf{J}_1 \approx \rho_{0,1} e^{-i\phi_{0,1}} \left(\sigma_0 + \frac{D_{H,1} - i\delta_{H,1}}{2} \sigma_1 + \frac{D_{45,1} - i\delta_{45,1}}{2} \sigma_2 \right), \tag{15.7}$$

$$\mathbf{J}_2 \approx \rho_{0,2} e^{-i\phi_{0,2}} \left(\sigma_0 + \frac{D_{H,2} - i\delta_{H,2}}{2} \sigma_1 + \frac{D_{45,2} - i\delta_{45,2}}{2} \sigma_2 \right). \tag{15.8}$$

Using the Pauli matrix identities in Section 14.2.1, $\sigma_0 \sigma_0 = \sigma_0$, $\sigma_0 \sigma_1 = \sigma_1 \sigma_0 = \sigma_1$, $\sigma_0 \sigma_2 = \sigma_2 \sigma_0 = \sigma_2$, and

$$\sigma_1 \sigma_2 = \begin{pmatrix} 1 & 0 \\ 0 & -1 \end{pmatrix} \begin{pmatrix} 0 & 1 \\ 1 & 0 \end{pmatrix} = \begin{pmatrix} 0 & 1 \\ -1 & 0 \end{pmatrix} = i \begin{pmatrix} 0 & -i \\ i & 0 \end{pmatrix} = -\sigma_2 \sigma_1 = i\sigma_3, \tag{15.9}$$

the product $\mathbf{J}_2 \mathbf{J}_1$ is

$$\rho_{0,1} \rho_{0,2} e^{-i(\phi_{0,1}+\phi_{0,2})} \begin{pmatrix} \sigma_0 + \dfrac{(D_{H,1}+D_{H,2}) - i(\delta_{H,1}+\delta_{H,2})}{2} \sigma_1 \\ + \dfrac{(D_{45,1}+D_{4,2}) - i(\delta_{45,1}+\delta_{45,2})}{2} \sigma_2 + \dfrac{\chi - iX}{2} \sigma_3 \end{pmatrix}, \tag{15.10}$$

where χ and X are higher order in the D's and δ's. To first order, the linear diattenuation of the matrix product $\mathbf{J}_2 \mathbf{J}_1$ is the sum of the individual diattenuation components: $D_{H,1} + D_{H,2}$ and $D_{45,1} + D_{45,2}$. Similarly, the linear retardance of $\mathbf{J}_2 \mathbf{J}_1$ is the sum of the linear retardance components $\delta_{H,1} + \delta_{H,2}$ and $\delta_{45,1} + \delta_{45,2}$. This is the main simplifying result for weak polarization elements.

The σ_3 (circular) components of $\mathbf{J}_2 \mathbf{J}_1$ arise from the following product terms:

$$\frac{D_{H,2} - i\delta_{H,2}}{2} \sigma_1 \frac{D_{45,1} - i\delta_{45,1}}{2} \sigma_2 + \frac{D_{45,2} - i\delta_{45,2}}{2} \sigma_2 \frac{D_{H,1} - i\delta_{H,1}}{2} \sigma_1 = \frac{\chi + iX}{2} \sigma_3, \tag{15.11}$$

the interaction of the σ_1-component of one ray intercept with the σ_2-component of the other. The real part of the σ_3-component is a circular diattenuation term,

$$\chi = (D_{H,2}\delta_{45,1} + D_{45,1}\delta_{H,2} - D_{45,2}\delta_{H,1} - D_{H,1}\delta_{45,2})/2, \tag{15.12}$$

which arises from the interaction of the diattenuation of one surface with the retardance of the other surface. The imaginary part of the σ_3-component is a circular retardance term

$$X = (D_{H,2}D_{45,1} - D_{45,2}D_{H,1} - \delta_{H,2}\delta_{45,1} + \delta_{45,2}\delta_{H,1})/2. \tag{15.13}$$

It is obvious that combinations of linear retardance 45° apart generate a circular retardance component (i.e., elliptical retardance), as seen in the latter two terms above. Less obvious is the first two terms where the interaction of two diattenuations also generates elliptical retardance, since they generate an overall rotation on the Poincaré sphere; see Example 14.3. Both $\mathcal{X}$ and X involve the products of two small D and δ coefficients and hence are second-order terms. Thus, if the diattenuations and retardances of a ray intercept are small, for example, of the order 10^{-3}, these circular terms are of the order 10^{-6}, which is negligible. Also note that when the order is reversed, the product $\mathbf{J}_1 \mathbf{J}_2$ is

$$\rho_{0,1}\, \rho_{0,2}\, e^{-i(\phi_{0,1}+\phi_{0,2})} \left(\begin{array}{l} \boldsymbol{\sigma}_0 + \dfrac{(D_{H,1}+D_{H,2})-i(\delta_{H,1}+\delta_{H,2})}{2}\boldsymbol{\sigma}_1 \\ + \dfrac{(D_{45,1}+D_{45,2})-i(\delta_{45,1}+\delta_{45,2})}{2}\boldsymbol{\sigma}_2 - \dfrac{\mathcal{X}-iX}{2}\boldsymbol{\sigma}_3 \end{array} \right), \qquad (15.14)$$

which is the same as $\mathbf{J}_2 \mathbf{J}_1$ except that the circular terms $\mathcal{X}$ and X changed sign. This is an important result. *To first order, the product of weakly polarizing Jones matrices is order independent, and the diattenuation and retardance contributions can just be added.*

In weakly polarizing optical systems, the circular terms are interesting but very small and thus are usually not important. However, as the strength of the interactions increases, these circular terms become more important.

Example 15.1 Weak Plarization Aberration in a Ray

A *meridional ray* propagates through an uncoated lens with two surfaces in a plane at 45° between x and y. At the first ray intercept, the ray has an amplitude change $\rho_{0,1} = 0.79889$ with diattenuation $D_{H,1} = -0.00020$ and $D_{45,1} = 0.00020$. At the second ray intercept, the ray has an amplitude change $\rho_{0,2} = 1.21583$ with diattenuation $D_{H,2} = -0.00186$ and $D_{45,2} = 0.00186$. The overall Jones matrix is

$$1.21583(\boldsymbol{\sigma}_0 - 0.00186\,\boldsymbol{\sigma}_1 + 0.00186\,\boldsymbol{\sigma}_2) \times 0.79889(\boldsymbol{\sigma}_0 - 0.00020\,\boldsymbol{\sigma}_1 + 0.00020\,\boldsymbol{\sigma}_2)$$
$$= 0.97131(\boldsymbol{\sigma}_0 - 0.00206\,\boldsymbol{\sigma}_1 + 0.00206\,\boldsymbol{\sigma}_2) \qquad (15.15)$$

Note that the diattenuation on the two surfaces are aligned; hence, the circular cross term is zero.

15.2.2 Polarization of a Sequence of Weakly Polarizing Ray Intercepts

Section 15.2.1 demonstrated how the linear diattenuation and retardance of two *weakly linearly polarizing interfaces*, to first order, just add. Next, this result is extended to an arbitrary number of weakly linearly polarizing interfaces, such as optical systems formed from uncoated interfaces, antireflection coatings, or metallic reflections at small angles of incidence. A geometrical picture of this summation is provided by adding the *complex Pauli coefficients* in a vector-like manner.

15.2 Polarization Aberrations

A light ray propagating through an optical system encounters a series of *weakly polarizing ray intercepts* numbered, $q = 1, 2, \ldots Q$, each with Jones matrix $\mathbf{J}_q$ expressed in the *normalized Pauli summation* form

$$\mathbf{J}_q = c_{0,q}(\boldsymbol{\sigma}_0 + d_{1,q}\boldsymbol{\sigma}_1 + d_{2,q}\boldsymbol{\sigma}_2 + d_{3,q}\boldsymbol{\sigma}_3)$$
$$= \rho_{0,q} e^{-i\phi_{0,q}} \left(\boldsymbol{\sigma}_0 + \frac{D_{H,q} - i\delta_{H,q}}{2}\boldsymbol{\sigma}_1 + \frac{D_{45,q} - i\delta_{45,q}}{2}\boldsymbol{\sigma}_2 + \frac{D_{L,q} - i\delta_{L,q}}{2}\boldsymbol{\sigma}_3 \right). \quad (15.16)$$

Although we are principally discussing optics with interfaces with linear eigenpolarizations (zero σ_3), σ_3 is included here for generality; its coefficients $D_{L,q}$ and $\delta_{L,q}$ will usually be zero. The Jones matrix $\mathbf{J}$ for the weakly polarizing sequence to first order is

$$\mathbf{J} = \mathbf{J}_Q \cdot \mathbf{J}_{Q-1} \cdot \ldots \cdot \mathbf{J}_2 \cdot \mathbf{J}_1$$
$$= \prod_{q=1}^{Q} \mathbf{J}_{Q-q+1} \quad (15.17)$$
$$\approx c(\boldsymbol{\sigma}_0 + d_1\boldsymbol{\sigma}_1 + d_2\boldsymbol{\sigma}_2 + d_3\boldsymbol{\sigma}_3).$$

The *complex Pauli coefficients* to first order are

$$c = \prod_{q=1}^{Q} c_{0,q}, \quad d_1 = \sum_{q=1}^{Q} d_{1,q}, \quad d_2 = \sum_{q=1}^{Q} d_{2,q}, \quad d_3 = \sum_{q=1}^{Q} d_{3,q}, \quad (15.18)$$

where

$$c = P e^{-iK}, \quad P = \prod_{q=1}^{Q} \rho_{0,q}, \quad K = \sum_{q=1}^{Q} \phi_{0,q}. \quad (15.19)$$

Thus, for sequences of weakly polarizing ray intercepts, the net *diattenuation* is approximately the sum of the individual diattenuation components, that is, a sum over the $\boldsymbol{\sigma}_1$ components, a sum over the $\boldsymbol{\sigma}_2$ components, and, if present, a sum over the $\boldsymbol{\sigma}_3$ components. Similarly, the net retardance is approximated by the sum of the individual $\boldsymbol{\sigma}_1$, $\boldsymbol{\sigma}_2$, and, if present, the $\boldsymbol{\sigma}_3$ retardance components. The diattenuations correspond to real parts and the retardances correspond to imaginary parts.

Equation 15.18 leads to a vector-like geometrical representation of the summation, not in x–y space but in ($\boldsymbol{\sigma}_1$, $\boldsymbol{\sigma}_2$, $\boldsymbol{\sigma}_3$) space as follows. Consider just the linear diattenuation contributions from a series of three ray intercepts in four steps as in Figure 15.2. (1) The top row represents the magnitudes and orientations of the three diattenuations in x–y space as lines or diattenuation orientors (see Section 15.5.2). Diattenuations repeat after a rotation of 180°. Remember, the Pauli bases $\boldsymbol{\sigma}_1$ and $\boldsymbol{\sigma}_2$ are only 45° apart. (2) To transform diattenuation lines into the Pauli basis, the angles from the x-axis are doubled, and the *lines* are converted to *vectors* as shown on the second row.

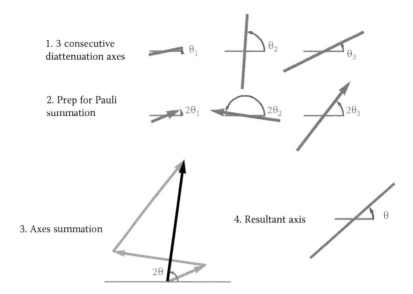

Figure 15.2 Geometrical view of the combination of weak diattenuations. (1) Three diattenuations with magnitudes indicated by line lengths and orientations θ_1, θ_2, and θ_3. (2) Converted into Pauli coefficient "vectors" by doubling the angles; magnitudes unchanged. (3) Vector addition of Pauli coefficients. (4) Resultant linear diattenuation obtained by halving the angle from the vector addition. The same construction applies to weak retarders.

(3) The "*Pauli vectors*" are added as vectors yielding the combined vector in black. (4) Finally, the angle is halved and the vector is converted back into a line, as it is returned to the *x–y* space. Note that this is an approximate calculation in the limit of weak diattenuators that ignores the σ_3 component, which would occur as a cross-product-like component out of the plane of the page. A separate and parallel Pauli-vector calculation would apply to the summation of the retardance components.

15.3 Paraxial Polarization Aberrations

A description of the *polarization aberrations* in the *paraxial region* of lens and mirror systems will be developed in this section in the form of an expansion in polarization aberration terms similar to the *Seidel aberrations*. A polarization aberration expansion to only second order in the diattenuation and retardance can provide an accurate polarization description of a large fraction of the object and pupil for many *radially symmetric systems*. A method is presented to calculate these coefficients from the *paraxial ray trace* combined with a Taylor series expansion description of the interface polarization, an expansion of either *Fresnel equations* for uncoated and metal interfaces, or thin film *amplitude coefficients*. A summary of paraxial ray tracing is included in the Appendix. This treatment includes the calculation of angles of incidence, planes of incidence, and the propagation vectors of skew rays, which are needed for polarization aberration calculation. These polarization aberration coefficients can also be determined by *polarization ray tracing* and fitting functions to the *Jones pupil*; an example of this is provided in Chapter 27 (Summary and Conclusions).

15.3.1 Paraxial Angle and Plane of Incidence

The optical system is described by normalized object H and pupil coordinates, for the entrance pupil ρ_E and exit pupil ρ_X as described in Section 10.7.1 and as shown in Figure 15.3.

15.3 Paraxial Polarization Aberrations

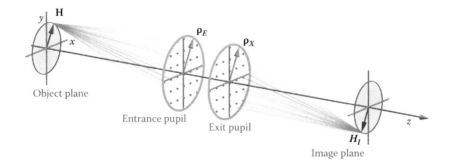

Figure 15.3 Normalized coordinates for the paraxial polarization expansion.

The polarization aberrations for lens and mirror surfaces depend on the *variation of angle of incidence* and the *orientation of the plane of incidence* across the beam. Figure 15.4 shows the wavefront from an on-axis object incident on a *spherical interface*. The angle of incidence over the pupil for this on-axis object is shown in Figure 15.4. The angle of incidence is zero for the ray in the center of the pupil ($\rho = 0$). For rays from the on-axis object point $H = 0$, the angle of incidence increases linearly to the marginal ray angle of incidence i_m at the edge of the pupil. The orientation Φ of the plane of incidence is radially oriented.

$$\theta(\mathbf{H}=0,\boldsymbol{\rho}) = |\boldsymbol{\rho}| i_m = \rho\, i_m, \quad \Phi(\mathbf{H}=0,\boldsymbol{\rho}) = \tan^{-1}\left(\frac{\rho_x}{\rho_y}\right). \tag{15.20}$$

Figure 15.5 shows the wavefront as the object moves away from the axis. Since the wavefront is spherical and the surface is spherical, the functional form remains like Figure 15.4, but is shifted and centered on whichever ray is at normal incidence, as graphed in Figure 15.6.

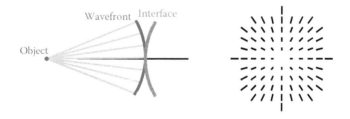

Figure 15.4 (Left) The wavefront from an on-axis object point incident on a spherical interface has an angle of incidence of zero at the center of the beam. (Right) The paraxial angle of incidence for an on-axis wavefront at a spherical surface increases linearly from the center and equals the marginal ray angle of incidence at the edge. Line lengths indicate the angle of incidence and the orientation indicates the plane of incidence.

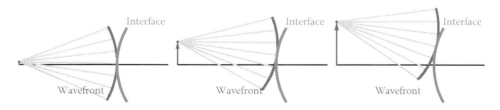

Figure 15.5 Wavefronts from an on-axis and two off-axis objects incident on a spherical surface. All three cases are a spherical wavefront tangent to a spherical interface.

Figure 15.6 Angle of incidence maps as the wavefronts of Figure 15.5 move off-axis. The patterns shift, but otherwise are unchanged. The ray through the center of each pattern (inside blue circle) is the chief ray for that field.

The angle of incidence, $\theta(\mathbf{H}, \boldsymbol{\rho})$ for marginal rays and skew rays at vector object position $\mathbf{H}$ and pupil position $\boldsymbol{\rho}$ is calculated using the Pythagorean theorem because $\theta(\mathbf{H}, \boldsymbol{\rho})$ has both x- and y-components:

$$\theta(\mathbf{H},\boldsymbol{\rho}) = \sqrt{\mathbf{H} \cdot \mathbf{H}\, i_c^2 + 2\, \mathbf{H} \cdot \boldsymbol{\rho}\, i_c i_m + \boldsymbol{\rho} \cdot \boldsymbol{\rho}\, i_m^2}, \qquad (15.21)$$

where i_c is the *chief ray angle of incidence*.

Consider an optical system with a series, $q = 1, 2, \ldots, Q$ surfaces. The paraxial angle of incidence at each surface q is a function of the chief $i_{c,q}$ and marginal ray angles $i_{m,q}$,

$$\begin{aligned}\theta_q &= \sqrt{\theta_{x,q}^2 + \theta_{y,q}^2} \\ &= \sqrt{(G^2 + H^2)\, i_{c,q}^2 + 2(G\, x + H\, y)\, i_{c,q} i_{m,q} + (x^2 + y^2)\, i_{m,q}^2}, \end{aligned} \qquad (15.22)$$

where (G, H) are the x- and y-components of $\mathbf{H}$, and (x, y) are the x- and y-components of $\boldsymbol{\rho}$. For simplicity, the object is preferably placed on the y-axis, $G = 0$, and the angle of incidence simplifies to

$$\theta_q = \sqrt{H^2\, i_{c,q}^2 + 2 H\, y\, i_{c,q} i_{m,q} + (x^2 + y^2)\, i_{m,q}^2}, \qquad (15.23)$$

or in polar pupil coordinates, ρ and ϕ,

$$\theta_q = \sqrt{H^2\, i_{c,q}^2 + 2 H \rho\, i_{c,q} i_{m,q} \sin\phi + \rho^2\, i_{m,q}^2}. \qquad (15.24)$$

The orientation of the plane of incidence, Φ, measured counterclockwise from the x-axis, is, as seen in Figure 15.6,

$$\tan\Phi = \frac{\theta_x}{\theta_y}. \qquad (15.25)$$

15.3 Paraxial Polarization Aberrations

If $\theta_x = 0$, the plane of incidence then intersects the x–y plane in a vertical line. The orientation of the plane of incidence is

$$\sin\Phi = \frac{\theta_x}{\sqrt{(\theta_x^2 + \theta_y^2)}} = \frac{x_e i_m}{|\theta|} = \frac{\rho \sin\phi \, i_m}{|\theta|}, \tag{15.26}$$

$$\cos\Phi = \frac{\theta_y}{\sqrt{(\theta_x^2 + \theta_y^2)}} = \frac{H i_c + \rho \cos\phi \, i_m}{|\theta|}, \tag{15.27}$$

so

$$\Phi = \arctan\left(\frac{\rho i_m \sin\phi}{H i_c + \rho i_m \cos\phi}\right). \tag{15.28}$$

15.3.2 Paraxial Diattenuation and Retardance

The paraxial polarization aberrations for a single lens or mirror surface are obtained by combining the paraxial angle of incidence with a quadratic expression for diattenuation or retardance as a function of angle of incidence. By combining the angle of incidence functions with approximations for the Jones matrix of a surface or coating, an approximate expression for the Jones matrix of a surface is developed. The diattenuation $D(\theta)$ and retardance $\delta(\theta)$ of interfaces are well approximated for all isotropic coatings and uncoated interfaces at small angles of incidence θ by simple quadratic equations,

$$D(\theta) \approx D_2 \theta^2, \quad \delta(\theta) \approx \delta_2 \theta^2, \tag{15.29}$$

where D_2 and δ_2 are coefficients of the diattenuation and retardance functions. For uncoated and coated interfaces, D_2 and δ_2 are found for each interface in the system from polynomial fits to the multilayer intensity reflection and transmission equations and associated diattenuation and retardance expressions; see Math Tip 13.1.

15.3.3 Diattenuation Defocus

Consider an uncoated refracting surface, which has no retardance, only diattenuation. A pupil map of the diattenuation for an on-axis object point, shown in Figure 15.7, is obtained by squaring the magnitude of the angle of incidence map of Figure 15.6 while leaving the orientation unchanged. The diattenuation magnitude at the edge of the pupil is $D_2 i_m^2$, the diattenuation of the marginal ray. This diattenuation pattern varies quadratically with pupil coordinate and so has been named *diattenuation defocus*. The *p*-Fresnel coefficient has a greater magnitude than the *s*-Fresnel coefficient; hence, the transmission axis is aligned with the plane of incidence. The uncoated interface has no

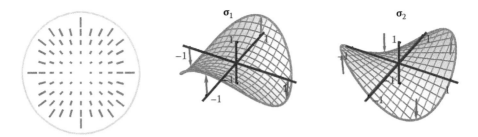

Figure 15.7 (Left) Pupil map for the diattenuation of a single refracting surface with an on-axis wavefront has a diattenuation that increases quadratically from the center and is radially oriented. The diattenuation magnitude around the edge of the pupil is the diattenuation for the interface evaluated at the marginal ray angle of incidence. The σ_1 (middle) and σ_2 (right) components of the diattenuation map are plotted. The red arrows points from the x–y plane to help visualize the shape σ1 has a cos 2ϕ form and σ2 has a sin 2ϕ form.

retardance since the refractive index is real on both sides of the interface. The corresponding Jones pupil equation is

$$\mathbf{J}(\rho,\phi) \approx (a_0+a_2\rho^2)e^{-i(\theta_0+\theta_2\rho^2)}\left(\sigma_0 + \frac{D_2}{2}\rho^2\cos 2\phi\,\sigma_1 + \frac{D_2}{2}\rho^2\sin 2\phi\,\sigma_2\right)$$

$$= (a_0+a_2\rho^2)e^{-i(\theta_0+\theta_2\rho^2)}\begin{pmatrix} 1+\dfrac{D_2}{2}\rho^2\cos 2\phi & \dfrac{D_2}{2}\rho^2\sin 2\phi \\ \dfrac{D_2}{2}\rho^2\sin 2\phi & 1-\dfrac{D_2}{2}\rho^2\cos 2\phi \end{pmatrix}. \qquad (15.30)$$

The Pauli coefficients for diattenuation defocus are plotted in Figure 15.7 (middle and right). The functional form of diattenuation defocus, for small D_2, is given by the matrix in Equation 15.30. The σ_1 is positive along the x-axis and negative along the y-axis while the σ_2 component is rotated by 45°. Figure 15.8 shows the output polarization state maps when 0°, 45°, 90°, and 135° linearly polarized light is incident. The edge of the pupil is brighter than the center along the axis aligned with the linear polarization and dimmer along the orthogonal axis. The polarization undergoes maximum polarization rotation at ±45° to the polarization axis. Figure 15.9 contains the corresponding maps when right and left circularly polarized light are incident. There is no polarization change at the center of the pupil. The change increases quadratically toward the edge with the light becoming elliptically polarized with a major axis parallel to the diattenuation.

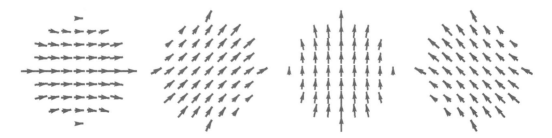

Figure 15.8 The transmitted polarization ellipse map for the diattenuation map of Figure 15.7 when, from left to right, 0°, 45°, 90°, and 135° linearly polarized light is incident. Diattenuation magnitude is 0.3 for the marginal ray in this example.

15.3 Paraxial Polarization Aberrations

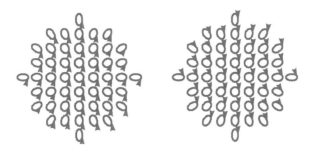

Figure 15.9 The transmitted polarization ellipse for the diattenuation map of Figure 15.7 when (left) left circular and (right) right circular polarized light is incident.

15.3.4 Diattenuation Defocus and Retardance Defocus

Next, compare a metal mirror on-axis with the uncoated lens surface of Section 15.3.3. The variation of the intensity reflectance and phase is quadratic near the origin for most interfaces. Figure 15.10 plots the Fresnel coefficients, diattenuation, and retardance of a typical metal mirror coated with aluminum. Such reflecting surfaces have non-zero retardance, and both diattenuation and retardance are nearly quadratic with incident angle.

The general spherical interface with a multilayer reflecting or refracting coating and illuminated on-axis has both diattenuation defocus and retardance defocus, and each can have either a radial or tangential axis. Thus, for an on-axis beam, there are four possibilities for the combined signs of the diattenuation and retardance for an interface, as shown in the four columns of Figure 15.11. Metal mirrors have the form of the third column.

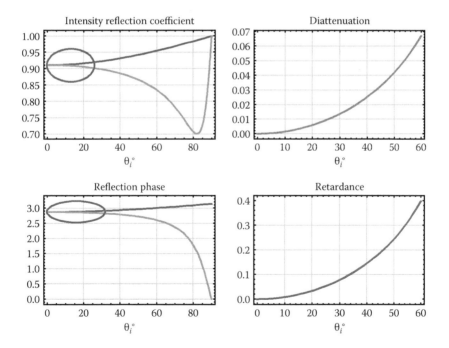

Figure 15.10 The Fresnel intensity reflection coefficients (upper left), phases (lower left), diattenuation (upper right), and retardance of an aluminum-coated mirror as a function of angle of incidence are all quadratic near the origin. Red indicates s-polarization and orange indicates p-polarization. These figures are for aluminum reflecting surface with $n = 1.262 + 7.185i$ at 600 nm.

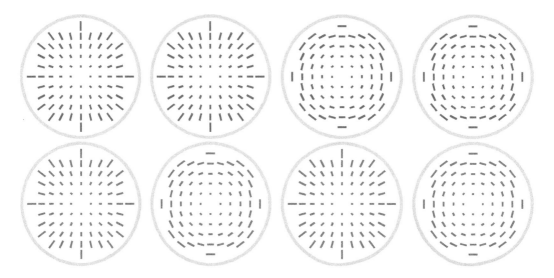

Figure 15.11 Combinations of retardance defocus (magenta) and diattenuation defocus (brown) can come with four combinations of signs: (left) positive retardance and positive diattenuation, (second) positive and negative, (third) negative and positive, and (right) negative and negative.

15.3.5 Diattenuation and Retardance across the Field of View

Figure 15.6 showed the variation of the paraxial polarization aberration, either diattenuation or retardance, as an object moves off-axis. The polarization aberration pattern translates in the direction **H** is moving. The magnitude of the σ_1 coefficient is a shifted quadratic, as shown in Figure 15.12. These patterns can be decomposed into quadratic, linear, and constant terms. The quadratic term maintains the same magnitude as the on-axis pattern as the pattern shifts. The Pauli coefficients at the edge of the field of view have the form shown in Figure 15.7.

Consider an object off axis in the y-direction. The polarization aberration pattern of Figure 15.13 (left) can be expressed as a combination of a quadratic diattenuation map (second), a linear

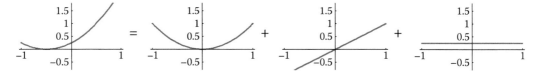

Figure 15.12 A decentered quadratic equation expressed as the sum of quadratic, linear, and constant components.

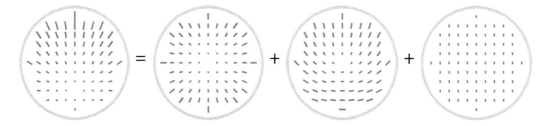

Figure 15.13 The diattenuation map for an off-axis beam (left) has a quadratic variation along the y-axis. This map can be expressed as the sum of a quadratic variation, linear variation, and constant diattenuation. These are diattenuation defocus, diattenuation tilt, and diattenuation piston aberrations. Parallel components add, crossed components subtract, and others combine as shown in Figure 15.2.

15.3 Paraxial Polarization Aberrations

diattenuation map (third), and a constant diattenuation map (right). These are the *second-order polarization aberrations*. These functional forms are given the names *polarization defocus*, *polarization tilt*, and *polarization piston*. When they refer to diattenuation, they become *diattenuation defocus*, *diattenuation tilt*, and *diattenuation piston*. Similarly, for retardance, they become *retardance defocus*, *retardance tilt*, and *retardance piston*.

15.3.6 Polarization Tilt and Piston

Diattenuation tilt and *retardance tilt* are linear variations in the pupil of $\boldsymbol{\sigma}_1$ and $\boldsymbol{\sigma}_2$. The $\boldsymbol{\sigma}_1$ component varies linearly in the meridional plane (here the *y*-axis) and the $\boldsymbol{\sigma}_2$ component varies linearly perpendicular to the meridional plane (here the *x*-axis), as shown in Figure 15.14. In a radially symmetric system, polarization tilt is zero on-axis and varies linearly with the object vector.

One way an optical system can generate pure polarization tilt is through the combination of two polarization defocus patterns, equal in magnitude and opposite in sign, shifted in opposite directions, as shown in Figure 15.15,

$$\mathbf{J}(\rho,\phi) \approx (a_0 + a_1 \rho \cos\phi) e^{-i(\theta_0 + \theta_1 \rho \cos\phi)} \left(\boldsymbol{\sigma}_0 + \frac{D_1 - i\delta_1}{2} \rho\cos\phi\, \boldsymbol{\sigma}_1 + \frac{D_1 - i\delta_1}{2} \rho\sin\phi\, \boldsymbol{\sigma}_2 \right). \quad (15.31)$$

The other second-order polarization aberration is polarization *piston*, a constant diattenuation or retardance shown in Figure 15.13 (right) and Figure 15.16. For a radially symmetric system, the piston is zero on axis and increases quadratically across the field as $\mathbf{H} \cdot \mathbf{H}$,

$$\mathbf{J}(\rho,\phi) \approx a_0 e^{-i\theta_0} \left(\boldsymbol{\sigma}_0 + \frac{D_0 - i\delta_0}{2} \boldsymbol{\sigma}_1 \right). \quad (15.32)$$

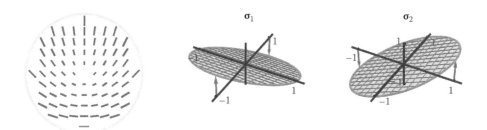

Figure 15.14 Polarization tilt, either diattenuation tilt or retardance tilt, has a linear variation of magnitude from the center and changes sign (90° rotation) passing through the center. The axis rotates by 180° around the edge of the pupil. The $\boldsymbol{\sigma}_1$ component varies linearly in the meridional plane, here the *y*-axis, and the $\boldsymbol{\sigma}_2$ component varies linearly in the orthogonal direction. The red arrows points from σ = 0.

Figure 15.15 Pure polarization tilt can be generated from two equal but opposite polarization defocus contributions, one shifted up, the other shifted down.

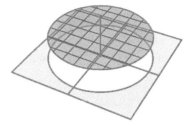

Figure 15.16 Polarization piston has a constant σ_1 component.

15.3.7 Binodal Polarization

One interesting pattern that occurs with second-order polarization aberration is the binodal polarization aberration shown in Figure 15.17 (left). Binodal indicates these are two zeros in the pupil, shown by the points on the *x*-axis. The axis rotates by 180° around each node. *Binodal polarization* can be generated by the combination of polarization defocus (center) and polarization piston (right), generating zeros where the two patterns are orthogonal. This distribution of polarization is very similar to the distribution of *astigmatism* in *binodal astigmatism*.[6,7]

15.3.8 Summation of Paraxial Polarization Aberrations over Surfaces

As a light ray propagates through a series of surfaces with *weak polarization aberration*, the aberration contributions at each surface, either retardance or diattenuation, can be summed to calculate the overall aberration. For example, Figure 15.18 overlays the polarization contributions from three

Figure 15.17 Binodal polarization aberration with two zeros or nodes with the polarization rotation by 180° around each node. In this example, the two nodes are located on the *x*-axis. Binodal polarization aberration can be created by the combination of polarization defocus (center) with polarization piston (right).

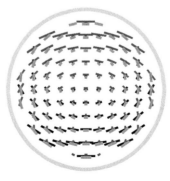

Figure 15.18 Polarization contributions from three surfaces, each with shifted polarization defocus, shown in black, purple, and orange, with small offsets for clarity.

15.3 Paraxial Polarization Aberrations

surfaces, all expressed in pupil coordinates. Here, the lines could represent either linear diattenuation or linear retardance contributions, which can be summed by the method of Section 15.2.2.

Figure 15.19 shows another example of *paraxial aberrations* for an off-axis beam cascaded over three surfaces. The first column shows the net polarization, retardance, or diattenuation, for each surface, and the total aberration from the three surfaces (bottom, left). The centers of the individual surface patterns are shifted due to the off-axis beam. Since the beam is off axis, the patterns for each surface can be decomposed into a polarization defocus (second column), polarization tilt (third column), and a polarization piston term (right column). The defocus terms can be separately added to yield the total defocus (bottom row, second column). Similarly, the tilt terms can be separately added, and the piston terms can be separately added. Hence, the net polarization aberration (bottom, left) is (1) the sum of the nine terms in the upper right, or (2) the sum of the three surface contributions in the first column, or (3) the sum of the total defocus, tilt, and piston in the bottom row. The sum is performed with the Pauli representation of the polarization (Section 15.2.2).

For the on-axis beam, the tilt and piston terms would be zero, so the net polarization aberration would be just the sum of the defocus terms (bottom row, second column). For a radially symmetric system, the tilt increases linearly with the field, the piston quadratically with the field, and the

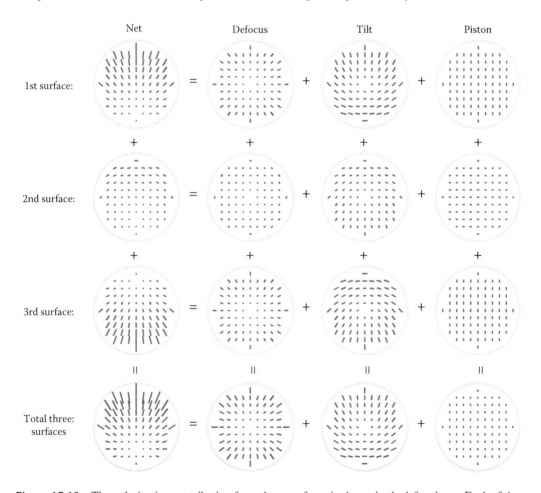

Figure 15.19 The polarization contribution from three surfaces is shown in the left column. Each of these polarization aberration maps can be decomposed into the summation of a defocus term (second column), a tilt term (third column), and a piston term (right column). These defocus, tilt, and piston columns can also be summed separately, by adding columns, to equal the bottom row's defocus, tilt, and piston terms, which sum to the overall polarization aberration map (lower left).

defocus is constant. Thus, a beam at twice the field angle of Figure 15.19 would have twice the tilt, four times the piston, but the same polarization defocus. Next, this paraxial polarization aberration method and these associated scaling rules will be tested with a high-etendué lens.

15.4 Paraxial Polarization Analysis of a Seven-Element Lens System

The *paraxial polarization aberration* method is demonstrated using the Polaris-M polarization analysis program with the *seven-element lens* shown in Figure 15.20. An exact polarization aberration calculation will be performed and compared to a paraxial calculation of the retardance and diattenuation defocus, tilt, and piston terms, showing a fit to within a few percent at the 10° field, and a fit off by only about 20% at the 30° field, a very large field angle. This example shows how the individual polarization aberration terms can be summed and also how large the paraxial region can be for polarization aberrations.

In the following calculation, each lens surface has a multilayer antireflection coating. The polarizations of the coated lenses are evaluated at 500 nm for objects at infinity. Figures 15.21 and 15.22

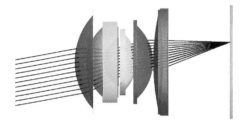

Figure 15.20 A seven-element lens (L1–L7) with several meridional ray paths drawn for the 10° field coming from infinity. The second lens L2 (orange) is cemented to the third lens L3 (green), and the sixth lens L6 (blue) is cemented to the seventh lens L7 (magenta). The stop is located between the third and fourth lenses.

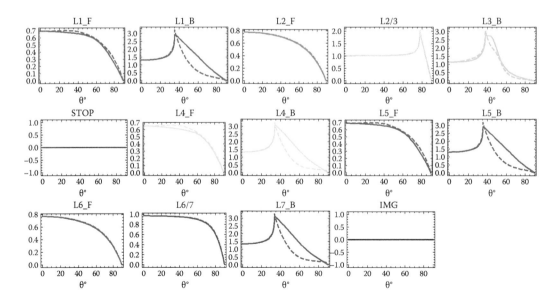

Figure 15.21 The magnitude of the *s* (solid) and *p* (dotted) amplitude coefficients are plotted for each interface for angles of incidence from 0° to 90°. L1, L2, and so on, refer to lens one, two, and so on. F refers to the front side toward the object and B refers to the back side toward the image. Beams exiting from glass into air show total internal reflection above their critical angles. F for front surface, and B for back surface.

15.4 Paraxial Polarization Analysis of a Seven-Element Lens System

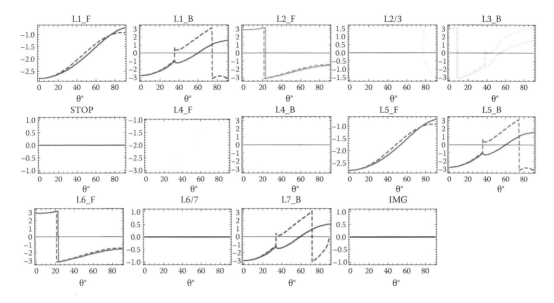

Figure 15.22 The *s* (solid) and *p* (dotted) *phase coefficients* are plotted in radians for each *antireflection* coated interface from 0° to 90°. Thin film program calculations often show 2π phase steps, such as on L3_B at 8°; these steps do not affect the *Fourier transforms* used for *point spread function* calculations, but can complicate *optical path length* calculations and *interferogram* calculation and interpretation. F for front surface, and B for back surface.

provide the coating performance, transmission amplitude, and phase for *s*- and *p*-polarizations, as a function of incident angle for each interface. Notice that the coatings have 2π discontinuities when the phase is $\pm\pi$ due to the arctan function.

Figure 15.23 shows the diattenuation calculated at a set of angles superposed on quadratic fits to the diattenuation at each interface; the *quadratic diattenuation coefficients* $D_{2,q}$, diattenuation per

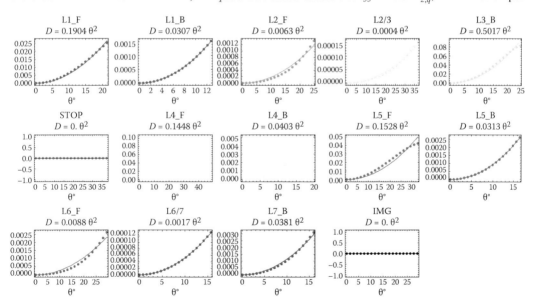

Figure 15.23 The diattenuation calculated for each coating is plotted as a set of points over the range of angles of incidence at each surface. The quadratic fits to the diattenuation are drawn as solid lines. Above each plot is an equation for the quadratic fit with θ in radians and the value of the quadratic diattenuation coefficients $D_{2,q}$ for each interface in numerical form. F for front surface, and B for back surface.

radian squared, are provided above each plot. Note how well quadratics fit over this range of angles. The $D_{2,q}$ are used in a *paraxial calculation* to determine the magnitudes of the *diattenuation defocus*, *diattenuation tilt*, and *diattenuation piston*. Similarly, Figure 15.24 shows the retardance quadratic fits at each interface with the *quadratic retardance coefficients* $\delta_{2,q}$. This provides the paraxial coating polarization for each interface, except L2/3 and L6/7 are cemented interfaces with no coatings.

Angle of incidence maps for the 10° field at each surface are shown in Figure 15.25. The ray at normal incidence in each map is located where the angle of incidence becomes zero. For some

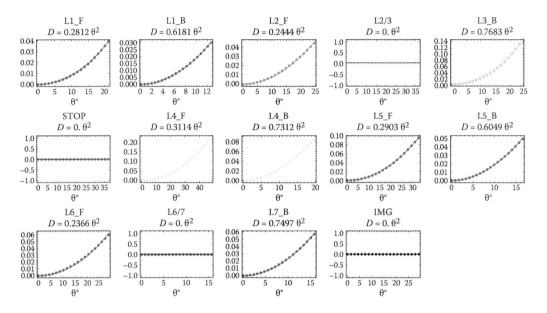

Figure 15.24 The retardance in radians calculated by the thin film algorithm (dots) for each coating is plotted as a set of points over the range of angles of incidence at each surface. The quadratic fits to the retardance are drawn as solid lines. Above each plot is an equation for the quadratic fit with θ in radians and the quadratic retardance coefficient $\delta_{2,q}$ in numerical form. F for front surface, and B for back surface.

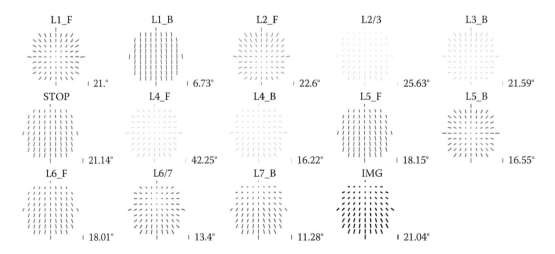

Figure 15.25 The angle of incidence maps for the lens of Figure 15.20 for the 10° field all have the form of the patterns of Figure 15.6, with the angles of incidence radially oriented about a ray at normal incidence and the magnitude increasing linearly from this node. F for front surface, and B for back surface.

15.4 Paraxial Polarization Analysis of a Seven-Element Lens System

surfaces such as L2/3, the normal incidence point is in the top of the pupil, while for others, including L4_F (F for front) and L4_B (B for back), the normal incidence point is in the lower part. For some surfaces such as lens 5_F and lens 6_F, the ray that would be at normal incidence is outside the aperture. The key at the lower right of each plot shows the length and value of the maximum angle of incidence in each plot. At each surface, the angle of incidence of the chief ray θ_C is the value in the center of the beam.

To evaluate the paraxial polarization aberration method, a comparison will be made between the result from an exact polarization ray trace and the paraxial polarization calculation. Figure 15.26 displays pupil maps of the surface-by-surface retardance contribution calculated by polarization ray tracing. These surface-by-surface retardance maps can also be calculated from the angle of incidence map. The retardance nodes and angle of incidence nodes on each surface are located in the same place. The values of all the retardances are small, less than 0.2 radians; thus, the retardances can be summed in Pauli coefficients by the method of Section 15.2.2.

Figure 15.27 (left) shows an exact calculation of the retardance for the marginal ray at each surface from a polarization ray trace. This is the magnitude of *retardance defocus* for each surface.

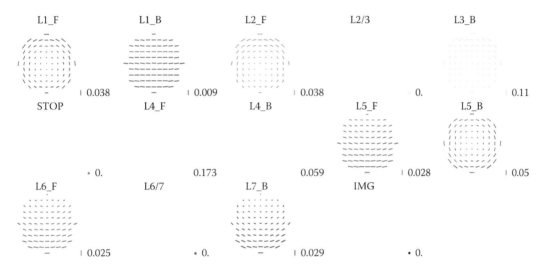

Figure 15.26 The retardance maps for the beams of Figure 15.25 for the 10° field all have the form of Figure 15.13 (left). Thus, the retardance map at each surface has a defocus, tilt, and piston component, like Figure 15.12. At each surface, the retardance of the chief ray, the value in the center of the beam, is given at the lower right. F indicates a lens front surface and B indicates a back surface.

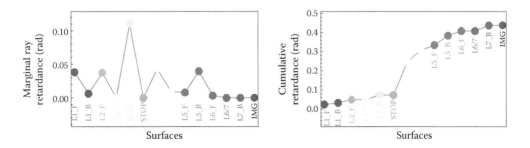

Figure 15.27 (Left) The surface-by-surface contributions to the retardance for the real marginal ray for the 10° field. (Right) The cumulative marginal ray retardance from object space through each surface increases monotonically to 0.4 radians, the exact value of the retardance defocus.

These values can be summed to yield the cumulative *retardance defocus* for the entire lens system. Figure 15.27 (right) shows the summation, representing the cumulative marginal ray retardance from object space through each surface. Similarly, Figure 15.28 shows an exact calculation of the retardance for the chief ray at each surface. This is the magnitude of *retardance piston* for each surface. Figure 15.29 (left) shows the total retardance (entrance pupil through exit pupil) for the 10° field ray paths. This retardance map can be decomposed into a retardance defocus term, calculated from the marginal ray, a retardance piston term, calculated from the chief ray, and a *retardance tilt* term, calculated from the *product of the chief and marginal rays at each surface*. Now, at this field and wavelength with these coatings, the paraxial approximation and exact polarization ray trace can be compared, and it is found that the three paraxial second-order retardance aberrations comprise more than 95% of the exact polarization aberration.

The exact and paraxial calculations of diattenuation are compared. Figure 15.30 maps the exact polarization ray tracing calculation of the diattenuation for each ray. The surface-by-surface diattenuation map is calculated from the angle of incidence map; the diattenuation nodes and angle of incidence nodes on each surface are located in the same place. The values of all the diattenuations are small, less than 0.1; hence, the diattenuations can be summed in Pauli coefficients by the method of Section 15.2.2.

The diattenuation is calculated by polarization ray tracing for comparison to the paraxial polarization aberration calculation. Figure 15.31 shows the exact calculations for the diattenuation for the marginal ray at each surface, the magnitude of the *diattenuation defocus* for each surface. These values can be summed to the cumulative diattenuation defocus for the entire lens system. Figure 15.32 shows the exact calculation for the diattenuation for the chief ray for each surface, the magnitude of *diattenuation piston* for each surface. Figure 15.33 (left) shows the end-to-end diattenuation map for the 10° field. This map can be decomposed into a diattenuation defocus term, calculated from the marginal ray, a diattenuation piston term, calculated from the chief ray, and a diattenuation tilt term, calculated from the product of the chief and marginal rays at each surface. At this field and

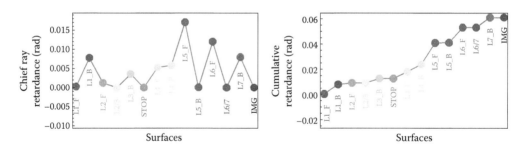

Figure 15.28 The surface-by-surface retardance contributions for the chief ray for the 10° field (left) and the accumulated values (right). The final accumulated value of 0.06 radians is the paraxial value for the retardance piston.

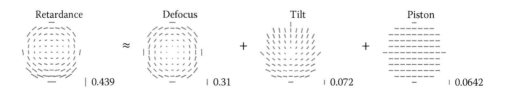

Figure 15.29 For the 10° field, the cumulative retardance map and its decomposition into a sum of retardance defocus, retardance tilt, and retardance piston are calculated from the paraxial ray trace. Comparing with the exact calculation values, the paraxial calculation is about 5% off for the chief ray and 2% off for the marginal ray.

15.4 Paraxial Polarization Analysis of a Seven-Element Lens System

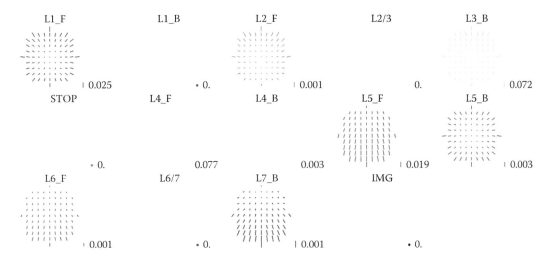

Figure 15.30 The exact diattenuation maps for the beam from the 10° field from a polarization ray trace. Since the contributions are quadratic in the angle of incidence, only a few surfaces with large marginal ray angles make substantial contributions.

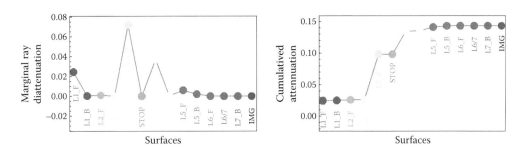

Figure 15.31 (Left) The marginal ray diattenuation is plotted for each surface for the 10° field. (Right) The cumulative marginal ray diattenuation from object space through each surface increases monotonically. The final value of 0.13 is the lenses' diattenuation defocus.

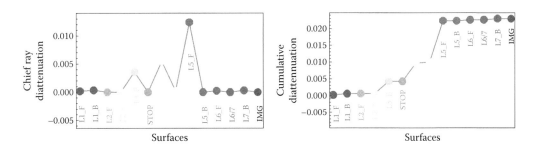

Figure 15.32 The surface-by-surface diattenuation contributions for the chief ray (left) and the accumulated values (right) for the 10° field. The final accumulated value of 0.02 is the value for the lenses' diattenuation piston.

Polarized Light and Optical Systems

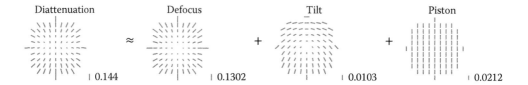

Figure 15.33 The cumulative diattenuation map for the 10° field (left) and its decomposition into a sum of diattenuation defocus, diattenuation tilt, and diattenuation piston (right three figures). Comparing with the exact calculation values shown on the left, the paraxial calculation is 8% off for the chief ray diattenuation, 2% off for the marginal ray at the top of the pupil, and 11% off for the marginal ray at the bottom of the pupil.

wavelength, with these coatings, the three second-order diattenuation aberrations comprise 87% of the polarization aberration.

For the on-axis field (Figure 15.34), the chief ray goes down the axis and the angle of incidence is zero at each interface. Thus, the diattenuation piston, retardance piston, diattenuation tilt, and retardance tilt are uniformly zero for each interface and for the entire lens, as shown in retardance and diattenuation maps in Figure 15.34 (middle and bottom rows). In the paraxial approximation, which is to second order, the defocus aberrations do not change over the field. The tilt terms increase linearly, while the piston terms increase quadratically.

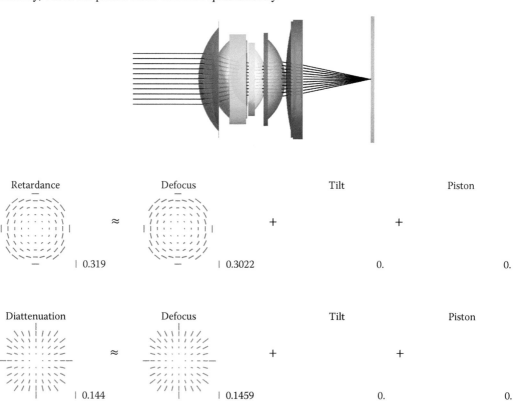

Figure 15.34 (Top) Ray paths for the on-axis field. (Middle row) The diattenuation map for the on-axis beam and the paraxial approximation, where only the retardance defocus term is non-zero. (Bottom row) The retardance map for the on-axis beam and the paraxial approximation. In radially symmetric systems, piston and tilt are always zero on-axis.

15.5 Higher-Order Polarization Aberrations

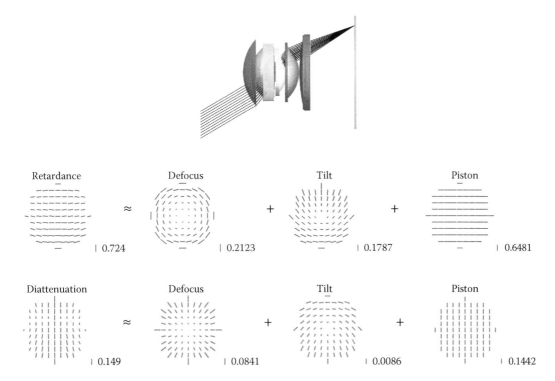

Figure 15.35 The example lens of Figure 15.20 with the beam traced from the 30° field. The paraxial calculation differs from the exact calculation by 28% for the retardance chief ray and 17% for the diattenuation chief ray.

Figure 15.35 shows the polarization of the ray paths through the lens for a 30° field, the cumulative diattenuation map, and the cumulative retardance map for the lens, which is now dominated by diattenuation piston and retardance piston, since this is the aberration that increases quadratically with field.

The paraxial polarization aberration method has been applied to a lens with large etendue to demonstrate the method of calculating the second-order polarization aberrations using quadratic fits for diattenuation and retardance. The particular numerical values for this example are not so important, but the method is powerful. The method and the resulting functional forms can form a template for other polarization analyses.

15.5 Higher-Order Polarization Aberrations

The paraxial aberration expansion of Section 15.3 is excellent for approximating the *Jones pupil* of radially symmetric lens and mirror systems and can also accurately describe many off-axis systems, such as *fold mirrors* and *off-axis telescopes*. In other cases, the variations of diattenuation and retardance are more complex than second-order terms can accurately describe. Then, a set of *higher-order basis functions* are useful to analyze such cases. First, *vector Zernike polynomials* are applied for describing higher-order variations of the electric field. Then, *orientors* are introduced for the expansion of angle of incidence, linear diattenuation, and linear retardance into a set of *basis functions*.

15.5.1 Electric Field Aberrations

Consider an arbitrary polarization aberrated monochromatic wavefront described on a reference sphere. The electric field distribution can be characterized by a *Jones vector* function **E** in normalized pupil coordinates, either polar coordinates, (ρ, ϕ), or Cartesian coordinates, (x, y),

$$\mathbf{E}(x,y) = \begin{pmatrix} E_x(\rho,\phi) \\ E_y(\rho,\phi) \end{pmatrix} = \begin{pmatrix} A_x(\rho,\phi)e^{i2\pi W_x(\rho,\phi)} \\ A_y(\rho,\phi)e^{i2\pi W_y(\rho,\phi)} \end{pmatrix}$$
$$= \begin{pmatrix} E_x(x,y) \\ E_y(x,y) \end{pmatrix} = \begin{pmatrix} A_x(x,y)e^{i2\pi W_x(x,y)} \\ A_y(x,y)e^{i2\pi W_y(x,y)} \end{pmatrix}. \tag{15.33}$$

Since **E** is a complex two-element vector function, four scalar functions, A_x, A_y, W_x, and W_y, are required for a full description. W is the wavefront's phase in waves.

Now, consider a simpler wavefront. Many wavefronts are linearly or nearly linearly polarized. For the description of such linear vector fields, the *Zernike polynomials* (Section 10.7.6) have been generalized into *vector Zernike polynomials*, $\mathbf{V}_{n,e}^m(\rho,\phi)$.[8] To construct the vector Zernike polynomials, the Zernike's $\cos(m\phi)$ are replaced by the vector, Θ_0^m, and terms $\sin(m\phi)$ are replaced by Θ_1^m, where the pair of orthogonally polarized basis vectors, Θ_0^m and Θ_1^m, are

$$\Theta_0^m = \begin{pmatrix} \cos(m\phi) \\ \sin(m\phi) \end{pmatrix}, \quad \Theta_1^m = \begin{pmatrix} -\sin(m\phi) \\ \cos(m\phi) \end{pmatrix}. \tag{15.34}$$

Index m is the order of the angular part of the Zernike polynomial. The vector Zernike polynomials through order $n = 4$ are listed in Table 15.1 and are graphed in Figure 15.36. For each vector Zernike polynomial in Figure 15.36, there is another term that is rotated by 90°; the first three are shown in the bottom row of Figure 15.37. The vector Zernike polynomials form an orthonormal basis set,

$$\int_0^1 \int_0^{2\pi} \mathbf{V}_{n,e}^m(\rho,\phi) \cdot \mathbf{V}_{n',e'}^{m'}(\rho,\phi) \rho \, d\phi \, d\rho = \Gamma \delta_{n,n'} \delta_{m,m'} \delta_{e,e'}, \tag{15.35}$$

where Γ is a normalization factor that here is chosen to equal one.

All the vector Zernike polynomials shown in Figures 15.36 and 15.37 have the same phase; therefore, the arrows are all at the end of the field line. Phase changes move the arrow around the polarization ellipse of a Jones vector, or in time, the arrow moves up and down along a linear Jones vector, as seen in Figure 15.38. Hence, the vector Zernike polynomials describe the linear polarization's amplitude and orientation but not the phase.

Consider an arbitrary linearly polarized vector function $\mathbf{E}_1$ of the form

$$\mathbf{E}_1(\rho,\phi) = \begin{pmatrix} A_x(\rho,\phi) \\ A_y(\rho,\phi) \end{pmatrix}, \tag{15.36}$$

15.5 Higher-Order Polarization Aberrations

Table 15.1 Vector Zernike Polynomials through Order $n = 4$

n	m	e	Vector Zernike Polar Form	Vector Zernike Cartesian Form
0	0	0	$\frac{1}{\sqrt{\pi}}\begin{pmatrix} 1 \\ 0 \end{pmatrix}$	$\frac{1}{\sqrt{\pi}}\begin{pmatrix} 1 \\ 0 \end{pmatrix}$
0	0	1	$\frac{1}{\sqrt{\pi}}\begin{pmatrix} 0 \\ 1 \end{pmatrix}$	$\frac{1}{\sqrt{\pi}}\begin{pmatrix} 0 \\ 1 \end{pmatrix}$
1	1	0	$\sqrt{\frac{3}{\pi}}\begin{pmatrix} \rho\cos(\phi) \\ \rho\sin(\phi) \end{pmatrix}$	$\sqrt{\frac{3}{\pi}}\begin{pmatrix} x \\ y \end{pmatrix}$
1	1	1	$\sqrt{\frac{3}{\pi}}\begin{pmatrix} -\rho\sin(\phi) \\ \rho\cos(\phi) \end{pmatrix}$	$\sqrt{\frac{3}{\pi}}\begin{pmatrix} -y \\ x \end{pmatrix}$
2	0	0	$\sqrt{\frac{5}{\pi}}\begin{pmatrix} 2\rho^2 - 1 \\ 0 \end{pmatrix}$	$\sqrt{\frac{5}{\pi}}\begin{pmatrix} 2(x^2 + y^2) - 1 \\ 0 \end{pmatrix}$
2	0	1	$\sqrt{\frac{5}{\pi}}\begin{pmatrix} 0 \\ 2\rho^2 - 1 \end{pmatrix}$	$\sqrt{\frac{5}{\pi}}\begin{pmatrix} 0 \\ 2(x^2 + y^2) - 1 \end{pmatrix}$
2	2	0	$\sqrt{\frac{5}{n}}\begin{pmatrix} \rho^2\cos(2\phi) \\ \rho^2\sin(2\phi) \end{pmatrix}$	$\sqrt{\frac{5}{\pi}}\begin{pmatrix} (x-y)(x+y) \\ 2xy \end{pmatrix}$
2	2	1	$\sqrt{\frac{5}{\pi}}\begin{pmatrix} -\rho^2\sin(2\phi) \\ \rho^2\cos(2\phi) \end{pmatrix}$	$\sqrt{\frac{5}{\pi}}\begin{pmatrix} -2xy \\ (x-y)(x+y) \end{pmatrix}$
2	2	2	$\sqrt{\frac{5}{\pi}}\begin{pmatrix} \rho^2\cos(2\phi) \\ -\rho^2\sin(2\phi) \end{pmatrix}$	$\sqrt{\frac{5}{\pi}}\begin{pmatrix} (x-y)(x+y) \\ -2xy \end{pmatrix}$
2	2	3	$\sqrt{\frac{5}{\pi}}\begin{pmatrix} -\rho^2\sin(2\phi) \\ -\rho^2\cos(2\phi) \end{pmatrix}$	$\sqrt{\frac{5}{\pi}}\begin{pmatrix} -2xy \\ y^2 - x^2 \end{pmatrix}$
3	1	0	$\sqrt{\frac{7}{\pi}}\begin{pmatrix} \rho(3\rho^2 - 2)\cos(\phi) \\ \rho(3\rho^2 - 2)\sin(\phi) \end{pmatrix}$	$\sqrt{\frac{7}{\pi}}\begin{pmatrix} x(3(x^2 + y^2) - 2) \\ y(3(x^2 + y^2) - 2) \end{pmatrix}$
3	1	1	$\sqrt{\frac{7}{\pi}}\begin{pmatrix} -\rho(3\rho^2 - 2)\sin(\phi) \\ \rho(3\rho^2 - 2)\cos(\phi) \end{pmatrix}$	$\sqrt{\frac{7}{\pi}}\begin{pmatrix} -y(3(x^2 + y^2) - 2) \\ x(3(x^2 + y^2) - 2) \end{pmatrix}$
3	3	0	$\sqrt{\frac{7}{\pi}}\begin{pmatrix} \rho^3\cos(3\phi) \\ \rho^3\sin(3\phi) \end{pmatrix}$	$\sqrt{\frac{7}{\pi}}\begin{pmatrix} x^3 - 3xy^2 \\ 3x^2y - y^3 \end{pmatrix}$

(*Continued*)

Table 15.1 (Continued) Vector Zernike Polynomials through Order $n = 4$

n	m	e	Vector Zernike Polar Form	Vector Zernike Cartesian Form
3	3	1	$\sqrt{\dfrac{7}{\pi}} \begin{pmatrix} -\rho^3 \sin(3\phi) \\ \rho^3 \cos(3\phi) \end{pmatrix}$	$\sqrt{\dfrac{7}{\pi}} \begin{pmatrix} y^3 - 3x^2 y \\ x^3 - 3xy^2 \end{pmatrix}$
3	3	2	$\sqrt{\dfrac{7}{\pi}} \begin{pmatrix} \rho^3 \cos(3\phi) \\ -\rho^3 \sin(3\phi) \end{pmatrix}$	$\sqrt{\dfrac{7}{\pi}} \begin{pmatrix} x^3 - 3xy^2 \\ y^3 - 3x^2 y \end{pmatrix}$
3	3	3	$\sqrt{\dfrac{7}{\pi}} \begin{pmatrix} -\rho^3 \sin(3\phi) \\ -\rho^3 \cos(3\phi) \end{pmatrix}$	$\sqrt{\dfrac{7}{\pi}} \begin{pmatrix} y^3 - 3x^2 y \\ 3xy^2 - x^3 \end{pmatrix}$
4	0	0	$\dfrac{3}{\sqrt{\pi}} \begin{pmatrix} 6\rho^4 - 6\rho^2 + 1 \\ 0 \end{pmatrix}$	$\dfrac{3}{\sqrt{\pi}} \begin{pmatrix} 6(x^2+y^2)^2 - 6(x^2+y^2) + 1 \\ 0 \end{pmatrix}$
4	0	1	$\dfrac{3}{\sqrt{\pi}} \begin{pmatrix} 0 \\ 6\rho^4 - 6\rho^2 + 1 \end{pmatrix}$	$\dfrac{3}{\sqrt{\pi}} \begin{pmatrix} 0 \\ 6(x^2+y^2)^2 - 6(x^2+y^2) + 1 \end{pmatrix}$
4	2	0	$\dfrac{3}{\sqrt{\pi}} \begin{pmatrix} \rho^2(4\rho^2 - 3)\cos(2\phi) \\ \rho^2(4\rho^2 - 3)\sin(2\phi) \end{pmatrix}$	$\dfrac{3}{\sqrt{\pi}} \begin{pmatrix} (x-y)(x+y)(4x^2 + 4y^2 - 3) \\ 2xy(4x^2 + 4y^2 - 3) \end{pmatrix}$
4	2	1	$\dfrac{3}{\sqrt{\pi}} \begin{pmatrix} -\rho^2(4\rho^2 - 3)\sin(2\phi) \\ \rho^2(4\rho^2 - 3)\cos(2\phi) \end{pmatrix}$	$\dfrac{3}{\sqrt{\pi}} \begin{pmatrix} -2xy(4x^2 + 4y^2 - 3) \\ (x-y)(x+y)(4x^2 + 4y^2 - 3) \end{pmatrix}$
4	2	2	$\dfrac{3}{\sqrt{\pi}} \begin{pmatrix} \rho^2(4\rho^2 - 3)\cos(2\phi) \\ -\rho^2(4\rho^2 - 3)\sin(2\phi) \end{pmatrix}$	$\dfrac{3}{\sqrt{\pi}} \begin{pmatrix} (x-y)(x+y)(4x^2 + 4y^2 - 3) \\ -2xy(4x^2 + 4y^2 - 3) \end{pmatrix}$
4	2	3	$\dfrac{3}{\sqrt{\pi}} \begin{pmatrix} -\rho^2(4\rho^2 - 3)\sin(2\phi) \\ -\rho^2(4\rho^2 - 3)\cos(2\phi) \end{pmatrix}$	$\dfrac{3}{\sqrt{\pi}} \begin{pmatrix} -2xy(4x^2 + 4y^2 - 3) \\ -4x^4 + 3x^2 + 4y^4 - 3y^2 \end{pmatrix}$
4	4	0	$\dfrac{3}{\sqrt{\pi}} \begin{pmatrix} \rho^4 \cos(4\phi) \\ \rho^4 \sin(4\phi) \end{pmatrix}$	$\dfrac{3}{\sqrt{\pi}} \begin{pmatrix} x^4 - 6y^2 x^2 + y^4 \\ 4x(x-y)y(x+y) \end{pmatrix}$
4	4	1	$\dfrac{3}{\sqrt{\pi}} \begin{pmatrix} -\rho^4 \sin(4\phi) \\ \rho^4 \cos(4\phi) \end{pmatrix}$	$\dfrac{3}{\sqrt{\pi}} \begin{pmatrix} -4x(x-y)y(x+y) \\ x^4 - 6y^2 x^2 + y^4 \end{pmatrix}$
4	4	2	$\dfrac{3}{\sqrt{\pi}} \begin{pmatrix} \rho^4 \cos(4\phi) \\ -\rho^4 \sin(4\phi) \end{pmatrix}$	$\dfrac{3}{\sqrt{\pi}} \begin{pmatrix} x^4 - 6y^2 x^2 + y^4 \\ -4x(x-y)y(x+y) \end{pmatrix}$
4	4	3	$\dfrac{3}{\sqrt{\pi}} \begin{pmatrix} -\rho^4 \sin(4\phi) \\ -\rho^4 \cos(4\phi) \end{pmatrix}$	$\dfrac{3}{\sqrt{\pi}} \begin{pmatrix} -4x(x-y)y(x+y) \\ -(x^2+y^2)^2 \cos(4\tan^{-1}(x,y)) \end{pmatrix}$

15.5 Higher-Order Polarization Aberrations

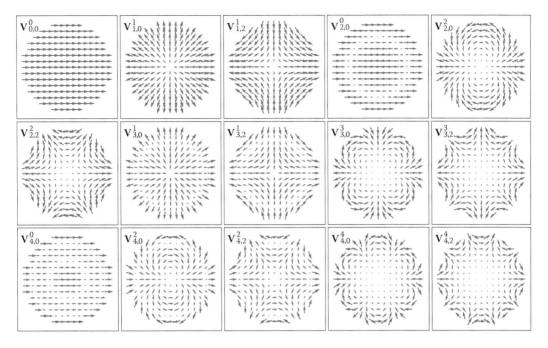

Figure 15.36 The vector Zernike polynomials $\mathbf{V}_{n,e}^m(x,y)$ for order $n = 0$ through 4 for terms $e = 0$ and 2.

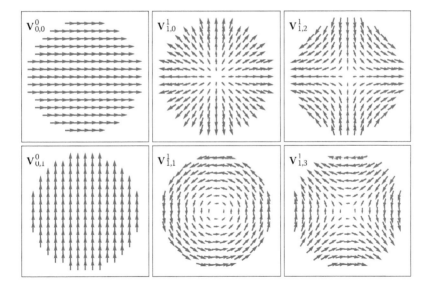

Figure 15.37 The first three vector Zernike polynomials $\mathbf{V}_{n,e}^m(x,y)$ for (top row) order $e = 0$ and 2, and (bottom row) order $e = 1$ and 3. For each vector Zernike polynomial in Figure 15.36, there is another term that is rotated by 90°; the first three are shown here.

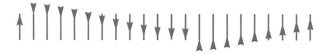

Figure 15.38 Phase change moves arrowheads up and down a linear polarization ellipse.

where the phases are equal to zero. Using the vector Zernike polynomials, Equation 15.36 can be expressed as the summation

$$\mathbf{E}_1(\rho,\phi) = \begin{pmatrix} A_x(\rho,\phi) \\ A_y(\rho,\phi) \end{pmatrix} = \sum_{n=1}^{\infty} \sum_{m=-n}^{n} \sum_{e=0}^{1} v_{n,e}^m \mathbf{V}_{n,e}^m(\rho,\phi). \tag{15.37}$$

The expansion coefficients $v_{n,e}^m$ are found from the inner product

$$\int_0^1 \int_0^{2\pi} \mathbf{V}_{n,e}^m(\rho,\phi) \cdot \mathbf{E}_1(\rho,\phi) \rho \, d\phi \, d\rho = v_{n,e}^m. \tag{15.38}$$

Equation 15.37 describes the variation of the amplitude and orientation of a linear polarization state with constant phase. To describe the phase, two additional functions are needed, either

(A) (1) an *x*-phase function and
 (2) a *y*-phase function, or
(B) (1) an average phase function (wavefront aberration function) and
 (2) a phase difference function (elliptical polarization).

(A) is straightforward, so (B) will be explored. First, the average of the phases

$$W(\rho,\phi) = \frac{W_x(\rho,\phi) + W_y(\rho,\phi)}{2} \tag{15.39}$$

describes a (polarization-independent) *wavefront aberration* contribution that can be expanded in its own set of Zernike polynomials to describe *defocus, tilt, spherical aberration, coma, astigmatism*, and so on. However, the light may not be uniformly linearly polarized. In this case, the vector Zernike polynomials describe the major axis of the *polarization ellipse*. The ellipticity of the light, $\varepsilon(\rho, \phi)$, varying from +1 for right circularly to −1 for left circularly polarized light, can then be expanded in an additional set of Zernike polynomials; for a linearly polarized field, these last Zernike coefficients would be all zero. Thus, in general, four sets of Zernike polynomials, counting the vector Zernike terms as two, are needed to fully describe a single polarized wavefront (i.e., from a single field point at a single wavelength) in an aberration expansion.

15.5.2 Orientors

In Section 15.5.1, the electric field in a circular pupil was expanded into *vector Zernike polynomials* because of the light's vector nature. *Vectors* repeat after a 360° rotation. *Angle of incidence, linear retardance*, and *linear diattenuation* are not vectors; their properties repeat after a 180° rotation. To account for this geometrical property of repetition after a 180° rotation, *orientors* are introduced, which provide basis functions for the expansion of angle of incidence, linear retardance, and linear diattenuation.[8] These orientor basis functions are derived from the vector Zernike polynomials.

15.5 Higher-Order Polarization Aberrations

Consider the behavior of *linear retardance*. Two retarders with retardances δ_1 and δ_2 with parallel fast axes have a net retardance $\delta_1 + \delta_2$. The retardances still add to $\delta_1 + \delta_2$ after one retarder is rotated by 180°, whereas two vectors would subtract when one vector is rotated by 180°. Consider the following geometrical construction that transforms angles. If the orientation angles (fast axis angles) of linear retarders are doubled, all θ are transformed to 2θ. Now, the transformed "orientation" 2θ repeats in 360° and a vector representation for "angle-doubled linear retardance" can be used. This "double the angle" property was seen earlier in Figure 15.2, where it followed from the *Pauli matrix* expressions for the combination of weak linear retardance or weak linear diattenuation. *Linear diattenuation* has the same behavior as retardance upon rotation. If one of two diattenuators is rotated by 180°, the two diattenuators combine in the same way. Equal magnitude diattenuators with axes 90° apart have a net diattenuation that cancels.

Orientors are defined using this "angle-doubled" approach and applied to representing *linear retardance, linear diattenuation*, and *angle of incidence* functions. Consider a pupil map in polar coordinates of linear retardance magnitude $\delta(\rho,\phi)$ and fast axis orientation $\psi(\rho,\phi)$, with ψ defined in the range $0 \le \psi \le 180°$, such as the arbitrary example retardance map in Figure 15.39. This pupil map is transformed into a distribution of vectors with the same magnitude $\delta(\rho,\phi)$ but with orientation $2 \times \psi(\rho, \phi)$. With this "double the angle" transformation, vector Zernike polynomials can now be used as a basis for linear diattenuation and linear retardance aberrations and angle of incidence maps providing a basis set for higher-order polarization aberrations. The orientors are thus vector Zernike polynomials but oriented at half the angle. Hence, an *orientor* is a line object with a magnitude and orientation ψ associated with a vector at twice the orientation $\phi = 2\psi$. Note that for linear retardance, linear diattenuation, and angle of incidence functions, there is no phase that needs to be described, as there was for the electric field in Section 15.5.1.

Next, the lowest-order orientor terms corresponding to the lowest-order vector Zernike polynomials will be considered. At zero order, the *orientor basis set* has two constant terms, $\mathbf{O}_{0,0}^0$ and $\mathbf{O}_{0,1}^0$, rotated by 45° from each other, as shown in Figure 15.40 (top row), and two corresponding vector basis functions, $\mathbf{V}_{0,0}^0$ and $\mathbf{V}_{0,1}^0$, rotated by 90° from each other and at twice the angle from the *x*-axis, as shown in Figure 15.40 (bottom row). Changing the sign of any orientor rotates its map by 90°.

At first order, there are four orientors shown in Figure 15.41. Two of the first-order orientor pupil maps, $\mathbf{O}_{1,0}^1$ and $\mathbf{O}_{1,1}^1$, rotate clockwise, moving clockwise around the pupil, and are associated with the angular distributions for positive values of the Zernike radial polynomial's coefficients,

$$\begin{pmatrix} \cos\left(\dfrac{\phi}{2}\right) \\ \sin\left(\dfrac{\phi}{2}\right) \end{pmatrix} \text{ and } \begin{pmatrix} \cos\left(\dfrac{\phi}{2}+\dfrac{\pi}{4}\right) \\ \sin\left(\dfrac{\phi}{2}+\dfrac{\pi}{4}\right) \end{pmatrix}. \tag{15.40}$$

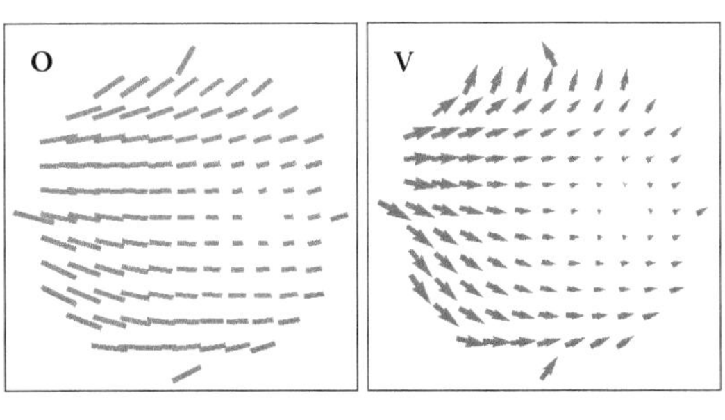

Figure 15.39 Example of the conversion of a map of orientors (left) into a map of vectors (right). The orientor distribution could be an angle of incidence map, diattenuation map, or a retardance map. To create the map of vectors, an arrowhead is added to the left (positive *x*) end of the orientor, and the angle from the *x*-axis is doubled.

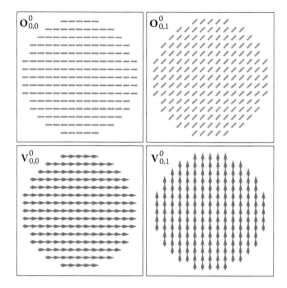

Figure 15.40 (Top row) The two zero-order orientor pupil maps, $\mathbf{O}_{0,0}^0$ and $\mathbf{O}_{0,1}^0$, are constant distributions of lines corresponding to piston. The corresponding vector Zernike maps, oriented at twice the angle of the orientors, are shown in dark red in the row below. For negative coefficient values, the orientors are rotated by 90° and the vectors are rotated by 180°.

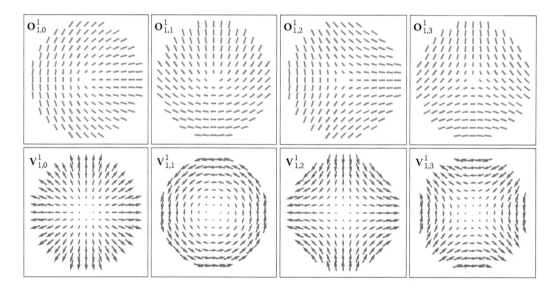

Figure 15.41 (Top) The four orientor maps at first order: $\mathbf{O}_{1,0}^0$, $\mathbf{O}_{1,1}^0$, $\mathbf{O}_{1,2}^1$, and $\mathbf{O}_{1,3}^1$. (Bottom) The corresponding Zernike vector polynomials, $\mathbf{V}_{1,0}^1$, $\mathbf{V}_{1,1}^1$, $\mathbf{V}_{1,2}^1$, and $\mathbf{V}_{1,3}^1$, are shown in magenta directly below. The $\mathbf{O}_{1,0}^0$, $\mathbf{O}_{1,1}^0$, $\mathbf{V}_{1,0}^1$, and $\mathbf{V}_{1,1}^1$ terms (left two columns) correspond to the diattenuation and retardance tilt terms. The orientors are shown for positive coefficients. For negative coefficient values, the orientors are rotated by 90°.

15.5 Higher-Order Polarization Aberrations

The other two first-order orientor pupil maps rotate counterclockwise, moving clockwise around the pupil, $\mathbf{O}_{1,2}^1$ and $\mathbf{O}_{1,3}^1$, associated with the angular distributions

$$\begin{pmatrix} \cos\left(\dfrac{\phi}{2}\right) \\ -\sin\left(\dfrac{\phi}{2}\right) \end{pmatrix} \quad \text{and} \quad \begin{pmatrix} \cos\left(\dfrac{\phi}{2}+\dfrac{\pi}{4}\right) \\ -\sin\left(\dfrac{\phi}{2}+\dfrac{\pi}{4}\right) \end{pmatrix}. \tag{15.41}$$

Six orientor basis functions are present at second order. Figure 15.42 shows the two terms with $m = 0$, $\mathbf{O}_{2,0}^0$ and $\mathbf{O}_{2,1}^0$. Note that the orientor's orientation changes sign as the corresponding *Zernike vector polynomial* (lower row) passes through zero. $\mathbf{O}_{2,0}^0$ and $\mathbf{O}_{2,1}^0$ describe quadratic magnitude variations with constant orientation. $\mathbf{O}_{2,0}^0$ and $\mathbf{O}_{2,1}^0$ pass through zero at a radius of $\rho = 1/\sqrt{2}$ to orthogonalize with the constant terms, $\mathbf{O}_{0,0}^0$ and $\mathbf{O}_{0,1}^0$. Figure 15.43 shows the four terms with $m = 2$, $\mathbf{O}_{2,0}^2$, $\mathbf{O}_{2,1}^2$, $\mathbf{O}_{2,2}^2$, and $\mathbf{O}_{2,3}^2$. $\mathbf{O}_{2,0}^2$ is our linear diattenuation defocus and linear retardance defocus aberration form and is ubiquitous in describing polarization aberrations of radially symmetric systems. The other three terms are present to a much lesser extent in typical pupil function expansions.

Figures 15.44 and 15.45 continue the expansion, showing the terms at third order, while Figures 15.46, 15.47, and 15.48 show the orientors and the corresponding Zernike vector polynomials at fourth order.

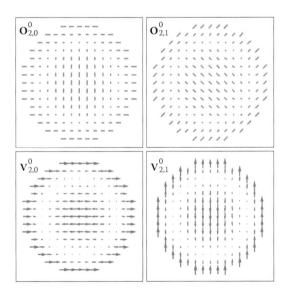

Figure 15.42 The second-order orientor pupil maps for $m = 0$, $\mathbf{O}_{0,0}^2$, and $\mathbf{O}_{0,1}^2$.

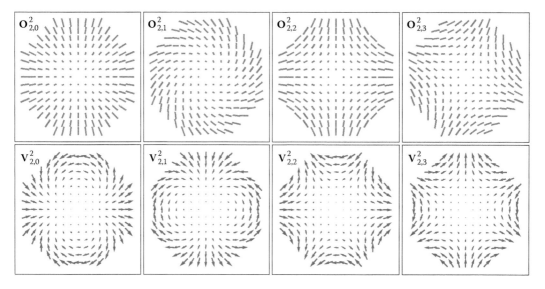

Figure 15.43 The second-order orientor pupil maps for $m = 2$, $\mathbf{O}^2_{2,0}$, $\mathbf{O}^2_{2,1}$, $\mathbf{O}^2_{2,2}$, and $\mathbf{O}^2_{2,3}$. The term $\mathbf{O}^2_{2,0}$ on the left corresponds to diattenuation defocus and retardance defocus.

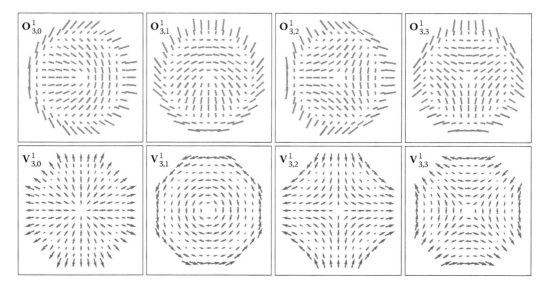

Figure 15.44 The third-order orientor maps (top) and vector Zernike polynomials (bottom) for $m = 1$.

15.5 Higher-Order Polarization Aberrations

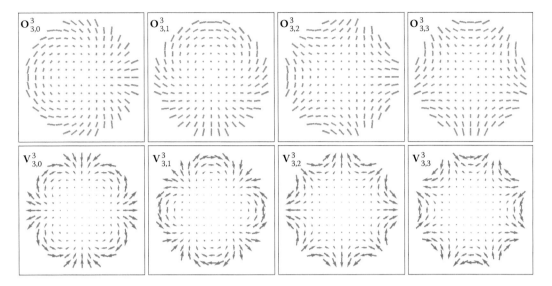

Figure 15.45 The third-order orientor maps (top) and vector Zernike polynomials (bottom) for $m = 3$.

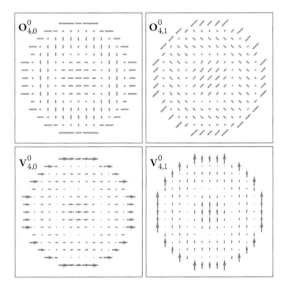

Figure 15.46 The fourth-order orientor maps (top) and vector Zernike polynomials (bottom) for $m = 0$.

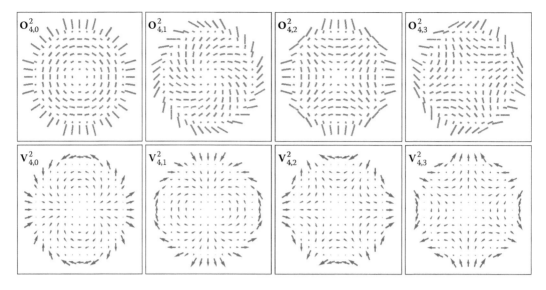

Figure 15.47 The fourth-order orientor maps (top) and vector Zernike polynomials (bottom) for $m = 2$.

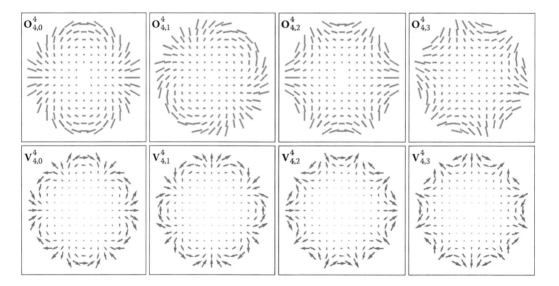

Figure 15.48 The fourth-order orientor maps (top) and vector Zernike polynomials (bottom) for $m = 4$.

15.5.3 Diattenuation and Retardance

Orientors were developed to describe linear diattenuation and linear retardance distributions in pupils with a series representation based on Zernike polynomials. Such a series representation can be constructed in different ways. One algorithm for describing an arbitrary Jones pupil's linear parts with orientors is as follows. Given a Jones pupil $\mathbf{J}(\rho, \phi)$, the function is divided into a hermitian (diattenuating) matrix function $\mathbf{H}(\rho, \phi)$ and a unitary (retarding) matrix function $\mathbf{U}(\rho, \phi)$ using the polar decomposition (Section 5.9.3),

$$\mathbf{J}(\rho,\phi) = \mathbf{H}(\rho,\phi) \cdot \mathbf{U}(\rho,\phi). \tag{15.42}$$

15.5 Higher-Order Polarization Aberrations

Both $\mathbf{H}(\rho, \phi)$ and $\mathbf{U}(\rho, \phi)$ have four degrees of freedom. The retardance is decomposed following Chapter 14, using the matrix logarithm to divide $\mathbf{U}(\rho, \phi)$ into Pauli components representing the average phase, ϕ, the 0° and 45° linear retardance components, δ_H and δ_{45}, and the circular retardance, δ_L,

$$2i \ln \mathbf{U} = \phi \boldsymbol{\sigma}_0 + \delta_H \boldsymbol{\sigma}_1 + \delta_{45} \boldsymbol{\sigma}_2 + \delta_L \boldsymbol{\sigma}_3. \tag{15.43}$$

This determines the linear retardance part of the Jones pupil, which has corresponding linear retarder Jones matrices

$$\begin{aligned}\mathbf{LR}(\delta_{linear}, \theta) &= \exp(-i(\delta_H \boldsymbol{\sigma}_1 + \delta_{45} \boldsymbol{\sigma}_2)/2) \\ &= \boldsymbol{\sigma}_0 \cos\left(\frac{\delta_{linear}}{2}\right) - i \sin\left(\frac{\delta_{linear}}{2}\right)\left(\frac{\delta_H \boldsymbol{\sigma}_1 + \delta_{45} \boldsymbol{\sigma}_2}{\delta_{linear}}\right). \end{aligned} \tag{15.44}$$

Next, $\mathbf{LR}(\delta_{linear}, \theta)$ has its angle doubled, ($\theta \to 2\theta$), and is treated as a vector function to be expanded into vector Zernike polynomials. The linear retardance is $\delta_{linear} = \sqrt{\delta_H^2 + \delta_{45}^2}$. $\mathbf{LR}$ corresponding to $\delta_H \boldsymbol{\sigma}_1 + \delta_{45} \boldsymbol{\sigma}_2$ is to be expanded in orientors, $\mathbf{O}(\delta_{linear}, 2\theta)$, by doubling the orientation angles and expanding $\mathbf{O}(\delta_{linear}, 2\theta)$ in the vector Zernike polynomials. The phase, the "scalar" wavefront aberration, will be expanded in "scalar" or ordinary Zernike polynomials, as is usual. The linear retardance normally comprises the majority of the retardance, but any significant circular retardance can also be expanded in its own set of scalar Zernike polynomials. Characterizing and understanding the linear retardance generally has a greater priority than the circular retardance.

Now, the linear retardance is given by a sum of vector Zernike polynomials,

$$\mathbf{O}(\delta_{Lin}, 2\theta)(\rho, \phi) = \sum_{n=1}^{\infty} \sum_{m=-n}^{n} \sum_{e=0}^{1} \Delta_{n,e}^m \mathbf{V}_{n,e}^m(\rho, \phi), \tag{15.45}$$

in a form similar to our Pauli matrix representation. The coefficients $\Delta_{n,e}^m$,

$$\int_0^1 \int_0^{2\pi} \mathbf{O}(\delta_{Lin}, 2\theta)(\rho, \phi) \cdot \mathbf{E}_1(\rho, \phi) \rho \, d\phi \, d\rho = \Delta_{n,e}^m, \tag{15.46}$$

describe the amount of each vector Zernike polynomial term, and the corresponding orientor term in the retardance pupil map, analogous to Zernike polynomial coefficients for a wavefront.

It was shown in Section 15.2.1 that for small values of linear retardance, $\delta_{linear} \ll 1$, retardances in the Pauli form add. Therefore, the linear retardance orientors expressed as vector Zernike polynomials also add in the weak retardance limit. If several parts of a weakly retarding system are expressed as vector Zernike polynomials,

$$\mathbf{O}_1(\rho, \phi) = \sum_{n=1}^{\infty} \sum_{m=-n}^{n} \sum_{e=0}^{1} \Delta_{n,e}^m \mathbf{V}_{n,e}^m(\rho, \phi), \quad \mathbf{O}_2(\rho, \phi) = \sum_{n=1}^{\infty} \sum_{m=-n}^{n} \sum_{e=0}^{1} E_{n,e}^m \mathbf{V}_{n,e}^m(\rho, \phi), \text{etc.}, \tag{15.47}$$

the resulting linear retardance distribution can be expressed approximately as the sum of the coefficients for each corresponding vector Zernike polynomial term, $\Delta_{n,e}^m + E_{n,e}^m + F_{n,e}^m + \ldots$. An example of the application of this method is found in Section VIII of the work of Ruoff and Totzeck.[8]

580 Chapter 15: Paraxial Polarization Aberrations

To express diattenuation maps in an expansion of orientors, the same procedure is applied as used above for retardance, except the linear diattenuation is obtained from the matrix logarithm of the hermitian part of the Jones matrix following the procedure of Section 14.4.5.

15.6 Polarization Aberration Measurements

Polarimetry is useful for measuring the *polarization aberrations* of optical systems and for characterizing optical and polarization components. Here, a few examples of polarization aberration measurements are provided. Optical system polarization aberrations can be measured by placing the system in the sample compartment of a *Mueller matrix imaging polarimeter* such as the Axometrics AxoStep Muller matrix imaging polarimeter. Usually, the exit pupil is imaged, measuring a Mueller matrix as a function of pupil coordinates. Maps of linear diattenuation, linear retardance, and other metrics are readily generated. Such a Mueller matrix pupil image is readily converted to a *Jones pupil*, but the *absolute phase*, the *wavefront aberration*, is not measured by a non-interferometric Mueller matrix image set.

Figure 15.49 (top) shows the polarimeter configuration to measure polarization aberrations for a pair of 0.55 numerical aperture microscope objectives. Collimated light from the polarization

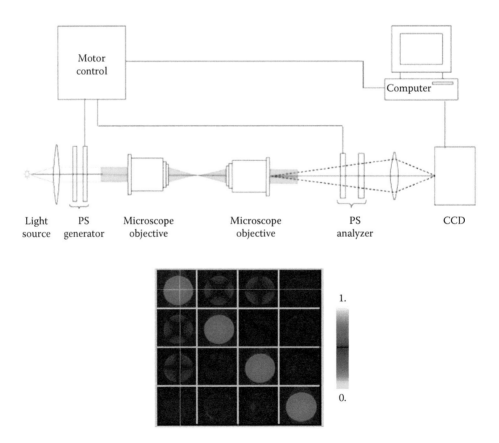

Figure 15.49 A schematic of a Mueller matrix imaging polarimeter measuring the polarization aberration of a pair of microscope objectives (top). Polarization states (PS) Generator is the polarization state generator and PS Analyzer is the polarization state analyzer. In this configuration, the camera is focused on the exit pupil of the microscope objectives. (Bottom) The Mueller matrix image for a microscope objective pair is close to the identity matrix with weak linear diattenuation evident in the top row and retardance in the off-diagonal elements of the lower right 3 × 3 elements.

15.6 Polarization Aberration Measurements

generator enters the pupil of the first objective, focuses at the joint focal point of the two objectives, is recollimated by the second objective, and measured in the polarization state analyzer. Figure 15.49 (bottom) shows a measured Mueller matrix pupil image for such a pair of objectives, specifically sold as low polarization objectives for polarization microscopes. Figure 15.50 plots the diattenuation and retardance maps calculated from the Mueller matrix image. This microscope objective pair has up to 5.4° of spatially varying retardance and 0.1 of spatially varying diattenuation. When placed between crossed linear polarizers, this pair of objectives will leak about 0.15% of the incident flux, averaged over the pupil.

Figure 15.51 shows the diattenuation and retardance aberrations of another microscope pair where the polarization aberrations were further reduced by thin film coating design. Figure 15.52

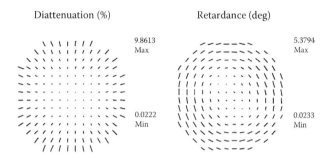

Figure 15.50 Linear diattenuation and linear retardance pupil maps for a pair of microscope objectives are nearly radially symmetric as expected. Deviations from radial symmetry are likely due to slight tilts and decenters.

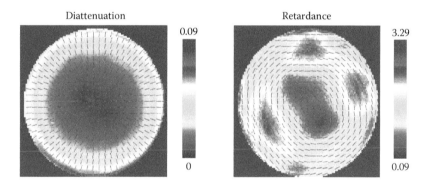

Figure 15.51 Diattenuation (left) and retardance (right) aberrations for another microscope objective pair with a reduced polarization coating design. Leakage through crossed polarizers for the low polarization microscope objective pair.

Figure 15.52 (Left) A microscope objective between crossed polarizers, and (right) the flux distribution in the exit pupil.

(left) schematically shows a microscope objective between crossed polarizers, while Figure 15.52 (right) shows the corresponding flux distribution in the exit pupil.

When significant polarization aberrations are present, an optical system illuminated with a uniform polarization state will have polarization variations within the *point spread function*. To characterize these variations and the dependence of the point spread function on the incident polarization state, a *Mueller matrix imaging polarimeter* focuses on the image of a point object and measures the *Mueller Point Spread Matrix*, **MPSM**, as a Mueller matrix image (Section 16.5). A measured **MPSM** with large polarization aberration is shown in Figure 15.53. A *vortex retarder* (Section 5.6.3) was placed in the pupil of an imaging system with a large F/# image on a camera focal plane, and a Mueller matrix image was acquired. This vortex retarder is a half wave linear retarder whose fast axis varies as a function of pupil angle.[9] The pupil image on the left side shows the retardance orientation varying by 360° around the pupil. The right side contains the **MPSM**. When the Stokes vector of the incident light is multiplied by the **MPSM**, the resulting Stokes vector function describes the flux (point spread function) and polarization state variations within the image as a *Stokes vector image*. Figure 15.54 shows the point spread function for a fixed incident polarization state and several analyzers, demonstrating the polarization variations within the point spread function.

Figure 15.55 shows the *depolarization aberration* measured from a lens whose coatings were damaged by heat and began flaking off.[10] The resulting Mueller matrix pupil image shows a few tenths of a percent depolarization in the damaged area. The undamaged area has a depolarization of only a few hundredths of a percent, more typical of coated lenses.

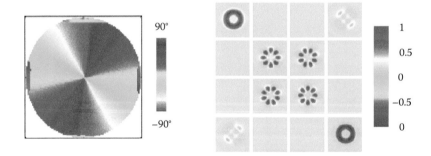

Figure 15.53 (Left) The orientation of the fast axis of the half wave vortex retarder rotates by 360° around the pupil. (Right) The Mueller point spread matrix (**MPSM**) describes the polarization dependence of the point spread function as a Mueller matrix image.

Figure 15.54 The measured point spread function of the vortex retarder completely changes with the analyzed polarization state: (left) no analyzer, (middle) horizontal linear analyzer, (right) vertical linear analyzer. Horizontal linearly polarized light is input.

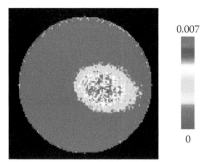

Figure 15.55 Depolarization index of lens with coating damage on the right center causing about 0.005 depolarization.

15.7 Summary and Conclusion

Paraxial optics provides straightforward and meaningful definitions of focal length, and the other "first order" properties of optical systems. Paraxial optics forms the underlying coordinate system for aberration theory. The Seidel wavefront aberrations are defined as deviations from paraxial performance. Similarly, for the derivation of polarization aberrations, paraxial optics forms an excellent basis for deriving the low-order forms of polarization aberration. In fact, the fraction of the *etendué* of an optical system that is well described by second-order polarization aberrations is generally far larger than the fraction of the etendué described by the *fourth-order wavefront aberrations*, that is, the region where the contributions of *spherical aberration*, *coma*, *astigmatism*, and *field curvature* is much less than one wave of optical path length.

The optical designer and optical engineer should be pleased to know that 95% of the polarization aberration of most optical systems can often be described with just three terms, polarization defocus, polarization tilt, and polarization piston.

15.8 Appendix

15.8.1 Paraxial Optics

Paraxial optics is the optics of ray paths in the vicinity of the optical axis through radially symmetric optical systems. As ray paths approach the optical axis, linear approximations to Snell's law and the location of ray intercepts become increasingly accurate. In paraxial optics, all the rays from an object point intersect at the same image point, forming a "perfect image." Thus, paraxial optics forms an excellent coordinate system for describing aberrations; aberrations are the deviations from paraxial behavior. Here, a brief summary of paraxial ray tracing is provided and augmented with the calculation of angles of incidence and propagation vectors for paraxial skew rays, a key result.

Paraxial optics is used to define focal length, nodal points, principal planes, pupil locations, magnification, and the other "first order" properties of optical systems. Our interest is primarily in the paraxial polarization aberrations; hence, these calculations are not treated here; the reader is referred to *Field Guide to Geometrical Optics* by John Greivenkamp,[11] whose notation has been adopted here.

The *paraxial region* of an optical system is a small region close to the *optical axis* where the ray paths are accurately calculated by applying the linearized form of Snell's law. At a refracting interface, *Snell's law* relates the angle of incidence θ_1 in an incident medium of refractive index n_1 to the angle of refraction θ_2 in a medium with refractive index n_2,

$$n_1 \sin\theta_1 = n_2 \sin\theta_2. \tag{15.48}$$

For rays propagating very near the optical axis, the angle of incidence is very small; thus, replacing $\sin\theta$ with its linear approximation θ yields the paraxial form of Snell's law

$$n_1 \theta_1 = n_2 \theta_2. \tag{15.49}$$

In calculating the *ray intercepts* of paraxial rays, a linear approximation to the intercept is all that is needed. Since the sag of spheres, parabolas, and other conics varies quadratically about the vertex,

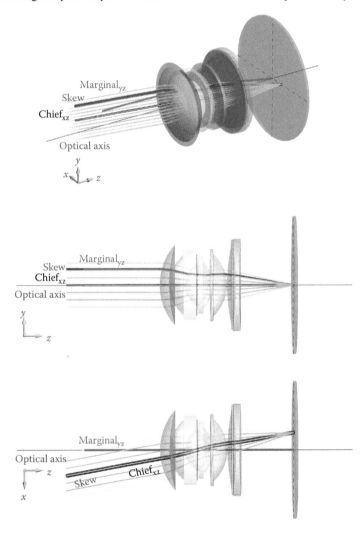

Figure 15.56 A grid of collimated rays (gray) at a 30° field angle in the xz-plane propagates through an example lens is shown in perspective view, in the yz-plane, and in the xz-plane. The chief ray of the grid of rays is shown in black propagating in the xz-plane. A marginal ray in the yz-plane is shown in red. A skew ray (purple) within the grid of collimated rays can be calculated by a combination of the marginal and chief rays (1 marginal ray height + 1 chief ray height).

the paraxial ray trace can ignore the sag; the paraxial ray intercept is the intersection of the ray with the vertex plane.

Because of the linearity of paraxial optics, all paraxial rays can be formed from linear combinations of two linearly independent paraxial rays. By convention, these two rays are chosen as the *marginal ray* and the *chief ray*. The marginal ray is chosen in the y–z plane from the center of the object through the top of the entrance pupil and edge of the aperture stop. The chief ray is a ray from a point on the edge of the field of view, through the center of the entrance pupil and center of the aperture stop. We distinguish between the paraxial marginal ray and the (real) marginal ray, and the paraxial chief ray and (real) chief ray. Figure 15.56 shows an example optical system with a marginal ray in the y–z plane, a chief ray in the x–z plane, and a skew ray formed from the addition of the yz marginal ray to the xz chief ray.

Using the results of a paraxial ray trace for the chief and marginal rays, together with quadratic Taylor series expansion coefficients for the Fresnel coefficients and amplitude reflection and transmission coefficients, yields simple approximate and easy-to-calculate polarization aberration expansions.

15.8.2 Setting Up the Optical System

The radially symmetric optical system at its reference wavelength is defined by a set of thicknesses, t_q, refractive indices, n_q, and curvatures, C_q. *Curvature* is the reciprocal of the radius of curvature R of a surface and has units of mm^{-1}. The index $q = 0, 1, 2, \ldots, Q-1, Q$ labels the surfaces. $q = 0$ indicates the object surface, also indicated by subscript O. $q = Q = I$ indicates the image surface. Subscript E indicates the entrance pupil; often, the entrance pupil is the first surface in the optical prescription. Similarly, subscript X indicates the exit pupil. Typically, $q = Q - 1$ will be the exit pupil. The *optical power*, Φ_q, the ability to focus light, of surface q is

$$\Phi_q = (n_q - n_{q-1})c_q, \qquad (15.50)$$

which is measured in inverse mm^{-1}.

The paraxial ray trace calculation can be organized into a standard table, the y–n–u ray tracing form (Table 15.2), or the information can be organized into an equivalent data structure in a computer. The optical system parameters, C_q, t_q, and n_q, are entered in the first three rows. The lens surface powers, Φ_q, and reduced thicknesses on rows four and five are calculated.

Table 15.2 y–n–u Paraxial Ray Tracing Form

	C_1		C_2		C_3		C_4	
t_0		t_1		t_2		t_3		t_4
n_0		n_1		n_2		n_3		n_4

	$-\Phi_1$		$-\Phi_2$		$-\Phi_3$		$-\Phi_4$	
t_0/n_0		t_1/n_1		t_2/n_2		t_3/n_3		t_4/n_4

	y_0		y_1		y_2		y_3		y_4
		nu_0		nu_1		nu_2		nu_3	
		u_0		u_1		u_2		u_3	

	$\bar{y}_0$		$\bar{y}_1$		$\bar{y}_2$		$\bar{y}_3$		$\bar{y}_4$
		$\overline{nu}_{c0}$		$\overline{nu}_{c1}$		$\overline{nu}_{c2}$		$\overline{nu}_{c3}$	
		$\bar{u}_{c0}$		$\bar{u}_{c1}$		$\bar{u}_{c2}$		$\bar{u}_{c3}$	

15.8.3 The Paraxial Ray Trace

The general paraxial ray trace procedure traces two rays, the *paraxial marginal ray* (from the center of the object to the edge of the entrance pupil) and full-field *paraxial chief ray* (from the top of the object to the center of the entrance pupil).[12] All paraxial rays can be calculated from linear combinations of these two rays. The *paraxial ray trace* will be performed in the y–z plane. The optical system is assumed fixed. Rays are started and their paths through the system are calculated. Consider an example meridional ray in the y–z plane. The ray is started at the object plane with y_0 and u_0. The intercept of the ray is found with the first surface's vertex plane, y_1. The *angle of incidence*, θ_1, is calculated from u_0 and the normal. Then, the ray is reflected or refracted determining u_1. The process is repeated for the second surface, the third surface, and so on, until the image is reached.

The paraxial marginal ray heights are indicated by $y_0, y_1, y_2, \ldots y_{Q-1}, y_Q$. The paraxial marginal ray angles, the angle with respect to the optical axis, are indicated by $u_0, u_1, u_2, \ldots u_{Q-1}, u_Q$. The slope of a ray, u, is positive if a counterclockwise rotation brings the axis to the ray. The paraxial angles are defined as the tangents of the actual angles; thus, the paraxial ray trace is linear with respect to ray angles and ray heights. The marginal ray angles of incidence are indicated by $\theta_0, \theta_1, \theta_2, \ldots \theta_{Q-1}, \theta_Q$. Chief ray quantities are indicated by the same letters and subscripts except with a bar over the letter, such as $\bar{y}_q, \bar{u}_q$, and $\bar{\theta}_q$.

The marginal ray is started on axis with $y_0 = 0$, and u_0 chosen to strike the edge of the entrance pupil. The chief ray is started at the edge of the object $\bar{y}_q$ with the angle $\bar{u}_q$ chosen to pass through the center of the entrance pupil. The starting values for the marginal ray $y_0 = 0$ and u_0 are entered on the left side of the next two rows, selected so that the marginal ray intersects the edge of the entrance pupil. The starting values for the chief ray, $\bar{y}_O$ and $\bar{u}_O$, are entered in the next two rows selected so that the chief ray intersects the center of the entrance pupil. With the marginal and chief ray starting values defined, the rays are transferred from each surface q to the next surface $q + 1$ with the paraxial ray transfer equations,

$$y_{q+1} = y_q + u_q t_q \text{ and } \bar{y}_q = \bar{y}_{q-1} + \bar{u}_q t_q. \tag{15.51}$$

Then, the paraxial refraction equation is applied with the revised values y_q and $\bar{y}_q$ to calculate the marginal ray angle

$$n_q u_q = n_{q-1} u_{q-1} - y_q \Phi_q \tag{15.52}$$

and chief ray angle

$$n_q \bar{u}_q = n_{q-1} \bar{u}_{q-1} - \bar{y}_q \Phi_q. \tag{15.53}$$

Transfer and refraction are repeatedly applied to complete these rows, systematically filling in the blank entries of the ray tracing form. The *angles of incidence*, θ_q, defined as the angle between a ray and the *normal* to its ray intercept, are then calculated for the marginal and chief ray intercepts,

$$\theta_q = u_q - \eta_{paraxial\ sphere} = u_q + y_q C_q. \tag{15.54}$$

θ is positive if a counterclockwise rotation brings the surface normal to the ray as shown in Figure 15.57. The paraxial ray trace algorithm presented in many references does not include θ;

15.8 Appendix

it is not needed for finding ray coordinates and cardinal points. For polarization analysis, calculating θ is an necessary objective since θ is needed for evaluating Fresnel and amplitude coefficients and calculating the polarization of interfaces.

For a reflecting surface, the power is set to $-\Phi_q$. A plane mirror has a power of -1. The surface normal $\boldsymbol{\eta}$ at a point $\mathbf{r}$ on a surface is a vector in the plane of incidence that is perpendicular to the tangent plane at $\mathbf{r}$. By convention, $\boldsymbol{\eta}$ points from the ray intercept away from the incident medium into the refracted medium. The slope of the surface normal, the scalar η, is likewise positive if clockwise rotation brings the optical axis to $\boldsymbol{\eta}$. The paraxial angle of $\boldsymbol{\eta}$ is calculated for a parabolic fit to the surface. For a spherical surface with curvature C, the sag in the y–z plane, and its second order approximation, is

$$z(y) = R - \sqrt{R^2 - y^2} \approx \frac{y^2}{2R} = \frac{C y^2}{2}. \tag{15.55}$$

As seen in Figure 15.58, to second order, the sphere is the same as the osculating parabola. The slope of the paraxial sphere's normal is

$$\eta_{paraxial\ sphere} = yC. \tag{15.56}$$

The definitions of u and $\eta_{paraxial\ sphere}$ are consistent with the conventional definition of slope as

$$m = \frac{df(y)}{dz}. \tag{15.57}$$

15.8.4 Reduced Thicknesses and Angles

The paraxial transfer and refractions can be simplified by incorporating the refractive index into angles and thicknesses. Reduced angles ω are paraxial angles times the refractive index,

$$\omega_q = n_q u_q \text{ and } \overline{\omega}_q = n_q \overline{u}_q. \tag{15.58}$$

Reduced thicknesses τ are physical thicknesses divided by the refractive index

$$\tau_q = t_q / n_q. \tag{15.59}$$

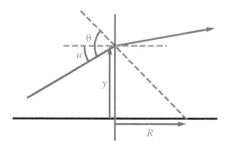

Figure 15.57 A ray (blue) with incident angle θ, paraxial angle u, and ray height y intercept at a surface with radius of curvature R.

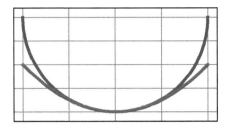

Figure 15.58 Sphere (blue) and its osculating parabola (red) at the vertex match shapes to second order.

With reduced quantities, the marginal and chief paraxial transfer and refraction equations take the following simplified form at the qth ray intercept:

$$\begin{aligned} y_{q+1} &= y_q + \omega_q \tau_q, \\ \bar{y}_{q+1} &= \bar{y}_q + \bar{\omega}_q \tau_q, \\ \omega_{q+1} &= \omega_q - y_q \phi_q, \\ \bar{\omega}_{q+1} &= \bar{\omega}_q - \bar{y}_q \phi_q. \end{aligned} \quad (15.60)$$

15.8.5 Paraxial Skew Rays

The skew aberration algorithm of Section 18.5 requires the propagation vector $\mathbf{k}_q$ for paraxial skew rays. Because of the linearity of paraxial optics, any meridional paraxial ray can be expressed as the linear combination of two linearly independent meridional rays, such as the marginal and chief rays. An arbitrary paraxial skew ray can be expressed as the linear combination of four linearly independent paraxial rays; the chief and marginal rays in the y–z plane and the chief and marginal rays rotated into the x–z plane are used as the basis ray set.

Thus, for the ray from the object point (G, H) through stop location (x, y), the ray intercept at the qth surface is

$$(x_q, y_q) = \left(G\,\bar{y}_q + x_e\,y_{m,q},\; H\,\bar{y}_q + y_e\,y_{m,q}\right), \quad (15.61)$$

where G and H are normalized object coordinates, G along the x-axis and H along the y-axis. Normalization is performed such that around the edge of a circular field of view, as defined by the chief ray height on the object plane, $\sqrt{G^2 + H^2} = 1$. Likewise, the ray slope after the qth interface is

$$(u_{x,q}, u_{y,q}) = (G\,u_{c,q} + x_e\,u_{m,q},\; H\,u_{c,q} + y_e\,u_{m,q}). \quad (15.62)$$

For skew rays, it is necessary to distinguish between quantities measured relative to the x-axis, subscript x, and y-axis, subscript y.

The skew aberration calculation uses the propagation vectors $\mathbf{k}_q$ specified after the qth ray intercept is along $\left(y_{q+1} - y_q, \bar{y}_{q+1} - \bar{y}_q, t_q\right)$. The normalized propagation vectors, $\mathbf{k}_q$, are

$$\mathbf{k}_q = \frac{(w_q, \bar{w}_q, 1)}{\sqrt{w_q^2 + \bar{w}_q^2 + 1}} \approx (w_q, \bar{w}_q, 1). \quad (15.63)$$

15.9 Problem Sets

15.1 Find the polarization aberration function for a paraxial silicon interface, $n = 4$, with a marginal ray angle of incidence of 0.2 radians for the on-axis beam.

15.2 Express the Jones matrices for the following polarization elements in the weak polarization element form:

$$\mathbf{J} = \rho_0 e^{-i\phi_0} \left[\begin{pmatrix} 1 & 0 \\ 0 & 1 \end{pmatrix} + \frac{d_1 - i\delta_1}{2} \begin{pmatrix} 1 & 0 \\ 0 & -1 \end{pmatrix} + \frac{d_2 - i\delta_2}{2} \begin{pmatrix} 0 & 1 \\ 1 & 0 \end{pmatrix} + \frac{d_3 - i\delta_3}{2} \begin{pmatrix} 0 & -i \\ i & 0 \end{pmatrix} \right]$$

a. $\mathbf{J}_1$: An ideal linear diattenuator with $D_1 = 0.02$, a transmission axis at 0°, and an average transmission of 1.
b. $\mathbf{J}_2$: An ideal linear retarder with $\delta_2 = 0.006$ and a fast axis at 45°, and an average transmission of 1.
c. $\mathbf{J}_3$: An ideal linear retarder with $\delta_3 = 0.01$ and a fast axis at $\theta_3 = \tan^{-1}(3/4)/2$, and an average transmission of 3/5.
d. Express $\mathbf{J}_1 \mathbf{J}_2$ and $\mathbf{J}_2 \mathbf{J}_1$ in weak polarization element form.
e. Which polarization effects are order independent?
f. Which polarization effects are order dependent?
g. Express $\mathbf{J}_1 \mathbf{J}_2 \mathbf{J}_3$ in weak polarization element form, keeping only first order terms.
h. Show how for $\mathbf{J}_1 \mathbf{J}_2 \mathbf{J}_3$ the first-order terms are order independent and are sums over the Pauli coefficients.

15.3 A beam of light goes through a series of five weak diattenuators each with diattenuation $D = 0.01$. The diattenuation transmission axes are oriented at angle 0°, 22.5°, 45°, 67.5°, and 90°.
a. Express each Jones matrix in the weak polarization element form. Find the σ_1 and σ_2 components for each element.
b. Add the components. What is the net diattenuation?
c. What is the orientation of the diattenuation transmission axis?

15.4 Row 1 contains eight polarization aberration maps, labeled A through H. Brown lines represent pupil maps of the magnitude and orientation of linear diattenuation. Pink lines represent linear retardance. These aberration maps are used to generate the ellipse maps in rows 2 and 3. Row 2 has been rearranged, as has row 3.
a. Match each ellipse map in row two, labeled 1 through 8, with the corresponding polarization aberration map and specify the incident polarization state.
b. Match each ellipse map in row three, labeled α through θ, with the corresponding polarization aberration map and specify the incident polarization state.

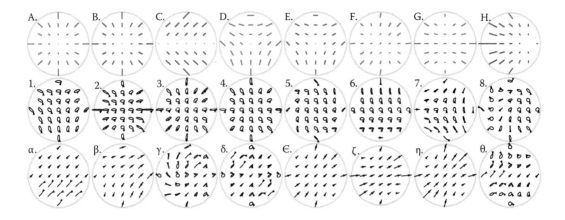

15.5 An optical system has retardance tilt for a particular object point. The diattenuation is zero everywhere in the pupil. The retardance magnitude increases linearly from the center of the pupil. The orientation ψ in degrees of the retardance fast depends on the angle θ in the pupil as $\psi(\theta) = 45 - \theta/2$. This retardance aberration pattern can be graphically represented as follows:

The position of the center of each line represents a location in the pupil. The length of the line represents the retardance magnitude. The orientation represents the orientation of the retardance fast axis. Assume that the maximum retardance at the edge of the pupil is much less than 1 radian. (Note: No calculations are requested or necessary in this problem.)

    a. Where in the pupil is there no change of polarization state for all polarization states?

    b. Where in the pupil is there no change of polarization state for 45° linearly polarized light?

    c. If the optical system is placed in a linear polariscope with an initial horizontal linear polarizer and final vertical polarizer, what will the distribution of flux be across the exit pupil? Does the flux vary linearly or quadratically?

    d. If the optical system is placed in a circular polariscope with an initial right circular polarizer and final left circular polarizer, describe the distribution of flux in the exit pupil.

    e. For what incident polarization states is the polarization change, integrated across the pupil, the largest, and why? This is the total leakage in the polariscope.

15.6 A four-element lens has eight surfaces. All elements are fabricated from the same glass. A coated lens assembly is fabricated and an uncoated lens assembly is fabricated. These are to be compared. The surfaces are described by the intensity transmittances shown below. The left graph is for light entering the lenses (external refraction, odd-numbered interfaces). The right graph is for light exiting the lenses (internal refraction).

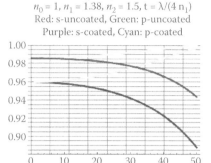

Uncoated/single-layer film comparison
$n_0 = 1$, $n_1 = 1.38$, $n_2 = 1.5$, $t = \lambda/(4\,n_1)$
Red: s-uncoated, Green: p-uncoated
Purple: s-coated, Cyan: p-coated

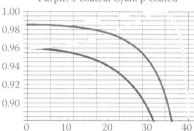

Uncoated/single-layer film comparison
$n_0 = 1.5$, $n_1 = 1.38$, $n_2 = 1$, $t = \lambda/(4\,n_1)$
Red: s-uncoated, Green: p-uncoated
Purple: s-coated, Cyan: p-coated

a. Estimate the intensity transmittance and the diattenuation down the axis for the coated and uncoated lenses.
b. If the marginal ray had angle of incidences of 0.8, 0.4, 0.6, 0.1, 0.2, 0.3, 0.7, 0.5 radians at each successive lens surface, estimate the diattenuation for the marginal ray path through the uncoated lens.
c. Repeat part (b) for antireflection-coated lens.

15.7 The following questions (a) to (d) assume a glass where diattenuation for small angles equals $D(AoI) = d_2\,AoI^2$. Write a polarization aberration function in terms of σ_0, σ_1, σ_2, and σ_3 for each question. An example would be the polarization aberration function for polarization defocus with a shift along the y-axis, Jones matrix $\mathbf{J}(x,y) = \sigma_0 + \dfrac{d_H - i\delta_H}{2}\left[\sigma_1(x^2 - (y - y_o)^2) + \sigma_2 2x(y - y_o)\right]$, where d_H is diattenuation magnitude at edge of pupil, δ_H is retardance magnitude at edge of pupil, y_o is pupil shift along the y-axis, and (x, y) are normalized pupil coordinates. (Feel free to make simplifying approximations.)

a. A collimated beam of light is deviated by 30° at a prism. The angles in air are equal at each surface as are the internal angles.
b. A collimated beam of diameter 2 mm enters a spherical surface of radius 10 mm, where the vertex of the spherical surface is shifted toward $+x$ by 1 mm.
c. Light enters a flat glass surface of diameter 2 mm at normal incidence and then exits through a cubic phase plate $z = 0.01\,(x + y)^3$.
d. A converging spherical beam enters and focuses at the center of a spherical marble and then refracts out the back side.

For the following, assume a metal reflector whose diattenuation for small angles equals $D(AoI) = d_2\,AoI^2$ and whose retardance for small angles equals $\delta(AoI) = \delta_2\,AoI^2$.

e. A beam of numerical aperture 0.2 with a central ray that reflects at normal incidence from the reflector.
f. A beam of numerical aperture 0.2 with a central ray with $\mathbf{k} = (0, 0, 1)$ that reflects at normal incidence from a reflector with normal $\boldsymbol{\eta} = \left(0.2, 0, \sqrt{1 - 0.2^2}\right)$.
g. A beam of numerical aperture 0.2 with a central ray with $\mathbf{k} = (0, 0, 1)$ that reflects at normal incidence from a reflector with normal $\boldsymbol{\eta}_1 = (\sin(0.2), 0, \cos(0.2))$, and then from a second mirror with normal $\boldsymbol{\eta}_1 = \left(\sin(0.4), \sin(0.2), \sqrt{1 - \sin^2(0.4) - \sin^2(0.2)}\right)$.
h. Reflection of a 2 mm diameter collimated beam from a toroid $z = 0.06\,x\,y$.

15.8 Two identical pieces of sheet retarder have a retardance of $\delta = 0.1 \approx 6°$. As a result of the fabrication process, stretching, the fast axis of retardance varies steadily from one side of the retarder to the other, as in the figure, by ±3°. One of the retarders is then rotated by 90°, so the fast axes are crossed at the center.

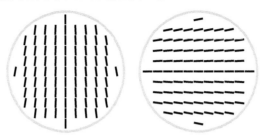

   a. Calculate the retardance of the composite retarder using the weak polarization approximation.
   b. What higher-order effects would be present, and about how small would they be?
   c. What would be the leakage between a horizontal and vertical polarizer?

15.9 For $H = (0, h_0)$, given the chief and marginal angles of incidence, i_c and i_m, which point in the pupil is at normal incidence, $\theta = 0$?

15.10 What is the condition on i_c and i_m for $\theta(H = 1, \rho = 1)$ to equal zero?

15.11 In Figure 15.8, assume the diattenuation is 0.3 for the marginal ray, if the four wavefronts are transmitted through a vertical analyzer, what will the flux $P(\rho,\phi)$ be?

References

1. J. Sasián, *Introduction to Aberrations in Optical Imaging Systems*, Figure 15.8, Cambridge University Press (2013).
2. H. Kuboda and S. Inoue, Diffraction images in the polarizing microscope, *J. Opt. Soc. Am* 49 (1959): 191–192.
3. R. A. Chipman, Polarization analysis of optical systems, *Opt. Eng.* 28(2) (1989): 90–99.
4. R. A. Chipman, Polarization aberrations, PhD dissertation, Optical Sciences Center, University of Arizona, Tucson, AZ (1987).
5. J. P. McGuire and R. A. Chipman, Polarization aberrations. 1. Rotationally symmetric optical systems, *Appl. Opt.* 33 (1994): 5080–5100.
6. R. V. Shack and K. Thompson, Influence of alignment errors of a telescope system on its aberration field, in *Proc. SPIE* 251, Optical Alignment I, 146 (1980).
7. K. Thompson, Description of the third-order optical aberrations of near-circular pupil optical systems without symmetry, *J. Opt. Soc. Am. A* 22 (2005): 1389–1401.
8. J. Ruoff and M. Totzeck, Orientation Zernike polynomials: A useful way to describe the polarization effects of optical imaging systems, *J. Micro/Nanolithogr. MEMS MOEMS* 8.3 (2009): 031404.
9. S. C. McEldowney, et al., Vortex retarders produced from photo-aligned liquid crystal polymers, *Opt. Exp.* 16(10) (2008): 7295–7308.
10. J. Wolfe and R. A. Chipman, Reducing symmetric polarization aberrations in a lens by annealing, *Opt. Exp.* 12(15) (2004): 3443–3451.
11. J. E. Greivenkamp, *Field Guide to Geometrical Optics*, SPIE Field Guides 1 (2004).
12. B. R. Irving, et al., *Code V, Introductory User's Guide*, Optical Research Associates (2001), p. 82.

16

Image Formation with Polarization Aberration

16.1 Introduction

A fundamental metric for optical imaging systems is the *point spread function* (PSF), which describes the image of a point object. To calculate a PSF, the methods of geometrical optics and ray tracing are augmented by physical optics and Fourier optics calculations. For systems with polarization aberration, the PSF depends on the incident polarization state.[1] In this chapter, the concept of a point spread function is generalized so that it describes imaging of arbitrary polarization states by introducing an *amplitude response matrix* (**ARM**) and *Mueller point spread matrix* (**MPSM**). It will be shown how the point spread function calculation of a conventional optical design program can be generalized through the use of Jones matrices and polarization ray tracing matrices. Next, the calculation of images for objects with spatially varying polarization states is considered. Polarization aberration affects the image quality of a system. Chapter 12 (Fresnel Aberrations) and Chapter 15 (Paraxial Polarization Aberrations) provided examples of polarization aberrations in optical systems and their effects on the PSF. Given a *wavefront aberration function* and *Jones pupil function*, the **MPSM** is calculated, which shows the polarization structure of the PSF and how the PSF varies with the incident polarization state. From the **MPSM**, an image of a point source with arbitrary polarization state can be calculated and one can vary incident polarization states and observe the resulting image polarization structure. Similarly, the *optical transfer function* (OTF) of conventional optics can be extended to *optical transfer matrix* (**OTM**) to show how the spatial filtering of an object during imaging depends on the incident polarization state.

The image of an extended object is calculated by replacing each point in the object with the corresponding response function. In lenses and other imaging systems, the response function varies

⟨Coherent imaging⟩

$$pupil = P$$
$$\downarrow \mathfrak{F}$$
$$cPSF = h = \mathfrak{F}[P]_{\xi=\frac{x}{\lambda f}}$$
$$\downarrow \mathfrak{F}$$
$$CTF = \frac{\mathfrak{F}[h]}{|\mathfrak{F}[h]|} = H \propto P$$

⟨Incoherent imaging⟩

$$pupil = P$$
$$\downarrow \mathfrak{F}$$
$$iPSF = |cPSF|^2 = |h|^2 = \left|\mathfrak{F}[P]_{\xi=\frac{x}{\lambda f}}\right|^2$$
$$\downarrow \mathfrak{F}$$
$$OTF = \frac{\mathfrak{F}[|h|^2]}{|\mathfrak{F}[|h|^2]|} = H \star H^*$$
$$= MTF \exp(iPTF)$$

Figure 16.1 Flowcharts for coherent and incoherent imaging calculations. The pupil function P represents the wavefront function at the exit pupil of an optical system. The spatial variable of the system is x; the wavelength of light is λ; the focal length of the imaging system is f. The coherent point spread function is $cPSF$ (i.e., amplitude response function, ARF); the coherent transfer function is CTF; the incoherent point spread function is $iPSF$; the optical transfer function is OTF; the modulation transfer function is MTF; and the phase transfer function is PTF. $\mathfrak{F}$ is the Fourier transform operation. $\star$ is the auto-correlation operation.

⟨Imaging polarized light⟩

$$Jones\,pupil = J_{pupil}$$
$$\downarrow \mathfrak{F}$$
$$\mathbf{ARM} = \mathfrak{F}[J_{pupil}] \rightarrow \mathbf{MPSM}$$
$$\downarrow \mathfrak{F}$$
$$\mathbf{OTM} = \mathbf{MTM} \exp(i\mathbf{PTM})$$

Figure 16.2 The imaging calculation for polarized light. The 2 × 2 amplitude response matrix is the **ARM**; the 4 × 4 Mueller point spread matrix is the **MPSM**; the 4 × 4 optical transfer matrix is the **OTM**; the 4 × 4 modulation transfer matrix is the **MTM**; and the 4 × 4 phase transfer matrix is the **PTM**.

across the field of view as the aberrations steadily change, moving from on-axis to other parts of the field. An *isoplanatic patch* is a region of the field over which the changes of the PSF are small. Then, simple linear system analysis methods (the **ARM** and **MPSM**) can be applied over space-invariant patches of the object plane, whose size depends on the rate of change of variation in wavefront aberration function and $JonesPupil(x, y)$.[2]

The flowcharts of the algorithms for coherent and incoherent imaging of scalar waves and for the imaging of systems with polarization aberrations are compared in Figures 16.1 and 16.2. Each operand and operation will be described in detail in this chapter.

16.2 Discrete Fourier Transformation

The calculation of diffraction in image formation involves Fourier transformation of the exit pupil functions. Most textbooks on diffraction and image formation provide examples that involve simple Fourier transform pairs such as Fourier transforming a *rectangle function* into a *sinc function* or a *Gaussian* into a *Gaussian function*. In optical design, it is necessary, however, to Fourier transform arbitrary functions such as *high-order aberrations* and *irregular pupil shapes* to calculate PSFs. Further, the results from ray tracing are also *sampled functions* and not continuous functions. For such pupil functions in optical design, evaluating the Fourier transforms as continuous functions is difficult and impractical. The *discrete Fourier transform* (DFT)[3–5] provides a straightforward algorithm to Fourier transform arbitrary functions on a regular sampled grid. The DFT is used in

16.2 Discrete Fourier Transformation

most optical design software diffraction calculations. All mathematical software packages such as MATLAB, Mathematica, and so on, provide built-in DFT algorithms; hence, only a brief summary is given in Math Tip 16.1.

Math Tip 16.1 1D Discrete Fourier Transformation

The one-dimensional (1D) discrete Fourier transformation of an array $u = (u_1, u_2,...)$ is

$$U_s = \frac{1}{\sqrt{n}} \sum_{r=1}^{n} u_r e^{2\pi i (r-1)(s-1)/n}, \quad (16.1)$$

where $r - 1$ and $s - 1$ occur since the index for the arrays u and U are counted from one. Equation 16.1 can be written as a matrix multiplication,

$$\begin{pmatrix} U_1 \\ U_2 \\ U_3 \\ U_4 \\ \vdots \end{pmatrix} = \frac{1}{\sqrt{n}} \begin{pmatrix} 1 & e^0 & e^0 & e^0 & \\ e^0 & e^\omega & e^{2\omega} & e^{3\omega} & \cdots \\ e^0 & e^{2\omega} & e^{4\omega} & e^{6\omega} & \\ e^0 & e^{3\omega} & e^{6\omega} & e^{9\omega} & \\ & \vdots & & & \ddots \end{pmatrix} \begin{pmatrix} u_1 \\ u_2 \\ u_3 \\ u_4 \\ \vdots \end{pmatrix} \quad (16.2)$$

where $\omega = i2\pi/n$. $\frac{1}{\sqrt{n}}$ is the normalization of the DFT.

The inverse discrete Fourier transformation of array $U = (U_1, U_2,...)$ is

$$u_r = \frac{1}{\sqrt{n}} \sum_{s=1}^{n} U_s e^{-2\pi i (r-1)(s-1)/n}. \quad (16.3)$$

Example 16.1 1D DFT Examples

Calculate the 1D DFT of the following arrays.

1. $\mathbf{u} = (\sqrt{2}, 0)$

$$U_1 = \frac{1}{\sqrt{2}} \sum_{r=1}^{2} u_r e^{i 2\pi (r-1)(1-1)/2} = \frac{1}{\sqrt{2}} (\sqrt{2} + 0) = 1.$$

$$U_2 = \frac{1}{\sqrt{2}} \sum_{r=1}^{2} u_r e^{i 2\pi (r-1)(2-1)/2} = \frac{1}{\sqrt{2}} (\sqrt{2} + 0) = 1.$$

Thus, $\mathbf{U} = (1, 1)$.

2. **u** = (0, 2, 0, 2)

$$\begin{pmatrix} U_1 \\ U_2 \\ U_3 \\ U_3 \end{pmatrix} = \frac{1}{\sqrt{4}} \begin{pmatrix} 1 & e^0 & e^0 & e^0 \\ e^0 & e^{i2\pi/4} & e^{i2\times 2\pi/4} & e^{i3\times 2\pi/4} \\ e^0 & e^{i2\times 2\pi/4} & e^{i4\times 2\pi/4} & e^{i6\times 2\pi/4} \\ e^0 & e^{i3\times 2\pi/4} & e^{i6\times 2\pi/4} & e^{i9\times 2\pi/4} \end{pmatrix} \begin{pmatrix} 0 \\ 2 \\ 0 \\ 2 \end{pmatrix} = \frac{1}{\sqrt{4}} \begin{pmatrix} 1 & 1 & 1 & 1 \\ 1 & i & -1 & -i \\ 1 & -1 & 1 & -1 \\ 1 & -i & -1 & i \end{pmatrix} \begin{pmatrix} 0 \\ 2 \\ 0 \\ 2 \end{pmatrix} = \begin{pmatrix} 2 \\ 0 \\ -2 \\ 0 \end{pmatrix}.$$

Thus, **U** = (2, 0, −2, 0).

The input function u can be written in terms of its DFT coefficients as

$$u = \frac{1}{\sqrt{4}} \left(U_1 e^{-i2\pi(r-1)0} + 0 + U_3 e^{-i2\pi(r-1)(3-1)/4} \right),$$

which expresses the input array as the sum of complex exponentials.

Math Tip 16.2 2D Discrete Fourier Transformation

The two-dimensional (2D) DFT transformation of an array $u_{q,r}$ is

$$U_{s,t} = \frac{1}{n} \sum_{q=1}^{n} \sum_{r=1}^{n} u_{q,r} e^{2\pi i[(r-1)(s-1)+(q-1)(t-1)]/n}. \tag{16.4}$$

This can be calculated by taking 1D DFTs of each row and then taking the DFTs of each column.

Math Tip 16.3 Shifting Functions

The discrete Fourier transform of 2D functions places the origin of the DFT, the constant component, at the (1, 1) element. This constant component is frequently referred to as the DC component, and is the mean of the input array. Because the "center of the DFT" is located in the corner, the DFT of most functions is divided between its four corners as in the left side of Figure 16.3. This makes viewing of most 2D DFTs difficult. Viewing DFT functions, such as the PSF, is easier when the DFT origin is at the center following a shifting operation; thus, throughout the chapter, DFT origins have been shifted to the center.

16.2 Discrete Fourier Transformation

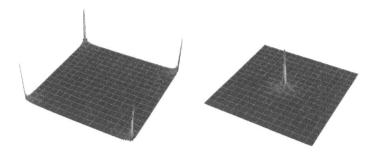

Figure 16.3 Fourier transform of a centered rectangular aperture (left) is concentrated in the four corners. After shifting the origin to the center, the transform has the expected appearance (right).

Math Tip 16.4 Padding for Higher Resolution

In order to increase the resolution of the DFT, the input array is often padded with zeros. The Jones pupil, a grid of Jones matrices at the exit pupil, can be padded with zero matrices $\begin{pmatrix} 0 & 0 \\ 0 & 0 \end{pmatrix}$, as shown in Figure 16.4, to achieve a higher resolution in Fourier transform domain. The padded grid should be at least twice the size of the original grid in order to avoid aliasing. Increasing padding by more than two times improves the resolution by having better interpolation in the Fourier transform domain.

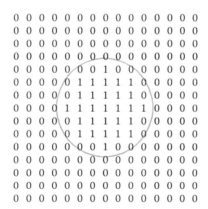

Figure 16.4 An element of an array is padded around the outside with zeros to provide higher resolution of its Fourier transform. The red circle indicates the exit pupil of the system.

16.3 Jones Exit Pupil and Jones Pupil Function

The polarization aberration of an imaging system can be described by a grid of Jones matrices at the exit pupil, that is, the Jones exit pupil described in Chapter 15 (Paraxial Polarization Aberrations). Figure 16.5 shows a Cassegrain telescope, which is used as this chapter's image formation example.

A grid of rays is traced through the Cassegrain telescope and the Jones pupil for an on-axis object is calculated. *Double pole local coordinates* were used to describe the Jones matrices on the spherical reference surface in the exit pupil (described in Chapter 11 [The Jones Pupil and Local Coordinate Systems]). Figure 16.6 shows the amplitude and phase of the Jones pupil. The system has zero wavefront aberration but some polarization aberration from its aluminum mirrors.

The full description of the wavefront and polarization at the exit pupil is divided into a combination of four functions: the *wavefront aberration function, amplitude function, aperture*

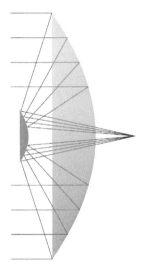

Figure 16.5 A side view of the Cassegrain telescope system with an on-axis grid of rays. The parabolic primary mirror is in blue and the hyperbolic secondary mirror is in pink.

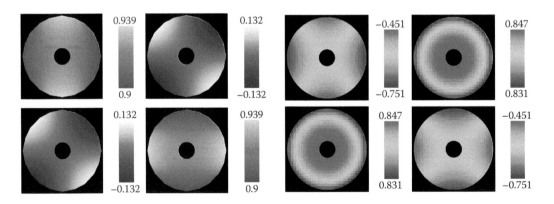

Figure 16.6 The amplitude (left) and phase (right) of the Cassegrain Jones pupil. A grid of on-axis rays was traced and each ray's Jones matrix at the exit pupil is plotted.

16.3 Jones Exit Pupil and Jones Pupil Function

function, and *polarization aberration function* at the exit pupil. This combination is called the *Jones pupil*,

$$JonesPupil(x,y) = aperture(x,y) \cdot a(x,y) \cdot \mathbf{J}(x,y) \cdot e^{-2\pi i W(x,y)}. \quad (16.5)$$

aperture(x, y) is an aperture function, which is one inside the aperture and zero outside the aperture. *a*(x, y) is an amplitude function that describes the transmission along the ray path, also known as the *apodization*. *W*(x, y) is the wavefront aberration function that characterizes the optical path difference between each ray and the chief ray. **J**(x, y) is the Jones matrix along ray paths from the entrance to exit pupil. (x, y) is the pupil coordinate. The *JonesPupil*(x, y) is spatially varying and is the starting point for the polarization image formation calculations. Two example Jones pupil functions are discussed in Examples 16.2 and 16.3.

Example 16.2 One Wave of Defocus with a Circular Aperture

Consider an optical system with a circular aperture with a radius of one, a wavefront aberration with one wave of defocus, uniform amplitude, and no polarization aberration. Since there is no polarization aberration, **J**(x, y) is an identity matrix and the Jones pupil is

$$aperture(x,y) = \text{If } (x^2 + y^2 \leq 1, 1, 0),$$

$$a(x,y) = 1, \quad \mathbf{J}(x,y) = \begin{pmatrix} 1 & 0 \\ 0 & 1 \end{pmatrix},$$

$$W(x,y) = W_{020}(x^2 + y^2), \quad W_{020} = 1,$$

$$JonesPupil(x,y) = aperture(x,y) \cdot a(x,y) \cdot \mathbf{J}(x,y) \cdot e^{-2\pi i W(x,y)}.$$

The amplitude and phase of the (1, 1) element of the *JonesPupil*(x, y) are shown in Figure 16.7. The amplitude is one inside the aperture and a quadratic phase is seen on the right.

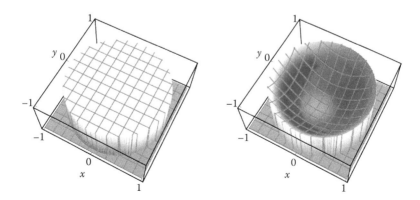

Figure 16.7 The amplitude (left) and phase (right) of the (1, 1) element of the *JonesPupil*(x, y) for one wave of defocus wavefront aberration. The Jones pupil has been sampled onto a square array of points.

Example 16.3 Quadratic Radially Oriented Retardance

One of the polarization aberrations described in Chapter 15 (Paraxial Polarization Aberrations) is retardance defocus with retardance magnitude that varies quadratically with the pupil radius $\delta_o (x^2 + y^2)$. Assume a retardance magnitude of $\delta_o = 1$ wave at the edge of the pupil, that is, one wave of retardance defocus. Further assume a tangentially varying fast axis orientation, $\tan^{-1}(y/x)$. Each ray path at the exit pupil has a retarder Jones matrix. Because of the retardance defocus aberration, each ray in the exit pupil experiences different magnitude and orientation of retardance as shown in Figure 16.8.

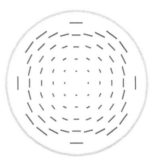

Figure 16.8 Retardance defocus aberration is plotted where the length of the line indicates the magnitude of the retardance and the orientation of the line indicates the orientation of the retardance. Mirror systems, such as Cassegrain telescopes, have tangentially oriented retardance of this form, but the retardance magnitude is much larger in this example.

Assume there is no wavefront aberration and the amplitude function is one. The Jones pupil function is

$$\text{aperture}(x, y) = \text{If } (x^2 + y^2 \leq 1, 1, 0),$$
$$a(x, y) = 1,$$
$$\mathbf{J}(x,y) = \begin{pmatrix} \dfrac{e^{-i\delta_o(x^2+y^2)/2}x^2 + e^{i\delta_o(x^2+y^2)/2}y^2}{x^2+y^2} & \dfrac{-2ixy\sin(\delta_o(x^2+y^2)/2)}{x^2+y^2} \\ \dfrac{-2ixy\sin(\delta_o(x^2+y^2)/2)}{x^2+y^2} & \dfrac{e^{i\delta_o(x^2+y^2)/2}x^2 + e^{-i\delta_o(x^2+y^2)/2}y^2}{x^2+y^2} \end{pmatrix},$$
$$W(x, y) = 0,$$
$$\text{JonesPupil}(x, y) = \text{aperture}(x, y) \cdot a(x, y) \cdot \mathbf{J}(x, y) \cdot e^{-2\pi i W(x,y)}.$$

Figure 16.9 shows the amplitude and phase of each of the four elements of the *JonesPupil*(*x*, *y*).

16.4 Amplitude Response Matrix (ARM)

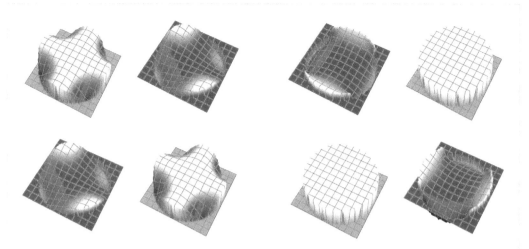

Figure 16.9 The amplitude (left) and phase (right) of the *JonesPupil(x,y)* for retardance defocus.

The polarization aberration of the Cassegrain telescope is a combination of retardance defocus and diattenuation defocus, with a larger retardance contribution.

16.4 Amplitude Response Matrix (ARM)

In conventional optics, the *amplitude response function* (ARF) at the image is the Fourier transform of the pupil function. To incorporate polarization aberration, the amplitude response function is generalized to the amplitude response matrix (**ARM**), which is the impulse response of the Jones pupil to coherent light. Using the Fraunhofer diffraction equation, the **ARM** describing the image of an optical system with a continuous Jones pupil can be calculated by 2D Fourier transformation of the Jones pupil function observed at z,

$$JonesARM(\xi,\eta) = \iint_{aperture} JonesPupil(x,y) \exp\left[\frac{i2\pi}{\lambda z}(x\xi + y\eta)\right] dx\, dy. \tag{16.6}$$

The **ARM** of a Jones pupil that has been regularly sampled by polarization ray tracing can be calculated by the discrete Fourier transform $(\Im)$ of each of the 2 × 2 elements of *JonesPupil(x, y)*, as stated earlier,

$$\begin{aligned}\mathbf{ARM}(r') &= \begin{pmatrix} ARM_{xx}(r') & ARM_{yx}(r') \\ ARM_{xy}(r') & ARM_{yy}(r') \end{pmatrix} \\ &= \begin{pmatrix} \Im[JonesPupil_{x,x}(x,y)] & \Im[JonesPupil_{y,x}(x,y)] \\ \Im[JonesPupil_{x,y}(x,y)] & \Im[JonesPupil_{y,y}(x,y)] \end{pmatrix}, \end{aligned} \tag{16.7}$$

where $\Im(JonesPupil_{l,m}(x, y)) = \Im(aperture(x,y) \cdot a(x, y) \cdot J_{l,m}(x, y) \cdot e^{-2\pi i W(x,y)})$ and $r' = (x', y')$ is the image plane coordinate. From the grid of ray parameters, the **ARM** is calculated by discrete Fourier transform from the Jones pupil. Each component of the *JonesPupil(x, y)* is a 2D array of electric field values cumulated at the pupil, and each component of the **ARM** is a 2D array characterizing the impulse response at the image plane.

For a given point source with Jones vector $\mathbf{E}_{object} = \begin{pmatrix} E_x \\ E_y \end{pmatrix}$, the amplitude response matrix multiplied by $\mathbf{E}_{object}$ yields an electric field distribution $\mathbf{E}_{image}(x, y)$ for the image, which characterizes the polarization variation. For coherent imaging, an x-polarized incident illumination produces an image with an x-polarized $\Im[JonesPupil_{x,x}(x, y)]$ and a y-polarized $\Im[JonesPupil_{x,y}(x, y)]$ amplitude response functions. The corresponding intensity point spread function is $\left|\Im[JonesPupil_{x,x}(x,y)]\right|^2 + \left|\Im[JonesPupil_{x,y}(x,y)]\right|^2$, which can also be calculated with the Mueller point spread matrix **MPSM** demonstrated in Section 16.5.

The four elements of the **ARM** describe the four ARFs seen when the imaging system is located between the four polarizer pairs (H&H, V&H, H&V, and V&V), where H stands for a horizontal polarizer and V stands for a vertical polarizer.

Example 16.4 ARM of an Aberration-Free System

For an imaging system with a circular aperture and no wavefront, amplitude, or polarization aberration, the Jones exit pupil is an identity matrix with circular apertures as shown in Figure 16.10 left. Thus, the diagonal elements of the **ARM** are real-valued Somb (sombrero or Airy) functions,[6] as shown in Figure 16.10 (right).

Thus, an image of a point object after this non-polarizing system has no polarization mixing since the **ARM** is diagonal; the polarization of the image is uniform and equal everywhere to the incident Jones vector.

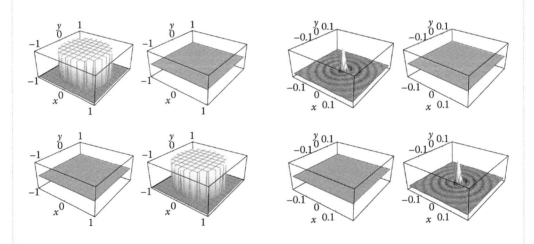

Figure 16.10 (Left) Jones pupil function and (right) amplitude response matrix of an aberration-free optical system. Both functions have had their origin shifted to the center of each array.

Figure 16.11 shows the amplitude of the Jones **ARM** for the example Cassegrain telescope of Figure 16.5 with aluminum mirrors ($n = 0.958 + 6.69i$) and marginal ray angles of incidence of 36° and 42° on the primary and secondary mirrors. The right figure zooms in to the center portion.

16.5 Mueller Point Spread Matrix (MPSM)

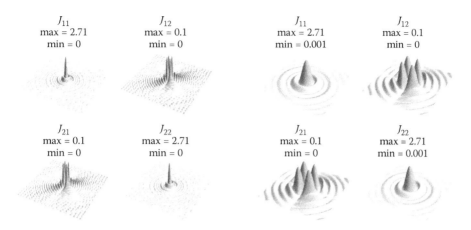

Figure 16.11 The amplitude response matrix (**ARM**) for the Cassegrain telescope (left) plots the entire array and (right) zooms into the center of the pupil.

This Cassegrain telescope **ARM** has off-diagonal components, which indicates a polarization mixing; for example, a horizontally polarized point source has a small, vertically polarized component in its image due to the four bumps in the off-diagonal components of the **ARM**. However, since the relative magnitudes of the diagonal elements are significantly larger than that of the off-diagonal elements (2.388: 0.101), the polarization mixing is small.

16.5 Mueller Point Spread Matrix (MPSM)

For incoherent light, a *point spread function* (PSF) describes the response of the system to a point source object in intensity space. Similarly, a *Mueller point spread matrix* (**MPSM**) is calculated by transforming the Jones **ARM** into an **MPSM**, a Mueller matrix representation using Equation 6.106 or 6.107. The **MPSM** relates the Stokes parameters of a point object to the Stokes parameter distribution of its image.

Simulating imaging within an isoplanatic patch simplifies the imaging calculation into a convolution.[6,7] The calculation of the images for coherent and incoherent objects involves the same *JonesPupil*(*x*, *y*) and discrete Fourier transform, but the calculation ends at different response functions, the amplitude response function for coherent objects or the point spread function for incoherent objects. Figure 16.12 illustrates this relationship between the object field and the diffracted field. The calculation of the scale of the **ARM** and **MPSM**, the distance between array points in the image, is explained in Section 16.6.

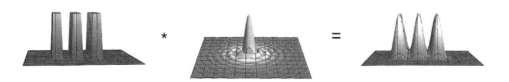

Figure 16.12 The convolution (*) of an example object with three bars (left) with a point spread function (PSF) yields a calculated image (right). In this case, the image is a blurred version of the three bars.

Math Tip 16.5 Convolution

A convolution of functions f and g is

$$(f * g)(t) = \int_{-\infty}^{\infty} f(\tau) g(t-\tau) \, d\tau. \qquad (16.8)$$

Example 16.5 MPSM of an Aberration-Free System

An **MPSM** of the aberration-free system with the **ARM** of Example 16.4 has an Airy disk along the diagonal elements and zeros in the off-diagonal elements as shown in Figure 16.13, calculated by converting the Jones matrices of Figure 16.11 into Mueller matrices.

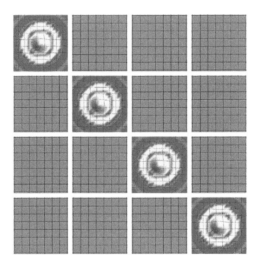

Figure 16.13 The **MPSM** of the polarization aberration-free optical system with a circular aperture is a diagonal matrix of sampled Airy functions.

Figure 16.14 shows the **MPSM** of the Cassegrain telescope system in a three-dimensional (3D) plot (left) and a raster plot (right). Note the relative magnitude difference between the diagonal elements and the off-diagonal elements; the presence of off-diagonal elements indicates polarization mixing, almost universally undesired, on the image plane, but here the small magnitude of the off-diagonal elements indicates only a small amount of polarization mixing. Note that the scale of the contour plot saturates the diagonal elements for a better visualization of the much smaller off-diagonal elements.

Padding the Jones exit pupil array with zeros (as shown in Math Tip 16.3) increases the resolution of the **MPSM** without changing the description of the wavefront at the image plane. To obtain the **ARM** and **MPSM** shown above with good resolution of structures, it was necessary to pad the Jones pupil array with zero matrices into an array eight times larger than the pupil. Without the

16.6 The Scale of the ARM and MPSM

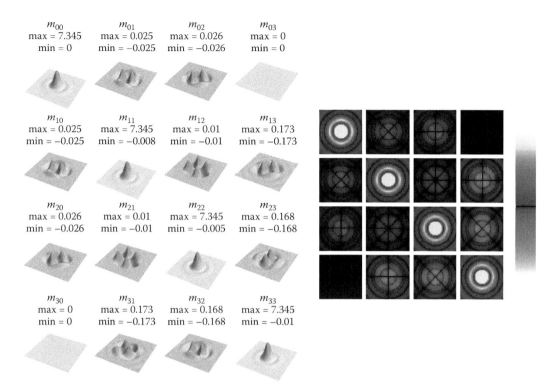

Figure 16.14 The **MPSM** of the Cassegrain telescope system in a 3D view (left) and the corresponding overexposed raster plot to show the dimmer feature (right).

padding, only a few points are calculated within the image such that the intensity distribution is difficult to visualize.

16.6 The Scale of the ARM and MPSM

The DFT of the Jones pupil provides an array of amplitude values for an image but does not provide the size of the image or the spacing between the elements of the **ARM** array. The Jones **ARM** and **MPSM** in this chapter are calculated using a discrete Fourier transformation of the *JonesPupil*(*x*, *y*). The scale of the **ARM** and **MPSM** is calculated by relating the Fraunhofer diffraction equation and DFT equation.

For simplicity, consider a 1D Fourier transformation. Then, Equation 16.6 becomes

$$JonesARM(\xi) = \int JonesPupil(x)\exp\left(\frac{i2\pi}{\lambda z}(x\xi)\right)dx, \quad (16.9)$$

where z is the location of the image plane from the exit pupil. The DFT equation is

$$U_s = \frac{1}{\sqrt{n}}\sum_{r=1}^{n} u_r e^{2\pi i(r-1)(s-1)/n}, \quad (16.10)$$

where n is the length of the u_r array. Comparing Equations 16.9 and 16.10, *JonesPupil*(*x*) = u_r and the exponent in Fourier transformation is equivalent to the exponent in DFT. Since s and r are discrete,

x and ξ can be written as $x = \Delta_x r$ and $\xi = \Delta_s s$, where Δ_x and Δ_s are the unit spacing in the spatial domain x and the spacing in the Fourier transform domain ξ. Thus,

$$\frac{2\pi i}{\lambda z} x\xi = \frac{2\pi i}{\lambda z} \Delta_x r \Delta_s s = \frac{2\pi i}{n} rs. \tag{16.11}$$

Therefore, the grid spacing in the **ARM** is

$$\Delta_s = \frac{\lambda z}{n \Delta_x}. \tag{16.12}$$

The denominator $n\Delta_x$ is the size of the Jones pupil array including the zero padding.

Example 16.6 Scale of the **ARM** and **MPSM**

A lens has a 20 mm exit pupil diameter ($XPD = 20$) and operates at a magnification of $M = 1/3$. The exit pupil is $z = L = 80$ mm from the image plane. A Jones pupil array has been calculated for $\lambda = 1.064$ μm by ray tracing a square grid with 41 rays across the exit pupil. The Jones pupil is then padded to a size of 200×200 Jones matrices ($n = 200$) with zero matrices, $\begin{pmatrix} 0 & 0 \\ 0 & 0 \end{pmatrix}$. The $2 \times 2 \times 200 \times 200$ amplitude response matrix is calculated by taking the 2D Fourier transforms of each 200×200 Jones pupil array. What is the spacing in millimeters between the elements of the **ARM**?

The spacing in the exit pupil is $\Delta_x = XPD/41 = 0.4878$ mm. Thus, the grid spacing in the **ARM** is $\Delta_s = \lambda L/(200 \cdot \Delta_x) = 0.00087$ mm. The spacing between the elements of the **MPSM** array is also Δ_s.

The spacing Δ_s can be interpreted as shown in Figure 16.15. Consider a plane wave exiting the exit pupil perpendicular to the z-axis. This plane wave maps to the center pixel in the image plane in Fourier domain. Then, a plane wave tilted by $\Delta\theta$ (as shown in dark green) maps to one

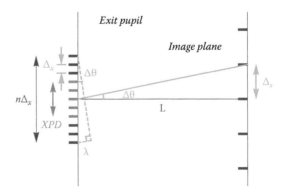

Figure 16.15 Consider a plane wave with one wave of tilt across the entire Jones grid (not just the exit pupil). This plane wave corresponds to the array spacing in the **ARM**.

pixel away from the center pixel. The optical path varies between the two plane waves by λ. Hence,

$$\Delta\theta = \frac{\lambda}{n\Delta_x}, \qquad (16.13)$$

and from the geometry,

$$\Delta_s = z\Delta\theta = \frac{\lambda L}{n\Delta_x}. \qquad (16.14)$$

16.7 Polarization Structure of Images

The polarization structure of images can be calculated using the **ARM** or **MPSM**. An example of the image of a linearly polarized point source, expressed as a Jones vector image, is shown in Figure 16.16. The Jones **ARM** multiplies an incident Jones vector, yielding a 3D plot of an image represented as a Jones vector. On the right side, a polarization state is plotted for each ray on the grid; the length of the line indicates the amplitude of the Jones vector and the orientation of the line indicates the orientation of the Jones vector. Since the off-diagonal elements of **ARM** are much smaller than the diagonal elements (as shown in Figure 16.11), the Jones vector image is mostly horizontally polarized with varying amplitudes.

A Stokes parameter image can be calculated from the **MPSM** with the object Stokes parameters; Figure 16.17 shows the Stokes parameter image when a horizontally polarized Stokes parameter is incident on the Cassegrain telescope. The Stokes image in this case is the sum of the first two columns of the **MPSM**. Note that the center of diagonal elements has been overexposed to reveal faint details. Because of the **MPSM**'s off-diagonal elements, the image of a horizontally polarized point object contains spatially varying S_2 and S_3 components.

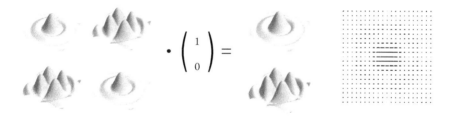

Figure 16.16 Jones vector image calculated from a Jones **ARM** when a horizontally polarized Jones vector point object is incident to the Cassegrain telescope.

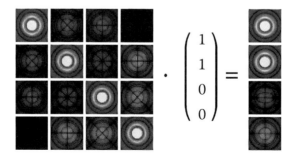

Figure 16.17 The **MPSM** operates on horizontally polarized Stokes parameters yielding a Stokes parameter image. The color scale is the same as in Figure 16.14.

16.8 Optical Transfer Matrix (OTM)

The *optical transfer function* (OTF) of conventional (scalar) optics describes imaging as a *spatial filtering operation*. In general, an imaging system accurately images the lowest frequencies of the object, but the amplitude of higher spatial frequencies is reduced until, at and beyond the cutoff frequency, the highest spatial frequency components are completely attenuated. As illustrated in Figure 16.18, the Fourier transform of an image is a Fourier transformation of an object multiplied by the optical system's OTF.[8] The OTF is usually specified in units of line pairs per millimeter. When imaging within an isoplanatic patch, the imaging operation is linear and linear systems theory applies. Within this isoplanatic patch, the image of a cosinusoidal object is always cosinusoidal.[6,7]

The extension of the OTF to *incoherent imaging systems* with polarization aberration is performed with the *Mueller optical transfer matrix* (**OTM**), a complex valued matrix function. A matrix is necessary because each of the object's Stokes parameters can couple into all four of the image's Stokes parameters. The Mueller **OTM** characterizes the spatial filtering of an object for each of the object's Stokes parameters. The Mueller **OTM** operates on the Fourier transform of the object's Stokes parameter. The modulus of the complex valued **OTM** (i.e., a Mueller modulation transfer matrix [**MTM**]) describes the change in modulation of each spatial frequency.

Figure 16.18 In the Fourier domain, the Fourier transform of the object of Figure 16.12 (left) times the OTF (center) is the Fourier transform of the image (right). The object of Figure 16.12 has a pair of strong frequency components about the central lobe, which become significantly reduced by the system's optical transfer function, as is seen in the Fourier transform of the image on the right, which corresponds to the blurred image on the right of Figure 16.12.

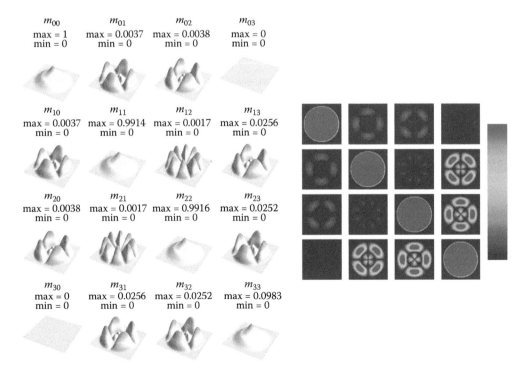

Figure 16.19 Amplitude of **OTM** of the Cassegrain telescope in 3D plot (left) and raster plot (right, overexposed).

16.8 Optical Transfer Matrix (OTM)

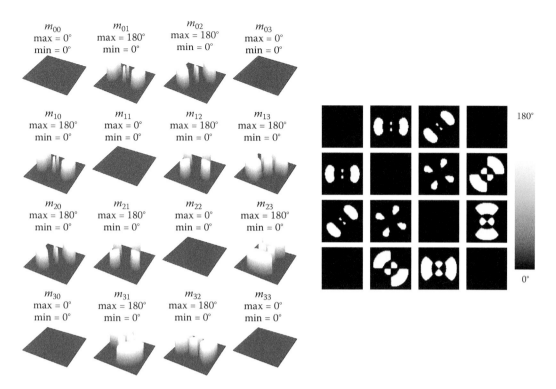

Figure 16.20 Phase of **OTM** of the Cassegrain telescope in 3D plot (left) and raster plot (right).

Figure 16.19 shows the Cassegrain telescope **MTM** as a 3D plot (left) and raster plot (right). Mueller matrices are *real*, but in general, the **OTM** is not real but *complex*, because the **MPSM** functions may have both *even* and *odd* components, and *real odd* functions have imaginary valued *Fourier transforms*. Physically, this means that the positions of the images of cosinusoidal objects are shifted from their geometrical images. Thus, the phase of the complex valued **OTM** (i.e., a Mueller phase transfer matrix [**PTM**]) describes the shift of each cosinusoidal spatial frequency from its geometrical image; a value of the **PTM** of π corresponds to a half period shift. Examples of the image degradation due to the *phase transfer function* are addressed in the work of Gaskill[7] beginning on page 236. Figure 16.20 shows the Cassegrain telescope **PTM** in 3D plot (left) and raster plot (right).

The raster plot of **MTM** is overexposed for the diagonal elements to show more details in the off-diagonal elements. The **MTM** is always normalized to one at the center of the pupil in the m_{00} element. This means that, for any incident Stokes parameters $\left(S_{0,obj} \ S_{1,obj} \ S_{2,obj} \ S_{3,obj} \right)^T$, the Mueller **MTM** provides image Stokes parameters that have a maximum of $S_{0,image} = S_{0,obj}$ at the center pixel of the array.

Example 16.7 Scale of **OTM**

Continuing from Example 16.6, calculate the spacing in line pairs per millimeters between the elements of the **OTM** $\Delta\xi$ and calculate the *cutoff frequency* in the **OTM**.

The size of the **MPSM** array is $n\Delta_s$. First, the lowest spatial frequency p_{image} of an **MPSM** is one period over $n\Delta_s$, $p_{image} = n\Delta_s = 0.174496$.

Since the system has a magnification of 1/3, the *fundamental period at the object*, that is, a spatial frequency in line pairs per millimeter at the object, is

$$p_{object} = p_{image}/M = 0.523488.$$

Thus, the spatial frequency interval $\Delta\xi$ in the **OTM** in line pairs per millimeter is

$$\Delta\xi = 1/p_{object} = 1.91026,$$

and the *cutoff frequency* $\Delta\xi_{Nyquist}$ is $\Delta\xi_{Nyquist} = \Delta\xi_n/2 = 191.026$ in line pairs per millimeter.

16.9 Example—Polarized Pupil with Unpolarized Object

As an incoherent imaging example with large and easily seen polarization effects, consider the following diffraction-limited imaging system with a square aperture. Let the center 2/3 of the aperture ($-2/3 \le x \le 2/3$) be filled with a vertical polarizer. The remainder ($-1 \le x \le -2/3$ and $2/3 \le x \le 1$) is filled with a horizontal polarizer as shown in Figure 16.21 (left). The *JonesPupil*(*x*, *y*) of this striped polarizer system is shown in Figure 16.21 (right), a square aperture with uniform amplitude and zero wavefront aberration. The non-zero part of the J_{11} component represents the horizontal polarizer at the edges of the pupil, while the J_{22} component represents the vertical polarizer over the center 2/3 of the pupil.

The Jones **ARM** of this striped polarizer system is calculated using Equation 16.6 and is shown in Figure 16.22 and the J_{11} component (left) and the J_{22} component (right) are shown in Figure 16.23. Since each Jones pupil element is an even function, each element of the Jones **ARM** is real. The J_{11} component resembles an interference pattern since the horizontal polarizer portion of the *JonesPupil*(*x*, *y*) is equivalent to a double slit separated along the *x*-axis. The J_{22} component of the

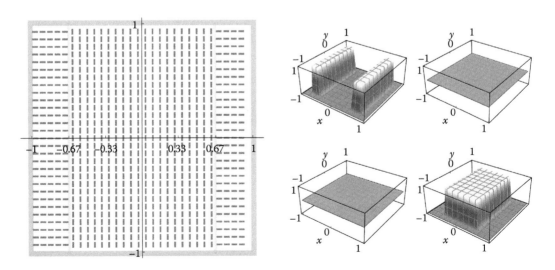

Figure 16.21 (Left) A square aperture where the center 2/3 is filled with a vertical polarizer and the remainder filled with a horizontal polarizer is shown. (Right) *JonesPupil*(*x*, *y*) for a square aperture with no wavefront aberration and uniform amplitude is shown. The center 2/3 of the square is a vertical polarizer and the remainder is a horizontal polarizer.

16.9 Example—Polarized Pupil with Unpolarized Object

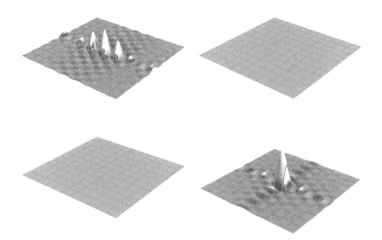

Figure 16.22 The Jones **ARM** of horizontal/vertical polarizer *JonesPupil*(x, y).

Figure 16.23 (Left) The (1, 1) component of the **ARM** is an interference pattern formed from the horizontal polarizer double slit *JonesPupil*(x, y). The amplitude varies between −4.40 and 4.71. (Right) The (2, 2) component of the **ARM** Fraunhofer diffraction pattern results from the vertical polarizer as a rectangular single slit *JonesPupil*(x, y). The amplitude varies between −2.04 and 9.42.

ARM is a sinc function with different widths in *x* and *y* since the vertically polarized aperture is equivalent to a single slit with wider dimension along the *y*-axis. Thus, the J_{22} component of the **ARM** is wider along the *x*-axis than along the *y*-axis.

As shown in Figure 16.1, the **MPSM** is calculated by converting the Jones **ARM** to a Mueller matrix. A scalar PSF of the system when a horizontally polarized object is viewed from a horizontal polarizer can be calculated by cascading the **MPSM** in between two horizontal polarizer Mueller matrices; this is shown in Figure 16.24 (left). This scalar PSF is called the horizontal–horizontal **MPSM**. Similarly, the scalar PSF of the system when a vertically polarized object is viewed through a vertical polarizer can be calculated and is shown in Figure 16.24 (right).

The Mueller **OTM** is calculated and plotted in Figure 16.25. Figure 16.26 shows the **OTM** (0, 0) element from two different viewpoints. The transfer function along the *x*-direction and *y*-direction exhibits drastically different profiles.

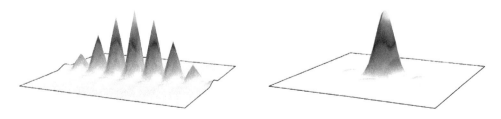

Figure 16.24 (Left) The PSF of the striped polarizer system viewed between two horizontal polarizers is a two-slit interference pattern with maximum and minimum intensity 22.18 and 0. (Right) The PSF viewed between two vertical polarizers has maximum and minimum intensity 88.66 and 0.

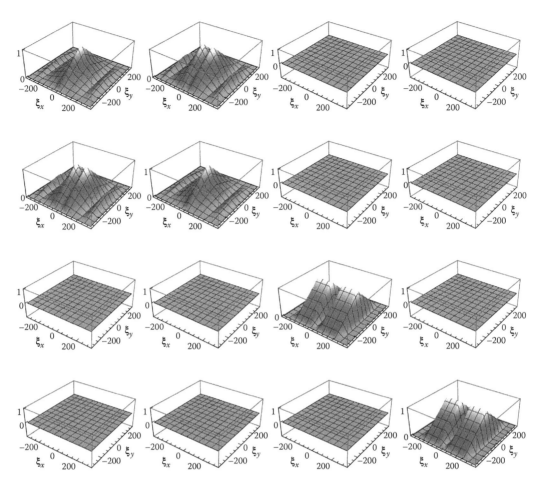

Figure 16.25 Mueller **OTM** of the example system.

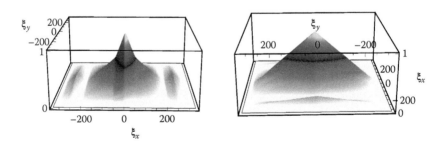

Figure 16.26 Two views of a scalar **OTF** when an unpolarized object enters the example system looking along the y-axis (left) and along the x-axis (right) showing different resolution. One band of spatial frequencies around 200 line pairs per millimeter (left) is completely removed where the **OTM** goes to zero. The resulting filtering is seen in Figure 16.29 (right).

16.9 Example—Polarized Pupil with Unpolarized Object

Figure 16.27 shows the Mueller **OTM** viewed in between two horizontal polarizers (horizontal–horizontal **OTM**) on the left and the **OTM** viewed in between two vertical polarizers (vertical–vertical **OTM**) on the right in two different views (top and bottom). For a horizontally polarized light, the **OTM** has three peaks since the *JonesPupil*(x, y) is a double slit. For a vertically polarized light, the **OTM** follows a familiar triangular shape since the *JonesPupil*(x, y) is a rectangular aperture with uniform amplitude.

Consider an unpolarized object consisting of three strips as shown in Figure 16.28. The Fourier transform of the image's Stokes parameters are calculated by matrix multiplying the Mueller **OTM** of the optical system to the Fourier transform of the object,

$$\Im \begin{pmatrix} S_{image,0} \\ S_{image,1} \\ S_{image,2} \\ S_{image,3} \end{pmatrix} = \mathbf{OTM} \cdot \Im \begin{pmatrix} S_{object,0} \\ S_{object,1} \\ S_{object,2} \\ S_{object,3} \end{pmatrix}, \qquad (16.15)$$

where the unpolarized object is $S_{object} = \begin{pmatrix} g(x,y) & 0 & 0 & 0 \end{pmatrix}^T$ and $g(x, y)$ has the object's intensity shown in Figure 16.28. The object $g(x, y)$ has a spatial frequency that is half the spatial frequency of the horizontal–horizontal **MPSM** of the system shown in Figure 16.24 (left).

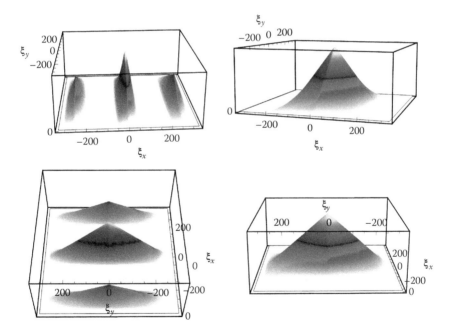

Figure 16.27 (Left) Mueller **OTM** viewed between two horizontal polarizers and (right) Mueller **OTM** viewed between two vertical polarizers. H–H **OTM** has a center peak value of 0.325 and V–V **OTM** has a center peak value of 0.675.

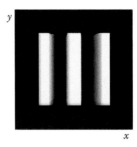

Figure 16.28 The s_0 Stokes parameter of the unpolarized object is $g(x, y)$ with three vertical strips is plotted.

Figure 16.29 (Left) Horizontally polarized component of the image and (right) vertically polarized component of the image drawn in the same scale.

Then, the inverse Fourier transform of Equation 16.15 provides the image of the object. Since the Mueller **OTM** is non-zero in m_{00} and m_{10} components, the unpolarized object has two orthogonal polarization components, horizontal and vertical, as shown in Figure 16.29.

The horizontal and vertical components of the image are completely different. For the vertical component, the image is a blurred version of the three strips object due to the loss in high frequency from a limited aperture size of the optical system. This is straightforward and can be observed from the scalar PSF/OTF calculations. The horizontal component of the image, however, requires more explanations. Since the object has a spatial frequency that is half of the horizontal–horizontal **MPSM** of the system, Fourier transform of the object has a frequency that is twice the horizontal–horizontal **OTM** of the system. This is why the horizontally polarized component of the image does not show any peaks (or spatial frequency of the object) but has an envelope that shows the blurred extent of the object. Thus, it is seen how a polarization-dependent image can be understood using the **MPSM/OTM** of the optical system.

16.10 Example—Solid Corner Cube Retroreflector

An interesting example of polarization image formation occurs with solid *corner cube retroreflectors* (CCRs), which have a large polarization aberration due to the total internal reflection (TIR) at the three back surfaces. First, a typical CCR is analyzed with a mid-refractive index glass, $n = 1.6$. Higher indices are preferred for solid corner cubes because the field of view, which is limited by the TIR failure, is larger. Then, a cube with a low refractive index chosen to operate at the critical angle $n = 1.227$ is considered for its theoretical interest.

A CCR consists of three mutually perpendicular reflective surfaces. As shown in Figure 16.30, a corner cube (shown in red) is literally the corner of a cube. For a solid glass cube, rays refracting into the front face undergo three TIRs from the back surfaces and exit the front face with their propagation vector reversed. For a normally incident ray, the angle of incidence (AOI) at each of the three reflecting faces is $\arccos(1/\sqrt{3}) = 54.74°$. For an example cube with $n = 1.6$, these rays experience three linear retardances of $\delta = 48.94°$, but each retardance is at a different orientation; hence they cascade into an elliptical retarder.

16.10 Example—Solid Corner Cube Retroreflector

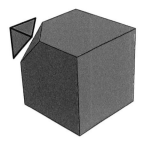

Figure 16.30 A corner cube retroreflector (red) has three mutually perpendicular reflecting surfaces. Rays refracting into the front face reflect three times and exit the front face with their propagation vector, thus retroreflecting the light.

Figure 16.31 shows how each surface of the CCR is numbered. There are six subapertures in the CCR; depending on the position of the entering ray in surface 1, six different ray paths exist. The retarder eigenstates are different for each of the six ray paths. Figure 16.32 lists the order of surface intercepts for the six subapertures viewed looking into surface 1. A Polaris-M polarization ray trace has been performed and Figure 16.33 follows the evolution of polarization for the six subapertures (each in a different column) for three different polarization states (different rows). The polarization ellipses are shown after each surface in 3D. For the linear states in the top two rows, the exiting light is always elliptical. Figure 16.34 shows polarization propagation for three more rays.

Figure 16.35 plots the amplitude and phase of the Jones pupil calculated by the polarization ray trace. Figure 16.36 shows eigenvectors of the 3D polarization ray tracing (PRT) matrix for each subaperture. Since these are for a retroreflection PRT, the polarization ellipses are associated with vectors propagating in opposite directions to the incident propagation vector. The ellipticities are the same in each subaperture. The ellipses' major axes rotate by 120° between subapertures.

Assuming a hexagonal aperture with an amplitude transmittance $a(x, y) = 1$ and free of wavefront aberration $W(x, y) = 0$, the far field diffraction pattern, the Jones **ARM** calculated by discrete Fourier transformation of the *JonesPupil*(x, y), is plotted in Figure 16.37. The ellipse map on the right is the Jones vector image for a horizontally polarized point source. Unlike Figure 16.16, the image's

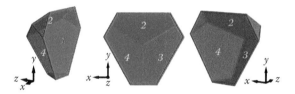

Figure 16.31 The front surface of the CCR where light enters and exits is numbered 1, and the three reflecting surfaces are numbered 2, 3, and 4.

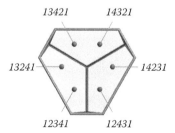

Figure 16.32 Depending on where light enters the front face, it encounters a different order of surface reflections as labeled in the figure. The corner cube exit pupil contains six distinct subapertures.

Polarized Light and Optical Systems

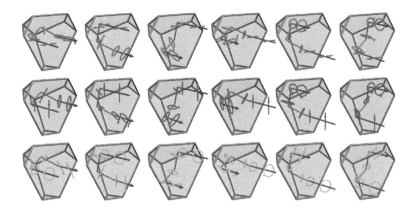

Figure 16.33 Polarization propagations through the CCR for horizontal (top row), vertical (middle), and left circularly polarized incident beams are shown in 3D following each surface. Each column follows one of the six subapertures.

Figure 16.34 The polarization change along the three different ray paths is shown for three different incident polarization states.

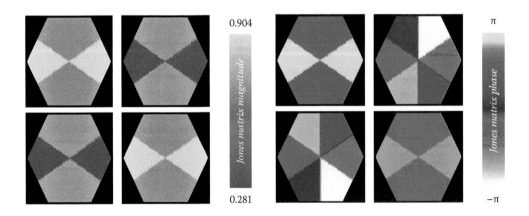

Figure 16.35 Amplitude (left) and phase (right) of the CCR's Jones pupil.

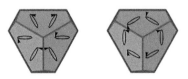

Figure 16.36 The two eigenpolarizations are plotted for each subaperture.

16.10 Example—Solid Corner Cube Retroreflector

polarization has substantial spatial variation in orientation and ellipticity since the off-diagonal elements of the Jones **ARM** are comparable in magnitude to the diagonal elements. Interestingly, at the radius of the blue dashed circle, Figure 16.37 (right) shows a 720° rotation of linear polarization.

The **MPSM** is calculated by transforming the Jones **ARM** into a Mueller matrix image, which is plotted in Figure 16.38. The first column contains the Stokes image for an unpolarized point source. The m_{00} element has six islands surrounding the central peak. The m_{10} and m_{20} elements show that the light at the same blue dashed line's radius has linear polarization that rotates (due to the six alternating lobes) by 540° around the center. Thus, even the PSF of unpolarized light has partially polarized regions.

The Mueller matrix optical transfer matrix (MOTM) is calculated by taking the discrete Fourier transformation of each element of the **MPSM**. The elements of the modulation transfer matrix and the phase transfer matrix are plotted in Figure 16.39. Because of this large polarization aberration, image formation is far from diffraction limited. The m_{00} element does not resemble a diffraction-limited transfer function for example.

Figure 16.37 The magnitude of the Jones **ARM**s for the $n = 1.6$ CCR (left) and a zoomed-in view of the center (middle) is shown. The Jones vector image for a horizontally polarized point source (right) is plotted with blue dashed lines indicating the orientation of the six subapertures. A 720° rotation of linear polarization can be observed following the blue dashed circle.

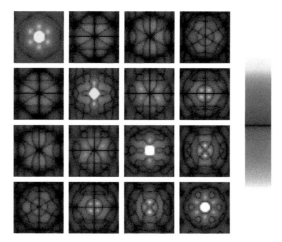

Figure 16.38 The Mueller point spread matrix of the $n = 1.6$ CCR shows large and complex amounts of polarization coupling between the incident and exiting Stokes parameters. The images are overexposed to show the structure of the dimmer regions.

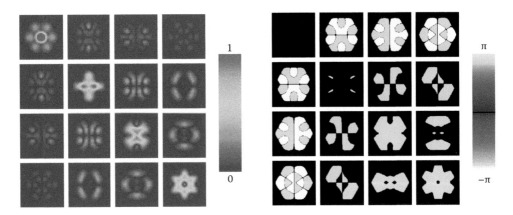

Figure 16.39 Amplitude (left) and phase (right) of the Mueller **OTM**.

16.11 Example—Critical Angle Corner Cube Retroreflector

A corner cube without retardance can be constructed by adjusting the refractive index to operate at the critical angle at the three reflections for normally incident light. This *critical angle corner cube retroreflector* has a very interesting property of rotating linear polarization in all six subapertures. Since the angle of incidence is 54.73°, the index of this cube is $n_{critical} = 1.2247$. TIR is associated with a retardance that varies rapidly with the angle of incidence, as plotted in Figure 16.40 for $n_{critical}$. At the critical angle, the retardance is zero, and the reflection is ideal and non-polarizing with $r_s = 1$ and $r_p = 1$. Since an on-axis ray experiences zero retardance at the three ray intercepts, it might seem that this cube is non-polarizing. However, by the time the beam changes direction three times and is retroreflected, the polarization states become rotated. This change of polarization is called *geometrical transformation* or *skew aberration*. This geometrical transformation is calculated using the **Q** matrix method of Chapter 17 (Parallel Transport and the Calculation of Retardance) and is understood as skew aberration as discussed in Chapter 18 (A Skew Aberration). In Figure 16.40, note that, at the critical angle, the slope of the retardance is *infinite*; hence, actually operating at the critical angle would require an impossibly perfectly collimated and perfectly aligned beam, rendering it impractical. The theoretical properties of this cube, however, provide a fascinating polarization aberration example.

Geometric transformation describes how the polarization changes as each ray bends or reflects at a non-polarizing surface (free of diattenuation and retardance). Geometric transformation does not change the ellipticity or the amplitude of the electric field. Since the retardance at each ray intercept is zero, the critical angle CCR has *no polarization aberration, only geometric transformation*. Figure 16.41 shows the Jones pupil for a grid of rays through the critical angle CCR.

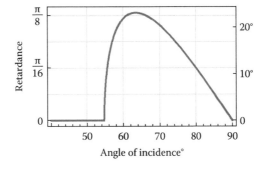

Figure 16.40 Retardance as a function of angles of incidence for internal reflection $n_{critical} = 1.2247$.

16.11 Example—Critical Angle Corner Cube Retroreflector

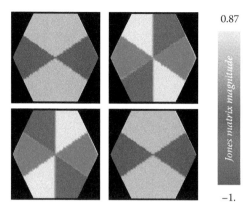

Figure 16.41 Jones exit pupil of the critical angle CCR* is real valued everywhere, corresponding to three pairs of different circular retarders.

Figure 16.42 Labels for the subapertures of the CCR.

This Jones pupil contains only real values since the phase of the Jones pupil elements is either π or $-\pi$. Each triangular subaperture (as labeled in Figure 16.42) has a Jones matrix, which is a rotation matrix along the z-axis with a rotation amount of 0°, 120°, and 240°. The Jones matrix of opposite pairs of subapertures is identical,

$$\mathbf{J}_A, \mathbf{J}_D = \begin{pmatrix} -1 & 0 \\ 0 & -1 \end{pmatrix}, \mathbf{J}_B, \mathbf{J}_E = \begin{pmatrix} 1/2 & -\sqrt{3}/2 \\ \sqrt{3}/2 & 1/2 \end{pmatrix}, \mathbf{J}_C, \mathbf{J}_F = \begin{pmatrix} 1/2 & \sqrt{3}/2 \\ -\sqrt{3}/2 & 1/2 \end{pmatrix}. \quad (16.16)$$

Thus, each subaperture of the Jones pupil has a *circular retarder* Jones matrix that rotates all polarization states by $\theta = 0°$, 120°, and 240° shown in Figure 16.43, demonstrating the polarization rotation after three reflections. All three reflections become perfect total internal reflections at the critical angle. Therefore, the resultant circular retardance of the reflected light is solely induced by the geometry of the ray paths. The polarization rotations around the exit surface are 0°, 120°, and 240°.

These properties and symmetries of the critical angle CCR yield a fascinating amplitude response matrix shown in Figure 16.44 (left), one of the most interesting and symmetric the authors have ever seen. An analysis shows that the **ARM** has circular retarder Jones matrices everywhere. The circular retardance magnitude of the **ARM** is plotted in Figure 16.44 (right). The center of the **ARM** is completely dark because the three components of the light, 0°, 120°, and 240° for example, destructively interfere. Around a small circle, at about 0.25 from the center to the edge of the figure, an 8π retardance change is observed circling around the pupil; this will cause a steady 4π rotation of linear

---

* The Jones pupil is calculated using local coordinates defined by the incident and exiting propagation vectors for the CCR.

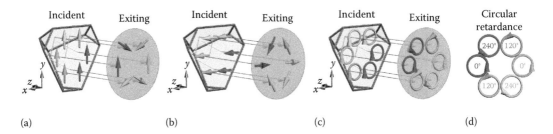

Figure 16.43 (a, b, and c) Six incident rays incident at different regions of the entrance surface follow different paths within the corner cube, and the corresponding exiting states have a different amount of polarization rotations for vertically, horizontally, and circularly polarized incident rays. (c) Circular retardance induced by the corner cube alters the absolute phase of circular incident light. The amount of circular retardance depends on the ray path inside the corner cube. (d) The rotation associated with the induced circular retardance are 0°, 120°, and 240° around the exit pupil.

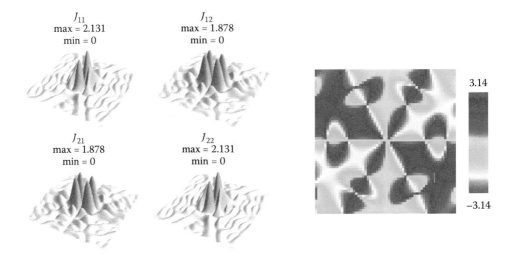

Figure 16.44 The amplitude of the amplitude response matrix (left) is purely real. A circular retardance map of the critical angle corner cube (right) is plotted in radians.

polarization around the pupil at this distance. The circular retardance is zero along the horizontal and vertical directions and has $2\pi/3$ retardance in the $+60°$ radial direction and $-2\pi/3$ retardance along the $-60°$ direction. Along each of these lines, two polarization singularities are found at zeros of the ARM; on the x-axis, the singularities are found at about 0.5 and 0.8 from the center to edge of the figure. Moving around a circle between these two zeros (about 0.65 on the x-axis), a continuous 16π retardance change is observed.

For a horizontally polarized incident plane wave, the Jones **ARM** yields a Jones vector image in the far field as shown in Figure 16.45. Because of the off-diagonal elements of the Jones **ARM**, vertically polarized light (lower) has been mixed into the horizontally polarized state and there is a spatial linear polarization variation on the image plane. The six subapertures that are visible in the circular retardance map in Figure 16.44 match the six islands in Figure 16.45. Different amounts of rotation of the horizontal polarization state are observed for each of the six islands.

The critical angle CCR's **MPSM** is shown in Figure 16.46. The three zero elements in the top row show that the intensity distribution is independent of the incident polarization state; there is no diattenuation. The first column shows that for unpolarized incident light, the image is unpolarized everywhere; there is no polarizance. The first and last columns show that a right circularly polarized

16.11 Example—Critical Angle Corner Cube Retroreflector

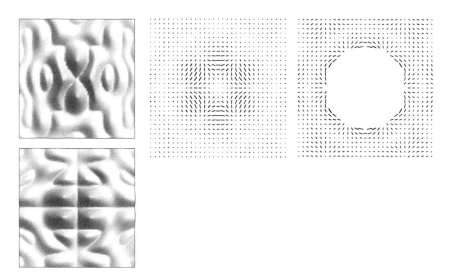

Figure 16.45 The diffraction pattern from the critical angle corner cube with horizontally polarized incident light shows a dark center and six inner islands of light at 0°, 120°, and −120° orientations. The *x*-amplitude (left top) and *y*-amplitude (left bottom) of the Jones **ARM** vector show a complex structure that, when represented as polarization ellipses (middle), clearly shows the six islands with varying polarization states. The 720° rotation of linear polarization can be observed by moving around the diffraction pattern at the radius of the six islands. (Right) The line lengths are magnified two times to show additional polarization structure further from the center; the lines in the center are suppressed. Since the **ARM** Jones matrices are pure circular retarders everywhere, this diffraction pattern is linearly polarized everywhere, free of elliptical polarization.

point source object generates a right circularly polarized image point and a left circularly polarized object forms a left circularly polarized image. The middle four elements are particularly interesting, having the form of a spatially varying circular retarder. For a linearly polarized source, the image is linearly polarized everywhere, with a rotating polarization state. For a circularly polarized object, the **MPSM** gives circularly polarized image Stokes parameters with the same handedness as the object, the opposite of the $n = 1.6$ CCR. The **MPSM** is dark at the center for all polarization states; a right circularly polarized point source object will have a right circularly polarized image point, and an unpolarized point source object will have an unpolarized image point.

Figure 16.47 shows the Mueller **OTM**'s magnitude (left) and the phase (right). Again, the **MPSM** is a circular retarder for all spatial frequencies and orientations. The **OTM** contains only real values with 0 or π phase. Figure 16.48 shows a cross section of the Mueller **MTM** along the horizontal direction.

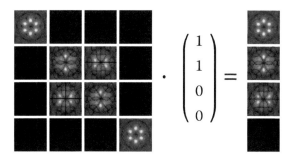

Figure 16.46 An **MPSM** for the critical angle CCR (left) operating on a horizontal linearly polarized set of Stokes parameters forms a diffraction pattern with spatially varying Stokes parameters corresponding to a varying polarization orientation.

The critical angle CCR provides a unique polarization imaging example. All it does is rotate polarization state between subapertures, which yields complex and highly symmetric PSFs and OTFs. This circular retardance-like change in polarization is present even though the critical angle CCR has no retardance or diattenuation. Such intrinsic geometric transformation is explored in more detail in Chapter 17 (Parallel Transport and the Calculation of Retardance) and Chapter 18 (A Skew Aberration).

16.12 Discussion and Conclusion

The analysis of the imaging system has been generalized to include polarization aberration. The optical system is described by an aperture function, amplitude function, wavefront aberration, and polarization aberration, which are combined into the *JonesPupil(x, y)*. The **ARM**, which describes image formation of a coherent optical system, can be calculated by applying the DFT to each of the four elements of the

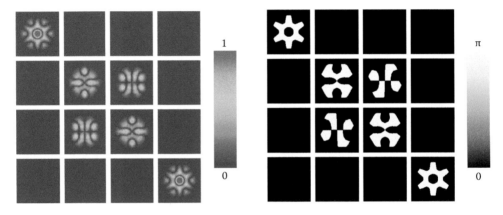

Figure 16.47 The Mueller **MTM** (left) and **PTM** (right) for the critical angle CCR, overexposed to reveal faint structures.

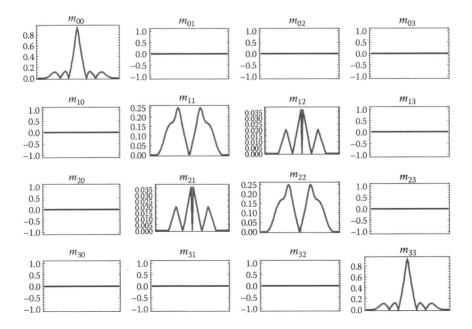

Figure 16.48 Cross section of the Mueller **MTM** along the horizontal direction.

JonesPupil(x, y). For an incoherent imaging system, the **MPSM** can be calculated by converting the Jones **ARM** into a Mueller matrix. An additional DFT results in modulation transfer functions between all the pairs of incident and exiting Stokes parameters in the form of the Mueller **OTM**.

The Jones **ARM**, **MPSM**, and Mueller **OTM** provide a full description of the image quality including polarization aberration. Often, these matrices need to be reduced to a scalar version of the corresponding metric. The Jones **ARM** can be reduced to an ARF when an object Jones vector is applied to the Jones **ARM** and viewed through a linear polarizer. Similarly, the **MPSM** or **OTM** can be reduced to the PSF and OTF by applying an object Stokes parameters to the corresponding Mueller matrix and viewing the image through a polarizer.

Off-diagonal elements in **ARM**, **MPSM**, or **OTM** describe the coupling between polarization components in the image. The relative magnitude of the off-diagonal elements with respect to the diagonal elements tells the amount of polarization mixing that occurs in the image plane. For example, the Cassegrain telescope system has small coupling between polarization states; thus, the image polarization was almost the same as the object polarization state. The two corner cube retroreflectors had significant off-diagonal elements in their **ARM**, **MPSM**, and **OTM**; hence, substantial spatially varying polarization coupling was observed in their images.

16.13 Problem Sets

16.1 Perform a 1D discrete Fourier transformation of the following arrays u. Identify the functions as sin, cos, constant, and so on.
   a. $u = \left(\sqrt{2}, 0\right)$
   b. $u = (0, 2, 0, 2)$
   c. $u = (0, 2i, 0, -2i)$
   d. $u = (0, 0, 4, 0)$
   e. $u = \left(0, 0, 2\sqrt{2}, 0, 0, 0, 0, 0\right)$
   f. $u = (4, 0, 0, 0, 2, 0, 0, 0, 0, 0, 0, 0, 2, 0, 0, 0,)$

16.2 Calculate the following 2D discrete Fourier transforms $U_{s,t}$ of the following sampled functions $u_{q,r}$. Identify the functions as sin, cos, constant, and so on.

a. $U = \begin{pmatrix} 4 & 0 & 0 & 0 \\ 0 & 0 & 0 & 0 \\ 0 & 0 & 0 & 0 \\ 0 & 0 & 0 & 0 \end{pmatrix}$, b. $U = \begin{pmatrix} 0 & 0 & 0 & 0 \\ 0 & 2 & 0 & 0 \\ 0 & 0 & 0 & 0 \\ 0 & 0 & 0 & 2 \end{pmatrix}$, c. $U = \begin{pmatrix} 0 & 0 & 0 & 0 \\ 2 & 0 & 0 & 0 \\ 0 & 0 & 0 & 0 \\ 2 & 0 & 0 & 0 \end{pmatrix}$,

d. $U = \begin{pmatrix} 0 & 2 & 0 & 2 \\ 2 & 0 & 0 & 0 \\ 0 & 0 & 0 & 0 \\ 2 & 0 & 0 & 0 \end{pmatrix}$, e. $U = \begin{pmatrix} 0 & 2 & 0 & 2 \\ 2 & 0 & 2 & 0 \\ 0 & 2 & 0 & 2 \\ 2 & 0 & 2 & 0 \end{pmatrix}$,

f. $U = \begin{pmatrix} 0 & 0 & 0 & 0 & 0 & 0 & 0 & 0 \\ 0 & 0 & 0 & 0 & 0 & 0 & 0 & 0 \\ 0 & 0 & 0 & 0 & 0 & 0 & 0 & 0 \\ 0 & 0 & 0 & 0 & 0 & 0 & 0 & 0 \\ 0 & 0 & 4 & 0 & 0 & 4 & 0 & 0 \\ 0 & 0 & 0 & 0 & 0 & 0 & 0 & 0 \\ 0 & 0 & 0 & 0 & 0 & 0 & 0 & 0 \\ 0 & 0 & 0 & 0 & 0 & 0 & 0 & 0 \end{pmatrix}$, and g. $U = \begin{pmatrix} 0 & 0 & 0 & 0 & 0 & 0 & 0 & 0 \\ 0 & 0 & 0 & 0 & 0 & 0 & 0 & 0 \\ 0 & 0 & 0 & 0 & 0 & 0 & 0 & 0 \\ 0 & 0 & 0 & 0 & 0 & 0 & 0 & 0 \\ 0 & 0 & 0 & 0 & 8 & 0 & 0 & 0 \\ 0 & 0 & 0 & 0 & 0 & 0 & 0 & 0 \\ 0 & 0 & 0 & 0 & 0 & 0 & 0 & 0 \\ 0 & 0 & 0 & 0 & 0 & 0 & 0 & 0 \end{pmatrix}$.

What is the highest frequency wave that can be described by a four-element sampled function? An eight-element sampled function?

16.3 Calculate the **PSM** for retardance defocus following the procedure below.

The point spread functions will be calculated for the polarization aberration "retardance defocus." A uniformly illuminated circular aperture is assumed with no wavefront aberration. The magnitude of the retardance defocus, δ_{mag}, is 1 radian. The exit pupil diameter is 2 mm. In $\mathbf{J}(x, y)$, the Jones matrices for a retarder have been expanded into linear and quadratic Taylor series terms.

$$a(x, y) = 1,$$

$$\mathbf{J}(x, y) = \begin{pmatrix} 1 - \dfrac{i\delta_{mag}(x^2 - y^2)}{2} & -i\delta_{mag} xy \\ -i\delta_{mag} xy & 1 - \dfrac{i\delta_{mag}(-x^2 + y^2)}{2} \end{pmatrix},$$

$$W(x, y = 0).$$

Use a computer program (MATLAB or Mathematica) to create a pupil array with odd dimensions such as 129 × 129.

   a. Create an exit pupil array with 33 rays across the circular exit pupil and padded with zeros out to 129 × 129. Graph the real and imaginary part of the 2 × 2 × 129 × 129 Jones Pupil.

   b. Calculate the 2 × 2 × 129 × 129 amplitude response matrix (**ARM**), the Jones matrix form of the coherent point spread function. Each 129 × 129 element of the 2 × 2 Jones pupil is separately Fourier transformed. Graph the real and imaginary part of the 2 × 2 × 129 × 129 **ARM**.

   c. Calculate and graph the 4 × 4 × 129 × 129 **MPSM**.

Based on the amplitude response matrix in parts a and b, answer the following questions:

   d. If the optic is illuminated with horizontal polarized light and viewed through a vertical polarizer, what form does the coherent point spread function have?

   e. What wavefront aberration is present in the $\mathbf{J}xx$ element? How does it compare to the aberration in the $\mathbf{J}_{yy}$ element?

   f. Describe and explain the difference between the horizontal-in horizontal-analyzed coherent point spread function and the vertical-in vertical-analyzed coherent point spread function.

Based on the **MPSM** in part c, answer the following questions:

   g. What is the polarization of the image of an unpolarized point source?

   h. How does the intensity distribution of the point spread function (the S_0 component) depend on the incident polarization state?

16.4 Refer to Figures 16.35 and 16.37. When right circularly polarized light is incident, is the center of the image right circularly polarized or left circularly polarized? Why? Will the six islands in the first diffraction ring have the same polarization as the center?

16.5 Figure 16.46 showed exiting Stokes parameters for the horizontally polarized light after the CC **MPSM**. Match the incident polarization state to the following exiting Stokes parameters.

a. $\mathbf{S} = \begin{pmatrix} 1 \\ -1 \\ 0 \\ 0 \end{pmatrix}$ b. $\mathbf{S} = \begin{pmatrix} 1 \\ 0 \\ 1 \\ 0 \end{pmatrix}$ c. $\mathbf{S} = \begin{pmatrix} 1 \\ 0 \\ -1 \\ 0 \end{pmatrix}$ d. $\mathbf{S} = \begin{pmatrix} 1 \\ 0 \\ 0 \\ 1 \end{pmatrix}$ e. $\mathbf{S} = \begin{pmatrix} 1 \\ 0 \\ 0 \\ -1 \end{pmatrix}$

(1) (2) (3) (4) (5)

16.6 What calculations are performed on the Jones pupil to calculate the ARM in regions near focus but not in focus?

16.7 Consider vortex retarders that are half wave linear retarders with their fast axes rotated $m/2$ times around the center. Their Jones matrix is

$$\mathbf{J}_m = \begin{pmatrix} \cos(m \cdot \arctan(x, y)) & \sin(m \cdot \arctan(x, y)) \\ \sin(m \cdot \arctan(x, y)) & -\cos(m \cdot \arctan(x, y)) \end{pmatrix}.$$

The retardance pupil map is shown below:

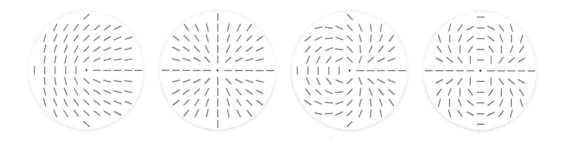

Calculate **PSM** of Jones matrices for $m = 1, 2, 3,$ and 4.

16.8 A collimated beam of 500 nm light is incident on a 1 mm × 1 mm square aperture over a Wollaston prism with the following polarization aberration function and pupil functions:

$$a(x,y) = 1,$$

$$\mathbf{J}(x,y) = \begin{pmatrix} \exp(-i\pi x/1) & 0 \\ 0 & \exp(i\pi x/1) \end{pmatrix},$$

$$W(x,y) = 0).$$

The prism is followed by a non-polarizing lens. The exit pupil is 100 mm from the image. Choose nine pixels across the aperture in a Jones pupil sized 2 × 2 × 199 × 199.

a. Calculate the 2 × 2 × 199 × 199 Jones pupil. Graph the magnitude and phase of a region just larger than the pupil.
b. Calculate the 2 × 2 × 199 × 199 amplitude response matrix. Graph the magnitude and phase of the center of the **ARM**.
c. Graph the magnitude and phase of the amplitude point spread function for horizontally polarized incident light, both E_x and E_y. Only graph the center where the majority of the light is located.
d. Graph the magnitude and phase of the center of the amplitude point spread function for vertically polarized incident light.
e. Graph the amplitude point spread function for 45° polarized incident light.
f. Calculate and graph the optical transfer matrix for horizontally polarized incident light.
g. Determine and graph the optical transfer matrix for 45° polarized incident light.

Based on the Jones pupil in part (a):
h. What aberrations are present?

Based on the amplitude point spread function for horizontal polarized incident light in part (c):
i. What is the polarization state of the amplitude point spread function for horizontally polarized incident light?

Based on the amplitude point spread function for vertical polarized incident light in part (d):
j. What is the polarization state of the amplitude point spread function for vertically polarized incident light?

Based on the amplitude point spread functions in parts (c) and (d):
k. How are horizontal point spread functions located relative to the vertical point spread functions? Explain.

Based on the amplitude point spread function for 45° incident light in part (e):
l. Describe the distribution of polarization states for 45° polarized point spread functions.

Based on the optical transfer matrix for horizontally polarized incident light in part (f):
m. How much is the image degraded by the polarization aberration? Can it be compared to a diffraction-limited optical transfer function?

Based on the optical transfer matrix for 45° polarized incident light in part (g):
n. Which spatial frequencies are filtered out (completely removed) by the system?
o. What is the separation between pixels in the **ARM** in millimeters?

References

1. N. Lindlein, S. Quabis, U. Peschel, and G. Leuchs, High numerical aperture imaging with different polarization patterns, *Opt. Express* 15(9) (2007).
2. J. P. McGuire and R. A. Chipman, Diffraction image formation in optical systems with polarization aberrations. I: Formulation and example, *J. Opt. Soc. Am. A* 7(9) (1990).
3. R. N. Bracewell, *The Fourier Transform and Its Applications*, Chapter 6, New York: McGraw-Hill Higher Education (2000).
4. R. W. Ramirez, *The FFT, Fundamentals and Concepts*, Hoboken, NJ: Prentice Hall (1985).
5. J. S. Walker, *Fast Fourier Transforms*, 2nd edition, Boca Raton, FL: CRC Press (1996).
6. J. W. Goodman, *Statistical Optics*, New York: Wiley (1985).
7. J. D. Gaskill, *Linear Systems, Fourier Transforms, and Optics*, New York: Wiley (1978).
8. E. Hecht, *Optics*, 4th edition, Boston, MA: Addison Wesley (2002).

17

Parallel Transport and the Calculation of Retardance

17.1 Introduction

Knowledge of an optical system's *retardance* provides important information on the polarization dependence of the exiting wavefronts.[1] This chapter develops an algorithm for the calculation of retardance, a subtle topic. It would be desirable, even expected, to have an algorithm that takes a *polarization ray tracing matrix* **P** as input, along with the associated input and output propagation vectors, $\mathbf{k}_{in}$ and $\mathbf{k}_{out}$, and returns the magnitude of the retardance, δ, and the fast and slow states. The actual situation is not so easy since the calculation of retardance for a ray path through an optical system depends on the entire sequence of **k** transiting the optical system.

When describing rays propagating through optical systems, the effects of coordinate system changes on refraction can masquerade as *circular retardance*; this is shown in Section 17.2.1. Similarly, *local coordinate* system changes on reflection can masquerade as a half wave of linear retardance. These two effects need to be calculated and corrected to obtain a correct retardance calculation. This chapter's objectives are the following: (1) critically consider several different definitions of retardance with the objective of finding a sufficiently robust definition generally applicable for optical design, (2) explore the local coordinate transformation associated with parallel transport of transverse vectors along ray paths through optical systems, and (3) present an algorithm for the calculation of *proper retardance* in *polarization ray tracing* using the 3 × 3 polarization ray tracing calculus.[2] This algorithm will separate the part of the polarization ray tracing matrix that describes proper retardance from the part that describes *non-polarizing rotations*, the *geometrical transformation*. The collection of geometrical transformations across a wavefront is the *skew aberration*, the subject of the next chapter. Examples are provided to highlight the associated subtleties.

The term *retardance* refers to a physical property by which *optical path length* accumulation depends on the incident polarization state. The classic *retarder* is a crystalline *waveplate* that divides a beam into two modes having two distinct polarization states and two optical path lengths.[3–5] The retardance is the phase difference that accrues corresponding to that optical path difference, a difference in transit time. For a combination of waveplates, the optical path length can be multivalued as shown in Figure 17.1. This waveplate picture of retardance needs to be generalized to incorporate the retardance contributions from metal reflections and multilayer thin films. A detailed discussion on multi-order compound retarders that have multivalued optical path lengths is presented in Chapter 26 (Multi-Order Retarders and the Mystery of Discontinuities).

The Jones matrix view of an ideal retarder is a device described by a *unitary matrix*. The unitary matrix has two orthogonal *eigenpolarizations*, which describe the two polarization states that exit in the same polarization state as the incident light. Depending on whether the *eigenvectors* are linear, circular, or elliptical, the retarder is termed a linear, circular, or elliptical retarder. The *unitary matrix* also has two unimodular eigenvalues, $e^{-i\phi_1}$ and $e^{-i\phi_2}$, which describe the change in phase of the two eigenpolarizations. The retardance is the difference in the phases of the eigenvalues, $\delta = |\phi_1 - \phi_2|$. Because of the unimodular eigenvalues, the *Jones matrix* is periodic in the retardance, repeating every wave of retardance, as in the matrix

$$\begin{pmatrix} 1 & 0 \\ 0 & e^{i\delta} \end{pmatrix}, \qquad (17.1)$$

where the *y*-component is delayed by δ radians. Note that this definition applies to ideal retarders and unitary matrices. Combinations of retardance and diattenuation, such as from polarization ray tracing, result in inhomogeneous matrices with non-orthogonal eigenpolarizations. One procedure is to decompose the Jones matrix into hermitian and unitary parts. This can be accomplished multiple ways. Three examples are the polar decomposition, the matrix logarithm, and the *Lu–Chipman decomposition* (for Mueller matrices), resulting in similar but not identical retardance magnitudes, fast and slow axes.

Another picture of retarders and retardance is devices or interactions that cause polarization states to change in a way described by rotations of the *Poincaré sphere*; the rotation angle is the retardance and the rotation axis of the Poincaré sphere identifies the fast and slow axes.[6] Retarder Mueller matrices are the associated rotation matrices for the Stokes parameters.

Retardance in optical systems is important because

  a. Retardance is measurable with *polarimeters* and *interferometers*.
  b. The separation of polarization properties into *diattenuation* and *retardance* helps simplify our description and understanding of *polarization aberrations*.
  c. When retardance aberration is present, *different incident polarization states exit with different wavefront aberrations; they have different interferograms, and they form different point spread functions.*

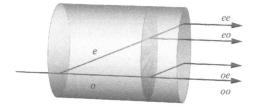

Figure 17.1 A series of two waveplates generates four exiting beams unless the two fast axes are exactly parallel or perpendicular. The beams are indicated as ordinary (*o*) or extraordinary (*e*) in each crystal. Therefore, four optical path lengths are associated with transmission through this two-element retarder.

17.2 Geometrical Transformations

There are complications in extending the concept of retardance to three-dimensional polarization ray trace matrices. Since the entering and exiting rays need not be collinear, the eigenvectors of the polarization ray tracing matrix usually are not in the incident and exiting transverse planes, and thus do not represent actual polarization states. Section 17.4 presents a well-defined retardance calculation algorithm for the 3 × 3 polarization ray tracing matrix; this algorithm requires the entire set of propagation vectors associated with all ray segments from object to image space to determine retardance. When dealing with a slippery concept like retardance, the precise algorithm used should be specified to aid communication and minimize misunderstanding.

Ray paths suffer retardance, diattenuation, amplitude, and phase changes. To keep the discussion focused, this chapter mostly deals with pure retarders and their matrices. These are *homogeneous matrices* with orthogonal eigenvectors. *Inhomogeneous matrices*, with non-orthogonal eigenvectors, involve combinations of diattenuation and retardance that are not aligned. These elements can be expressed as a product of a pure retarder with a pure diattenuator, and the retardance is well defined to be that of the pure retarder, as described by Lu and Chipman.[7] Often, skew rays through optical systems are slightly inhomogeneous; hence, this is an important concern in polarization ray tracing, but it does not complicate our treatment here.

17.1.1 Purpose of the Proper Retardance Calculation

The polarization-dependent phase change associated with a ray path through an optical system has two components:

1. The *proper retardance*; the *phase retardation* (*optical path difference*) arising from physical processes, such as propagation through birefringent materials or polarization-dependent phase changes occurring upon reflection or refraction from a surface
2. A *geometric transformation* due to the coordinate selection used for determining the phase

Often, *ideal reflection at normal incident* is represented as the Jones matrix $\begin{pmatrix} -1 & 0 \\ 0 & 1 \end{pmatrix}$, see Section 17.6.1. However, at normal incidence, how does the mirror differentiate between x- and y-polarized light? The rationale behind this Jones matrix and the resolution of this paradox is the subject of Section 17.6.1.

The retardance calculation algorithm needs to separate the geometric transformation, which is an "optical activity-like" geometric rotation and/or inversion, from the proper retardance. A *parallel transport matrix*, **Q**, described in Section 17.2.5, identifies canonical pairs of local coordinate systems for general sequences of ray paths, thus characterizing the geometric transformation. **Q** is the intermediate calculation to be used to separate retardance from geometric transformation.

17.2 Geometrical Transformations

The *local coordinates* that are necessary to specify *Jones vectors* propagating in arbitrary directions may be rotated and/or inverted between object and image space[8] in ways that are not associated with any retardance. To describe these coordinate effects, this section defines the parallel transport matrix and applies it to the description of geometric transformations of local coordinates. This will result in degenerate *s*- and *p*-polarizations at normal incidence.

17.2.1 Rotation of Local Coordinates: Polarimeter Viewpoint

Our philosophy is to perform as many calculations as possible in *global coordinates*. But the understanding of geometric transformations and proper retardance is one topic that benefits from a parallel discussion within the context of *local coordinates*, because it provides additional insight into the underlying issue. Jones matrices are defined with respect to a pair of *right-handed local coordinates*; one set is associated with the incident Jones vector and another set is associated with the exiting Jones vector. The retardance of a *homogeneous Jones matrix* **J**, with orthogonal eigenpolarizations $\mathbf{w}_1$ and $\mathbf{w}_2$, is calculated from a phase difference of the matrix eigenvalues, ξ_1 and ξ_2, as

$$\delta = |\arg(\xi_1) - \arg(\xi_2)|, \tag{17.2}$$

where $\mathbf{J}\mathbf{w}_1 = \xi_1 \mathbf{w}_1$ and $\mathbf{J}\mathbf{w}_2 = \xi_2 \mathbf{w}_2$.[9]

Consider a Jones or Mueller matrix polarimeter measurement of an empty sample compartment that has an identity Jones matrix as shown in Figure 17.2 (left). A Jones matrix polarimeter performing a measurement in air should result in an identity matrix.

By rotating the PSA by θ as shown in Figure 17.2 (right), the exiting local coordinates are rotated by an angle θ with respect to the incident local coordinates. Then, the measured Jones matrix becomes a *rotation matrix* instead of the identity matrix,

$$\mathbf{J}(\theta) = \mathbf{R}(\theta)\mathbf{I} = \begin{pmatrix} \cos\theta & -\sin\theta \\ \sin\theta & \cos\theta \end{pmatrix} \begin{pmatrix} 1 & 0 \\ 0 & 1 \end{pmatrix} = \begin{pmatrix} \cos\theta & -\sin\theta \\ \sin\theta & \cos\theta \end{pmatrix}, \tag{17.3}$$

with eigenvalues

$$\xi_1 = \exp(i\theta), \quad \xi_2 = \exp(-i\theta) \tag{17.4}$$

and right and left circularly polarized eigenvectors or eigenpolarizations

$$\mathbf{w}_1 = \begin{pmatrix} 1 \\ -i \end{pmatrix}, \quad \mathbf{w}_2 = \begin{pmatrix} 1 \\ i \end{pmatrix}. \tag{17.5}$$

Note the similarity of $\mathbf{J}(\theta)$ to the form of a *circular retarder* (Equation 5.48). Unless the exiting local coordinate orientation is parallel to the incident local coordinate orientation, a non-polarizing element appears to have a "circular retardance" of

$$\delta = |\arg(\xi_1) - \arg(\xi_2)| = 2\theta, \tag{17.6}$$

where θ is the local coordinate rotation. However, it should be obvious that rotating the PSA coordinates cannot introduce an optical path difference between right and left circularly polarized light in

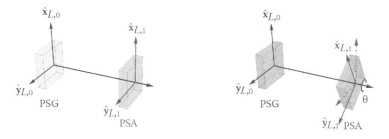

Figure 17.2 A polarimeter performing a measurement in air with the polarization state analyzer (PSA) (left) aligned with the polarization state generator (PSG) and (right) rotated to an arbitrary orientation. By rotating the PSA, the exiting local coordinates for the Jones matrix are also rotated. The measured retardance of the empty sample has a "circular retardance" of 2θ.

17.2 Geometrical Transformations

an empty sample compartment. Here, the "proper" retardance is zero; no optical path difference is present. The geometrical transformation is θ. Thus, it is seen how the retardance measured by polarimeters depends on the relative choice of incident and exiting local coordinates. The main objective of this chapter is to understand the similar issues that apply to polarization ray tracing calculations.

Now, consider the Jones matrix measurement of a retarder **U** (for unitary) using this polarimeter with a rotated PSA,

$$\mathbf{J}(\theta) = \begin{pmatrix} \cos\theta & -\sin\theta \\ \sin\theta & \cos\theta \end{pmatrix} \mathbf{U} = \begin{pmatrix} \cos\theta & -\sin\theta \\ \sin\theta & \cos\theta \end{pmatrix} \begin{pmatrix} j_{11} & j_{12} \\ j_{21} & j_{22} \end{pmatrix}. \tag{17.7}$$

The retardance calculated from Equations 17.2 and 17.7 depends on the rotation angle θ. For circular retarders with a retardance of δ, the measured retardance from the polarimeter as a function of the PSA angle θ is

$$\delta_{Measured} = 2\arg\left[\cos\left(\frac{\delta}{2}+\theta\right) + i\left|\sin\left(\frac{\delta}{2}+\theta\right)\right|\right] = \delta + 2\theta. \tag{17.8}$$

For a linear retarder with a retardance of δ, the measured retardance as a function of the PSA angle θ is

$$\delta_{Measured} = 2\arctan\left(\frac{\sqrt{1-\left(\cos\frac{\delta}{2}\cdot\cos\theta\right)^2}}{\cos\frac{\delta}{2}\cdot\cos\theta}\right). \tag{17.9}$$

17.2.2 Non-Polarizing Optical Systems

To identify the physical retardance or *proper retardance*, a coordinate system is needed to identify the relationship between incident and exiting polarization states in the absence of retardance. The concept of a *non-polarizing optical system* is introduced to define polarization relationships in the absence of diattenuation and retardance; then, the diattenuation and retardance can be defined with respect to the non-polarizing relationships. This coordinate system serves a role in polarization similar to the *paraxial ray trace* in geometrical optics. In paraxial optics, point sources generate spherical waves in exit pupils and all rays come to ideal image points. Then, aberrations are defined as deviations of real ray paths from paraxial optics. Paraxial optics provides an underlying coordinate system for aberrations.

In a non-polarizing optical system, at reflections and refractions, no polarization state change occurs—only a change of direction. It is equivalent to the Fresnel amplitude coefficients being equal to one, $a_s = a_p = 1$. The incident polarization ellipse, which is arbitrary, folds about the s-basis vector and continues with the same ellipse with the same major axis rotated. For refraction, the helicity of the state remains the same, right circularly polarized light refracts as right circularly polarized light. For reflection, there is an inversion of helicity, right circularly or elliptically polarized light reflects as left circularly or elliptically polarized light, associated with the change of the propagation direction. Thus, for the non-polarizing optical system, the polarization state only undergoes a series of rotations at reflections and changes of helicity. Figure 17.3 shows left elliptically polarized light propagating through a Dove prism, which has been ray traced as a non-polarizing optical system. At the entrance face, the polarization ellipse rotates with the propagation direction. Without other

Figure 17.3 Three views of polarized light propagating through a non-polarizing optical system, a Dove prism, undergoes a series of folds, or rotations and inversions, but is otherwise unaltered.

changes, it propagates to the bottom surface where it reflects as right elliptically polarized light. Finally, it rotates again following the refraction direction out of the Dove prism. This would be the evolution of the polarization state if all the Fresnel coefficients equaled one.

17.2.3 Parallel Transport of Vectors

In this section, the geometric transformations for ray paths through optical systems are explored using parallel transport where skew rays are of particular interest. *Parallel transport* of a vector over a sphere is a process of moving the vector along a series of great circle arcs such that the angle between the vector and each arc is constant,[10,11] as shown in Figure 17.4. At a vertex where the path transitions from the first to the second arc (and so on), the angle at the vertex is maintained along the second arc, and so on, for the sequence of arcs as shown in arrows. This section introduces a *propagation sphere* or *k-sphere* for representing sets of propagation vectors for a sequence of ray segments.

Consider tracing a single ray through a lens system with N interfaces for the special case where the incident propagation vector $\hat{\mathbf{k}}_{incident}$ (first ray segment) is parallel to the exiting propagation vector $\hat{\mathbf{k}}_{exit}$ (last ray segment), but the ray has changed its propagation direction $\hat{\mathbf{k}}_q$ many times while refracting through the optical system. This set of $\hat{\mathbf{k}}_q$ can be represented as points on the unit **k**-sphere connected by great circle arcs. Since $\hat{\mathbf{k}}_{incident} = \hat{\mathbf{k}}_{exit}$, the arcs form a closed spherical polygon. Let a pair of orthogonal local coordinate vectors be arbitrarily chosen in the transverse plane for the first ray segment. When this pair of local coordinates is carried through the system by parallel transport as shown in Figure 17.4, the parallel-transformed local coordinates at the exiting propagation vector have rotated with respect to the initial local coordinates by an angle in radians equal to the spherical polygon's solid angle.[12]

Consider the example of a sequence of **k** vectors on a unit **k**-sphere shown in Figure 17.4, with a ray propagating along $\hat{\mathbf{k}}_1$, refracting to $\hat{\mathbf{k}}_2$, then refracting to $\hat{\mathbf{k}}_3$, and exiting along $\hat{\mathbf{k}}_4 = \hat{\mathbf{k}}_1$. In the plane transverse to $\hat{\mathbf{k}}_1$, an arbitrary pair of orthogonal right-handed local basis vectors is defined and can be used for describing local coordinates for the incident Jones vector. This pair of local coordinates can then be moved through the optical system by parallel transport as shown in Figure 17.4. Vectors $(\mathbf{x}_A, \mathbf{y}_A)$ in the tangent plane of point A are parallel transported from point A with

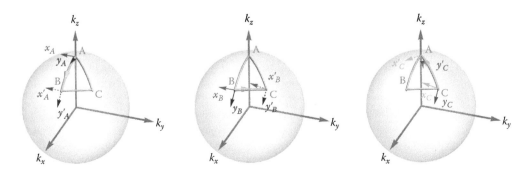

Figure 17.4 Parallel transport of the incident local basis vectors through point $A \rightarrow B \rightarrow C \rightarrow A$.

17.2 Geometrical Transformations

$\hat{\mathbf{k}}_{incident} = \hat{\mathbf{k}}_1 = (0,0,1)$ to point B with $\hat{\mathbf{k}}_2 = (1,0,1)/\sqrt{2}$, and $(\mathbf{x}_A, \mathbf{y}_A)$ becomes $(\mathbf{x}'_A, \mathbf{y}'_A) = (\mathbf{x}_B, \mathbf{y}_B)$ which is in the tangent plane at point B. Then, vectors $(\mathbf{x}_B, \mathbf{y}_B)$ parallel transports from point B to point C with $\hat{\mathbf{k}}_3 = (0.5, 0.5, 1/\sqrt{2})$, and $(\mathbf{x}_B, \mathbf{y}_B)$ becomes $(\mathbf{x}'_B, \mathbf{y}'_B) = (\mathbf{x}_C, \mathbf{y}_C)$. Finally, the vectors transport back to point A where $\hat{\mathbf{k}}_{exit} = \hat{\mathbf{k}}_4 = (0,0,1)$, where $(\mathbf{x}_C, \mathbf{y}_C)$ becomes $(\mathbf{x}'_C, \mathbf{y}'_C)$, which does not equal to $(\mathbf{x}_A, \mathbf{y}_A)$. As the propagation vector changes its direction through an optical system, the total rotation of the vector in its transverse plane is equivalent to the *Pancharatnam phase*[13,14] or the *Berry phase*.[15]

The incident propagation vector $\hat{\mathbf{k}}_{incident} = \hat{\mathbf{k}}_1$ is (0, 0, 1), and after three refractions, the exiting propagation vector $\hat{\mathbf{k}}_{exit} = \hat{\mathbf{k}}_4$ is also (0, 0, 1). Thus, the propagation vectors are mapped to points A, B, and C and back to A on the **k**-sphere. One might naively select the same local coordinates for the incident space and exiting space since $\hat{\mathbf{k}}_1 = \hat{\mathbf{k}}_4$. However, the geometric transformation for this ray path is a 12.35° rotation, which is a circular retardance-like effect. As shown in Figure 17.5, the solid angle subtended by the *spherical triangle ABC* is the 12.35° geometrical rotation. Any incident linear polarization will exit propagating parallel but with its polarization rotated by 12.35° in the transverse plane. Right and left circular polarized light will have a phase shift of 12.35/360 waves between them, in addition to their optical path length. Since the state rotates counterclockwise looking into the beam, the phase of left circularly polarized light advances and right circularly polarized light is delayed.

A simple example of a similar sequence of refractions through a three-prism system is shown in Figure 17.6. The set of propagation vectors through the prisms are very similar to the ones in Figure 17.4 but subtend a much smaller spherical triangle with a solid angle of 2.8°. Note that the edges of lenses resemble prisms. Skew rays in lenses do propagate in a similar manner, as shown in Figure 17.4, where the succession of **k** vectors move in either a clockwise or a counterclockwise direction.

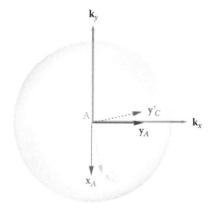

Figure 17.5 The rotation amount between $(\mathbf{x}_A, \mathbf{y}_A)$ and $(\mathbf{x}'_C, \mathbf{y}'_C)$ is equal to the solid angle of the spherical triangle ABC.

Figure 17.6 Three-prism example that refracts propagation vectors and forms a similar spherical polygon on the **k**-sphere as the one shown in Figure 17.4.

If the initial local coordinates and the exiting local coordinates for a Jones matrix or polarimeter are chosen to be parallel to each other, the ray path appears to have "circular retardance." However, if a polarimeter is set up with a 12.35° rotation counterclockwise, as in Figure 17.2 (right), the Jones matrix, calculated or measured with our earlier Jones polarimeter, will be an identity matrix, as expected for a non-polarizing system with no physical retardance.

17.2.4 Parallel Transport of Vectors with Reflection

The modification of parallel transports for reflection is described in this section. *Reflection* differs from refraction in that, during reflection, *circular polarized light* changes helicity upon reflection.* Consider a reflection example of the three-fold mirror system shown in Figure 17.7. The three mirrors are arranged so that the angle of incidence for a collimated beam at each mirror is 45°. Let each reflection be an ideal non-polarizing reflection so that the incident polarization ellipse enters and exits each reflection unchanged, and thus exits the optical system with the same ellipticity.

The incident propagation vector $\hat{\mathbf{k}}_{incident} = \hat{\mathbf{k}}_1$ is (0, 0, 1), and after three reflections, the exiting propagation vector $\hat{\mathbf{k}}_4$ is also (0, 0, 1) with $\hat{\mathbf{k}}_2 = (1,0,0)$ and $\hat{\mathbf{k}}_3 = (0,1,0)$. Thus, the propagation vectors are mapped to points *A*, *B*, and *C* on the *k-sphere* as shown in Figure 17.8. The arbitrary right-handed local basis vectors along the first ray segment have been chosen with x_A aligned with *s*-polarization and y_A aligned with *p*-polarization. The incident *local basis vectors* (blue arrows) are parallel transported along the great circle arc from points *A* to *B*. At point *B*, upon reflection, $\hat{\mathbf{k}}_1$ reflects to $\hat{\mathbf{k}}_2$, and one of the basis vectors $\left(\mathbf{x}'_A \text{ or } \mathbf{y}'_A\right)$ is inverted due to reflection. Our convention inverts the *p*-polarization; thus, $\mathbf{y}'_A$ is inverted to $\mathbf{y}_B$ from the left panel to the middle panel of Figure 17.8. The vectors pair $\mathbf{x}_B$ and $\mathbf{y}_B$ (red arrows) are parallel transported to point *C* where the *p*-polarization vector $\mathbf{x}'_B$ is inverted to $\mathbf{x}_C$ (Figure 17.8, middle to right panels). Then, the vectors pair $\mathbf{x}_C$ and $\mathbf{y}_C$ (green arrows) are parallel transported back to point *A* where the final reflection occurs. At point *A*, the *p*-polarization $\left(\mathbf{y}'_C\right)$ is inverted to $\mathbf{y}''_A$, as shown in Figure 17.9.

Thus, the geometric transformation for this ray path is a 90° rotation, which involves the solid angle subtended by the spherical triangle *ABC*, and also an inversion from the odd number of reflections. As shown in Figure 17.9, between $(\mathbf{x}_A, \mathbf{y}_A)$ and $\left(\mathbf{x}'_A, \mathbf{y}'_A\right)$, there is a 90° rotation and an inversion.

Associated with this rotation of the local coordinates, incident *y*-polarized light (blue) in Figure 17.7 exits the system as *x*-polarized light (orange), even though the mirrors are non-polarizing. If this non-polarizing system is measured by a polarimeter with parallel polarization state generator and analyzer, a 180° circular retardance will be measured due to inversion resulting from the odd number of reflections. But the retardance measured here is purely from the local coordinate transformation. To measure the *proper retardance*, the analyzer should be rotated 90° from the generator and its local coordinates need to be corrected for the inversion. This 90° is the Berry phase for this sequence of propagation vectors.

17.2.5 Parallel Transport Matrix, Q

To calculate the *Berry phase*, a matrix algorithm is applied to handle the effect of parallel transport on ray propagation through an optical system with *N* ray intercepts. The *parallel transport matrix* $\mathbf{Q}_q$ at *q*th ray intercepts is defined as a real unitary 3 × 3 ray tracing matrix, calculated by assuming that each ray intercept is non-polarizing.

---

* This is considered at around normal incidence and below the Brewster angle. Beyond the Brewster angle, the helicity change is opposite.

17.2 Geometrical Transformations

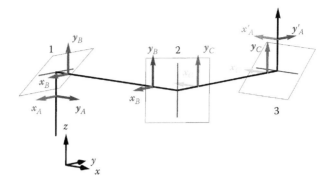

Figure 17.7 A *three-fold mirror* system with the incident local basis vectors (blue arrows) and its propagation through the non-polarizing optical system as red, green, and orange arrows. When a collimated beam enters the system along the z-axis, the beam exits along the z-axis.

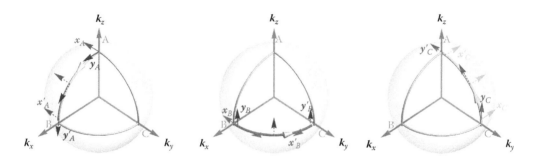

Figure 17.8 The evolution of a local coordinate pair $(\mathbf{x}_A, \mathbf{y}_A)$ after the first reflection (from left to middle), the second reflection (from middle to right), and right before the last reflection are shown through a system of three-fold mirrors in Figure 17.7.

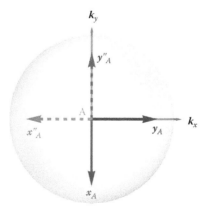

Figure 17.9 The exiting local coordinates $\left(\mathbf{x}''_A, \mathbf{y}''_A\right)$ (dashed arrows) undergo a 90° rotation plus an inversion from the initial local coordinates $(\mathbf{x}_A, \mathbf{y}_A)$ (solid arrows).

$\mathbf{Q}_q$ for *refraction* is a rotation about $\hat{\mathbf{k}}_{q-1} \times \hat{\boldsymbol{\eta}}_q$ (the $\hat{\mathbf{s}}$ vector) by the ray deviation angle χ from Snell's law

$$n \sin \theta = n' \sin \theta', \quad \chi = |\theta' - \theta|, \qquad (17.10)$$

where (n, n') are refractive indices, and (θ, θ') are angles of incidence before and after the refraction. Let $\mathbf{E}_{q-1}$ be the polarization state exiting surface $q-1$ and incident on surface q. Then, for a non-polarizing interface, the refracted polarization state is $\mathbf{E}_q = \mathbf{Q}_q \mathbf{E}_{q-1}$. $\mathbf{Q}_q$ for refraction rotates the incident polarization ellipse $\mathbf{E}_{q-1}$ to the same ellipse relative to the propagation vector $\mathbf{E}_q$, which is in the plane transverse to $\hat{\mathbf{k}}_q$. Similarly, $\mathbf{Q}_q$ for *reflection* is a rotation about the $\hat{\mathbf{s}}$ vector, which brings $\hat{\mathbf{k}}_{q-1}$ to $\hat{\mathbf{k}}_q$ plus an inversion that transforms a right-handed incident coordinate system to a left-handed coordinate system.

A sequence of $\mathbf{Q}_q$ matrices performs the vector parallel transport from a given incident s–p basis pair into the corresponding exiting s–p basis pair. Other than this geometric transformation, $\mathbf{Q}_q$ incorporates no polarization effects; it is purely non-polarizing.

The cumulative parallel transport ray tracing matrix for a ray through a system with N interfaces is

$$\mathbf{Q}_{Total} = \prod_{q=1}^{N} \mathbf{Q}_{N-q+1} = \mathbf{Q}_N \mathbf{Q}_{N-1} \cdots \mathbf{Q}_q \cdots \mathbf{Q}_2 \mathbf{Q}_1. \qquad (17.11)$$

This is equivalent to sliding basis vectors around a unit sphere via parallel transport as in Figure 17.4. For an incident polarization state $\mathbf{E}_0$, the state exiting the system in the absence of any diattenuation or retardance is

$$\mathbf{E}_N = \mathbf{Q}_{Total} \mathbf{E}_0. \qquad (17.12)$$

$\mathbf{E}_N$ provides a reference polarization state for the actual exiting polarization state,

$$\mathbf{E}'_N = \mathbf{P}_{Total} \mathbf{E}_0. \qquad (17.13)$$

The refraction $\mathbf{Q}_q$ is only a function of $\hat{\mathbf{k}}_{q-1}$ and $\hat{\mathbf{k}}_q$

$$\mathbf{Q}_q = \mathbf{O}_{out,q} \mathbf{I} \mathbf{O}_{in,q}^{-1} = \begin{pmatrix} \hat{s}_{x,q} & \hat{p}'_{x,q} & \hat{k}_{x,q} \\ \hat{s}_{y,q} & \hat{p}'_{y,q} & \hat{k}_{y,q} \\ \hat{s}_{z,q} & \hat{p}'_{z,q} & \hat{k}_{z,q} \end{pmatrix} \begin{pmatrix} 1 & 0 & 0 \\ 0 & 1 & 0 \\ 0 & 0 & 1 \end{pmatrix} \begin{pmatrix} \hat{s}_{x,q} & \hat{s}_{y,q} & \hat{s}_{z,q} \\ \hat{p}_{x,q} & \hat{p}_{y,q} & \hat{p}_{z,q} \\ \hat{k}_{x,q-1} & \hat{k}_{y,q-1} & \hat{k}_{z,q-1} \end{pmatrix}, \qquad (17.14)$$

$$= \left(\hat{\mathbf{s}}_q \ \hat{\mathbf{p}}'_q \ \hat{\mathbf{k}}_q \right) \left(\hat{\mathbf{s}}_q \ \hat{\mathbf{p}}_q \ \hat{\mathbf{k}}_{q-1} \right)^T,$$

where $\mathbf{I}$ is the 3×3 identity matrix and the column vectors are defined as

$$\hat{\mathbf{s}}_q = \frac{\hat{\mathbf{k}}_{q-1} \times \hat{\boldsymbol{\eta}}_q}{|\hat{\mathbf{k}}_{q-1} \times \hat{\boldsymbol{\eta}}_q|}, \quad \hat{\mathbf{p}}_q = \hat{\mathbf{k}}_{q-1} \times \hat{\mathbf{s}}_q, \quad \hat{\mathbf{p}}'_q = \hat{\mathbf{k}}_q \times \hat{\mathbf{s}}_q. \qquad (17.15)$$

17.2 Geometrical Transformations

Note that Equation 17.14 is essentially a polarization ray tracing matrix **P** (introduced in Chapter 9 [Polarization Ray Tracing Calculus]) when the Jones matrix is an identity matrix (i.e., non-polarizing matrix). Similarly, $\mathbf{Q}_q$ for reflection is

$$\mathbf{Q}_q = \begin{pmatrix} \hat{s}_{x,q} & \hat{p}'_{x,q} & \hat{k}_{x,q} \\ \hat{s}_{y,q} & \hat{p}'_{y,q} & \hat{k}_{y,q} \\ \hat{s}_{z,q} & \hat{p}'_{z,q} & \hat{k}_{z,q} \end{pmatrix} \begin{pmatrix} 1 & 0 & 0 \\ 0 & -1 & 0 \\ 0 & 0 & 1 \end{pmatrix} \begin{pmatrix} \hat{s}_{x,q} & \hat{s}_{y,q} & \hat{s}_{z,q} \\ \hat{p}_{x,q} & \hat{p}_{y,q} & \hat{p}_{z,q} \\ \hat{k}_{x,q-1} & \hat{k}_{y,q-1} & \hat{k}_{z,q-1} \end{pmatrix} \qquad (17.16)$$

$$= \begin{pmatrix} \hat{\mathbf{s}}_q & -\hat{\mathbf{p}}'_q & \hat{\mathbf{k}}_q \end{pmatrix} \begin{pmatrix} \hat{\mathbf{s}}_q & \hat{\mathbf{p}}_q & \hat{\mathbf{k}}_{q-1} \end{pmatrix}^T,$$

where the negative sign in the diagonal matrix indicates that the $\mathbf{Q}_q$ for reflection introduces inversion.

Example 17.1 Q Matrix for a Fold Mirror

Consider a fold mirror with a surface normal along $\hat{\boldsymbol{\eta}} = (-1,0,1)/\sqrt{2}$ and a ray entering the mirror along $\hat{\mathbf{k}}_0 = (0,0,1)$. Calculate **Q** for the mirror.

Using $\hat{\mathbf{s}} = (0,-1,0)$, $\hat{\mathbf{p}} = (1,0,0)$, $\hat{\mathbf{p}}' = (0,0,-1)$, $\hat{\mathbf{k}} = (1,0,0)$,

$$\mathbf{Q} = \begin{pmatrix} -\hat{\mathbf{p}}' & -\hat{\mathbf{s}} & \hat{\mathbf{k}} \end{pmatrix} = \begin{pmatrix} 0 & 0 & 1 \\ 0 & 1 & 0 \\ 1 & 0 & 0 \end{pmatrix}.$$

Example 17.2 Q Matrix for a Dove Prism

Another example with geometrical transformation is seen in the *Dove prism*. Figure 17.10 shows a Dove prism with *non-polarizing reflection*. When a Dove prism rotates about the direction of the incident light, the exiting x-polarization rotates twice as much as the prism. Since the internal reflection is beyond critical angle, the incident y-polarization has an inversion, as shown in Figure 17.11. The Dove prism induces a half wave of retardance with a fast axis lying along the bottom surface of the prism.

Figure 17.10 Horizontally polarized light propagates through a Dove prism that rotates about the incident direction by 0°, 30°, and 60°. Then, the resultant polarization rotates 0°, 60°, and 120° about the exiting direction.

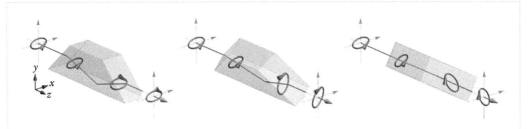

Figure 17.11 Elliptically polarized light propagates through a Dove prism, which is rotated about the incident direction by 0°, 30°, and 60°. The major axis of the resultant polarization ellipse rotates by 0°, 60°, and 120° with a change of helicity.

The **P** matrix derived in Chapter 9 (Polarization Ray Tracing Calculus) or measured from a polarimeter contains both the physical retardance and geometrical transformation. To extract the physical retardance from **P**, the geometrical transformation is undone by the inverse of **Q**. For an optical system with N interfaces, and arbitrary incident local coordinates $\hat{\mathbf{x}}_{L,0}$ and $\hat{\mathbf{y}}_{L,0}$,

$$\begin{aligned} \mathbf{Q}_{Total}\,\hat{\mathbf{x}}_{L,0} &= \hat{\mathbf{x}}_{L,N} & \mathbf{Q}_{Total}^{-1}\,\hat{\mathbf{x}}_{L,N} &= \hat{x}_{L,0} \\ \mathbf{Q}_{Total}\,\hat{\mathbf{y}}_{L,0} &= \hat{\mathbf{y}}_{L,N} \quad\Rightarrow\quad & \mathbf{Q}_{Total}^{-1}\,\hat{\mathbf{y}}_{L,N} &= \hat{y}_{L,0} \,, \\ \mathbf{Q}_{Total}\,\hat{\mathbf{k}}_{0} &= \hat{\mathbf{k}}_{N} & \mathbf{Q}_{Total}^{-1}\,\hat{\mathbf{k}}_{N} &= \hat{\mathbf{k}}_{0} \end{aligned} \qquad (17.17)$$

where $\hat{\mathbf{x}}_{L,N}$ and $\hat{\mathbf{y}}_{L,N}$ are the geometrically transformed coordinates in the exit space after operation by $\mathbf{Q}_{Total}$. The vectors $\hat{\mathbf{x}}_{L,0}$ and $\hat{\mathbf{y}}_{L,0}$ here are assumed to be an arbitrary pair of orthogonal vectors in the transverse plane of the first ray segment. Equation 17.17 also relates the incident and exiting propagation vectors by $\mathbf{Q}_{Total}$. For a given set of right-handed incident local coordinates $(\hat{\mathbf{x}}_{L,0},\hat{\mathbf{y}}_{L,0},\hat{\mathbf{k}}_0)$, the geometrically transformed coordinates $(\hat{\mathbf{x}}_{L,N},\hat{\mathbf{y}}_{L,N},\hat{\mathbf{k}}_N)$ forms right-handed coordinates unless the ray had an odd number of reflections in its ray path.

$\mathbf{Q}_{Total}$ represents a polarization ray trace through a non-polarizing optical system for a ray incident along $\hat{\mathbf{k}}_0$, which exits along $\hat{\mathbf{k}}_N$; since $\mathbf{Q}_{Total}$ only keeps track of the geometric transformation during ray propagation without any polarization changes, it is equivalent to ray tracing through a non-polarizing optical system.

$\mathbf{Q}_{Total}$ provides the much needed coordinate system with which to measure polarization change in optical systems, comparable to the role of paraxial optics for defining wavefront aberrations as discussed at the beginning of Section 17.2.2. Besides reflection and refraction, $\mathbf{Q}_q$ can be calculated for gratings, holograms, and scattering as well. If both $\hat{\mathbf{k}}_{q-1}$ and $\hat{\mathbf{k}}_q$ are on the same side of the interface, the reflection algorithm is used to calculate $\mathbf{Q}_q$. If the two propagation vectors are on opposite sides, the refraction algorithm for $\mathbf{Q}_q$ is used. In either case, the objective is to carry the polarization ellipse through the interface with no polarization change, only change of direction.

17.3 Canonical Local Coordinates

The sequence of $\mathbf{Q}_q$ provides a means to define a consistent set of *coordinate vectors** along a ray path from object to image space. The initial orthogonal right-handed coordinate vectors $(\hat{\mathbf{x}}_{L,0},\hat{\mathbf{y}}_{L,0})$

---

* Coordinate vectors may not be the local coordinate $(\hat{\mathbf{s}},\hat{\mathbf{p}},\hat{\mathbf{k}})$ vectors that are used to construct **P** or **Q** matrices. In this book, $(\hat{\mathbf{s}},\hat{\mathbf{p}},\hat{\mathbf{k}})$ is always a right-handed system to be consistent with how Jones matrices are defined. The coordinate vectors can be left-handed since they represent how given coordinate vectors transform via a non-polarizing optical system calculated by **Q** matrices.

17.3 Canonical Local Coordinates

are chosen arbitrarily in object space in the transverse plane. After the first interface, the next set of coordinate vectors is

$$\left(\hat{\mathbf{x}}_{L,1}, \hat{\mathbf{y}}_{L,1}\right) = \left(\mathbf{Q}_1 \hat{\mathbf{x}}_{L,0}, \mathbf{Q}_1 \hat{\mathbf{y}}_{L,1}\right). \tag{17.18}$$

With successive matrix multiplications by $\mathbf{Q}_q$, $q = 2, 3, 4, \ldots$, the coordinate vectors propagate along the ray path defining coordinate vectors for each ray segment. The coordinate vectors are canonical because in the absence of diattenuation and retardance, an incident polarization state $\hat{\mathbf{x}}_{L,0}$ would pass through the sequence of states $\hat{\mathbf{x}}_{L,2}, \hat{\mathbf{x}}_{L,3}, \hat{\mathbf{x}}_{L,4}, \ldots$, as it propagates through the system, providing a well-defined and meaningful relationship between the sets.

Depending on the number of reflections in the system, $\left(\hat{\mathbf{x}}_{L,q}, \hat{\mathbf{y}}_{L,q}, \hat{\mathbf{k}}_q\right)$ can either form a right-handed or a *left-handed coordinate system*. Figure 17.12 shows the incident $\left(\hat{\mathbf{x}}_{L,0}, \hat{\mathbf{y}}_{L,0}\right)$, reflected $\left(\hat{\mathbf{x}}_{L,r,1}, \hat{\mathbf{y}}_{L,r,1}\right)$, and transmitted $\left(\hat{\mathbf{x}}_{L,t,1}, \hat{\mathbf{y}}_{L,t,1}\right)$ coordinate vector pairs calculated from $\mathbf{Q}_r$ and $\mathbf{Q}_t$, where the incident coordinate vectors are

$$\hat{\mathbf{x}}_{L,0} = \frac{\hat{\mathbf{k}}_0 \times \hat{\boldsymbol{\eta}}_1}{\left|\hat{\mathbf{k}}_0 \times \hat{\boldsymbol{\eta}}_1\right|}, \; \hat{\mathbf{y}}_{L,0} = \hat{\mathbf{k}}_0 \times \hat{\mathbf{x}}_{L,0}. \tag{17.19}$$

Note that $\left(\hat{\mathbf{x}}_{L,0}, \hat{\mathbf{y}}_{L,0}, \hat{\mathbf{k}}_0\right)$ and $\left(\hat{\mathbf{x}}_{L,t,1}, \hat{\mathbf{y}}_{L,t,1}, \hat{\mathbf{k}}_{t,1}\right)$ form a right-handed coordinate vector set while $\left(\hat{\mathbf{x}}_{L,r,1}, \hat{\mathbf{y}}_{L,r,1}, \hat{\mathbf{k}}_{r,1}\right)$ forms a left-handed set. The subscript r stands for reflection and the subscript t stands for transmission.

A two-mirror example in Figure 17.13 shows the $\left(\hat{\mathbf{x}}_{L,i}, \hat{\mathbf{y}}_{L,i}\right)$ in object space ($i = 0$) and their geometric transformation along each ray segment using a set of $\mathbf{Q}_q$. Note that $\left(\hat{\mathbf{x}}_{L,0}, \hat{\mathbf{y}}_{L,0}\right) = \left(\hat{\mathbf{s}}_1, \hat{\mathbf{p}}_1\right)$ for this example. It is seen that the right-handed coordinate vectors in object space becomes left handed after the first reflection and then becomes right handed after the second reflection.

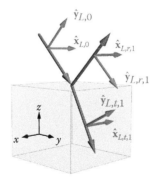

Figure 17.12 Incident, reflected, and transmitted coordinate vectors calculated from a parallel transport matrix. $\left(\hat{\mathbf{x}}_{L,0}, \hat{\mathbf{y}}_{L,0}, \hat{\mathbf{k}}_0\right)$ are right-handed incident coordinate vectors, $\left(\hat{\mathbf{x}}_{L,r,1}, \hat{\mathbf{y}}_{L,r,1}, \hat{\mathbf{k}}_{r,1}\right)$ are left-handed reflected coordinate vectors, and $\left(\hat{\mathbf{x}}_{L,t,1}, \hat{\mathbf{y}}_{L,t,1}, \hat{\mathbf{k}}_{t1}\right)$ are right-handed transmitted coordinate vectors.

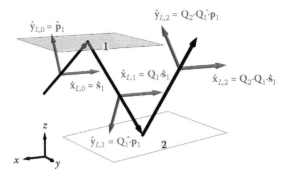

Figure 17.13 A two-mirror system. The red solid lines show the *s*-vector at the first mirror and its geometric transformation along each ray segment using **Q**. The blue lines show the *p*-vector in object space and its geometric transformations.

17.4 Proper Retardance Calculations

This section presents retardance algorithms for calculating the proper retardance for ray paths represented by **P** matrices. The proper retardance contains no geometric transformation, where geometric transformation is described by the **Q** matrices.

17.4.1 Definition of the Proper Retardance

Proper retardance or just *retardance* is the accumulation of the polarization-dependent *optical path difference* from physical processes associated with the ray path. Retardance is generated by mechanisms that cause a polarization-dependent phase change, such as *s*- and *p*-phase differences in reflection or refraction, propagation through a waveplate, birefringent material, or interaction with a diffraction grating. *Retardance is invariant with respect to the selection of local or global coordinates.*

17.5 Separating Geometric Transformations from P

The *parallel transport matrix* allows the retardance to be uniquely defined for *ray paths* despite the incident and exiting propagation vectors being different. $\mathbf{Q}_{Total}$ in Equation 17.11 provides a relationship between canonical local coordinates in the two transverse planes; it tracks the coordinate vector transformations through an optical system. When an arbitrary polarization state **v** is specified in the transverse plane along the first ray segment, the corresponding polarization state propagates through a non-polarizing system and exits as $\mathbf{Q}_{Total}\mathbf{v}$. $\mathbf{Q}_{Total}^{-1}$ will reverse the *geometric transformation* by mapping the coordinate vectors in the exit space back to the initial coordinate vectors. Performing this operation on $\mathbf{P}_{Total}$ maps the actual exiting polarization state back onto the incident transverse plane, in the incident state's coordinates,

$$\mathbf{M}_{Total} = \mathbf{Q}_{Total}^{-1}\mathbf{P}_{Total}. \quad (17.20)$$

Thus, $\mathbf{M}_{Total}$ is a *polarization ray tracing matrix* that has its exiting coordinate vectors straightened and aligned with the incident coordinate vectors, because the exiting polarization state has been unfolded (with a possible inversion from reflections). The incident and exiting transverse planes for

17.5 Separating Geometric Transformations from P

$\mathbf{M}_{Total}$ are parallel and both are orthogonal to $\hat{\mathbf{k}}_0$ since $\mathbf{M}_{Total}\hat{\mathbf{k}}_0 = \hat{\mathbf{k}}_0$. $\mathbf{M}_{Total}$ contains the diattenuation and the proper retardance of the ray path, the polarization contributions from all the thin films and other interactions. Because calculation of $\mathbf{Q}$ requires knowledge of $\hat{\mathbf{k}}_i$ for all ray segments, the proper retardance cannot be separated from $\mathbf{P}$ for an unknown "black box" optical system. *If only incident and exiting ray properties are known without the internal ray directions, $\mathbf{Q}$ and the proper retardance cannot be calculated.*

Math Tip 17.1 Polar Decomposition of a Matrix

The *polar decomposition* of a matrix $\mathbf{M}$ calculates a *Hermitian* matrix $\mathbf{H}$ and *unitary* matrix $\mathbf{U}$, which, when multiplied, equal $\mathbf{M}$. The Hermitian matrix has nonnegative real eigenvalues and represents an ideal diattenuator. The unitary matrix has unimodular eigenvalues and represents an ideal retarder. Since matrices in general do not commute, the polar decomposition can be performed in two orders: unitary followed by hermitian and hermitian followed by unitary. The matrices calculated depend on the order of the decomposition selected,

$$\mathbf{M} = \mathbf{UH} = \mathbf{H'U}. \tag{17.21}$$

Note that in the two polar decompositions, $\mathbf{U}$ is the same but the hermitian component changes. The polar decompositions are calculated from the singular value decomposition of $\mathbf{M}$,

$$\mathbf{M} = \mathbf{W}\mathbf{\Lambda}\mathbf{V}^{-1}. \tag{17.22}$$

The first decomposition is

$$\mathbf{M} = \mathbf{UH} = \mathbf{WV}^{-1}\mathbf{V}\mathbf{\Lambda}\mathbf{V}^{-1}, \quad \mathbf{U} = \mathbf{WV}^{-1}, \quad \mathbf{H} = \mathbf{V}\mathbf{\Lambda}\mathbf{V}^{-1}. \tag{17.23}$$

The second is

$$\mathbf{M} = \mathbf{H'U} = \mathbf{W}\mathbf{\Lambda}\mathbf{W}^{-1}\mathbf{WV}^{-1}, \quad \mathbf{H'} = \mathbf{W}\mathbf{\Lambda}\mathbf{W}^{-1}, \quad \mathbf{U} = \mathbf{WV}^{-1}. \tag{17.24}$$

To calculate the proper retardance, the next step is to separate the retardance and diattenuation of $\mathbf{M}_{Total}$. Applying the polar decomposition to $\mathbf{M}_{Total}$ yields the product of pure retarder and pure diattenuator matrices; then, data reduction can be performed separately on each. Two methods are shown: (1) the polar decomposition can be applied directly to the 3 × 3 $\mathbf{M}_{Total}$, or (2) a 2 × 2 Jones matrix is calculated from $\mathbf{M}_{Total}$ and the methods in Ref. [16] are used. Section 17.5.1 develops the first approach and the second approach is shown in Section 17.5.2.

17.5.1 The Proper Retardance Algorithm for P, Method 1

Applying the polar decomposition to the straightened-out polarization ray tracing matrix $\mathbf{M}_{Total}$ yields the desired diattenuation and retardance matrices for the ray path,

$$\mathbf{M}_{Total} = \mathbf{M}_{Total,R}\mathbf{M}_{Total,D} = \mathbf{M}'_{Total,D}\mathbf{M}_{Total,R}, \tag{17.25}$$

where $\mathbf{M}_{Total,R}$ is a retarder (unitary) matrix and $\mathbf{M}_{Total,D}$ and $\mathbf{M}'_{Total,D}$ are the diattenuator (nonnegative definite hermitian) matrices expressed in object space. The retardance of $\mathbf{M}_{Total}$ is the retardance of $\mathbf{M}_{Total,R}$. $\mathbf{M}_{Total,R}$ has three eigenvectors

$$\left(\mathbf{v}_1, \mathbf{v}_2, \hat{\mathbf{k}}_0\right), \tag{17.26}$$

and three associated eigenvalues

$$\left(\xi_1, \xi_2, \xi_3\right). \tag{17.27}$$

At least one of the eigenvalues is unity, here chosen as $\xi_3 = 1$, the eigenvalue that relates the incident propagation vector $\hat{\mathbf{k}}_0$ to the rotated exiting propagation vector, which is now parallel to $\hat{\mathbf{k}}_0$.

As expected, the retardance (δ) is calculated from the two eigenvalues associated with the transverse plane, ξ_1 and ξ_2, as the difference in their phases,

$$\delta = \arg(\xi_2) - \arg(\xi_1), \tag{17.28}$$

assuming $\arg(\xi_2) > \arg(\xi_1)$. If δ is less than $\pi/2$, the fast axis is associated with the smaller eigenvalue. If δ is greater than $\pi/2$, the matrix multiplication of complex numbers has lost track of which state is fast and slow, because the product of a series of complex numbers in Cartesian form $(x + i\, y)$ has its phase limited to the range $-\pi < \phi \le \pi$.

When $\mathbf{P}_{Total}$ is homogeneous, applying the polar decomposition to $\mathbf{M}_{Total}$ is unnecessary; $\mathbf{M}_{Total}$ and $\mathbf{M}_{Total,R}$ have the same eigenvalues and eigenpolarizations. Then, Equation 17.28 gives the retardance of $\mathbf{P}_{Total}$, where ξ_1 and ξ_2 are the eigenvalues of $\mathbf{M}_{Total}$.

17.5.2 The Proper Retardance Algorithm for P, Method 2

A second algorithm for *proper retardance* is developed in this section. A 2 × 2 Jones matrix $\mathbf{J}$ is retrieved from an inhomogeneous $\mathbf{M}_{Total}$, and the retardance is calculated from $\mathbf{J}$. First, a unitary change of basis is applied to $\mathbf{M}_{Total}$ so that among the elements of the last row and last column, only the (3, 3) element is nonzero,

$$\mathbf{S}_R = \mathbf{U}\mathbf{M}_{Total}\mathbf{U}^\dagger = \begin{pmatrix} & & 0 \\ & \mathbf{J} & 0 \\ 0 & 0 & 1 \end{pmatrix}. \tag{17.29}$$

$\mathbf{U}$ rotates by $\theta = \cos^{-1}(\hat{\mathbf{k}}_0 \cdot \hat{\mathbf{z}})$ counterclockwise about the $\hat{\mathbf{k}}_0 \times \hat{\mathbf{z}}$ axis so that $\hat{\mathbf{k}}_0$ is rotated to $\hat{\mathbf{z}}$. For $\hat{\mathbf{k}}_0 = \left(\hat{k}_x, \hat{k}_y, \hat{k}_z\right)$,

$$\mathbf{U} = \frac{1}{H}\begin{pmatrix} k_x^2 \cos\theta + k_y^2 & k_x(\cos\theta - k_y) & -\sqrt{H}\,k_x \sin\theta \\ k_x k_y(\cos\theta - 1) & k_x^2 + k_y^2 \cos\theta & -\sqrt{H}\,k_y \sin\theta \\ \sqrt{H}\,k_x \sin\theta & \sqrt{H}\,k_y \sin\theta & H\cos\theta \end{pmatrix}, \tag{17.30}$$

where $H = k_x^2 + k_y^2$. The upper left 2×2 submatrix of $\mathbf{S}_R$ is a Jones matrix $\mathbf{J}$. The retardance of $\mathbf{J}$, which is also the retardance of $\mathbf{P}_{Total}$, is given by the rather complicated equation:

$$\delta = 2\cos^{-1}\left(\frac{\left|tr(\mathbf{J}) + \frac{\det(\mathbf{J})}{|\det(\mathbf{J})|}tr(\mathbf{J}^\dagger)\right|}{2\sqrt{tr(\mathbf{J}^\dagger\mathbf{J}) + 2|\det(\mathbf{J})|}}\right). \tag{17.31}$$

Equation 17.31 is complicated because the diattenuation of $\mathbf{J}$ needs to be removed before applying the arccosine to calculate retardance. The unitary matrix (retarder) of polar decomposed $\mathbf{J}$ ($\mathbf{J}_R$ in Ref. [16]) has two eigenpolarizations ($\mathbf{w}_1$, $\mathbf{w}_2$). These eigenpolarizations can be written as three-element electric field vectors, which provide a canonical basis set in the incident space,

$$\mathbf{v}_1 = \mathbf{U}^\dagger\mathbf{w}_1', \quad \mathbf{v}_2 = \mathbf{U}^\dagger\mathbf{w}_2', \quad \hat{\mathbf{k}}_0, \tag{17.32}$$

where $\mathbf{w}_1' = (\mathbf{w}_{x,1}, \mathbf{w}_{y,1}, 0)$ and $\mathbf{w}_2' = (\mathbf{w}_{x,2}, \mathbf{w}_{y,2}, 0)$. In exit space, the corresponding canonical basis set is

$$\mathbf{v}_1' = \mathbf{Q}\mathbf{v}_1 = \mathbf{Q}\mathbf{U}^\dagger\mathbf{w}_1', \quad \mathbf{v}_2' = \mathbf{Q}\mathbf{v}_2 = \mathbf{Q}\mathbf{U}^\dagger\mathbf{w}_2', \quad \hat{\mathbf{k}}_N. \tag{17.33}$$

In the special case when the polarization of the ray path corresponds to a polarizer, the retardance is undefined; there is no second beam; hence, the second beam's phase is undefined. If $\mathbf{J}$ and $\mathbf{P}_{Total}$ describe a polarizer, then $\frac{\det(\mathbf{J})}{|\det(\mathbf{J})|}$ in Equation 17.31 is undefined.

17.5.3 Retardance Range

Optical path difference and *retardance* may assume any value between 0 and infinity. However, in Jones calculus and Mueller calculus, the retardance algorithms in the previous sections perform an arccosine to calculate the retardance and return a value with a modulo of π (a half wave). This situation is similar to the phase of the electric field, which is usually represented by a modulo of 2π, while the optical path length can assume any value. Therefore, it is frequently desired to know the *order of the retarder*, the number of waves of the optical path difference. Unfortunately, the matrix multiplication of complex numbers does not preserve the order. Further discussion of methods to extend the retardance calculation beyond 2π by *retardance unwrapping* or other methods is beyond the scope of this section,[17,18] but is addressed in Chapter 26 (Multi-Order Retarders and the Mystery of Discontinuities).

17.6 Examples

In this section, the *paradox* of the Jones matrix for *normal incidence reflection*, introduced in Chapter 9 (Polarization Ray Tracing Calculus), is resolved. This problem with Jones matrices is one of the arguments for the use of 3×3 versus 2×2 matrices for polarization ray tracing. Simple examples using homogeneous matrices are presented in the following; thus, the eigenvalues of $\mathbf{M}_{Total}$ can be used directly to calculate retardance.

17.6.1 Ideal Reflection at Normal Incidence

Consider ideal (100%) *reflection at normal incidence* from a mirror. Since mirrors are non-polarizing at normal incidence, the retardance will be zero; there is no difference between s- and p-polarizations; they are degenerate. As different linearly polarized states reflect, the phase delay due to reflection must stay the same. The **P** matrix for $\hat{\mathbf{k}}_0 = (0,0,1)$ and $\boldsymbol{\eta} = (0, 0, 1)$ is

$$\mathbf{P} = \begin{pmatrix} -1 & 0 & 0 \\ 0 & -1 & 0 \\ 0 & 0 & -1 \end{pmatrix}, \tag{17.34}$$

where fields of x- and y-polarized light reflect without a differential phase change, while the direction of the propagation vector flips from z to −z. The upper diagonal elements are −1 due to a π phase shift upon external reflection. For right circularly polarized incident light,

$$\begin{pmatrix} -1 & 0 & 0 \\ 0 & -1 & 0 \\ 0 & 0 & -1 \end{pmatrix} \begin{pmatrix} 1 \\ -i \\ 0 \end{pmatrix} = e^{i\pi} \begin{pmatrix} 1 \\ -i \\ 0 \end{pmatrix}, \tag{17.35}$$

which yields the same electric field vector. However, since the propagation vector changed to (0, 0, −1), the reflected light is left circularly polarized. Similarly, for linearly polarized incident light,

$$\begin{pmatrix} -1 & 0 & 0 \\ 0 & -1 & 0 \\ 0 & 0 & -1 \end{pmatrix} \begin{pmatrix} \cos\theta \\ \sin\theta \\ 0 \end{pmatrix} = e^{i\pi} \begin{pmatrix} \cos\theta \\ \sin\theta \\ 0 \end{pmatrix}. \tag{17.36}$$

It yields the same electric field vector oscillating in the same plane in the global perspective. However, in the corresponding Jones matrix, the incident angle θ is mapped into −θ because of right-handed local coordinates. Note that in Equation 17.36, no relative phase change is introduced between the x- and y-components of the electric field.

This is very different from the standard Jones matrix $\mathbf{J}_f$ for *reflection*,[19,20]

$$\mathbf{J}_f = \begin{pmatrix} -1 & 0 \\ 0 & 1 \end{pmatrix}. \tag{17.37}$$

$\mathbf{J}_f$ appears to include a π phase shift (−1) between the x- and y-polarization components. In $\mathbf{J}_f$, this phase shift serves two purposes: (1) it reflects right circularly polarized light into left circularly polarized and vice versa, and (2) it changes the orientation of incident linearly polarized light from θ to −θ such as 45° into 135°, which is appropriate when maintaining right-handed local coordinates after reflection.

In order to keep all the Jones matrix local coordinates right-handed (as shown in Figure 17.14), the Jones matrix for reflection must contain −1 in one of the diagonal elements. Thus, the Jones reflection matrix has the same form as half wave linear retarders! This is the paradox of the reflection Jones matrix. The minus sign *does not* indicate a physical half wave linear retardance; it indicates a local coordinate flip.

17.6 Examples

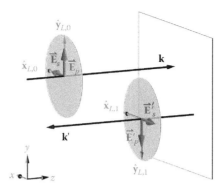

Figure 17.14 Incident and exiting right-handed local coordinates, $(\hat{\mathbf{x}}_{L,0}, \hat{\mathbf{y}}_{L,0})$ and $(\hat{\mathbf{x}}_{L,1}, \hat{\mathbf{y}}_{L,1})$, and s- and p-vectors, $(\vec{\mathbf{E}}_s, \vec{\mathbf{E}}_p)$ and $(\vec{\mathbf{E}}'_s, \vec{\mathbf{E}}'_p)$, for an ideal reflection at normal incidence. In this particular choice of local coordinates, the $\hat{\mathbf{x}}_L$ vector was flipped after the reflection.

The local coordinate transformation for this normal incidence reflection can be clearly revealed with the **Q** matrix,

$$\mathbf{Q} = \begin{pmatrix} 1 & 0 & 0 \\ 0 & 1 & 0 \\ 0 & 0 & -1 \end{pmatrix}. \tag{17.38}$$

The minus sign is now associated with the propagation vector, where it properly belongs, not with one of the electric field components. Applying Equations 17.28 and 17.38, the ideal retroreflection is calculated to have zero retardance, as it must.

17.6.2 An Aluminum-Coated Three-Fold Mirror System Example

The three-fold mirror system of Figure 17.7 is analyzed as a real, not as a non-polarizing, optical system incorporating the effects of aluminum-coated mirrors with a refractive index of $0.77 + 6.06i$. Light enters at each mirror at a 45° angle of incidence. Table 17.1 lists the propagation vectors, the **P** matrices, and the **Q** matrices for each surface calculated from methods described in Chapter 9 (Polarization Ray Tracing Calculus). The exiting propagation vector $\hat{\mathbf{k}}_4$ is the same as the incident propagation vector $\hat{\mathbf{k}}_1$; both are along the z-axis.

The system's **P** matrix is $\mathbf{P}_{Total} = \begin{pmatrix} 0 & -0.549 + 0.705i & 0 \\ -0.365 + 0.788i & 0 & 0 \\ 0 & 0 & 1 \end{pmatrix}$. This shows how x-polarized incident light exits as y-polarized light and y-polarized incident light exits as x-polarized light despite the incident and exiting propagation vectors being the same. The diattenuation for this ray path is 0.0285. The diattenuation magnitudes of the second and third mirrors are equal, but their orientations are 90° apart, and therefore they cancel to zero. Therefore, the total diattenuation is equal to the first mirror's contribution. For the diattenuation calculation algorithm, see Section 9.4 (Diattenuation Calculation Using Singular Value Decomposition).

The corresponding **Q** matrix is $\mathbf{Q}_{Total} = \begin{pmatrix} 0 & 1 & 0 \\ 1 & 0 & 0 \\ 0 & 0 & 1 \end{pmatrix}$. Figure 17.7 showed how each ideal reflection transforms the incident coordinate vectors, $(\hat{\mathbf{x}}_A, \hat{\mathbf{y}}_A)$ via **Q** matrices. As shown in Section 17.2.4,

Table 17.1 Propagation Vectors, **P**, and **Q** for a Ray Propagating through the Aluminum-Coated Three-Fold Mirror System

q	$\hat{\mathbf{k}}$	P	Q
1	$\begin{pmatrix} 1 \\ 0 \\ 0 \end{pmatrix}$	$\begin{pmatrix} 0 & 0 & 1 \\ 0 & -0.947+0.219i & 0 \\ -0.849+0.415i & 0 & 0 \end{pmatrix}$	$\begin{pmatrix} 0 & 0 & 1 \\ 0 & 1 & 0 \\ 1 & 0 & 0 \end{pmatrix}$
2	$\begin{pmatrix} 0 \\ 1 \\ 0 \end{pmatrix}$	$\begin{pmatrix} 0 & -0.849+0.415i & 0 \\ 1 & 0 & 0 \\ 0 & 0 & -0.947+0.219i \end{pmatrix}$	$\begin{pmatrix} 0 & 1 & 0 \\ 1 & 0 & 0 \\ 0 & 0 & 1 \end{pmatrix}$
3	$\begin{pmatrix} 0 \\ 0 \\ 1 \end{pmatrix}$	$\begin{pmatrix} -0.947+0.219i & 0 & 0 \\ 0 & 0 & -0.849+0.415i \\ 0 & 1 & 0 \end{pmatrix}$	$\begin{pmatrix} 1 & 0 & 0 \\ 0 & 0 & 1 \\ 0 & 1 & 0 \end{pmatrix}$

the incident coordinate vectors $(\hat{\mathbf{x}}_A, \hat{\mathbf{y}}_A, \hat{\mathbf{k}}_1)$ are rotated by 90° and inverted to $(\hat{\mathbf{x}}''_A, \hat{\mathbf{y}}''_A, \hat{\mathbf{k}}_4)$. Thus, one canonical pairing of Jones matrix basis vectors between entrance and exit space would be $(\hat{\mathbf{x}}_A, \hat{\mathbf{y}}_A) = (-\hat{\mathbf{y}}, \hat{\mathbf{x}})$ and $(\hat{\mathbf{x}}''_A, \hat{\mathbf{y}}''_A) = (-\hat{\mathbf{x}}, \hat{\mathbf{y}})$. This pairing is not unique; other canonical pairings are obtained by rotating both spaces' basis sets. Because of inversion in reflection, the coordinate vectors transformed by the **Q** matrix change handedness if a system has an odd number of reflections and maintain their handedness for an even number of reflections. $(\hat{\mathbf{x}}''_A, \hat{\mathbf{y}}''_A, \hat{\mathbf{k}}_4)$ are the canonical set of coordinate vectors for the polarization state analyzer (as described in Section 17.2.1) for measuring the proper retardance of the system.

Multiplying $\mathbf{P}_{Total}$ by $\mathbf{Q}_{Total}^{-1}$ cancels the geometric transformation. $\mathbf{M}_{Total}$ of the system is

$$\mathbf{M}_{Total} = \mathbf{Q}_{Total}^{-1} \mathbf{P}_{Total} = \begin{pmatrix} -0.365+0.788i & 0 & 0 \\ 0 & -0.549+0.705i & 0 \\ 0 & 0 & 1 \end{pmatrix}. \quad (17.39)$$

Since $\mathbf{P}_{Total}$ is homogeneous, the retardance of the system is found by calculating eigenvalues of $\mathbf{M}_{Total}$,

$$\xi_1 = 0.868 e^{i2.005}, \quad \xi_2 = 0.8938 e^{i2.232}, \quad \xi_3 = 1, \quad (17.40)$$

and the eigenpolarization states associated with these eigenvalues are

$$\mathbf{v}_1 = (1,0,0), \quad \mathbf{v}_2 = (0,1,0), \quad \mathbf{v}_3 = \hat{\mathbf{k}}_0 = (0,0,1). \quad (17.41)$$

The retardance of the system is

$$\delta = \arg(\xi_2) - \arg(\xi_1) = 0.227, \quad (17.42)$$

with the fast axis orientation along the global $\hat{\mathbf{x}}$-axis.

The retardance calculated from this method does not contain any effects from the geometric transformation. Similar to the cancelation of diattenuation described above, this proper retardance is equal to just the first mirror's contribution since the retardance of the last two crossed mirrors cancel.

The retardance calculated from the Jones matrix of the first mirror is

$$\xi_1 = 0.945 e^{-i 0.455}, \; \xi_2 = 0.972 e^{i 2.914} \Rightarrow \delta = \arg(\xi_2) - \arg(\xi_1) = 3.369 = 193.0°, \quad (17.43)$$

with the fast axis orientation along the $\mathbf{y}_{L,0}$, which is the global $\hat{\mathbf{x}}$. The retardance from the Jones calculus and the one from the polarization ray tracing matrix differ by π since Jones calculus uses right-handed local coordinates for data reduction.

17.7 Conclusion

This chapter began with a critical analysis of *retardance*, a concept simple to define for a waveplate, but difficult to generalize to ray paths through optical systems. Retardance is a polarization-dependent optical path length through an optical system that creates polarization transformations well described as rotation of polarization states on the Poincaré sphere, a concept developed further in Chapter 6 (Mueller Matrices). The *polarization ray tracing matrix* for a ray path **P** describes the polarization state changes due to *diattenuation*, *retardance*, and *geometric transformations*. The ray's parallel transport matrix **Q** describes the associated non-polarizing optical system and thus keeps track of just the geometric transformation. Further discussion of non-polarizing systems is found in Chapter 18 (A Skew Aberration), which is the functional form of the geometric transport from the entrance pupil to the exit pupil for an entire wavefront.

To calculate the *proper retardance*, the geometric transformation needs to be removed, $\mathbf{M} = \mathbf{Q}^{-1} \mathbf{P}$. **M** is the fundamental equation for calculating retardance without spurious *circular retardance* arising from a poor choice of coordinate vectors. **M** also tracks the inversions of the polarization state due to reflections and thus clarifies the meaning of the troublesome minus sign in the Jones matrix for *reflection*. The difference in eigenvalue arguments of $\mathbf{M}_R$, the unitary part of the polar decomposed **M**, gives the proper retardance, which is not calculated from **P** alone. It is important to emphasize that the proper retardance cannot be assigned to a ray through an optical system in a black box whose internal ray path is unknown.

Minus signs associated with reflections and *local coordinates* are treacherous! We have explained each "minus sign," at the expense of length and some repetition, and the reader is advised to be very careful in this regard.

17.8 Problem Sets

17.1 What is the difference between the retardance and proper retardance?

17.2 Why are sets of left-handed coordinate basis vectors useful? (See Section 17.3.)

17.3 Simulate a Jones matrix measurement with a polarimeter that has its polarization state generator rotated by $\pi/6$ for the following elements: (a) a linear retarder **LR**(δ, 0), and (b) a circular retarder **CR**(δ). How does the measured retardance differ from the physical retardance?

17.4 Consider a Mach–Zehnder interferometer where the beam takes a path through the points (0, 0, −1), (0, 0, 0), (0, 0, 1), (0, 1, 1), (1, 1, 1), and (2, 1, 1), and the second beam passes through (0, 0, 0), (1, 0, 0), (1, 1, 0), and (1, 1, 1), as shown below. Find the parallel transport matrices for the two paths. When linearly polarized light is incident, are the two exiting parallel or orthogonal? Use Equation 17.11 ($\mathbf{Q}_{Total}$) to compare two beam paths.

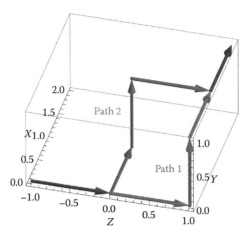

17.5 Using Equation 17.31, calculate the retardance of the following Jones matrices:

a. $\dfrac{1}{2}\begin{pmatrix} \sqrt{3}-i & 0 \\ 0 & \dfrac{1}{\sqrt{3}}+\dfrac{i}{6} \end{pmatrix}$

b. $\begin{pmatrix} \cos\dfrac{\pi}{8} & \dfrac{1}{2}\sin\dfrac{\pi}{8} \\ -\sin\dfrac{\pi}{8} & \dfrac{1}{2}\cos\dfrac{\pi}{8} \end{pmatrix}$

c. $\dfrac{1}{6}\begin{pmatrix} -3-\sqrt{3}-i & -3+\sqrt{3}+i \\ -3+\sqrt{3}+i & -3-\sqrt{3}-i \end{pmatrix}$

17.6 Explain the difference between the Jones matrix and polarization ray tracing matrix for ideal reflection at normal incidence.

17.7 (a) If a beam went through only the first two of the three mirror systems shown in Section 17.2.4, what would be the $\mathbf{Q}_{Total}$? (b) What if we add a retro-reflecting fourth mirror after the three-mirror system so that the beam goes from mirrors 1, 2, 3, 4 and back to mirrors 3, 2, 1?

17.8 Given $\mathbf{k}_1$, $\mathbf{k}_2$, $\mathbf{P}$, and $\mathbf{Q}$ for a ray, what are $\mathbf{P}'$ and $\mathbf{Q}'$ if the light is reversed so that $-\mathbf{k}_2$ becomes the incident direction and $-\mathbf{k}_1$ is the exiting direction? Also, explain how they are related to $\mathbf{P}$ and $\mathbf{Q}$.

17.9 Find the singular valued decomposition and the two polar decompositions for the following matrices. Show that the retardance is the same in both polar decompositions.

a. $\mathbf{M}_a = \begin{pmatrix} 1 & 0 \\ 0 & i/2 \end{pmatrix}$

b. $\mathbf{M}_b = \begin{pmatrix} 0 & 1 \\ 0 & 0 \end{pmatrix}$

c. $\mathbf{M}_c = \begin{pmatrix} \cos 1 & \dfrac{-\sin 1}{3} \\ \sin 1 & \dfrac{\cos 1}{3} \end{pmatrix}$

17.10 Consider the Jones matrix measurement of a retarder **LR**(δ,0) using a polarimeter with a PSA rotated by θ. The measured Jones matrix is $\begin{pmatrix} \cos\theta & -\sin\theta \\ \sin\theta & \cos\theta \end{pmatrix}\begin{pmatrix} e^{-i\delta/2} & 0 \\ 0 & e^{i\delta/2} \end{pmatrix}$. Show that the measured retardance is $\delta_{measured} = 2\arctan\dfrac{\sqrt{1-\cos^2\theta\cos^2(\delta/2)}}{\cos\theta\cos(\delta/2)}$. It is recommended to work the problem by putting the Jones matrices into Pauli matrix form.

References

1. R. A. Chipman, Polarization analysis of optical systems, *Proc. SPIE* 891 (1988):10.
2. G. Yun, K. Crabtree, and R. A. Chipman, Three-dimensional polarization ray-tracing calculus I. Definition and diattenuation, *Appl. Opt.* 50 (2011): 2855–2865.
3. R. C. Jones, A new calculus for the treatment of optical systems, *J. Opt. Soc. Am.* 31 (1941): 488–493, 493–499, 500–503; 32 (1942): 486–493; 37 (1947): 107–110, 110–112; 38 (1948): 671–685; 46 (1956): 126–131.
4. W. A. Shurcliff, *Polarized Light*, Harvard University Press (1962).
5. E. Hecht, *Optics*, Addison-Wesley (2002).
6. W. A. Shurcliff, *Polarized Light*, Harvard University Press (1962).
7. S. Lu and R. A. Chipman, Homogeneous and inhomogeneous Jones matrices, *J. Opt. Soc. Am. A* 11 (1994): 766–773.
8. P. Torok, P. Varga, Z. Laczik, and G. R. Booker, Electromagnetic diffraction of light focused through a planar interface between materials of mismatched refractive indices: An integral representation, *J. Opt. Soc. Am. A* 12 (1995): 325–332.
9. C. Brosseau, *Fundamentals of polarized Light, A Statistical Optics Approach*, New York: John Wiley & Sons (1998).
10. D. W. Henderson and D. Taimina, *Experiencing Geometry*, 3rd edition, Chapter 8, NJ: Pearson (2004).
11. R. Penrose, *The Road to Reality*, section 14.2, NY: Knopf (2005).
12. J. M. Leinaas and J. Myrheim, On the theory of identical particles, *Il Nuovo Cimento B* 37(1) (1977): 1–23.
13. S. Pancharatnam, Generalized theory of interference, and its applications, *Proc. Indian Acad. Sci. A* 44 (1956): 247.
14. R. Bhandari and J. Samuel, Observation of topological phase by use of a laser interferometer, *Phys. Review Lett.* 60 (1988): 1211.
15. M. V. Berry, The adiabatic phase and Pancharatnam's phase for polarized light, *J. Mod. Opt.* 34 (1987): 1401.
16. S. Lu and R. A. Chipman, Interpretation of Mueller matrices based on polar decomposition, *J. Opt. Soc. Am. A* 13 (1996): 1106–1113.
17. D. Bone, Fourier fringe analysis: The two-dimensional phase unwrapping problem, *Appl. Opt.* 30 (1991): 3627–3632.
18. A. Collaro, G. Franceschetti, F. Palmieri and M. S. Ferreiro, Phase unwrapping by means of genetic algorithms, *J. Opt. Soc. Am. A* 15 (1998): 407–418.
19. G. R. Fowles, *Introduction to Modern Optics*, Dover Publications (1975).
20. A. Macleod, Phase matters, *SPIE's OE Magazine*, June/July 29–31 (2005).

18

A Skew Aberration

18.1 Introduction

This chapter considers the polarization aberrations of non-polarizing optical systems. If all the diattenuation and retardance are removed from an optical system, systematic polarization changes remain, named *skew aberration*.

Aberrations can be considered as deviations from the ideal behavior of imaging optical systems, that is, deviations from the mapping of spherical waves with uniform amplitude and polarization into spherical waves with uniform amplitude and polarization. The main aberration categories are *wavefront aberration*,[1] *apodization* (*amplitude aberration*),[2–4] and *polarization aberration*.[5] Wavefront aberration is the variation in optical path length, which is calculated in all the commercial ray tracing programs. Apodization is an amplitude aberration; different rays have different transmittances due to reflection losses and absorption. Polarization aberration, a non-uniform polarization change across wavefronts, is divided into (1) *diattenuation aberration*, which is polarization-dependent transmission or reflection, (2) *retardance aberration*, which is polarization-dependent optical path difference, and (3) *skew aberration*, the polarization change in the absence of diattenuation and retardance aberration.

One example of skew aberration occurs in a *corner cube* system where there is polarization change even when the diattenuation and retardance at three reflections are set to zero. This example is explained in detail in Section 16.11. As shown in Figure 18.1, the incident polarization (*y*-polarized) rotates by 120° as it propagates through the corner cube solely due to geometric transformation, that is, skew aberration. Further, the rotation is different in the three pairs of subapertures. This variation

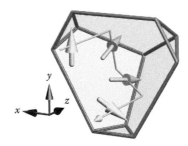

Figure 18.1 Polarization state rotates due to skew aberration as the ray propagates through the non-polarizing critical angle corner cube system.

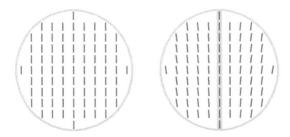

Figure 18.2 The effect of linear skew aberration transforming a uniform incident linearly polarized state (left) into a skew aberrated state (right). This linear form of skew aberration occurs in most lenses and other rotationally symmetric systems in addition to higher-order variations of skew aberration-induced polarization rotation.

of the geometrical transformation across the pupil has a profound effect on the image formation with corner cubes as shown in Figure 16.44.

The effect of skew aberration on point spread functions (PSFs) is examined using the *Mueller point spread matrix* (**MPSM**) (see Chapter 16 [Image Formation with Polarization Aberration]) to show how non-polarizing and ideal optical systems can have undesired polarization mixing due to skew aberration. Skew aberration's separate origin and behavior is fascinating. In radially symmetric systems, *skew rays* have skew aberration but *meridional rays* do not. Thus, the name skew aberration is applied.

In this chapter, skew aberration is defined and algorithms are provided for its calculation. The skew aberration of a high numerical aperture (NA) lens with a large field of view (FOV) is analyzed in detail. The effect of a linear variation of skew aberration, as shown in Figure 18.2, the most common form, on a diffraction-limited **MPSM** is treated. Finally, the skew aberration is calculated for 2383 lens systems in the CODE V^6 patent library and their statistics provide perspective on the role and importance of skew aberration.

18.2 Definition of Skew Aberration

Skew aberration is the rotation of each ray's polarization state between the *entrance pupil* and the *exit pupil* due to the intrinsic *geometric transformation* of polarization states. Skew aberration occurs even for *non-polarizing optical systems*. It seems odd at first that a non-polarizing optical system should have polarization aberration. As shown in Chapter 17, the propagation of polarized light in a non-polarizing optical system can be simulated with the parallel transport of vectors on a sphere, and skew aberration is related to *Berry phase* and *Pancharatnam phase* calculated using the *parallel transport matrices* **Q** in Chapter 17. Thus, if a ray has non-zero geometric transformation, its skew aberration is also non-zero.

Polarized Light and Optical Systems

18.3 Skew Aberration Algorithm

Skew aberration is independent of the incident polarization state or polarization properties of optical elements such as polarizers, retarders, and coatings. As was shown in the *critical angle corner cube reflector* example in Chapter 16, non-polarizing optical system with zero diattenuation or retardance can have a significant amount of spatially varying polarization aberration. Although there is no optical path difference in the wavefront exiting the non-polarizing optical system, the wavefront may have varying optical phase similar to the six subapertures of critical angle corner cube. This phase difference can be measured by *interferometers* and may cause large polarization aberration.

Skew aberration is a *polarization rotation*, rotating the plane of polarization of linearly polarized light or the axis of elliptically polarized light. The rotation of circularly polarized light causes a phase change but not a change of polarization state. Skew aberration is distinct from diattenuation aberration and retardance aberration since its origin arises from purely geometric effects.

Since skew aberration occurs even for rays propagating through ideal, aberration-free, and non-polarizing optical systems, the example systems chosen for this chapter are non-polarizing. Such a system refracts and reflects the incident polarization ellipses into image space without changing its ellipticity.[7] For example, a refraction through a surface with $t_s = t_p = 1$, where t_s and t_p are Fresnel transmission coefficients for *s*- and *p*-polarization, is ideal and non-polarizing. The effect of skew aberration in radially symmetric optical systems, such as a lens or Cassegrain telescope, occurs for the off-axis fields. The polarization of the exiting wavefront relative to the incident wavefront has a steady rotation of its polarization state, which increases linearly with the distance from the meridional plane; this linear rotation continues toward the origin and is present even for paraxial ray paths. Figure 18.2 shows an example of this generic *skew aberration*. The polarization is not rotated along the center of the pupil, rotated clockwise on one side of the pupil and counterclockwise on the other side. The magnitude of rotation shown in here is greater than commonly encountered skew aberration. The skew aberration shown is linear but may also have quadratic, cubic, or contributions of other functional forms; Figure 18.2 shows the most common manifestation.

The skew aberration is determined solely by the ray's propagation path, that is, its sequence of normalized *propagation vectors* ($\mathbf{k}_{In}$, $\mathbf{k}_1$, $\mathbf{k}_2$, ...$\mathbf{k}_j$, ...$\mathbf{k}_{Exit}$). $\mathbf{k}_{In}$, $\mathbf{k}_j$, and $\mathbf{k}_{Exit}$ correspond to the propagation vector at the entrance pupil, after the *j*th surface, and at the exit pupil, respectively.

18.3 Skew Aberration Algorithm

An algorithm is presented to define the polarization changes between the entrance and exit pupil of a non polarizing optical system. Consider Figure 18.3 showing a grid of polarization states in the entrance pupil and the corresponding grid in the exit pupil. These pupils have different NAs and propagation directions. To compare two different wavefronts, a scale invariant reference polarization grid, the double pole basis vector, is selected; a grid of double pole ($\mathbf{g}_{In,i}$) are defined on the entrance pupil, traced via parallel transport of vectors, and compared with an exit pupil double

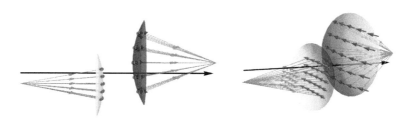

Figure 18.3 A grid of reference vectors ($\mathbf{g}_{In,i}$) on the entrance pupil (green) and another grid of reference vectors ($\mathbf{g}_{Exit,i}$) on the exit pupil (purple) are shown in two different views.

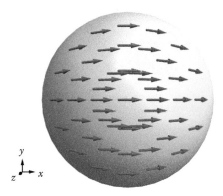

Figure 18.4 An array of double pole reference vectors on a unit wavefront sphere viewed along the chief ray's propagation vector. These are reference vectors for defining the *x*-component of Jones vectors over a spherical wavefront.

pole ($\mathbf{g}_{\text{Exit},i}$). As shown in Figure 18.4, this function does not change when the function is radially enlarged or shrunk about its center. If a *double pole* polarization pattern is present in the entrance and exit pupil, it is justified that there is no skew aberration and the resulting PSF would not be degraded by polarization aberration.

The following steps are a recap of calculating the double pole reference *basis vectors*, which were introduced in Chapter 11, since they are used as reference vectors in this chapter. First, define a vector ($\mathbf{g}_C$) that is perpendicular to the center ray's propagation vectors on the entrance and exit pupil

$$\mathbf{g}_C = \mathbf{k}_{\text{In},C} \times \mathbf{k}_{\text{Exit},C}. \tag{18.1}$$

The grid of reference vectors $\mathbf{g}_{\text{In},i}$ are generated by a counterclockwise rotation of $\mathbf{g}_C$ along $\mathbf{axis}_{\text{In},i}$ by $\theta_{\text{In},i}$. Similarly, the $\mathbf{g}_{\text{Exit},i}$ grid is obtained by a counterclockwise rotation of $\mathbf{g}_C$ along $\mathbf{axis}_{\text{Exit},i}$ by $\theta_{\text{Exit},i}$,

$$\begin{cases} \mathbf{g}_{\text{In},i} = \mathbf{R}(\theta_{\text{In},i}, \mathbf{axis}_{\text{In},i}) \mathbf{g}_C \\ \mathbf{g}_{\text{Exit},i} = \mathbf{R}(\theta_{\text{Exit},i}, \mathbf{axis}_{\text{Exit},i}) \mathbf{g}_C \end{cases} \tag{18.2}$$

where index *i* indicates the *i*th ray,

$$\begin{cases} \theta_{\text{In},i} = \cos^{-1}(\mathbf{k}_{\text{In},i} \cdot \mathbf{k}_{\text{In},C}), \ \mathbf{axis}_{\text{In},i} = \mathbf{k}_{\text{In},C} \times \mathbf{k}_{\text{In},i}, \\ \theta_{\text{Exit},i} = \cos^{-1}(\mathbf{k}_{\text{Exit},i} \cdot \mathbf{k}_{\text{Exit},C}), \ \mathbf{axis}_{\text{Exit},i} = \mathbf{k}_{\text{Exit},C} \times \mathbf{k}_{\text{Exit},i}, \end{cases} \tag{18.3}$$

and $\mathbf{R}(\theta, \mathbf{axis})$ is the 3D rotation matrix for a counterclockwise rotation around $\mathbf{axis}$ by θ.

The pair of $\mathbf{g}_{\text{In},i}$ and $\mathbf{g}_{\text{Exit},i}$ are invariant to magnification of the system and allows the polarization change from the entrance pupil to the exit pupil to be readily visible since it is defined by an angle relative to a radially symmetric set of lines. This rotation method described here is analogous to parallel transporting $\mathbf{g}_C$ along a great circle arc on a unit $\mathbf{k}$-sphere that connects points $\mathbf{k}_{\text{In},C}$ and $\mathbf{k}_{\text{In},i}$. This is the "double pole grid" described in Chapter 11, which defines the "standard" to describe a linearly polarized spherical wavefront, as shown in Figure 18.4.

Once the reference vectors $\mathbf{g}_{\text{In},i}$ and $\mathbf{g}_{\text{Exit},i}$ are established for an optical system, the system's geometric transformation can be calculated. The geometric transformation of each ray intercept due to the change in ray propagation direction from $\mathbf{k}_{j-1}$ to $\mathbf{k}_j$ is described by the parallel transport matrix $\mathbf{Q}_j$ (see Chapter 17 [Parallel Transport and the Calculation of Retardance]). $\mathbf{Q}_j$ for refracting

18.3 Skew Aberration Algorithm

surface j is equivalent to sliding vectors from a point $\mathbf{k}_{j-1}$ to a point $\mathbf{k}_j$ on a unit $\mathbf{k}$-sphere following the great circle arc that connects two points as shown in Figure 17.4. $\mathbf{Q}_j$ for reflecting surface j is equivalent to inverting vectors on a point $\mathbf{k}_{j-1}$ about $\mathbf{k}_{j-1} - \mathbf{k}_j$ and then moving them to a point $\mathbf{k}_j$ by parallel transport. The cumulative geometric transformation through the system along a single ray is $\mathbf{Q}_{Total,i}$,

$$\mathbf{Q}_{Total,i} = \prod_{j=\text{Exit}}^{\text{In}} \mathbf{Q}_j = \mathbf{Q}_{\text{Exit}} \cdots \mathbf{Q}_j \cdots \mathbf{Q}_1 \mathbf{Q}_{\text{In}}. \tag{18.4}$$

This maps the reference input polarization into $\mathbf{g}'_{\text{Exit},i}$,

$$\mathbf{g}'_{\text{Exit},i} = \mathbf{Q}_{Total,i} \mathbf{g}_{\text{In},i}. \tag{18.5}$$

The ith ray's skew aberration is defined as the angle between the ideal vector $\mathbf{g}_{\text{Exit},i}$ and the non-polarizing system's vevtor $\mathbf{g}'_{\text{Exit},i}$. If $\mathbf{g}'_{\text{Exit},i}$ results from a counterclockwise rotation from the $\mathbf{g}_{\text{Exit},i}$ looking into the beam, the ray has a positive skew aberration.

$\mathbf{Q}_{Total,i}$ serves a role for polarization evolution similar to the role of paraxial optics. *Paraxial optics* describes ray paths in an aberration-free system; it describes ideal systems. When a spherical wave is traced via paraxial optics, the exiting spherical wave forms an image at the ideal location free of distortion, astigmatism, and other aberrations. When real rays are traced through the same system, the results are quite complex, containing large sets of ray intercepts, optical path lengths, and so on. Optical designers can compare these real ray trace results with paraxial optics to calculate wavefront aberrations, since aberration is often defined as the deviation from the paraxial optics. These wavefront aberrations are convenient for communication with other optical designers. Similarly, $\mathbf{Q}_{Total,i}$ is "paraxial optics-like" since it provides $\mathbf{g}'_{\text{Exit},i}$, which is the "natural coordinate system" for a non-polarizing optical system; other polarization changes that deviate from $\mathbf{g}'_{\text{Exit},i}$ due to polarizing elements or interactions such as coatings, Fresnel coefficients, and so on, are the polarization aberration. The skew aberration calculates a change in polarization that is natural (i.e., intrinsic) to the system, and all other polarization changes are aberrations due to light–matter interactions.

Figure 18.5 shows the form of a typical skew aberrated polarization state, $\mathbf{g}'_{\text{Exit},i}$, for a rotationally symmetric refractive optical systems of an off-axis source along y-axis with $\mathbf{g}_{\text{in},i}$ shown in Figure 18.4. Note that in the y–z plane, $\mathbf{g}'_{\text{Exit},t} = \mathbf{g}_{\text{Exit},t}$, while the skew rays' $\mathbf{g}'_{\text{Exit},t}$ are rotated from $\mathbf{g}_{\text{Exit},t}$.

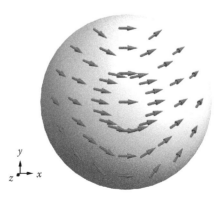

Figure 18.5 The form of the skew aberration, greatly exaggerated, $\mathbf{g}'_{\text{Exit},i}$ for a rotationally symmetric refractive optical systems for $\mathbf{g}_{\text{Exit},i}$ shown in Figure 18.4.

18.4 Lens Example—U.S. Patent 2,896,506

In a *radially symmetric optical system*, *meridional rays* lie on a plane containing the optical axis of the system. The *chief ray* and *marginal ray* are examples of meridional rays. They remain in one plane as they reflect and refract through the system from object space to image space. *Skew rays*, on the other hand, do not stay in one plane. In a long lens, skew rays will spiral around the optical axis in either a purely clockwise or counterclockwise sense. *Skew aberration* only occurs for skew rays, not for meridional rays. When the optical system is not radially symmetric, definitions of meridional rays are no longer clearly defined but the skew aberration algorithm still applies.

Skew aberration naturally increases with NA and FOV; hence, systems with high NA and wide FOV tend to have larger skew aberration. For example, U.S. Patent 2,896,506[8] has a comparatively large skew aberration. The system is rotationally symmetric with F/1.494 and a maximum FOV of 32°. Figure 18.6 shows the layout of this seven-lens system with two field angles, 0° and 20°.

Figure 18.7 (left) shows the skew aberration at the exit pupil calculated for a grid of rays from the 32° field angle. The skew aberration is generally largest at the edge of the pupil of the maximum field angle. For this example, the skew ray formed from the chief ray in the y–z plane plus the marginal ray in the x-direction at the 32° field angle has the largest skew aberration of 7.01°, shown in Figure 18.7 (left) as a gray point (point B). The skew ray at the opposite side of the pupil (point A) has −7.01° of skew aberration. The pupil is elliptical due to vignetting and pupil distortion. A meridional fan of rays through the center of the exit pupil has zero skew aberration.

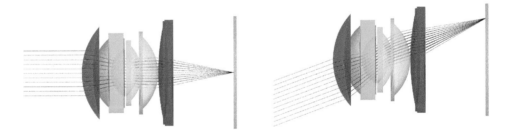

Figure 18.6 Optical system layout of U.S. Patent 2896506 from the CODE V lens patent library with seven lenses. Two field angles at 0° (right) and 20° (left) are shown.

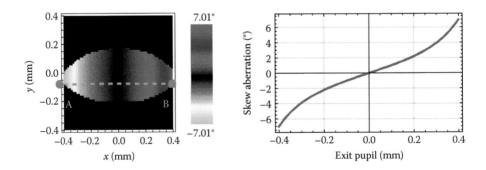

Figure 18.7 (Left) Skew aberration of a ray grid from the 32° field is evaluated at the exit pupil of the US lens patent 2896506 with skew aberration varying from −7.01° to +7.01°. (Right) The skew aberration along a horizontal cross section (orange dashed line in left figure) perpendicular to the meridional plane. It shows zero skew aberration for the center chief ray with a linear variation at the center of the pupil and higher-order variations toward the edge. In the meridional plane along the y-axis, the skew aberration is zero.

18.4 Lens Example—U.S. Patent 2,896,506

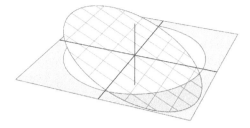

Figure 18.8 Skew aberration map showing linearly varying skew aberration.

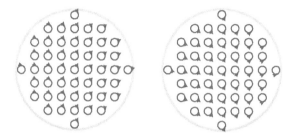

Figure 18.9 (Left) The effect of linear skew aberration on a right circularly polarized beam and (right) a left circularly beam that acquire opposite linear phase shift across the pupil due to skew aberration.

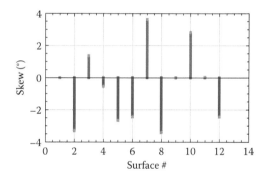

Figure 18.10 Skew ray A's skew aberration contribution from each lens surface sums to -7.01° through the system.

Figure 18.7 (right) shows that the skew aberration point A to point B has a form of a linear variation in x (*circular retardance tilt**) plus a coma-like cubic variation in x. The linearly varying circular retardance causes polarization-dependent tilt in the wavefront as shown in Figure 18.8.

Consider a circularly polarized plane wave incident on a non-polarizing system with a linear skew aberration, such as Figure 18.8. Along the y-axis, the light's phase stays unchanged. On the +x side of the pupil, the phase advances; on the −x side of the pupil, the phase retards, as shown in Figure 18.9 for left and right circularly polarized light. Thus, they focus at two shifted image points, by a fraction of an Airy disk diameter, enlarging the PSF in a way similar to astigmatism.

For lens systems, skew aberration tends to be largest from the edge of the object and at the side of the pupil furthest from the meridional plane. The skew aberration contribution for patent lens 2,896,506 for the ray corresponding to point A in Figure 18.7 at each lens surface is shown in Figure 18.10. Such surface contribution plots identify surfaces with the largest skew contributions.

---

* The linear variation of skew aberration adds a linear phase to the wavefront, which results in a tilt-like wavefront aberration.

18.5 Skew Aberration in Paraxial Ray Trace

Next, the skew aberration is studied in the *paraxial limit*. The *linear skew aberration* that appeared through the center of the pupil in Figure 18.7 clearly shows that skew aberration is present for paraxial rays. *Paraxial optics* is a method of determining the first-order properties of a radially symmetric optical system. It assumes that all ray angles and angles of incidence are small.[9] More information on paraxial optics and ray tracing is found in Chapter 15.

The general paraxial ray trace procedure traces two rays, the *paraxial marginal ray* (from the center of the object to the edge of entrance pupil) and the full-field *paraxial chief ray* (from the top of the object to the center of the entrance pupil).[10] Then, all other paraxial rays can be calculated from linear combinations of these two rays. For the *paraxial skew ray* from the top of the object and the edge of the pupil, the paraxial marginal ray height at each surface is the *x*-coordinate of the skew ray, the paraxial chief ray height is the *y*-coordinate of the skew ray, and the vertex of each surface is the *z*-coordinate of the skew ray.

The skew aberration calculation uses the propagation vector specified after the qth ray intercept, which is along $(y_{q+1} - y_q, \bar{y}_{q+1} - \bar{y}_q, t_q)$, where y_q is the marginal ray height, $\bar{y}_q$ is the chief ray height, and t_q is the distance between the qth and the $(q+1)$th surface vertices along the axis. The normalized *propagation vectors* $\mathbf{k}_q$ are

$$\mathbf{k}_q = \frac{(w_q, \bar{w}_q, 1)}{\sqrt{w_q^2 + \bar{w}_q^2 + 1}}, \tag{18.6}$$

where u_q is the *marginal ray angle*, $\bar{u}_q$ is the *chief ray angle*, $w_q = n_q u_q, t_q$ and $\bar{w}_q = n_q \bar{u}_q, t_q$ are reduced angles, and n_q is the *refractive index* following the qth surface.

In paraxial ray trace, the calculation of the *spherical polygon*'s area associated with the *parallel transport* of skew rays, as shown in Section 17.2, reduces to a polygon on the plane perpendicular to the optical axis. Thus, the skew aberration in paraxial ray trace is proportional to the area of the polygon. By dropping the *z*-component of the propagation vectors, the remaining 2D propagation vectors that form the polygon can be calculated as

$$\mathbf{k}_{2D,q} = \frac{(w_q, \bar{w}_q)}{\sqrt{w_q^2 + \bar{w}_q^2 + 1}} \approx (w_q, \bar{w}_q). \tag{18.7}$$

The area of the triangle shown in Figure 18.11 which connects the origin, $\mathbf{k}_{2D,q}$, and $\mathbf{k}_{2D,q+1}$ is

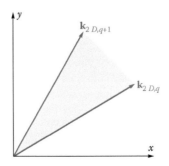

Figure 18.11 A triangle that connects the origin and two-dimensional propagation vectors.

18.5 Skew Aberration in Paraxial Ray Trace

$$\text{Area}_q = \frac{1}{2} \frac{w_q \bar{w}_{q+1} - w_{q+1} \bar{w}_q}{\sqrt{w_q^2 + \bar{w}_q^2 + 1} \sqrt{w_{q+1}^2 + \bar{w}_{q+1}^2 + 1}}. \tag{18.8}$$

Further manipulations using $w_{q+1} = w_q - y_q \phi_q$, $\bar{w}_{q+1} = \bar{w}_q - \bar{y}_q \phi_q$ results in a relationship for the area

$$\begin{aligned}\text{Area}_q &= \frac{\phi_q}{2} \frac{\bar{w}_q y_q - w_q \bar{y}_q}{\sqrt{w_q^2 + \bar{w}_q^2 + 1}\sqrt{w_{q+1}^2 + \bar{w}_{q+1}^2 + 1}} \\ &= \frac{H}{2} \frac{\phi_q}{\sqrt{w_q^2 + \bar{w}_q^2 + 1}\sqrt{w_{q+1}^2 + \bar{w}_{q+1}^2 + 1}},\end{aligned} \tag{18.9}$$

where $H = \bar{w}_q y_q - w_q \bar{y}_q$ is the *Lagrange invariant* of the system.

Therefore, the paraxial skew aberration of the system is proportional to the Lagrange invariant and is closely related to the sum of the individual surface powers in Equation 18.9,

$$\text{Total Area} = \frac{H}{2} \sum_q \frac{\phi_q}{\sqrt{w_q^2 + \bar{w}_q^2 + 1}\sqrt{w_{q+1}^2 + \bar{w}_{q+1}^2 + 1}}. \tag{18.10}$$

Frequently, paraxial rays with large y, $\bar{y}$, w, $\bar{w}$ are traced instead of small values. Since paraxial values are small, $\sqrt{w_q^2 + \bar{w}_q^2 + 1} \approx 1$ and $\sqrt{w_{q+1}^2 + \bar{w}_{q+1}^2 + 1} \approx 1$. Then, Equation 18.10 simplifies to

$$\text{Total Area} = \frac{H}{2} \sum_q \phi_q. \tag{18.11}$$

The example lens in Section 18.4 demonstrates the calculation of paraxial skew aberration. A paraxial skew ray (point A in Figure 18.7) is created by adding a paraxial marginal ray in the x–z plane to a paraxial chief ray in the y–z plane, yielding the paraxial skew aberration of $-4.49° \approx 0.078$. The surface-by-surface skew aberration contribution for ray A is shown in Figure 18.12.

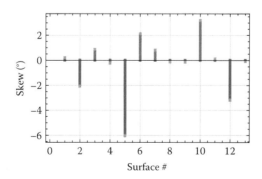

Figure 18.12 Paraxial skew ray at point A's skew aberration contribution from each lens surface sums to $-4.49°$ through the system.

18.6 Example of Paraxial Skew Aberration

In this section, an example of *paraxial skew aberration* is calculated using a four-element lens relay system that is chosen such that the exiting rays are parallel to the incident rays to simplify the understanding of the skew aberration concept.

Four thin lenses with the same effective focal length f are spaced $2f$ from each other as shown in Figure 18.13. The object plane is $2f$ in front of the first lens. The first lens is the entrance pupil. The first lens images the object plane with magnification −1 at the second lens. The second lens, a field lens, images the entrance pupil onto the third lens with magnification −1, such that the third lens is a pupil image. The third lens images the object onto the fourth lens, with a magnification of one. The fourth lens is another field lens, such that all paraxial rays in the image space after the fourth lens are parallel to the corresponding incident rays.

Table 18.1 contains the paraxial ray trace where each lens is specified by power, not by curvature and index. Figure 18.14 (left) shows the x–y components of the *propagation vectors* forming

Figure 18.13 Four thin lenses (blue) with effective focal length of 100 mm are spaced by 200 mm. The object is shown in green. Paraxial chief ray is shown in red and the marginal ray is shown in blue.

Table 18.1 Paraxial Ray Tracing Form for Four-Element Relay Lens

	0	1	2	3	4
$-\phi_q$		−0.01	−0.01	−0.01	−0.01
τ_q	200	200	200	200	
y_0	0	30	0	−30	0
u_0	0.15	−0.15	−0.15	−0.15	−0.15
$\bar{y}_0$	−30	0	30	0	−30
$\bar{u}_0$	0.15	0.15	−0.15	−0.15	0.15
$\mathbf{k}_0$	$\begin{pmatrix} 0.15 \\ -0.15 \\ 0.98 \end{pmatrix}$	$\begin{pmatrix} -0.15 \\ -0.15 \\ 0.98 \end{pmatrix}$	$\begin{pmatrix} -0.15 \\ 0.15 \\ 0.98 \end{pmatrix}$	$\begin{pmatrix} 0.15 \\ 0.15 \\ 0.98 \end{pmatrix}$	$\begin{pmatrix} 0.15 \\ -0.15 \\ 0.98 \end{pmatrix}$

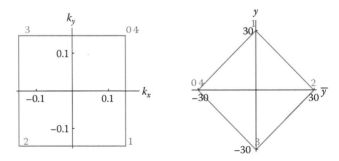

Figure 18.14 (Left) Propagation vector sequence near the z-axis for the relay lens and (right) the $y - \bar{y}$ diagram.

18.7 Skew Aberration's Effect on PSF

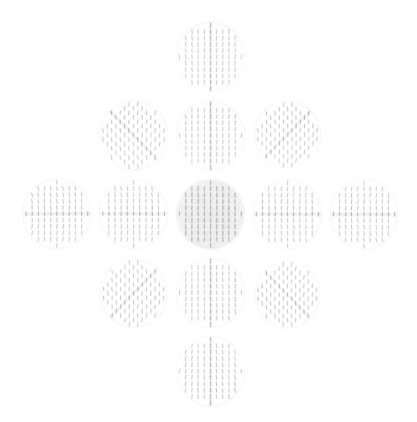

Figure 18.15 Functional form for the variation of paraxial skew aberration for a set of 13 fields for vertically polarized incident light. The skew aberration is the rotation from vertical. Meridional rays, shown over a gray background, have no skew aberration. In paraxial systems, the skew aberration increases linearly from the meridional plane and increases linearly with field.

a four-sided *spherical polygon* about the z-axis on the *propagation vector sphere*, which would be a square if the surface were flat. The four-sided spherical polygon subtends about $0.3 \times 0.3 = 0.09$ steradians. Thus, the *skew aberration* for this extreme skew ray is 0.09 radians $\cong 5°$. The $y - \bar{y}$ diagram shows surfaces 0, 2, and 4 are where the object and its images are located since $y = 0$, and due to the field lens at 4, each ray exits the system parallel to the incident light.

The variation of the skew aberration is shown for a set of fields in Figure 18.15. All meridional rays are shown over a gray background; other rays in the white background are skew rays. There is no skew aberration on-axis since all on-axis rays in radially symmetric systems are meridional rays. For fields along the y-axis where the meridional rays are down the vertical axis, there is no rotation, and along the horizontal axis for the x fields, the skew aberration is zero.

18.7 Skew Aberration's Effect on PSF

Skew aberration creates undesired polarization components in the *exit pupil* and the *image*. Skew aberration modifies the *PSF* and thus image quality can be degraded, even in the absence of wavefront, retardance, and diattenuation aberrations. Typically, cross-polarized satellites form around the PSF.[11–13] This section applies the methods in Chapter 16 to show the effect of *skew tilt* on the PSF.[13]

Skew aberration causes a spatially varying "*circular retardance*–like" behavior across the pupil. Skew aberration occurs with different functional forms, such as constant, linear, and quadratic variations, just like *wavefront aberrations*. Section 18.6 showed that the linearly varying skew aberration is the expected form in *paraxial optics* and is likely to be the dominant skew aberration component in many other systems. For example, linear variation is a significant component in Figure 18.7. At the lowest order, skew aberration has a linearly varying component, like the wavefront aberration named tilt; thus, this linear skew component is named *skew tilt*. Skew tilt has a *Jones pupil* function of the form

$$\mathbf{J}_{pupil}(u,v) = p(u,v)\begin{pmatrix} \cos(u\Delta) & \sin(u\Delta) \\ -\sin(u\Delta) & \cos(u\Delta) \end{pmatrix}, \quad (18.12)$$

where u is the x-pupil coordinate in the direction perpendicular to the meridional plane, v is the y-coordinate, and Δ is the magnitude of skew aberration at the edge of the pupil in the u-direction. $p(u, v)$ is the pupil function of the system.

To evaluate the effect of skew aberration on the PSF, an example with $\Delta = \pi$ is analyzed, a value far larger than expected, but good for tutorial purposes. Let's consider a circular aperture for

$$p(u,v) = \begin{cases} 1 & u^2 + v^2 \leq 1 \\ 0 & \text{otherwise} \end{cases}. \quad (18.13)$$

The Jones pupil is plotted in Figure 18.16 as a density plot (left) and a cross-section plot (right). This Jones pupil has purely real components. Note that the sine and cosine functions are plotted for $-\pi \leq u\Delta \leq \pi$.

A two-dimensional *Fourier transform* of $\mathbf{J}_{pupil}(u,v)$ gives a 2 × 2 Jones *amplitude response matrix* (**ARM**) of the system, which is then converted into a 4 × 4 **MPSM** as shown in Figure 18.17. For more details of these calculations, see Chapter 16.

The m_{03} component of the **MPSM** shows two peaks. Referring to Figures 18.8 and 18.9, the right circularly polarized light has been tilted to one side and the left circularly polarized light has been tilted to the other. This is seen in the two peaks in the m_{03}, one positive (right circular) and one negative (left circular). This means *unpolarized* incident light produces a *double image* with opposite circular components. Considering the m_{00}, m_{03}, m_{30}, and m_{33} elements, right circularly polarized incident light produces a single shifted peak to the left, and left circularly polarized incident light produces a single shifted peak to the right. Thus, the skew aberration divides the incident light into left and right circularly polarized states and shifts the peaks in opposite direction depending on the incident polarization state.

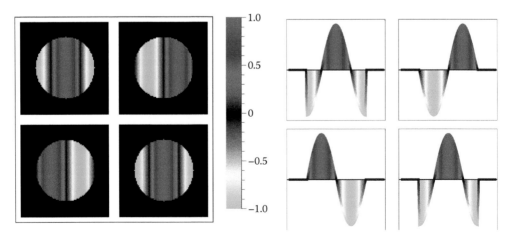

Figure 18.16 A density plot (left) and a cross-section plot (right) of the Jones pupil for skew tilt with a circular aperture.

18.8 PSM for U.S. Patent 2,896,506

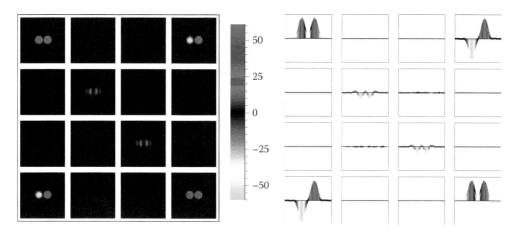

Figure 18.17 A density plot (left) and a cross-section plot (right) of the MPSM of the example system.

This example chose a half wave ($\Delta = \pi$) of skew aberration. At this level, the two peaks are entirely shifted from each other to visually show the effect of skew tilt on the **MPSM**. A typical skew tilt is in order of 1/100 waves; hence, the two images would overlap with just a small shear. Similar analyses are easily performed for different aperture functions such as rectangular apertures and higher-order skew aberration than linear skew aberration.

18.8 PSM for U.S. Patent 2,896,506

In this section, **MPSM** of the example optical system shown in Section 18.4 is further analyzed. The *Jones pupil* has been calculated by *polarization ray tracing* determining the *geometric transformation* for each ray using $\mathbf{Q}_{\text{Total},i}$. Since Jones matrices are defined in *local coordinates*, the exit pupil reference vectors ($\mathbf{g}_{\text{Exit},i}$) are used as local coordinate (u, v) when converting each $\mathbf{Q}_{\text{Total},i}$ to a Jones matrix. The choice of local coordinates is critical since the geometric transformation of each ray can be obscured by choosing local coordinates that already contain skew aberration, and in that basis, each Jones matrix would not show any skew aberration. For example, if $\mathbf{g}'_{\text{Exit},i}$ in Equation 18.5 is used to calculate ith ray's Jones matrix, the matrix will not show any skew aberration since $\mathbf{g}'_{\text{Exit},i}$ is already a *skew aberrated local coordinate*.

The Jones matrix pupil of the example high-NA lens, shown in Figure 18.18, becomes elliptical as the object moves far off-axis, 32° in this case, due to pupil distortion.

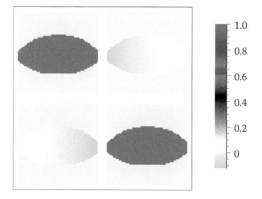

Figure 18.18 Jones matrix pupil at the exit pupil of the patent lens system calculated from the parallel transport matrix of the system at the exit pupil.

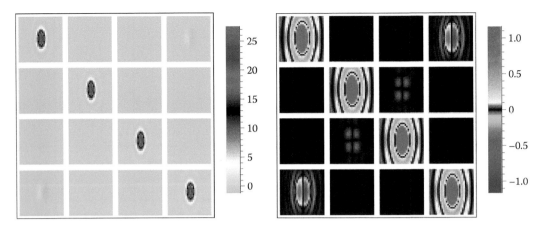

Figure 18.19 A **MPSM** of the patent lens system calculated from a discrete Fourier transform of the Jones matrix pupil. On the left, green indicates zero, and a few nonzero regions, most obviously in elements m_{03} and m_{30}, are shown in darker green and yellow. On the right, the scale has been enlarged from −1 to 1 with zero mapped to black to show nonzero regions of m_{03}, m_{12}, m_{21}, and m_{30}.

As shown in Figure 18.7, the system has *circular retardance*-like skew aberration described in Equation 18.12, which has linear and cubic variation along the pupil u-axis. Since the skew aberration is small, $\cos(u\Delta) \approx 1$, $\sin(u\Delta) \approx u\Delta$, the $j_{1,1}$ and $j_{2,2}$ components of the Jones matrix pupil are constant, and the magnitudes of $j_{1,2}$ and $j_{2,1}$ components are varying along the pupil u-axis.

The **MPSM** is plotted in two different scales in Figure 18.19; the legend of the left plot shows the full range of the **MPSM** values while the legend of the right plot is limited to the minimum and maximum values of the m_{03} component to highlight the m_{03} and m_{30} components. Note that the shape of the **MPSM** elements is elongated in the vertical direction and shortened along the horizontal direction due to the *discrete Fourier transform* of an elliptical aperture becoming elongated in the opposite direction.

The non-zero m_{03} and m_{30} components show, for *unpolarized light*, that the right circularly polarized component is tilted one way and the left circularly polarized component is tilted the opposite way. For this example, the shift between right and left circularly polarized light is quite small. However, the presence of off-diagonal components shows the degradation to the image quality arising from the skew aberration.

18.9 Statistics—CODE V Patent Library

Skew aberration is typically a small effect, but how small can only be addressed in the context of the magnitudes of the other families of aberrations. To understand the typical magnitude of skew aberration and its importance relative to other aberrations, the skew aberration of 2382 non-reflecting optical systems in CODE V's U.S. patent library was calculated. Figure 18.20 (left) shows the skew aberration for a non-paraxial ray from the largest defined field through the edge of the pupil along an axis perpendicular to the field point. Figure 18.20 (right) zooms in to the range from 2° to 6.5°, indicating the number of optical systems with skew aberration in each range. The mean of the overall skew aberration is 0.28° with a standard deviation of 0.47°. The maximum skew aberration in this set of lenses is 6.44°. Thus, for more than 90% of these lenses, the skew aberration corresponds to less than a hundredth of a wave of circular retardance and is insignificant.

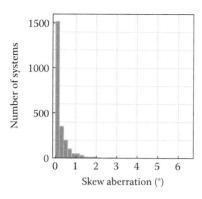
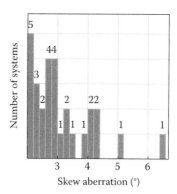

Figure 18.20 (Left) Histogram of the maximum skew aberration evaluated for 2382 non-reflecting optical systems in CODE V's library of patented lenses. (Right) Zoomed-in view of the histogram for skew aberration of 2° to 6.5°.

18.10 Conclusion

Skew aberration is the component of *polarization aberration* that originates from pure geometric effects associated with the sequence of *propagation vectors* through an optical system. Skew aberration is calculated by following the *parallel transport* of polarization states through *non-polarizing optical systems*. Skew aberration depends only on the sequence of **k** vectors; it does not change by varying the *coatings* applied to an optical system. The variation of skew aberration across the pupil often has a linear component, *skew tilt*, orthogonal to the *meridional plane*. A constant skew, similar to *wavefront piston*, would uniformly change the polarization and would not affect the PSF. It is the variation of skew (i.e., skew aberration) that affects the **MPSM** and, thus, degrades image quality. Skew aberration is typically a small effect in lenses but is a large effect in *corner cubes*. Skew aberration was calculated for two example systems to demonstrate how to separate skew aberration from other retardance effects and its effect on the **MPSM**. Greater skew aberration is expected in systems with high NA and large FOV such as *microlithography* optics and other polarization-sensitive systems.

As we understand it, the polarization ray tracing algorithms in most of the optical design and ray tracing software account for polarization aberrations but do not identify skew aberration as a separate component. Thus, we expect that these PSF calculations will incorporate the effects of skew aberration.

18.11 Problem Sets

18.1 A skew ray propagates through a sequence of four thin lenses with the following propagation vectors:
$\mathbf{k}_0 = (3, 4, 12)/13$, $\mathbf{k}_1 = (4, -3, 12)/13$, $\mathbf{k}_2 = (-3, -4, 12)/13$,
$\mathbf{k}_3 = (-4, 3, 12)/13$, and $\mathbf{k}_4 = (3, 4, 12)/13$.
   a. Calculate **Q** matrices for the four refractions: $\mathbf{Q}_1$, $\mathbf{Q}_2$, $\mathbf{Q}_3$, and $\mathbf{Q}_4$.
   b. Find the cumulative **Q** matrix for the ray path. Rational or floating point numbers are acceptable.
   c. What is the skew aberration in radians?
   d. Derive a set of local coordinates for the planes transverse to $\mathbf{k}_0$ and $\mathbf{k}_4$ using $\mathbf{y}_{Local} = (4/5, -3/5, 0)$.

e. Convert **Q** into a Jones matrix **J** using these local coordinates.
f. Find the eigenvalues and eigenvectors for **J**. What type of polarization element does **J** behave as?

18.2 A 1000 mm focal length lens with a 2 cm × 2 cm square aperture has $\xi_o = 0.3$ radians of skew aberration (polarization rotation varying linearly across the pupil) such that its polarization aberration function is $\mathbf{J}_{skew} = \begin{pmatrix} \cos\xi_o x & -\sin\xi_o x \\ \sin\xi_o x & -\cos\xi_o x \end{pmatrix}$, where $-1 < (x, y) < 1$ are normalized pupil coordinates. A collimated beam of 400 nm light is incident on-axis.

a. How much amplitude and intensity are transmitted when the lens is located between a horizontal and a vertical linear polarizer?
b. What is the associated wavefront aberration and apodization of the cross-polarized component?
c. Calculate the 2 × 2 amplitude response function at the image through suitable Fourier transforms of rectangle functions. Use the Fourier transform, not discrete Fourier transform, no approximations.

References

1. H. H. Hopkins, *Wave Theory of Aberrations*, London: Clarendon (1950).
2. E. Sklar, Effects of small rotationally symmetrical aberrations on the irradiance spread function of a system with Gaussian apodization over the pupil, *J. Opt. Soc. Am.* 65 (1975): 1520–1521.
3. J. P. Mills and B. J. Thompson, Effect of aberrations and apodization on the performance of coherent optical systems. I. The amplitude impulse response, *J. Opt. Soc. Am. A* 3 (1986): 694–703
4. J. P. Mills and B. J. Thompson, Effect of aberrations and apodization on the performance of coherent optical systems. II. Imaging, *J. Opt. Soc. Am. A* 3 (1986): 704–716.
5. R. A. Chipman, Polarization analysis of optical systems, *Opt. Eng.* 28 (1989): 90–99.
6. CODE V Version 10.3, Synopsys, Inc. (http://www.opticalres.com/cv/cvprodds_f.html).
7. G. Yun, S. McClain, and R. A. Chipman, Three-dimensional polarization ray-tracing calculus II. Retardance, *Appl. Opt.* 50 (2011): 2866–2874.
8. H. Azuma, High aperture wide-angle objective lens, U.S. Patent 2,896,506 (July 28, 1959).
9. J. E. Greivenkamp, *Field Guide to Geometrical Optics*, SPIE Press (2004).
10. B. R. Irving et al., *Code V, Introductory User's Guide*, Optical Research Associates (2001).
11. J. Ruoff and M. Totzeck, Orientation Zernike polynomials: A useful way to describe the polarization effects of optical imaging systems, *J. Micro/Nanolithogr. MEMS MOEMS* 8(3) (2009): 031404.
12. M. Mansuripur, Effects of high-numerical-aperture focusing on the state of polarization in optical and magneto-optic data storage systems, *Appl. Opt.* 50 (1991): 3154–3162.
13. J. P. McGuire and R. A. Chipman, Diffraction image formation in optical systems with polarization aberrations I: Formulation and example, *J. Opt. Soc. Am. A* 7 (1990): 1614–1626.

19

Birefringent Ray Trace

19.1 Ray Tracing in Birefringent Materials

The optical properties of *birefringent materials* depend on the direction of the light's polarization, as opposed to isotropic materials, which have identical properties in all directions. Isotropic and anisotropic materials are characterized by 3 × 3 *dielectric tensors* and 3 × 3 *gyrotropic tensors*. Ray tracing through birefringent materials is different from tracing through isotropic materials. Rays refracting into anisotropic media are decomposed into two rays with different propagation directions and orthogonal polarizations. These two rays are eigen-*modes* and propagate without change of polarization state. The ray tracing details are different for each type of birefringent materials: uniaxial, biaxial, and optically active materials (Figure 19.1).

This chapter describes the interaction of light with birefringent materials, with algorithms relating the light fields before and after birefringent interfaces and a method to track the multiple rays generated due to double refraction/ray doubling. The ray tracing algorithm for birefringent ray intercepts tracks the light field, amplitude, and direction change through birefringent interfaces using the polarization ray tracing matrix. This integrates the birefringent ray trace with the polarization ray tracing methods of Chapter 10, maintaining a global three-dimensional matrix representation. Additional details on light propagation in uniaxial materials such as calcite and uniaxial devices such as waveplates are found in Chapter 21. An example polarization aberration analysis of a common uniaxial optical element, the Glan–Taylor polarizer, an application of this chapter's algorithms, is presented in Chapter 22 (Crystal Polarizers).

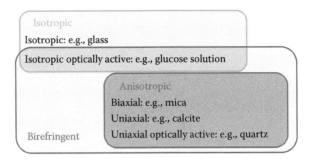

Figure 19.1 Classes of isotropic and anisotropic media. Isotropic optically active materials are both isotropic and birefringent.

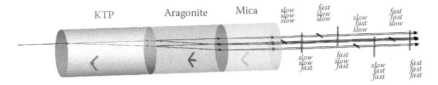

Figure 19.2 A normally incident ray propagates through three blocks of anisotropic materials (potassium titanyl phosphate [KTP], aragonite, and mica), each with different crystal axis orientations shown as three lines inside each block. One incident ray results in eight exiting rays each with different sequences of polarizations and different *OPL*.

To see why ray tracing birefringent materials is complicated, consider a real ray trace with the Polaris-M software* for a ray propagating through an example anisotropic system shown in Figure 19.2. First, a ray refracts from air into a biaxial KTP (potassium titanyl phosphate, $KTiOPO_4$) crystal, where the light divides into two modes, labeled *fast* (f_1) and *slow* (s_1), due to *double refraction*. These two modes then refract into a crystal of aragonite. The aragonite's crystal axes (CA) are not aligned parallel to the KTP's axes. Therefore, the f_1 ray couples into two modes, f_2 and s_2, and similarly, the s_1-mode couples into f_2- and s_2-modes. The collective mode labels after propagating through KTP and aragonite are *fast–fast* ($f_1 f_2$), *fast–slow* ($f_1 s_2$), *slow–fast* ($s_1 f_2$), and *slow–slow* ($s_1 s_2$). This ray doubling continues into the third biaxial crystal, mica. When the incident light exits the three crystals, eight modes emerge, labeled as *fff, ffs, fsf, fss, sff, sfs, ssf,* and *sss*. Each *f* and *s* represent distinct electric field orientations along separate ray segments. The exiting polarization state and phase are found from the addition (superposition) of eight waves. The optical path length (*OPL*) has eight different values for the eight *partial waves*, a term for the division of an incident wave into multiple waves.

Depending on the type of anisotropic material, different symbols and subscripts label the types eigenmodes as tabulated in Tables 19.1 and 19.2. Isotropic materials are a special case with degenerate modes. When refracting into an isotropic material, both refracted modes, *s* and *p*, share the same Poynting vector direction $\hat{S}$, the same propagation direction $\hat{k}$, and the same refractive index; hence, these modes are *degenerate*. Thus, for isotropic refraction, the *s*- and *p*-modes can be combined and treated as a single mode labeled *i*, denoting an *isotropic mode*. In uniaxial materials, the two modes are labeled *o* for the *ordinary* and *e* for the *extraordinary* modes. In biaxial materials, the two modes are distinguished by the associated refractive index of the ray, the mode with the higher index being the *slow*-mode and the other mode being the *fast*-mode. In isotropic optically active materials, the two modes are the *right* and *left* circularly polarized modes. Note the symbols

---

* See Preface *xxvii* for Polaris-M.

19.1 Ray Tracing in Birefringent Materials

Table 19.1 Mode Labeling for Rays in Different Types of Birefringent Interfaces

Isotropic/Anisotropic Material	Descriptions of Eigenmodes	Mode Label
Biaxial	Mode with smaller n	f-mode; *fast*-mode
	Mode with larger n	s-mode; *slow*-mode
Uniaxial	Ordinary ray	o-mode
	Extraordinary ray	e-mode
Optically active	Left circularly polarized	l-mode; *left*-mode
	Right circularly polarized	r-mode; *right*-mode
Isotropic	Polarized in plane of incidence	p-polarization
	Polarized out of the plane of incidence	s-polarization
	Combined s- and p-states	i-mode

Eigenpolarizations in isotropic/anisotropic material		Labeling
Incidence; *inc*	Two incident modes	m, n
Exiting	Two exiting modes	v, w
Transmission; t	Two transmitted modes	ta, tb
Reflection; r	Two reflected modes	rc, rd

Note: The s- and p-polarizations in isotropic material have the same propagation directions and are grouped to one mode, *isotropic*.

Table 19.2 Parameters to Be Calculated for Each Birefringent Ray Intercept to Characterize Each Exiting Ray

Parameter	Symbol
Ray intercept coordinates	**r**
Propagation vector	$\hat{\mathbf{k}}$
Poynting vector	$\hat{\mathbf{S}}$
Normal to surface	$\hat{\boldsymbol{\eta}}$
Mode label	f, s, o, e, l, r, i
Mode refractive index	$n_f, n_s, n_o, n_e, n_l, n_r, n$
Optical path length	OPL
Electric field vector	**E**
Magnetic field vector	**H**
Ray status	e.g., active or missed aperture
Surface order	e.g., (1, 2, 3, 4)
Polarization ray tracing matrix for interface	**P**
Polarization ray tracing matrix from object space	$\mathbf{P}_{cumulative}$
Geometrical transformation for interface	**Q**
Geometrical transformation from object space	$\mathbf{Q}_{cumulative}$
Amplitude reflection or transmission coefficient	a

for modes are in lowercase. A list of parameters needed by the polarization ray trace for each ray segment is presented in Table 19.2. Many of these parameters were first introduced in Chapters 9 and 10, but birefringent interfaces need additional parameters.

To simulate the propagation of a wavefront through optical systems with birefringent optical elements, a large number of rays are usually traced, as shown in Figure 19.3. For one

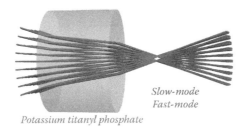

Slow-mode
Fast-mode
Potassium titanyl phosphate

Figure 19.3 A converging beam focuses through a KTP plate. Because of ray doubling, two foci are observed, one for the fast mode and another one for the slow mode.

birefringent element, one incident wavefront yields two exiting wavefronts. Each of these wavefronts focuses in different locations and has differing amounts of astigmatism and other aberrations.

19.2 Description of Electromagnetic Waves in Anisotropic Media

The description of light fields of a ray used by the polarization ray tracing algorithm is provided in this section. Light is a transverse electromagnetic wave characterized by its electric field **E**, its magnetic field **H**, its displacement field **D**, and its induction field **B** by Maxwell's equation.[1] The electromagnetic fields of a monochromatic plane wave in space **r** and time t with wavelength λ are

$$\mathbf{E}(\mathbf{r},t) = \mathcal{R}e\left\{\mathbf{E}\exp\left[i\left(\frac{2\pi n}{\lambda}\hat{\mathbf{k}}\cdot\mathbf{r} - \omega t\right)\right]\right\},$$

$$\mathbf{H}(\mathbf{r},t) = \mathcal{R}e\left\{\mathbf{H}\exp\left[i\left(\frac{2\pi n}{\lambda}\hat{\mathbf{k}}\cdot\mathbf{r} - \omega t\right)\right]\right\},$$

$$\mathbf{D}(\mathbf{r},t) = \mathcal{R}e\left\{\mathbf{D}\exp\left[i\left(\frac{2\pi n}{\lambda}\hat{\mathbf{k}}\cdot\mathbf{r} - \omega t\right)\right]\right\}, \text{ and}$$

$$\mathbf{B}(\mathbf{r},t) = \mathcal{R}e\left\{\mathbf{B}\exp\left[i\left(\frac{2\pi n}{\lambda}\hat{\mathbf{k}}\cdot\mathbf{r} - \omega t\right)\right]\right\}.$$

(19.1)

The normalized *propagation vector* in a medium with refractive index n is $\hat{\mathbf{k}}$ with wavenumber $\frac{2\pi n}{\lambda}$. In an absorptive material, the complex refractive index is $n + i\kappa$, so the magnitude of the fields in an absorptive material decays exponentially as light propagates.

The *polarization vector* describes the light's polarization state in 3D as a 3×1 vector*

$$\mathbf{E} = E_o e^{i\phi_o}\hat{\mathbf{E}} = E_o e^{i\phi_o}\begin{pmatrix} E_x \\ E_y \\ E_z \end{pmatrix} = E_o e^{i\phi_o}\begin{pmatrix} |E_x|e^{i\phi_x} \\ |E_y|e^{i\phi_y} \\ |E_z|e^{i\phi_z} \end{pmatrix}.$$

(19.2)

---

* Polarization vector is described in Chapter 2.

The field has an absolute complex magnitude $E_0 e^{i\phi_0}$ and complex components (E_x, E_y, E_z), where $\hat{\mathbf{E}} \cdot \hat{\mathbf{E}}^* = |E_x|^2 + |E_y|^2 + |E_z|^2 = 1$.

19.3 Defining Birefringent Materials

For the purpose of polarization ray tracing, birefringent materials, including biaxial, uniaxial, and optically active materials, are described by a *dielectric tensor* $\boldsymbol{\varepsilon}$ and a *gyrotropic tensor* $\mathbf{G}$.

In an isotropic material, light with wavelength λ experiences the same refractive index regardless of propagation direction and polarization state. Optical glasses are isotropic, as are air, water, and vacuum. In birefringent materials, the refractive index experienced by the light varies with the direction of the light's electric field. Many crystals, such as calcite and rutile, are anisotropic. Materials also become birefringent due to stress, strain, or applied electric or magnetic fields. In an anisotropic material, an *optic axis* is a direction of propagation in which light experiences zero birefringence. When light propagates along the optic axis, the refractive index is the same for all electric field components in the transverse plane. For propagation near the optic axis, the birefringence, the difference in refractive indices between two modes with the same $\mathbf{k}$ vector, is small. A *biaxial material* has three distinct principal indices, and there are four directions, plus and minus along two lines within the material having degenerate eigenpolarizations, as is explained in Section 19.5. Unlike uniaxial material, biaxial material has two optic axes, thus the label biaxial.

The *dielectric tensor* $\boldsymbol{\varepsilon}$ relates the variation of refractive index with the light's polarization state by relating $\mathbf{E}$ to $\mathbf{D}$[1–3]:

$$\mathbf{D} = \boldsymbol{\varepsilon} \mathbf{E} = \begin{pmatrix} D_x \\ D_y \\ D_z \end{pmatrix} = \begin{pmatrix} \varepsilon_{XX} & \varepsilon_{XY} & \varepsilon_{XZ} \\ \varepsilon_{YX} & \varepsilon_{YY} & \varepsilon_{YZ} \\ \varepsilon_{ZX} & \varepsilon_{ZY} & \varepsilon_{ZZ} \end{pmatrix} \begin{pmatrix} E_x \\ E_y \\ E_z \end{pmatrix}. \quad (19.3)$$

When light's oscillating $\mathbf{E}$ field is propagating through a crystal, the response of the crystal changes its orientation with respect to atomic configuration of the crystal and directions of the different molecular bonds. Under the influence of the light field, the charges oscillate at optical frequencies that contribute to the $\mathbf{E}$ field. The result is the $\mathbf{D}$ field, which includes the light's field and a contribution from dipoles induced in the material. This relationship is described by the 3×3 dielectric tensor. The tensor $\boldsymbol{\varepsilon}$ can always be rotated into a diagonal form

$$\boldsymbol{\varepsilon} = \begin{pmatrix} \varepsilon_X & 0 & 0 \\ 0 & \varepsilon_Y & 0 \\ 0 & 0 & \varepsilon_Z \end{pmatrix} = \begin{pmatrix} (n_X + i\kappa_X)^2 & 0 & 0 \\ 0 & (n_Y + i\kappa_Y)^2 & 0 \\ 0 & 0 & (n_Z + i\kappa_Z)^2 \end{pmatrix}, \quad (19.4)$$

where n_X, n_Y, and n_Z in uppercase subscripts are the *principal refractive indices* associated with three orthogonal *principal axes* or *crystal axes* (CA) and κ_X, κ_Y, and κ_Z are the associated absorption coefficients along those three axes. Biaxial materials, such as mica and topaz, have three different principal refractive indices—n_S for the largest index, n_M, and n_F for the smallest index.

An isotropic material can be considered a special case of an anisotropic material with $\varepsilon_X = \varepsilon_Y = \varepsilon_Z = \varepsilon$,

$$\boldsymbol{\varepsilon} = \begin{pmatrix} \varepsilon & 0 & 0 \\ 0 & \varepsilon & 0 \\ 0 & 0 & \varepsilon \end{pmatrix} = \begin{pmatrix} (n + i\kappa)^2 & 0 & 0 \\ 0 & (n + i\kappa)^2 & 0 \\ 0 & 0 & (n + i\kappa)^2 \end{pmatrix}. \quad (19.5)$$

The dielectric tensor of an isotropic material is proportional to the identity matrix; light experiences the same refractive index $n + i\kappa$ regardless of propagation direction and polarization state. A *uniaxial material* has two equal principal refractive indices on the diagonal, n_O as the principal ordinary index, and n_E as the principal extraordinary index,

$$\varepsilon = \begin{pmatrix} \varepsilon_O & 0 & 0 \\ 0 & \varepsilon_O & 0 \\ 0 & 0 & \varepsilon_E \end{pmatrix} = \begin{pmatrix} (n_O + i\kappa_O)^2 & 0 & 0 \\ 0 & (n_O + i\kappa_O)^2 & 0 \\ 0 & 0 & (n_E + i\kappa_E)^2 \end{pmatrix}. \tag{19.6}$$

The principal index n_O is associated with a plane and the principal index n_E is associated with the principal axis orthogonal to that plane as shown in the middle of Table 19.3. The principal axis related to n_E is the *optic axis*, which specifies the orientation of a uniaxial crystal. By definition, a *negative uniaxial crystal*, such as calcite, has $n_O > n_E$, while a *positive uniaxial crystal* has $n_O < n_E$.

Table 19.3 Properties of Biaxial, Uniaxial, and Isotropic Materials When the Principal Axes Are Aligned to the Global Coordinate System

Material	Principal Label	Principal Refractive Index	Diagonal Dielectric Tensor ε_D
Biaxial	Slow / Medium / Fast	(n_S, n_M, n_F)	$\begin{pmatrix} \varepsilon_S & 0 & 0 \\ 0 & \varepsilon_M & 0 \\ 0 & 0 & \varepsilon_F \end{pmatrix} = \begin{pmatrix} n_S^2 & 0 & 0 \\ 0 & n_M^2 & 0 \\ 0 & 0 & n_F^2 \end{pmatrix}$
Uniaxial	Ordinary / Extraordinary	(n_O, n_E)	$\begin{pmatrix} \varepsilon_O & 0 & 0 \\ 0 & \varepsilon_O & 0 \\ 0 & 0 & \varepsilon_E \end{pmatrix} = \begin{pmatrix} n_O^2 & 0 & 0 \\ 0 & n_O^2 & 0 \\ 0 & 0 & n_E^2 \end{pmatrix}$
Isotropic	Isotropic	n	$\begin{pmatrix} \varepsilon & 0 & 0 \\ 0 & \varepsilon & 0 \\ 0 & 0 & \varepsilon \end{pmatrix} = \begin{pmatrix} n^2 & 0 & 0 \\ 0 & n^2 & 0 \\ 0 & 0 & n^2 \end{pmatrix}$

Note: A biaxial material has three principal refractive indices (n_F, n_M, n_S). Their associated principal axes are oriented orthogonally. The uniaxial material has two principal refractive indices, the ordinary refractive index n_O and extraordinary index n_E. The isotropic material has one refractive index n.

19.3 Defining Birefringent Materials

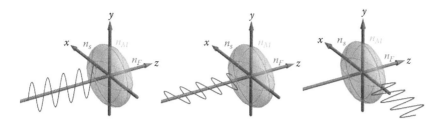

Figure 19.4 (Left) Light propagating along z and polarized along x experiences the refractive index n_S. (Middle) Light propagating along z and polarized along y experiences refractive index n_M. (Right) Light propagating along x and polarized along z experiences refractive index n_F. Thus, the refractive index depends on the light polarization, not the direction of propagation.

The refractive index experienced by a light ray depends on the electric field orientation (the polarization state) relative to the principal axes of the material. Light linearly polarized with its electric field along each of the three principal axes is depicted in Figure 19.4, which demonstrates how the refractive index of the light depends on its polarization, not the direction of propagation. The refractive index characterizes how strongly the electrons in a material oscillate in response to an electromagnetic wave, which governs how fast the mode propagates.[4]

The *birefringence* Δn is the refractive index difference of two eigenpolarizations propagating in the same direction in a birefringent material. The maximum Δn of biaxial and uniaxial materials are $n_S - n_F$ and $n_E - n_O$. Birefringence is dispersive and changes with wavelength. The maximum birefringence of various biaxial and uniaxial material as a function of wavelength is shown in Figures 19.5 through 19.7.[5-9]

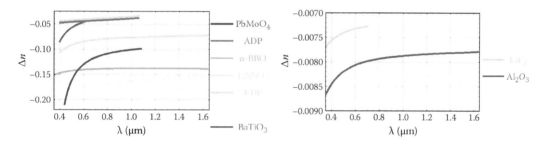

Figure 19.5 Birefringence spectra of common negative uniaxial materials.

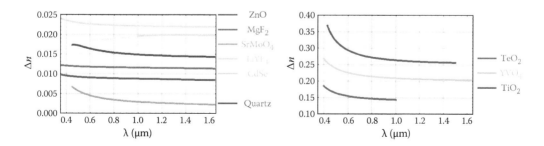

Figure 19.6 Birefringence spectra of common positive uniaxial materials.

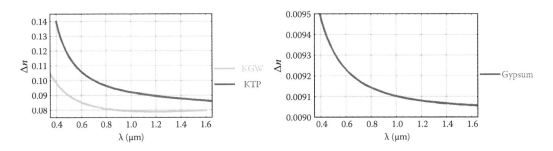

Figure 19.7 Birefringence spectra of common biaxial materials.

Optically active materials have a molecular structure, typically helical, that induces a rotation of the plane of the electric field oscillations as light passes through the material. The effect of optical activity is described by a *gyrotropic tensor* **G** in the *constitutive relation*:

$$\mathbf{D} = \varepsilon \mathbf{E} + i\mathbf{G}\mathbf{H},$$
$$\mathbf{B} = \mu \mathbf{H} - i\mathbf{G}\mathbf{E}, \tag{19.7}$$

where **μ** is the *magnetic permeability tensor* and **B** is the *magnetic induction*.[2,10] Organic liquids such as glucose and sucrose solutions are common examples of isotropic optically active liquids that induce birefringence, or a phase shift, between left and right circularly polarized light. The two circularly polarized eigenmodes have slightly different refractive index n_R and n_L. Thus, optical activity is a source of circular birefringence. The difference between these two indices is often characterized by the *optical rotatory power* α, which is related to the *gyrotropic constant* g,[11–14]

$$\alpha = \frac{2\pi}{\lambda} g = \frac{\pi}{\lambda} |n_R - n_L|. \tag{19.8}$$

In general, **G** is a symmetric tensor with six independent coefficients,

$$\mathbf{G} = \begin{pmatrix} g_{11} & g_{12} & g_{13} \\ g_{12} & g_{22} & g_{23} \\ g_{13} & g_{23} & g_{33} \end{pmatrix}. \tag{19.9}$$

For an isotropic optically active liquid, such as a sugar solution, **G** is a diagonal tensor with only one parameter,

$$\mathbf{G} = \begin{pmatrix} g & 0 & 0 \\ 0 & g & 0 \\ 0 & 0 & g \end{pmatrix}. \tag{19.10}$$

The majority of biaxial and uniaxial materials have no optical activity, so **G** is **0**. A few crystals combine uniaxial or biaxial properties with optical activity such as mercury sulfide. In general, molecules that lack mirror symmetry are optically active, that is, a molecule that cannot be superposed on its mirror image, similar to a left shoe and a right shoe.

19.3 Defining Birefringent Materials

Crystalline quartz has both uniaxial and optically active characteristics. In quartz, the optical activity is only significant when the light propagates near the uniaxial optical axis, so **G** for quartz has two dependent values,[15]

$$\mathbf{G} = \begin{pmatrix} g_O & 0 & 0 \\ 0 & g_O & 0 \\ 0 & 0 & g_E \end{pmatrix}, \qquad (19.11)$$

where $g_O = \tfrac{1}{2}(n_R - n_L) = 3 \times 10^{-5}$ and $g_E = -1.92 g_O$ at 589 nm. **G** is also a function of externally applied magnetic fields.[16] Magnetically induced circular birefringence is known as the Faraday effect.

A schematic of the electric field for the two circularly polarized modes propagating in an optically active material is shown in Figure 19.8. When linearly polarized light passes through the optically active material, its plane of polarization rotates steadily through the medium, as shown in Figure 19.9. When it centers the medium, it decomposes into left and right circular components with equal amplitudes that propagate at different speeds, and it exits the material with rotated orientation but still linearly polarized. Optical activity is readily observed under a polariscope. In Figure 19.10, a bottle of concentrated sugar solution is placed between a pair of linear polarizers. The transmission between the polarizers depends on the amount of optical activity ($n_R - n_L$), the orientation of the polarizers, and the wavelength.

To trace rays through optical elements formed from birefringent materials, the dielectric tensor and gyrotropic tensor are expressed in the optical system's global xyz Cartesian coordinates. Arbitrarily oriented tensors are obtained by rotating the material's diagonal tensor, which is

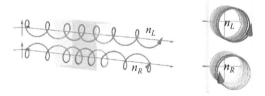

Figure 19.8 The side and front view of the left (red) and right (blue) circularly polarized electric field propagating through an optically active material. The left circularly polarized beam propagates for three wavelengths, and the right propagates for three and a half wavelengths through the material, yielding one-half wave of circular retardance.

Figure 19.9 The plane of polarization of linearly polarized light rotates at a uniform rate when propagating in an optically active medium. This one-and-a-half wave circular retarder generates 270° of optical rotation.

Figure 19.10 Light propagating through corn syrup (concentrated sugar water solution) in a polariscope. The angle between the polarizer axes rotates 180° from the left figure to the right figure. The polarizer axes are crossed in the left-hand image, parallel in the middle image, and crossed again in the right-hand image. The colors arise from the dispersion of the polarization state, being rotated by a larger angle in the blue and a smaller angle in the red. For this specific jar, the blue–purple light has been rotated by more than 180° while the red light has been rotated by about 135°.

tabulated in optical materials tables.[17] For principal refractive indices (n_A, n_B, n_C) with principal axis orientations specified by the unit vectors (v_A, v_B, v_C), the diagonal dielectric tensor is

$$\boldsymbol{\varepsilon}_D = \begin{pmatrix} n_A^2 & 0 & 0 \\ 0 & n_B^2 & 0 \\ 0 & 0 & n_C^2 \end{pmatrix}, \tag{19.12}$$

and the dielectric tensor in the optical system's global coordinate is

$$\boldsymbol{\varepsilon} = \begin{pmatrix} v_{Ax} & v_{Bx} & v_{Cx} \\ v_{Ay} & v_{By} & v_{Cy} \\ v_{Az} & v_{By} & v_{Cy} \end{pmatrix} \cdot \boldsymbol{\varepsilon}_D \cdot \begin{pmatrix} v_{Ax} & v_{Bx} & v_{Cx} \\ v_{Ay} & v_{By} & v_{Cy} \\ v_{Az} & v_{By} & v_{Cy} \end{pmatrix}^{-1}. \tag{19.13}$$

This rotation is another example of the orthogonal transformation matrices introduced in Chapter 9. Gyrotropic tensors are rotated in the same manner.

Example 19.1 Rotation from Diagonal $\boldsymbol{\varepsilon}$ to Global $\boldsymbol{\varepsilon}$

Rotate the diagonal dielectric tensor $\boldsymbol{\varepsilon}_D = \begin{pmatrix} \varepsilon_x & 0 & 0 \\ 0 & \varepsilon_y & 0 \\ 0 & 0 & \varepsilon_z \end{pmatrix}$ with principal axes $v_x = \begin{pmatrix} 0 \\ 0 \\ 1 \end{pmatrix}$, $v_y = \begin{pmatrix} 0 \\ 1 \\ 0 \end{pmatrix}$, and $v_z = \begin{pmatrix} -1 \\ 0 \\ 0 \end{pmatrix}$ into global coordinates (Figure 19.11).

19.4 Eigenmodes of Birefringent Materials

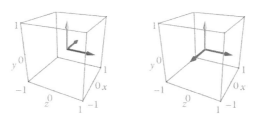

Figure 19.11 Three orthogonal principal axes shown in red, green, and blue on the left in its principal coordinate are rotated to the global coordinate shown on the right.

Using Equation 19.13,

$$\varepsilon = \begin{pmatrix} 0 & 0 & -1 \\ 0 & 1 & 0 \\ 1 & 0 & 0 \end{pmatrix} \cdot \varepsilon_D \cdot \begin{pmatrix} 0 & 0 & -1 \\ 0 & 1 & 0 \\ 1 & 0 & 0 \end{pmatrix}^{-1} = \begin{pmatrix} \varepsilon_z & 0 & 0 \\ 0 & \varepsilon_y & 0 \\ 0 & 0 & \varepsilon_x \end{pmatrix}.$$

This operation, for example, can rotate a C-plate, a waveplate with the optic axis normal to the surfaces and no retardance on-axis, to a conventional A-plate waveplate (Figure 19.12).*

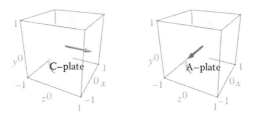

Figure 19.12 The optic axis of a C-plate can rotate to the optic axis of an A-plate via the rotation operation of Figure 19.11.

---

* Uniaxial waveplates are described in Chapter 21 (Uniaxial Materials and Components).

19.4 Eigenmodes of Birefringent Materials

Ray tracing through birefringent materials results in multiple rays with different polarizations. These polarization states are the *eigenpolarizations* or *eigenmodes*. When light refracts into a birefringent material, its energy divides into orthogonal polarized eigenmodes in a process called *double refraction* or *ray splitting*. These eigenmodes are the only polarization states that can propagate in a given direction without change of polarization. Figure 19.13 shows the image of the text "POLARIS" seen

Figure 19.13 Double refraction through a calcite rhomb.

through a calcite crystal. The two images have orthogonal linear polarization states, so either image can be selected by rotating a linear polarizer in front of the crystal.

Figure 19.14 shows the schematic of a ray refracting into a plate of the biaxial crystal ulexite, where it splits into the *fast-* and *slow-*modes, which refract into different directions due to their higher and lower refractive indices, propagating to the back face, and exiting the crystal. Here, the *fast* ray has its polarization coming out of the page and the *slow* ray has its polarization in the plane of the page. Any incident ray divides into two orthogonal polarization modes within the crystal. Unpolarized incident light divides its flux equally into both the *fast-* and *slow-*modes; other states divide unequally. The energy distribution between the two modes depends on the polarization state of the incident light, that is, electric field oscillation with respect to the three crystal axes. Similarly, for refraction into uniaxial materials, two eigenmodes emerge after the uniaxial interface with orthogonal polarizations labeled *ordinary* and *extraordinary*, while for optically active materials, the modes are labeled *left* and *right*.

In general, ray doubling occurs each time a beam enters into or reflects toward a birefringent medium, unless the polarization is aligned exactly such that only one mode is excited with energy. An incident ray refracting through N birefringent interfaces results in a potential of 2^N separate exiting rays with 2^N different mode sequences. Each of these modes takes a different path and has its own amplitude, polarization, and *OPL*. To retrieve the properties of each of these modes, a list of ray parameters, presented in Table 19.2, are calculated for each ray segment during the polarization ray trace through an optical system. Mode label is an additional ray parameter needed for tracking ray doubling at each birefringent interface. The collective mode label of a resultant ray at the exit pupil describes the evolution of polarization along that specific ray path. Polarization ray tracing a grid of incident rays emerging from an incident wavefront through N birefringent surfaces produces 2^N separate wavefronts at the exit pupil. The exiting rays with the same mode label represent one of these wavefronts. By studying the properties of these rays at the exit pupil, one mode at a time,

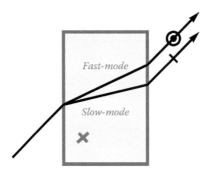

Figure 19.14 A ray splits into two modes through a ulexite biaxial block. Two of the crystal axes are shown in the bottom left corner and the third is out of the plane of the page. In this case, the *fast-*mode oscillates in and out of the page and the *slow-*mode oscillates within the plane of the page.

Polarized Light and Optical Systems

the wavefront aberrations (amplitude aberration, defocus, spherical aberration, coma, astigmatism, etc.) of each mode sequence can be analyzed and the effects of overlapping these wavefronts can be calculated.

19.5 Reflections and Refractions at Birefringent Interface

When light is incident at an optical interface, part of its energy refracts and another part reflects. At a birefringent intercept, the energy of the light divides into eigenmodes. The exiting ray parameters and the beam's amplitudes are shown in Table 19.2 before and after the intercept are calculated from the incident ray parameters, as well as the material properties before and after the intercept. The algorithms to calculate the exiting ray parameters for polarization ray tracing in refraction and reflection for each exiting modes at an uncoated birefringent intercept are described in this section. These resultant ray parameters will be used in Section 19.7 to calculate their polarization ray tracing matrices, **P**, which represent the polarization properties of one or a series of ray intercepts. The type of uncoated birefringent interfaces described in this section includes isotropic/birefringent, birefringent/isotropic, and birefringent/birefringent interfaces. Depending on the type of the interface, the number of exiting rays varies, the resultant fields behave differently, but the calculations are similar and can be generalized. The incident, refracted, and reflected ray parameters are distinguished by their subscripts *inc*, *t*, and *r*, respectively, in the following discussion.

An isotropic interface is a special case of birefringent interfaces. In isotropic materials, **k** and **S** are aligned in the same direction, and the polarization of light describes the oscillation direction of **E** field. At an isotropic intercept, **E** is decomposed into its *s*- and *p*-components, $\mathbf{E}_s$ and $\mathbf{E}_p$. These are defined by the surface normal $\hat{\boldsymbol{\eta}}$ and $\hat{\mathbf{k}}$, where $\mathbf{E}_s$ is in the direction of $(\hat{\mathbf{k}} \times \hat{\boldsymbol{\eta}})$ and $\mathbf{E}_p$ is in the direction of $(\hat{\mathbf{k}} \times \hat{\mathbf{E}}_s)$,* as shown in Figure 19.15. The transmitted electric field inside the isotropic material is the superposition of the transmitted *s*- and *p*-components,[2]

$$\mathbf{E}_t(\mathbf{r},t) = \mathcal{R}e\left[E_{inc}\left(a_{ts}\hat{\mathbf{E}}_{ts} + a_{tp}\hat{\mathbf{E}}_{tp} \right) e^{i\frac{2\pi}{\lambda}n\hat{\mathbf{k}}_t \cdot \mathbf{r}} e^{-i\omega t} \right], \tag{19.14}$$

where E_{inc} is the incident electric field amplitude, and a_{ts} and a_{tp} are the complex electric field amplitude transmission coefficients. Regardless of the propagation direction, the two refracted components experience the same refractive index n, and $\hat{\mathbf{k}}_t = \hat{\mathbf{S}}_t$.

Inside birefringent materials, however, the refracted rays experience different refractive indices depending on the incident polarization and the direction of $\mathbf{k}_{inc}$ relative to the crystal axis orientations. $\mathbf{E}_{inc}$ splits into two orthogonal eigenpolarizations and propagate in different directions. As an example, consider a ray normally incident onto a biaxial crystal block as shown in Figure 19.16. In transmission, the incident ray splits into two orthogonal modes with different refractive indices that lie between the largest and smallest of the three principal indices of the biaxial material. Inside the crystal, each mode's constant phase wavefront is not perpendicular to the energy propagation direction; $\mathbf{k}_t$ is not aligned with $\mathbf{S}_t$. $\mathbf{E}_t$ is oscillating perpendicular to $\mathbf{S}_t$, but not $\mathbf{k}_t$. The energy of the two modes propagates along different paths; the *fast*-mode's energy propagates along $\mathbf{S}_{tf}$ and the *slow*-mode's energy propagates along $\mathbf{S}_{ts}$. The calculation of these ray parameters is shown later in this section. After the first interface and assuming plane waves, the refracted electric field in the biaxial material is the sum of the two modes having different propagation directions:

$$\mathbf{E}_t(\mathbf{r},t) = \mathcal{R}e\left\{ E_{inc}\left[a_{tf}\hat{\mathbf{E}}_{tf} e^{i\frac{2\pi}{\lambda}n_f \hat{\mathbf{k}}_{tf} \cdot \mathbf{r}} + a_{ts}\hat{\mathbf{E}}_{ts} e^{i\frac{2\pi}{\lambda}n_s \hat{\mathbf{k}}_{ts} \cdot \mathbf{r}} \right] e^{-i\omega t} \right\}, \tag{19.15}$$

---

* Convention in Chapter 10 (Optical Ray Tracing).

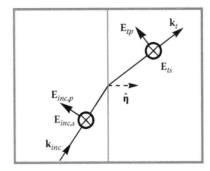

Figure 19.15 At isotropic interface, $\mathbf{E}_s$ is in the direction of $(\hat{\mathbf{k}}_{inc} \times \hat{\boldsymbol{\eta}})$ and $\mathbf{E}_{inc,p}$ is in the direction of $(\hat{\mathbf{k}} \times \hat{\mathbf{E}}_{inc,s})$. They form a right-handed (**s**, **p**, **k**) basis.

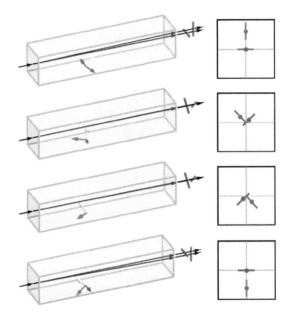

Figure 19.16 A ray at normal incidence propagates into a block of ulexite. The direction of energy flow is along the Poynting vector **S**, which is shown as the black arrows passing through the crystal block. The ulexite's principal axes associated with n_F, n_M, and n_S are represented by the yellow, blue, and red arrows inside the block and change in each cases. The *fast*-mode is light blue and the *slow*-mode is magenta. The boxes in the right column show the locations of the two exiting modes at the exit surface; the gray axes mark the center of the exit surface in-line with the incident ray. One mode generally has a larger shift than the other as the crystal axes rotate, but neither of them stays fixed in a biaxial material.

where a_{tf} and a_{ts} are the complex electric field amplitude coefficients, and n_f and n_s are the refractive indices encountered by each mode.

In the following discussion, the transmitted and reflected modes are identified as *ta*, *tb*, *rc*, and *rd* for all types of materials. The four combinations of isotropic and birefringent interfaces along with the corresponding ray splitting configurations are depicted in Figure 19.17.[15]

In an isotropic material, the refractive index remains constant and is independent of the polarization of the ray. The **k** and **S** vectors of the *s*- and *p*-polarizations are parallel and one incident ray produces one reflected and one refracted ray. In this case, the reflected and refracted constant phase wavefronts are perpendicular to the energy propagation direction. In birefringent materials,

19.5 Reflections and Refractions at Birefringent Interface

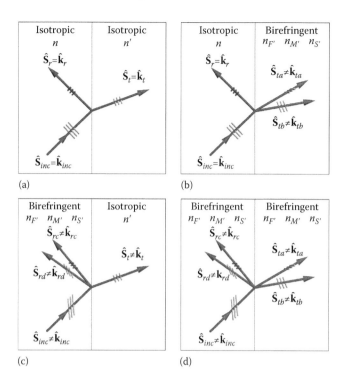

Figure 19.17 The four configurations of birefringent interfaces are shown with corresponding reflected rays toward the incident medium and refracted rays into the transmitting medium for a given incident ray. The black arrow represents **S**, the Poynting vector direction, which is the direction of energy flow and is not necessarily the same as **k**, the propagation vector direction. The transmitted and reflected modes are identified by subscripts *ta*, *tb*, *rc*, and *rd*. The three gray parallel lines along each ray represent the wavefronts (direction of **D**), which are perpendicular to **k**, but not necessarily perpendicular to **S**. (a) A ray propagating from an isotropic medium to another isotropic medium results in one reflected ray and one refracted ray. (b–d) If the incident medium or/and the transmitted medium is/are birefringent, ray splitting occurs and the two resultant rays in each birefringent material have different **k** and **S** directions.

the wavefronts are not generally perpendicular to the direction of energy propagation; **k** and **S** are not aligned in the same direction. At birefringent intercepts, one incident ray may result in up to four exiting rays, two reflected and two refracted.

The ray tracing algorithm used with isotropic materials derived from Snell's law must be generalized to compute refraction and reflection at birefringent interfaces. Also, the corresponding Fresnel amplitude transmission and reflection coefficients are more complex for birefringent intercept, since the refractive index varies with **E**. The necessary set of parameters for ray tracing comprises six 3×1 vectors: (**k**, **S**, **E**, **D**, **B**, and **H**) and the associated refractive index of the ray. These are calculated by solving Maxwell's equations with appropriate boundary values.[15] The normalized refracted or reflected vector **k** at a birefringent material is

$$\hat{\mathbf{k}} = \frac{n\,\hat{\mathbf{k}}_{inc} + \left(-n\,\hat{\mathbf{k}}_{inc} \cdot \hat{\boldsymbol{\eta}} \pm \sqrt{n^2(\hat{\mathbf{k}}_{inc} \cdot \hat{\boldsymbol{\eta}})^2 + (n'^2 - n^2)}\right)\hat{\boldsymbol{\eta}}}{\left\|n\,\hat{\mathbf{k}}_{inc} + \left(-n\,\hat{\mathbf{k}}_{inc} \cdot \hat{\boldsymbol{\eta}} \pm \sqrt{n^2(\hat{\mathbf{k}}_{inc} \cdot \hat{\boldsymbol{\eta}})^2 + (n'^2 - n^2)}\right)\hat{\boldsymbol{\eta}}\right\|}, \quad (19.16)$$

where n is the refractive index for the incident ray, n' is the refractive index of the exiting ray, and the sign of the square root is + for refraction and − for reflection. Equation 19.16 is the general form of the refraction and reflection equations in Chapter 10. The solution is complicated by the fact that $\hat{\mathbf{k}}$ is a function of n', which is not specified at the beginning of the calculation. Hence, the reflection and refraction algorithm must simultaneously solve for $\hat{\mathbf{k}}$ and n'. By combining the constitutive relations in Equation 19.7 with Maxwell's equations, the eigenvalue equations for the **E** fields are formed:

$$\left[\boldsymbol{\varepsilon}' + (n'\mathbf{K}_t + i\mathbf{G}')^2\right]\mathbf{E}_t = 0 \text{ and } \left[\boldsymbol{\varepsilon} + (n\mathbf{K}_r + i\mathbf{G})^2\right]\mathbf{E}_r = 0 \tag{19.17}$$

for refraction and reflection, respectively, where

$$\mathbf{K} = \begin{pmatrix} 0 & -k_z & k_y \\ k_z & 0 & -k_x \\ -k_y & k_x & 0 \end{pmatrix} \tag{19.18}$$

and $\hat{\mathbf{k}} = (k_x, k_y, k_z)$. For a non-zero exiting $\mathbf{E}_t$ and $\mathbf{E}_r$, the determinant of Equation 19.19 has to be zero,

$$\left|\boldsymbol{\varepsilon}' + (n'\,\mathbf{K}_t + i\,\mathbf{G}')^2\right| = 0 \text{ and } \left|\boldsymbol{\varepsilon} + (n'\,\mathbf{K}_r + i\,\mathbf{G})^2\right| = 0. \tag{19.19}$$

Equations 19.16 and 19.19 are solved simultaneously for n' and $\mathbf{k}$. The exiting **E** fields are calculated by Equation 19.17 through singular value decomposition. The exiting **H** and **S** fields are calculated with Equations 19.20 and 19.21.

$$\mathbf{H}_t = (n_t\mathbf{K}_t + i\mathbf{G}')\mathbf{E}_t \text{ and } \mathbf{H}_r = (n_r\mathbf{K}_r + i\mathbf{G})\mathbf{E}_r \tag{19.20}$$

$$\mathbf{S}_t = \frac{\mathcal{R}e\left[\mathbf{E}_t \times \mathbf{H}_t^*\right]}{\left|\mathcal{R}e\left[\mathbf{E}_t \times \mathbf{H}_t^*\right]\right|} \text{ and } \mathbf{S}_r = \frac{\mathcal{R}e\left[\mathbf{E}_r \times \mathbf{H}_r^*\right]}{\left|\mathcal{R}e\left[\mathbf{E}_r \times \mathbf{H}_r^*\right]\right|}. \tag{19.21}$$

Then, **E**, **D**, **B**, **H**, and **S** fields are calculated for all exiting modes. Since highly transparent materials are preferred in optical systems, the absorption is assumed negligibly small. Extension of these methods to absorbing and dichroic materials is included in Refs. [18–21].

The equation for *optical path length* (OPL) in birefringent materials is generalized from its definition for isotropic materials in Equation 19.22. OPL describes the phase accumulated along a ray path between optical element interfaces and is calculated separately for each mode. The physical ray path ℓ is along **S**, which is the direction of energy flow, while the phase increases along **k**. Therefore, the *OPL* for the ray segment is

$$OPL = n\,\ell\,\hat{\mathbf{k}}\cdot\hat{\mathbf{S}} \tag{19.22}$$

as shown in Figure 19.18.

The fraction of the incident energy that couples into each of the four modes is described by the amplitude transmission coefficients (a_{ta} and a_{tb}) and the amplitude reflection coefficients (a_{rc} and a_{rd}). These coefficients at an isotropic-to-isotropic intercept are the conventional Fresnel coefficients for the *s*- and *p*-polarizations. By matching the boundary conditions at the interface for **E**

19.5 Reflections and Refractions at Birefringent Interface

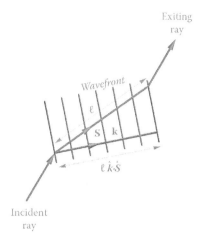

Figure 19.18 The calculation of the *OPL* for a ray propagates through a birefringent material where **S** (orange) and **k** (gray) are not aligned. The wavefronts (gray parallel lines) are perpendicular to the **k** vector. The energy propagation direction is along **S**, which determines the location of ray intercept at the next surface. The *OPL* is the number of wavelengths between the two intercepts, multiplied by the projection of the ray path ℓ onto **k**, the refractive index of the mode, and the wavelength.[15]

and **H** fields, all four exiting amplitude coefficients at intercept are calculated using Equations 19.23 through 19.26,[2,15,18,19]

$$\mathbf{A} = \mathbf{F}^{-1} \cdot \mathbf{C}, \tag{19.23}$$

where

$$\mathbf{A} = (a_{ta}, a_{tb}, a_{rc}, a_{rd})^T, \tag{19.24}$$

$$\mathbf{F} = \begin{pmatrix} \mathbf{s}_1 \cdot \hat{\mathbf{E}}_{ta} & \mathbf{s}_1 \cdot \hat{\mathbf{E}}_{tb} & -\mathbf{s}_1 \cdot \hat{\mathbf{E}}_{rc} & -\mathbf{s}_1 \cdot \hat{\mathbf{E}}_{rd} \\ \mathbf{s}_2 \cdot \hat{\mathbf{E}}_{ta} & \mathbf{s}_2 \cdot \hat{\mathbf{E}}_{tb} & -\mathbf{s}_2 \cdot \hat{\mathbf{E}}_{rc} & -\mathbf{s}_2 \cdot \hat{\mathbf{E}}_{rd} \\ \mathbf{s}_1 \cdot \vec{\mathbf{H}}_{ta} & \mathbf{s}_1 \cdot \vec{\mathbf{H}}_{tb} & -\mathbf{s}_1 \cdot \vec{\mathbf{H}}_{rc} & -\mathbf{s}_1 \cdot \vec{\mathbf{H}}_{rd} \\ \mathbf{s}_2 \cdot \vec{\mathbf{H}}_{ta} & \mathbf{s}_2 \cdot \vec{\mathbf{H}}_{tb} & -\mathbf{s}_2 \cdot \vec{\mathbf{H}}_{rc} & -\mathbf{s}_2 \cdot \vec{\mathbf{H}}_{rd} \end{pmatrix}, \tag{19.25}$$

$$\mathbf{C} = \left(\mathbf{s}_1 \cdot \hat{\mathbf{E}}_{inc}, \mathbf{s}_2 \cdot \hat{\mathbf{E}}_{inc}, \mathbf{s}_1 \cdot \vec{\mathbf{H}}_{inc}, \mathbf{s}_2 \cdot \vec{\mathbf{H}}_{inc} \right)^T, \tag{19.26}$$

$\mathbf{s}_1 = \hat{\mathbf{k}}_{inc} \times \hat{\boldsymbol{\eta}}$, and $\mathbf{s}_2 = \hat{\boldsymbol{\eta}} \times \mathbf{s}_1$. $\mathbf{s}_1 \cdot \mathbf{V}$ and $\mathbf{s}_2 \cdot \mathbf{V}$ operate on vector **V** to extract the tangential and normal components of **V**. Therefore, the transmitted electric field is the superposition of the two transmitted modes:

$$\mathbf{E}_t(\mathbf{r},t) = \mathcal{R}e \left[E_{inc} \left(a_{ta} \hat{\mathbf{E}}_{ta} e^{i\frac{2\pi}{\lambda} n_{ta} \hat{\mathbf{k}}_{ta} \cdot \mathbf{r}} + a_{tb} \hat{\mathbf{E}}_{tb} e^{i\frac{2\pi}{\lambda} n_{tb} \hat{\mathbf{k}}_{tb} \cdot \mathbf{r}} \right) e^{-i\omega t} \right], \tag{19.27}$$

and the reflected electric field is

$$\mathbf{E}_r(\mathbf{r},t) = \mathcal{R}e\left[E_{inc}\left(a_{rc}\hat{\mathbf{E}}_{rc}e^{i\frac{2\pi}{\lambda}n_{rc}\hat{\mathbf{k}}_{rc}\cdot\mathbf{r}} + a_{rd}\hat{\mathbf{E}}_{rd}e^{i\frac{2\pi}{\lambda}n_{rd}\hat{\mathbf{k}}_{rd}\cdot\mathbf{r}}\right)e^{-i\omega t}\right]. \quad (19.28)$$

The algorithms above explain the calculation for uncoated birefringent interfaces. The calculations of the amplitude coefficients for layered birefringent slabs and birefringent coatings are discussed by Mansuripur,[22] Abdulhalim,[23,24] and others.[25,26]

The *light intensity* I of a given $\mathbf{E}$ is calculated by multiplying $|\mathbf{E}|^2$ with the cross section scaling factor, $n_2\cos\theta_{s2}/n_1\cos\theta_{s1}$, at a ray intercept as

$$I = \frac{n_2\cos\theta_{s2}}{n_1\cos\theta_{s1}}|\mathbf{E}|^2, \quad (19.29)$$

where θ_s is the angle of the Poynting vector, subscript 1 for parameters before the interface, and subscript 2 for parameters after the interface.

For rays refracting into a uniaxial medium, the *a*- and *b*-modes are the *o*- and *e*-modes in Equation 19.27. For rays refracting into an optically active medium, the resultant two modes are *left*- and *right*-modes. It is important to note that Equations 19.27 and 19.28 must be repeated for the orthogonally polarized incident modes to yield the four refracting or four reflecting modes, as shown in Section 19.7. For biaxial-to-biaxial interface, there are *fast–fast*-, *fast–slow*-, *slow–fast*-, and *slow–slow*-modes; for uniaxial-to-uniaxial interface, there are *oo*-, *oe*-, *eo*-, and *ee*-modes. Further, at birefringent-to-birefringent interfaces, *s* is no longer refracting into *s*, or *p* into *p*, since *s* and *p* are not eigenpolarizations. At an uncoated isotropic interface, refraction (or reflection) can be represented in *s–p* coordinates as a diagonal Jones matrix. At a birefringent interface, the Jones matrix is not diagonal; *s* couples into *p*, and *p* also couples into *s*.

Example 19.2 Calculate Ray Parameters at an Air/KTP Ray Intercept

This example shows the step-by-step procedure to propagate a ray through the birefringent interface of Figure 19.19. Given the incident ray parameters, the exiting ray parameters are calculated. A light ray propagates from an isotropic medium (air, $n = 1$) into a biaxial material (KTP), $(n_F, n_M, n_S) = (1.78559, 1.79718, 1.90206)$ with crystal axes along $\begin{pmatrix}1\\0\\0\end{pmatrix}, \begin{pmatrix}0\\1\\0\end{pmatrix}, \begin{pmatrix}0\\0\\1\end{pmatrix}$.

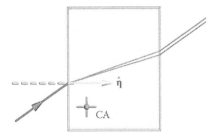

Figure 19.19 A ray propagates into a KTP block of crystal and splits into two rays. Note the normal points away from the incident interface.

19.5 Reflections and Refractions at Birefringent Interface

In air, the light direction is $\mathbf{k}_{inc} = \begin{pmatrix} 0 \\ \sin\theta \\ \cos\theta \end{pmatrix} = \begin{pmatrix} 0 \\ \sin 35° \\ \cos 35° \end{pmatrix}$ and $\mathbf{S}_{inc} = \begin{pmatrix} 0 \\ \sin\theta \\ \cos\theta \end{pmatrix} = \begin{pmatrix} 0 \\ \sin 35° \\ \cos 35° \end{pmatrix}$.

The interface has $\boldsymbol{\eta} = \begin{pmatrix} 0 \\ 0 \\ 1 \end{pmatrix}$, $\boldsymbol{\varepsilon} = \begin{pmatrix} 1 & 0 & 0 \\ 0 & 1 & 0 \\ 0 & 0 & 1 \end{pmatrix}$, $\boldsymbol{\varepsilon}' = \begin{pmatrix} n_F^2 & 0 & 0 \\ 0 & n_M^2 & 0 \\ 0 & 0 & n_S^2 \end{pmatrix}$, and $\mathbf{G} = \mathbf{G}' = \begin{pmatrix} 0 & 0 & 0 \\ 0 & 0 & 0 \\ 0 & 0 & 0 \end{pmatrix}$.

Note the normal points away from the incident medium.

With Equations 19.16 and 19.18, the resultant propagation parameters are

$$\hat{\mathbf{k}} = \frac{1}{n'}\begin{pmatrix} 0 \\ \sin\theta \\ \pm\sqrt{n'^2 - \sin^2\theta} \end{pmatrix} \text{ and } \mathbf{K} = \frac{1}{n'}\begin{pmatrix} 0 & \mp\sqrt{n'^2 - \sin^2\theta} & \sin\theta \\ \pm\sqrt{n'^2 - \sin^2\theta} & 0 & 0 \\ -\sin\theta & 0 & 0 \end{pmatrix}$$

for transmission and reflection.
In transmission, based on Equation 19.17,

$$\begin{pmatrix} n_F^2 - n'^2 & 0 & 0 \\ 0 & \sin^2\theta + n_M^2 - n'^2 & \sqrt{n'^2 - \sin^2\theta}\sin\theta \\ 0 & \sqrt{n'^2 - \sin^2\theta}\sin\theta & n_S^2 - \sin^2\theta \end{pmatrix}\mathbf{E}_t = 0.$$

For non-zero $\mathbf{E}_t$, the determinant of the matrix is zero:

$$\frac{1}{2}(n_F^2 - n'^2)\left[n_M^2 + n_S^2(-1 - 2n_M^2 + 2n'^2) + (n_S^2 - n_M^2)\cos 2\theta\right] = 0.$$

Since $n' > 0$, $n' = n_F$, or $n' = \sqrt{n_M^2 + \left(1 - \frac{n_M^2}{n_S^2}\right)\sin^2\theta}$, $n_f = 1.78559$ and $n_s = 1.80697$.

- The transmitted *fast*-mode satisfies

$$\begin{pmatrix} n_F^2 - n_f^2 & 0 & 0 \\ 0 & \sin^2\theta - n_f^2 + n_M^2 & \sqrt{n_f^2 - \sin^2\theta}\sin\theta \\ 0 & \sqrt{n_f^2 - \sin^2\theta}\sin\theta & n_S^2 - \sin^2\theta \end{pmatrix}\mathbf{E}_{tf} = 0,$$

$$\begin{pmatrix} 0 & 0 & 0 \\ 0 & 0.371 & 0.970 \\ 0 & 0.970 & 3.289 \end{pmatrix} \mathbf{E}_{tf} = 0.$$

By singular value decomposition,

$$\begin{pmatrix} 0 & 0 & 0 \\ 0 & 0.371 & 0.970 \\ 0 & 0.970 & 3.289 \end{pmatrix} = \begin{pmatrix} 0 & 0 & 1 \\ -0.289 & -0.957 & 0 \\ -0.957 & 0.289 & 0 \end{pmatrix} \begin{pmatrix} 3.582 & 0 & 0 \\ 0. & 0.078 & 0 \\ 0 & 0 & 0 \end{pmatrix} \begin{pmatrix} 0 & 0 & 1 \\ -0.289 & -0.957 & 0 \\ -0.957 & 0.289 & 0 \end{pmatrix}^{\dagger}.$$

The exiting state corresponds to the singular value of zero with electric field:

$$\mathbf{E}_{tf} = \begin{pmatrix} 1 \\ 0 \\ 0 \end{pmatrix}.$$

The **H** and **S** fields are

$$\mathbf{H}_{tf} = (n_f \mathbf{K}_{tf} + i\,\mathbf{G}')\hat{\mathbf{E}}_{tf} = \begin{pmatrix} 0 & -1.691 & 0.574 \\ 1.691 & 0 & 0 \\ -0.574 & 0 & 0 \end{pmatrix} \begin{pmatrix} 1 \\ 0 \\ 0 \end{pmatrix} = \begin{pmatrix} 0 \\ 1.691 \\ -0.574 \end{pmatrix} \text{ and}$$

$$\hat{\mathbf{S}}_{tf} = \frac{\mathcal{R}e[\hat{\mathbf{E}}_{tf} \times \hat{\mathbf{H}}_{tf}{}^*]}{|\mathcal{R}e[\hat{\mathbf{E}}_{tf} \times \hat{\mathbf{H}}_{tf}{}^*]|} = \begin{pmatrix} 0 \\ 0.321 \\ 0.947 \end{pmatrix}.$$

- The transmitted *slow*-mode satisfies

$$\begin{pmatrix} n_F^2 - n_s^2 & 0 & 0 \\ 0 & \sin^2\theta + n_M^2 - n_s^2 & \sqrt{n_s^2 - \sin^2\theta}\sin\theta \\ 0 & \sqrt{n_s^2 - \sin^2\theta}\sin\theta & n_S^2 - \sin^2\theta \end{pmatrix} \mathbf{E}_{ts} = 0$$

$$\begin{pmatrix} -0.077 & 0 & 0 \\ 0 & 0.294 & 0.983 \\ 0 & 0.983 & 3.289 \end{pmatrix} \mathbf{E}_{ts} = 0.$$

19.5 Reflections and Refractions at Birefringent Interface

By singular value decomposition,

$$\begin{pmatrix} -0.077 & 0 & 0 \\ 0 & 0.294 & 0.983 \\ 0 & 0.983 & 3.289 \end{pmatrix} = \begin{pmatrix} 0 & -1 & 0 \\ -0.286 & 0 & -0.958 \\ -0.958 & 0 & 0.286 \end{pmatrix} \begin{pmatrix} 3.583 & 0 & 0 \\ 0 & 0.077 & 0 \\ 0 & 0 & 0 \end{pmatrix} \begin{pmatrix} 0 & 1 & 0 \\ -0.286 & 0 & -0.958 \\ -0.958 & 0 & 0.286 \end{pmatrix}^{\dagger}.$$

The exiting state corresponds to the singular value of zero with electric field:

$$\mathbf{E}_{ts} = \begin{pmatrix} 0 \\ -0.958 \\ 0.286 \end{pmatrix}.$$

The **H** and **S** fields are

$$\mathbf{H}_{ts} = (n_s \mathbf{K}_{ts} + i\mathbf{G}')\hat{\mathbf{E}}_{ts} = \begin{pmatrix} 0 & -1.714 & 0.574 \\ 1.714 & 0 & 0 \\ -0.574 & 0 & 0 \end{pmatrix} \begin{pmatrix} 0 \\ -0.958 \\ 0.286 \end{pmatrix} = \begin{pmatrix} 1.806 \\ 0 \\ 0 \end{pmatrix}$$

and

$$\hat{\mathbf{S}}_{ts} = \frac{\mathcal{R}e[\hat{\mathbf{E}}_{ts} \times \hat{\mathbf{H}}_{ts}{}^*]}{|\mathcal{R}e[\hat{\mathbf{E}}_{ts} \times \hat{\mathbf{H}}_{ts}{}^*]|} = \begin{pmatrix} 0 \\ 0.286 \\ 0.958 \end{pmatrix}.$$

- In reflection, $[\boldsymbol{\varepsilon} + (n_r \mathbf{K}_r + i\mathbf{G})^2] \mathbf{E}_r = 0$, where $n_r = 1$:

$$\begin{pmatrix} 1 - n_r^2 & 0 & 0 \\ 0 & 1 - n_r^2 + \sin^2\theta & -\sqrt{n_r^2 - \sin^2\theta}\sin\theta \\ 0 & -\sqrt{n_r^2 - \sin^2\theta}\sin\theta & \cos^2\theta \end{pmatrix} \mathbf{E}_r = 0$$

$$\begin{pmatrix} 0 & 0 & 0 \\ 0 & 0.329 & -0.470 \\ 0 & -0.470 & 0.671 \end{pmatrix} \mathbf{E}_r = 0.$$

By singular value decomposition,

$$\begin{pmatrix} 0 & 0 & 0 \\ 0 & 0.329 & -0.470 \\ 0 & -0.470 & 0.671 \end{pmatrix} = \begin{pmatrix} 0 & 0 & 1 \\ -0.574 & 0.819 & 0 \\ 0.819 & 0.574 & 0 \end{pmatrix} \begin{pmatrix} 1 & 0 & 0 \\ 0 & 0 & 0 \\ 0 & 0 & 0 \end{pmatrix} \begin{pmatrix} 0 & 0 & 1 \\ -0.574 & 0.819 & 0 \\ 0.819 & 0.574 & 0 \end{pmatrix}^\dagger.$$

The exiting states correspond to the singular value of zero, which are $\mathbf{E}_{rs} = \begin{pmatrix} 1 \\ 0 \\ 0 \end{pmatrix}$ and $\mathbf{E}_{rp} = \begin{pmatrix} 0 \\ 0.819 \\ 0.574 \end{pmatrix}$.

Their **H** and **S** fields are

$$\mathbf{H}_{rs} = (n_r \mathbf{K}_r + i\mathbf{G})\hat{\mathbf{E}}_{rs} = \begin{pmatrix} 0 & 0.819 & 0.574 \\ -0.819 & 0 & 0 \\ -0.574 & 0 & 0 \end{pmatrix} \begin{pmatrix} 1 \\ 0 \\ 0 \end{pmatrix} = \begin{pmatrix} 0 \\ -0.819 \\ -0.574 \end{pmatrix},$$

$$\mathbf{H}_{rp} = (n_r \mathbf{K}_r + i\mathbf{G})\hat{\mathbf{E}}_{rp} = \begin{pmatrix} 0 & 0.819 & 0.574 \\ -0.819 & 0 & 0 \\ -0.574 & 0 & 0 \end{pmatrix} \begin{pmatrix} 0 \\ 0.819 \\ 0.574 \end{pmatrix} = \begin{pmatrix} 1 \\ 0 \\ 0 \end{pmatrix},$$

and

$$\hat{\mathbf{S}}_r = \frac{\mathcal{R}e[\hat{\mathbf{E}}_r \times \hat{\mathbf{H}}_r{}^*]}{\left|\mathcal{R}e[\hat{\mathbf{E}}_r \times \hat{\mathbf{H}}_r{}^*]\right|} = \begin{pmatrix} 0 \\ 0.574 \\ -0.819 \end{pmatrix}.$$

Now, we will construct the **F** matrix with $\mathbf{s}_1 = \hat{\mathbf{k}}_{inc} \times \hat{\boldsymbol{\eta}} = \begin{pmatrix} \sin 35° \\ 0 \\ 0 \end{pmatrix}$ and $\mathbf{s}_2 = \hat{\boldsymbol{\eta}} \times \mathbf{s}_1 = \begin{pmatrix} 0 \\ \sin 35° \\ 0 \end{pmatrix}$:

19.5 Reflections and Refractions at Birefringent Interface

$$\mathbf{F} = \begin{pmatrix} \mathbf{s}_1 \cdot \hat{\mathbf{E}}_{tf} & \mathbf{s}_1 \cdot \hat{\mathbf{E}}_{ts} & -\mathbf{s}_1 \cdot \hat{\mathbf{E}}_{rs} & -\mathbf{s}_1 \cdot \hat{\mathbf{E}}_{rp} \\ \mathbf{s}_2 \cdot \hat{\mathbf{E}}_{tf} & \mathbf{s}_2 \cdot \hat{\mathbf{E}}_{ts} & -\mathbf{s}_2 \cdot \hat{\mathbf{E}}_{rs} & -\mathbf{s}_2 \cdot \hat{\mathbf{E}}_{rp} \\ \mathbf{s}_1 \cdot \vec{\mathbf{H}}_{tf} & \mathbf{s}_1 \cdot \vec{\mathbf{H}}_{ts} & -\mathbf{s}_1 \cdot \vec{\mathbf{H}}_{rs} & -\mathbf{s}_1 \cdot \vec{\mathbf{H}}_{rp} \\ \mathbf{s}_2 \cdot \vec{\mathbf{H}}_{tf} & \mathbf{s}_2 \cdot \vec{\mathbf{H}}_{ts} & -\mathbf{s}_2 \cdot \vec{\mathbf{H}}_{rs} & -\mathbf{s}_2 \cdot \vec{\mathbf{H}}_{rp} \end{pmatrix} = \begin{pmatrix} 0.574 & 0 & -0.574 & 0 \\ 0 & -0.550 & 0 & -0.470 \\ 0 & 1.036 & 0 & -0.574 \\ 0.970 & 0 & 0.470 & 0 \end{pmatrix},$$

$$\mathbf{F}^{-1} = \begin{pmatrix} 0.569 & 0 & 0 & 0.695 \\ 0 & -0.715 & 0.586 & 0 \\ -1.174 & 0 & 0. & 0.695 \\ 0 & -1.292 & -0.685 & 0. \end{pmatrix}.$$

With s-polarized incident light,

$$\hat{\mathbf{E}}_{inc,s} = \frac{\hat{\mathbf{k}}_{inc} \times \hat{\mathbf{\eta}}}{\|\hat{\mathbf{k}}_{inc} \times \hat{\mathbf{\eta}}\|} = \begin{pmatrix} 1 \\ 0 \\ 0 \end{pmatrix}, \quad \mathbf{H}_{inc,s} = (n\,\mathbf{K}_{inc} + i\,\mathbf{G})\hat{\mathbf{E}}_{inc,s} = \begin{pmatrix} 0 \\ \cos 35° \\ -\sin 35° \end{pmatrix},$$

$$\mathbf{C}_s = \begin{pmatrix} \mathbf{s}_1 \cdot \hat{\mathbf{E}}_{inc,s} \\ \mathbf{s}_2 \cdot \hat{\mathbf{E}}_{inc,s} \\ \mathbf{s}_1 \cdot \vec{\mathbf{H}}_{inc,s} \\ \mathbf{s}_2 \cdot \vec{\mathbf{H}}_{inc,s} \end{pmatrix} = \begin{pmatrix} \sin 35° \\ 0 \\ 0 \\ \cos 35° \sin 35° \end{pmatrix},$$

and

$$\mathbf{A}_s = \mathbf{F}^{-1} \cdot \mathbf{C}_s = \begin{pmatrix} 0.569 & 0 & 0 & 0.695 \\ 0 & -0.715 & 0.586 & 0 \\ -1.174 & 0 & 0. & 0.695 \\ 0 & -1.292 & -0.685 & 0. \end{pmatrix} \begin{pmatrix} \sin 35° \\ 0 \\ 0 \\ \cos 35° \sin 35° \end{pmatrix} = \begin{pmatrix} 0.653 \\ 0 \\ -0.347 \\ 0 \end{pmatrix} = \begin{pmatrix} a_{s \to tf} \\ a_{s \to ts} \\ a_{s \to rs} \\ a_{s \to rp} \end{pmatrix}.$$

With *p*-polarized incident light,

$$\hat{\mathbf{E}}_{inc,p} = \frac{\hat{\mathbf{k}}_{inc} \times \hat{\mathbf{E}}_{inc,s}}{\|\hat{\mathbf{k}}_{inc} \times \hat{\mathbf{E}}_{inc,s}\|} = \begin{pmatrix} 0 \\ \cos 35° \\ -\sin 35° \end{pmatrix}, \quad \mathbf{H}_{inc,p} = (n\,\mathbf{K}_{inc} + i\,\mathbf{G})\hat{\mathbf{E}}_{inc,p} = \begin{pmatrix} -1 \\ 0 \\ 0 \end{pmatrix},$$

$$\mathbf{C}_p = \begin{pmatrix} \mathbf{s}_1 \cdot \hat{\mathbf{E}}_{inc,p} \\ \mathbf{s}_2 \cdot \hat{\mathbf{E}}_{inc,p} \\ \mathbf{s}_1 \cdot \vec{\mathbf{H}}_{inc,p} \\ \mathbf{s}_2 \cdot \vec{\mathbf{H}}_{inc,p} \end{pmatrix} = \begin{pmatrix} 0 \\ 0.470 \\ -0.574 \\ 0 \end{pmatrix},$$

and

$$\mathbf{A}_p = \mathbf{F}^{-1} \cdot \mathbf{C}_p = \begin{pmatrix} 0.569 & 0 & 0 & 0.695 \\ 0 & -0.715 & 0.586 & 0 \\ -1.174 & 0 & 0 & 0.695 \\ 0 & -1.292 & -0.685 & 0 \end{pmatrix} \begin{pmatrix} 0 \\ 0.470 \\ -0.574 \\ 0 \end{pmatrix} = \begin{pmatrix} 0 \\ -0.672 \\ 0 \\ -0.214 \end{pmatrix} = \begin{pmatrix} a_{p \to tf} \\ a_{p \to ts} \\ a_{p \to rs} \\ a_{p \to rp} \end{pmatrix}.$$

In summary, the amplitude coefficients are $\begin{pmatrix} a_{s \to tf} \\ a_{s \to ts} \\ a_{s \to rs} \\ a_{s \to rp} \end{pmatrix} = \begin{pmatrix} 0.653 \\ 0 \\ -0.347 \\ 0 \end{pmatrix}$ and $\begin{pmatrix} a_{p \to tf} \\ a_{p \to ts} \\ a_{p \to rs} \\ a_{p \to rp} \end{pmatrix} = \begin{pmatrix} 0 \\ -0.672 \\ 0 \\ -0.214 \end{pmatrix}.$

When Equation 19.16 yields a ray with complex **k**, the corresponding mode is an evanescent wave. The mode is *total internally reflected* for a complex $\mathbf{k}_t$ and all the energy is reflected with zero transmission.[27] In a birefringent material, *inhibited reflection* occurs when the reflected ray has a complex $\mathbf{k}_r$; then, all the energy transmits with zero reflection.[28]

Conical refraction is a complex phenomenon in biaxial materials where light refracts into a continuous cone of light, not just into two modes in two directions. It only happens when light propagates along one of the two optic axes[2,29,30] in a biaxial material and the two orthogonal modes experience the same refractive index.[31] Biaxial materials have two optic axes, as shown in Figure 19.20 (left); the optic axes do not correspond to any of the principal axes. When conical refraction occurs, the solution to Maxwell's equations becomes degenerate with a family of **k** and **E** pairs. The associated refracted energy forms a hollow cone of light as shown in Figure 19.20 (right). The distribution of the energy depends on the incident polarization; the corresponding polarization state rotates around the cone through 180° and forms a ring on the next interface.[32,33]

Figure 19.21 shows measurements of conical refraction in a KTP crystal. The distribution of the energy depends on the incident polarization; the corresponding polarization state rotates around the cone through 180° and forms a ring on the next interface. Thus, because of conical refraction, special care is required when ray tracing near the optic axes. For example, the cone of light can be modeled as a cone of a large number of rays.

Under conoscope (described in Section 7.7.1.5), the two optic axes of the biaxial material are easily visible as shown in Figures 19.22 and 19.23.

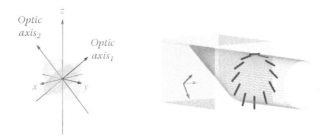

Figure 19.20 (Left) Orientations of the two optic axes of a biaxial material are perpendicular to the two circular cross sections through the index ellipsoid. (Right) A ray refracts into the direction of the biaxial optic axis. The incident polarizations distribute their Poynting vectors into a cone (the solid angle of the cone shown is exaggerated) and propagate as a cone through the biaxial crystal. The associated polarization (shown as the purple lines) rotates by π around the refracted cone. The distribution of flux around the cone depends on the distribution of incident polarization states. The *fast*, *medium*, and *slow* crystal axes are shown as the red, green, and blue arrows.

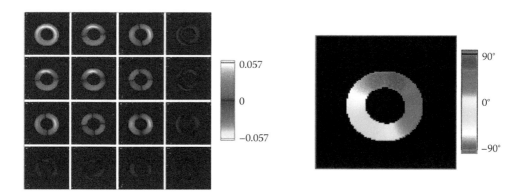

Figure 19.21 A biaxial KTP crystal is set up to display conical refraction and its refracted cone is measured in an imaging polarimeter. The Mueller matrix image (left) and diattenuation orientation image (right) of the refracted cone are shown.

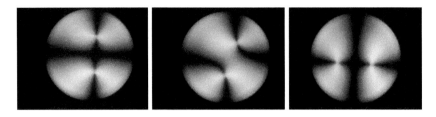

Figure 19.22 An aragonite rotates under conoscope.

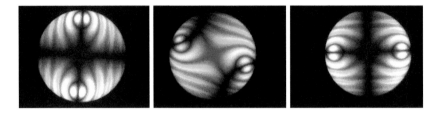

Figure 19.23 A piece of muscovite rotates under conoscope.

19.6 Data Structure for Ray Doubling

A data structure in the form of a tree is developed to keep track of the multiplicity of rays generated by birefringent elements when ray tracing optical systems. When a ray propagates through several birefringent interfaces, the number of resultant rays or modes typically doubles at each birefringent interface. In general, a polarization ray tracing program should not impose any assumptions and should automate the handling of ray doubling in order to properly simulate multiple wavefronts exiting the optical system. The optical designer may request a particular subset of partial rays to be traced, but otherwise, the program should generate and keep track of the entire tree of rays generated. Often, reflected rays have such a small effect that only refracted rays are traced.

Treating each split ray separately without assumptions allows us to observe polarization aberration easily by tracing a grid of rays with different positions or incident angles. In systems with only isotropic components, one incident wavefront produces one exiting wavefront. When birefringent components are involved, however, multiple wavefronts are produced as seen in Figure 19.3. The resultant wavefronts can be assembled after the ray trace by collecting and sorting rays using their sequence of eigenmodes, such as shown in the detailed analysis of a Glan-type polarizer in Chapter 22 (Crystal Polarizers). One ray passing through two Glan-type polarizers results in as many as 16 exiting rays, due to the sequence of four birefringent interfaces. Manually setting up and keeping track of 16 ray paths is tedious, so the birefringent ray tracing program should automate this.

Consider a ray propagating from one uniaxial material into another uniaxial material whose optic axes are not aligned, as shown in Figure 19.24. The eigenmodes (o_1 and e_1) in the first uniaxial material have different polarization orientations compared to the second material's eigenmodes (o_2 and e_2). The o_1 ray from the first uniaxial material splits into o_2- and e_2-modes in the second uniaxial material; these resultant modes will be labeled the *oo*- and *oe*-modes. This mode labeling provides a method of useful abbreviation for tracking the history of each partial ray.

Figure 19.25 shows a ray incident on a uniaxial plate and an air space followed by a uniaxial block with an angled back surface. The two optic axes are configured, such that the normal incident ray produces four resultant rays with *eie*-, *eio*-, *oie*-, and *oio*-modes incident on the angled back surface. At each interface entering a uniaxial medium, considering only refraction, a maximum of two rays

19.7 Polarization Ray Tracing Matrices for Birefringent Interfaces

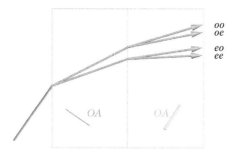

Figure 19.24 A ray splits into four modes after refracting through two uniaxial blocks with unaligned optic axes.

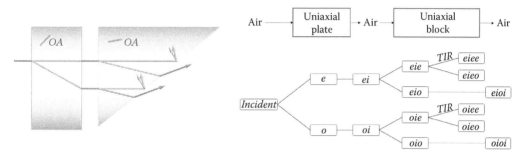

Figure 19.25 (Left) A normal incident ray propagates through two uniaxial components with different optic axis (purple lines) as shown. (Right) A ray tree describes the resultant modes from each interface interaction through the system.

are produced due to the ray doubling inside the uniaxial material. A ray trace program traces all the ray branches in the ray tree shown in Figure 19.25, from left to right and top to bottom, to ensure that all *daughter rays* are accounted for. It starts at the first interface, calculates the *e*-mode, exits the first plate to air, calculates the *ei*-mode, entering the uniaxial block, and calculates the *eie*-mode. The *eie*-mode enters the last surface at an incidence angle that is greater than the critical angle for the *e*-mode. Hence, the *eiei*-mode is evanescent and the refraction branch stops with its energy being totally internally reflected back to the second uniaxial block as *eiee*- and *eieo*-modes. After the first exiting ray stopped, the program moves to the next branch where the *eio*-mode enters the last surface less than the critical angle for the *o*-mode. The *eio*-mode refracts out of the second plate and stops as the *eioi*-mode, since there are no more possible ray intercepts for this ray. The program repeats the process, moving through all the branches until all the possible ray splitting paths are calculated and all the rays terminate. The user of the ray trace program can set up *ray stop criteria*, such as (1) all rays stop at a certain surface, (2) all rays stop below a certain energy level, or (3) all rays stop after a certain number of surface interactions. In the system of Figure 19.25, one incident ray couples to two refracted rays through the first uniaxial plate, and four refracted rays after the first surface of the second uniaxial block. The orientation of the back surface selects the two *o*-modes out of four modes to refract out into air. The other two modes, *eie* and *oie*, are lost due to the evanescence. Section 19.9 contains a complex example of a ray splitting calculation and data structure, where the number of daughter rays is not simply 2^N due to evanescence in reflection as well as refraction.

19.7 Polarization Ray Tracing Matrices for Birefringent Interfaces

This section develops the 3 × 3 polarization ray tracing matrix **P** for each resultant mode. The **P** matrix keeps track of the resultant electric field direction, amplitude coefficients, and the mode direction in the global coordinates that are preferred for polarization ray tracing. As discussed in earlier chapters, Jones matrices with associated local coordinate systems can be used for ray tracing birefringent materials as well. However, our experience has convinced us that keeping the calculation in global coordinates has many advantages, and the **P** matrix described in Chapter 9 (Polarization Ray Tracing Calculus) is the ideal tool to accurately keep track of the polarization changes throughout complex systems.

The **P** matrix describes polarization interactions by generalizing Jones matrices into a 3 × 3 matrix formalism to handle arbitrary propagation directions in 3D coordinates; this avoids carrying along local coordinates, which is necessary with Jones matrices. A formalism that calculates the **P** matrix directly from the 3D orthonormal basis is presented in this section. This formalism is the basis for calculating the **P** matrix through birefringent components using only the ray parameters (**S**, **E**, *a*, and *OPL*) calculated from the birefringent ray tracing algorithms shown in Section 19.5. The resultant **P** matrices are used to study the polarization aberrations of birefringent elements, such as the angle dependence of crystal retarders and polarizers. For a pair of incident and exiting modes through a birefringent intercept, the **P** matrix relates the electric field amplitude and phase across the interface as well as the change of propagation direction. For example, part of the energy of an *e*-mode refracts to a *fast*-mode at a uniaxial/biaxial interface, or the *p*-polarized component couples to right circularly polarized light through an isotropic/optically active interface.

The electromagnetic fields of a pair of incident and exiting modes propagating through an isotropic interface and a birefringent interface are depicted in Figure 19.26. $\hat{\mathbf{E}}_s \perp \hat{\mathbf{E}}_p \perp \hat{\mathbf{S}} = \hat{\mathbf{k}}$ within the isotropic medium, and $\hat{\mathbf{E}}_m \perp \hat{\mathbf{E}}_n \perp \hat{\mathbf{S}} \neq \hat{\mathbf{k}}$ within the birefringent medium, where $\hat{\mathbf{E}}_m$ and $\hat{\mathbf{E}}_n$ are orthogonal electric fields in the transverse plane of $\hat{\mathbf{S}}$.

In a birefringent ray trace, two **P** matrices are needed to represent the two refracted or reflected modes at a birefringent ray intercept because the two exiting rays take different paths. Since the incident and exiting eigenmodes are generally not aligned, one of the incident eigenmodes can couple light into both exiting eigenmodes. The **P** matrix maps the three incident orthonormal basis vectors $(\hat{\mathbf{E}}_m, \hat{\mathbf{E}}_n, \hat{\mathbf{S}})$ into three exiting vectors $(\mathbf{E}'_m, \mathbf{E}'_n, \hat{\mathbf{S}}')$ associated with one eigenstate in the exiting medium along $\hat{\mathbf{S}}'$. Thus, the conditions defining the two exiting **P** matrices at a birefringent ray intercept are

$$\begin{cases} \mathbf{P}_v \, \hat{\mathbf{E}}_m = \mathbf{E}'_{mv} = a_{mv}\hat{\mathbf{E}}_v, \\ \mathbf{P}_v \, \hat{\mathbf{E}}_n = \mathbf{E}'_{nv} = a_{nv}\hat{\mathbf{E}}_v, \\ \mathbf{P}_v \, \hat{\mathbf{S}} = \hat{\mathbf{S}}'_v, \end{cases} \text{ and } \begin{cases} \mathbf{P}_w \, \hat{\mathbf{E}}_m = \mathbf{E}'_{mw} = a_{mw}\hat{\mathbf{E}}_w, \\ \mathbf{P}_w \, \hat{\mathbf{E}}_n = \mathbf{E}'_{nw} = a_{nw}\hat{\mathbf{E}}_w, \\ \mathbf{P}_w \, \hat{\mathbf{S}} = \hat{\mathbf{S}}'_w, \end{cases} \quad (19.30)$$

for refraction/reflection, where $(a_{mv}, a_{mw}, a_{nv}, a_{nw})$ are the complex amplitude coefficients associated with the **E** field coupling between each pair of incident and exiting states. These coefficients are calculated in Section 19.5.

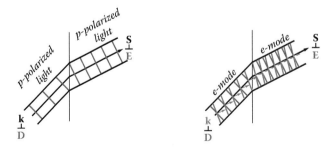

Figure 19.26 The incident and refracted **E**, **D**, **k**, and **S** field orientations of (left) an isotropic-to-isotropic and (right) a birefringent-to-birefringent interface. Only *p*-polarization coupling to *p*-polarization is shown refracting through the isotropic interface. Only *e*-mode coupling to *e*-mode is shown refracting through the birefringent interface. In isotropic media, the **k** and **S** are in the same direction, and the **E** and **D** are in the same direction. These fields are no longer aligned in anisotropic materials, but **E** stays orthogonal to **S**, and **k** stays orthogonal to **D**.

19.7 Polarization Ray Tracing Matrices for Birefringent Interfaces

The exiting v- and w-modes of Equation 19.30 are the transmitted ta- and tb-modes or the reflected rc- and rd-modes in Figure 19.17. In general, the incident $\hat{\mathbf{E}}_m$ couples to both $\mathbf{E}'_{mv}$ and $\mathbf{E}'_{mw}$, and $\hat{\mathbf{E}}_n$ couples to both $\mathbf{E}'_{nv}$ and $\mathbf{E}'_{nw}$. These exiting $\mathbf{E}$ fields $\mathbf{E}'_{mv}$ and $\mathbf{E}'_{nv}$ are in the transverse plane of $\hat{\mathbf{S}}'_v$; $\mathbf{E}'_{mw}$ and $\mathbf{E}'_{nw}$ are in the transverse plane of $\hat{\mathbf{S}}'_w$ in the exiting medium. The refracted/reflected exiting electric fields resulting from $\hat{\mathbf{E}}_m$ and $\hat{\mathbf{E}}_n$ at the ray intercept are

$$\mathbf{E}'_m(\mathbf{r},t) = \mathcal{R}e\left\{E_{inc}\left[a_{mv}\hat{\mathbf{E}}_v e^{i\frac{2\pi}{\lambda}n_{mv}\hat{\mathbf{k}}_{mv}\cdot\mathbf{r}} + a_{mw}\hat{\mathbf{E}}_w e^{i\frac{2\pi}{\lambda}n_{mw}\hat{\mathbf{k}}_{mw}\cdot\mathbf{r}}\right]e^{-i\omega t}\right\} \quad (19.31)$$

and

$$\mathbf{E}'_n(\mathbf{r},t) = \mathcal{R}e\left\{E_{inc}\left[a_{nv}\hat{\mathbf{E}}_v e^{i\frac{2\pi}{\lambda}n_{nv}\hat{\mathbf{k}}_{nv}\cdot\mathbf{r}} + a_{nw}\hat{\mathbf{E}}_w e^{i\frac{2\pi}{\lambda}n_{nw}\hat{\mathbf{k}}_{nw}\cdot\mathbf{r}}\right]e^{-i\omega t}\right\}. \quad (19.32)$$

Consider the o- and e-modes with two different $\hat{\mathbf{S}}'$ emerging from an isotropic/uniaxial intercept: they each have their own $\mathbf{P}$ matrix because $\hat{\mathbf{S}}'_o \neq \hat{\mathbf{S}}'_e$. In this case, (m, n, v, w) in Equation 19.30 are (s, p, o, e); the detailed calculations of $\mathbf{P}_o$ and $\mathbf{P}_e$ are described in Section 19.7.2. In general, $\hat{\mathbf{E}}_m$ can couple to both $\hat{\mathbf{E}}_v$ and $\hat{\mathbf{E}}_w$, and $(a_{mv}, a_{mw}, a_{nv}, a_{nw})$ are all non-zero. In some situations, the amplitude coefficients are set to zero for certain properties of the interface. For example, at an uncoated isotropic ray intercept, the coupling between s- and p-polarizations is zero; $(m, n, v, w) = (s, p, s', p')$ and $(a_{mw}, a_{nv}) = (a_{sp'}, a_{ps'}) = (0, 0)$. Also, if a ray is incident at a uniaxial/isotropic intercept polarized in only one eigenmode for a given incident $\hat{\mathbf{S}}$, then $(m, n, v, w) = (e, e_\perp, s, p)$ and $(a_{nv}, a_{nw}) = (a_{e_\perp s}, a_{e_\perp p}) = (0, 0)$, because $e_\perp$-mode carries zero energy to begin with. This will be further explained in Section 19.7.3. Since each exiting mode is in a particular eigenpolarization, the corresponding $\mathbf{P}$ matrix of the exiting mode is a singular matrix; that is, it has the form of a polarizer.

By placing the three pairs of incident and exiting 3 × 1 vectors in matrix form, the $\mathbf{P}$ matrix is calculated as

$$\begin{aligned}
\mathbf{P} &= \begin{pmatrix} \mathbf{E}'_m & \mathbf{E}'_n & \hat{\mathbf{S}}' \end{pmatrix} \cdot \begin{pmatrix} \hat{\mathbf{E}}_m & \hat{\mathbf{E}}_n & \hat{\mathbf{S}} \end{pmatrix}^{-1} \\
&= \begin{pmatrix} E'_{m,x} & E'_{n,x} & S'_x \\ E'_{m,y} & E'_{n,y} & S'_y \\ E'_{m,z} & E'_{n,z} & S'_z \end{pmatrix} \cdot \begin{pmatrix} E_{m,x} & E_{n,x} & S_x \\ E_{m,y} & E_{n,y} & S_y \\ E_{m,z} & E_{n,z} & S_z \end{pmatrix}^{-1} \\
&= \begin{pmatrix} E'_{m,x} & E'_{n,x} & S'_x \\ E'_{m,y} & E'_{n,y} & S'_y \\ E'_{m,z} & E'_{n,z} & S'_z \end{pmatrix} \cdot \begin{pmatrix} E_{m,x} & E_{n,x} & S_x \\ E_{m,y} & E_{n,y} & S_y \\ E_{m,z} & E_{n,z} & S_z \end{pmatrix}^T,
\end{aligned} \quad (19.33)$$

where $\begin{pmatrix} \hat{\mathbf{E}}_m & \hat{\mathbf{E}}_n & \hat{\mathbf{S}} \end{pmatrix}$ is a real unitary matrix; hence, its inverse equals its transpose. The amplitude coefficients for the interface are contained in $\begin{pmatrix} \mathbf{E}'_m & \mathbf{E}'_n & \hat{\mathbf{S}}' \end{pmatrix}$.

The multiple exiting modes from birefringent interfaces need to be described by multiple $\mathbf{P}$ matrices. The derivations of the $\mathbf{P}$ matrices for the four cases of uncoated birefringent interfaces—(1) isotropic/isotropic, (2) isotropic/birefringent, (3) birefringent/birefringent, and (4) birefringent/birefringent interfaces in refraction and reflection—are shown in the following subsections. In the derivations, the (m, n, v, w) states used in Equation 19.30 for each of the cases are shown in Table 19.4. In an exiting isotropic medium, the exiting s'- and p'-modes share the same $\mathbf{S}$; thus, the

Table 19.4 The Polarization States for Calculating the **P** Matrices for an Uncoated Interface

Interface	Reflected $(m, n) \rightarrow v=rc$ and $(m, n) \rightarrow w=rd$		Refracted $(m, n) \rightarrow v=ta$ and $(m, n) \rightarrow w=tb$	
Isotropic/ isotropic	$(s, p) \rightarrow s'$ $(s, p) \rightarrow p'$	$\Rightarrow (s, p) \rightarrow (s', p')$	$(s, p) \rightarrow s'$ $(s, p) \rightarrow p'$	$\Rightarrow (s, p) \rightarrow (s', p')$
Isotropic/ birefringent	$(s, p) \rightarrow s'$ $(s, p) \rightarrow p'$	$\Rightarrow (s, p) \rightarrow (s', p')$	$(s, p) \rightarrow v$ $(s, p) \rightarrow w$	
Birefringent/ isotropic	$(m, n) \rightarrow v$ $(m, n) \rightarrow w$		$(m, n) \rightarrow s'$ $(m, n) \rightarrow p'$	$\Rightarrow (m, n) \rightarrow (s', p')$
Birefringent/ birefringent	$(m, n) \rightarrow v$ $(m, n) \rightarrow w$		$(m, n) \rightarrow v$ $(m, n) \rightarrow w$	

Note: (m, n, v, w) are defined in Equation 19.30. $'$ indicates an exiting mode. The split eigenmodes in a birefringent material are described by two **P** matrices. However, the two **P** matrices for s' and p' exiting states can be combined ($\Rightarrow$) to one **P** matrix.

associated **P** matrices can be reduced to one **P** matrix (see Sections 19.7.1 through 19.7.3). In an incident birefringent medium, the incident (m, n) modes are $(o, o_\perp)$, $(e, e_\perp)$, $(fast, fast_\perp)$, $(slow, slow_\perp)$, $(right, right_\perp)$, or $(left, left_\perp)$ modes. In an exiting birefringent medium, the exiting (v, w) are (o, e), $(fast, slow)$, or $(right, left)$ modes.

19.7.1 Case I: Isotropic-to-Isotropic Intercept

For the isotropic-to-isotropic interface, the incident eigenstates are s- and p-polarizations,

$$\hat{\mathbf{E}}_{inc,s} = \frac{\hat{\mathbf{S}}_{inc} \times \hat{\boldsymbol{\eta}}}{|\hat{\mathbf{S}}_{inc} \times \hat{\boldsymbol{\eta}}|} \quad \text{and} \quad \hat{\mathbf{E}}_{inc,p} = \hat{\mathbf{S}}_{inc} \times \hat{\mathbf{E}}_{inc,s}, \tag{19.34}$$

as shown in Figure 19.27, where s and p are m and n in Equation 19.33. In this case, the four exiting modes are ts, tp, rs, and rp associated with four **P** matrices, $\mathbf{P}_{ts}$, $\mathbf{P}_{tp}$, $\mathbf{P}_{rs}$, and $\mathbf{P}_{rp}$.

Equation 19.34 is for the isotropic interface, where $\hat{\mathbf{k}} = \hat{\mathbf{S}}$. No ray splitting occurs; the two reflected rays are degenerate, as are the two transmitted rays. Although the figure shows the amplitude coefficients for a reflected and transmitted p-polarized ray for the incident s-polarized ray to provide the most general description of the **P** matrix calculation, the coupling between s and p is zero for an uncoated isotropic interface.

Their electric fields $\hat{\mathbf{E}}_{ts}$, $\hat{\mathbf{E}}_{tp}$, $\hat{\mathbf{E}}_{rs}$, and $\hat{\mathbf{E}}_{rp}$, and propagation vectors are calculated by the method of Section 19.5. The amplitude coefficients a_s and a_p are calculated by Equation 19.23 as

$$\begin{pmatrix} a_{inc,s \rightarrow ts} \\ a_{inc,s \rightarrow tp} \\ a_{inc,s \rightarrow rs} \\ a_{inc,s \rightarrow rp} \end{pmatrix} = \mathbf{F}^{-1} \cdot \begin{pmatrix} \vec{\mathbf{s}}_1 \cdot \hat{\mathbf{E}}_{inc,s} \\ \vec{\mathbf{s}}_2 \cdot \hat{\mathbf{E}}_{inc,s} \\ \vec{\mathbf{s}}_1 \cdot \vec{\mathbf{H}}_{inc,s} \\ \vec{\mathbf{s}}_2 \cdot \vec{\mathbf{H}}_{inc,s} \end{pmatrix} \quad \text{and} \quad \begin{pmatrix} a_{inc,p \rightarrow ts} \\ a_{inc,p \rightarrow tp} \\ a_{inc,p \rightarrow rs} \\ a_{inc,p \rightarrow rp} \end{pmatrix} = \mathbf{F}^{-1} \cdot \begin{pmatrix} \vec{\mathbf{s}}_1 \cdot \hat{\mathbf{E}}_{inc,p} \\ \vec{\mathbf{s}}_2 \cdot \hat{\mathbf{E}}_{inc,p} \\ \vec{\mathbf{s}}_1 \cdot \vec{\mathbf{H}}_{inc,p} \\ \vec{\mathbf{s}}_2 \cdot \vec{\mathbf{H}}_{inc,p} \end{pmatrix}, \tag{19.35}$$

19.7 Polarization Ray Tracing Matrices for Birefringent Interfaces

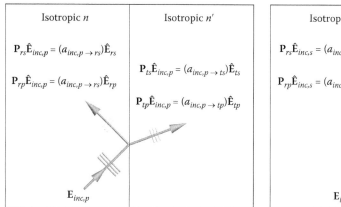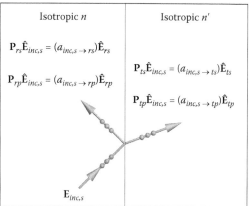

Figure 19.27 A *p*-polarized (left) and an *s*-polarized (right) incident ray are launched into the interface to calculate **E**, **S**, *a*, and the **P** matrix of all exiting modes. The planes of incidence are shown in both cases. The arrows indicate the **S** directions. (Left) The triple parallel lines (blue and green) shown in the plane of incidence are the direction of $\mathbf{E}_p$. (Right) The triple parallel lines (red and magenta) shown orthogonal to the plane of incidence are the direction of $\mathbf{E}_s$.

where the electric field amplitude from incident *s*-polarization coupled into the transmitted *s*-polarization is denoted as $a_{inc,s \to ts}$. For uncoated dielectric interfaces, this yields the Fresnel amplitude coefficients. With these amplitude coefficients and applying Equation 19.30, the **P** matrices for the exiting *s*- and *p*-modes should satisfy

$$\begin{cases} \mathbf{P}_{rs}\,\hat{\mathbf{E}}_{inc,s} = a_{inc,s\to rs}\hat{\mathbf{E}}_{rs} \\ \mathbf{P}_{rs}\,\hat{\mathbf{E}}_{inc,p} = a_{inc,p\to rs}\hat{\mathbf{E}}_{rs}, \\ \mathbf{P}_{rs}\,\hat{\mathbf{S}}_{inc} = \hat{\mathbf{S}}_{rs} \end{cases} \begin{cases} \mathbf{P}_{rp}\,\hat{\mathbf{E}}_{inc,s} = a_{inc,s\to rp}\hat{\mathbf{E}}_{rp} \\ \mathbf{P}_{rp}\,\hat{\mathbf{E}}_{inc,p} = a_{inc,p\to rp}\hat{\mathbf{E}}_{rp}, \\ \mathbf{P}_{rp}\,\hat{\mathbf{S}}_{inc} = \hat{\mathbf{S}}_{rp} \end{cases}$$
$$\begin{cases} \mathbf{P}_{ts}\,\hat{\mathbf{E}}_{inc,s} = a_{inc,s\to ts}\hat{\mathbf{E}}_{ts} \\ \mathbf{P}_{ts}\,\hat{\mathbf{E}}_{inc,p} = a_{inc,p\to ts}\hat{\mathbf{E}}_{ts}, \\ \mathbf{P}_{ts}\,\hat{\mathbf{S}}_{inc} = \hat{\mathbf{S}}_{ts} \end{cases} \begin{cases} \mathbf{P}_{tp}\,\hat{\mathbf{E}}_{inc,s} = a_{inc,s\to tp}\hat{\mathbf{E}}_{tp} \\ \mathbf{P}_{tp}\,\hat{\mathbf{E}}_{inc,p} = a_{inc,p\to tp}\hat{\mathbf{E}}_{tp}. \\ \mathbf{P}_{tp}\,\hat{\mathbf{S}}_{inc} = \hat{\mathbf{S}}_{tp} \end{cases} \quad (19.36)$$

Using Equation 19.33, the four **P** matrices are calculated as

$$\begin{aligned} \mathbf{P}_{ts} &= \left(a_{inc,s\to ts}\hat{\mathbf{E}}_{ts} \quad a_{inc,p\to ts}\hat{\mathbf{E}}_{ts} \quad \hat{\mathbf{S}}_{ts} \right) \cdot \left(\hat{\mathbf{E}}_{inc,s} \quad \hat{\mathbf{E}}_{inc,p} \quad \hat{\mathbf{S}}_{inc} \right)^T, \\ \mathbf{P}_{tp} &= \left(a_{inc,s\to tp}\hat{\mathbf{E}}_{tp} \quad a_{inc,p\to tp}\hat{\mathbf{E}}_{tp} \quad \hat{\mathbf{S}}_{tp} \right) \cdot \left(\hat{\mathbf{E}}_{inc,s} \quad \hat{\mathbf{E}}_{inc,p} \quad \hat{\mathbf{S}}_{inc} \right)^T, \\ \mathbf{P}_{rs} &= \left(a_{inc,s\to rs}\hat{\mathbf{E}}_{rs} \quad a_{inc,p\to rs}\hat{\mathbf{E}}_{rs} \quad \hat{\mathbf{S}}_{rs} \right) \cdot \left(\hat{\mathbf{E}}_{inc,s} \quad \hat{\mathbf{E}}_{inc,p} \quad \hat{\mathbf{S}}_{inc} \right)^T, \text{ and} \\ \mathbf{P}_{rp} &= \left(a_{inc,s\to rp}\hat{\mathbf{E}}_{rp} \quad a_{inc,p\to rp}\hat{\mathbf{E}}_{rp} \quad \hat{\mathbf{S}}_{rp} \right) \cdot \left(\hat{\mathbf{E}}_{inc,s} \quad \hat{\mathbf{E}}_{inc,p} \quad \hat{\mathbf{S}}_{inc} \right)^T. \end{aligned} \quad (19.37)$$

As depicted in Figure 19.17a, inside isotropic materials, $\hat{\mathbf{S}}_{ts} = \hat{\mathbf{S}}_{tp}$ and $\hat{\mathbf{S}}_{rs} = \hat{\mathbf{S}}_{rp}$. The result can be obtained by combining the two exiting modes. The electric fields can always be added, but care is needed in adding **P** matrices.* Since the condition of $\hat{\mathbf{S}}_t = \mathbf{P}_t \hat{\mathbf{S}}_{inc}$ is present for both modes, $\mathbf{P}_{ts}$ and $\mathbf{P}_{tp}$ can be combined as follows:

$$\mathbf{P}_t = \left(a_{inc,s \to ts}\hat{\mathbf{E}}_{ts} + a_{inc,s \to tp}\hat{\mathbf{E}}_{tp} \quad a_{inc,p \to ts}\hat{\mathbf{E}}_{ts} + a_{inc,p \to tp}\hat{\mathbf{E}}_{tp} \quad \hat{\mathbf{S}}_t \right) \cdot \left(\hat{\mathbf{E}}_{inc,s} \quad \hat{\mathbf{E}}_{inc,p} \quad \hat{\mathbf{S}}_{inc} \right)^T. \tag{19.38}$$

Similarly, $\mathbf{P}_{rs}$ and $\mathbf{P}_{rp}$ can be combined as

$$\mathbf{P}_r = \left(a_{inc,s \to rs}\hat{\mathbf{E}}_{rs} + a_{inc,s \to rp}\hat{\mathbf{E}}_{rp} \quad a_{inc,p \to rs}\hat{\mathbf{E}}_{rs} + a_{inc,p \to rp}\hat{\mathbf{E}}_{rp} \quad \hat{\mathbf{S}}_r \right) \cdot \left(\hat{\mathbf{E}}_{inc,s} \quad \hat{\mathbf{E}}_{inc,p} \quad \hat{\mathbf{S}}_{inc} \right)^T. \tag{19.39}$$

Equation 19.37 shows the most general derivation for **P** matrices, which aids in understanding the procedures for non-isotropic interfaces introduced in Section 19.7.2 through 19.7.4. For an isotropic interface, $a_{inc,s \to tp}$, $a_{inc,s \to rp}$, $a_{inc,p \to ts}$, and $a_{inc,p \to rs}$ are zero since the s-component only couples to s-polarization and the p-component only couples to p-polarization. Therefore, Equations 19.38 and 19.39 become

$$\mathbf{P}_t = \left(a_{inc,s \to ts}\hat{\mathbf{E}}_{ts} \quad a_{inc,p \to tp}\hat{\mathbf{E}}_{tp} \quad \hat{\mathbf{S}}_t \right) \cdot \left(\hat{\mathbf{E}}_{inc,s} \quad \hat{\mathbf{E}}_{inc,p} \quad \hat{\mathbf{S}}_{inc} \right)^T \text{ and}$$

$$\mathbf{P}_r = \left(a_{inc,s \to rs}\hat{\mathbf{E}}_{rs} \quad a_{inc,p \to rp}\hat{\mathbf{E}}_{rp} \quad \hat{\mathbf{S}}_r \right) \cdot \left(\hat{\mathbf{E}}_{inc,s} \quad \hat{\mathbf{E}}_{inc,p} \quad \hat{\mathbf{S}}_{inc} \right)^T. \tag{19.40}$$

19.7.2 Case II: Isotropic-to-Birefringent Interface

The isotropic-to-birefringent interface case uses the same s- and p-incident basis as in the isotropic interface case in Equation 19.34. As shown in Figure 19.17b, the four exiting modes are rs, rp, ta, and tb; the two reflected modes reflect back to the incident isotropic medium and two refracted modes refract into the birefringent medium. The two reflected s- and p-modes share the same $\hat{\mathbf{S}}'$, while the two refracted birefringent modes, labeled with subscripts v and w in Equation 19.30, split into two directions. Hence, $(m, n, v, w) = (s, p, s', p')$ in reflection and (s, p, v, w) for refraction. If the refracting medium is biaxial, the two refracted modes are *fast*- and *slow*-modes; $(m, n, v, w) = (s, p,$ *fast, slow*$)$. The refraction of this case is depicted in Figure 19.28.

The two refracted modes propagate in two different directions with $\hat{\mathbf{k}}_{ta} \ne \hat{\mathbf{k}}_{tb} \ne \hat{\mathbf{S}}_{ta} \ne \hat{\mathbf{S}}_{tb}$. From Equation 19.35, the amplitude coefficients of the exiting modes from each incident mode are

$$\begin{pmatrix} a_{inc,s \to ta} \\ a_{inc,s \to tb} \\ a_{inc,s \to rs} \\ a_{inc,s \to rp} \end{pmatrix} = \mathbf{F}^{-1} \cdot \begin{pmatrix} \vec{\mathbf{s}}_1 \cdot \hat{\mathbf{E}}_{inc,s} \\ \vec{\mathbf{s}}_2 \cdot \hat{\mathbf{E}}_{inc,s} \\ \vec{\mathbf{s}}_1 \cdot \vec{\mathbf{H}}_{inc,s} \\ \vec{\mathbf{s}}_2 \cdot \vec{\mathbf{H}}_{inc,s} \end{pmatrix} \text{ and } \begin{pmatrix} a_{inc,p \to ta} \\ a_{inc,p \to tb} \\ a_{inc,p \to rs} \\ a_{inc,p \to rp} \end{pmatrix} = \mathbf{F}^{-1} \cdot \begin{pmatrix} \vec{\mathbf{s}}_1 \cdot \hat{\mathbf{E}}_{inc,p} \\ \vec{\mathbf{s}}_2 \cdot \hat{\mathbf{E}}_{inc,p} \\ \vec{\mathbf{s}}_1 \cdot \vec{\mathbf{H}}_{inc,p} \\ \vec{\mathbf{s}}_2 \cdot \vec{\mathbf{H}}_{inc,p} \end{pmatrix}. \tag{19.41}$$

---

* A combination of the **P** matrix is explained in Section 9.6 (The Addition Form of Polarization Ray Tracing Matrices), where $\hat{\mathbf{S}}' = \mathbf{P}\hat{\mathbf{S}}_{inc}$ should not be double counted.

19.7 Polarization Ray Tracing Matrices for Birefringent Interfaces

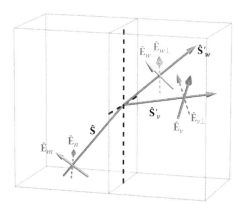

Figure 19.28 Mode coupling in refraction through an isotropic-to-birefringent interface. The incident ray with orthogonal modes, labeled as n (red) and m (blue), splits into two exiting modes as $v = ta$ (pink) and $w = tb$ (green) in two directions. In biaxial and uniaxial materials, $\mathbf{E}_v$ and $\mathbf{E}_w$ are linearly polarized. In optically active materials, $\mathbf{E}_v$ and $\mathbf{E}_w$ are circularly polarized. In birefringent and optically active materials like quartz, $\mathbf{E}_v$ and $\mathbf{E}_w$ are elliptically polarized. Given a ray with a specific pair of $\mathbf{k}'_v$ and $\mathbf{S}'_v$ in a birefringent material, the ray can only be polarized in $\mathbf{E}_v$; thus, the orthogonal state $\mathbf{E}_{v\perp}$ (dashed arrow) has zero amplitude.

Using Equations 19.16 and 19.19, $\hat{\mathbf{E}}_{ta}$, $\hat{\mathbf{E}}_{tb}$, $\hat{\mathbf{E}}_{rs}$, and $\hat{\mathbf{E}}_{rp}$ can be calculated. The two $\mathbf{P}$ matrices for refraction are

$$\mathbf{P}_{ta} = \left(a_{inc,s \to ta} \hat{\mathbf{E}}_{ta} \quad a_{inc,p \to ta} \hat{\mathbf{E}}_{ta} \quad \hat{\mathbf{S}}_{ta} \right) \cdot \left(\hat{\mathbf{E}}_{inc,s} \quad \hat{\mathbf{E}}_{inc,p} \quad \hat{\mathbf{S}}_{inc} \right)^T \text{ and}$$

$$\mathbf{P}_{tb} = \left(a_{inc,s \to tb} \hat{\mathbf{E}}_{tb} \quad a_{inc,p \to tb} \hat{\mathbf{E}}_{tb} \quad \hat{\mathbf{S}}_{tb} \right) \cdot \left(\hat{\mathbf{E}}_{inc,s} \quad \hat{\mathbf{E}}_{inc,p} \quad \hat{\mathbf{S}}_{inc} \right)^T, \qquad (19.42)$$

and the two $\mathbf{P}$ matrices for reflection are

$$\mathbf{P}_{rs} = \left(a_{inc,s \to rs} \hat{\mathbf{E}}_{rs} \quad a_{inc,p \to rs} \hat{\mathbf{E}}_{rs} \quad \hat{\mathbf{S}}_{rs} \right) \cdot \left(\hat{\mathbf{E}}_{inc,s} \quad \hat{\mathbf{E}}_{inc,p} \quad \hat{\mathbf{S}}_{inc} \right)^T \text{ and}$$

$$\mathbf{P}_{rp} = \left(a_{inc,s \to rp} \hat{\mathbf{E}}_{rp} \quad a_{inc,p \to rp} \hat{\mathbf{E}}_{rp} \quad \hat{\mathbf{S}}_{rp} \right) \cdot \left(\hat{\mathbf{E}}_{inc,s} \quad \hat{\mathbf{E}}_{inc,p} \quad \hat{\mathbf{S}}_{inc} \right)^T. \qquad (19.43)$$

The two reflected modes share the same pair of $\hat{\mathbf{S}}_{rs} = \hat{\mathbf{S}}_{rp}$, and the couplings between s- and p-states are zero for an uncoated surface ($a_{inc,s \to rp} = a_{inc,p \to rs} = 0$). Therefore, $\mathbf{P}_{rs}$ and $\mathbf{P}_{rp}$ are combined to

$$\mathbf{P}_r = \left(a_{inc,s \to rs} \hat{\mathbf{E}}_{rs} \quad a_{inc,p \to rp} \hat{\mathbf{E}}_{rp} \quad \hat{\mathbf{S}}_r \right) \cdot \left(\hat{\mathbf{E}}_{inc,s} \quad \hat{\mathbf{E}}_{inc,p} \quad \hat{\mathbf{S}}_{inc} \right)^T, \qquad (19.44)$$

the same as in Equation 19.40. When light refracts or reflects into an isotropic medium, the two modes combine to one $\mathbf{P}$ matrix because they have the same $\mathbf{S}$ direction. However, when light propagates into a birefringent medium, the incident ray splits into two directions and the exiting modes have two different $\mathbf{S}$. In this case, two $\mathbf{P}$ matrices are needed to describe the two modes, which cannot be combined.

19.7.3 Case III: Birefringent-to-Isotropic Interface

A ray in a birefringent incident medium with a specified $\hat{\mathbf{k}}_{inc}$ and $\hat{\mathbf{S}}_{inc}$ on an interface is constrained to be one of the two eigenmodes that is calculated from the previous ray intercept by Equations

19.17 through 19.19. The eigenmode is *o*- or *e*-modes for uniaxial materials, *fast*- or *slow*-modes for biaxial materials, and *right*- or *left*-modes for optically active materials. The electric field and magnetic field for this incident ray are $\hat{\mathbf{E}}_m$ and $\hat{\mathbf{H}}_m$ with index n_m. To construct the $\mathbf{P}$ matrix, a *pseudo electric field* or *absent mode* $\hat{\mathbf{E}}_n = \hat{\mathbf{E}}_{m\perp}$ conveying no power and orthogonal to $\hat{\mathbf{E}}_m$ is calculated as

$$\hat{\mathbf{E}}_n = \frac{\hat{\mathbf{S}}_{inc} \times \hat{\mathbf{E}}_m}{\left|\hat{\mathbf{S}}_{inc} \times \hat{\mathbf{E}}_m\right|}. \tag{19.45}$$

This absent mode has no power because the orthogonal polarization has refracted into another direction and is described by the other $\mathbf{P}$ matrix. This state needs to be defined, however, to properly set up our 3 × 3 mode matrices.

The exiting modes for $\hat{\mathbf{E}}_m$ are the *s*- and *p*-polarized transmitted states and the two bi-reflected rays with $\hat{\mathbf{E}}_v$ and $\hat{\mathbf{E}}_w$. The refraction is depicted in Figure 19.29.

The four exiting electric fields in transmission and reflection are $\hat{\mathbf{E}}_{ts}$, $\hat{\mathbf{E}}_{tp}$, $\hat{\mathbf{E}}_{rc}$, and $\hat{\mathbf{E}}_{rd}$. Inside the incident birefringent medium, only $\hat{\mathbf{E}}_m$ carries non-zero amplitude,

$$\begin{pmatrix} a_{inc,m \to ts} \\ a_{inc,m \to tp} \\ a_{inc,m \to rc} \\ a_{inc,m \to rd} \end{pmatrix} = \mathbf{F}^{-1} \cdot \begin{pmatrix} \vec{\mathbf{s}}_1 \cdot \hat{\mathbf{E}}_m \\ \vec{\mathbf{s}}_2 \cdot \hat{\mathbf{E}}_m \\ \vec{\mathbf{s}}_1 \cdot \vec{\mathbf{H}}_m \\ \vec{\mathbf{s}}_2 \cdot \vec{\mathbf{H}}_m \end{pmatrix}, \tag{19.46}$$

and the amplitude coefficients from $\hat{\mathbf{E}}_n$, ($a_{inc,n \to ts}$, $a_{inc,n \to tp}$, $a_{inc,n \to ts}$, $a_{inc,n \to ts}$) are zeros. As depicted in Figure 19.29, the three pairs of conditions for each transmitted $\mathbf{P}$ matrix are

$$\begin{cases} \mathbf{P}_{ts} \hat{\mathbf{E}}_m = a_{inc,ms \to ts} \hat{\mathbf{E}}_{ts}, \\ \mathbf{P}_{ts} \hat{\mathbf{E}}_n = a_{inc,ns \to ts} \hat{\mathbf{E}}_{ts} = \mathbf{0}, \quad \text{and} \quad \\ \mathbf{P}_{ts} \hat{\mathbf{S}}_{inc} = \hat{\mathbf{S}}_{ts}, \end{cases} \begin{cases} \mathbf{P}_{tp} \hat{\mathbf{E}}_m = a_{inc,ms \to tp} \hat{\mathbf{E}}_{tp}, \\ \mathbf{P}_{tp} \hat{\mathbf{E}}_n = a_{inc,ns \to tp} \hat{\mathbf{E}}_{tp} = \mathbf{0}, \\ \mathbf{P}_{tp} \hat{\mathbf{S}}_{inc} = \hat{\mathbf{S}}_{tp}, \end{cases} \tag{19.47}$$

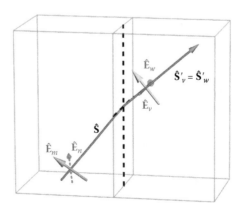

Figure 19.29 Mode coupling in refraction through a birefringent-to-isotropic interface. The incident ray is polarized along $\mathbf{E}_m$ (blue) and has a zero amplitude component $\mathbf{E}_n$ (red, dashed arrow). The incident polarization couples to the *s*- and *p*-states in the isotropic medium, which are $\mathbf{E}_v$ (pink) and $\mathbf{E}_w$ (green) and propagate in the same direction $\mathbf{S}'$.

19.7 Polarization Ray Tracing Matrices for Birefringent Interfaces

where **0** is a 3 × 1 zero vector. Since the transmitted medium is isotropic, $\hat{\mathbf{S}}_{ts} = \hat{\mathbf{S}}_{tp} = \hat{\mathbf{S}}_t$, and the refracted **P** matrix is

$$\mathbf{P}_t = \left(a_{inc,m \to ts} \hat{\mathbf{E}}_{ts} + a_{inc,m \to tp} \hat{\mathbf{E}}_{tp} \quad \mathbf{0} \quad \hat{\mathbf{S}}_t \right) \cdot \left(\hat{\mathbf{E}}_m \quad \hat{\mathbf{E}}_n \quad \hat{\mathbf{S}}_{inc} \right)^T. \tag{19.48}$$

The two reflected **P** matrices for the two rays reflecting back to the birefringent medium are

$$\mathbf{P}_{rc} = \left(a_{inc,m \to rc} \hat{\mathbf{E}}_{rc} \quad \mathbf{0} \quad \hat{\mathbf{S}}_{rc} \right) \cdot \left(\hat{\mathbf{E}}_{inc,m} \quad \hat{\mathbf{E}}_{inc,n} \quad \hat{\mathbf{S}}_{inc} \right)^T \text{ and}$$
$$\mathbf{P}_{rd} = \left(a_{inc,m \to rd} \hat{\mathbf{E}}_{rd} \quad \mathbf{0} \quad \hat{\mathbf{S}}_{rd} \right) \cdot \left(\hat{\mathbf{E}}_{inc,m} \quad \hat{\mathbf{E}}_{inc,n} \quad \hat{\mathbf{S}}_{inc} \right)^T. \tag{19.49}$$

The calculation shown in this section is for one incident mode only. In general, each incident mode has its own associated calculations of **P** matrices for transmission and reflection.

19.7.4 Case IV: Birefringent-to-Birefringent Interface

Similar to Section 19.7.3, the basis for the incident fields is chosen as $\hat{\mathbf{E}}_m$ and $\hat{\mathbf{E}}_n$, where $\hat{\mathbf{E}}_n$ is an absent mode with zero energy constructed orthogonal to $\hat{\mathbf{E}}_m$. The four exiting modes all propagate in different directions. The refraction is depicted in Figure 19.30.

The amplitude coefficients associated with $\hat{\mathbf{E}}_m$ are calculated by Equation 19.45, and the amplitude coefficients associated with $\hat{\mathbf{E}}_n$ are zeros. The **P** matrices for the two transmitted rays are

$$\mathbf{P}_{ta} = \left(a_{inc,m \to ta} \hat{\mathbf{E}}_{ta} \quad \mathbf{0} \quad \hat{\mathbf{S}}_{ta} \right) \cdot \left(\hat{\mathbf{E}}_{inc,m} \quad \hat{\mathbf{E}}_{inc,n} \quad \hat{\mathbf{S}}_{inc} \right)^T \text{ and}$$
$$\mathbf{P}_{tb} = \left(a_{inc,m \to tb} \hat{\mathbf{E}}_{tb} \quad \mathbf{0} \quad \hat{\mathbf{S}}_{tb} \right) \cdot \left(\hat{\mathbf{E}}_{inc,m} \quad \hat{\mathbf{E}}_{inc,n} \quad \hat{\mathbf{S}}_{inc} \right)^T. \tag{19.50}$$

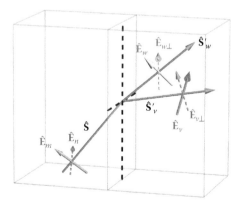

Figure 19.30 Mode coupling in refraction through a birefringent-to-birefringent interface. The incident ray propagating in **S** is polarized along $\mathbf{E}_m$ (blue) and has a zero amplitude component $\mathbf{E}_n$ (red, dashed arrow). It splits into two exiting modes as v (green) and w (pink) in two directions. Their orthogonal states $\mathbf{E}_{v\perp}$ and $\mathbf{E}_{w\perp}$ (dashed arrows) both have zero amplitude.

Similarly, the **P** matrices for the two reflected rays are

$$\mathbf{P}_{rc} = \left(a_{inc,m \to rc} \hat{\mathbf{E}}_{rc} \quad \mathbf{0} \quad \hat{\mathbf{S}}_{rc} \right) \cdot \left(\hat{\mathbf{E}}_{inc,m} \quad \hat{\mathbf{E}}_{inc,n} \quad \hat{\mathbf{S}}_{inc} \right)^T \text{ and}$$

$$\mathbf{P}_{rd} = \left(a_{inc,m \to rd} \hat{\mathbf{E}}_{rd} \quad \mathbf{0} \quad \hat{\mathbf{S}}_{rd} \right) \cdot \left(\hat{\mathbf{E}}_{inc,m} \quad \hat{\mathbf{E}}_{inc,n} \quad \hat{\mathbf{S}}_{inc} \right)^T. \quad (19.51)$$

Example 19.3 Construct the **P** Matrix for a Uniaxial-to-Isotropic Interface

Consider the ray doublings shown in Figure 19.31. The o–i and e–i couplings at the second surface are described by two **P** matrices.

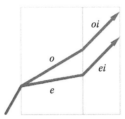

Figure 19.31 An incident ray refracts through two uniaxial interfaces to two rays. A ray starts from an isotropic medium, splits into o- and e-modes, and refracts into oi- and ei-modes.

Following the one ray path from o to oi, the Poynting vector changes from $\hat{\mathbf{S}}_o$ to $\hat{\mathbf{S}}_i$; the $o_\perp$-mode (orthogonal states of o-mode) along $\hat{\mathbf{S}}_o$ has zero energy, so the coupling from $o_\perp$ to $o_\perp$–i is zero. Then, applying Equation 19.30,

$$\begin{cases} \mathbf{P}_{oi} \hat{\mathbf{E}}_o = a_{o \to ts} \hat{\mathbf{E}}_{ts} + t_{o \to tp} \hat{\mathbf{E}}_{tp}, \\ \mathbf{P}_{oi} \hat{\mathbf{E}}_{o\perp} = \mathbf{0}, \text{ and} \\ \mathbf{P}_{oi} \hat{\mathbf{S}}_o = \hat{\mathbf{S}}_i. \end{cases}$$

Then,

$$\mathbf{P}_{oi} = \left(a_{o \to ts} \hat{\mathbf{E}}_{ts} + a_{o \to tp} \hat{\mathbf{E}}_{tp} \quad \mathbf{0} \quad \hat{\mathbf{S}}_i \right) \cdot \left(\hat{\mathbf{E}}_o \quad \hat{\mathbf{E}}_{o\perp} \quad \hat{\mathbf{S}}_o \right)^T.$$

Similarly,

$$\mathbf{P}_{ei} = \left(a_{e \to ts} \hat{\mathbf{E}}_{ts} + a_{e \to tp} \hat{\mathbf{E}}_{tp} \quad \mathbf{0} \quad \hat{\mathbf{S}}_i \right) \cdot \left(\hat{\mathbf{E}}_e \quad \hat{\mathbf{E}}_{e\perp} \quad \hat{\mathbf{S}}_e \right)^T.$$

19.7 Polarization Ray Tracing Matrices for Birefringent Interfaces

Example 19.4 Calculate the **P** Matrix for an Isotropic-to-Biaxial Interface

This example uses the calculation results (**E**, **S**, and *a*'s) from Example 19.2 to construct reflection and transmission **P** matrices. With Equations 19.42 and 19.44,

$$\mathbf{P}_{tf} = \begin{pmatrix} 0.653 & 0 & 0 \\ 0 & 0 & 0.321 \\ 0 & 0 & 0.947 \end{pmatrix} \begin{pmatrix} 1 & 0 & 0 \\ 0 & 0.819 & 0.574 \\ 0 & -0.574 & 0.819 \end{pmatrix}^{-1} = \begin{pmatrix} 0.653 & 0 & 0 \\ 0 & 0.184 & 0.263 \\ 0 & 0.543 & 0.776 \end{pmatrix},$$

$$\mathbf{P}_{ts} = \begin{pmatrix} 0 & 0 & 0 \\ 0 & 0.644 & 0.286 \\ 0 & -0.192 & 0.958 \end{pmatrix} \begin{pmatrix} 1 & 0 & 0 \\ 0 & 0.819 & 0.574 \\ 0 & -0.574 & 0.819 \end{pmatrix}^{-1} = \begin{pmatrix} 0 & 0 & 0 \\ 0 & 0.692 & -0.135 \\ 0 & 0.392 & 0.895 \end{pmatrix}, \text{ and}$$

$$\mathbf{P}_r = \begin{pmatrix} -0.347 & 0 & 0 \\ 0 & -0.175 & 0.574 \\ 0 & -0.123 & -0.819 \end{pmatrix} \begin{pmatrix} 1 & 0 & 0 \\ 0 & 0.819 & 0.574 \\ 0 & -0.574 & 0.819 \end{pmatrix}^{-1} = \begin{pmatrix} -0.347 & 0 & 0 \\ 0 & 0.185 & 0.570 \\ 0 & -0.570 & -0.601 \end{pmatrix}.$$

The singular value decomposition of each exiting **P** matrix gives the incident and exiting **E** fields and **S** vectors. The singular value of 1 corresponds to the **S** vector. The other two singular values represent the magnitude of the amplitude coefficients of its two exiting modes. The **P** matrix for the transmitting *slow*-mode,

$$\mathbf{P}_{ts} = \begin{pmatrix} 0 & 0 & 0 \\ 0 & 0.692 & -0.135 \\ 0 & 0.392 & 0.895 \end{pmatrix} = \begin{pmatrix} 0 & 0 & 1 \\ 0.286 & 0.958 & 0 \\ 0.958 & -0.286 & 0 \end{pmatrix} \begin{pmatrix} 1 & 0 & 0 \\ 0 & 0.672 & 0 \\ 0 & 0 & 0 \end{pmatrix} \begin{pmatrix} 0 & 0 & 1 \\ 0.574 & 0.819 & 0 \\ 0.819 & -0.574 & 0 \end{pmatrix}^{\dagger},$$

shows that the incident $\mathbf{S}_{inc}$ maps to $\mathbf{S}_{ts}$ and $\mathbf{E}_{inc,p}$ maps to $\mathbf{E}_{ts}$ with 0.672 attenuation.

The reflection matrix,

$$\mathbf{P}_r = \begin{pmatrix} -0.347 & 0 & 0 \\ 0 & 0.185 & 0.570 \\ 0 & -0.570 & -0.601 \end{pmatrix} = \begin{pmatrix} 0 & -1 & 0 \\ -0.574 & 0 & 0.819 \\ 0.819 & 0 & 0.574 \end{pmatrix} \begin{pmatrix} 1 & 0 & 0 \\ 0 & 0.347 & 0 \\ 0 & 0 & 0.214 \end{pmatrix} \begin{pmatrix} 0 & 1 & 0 \\ -0.574 & 0 & -0.819 \\ -0.819 & 0 & 0.574 \end{pmatrix}^{\dagger},$$

shows that the incident $\mathbf{S}_{inc}$ maps to $\mathbf{S}_r$, $\mathbf{E}_{inc,s}$ maps to $\mathbf{E}_{rs}$ with −0.347 amplitude coefficient, and $\mathbf{E}_{inc,p}$ maps to $\mathbf{E}_{rp}$ with 0.214 attenuation.

19.8 Example: Ray Splitting through Three Biaxial Crystal Blocks

The ray tracing example of a normal incident ray propagating through a sequence of anisotropic materials in Figure 19.2 will help explain the calculation of the **P** matrix. The principal refractive indices and orientations of the three biaxial plane parallel plates, KTP, aragonite, and mica, are given in Table 19.5 for λ = 500 nm.

The three biaxial blocks produce $2^3 = 8$ exiting modes as shown in the ray tree in Figure 19.32. The directions of ray doubling at each of the three interfaces are different depending on the principal indices and orientations relative to the ray's electric field and propagation direction. The ray locations at the exit surface of each biaxial block are shown in Figure 19.33. The first crystal (KTP) splits the two rays up and down. The second crystal (aragonite) splits the rays diagonally. Looking straight onto the interface, these four rays form a parallelogram. The last crystal (mica) splits the rays up and down with a slight shift and the resulting rays form a double parallelogram.

For the normal incident ray down the z-axis, all the intermediate **k** vectors and the final **k** vectors remains the same as the incident **k**, while the orientation of **S** vectors changes along each ray

Table 19.5 Principal Refractive Indices and Orientation for the KTP, Aragonite, and Mica Example

Biaxial Materials	Principal Refractive Indices (n_F, n_M, n_S) at Wavelength 500 nm	Principal Axes Orientations Unit Vector of n_F Axis, Unit Vector of n_S Axis
KTP	(1.786, 1.797, 1.902)	(0.00, 0.64, 0.77), (0.00, −0.77, 0.64)
Aragonite	(1.530, 1.681, 1.685)	(0.38, 0.64, 0.66), (0.32, −0.77, 0.56)
Mica	(1.563, 1.596, 1.601)	(−0.12, 0.74, 0.66), (0.74, −0.38, 0.56)

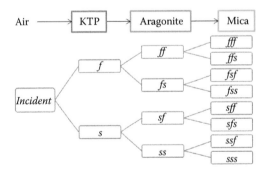

Figure 19.32 The ray tree shows that one incident ray results in eight exiting modes after propagating through three blocks of biaxial materials.

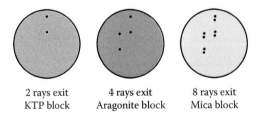

Figure 19.33 The ray locations at the end of each biaxial block.

Table 19.6 Resultant Ray Trace Parameter for *Fast–Slow–Fast*-Mode

Mode	*n* for the Mode	P	Normalized E of the Mode
f	1.797	$\begin{pmatrix} 0.715 & 0 & 0 \\ 0 & 0 & 0 \\ 0 & 0 & 1 \end{pmatrix}$	$\begin{pmatrix} 1 \\ 0 \\ 0 \end{pmatrix}$
fs	1.683	$\begin{pmatrix} 0.750 & 0 & 0.003 \\ -0.461 & 0 & -0.002 \\ -0.003 & 0 & 1 \end{pmatrix}$	$\begin{pmatrix} 0.852 \\ -0.523 \\ -0.003 \end{pmatrix}$
fsf	1.578	$\begin{pmatrix} 0.041 & -0.025 & 0.002 \\ -0.517 & 0.317 & -0.020 \\ -0.009 & 0.005 & 1 \end{pmatrix}$	$\begin{pmatrix} 0.080 \\ -0.997 \\ -0.022 \end{pmatrix}$
fsfi exiting the surface	1	$\begin{pmatrix} 0.008 & -0.097 & -0.002 \\ -0.097 & 1.216 & 0.027 \\ 0.002 & -0.022 & 1 \end{pmatrix}$	$\begin{pmatrix} 0.080 \\ -0.997 \\ 0 \end{pmatrix}$

segment. Each exiting mode has a unique **P** matrix that tracks the polarization and electric field amplitude. The **P** matrix for each ray intercept for the *fast–slow–fast*-mode is shown as an example in Table 19.6.

The cumulative **P** matrix from the entrance to beyond the exit surface is

$$\mathbf{P}_{fsfi,total} = \mathbf{P}_{fsfi}\mathbf{P}_{fsf}\mathbf{P}_{fs}\mathbf{P}_f = \begin{pmatrix} 0.037 & 0 & 0 \\ -0.467 & 0 & 0 \\ 0 & 0 & 1 \end{pmatrix},$$

which is calculated by matrix multiplication. The resultant **E** is along the unit direction (0.080, −0.997, 0) and its amplitude depends on the incident polarization state. For an incident **E** = (1, 0, 0), the exiting **E** is (0.037, −0.467, 0). For an incident **E** of (0, 1, 0), the exiting **E** is (0, 0, 0). Therefore, this mode sequence acts as a linear polarizer with a transmission axis along the *fsf*-mode. In fact, all the exiting modes are linearly polarized and all the associated **P** matrices have the form of linear polarizers.

19.9 Example: Reflections Inside a Biaxial Cube

An example of a non-sequential ray trace in a biaxial crystal involving evanescent waves is considered in this section. Aragonite is a natural form of calcium carbonate $CaCO_3$, different from calcite, which is biaxial with principal indices (1.530, 1.681, 1.685) at 500 nm. Figure 19.34 shows a cube of aragonite with its crystal axes aligned along the edge of the cube. When a laser shines into this cube at a certain range of angles, part of the light will refract into the crystal, then reflect within the crystal, and eventually refract out through the entrance surface. The example ray enters the aragonite block, reflects three times within the crystal, and exits through the front surface. There are four birefringent interfaces along this ray path; thus, a maximum of $2^4 = 16$ modes can potentially exit

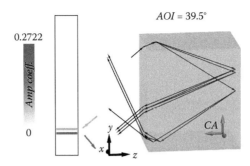

Figure 19.34 Ray paths through a cube of aragonite arising from one incident ray (top left corner of block) at an angle of incidence in the y–z plane of 39.5°. Multiple rays are generated at each internal interface and six modes exit back out the entrance surface. The crystal axis orientations (shown below the CA label) are aligned with the edges of the block; $(\mathbf{v}_F, \mathbf{v}_M, \mathbf{v}_S) = (\mathbf{y}, \mathbf{z}, \mathbf{x})$. Note that not all reflected modes carry significant energy. The white bar shows the location of the exiting rays. The shade of the color red shows the amplitude associated with each exiting mode; four modes have negligible flux. The red arrows to the left of the white bar show the polarization ellipse, all linear, for each exiting mode.

the front surface. For some incident directions, due to total internal reflections and inhibited reflections, the number of modes decreases. Since the number of exiting modes depends on the crystal axis orientations, ray tracing results will be compared with two sets of crystal axis orientations for the same set of incident rays.

When the crystal axes are aligned with the edge of the cube, rays incident in the y–z plane remain propagating within the y–z plane of Figure 19.34, and each mode couples entirely into only one mode at the next surface. In general, during refraction, the energy distributes between *fast*- and *slow*-modes. In this case, all the energy from the *fast*-mode couples to the *fast–fast*-mode and all the energy from the *slow*-mode couples to the *slow–slow*-mode; hence, the *fast–slow*-mode and the *slow–fast*-mode carry no energy. This behavior applies to all of the reflections within the crystal. Eventually, only two exiting rays carry energy, the purely *fast*-mode and the purely *slow*-mode as shown in Figure 19.35. The exiting intensities from these two modes are shown in Figure 19.36.

When the incidence angle is small, the two modes reflect twice inside the cube and most energy is lost due to inhibited reflection at the second reflection since one mode is evanescent at this steep incident angle. As the angle of incidence increases, the top surface inhibited reflection ceases, and

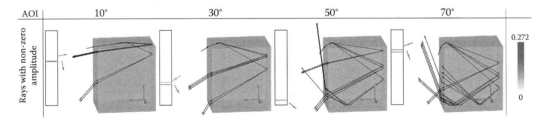

Figure 19.35 Rays are traces through a block of aragonite with crystal axes aligned with the edges of the block for incident angles of 10°, 30°, 50°, and 70°. The figures show only the exiting rays with non-zero amplitude, which are the purely *slow*-mode and the purely *fast*-mode. The amplitude coefficient distribution at the exiting surface is shown in the white bar, where red denotes high amplitude and white denotes zero amplitude. The red arrows on the left of the white bar represent the polarization ellipses of the exiting modes.

19.9 Example: Reflections Inside a Biaxial Cube

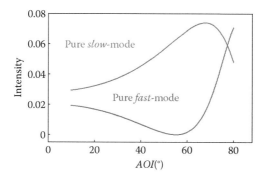

Figure 19.36 Exiting intensity from the pure *fast*-mode and pure *slow*-mode versus angle of incidence in air for unit incident intensity.

total internal reflection occurs after the first reflection. For incident angles greater than 56° for the pure *fast*-mode and 63° for the pure *slow*-mode, there are three reflections instead of two reflections inside the cube. The ray path changes rapidly with incident angle and the crystal axis orientation, so non-sequential ray tracing is used. The ray tree for the 39.5° ray is shown in Figure 19.37, which has inhibited reflection in the *slow*-mode paths and total internal reflection in the end of the *fast*-mode paths.

Next, consider the case with crystal axis orientations rotated from *xyz* (alignment with the cube edges). Now the *slow*- and *fast*-modes couple at each interface, modes that previously had zero amplitude acquire flux, and propagation within the crystal is no longer confined to a plane. As shown in Figure 19.38, many exiting rays have amplitudes near zero. The exiting linear polarization and amplitude distribution also change with the incident angle. By such methods, tolerance analysis can be performed on the crystal axis orientations, fabrication angles, thicknesses, and other parameters of birefringent devices.

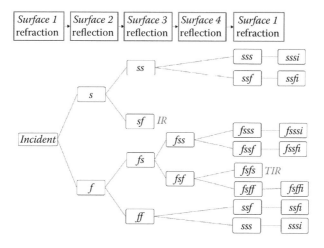

Figure 19.37 The ray tree showing mode splitting for 39.5°. The entrance surface is *surface 1*, the top surface is *surface 2*, the right side surface is *surface 3*, and the bottom surface is *surface 4*. *IR* denotes inhibited refraction; *TIR* denotes total internal reflection.

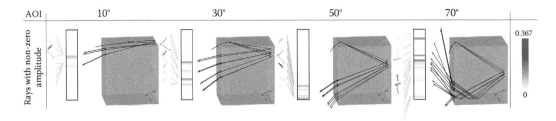

Figure 19.38 A block of aragonite with crystal axes not aligned to the block edges is simulated with various incident angles. The *fast* crystal axis orientation is (−0.36, 0.39, 0.85), and the *slow* crystal axis orientation is (0.15, 0.92, −0.36).

19.10 Conclusion

The definition of the 3D polarization ray tracing matrix has been extended to incorporate birefringent ray tracing. The calculations of the **P** matrix shown in this chapter provide the basis to perform systematic ray tracing through complex sequences of birefringent interfaces. Propagation through systems with birefringent materials generates multiple exiting rays from each incident ray. Tracing a single ray samples a single point on the incident wavefront. This wavefront splits into multiple wavefronts propagating through a birefringent assembly with a potential of 2^N resultant wavefronts after N birefringent interface interactions. With birefringent components, accounting for ray doubling is only the first step for accurate analysis. Further calculations may be required to manipulate the ray tracing results of all the bifurcated modes exiting the system. Each of these **P** matrices represents polarization coupling between a single pair of eigenmodes in the incident and exiting materials. Optical system performance is usually evaluated through recreating the exiting wavefront, after tracing a grid of rays from the incident wavefront. The multiple exiting wavefronts require algorithms such as discussed in Chapter 20 (Beam Combination with Polarization Ray Tracing Matrices) to combine them appropriately in the image space. When the exiting rays are propagating in the same direction, the $\breve{\mathbf{P}}$ (defined in Section 9.6) form of the **P** matrices can be added. When the exiting rays have different **S**, the $\breve{\mathbf{P}}$'s don't add and the resultant **E**'s must be added instead. Although the ray parameter amplitude calculations in Section 19.5 are for uncoated birefringent interfaces, for coated birefringent interfaces, the amplitude coefficients calculated for these coated interfaces[22–26] can be substituted into the **P** matrix calculations in Section 19.7.

Often, assumptions can simplify the analysis of the multiple exiting modes, such as cases with small shear, small ray separation, or parallel exiting rays. Simple systems such as plane parallel waveplates and birefringent crystal polarizers are designed for a small range of incident angles; hence, the shear will often be small. In the case of a quarter waveplate, a normally incident beam produces two orthogonally polarized modes propagating in the same direction with a quarter wave phase delay, and a circularly polarized incident beam yields a linearly polarized exiting beam. With non-normal incident beams, the exiting modes have a slight displacement; the optical path lengths of the two modes may increase or decrease, and the result is an elliptically polarized exiting beam. The larger the incident angle is, the more elliptical the exiting polarization state becomes. Because of the different angles and ray paths, there are two crescent-shaped regions of light around an exiting circular beam area with only one mode present; thus, the majority of the beam may be circularly polarized with thin strips of horizontally polarized light around one side, and vertically polarized light around the other side, as shown in Figure 19.39. Further analysis on waveplates is included in Chapter 21. A detailed analysis of the Glan–Taylor crystal polarizer is included in Chapter 22.

The simple meaning of optical path length and optical path difference becomes complicated when more than two exiting rays emerge from each incident ray. Many modes might propagate close to each other in the same direction; thus, the concept of optical path length from conventional

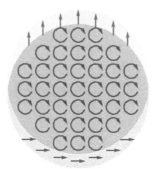

Figure 19.39 The modes exiting a birefringent plate such as a waveplate with off-axis illumination are slightly shifted (shear). Thus, for a quarter waveplate illuminated with 45° light, the majority of the exiting beam is circularly polarized, but two crescent-shaped areas are illuminated by only one mode, in this case horizontally and vertically polarized light.

optical design must be generalized in polarization optical design. To simulate a measurement, all the partial waves need to be added correctly, and the resultant amplitude, phase, and polarization state of the exiting wave need to be calculated at the exit pupil or terminal surface of the optical system. After this beam combination, the phase of the light remains well defined; it is the optical path length that becomes *multivalued*. For example, in a multibeam interferometer with monochromatic light, the phase of the light can always be measured despite a large number of overlapping beams. For birefringent systems illuminated by short laser pulses, the addition of partial waves needs to account for multiple pulses that can arrive in the exit pupil at different times.

In some components such as lenses with stress birefringence or electro-optic devices, the magnitude of the birefringence is small enough that the ray splitting angle is small and safely ignored because the deviation between the two ray paths is negligible. However, the ray's polarization changes may still be significant if the accumulated retardance is large enough. These close ray paths are typically handled as follows in polarization ray tracing. Instead of tracing two rays through the remainder of the system, a retardance matrix is associated with the ray segment, and then the ray segment can be handled as a single ray. Stress birefringence is further discussed in Chapter 25. Another example of close ray paths is liquid crystal cells. The liquid crystal interfaces are parallel, so all the modes exit in the same direction. Because the cell is so thin, 1 to 7 μm, the ray paths do not separate by a significant distance; the shear is small. Depending on the objective of the calculation, a retardance matrix can usually be used to describe each ray segment through the cells, and the light propagation can be handled as a single ray. The simulation of liquid crystal is described in Chapter 24.

19.11 Problem Sets

19.1 Explain the difference between ray tracing the combinations of isotropic, uniaxial, biaxial, and optically active interfaces.

19.2 Calculate the dielectric tensor in global coordinates for a material with principal refractive indices (1.3, 1.4, 1.5) and principal axes oriented at $\begin{pmatrix} 0.66 \\ -0.24 \\ 0.71 \end{pmatrix}$, $\begin{pmatrix} 0.34 \\ 0.94 \\ 0 \end{pmatrix}$, and $\begin{pmatrix} -0.66 \\ 0.24 \\ 0.71 \end{pmatrix}$ in global coordinates.

19.3 The refractive indices for the biaxial material sulfoiodide SbSI are (2.7, 3.2, 3.8). What is the principal dielectric tensor? What is the largest angle obtainable between **D** and **E**? What is the corresponding direction of propagation?

19.4 A sample is measured to have the dielectric tensor $\boldsymbol{\varepsilon} = \begin{pmatrix} 1.906 & -0.076 & 0.199 \\ -0.076 & 1.971 & -0.023 \\ 0.199 & -0.023 & 2.355 \end{pmatrix}$.

Find the unitary transformation that diagonalizes this matrix. How are the crystal axes oriented with respect to x, y, and z? What are n_x, n_y, n_z?

19.5 Given a compound retarder formed from two materials with fast axes 45° apart, fast mode optical path lengths OPL_1 and OPL_2, and retardances δ_1 and δ_2, find the four optical path lengths associated with the four resultant modes. Combine the four modes into a Jones matrix.

19.6 Consider a collimated beam incident on a tilted plane parallel plate of a uniaxial material. Is it possible to separate two polarizations by total internal reflecting one mode at the back face and transmitting the orthogonal mode at Brewster's angle?

19.7 What is the number of potential modes from the aragonite block example in Figure 19.34, including all the modes with zero energy? What is the polarization of the two modes from Figure 19.35 relative to each other? According to Figure 19.36, what incident angle for the aragonite block gives the highest diattenuation?

19.8 Given a 10 mm thick plane parallel anisotropic plate oriented along the z-axis (0,0,1), what is the OPL for a normal incident ray, where the two modes have indices $n_s = 1.85124$ and $n_f = 1.79718$, propagation vectors $\mathbf{k}_s = \mathbf{k}_f = \begin{pmatrix} 0 \\ 0 \\ 1 \end{pmatrix}$, and Poynting vectors $\mathbf{S}_s = \begin{pmatrix} 0 \\ 0 \\ 1 \end{pmatrix}$ and $\mathbf{S}_f = \begin{pmatrix} 0 \\ -0.0627 \\ 0.998 \end{pmatrix}$?

19.9 Consider a biaxial material whose refractive indices are described by its index ellipsoid: $\frac{x^2}{n_x^2} + \frac{y^2}{n_y^2} + \frac{z^2}{n_z^2} = 1$. The only two circular cross sections through this ellipsoid are shown in Figure 19.20, associated with propagation directions $\mathbf{k}_1$ and $\mathbf{k}_2$. These directions have no birefringence, because the electric field in any transverse direction has the same refractive index. Show the angle, θ, of these special directions known as two optic axes of a biaxial material, from the n_z axis given by $\tan(\theta) = \sqrt{\frac{\frac{1}{n_y^2} - \frac{1}{n_x^2}}{\frac{1}{n_z^2} - \frac{1}{n_y^2}}}$, where $n_y < n_x < n_z$.

How will the polarization evolve for propagation along the optic axis?

19.10 Build a ray tree for an off-axis ray refracting into a biaxial plate with two internal reflections. Each ray intercept produces multiple reflected and refracted rays. Figure 19.40 shows the ray splitting at the first internal reflection, which generates six rays.

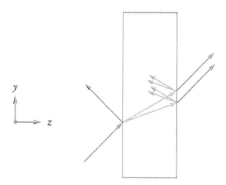

Figure 19.40 An off-axis ray refracts into a biaxial plate.

References

1. M. Born and E. Wolf, *Principles of Optics*, 6th edition, Pergamon (1980).
2. A. Yariv and P. Yeh, *Optical Waves in Crystals*, New York: Wiley (1984).
3. M. Mansuripur, *Field, Force, Energy and Momentum in Classical Electrodynamics*, Bentham e-Books, Bentham Science Publishers (2011).
4. M. Mansuripur, The Ewald–Oseen extinction theorem, in *Classical Optics and Its Applications*, Chapter 16, 2nd edition, Cambridge University Press (2009).
5. W. J. Tropf, M. E. Thomas, and E. W. Rogala, Properties of crystal and glasses, in *Handbook of Optics*, Chapter 2, 3rd edition, Vol. 4, ed. M. Bass, McGraw-Hill (2009).
6. M. C. Pujol et al. Crystalline structure and optical spectroscopy of Er^{3+}-doped $KGd(WO_4)_2$ single crystals, *Appl. Phys. B* 68(2) (1999): 187–197.
7. V. G. Dmitriev, G. G. Gurzadyan, and D. N. Nikogosyan, *Handbook of Nonlinear Optical Crystals*, 2nd edition.
8. V. A. Dyakov et al. Sellmeier equation and tuning characteristics of KTP crystal frequency converters in the 0.4–4.0 μm range, *Sov. J. Quant. Electron.* 18 (1988): 1059.
9. M. Emam-Ismail, Spectral variation of the birefringence, group birefringence and retardance of a gypsum plate measured using the interference of polarized light, *Opt. Laser Technol.* 41(5) (2009): 615–621.
10. E. U. Condon, Theories of optical rotatory power, *Rev. Mod. Phys.* 9 (1937): 432–457.
11. F. I. Fedorov, On the theory of optical activity in crystals, *Opt. Spektrosk.* 6 (1959): 49–53.
12. E. J. Post, *Formal Structure of Electromagnetics*, Amsterdam: North Holland (1962).
13. A. Lakhtakia, V. K. Varadan, and V. V. Varadan, *Time Harmonic Electromagnetic Fields in Chiral Media*, Berlin: Springer-Verlag (1989), pp. 13–18.
14. L. D. Landau and E. M. Lifshitz, *Electrodynamics of Continuous Media*, 2nd edition, Oxford: Pergmon (1987), pp. 54, 331–357.
15. S. C. McClain, L. W. Hillman, and R. A. Chipman, Polarization ray tracing in anisotropic optically active media. II. Theory and physics, *J. Opt. Soc. Am. A* 10 (1993): 2383–2393.
16. R. Vlokh, Partial reciprocity of Faraday rotation in gyrotropic crystals, *Ferroelectrics* 414(1), 70–76, 2011.
17. E. E. Palik (ed.), *Handbook of Optical Constants of Solids*, Elsevier Inc., (1997).
18. Y. Wang, L. Liang, H. Xin, and L. Wu, Complex ray tracing in uniaxial absorbing media, *J. Opt. Soc. Am. A* 25 (2008): 653–657.
19. Y. Wang, P. Shi, H. Xin, and L. Wu, Complex ray tracing in biaxial anisotropic absorbing media, *J. Opt. A: Pure Appl. Opt.* 10 (2008).

20. G. D. Landry and T. A. Maldonado, Complete method to determine transmission and reflection characteristics at a planar interface between arbitrarily oriented biaxial media, *J. Opt. Soc. Am. A* 12 (1995): 2048–2063.
21. W.-Q. Zhang, General ray-tracing formulas for crystal, *Appl. Opt.* 31 (1992): 7328–7331.
22. M. Mansuripur, Analysis of multilayer thin-film structures containing magneto-optic and anisotropic media at oblique incidence using 2×2 matrices, *J. Appl. Phys.* 67(10) (1990): 6466–6475.
23. I. Abdulhalim, 2×2 Matrix summation method for multiple reflections and transmissions in a biaxial slab between two anisotropic media, *Opt. Commun.* 163 (1999): 9–14.
24. I. Abdulhalim, Analytic propagation matrix method for linear optics of arbitrary biaxial layered media, *J. Opt. A: Pure Appl. Opt.* 1 (1999): 646.
25. K. Mehrany and S. Khorasani, Analytical solution of non-homogeneous anisotropic wave equations based on differential transfer matrices, *J. Opt. A: Pure Appl. Opt.* 4 (2002): 624.
26. K. Postava, T. Yamaguchi and R. Kantor, Matrix description of coherent and incoherent light reflection and transmission by anisotropic multilayer structures, *Appl. Opt.* 41 (2002): 2521–2531.
27. M. C. Simon, Internal total reflection in monoaxial crystals, *Appl. Opt.* 26 (1987): 3878–3883.
28. M. C. Simon and R. M. Echarri, Inhibited reflection in uniaxial crystal, *Opt. Lett.* 14 (1989): 257–259.
29. W. R. Hamilton, Third supplement to an essay on the theory of systems of rays, *Trans. Roy. Irish Acad.* 17 (1833): 1.
30. H. Lloyd, On the phenomenon presented by light in its passage along the axis of biaxial crystals, *Trans. R. Irish Acad.* 17 (1833): 145–158.
31. M. Mansuripur, *Classical Optics & Its Applications*, Chapter 21, Cambridge University Press, (2002).
32. D. L. Portigal and E. Burstein, Internal conical refraction, *J. Opt. Soc. Am.* 59 (1969): 1567–1573.
33. E. Cojocaru, Characteristics of ray traces at the back of biaxial crystals at normal incidence, *Appl. Opt.* 38 (1999): 4004–4010.

20

Beam Combination with Polarization Ray Tracing Matrices

20.1 Introduction

Many optical systems divide a light beam into two or more partial waves, operate on each of these beams separately, and interfere the beams at an output plane. Such systems include beam splitters, interferometers, achromatic retarders, Lyot filters, optical isolators, crystal polarizers, and many others. For example, all birefringent components generate multiple wavefronts due to double refraction. The exiting wavefronts can completely or partially overlap in case of a retarder, or split and not overlap in case of beam splitters. When these wavefronts do overlap at an output plane, the resultant wavefront is the interference between all the partial waves. The output plane could be a detector such as a CCD detector, an exit pupil, screen for viewing interferograms,[1-4] a device to record an interference pattern or hologram,[5,6] or a surface to be illuminated, perhaps with structured light.

This chapter addresses methods to simulate the combination of overlapping wavefronts. Chapter 4 (Interference of Polarized Light) considered the interference between different polarization states. This chapter is preparation for Chapter 22 (Crystal Polarizers) and Chapter 26 (Multi-Order Retarders and the Mystery of Discontinuities), where the polarization aberrations of sophisticated devices are simulated for multiple overlapping beams of polarized light.

The algorithms for calculating the set of wavefronts from sequences of anisotropic materials, the set of all combinations of eigenmodes generated from ray doubling, were presented in Chapter 19. In general, the electric fields (**E**) of overlapping wavefronts (represented by sets of rays) add, but their associated polarization ray tracing matrices (**P**) do not, because the **P** matrices for different rays can be associated with different incident propagation directions and thus different **E**'s. Each **P** relates a

Table 20.1 Summary of Beam Combination Principles

Case 1	The **P** matrices of the overlapping wavefronts can only be combined when the following entities are the same: • Incident $\hat{\mathbf{S}}$ vectors • Exiting $\hat{\mathbf{S}}'$ vectors • Incident polarization states
Case 2	Given the incident polarization states, the resultant **E** fields of the overlapping wavefronts can always be added.

unique pair of incident and exiting Poynting vectors (**S**). Therefore, the **P** matrices at a given point in image space that may originate at different points in object space cannot be added as matrices, since they will no longer obey the property $\mathbf{P}\hat{\mathbf{k}} = \hat{\mathbf{k}}'$ or $\mathbf{P}\hat{\mathbf{S}} = \hat{\mathbf{S}}'$. Instead, the modified *polarization ray tracing matrix for addition* of Section 9.6:

$$\breve{\mathbf{P}}\hat{\mathbf{k}} = 0\hat{\mathbf{k}}' \text{ or } \breve{\mathbf{P}}\hat{\mathbf{S}} = 0\hat{\mathbf{S}}' \tag{20.1}$$

is used.

Wavefront combination procedures have two subtle issues. (1) The exiting rays of each partial wavefront calculated from ray tracing are not on the same exiting grid. This is addressed by interpolating the partial waves into continuous functions, which can then be resampling onto a uniform grid before simulating the mode combination. (2) When the rays are converging or diverging close to focus, caustics form, the wavefront folds over on itself, and the optical path length (*OPL*) becomes multivalued at parts of the wavefront.[7,8] To avoid this complication, it is much easier, when possible, to perform the wavefront resampling at the exit pupil where the rays are spatially distributed in a well-spaced grid or at least resample in regions where caustics can be avoided.

In the process of combining wavefronts, functions of ray parameters, such as the *OPL*s or **P** matrices evaluated from ray tracing, on a grid of values are combined. Because of the shear between the rays in different partial wavefronts, it is best to convert the ray tracing result from the ray-by-ray description to a function of **E** field, since **E** fields can be added as vectors. Interpolation is a necessary tool to construct the **E** field functions from ray data. *Interpolation* refers to methods that construct new values of a sampled function at intermediate locations. Interpolating functions over pupils allows estimating the associated values anywhere within the pupil from the ray trace data. In particular, when ray tracing a single incident ray into multiple modes through interferometers or anisotropic materials, the exiting rays for the different modes intersect the pupil or other surfaces at different locations. By interpolating the ray trace data for each mode or partial wavefront, the functions can be interpolated onto a common grid and then be readily combined.

Table 20.1 summarizes the important points on beam combination. The first half of this chapter considers the simpler Case 1 in detail, where the incident **S**, incident **E**, and exiting **S** vectors for all the beams are the same. This is the typical case for waveplates, crystal polarizers, and many other birefringent devices with plane parallel surfaces. The second half of the chapter treats the more general Case 2.

20.2 Wavefronts and Ray Grids

The optical ray trace commonly starts by tracing a grid of rays emerging from a point on the object. This grid of rays is usually evenly distributed over the incident wavefront, as shown in Figure 20.1.

When these wavefronts encounter beam splitters or birefringent components, they split into multiple wavefronts, eventually reaching an image plane or exit surface where the interference of the

20.2 Wavefronts and Ray Grids

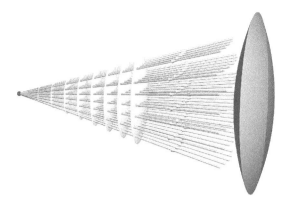

Figure 20.1 The point source on the left emits spherical wavefronts (yellow surfaces) that are represented by a grid of rays (orange arrows) propagating toward the lens (red).

wavefronts may need to be evaluated. Depending on the system, the interfering wavefronts could be as simple as two overlapping collimated wavefronts exiting a retarder or as complicated as hundreds of partially overlapping converging or diverging wavefronts from systems such as Lyot filters and Fabry–Perot interferometers. These resultant rays at the output surface are most likely not evenly spaced due to aberration differences between modes. In Figure 20.2, an incident wavefront sampled by a grid of rays converges through a birefringent material and exits as two grids of rays representing the two polarized exiting wavefronts. These two sets of rays are traced separately after splitting, and the effect of the birefringent plate is contained in the combination of these two wavefronts. Therefore, accurate analysis needs to encompass both ray grids where the wavefronts overlap.

There are two broad classes of optical systems: imaging and non-imaging. Illumination systems that shine light onto surfaces are a common example of a non-imaging system. They are often intended to provide uniform illumination on an object, such as illuminating a hologram with a uniform spherical wavefront or illuminating an LCD device screen with a polarized wavefront inside a liquid crystal projector.[9,10] On the other hand, imaging systems are usually designed to take spherical wavefronts from a light source and image them through a series of optical components onto exiting spherical wavefront centered at an image plane. Because of aberrations, the output wavefronts will deviate from the ideal spherical shape. As the aberrated wavefront approaches to its focus, it tends to have a complex **E** field distribution. Figure 20.3 shows light focusing through a

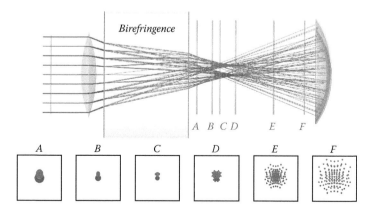

Figure 20.2 A collimated wavefront passes through a lens and converges through a birefringent plate. Two aberrated wavefronts (red and blue) exit the birefringent plate and pass a series of planes labeled from *A* to *F*. The ray positions from the two modes are spaced differently and only partially overlap or, near these images, don't overlap at all.

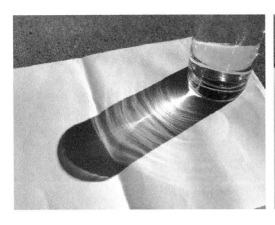
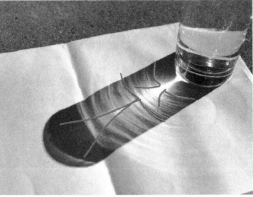

Figure 20.3 (Left) Illumination pattern created from a light focusing through a glass of water. The wavefront contains many bright lines, which are the caustics. (Right) The caustics where the wavefront folds over on itself are highlighted in red.

glass of water and forming a complex light distribution. The caustics (nephroid-shaped bright areas) are locations where the wavefront folds over on itself and neighboring rays intersect. We try to avoid such caustics when simulating wavefront combination.

Instead, the wavefronts often best combine at the exit pupil of imaging systems where the ray grid tends to be relatively well spaced. For example, for a beam with spherical aberration, the region containing the caustic would be avoided. Then, the **E** field at the image plane can be evaluated by diffraction theory applied to the exit pupil, as described in Chapter 16.

20.3 Co-Propagating Wavefront Combination

The simplest beam combining configuration occurs when combining two collimated wavefronts propagating in the same direction (Case 1 of Table 20.1). Many retarders, polarizers, and other anisotropic devices consist of planar surfaces; thus, it is a common occurrence that a collimated incident beam yields co-propagating collimated exiting beams. In general, **P** matrices do not combine, but in this case of co-propagating wavefronts, the **P**'s can be combined.

The *OPL*s of all the rays and ray segments and the **P** matrices are needed to accurately account for the phase changes through the optical system. For the combination, the mapping of two orthogonal basis vectors and the Poynting vector in the entrance pupil, $(\hat{\mathbf{E}}_m, \hat{\mathbf{E}}_n, \hat{\mathbf{S}})$, into the exit pupil is $\left(a_m e^{i\frac{2\pi}{\lambda} OPL} \hat{\mathbf{E}}'_m, a_n e^{i\frac{2\pi}{\lambda} OPL} \hat{\mathbf{E}}'_n, \hat{\mathbf{S}}'\right)$, where the *OPL* of an eigenmode propagating in anisotropic material is defined in Equation 19.22.

In some applications, such as Michelson interferometers, optical coherence tomography, and fiber optic pulse measurement, the cumulative *OPL* is desired for accurate simulation, and the absolute *OPL* should be kept separately from the **P** matrix. It is also important to be able to describe the exiting wavefront's peak-to-valley wavefront aberration and show phase maps that are unwrapped. These are handled by a separate calculation of *OPL* (Equation 19.22). The absolute *OPL* is also needed in calculations such as determining where the white light fringe will occur in Michelson interferometers, Fourier transform spectrometers, and optical coherence tomography, where the differences in *OPL* between wavelengths are required.

In other applications, such as utilizing retarders to generate elliptically polarized light, only the relative phase, modulus 2π, is needed, and the *OPL* can be included in the **P** matrix. The phase of all the light components exiting an optical system, usually in the exit pupil, is needed for image formation and optical transfer function calculations. These pupil functions, the wavefront aberration

20.3 Co-Propagating Wavefront Combination

function and the Jones pupil, are inputs to Fourier transform algorithms, which only need the input complex numbers modulo 2π. This phase contribution is contained in the $\mathbf{P}_{segment}$ matrices as modulo 2π.

This section presents two algorithms that combine the *OPL* into the **P** matrix. Depending on the goal, one method might be better suited than the other, but both produce the same overall **P** matrix. These algorithms will be applied in Chapter 21 in the analysis of a waveplate.

In the first algorithm, the **P** matrix of the light–surface interaction and the **P** matrix related to the propagation are considered as two separate polarization changes happening sequentially along a ray path. At a ray intercept, $\mathbf{P}_{intercept}$ maps $(\hat{\mathbf{E}}_m, \hat{\mathbf{E}}_n, \hat{\mathbf{S}})$ to $(a_m \hat{\mathbf{E}}'_m, a_n \hat{\mathbf{E}}'_n, \hat{\mathbf{S}}')$. With Equation 19.33,

$$\mathbf{P}_{intercept} = \begin{pmatrix} a_m \hat{\mathbf{E}}'_m & a_n \hat{\mathbf{E}}'_n & \hat{\mathbf{S}}' \end{pmatrix} \begin{pmatrix} \hat{\mathbf{E}}_m & \hat{\mathbf{E}}_n & \hat{\mathbf{S}} \end{pmatrix}^T. \tag{20.2}$$

As this mode propagates, the absolute phase of the mode increases through the material, which is represented by another **P** matrix, $\mathbf{P}_{segment}$, which maps $(\hat{\mathbf{E}}'_m, \hat{\mathbf{E}}'_n, \hat{\mathbf{S}}')$ onto $\left(e^{i\frac{2\pi}{\lambda}OPL} \hat{\mathbf{E}}'_m, e^{i\frac{2\pi}{\lambda}OPL} \hat{\mathbf{E}}'_n, \hat{\mathbf{S}}' \right)$;

$$\mathbf{P}_{segment} = \begin{pmatrix} e^{i\frac{2\pi}{\lambda}OPL} \hat{\mathbf{E}}'_m & e^{i\frac{2\pi}{\lambda}OPL} \hat{\mathbf{E}}'_n & \hat{\mathbf{S}}' \end{pmatrix} \begin{pmatrix} \hat{\mathbf{E}}'_m & \hat{\mathbf{E}}'_n & \hat{\mathbf{S}}' \end{pmatrix}^T. \tag{20.3}$$

Then, net **P** matrix just before the next ray intercept is

$$\begin{aligned}
&\mathbf{P}_{segment} \cdot \mathbf{P}_{intercept} \\
&= \begin{pmatrix} e^{i\frac{2\pi}{\lambda}OPL} \hat{\mathbf{E}}'_m & e^{i\frac{2\pi}{\lambda}OPL} \hat{\mathbf{E}}'_n & \hat{\mathbf{S}}' \end{pmatrix} \begin{pmatrix} \hat{\mathbf{E}}'_m & \hat{\mathbf{E}}'_n & \hat{\mathbf{S}}' \end{pmatrix}^T \begin{pmatrix} a_m \hat{\mathbf{E}}'_m & a_n \hat{\mathbf{E}}'_n & \hat{\mathbf{S}}' \end{pmatrix} \begin{pmatrix} \hat{\mathbf{E}}_m & \hat{\mathbf{E}}_n & \hat{\mathbf{S}} \end{pmatrix}^T \\
&= \begin{pmatrix} a_m e^{i\frac{2\pi}{\lambda}OPL} \hat{\mathbf{E}}'_m & a_n e^{i\frac{2\pi}{\lambda}OPL} \hat{\mathbf{E}}'_n & \hat{\mathbf{S}}' \end{pmatrix} \begin{pmatrix} \hat{\mathbf{E}}_m & \hat{\mathbf{E}}_n & \hat{\mathbf{S}} \end{pmatrix}^T,
\end{aligned} \tag{20.4}$$

where $\hat{\mathbf{E}}_m \cdot \hat{\mathbf{E}}_n = 0$, $\hat{\mathbf{S}}' \cdot \hat{\mathbf{E}}'_m = 0$, and $\hat{\mathbf{S}}' \cdot \hat{\mathbf{E}}'_n = 0$. Hence, $(\hat{\mathbf{E}}_m, \hat{\mathbf{E}}_n, \hat{\mathbf{S}})$ gets mapped into $\left(a_m e^{i\frac{2\pi}{\lambda}OPL} \hat{\mathbf{E}}'_m, a_n e^{i\frac{2\pi}{\lambda}OPL} \hat{\mathbf{E}}'_n, \hat{\mathbf{S}}' \right)$.

When the ray propagation and the surface interaction to be combined into one **P** matrix are not in sequence, as shown in the example in Figure 20.6, the mapping of the Poynting vectors $(\hat{\mathbf{S}} \rightarrow \hat{\mathbf{S}}')$ needs to be completely removed from $\mathbf{P}_{intercept}$ before incorporating $e^{i\frac{2\pi}{\lambda}OPL}$. By removing $(\hat{\mathbf{S}} \rightarrow \hat{\mathbf{S}}')$, the operations affect only the **E** field component. After that, the original mapping of **S** is added back into the final **P** matrix. The step-by-step procedures are as follows:

1. The mapping of $(\hat{\mathbf{S}} \rightarrow \hat{\mathbf{S}}')$ is represented by the 3×3 matrix $\mathbf{S_D}$ (**S** dyad), which is the outer product of $\hat{\mathbf{S}}$ and $\hat{\mathbf{S}}'$,

$$\mathbf{S_D} = \hat{\mathbf{S}}' \cdot \hat{\mathbf{S}}^T. \tag{20.5}$$

$\mathbf{S_D}$ maps $(\hat{\mathbf{E}}_m, \hat{\mathbf{E}}_n, \hat{\mathbf{S}})$ to $(\mathbf{0}, \mathbf{0}, \hat{\mathbf{S}}')$, and all other vectors orthogonal to $\hat{\mathbf{S}}$ into **0**.

2. The mapping of the Poynting vector is removed from $\mathbf{P}_{intercept}$ by subtracting $\mathbf{S_D}$. Then,

$$\breve{\mathbf{P}} = \mathbf{P}_{intercept} - \mathbf{S_D}, \tag{20.6}$$

which maps $(\hat{\mathbf{E}}_m, \hat{\mathbf{E}}_n, \hat{\mathbf{S}})$ to $(a_m \hat{\mathbf{E}}'_m, a_n \hat{\mathbf{E}}'_n, \mathbf{0})$. Thus, $\breve{\mathbf{P}} \cdot \mathbf{S} = \mathbf{0}$.

3. By multiplying $e^{i\frac{2\pi}{\lambda}OPL}$ to $\breve{\mathbf{P}}$,

$$\bar{\mathbf{P}} = \breve{\mathbf{P}} e^{i\frac{2\pi}{\lambda}OPL}, \tag{20.7}$$

which maps $(\hat{\mathbf{E}}_m, \hat{\mathbf{E}}_n, \hat{\mathbf{S}})$ to $\left(a_m e^{i\frac{2\pi}{\lambda}OPL} \hat{\mathbf{E}}'_m, a_n e^{i\frac{2\pi}{\lambda}OPL} \hat{\mathbf{E}}'_n, \mathbf{0}\right)$.

4. Then, by adding $\mathbf{S_D}$ back into the $\mathbf{P}$ matrix, the mapping of $\hat{\mathbf{S}}$ is restored. Hence,

$$\bar{\bar{\mathbf{P}}} = \bar{\mathbf{P}} + \mathbf{S_D}, \tag{20.8}$$

which maps $(\hat{\mathbf{E}}_m, \hat{\mathbf{E}}_n, \hat{\mathbf{S}})$ to $\left(a_m e^{i\frac{2\pi}{\lambda}OPL} \hat{\mathbf{E}}'_m, a_n e^{i\frac{2\pi}{\lambda}OPL} \hat{\mathbf{E}}'_n, \hat{\mathbf{S}}'\right)$.

Finally, the overall $\mathbf{P}$ matrix with both surface interactions and propagation effects is

$$\bar{\bar{\mathbf{P}}} = (\mathbf{P}_{intercept} - \mathbf{S_D}) e^{i\frac{2\pi}{\lambda}OPL} + \mathbf{S_D}, \tag{20.9}$$

which keeps the mapping of $\hat{\mathbf{S}}$ unchanged and incorporates the *OPL* effect to the electric fields.

Many intermediate $\mathbf{P}$ matrices are used in these steps to incorporate *OPL* into the overall $\mathbf{P}$ matrix. These intermediate $\mathbf{P}$ matrices are summarized in Table 20.2.

Consider a collimated beam normally incident onto a birefringent plate, as shown in Figure 20.4. Two collimated and completely overlapping beams are generated. They are considered as two

Table 20.2 Summary of the Intermediate $\mathbf{P}$ Matrices for Incorporating *OPL*

$\mathbf{P}_{intercept}$	$(\hat{\mathbf{E}}_m, \hat{\mathbf{E}}_n, \hat{\mathbf{S}})$	$\rightarrow$	$\left(a_m \hat{\mathbf{E}}'_m,\ a_n \hat{\mathbf{E}}'_n,\ \hat{\mathbf{S}}'\right)$
$\mathbf{P}_{segment}$	$(\hat{\mathbf{E}}_m, \hat{\mathbf{E}}_n, \hat{\mathbf{S}})$	$\rightarrow$	$\left(e^{i\frac{2\pi}{\lambda}OPL} \hat{\mathbf{E}}_m,\ e^{i\frac{2\pi}{\lambda}OPL} \hat{\mathbf{E}}_n,\ \hat{\mathbf{S}}\right)$
$\breve{\mathbf{P}}$	$(\hat{\mathbf{E}}_m, \hat{\mathbf{E}}_n, \hat{\mathbf{S}})$	$\rightarrow$	$\left(a_m \hat{\mathbf{E}}'_m,\ a_n \hat{\mathbf{E}}'_n,\ \mathbf{0}\right)$
$\bar{\mathbf{P}}$	$(\hat{\mathbf{E}}_m, \hat{\mathbf{E}}_n, \hat{\mathbf{S}})$	$\rightarrow$	$\left(a_m e^{i\frac{2\pi}{\lambda}OPL} \hat{\mathbf{E}}'_m,\ a_n e^{i\frac{2\pi}{\lambda}OPL} \hat{\mathbf{E}}'_n,\ \mathbf{0}\right)$
$\bar{\bar{\mathbf{P}}}$	$(\hat{\mathbf{E}}_m, \hat{\mathbf{E}}_n, \hat{\mathbf{S}})$	$\rightarrow$	$\left(a_m e^{i\frac{2\pi}{\lambda}OPL} \hat{\mathbf{E}}'_m,\ a_n e^{i\frac{2\pi}{\lambda}OPL} \hat{\mathbf{E}}'_n,\ \hat{\mathbf{S}}'\right)$

20.3 Co-Propagating Wavefront Combination

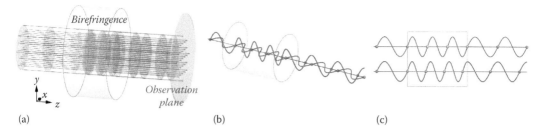

Figure 20.4 (a) A normally incident collimated beam with a planar wavefront (purple) incident onto a birefringent plate. The disks in the beam indicate the start of each period of the oscillating **E** fields. The incident beam splits into two wavefronts (red, higher index; blue, lower index) exiting with a phase delay and then reaching an observation plane. (b) Schematic of the **E** fields of the two orthogonal eigenstates, red for *y*-polarized and blue for *x*-polarized. (c) Both **E** field oscillations are plotted in one plane to show that the *x*-polarized mode has 3 periods while the *y*-polarized mode has 3.25 periods traversing the plate.

separate wavefronts in a sense that the birefringent plate delayed the phase of one mode relative to the other mode.

The orthogonally polarized eigenstates of the plate are chosen to be *x*- and *y*-polarized light with the **P** matrices for the two wavefronts, $\mathbf{P}_1 = \begin{pmatrix} \alpha_1 & 0 & 0 \\ 0 & 0 & 0 \\ 0 & 0 & 1 \end{pmatrix}$ and $\mathbf{P}_2 = \begin{pmatrix} 0 & 0 & 0 \\ 0 & \alpha_2 & 0 \\ 0 & 0 & 1 \end{pmatrix}$, where α_1 and α_2 are complex amplitude transmittances. These two wavefronts are the *o*- and *e*-modes for a uniaxial material, or the *fast*- and *slow*-modes for a biaxial material. Both **P** matrices perform the same transformation for **S**; $\mathbf{S}' = \mathbf{P}_1 \cdot \mathbf{S}$ and $\mathbf{S}' = \mathbf{P}_2 \cdot \mathbf{S}$. Thus, $\mathbf{P}_1$ and $\mathbf{P}_2$ contain within them the same dyad for operating on **S**, $\mathbf{S_D} = \begin{pmatrix} 0 & 0 & 0 \\ 0 & 0 & 0 \\ 0 & 0 & 1 \end{pmatrix}$, defined in Equation 20.5. By subtracting $\mathbf{S_D}$ from **P**, as shown in Equation 20.6, the resultant $\breve{\mathbf{P}} = \mathbf{P} - \mathbf{S_D}$, the *addition form of the polarization ray tracing matrix* (see Section 9.6), contains the **E** field transformation, but not the **S** transformation; thus, the *OPL* accumulated from each mode can be coherently incorporated into these **E** field transformations to $\overline{\mathbf{P}} = \breve{\mathbf{P}} e^{i\frac{2\pi}{\lambda}OPL} = (\mathbf{P} - \mathbf{S_D}) e^{i\frac{2\pi}{\lambda}OPL}$, as shown in Equation 20.7. The overbar in $\overline{\mathbf{P}}$ indicates the incorporation of the *OPL*. For an incident **E**, the combined exiting **E** of two overlapping beams is

$$\begin{aligned}\mathbf{E}' &= \overline{\mathbf{P}}_1 \cdot \mathbf{E} + \overline{\mathbf{P}}_2 \cdot \mathbf{E} \\ &= (\overline{\mathbf{P}}_1 + \overline{\mathbf{P}}_2) \cdot \mathbf{E} \\ &= \overline{\mathbf{P}}_{combine} \cdot \mathbf{E},\end{aligned} \quad (20.10)$$

hence, the $\overline{\mathbf{P}}$ matrices can be combined. The combination is done by adding the $\overline{\mathbf{P}}$ matrices of each mode with one component of $\mathbf{S_D}$, so $\mathbf{P}_{combine} \cdot \mathbf{S} = \mathbf{S}'$. The combined **P** matrix is

$$\begin{aligned}\mathbf{P}_{combine} &= (\mathbf{P}_1 - \mathbf{S_D}) e^{i\frac{2\pi}{\lambda}OPL_1} + (\mathbf{P}_2 - \mathbf{S_D}) e^{i\frac{2\pi}{\lambda}OPL_2} + \mathbf{S_D} \\ &= \left(\breve{\mathbf{P}}_1 e^{i\frac{2\pi}{\lambda}OPL_1} + \breve{\mathbf{P}}_2 e^{i\frac{2\pi}{\lambda}OPL_2} \right) + \mathbf{S_D} \\ &= (\overline{\mathbf{P}}_1 + \overline{\mathbf{P}}_2) + \mathbf{S_D},\end{aligned} \quad (20.11)$$

as shown in Equations 20.8 and 20.9.

For the example in Figure 20.4, the waveplate induces $OPL_1 = 3\lambda$ and $OPL_2 = 3.25\lambda$ and assumes that the amplitude transmittances t_1 and t_2 are both 1. The combined **P** matrix is

$$\mathbf{P}_{combine} = \begin{pmatrix} 1 & 0 & 0 \\ 0 & 0 & 0 \\ 0 & 0 & 0 \end{pmatrix} e^{i6\pi} + \begin{pmatrix} 0 & 0 & 0 \\ 0 & 1 & 0 \\ 0 & 0 & 0 \end{pmatrix} e^{i6.5\pi} + \begin{pmatrix} 0 & 0 & 0 \\ 0 & 0 & 0 \\ 0 & 0 & 1 \end{pmatrix}$$

$$= \begin{pmatrix} e^{i6\pi} & 0 & 0 \\ 0 & e^{i6.5\pi} & 0 \\ 0 & 0 & 1 \end{pmatrix} = \begin{pmatrix} 1 & 0 & 0 \\ 0 & i & 0 \\ 0 & 0 & 1 \end{pmatrix},$$

which describes a linear quarter wave retarder as expected and conveys absolute phase and information on the relative phase induced between the two wavefronts. In general, the combined **P** for M co-propagating overlapped wavefronts is

$$\begin{aligned} \mathbf{P}_{combine} &= \left(\sum_{m}^{M} (\mathbf{P}_m - \mathbf{S}_\mathbf{D}) e^{i\frac{2\pi}{\lambda} OPL_m} \right) + \mathbf{S}_\mathbf{D} \\ &= \left(\sum_{m}^{M} \breve{\mathbf{P}}_m e^{i\frac{2\pi}{\lambda} OPL_m} \right) + \mathbf{S}_\mathbf{D} \\ &= \left(\sum_{m}^{M} \bar{\mathbf{P}}_m \right) + \mathbf{S}_\mathbf{D}. \end{aligned} \qquad (20.12)$$

Consider the 45° polarized incident beam in the system in Figure 20.4,

$$\begin{aligned} \mathbf{E}' &= \mathbf{P}_{combine} \cdot \mathbf{E} \\ &= \begin{pmatrix} e^{i\frac{2\pi}{\lambda} OPL_1} & 0 & 0 \\ 0 & e^{i\frac{2\pi}{\lambda} OPL_2} & 0 \\ 0 & 0 & 1 \end{pmatrix} \begin{pmatrix} 1 \\ 1 \\ 0 \end{pmatrix} = \begin{pmatrix} e^{i\frac{2\pi}{\lambda} OPL_1} \\ e^{i\frac{2\pi}{\lambda} OPL_2} \\ 0 \end{pmatrix} \\ &= e^{i\frac{2\pi}{\lambda} OPL_1} \begin{pmatrix} 1 \\ e^{i\frac{2\pi}{\lambda}(OPL_2 - OPL_1)} \\ 0 \end{pmatrix} = e^{i\frac{2\pi}{\lambda} OPL_1} \begin{pmatrix} 1 \\ e^{i\frac{2\pi}{\lambda} \Delta OPL} \\ 0 \end{pmatrix}. \end{aligned}$$

The resultant polarizations after the waveplate are a function of relative OPL between the two eigenstates, ΔOPL, as shown in Figure 20.5.

20.3 Co-Propagating Wavefront Combination

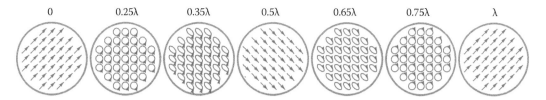

Figure 20.5 The resultant polarization states as a function of ΔOPL.

For the Figure 20.4 example, $\Delta OPL = OPL_2 - OPL_1 = 6.5\pi - 6\pi = 0.5\pi$, thus, $\mathbf{E}' = e^{i6\pi}(1, i, 0)$, which is a left circularly polarized ray propagating in the z-direction.

Math Tip 20.1 The Singular Values of $\mathbf{P}$ and $\breve{\mathbf{P}}$

The singular values of the *polarization ray tracing matrix* $\mathbf{P}$ and the *addition form of the polarization ray tracing matrix* $\breve{\mathbf{P}}$ are the same, except for the singular value corresponding to $\hat{\mathbf{S}}$:

$$\begin{cases} \mathbf{P} \cdot \mathbf{v}_1 = \Lambda_1 \mathbf{u}_1 \\ \mathbf{P} \cdot \mathbf{v}_2 = \Lambda_2 \mathbf{u}_2 \\ \mathbf{P} \cdot \hat{\mathbf{S}} = \hat{\mathbf{S}}' \end{cases} \text{ and } \begin{cases} \breve{\mathbf{P}} \cdot \mathbf{v}_1 = (\mathbf{P} - \mathbf{S}_D) \cdot \mathbf{v}_1 = \Lambda_1 \mathbf{u}_1 \\ \breve{\mathbf{P}} \cdot \mathbf{v}_2 = (\mathbf{P} - \mathbf{S}_D) \cdot \mathbf{v}_2 = \Lambda_2 \mathbf{u}_2, \\ \breve{\mathbf{P}} \cdot \hat{\mathbf{S}} = (\mathbf{P} - \mathbf{S}_D) \cdot \hat{\mathbf{S}}' = 0 \end{cases} \qquad (20.13)$$

$$\mathbf{P} = \mathbf{U}\mathbf{D}\mathbf{V}^\dagger = \begin{pmatrix} \hat{S}_{x,Q} & u_{x,1} & u_{x,2} \\ \hat{S}_{y,Q} & u_{y,1} & u_{y,2} \\ \hat{S}_{z,Q} & u_{z,1} & u_{z,2} \end{pmatrix} \begin{pmatrix} 1 & 0 & 0 \\ 0 & \Lambda_1 & 0 \\ 0 & 0 & \Lambda_2 \end{pmatrix} \begin{pmatrix} \hat{S}^*_{x,0} & \hat{S}^*_{y,0} & \hat{S}^*_{z,0} \\ v^*_{x,1} & v^*_{y,1} & v^*_{z,1} \\ v^*_{x,2} & v^*_{y,2} & v^*_{z,2} \end{pmatrix}, \qquad (20.14)$$

$$\breve{\mathbf{P}} = \breve{\mathbf{U}}\breve{\mathbf{D}}\breve{\mathbf{V}}^\dagger = \begin{pmatrix} \hat{S}_{x,Q} & u_{x,1} & u_{x,2} \\ \hat{S}_{y,Q} & u_{y,1} & u_{y,2} \\ \hat{S}_{z,Q} & u_{z,1} & u_{z,2} \end{pmatrix} \begin{pmatrix} 0 & 0 & 0 \\ 0 & \Lambda_1 & 0 \\ 0 & 0 & \Lambda_2 \end{pmatrix} \begin{pmatrix} \hat{S}^*_{x,0} & \hat{S}^*_{y,0} & \hat{S}^*_{z,0} \\ v^*_{x,1} & v^*_{y,1} & v^*_{z,1} \\ v^*_{x,2} & v^*_{y,2} & v^*_{z,2} \end{pmatrix}. \qquad (20.15)$$

The wavefront exiting systems of multiple birefringent plates or wedges do not always perfectly overlap. The difference in position between two wavefronts is referred to as their *shear*. They may be laterally sheared, where one is translated relative to the other, as in the case of lateral shearing interferometers.[11] The wavefronts may be rotationally sheared, where one wavefront is rotated. Or the wavefronts may have a different size or magnification, a radial shear.

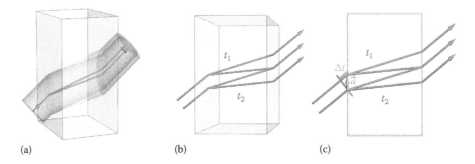

Figure 20.6 (a) An off-axis collimated wavefront refracts through a birefringent plate. Two partially overlapping collimated beams exit the plate propagating in the same direction. (b) Two rays from the incident wavefront are traced through the plate; note how the mode 2 ray (red) from the upper incident ray exits on top of the mode 1 ray (blue) from the lower ray, so these are the rays to be combined. (c) The resultant shear between the incident rays of $\vec{a}$ causes a contribution to the *OPL* difference Δt between the two rays.

Figure 20.6 contains an example of two partially overlapping wavefronts propagating in the same direction with a lateral shear. The ray in the center of the incident beam shown in Figure 20.6a can represent the propagation of a collimated plane wave in air passing through the birefringent waveplate, and the two exiting rays (modes 1 and 2) represent the two partially overlapping exiting plane waves. All rays of mode 1 have the same cumulative $\mathbf{P}_{t1}$ and all rays of mode 2 have the same cumulative $\mathbf{P}_{t2}$. In the region where the exiting wavefronts overlap, the $\mathbf{P}$ matrices of the two modes with the same incident and exiting $\hat{\mathbf{S}}$ can be combined.

For the rays in the partially overlapping region, as shown by the middle rays in Figure 20.6b, due to the shear between the two exiting beams, two different exiting rays, one from each mode, appear to exit as one ray. In the polarization ray trace, both of the superposed rays must be traced to explicitly show the *optical path difference* contribution due to the shear between the modes. However, due to the waveplate's parallel surfaces, the behavior of all collimated rays is the same, and the ray tracing results from tracing all the modes of one incident ray provide sufficient information to analyze the overall effect of the waveplate for this angle and wavelength.

The separation between the two incident rays whose exit overlaps depends on the thickness of the plate, such that the rays are closer together with thinner plates. As illustrated in Figure 20.6c, there is an optical path Δt associated with propagation outside of the waveplate that needs to be accounted for in addition to the *OPL* within the birefringent plate, where

$$\Delta t = \vec{a} \cdot \hat{\mathbf{S}}. \tag{20.16}$$

The waveplate induces an OPL_1 to mode 1 (shown in blue). At the exit surface where mode 1 and mode 2 meet, mode 2 has a corresponding $OPL_2 + \Delta t$, which includes the extra path (shown in green) outside the plate, where $\vec{a} = \mathbf{r}_1 - \mathbf{r}_2$ is the lateral shear between two rays measured on the exit face. Therefore, the retardance induced by the waveplate for an off-axis beam is as follows:

$$\begin{aligned}
\delta &= \frac{2\pi}{\lambda}\left[(OPL_2 + \Delta t) - OPL_1\right] \\
&= \frac{2\pi}{\lambda}\left[(OPL_2 + \vec{a}\cdot\hat{\mathbf{S}}) - OPL_1\right] \\
&= \frac{2\pi}{\lambda}\left[\left(OPL_2 + (\mathbf{r}_1 - \mathbf{r}_2)\cdot\hat{\mathbf{S}}\right) - OPL_1\right].
\end{aligned} \tag{20.17}$$

20.3 Co-Propagating Wavefront Combination

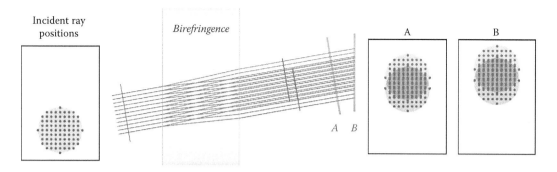

Figure 20.7 One off-axis collimated ray grid propagates through a birefringent plate dividing into two parallel and collimated wavefronts. They partially overlap each other on plane *A* and plane *B*.

Consider an evenly spaced incident grid of rays propagating through the birefringent plate at an angle, as shown in Figure 20.7. This results in exiting ray grids representing the two wavefronts. Depending on the system, the exiting rays may be analyzed on plane *A*, orthogonal to the ray direction, plane *B*, which is parallel to the plate, or some other surface. These two wavefronts are determined from the ray trace information for the two ray sets. Since the ray locations of these two exiting modes do not coincide, the individual wavefront can be interpolated onto the same grid before combining the **P** matrices or before combining the **E** fields. The incident grid of rays should be dense enough to resolve the structure of the beams and accurately represent their boundaries. Although these rays are individual points on an exit plane (e.g., the exit pupil), in this case, they represent a smooth and predictable wavefront. Further considerations of interpolating wavefronts are given in Section 20.5.

The two ray sets in Figure 20.7 are assumed to have a uniform **E** field across the exit plane. One exiting ray set is horizontally polarized and the other ray set is vertically polarized, as shown in Figure 20.8. The result of combining these two modes gives an overlapping area of interference surrounded by two crescents with the polarization of the two individual modes.

The polarization of the overlapping area depends on the relative phase between the two modes. Figure 20.9 shows the various polarization ellipse patterns for different relative phases.

In general, ray doubling occurs each time a beam enters an anisotropic medium. An optical system containing *N* anisotropic interfaces can have 2^N separate exiting modes in transmission for one incident ray. To combine cumulative **P** matrices of multiple off-axis modes from a system that consists of *N* plane parallel anisotropic plates, Equation 20.12 becomes Equation 20.18. Instead of 2 exiting modes, 2^N modes are combined within the beam overlapping region as

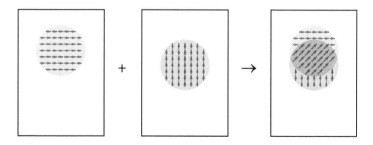

Figure 20.8 The two resultant wavefronts are horizontally (blue) and vertically (red) polarized. The overlapping area of the two orthogonal modes is 45° polarized.

Figure 20.9 Two individual wavefronts, one horizontally and one vertically polarized, are laterally sheared by the waveplate. They partially overlap each other. The polarization of the overlapping region depends on the phase difference of the individual wavefront. Showing from left to right, the relative phases are (0, 0.15, 0.25, 0.35, 0.5) waves.

$$\begin{aligned}
\mathbf{P}_{overlap} &= \left(\sum_{m=1}^{M} \bar{\mathbf{P}}_{m,total}\right) + \mathbf{S}_\mathbf{D} = \sum_{m=1}^{M}\left(\breve{\mathbf{P}}_{m,total} e^{i\frac{2\pi}{\lambda}(OPL_m + \Delta t_m)}\right) + \mathbf{S}_\mathbf{D} \\
&= \sum_{m=1}^{M}\left[(\mathbf{P}_{m,total} - \mathbf{S}_\mathbf{D})e^{i\frac{2\pi}{\lambda}(OPL_m + \Delta t_m)}\right] + \mathbf{S}_\mathbf{D} \\
&= \left\{\sum_{m=1}^{M}\left[\left(\prod_{n=1}^{N}\mathbf{P}_{m,N-n+1}\right) - \mathbf{S}_\mathbf{D}\right]e^{i\frac{2\pi}{\lambda}(OPL_m + \Delta t_m)}\right\} + \mathbf{S}_\mathbf{D},
\end{aligned} \qquad (20.18)$$

where $M = 2^N$. $\mathbf{P}_{m,total} = \prod_{n=1}^{N}\mathbf{P}_{m,N-n+1}$ is the cumulative $\mathbf{P}$ matrix for the effects from all of the interfaces, $\Delta t_m = \vec{\mathbf{a}}_m \cdot \hat{\mathbf{S}} = (\vec{\mathbf{r}}_1 - \vec{\mathbf{r}}_m)\cdot \hat{\mathbf{S}}$, and $\vec{\mathbf{r}}_m$ is the ray intercept of mode m at the last surface of the anisotropic plate assembly. Different parts of the exiting beam are expected to have different numbers of modes present.

Example 20.1 Polarization Aberrations of a Calcite Quarter Waveplate

For sheared non-planar wavefronts, the $\mathbf{E}$ fields can be added, but the $\mathbf{P}$ matrices cannot. The method of Section 20.3 will now be applied to calculate the variation of a waveplate's retardance with angle. Consider a first-order calcite quarter waveplate with ordinary and extraordinary refractive indices $n_O = 1.656$ and $n_E = 1.485$ at $\lambda = 633$ nm. The waveplate thickness is $(1 + \frac{1}{4})\lambda/(n_O - n_E) = 4.639$ µm and the fast axis is oriented at 45° counterclockwise from the x-axis. A square grid of vertically polarized incident rays spanning a ±30° × 30° field propagates through the plate as shown in Figure 20.10, simulated by polarization ray tracing $\mathbf{P}$ matrices. Each incident ray has unit $\mathbf{E}$ field amplitude. After passing through the waveplate, a vertically polarized ray becomes a left circularly polarized light at the center of the field. For the converging field, the two orthogonally polarized exiting modes are combined as in Equation 20.18, to calculate an angularly varying retardance. The orientations of their major axis are oriented vertically along one diagonal and horizontally along the other diagonal. The ellipticity changes from circularly polarized moving away from both the x- and y-axes.

20.3 Co-Propagating Wavefront Combination

The retardance of the 1¼ λ plate, shown in Figure 20.10, can be calculated by Equation 20.17 or extracted from the **P** matrix using the retardance algorithm in Chapter 17. The center of the field, corresponding to normal incidence, has retardance $3\pi/2$. As the angle of incidence increases along the 45° plane containing the optic axis, the retardance increases about normal incidence. As the angle increases in the direction orthogonal to the optic axis, the retardance decreases. The retardance for this waveplate has a saddle shape symmetric about the optic axis seen in Figure 20.11 (left).

The waveplate has a small diattenuation, shown in Figure 20.11 (right), arising from the difference in Fresnel transmission coefficients due to the difference in ordinary and extraordinary refractive indices. The diattenuation is low for this crystal waveplate. Two incident angles with zero diattenuation, *nodes*, occur in the plane containing the optic axis, where $T_{p\to e} = T_{s\to o}$.

For the vertically polarized incident light shown in Figure 20.10, the resultant **E** field amplitude and the major axis orientation and ellipticity of the polarization ellipses are mapped in Figure 20.12. The exiting amplitude has a very small variation, an average of 0.95, due to a small variation in Fresnel transmission in a skewed saddle pattern. Higher ellipticity is observed along the *x*- and *y*-directions. Ellipticities of 0 and 1 denote linearly and circularly polarized light, respectively. The exiting light is not circularly polarized at normal incident due to the slight transmission difference of the two modes. A circularly polarized beam is possible by finding the angle, along the *y*-axis in this case, which balances the diattenuation and retardance. As the incident ray scans across the *x*- or *y*-direction, the orientation of the ellipse switches from 0° to 90°.

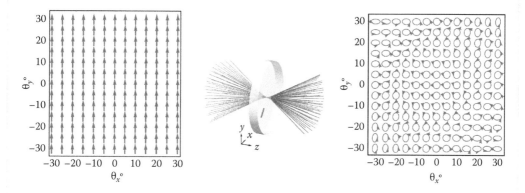

Figure 20.10 (Middle) The angular variation of retardance is seen for a square grid of rays with incident angle ±30° × 30° propagating through a first-order quarter waveplate whose optic axis is at 45° in the *xy* plane. (Left) The ellipse maps show the incident states and (right) exiting polarizations in double pole coordinates. The size of the ellipses is proportional to the electric field amplitude.

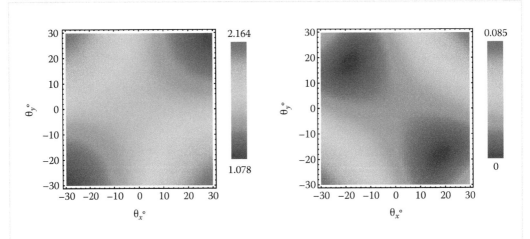

Figure 20.11 Retardance in radian (left) and diattenuation (right) as a function of incident angle θ_x and θ_y for 1¼λ plate. Here, retardance is constant along the x- and y-axes.

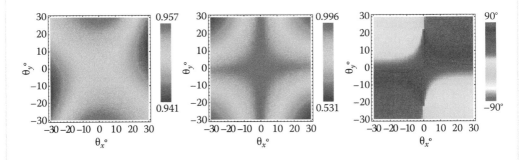

Figure 20.12 Electric field amplitude, the ellipticity, and the major axis of the polarization ellipses for vertically polarized incident light are shown from left to right.

20.4 Non-Co-Propagating Wavefront Combination

For spherical and other nonplanar wavefronts, the **E** fields can be added, but their corresponding **P** matrices cannot. The discussions in previous sections only address collimated beams, which can always be represented by a single **P** matrix. This section addresses the combination of sets of **E** fields using spatially overlapping ray grids propagating in different directions.

In the simulation of optical systems, a frequent task is evaluating the light distribution on a surface of interest. Examples include the calculation of flux and wavefront aberrations in an exit pupil, the point spread function on a detector, or the illumination pattern on some other arbitrary surface. When light rays from different parts of the object or pupil with different ray paths overlap on a surface, all these light rays are associated with **P** matrices with different $\mathbf{S_D}$. To properly simulate the beam combination, a conversion needs to be made from rays to wavefronts and **E** fields. When M modes overlap, for example, when rays from M modes span the same area, the **E** fields can be combined coherently as

$$\mathbf{E}_{total} = \sum_{m=1}^{M} \mathbf{E}_m e^{i\frac{2\pi}{\lambda}OPL_m}. \tag{20.19}$$

20.4 Non-Co-Propagating Wavefront Combination

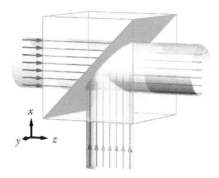

Figure 20.13 Two collimated wavefronts (red and blue) exiting a PBS into the same direction.

A simple example is merging two grids of polarized rays with a polarizing beam splitter (PBS) onto the same plane as is shown in Figure 20.13. In this case, the two wavefronts come from different directions, may have the same amplitude, and may be polarized orthogonally and merge together after the beam combiner.

In one path, incident *x*-polarization with $\mathbf{P}_1 = \begin{pmatrix} 1 & 0 & 0 \\ 0 & 0 & 0 \\ 0 & 0 & 1 \end{pmatrix}$ propagates in +*z*-direction, transmits through the PBS, and exits the PBS as $\mathbf{E}_1 = (1, 0, 0)$. For the other path, rays reflect from *x*- to *z*-direction, and incident *y*-polarization with $\mathbf{P}_2 = \begin{pmatrix} 0 & 0 & 0 \\ 0 & 1 & 0 \\ 1 & 0 & 0 \end{pmatrix}$ reflects into $\mathbf{E}_2 = (0, 1, 0)$. Since the two grids of rays come from different directions, it does not make sense to add their **P** matrices. Instead, the resultant **E** fields are combined. These two grids are shown coinciding with each other, but in general, they do not share the exact same ray coordinates. The method to combine such overlapping beams with misaligned ray grids will be explained in Section 20.5.

The ray tracing algorithms will calculate the optical path lengths OPL_1 and OPL_2 for the two ray paths. The exiting polarization state depends on the optical path difference ΔOPL between the two beams,

$$\mathbf{E}_{total} = \mathbf{E}_1 e^{i\frac{2\pi}{\lambda}OPL_1} + \mathbf{E}_2 e^{i\frac{2\pi}{\lambda}OPL_2} = \begin{pmatrix} 1 \\ e^{i\frac{2\pi}{\lambda}(OPL_2-OPL_1)} \\ 0 \end{pmatrix} e^{i\frac{2\pi}{\lambda}OPL_1} = \begin{pmatrix} 1 \\ e^{i\frac{2\pi}{\lambda}\Delta OPL} \\ 0 \end{pmatrix} e^{i\phi}. \quad (20.20)$$

If the two beams are perfectly collimated with no deviation, and the ray grids for the two wavefronts overlap each other perfectly, the resultant **E** will have a constant ΔOPL across the resultant beam and produce a uniformly polarized beam, as shown in Figure 20.5. If the beams are aberrated or tilted with respect to each other, the ΔOPL is spatially varying and the resultant polarization ellipses will also be spatially varying across the beam, as shown in Figure 20.14.

In this case, the ray grids provide sampled information on the variation in ΔOPL between wavefronts. An accurate calculation of these patterns requires sufficient ray density to resolve the ΔOPL variation across the exit pupil and avoid aliasing.

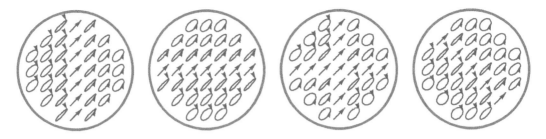

Figure 20.14 Example polarization state distributions with spatially varying ΔOPL across the beam.

20.5 Combining Irregular Ray Grids

In general, the exiting ray locations corresponding to multiple wavefronts do not coincide. When a spherical wave propagates through a retarder, the two sets of rays for the two split modes are sheared with respect to each other. The two modes have different spacing between rays, as shown in plane F in Figure 20.2, as well as Figure 20.15. Although the rays do not overlap exactly on top of each other, their wavefronts do substantially overlap, so the ray parameters carried by their ray sets need to be combined appropriately to reveal their overall polarization.

This section describes the procedures to reconstruct and combine the wavefronts of these misaligned ray grids. These procedures involve interpolating data, and an example interpolation algorithm is also provided.

20.5.1 General Steps to Combine Misaligned Ray Data

To combine **E** field functions by vector addition, the discrete ray data for the **E** fields needs to be interpolated into continuous functions. The discrete set of **E** fields calculated for each wavefront from the ray trace is spatially distributed over a surface. The discrete **E** fields on the irregular ray grid can be reconstructed into continuous function by interpolation. Interpolation is a method to estimate intermediate values between known data, so resampling can be done on the interpolated function. There are many algorithms to interpolate data, such as bilinear interpolation, spline interpolation, Kriging interpolation, and many others.[12-16] Values estimated by interpolation contain interpolation errors. However, with sufficient sample points and smooth input ray data, these errors can be minimized. By interpolating the ray data of the ray grids and resampling the interpolated functions to a common ray grid, the resultant resampled **E** fields can be added to simulate multiple interfering beams of light. Since the **E** field is summed on a mode-by-mode basis, the ray trace data are first grouped into individual modes. Then, the **E** field, as shown in Equation 20.21 with seven components (three orthogonal components of the complex **E** field and an *OPL*) for each mode, is

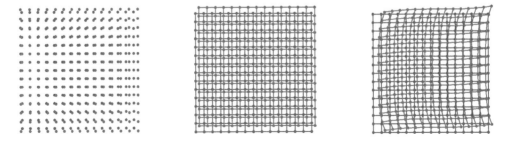

Figure 20.15 (Left) The ray positions of the exiting rays on a detector plane with two exiting modes (blue and red). (Middle) Exiting rays on regular grids. (Right) Exiting rays on irregular grids with uneven spacing.

20.5 Combining Irregular Ray Grids

Table 20.3 Steps to Reconstruct Wavefront from Wavefront Combination

Steps	Operation	Output
1	Group exiting ray trace data by mode	Grids of ray data
2	Calculate exiting **E** for the grids of rays	Grids of **E**
3	Interpolate each grid of **E**	**E** functions for each mode
4	Resample **E** from each **E** function onto a common grid	Grids of resampled **E**
5	Add the resampled grids of **E**	Final grid of **E**

interpolated using the discrete ray data yielding a continuous function. The resultant functions of **E** of each mode are added to represent the combined **E** field. The steps of constructing the overall **E** field are summarized in Table 20.3.

$$\mathbf{E} = \begin{pmatrix} \mathcal{R}e[E_x] + i\,\mathcal{J}m[E_x] \\ \mathcal{R}e[E_y] + i\,\mathcal{J}m[E_y] \\ \mathcal{R}e[E_z] + i\,\mathcal{J}m[E_z] \end{pmatrix} e^{i\frac{2\pi}{\lambda}OPL} = \begin{pmatrix} |E_x|e^{i\phi_x} \\ |E_y|e^{i\phi_y} \\ |E_z|e^{i\phi_z} \end{pmatrix} e^{i\frac{2\pi}{\lambda}OPL}. \tag{20.21}$$

An example of three overlapping grids of rays whose ray positions are unequally spaced and do not coincide is shown in Figure 20.16 (left). Rectangular grids are preferred so that matrix data structures and operations can be used. Therefore, the regions without data are usually padded with zeros. In Figure 20.16 (right), the three ray grids have been resampled using interpolation onto a new grid with zeros around the data. This new grid is evenly spaced, which is useful for further analysis using diffraction theory. For M individual modes, the resultant **E** field is calculated by adding M interpolated **E** fields as

$$\mathbf{E}(x,y) = \sum_m^M \left[e^{i\frac{2\pi}{\lambda}OPL_m(x,y)} \begin{pmatrix} E_{x,m}(x,y) \\ E_{y,m}(x,y) \\ E_{z,m}(x,y) \end{pmatrix} \right], \tag{20.22}$$

where (x, y) is the coordinate system at the surface of mode combination.

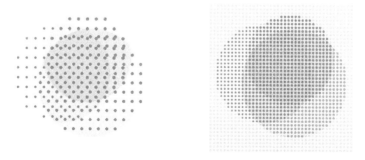

Figure 20.16 Three grids of rays corresponding to three modes reach a 2D plane. (Left) The location of the rays is shown as dots. (Right) The ray data are resampled in a regular grid with zero values (gray dots) outside the region of the three beams.

20.5.2 Inverse-Distance Weighted Interpolation

This section describes one interpolation algorithm, the inverse-distance weighted interpolation algorithm,[17] which is used in this chapter. This method uses the weighted averages of data from nearby points.[15] This interpolation method is versatile and fairly accurate for a dense grid of sample points with smooth values. Comparison between different interpolation methods are discussed in Ref. 18.

Consider A samples of electric field ($\mathbf{E}_1, \mathbf{E}_2, \ldots \mathbf{E}_a, \ldots \mathbf{E}_A$) at A irregularly spaced locations ($\mathbf{r}_{E1}, \mathbf{r}_{E2}, \ldots \mathbf{r}_{Ea}, \ldots \mathbf{r}_{EA}$), which will be resampled to a new evenly spaced grid at B locations ($\mathbf{r}_1, \mathbf{r}_2, \ldots \mathbf{r}_b, \ldots \mathbf{r}_B$), as shown in Figure 20.17 (left). The Q closest data points from $\mathbf{r}_b$, as shown in Figure 20.17 (right), will be used to estimate the value at $\mathbf{r}_b$. At point $\mathbf{r}_b$, the Q closest electric field samples are ($\mathbf{E}_{c1}, \mathbf{E}_{c2}, \ldots \mathbf{E}_{cq} \ldots \mathbf{E}_{cQ}$) at ($\mathbf{r}_{c1}, \mathbf{r}_{c2}, \ldots \mathbf{r}_{cq}, \ldots \mathbf{r}_{cQ}$). Q is chosen to produce a reasonable electric field estimation at intermediate locations. The distance of these Q data points from point $\mathbf{r}_b$ is $\left(|\bar{r}_{c1}|, |\bar{r}_{c2}|, \ldots |\bar{r}_{cq}|, \ldots |\bar{r}_{cQ}|\right) = (|\mathbf{r}_{c1} - \mathbf{r}_b|, |\mathbf{r}_{c2} - \mathbf{r}_b|, \ldots |\mathbf{r}_{cq} - \mathbf{r}_b|, \ldots |\mathbf{r}_{cQ} - \mathbf{r}_b|)$. Then, the interpolated value of the $\mathbf{E}$ field at $\mathbf{r}_b$ weighted by the Q data points is

$$\mathbf{E}_b = \sum_{q=1}^{Q} S_q \mathbf{E}_{cq}, \quad (20.23)$$

where the individual weights S_q are

$$S_q = \frac{\left(|\bar{r}_{cq}| + \varepsilon\right)^{-p}}{\sum_{q=1}^{Q}\left[\left(|\bar{r}_{cq}| + \varepsilon\right)^{-p}\right]}. \quad (20.24)$$

S_q is a scaling factor based on distance $|\bar{r}_{cq}|$ and has a maximum value of 1. ε is a small number chosen to avoid the situation of "divide by zero" in computer programs, when the point to be interpolated falls exactly on one of the set of data points; then, $|\bar{r}_{cq}| = 0$. $\varepsilon = 10^{-17}$ is a typical value. p is the inverse-distance weighting power that controls the region of influence of each of the data locations. As p increases, the region of influence decreases. When $p = 0$, Equation 20.24 simply averages the sampled values. When $\mathbf{r}_{cq}$ approaches $\mathbf{r}_b$, S_q emphasizes $\mathbf{E}_{cq}$. If $\mathbf{r}_b$ is exactly on top of $\mathbf{r}_{cq}$, $|\bar{r}_{cq}| = 0$, then $S_q \approx 1$. When $\mathbf{r}_b$ is far away from any one of the A data points, $S_q \to 0$. The abrupt changes in magnitude at the edges of apertures and at other discontinuities always result in interpolation

Figure 20.17 (Left) An unevenly spaced ray grid at locations ($\mathbf{r}_{E1}, \mathbf{r}_{E2}, \ldots \mathbf{r}_{Ea}, \ldots \mathbf{r}_{EA}$) is shown in red. The evenly spaced grid ($\mathbf{r}_1, \mathbf{r}_2, \ldots \mathbf{r}_b, \ldots \mathbf{r}_B$) for resampling is shown in black. (Right) Interpolation involves weighting contributions from the nearest neighbors. The five closest data locations ($\mathbf{r}_{c1}, \mathbf{r}_{c2}, \mathbf{r}_{c3}, \mathbf{r}_{c4}, \mathbf{r}_{c5}$) to $\mathbf{r}_b$ are shown.

20.5 Combining Irregular Ray Grids

artifacts, and such artifacts can be minimized by using a small number of Q and limiting the area where the sample can be accounted for in Equation 20.23 during resampling.

Two interpolation examples are shown in Figures 20.18 and 20.19 for a set of 1D data and a set of 2D data. The interpolation algorithms produce a smooth interpolation function with $p = 3$. p should be chosen depending on the physical properties of the data. For example, Kelway[19] and NOAA[20] use $p = 1.65$ and $p = 2$ for interpolating rainfall. The ARMOS model[21] suggests that p ranges from 4 to 8 for interpolating oil pressure heads.

Variations of the inverse-distance weighted interpolation algorithm can reduce the artifacts of the interpolation function, such as limiting the radius of the samples being used or taking the slope

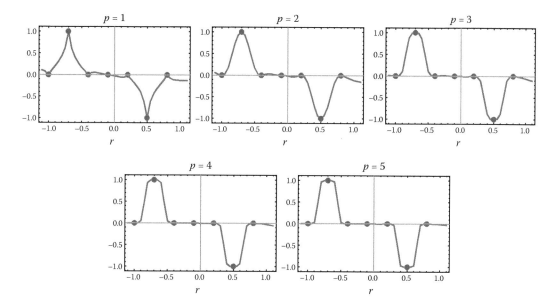

Figure 20.18 Seven data points (red) are interpolated using Equation 20.23. The resultant interpolation functions are shown in blue for five different p.

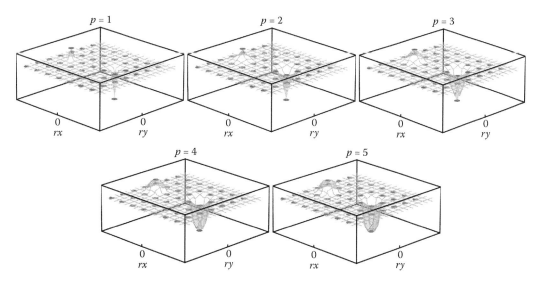

Figure 20.19 Two non-zero data points (red) are interpolated using Equation 20.23. The resultant 2D interpolation functions shown in orange for the different p reveal the influence function for the different p.

of the samples into account.[22] Other interpolation algorithms, such as Kriging interpolation[23] and thin plate spline,[24] provide more sophisticated estimations and require more involved computer programming. These interpolation functions are usually built into programs such as Mathematica and MATLAB.

Example 20.2 Polarization Variations of a Combined Wavefront

Consider combining the two wavefronts exiting from a converging beam focused through the calcite quarter waveplate, shown in Figure 20.2. The crystal's optic axis is oriented at (sin 45°, cos 45°, 0). The resultant wavefronts are to be combined on a plane after the crystal plate. The **E** field components for each mode are fit to the same grid as shown in Figure 20.20. These sample points are nearly equally spaced on a sphere and thus look bunched together toward the circumference. For this waveplate, the shear between the exiting modes is very small since the quarter waveplate is thin. The incident **E**'s are chosen as 45° linearly polarized. At each of the exiting locations, each mode has a **P** matrix calculated from the polarization ray trace. These **P** multiply $\mathbf{E}_{45°}$ and two sets of exiting **E** are obtained—$\mathbf{E}_o$ and $\mathbf{E}_e$. There are four real functions to be fit for each mode, as shown in Figure 20.21.

Then, the resultant $\mathbf{E}_{total}$ is calculated by Equation 20.22 where m is o and e. The resultant polarization ellipses of $\mathbf{E}_{total}$ for 1¼λ, 2¼λ, and 5¼λ waveplates are shown in Figure 20.22. As the plate gets thicker, the ellipticity changes faster with angle, but always remains nearly circular along the x- and y-axes. For all three cases, the elliptical polarization's major axis orientation is at 0° in two diagonal quadrants and is at 90° in the other two diagonal quadrants. As the plate gets thicker, polarization fringes, periodic modulations of polarization state, are observed. This can be seen, for example, in Figure 20.23, where the plate has been thickened to 50¼λ plate to produce about seven polarization fringes across each diagonal. Now, by displacing the individual field distributions of o and e modes spatially, the fringe distribution is also displaced, as shown in Figure 20.23.

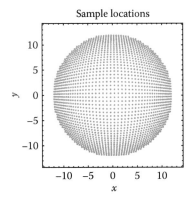

Figure 20.20 The grid locations on a spherical surface where the fields and *OPLs* will be resampled.

20.5 Combining Irregular Ray Grids

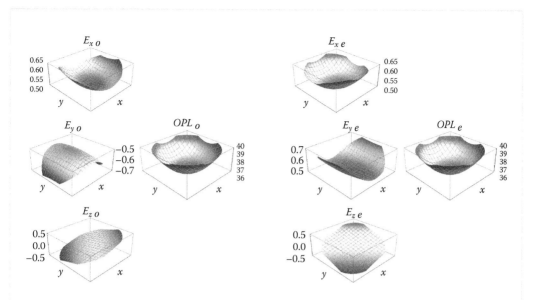

Figure 20.21 Interpolated function for the three **E** field components (real) and the *OPL* for the two exiting modes.

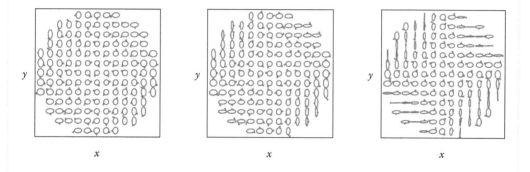

Figure 20.22 The polarization ellipse on a plane after the 1¼ λ, 2¼ λ, and 5¼ λ waveplates.

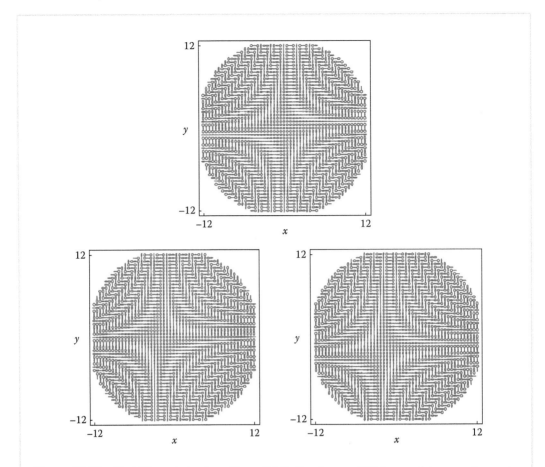

Figure 20.23 (Top) The polarization ellipse distribution from a 50¼ λ waveplate generates many polarization fringes. (Bottom left) Additional spatial shear between modes in *y* direction, and (bottom right) additional shear in diagonal direction of the field.

20.6 Conclusion

The results of a birefringent ray trace are grids of rays associated with multiple modes and multiple wavefronts. Each wavefront takes a different path and has its own amplitude, polarization, *OPL*, and aberration functions. When these wavefronts overlap, their combined effects are calculated from interfering their wavefronts. The optical analysis of interferometers and anisotropic optical elements requires combining these multiple wavefronts after the ray trace. This is done by (1) sampling each wavefront through polarization ray tracing, (2) fitting them to interpolating functions, one for each mode, and then (3a) adding these functions of **P** or **E** to represent the interference at that plane, or (3b) resampling **P** or **E** from their interpolation functions on an evenly spaced grid. Then, this recombined wavefront is used to calculate overall aberrations, point spread functions, and other metrics.

It is preferred to combine the **P** matrices rather than the **E** fields from each mode when the beams are collimated. The advantages of combining **P** matrices instead of **E** fields are as follows: (1) it preserves the polarization information along the ray path for any incident polarization, and (2) the calculations of diattenuation and retardance described in Chapters 9 and 17 still apply. However,

20.7 Problem Sets

when the S_D of the modes being combined is not the same (rays with different entering directions to be added), the **P** matrices cannot be combined, so the **E** fields are combined instead.

The calculations of wavefront combination are usually done at the exit pupil of an imaging system or at the plane of illumination, where the **P** matrices, **E** fields, and their *OPLs* sampled from ray tracing are located on a smooth and well-spaced exiting grid, and where caustics can be avoided. These sampled wavefront parameters are fitted and interpolated to wavefront functions. Interpolation over the pupil allows estimation of the associated values anywhere within the pupil using the ray trace data and constructs new values at intermediate locations. Then, the combining effect of each mode is calculated from the summation of these wavefront functions.

Assumptions are often made to reduce complex calculations, such as ignoring the small wavefront shear from a thin waveplate. However, for thick biaxial components or birefringent crystals that split an incident beam into different paths and then redirect the beams to be combined to a plane of interest, care and caution are needed. When the calculation is done without assumptions, subtle effects of misalignments and fabrication defects in the system can be simulated and yield accurate results. Large amounts of ray tracing and interpolation may be required for wavefronts that are not smooth.

Detailed analysis of a crystal waveplate and a Glan-type polarizer, using the ray tracing algorithms in Chapter 19 with the method of mode combination in this chapter, will be covered in Chapters 21 and 22. These simulations show the aberrations, undesired modes, and angle of incidence effects observed in laboratory measurements, highlighting optical effects that limit these high-performance birefringent components.

20.7 Problem Sets

20.1 Two uniformly polarized nearly co-propagating plane waves of monochromatic light are interfered, yielding the following interference patterns. Describe the two interfering polarization states and provide example Jones vectors.

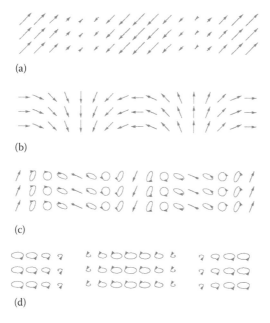

(a)

(b)

(c)

(d)

20.2 A 500-nm-wavelength normal incident beam with propagation vector $\mathbf{k} = (0, 0, 1)$ passes through a waveplate and produces two exiting modes with the **P** matrices shown below:

$$\mathbf{P}_e = \begin{pmatrix} 0.48063 & 0.48063 & 0 \\ 0.48063 & 0.48063 & 0 \\ 0 & 0 & 1 \end{pmatrix} \text{ and } \mathbf{P}_o = \begin{pmatrix} 0.46877 & -0.46877 & 0 \\ -0.46877 & 0.46877 & 0 \\ 0 & 0 & 1 \end{pmatrix}$$

Their cumulative *OPL*s are $OPL_e = 0.00316754$ mm and $OPL_o = 0.00354254$ mm. For this simple waveplate, $\mathbf{k} = \mathbf{k}'$.

a. Combine the two exiting modes using

$$\mathbf{P}' = (\mathbf{P}_o - \mathbf{k_D})e^{i\frac{2\pi}{\lambda}OPL_o} + (\mathbf{P}_e - \mathbf{k_D})e^{i\frac{2\pi}{\lambda}OPL_e} + \mathbf{k_D},$$

where $\mathbf{k_D}$ is the outer product of the $\mathbf{k}$ and $\mathbf{k}'$.

The outer product of (a_1, a_2, a_3) and (b_1, b_2, b_3) is $\begin{pmatrix} a_1 b_1 & a_1 b_2 & a_1 b_3 \\ a_2 b_1 & a_2 b_2 & a_2 b_3 \\ a_3 b_1 & a_3 b_2 & a_3 b_3 \end{pmatrix}$.

b. What incident **E** field will give a circularly polarized output **E** field $(1, i, 0)$?

20.3 Consider the interferometer setup:

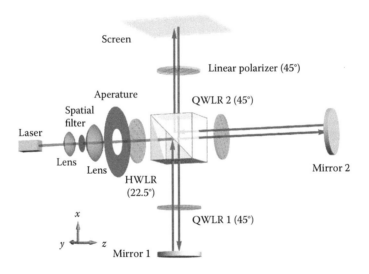

a. Is it valid to add the resultant **P** matrices? Why?
b. How does the overall **P** change as Mirror 2 moves along the *z*-axis by 0.25 waves?

20.4 Find the dyads $\mathbf{S}_D$ for the following vector fields (see Equation 20.5):
a. $(1, 0, 0)$ and $(1, 0, 0)$
b. $(1, 0, 0)$ and $(0, 1, 0)$
c. $(3, 4, 5)$ and $(3, 4, 5)$
d. $(3, 4, 5)$ and $(-3, -4, -5)$

20.5 Four rays exiting a system of birefringent elements with **P** matrices

$$\begin{pmatrix} 0.0853 & 0 & 0 \\ 0.2704 & 0 & 0 \\ 0 & 0 & 1 \end{pmatrix}, \begin{pmatrix} 0.8527 & 0 & 0 \\ -0.2689 & 0 & 0 \\ 0 & 0 & 1 \end{pmatrix}, \begin{pmatrix} 0 & 0.2720 & 0 \\ 0 & 0.8627 & 0 \\ 0 & 0 & 1 \end{pmatrix}, \text{ and } \begin{pmatrix} 0 & -0.2707 & 0 \\ 0 & 0.0853 & 0 \\ 0 & 0 & 1 \end{pmatrix}, \text{ and}$$

associated *OPL* 45.824 mm, 44.330 mm, 44.960 mm, and 45.466 mm, respectively. Calculate $\breve{\mathbf{P}}$ for each **P** matrix, and find $\mathbf{P}_{total}$ for the combination of the four exiting beams.

20.6 A Fabry–Perot cavity is formed with a birefringent crystal with thickness $t = 0.06$ mm between two partially reflecting mirrors. Each mirror transmits 0.64 of the flux with no phase shift and reflects 0.36 of the flux with a π phase shift. The dispersion of the ordinary refractive index is $n_O(\lambda) = 1.5 \times 0.6/\lambda$, and for the extraordinary index, $n_E(\lambda) = 1.6 \times 0.6/\lambda$. Find the $\mathbf{P}_0(\lambda)$ matrix for the directly transmitted beam, $\mathbf{P}_1(\lambda)$, the beam that makes one reflection at each surface and exits, and $\mathbf{P}_2(\lambda)$, which makes two reflections at each surface. Higher-order beams make a contribution of less than 0.2% of the total flux. At which wavelengths does the extraordinary transmitted flux peak? Ordinary?

References

1. A. Y. Karasik and V. A. Zubov. Laser Interferometry Principles, CRC Press (1995).
2. E. P. Goodwin and J. C. Wyant, *Field Guide to Interferometric Optical Testing*, SPIE (2006).
3. P. L. Polavarapu (ed.), *Principles and Applications of Polarization-Division Interferometry*, John Wiley & Sons (1998).
4. M. P. Rimmer and J. C. Wyant, Evaluation of large aberrations using a lateral-shear interferometer having variable shear, *Appl. Opt.* 14 (1975): 142–150.
5. C. M. Vest, *Holographic Interferometry*, New York: John Wiley & Sons (1979).
6. P. K. Rastogi, *Holographic Interferometry*, Vol. 1, Springer (1994).
7. C. L. Adler and J. A. Lock, Caustics due to complex water menisci, *Appl. Opt.* 54 (2015): B207–B221.
8. M. Avendaño-Alejo, Caustics in a meridional plane produced by plano-convex aspheric lenses, *J. Opt. Soc. Am. A* 30 (2013): 501–508.
9. H. Noble, E. Ford, W. Dallas, R. A. Chipman, I. Matsubara, Y. Unno, S. McClain, P. Khulbe, D. Hansen, and T. D. Milster, Polarization synthesis by computer-generated holography using orthogonally polarized and correlated speckle patterns, *Opt. Lett.* 35 (2010): 3423–3425.
10. H. Noble, W. Lam, W. Dallas, R. Chipman, I. Matsubara, Y. Unno, S. McClain, P. Khulbe, D. Hansen, and T. Milster, Square-wave retarder for polarization computer-generated holography, *Appl. Opt.* 50 (2011): 3703–3710.
11. D. Malacara (ed.), *Optical Shop Testing*, 2nd edition, John Wiley & Sons (1992).
12. X. Geng and P. Hu, Spatial interpolation method of scalar data based on raster distance transformation of map algebra, in *IPTC, Intelligence Information Processing and Trusted Computing, International Symposium*, 188–191 (2010).
13. I. Amidror, Scattered data interpolation methods for electronic imaging systems: A survey, *J. Electronic Imaging* 2(11) (2002): 157–176.
14. S. J. Owen, An implementation of natural neighbor interpolation in three dimensions, Master's thesis, Brigham Young University (1992).
15. D. Shepard, A two-dimensional interpolation function for irregularly-spaced data, *Proc. 23rd Natl. Conf. ACM*, 517–524 (1968).
16. W. T. Vetterling, W. H. Press, S. A. Teukolsky, and B. P. Flannery, *Numerical Recipes Example Book C (The Art of Scientific Computing)*, Cambridge University Press (1993).
17. N. S. Lam, Spatial interpolation methods review, *Am. Cartogr.* 10 (1983): 129–149.
18. W. H. Press, B. P. Flannery, S. A. Teukolsky, and W. T. Vetterling, *Numerical Recipes in C: The Art of Scientific Computing*, 2nd revised edition, Cambridge University Press (1992).

19. P. S. Kelway, A scheme for assessing the reliability of interpolated rainfall estimates, *J. Hydrol.* 21 (1974): 247–267.
20. NOAA, National Weather Service River Forecast System Forecast Procedures, TM NWS HYDRO-14, U.S. Department of Commerce, Washington, DC (1972).
21. J. J. Kaluarachchi, J. C. Parker, and R. J. Lenhard, A numerical model for areal migration of water and light hydrocarbon in unconfined aquifers, *Adv. Water Resour.* 13 (1990): 29–40.
22. D. F. Watson and G. M. Philip, A refinement of inverse distance weighted interpolation, *Geo-Processing*, 2 (1985): 315–327.
23. D. G. Krige, A statistical approach to some mine valuations and allied problems at the Witwatersrand, Master's thesis, the University of Witwatersrand (1951).
24. J. Duchon, Splines minimizing rotation invariant semi-norms in Sobolev spaces, in *Constructive Theory of Functions of Several Variables*, Oberwolfach (1976), pp. 85–100.

21

Uniaxial Materials and Components

21.1 Optical Design Issues in Uniaxial Materials

Uniaxial materials are the most common anisotropic materials used as optical components. They are a special case of biaxial materials with two equal principal refractive indices; thus, the two corresponding principal axes become degenerate. The wavefront aberrations induced by the birefringent material vary with the polarization of the light and the direction of propagation. Uniaxial materials, such as calcite and quartz, are frequently used to generate phase delays between eigenpolarizations or to divide wavefronts based on polarization and direct these components into different directions.

The objectives of this chapter are the following:

- Understanding light propagation through uniaxial optical components
- Finding the shape of the *extraordinary wavefront*
- Determining how light divides based on polarization into *eigenmodes*
- Calculating the resultant phase and polarization state
- Understanding the polarization aberrations of example uniaxial optical components

The special properties of uniaxial crystals were first described by Erasmus Bartholin who in 1669 discovered *double refraction* when light transmitted through the *crystal Iceland spar*, which is one of the names for calcite.[1] After Bartholin first observed double refraction (Figure 21.1), he could not explain the phenomenon, but he did obtain the *ordinary refractive index* of the Iceland crystal. Ten years later in 1690, Christiaan Huygens developed his revolutionary wave optics, which

Figure 21.1 Imaging doubling through a piece of calcite.

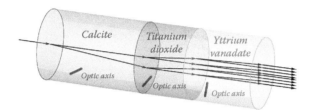

Figure 21.2 A real ray trace through three blocks of uniaxial material with arbitrarily oriented optic axes. Entering the first crystal, calcite, the incident ray splits into two rays. Each of these splits again at the second interface, and they further divide into eight rays at the third interface.

included a law of refraction for the *extraordinary ray*. E. L. Malus in 1808 described the image doubling through calcite with polarized light and confirmed Huygens' extraordinary refraction result.[2,3]

Common uniaxial crystalline polarization elements include crystal waveplates and crystal polarizers, such as Glan–Taylor polarizers,* Glan–Thompson polarizers, and Wollaston prisms. Many birefringent devices contain more than one uniaxial components for athermalization or achromatization.[4,5] Ray tracing through such birefringent components, such as Lyot filters, becomes complicated. Consider a ray propagating through a crystal assembly shown in Figure 21.2, where ray doubling at every birefringent interface results in multiple exiting rays with different polarizations. The one incident ray shown is one of many rays from a spherical wavefront, and the multiple exiting rays correspond to multiple exiting wavefronts. Each of these polarized wavefronts has a different sequence of modes, a combination of ordinary and extraordinary eigenmodes inside the uniaxial media. And each takes a completely different path, focuses in different locations, and has a different amount of aberration, as described in Chapter 19.

Understanding the resultant wavefronts and their aberrations aids in finding suitable configurations of optical elements and balancing aberrations. Chapter 19 developed the general method for polarization ray tracing through all types of birefringent materials, including uniaxial materials, where each material is represented by its dielectric tensor. The algorithm allows optical design software to trace rays through general systems of birefringent materials. This chapter presents the index ellipsoid or the optical indicatrix, which represents the dielectric tensor in the shape of an ellipsoid to help visualize the birefringent wavefronts.

The wavefront aberrations of optical waveplates are evaluated in this chapter by simulating the propagation of the wavefront through the uniaxial crystal assembly with a large number of rays in a grid, as shown in Figure 21.3 for one uniaxial calcite crystal. The combined effect of the multiple exiting polarized wavefronts is analyzed in Chapter 22 (Crystal Polarizers).

---

* A detailed analysis on Glan–Taylor polarizers is included in Chapter 22 (Crystal Polarizers).

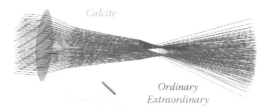

Figure 21.3 A converging beam is focused through a thick plate of calcite with optic axis (green line) in the plane of the page. Because of ray doubling, two separated foci are observed, one for the ordinary (*o*) mode and another one for the extraordinary (*e*) mode.

21.2 Descriptions of Uniaxial Materials

In general, when a plane wave is incident onto a uniaxial crystal interface, it refracts into two plane waves with mutually orthogonal polarizations in different propagation directions. These two waves are the two *eigenmodes*, *ordinary o*-mode and *extraordinary e*-mode. The refractive index of one of these eigenmodes, the *e*-mode, varies with the direction of the polarization state, due to the electric and magnetic field's interaction with the varying atomic arrangement in different crystal directions.

Uniaxial materials include the tetragonal, hexagonal, and trigonal crystallized minerals.[6] Inside the crystal, the response of the material changes the light field oscillation depending on the field orientation relative to the different chemical bonds. In each uniaxial crystal calcite, $CaCO_3$, all the calcium–carbon bonds are aligned in one direction, which is the *extraordinary principal axis*. The three carbon–oxygen bonds in the carbonate radical are oriented in the perpendicular plane, as shown in Figure 21.4. The response of the ionic calcium–carbon bond to the light's electric field is different from the response of the covalent bonds in the carbonate group, therefore causing charges to oscillate along the calcium–carbon bond, which generates a refractive index different from driving charges in the plane of the carbonate bonds.

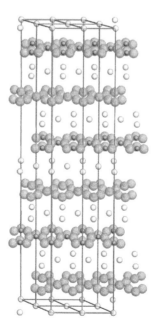

Figure 21.4 Calcite $CaCO_3$ crystal structure with the vertical optic axis, connecting the calcium (blue) and carbon (purple) atoms.

The *optic axis* of a birefringent material is a direction of propagation in which the light experiences zero birefringence. Uniaxial materials only have a single optic axis, hence the name uniaxial. When light propagates in either direction along the optic axis of uniaxial calcite, the electric field components in the transverse plane with the threefold symmetry of the carbon–oxygen bonds can only experience the ordinary refractive index n_O; the two eigenmodes are degenerate. Therefore, the optic axis of a uniaxial material is the extraordinary principal axis.

In a uniaxial medium, under the influence of the light field, the charges oscillate and contribute to the electric field of the light propagating through the medium. The relationship of the displacement field **D** to electric field **E** is described by a tensor ε in Equation 21.1,[6–8] which shows that the bound charges of uniaxial material respond in a different direction from the direction of **E**.

$$\mathbf{D} = \boldsymbol{\varepsilon}\,\mathbf{E}$$

$$\begin{pmatrix} D_x \\ D_y \\ D_z \end{pmatrix} = \begin{pmatrix} \varepsilon_O & 0 & 0 \\ 0 & \varepsilon_O & 0 \\ 0 & 0 & \varepsilon_E \end{pmatrix} \begin{pmatrix} E_x \\ E_y \\ E_z \end{pmatrix}$$

$$\begin{pmatrix} D_x \\ D_y \\ D_z \end{pmatrix} = \begin{pmatrix} (n_O + i\kappa_O)^2 & 0 & 0 \\ 0 & (n_O + i\kappa_O)^2 & 0 \\ 0 & 0 & (n_E + i\kappa_E)^2 \end{pmatrix} \begin{pmatrix} E_x \\ E_y \\ E_z \end{pmatrix}, \quad (21.1)$$

where ε is the dielectric tensor for uniaxial birefringent materials. The dielectric tensor components, $\varepsilon_O = (n_O + i\kappa_O)^2$ and $\varepsilon_E = (n_E + i\kappa_E)^2$, are the ordinary and extraordinary delectric constants. Absorption is represented by the imaginary refractive index components κ, the restoring force of electrons. For a transparent crystal, κ is negligibly small. The two principal refractive indices of a uniaxial material are n_O and n_E.

Equation 21.1 describes the dielectric tensor of a material whose optic axis is oriented along the z-axis. To represent a uniaxial medium with arbitrary orientation, the uniaxial dielectric tensor is rotated from its diagonal form. For an optic axis oriented along unit vector $\mathbf{v_E}$ and two mutually orthogonal unit vectors $\mathbf{v_{O1}}$ and $\mathbf{v_{O2}}$, both perpendicular to $\mathbf{v_E}$, the resultant dielectric tensor is

$$\boldsymbol{\varepsilon} = \begin{pmatrix} v_{O1x} & v_{O2x} & v_{Ex} \\ v_{O1y} & v_{O2y} & v_{Ey} \\ v_{O1z} & v_{O2z} & v_{Ez} \end{pmatrix} \cdot \boldsymbol{\varepsilon_D} \cdot \begin{pmatrix} v_{O1x} & v_{O2x} & v_{Ex} \\ v_{O1y} & v_{O2y} & v_{Ey} \\ v_{O1z} & v_{O2z} & v_{Ez} \end{pmatrix}^{-1}, \text{ where } \boldsymbol{\varepsilon_D} = \begin{pmatrix} n_O^2 & 0 & 0 \\ 0 & n_O^2 & 0 \\ 0 & 0 & n_E^2 \end{pmatrix}^{-1}. \quad (21.2)$$

This is a matrix rotation operation, a unitary change of basis.

Uniaxial materials are categorized by their values of n_O and n_E. *Negative uniaxial crystals* have $n_O > n_E$ and include calcite and sapphire. *Positive uniaxial crystals*, such as magnesium fluoride, have $n_O < n_E$. The principal refractive indices for some common uniaxial crystals are tabulated in Table 21.1.

The maximum *birefringence* of a uniaxial medium is the difference between n_O and n_E, which is the maximum birefringence a ray can experience. Since the refractive index changes with wavelength, the birefringence is also a function of wavelength. The birefringence spectra Δn of several uniaxial materials are plotted in Figure 21.5, where

$$\Delta n = n_E - n_O. \quad (21.3)$$

Positive and negative uniaxial materials have positive and negative birefringence, respectively.

21.3 Eigenmodes of Uniaxial Materials

Table 21.1 Uniaxial Material Principal Refractive Index at 587.56 nm

Material		n_O	n_E
Negative Uniaxial Crystals[9]			
Sapphire	α-Al_2O_3	1.76817	1.76009
Calcite	$CaCO_3$	1.65864	1.48649
Lithium niobate	$LiNbO_3$	2.30014	2.21453
Positive Uniaxial Crystals[9,10]			
Magnesium fluoride	MgF_2	1.37775	1.38957
Quartz[a]	SiO_2	1.54431	1.55343
Rutile	TiO_2	2.61423	2.91031
Yttrium vanadate	YVO_4	2.00269	2.22940
Zinc oxide	ZnO	2.00337	2.01986
Titanium oxide	TiO_2	2.61605	2.91260

[a] Quartz has small optical activity within a few degrees of the optic axis.

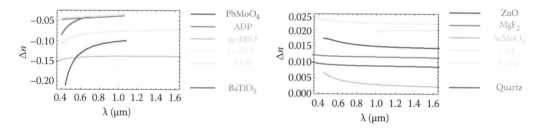

Figure 21.5 Birefringence spectra of common negative and positive uniaxial materials.[9,10]

21.3 Eigenmodes of Uniaxial Materials

When light refracts into a birefringent material, its energy divides into two eigenmodes in the process of *double refraction* or *ray splitting*. The refractive indices associated with these two eigenmodes are n_o and n_e. The *principal section* of the crystal is the plane of ray splitting. Light polarized orthogonal to the optic axis, for example, in the carbonate plane of calcite, is the *ordinary o-mode* and has the ordinary refractive index n_o, which is always equal to the material principal index n_O. Light polarized perpendicular to the o-mode, that is, with a component along the optic axis, is the *extraordinary e-mode* and experiences the extraordinary refractive index n_e, which has a value between n_E and n_O. The *birefringence of a ray* is the refractive index difference between the eigenmodes, $|n_o - n_e|$, which is equal to or less than the birefringence of the medium.

Ray doubling is observed in Figure 21.6, which shows the double refraction when imaging the text "POLARIS" through a calcite crystal. The direction of image doubling rotates as the crystal rotates, since the principal section is fixed with the crystal. As the crystal rotates, the principal section also rotates; the o-mode image remains fixed while the e-mode image rotates around the o-image. One can demonstrate that the two modes have orthogonal linear polarization states by rotating a linear polarizer above the crystal.

Figure 21.7 (right) shows the schematic of a ray refracting into a uniaxial plate, which has an optic axis in the plane of the page 42° to the surface normal. The optic axis and the incident ray are both in the plane of the page, the principal section of the crystal that is also the plane of ray splitting.

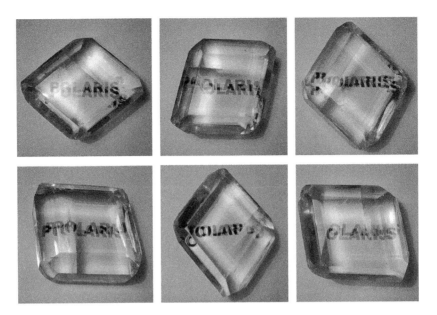

Figure 21.6 Double refraction through a calcite rhomb. When rotating the calcite, the ordinary image remains fixed while the extraordinary image rotates around it.

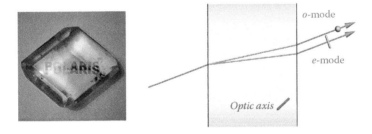

Figure 21.7 (Left) Double refraction through a calcite rhomb. (Right) A schematic of a ray splitting into two modes through a calcite block. The optic axis shown is 42° from horizontal. In this case, the e-state is in the plane of the page; the o-state is oriented out of the plane of the page.

The refracted polarization state in the principal section is the e-mode, and the other refracted ray polarized perpendicular to the principal section is the o-mode. For unpolarized incident ray, the incident flux divides between the o- and e-modes in refraction. For light polarized in the plane of the page, all energy refracts into the e-mode. Similarly, for light polarized perpendicular to the plane of the page, all the light refracts into the o-mode. Thus, the amount of flux coupled into the two modes is a function of the incident polarization state.

21.4 Reflections and Refractions at a Uniaxial Interface

When light strikes a uniaxial interface, its flux divides between refracted and reflected beams. The portion of the flux refracted and reflected depends on the incident light's polarization and the materials involved. When light propagates into a uniaxial medium, its energy splits into two directions between the o- and the e-modes. Because of the uniaxial crystal symmetry, the e-mode electric field

21.4 Reflections and Refractions at a Uniaxial Interface

is always in the plane containing the optic axis, and the *o*-mode electric field is always in the plane perpendicular to the optic axis.

Example 21.1 shows the different orientations and refractive indices associated with the two orthogonally polarized modes. The *o*-mode always follows Snell's law and has index $n_o = n_O$. When the *e*-mode electric field is orthogonal to the optic axis, it has the same index as the *o*-mode, $n_e = n_O$. As soon as the direction of the *e*-mode electric field deviates from being perpendicular to the optic axis, n_e departs from n_O toward n_E. This *e*-ray index becomes closer to n_E as the *e*-mode electric field oscillates closer to the orientation of the optic axis. When the *e*-mode electric field is aligned with the optic axis, $n_e = n_E$.

Example 21.1 Refractive Index of *o*- and *e*-Modes

Figure 21.8 depicts a ray propagating inside a uniaxial medium in four different directions relative to the optic axis. What are the refractive indices experienced by the electric field components in each case relative to the principal indices n_O and n_E?

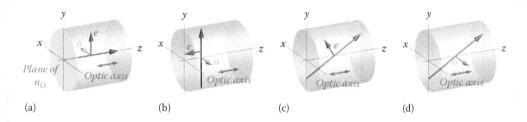

Figure 21.8 A ray (black arrow) propagates in four different directions inside a uniaxial material whose optic axis (red arrow) lies along the *z*-axis. The plane corresponding to the ordinary index n_O, shown as a green plane in (a), is orthogonal to the optic axis. The *e* and *o* electric fields of the ray are represented by the blue and green arrows, respectively. The ray is propagating along the *z*-direction in (a), along the *y*-direction in (b), and in the *y*–*z* plane in (c) and (d). The yellow plane shown in (c) and (d) is the *y*–*z* plane in which the ray is propagating.

A uniaxial material is characterized by two principal refractive indices and an optic axis direction. The *extraordinary principal refractive index* n_E applies to light polarized along the optics axis; the *ordinary principal refractive index* n_O applies to light polarized in the plane perpendicular to the optic axis.

In Figure 21.8a, a ray propagates in the direction of the optic axis. Both components of the electric field are oscillating orthogonal to the optic axis and in the plane of n_O. Therefore, they both experience refractive index n_O; $n_o = n_e = n_O$.

In Figure 21.8b, a ray propagates perpendicular to the optic axis with its *e*-component electric field oscillating along the optic axis with an index n_e equal to the principal index n_E. The *o*-component is orthogonal to the optic axis and has an index n_o equal to the ordinary principal index n_O.

When a ray propagates neither parallel nor perpendicular to the optic axis, as in Figure 21.8c and d, the mode refractive index depends on the angle between its electric field and the optic axis. The ray of the *e*-mode in Figure 21.8c lies in the *y*–*z* plane and has an electric field oscillating in the *y*–*z* plane orthogonal to the ray direction. The refractive index of the *e*-mode is between n_O and n_E. For the *o*-mode shown in Figure 21.8d, the electric field is orthogonal to the optic axis and it has index n_O.

In refraction or reflection, the energy of the two modes propagates in two separate paths with directions $\hat{\mathbf{S}}_o$ and $\hat{\mathbf{S}}_e$ in uniaxial material. Consider a ray is normally incident onto several negative uniaxial crystal blocks with different optic axis orientations in the plane of the page, as shown in Figure 21.9. The unpolarized incident ray divides into o- and e-rays in two directions. The o-mode always has its Poynting vector aligned with its propagation vector; $\hat{\mathbf{S}}_o = \hat{\mathbf{k}}_o$. For all the cases shown in Figure 21.9, the o-mode electric field $\mathbf{E}_o$ oscillates along the x-axis; hence, it is always orthogonal to the optic axis. For normal incidence, the propagation vectors of both modes, $\hat{\mathbf{k}}_o$ and $\hat{\mathbf{k}}_e$, are along the z-direction. However, the wavefront of the e-mode (the surfaces of constant phase) is not orthogonal to the Poynting vector; thus, $\hat{\mathbf{S}}_e$ is not aligned with $\hat{\mathbf{k}}_e$; the electric field $\mathbf{E}_e$ oscillates perpendicular to $\hat{\mathbf{S}}_e$, but not $\hat{\mathbf{k}}_e$. Their direction changes with the optic axis orientation and the corresponding n_e varies quadratically through n_E. To be orthogonal to both the optic axis and the o-mode electric field in this example, $\mathbf{E}_e$ oscillates in the y–z plane. The refracted electric field inside the uniaxial material is the sum of the two modes:

$$\begin{aligned}\mathbf{E}(\mathbf{r},t) &= \mathcal{R}e\left\{E_{inc}\left[a_o\hat{\mathbf{E}}_o e^{i\frac{2\pi}{\lambda}n_o\hat{\mathbf{k}}_o\cdot\mathbf{r}} + a_e\hat{\mathbf{E}}_e e^{i\frac{2\pi}{\lambda}n_e\hat{\mathbf{k}}_e\cdot\mathbf{r}}\right]e^{-i\omega t}\right\} \\ \mathbf{E}(z,t) &= \mathcal{R}e\left\{E_{inc}\left[a_o\hat{\mathbf{E}}_o e^{i\frac{2\pi}{\lambda}n_o k_{oz}z} + a_e\hat{\mathbf{E}}_e e^{i\frac{2\pi}{\lambda}n_e k_{ez}z}\right]e^{-i\omega t}\right\},\end{aligned} \qquad (21.4)$$

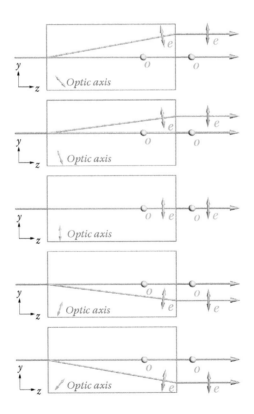

Figure 21.9 A ray at normal incidence refracts into a negative uniaxial crystal block. The direction of energy flow is along the Poynting vectors **S**, shown as arrows passing through the crystal block. The optic axis is shown for several orientations in the y–z plane. Incident flux divides into the o-mode and the e-mode. Only the propagation of the e-mode depends on the optic axis orientation. The electric field of the o- and e-modes is shown orthogonal to **S** perpendicular to the page and in the plane of the page, respectively.

where a_o and a_e are the complex electric field amplitude transmission coefficients. When the normal incident $\mathbf{k}_{inc}$ is perpendicular to the optic axis, as shown in the middle of Figure 21.9, the two modes follow the same path, $\hat{\mathbf{k}}_o = \hat{\mathbf{k}}_e$ and $\hat{\mathbf{S}}_O = \hat{\mathbf{S}}_e$, but the *e*-mode has a lower refractive index than the *o*-mode in the negative uniaxial material and thus results in a linearly increasing phase delay between the two eigenmodes as they propagate.

21.5 Index Ellipsoid, Optical Indicatrix, and *K*- and *S*-Surfaces

The refraction and reflection at birefringent interface can be understood by a geometrical construction as well as the algebraic algorithm in Chapter 19. In this geometrical approach, the calculation of ray parameters is visualized using the *index ellipsoid* and *optical indicatrix*, which are very helpful for understanding wavefront shapes and aberrations.

The **E** and **D** fields are related by the dielectric tensor, and the *energy density u* (in volts per meter) in the electromagnetic wave,

$$\mathbf{D} = \varepsilon\, \mathbf{E} \quad \text{and} \quad u = \frac{1}{2} \mathbf{E} \cdot \mathbf{D}. \tag{21.5}$$

In the principal coordinates, the dielectric tensor is diagonal and the three principal axes are aligned with the *xyz*-axes,

$$\begin{pmatrix} D_x \\ D_y \\ D_z \end{pmatrix} = \begin{pmatrix} n_x^2 & 0 & 0 \\ 0 & n_y^2 & 0 \\ 0 & 0 & n_z^2 \end{pmatrix} \begin{pmatrix} E_x \\ E_y \\ E_z \end{pmatrix}; \tag{21.6}$$

then,

$$u = \frac{1}{2}\left(n_x^2 E_x + n_y^2 E_y + n_z^2 E_z\right). \tag{21.7}$$

Hence,

$$1 = \frac{\left(\dfrac{E_x}{\sqrt{2u}}\right)^2}{(1/n_x)^2} + \frac{\left(\dfrac{E_y}{\sqrt{2u}}\right)^2}{(1/n_y)^2} + \frac{\left(\dfrac{E_z}{\sqrt{2u}}\right)^2}{(1/n_z)^2}. \tag{21.8}$$

The solutions for the electric field vector in Equation 21.8 can be represented as a *ray ellipsoid* with semi-axes $1/n_x$, $1/n_y$, and $1/n_z$, as shown in Figure 21.10. For rays emanating from the origin, the distance to the ellipsoid is proportional to the propagation distance for a ray with the corresponding **E** field. For the uniaxial example with propagation along the *z*-axis, the field can be located anywhere in the *x*–*y* plane, and all those rays propagate with equal speed; thus, that cross section of the ellipse is circular.

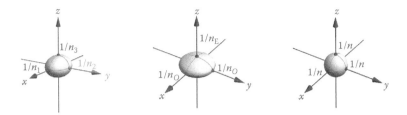

Figure 21.10 The ray ellipsoid of biaxial (left), positive uniaxial (middle), and isotropic materials (right). The principal axes are aligned with the coordinate axis with $(x, y, z) = (E_x, E_y, E_z)/\sqrt{2u}$.

On the other hand, the electric field can be expressed in terms of the displacement field as

$$\mathbf{E} = \boldsymbol{\varepsilon}^{-1}\mathbf{D} \text{ or } \begin{pmatrix} E_x \\ E_y \\ E_z \end{pmatrix} = \begin{pmatrix} 1/n_x^2 & 0 & 0 \\ 0 & 1/n_y^2 & 0 \\ 0 & 0 & 1/n_z^2 \end{pmatrix} \begin{pmatrix} D_x \\ D_y \\ D_z \end{pmatrix}, \quad (21.9)$$

where the inverse of $\boldsymbol{\varepsilon}$ is the *impermeability tensor*

$$\boldsymbol{\varepsilon}^{-1} = \begin{pmatrix} 1/n_x^2 & 0 & 0 \\ 0 & 1/n_y^2 & 0 \\ 0 & 0 & 1/n_z^2 \end{pmatrix}. \quad (21.10)$$

Then, the energy density is

$$u = \frac{1}{2}\mathbf{E} \cdot \mathbf{D} = \frac{1}{2}\left(\frac{D_x^2}{n_x^2} + \frac{D_y^2}{n_y^2} + \frac{D_z^2}{n_z^2}\right) \quad (21.11)$$

and

$$1 = \frac{\left(\frac{D_x}{\sqrt{2u}}\right)^2}{n_x^2} + \frac{\left(\frac{D_y}{\sqrt{2u}}\right)^2}{n_y^2} + \frac{\left(\frac{D_z}{\sqrt{2u}}\right)^2}{n_z^2}. \quad (21.12)$$

With Equation 21.12, the dielectric tensor is visualized as an *index ellipsoid* shown in Figure 21.11, whose semi-axes are the principal refractive indices in the principal coordinates. The index ellipsoid is variously called the *ellipsoid of wave normal*, *reciprocal ellipsoid*, or *index indicatrix*.[7,9,11–13] The distance from the origin corresponds to the refractive index for a mode with the corresponding **D** field direction.

Consider a ray normally incident onto a uniaxial crystal in Figure 21.12 (left), which, as the index ellipsoid, rotated from the z-axis by α. At normal incidence, the refracted o- and e-rays have the same $\mathbf{k} = (0, 0, 1)$. The **D** field and their corresponding refractive indices can be calculated geometrically using the ellipsoid. In Figure 21.12 (right), the index ellipsoid has been rotated with its optic axis at an angle α from z.

21.5 Index Ellipsoid, Optical Indicatrix, and K- and S-Surfaces

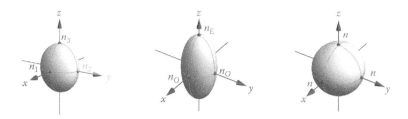

Figure 21.11 Index ellipsoid of biaxial (left), positive uniaxial (middle), and isotropic materials (right). Their principal axes are aligned with the coordinate axes with $(x, y, z) = \left(D_x, D_y, D_z\right)/\sqrt{2u}$.

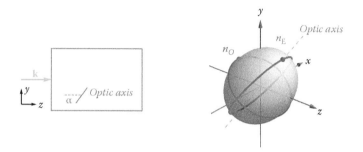

Figure 21.12 (Left) A ray normally incident to a block of uniaxial material. (Right) The index ellipsoid in the system coordinates where the optic axis is shown rotated by α from the z-axis in the y–z plane. The blue ellipse goes through the optic axis. The red circle is perpendicular to the optic axis.

The eigenmodes in the uniaxial material follow from Maxwell's equations,

$$\mathbf{k} \times \mathbf{H} = -\omega\, \mathbf{D} \quad \text{and} \quad \mathbf{k} \times \mathbf{E} = \omega\, \mu_o \mathbf{H}, \tag{21.13}$$

where ω is the light frequency in radians per second. Therefore,

$$\mathbf{k} \times (\mathbf{k} \times \mathbf{E}) = -\omega^2 \mu_o \mathbf{D}. \tag{21.14}$$

With Equation 21.6,

$$-\frac{\mathbf{k}}{k} \times \left(\frac{\mathbf{k}}{k} \times \boldsymbol{\varepsilon}^{-1} \mathbf{D}\right) = \frac{1}{n^2} \mathbf{D}, \tag{21.15}$$

where $k = n\omega/c$ and $k = |\mathbf{k}|$. This reveals that the projection of $\boldsymbol{\varepsilon}^{-1}\mathbf{D}$ onto a plane perpendicular to $\mathbf{k}$ is the $\mathbf{D}$ field direction of the e-mode. Within the index ellipsoid, the surface orthogonal to $\mathbf{k}$ including the origin is the *index ellipse* shown in Figure 21.13. For the example in Figure 21.12, the index ellipse is the intersection of the x–y plane with the crystal's index ellipsoid. The $\mathbf{D}$ fields of the two eigenmodes are along the major and minor axes of the index ellipse. The lengths of these two axes correspond to the eigenmode refractive indices. For the e-mode, $\mathbf{D}_e$ is in the plane of the optic axis and orthogonal to $\mathbf{k}$, as shown in Figure 21.13, and $\mathbf{D}_o$ for the o-mode is in the plane orthogonal to $\mathbf{D}_e$ and $\mathbf{k}$.

A general uniaxial index ellipse for all possible $\mathbf{k}$ always has at least one axis with length n_O, as shown in Figure 21.14 corresponding to the o-ray index n_o. For a positive uniaxial material, the

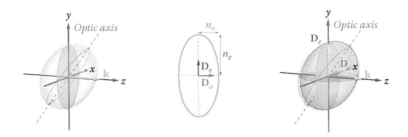

Figure 21.13 (Left) For a ray with propagation vector **k** (green) propagating through a negative uniaxial crystal, the ray index ellipse (green disk) in the x–y plane is the intersection of the plane normal to **k** and the optical indicatrix. The brown ellipse is the great ellipse containing the optic axis. (Middle) The minor and major axes of the index ellipse correspond to the **D** field directions of the o- and e-modes, and the lengths of these two axes are the refractive indices. (Right) $\mathbf{D}_e$ (pink arrow) is on the pink plane in the y–z plane containing both the optic axis (dashed brown line) and the **k** vector. $\mathbf{D}_o$ (blue arrow) is on the blue plane in the x–z plane orthogonal to the pink $\mathbf{D}_e$ plane. The three planes are orthogonal to each other, and both $\mathbf{D}_o$ and $\mathbf{D}_e$ are lying on the index ellipse in the x–y plane.

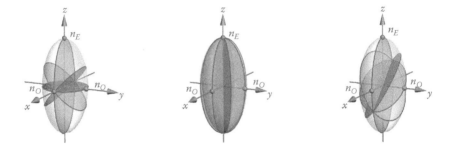

Figure 21.14 Several possible index ellipses (blue) in the positive uniaxial index ellipsoid are shown. (Left) The index ellipses shown contain the x-axis. (Middle) The index ellipses shown contain the z-axis. (Right) The index ellipses shown contain an axis 45° from the x-axis in the x–z plane.

length of the major axis is the e-mode index n_e. The equation for n_e is obtained from Figure 21.15 by projecting the index ellipsoid and the index ellipse onto a 2D plane:

$$\frac{1}{n_e(\theta)^2} = \frac{\cos^2\theta}{n_O^2} + \frac{\sin^2\theta}{n_E^2}, \tag{21.16}$$

where θ is the angle between **k** and the optic axis. n_e becomes n_O in the limit when **k** is along the optic axis, and n_e is n_E when **k** lies in the plane orthogonal to the optic axis.

The direction of energy propagation, the Poynting vector **S**, and its corresponding **E** field are calculated using another ellipsoid derived from Equation 21.14, which reduces to

$$\mathbf{k} \times (\mathbf{k} \times \mathbf{E}) + \mu\varepsilon\omega^2 \mathbf{E} = 0$$

$$\begin{pmatrix} n_1^2 k_0^2 - k_y^2 - k_z^2 & k_x k_y & k_x k_z \\ k_y k_x & n_2^2 k_0^2 - k_x^2 - k_z^2 & k_y k_z \\ k_z k_x & k_z k_y & n_3^2 k_0^2 - k_x^2 - k_y^2 \end{pmatrix} \begin{pmatrix} E_x \\ E_y \\ E_z \end{pmatrix} = \begin{pmatrix} 0 \\ 0 \\ 0 \end{pmatrix}, \tag{21.17}$$

21.5 Index Ellipsoid, Optical Indicatrix, and K- and S-Surfaces

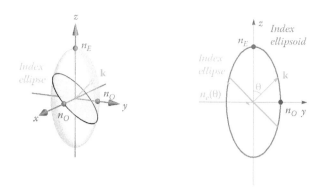

Figure 21.15 (Left) A ray propagating in the y–z plane has an index ellipse (green disk) lying in a plane orthogonal to **k** and containing the x-axis. **k** lies at an angle θ from the optic axis along the z-axis. (Right) The projected length of the index ellipse is the magnitude of $n_e(\theta)$.

where $\mathbf{k} = (k_x, k_y, k_z)$, $k_0 = \omega/c$ and ω is the angular frequency. For a nontrivial solution of **E**, the matrix determinant in Equation 21.17 must be zero, which results in the dispersion relation and defines an ellipsoid called the *K-surface*, also called the *normal surface*.[6,14] For a uniaxial material, this is a function of k_0:

$$\left(\frac{k_x^2/k_0^2}{n_O^2} + \frac{k_y^2/k_0^2}{n_O^2} + \frac{k_z^2/k_0^2}{n_O^2} - 1 \right)\left(\frac{k_x^2/k_0^2}{n_E^2} + \frac{k_y^2/k_0^2}{n_E^2} + \frac{k_z^2/k_0^2}{n_O^2} - 1 \right) = 0. \qquad (21.18)$$

The left part of the equation is the solution for the *o*-mode with a spherical *K*-surface, while the right part is the solution for the *e*-mode with an ellipsoid *K*-surface. Both of these *K*-surfaces are plotted in the frame of their principal axes $(k_x/k_0, k_y/k_0, k_z/k_0)$ in Figure 21.16.

For the example shown in Figure 21.12, the *K*-surfaces are rotated by α; hence, Figure 21.16 becomes Figure 21.17. The Poynting vector **S** is normal to the *K*-surface at the intersection of **k**, as shown in Figures 21.17 and 21.18. The ordinary $\mathbf{k}_o$ is always in the same direction as $\mathbf{S}_o$; the ordinary wavefront is spherical just like an isotropic wavefront. On the other hand, the extraordinary $\mathbf{k}_e$ is not parallel with $\mathbf{S}_e$, unless $\mathbf{k}_e$ is along the optic axis or orthogonal to it. As the energy follows $\mathbf{S}_e$, the constant phase fronts along the $\mathbf{D}_e$ field that are perpendicular to $\mathbf{k}_e$ are not orthogonal to $\mathbf{S}_e$, and the $\mathbf{E}_e$ field orthogonal to $\mathbf{S}_e$ lies tangent to the *K*-surface.

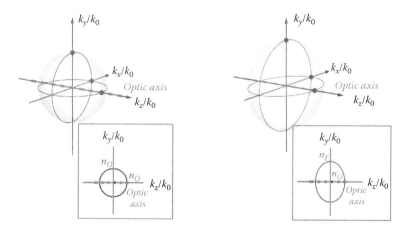

Figure 21.16 *K*-surface and its cross section for the *o*-mode (red sphere) and the *e*-mode (blue ellipsoid). The optic axis lies on the k_z/k_0 axis.

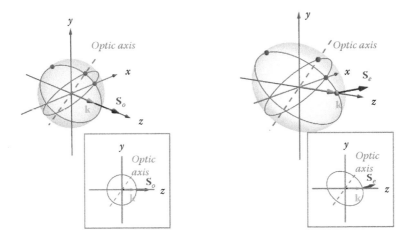

Figure 21.17 The *K*-surfaces are oriented for the example in Figure 21.12, where the optic axis is rotated by α from the *z*-axis. The Poynting vectors are normal to the *K*-surface.

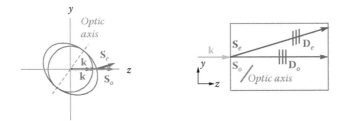

Figure 21.18 (Left) The Poynting vectors are normal to the *K*-surface. (Right) The incident energy divides into S_e and S_o directions. Their wavefronts **D**, shown as the three parallel lines, are orthogonal to **k** and not aligned with **S**.

When a ray refracts through a birefringent interface, the refraction directions are fixed by the phase matching condition across the boundary for both eigenstates. The ordinary k_o is calculated as in an isotropic medium. The extraordinary k_e still obeys Snell's law with consideration for its varying refractive index as a function of direction:

$$n \sin\theta_i = n_o \sin\theta_o$$
$$n \sin\theta_i = n_e(\theta_a + \theta_e)\sin\theta_e, \qquad (21.19)$$

where θ_a is the angle between the surface normal and the optic axis, θ_e is the refraction angle, and n_e is a function of θ_a and θ_e. Equations 21.16 and 21.19 must be solved simultaneously for the *e*-mode index and refraction angle. For the normal incidence example, the incident and exiting *K*-surfaces are shown in Figure 21.19, where $\theta_i = \theta_o = \theta_e = 0$.

Trace ray using this geometrical method requires the crystal's principal indices, the principal axis orientation, and the incident **k** direction. The ray trace procedure is as follows:

1. Calculate the refractive indices of the two eigenstates, and their propagation directions using Equations 21.16 and 21.19.
2. Use the index ellipsoid to determine **D**, from which **E** is calculated.
3. Use the *K*-surfaces to determine **S**.

21.5 Index Ellipsoid, Optical Indicatrix, and K- and S-Surfaces

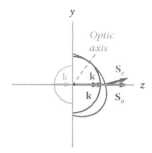

Figure 21.19 The K-surface of refraction through isotropic material to uniaxial material for normal incidence.

The method works for both on-axis and off-axis rays as shown in Example 21.2. The ray parameters determined from the index ellipsoid and its K-surface are the same as calculated using the general anisotropic interface refraction method in Chapter 19. A detailed calculation of the necessary ray parameters described in Section 19.4, including the **P** matrices, through a birefringent interface is shown in Example 21.3, which traces the same ray as in Example 21.2.

Example 21.2 Off-Axis Ray Trace from Air into a Uniaxial Medium

In this example, an off-axis ray propagates from air into a uniaxial material. The ray tracing parameters are calculated using the optical ellipsoid and K-surface described in Section 21.5. The uniaxial material has principal indices $n_O = 1.3$ and $n_E = 2.5$ with its optic axis oriented in the y–z plane 10° from the z-axis. The off-axis incident ray propagates in the y–z plane 50° from the z-axis.

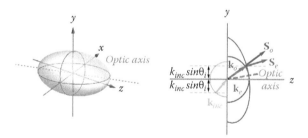

Figure 21.20 (Left) Index ellipsoid for a uniaxial medium. (Right) Refraction through the K-surfaces of the isotropic material into the uniaxial material for off-axis incidence. **k** vectors in both media are phase matched through the interface.

With Equation 21.19, the refraction angle for o-mode is calculated as

$$1 \times \sin 50° = 1.3 \times \sin \theta_o$$
$$\theta_o = 36.10°.$$

$\hat{\mathbf{S}}_o = \hat{\mathbf{k}}_o = (0, \sin 36.10°, \cos 36.10°)$. For the extraordinary mode, using Equations 21.19 and 21.16,

$$1 \times \sin 50° = n_e \sin \theta_e, \text{ and}$$

$$\frac{1}{n_e^2} = \frac{\cos^2(\theta_e + \theta_a)}{1.3^2} + \frac{\sin^2(\theta_e + \theta_a)}{2.5^2},$$

where $\theta_a = -10°$. Thus, the refraction angle for the e-mode is $33.62°$ and n_e is 1.38355.
The K-surface for the e-mode intercepts $\mathbf{k}_e$, as in Figure 21.20 (right):

$$\begin{pmatrix} 1 & 0 & 0 \\ 0 & \cos(-10°) & -\sin(-10°) \\ 0 & \sin(-10°) & \cos(-10°) \end{pmatrix} \begin{pmatrix} 2.5\cos(u)\cos(v) \\ 2.5\cos(u)\sin(v) \\ 1.3\sin(u) \end{pmatrix} = \begin{pmatrix} 0 \\ k_{ey} \\ k_{ez} \end{pmatrix} t = \begin{pmatrix} 0 \\ \sin(33.62°) \\ \cos(33.62°) \end{pmatrix} t,$$

where $u = 1.794$, $v = -1.571$, and $t = 1.384$. Hence, the intercept is $(0, 0.766, 1.152)$. The surface normal at that intercept is $\hat{\mathbf{S}}_e = (0, 0.288, 0.958)$.

The refractive index of the extraordinary mode in Equation 21.16 is related to the *phase velocity* as $v_{p,e} = c/n_e$. The surface of equal phase, the *S-surface*, originates from a point source and expands at the speed of v_p along $\mathbf{k}$. Thus, the ordinary wave surface is a sphere while the extraordinary wave surface is proportional to $1/n_e$. The energy of the light flows along $\mathbf{S}$ at *ray velocity* $v_r = v_p/\cos \beta$, where β is the angle between $\mathbf{k}$ and $\mathbf{S}$.[7,15,16] Figure 21.21 shows the *S*-surfaces in the principal plane for positive and negative uniaxial materials. Consider a negative uniaxial material, when $\mathbf{k}$ is along the optic axis; then, $v_{p,o} = v_{p,e}$. At other propagation directions, $n_e < n_o$, so $v_{p,o} < v_{p,e}$.

The K-surface and the S-surface of the e-mode are related to each other in Figure 21.22. Considering a $\mathbf{k}$ vector originating from the origin, its intercept at the K-surface reveals the $\mathbf{E}$ and $\mathbf{S}$ directions to be the tangent and normal of the K-surface. For this $\mathbf{S}$ direction originating from the origin, the intercept at the S-surface reveals the $\mathbf{D}$ and $\mathbf{k}$ directions to be the tangent and normal of the S-surface. The e-wavefront has the ellipsoidal shape of the S-surface, which extends along $\mathbf{S}$ at the ray velocity.

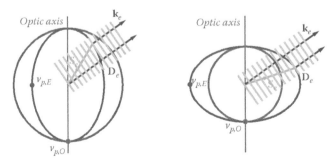

Figure 21.21 *S*-surface for positive (left) and negative (right) uniaxial material. Blue for e-mode, and red for o-mode. The optic axis (brown) is oriented vertically.

21.5 Index Ellipsoid, Optical Indicatrix, and K- and S-Surfaces

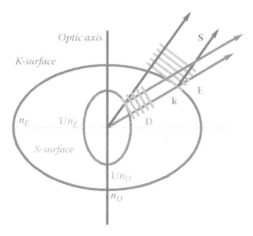

Figure 21.22 The K-surface and S-surface of extraordinary mode in positive uniaxial material.

Example 21.3 Ray Parameters Refract and Reflect from Air to Calcite

This example shows the step-by-step procedure presented in Chapter 19 to propagate a ray through a uniaxial interface (Figure 21.23). Calculate the exiting ray parameters (**E**, **k**, **H**, **S**, *a*, and **P** matrices) for a ray propagating from air ($n = 1$) to a uniaxial medium with $(n_O, n_E) = (1.3, 2.5)$ with optic axis along $(0, \sin 10°, \cos 10°)$. The results of this example and Example 21.2 show that both methods produce the same results.

In air, the light direction is $\mathbf{k}_{inc} = \begin{pmatrix} 0 \\ \sin\theta \\ \cos\theta \end{pmatrix} = \begin{pmatrix} 0 \\ \sin 50° \\ \cos 50° \end{pmatrix}$ and $\mathbf{S}_{inc} = \begin{pmatrix} 0 \\ \sin\theta \\ \cos\theta \end{pmatrix} = \begin{pmatrix} 0 \\ \sin 50° \\ \cos 50° \end{pmatrix}$.

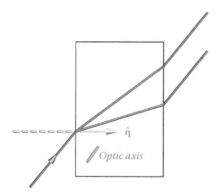

Figure 21.23 A ray propagates into a uniaxial crystal and splits into two rays. The plane of incident and the plane of ray doubling are both in the plane of the page.

The interface has $\boldsymbol{\eta} = \begin{pmatrix} 0 \\ 0 \\ 1 \end{pmatrix}$, the material before the interface has $\boldsymbol{\varepsilon} = \begin{pmatrix} 1 & 0 & 0 \\ 0 & 1 & 0 \\ 0 & 0 & 1 \end{pmatrix}$, the material after the interface has

$$\boldsymbol{\varepsilon}' = \begin{pmatrix} n_o^2 & 0 & 0 \\ 0 & n_o^2 \cos^2 \alpha + n_E^2 \sin^2 \alpha & \frac{1}{2}(n_E^2 - n_o^2)\sin 2\alpha \\ 0 & \frac{1}{2}(n_E^2 - n_o^2)\sin 2\alpha & n_E^2 \cos^2 \alpha + n_o^2 \sin^2 \alpha \end{pmatrix} = \begin{pmatrix} 1.690 & 0 & 0 \\ 0 & 1.828 & 0.780 \\ 0 & 0.780 & 6.112 \end{pmatrix}, \text{ and}$$

$$\mathbf{G} = \mathbf{G}' = \begin{pmatrix} 0 & 0 & 0 \\ 0 & 0 & 0 \\ 0 & 0 & 0 \end{pmatrix} \text{ for both materials before and after the interface.}$$

The resultant propagation parameters inside the crystal are

$$\hat{\mathbf{k}} = \frac{1}{n'}\begin{pmatrix} 0 \\ \sin\theta \\ \pm\sqrt{n'^2 - \sin^2\theta} \end{pmatrix} \text{ and } \mathbf{K} = \frac{1}{n'}\begin{pmatrix} 0 & \mp\sqrt{n'^2 - \sin^2\theta} & \sin\theta \\ \pm\sqrt{n'^2 - \sin^2\theta} & 0 & 0 \\ -\sin\theta & 0 & 0 \end{pmatrix}$$

for transmission and reflection.
In transmission, based on Equation 19.17,

$$\begin{pmatrix} n_o^2 - n'^2 & 0 & 0 \\ 0 & \sin^2\theta - n'^2 + n_o^2\cos^2\alpha + n_E^2\sin^2\alpha & \frac{1}{2}(n_E^2 - n_o^2)\sin 2\alpha + \sin\theta\sqrt{n'^2 - \sin^2\theta} \\ 0 & \frac{1}{2}(n_E^2 - n_o^2)\sin 2\alpha + \sin\theta\sqrt{n'^2 - \sin^2\theta} & n_E^2\cos^2\alpha + n_o^2\sin^2\alpha - \sin^2\theta \end{pmatrix} \mathbf{E}_t = 0.$$

For non-zero $\mathbf{E}_t$, the determinant of the matrix is zero:

$$\frac{1}{2}(n_O - n')(n_O + n')\left\{ \begin{array}{l} 2n_E^2 n_O^2 - (n_E^2 + n_O^2)n'^2 - \\ (n_E - n_O)(n_E + n_O)\left[\cos 2\alpha(-1 + n'^2 + \cos 2\theta) + 4\cos\alpha\sqrt{n'^2 - \sin 2\theta}\sin\alpha\sin\theta\right] \end{array} \right\} = 0.$$

21.5 Index Ellipsoid, Optical Indicatrix, and K- and S-Surfaces

Since $n' > 0$,

$n' = n_O$, or

$$n' = \frac{\sqrt{\begin{array}{c}2n_E^4\cos^2\alpha\left(1+2n_O^2-\cos 2\theta\right)+2n_E^2n_O^2\left(n_O^2-n_O^2\cos 2\alpha+\cos 2\theta-1\right)+4n_O^4\sin^2\alpha\sin^2\theta- \\ 2\sqrt{2}n_O\sqrt{n_E^2\left(n_E^2-n_O^2\right)^2\left(n_E^2+n_O^2+\left(n_E^2-n_O^2\right)\cos 2\alpha+\cos 2\theta-1\right)\sin^2 2\alpha\sin^2\theta}\end{array}}}{n_E^2+\left(n_E^2-n_O^2\right)\cos 2\alpha+no^2},$$

so $n_o = 1.3$ and $n_e = 1.38355$. The corresponding two propagation vectors are $\mathbf{k}_{to} = \begin{pmatrix} 0 \\ 0.5893 \\ 0.8079 \end{pmatrix}$

and $\mathbf{k}_{te} = \begin{pmatrix} 0 \\ 0.5537 \\ 0.8323 \end{pmatrix}$.

- The transmitted *o*-mode satisfies

$$\begin{pmatrix} 0 & 0 & 0 \\ 0 & \sin^2\theta+\left(n_E^2-n_O^2\right)\sin^2\alpha & \frac{1}{2}\left(n_E^2-n_O^2\right)\sin 2\alpha+\sin\theta\sqrt{n_O^2-\sin^2\theta} \\ 0 & \frac{1}{2}\left(n_E^2-n_O^2\right)\sin^2\alpha+\sin\theta\sqrt{n_O^2-\sin^2\theta} & n_E^2\cos^2\alpha+n_O^2\sin^2\alpha-\sin^2\theta \end{pmatrix}\mathbf{E}_{tO}=0,$$

$$\begin{pmatrix} 0 & 0 & 0 \\ 0 & 0.7243 & 1.5844 \\ 0 & 1.5844 & 5.5257 \end{pmatrix}\mathbf{E}_{tO}=0.$$

By singular value decomposition,

$$\begin{pmatrix} 0 & 0 & 0 \\ 0 & 0.7243 & 1.5844 \\ 0 & 1.5844 & 5.5257 \end{pmatrix} = \begin{pmatrix} 0 & 0 & 1 \\ -0.2876 & -0.9578 & 0 \\ -0.9578 & 0.2876 & 0 \end{pmatrix}\begin{pmatrix} 6.0014 & 0 & 0 \\ 0 & 0.2486 & 0 \\ 0 & 0 & 0 \end{pmatrix}\begin{pmatrix} 0 & 0 & 1 \\ -0.2876 & -0.9578 & 0 \\ -0.9578 & 0.2876 & 0 \end{pmatrix}^\dagger.$$

The exiting state corresponding to the singular value of zero has electric field $\mathbf{E}_{tO} = \begin{pmatrix} 1 \\ 0 \\ 0 \end{pmatrix}$.

The **H** and **S** fields are

$$\mathbf{H}_{tO} = (n_O \mathbf{K}_{tO} + i\mathbf{G}) \hat{\mathbf{E}}_{tO} = \begin{pmatrix} 0 & -1.0503 & 0.7660 \\ 1.0503 & 0 & 0 \\ -0.7660 & 0 & 0 \end{pmatrix} \begin{pmatrix} 1 \\ 0 \\ 0 \end{pmatrix} = \begin{pmatrix} 0 \\ 1.0503 \\ -0.7660 \end{pmatrix}$$

and

$$\hat{\mathbf{S}}_{tO} = \frac{\mathcal{Re}[\hat{\mathbf{E}}_{tO} \times \hat{\mathbf{H}}_{tO}{}^*]}{\|\mathcal{Re}[\hat{\mathbf{E}}_{tO} \times \hat{\mathbf{H}}_{tO}{}^*]\|} = \begin{pmatrix} 0 \\ 0.5893 \\ 0.8079 \end{pmatrix}.$$

- The transmitted *e*-mode satisfies

$$\begin{pmatrix} n_O^2 - n_O^2 & 0 & 0 \\ 0 & \sin^2\theta - n_e^2 + n_O^2 \cos^2\alpha + n_E^2 \sin^2\alpha & \frac{1}{2}(n_E^2 - n_O^2)\sin 2\alpha + \sin\theta\sqrt{n_e^2 - \sin^2\theta} \\ 0 & \frac{1}{2}(n_E^2 - n_O^2)\sin 2\alpha + \sin\theta\sqrt{n_e^2 - \sin^2\theta} & n_E^2 \cos^2\alpha + n_O^2 \sin^2\alpha - \sin^2\theta \end{pmatrix} \mathbf{E}_{te} = 0,$$

$$\begin{pmatrix} -0.2242 & 0 & 0 \\ 0 & 0.5001 & 1.6624 \\ 0 & 1.6624 & 5.5257 \end{pmatrix} \mathbf{E}_{te} = 0.$$

By singular value decomposition,

$$\begin{pmatrix} -0.2242 & 0 & 0 \\ 0 & 0.5001 & 1.6624 \\ 0 & 1.6624 & 5.5257 \end{pmatrix} = \begin{pmatrix} 0 & -1 & 0 \\ -0.2881 & 0 & -0.9576 \\ -0.9576 & 0 & 0.2881 \end{pmatrix} \begin{pmatrix} 6.0258 & 0 & 0 \\ 0 & 0.2242 & 0 \\ 0 & 0 & 0 \end{pmatrix} \begin{pmatrix} 0 & 1 & 0 \\ -0.2881 & 0 & -0.9576 \\ -0.9576 & 0 & 0.2881 \end{pmatrix}^\dagger.$$

The exiting state corresponding to the singular value of zero has electric field

$$\mathbf{E}_{te} = \begin{pmatrix} 0 \\ -0.9576 \\ 0.2881 \end{pmatrix}.$$

21.5 Index Ellipsoid, Optical Indicatrix, and K- and S-Surfaces

The **H** and **S** fields are

$$\mathbf{H}_{te} = (n_e \mathbf{K}_{te} + i\mathbf{G}')\hat{\mathbf{E}}_{te} = \begin{pmatrix} 0 & -1.1521 & 0.7660 \\ 1.1521 & 0 & 0 \\ -0.7660 & 0 & 0 \end{pmatrix} \begin{pmatrix} 0 \\ -09576 \\ 0.2881 \end{pmatrix} = \begin{pmatrix} 1.3240 \\ 0 \\ 0 \end{pmatrix}$$

and

$$\hat{\mathbf{S}}_{te} = \frac{\mathcal{R}e[\hat{\mathbf{E}}_{te} \times \hat{\mathbf{H}}_{te}{}^*]}{\left|\mathcal{R}e[\hat{\mathbf{E}}_{te} \times \hat{\mathbf{H}}_{te}{}^*]\right|} = \begin{pmatrix} 0 \\ 0.2881 \\ 0.9576 \end{pmatrix}.$$

- In reflection, $[\varepsilon + (n_r \mathbf{K}_r + i\mathbf{G})^2]\mathbf{E}_r = 0$, where $n_r = 1$:

$$\begin{pmatrix} 1 - n_r^2 & 0 & 0 \\ 0 & 1 - n_r^2 + \sin^2\alpha & -\sqrt{n_r^2 - \sin^2\theta}\sin\theta \\ 0 & -\sqrt{n_r^2 - \sin^2\theta}\sin\theta & \cos^2\theta \end{pmatrix} \mathbf{E}_r = 0$$

$$\begin{pmatrix} 0 & 0 & 0 \\ 0 & 0.5868 & -0.4924 \\ 0 & -1.4924 & 0.4132 \end{pmatrix} \mathbf{E}_{tr} = 0.$$

By singular value decomposition,

$$\begin{pmatrix} 0 & 0 & 0 \\ 0 & 0.5868 & -0.4924 \\ 0 & -0.4924 & 0.4132 \end{pmatrix} = \begin{pmatrix} 0 & 0 & 1 \\ -0.7660 & 0.6428 & 0 \\ 0.6428 & 0.7660 & 0 \end{pmatrix} \begin{pmatrix} 1 & 0 & 0 \\ 0 & 0 & 0 \\ 0 & 0 & 0 \end{pmatrix} \begin{pmatrix} 0 & 0 & 1 \\ -0.7660 & 0.6428 & 0 \\ 0.6428 & 0.7660 & 0 \end{pmatrix}^\dagger.$$

The exiting states correspond to the singular value of zero, which are $\mathbf{E}_{rs} = \begin{pmatrix} 1 \\ 0 \\ 0 \end{pmatrix}$ and $\mathbf{E}_{rp} = \begin{pmatrix} 0 \\ 0.6428 \\ 0.7660 \end{pmatrix}.$

Their **H** and **S** fields are

$$\mathbf{H}_{rs} = (n_r\mathbf{K}_r + i\mathbf{G})\hat{\mathbf{E}}_{rs} = \begin{pmatrix} 0 & 0.6428 & 0.7660 \\ -0.6428 & 0 & 0 \\ -0.7660 & 0 & 0 \end{pmatrix}\begin{pmatrix} 1 \\ 0 \\ 0 \end{pmatrix} = \begin{pmatrix} 0 \\ -0.6428 \\ -0.7660 \end{pmatrix},$$

$$\mathbf{H}_{rp} = (n_r\mathbf{K}_r + i\mathbf{G})\hat{\mathbf{E}}_{rp} = \begin{pmatrix} 0 & 0.6428 & 0.7660 \\ -0.6428 & 0 & 0 \\ -0.7660 & 0 & 0 \end{pmatrix}\begin{pmatrix} 0 \\ 0.6428 \\ 0.7660 \end{pmatrix} = \begin{pmatrix} 1 \\ 0 \\ 0 \end{pmatrix}, \text{and}$$

$$\hat{\mathbf{S}}_r = \frac{\mathcal{R}e\left[\hat{\mathbf{E}}_r \times \hat{\mathbf{H}}_r{}^*\right]}{\left\|\mathcal{R}e\left[\hat{\mathbf{E}}_r \times \hat{\mathbf{H}}_r{}^*\right]\right\|} = \begin{pmatrix} 0 \\ 0.7660 \\ -0.6428 \end{pmatrix}.$$

Now, we will construct the **F** matrix with $\mathbf{s}_1 = \hat{\mathbf{k}}_{inc} \times \hat{\boldsymbol{\eta}} = \begin{pmatrix} \sin 50° \\ 0 \\ 0 \end{pmatrix}$ and $\mathbf{s}_2 = \hat{\boldsymbol{\eta}} \times \mathbf{s}_1 = \begin{pmatrix} 0 \\ \sin 50° \\ 0 \end{pmatrix}$:

$$\mathbf{F} = \begin{pmatrix} \mathbf{s}_1\cdot\hat{\mathbf{E}}_{to} & \mathbf{s}_1\cdot\hat{\mathbf{E}}_{te} & -\mathbf{s}_1\cdot\hat{\mathbf{E}}_{rs} & -\mathbf{s}_1\cdot\hat{\mathbf{E}}_{rp} \\ \mathbf{s}_2\cdot\hat{\mathbf{E}}_{to} & \mathbf{s}_2\cdot\hat{\mathbf{E}}_{te} & -\mathbf{s}_2\cdot\hat{\mathbf{E}}_{rs} & -\mathbf{s}_2\cdot\hat{\mathbf{E}}_{rp} \\ \mathbf{s}_1\cdot\hat{\mathbf{H}}_{to} & \mathbf{s}_1\cdot\hat{\mathbf{H}}_{te} & -\mathbf{s}_1\cdot\hat{\mathbf{H}}_{rs} & -\mathbf{s}_1\cdot\hat{\mathbf{H}}_{rp} \\ \mathbf{s}_2\cdot\hat{\mathbf{H}}_{to} & \mathbf{s}_2\cdot\hat{\mathbf{H}}_{te} & -\mathbf{s}_2\cdot\hat{\mathbf{H}}_{rs} & -\mathbf{s}_2\cdot\hat{\mathbf{H}}_{rp} \end{pmatrix} = \begin{pmatrix} 0.7660 & 0 & -0.7660 & 0 \\ 0 & -0.7336 & 0 & -0.4924 \\ 0 & 1.0142 & 0 & -0.7660 \\ 0.8046 & 0 & 0.4924 & 0 \end{pmatrix},$$

$$\mathbf{F}^{-1} = \begin{pmatrix} 0.4956 & 0 & 0 & 0.7710 \\ 0 & -0.7218 & 0.4639 & 0 \\ -0.8098 & 0 & 0 & 0.7710 \\ 0 & -0.9556 & -0.6912 & 0 \end{pmatrix}.$$

With *s*-polarized incident light,

$$\hat{\mathbf{E}}_{inc,S} = \frac{\hat{\mathbf{S}}_{inc} \times \hat{\boldsymbol{\eta}}}{\left\|\hat{\mathbf{S}}_{inc} \times \hat{\boldsymbol{\eta}}\right\|} = \begin{pmatrix} 1 \\ 0 \\ 0 \end{pmatrix}, \mathbf{H}_{inc,S} = (n\mathbf{K}_{inc} + i\mathbf{G})\hat{\mathbf{E}}_{inc,S} = \begin{pmatrix} 0 \\ \cos 50° \\ -\sin 50° \end{pmatrix},$$

21.5 Index Ellipsoid, Optical Indicatrix, and K- and S-Surfaces

$$C_s = \begin{pmatrix} \mathbf{s}_1 \cdot \hat{\mathbf{E}}_{inc,s} \\ \mathbf{s}_2 \cdot \hat{\mathbf{E}}_{inc,s} \\ \mathbf{s}_1 \cdot \hat{\mathbf{H}}_{inc,s} \\ \mathbf{s}_2 \cdot \hat{\mathbf{H}}_{inc,s} \end{pmatrix} = \begin{pmatrix} \sin 50° \\ 0 \\ 0 \\ \cos 50° \sin 50° \end{pmatrix}, \text{ and}$$

$$\mathbf{A}_s = \mathbf{F}^{-1} \cdot \mathbf{C}_s = \begin{pmatrix} 0.496 & 0 & 0 & 0.771 \\ 0 & -0.722 & 0.464 & 0 \\ -0.810 & 0 & 0 & 0.771 \\ 0 & -0.956 & -0.691 & 0 \end{pmatrix} \begin{pmatrix} 0.766 \\ 0 \\ 0 \\ 0.492 \end{pmatrix} = \begin{pmatrix} 0.759 \\ 0 \\ -0.241 \\ 0 \end{pmatrix} = \begin{pmatrix} a_{s \to to} \\ a_{s \to te} \\ a_{s \to rs} \\ a_{s \to rp} \end{pmatrix}.$$

With p-polarized incident light,

$$\hat{\mathbf{E}}_{inc,p} = \frac{\hat{\mathbf{S}}_{inc} \times \hat{\boldsymbol{\eta}}}{\|\hat{\mathbf{S}}_{inc} \times \hat{\boldsymbol{\eta}}\|} = \begin{pmatrix} 0 \\ \cos 50° \\ -\sin 50° \end{pmatrix}, \mathbf{H}_{inc,p} = (n\mathbf{K}_{inc} + i\mathbf{G})\hat{\mathbf{E}}_{inc,p} = \begin{pmatrix} -1 \\ 0 \\ 0 \end{pmatrix},$$

$$C_p = \begin{pmatrix} \mathbf{s}_1 \cdot \hat{\mathbf{E}}_{inc,s} \\ \mathbf{s}_2 \cdot \hat{\mathbf{E}}_{inc,s} \\ \mathbf{s}_1 \cdot \hat{\mathbf{H}}_{inc,s} \\ \mathbf{s}_2 \cdot \hat{\mathbf{H}}_{inc,s} \end{pmatrix} = \begin{pmatrix} 0 \\ \cos 50° \sin 50° \\ -\sin 50° \\ 0 \end{pmatrix}, \text{ and}$$

$$\mathbf{A}_p = \mathbf{F}^{-1} \cdot \mathbf{C}_p = \begin{pmatrix} 0.496 & 0 & 0 & 0.771 \\ 0 & -0.722 & 0.464 & 0 \\ -0.810 & 0 & 0 & 0.771 \\ 0 & -0.956 & -0.691 & 0 \end{pmatrix} \begin{pmatrix} 0 \\ 0.492 \\ -0.766 \\ 0 \end{pmatrix} = \begin{pmatrix} 0 \\ -0.711 \\ 0 \\ 0.0589 \end{pmatrix} = \begin{pmatrix} a_{p \to to} \\ a_{p \to te} \\ a_{p \to rs} \\ a_{p \to rp} \end{pmatrix}.$$

The **P** matrices of the exiting rays are calculated by the resultant **E**, **S**, and a's, as shown in Section 19.7.2:

$$\mathbf{P}_{to} = \begin{pmatrix} 0.759 & 0 & 0 \\ 0 & 0 & 0.589 \\ 0 & 0 & 0.808 \end{pmatrix} \begin{pmatrix} 1 & 0 & 0 \\ 0 & 0.643 & 0.766 \\ 0 & -0.766 & 0.643 \end{pmatrix}^{-1} = \begin{pmatrix} 0.759 & 0 & 0 \\ 0 & 0.451 & 0.379 \\ 0 & 0.619 & 0.519 \end{pmatrix}$$

$$\mathbf{P}_{te} = \begin{pmatrix} 0 & 0 & 0 \\ 0 & 0.681 & 0.288 \\ 0 & -0.205 & 0.958 \end{pmatrix} \begin{pmatrix} 1 & 0 & 0 \\ 0 & 0.643 & 0.766 \\ 0 & -0.766 & 0.643 \end{pmatrix}^{-1} = \begin{pmatrix} 0 & 0 & 0 \\ 0 & 0.658 & -0.336 \\ 0 & 0.602 & 0.772 \end{pmatrix}$$

$$\mathbf{P}_r = \begin{pmatrix} -0.241 & 0 & 0 \\ 0 & 0.038 & 0.766 \\ 0 & 0.045 & -0.643 \end{pmatrix} \begin{pmatrix} 1 & 0 & 0 \\ 0 & 0.643 & 0.766 \\ 0 & -0.766 & 0.643 \end{pmatrix}^{-1} = \begin{pmatrix} -0.241 & 0 & 0 \\ 0 & 0.611 & 0.463 \\ 0 & -0.463 & -0.448 \end{pmatrix}$$

The singular value decomposition of each exiting $\mathbf{P}$ matrix gives the incident and exiting $\mathbf{E}$ fields and $\mathbf{S}$ vectors as

$$\mathbf{P} = \begin{pmatrix} S_{inc,x} & E_{inc,x} & E_{inc\perp,x} \\ S_{inc,y} & E_{inc,y} & E_{inc\perp,y} \\ S_{inc,z} & E_{inc,z} & E_{inc\perp,z} \end{pmatrix} \begin{pmatrix} 1 & 0 & 0 \\ 0 & |a| & 0 \\ 0 & 0 & 0 \end{pmatrix} \begin{pmatrix} S_{out,x} & E_{out,x} & E_{out\perp,x} \\ S_{out,y} & E_{out,y} & E_{out\perp,y} \\ S_{out,z} & E_{out,z} & E_{out\perp,z} \end{pmatrix}^{-1},$$

where $|a|$ is the amplitude coefficients of the transmitted mode; thus,

$$\mathbf{P} \cdot \begin{pmatrix} E_{inc,x} \\ E_{inc,y} \\ E_{inc,z} \end{pmatrix} = |a| \begin{pmatrix} E_{out,x} \\ E_{out,y} \\ E_{out,y} \end{pmatrix} \text{ or } \mathbf{P} \cdot \hat{\mathbf{E}}_{inc} = |a| \hat{\mathbf{E}}_{out}.$$

The singular value of one corresponds to the $\mathbf{S}$ vector,

$$\mathbf{P} \cdot \begin{pmatrix} S_{inc,x} \\ S_{inc,y} \\ S_{inc,z} \end{pmatrix} = \begin{pmatrix} S_{out,x} \\ S_{out,y} \\ S_{out,y} \end{pmatrix} \text{ or } \mathbf{P} \cdot \mathbf{S}_{inc} = \mathbf{S}_{out}.$$

The other two singular values represent the magnitude of the amplitude coefficients of its two exiting modes; hence,

$$\mathbf{P} \begin{pmatrix} E_{inc,x} \\ E_{inc,y} \\ E_{inc,z} \end{pmatrix} = a \begin{pmatrix} E_{out,x} \\ E_{out,y} \\ E_{out,y} \end{pmatrix} \text{ or } \mathbf{P} \cdot \mathbf{E}_{inc} = \mathbf{E}_{out}.$$

The $\mathbf{P}$ matrix for the extraordinary mode in transmission,

$$\mathbf{P}_{te} = \begin{pmatrix} 0 & 0 & 0 \\ 0 & 0.658 & -0.336 \\ 0 & 0.602 & 0.772 \end{pmatrix}$$

$$= \begin{pmatrix} 0 & 0 & 1 \\ \sin 16.74° & \cos 16.74° & 0 \\ \cos 16.74° & -\sin 16.74° & 0 \end{pmatrix} \begin{pmatrix} 1 & 0 & 0 \\ 0 & 0.711 & 0 \\ 0 & 0 & 0 \end{pmatrix} \begin{pmatrix} 0 & 0 & 1 \\ -\sin 50° & -\cos 50° & 0 \\ -\cos 50° & \sin 50° & 0 \end{pmatrix}^{\dagger},$$

shows the incident $\mathbf{S}_{inc} = \begin{pmatrix} 0 \\ \sin 50° \\ \cos 50° \end{pmatrix}$ maps to $\mathbf{S}_{te} = \begin{pmatrix} 0 \\ \sin 16.74° \\ \cos 16.74° \end{pmatrix}$ in the first column and

$\mathbf{E}_{inc} = \begin{pmatrix} 0 \\ \cos 50° \\ -\sin 50° \end{pmatrix}$ maps to $\mathbf{E}_{te} = \begin{pmatrix} 0 \\ -\cos 16.74° \\ \sin 16.74° \end{pmatrix}$ with 0.711 amplitude relative to the incident field.

The reflection matrix,

$$\mathbf{P}_r = \begin{pmatrix} -0.241 & 0 & 0 \\ 0 & 0.6112 & 0.463 \\ 0 & -0.463 & -0.448 \end{pmatrix}$$

$$= \begin{pmatrix} 0 & -1 & 0 \\ -\sin 50° & 0 & -\cos 50° \\ \cos 50° & 0 & -\sin 50° \end{pmatrix} \begin{pmatrix} 1 & 0 & 0 \\ 0 & 0.241 & 0 \\ 0 & 0 & 0.059 \end{pmatrix} \begin{pmatrix} 0 & 1 & 0 \\ -\sin 50° & 0 & -\cos 50° \\ -\cos 50° & 0 & \sin 50° \end{pmatrix}^\dagger,$$

shows the incident $\mathbf{S}_{inc} = \begin{pmatrix} 0 \\ \sin 50° \\ \cos 50° \end{pmatrix}$ maps to $\mathbf{S}_r = \begin{pmatrix} 0 \\ \sin 50° \\ -\cos 50° \end{pmatrix}$ in the first column,

$\mathbf{E}_{inc,s} = \begin{pmatrix} 1 \\ 0 \\ 0 \end{pmatrix}$ maps to $\mathbf{E}_{rs} = \begin{pmatrix} -1 \\ 0 \\ 0 \end{pmatrix}$ with 0.241 amplitude coefficient in the second column,

and $\mathbf{E}_{inc,p} = \begin{pmatrix} 0 \\ \cos 50° \\ -\sin 50° \end{pmatrix}$ maps to $\mathbf{E}_{rp} = \begin{pmatrix} 0 \\ \cos 50° \\ \sin 50° \end{pmatrix}$ with 0.059 amplitude coefficient in the third column.

21.6 Aberrations of Crystal Waveplates

A *waveplate* is a common type of retarder formed from a plane parallel plate of birefringent material. Its function is to introduce retardance between its two orthogonal eigenmodes in the transmitted light. They are utilized to control polarized light, adjust ellipticity, rotate polarization, and even fine-tune laser wavelengths when used inside laser cavities. They are widely used in polarimetry, medical imaging, microscopy, telecom, and the laser cutting industry. Simple isotropic plane parallel plates cause focus shift and spherical aberration on-axis and add coma and astigmatism off-axis due to the variation of the optical path length with angle.[17] The aberrations of the retarder's ordinary wave are the same as for the isotropic plate. The aberrations of the extraordinary wave are more complicated because they include all the isotropic aberrations as well as additional aberrations arising from the variation of refractive index with direction. These aberrations and their effect on image formation will be calculated from polarization ray tracing.

Waveplates are characterized by the plate thickness t and the on-axis retardance δ at a reference wavelength λ_{ref}.

$$\delta = \frac{2\pi}{\lambda_{ref}} \Delta n t, \qquad (21.20)$$

where δ is the on-axis retardance in radians and $\Delta n = |n_e - n_o|$ is the birefringence. The *optical path difference* of the two modes is $\Delta OPL = \Delta n t$. Common waveplates include *quarter waveplates* and *half waveplates*. A zero-order quarter waveplate with a thickness of $\lambda_{ref}/(4\Delta n)$ is a *true zero-order quarter waveplate*. It is most commonly made of quartz or magnesium fluoride rather than calcite, because the smaller Δn translates into a more practical thickness. Calcite would require a thickness of less than 10 μm to achieve 3½ waves or retardance (a third-order waveplate) for 0.5 μm, which is impractically thin to manufacture. A 1¼-wave-thick waveplate generates the same polarization changes on-axis at λ_{ref} as a zero-order quarter waveplate and is labeled a *first-order quarter waveplate*. Similarly, a 2¼ wave waveplate is a *second-order quarter waveplate*. The waveplates of different orders have the same effect on normal incident light at λ_{ref}. The effects are different when the optical path length *OPL* is different, that is, at other wavelengths or incident angles. A waveplate's retardance changes with incident angles and with changes in refractive index due to dispersion. It is common to stack together multiple waveplates together with different dispersion properties in order to produce a nearly constant retardance over multiple wavelengths. Such a waveplate is called an *achromatic waveplate*.[4,5]

Algorithms for ray paths through birefringent elements were described in Section 19.7 where the flux of an incident ray divides into eigenmodes at the birefringent intercept. These algorithms are applied in Section 20.3 to study the general ray path of a ray refracting through a waveplate where the ΔOPL and the retardance are calculated between the two eigenmodes for on- and off-axis incident rays. In this section, the aberrations of two common waveplate configurations, the A-plate and the C-plate, are studied. Consider a collimated beam passing through a uniaxial plate shown in Figure 21.24. The lateral shear between the two wavefronts is Δr and the two beams partially overlap after the plate. Depending on the incident angle, the incident polarization, and the retardance of the plate, the two exiting states can combine to linear, circular, or elliptical polarization.

In the following discussion, the incident propagation vector $\mathbf{k} = (k_x, k_y, k_z)$ is defined similar to the double pole coordinates in Chapter 11 as

Figure 21.24 A ray of an incident beam refracts into a uniaxial plane parallel plate and splits into two rays in two directions. After these two rays refract out of the crystal, they both propagate in the same direction as the incident ray, but are displaced by Δr.

21.6 Aberrations of Crystal Waveplates

$$\hat{\mathbf{k}} = \left(\frac{\tan\theta_x}{\sqrt{\tan^2\theta_x + \tan^2\theta_y + 1}}, \frac{\tan\theta_y}{\sqrt{\tan^2\theta_x + \tan^2\theta_y + 1}}, \frac{1}{\sqrt{\tan^2\theta_x + \tan^2\theta_y + 1}} \right) \quad (21.21)$$

where $\tan\theta_x = k_x/k_z$, $\tan\theta_y = k_y/k_z$, and θ is the incident angle in air.

21.6.1 A-Plate Aberrations

An *A-plate* is a uniaxial waveplate whose optic axis lies on the plate surface, as shown in Figure 21.25. In this configuration, the maximum birefringence is obtained at normal incidence.

Consider a zero-order quarter wave calcite A-plate with an optic axis along the y-axis. Ray splitting occurs for all off-axis rays. The index variation of the *e*-mode, n_e, for wavelength 0.5 μm is shown in Figure 21.26 (left) over a ±40° square field of view (FOV) in air. $n_e(\theta_x, 0)$ is constant for light incident in the *x–z* plane and decreases quadratically with respect to angle θ_y above and below that plane for any negative uniaxial material. All *e*-rays propagating in the *x–z* plane are linearly polarized along the optic axis, with the same index n_E. For *e*-rays in the *y–z* plane, its polarization increasingly deviates from the optic axis as they move off-axis, changing n_e toward n_O.

The high birefringence of the A-plate produces small lateral shear Δr between the two parallel exiting modes, quadratic with angle and slightly larger along the *x*-axis as plotted in Figure 21.26 (right). This shear is a function of Δn and increases with the plate's thickness.

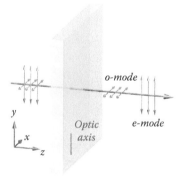

Figure 21.25 Negative uniaxial A-plate with *y* oriented optic axis delaying the *o*-mode with respect to the *e*-mode.

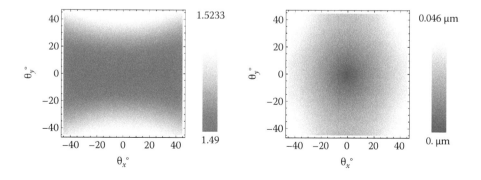

Figure 21.26 (Left) n_e at wavelength 0.5 μm in a *y*-oriented calcite A-plate as a function of incident angle (θ_x, θ_y) in air. (Right) The shear displacement Δr between the two exiting modes through the quarter wave calcite A-plate, measured transverse to the **k** vector.

The retardance induced by a waveplate depends not only on Δn but also the physical ray paths of the two eigenstates. The ray paths of three rays at 0.5 μm with increasing angle of incidence in the y–z plane passing through a quarter wave A-plate with thickness 0.0007 mm are shown in Figure 21.27. The path length increases off-axis, but the birefringence reduces faster with angle, described in Section 20.3, yielding the net decrease in retardance.

Because of the varying refractive index of the e-mode with angle, the angular behavior of the e-wavefront is very different from that of the o-wavefront. Consider a converging beam focused through an A-plate. The wavefronts of the individual o- and e-modes take different paths, experience different refractive indices, and have different aberrations. Figure 21.28 shows the ray bundles and caustic formed near focus in both sagittal and tangential planes. The e-mode converges before the o-mode in the x–z plane and after the o-mode in the y–z plane.

The *OPL*s of the two modes exiting the A-plate as a function of angle are calculated and shown in Figure 21.29. For a spherical wavefront, the resultant o-mode's *OPL* is axially symmetric, which produces circularly symmetric aberration about the z-axis, comprising a focus shift and some spherical aberration. In contrast, the e-mode ray path is not rotationally symmetric; the tangential e-rays have a larger *OPL* than the sagittal e-rays. The difference in extraordinary *OPL* between the tangential and sagittal planes demonstrates that the e-wavefront has on-axis astigmatism when the input is not collimated. The exiting e-beam has an ellipsoidal wavefront due to this extraordinary astigmatism.

The orientation of the exiting polarization states for the two modes is shown in Figure 21.30 (middle and right). The *OPL* difference between the o- and e-rays is the retardance, ΔOPL, which is shown in Figure 21.30 (left). The variation of retardance with angle displays a saddle shape, increasing perpendicular to the optic axis (x–z plane) and decreasing along the y–z plane.

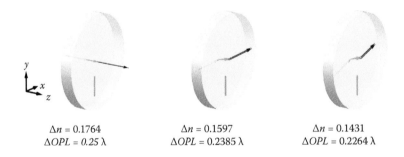

$\Delta n = 0.1764$ $\Delta n = 0.1597$ $\Delta n = 0.1431$
$\Delta OPL = 0.25\ \lambda$ $\Delta OPL = 0.2385\ \lambda$ $\Delta OPL = 0.2264\ \lambda$

Figure 21.27 Three incident rays at 0.5 μm along the principal section with incident angles 0°, 30°, and 45° pass through a quarter wave A-plate and experience different birefringence Δn and ΔOPL.

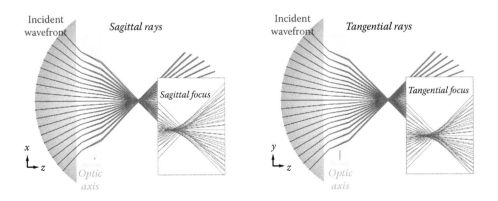

Figure 21.28 (Left) Sagittal and (right) tangential ray bundles at the focus are shown with o-rays in red and e-rays in blue.

21.6 Aberrations of Crystal Waveplates

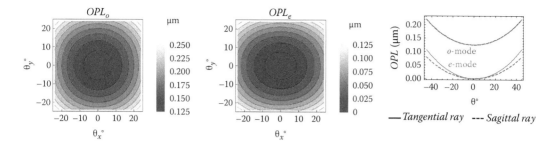

Figure 21.29 Left and middle panels show the *OPL* of the *o*- and *e*-wavefronts as a function of angle of incidence at 0.5 μm wavelength after the A-plate. The *OPL* shown is relative to the minimum OPL_e. The right panel shows the *OPL* values in the tangential (*y–z* plane) and sagittal (*x–z* plane) planes.

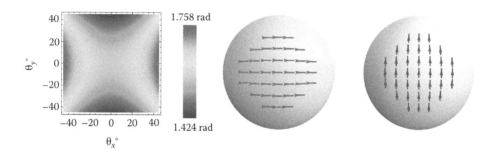

Figure 21.30 (Left) Retardance magnitude through the quarter wave A-plate. (Middle and right) *o*- and *e*-polarizations are shown on the exiting wavefronts in a three-dimensional view, thus the slight rotation. The arrow locations indicate phase.

The retardance magnitude in the retardance map is scaled as Equation 21.20, linear with the plate thickness and Δ*n*, and inversely with wavelength. The shape of the retardance map is symmetric about the optic axis. If the optic axis lies orthogonal to the waveplate (as shown in the next section for the C-plate configuration), the retardance map will become rotationally symmetric, zero retardance at normal incidence and increases quadratically with angle.

21.6.2 C-Plate Aberrations

A *C-plate* is a waveplate whose optic axis is perpendicular to the plate surface. There is no retardance at normal incidence, but the retardance increases quadratically off-axis. As the angle of incidence approaches zero, the *o*- and *e*-rays become degenerate; there is no phase difference and Δ*OPL* = 0. A quarter wave calcite C-plate with thickness $\lambda_{ref}/(4|n_O - n_E|)$ is shown in Figure 21.31, where ray splitting is observed for off-axis rays.

The *e*-mode refractive index distribution is shown in Figure 21.32, which is rotationally symmetric about the optic axis along *z*. Zero birefringence is obtained on-axis where the two eigenmodes experience the same refractive index: $n_o = n_e = n_O$, the modes are degenerate but are defined as orthogonal to each other. As the angle of incidence increases, the *e*-mode refractive index n_e decreases from n_O toward n_E, and the birefringence increases as does the retardance. The shear Δ*r* between the two modes is shown in Figure 21.32, which is similar to the A-plate's but with zero shear at normal incidence. Its overall shape is rotationally symmetric about the optical axis.

Unlike the A-plate, a C-plate gives a rotationally symmetric *OPL* for both exiting modes about the optical axis as shown in Figure 21.33, which can be useful for some devices. Therefore, both

770 Chapter 21: Uniaxial Materials and Components

$\Delta n = 0$
$\Delta OPL = 0\,\lambda$

$\Delta n = 0.0189$
$\Delta OPL = 0.0281\,\lambda$

$\Delta n = 0.0381$
$\Delta OPL = 0.0597\,\lambda$

Figure 21.31 Three incident rays at 0.5 μm with incident angles 0°, 30°, and 45° pass through a quarter wave C-plate and experience different birefringence Δn and ΔOPL. An exaggerated thickness is shown to demonstrate the ray splitting.

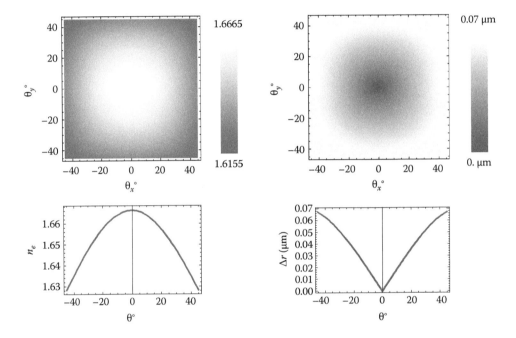

Figure 21.32 (Left) The refractive index of the exiting e-mode inside a quarter wave calcite C-plate as a function of incident angle in air. (Right) The shear displacement between the two exiting modes after the C-plate is also rotationally symmetric.

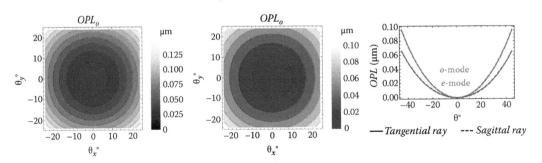

Figure 21.33 (Left and middle) The OPL relative to the axial ray for the o- and e-modes through the quarter wave C-plate are both circularly symmetric, shown on the left and right side, respectively. (Right) The OPL of o-mode (red) and e-mode (blue) rays for both tangential (solid line) and sagittal (dotted line) rays are the same.

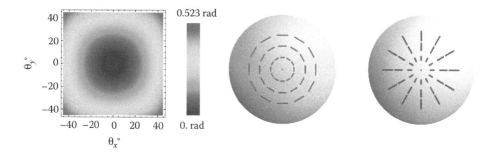

Figure 21.34 (Left) The retardance magnitude of a spherical wavefront passing through a C-plate is rotationally symmetric about normal incidence. The polarization orientation of the *o*-mode (middle) and the *e*-mode (right) is tangentially and radially oriented, respectively. The eigenmodes at normal incidence are degenerate.

Figure 21.35 The exiting polarization ellipses from a 6.3 μm C-plate for horizontally (left), vertically (middle), and circular (right) polarized spherical incident wavefronts.

tangential and sagittal rays have the same *OPL*, with the *e*-mode having a smaller *OPL* than the *o*-mode for the negative uniaxial calcite. Also, the *OPL* difference becomes zero on-axis due to the mode degeneracy.

The retardance of the quarter wave C-plate is plotted at wavelength 0.5 μm in Figure 21.34 (left), showing that the retardance is rotationally symmetric about the optical axis with zero retardance at normal incidence. The polarization orientation of the modes is very different from the A-plate. In Figure 21.34 (middle), the *o*-mode has tangentially oriented polarization around the exit pupil while the *e*-mode in Figure 21.34 (right) is radially oriented.

The polarization states through a thicker C-plate for vertically, horizontally, and circularly polarized spherical incident wavefronts are shown in Figure 21.35. The method for combining separate *o*- and *e*-modes is presented in Chapter 20. For the two linear wavefronts, the waveplate induces ellipticity off the *x*- and *y*-axes. For a circularly polarized wavefront, the ellipticity decreases and the orientation of the major axis becomes 45° to the radial direction.

21.7 Image Formation through an A-Plate

This section follows the methods in Chapter 16 to analyze the effects of aberrations of a crystal A-plate on image formation. A polarization ray trace of a converging beam with 0.5 NA at 0.5 μm focusing through a quarter wave uniaxial calcite A-plate (Figure 21.36) is performed. Although this zero-order quarter waveplate is very thin and induces small and arbitrary forms of aberrations, every uniaxial A-plate has similar polarization aberrations and the form and interpretation are similar to other waveplates.

From the ray tracing results, the wavefront aberrations at the exit pupil of the two modes are shown in Figure 21.37. The *o*-wavefront is rotationally symmetric with spherical aberration, while

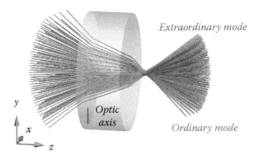

Figure 21.36 A beam with 0.5 NA is simulated by a grid of rays that focuses through a quarter wave calcite A-plate with optic axis oriented along *y*. The incident ray grid splits into two ray grids, one for the ordinary mode (red) and one for the extraordinary mode (blue).

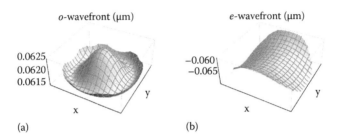

Figure 21.37 The *o*- and *e*-wavefronts at the exit pupil of the quarter wave calcite A-plate for 0.5 NA at 0.5 μm. (a) *o*-wavefront with spherical aberration. (b) *e*-wavefront with on-axis astigmatism.

the *e*-wavefront has astigmatism. Although the magnitude of the aberrations are small due to the example thin plate, they scale linearly with the plate thickness t and birefringence Δn and changes form with the optic axis orientation.

The **P** matrices at the spherical exit pupil for the 0.5 NA are calculated using the method in Chapter 19 and are shown in Figure 21.38. To examine the polarization aberrations, the **P** matrices are extracted from the 3D representation and converted into a Jones pupil, as shown in Figure 21.39.

The *o*-wavefront is an *x*-polarizer with aberrations; the *xx*-component with the majority of the light has apodization, and its phase shows the spherical aberration. The *e*-wavefront is a *y*-polarizer with aberrations; the *yy*-component has the most amplitude with apodization, and its phase shows astigmatism.

Next, the form and polarization structure of these 0.5 NA images are examined. The focus of the two modes is evaluated by the response matrix of the Jones pupil, which is the amplitude response matrix **ARM** calculated by diffraction theory for coherent imaging. The magnitude of the pair of

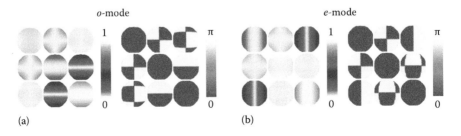

Figure 21.38 The **P** matrices of the *o*- and *e*-wavefronts at the exit pupil. The magnitude (left 3 × 3 panels) and phase in radians (right 3 × 3 panels) of the (a) *e*-mode's and (b) *o*-mode's **P** matrices.

21.7 Image Formation through an A-Plate

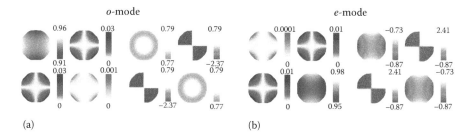

Figure 21.39 The Jones pupil of the o- and e-wavefronts at the exit pupil are shown in (a) and (b), respectively. The magnitude (left 2 × 2 panels) and phase in radians (right 2 × 2 panels) of the (a) e-mode's and (b) o-mode's Jones pupils.

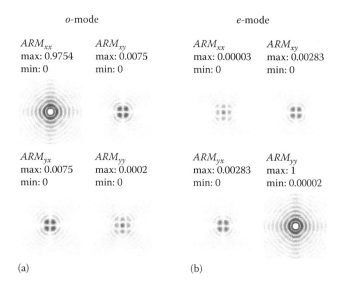

Figure 21.40 The amplitude of the 2 × 2 **ARM**s of the o- and e-wavefronts are shown in (a) and (b), respectively. Zero amplitude in gray, medium amplitude in red, and highest amplitude in yellow.

ARMs, one for each mode, is plotted in Figure 21.40. With the small aberration of the thin quarter wave plate, the major polarization (ARM_{xx} of the o-mode and ARM_{yy} of the e-mode) forms an Airy disk. The cross coupling (ARM_{xy} and ARM_{yx}) forms a dark center and four islands, and the orthogonal polarization (ARM_{yy} of the o-mode and ARM_{xx} of the e-mode) forms nine island leakages with extremely low amplitude.

Since the magnitude of the aberration scales with the plate thickness, the effects of the aberration on the image are more visible with a thicker A-plate. Two more A-plates, 3rd-order and 17th-order half waveplates, are analyzed with the same procedures. As the thickness of the A-plate increases, the **ARM** of the ordinary focus remains about the same, while the structure of the extraordinary focus diffuses from the Airy disk due to aberrations, as shown in Figure 21.41. For the thicker plate, the e-ARM_{yy} has a higher energy in the y-direction than in the x-direction, which is formed from the astigmatism residual to the uniaxial plate.

The overall **ARM** at the image is the sum of complex **ARM**$_o$ and **ARM**$_e$. In the absence of polarization aberration, the **ARM**s are $\begin{pmatrix} e^{i\pi/4} & 0 \\ 0 & e^{-i\pi/4} \end{pmatrix}$ and $\begin{pmatrix} 1 & 0 \\ 0 & -1 \end{pmatrix}$ modulated by the Airy disk patterns for the quarter and half waveplate, respectively. The thickness of the waveplate induces aberrations

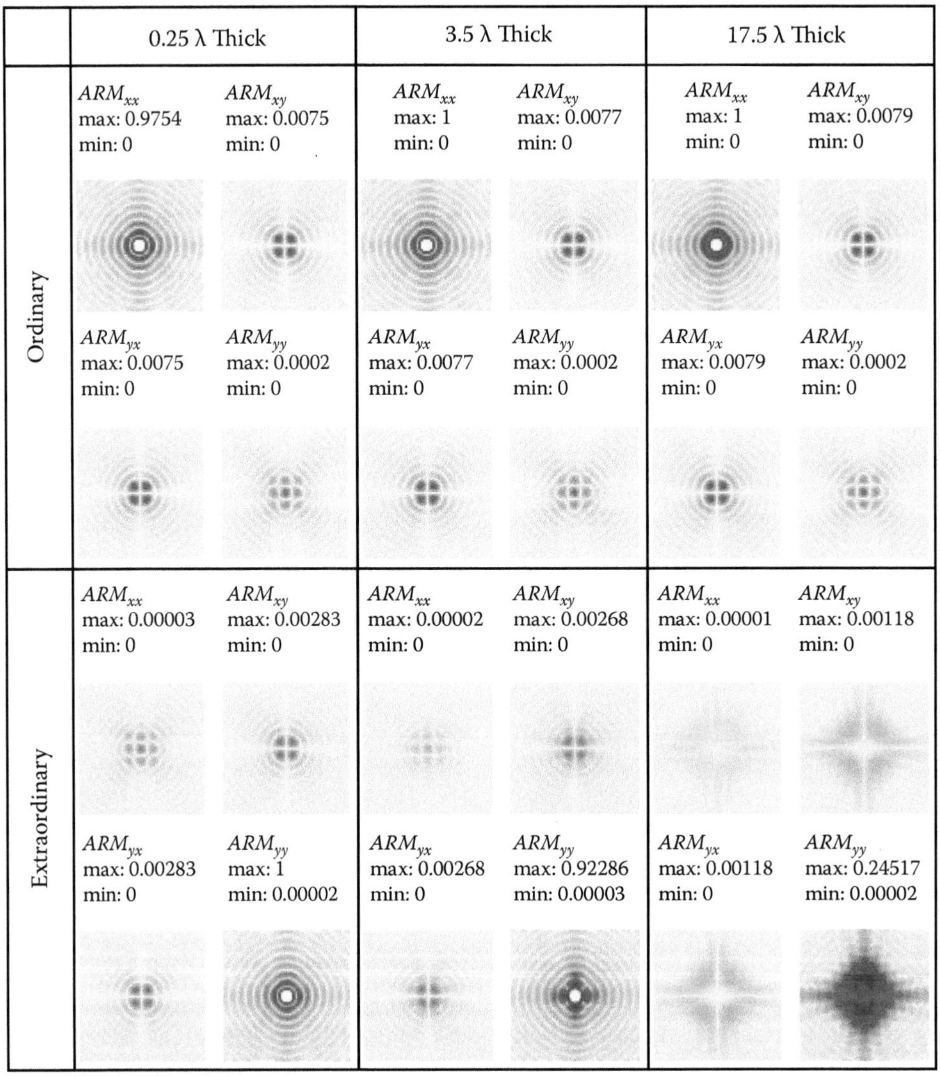

	0.25 λ Thick		3.5 λ Thick		17.5 λ Thick	
Ordinary	ARM_{xx} max: 0.9754 min: 0	ARM_{xy} max: 0.0075 min: 0	ARM_{xx} max: 1 min: 0	ARM_{xy} max: 0.0077 min: 0	ARM_{xx} max: 1 min: 0	ARM_{xy} max: 0.0079 min: 0
	ARM_{yx} max: 0.0075 min: 0	ARM_{yy} max: 0.0002 min: 0	ARM_{yx} max: 0.0077 min: 0	ARM_{yy} max: 0.0002 min: 0	ARM_{yx} max: 0.0079 min: 0	ARM_{yy} max: 0.0002 min: 0
Extraordinary	ARM_{xx} max: 0.00003 min: 0	ARM_{xy} max: 0.00283 min: 0	ARM_{xx} max: 0.00002 min: 0	ARM_{xy} max: 0.00268 min: 0	ARM_{xx} max: 0.00001 min: 0	ARM_{xy} max: 0.00118 min: 0
	ARM_{yx} max: 0.00283 min: 0	ARM_{yy} max: 1 min: 0.00002	ARM_{yx} max: 0.00268 min: 0	ARM_{yy} max: 0.92286 min: 0.00003	ARM_{yx} max: 0.00118 min: 0	ARM_{yy} max: 0.24517 min: 0.00002

Figure 21.41 The **ARM** amplitude of the o- and e-wavefronts with 0.5 NA for various A-plate thicknesses.

and increases the spatial extent of the yy-component of the overall **ARM**. The magnitudes of the overall **ARM**s are shown in Figure 21.42. The aberrations also increase the spatial extent of the xy- and yz-leakages. These aberrations originate from the aberrated e-wavefront. The amplitude of the overall ARM_{yy} decreases and its spatial extent increases approximately quadratically with thickness, as shown in Figure 21.43.

For the analysis of incoherent light, the **ARM** converts to the 4×4 Mueller point spread matrix **MPSM** with the same algorithm converting Jones matrix into Mueller matrix, as described in Chapter 16. The **MPSM**s for the different plate thicknesses are shown in Figure 21.44. The resulting **MPSM**s are consistent with the **ARM**s. Without aberrations, the **MPSM**s are the Mueller matrices

$$\begin{pmatrix} 1 & 0 & 0 & 0 \\ 0 & 1 & 0 & 0 \\ 0 & 0 & 0 & -1 \\ 0 & 0 & 1 & 0 \end{pmatrix} \text{ and } \begin{pmatrix} 1 & 0 & 0 & 0 \\ 0 & 1 & 0 & 0 \\ 0 & 0 & -1 & 0 \\ 0 & 0 & 0 & -1 \end{pmatrix}$$

modulated by the Airy disk pattern for the quarter and half

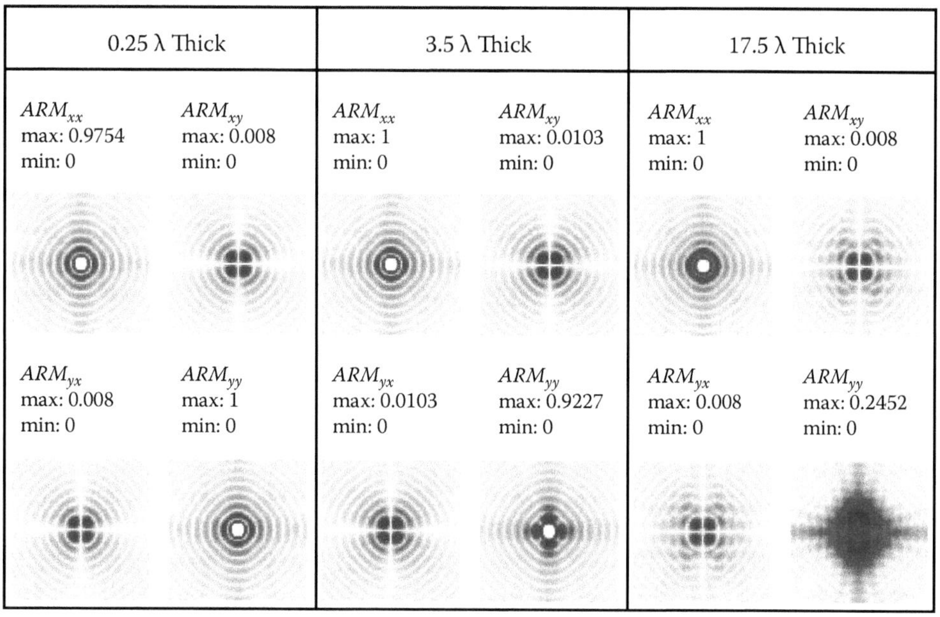

21.7 Image Formation through an A-Plate

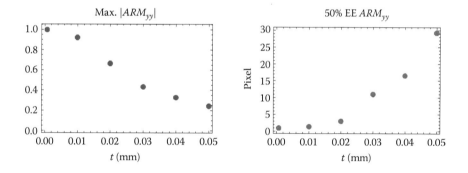

Figure 21.42 The amplitude of **ARM** of the combined o- and e-wavefronts with 0.5 NA for various A-plate thicknesses.

Figure 21.43 (Left) The maximum amplitude and (right) the 50% encircled energy (EE) for $ARM_{e,yy}$ calculated from ray tracing results at six different plate thicknesses t.

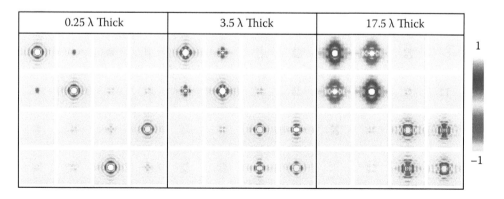

Figure 21.44 **MPSM** of the combined o- and e-wavefronts with 0.5 NA for various A-plate thicknesses.

waveplates, respectively. The aberrated **MPSM**s for the thicker A-plates, as shown in Figure 21.44, contain large aberrations, produce polarization leakages, and induce non-Airy disk image.

The image of an unpolarized or partially polarized point object is calculated by multiplying the system **MPSM** to the incident Stokes parameters. The Stokes images of a point source for seven different incident polarization states and for three different A-plate thicknesses are shown in Figures 21.45 through 21.47.

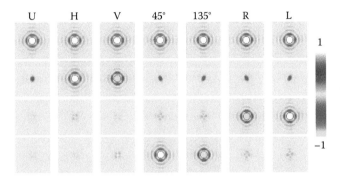

Figure 21.45 Exiting Stokes parameters for unpolarized, horizontally polarized, vertically polarized, 45° linearly polarized, 135° linearly polarized, right circularly polarized, and left circularly polarized incident light for the quarter wave A-plate.

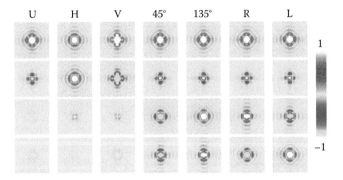

Figure 21.46 Exiting Stokes parameters for unpolarized, horizontally polarized, vertically polarized, 45° linearly polarized, 135° linearly polarized, right circularly polarized, and left circularly polarized incident light for the 3.5 wave A-plate.

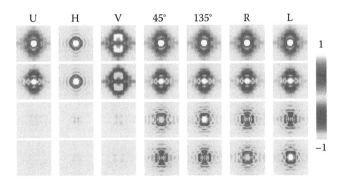

Figure 21.47 Exiting Stokes parameters for unpolarized, horizontally polarized, vertically polarized, 45° linearly polarized, 135° linearly polarized, right circularly polarized, and left circularly polarized incident light for the 17.5 wave A-plate.

21.8 Walk-Off Plate

In summary, the *o*-wavefront demonstrates isotropic aberrations while the *e*-wavefront demonstrates the more complex aberrations of extraordinary rays. The *e*-wavefront has larger wavefront aberrations than the *o*-wavefront in the exit pupil and in the image plane. The combined wavefront at the exit pupil contains polarization aberrations that induce various polarization states across the pupil as a function of the plate thickness, material's birefringence, the NA of the beam, and the orientation of the optic axis.

21.8 Walk-Off Plate

A *walk-off plate* is a birefringent plate that spatially separates one incident ray into two parallel exiting rays with orthogonal polarizations. The orientation of the optic axis is usually selected to maximize the ray separation angle as shown in the middle example of Figure 21.48.

The mode separation depends on the plate birefringence and the optic axis orientation. By adjusting the optic axis, the plate can yield a useful separation between the exiting *o*- and *e*-beams, for example, forming a polarizer in Figure 21.49 (left) or a polarizing interferometer in Figure 21.49 (right).

The two mode refractive indices as a function of the optic axis orientation of a calcite piece are plotted in Figure 21.50 (left) for a normal incidence ray at wavelength 0.5 μm. The orientation of the optic axis α is measured from the ray's propagation vector. The *walk-off angle* γ, the angular separation of the *o*- and *e*-modes' Poynting vectors in the uniaxial medium, is zero when α is 0° or 90° and has a maximum in between the two zero γ.

Figure 21.48 Polarization ray trace of a normal incident beam into a negative uniaxial block with varying optic axis orientation in the plane of the page, showing an optimum separation near 45°.

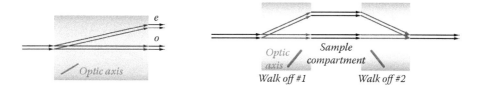

Figure 21.49 Walk-off plate as (left) polarizer and (right) polarizing interferometer.

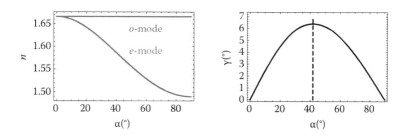

Figure 21.50 (Left) Mode index n_o and n_e for calcite versus optic axis angle α. (Right) Calcite walk-off angle as a function of optic axis orientation at 500 nm. The maximum walk-off angle of 6.42° occurs when α = 41.8°.

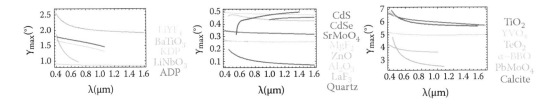

Figure 21.51 The maximum walk-off angle versus wavelength for various uniaxial materials.

The condition for maximum γ is obtained when[8,*]

$$\alpha = \tan^{-1}(n_E/n_O). \tag{21.22}$$

This walk-off angle versus optic axis orientation is plotted for calcite in Figure 21.50 (right) where the maximum beam separation is 6.42°.

The maximum walk-off angle varies with wavelength because the principal refractive index varies with wavelength. Several γ_{max} spectra are plotted in Figure 21.51.

21.9 Crystal Prisms

Combinations of uniaxial crystal blocks have been developed and applied since the eighteenth century as a technique to polarize light or to separate beams of polarized light. Several crystal polarizers operate by total internal reflection (TIR) of one polarization and transmission of the orthogonal polarization. This very efficient polarizing mechanism can produce far higher extinction ratios than dichroic sheet polarizers or polarizing beam splitter cubes.

The *Nicol prism* historically was one of the most common crystal polarizers, so common that polarizers were often called "Nicols." With the Nicol prism[18] geometry, formed by cementing two calcite prisms together, the o-mode total internally reflects and the e-mode passes through the prism as depicted in Figure 21.52. The calcite prisms are cut from a calcite rhomb in its preferred direction due to the atomic structure of calcite. For simple fabrication, this optimum cut angle leaves the tilted front and rear faces.

The FOV for the Nichol prism is limited by two factors: the TIR condition of the o-mode at the balsam interface and the transmission of the e-mode through the interface. Since $n_O > n_{balsam} > n_E$, TIR can happen to both modes when they enter the balsam layer. At 589.3 nm, $n_O = 1.658$, $n_E = 1.486$, and $n_{balsam} = 1.54$. The critical angles of o-mode and e-mode are at 16.03° below horizontal and 10.33° above horizontal, respectively, at the entrance surface to produce evanescence at the balsam interface. For o-mode TIR, the incident angle $\theta_{in} < 16.03°$; for e-mode to avoid TIR and transmit through, $\theta_{in} > -10.33°$. Thus, the prism has a total of 26.36° asymmetric full FOV. Since the Nichol prism is a rather long device, the FOV can also be limited by vignetting.

Other types of crystal polarizer have been developed after the Nicol prism, for example, Glan–Foucault prisms, Glan–Taylor prisms, Glan–Thompson prisms, Wollaston prisms, Rochon prisms, and so on.[19] All of them utilize double refraction or TIR to separate the o- and e-modes. Their differing geometries favor different applications. In Chapter 22, detailed analyses are performed on the Glan-type polarizers, which have exceptional performance but severe limitations in their FOV.

---

* See Problem Set 21.11.

Figure 21.52 (Left) Nicol prism consists of two pieces of calcite with a layer of balsam adhesive (refractive index 1.54) in between. The geometry of the two crystal blocks is identical. A = 68°, B = 90°, and both crystal axes are at C = 63.73° from the prism base. (Right) A normal incidence unpolarized beam incident onto the Nicol prism. The ordinary polarization (red dot represents polarization oscillating in and out of the page) undergoes TIR at the balsam interface while the vertically polarized extraordinary polarization (blue) passes to the exit surface.

21.10 Problem Sets

21.1 Describe the meaning of (a) positive uniaxial, (b) negative uniaxial, (c) birefringence, (d) ordinary mode, (e) extraordinary mode, and (f) optic axis.

21.2 In Figure 21.2, how many rays exit the series of three uniaxial plates, considering only refraction? How many different polarization states are exiting? Draw a picture showing the propagation of the ordinary *ooo*-mode and the extraordinary *eee*-mode through the assembly.

21.3 Consider a six-element cascaded crystal plate (Figure 21.53).
   a. Considering only refraction, how many rays can exit from six birefringent crystals when one ray is incident?
   b. Build a ray tree for a normal incident ray propagating from air through a stack of six uniaxial crystal plates. The front and back plates have identical crystal axis with *x*-, *y*-, and *z*-components. The four middle plates have crystal axes in the *y*–*z* plane, as shown in the figure on the right.
   c. In this system, how many of the modes would have zero power, because *o* and *e* will not generate two modes on refraction?
   d. List the transmitted modes, like *oooooo*.

21.4 a. In Figure 21.3, explain why one mode converges before than the other. If the plate is formed from MgF_2, which mode focuses closer to the plate?
   b. An unpolarized wavefront focuses through a C-plate that is made of negative uniaxial crystal. Which resultant wavefront, ordinary or extraordinary, focuses closer to the plate?

21.5 Consider light incident from air to a plane parallel plate of uniaxial material such that the ordinary beam is at the critical angle at the exiting (back) surface. Show that the

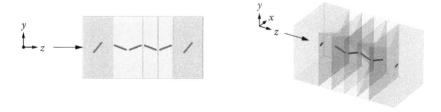

Figure 21.53 Six blocks of uniaxial material. The optic axis orientations are (−0.766, 0.766, 0.643) (the red lines), (0, −0.342, 0.940) (the first pair of blue lines), and (0, 0.342, 0.940) (the second pair of blue lines).

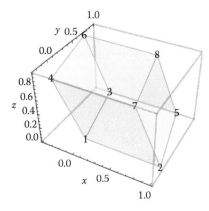

Figure 21.54 Schematics of a calcite rhomb.

extraordinary ray inside the anisotropic material is also incident on the exiting surface at the critical angle.

21.6   a.   Given a sphere of a uniaxial material, how can the optic axis be determined?
       b.   In the upper left plate of Figure 21.6, does the optic axis lie in the horizontal or vertical plane?

21.7   A calcite rhomb (cut along planes of crystal symmetry) has 12 edges each of which is one unit long (Figure 21.54). The angles for each rhombus-shaped face are 102°, 78°, 102°, and 78°. The eight vertices are located as follows:

v1	v2	v3	v4	v5	v6	v7	v8
$\begin{pmatrix} 0 \\ 0 \\ 0 \end{pmatrix}$	$\begin{pmatrix} 1 \\ 0 \\ 0 \end{pmatrix}$	$\begin{pmatrix} -0.208 \\ 0.978 \\ 0 \end{pmatrix}$	$\begin{pmatrix} -0.208 \\ 0.257 \\ 0.944 \end{pmatrix}$	$\begin{pmatrix} 0.792 \\ 0.978 \\ 0 \end{pmatrix}$	$\begin{pmatrix} -0.416 \\ 0.721 \\ 0.944 \end{pmatrix}$	$\begin{pmatrix} 0.792 \\ -0.257 \\ 0.944 \end{pmatrix}$	$\begin{pmatrix} 0.584 \\ 0.721 \\ 0.944 \end{pmatrix}$

The vertices of the six faces are defined as follows:

Face 1	(v1, v2, v5, v3)
Face 2	(v1, v3, v6, v4)
Face 3	(v1, v4, v7, v2)
Face 4	(v6, v8, v7, v4)
Face 5	(v7, v8, v5, v2)
Face 6	(v6, v3, v5, v8)

       a.   Between which two vertices does the optic axis run?
       b.   To cut a C-plate (no birefringence normal to plate), how should the C-plate be oriented?
       c.   Rotate the 3D view to look right down the optical axis. Find the largest area C-plate that can be cut; it is an equilateral triangle and can be cut two different ways. Calculate the length of the edges of the triangle and the area of the triangle.
       d.   To cut an A-plate (maximum birefringence), how should the plate be oriented?
       e.   Would a plate cut through points 1, 2, 3, and 4 be an A-plate? 1, 2, 7, and 8? 1, 5, 6, and 8?

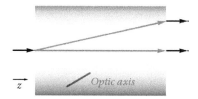

Figure 21.55 A ray normally incident onto a block of uniaxial material.

21.8 A normally incident ray propagates through a 6 mm thick uniaxial plate, optic axis 30° from z-axis, and refractive index $(n_O, n_E) = (1.7, 2.3)$ at wavelength 0.5 μm (Figure 21.55).
   a. In the crystal, are the o- and e-ray propagation vectors parallel to Poynting vectors? How about after exiting the crystal?
   b. What is the walk-off angle between the ordinary and extraordinary modes?
   c. What are the optical path lengths for the two modes?
   d. What is the retardance for this normally incident ray?
   e. What are the reflection losses per surface, and for the combination of entrance and exit surfaces for each mode? Given the **P** matrices through the plate $(\mathbf{P}_2 \cdot \mathbf{P}_1 = \mathbf{P})$:
   At the first interface,

$$\mathbf{P}_o = \begin{pmatrix} 0.7407 & 0 & 0 \\ 0 & 0 & 0 \\ 0 & 0 & 1 \end{pmatrix} \text{ and } \mathbf{P}_e = \begin{pmatrix} 0 & 0 & 0 \\ 0 & 0.7129 & 0.2163 \\ 0 & -0.1580 & 0.9763 \end{pmatrix}.$$

   At the exit interface,

$$\mathbf{P}_o = \begin{pmatrix} 1.2593 & 0 & 0 \\ 0 & 0 & 0 \\ 0 & 0 & 1 \end{pmatrix} \text{ and } \mathbf{P}_e = \begin{pmatrix} 0 & 0 & 0 \\ 0 & 1.2269 & -0.2719 \\ 0 & 0.2163 & 0.9763 \end{pmatrix}.$$

   f. If this retarder was five times thicker, how would the retardance change?

21.9 Consider two uniaxial blocks with refractive index $(n_O, n_E) = (1.6, 2.5)$ assembled along the z-axis, with their optic axes oriented at 50° and −50° from the z-direction (Figure 21.56). Draw the ray tree of all the resultant modes through the system. What are the propagation vectors and Poynting vectors of each exiting ray?

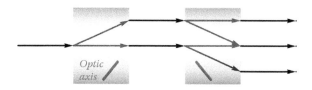

Figure 21.56 Normal incident onto two uniaxial blocks with optic axis oriented at 50° and −50° from the z-direction.

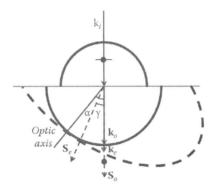

Figure 21.57 K-surfaces of a ray incident from an isotropic material to a uniaxial material, and refracts into o- and e-modes.

21.10 Consider the index indicatrix shown in Figure 21.57.
   a. Given the separation between the Poynting vector and the propagation vector for the extraordinary mode in uniaxial material is γ, where

   $$\tan \gamma = -\frac{\left(n_E^2 - n_O^2\right)\sin\alpha \cos\alpha}{n_E^2 \cos^2\alpha + n_O^2 \sin^2\alpha}$$

   and α is the angle between the optic axis and the propagation vector. Calculate the maximum γ and its corresponding α of a given n_O and n_E.
   b. The effective refraction index for extraordinary ray is

   $$n_{eff} = \frac{n_E n_O}{\sqrt{n_E^2 \cos^2(\theta_e + \alpha) + n_O^2 \sin^2(\theta_e + \alpha)}},$$

   where θ_e is the angle of refraction. Calculate n_{eff} for maximum walk-off in terms of n_O and n_E corresponding to normal incidence.

21.11 In Example 21.2, calculate the Poynting vector direction for both o- and e-modes.

21.12 For propagation inside a uniaxial material near the optical axis, $\theta \ll 1$ radian, the extraordinary refractive index varies approximately quadratically:

$$n_e(\theta) \approx n_E + n_2 \theta^2.$$

Expand $n_e(\theta)$ from Equation 21.16 in a Taylor series to find n_2 in this approximation. How large is n_2 for calcite? How will this affect the focal position of a spherical wave focusing through a calcite C-plate of thickness t?

21.13 Consider the equation for the extraordinary refractive index as a function of propagation angle within the material θ, $\frac{1}{n_e^2(\theta)} = \frac{\cos^2\theta}{1.6^2} + \frac{\sin^2\theta}{1.5^2}$. Plotting this function (in blue), it closely resembles a cosinusoidal function (Figure 21.58).

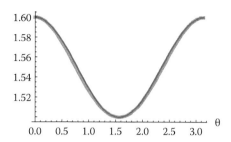

Figure 21.58 n_e versus θ.

When a cosinusoidal function $\dfrac{n_o + n_e}{2} + \dfrac{n_o - n_e}{2}\cos\theta$ is overlaid (in red), a small deviation is seen. About $\theta = 0$, $\dfrac{n_o + n_e}{2} + \dfrac{n_o - n_e}{2}\cos\theta$ has a Taylor series expansion to second order of $n_o + \dfrac{1}{4}(n_e - n_o)\theta^2$ and varies as $\dfrac{1}{4}(n_e - n_o)\theta^2$.

a. Find the second-order Taylor series for $n_e(\theta)$ and compare.

b. What is the difference between $\dfrac{n_o + n_e}{2} + \dfrac{n_o - n_e}{2}\cos\theta$ and $\left(\dfrac{\cos^2\theta}{n_o^2} + \dfrac{\sin^2\theta}{n_e^2}\right)^{-1/2}$ at $\theta = \pi/2$?

21.14 For the Nicol prism described in Section 21.10, what is the polarization orientation of the extraordinary wave? Does the extraordinary beam have astigmatism? How do the two mode orientations vary over the beam? What is the horizontal angular FOV?

21.15 Considering the geometry of the Nicol prism shown in Figure 21.53, what is the refractive index n_e for TIR at the interface for the e-mode?

21.16 Ahren's prism polarizer consists of three polished prisms of calcite cemented together (Figure 21.59). Calcite has refractive indices $n_O = 1.658$ and $n_E = 1.486$. For collimated incident light normal to the front face, one of the modes is transmitted, deviated, and translated. The other state is totally internally reflected and does not reach the final surface.

a. How is the optic axis oriented in crystals A, B and C, along x or y?
b. Which state is transmitted, o or e? What is its mode in the first and second crystals?
c. What is the minimum prism angle, α_{min}?
d. What is the angle of the totally internally reflected ray from the z-axis?
e. What is the transmittance of Ahren's prism at α_{min}?

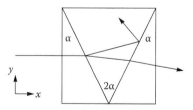

Figure 21.59 A ray normally incident onto an Ahren's prism.

References

1. E. Bartholin, *Experiments on Birefringent Icelandic Crystal (Acta Historica Scientiarum Naturalium et Medicinalium)*, Translated by T. Archibald, Danish National Library of Science and Medicine (1991).
2. J. Z. Buchwald and M. Feingold, *Newton and the Origin of Civilization*, Princeton University Press (2013).
3. J. Z. Buchwald, *The Rise of the Wave Theory of Light: Optical Theory and Experiment in the Early Nineteenth Century*, University of Chicago Press (1989).
4. A.-B. Mahler, S. McClain, and R. Chipman, Achromatic athermalized retarder fabrication, *Appl. Opt.* 50 (2011): 755–765.
5. J. L. Vilas, L. M. Sanchez-Brea, and E. Bernabeu, Optimal achromatic wave retarders using two birefringent wave plates, *Appl. Opt.* 52 (2013): 1892–1896.
6. A. Yariv, and P. Yeh, *Optical Waves in Crystals*, New York: Wiley (1984).
7. M. Born and E. Wolf, *Principles of Optics*, 6th edition, Pergamon (1980).
8. J. N. Damask, Interaction of light and dielectric media, in *Polarization Optics in Telecommunications*, Chapter 3, Springer Series in Optical Sciences (2004).
9. W. J. Tropf, M. E. Thomas, and E. W. Rogala, Properties of crystal and glasses, in *Handbook of Optics*, Vol. 4, Chapter 2, 3rd edition, ed. M. Bass, McGraw-Hill (2009).
10. Refractive index provided by Karl Lambrecht Corporation (http://www.klccgo.com/) accessed in 2014.
11. W. D. Nesse, *Introduction to Optical Mineralogy*, 3rd edition Oxford University Press (2004).
12. E. E. Wahlstrom, *Optical Crystallography*, 3rd edition, Wiley (1960).
13. B. Saleh and M. Teich, *Fundamentals of Photonics*, 2nd edition, Wiley (2007).
14. J. Wilson and J. Hawkes, *Optoelectronics: An Introduction*, 3rd edition, Prentice Hall Europe (1998).
15. M. C. Simon and K. V. Gottschalk, Waves and rays in uniaxial birefringent crystal, *Optik* 118 (2007): 457–470.
16. D. H. Goldstein, Anisotropic materials, in *Handbook of Optical Engineering*, Chapter 24, eds. D. Malacara and B. J. Thompson, New York, NY: CRC Press (2001).
17. W. J. Smith, *Modern Optical Engineering*, 3rd edition, McGraw-Hill (2000).
18. M. C. Simon and R. M. Echarri, Internal total reflection in monoaxial crystal, *App. Opt.* 26(18) (1987).
19. J. M. Bennett, Polarizers, in *Handbook of Optics*, ed. M. Bass, New York: McGraw-Hill (1995).

22

Crystal Polarizers

22.1 Introduction to Crystal Polarizers

Crystal polarizers fabricated from anisotropic materials utilize double refraction* to obtain highly polarized exiting beams of light. The highest-performing polarizers with the greatest extinction ratio are crystal polarizers, including the Glan–Taylor (Figure 22.1, right), Glan–Thompson, and the Nichol prism. Our analysis focuses on several critical performance issues, the small field of view (FOV) of crystal polarizers, transmission loss and large apodization, and the transmission patterns for modes other than the desired mode. These interfering modes can significantly reduce the performance. The crystal polarizers operate by total internally reflecting (TIR) one eigenmode while transmitting most of the orthogonal mode. Since TIR reflects 100% of the incident beam, the degree of polarization of the transmitted beam can be very close to one, leading to exceptional extinction ratio and diattenuation. Polarization states are separated in angle by double refraction, such that two orthogonal polarization states exit in different directions, as with the Wollaston prism shown in Figure 22.1 (left). A comprehensive review of crystal polarizers is contained in Refs. [1–3] written by J. M. Bennett and H. E. Bennett in *OSA Handbook of Optics*.

Crystal polarizers provide high performance in optical systems at a high penalty: limited aperture and significant length. Total internal reflection polarizers tend to be long when compared to sheet and wire grid polarizers, polarizers that are particularly thin. Since the light needs to intersect the crystal surface at a large angle for TIR, the ratio of polarizer length to aperture size is large, about 1 for Glan–Taylor polarizers and 3–5 for Glan–Thompson polarizers. The common issues when

---
* See Chapter 19.

Figure 22.1 (Left) A Wollaston prism is made of two blocks of calcite that have orthogonal optic axes (green lines inside crystals). It directs two orthogonal modes into two directions, due to the different refractive index associated to each mode. (Right) A Glan–Taylor polarizer is also made of two blocks of calcite with the same optic axis orientation. It transmits the *e*-mode (blue arrows) while redirecting the *o*-mode (red arrows) at the interface.

incorporating Glan types of polarizers into imaging systems are small FOV, restricted etendué, and vignetting. Crystal polarizers are commonly used in specialized applications, such as ellipsometry and polarimetry, which require high performance, as well as laser systems where lengthy components are less of an issue for collimated beams. High-power laser systems are another application for Glan–Taylor polarizers, because their ability to handle much higher powers than sheet and wire grid polarizers is compelling. Such high-power applications require highly transparent optical grade crystals, typically calcite and rutile, fabricated to tight specifications.

The complex behavior of crystal polarizers should be considered when designing or using these polarizers. The intended exiting polarization is the extraordinary mode, since the refractive index of the extraordinary mode varies with angle. This introduces astigmatism, apodization, complex wavefront aberrations, and polarization aberrations, which are often daunting for these excellent crystal polarizers. Additionally, crystal polarizers have a small FOV and substantial chromatic aberration. At the edge of the FOV, the reflected ordinary rays begin to leak through, rapidly decreasing the degree of polarization of the transmitted light and varying the polarization state in a complex manner across the pupil.

22.2 Materials for Crystal Polarizers

The ideal material for crystal polarizers should be highly transparent over a substantial spectral range and easy to polish to obtain high-quality optical surfaces and should have a large birefringence. In practice, almost all crystal polarizers are fabricated from calcite, which is a soft crystal with a Mohr hardness of 3, can be polished to very smooth surfaces, and has a large birefringence over the 400 to 1600 nm wavelength range. Optical calcite is a natural material mined most commonly in Mexico and Brazil. Although laboratory growth of calcite has been reported,[4–7] such processes have not been commercialized. As a natural material, it frequently contains stria, small refractive index inhomogeneities that often resemble curtains of weakly scattering inclusions, bubbles, and other scattering defects. These defects can be assessed by shining a laser through calcite and observing scattering sites. Because of its crystal structure, small pyramid-shaped voids may occur on the polished surface if proper care is not taken. Also, some calcite can fluoresce;[8] it fluoresces red, blue, white, pink, green, or orange, depending on impurities.

In contrast, quartz is another highly desirable birefringent material because it can be synthetically grown and is inexpensive. However, the birefringence of quartz is too small for effective Glan–Thompson and Glan–Taylor polarizers; it is more useful for waveplate retarders. Rutile has high refractive index and is a strong birefringent crystal that can also be synthetically grown to manufacture polarizing cubes and coupling prisms.[9,10] Many other minerals, such as sapphire and magnesium fluoride, are also used to make prism and vacuum viewing ports for specific wavefront separation and combining purposes.

22.3 Glan–Taylor Polarizer

A Glan–Taylor polarizer is an air-spaced crystal polarizer constructed from two right triangular pieces of calcite as shown in Figure 22.2a. Its design was first described in 1948.[11] Light propagation in an example Glan–Taylor polarizer, shown in Figure 22.2b, is modeled in this chapter to study its wavefront and polarization aberrations. The optical axes of both calcite pieces are oriented vertically along the y-axis. In the following simulation, the hypotenuse is inclined at an angle $\theta_A = 40°$ with respect to the entrance and exit surfaces.

At 589.3 nm, the refractive indices of calcite are $n_O = 1.659$ and $n_E = 1.486$. Consider the ray paths for internal refraction at a calcite/air interface as depicted in Figure 22.3a and b. When the plane of incidence contained the optic axis, the o-ray refracts as s-polarized light and the e-ray refracts as p-polarized light. Since the two modes have different refractive indices, they have different critical angles. The critical angle propagating from calcite to air for the o-ray is $\psi_o = 37.08°$ and the critical angle for e-ray, after accounting for the varying n_e, is $\psi_e = 40.22°$ in the y–z plane of Figure 22.2, as shown in Figure 22.3d. The difference between the critical angle of the o- and e-modes is the key property for the polarizer operation.

As shown in Figure 22.3e, the refracted o-ray has decreasing intensity with incident angle and zero transmission above critical angle. On the other hand, the refracted e-ray has increasing intensity with angle, and 100% transmission at Brewster's angle around 35°; thus, a smaller θ_A is desirable. As the incident angle continues to increase, the transmission of e-ray decreases rapidly to zero at a critical angle.

For light incident on the hypotenuse calcite/air interface of the Glan–Taylor with angles in the range between ψ_o and ψ_e, the o-ray (above its critical angle) undergoes TIR and reflects toward the top surface, while the e-ray divides into reflected and refracted rays. For light normally incident on the front face, as shown in Figure 22.4, only the e-mode is transmitted. The e-mode transmitted into the air gap couples entirely to the e-mode in the second crystal and exits the polarizer as y-polarized light.

22.3.1 Limited FOV

The Glan–Taylor polarizer is an excellent polarizer within the specified 3° FOV, where the undesired polarization component is efficiently redirected by TIR. For incident angles further off normal incidence, other undesired modes can be present. The desired light path through the polarizer is the eie-mode by the convention of Chapter 19: (e)xtraordinary in the first crystal, (i)sotropic in air gap, and then (e)xtraordinary in the second crystal. Ray doubling can occur in the second crystal, but in case of normal incidence, when the orientations of the optic axes are perfectly aligned, the undesired eio-mode has zero amplitude.

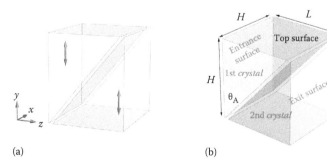

(a)   (b)

Figure 22.2 (a) A Glan–Taylor polarizer consists of two calcite right-triangular pieces separated by an air gap with vertically oriented optic axes (black arrows inside the two blocks). (b) The geometry of the crystal polarizer is defined by θ_A, H, L, and C. The combination of the two crystal blocks forms a rectangular cube. The entrance surface has an area of $H \times H$. The length of the crystal is $L = H \tan \theta_A$. The air gap has a thickness of C.

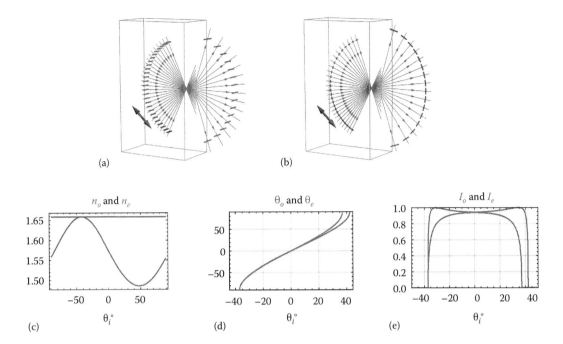

Figure 22.3 *o*- and *e*-rays internally refract through a calcite/air interface. The incident and exiting polarization orientations are shown at the start and at the end of each ray. (a) The *o*-mode is orthogonal to the optic axis (black arrow inside the block) and (b) the *e*-mode is in the plane of the optic axis. (c) Refractive index as a function of incident angle in calcite θ_i. (d) Refraction angle as a function of incident angle; the *o*-mode critical angle is 37.08° and the *e*-mode critical angle is 42.22°. (e) The intensity transmission as a function of incident angle, where the maximum *e*-transmission occurs at 34.54°.

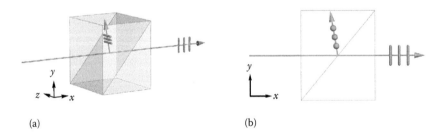

Figure 22.4 For a normal incident ray, the *o*-mode (red) undergoes TIR at the air gap, while the *e*-mode (blue) transmits to the exit surface as vertically polarized light. (The reflected *e*-mode in the first crystal is not shown here.) The *o*-mode either exits through the top surface or is absorbed by a black coating on the top surface. The partial reflections with small amplitude are not shown here, for example, the partial reflection of the *e*-mode at the hypotenuse.

The FOV of the crystal polarizer is defined as the solid angle within which only the *eie*-mode is present in the transmitted beam, and thus the degree of polarization of the exiting light is one. In practice, (1) the TIR of the *o*-mode and (2) the transmission of the *e*-mode for the Glan–Taylor polarizer only occur within a few degrees around normal incidence, as shown in Figure 22.5, leading to a small FOV which is about ±3.5° in the *y*-direction.

Outside of the FOV, some *o*-mode transmits to the second crystal (below purple band in Figure 22.5b), or the *e*-mode undergoes TIR at the air gap interface (above purple band) and no light reaches the second crystal. As the angle of incidence on the front surface varies, the corresponding

22.3 Glan–Taylor Polarizer

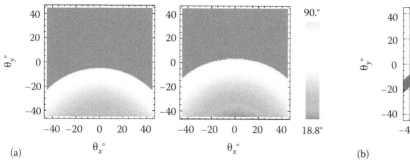
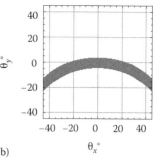

Figure 22.5 (a) The angle of refraction for the o- and the e-modes at the first air gap interface as a function of the incident angle (θ_x, θ_y) in air at the polarizer's entrance surface. The false color scale indicates the refraction angle, where gray denotes TIR. At the critical angle, light refracts at 90° (light blue). The vertical shift between these two functions for oi- and ei-modes is the basis of operation for the Glan–Taylor polarizer, indicating the incident angles where the o-mode TIRs while the e-mode partially transmits. This shift is responsible for the highly efficient performance of the Glan–Taylor. (b) The intended field for the Glan–Taylor polarizer shown ensures that the o-mode will TIR at the interface, while the e-mode transmits through the polarizer. At $\theta_x = 0$, the polarizers field along θ_y ranges from −4.5° to +3.5°.

angle of incidence at the first air gap interface changes. When the angle of incidence at the air gap is below ψ_o, the o-mode partially transmits to the second crystal and generates oie- and oio-modes, thus spoiling polarizer performance. When the angle of incidence at the air gap is greater than ψ_e, the e-mode switches from partially transmitting to TIR, no light reaches the second crystal, and the polarizer goes dark. Also, for an incident ray outside of the y–z plane (θ_x is non-zero), the e-mode generates a small amplitude of eio-mode that leaks through the polarizer as an undesired polarization state in the transmitted light.

The Glan–Taylor polarizer is a complicated device outside its limited FOV, and users should be prepared for these additional undesired beams if the incident beam's angular range is not properly limited. Experiments with laser pointers and crystal polarizers easily demonstrate these additional beams.

22.3.2 Multiple Potential Ray Paths

A polarization ray trace of a Glan–Taylor polarizer over a large range of angles is performed to analyze the paths of all the modes, taking into account the TIR at the air gap. The calcite is simulated as uncoated and the Fresnel losses at all surfaces are included. A ray tree depicting all the possible ray paths (coupling between modes) through the polarizer is shown in Figure 22.6.

The focus of this analysis is refractions, not reflections, unless TIR occurs, which prevents a ray from refracting through the system. The incident light (inc) first divides into o- and e-states entering the polarizer. At the first crystal's hypotenuse (the birefringent/isotropic interface), each mode reflects into both o- and e-modes; the coupling of o into e and e into o is small. However, this coupling is only zero in the vertical plane, the plane containing the surface normal and the optic axis. Within the air gap between the two crystals, the transmitted rays are in an *isotropic* mode. These rays refract part of their flux into the second crystal, generating o- and e-rays. Again, coupling from e in the first crystal into o in the second crystal, or vice versa, is zero only for rays propagating in the vertical plane. The light reflecting at the first hypotenuse is undesired light, which will reach the top surface and preferably either escape or be absorbed by a top surface coating. As a ray passes through the air gap (ei-mode or oi-mode) into the second crystal, the light splits again into o- and e-modes. For each incident ray, a maximum of four modes (eie, eio, oie,

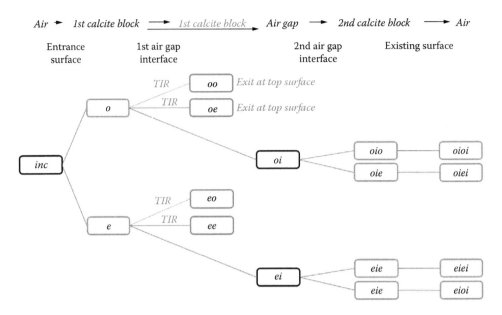

Figure 22.6 A ray tree of all possible ray paths with different mode combinations propagating through a Glan–Taylor polarizer. The incident ray splits into *o*- and *e*-modes refracting at the entrance surface. Without loss of generality, at the first surface of the air gap, both modes can refract through the gap or TIR back to the first crystal. For light reflected from the hypotenuse, the rays split further into *oo*- and *oe*-modes or *eo*- and *ee*-modes. These modes may escape through the top surface. Rays transmitted through the air gap further split into *oio*- and *oie*- or *eio*- and *eie*-modes, and exit the polarizer. Up to four exiting modes may occur for a given incident ray, but depending on the incident ray propagation direction, some of these modes have zero amplitude.

and *oio*) may emerge from the exit surface. Small amounts of multiply reflected light also occur within the air gap.

The polarization ray tracing matrix $\mathbf{P}_{eie}$ of the *eie*-mode for normal incidence is calculated by matrix multiplication of the $\mathbf{P}$ matrices from each of the four interfaces ($\mathbf{P}_1$, $\mathbf{P}_2$, $\mathbf{P}_3$, and $\mathbf{P}_4$). Table 22.1 shows the $\mathbf{P}$ matrices and cumulative $\mathbf{E}$ field after each interface.

$$\mathbf{P}_{eie} = \mathbf{P}_4\mathbf{P}_3\mathbf{P}_2\mathbf{P}_1 = \begin{pmatrix} 0 & 0 & 0 \\ 0 & 0.891 & 0 \\ 0 & 0 & 1 \end{pmatrix}, \qquad (22.1)$$

The $\mathbf{P}_{eie}$ corresponds to a *y*-polarizer; the incident *x*-polarized light is completely extinguished. For the incident polarization vector $\mathbf{E}_i = (0, 1, 0)$, the amplitude of the exiting $\mathbf{E}$ field is 0.891, which corresponds to an intensity transmission of 0.793 as calculated in Table 22.2.

Total internal reflection from the crystal/air interface is a clever and effective mechanism to redirect the unwanted polarization state out of the beam. However, this mechanism is strongly angle of incidence dependent. As the incident angle increases further from normal incidence, as shown in Figure 22.7, undesired modes leak to the exiting surface.

Incident rays in the *y–z* plane induce only *oo*-, *ee*-, *oio*-, and *eie*-modes; the *oe*-, *eo*-, *oie*-, and *eio*-modes have zero amplitude since the optic axis is along the *y*-axis. As the incident ray is tilted away from the vertical plane in the *x*-direction, the cross-coupling modes, *oe*-, *eo*-, *oie*-, and *eio*-modes, start to linearly gain amplitude. These leakages from the undesired modes cause reduction of the extinction ratio and other aberrations and affect image quality.

22.3 Glan–Taylor Polarizer

Table 22.1 Resultant e-Mode **P** Matrices and **E** Fields for a Normally Incident Ray through a Glan–Taylor Polarizer

Ray Path	**P** for Each Individual Surface	Cumulative **E** Field after Each Surface	**E** Field Amplitude after Each Surface
Entrance surface	$\mathbf{P}_1 = \begin{pmatrix} 1 & 0 & 0 \\ 0 & 0.804 & 0 \\ 0 & 0 & 1 \end{pmatrix}$	$\mathbf{E}_1 = \begin{pmatrix} 0 \\ 0.804 \\ 0 \end{pmatrix}$	0.804
First air gap interface	$\mathbf{P}_2 = \begin{pmatrix} 1 & 0 & 0 \\ 0 & 1.588 & 0.542 \\ 0 & -1.025 & 0.840 \end{pmatrix}$	$\mathbf{E}_2 = \begin{pmatrix} 0 \\ 1.278 \\ -0.825 \end{pmatrix}$	1.521
Second air gap interface	$\mathbf{P}_3 = \begin{pmatrix} 1 & 0 & 0 \\ 0 & 0.412 & -0.266 \\ 0 & 0.542 & 0.840 \end{pmatrix}$	$\mathbf{E}_3 = \begin{pmatrix} 0 \\ 0.745 \\ 0 \end{pmatrix}$	0.745
Exit surface	$\mathbf{P}_4 = \begin{pmatrix} 1 & 0 & 0 \\ 0 & 1.196 & 0 \\ 0 & 0 & 1 \end{pmatrix}$	$\mathbf{E}_4 = \begin{pmatrix} 0 \\ 0.891 \\ 0 \end{pmatrix}$	0.891

Table 22.2 Resultant Transmittance and Transmission of the eie-Mode from Normal Incident Are Calculated

Ray Path	Incident and Refraction Angles at Each Surface		Refractive Index	t_i	$T_i = \dfrac{n_{out} \cos\theta_{out}}{n_{in} \cos\theta_{in}} t_i^2$
Entrance surface	0°	0°	1	0.8044	$1.486 \times 0.8044^2 = 0.9617$
First air gap interface	40°	72.836°	1.486	1.8904	$\dfrac{\cos 72.836°}{1.486 \cos 40°} \times 1.8904^2 = 0.926$
Second air gap interface	72.836°	40°	1	0.4899	$\dfrac{1.486 \cos 40°}{\cos 72.836°} \times 0.4899^2 = 0.926$
Exit surface	0°	0°	1.486	1.1956	$\dfrac{1}{1.486} \times 1.1956^2 = 0.9617$
			Total:	$\prod_i t_i = 0.8907$	$\prod_i T_i = 0.793$

t_i is the Fresnel amplitude coefficient and T_i is the Fresnel transmission coefficient.

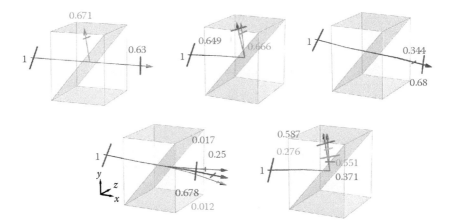

Figure 22.7 Ray traces of several incident angles. Only rays associated with TIR and transmitted through the air gap are shown. The top left is normal incidence. The incident rays for the top two ray traces on the right are tilted along the y-axis with incident **k** = (0, 0.087, 0.996) in air and (0, −0.122, 0.993). The incident rays in the third row are tilted in both the x- and y-directions with incident **k** = (0.104, −0.104, 0.989) and (0.104, 0.104, 0.989). All incident beams are 45° linearly polarized with an amplitude of 1. The resultant *eie*- (blue), *eio*- (green), *oie*- (magenta), *oio*- (red), *ee*- (light blue), *oo*- (orange), *eo*- (green), and *oe*- (coral) polarizations are shown with their corresponding amplitude.

A spherical wavefront converging through the polarizer becomes quite distorted. Figure 22.8 shows a polarization ray trace for a ±45° fan of rays in the y–z plane. Part of the rays is redirected by TIR, while the rest of the *eie*- and *oio*-rays transmit through with an uneven spatial distribution. Each ray exits parallel to the incident ray, since the refractive indices experienced by the ray in the first and second crystals are equal in this case, but with large sideways

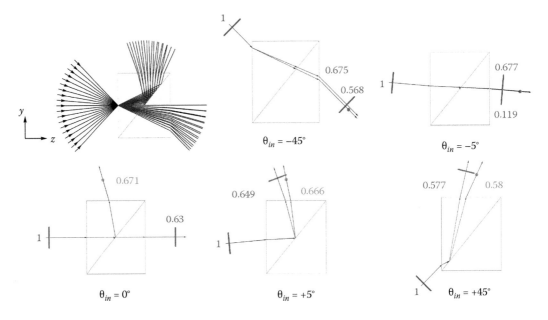

Figure 22.8 Ray paths for a ±45° fan of rays in the y–z plane. Details of ray paths for five circularly polarized incident rays in the y–z plane are shown for refraction and TIR. (To keep the figures simple, the partial reflections are not shown.) The resultant *eie*- (blue), *oio*- (red), *ee*- (light blue), and *oo*- (orange) polarizations are shown with their corresponding electric field amplitude.

22.3 Glan–Taylor Polarizer

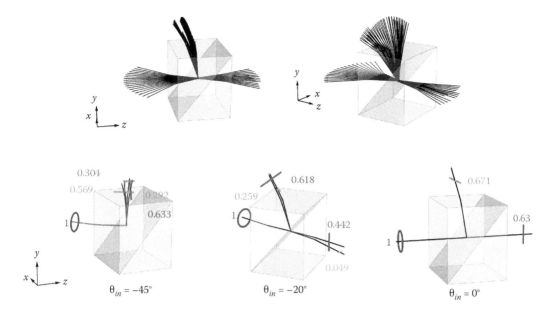

Figure 22.9 A ±45° x–z fan of rays through a Glan–Taylor polarizer. Individual rays in the x–z plane at different incident angles with circularly polarized incident ray (purple) with incident angle θ_{in} and an amplitude of 1. The resultant *eie*- (blue), *eio*- (green), *ee*- (light blue), *oo*- (orange), *eo*- (light green), and *oe*- (coral) polarizations are shown with their corresponding **E** field amplitudes.

displacements, indicating substantial pupil aberration. The individual ray plots for five of the rays in the ray fan are shown, producing varying ray propagation across the field. Because of the asymmetric geometry of the polarizer in the y–z plane, the exiting wavefronts are also non-symmetric.

Figure 22.9 contains a similar ray trace to Figure 22.8 but in the x–z plane. The polarizer geometry is symmetric in this plane and therefore the resultant wavefront is symmetric.

These calculations show that the desired *eie*-mode is highly apodized due to the large variation of the Fresnel transmission close to critical angle at the hypotenuse, as shown in Figure 22.3e. Precision radiometry is thus extremely difficult using this *eie*-beam!

22.3.3 Multiple Polarized Wavefronts

Understanding crystal polarizers, especially in the regions where undesired modes are present, involves calculations to combine the effects of multiple polarized wavefronts. To do this, a spherical wavefront, simulated by a grid of rays, is sent through the polarizer, shown in Figure 22.10 (top), to study the off-axis effects. According to the ray tree in Figure 22.6, four groups of rays are expected for the four exiting modes. These modes are examined individually before combining the wavefronts to determine the overall **E** field. The exiting polarization states for each mode are shown in Figure 22.10 (middle and bottom).

The exiting amplitude distributions for the three incident beams with $0.088 = \sin(5°)$, $0.259 = \sin(15°)$, and $0.707 = \sin(45°)$ numerical aperture (NA) are shown in Figure 22.11. These incident ray grids are an evenly spaced square grid of rays. None of the four exiting modes maintain the original square beam footprint, and all have large amplitude variations across their fields. The exiting *eie*- and *eio*-modes have the same maximum exiting angle because they have the same critical angle at the air gap. Similarly, all the *eio*- and *oio*-modes have the same maximum exiting angle.

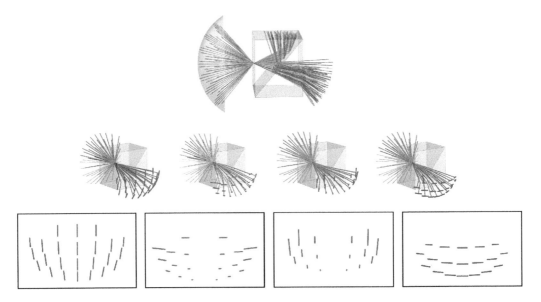

Figure 22.10 (Top) An incident spherical wavefront (yellow) is ray traced for a grid of rays (gray lines) shown within a ±45° cone. The wavefront focuses at the entrance of the Glan–Taylor polarizer and diverges into the crystal. Because of ray doubling, four resultant modes, shown in separate colors, pass through the polarizer, exit, and overlap unevenly. The *eie*-mode is blue, *eio*-mode is green, *oie*-mode is red, and *oio*-mode is magenta. (Middle row) The grid of rays for each refracted mode is plotted separately. The exiting polarization states are determined in 3D. Note that only the bottom half of the rays transmit through. The polarization ellipses of each ray are shown at the end of the ray. (Bottom row) The 3D polarization states scaled by their amplitudes viewed after the crystal.

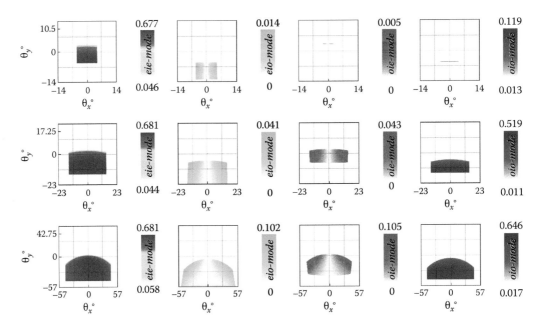

Figure 22.11 The distribution of the **E** field amplitudes for each exiting modes in Figure 22.10 as a function of exiting angle. The NA of the incident beam for each row is (top) $0.088 = \sin(5°)$, (middle) $0.259 = \sin(15°)$, and (bottom) $0.707 = \sin(45°)$.

22.3 Glan–Taylor Polarizer

Next, the exiting electric field amplitudes are calculated for 45° linearly polarized incident rays. The desired *eie*-mode is vertically polarized and has higher amplitude toward the bottom of the field where it is approaching Brewster's angle, with maximum transmitted amplitude 0.68. The undesired *oio*-mode is horizontally polarized and has increasing amplitude with increasing NA. For a cone of ±45° incident beam, the maximum amplitude of the *oio*-mode reaches 0.65. The amplitudes of both of these modes decrease rapidly toward the TIR boundary, and their ray angles are unchanged after the polarizer because the refractive indices in the first and second crystals are the same. On the other hand, the *oie*- and *eio*-modes can be classified as cross-coupling leakages. For rays out of the *y–z* plane, crossing the air gap, small amounts of *o* couple into *e* and vice versa. They have small amplitudes, less than 10%, which increase approximately linearly with increasing θ_x. Their angles are different in air, in each crystal, and in air again due to the change of refractive indices, from n_o to n_e and vice versa. For the 0.088 NA beam, these two cross-coupled modes with undesired polarization should be blocked by aperture. For larger NA, these four modes overlap each other after the polarizer, as shown in Figure 22.12.

The optical path lengths (*OPLs*) evaluated on a best-fit sphere of the *eie*-mode are shown in Figure 22.13. Because of the large crystal block, the *OPL* contains large values.

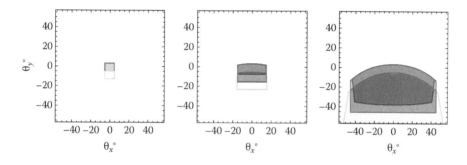

Figure 22.12 The overlap between the four exiting modes for diverging ±5°, ±15°, and ±45° wavefronts with the *eie*-mode in blue, *eio*-mode in green, *oie*-mode in red, and *oio*-mode in magenta.

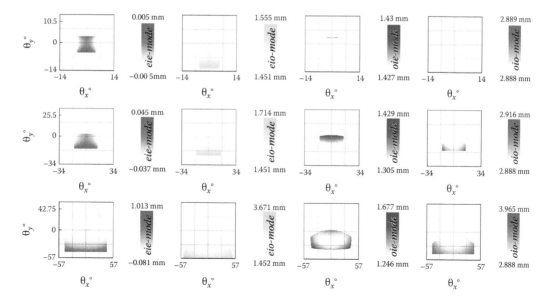

Figure 22.13 The *OPL* of each mode corresponds to the amplitude shown in Figure 22.11 and relative to the *OPL* of the on-axis *eie*-ray.

22.3.4 Polarized Wavefronts Exiting from the Polarizer

To understand the polarizer's performance, the fields of the four exiting modes are combined. The method of Chapter 20 for combining polarized wavefronts represented by irregular grids of rays is applied. The resultant wavefronts from each mode is approximated by resampling each grid of rays onto the same grid. The overall **E** field is

$$\mathbf{E}(\theta_x, \theta_y) = \sum_m^M e^{i\frac{2\pi}{\lambda}OPL_m(\theta_x,\theta_y)} \begin{pmatrix} E_{mx}(\theta_x,\theta_y) \\ E_{my}(\theta_x,\theta_y) \\ E_{mz}(\theta_x,\theta_y) \end{pmatrix}, \quad (22.2)$$

where OPL_m and $\mathbf{E}_m$ are generated from interpolating the ray parameters in terms of θ_x and θ_y for mode m.

The resultant far field intensity distributions in angle space of the four exiting modes are shown in Figure 22.14 for three incident NAs. The resultant intensity drops off toward the top of the field due to the decreasing Fresnel coefficients toward the critical angle of the pure e-mode and pure o-mode. Both of these modes eventually become zero due to TIR. A faint ghost mode (pink) is observed at the bottom of the field from the eio-mode.

These simulation results can be compared with the polarimetric measurement of a Glan–Taylor crystal polarizer shown in Figure 22.15. The measurement shows the angular behavior of the

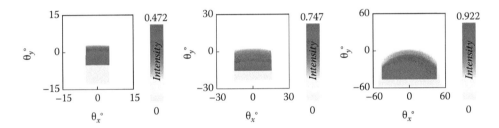

Figure 22.14 The resultant intensity distribution for Figures 22.11 and 22.13. The incident NA for each of them is (left) 0.088, (middle) 0.259, and (right) 0.707. Light pink is chosen to represent the lower intensity.

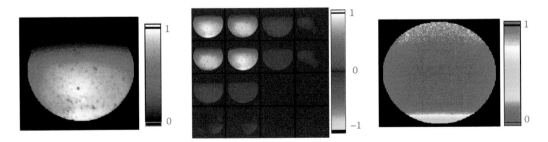

Figure 22.15 Polarimeter measurement of a Glan–Taylor polarizer with a 0.1 NA microscope objective pair. (Left) Measured irradiance, (middle) Mueller matrix image, and (right) diattenuation image.

22.4 Aberrations of the Glan–Taylor Polarizer

polarizer in the form of a Mueller matrix image measured by a Mueller matrix imaging polarimeter in the Polarization Lab at the University of Arizona.[12] The image is acquired over a range of incident angles by illuminating the crystal polarizer with a 0.1 NA focused beam from a polarization state generator. After the beam refracts through the crystal polarizer, the transmitted light is collected by a polarization state analyzer, and an image of the collection objective lens' exit pupil is measured with a CCD detector. Each pixel of the image represents light from a different angle exiting from the crystal polarizer.

The measured irradiance (Figure 22.15, left) shows that the polarizer has a decreasing transmittance toward the top of the field, and no light in the top 1/3 of the 0.1 NA. This region with zero transmittance corresponds to the TIR of all modes. The measured Mueller matrix image (Figure 22.15, middle) contains the polarization properties of the crystal polarizer, principally vertical diattenuation. Except for m_{00}, m_{01}, m_{10}, and m_{11}, the other elements are nearly zero. The Mueller matrix image has a form close to $\begin{pmatrix} 1 & -1 & 0 & 0 \\ -1 & 1 & 0 & 0 \\ 0 & 0 & 0 & 0 \\ 0 & 0 & 0 & 0 \end{pmatrix}$ in 2/3 of the 0.1 NA, which corresponds to a vertical polarizer. The diattenuation image (Figure 22.15, right) is calculated from the Mueller matrix image. It shows high diattenuation within the middle region and about 0.5 diattenuation in a small region at the bottom of the field where the *o*-mode has begun leakage. This low diattenuation corresponds to the multiple modes overlapping as in the simulation. Because of the aperture of the microscope objective, further leakages from the undesired modes are cropped, and the ghost shadow in the simulation does not reach the detector.

22.4 Aberrations of the Glan–Taylor Polarizer

The aberrations of the *eie*-mode (the principal and the only desired mode) and the extinction ratio of the Glan–Taylor polarizer are examined in this section. For the 0.088 NA, Figure 22.11 shows how the *oio*-mode hardly overlaps the *eie*-mode at the center ±5°. The wavefront aberrations of the *eie*-mode are calculated from the *OPL* distribution shown in Figure 22.13 (top left). With a Zernike polynomial fitting to the second- and fourth-order terms,

$$\begin{aligned} OPL(\rho,\varphi) &= P + T_1 + T_2 + D + A_1 + A_2 + C_1 + C_2 + S \\ &= a_0 + a_1(\rho\cos\varphi) + a_2(\rho\sin\varphi) + a_3(2\rho^2 - 1) \\ &\quad + a_4(\rho^2\sin 2\varphi) + a_5(\rho^2\cos 2\varphi) \\ &\quad + a_6\rho(3\rho^2 - 2)\sin\varphi + a_7\rho(3\rho^2 - 2)\cos\varphi + a_8(6\rho^4 - 6\rho^2 + 1). \end{aligned} \qquad (22.3)$$

The forms of the (P, T_1, T_2, D, A_1, A_2, C_1, C_2, S) aberrations are tabulated in Table 22.3, a's are the aberration coefficients, ρ is normalized pupil coordinate, and $\tan\varphi = \varphi_x/\varphi_y$ is the pupil angle measured counterclockwise from the *x*-axis. This Zernike fit to nine terms has a 0.015 μm fit residual due to a small amount of higher-order aberrations.

After subtracting the piston, tilt, and defocus, the residual wavefront of the *eie*-mode, shown in Figure 22.16, is primarily 7.8 waves astigmatism.

Table 22.3 Zernike Coefficients for the *eie*-Wavefront for the System with 0.088 NA

Zernike Aberration			a Coefficients of the *eie*-Wavefront (Waves)
Piston, P	a_0		0
Tip, T_1	$a_1 (\rho \cos\varphi)$		0
Tilt, T_2	$a_2 (\rho \sin\varphi)$		0
Defocus, D	$a_3 (2\rho^2 - 1)$		0
Astigmatism 1, A_1	$a_4 (\rho^2 \sin 2\varphi)$		0
Astigmatism 2, A_2	$a_5 (\rho^2 \cos 2\varphi)$		7.801
Coma 1, C_1	$a_6 \rho (3\rho^2 - 2)\sin\varphi$		0

(Continued)

Table 22.3 (Continued) Zernike Coefficients for the *eie*-Wavefront for the System with 0.088 NA

	Zernike Aberration		a Coefficients of the eie-Wavefront (Waves)
Coma 2, C_2	$a_7 \rho (3\rho^2 - 2)\cos\varphi$		−0.001
Spherical, S	$a_8 (6\rho^4 - 6\rho^2 + 1)$		0.023

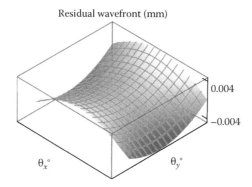

Figure 22.16 Wavefront of the *eie*-mode with zero piston, tilt, and defocus.

22.5 Pairs of Glan–Taylor Polarizers

A figure of merit for polarizer performance is *extinction ratio*:

$$\text{Extinction Ratio} = \frac{\text{Intensity}_{co-polarizer}}{\text{Intensity}_{cross-polarizer}}, \quad (22.4)$$

the ratio of the transmission of the co-polarized system to the cross-polarized system for unpolarized incident light. For sheet polarizers, the cross-polarized system is two polarizers with their transmission axes crossed, and the co-polarized system is two polarizers with their transmission axes in the same orientation.

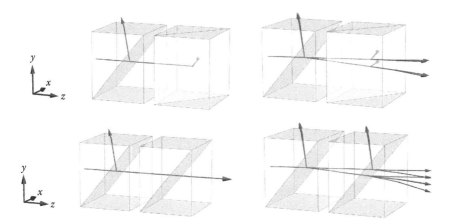

Figure 22.17 Polarizer configurations for calculating extinction of Glan–Taylor polarizers. (Left) A normal incident ray propagates through the (top) cross and (bottom) co-polarized systems. (Right) An off-axis ray with $\theta_y = 3°$ incident angle propagates through the two systems, where light leaks through the cross-polarized system while more rays split through the co-polarized system.

Figure 22.17 shows the co- and cross-polarized configurations. At normal incidence, the co-polarized system produces a vertically polarized ray, and the cross-polarized system blocks all incident light. Therefore, the extinction ratio at normal incidence is infinite.

The advertised FOV of commercial Glan–Taylor polarizer is stated to be around 3° to 5° at 0.5 μm.[13] A $\theta_y = 3°$ incident ray shown in Figure 22.17 (right) creates four exiting modes in the cross-polarized configuration (extinction) and five exiting modes in the co-polarized configuration (adding the desired mode). Therefore, the cross-polarized system leaks light at this angle. The leaked light has a very small flux that decreases the extinction ratio.

Both configurations have four calcite blocks with eight birefringent interfaces; hence, they potentially yield 2^4 or a total of 16 possible exiting modes, as shown in Figure 22.18. These 16 rays involve all possible couplings between *o*- and *e*-modes. When the ray enters the air gap between the crystals, it does not split, and this mode is represented by *i*, for isotropic air. Thus, the 16 possible exiting modes are: *eieieie, eieieio, eieioie, eieioio, eioieie, eioieio, eioioie, ieoioio, oieieie, oieieio, oieioie, oieioio, oioieie, oioieio, oioioie,* and *oioioio.*

The performance of the Glan–Taylor polarizer is first examined for a square grid of incident rays with 0.05 NA. At this small NA, there are four transmitted modes for the cross-polarized system and six transmitted modes for the co-polarized system. Figure 22.19 shows the locations for all the resultant modes at the far field. The number of exiting modes increases with increasing field for this system.

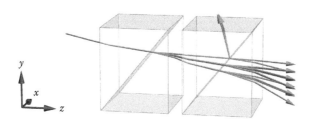

Figure 22.18 A ray propagates through a co-polarized system with four crystals. Beyond $\theta_y = -10°$, one incident ray produces 16 exiting rays.

22.5 Pairs of Glan–Taylor Polarizers

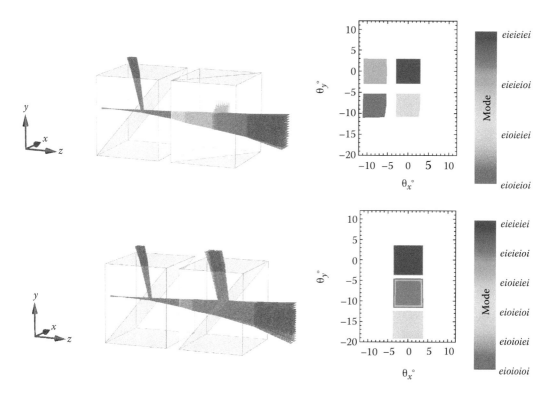

Figure 22.19 (Left) A grid of rays with 0.05 NA propagates through the cross- and co-polarized systems, where rays leak through for non-normal incident rays. (Right) The exiting modes are divided into groups by different colors and are plotted as a function of exiting angles. In the co-polarized system, the *eieieie*-mode dominates the center of the field.

The flux of the exiting beam near the optical axis (z-axis) is dominated by the pure e-modes for co- and crossed Glan–Taylor prisms. The four exiting modes from the cross-polarized system propagates in four distinct directions and results in four wavefront patches that do not overlap. On the other hand, three of the six modes from the co-polarized system have the same exiting propagation vector and overlap right below the dominant *eieieiei*-wavefront, as shown in Figure 22.19. Exiting rays from the same mode produce a smooth wavefront. Therefore, as in the last section, each mode is interpolated individually into a continuous function before combining them.

The polarization amplitude of each mode exiting the cross- and co-polarized system is shown in color in Figures 22.20 and 22.21 showing their exiting angles. In the cross-polarized system, a Maltese cross-like pattern is observed for both the *eieieie*- and *eieieio*-modes, with the *eieieioi*-mode

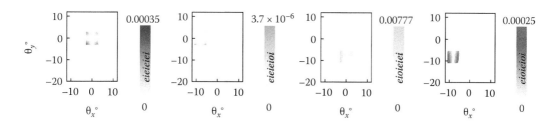

Figure 22.20 Polarization amplitudes through the cross-polarized system with 0.05 NA for each resultant mode expressed in the exiting angle coordinates.

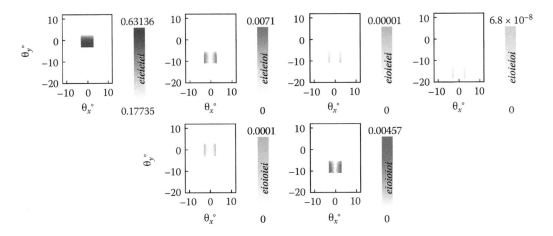

Figure 22.21 Polarization amplitudes through the co-polarized system with 0.05 NA for each resultant mode in exiting angle coordinates.

carrying the least energy. The *eioieie*- and *eioieio*-modes have zero amplitude for incident rays along the y–z plane. These amplitudes increase with incident angle away from the y–z plane. These four modes appear at distinct areas of the exit surface, where the *eieieie*-mode stays at the center of the optical axis, and all the other modes are displaced. From the co-polarized system, the dominant extraordinary mode has rapidly decreasing amplitude, apodization, with increasing θ_y. The rest of the five exiting modes displaced from the center have minimum amplitude with zero θ_x component and increase with θ_y.

The combined intensity distributions at the far field for both of the systems are shown in Figure 22.22, where most of the undesired modes are directed away from the center of the field, except for the pure e-mode. The highest-intensity leak in the cross-polarized system is the *eioieie*-mode, which is just below the centered pure e-mode. Special care is necessary to prevent this significant mode from interfering with an optical system's performance.

The extinction ratio will be calculated for a fully illuminated ±3° square FOV, as shown in Figure 22.23 as a function of the exiting angle.

The extinction ratio distribution is the ratio of these two intensity distributions shown in Figure 22.24. The highest extinction ratio, essentially infinite in simulation, occurs at normal incidence

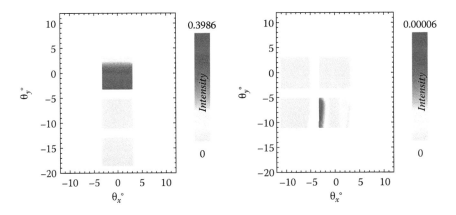

Figure 22.22 Resultant electric field intensity at the exit surface of the (left) co-polarized and (right) cross-polarized systems with 0.05 NA.

22.5 Pairs of Glan–Taylor Polarizers

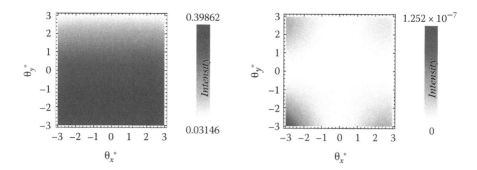

Figure 22.23 The intensity distribution at the center ±3° in the far field for the (left) co-polarized and (right) cross-polarized systems for the ±3° square illumination. The light leakage from off-axis rays in the cross-polarized system has a Maltese cross pattern. The co-polarized system has decreasing amplitude toward the +y-direction, going to zero where TIR begins.

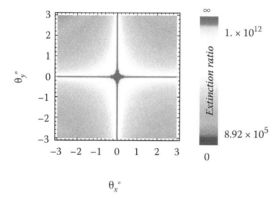

Figure 22.24 Extinction ratio map for ±3° square illumination. The highest extinction ratio is at normal incidence, and along the x- and y-directions.

(shown by the center point) and along the x- and y-direction (zero θ_x and/or θ_y), because the intensity in the Maltese cross leakage of the cross-polarized system is zero. Outside those regions, the extinction ratio drops off rapidly with increasing angles to about 10^5.

Next, a similar analysis is performed for a 0.1 NA. The exiting modes have more overlap than the 0.05 NA, as shown in Figure 22.25. The co-polarized system has a decreasing transmission toward the +y-direction and only transmits light up to +3° due to TIR. In the cross-polarized system, the center ±6° region is highly affected by the overlapping modes. The low Maltese cross leakage is boosted by the light couplings from the overlapped modes; the abrupt changes of intensity is the sign of overlapped modes. Also, the cross-polarized system does not transmit light beyond +3° in both x- and y-directions. The extinction ratio is undefined when the co-polarized system has zero intensity, which is when $\theta_y > +3°$. It is infinity when the cross-polarized system has zero intensity at normal incidence and $\theta_x > +3°$. At the mode overlapping regions ($\theta_x \approx -6$, $\theta_x \approx +3$, and $\theta_y \approx -6$) of the cross-polarized system, the extinction ratio drops to 0.13. The minimum extinction ratio of the remaining area is ~3000, which is lower than that for the 0.05 NA in Figure 22.24.

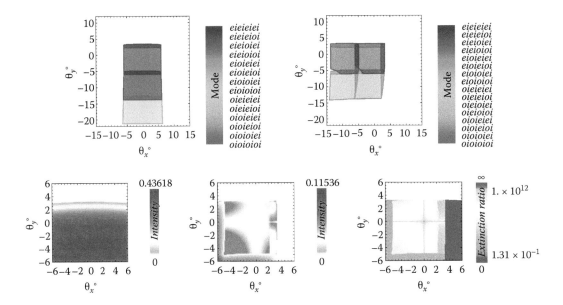

Figure 22.25 Extinction ratio analysis for a ±6° incident beam. (Top) The mode overlapped after the co- and cross-polarized systems. (Bottom, left, and middle) The exiting intensity within a ±6° exiting angle for the co- and cross-polarized systems. (Bottom, right) Extinction ratio plot with regions of overlapping modes.

22.6 Conclusion

This chapter presented methods to analyze complex anisotropic optical components by calculating the significant modes generated from ray doubling. The analysis shows how the high-performance crystal polarizer—a Glan–Taylor prism—is highly sensitive to incident angle. To see the important effects, incident beams with various NAs were traced through the polarizer, and the multiple exiting modes were simulated. The light leakage from undesired modes has an increasing effect with increasing FOV.

The Glan–Taylor polarizer shows interesting wavefront and polarization aberrations due to the non-rotationally symmetric geometry of the component. For small FOV, the center region is dominated by the extraordinary mode. When focusing through the Glan–Taylor, the *eie*-mode has large residual astigmatism due to the crystal birefringence as well as the strong apodization, a rapidly decreasing intensity toward critical angle. The undesired modes propagating through the birefringent materials should be avoided. Also, the suppression of the TIR beam at the top surface is an issue and a potential source of scattered light.

Additionally, two Glan–Taylor configurations, co- and cross-crystal polarized systems, were studied. The performance of the Glan–Taylor crystal polarizer decreases drastically when operating outside of the intended FOV. Such configurations can be found in interferometers with co- and cross-beam splitters, such as heterodyne interferometers[14] and other multi-channel interferometers,[15–17] where similar light leakage could be observed and affect the resultant fringe visibility.

22.7 Problem Sets

The performance of crystal polarizers also depends on the proper alignment of the crystal axes. One common defect is that the crystal axes of the two crystals are not parallel; then, the diattenuation and polarizance are not aligned. Such misalignment can distort the FOV of the polarizer.

22.7 Problem Sets

22.1 What are the advantages of crystal polarizers over sheet polarizers?

22.2 A Glan–Taylor polarizer uses two calcite crystal pieces. Explain how and why the second piece of calcite can be replaced by glass. What disadvantages could occur using two different materials, that is, what useful characteristics of the Glan–Taylor polarizer would change?

22.3 Why is the on-axis transmission of the Glan–Taylor polarizer only 0.793? What could be done to increase this value? Why is the transmitted *eie*-mode not uniform across the field?

22.4 Consider a 500 nm wavelength normal incident ray propagating through the first surface of a crystal polarizer as shown in Figure 22.4. The two crystals are made up of two calcite blocks with $n_o = 1.666$ and $n_e = 1.490$ at 500 nm, with an air gap (index = 1). The surface normal η of the interface is $(0, -\sin\theta_A, \cos\theta_A)$, where $\theta_A = 40°$. Calculate the Fresnel coefficients for all exiting modes at the first calcite/air interface in (a) and (b) for different incident modes, using

$$\mathbf{F} \cdot \mathbf{A} = \mathbf{B},$$

where

$$\mathbf{F} = \begin{pmatrix} \vec{s}_1 \cdot \hat{\mathbf{E}}_{tA} & \vec{s}_1 \cdot \hat{\mathbf{E}}_{tB} & -\vec{s}_1 \cdot \hat{\mathbf{E}}_{rC} & -\vec{s}_1 \cdot \hat{\mathbf{E}}_{rD} \\ \vec{s}_2 \cdot \hat{\mathbf{E}}_{tA} & \vec{s}_2 \cdot \hat{\mathbf{E}}_{tB} & -\vec{s}_2 \cdot \hat{\mathbf{E}}_{rC} & -\vec{s}_2 \cdot \hat{\mathbf{E}}_{rD} \\ \vec{s}_1 \cdot \vec{\mathbf{H}}_{tA} & \vec{s}_1 \cdot \vec{\mathbf{H}}_{tB} & -\vec{s}_1 \cdot \vec{\mathbf{H}}_{rC} & -\vec{s}_1 \cdot \vec{\mathbf{H}}_{rD} \\ \vec{s}_2 \cdot \vec{\mathbf{H}}_{tA} & \vec{s}_2 \cdot \vec{\mathbf{H}}_{tB} & -\vec{s}_2 \cdot \vec{\mathbf{H}}_{rD} & -\vec{s}_2 \cdot \vec{\mathbf{H}}_{rD} \end{pmatrix},$$

$$\mathbf{A} = \begin{pmatrix} a_{tA} \\ a_{tB} \\ a_{rC} \\ a_{rD} \end{pmatrix}, \mathbf{B} = \begin{pmatrix} \vec{s}_1 \cdot \hat{\mathbf{E}}_i \\ \vec{s}_2 \cdot \hat{\mathbf{E}}_i \\ \vec{s}_1 \cdot \vec{\mathbf{H}}_i \\ \vec{s}_2 \cdot \vec{\mathbf{H}}_i \end{pmatrix},$$

$\vec{s}_1 = \hat{k}_i \times \hat{\eta}$, and $\vec{s}_2 = \hat{\eta} \times \vec{s}_1$, as shown in Chapter 19.

a. Consider the horizontally polarized ordinary ray propagating toward the calcite/air interface,

$$\begin{cases} n_o = 1.666 \\ \mathbf{k}_{i,o} = (0,0,1) \\ \mathbf{E}_{i,o} = (1,0,0) \end{cases}$$

By using the "reflecting to birefringent material algorithm" for the reflecting modes and the "transmitting to isotropic material algorithm" for the transmitting modes:

Mode	n	$\hat{\mathbf{k}}$	$\hat{\mathbf{E}}$	$\hat{\mathbf{H}}$
Reflected oo	1.666	$\begin{pmatrix} 0 \\ 0.98481 \\ -0.17365 \end{pmatrix}$	$\begin{pmatrix} 1 \\ 0 \\ 0 \end{pmatrix}$	$\begin{pmatrix} 0 \\ -0.28938 \\ -1.64113 \end{pmatrix}$
Reflected oe	1.660	$\begin{pmatrix} 0 \\ 0.98534 \\ -0.17063 \end{pmatrix}$	$\begin{pmatrix} 0 \\ 0.21169 \\ 0.97734 \end{pmatrix}$	$\begin{pmatrix} 1.65895 \\ 0 \\ 0 \end{pmatrix}$
Transmitted os	1	$\begin{pmatrix} 0 \\ 0.72223 - 0.21689i \\ 0.60509 + 0.25847i \end{pmatrix}$	$\begin{pmatrix} 1 \\ 0 \\ 0 \end{pmatrix}$	$\begin{pmatrix} 0 \\ 0.60509 + 0.25847i \\ -0.72112 + 0.21689i \end{pmatrix}$
Transmitted op	1	$\begin{pmatrix} 0 \\ 0.72223 - 0.21689i \\ 0.60509 + 0.25847i \end{pmatrix}$	$\begin{pmatrix} 0 \\ 0.60509 + 0.25847i \\ -0.72112 + 0.21689i \end{pmatrix}$	$\begin{pmatrix} -0.7723 \\ 0 \\ 0 \end{pmatrix}$

There are two reflected rays and two transmitted rays. The s- and p-modes can be combined to a single i-mode in the isotropic air gap. One of the reflected modes and one of the transmitted modes will have zero energy. The transmitted ray has a complex $\mathbf{k}$ indicating TIR.

b. Consider the vertically polarized extraordinary ray propagating toward the calcite/air interface,

$$\begin{cases} n_{eo} = 1.490 \\ \mathbf{k}_{i,o} = (0,0,1) \\ \mathbf{E}_{i,o} = (1,0,0) \end{cases}$$

By using the "reflecting to birefringent material algorithm" for the reflecting modes and the "transmitting to isotropic material algorithm" for the transmitting modes:

Mode	n	$\hat{\mathbf{k}}$	$\hat{\mathbf{E}}$	$\hat{\mathbf{H}}$
Reflected eo	1.666	$\begin{pmatrix} 0 \\ 0.96629 \\ -0.25744 \end{pmatrix}$	$\begin{pmatrix} 1 \\ 0 \\ 0 \end{pmatrix}$	$\begin{pmatrix} 0 \\ -0.42901 \\ -1.61028 \end{pmatrix}$
Reflected ee	1.653	$\begin{pmatrix} 0 \\ 0.96772 \\ -0.25204 \end{pmatrix}$	$\begin{pmatrix} 0 \\ 0.30975 \\ 0.95082 \end{pmatrix}$	$\begin{pmatrix} 1.65034 \\ 0 \\ 0 \end{pmatrix}$
Transmitted es	1	$\begin{pmatrix} 0 \\ 0.54891 \\ 0.83588 \end{pmatrix}$	$\begin{pmatrix} 1 \\ 0 \\ 0 \end{pmatrix}$	$\begin{pmatrix} 0 \\ 0.83588 \\ -0.54891 \end{pmatrix}$
Transmitted ep	1	$\begin{pmatrix} 0 \\ 0.54891 \\ 0.83588 \end{pmatrix}$	$\begin{pmatrix} 0 \\ 0.83588 \\ -0.54891 \end{pmatrix}$	$\begin{pmatrix} -1 \\ 0 \\ 0 \end{pmatrix}$

There are two reflected rays and two transmitted rays. One of the reflected modes and one of the transmitted modes will have zero energy. Note that the transmission coefficient could be larger than one when ray passes from high index to low index.

22.5 In Table 22.1, the amplitude transmission coefficient from the first crystal into air is 1.5. Why is this greater than 1?

22.6 At wavelength 589.3 nm, the calcite-made Glan–Taylor polarizer has $n_O = 1.65852$ and $n_E = 1.48644$.
   a. At normal incidence, calculate the refraction angles at all interface corresponding to e-, ei-, eie-, and eiei-modes.

   $$\frac{1}{n_e(\theta)} = \frac{\cos^2\theta}{n_O^2} + \frac{\sin^2\theta}{n_E^2},$$ where θ is the angle between the **k** vector and the optic axis.

   $n_i \sin\theta_i = n_o \sin\theta_o = n_e(\theta_a + \theta_e)\sin\theta_e$, where θ_a is the angle between the optic axis and the surface normal, and θ_e is the angle between the surface normal and **k** vector.

   b. What are the corresponding refractive indices of the modes calculated in part (a)?
   c. Calculate the Fresnel coefficients of these modes. Notice at the second surface that e-mode is completely coupled to p-polarization. $t_p = \dfrac{2n\cos\theta}{n'\cos\theta + n\cos\theta'}$
   d. Calculate the Fresnel intensity of these modes. $T_p = \dfrac{n'\cos\theta'}{n\cos\theta}t_p^2$
   e. What is the intensity transmission of vertically polarized light?

22.7 What is the **P** matrix for the TIR reflected beam in the Glan–Taylor polarizer for normal incident light?

22.8 Consider a normal incident ray at 500 nm propagating through the first surface of a crystal polarizer with geometry similar to Figure 22.2. For normal incident, the two calcite crystal blocks have $n_o = 1.666$ and $n_e = 1.490$ at 500 nm. Let $\theta_A = 14°$ and the optic axis be oriented in the *x*-direction. The cement between the two crystals has an index of $n_c = 1.540$.

a. What are the Brewster angles at the first calcite/air interface for the ordinary mode and the extraordinary mode?

$$\text{Brewster angle equation: } \theta_B = \tan^{-1}\left(\frac{n_{ext}}{n_{inc}}\right)$$

b. What is the critical angle at the first calcite/air interface for the ordinary mode? The extraordinary mode never experiences TIR, since $n_e < n_c$.

$$\text{Critical angle equation: } \theta_C = \sin^{-1}\left(\frac{n_{ext}}{n_{inc}}\right)$$

22.9 What are the effects of rotating the optic axis of one of the crystals in the crystal polarizer in the *x*–*y* plane? In the *y*–*z* plane? Which would be worse?

22.10 Based on Figure 22.11, for a square beam with ±45° incident angle, where in the exiting beam is the light linearly polarized, and how is the polarization oriented?

22.11 How much volume of the calcite rhombs are actually used in Glan–Taylor and Glan–Thompson polarizers?

22.12 What is the accepted cone angle for both Glan–Taylor and Glan–Thompson polarizers? What are the etendué for the intended use of both polarizers?

22.13 Which polarizer should be used with high-power lasers, Glan–Taylor or Glan–Thompson polarizer?

22.14 What would happen if the second optic axis in a Glan-type polarizer is rotated by 1°?

22.15 How does the birefringence of calcite compare to quartz? Can a Glan–Taylor crystal polarizer be fabricated from quartz? What would the advantages or disadvantages be?

22.16 Generate two ray trees to depict all the possible modes for the systems shown in Figure 22.17.

22.17 The Glan–Taylor and Glan–Thompson crystal polarizers are both fabricated by calcite, which is a negative uniaxial crystal. What is the modification for using a positive uniaxial crystal?

References

1. H. E. Bennett and J. M. Bennett, Polarization, in *Handbook of Optics*, 1st edition, Chapter 10, eds. W. G. Driscoll and W. Vaughan, New York: McGraw-Hill (1978).
2. J. M. Bennett, Polarizers, in *Handbook of Optics Vol. II*, 2nd edition, Chapter 3, ed. M. Bass. New York: McGraw-Hill (1995).
3. J. M. Bennett, Polarizers, in *Handbook of Optics Vol. I*, 3rd edition, Chapter 13, ed. M. Bass, New York: McGraw-Hill (2010).
4. K. Yanagisawa, Q. Feng, K. Ioku and N. Yamasaki, Hydrothermal single crystal growth of calcite in ammonium acetate solution, *J. Cryst. Growth*, 163 (1996): 285–294.
5. K. Yanagisawa, Preparation of single crystals under hydrothermal conditions, *J. Cer. Soc. Jpn.* 163 (1996): 285–294.

References

6. K. Yanagisawa, K. Ioku, and N. Yamasaki, Solubility and crystal growth of calcite in organic salt solutions under hydrothermal conditions, *J. Mater. Sci. Lett.* 14 (1995): 256–257.
7. Y. K. Lee and S. J. Chung, Hydrothermal growth of calcite single crystal in NH_4Cl solution, *J. Cryst. Growth*, 192 (1998): 350–353.
8. Fluorescent Minerals (http://geology.com/articles/fluorescent-minerals/, accessed July 2015).
9. Del Mar Photonics Newsletter (http://www.dmphotonics.com/rutile_coupling_prism.htm, accessed January 2015).
10. Greyhawk Optics (http://greyhawkoptics.com, accessed January 2015).
11. J. F. Archard and A. M. Taylor, Improved Glan-Foucault prism, *J. Sci. Instrum.* 25(12) (1948): 407–409.
12. N. A. Beaudry, Y. Zhao, and R. Chipman, Dielectric tensor measurement from a single Mueller matrix image, *J. Opt. Soc. Am. A* 24 (2007): 814–824.
13. Karl Lambrecht Corporation (http://www.klccgo.com/glantaylor.htm, accessed March 2015).
14. S. F. Jacobs and D. Shough, Thermal expansion uniformity of Heraeus-Amersil TO8E fused silica, *Appl. Opt.* 20 (1981): 3461–3463.
15. C.-C. Chen, H.-D. Chien, and P.-G. Luan, Photonic crystal beam splitters, *Appl. Opt.* 43 (2004): 6187–6190.
16. M. Pavičić, Spin-correlated interferometry with beam splitters: Preselection of spin-correlated photons, *J. Opt. Soc. Am. B* 12 (1995): 821–828.
17. L. Kaiser, E. Frins, B. Hils, L. Beresnev, W. Dultz, and H. Schmitzer, Polarization analyzer for all the states of polarization of light using a structured polarizer, *J. Opt. Soc. Am. A* 30 (2013): 1256–1260.

23

Diffractive Optical Elements

23.1 Introduction

Diffractive optical elements (DOEs) are optical elements incorporating fine structures to diffract light, as opposed to reflecting or refracting light. In general, DOEs diffract an incident plane wave into a number of reflected and/or transmitted diffraction orders as shown in Figure 23.1. The theory of diffraction grating was first described by James Gregory in 1673.[1] David Rittenhouse made the first diffraction grating in 1785 by placing hairs between finely threaded screws.[2] In the late nineteenth century, Henry Rowland made major advancements in diffraction grating technology, including concave gratings development, enabling the rapid development of spectroscopic technology.[3,4] The diffraction operation is described by a set of amplitude coefficients, like Fresnel coefficients, for each diffraction order. Since each diffraction order diffracts a different amount of *transverse electric* (TE) and *transverse magnetic* (TM) modes, each order has different diattenuation and retardance. Often, diffraction orders are quite polarizing.

The basic theory of diffractive optics is presented to understand the in-plane and out-of plane propagation of diffracted rays. The calculation of these polarization-dependent amplitude coefficients is generally performed by the *rigorous coupled wave analysis* (RCWA) algorithm, which is briefly summarized at the end of the chapter.

The oldest application of DOE is reflection diffraction gratings, commonly used in monochromators and spectrometers. Gratings are the source of many polarization problems in monochromators and other optical systems. A typical commercial reflection grating is analyzed in Section 23.3. Another DOE common application, wire grid polarizers, is analyzed by RCWA in Section 23.3.2, where the effect of the depth of the wires on the polarizer performance is considered. Subwavelength

812   Chapter 23: Diffractive Optical Elements

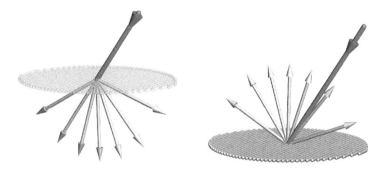

Figure 23.1 A ray diffracts into multiple diffraction orders through a transmission (left) and reflection (right) diffractive grating.

phase gratings have retardance, but little diattenuation; hence, the application of DOEs to retarders is considered in Section 23.3.3, where it is shown that the retardance is too small for most applications. Finally, subwavelength gratings used as antireflection coatings on a lens are analyzed in Section 23.3.4 and the resulting polarization aberrations are studied.

The different classes of DOEs are briefly outlined:

- An *amplitude grating* has alternating opaque and transparent structures imposing amplitude modulation on the reflected and transmitted light. Figure 23.2 shows a periodic amplitude grating.
- A *phase grating* is a transparent structure with phase variations that produce phase modulation. The phase variations can be created by thickness variations, refractive index variations, and similar means. Figure 23.3 shows a periodic square wave phase grating.
- A *surface grating* is a thin grating where all diffractions happen at one surface.
- A *volume grating* is a thick grating, usually thicker than the grating's period, formed from refractive index variations. Volume gratings can have high diffraction efficiency when operating near the Bragg condition. Because of the thickness, an appreciable interaction can occur, even for small refractive index differences. Volume gratings are not always formed from refractive index variation. For example, a volume hologram can be made on a photographic film, where silver halide particles scatter the incident light and no index variation can be observed.
- A *periodic grating* has periodic structure, either in amplitude, phase, and/or both. Two example gratings of this type are shown in Figures 23.2 and 23.3.

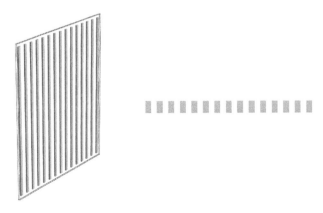

Figure 23.2 Amplitude grating structure, where the gray area represents the opaque region and the white area in between is the transparent region.

Polarized Light and Optical Systems

23.1 Introduction

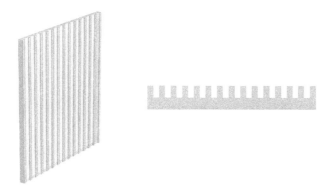

Figure 23.3 Phase grating structure, where the (pink) transparent material introduces a spatially alternating phase change.

- A *focusing grating* uses a grating structure to direct light to certain locations, similar to a lens focusing light. A *Fresnel zone plate* is an amplitude grating with non-periodic binary circular rings as shown in Figure 23.4a. A *Fresnel lens* has a structured surface with non-periodic circular blazed structure as shown in Figure 23.4b.
- A *non-periodic grating* diffracts plane or spherical waves into more complex wavefronts. A hologram is an image recorded as a non-periodic grating; when an appropriate wavefront illuminates the hologram, the resultant diffracted light recreates the image.
- A *subwavelength grating* (SWG) has a period less than half the illumination wavelength at normal incidence. With a sufficiently small period, all higher diffraction orders are evanescent, leaving only the transmitted and reflected 0th orders. Because light is not lost to the ±1st, ±2nd, and other orders, the efficiency is often high. Later sections consider subwavelength gratings as polarizers, retarders, and antireflection coatings.

Two common periodic gratings are the rectangular grating and the blazed grating depicted in Figure 23.5. As shown in Figure 23.5a, a *rectangular grating* contains a repeating rectangular profile. A *blazed grating* has a triangular profile. Consider light reflecting from a grating facet. A blazed grating will tend to have high efficiency in orders near the reflected direction. Thus, blazing is used to maximize the diffraction efficiency into a desired order. *Arbitrary shape gratings* are those with the arbitrary profiles. Several terminologies and notations of these periodic grating structures used in the following sections are highlighted in Table 23.1.

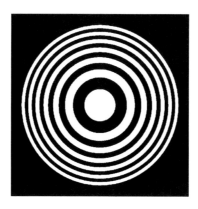

Figure 23.4 Fresnel zone patterns where white has a transmission of 1 and black has 0 transmission.

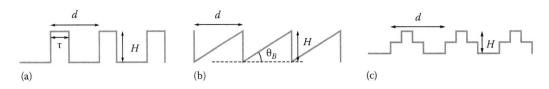

Figure 23.5 Structure of a periodic (a) rectangular grating, (b) blazed grating, and (c) multi-step grating.

Table 23.1 Grating Parameters and Notation Definitions

Grating Parameters	Definitions	Notation	
Grating period		d	
Grating height		H	
Duty cycle	The positive fraction of the square wave in one period	$D = \dfrac{\tau}{d}$	(23.1)
Aspect ratio	The height of the grating relative to its period	$A = \dfrac{H}{d}$	(23.2)
Blaze angle	The base angle of a triangular grating	θ_B	
Amplitude coefficient	A complex coefficient relating the amplitude a and phase ϕ of dth diffractive order to the incident light	$\dfrac{a_d e^{-i\phi_d}}{a_{in} e^{-i\phi_{in}}}$	(23.3)
Diffraction efficiency	The ratio between the energy flux in a particular order and the energy flux in the incident beam for incident angle θ_{in} and diffractive angle θ_d	$\dfrac{|a_d|^2 \cos\theta_d}{|a_{in}|^2 \cos\theta_{in}}$	(23.4)

Diffractive structures are manufactured in many different ways. Originally, they were produced as large numbers of parallel wires; this is still done in millimeter and radio frequencies. Conventionally, diffraction gratings have been produced by scribing a diamond tool into a soft metal like aluminum. A replica grating is copied from a master grating into epoxy, allowing many gratings to be produced from the master ruling. Grating copies can also be electroformed. Interference patterns can be generated with interferometers and then exposed in photoresist, dichromated gelatin, photographic film, and other materials to create DOEs. Microlithography is a flexible tool for producing very fine gratings in photoresist, which can then be processed into many materials through etching, coating, and so on. The computer-controlled electron beam etching allows patterning gratings with a resolution of a few nanometers, with a trade-off of long processing time. The diffraction pattern can be fabricated on a thin sheet, flat or curved surface.

23.2 The Grating Equation

To ray trace a DOE, the grating period and orientation must be specified along with the direction of the incident light. A grating is operated *in-plane* when the plane perpendicular to the rulings contains the incident light's **k** vector. All the diffracted orders will lie in the same plane, and the mathematics is easier than the general case of *out-of-plane* diffraction, where the light has an arbitrary direction. The directions of diffracted light are given by the grating equation in one of these two forms, Equation 23.6 or 23.9. Either equation describes both transmission and reflection gratings.

23.2 The Grating Equation

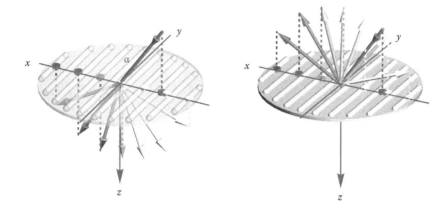

Figure 23.6 An in-plane incident light ray is diffracting from a refractive grating (left) and a reflective grating (right). The colors of the arrows distinguish the incident and diffractive directions for one wavelength. The incident ray shown in blue results in multiple diffraction orders shown in different colors. The projections from the unit sphere are equally spaced on the x–y plane. Red indicates the zero order, with orange, yellow, and green arrows indicating negative orders and red–purple is the only positive order present.

These equations specify the diffracted light's direction for all diffracted orders, since this only depends on the grating period d, but say nothing about the amount of energy carried by each order, which depends on the grating profile. This can be calculated using the RCWA algorithm.

For a grating in the x–y plane with a surface normal along the +z-direction and the grating's grooves along the y-direction, as shown in Figure 23.6, an in-plane incident ray at an angle of incidence α diffracts into the x–z plane with diffraction angles β_m for each diffraction order m as

$$n_m d \sin \beta_m - n_i d \sin \alpha = m\lambda. \tag{23.5}$$

With media refractive index $n_i = n_m = 1$,*

$$\sin \beta_m = \sin \alpha + \frac{m\lambda}{d}, \tag{23.6}$$

where λ is the light's wavelength.

In the 0th order ($m = 0$), all wavelengths are diffracted in the same direction; this corresponds to the ordinary reflected and transmitted beams. In this 0th order, white light diffracts into white light. The other orders disperse the light, with shorter wavelengths diffracting nearer to the 0th order and longer wavelengths diffracting further from the 0th order,[†] as shown in Figure 23.7. From the incident beam, the positive orders diffract beyond the 0th order; negative orders lie on the same side of the 0th order as the incident light. The normal always lies in the region of the negative orders.

The *angular dispersion* is the rate of change of the diffracted angle with wavelength,

$$\frac{d\beta}{d\lambda} = \frac{m}{d \cos \beta}. \tag{23.7}$$

---

* A sign convention can be defined to have positive diffraction orders refract/reflect away from surface normal while negative orders refract/reflect toward the surface normal. In this convention, Equation 23.6 becomes $\sin \beta_m = \sin \alpha + \frac{\text{Sign}[\alpha] \times m\lambda}{d}$, where $\text{Sign}[\alpha] = +1$ for positive α and $\text{Sign}[\alpha] = -1$ for negative α.

† Each photon that diffracts from a diffraction grating transfers mh/λ of momentum to the grating in the x–y plane where h is Plank's constant.

Figure 23.7 An incident beam diffracts into multiple orders. The color of the arrows represents different wavelengths. Because of dispersion, the diffraction angles are changed with wavelength except for the 0th diffraction order, which diffracts white light. The shorter wavelength (blue) diffracts closer to the 0th order, while the longer wavelength (red) diffracts further from the 0th order.

The more general *out-of-plane diffraction grating equation* is derived from the three-dimensional grating equation $\mathbf{k}_i - \mathbf{k}_m = \mathbf{K}_g$ depicted in Figure 23.8 where $|\mathbf{k}_m| = \dfrac{2\pi n_m}{\lambda}$ and $|\mathbf{K}_g| = \dfrac{2\pi}{d}$.

Consider a plane grating surface in the x–y plane with an incident medium refractive index of n_I, exiting medium index n_{II}, and incident wavevector

$$\mathbf{k}_i = \begin{pmatrix} k_{xo} \\ k_{yo} \\ k_{I,zo} \end{pmatrix} = \begin{pmatrix} k_o n_I \sin\theta \cos\phi \\ k_o n_I \sin\theta \sin\phi \\ k_o n_I \cos\theta \end{pmatrix}, \tag{23.8}$$

with $k_o = \dfrac{2\pi}{\lambda}$. The *diffracted wavevectors* are

$$\mathbf{k}_{I,m} = \begin{pmatrix} k_{xm} \\ k_{yo} \\ -k_{I,zm} \end{pmatrix} \text{ and } \mathbf{k}_{II,m} = \begin{pmatrix} k_{xm} \\ k_{yo} \\ k_{II,zm} \end{pmatrix} \tag{23.9}$$

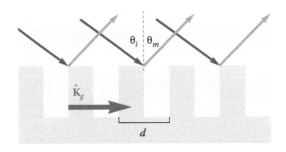

Figure 23.8 Grating vector $\mathbf{K}_g$ describes the period and orientation of the grating grooves.

23.2 The Grating Equation

for mth reflected $\mathbf{k}_{I,m}$ and transmitted $\mathbf{k}_{II,m}$ diffraction order wavevectors, respectively,[5-7] where

$$k_{xm} = k_o(n_I \sin\theta\cos\phi + m\lambda/d), \qquad (23.10)$$

$$k_{I,zm} = \begin{cases} \sqrt{(k_o n_I)^2 - (k_{xm}^2 + k_{yo}^2)} & \text{for } \sqrt{k_{xm}^2 + k_{yo}^2} \le k_o n_I, \\ -i\sqrt{(k_{xm}^2 + k_{yo}^2) - (k_o n_I)^2} & \text{for } \sqrt{k_{xm}^2 + k_{yo}^2} > k_o n_I, \end{cases} \qquad (23.11)$$

and

$$k_{II,zm} = \begin{cases} \sqrt{(k_o n_{II})^2 - (k_{xm}^2 + k_{yo}^2)} & \text{for } \sqrt{k_{xm}^2 + k_{yo}^2} \le k_o n_{II} \\ -i\sqrt{(k_{xm}^2 + k_{yo}^2) - (k_o n_{II})^2} & \text{for } \sqrt{k_{xm}^2 + k_{yo}^2} > k_o n_{II}. \end{cases} \qquad (23.12)$$

Equations 23.6 and 23.9 describe the directions of the diffracted orders for gratings in the x–y plane. For other grating orientations, these results must be rotated. This *out-of-plane grating equation* is not so widely presented, but is essential for ray tracing (Figure 23.9).

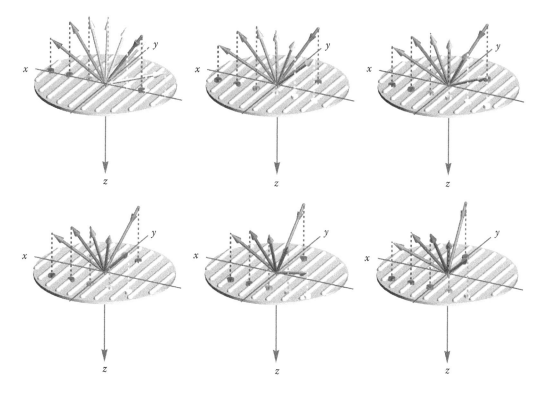

Figure 23.9 A graphic depiction of the out-of-plane grating equation. The blue vectors pointing into the origin are the incident light. In the six figures, the incident vector is rotated about the normal, starting in the x–z plane and ending in the y–z plane. The upper left figure shows the in-plane case, where multiple diffraction orders (red, orange, and yellow vectors), extending to a unit hemisphere, have equally spaced projections onto the x-axis. As the incident vector rotates toward the y-axis, the diffraction orders lie on the surface of a cone, but the projection onto the x–y plane remains equally spaced.

23.3 Ray Tracing DOEs

Geometrical ray tracing through a DOE calculates the propagation vectors of the diffraction orders, but not, the amplitude coefficients, and the state of the electric fields resulting from the light–DOE interaction. The wavevectors are obtained from Equation 23.9 to Equation 23.12 for all diffracted rays. The derivation of the amplitude coefficients for all the diffraction orders is calculated from Maxwell's equations for periodic boundary conditions. This is most often performed by the RCWA calculation summarized in Section 23.4. The eigenpolarizations of DOEs are not necessarily the s- and p-polarizations. But the changes to the light's amplitude and phase can be expressed with Jones matrices or **P** matrices for each reflected and transmitted diffraction order. This section includes four examples of DOEs showing their polarization characteristics simulated by RCWA.

23.3.1 Reflection Diffractive Gratings

Diffraction gratings are essential elements in spectrometers and monochromators. However, they are also the root cause of many serious polarization problems in these systems. A reflection grating example is presented to assist in understanding how to interpret the manufacturer's specifications and measurement data as well as using the outputs from RCWA grating simulations in polarization ray tracing.

Reflection gratings are in widespread use in spectroscopy to disperse polychromatic light for the purpose of measuring spectra or producing monochromatic light.* For example, a Czerny–Turner monochromator, shown in Figure 23.10 (left), uses a mirror to collimate light from the incident slit onto the grating. The dispersed light is focused by a second mirror and forms an image, the spectrum. An exit slit selects and passes a small spectral band. By rotating the grating, the monochromator scans the exiting wavelength. Another common monochromator is the Littrow monochromator, Figure 23.10 (right), which uses one focusing mirror twice.

Reflective diffraction gratings commonly have triangular profiles, where one facet receives most of the illumination. Aluminum grating surfaces are common for visible applications while gold surfaces are common for infrared applications. Light is distributed among all the diffraction orders in a complex fashion and can be calculated by RCWA. Despite the diffraction, light has a natural tendency to reflect from the facet. The diffraction orders near this reflection direction are typically the brightest. The term *blazing* means choosing the reflective facet angle to reflect one wavelength, the *blaze wavelength* λ_B, into the direction of a desired diffraction order. Thus, blazing

---

* The *linear dispersion* of a monochromator or spectrometer is the rate of change of length per wavelength at the exit slit,

$$\frac{dx}{d\lambda} = \frac{f\,m}{d\cos\beta},$$

where f is the focal length of the focusing mirror. The *grating resolving power* is the ability to discern closely spaced spectral lines. It is proportional to the number of rulings N being illuminated. The *grating free spectral range* is the maximum wavelength range detected without overlapping diffraction orders. When a monochromator is illuminated with monochromatic light, the flux through the exit slit as a function of wavelength is the *line spread function*, the spectral profile of a monochromatic input. The line spread function is calculated as the convolution of the entrance slit (a rectangle function) with the point spread function (PSF) of the imaging optics in the monochromator, which is then convolved with the exit slit,

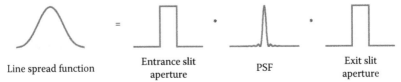

| Line spread function | Entrance slit aperture | PSF | Exit slit aperture |

The convolution operation is indicated by *. The spectrum measured by a monochromator is the convolution of the input light's spectrum with the line spread function. The *spectral resolution* describes the minimum separation in wavelength, which can be identified as two separate spectral lines. The spectral resolution can be estimated as the width of the line spread function in millimeters multiplied by the linear dispersion in nanometers per millimeter.

23.3 Ray Tracing DOEs

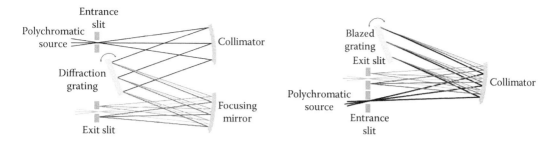

Figure 23.10 (Left) The Czerny–Turner monochromator and (right) the Littrow monochromator.

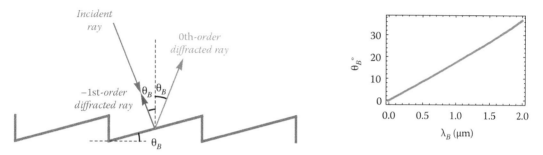

Figure 23.11 (Left) The Littrow configuration where the −1st diffraction order shown is optimized for maximum efficiency by setting the incident angle and the blaze angle equal to θ_B. Gratings can be blazed for other negative orders. (Right) The Littrow blaze angle as a function of wavelength for a grating period of 1.67 μm.

will provide high diffraction efficiency for wavelengths near the blaze wavelength. By convention, the blaze wavelength and *blaze angle* θ_B are defined as the wavelength and facet angle where, in a Littrow monochromator, the incident light and the −1st order lie along the same line, as depicted in Figure 23.11. The blaze wavelength satisfies the equation

$$\lambda_B = 2d \sin \theta_B. \tag{23.13}$$

Example 23.1 Reflection Diffractive Grating Example

A reflection diffractive grating is simulated, made of aluminum, with 600 grooves per millimeter, a blaze wavelength $\lambda_B = 500$ nm, and a blaze angle $\theta_B = 8.6°$. The triangular blaze structure is simulated by 30 layers of rectangular steps as shown in Figure 23.12 to mimic the triangular shape.

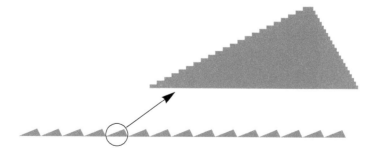

Figure 23.12 The profile of a blazed diffraction grating simulated with 30 rectangular steps.

The diffraction efficiency of this grating is calculated by RCWA[5,6] for in-plane illumination near the Littrow configuration for both TE and TM modes, where TE field is parallel to the grating grooves and TM field is perpendicular to the grating grooves. In our convention, TE mode is s-polarization and TM mode is p-polarization. Figure 23.13 shows measured and simulated grating efficiencies for the −1st diffraction order. Note that the diffraction efficiencies peak around 500 nm at the blaze wavelength.

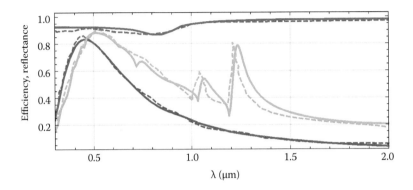

Figure 23.13 Data for a 600 line/mm aluminum grating in the −1st order. The reflectance of aluminum at normal incidence and the measured diffraction efficiency data are shown in dashed lines. The blue line is the reflectance dispersion of the aluminum used in the simulation. Red and green lines are the diffraction efficiency for s- and p-polarizations (TE and TM modes), respectively. Woods anomalies can be seen at 1.05 and 1.21 μm. Very good agreement was found between the measured and the simulated grating efficiencies for the −1st order.

The abrupt variations of the TM mode's diffraction efficiency at 1.02 and 1.22 μm (Figure 23.13) are examples of *Woods anomalies*.[8,9] Woods discovered in 1902 that abrupt changes in diffraction efficiency from metal gratings are observed at certain wavelengths and angles. In 1907, Rayleigh[10,11] explained that the anomalies occur when the diffracted light in one order is diffracting at $\beta_m = 90°$ (i.e., tangential to the grating's surface) and is transitioning from a real into an evanescent wave. In this situation, the energy of the diffracted light is redistributed to lower diffraction orders, which causes abrupt changes in the spectrum. In 1941, Fano[12] and others[13,14] described the anomalies as the results of surface plasmon resonance effects, and the characteristics of the materials close to the interface also affect the sharp changes to the spectrum. The anomalies are generally observed for both s-polarization and p-polarization, with the p-polarization showing larger anomalies.[15–18]

The diffraction efficiencies of the TE and TM modes are very different and change rapidly with wavelength. These lead to large grating polarization effects. The diattenuation of the example grating, in Figure 23.14 (right), is significant, rapidly changing, and not described by any simple linear or quadratic function. As a result, when a polarized source is measured by a spectrometer using these gratings, very different spectra are obtained if the incident light is in the TE or TM mode. Gratings also have retardance and the retardance of the −1st diffraction order for this example grating is plotted in Figure 23.14 (left) with an interesting shape.

23.3 Ray Tracing DOEs

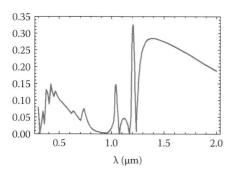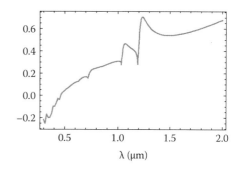

Figure 23.14 The simulated retardance in radian (left) and diattenuation (right) of −1st diffraction order.

Figure 23.13 showed the grating efficiency as measured in a "Littrow" monochromator. But a Littrow can never operate at exactly the Littrow condition, the entrance and exit slit would be on top of each other. In this case the angle between the slits is not specified, making matching calculations with measurements difficult. To obtain a close match between the simulation and the measurement, the grating was simulated for a series of incident and diffracted angles about the Littrow condition. The best match, shown in Figure 23.15, was found when the incident beam was about 0.15 radians from Littrow, and the diffracted beam is collected at about 0.15 radians on the other side. In particular, the locations and the shapes of the two Woods anomalies at 1.02 and 1.22 μm were very sensitive to the incident angle and provided an accurate method to estimate the angle from Littrow in the monochromator configuration used for the measurements. The measured and simulated diffraction efficiency agree well, particularly since real gratings have many defects, including not perfectly rectangular or triangular profiles and small variability in the grating's spacing.

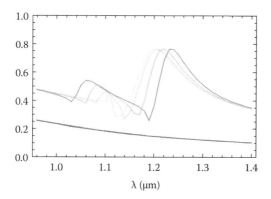

Figure 23.15 TE and TM's diffraction efficiencies calculated with RCWA for the −1st order at several angles of incidence. The incident angles are specified in radians from the Littrow condition: 0.05 offset is light green, 0.1 and 0.15 in darker greens, and 0.2 offset is deep green. The best fit to the diffraction efficiencies occurred for an offset from the Littrow condition of 0.15 radians.

23.3.2 Wire Grid Polarizers

A *wire grid polarizer* is formed as a series of fine parallel metal strips on a transparent substrate. To be an effective polarizer, the wire spacing should be less than half of the wavelength of the light. This group of wires forms the diffraction grating as shown in Figure 23.2. When an electromagnetic field is incident onto the metal wires, it easily moves the free electrons along the wires, and this current generates a strong reflected beam polarized along the wires. The orthogonal components of the field, oscillating perpendicular to the metal wires, have a much smaller interaction and thus nearly all the light passes through the polarizer. Therefore, a wire grid polarizer with wires oriented in the y-direction transmits mostly x-polarized light and reflects mostly y-polarized light.

Example 23.2 An Aluminum Wire Grid Polarizer

An aluminum wire grid polarizer with 0.2 µm period, 30% duty cycle, and 0.225 µm wire height is simulated at 550 nm wavelength. A circularly polarized converging wavefront passes through the vertically oriented wire grid shown in Figure 23.16a. The horizontal component transmits and forms the wavefront shown in Figure 23.16b, while the vertical component reflects as shown in Figure 23.16c. Thus, this wire grid acts as a horizontal polarizer in transmission.

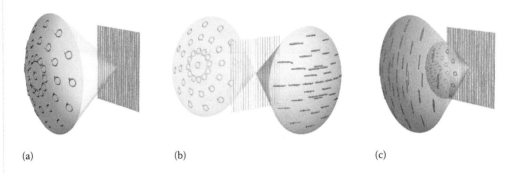

(a)   (b)   (c)

Figure 23.16 (a) A circularly polarized converging incident ray grid (green) focuses on a wire grid polarizer with vertically oriented wires. This has a horizontal transmission axis. (b) A horizontally polarized wavefront (blue) transmits through the polarizer. (c) A vertically polarized wavefront (red) reflects from the polarizer for incident circularly polarized light.

The wires are oriented vertically in a 2D plane on the polarizer's surface, while the incident light rays are converging. For the on-axis ray, its transverse plane is in the plane of the polarizer's surface, and thus transmitted ray has horizontal linear polarization. For the other off-axis rays, their tranverse planes do not coincide with the polarizer's surface. The higher the incident angle, the further its transverse plane is tilted from the polarizer's surface, and the more light in vertical polarization leaks through, rotating the planes of polarization, and yielding elliptically polarized transmitted rays.

To view the polarization states of the incident, transmitted, and reflected beams, *the dipole local coordinate system** with dipole axis along the wires was selected, since this closely matches the physics. The polarization of the incident, reflected, and transmitted beams is plotted in Figure 23.17. The exiting states are predominantly linear with a small amount of ellipiticity visible at some off-axis rays.

---

* See Chapter 11 (The Jones Pupil and Local Coordinate Systems).

23.3 Ray Tracing DOEs

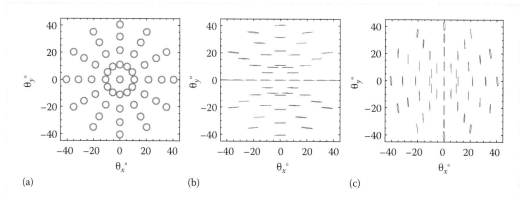

Figure 23.17 The local polarization ellipses of the (a) incident, (b) transmitted, and (c) reflected rays in Figure 23.16 over a 45° spherical wavefront.

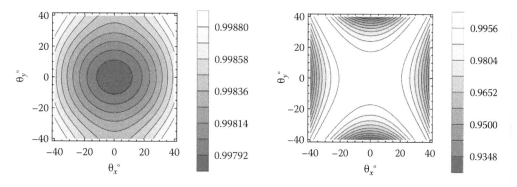

Figure 23.18 The transmitted (left) and reflected (right) diattenuation magnitude of the polarizer.

Figure 23.18 plots the diattenuation of the reflected and transmitted light; both are above 90% diattenuation, but are never 100%, such as achieved by crystal polarizers. The reflected light has a maximum diattenuation on-axis, while the transmitted diattenuation is not at maximum on-axis. The extinction ratio shown in Figure 23.19 for this wire grid polarizer in transmission has a maximum of more than 1000 for the on-axis beam and decreasing to around 20 at the corner of the field (~56°).

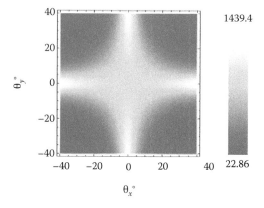

Figure 23.19 Angle of incidence variation of the extinction ratio for the simulated wire grid polarizer.

When designing a wire grid polarizer, the design parameters include the grating's metal layer thickness, cross-sectional shape, period, duty cycle, and the choice of metal. In most cases, the transmission and/or extinction ratio are optimized. As a rule of thumb, a metal with a large real index and an absorption index in the range of 1 to 2 works well. A smaller duty cycle with a thicker metal layer is preferred to a larger duty cycle and thinner metal layer since it can yield better extinction with smaller absorption. The wire grids produce substantial retardance but since the flux of orthogonal polarization is generally very small, the effect is barely noticeable. Equation 23.14 is an example merit function for wire grid polarizer optimization,

$$\text{Merit} = \sqrt{c_1(1-T_p)^2 + c_2 T_s^2 + c_3(1-R_s)^2 + c_4 R_p^2}, \tag{23.14}$$

where (c_1, c_2, c_3, c_4) are weighted to trade off the transmission and reflection coefficients for s- and p-polarizations (T_s, T_p, R_s, R_p). For high contrast ratio, c_1 should be large. For better transmitted flux, c_2 should increase. The ideal wire grid polarizer would have a merit function of zero. Wire grid polarizers are somewhat angle sensitive, with more variation for the polarization orthogonal to the wire direction or along the transmission axis, than along the wire direction.

Example 23.3 Performance as a Function of Aluminum Wire Thickness

To understand these trade-offs in Equation 23.14, consider the following simulation (Figure 23.20) showing the TE and TM mode transmissions as a function of aluminum layer thickness at various wavelengths. The TE transmission, parallel to the wires, is not well extinguished until the thickness is greater than 0.2 μm. The TM transmission has a rather small oscillation with thickness.

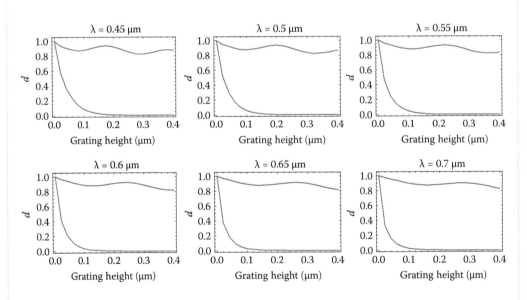

Figure 23.20 Wire grid polarizer transmission of TM polarization (red) and TE polarization (blue) versus aluminum wire height at six wavelengths.

23.3 Ray Tracing DOEs

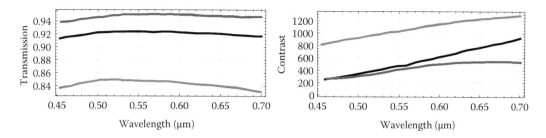

Figure 23.21 Performance comparison of three wire grid polarizers: (left) transmission of *p*-state and (right) contrast ratio. Red for ultra-high transmission polarizer, black for high transmission polarizer, and blue for high contrast polarizer.

When selecting commercial wire grid polarizers, it is helpful to understand that these polarizers can be optimized for high extinction, transmission efficiency, or a balance between the two, as shown in Figure 23.21. These graphs make clear the design trade-off between contrast and transmission; higher contrast polarizers have a lower transmission.

23.3.3 Diffractive Retarders

Diffractive polarizers, wire grid polarizers, are very common optical components, so one might wonder why diffractive retarders are not as common. Light transmitted through grating structures experiences retardance; thus, it is possible to fabricate retarders from DOEs. However, it is difficult to obtain substantial retardance from such DOEs. This section explain why that is the case.

An SWG only reflects and transmits the 0th order; the angle of the first diffraction order would be greater than 90°, if that was possible. In our terminology, all of the orders for a subwavelength grating except the 0th order are evanescent. Thus, SWGs are more efficient than gratings that are not subwavelength, which must divide their flux between orders. Example 23.4 explains the difficulty with diffractive retarders.

Example 23.4 A Family of Fused Silica Diffractive Retarders

An etched fused silica grating is simulated with a period of 170 nm and various aspect ratios correspond to grating heights of 42.5, 85, 170, and 340 nm. Figure 23.22 shows the retardance spectra for different aspect ratios. Higher retardance is obtained with a larger aspect ratio, though these SWGs are difficult to make, especially when the aspect ratio increases! Even for an aspect ratio of 2, these diffractive retarders have such weak retardance that it is hardly practical to make a quarter wave retarder, much less for half wave retarders.

The retardance dispersions of the SWG with an aspect ratio of 1 from Example 23.4 is compared in Figure 23.22 (right) with three waveplates made of different crystals: a 2.198-μm-thick sapphire A-plate, a 0.103-μm-thick calcite A-plate, and a 1.95-μm-thick quartz A-plate. The thicknesses are chosen for a 0.186-rad retardance at 600 nm. Note the similarity of the retardance spectra! If the shape of the diffractive retardance dispersion was different from the crystal's dispersion, the diffractive retarders could play a valuable role in achromatizing retarders even with small retardance, but this is not the case. Thus, diffractive retarders appear to be difficult to fabricate due to the need for very large aspect ratios and do not significantly enlarge the design space for controlling the dispersion of retardance.

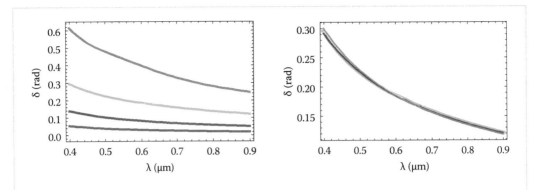

Figure 23.22 (Left) The retardance of subwavelength fused silica gratings with a 170-nm period and aspect ratios of 0.25 (blue), 0.5 (red), 1 (green), and 2 (orange). (Right) A comparison of the retardance of the aspect ratio of 1 SWG (green) with comparable birefringent waveplates with thicknesses chosen to match the 600-nm retardance: sapphire (purple), calcite (pink), and quartz (cyan).

23.3.4 Diffractive Subwavelength Antireflection Coatings

The SWG can be used to replace thin film coatings, such as anitireflection coatings and polarizing beam splitters. An ideal antireflection SWG can be structured to create an effective medium layer that has a gradient index, from the substrate index to the index of the surrounding medium.[19] An antireflection SWG[20,21] shown in Figure 23.23 is simulated with the RCWA, and the associated wavefront aberrations and polarization aberrations are studied. These aberrations depend on the direction of the incident light, the polarization state, and the orientation of the grating's grooves.

Figure 23.24 compares the transmission of the SWG with the transmission of a standard quarter wave antireflection MgF_2 coating. The SWG shows very high transmission up to 99.5%. In the case where the plane of incidence is parallel to the grooves, it has less polarization dependence than when the plane of incidence is perpendicular to the grooves. For light propagating in the plane of incidence perpendicular to the grooves, as the incident angle increases, the apparent grating period shrinks, and resonant interactions between the grating and the incident field occurs. As the angle of incidence increases to 28°, a 1st diffraction order comes into existence (the first grating order ceases to be evanescent) and a rapid reduction in transmission occurs as the 1st order gains energy. The corresponding diattenuation for the 0th order rapidly changes. This sets a limit on the practical angle of incidence range of the antireflection SWG.

The SWG introduces retardance even at normal incidence, unlike a typical antireflection coating. Figure 23.25 shows the phase delay in transmitted light for TE and TM polarized light in two orthogonal planes. As the plane of incidence switches from being parallel to perpendicular

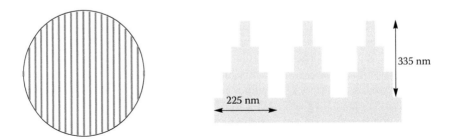

Figure 23.23 (Left) An antireflection grating's groove orientation. (Right) The subwavelength structure[21] of the grating acts like a gradient material since the duty cycle decreases moving from the substrate.

23.3 Ray Tracing DOEs

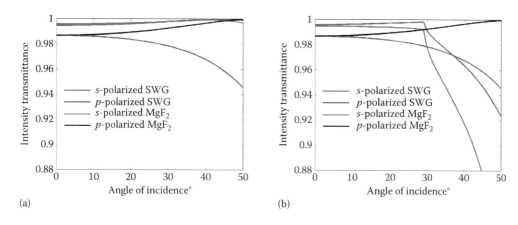

Figure 23.24 The intensity transmission of the antireflection SWG compared with a standard quarter wave MgF$_2$ antireflection coating. The SWG data correspond to the grating's grooves parallel (a) and perpendicular (b) to the plane of incidence.

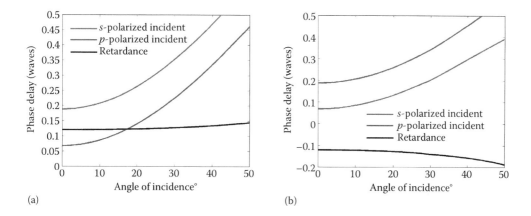

Figure 23.25 The transmitted phase and retardance of the SWG for the plane of incidence parallel (a) and perpendicular (b) to the grating's groves.

to the grating's groves, the resultant retardance switches sign, because the polarization orientation along the grating's groove has switched. As the angle of incidence increases, the two retardance functions also diverge. The phase shifts and retardance at 24° incident angle for each polarization and grating orientation are given in Table 23.2. The phases are given in terms of the s- and p-components, but which of these is parallel to the grating lines depends on the plane of incidence. Therefore, the s-component parallel to the grating should be compared against the p-component perpendicular to the grating. This shows the phase in the relative orientation of the plane of incidence and the gratings produce a 0.005 wave change in the electric field component perpendicular to the gratings and a 0.012 wave change in the component parallel to the gratings. This causes a 0.007 wave difference in the retardance magnitude between the two grating orientations at 24°.

The polarization aberration function of the example SWG is analyzed on an ideal spherical surface using the Polaris-M polarization ray tracing program (Airy Optics, Inc.). Since the plane of incidence is radially oriented on a spherical surface, the groove's orientation plays a role in the resultant polarization aberration as well. The polarization aberration functions are shown in Figures 23.26 through 23.28 for an on-axis beam with the grating lines oriented horizontally. Both co-polarized

Table 23.2 Phase Shift of SWG at 24° Incident Angle for Plane of Incidence (POI) Parallel and Perpendicular to the Grating's Grooves

	s-Phase (Wave)	p-Phase (Wave)	Retardance (Wave)
POI // to the groove	0.3037	0.1785	0.1252
POI ⊥ to the groove	0.1662	0.2986	−0.1324

terms with subscripts xx and yy have a small quadratic apodization due to incident angle changes across the field. The cross-coupled terms xy and yx have the Maltese cross pattern with a 3% maximum amplitude leakage. The abrupt amplitude and phase change toward the top and bottom of the field indicates where the subwavelength condition fails and the 1st order begins robbing energy. The resultant retardance pattern contains elements of piston, astigmatism, and defocus. The retardance orientation has a very small 3° variation toward the edge of the field. Overall, the resulting wavefront aberrations are dominated by astigmatism, with a magnitude less than 0.1 waves.

The polarization-dependent aberrations from SWG antireflection coatings give an additional complexity to aberration compensation compared to the traditional geometrical wavefront optimization. In the example described here, using two identical gratings in orthogonal orientation on two similar surfaces can compensate for the on-axis retardance. The higher-order aberrations remain, but with very small magnitudes. Using the RCWA simulation capability to fine-tune the shape of an SWG in a point-by-point basis can potentially result in high levels of polarization correction.

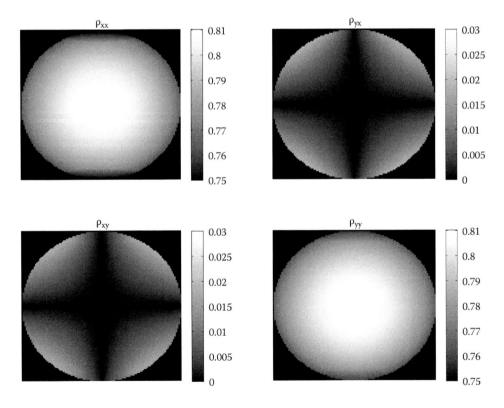

Figure 23.26 The amplitude ρ of the Jones pupil for the example lens with two SWG antireflection-coated surfaces. Dark bands across the tops and bottoms of the xx and yy amplitudes are due to increasing amounts of light lost into the first diffraction order.

23.4 Summary of the RCWA Algorithm

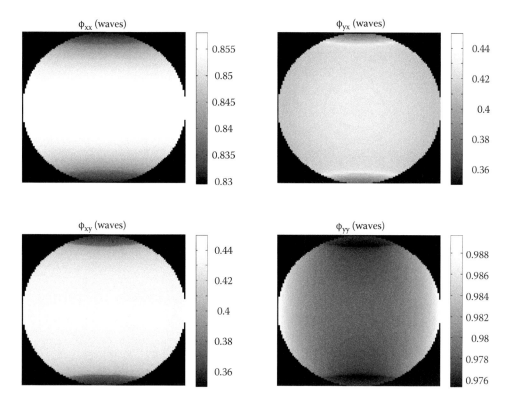

Figure 23.27 The wavefront aberration, the phase ϕ of the Jones pupil, for the example lens with two SWG antireflection-coated surfaces.

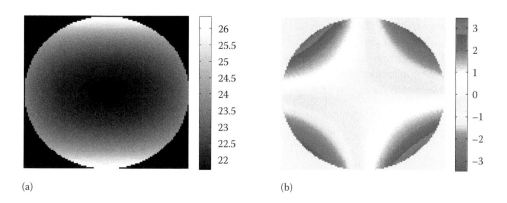

(a) (b)

Figure 23.28 Retardance magnitude in waves (a) and retardance orientation in degrees (b).

23.4 Summary of the RCWA Algorithm

The RCWA algorithm is widely used for analyzing periodic diffractive structures, such as diffraction gratings, wire grid polarizers, holograms, and all the examples presented in this chapter. RCWA calculates exact solutions for the reflection from and transmission through periodic structures. It is a relatively straightforward, non-iterative, and deterministic technique that calculates the amplitude coefficients for each diffraction order. The accuracy depends solely on the number of Fourier terms retained in the Floquet–Fourier expansions of the grating structure. The algorithm

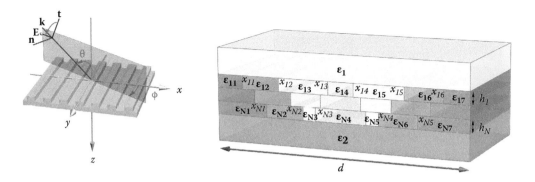

Figure 23.29 (Left) Coordinate system for the incident light onto a grating structure. (Right) The specification for a grating or diffractive optical element in RCWA divides the grating into many layers, and each layer into rectangles of uniform refractive index or dielectric tensor. Only one period of the grating is specified as input.

was initially developed for modeling volume holographic gratings and then extended to surface-relief and multilevel grating structures. RCWA has been successfully applied to transmission and reflection planar dielectric and/or absorption holographic gratings, arbitrarily profiled dielectric and/or metallic surface-relief gratings, multiplexed holographic gratings, two-dimensional surface-relief gratings, and anisotropic gratings, for both planar and conical diffraction.[5,6,22–24] This section will provide a summary of the RCWA algorithm.

In the RCWA algorithm, the grating is divided into a large number of sufficiently thin planar slabs, as shown in Figure 23.29, to approximate the grating profile to an arbitrary degree of accuracy. The information required is listed in Table 23.3 and Figure 23.29.

The basic algorithm for RCWA is similar to the algorithms for Fresnel equations and thin film–coated interfaces (Figure 23.30). The principal difference is the superposition of many diffracted orders of reflected, transmitted, forward, and backward waves in the grating region because of the periodic grating structure. The explicit form of the incident, reflected, and transmitted waves given by the superposition of individual modes in the grating region is obtained from Floquet's condition, described in Figure 23.8, which relates the wavevectors of the diffracted waves with the grating vector and the dispersion relationships. The fields propagating in each grating layer are calculated via a coupled wave approach. In the lth grating layer, the fields can be expanded to

$$\mathbf{E}_{g,l} = \sum_n \mathbf{S}_{l,n}(z)\, e^{-i(k_{xn}x + k_{yn}y)} \quad \text{and} \quad \mathbf{H}_{g,l} = \sqrt{\varepsilon_o/\mu_o} \sum_n \mathbf{U}_{l,n}(z) e^{-i(k_{xn}x + k_{yn}y)}, \qquad (23.15)$$

where $\mathbf{S}_n$ and $\mathbf{U}_n$ are the nth components of the vector Fourier coefficients function. The coupled wave equations are obtained by substituting the constitutive relations and electromagnetic fields in their Fourier series forms into Maxwell's equations.

After the substitutions, $\mathbf{S}_z$ and $\mathbf{U}_z$ are eliminated and the coupled wave equations are obtained:

$$\frac{1}{k_o}\frac{\partial}{\partial z}\begin{pmatrix} \mathbf{S}_{l,x}(z) \\ \mathbf{S}_{l,y}(z) \\ \mathbf{U}_{l,x}(z) \\ \mathbf{U}_{l,y}(z) \end{pmatrix} = i\mathbf{\Gamma}\begin{pmatrix} \mathbf{S}_{l,x}(z) \\ \mathbf{S}_{l,y}(z) \\ \mathbf{U}_{l,x}(z) \\ \mathbf{U}_{l,y}(z) \end{pmatrix}, \qquad (23.16)$$

23.4 Summary of the RCWA Algorithm

Table 23.3 Rigorous Coupled Wave Analysis Input and Output Parameters

Incident Light Descriptions	Incident Light Parameters
Incident wavelength	λ
Angle of incidence	θ
Azimuthal angle	ϕ
Incident polarization	**E**

Grating Descriptions	Grating Parameters
Dielectric constant or tensor of incident medium[a]	ε_1
Dielectric constant or tensor of grating substrate[a]	ε_2
Layers of dielectric constant or tensor of grating structure[a]	$\{\{\varepsilon_{11}, \varepsilon_{12},\ldots, \varepsilon_{1M}\},\ldots\{\varepsilon_{M1}, \varepsilon_{M2}, \ldots, \varepsilon_{MN}\}\}$
Subperiod dimensions for each layer	$\{\{x_{11}, x_{12},\ldots, x_{1M}\},\ldots\{x_{N1}, x_{N2}, \ldots, x_{NM}\}\}$
Layer thicknesses	$\{h_1, \ldots, h_N\}$
Grating period	d

RCWA Parameter Descriptions	Algorithm's Parameters
Number of Fourier orders retained in calculations	$(2f + 1)$ orders: $-f$th, $-(f-1)$th, $\ldots$, -2nd, -1st, 0, $+1$st, $+2$nd, $\ldots$, $(f-1)$th, fth

Diffracted Light Parameters	
$\mathbf{k}_m$	mth diffraction order propagation vector
$\mathbf{S}_m$	mth diffraction order Poynting vector
$\mathbf{D}_m$	mth diffraction order **D**-field
$\mathbf{E}_m$	mth diffraction order polarization state
$\mathbf{H}_m$	mth diffraction order **H**-field
a_m	mth diffraction order complex amplitude coefficients
R_m and T_m	mth diffraction order diffraction efficiencies

[a] Similarly for gyrotropic tensor, when optical activity is present.

where $\mathbf{\Gamma}$ is a block matrix described in Refs. [22] and [25]. The coupled wave equation is a 1st-order ordinary differential equation, and its solution for the field in the z-component (orthogonal to interface) is analytically obtained as

$$\mathbf{V}_l(z) = \sum_m c_{l,m} \mathbf{w}_{l,m} e^{ik_o \lambda_{l,m}(z-z_{l-1})} \quad \text{and} \quad \mathbf{V}_l = \begin{pmatrix} \mathbf{S}_{l,x} \\ \mathbf{S}_{l,y} \\ \mathbf{U}_{l,x} \\ \mathbf{U}_{l,y} \end{pmatrix}. \tag{23.17}$$

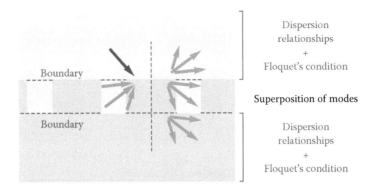

Figure 23.30 The basic algorithm of RCWA solves the boundary value problem for forward and backward propagating waves for all diffraction orders. Each wave interacts with all the components of a three-dimensional refractive structure described by its Fourier series. The incident, reflected, and transmitted waves are obtained by Floquet's condition and the dispersion relations.

$\lambda_{l,m}$ are the set of eigenvalues and $\mathbf{w}_{l,m}$ are the eigenvectors of Γ_l for the lth layer. This yields the amplitude coefficients $c_{l,m}$, determined by the boundary conditions. By applying the electromagnetic boundary conditions—continuity of the tangential electric and magnetic field components—to the interfaces at the output region, the individual grating layers, and finally the input region, the Fresnel coefficients of the reflected and transmitted diffractive fields are obtained for each order. More detailed description of the RCWA algorithm and its solution for isotropic material is provided in Glytsis and Gaylord,[26] section 2, Appendix C. For anisotropic and optically active gratings, the dielectric and gyrotropic tensors are needed in the grating description, and the scattering matrix method is employed to solve the boundary condition equations for numerical stability.[27,28] The diffraction efficiencies, complex amplitude reflectance and transmittance coefficients, wavevectors and the electromagnetic field vectors from RCWA, Jones matrices, Mueller matrices, or polarization ray tracing matrices can be constructed for each order for polarization analysis.

For an accurate calculation from RCWA, not just the reflected and diffracted orders are calculated, but also a large number of evanescent orders. This depends on the number of Fourier terms needed to accurately express the discontinuous piecewise electric permittivity ε, electric field $\mathbf{E}$, and magnetic field $\mathbf{B}$ at the edge of the grating's grooves.[29–32] In general, more Fourier terms yield greater accuracy, but with a longer simulation time. As the number of Fourier terms is increased, the diffraction efficiency oscillates and converges. Usually, the RCWA calculation for metal gratings converges slower than that for dielectric gratings. Also, RCWA calculation for binary gratings requires more Fourier terms and takes longer to compute than that for sinusoidal and/or continuous gradient index gratings due to the convergence of the Fourier series of the grating index function. For conventional one-dimensional dielectric gratings, 10 to 20 Fourier terms are usually sufficient for accurate RCWT calculation. As an example, the convergence of the RCWA algorithm for the wire grid polarizer example of Section 23.3.2 is shown in Figure 23.31. The resultant 0th-order diffraction efficiency converges nicely after 100 terms while the p-polarization efficiency fluctuates more and converges slowly.

23.5 Problem Sets

23.1 Derive the angular dispersion equation (Equation 23.7) from Equation 23.6 assuming $\alpha > 0$.

23.2 How many transmitted diffraction orders are presented from a fused silica grating at normal incident with wavelength 0.5 m for periods d = 800 nm, 1600 nm, 2000 nm, and 2800 nm?

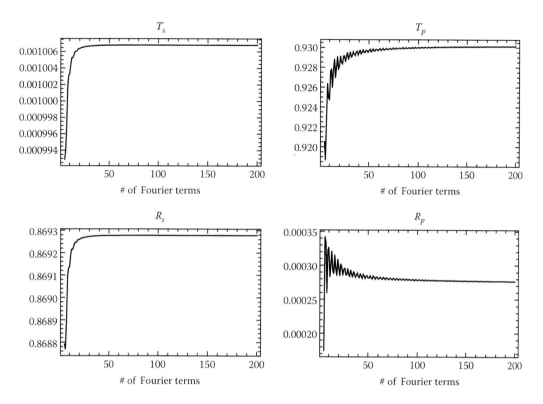

Figure 23.31 The convergence of the 0th-order transmitted and reflected diffraction efficiencies for the wire grid polarizer as the number of Fourier terms increases.

23.3 How many reflected diffraction orders are presented from an aluminum grating for periods d = 950 nm, 1600 nm, and 2400 nm with wavelength 0.5 μm for out-of-plane angle of incidence θ = 40° and ϕ being (a) 10°, (b) 30°, (c) 60°, and (d) 80°?

23.4 Calculate the Littrow angle for blaze grating with 0.5 μm period.
    a. Diffraction order m = 1 and wavelength λ = 0.5, 0.6, and 0.7 μm.
       The Littrow condition is not limited to the 1st order. What is the Littrow blaze angle for the following configurations?
    b. Diffraction order m = 2 and wavelength λ = 0.5, 0.6, and 0.7 μm.

23.5 Calculate the Woods anomaly locations for a 600 line/mm diffraction grating in the Littrow condition and compare the result to Figure 23.13.

23.6 In Figure 23.22, what would be the aspect ratio needed for a half wave and quarter wave retarder?

23.7 At what wavelength is the diattenuation of the example grating in Figure 23.13 zero? Approximate how fast is the diattenuation varying at nearby wavelengths, that is, estimate $\frac{dD}{d\lambda}$.

23.8 A diffraction grating has the following diffraction efficiencies (output flux/input flux) as a function of wavelength for its −1st order. Green is for light polarized parallel to the rulings, and red is for light polarized perpendicular to the rulings (Figure 23.32).

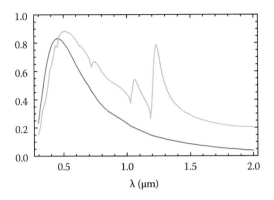

Figure 23.32 Diffraction efficiencies of diffracted light with polarization parallel (green) and perpendicular (red) to the grating grooves.

a. For an unpolarized incident light, at what wavelengths would the output light have a degree of linear polarization (DoLP) of about 1/4?
b. Estimate in which spectral regions is the variation of DoP with wavelength the smallest?

Acknowledgments

This chapter incorporates the work of several colleagues including Karlton Crabtree, who performed the analysis of the SWG antireflection-coated lens, and Michihisa Onishi, who wrote the RCWA code used for many of these examples.

References

1. *Correspondence of Scientific Men of the Seventeenth Century: Including Letters of Barrow, Flamsteed, Wallis, and Newton, Printed from the Originals in the Collection of the Right Honourable the Earl of Macclesfield*, Vol. 2, ed. S. J. Rigaud, England: Oxford University Press (1841), pp. 251–255.
2. *The Scientific Writings of David Rittenhouse*, ed. B. Hindle, New York: Arno Press (1980), pp. 377–382.
3. H. Rowland, Preliminary notice of results accomplished on the manufacture and theory of gratings for optical purposes, *Phil. Mag. Suppl.* 13 (1882): 469–474.
4. G. R. Harrison and E. G. Loewen, Ruled gratings and wavelength tables, *Appl. Opt.* 15 (1976): 1744–1747.
5. W. T. Welford, Aberrations of optical systems, Chapter 5, New York: Taylor & Francis Group (1986), pp. 75–78.
6. M. G. Moharam and T. K. Gaylord, Rigorous coupled-wave analysis of grating diffraction E-mode polarization and losses, *J. Opt. Soc. Am.* 73 (1983): 451–455.
7. M. G. Moharam and T. K. Gaylord, Rigorous coupled-wave analysis of metallic surface-relief gratings, *J. Opt. Soc. Am.* A 3 (1986): 1780–1787.
8. R. W. Wood, On the remarkable case of uneven distribution of light in a diffraction grating spectrum, *Philos. Mag.* 4 (1902): 396–402.
9. R. W. Wood, Anomalous diffraction gratings, *Phys. Rev.* 48 (1935): 928–936.
10. L. Rayleigh, Note on the remarkable case of diffraction spectra described by Prof. Wood, *Philos. Mag.* 14 (1907): 60–65.
11. L. Rayleigh, On the dynamical theory of gratings, *Proc. R. Soc. Lond.* 79 (1907): 399–416.
12. U. Fano, The theory of anomalous diffraction gratings and of quasi-stationary waves on metallic surfaces (Sommerfeld's waves), *J. Opt. Soc. Am.* 31 (1941): 213–222.
13. A. Hessel and A. A. Oliner, A new theory of Wood's anomalies on optical gratings, *Appl. Opt.* 4 (1965): 1275–1297.

14. R. H. Ritchie, E. T. Arakawa, J. J. Cowan, and R. N. Hamm, Surface-plasmon resonance effect in grating diffraction, *Phys. Rev. Lett.* 21 (1968): 1530–1533.
15. C. H. Palmer Jr., Parallel diffraction grating anomalies, *J. Opt. Soc. Am.* 42 (1952): 269–276.
16. C. H. Palmer Jr., Diffraction grating anomalies, II, coarse gratings, *J. Opt. Soc. Am.* 46 (1956): 50–53.
17. G. P. Bryan-Brown, J. R. Sambles, and M. C. Hutley, Polarisation conversion through the excitation of surface plasmons on a metallic grating, *J. Mod. Opt.* 37 (1990): 1227–1232.
18. S. J. Elston, G. P. Bryan-Brown, and J. R. Sambles, Polarization conversion from diffraction gratings, *Phys. Rev.* B 44 (1991): 6393–6400.
19. W. C. Sweatt, S. A. Kemme, and M. E. Warren, *Diffractive Optical Elements Optical Engineer's Desk Reference: Ch. 17*, ed. W. L. Wolfe, in SPIE Press Monograph Vol. PM131, SPIE Publications (2003).
20. J. dos Santos and L. Bernardo, Antireflection structures with use of multilevel subwavelength zero-order gratings, *Appl. Opt.* 36 (1997): 8935–8938.
21. K. Crabtree and R. A. Chipman, Subwavelength-grating-induced wavefront aberrations: A case study, *Appl. Opt.* 46 (2007): 4549–4554.
22. M. Onishi, Rigorous Coupled Wave Analysis for Gyrotropic Materials, PhD dissertation, College of Optical Sciences, University of Arizona (2011).
23. M. G. Moharam, E. B. Grann, D. A. Pommet, and T. K. Gaylord, Formulation for stable and efficient implementation of the rigorous coupled-wave analysis of binary gratings, *J. Opt. Soc. Am. A* 12 (1995): 1068–1076.
24. M. G. Moharam, D. A. Pommet, E. B. Grann, and T. K. Gaylord, Stable implementation of the rigorous coupled-wave analysis for surface-relief gratings: Enhanced transmittance matrix approach, *J. Opt. Soc. Am. A* 12 (1995): 1077–1086.
25. L. Li and C. W. Haggans, Convergence of the coupled-wave method for metallic lamellar diffraction gratings, *J. Opt. Soc. Am. A* 10 (1993): 1184–1189.
26. E. N. Glytsis and T. K. Gaylord, Rigorous three-dimensional coupled-wave diffraction analysis of single cascaded anisotropic gratings, *J. Opt. Soc. Am. A* 4(11) (1987): 2061–2080.
27. E. N. Glytsis and T. K. Gaylord, Three-dimensional (vector) rigorous coupled-wave analysis of anisotropic grating diffraction, *J. Opt. Soc. Am. A* 7 (1990): 1399–1420.
28. M. Onishi, K. Crabtree, and R. Chipman, Formulation of rigorous coupled-wave theory for gratings in bianisotropic media, *J. Opt. Soc. Am. A*, 28(8) (2011): 1747–1758.
29. L. Li, Use of Fourier series in the analysis of discontinuous periodic structures, *J. Opt. Soc. Am. A* 13 (1996): 1870–1876.
30. L. Li, Reformulation of the Fourier modal method for surface-relief gratings made with anisotropic materials, *J. Mod. Opt.* 45 (1998): 1313–1334.
31. E. Popov and M. Nevière, Maxwell equations in Fourier space: Fast-converging formulation for diffraction by arbitrary shaped, periodic, anisotropic media, *J. Opt. Soc. Am. A* 18 (2001): 2886–2894.
32. R. Antos, Fourier factorization with complex polarization bases in modeling optics of discontinuous bi-periodic structures, *Opt. Express* 17 (2009): 7269–7274.

24

Liquid Crystal Cells

24.1 Introduction

Liquid crystal cells are optical elements that manipulate polarized light by rotating liquid crystal molecules. Liquid crystal cells are used as electrically controllable retarders, polarization controllers, and spatial light modulators. *Liquid crystal displays* (LCDs) incorporate liquid crystal cells with illumination and electronics for displaying information. LCDs have traditionally been utilized in flat screen televisions, mobile phone screens, and computer screens.

LCDs comprise a large fraction of the *polarization economy*. LCDs were introduced in the 1970s and have dominated the display business since the late 1980s for many reasons.

- LCDs operate at *low voltages*, typically less than 5 V.
- LCDs have *large etendue*, a combination of large area and large numerical aperture; thus, LCDs can utilize a large fraction of the light from an incandescent, fluorescent, or LED illumination system.
- LCDs can be fabricated in arrays with *large numbers of small pixels*; pixel sizes below 2 μm are common.
- LCDs can be produced at *low cost* through the use of complex foundries.

The image quality of early LCDs was poor. To attain their position of dominance, LCDs had to overcome many obstacles, including absorption, scattering, low contrast, muddy colors, long switching times, uniformity for larger display, limited viewing angle, disclinations within pixels, and polarization aberration. LCDs had to compete with the installed base of cathode ray tube

Table 24.1 Common Types of Liquid Crystal Cells

1. Fréedericksz, untwisted nematic cell
   - Retardance and phase modulators, polarimetry, signal processing
2. Twisted nematic (TN)
   - Most common display technology
3. Super twisted nematic (STN)
   - Displays
4. Vertically aligned mode (VAN mode)
   - Displays
5. Hybrid aligned
   - Displays
6. In-plane switching, fast field switching (FFS), and fringe-field-switching (FFS)
   - iPhone/iPad/high-end monitors/televisions
7. Multi-domain vertically aligned (MVA), PSVA, UV2A, etc.
   - Televisions

televisions and displays. Other new technologies also threatened the early LCD industry. Over time, billions of dollars of investment in production technologies has addressed these issues and LCDs became the dominant display technology. This chapter discusses the construction, operation, and polarization aberrations of common liquid crystal devices.

Several common types of liquid crystal cells are listed and described in Table 24.1. The construction of these liquid crystal cells is explained and fabrication issues are discussed, along with more detailed discussions of the types of liquid crystal cells and their evolution. The testing of liquid crystals is described using imaging polarimetry. Finally, the modeling of cells' polarization aberrations and their compensation with multilayer biaxial films are shown.

24.2 Liquid Crystals

Liquid crystals (LCs) are a state of matter with properties between solid and liquid. When typical substances are heated and melt, they transform directly from a highly ordered crystalline solid structure (Figure 24.1, left) into a disordered isotropic liquid form (Figure 24.1, right). In contrast, when a solid LC material melts, it transitions through a partially ordered liquid state (Figure 24.1, middle) before becoming an isotropic liquid. Thus, a material in a liquid crystal state is an ordered

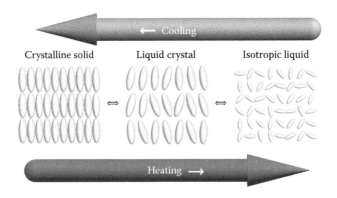

Figure 24.1 State of matter between solid and liquid.

24.2 Liquid Crystals

fluid that exhibits less order than found in solids, but more order than found in liquids. LC molecules are not ordered by position, but the orientation of the molecule is correlated to the orientation of neighboring molecules. The shape of the LC molecule is highly anisotropic, which affects the preferred orientation of its nearby molecules. Thus, it has physical properties of a liquid but with a short range of crystal-like order.

A botanist in Austria, Friedrich Reinitzer, first described liquid crystals in 1888. He encountered a material, cholesteryl benzoate, that exhibited a mesophase between the solid and liquid states. When heating through 145°C, it melted into a viscous white fluid. At 179°C, it transitioned into a clear isotropic liquid. He shared his discovery with a physics professor at the Technical University Karlsruhe, Otto Lehmann, describing the two melting points. Lehmann observed that in the mesophase, the liquid demonstrated the double refraction effect, a characteristic of a crystal. Since cholesteryl benzoate showed both liquid and crystal characteristics, he named it a "fliessende crystal." This translated into the English name "liquid crystal." Hence, the molecules themselves are not the liquid crystals. Liquid crystal refers to the phase, the state between solid and liquid.

The most common molecules used in LC cells are the *nematic liquid crystals*, which are rod-like positive uniaxial molecules. The orientation of the LC molecules is specified by the *director*, which is the direction of the extraordinary axis, also called the optic axis. An example of a typical molecule used in LC cells is shown in Figure 24.2. This molecule is elongated or rod-like, and has a large permanent dipole moment. Note the three fluorine atoms attached to a benzene ring on one end. These three highly electronegative fluorine atoms draw electrons from the benzene ring and beyond toward their end of the molecule, leaving the other end positively charged, setting up a significant dipole moment.

When LC cells are drawn, lines or ellipsoids are drawn to show the variation of the director throughout the cell, as seen in Figure 24.3. The sign of the director is not used; the director has unit length, since its magnitude is not important. Because the material is liquid, the director can vary throughout the cell. The molecules locally strive to remain aligned with each other to reduce their

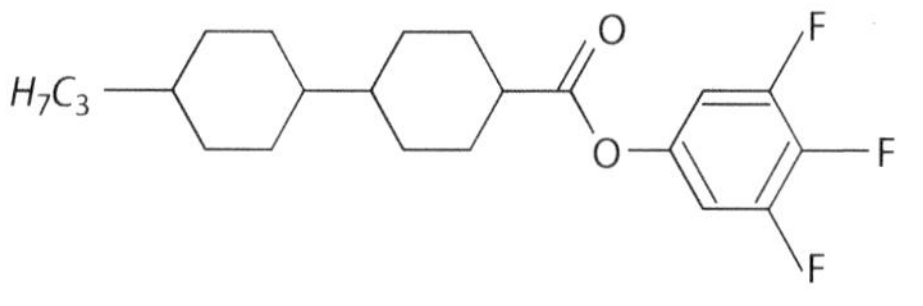

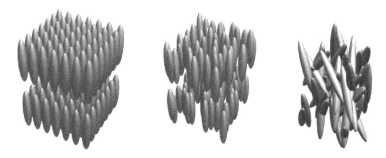

Figure 24.2 Chemical structure of 4′-Propyl-bicyclohexyl-4-carboxylic acid 3,4,5-trifluoro-phenyl ester ($C_{22}H_{29}F_3O_2$), one of the common liquid crystal molecules used in displays. The director of this molecule, indicating the axis along which the dipole is oriented, is horizontally oriented.

Figure 24.3 Temperature influences the order in nematic LC molecules. At high temperature, (right) the LCs are in an isotropic phase where no order is present. (Middle) Lowering the temperature leads to orientation order in the nematic phase, and when cooled to crystallization, (left) the molecules assume long-range positional and orientation order.

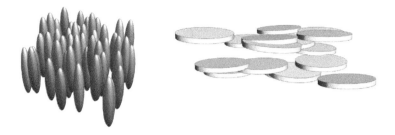

Figure 24.4 Schematics for nematic molecules (left) and discotic molecules (right).

energy, but mechanical and electrical forces cause a change in the director orientation. Rotating the LC molecules provides a tunable birefringence. For nematic LCs, the ordinary refractive index is usually about $n_O = 1.5$ and the birefringence Δn varies from 0.05 to 0.5.

Molecules used in LC devices are fabricated in a great variety of shapes. For LC displays, two shape families dominate; the nematic LC, with a rod-shaped molecule as shown in Figure 24.4, and the discotic LC, with a disk-like or pancake-like molecule.

24.2.1 Dielectric Anisotropy

LC molecules have an intrinsic *dipole moment*; the center of a molecule's positive charge is displaced from the center of the negative charge. One end of the molecule has an excess of electrons. When an electric field is applied, a *torque* is induced, which rotates the molecules. For molecules with *positive dielectric anisotropy*, the directors rotate toward aligning *parallel* with the electric field, as shown in Figure 24.5. Rod-like molecules with positive dielectric anisotropy have an excess of charge at one end of the rod. Such molecules have a larger retardance without a field and near-zero retardance when large voltage is applied. For *negative dielectric anisotropy*, the torque tends to orient the directors perpendicular to the field. These molecules have an excess of electrons near the middle of the rod on one side.

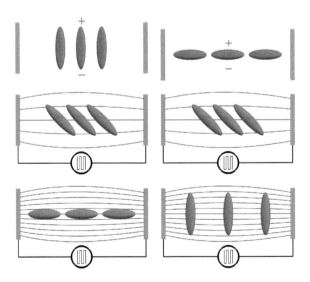

Figure 24.5 Molecules with (left) positive dielectric anisotropy and (right) negative dielectric anisotropy are shown. The LCs are at their rest state at 0 V at the top row. The electric field lines are shown in blue, with number of lines indicating field strength.

24.3 Liquid Crystal Cells

A *liquid crystal cell* is a device that packages a layer of liquid crystal with *electrodes* in order to *modulate* the distribution of *directors* and thus modulate the phase and polarization of the transmitted light. Liquid crystal cells allow the electrical control of light at low voltages, over large areas, and with large numbers of pixels. The modulation schemes are divided into three classes:

1. Intensity modulators, commonly used for the display of information
2. Polarization modulators, for polarization manipulation and control
3. Phase modulators, for wavefront control, interferometry, diffractive optical elements, and so on

Consider the earliest type of liquid crystal cell, the Fréedericksz cell shown in Figure 24.6. The space between two glass plates is filled with a thin layer of nematic liquid crystal with positive anisotropic anisotropy, usually between 0.5 and 7 μm thick depending on the application. When no voltage is applied (Figure 24.6, left), the directors are all oriented parallel to the glass plates (*x*-axis in figure). In this orientation, the LC cell acts like an *A-plate*, a conventional waveplate, and each layer contributes the same retardance to the cells overall retardance. The electric field across the cell is adjusted by charging electrodes outside the glass, and because of their positive dielectric anisotropy, the rod-like LC molecules rotate toward the *z*-axis, the normal to the glass and the nominal direction of light propagation. As the LC molecules rotate, the birefringence along the light propagation direction is reduced, and the overall retardance (along the optical axis) is reduced. As the voltage increases (toward the right side of Figure 24.6), the cell's retardance is minimized. At higher voltages, most of the molecules are aligned along the *z*-axis, except for the thin layers along the glass that cannot rotate as these are attached or stuck to the cell faces. The distributions of directors will be described by a twist angle ϕ and a tilt angle θ as shown in Figure 24.7.

Figure 24.8 shows a typical retardance versus voltage curve for the Fréedericksz cell. Thus, this cell can operate as a linear *retardance modulator* with axes along *x* and *y*. As the cell modulates, the refractive index for the *x*-component of the light changes; the molecules are rotating in the *x–z* plane, but the *y*-refractive index remains nearly constant. If a polarizer is placed above the cell with the transmission axis along *x*, the light that is transmitted through the cell is in an *eigenpolarization*

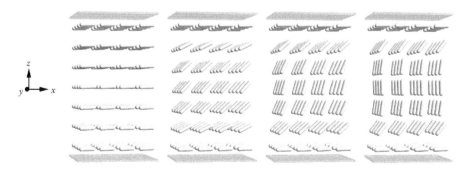

Figure 24.6 Variation of director orientation (purple rods) in a Fréedericksz cell with light propagating from top, through a transparent electrode and glass plate (orange sheet), through the liquid crystal layer, and out the bottom glass and electrode (orange sheet). The applied voltage increases from 0 V on the left to the maximum voltage on the right, and the nematic liquid crystals with positive dielectric anisotropy rotate closer to the field as the electric field increases. The LC liquid crystal cell modulates from an *A-plate* retarder (like a waveplate) at low voltages to almost a *C-plate* (optic axis along the direction of light propagation) as the voltage increases. Thus, this LC cell is a retardance modulator.

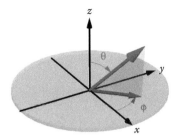

Figure 24.7 Tilt θ is the angle measured from the z-axis to the vector. Twist ϕ is the angle measured from the x-axis to the projected vector on the xy plane.

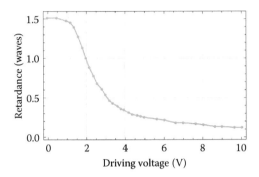

Figure 24.8 A typical retardance versus voltage curve for a Fréedericksz cell. Below a threshold voltage of around 2 V, the LC molecules in this cell barely rotate. Further increasing the voltage reduces the retardance, which approaches but does not reach zero at higher voltages.

at every layer of the LC and thus does not change polarization. The refractive index is however changing, so the phase will vary with voltage. Thus, this liquid crystal cell can be used as a *phase modulator*. Finally, an *intensity modulator* can be constructed by placing a polarizer at 45° before the cell and a polarizer at 135° after the cell. The cell thickness and liquid crystal birefringence are selected; thus, the cell's retardance modulates from about ½ wave of retardance to close to 0 retardance. At high voltages, the LC retardance is small, the polarization is not changed much since the cell is acting like a *C-plate*, and the transmission through the cell and crossed polarizers is nearly dark. At low voltages, the LC retardance is about 180° and the 45° light transmitted through the first polarizer is rotated to 135°, aligned with the second polarizer creating a bright state. Intermediate voltages provide light levels from light to dark.

This concept of varying the output polarization state from being *aligned* with the output polarizer to being *orthogonal* to the output polarizer as a function of voltage is the basis of all polarization-based LCDs, from calculators, to watches, ATMs, computer displays, projectors, and televisions. Almost all liquid crystal cells are built for intensity modulation since the human eye is sensitive to changes in intensity and blind to changes in light polarization and phase. LC cells for polarization modulation and phase modulation, while important as optical components, comprise much less than 1% of the LC cell market, which is dominated by displays.

24.3.1 Construction of Liquid Crystal Cells

Figure 24.9 shows a typical LC cell configuration. The LC cell provides a means for transmitting light through a thin layer of LC and controlling the orientation of the layers with an applied electric field. The LC layer is 1 to 7 μm thick and contained between two parallel thin glass plates.

24.3 Liquid Crystal Cells

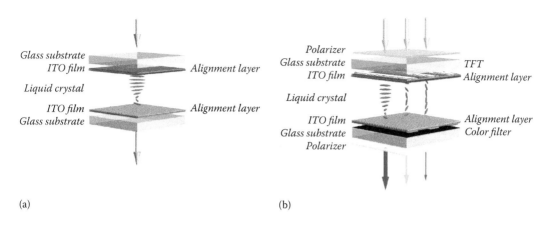

Figure 24.9 (a) A typical LC cell configuration and an LCD configuration are shown. The LC cells are illuminated from the top. The modulated light exits at the bottom of the cell. (b) For a color display, adjacent pixels incorporate red, green, and blue color filters. Intensity modulation is obtained by the placement of polarizers on top and bottom.

These glass plates, which mechanically hold the LC cell, must have very low birefringence so they do not change the polarization modulation due to the LC layer. The inside layer of each glass plate, the surface against the LC, has a thin layer of transparent conductor such as indium tin oxide (ITO) about 100 nm thick. Over the electrode is a thin *alignment layer*, a soft plastic layer in which grooves can easily be formed, typically made of *polyimide*, which controls the orientation of the LC layers adjacent to the glass substrate. For *retardance modulators* (polarization controller) and *phase modulators*, such as Figure 24.9a, this is all that is required; no polarizers are included. A *liquid crystal display* (LCD) includes several additional components to perform *intensity modulation* and *color modulation*. Outside the glass substrates, additional retardation films, often called *biaxial multilayer films*, are usually applied to optimize the color of the display and minimize the variation of color with angle. Over these films, film polarizers are glued. The film polarizers are needed to change the cell from a polarization modulator into an intensity modulator. Electrodes are connected to the transparent electrodes to apply and hold the voltages. One or both electrodes will be pixelated with associated transistors and capacitors to set the voltage per pixel each using an individual patch of ITO film. For *color displays*, an array of color filters, red, green and blue, will be placed outside the glass plates.

24.3.2 Restoring Forces

The intermolecular forces between nearby LC molecules can be visualized as tiny springs. The *elastic constant* k of the springs defines how hard the LC molecules "push" against each other. There are three types of k values: the "splay" direction push, the" twist" direction push, and the "bend" direction push as shown in Figure 24.10.

LC cells are voltage controlled retardance modulators that change their retardance by rotating anisotropic LC molecules in an electric field. The LC cell is analyzed and understood as a spatially varying anisotropic material with spatially varying retardance. The spatial variation of the director is easily perturbed by electric fields and magnetic fields and by the shape of the bounding surfaces. In the absence of voltage, the liquid crystal seeks its lowest free energy configuration; it mechanically relaxes into the lowest energy state, the ground state. Then, applied fields cause the molecules to rotate, increasing the volume somewhat and increasing the spring-like energy between the molecules. When the field is removed, the molecules push each other back into the ground state. The type of cell is typically named after the configuration of directors in this ground state.

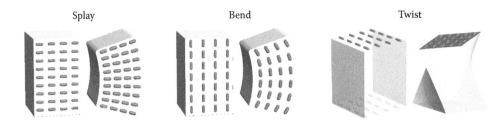

Figure 24.10 (Top) Director distributions with splay, (middle) bend, (bottom) and twist.

Figure 24.11 shows a vertically aligned nematic (VAN) LC cell with negative dielectric anisotropy sandwiched between two transparent electrodes. The LC molecules against the outer surfaces are attached perpendicular to the glass plates, which set the director distribution throughout the cell parallel to normal incident light at zero voltage. Thus, the directors line up with the electric field. As the voltage increases, the negative dielectric anisotropy LCs rotate perpendicular to the applied electric field as a function of the applied voltage. At the highest applied voltage shown, the directors at the center of the cell are twisted 90°. The top and bottom layers of the LCs do not rotate, since the ends of the molecules have been anchored to the surfaces of the cell. Figure 24.11 (Bottom) The typical transmittance for a VAN mode cell between crossed polarizers.

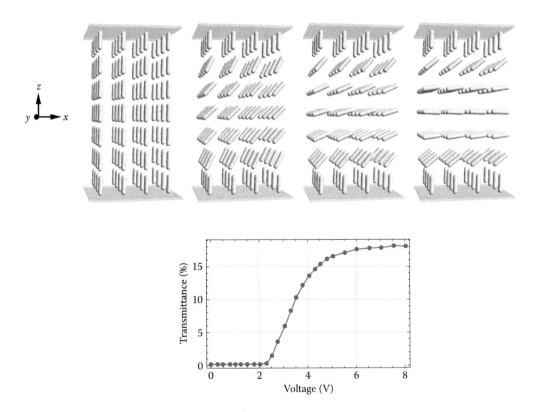

Figure 24.11 A VAN mode LC cell with negative dielectric anisotropy is shown. (Top left) All LC's layers are aligned at zero voltage. As the voltage increases, (toward the right) the LC's molecules rotation increases. (Top right) At the highest voltage, most of the directors are nearly perpendicular to the field. (Bottom) Typical transmittance vs. voltage.

Dozens of different "modes" of LC cells have been invented for different purposes such as the following:

- Wider viewing angle
- Lower electrical power
- Higher contrast
- Brighter
- Blacker "blacks"
- Better color
- Lower manufacturing cost
- Unaffected by squeezing for the purpose of touch screen capability

24.3.3 Liquid Crystal Display: High Contrast Ratio Intensity Modulation

Most LC devices are operated as intensity modulators, not as phase or polarization modulators. The liquid crystal cell is placed between two polarizers, usually linear polarizers. When the voltage applied to the cell changes, the cell's retardance and exiting polarization state changes, modulating the intensity. Retarders are often used in conjunction with the linear polarizers if they improve performance. However, cost is a driving factor for high volume production, and simplicity is highly valued.

One metric for evaluating display performance is the *contrast ratio* C, defined as the maximum flux I_{max} divided by the minimum flux I_{min},

$$C = \frac{I_{max}}{I_{min}}. \quad (24.1)$$

A high I_{max} is desirable for a bright display, but usually a small I_{min} and a good black is far more important and more difficult to achieve. Thus, achieving a very low I_{min} often is a driving factor in the display design. Some of the reasons why a cell might have a poor I_{min} are the following:

- Misalignment between analyzer and exiting polarization state
- Spectral variation of exiting polarization state
- Variation of polarization state with angle through the LC cell
- Scattering and depolarization
- Polarizer leakage, poor-quality polarizers

Typical contrast from an LC projector is about 100 and an average LC computer screen has a contrast ratio of about 1000. The contrast of most LC-based TV sets is much greater than 1000, and some models can reach a contrast ratio of 10,000 to 30,000. In general, the more expensive the TV, the better the black, and the higher the contrast.

One ideal configuration consists of a liquid crystal cell as a variable linear retarder oriented at 45° to the first polarizer, which can modulate over a range of at least a half of a wavelength. This variable retarder brings 0° incident light through a series of elliptical polarizations, through circularly polarized light when $\delta = \pi/2$ into vertically linearly polarized light. This polarization evolution is well described by a spectral trajectory on the Poincarè sphere along a circle of longitude about the variable retarders' fast and slow axes. The analyzer can be oriented at either 0° for a *dark state* at low retardance or 90° for a *dark state* at high retardance. The choice is usually driven by which configuration produces the better *dark state*. Retardance modulation of greater than half a wavelength generally does not benefit the performance of an intensity modulator, since the device has already spanned its entire intensity range.

24.4 Configurations of Liquid Crystal Cells

Most transmissive LC displays share a basic configuration: a pair of polarizers with an adjustable LC retarder in between them. The LC retarder varies between *on* and *off states*. At the bright state, the LC retarder transforms light exiting the incident polarizer so it aligns with the exiting polarizer, thus allowing maximum light to pass. At the *dark state*, the light is orthogonal to the exiting polarizer. Most commercial LCD designs are variants of this approach. Throughout the evolution of LC cells, complexity has continually increased.[1]

Many different LC configurations are manufactured. Configurations differ in the director distribution functions, electrode locations, and subpixel structures. One of the oldest LC cell designs is the untwisted Fréedericksz cell, which is preferred for polarization control and acts well as an adjustable retarder. The twisted nematic cells are very popular in the LCD industry; the three major families of commercial LC cell technologies used are the twisted nematic, the in-plane switching, and the vertical-aligned nematic. Many more types of LC cells have been developed. The following are three of the most important families for displays:

- Twisted nematic cell
  - Inexpensive
  - Most common
  - Significant polarization and color variation with angle
- Vertically aligned [nematic] (VA, VAN, MVA, PVA, S-PVA, etc.)
  - Good switching speed, off-axis viewing (wide viewing angle), intrinsic color gamut (range), relative to TN; black level overall
  - Intrinsic image quality not as good as IPS, but many incremental improvements have been introduced
  - Popular for TVs
- In-plane switching (IPS, S-IPS, AS-IPS, H-IPS, A-TW-IPS, etc.)
  - High intrinsic image quality; reference standard for color output, and off-axis performance (wide viewing angle)
  - Initial obscuration and switching speed handicap, largely overcome by design and process evolution
  - Costly

24.4.1 The Fréedericksz Cell

The first LC cell developed was the *Fréedericksz cell*, also known as the *untwisted nematic cell*, or *planar aligned nematic cell*, as shown in Figure 24.12. The directors align and rotate in a single plane; thus, the label untwisted. At 0 V in the *off state*, all directors are parallel and aligned to the electrodes, like an *A-plate* retarder, with a small pretilt. As the voltage increases, the LCs rotate in the x–z plane toward the direction of light propagation, thus reducing the retardance. At high voltages, the directors rotate to almost vertical at the middle layer of the cell, while the molecules at either end are anchored at the glass plates. Figure 24.13 shows a small single-pixel Fréedericksz cell retardance modulator with a pair of wires for the driving voltage.

Since the directors are untwisted and only rotate in one plane, the Fréedericksz cell is always a linear retarder with constant orientation; the retardance axis usually oriented at 45° to the edge of the cell. The cell is often driven by 500–10,000 Hz square wave at low voltage (1–13 V). It is a desirable 180° phase modulator. The Fréedericksz cells are small and inexpensive compared to other retardance modulators, such as electro-optical modulators, photo elastic modulators, magneto-optical modulators, or retarders mounted on rotating motors. However, the parallel aligned LC cells

24.4 Configurations of Liquid Crystal Cells

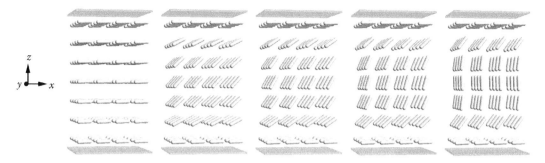

Figure 24.12 Director orientations in a Fréedericksz cell. The applied voltage increases from (left) 0 to 5 V (right). The LCs modulate from an *A-plate*-like retarder configuration to almost a *C-plate*.

Figure 24.13 A small Fréedericksz cell retardance modulator for adjustable linear retardance. The LC is sealed between glass plates.

have many issues; they are often too slow and have large variation of color and retardance with field of view.

24.4.2 90° Twisted Nematic Cell

The *twisted nematic cell* (TN cell) was the first LC technology in production for LC displays, introduced by Schadt and Helfrich and Fergason in 1971. The TN technology still has a large market share. It is often found in notebook PC screens, desktop LCD monitors, and some low-end cell phones. Because of decades of experience of manufacturing TN cells, the production cost is relatively low. The TN cell operates at a lower voltage and switches considerably faster than the Fréedericksz cell.

The TN directors twist steadily from window to window at zero voltage, while remaining parallel to the glass plates (no tilt). In the *off state*, the directors are twisted by 90° around the direction of light propagation, as shown in Figure 24.14, and the pixel has an approximately 180° of retardance. That pixel becomes *bright* when placed between crossed polarizers. When a TN pixel goes dead, the transmission gets stuck *on* and appears white. As shown in Figure 24.15, when voltage is applied, the LC molecules rotate toward the z-axis, becoming normal to the cell, and the retardance reduces, since birefringence goes to zero for propagation along the director. At the high range of applied voltage, the retardance becomes almost zero, as shown in Figures 24.15 (right) and 24.16. Then, the pixel is *dark* when placed between crossed polarizers. At the *dark* state, the retardance at the top and the bottom of the cell cancels each other, thus providing a larger FOV.

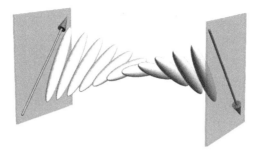

Figure 24.14 The helix distribution of the liquid crystal directors in a 90° TN-LC cell with a left-handed twist is indicated by the distribution of the rods. The arrows on the ends indicate the rubbing directions in the alignment layer.

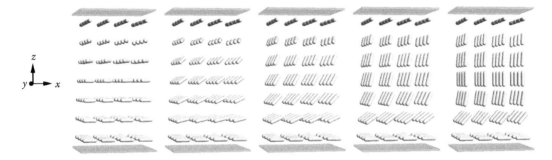

Figure 24.15 Director distributions for a 90° twisted nematic LC cell and the variation of the directors with voltage. The entering and exiting windows are indicated in orange. The leftmost image is at 0 V, where all molecules are parallel to the windows but twist through 90°. The voltage increases from left to right in arbitrary increments. At high voltages, above 5 V, the molecules have mostly rotated to the vertical direction except for the thin layers near the top and bottom alignment layers.

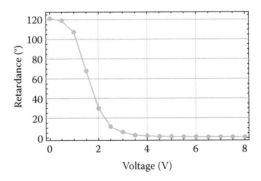

Figure 24.16 Resultant retardance as a function of voltage across the 90° TN cell.

For a TN cell, as the voltage begins increasing from 0 V, very little LC molecular motion occurs, until suddenly at around 1 V, enough torque is applied for the molecules to overcome their intermolecular attraction, and they begin to rotate. As the voltage increases, the central layer of the molecules rotates to vertical and then more layers become vertical; most of the cell is approaching a *C-plate*, until finally most of the cell volume has a vertical orientation except for the layers at the very top and bottom, which are not free to rotate. The elliptical eigenpolarizations of the LC cell are depicted in Figure 24.17. The eigenpolarizations are the states that exit the cell in the incident

24.4 Configurations of Liquid Crystal Cells

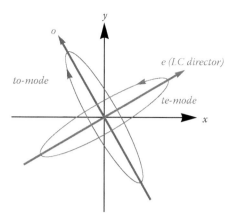

Figure 24.17 The polarization ellipses of the normal modes in a TN-LC.

polarization state. The phase delay between the two orthogonal eigenpolarizations is the retardance. The structure of the LC cell is like slices of linear retarders rotating through 90°. Since misaligned linear retarders give circular retardance, the eigenpolarizations are elliptically polarized. As the wavelength and the incident angle vary, these eigenpolarizations and the retardance also change in magnitude and orientation. The minimum retardance is asymptotically approached at high voltage state, above 5 V as shown in Figure 24.16, since most of the directors are vertical. They are like the *C-plates* with no retardance. Since the retardance does not decrease all the way down to zero due to small residual retardance from layers against the cell windows, sometimes a weak retarder is added in series, a *trim retarder* perhaps with about 5° of retardance, to bring the retardance down to zero. Because of retardance dispersion, a shorter wavelength produces a larger retardance modulation while a longer wavelength gives a smaller retardance modulation.

The shortcomings of the TN cell motivated the development of other cell designs to provide higher contrast and better display. A common defect of TN cells is crossed-polarizer leakage due to linear state obliquity. Another issue is that for the *dark state*, the TN cell converts the input linear state to a slightly elliptical state that the output polarizer cannot completely block.

The retardance of TN liquid crystal cells has a large variation with angle of incidence, which is simulated in Section 24.5.4. Figure 24.35 maps the retardance versus angle of incidence for a typical TN cell with increasing voltage.

24.4.3 Super Twisted Nematic Cell

The *super twisted nematic cell* (STN) is a variation of the TN cell, which was commonly found in displays in the 1990s and 2000s.[2] The term *super* refers to a twist angle larger than the 90° twist typical of TN cells. This lets the STN cell switch faster at lower voltages. The angle of incidence characteristics are better but the cells are more expensive to fabricate. The schematics of the directors for the STN cell with 180° and 270° twist at 0 V are shown in Figure 24.18.

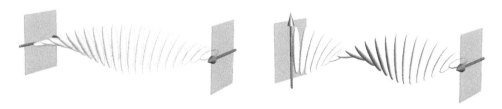

Figure 24.18 Director orientations through the cell for STN cells with (left) 180° and (right) 270° rotations.

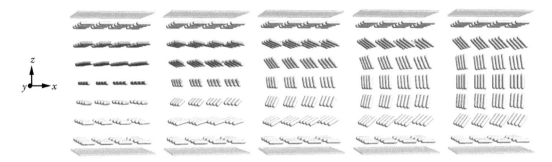

Figure 24.19 The 180° supertwisted nematic liquid crystal cell's directors at several voltages. At low voltage (left), the director twists in the plane of the cell, in this example by 180°. At high voltages (right), the directors tilt by almost 90° and are aligned along the direction of light propagation except for thin layers against the alignment layers (orange planes).

In Figure 24.19 (left), with no applied voltage, the directors twist by 180°. At maximum voltage (right), the director orientation approaches a *C-plate* configuration, with the directors substantially aligned along the direction of light propagation.

24.4.4 Vertically Aligned Nematic Cell

The *vertically aligned nematic* (VAN) LC was a common configuration for TVs in the early 2000s, often replacing the TN cell in this application. The VAN LC cell was preferred for its high contrast in the *off* state and good viewing angle performance.[3] The VAN mode cell uses a negative dielectric anisotropy molecule. In the *off* state, the directors are tilted vertically, called *homeotropic alignment*, perpendicular to the electrodes, like a *C-plate* with small pretilt. As shown in Figure 24.20, in the ground state with 0 V, the directors are near vertical. The applied voltage then rotates the directors in the *x–z* plane.

At 0 V, the VAN cell is basically a *C-plate* providing an extremely good *dark state* between crossed polarizers with a wide field of view. As a function of angle of incidence, the ordinary and extraordinary modes are shown in Figure 24.21. At normal incidence, the two modes are degenerate, and the LC cell has zero retardance.

The VAN mode cell functions well as a phase modulator. The directors of the VAN cell remain in one plane as voltage is applied as is seen in Figure 24.20. If the cell is illuminated with light in this plane of polarization, the light remains in an eigenpolarization as it propagates through the cell; hence, the polarization state emerges unchanged. But the refractive index changes from n_O toward n_E as the molecules rotate, thus changing the phase without changing the polarization.

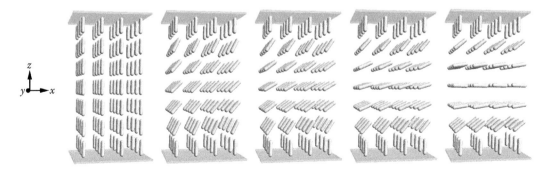

Figure 24.20 The VAN cell's directors at several voltages. It gives a *dark state* at 0 V shown on the left, and a bright state at high voltage shown on the right.

24.4 Configurations of Liquid Crystal Cells

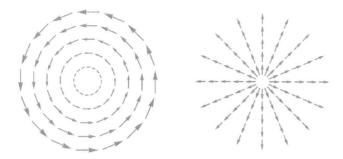

Figure 24.21 VAN-mode eigenpolarizations in a cell without pretilt, (left) ordinary mode is tangentially polarized, and (right) extraordinary mode is radially polarized. At normal incidence, the modes are degenerate; the ordinary and extraordinary modes have the same refractive index.

The *dark state* of the VAN cell is an improvement from the TN and STN cells. The wide-angle contrast is still limiting by the leakage in the *dark state*. But this can be well compensated by additional retarders before and after the LC cell. The nematic LC molecules are normal to the glass interfaces, which can produce stability issues—moving picture image sticking (MPIS). In some severe cases, the image sticking remains for a short time after the image has changed, as shown in Figure 24.22. Because of disclination, the LC director orientations are undefined and can easily flip the wrong way. These lead to transmission variability and slower response time.

The *patterned vertical-aligned cell* (PVA) was developed by Samsung in 1996. The PVA mode tackles disclination by having a non-normal field applied by patterning the electrodes on planar layers bounding the LC so they are offset from each other, as shown in Figure 24.23. This configuration produces multiple domains, which can address the asymmetric behavior of the overall display when arranged correctly and give excellent viewing angle. In PVA, the electrodes are arranged as zigzag chevron geometry (Figure 24.23, right).

Another method to treat disclinations in VAN mode cells is the multi-domain vertically aligned (MVA) cell. In the MVA mode, structures with protrusions are formed in the layers bounding the LC, which makes the director non-normal to the display axis in the *off* state, as shown in Figure 24.24 with pyramids on two sides or in Figure 24.25. Multiple domains are produced in a symmetric pattern, which overcomes any asymmetry in transmission characteristics. Another variation of the MVA configuration replaces the protrusions on one substrate with patterned ITO slits. This reduces

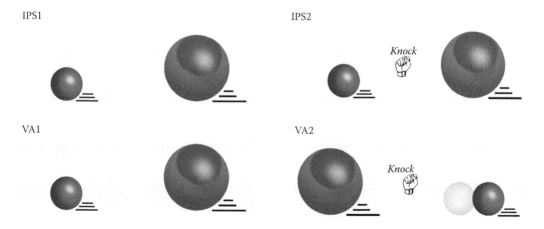

Figure 24.22 Example of motion picture image sticking that appears in VA panels and does not appear in S-IPS mode panels.

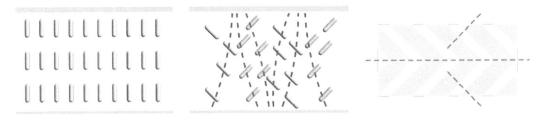

Figure 24.23 (Left) The director distribution in a PVA cell in its *off* state resembles a VAN mode cell. (Middle) The director distribution in the *on* state is perpendicular to the electric field lines (red). (Right) Each surface's electrode distribution within each pixel has two regions (indicated by the blue line) with electrodes patterned in a different direction.

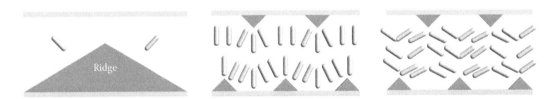

Figure 24.24 The configuration for MVA panels uses arrays of pyramids or ridges (left) for alignment and anchoring. (Middle) Directors in the *off* state. (Right) Directors in the *on* state.

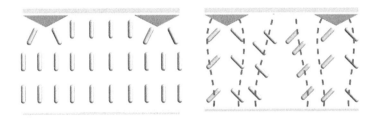

Figure 24.25 A variation of the MVA concept with pyramids on only one surface in (left) *off* and (right) *on* state.

the number of manufacturing steps and increases the contrast ratio because the residual birefringence around the protrusions on one substrate is removed.

24.4.5 In-Plane Switching Cell

The *in-plane switching* (IPS) LC cell was developed by Hitachi in 1996 to increase the viewing angle, improve the color reproduction, and provide stable image quality.[4,5] The IPS mode is most commonly found in high-end TVs and cell phone displays, including the Apple Cinema and Thunderbolt Display.

The IPS cell configuration has both electrodes on the same substrate, as shown in Figure 24.26. Therefore, the electric field is predominantly parallel to the glass plates. The molecules are not anchored at the boundaries. The applied voltage rotates the directors along the applied field direction, which is in the plane of the glass plates. The directors always remain perpendicular to the display normal, so the LC molecules change very little with switching.[6] The IPS cell acts like a rotating half-waveplate. At low voltage, the directors are aligned with one of the polarizers to give a dark state. The applied voltage twists the directors about the z-axes. At the highest voltage shown, the director is 45° from the polarizers, providing a bright white state. Unlike the TN cell, when a

24.4 Configurations of Liquid Crystal Cells

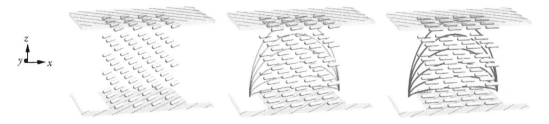

Figure 24.26 (Left) The in-plane switching (IPS) LC cell has two electrodes, shown in yellow, both on one of the glass plates. The applied voltage increases from the left figure to the right. The off state of the IPS cell is shown on the left, where, with crossed polarizers, indicated by gray lines, no light transmits through the cell. As the applied voltage increases, the directors rotate in the x–y plane. As the directors twist, the brightness increases. (Right) At the highest voltage, the directors are rotated 45° to the polarizers, the cell thickness chosen for a half wave of retardance, and the highest brightness is observed.

pixel fails, IPS gives a dark dead pixel. The dark and white states could be reversed by orienting the polarizers parallel to the directors in the *off* state.

Pressure applied through touching the screen causes little change to the transmission. This provides an easy way to identify an IPS display, shown in Figure 24.27, and has contributed to IPS becoming the preferred technology for touch screens.

The IPA technology rapidly became a mainstream technology. Because of the director arrangement, the IPA mode is inherently robust at large viewing angle. The initial production of Super-IPS (S-IPS) mode in 1998 provided a minimal color shift and has a truly wide viewing angle 178° in all directions. The IPS color reproduction and accuracy have always surpassed other LC modes. The S-IPS shows no image sticking even when touching a moving image.

One of the major drawbacks in the original IPS was slow response times. It was originally so slow, at around 60 ms, that it was unsuitable for viewing motion pictures. In 2005, LG. Phillips (now LG. Display) adapted the overdrive circuitry technology and produced the enhanced-IPS (E-IPS), often called the advanced S-IPS (AS-IPS), which improved response times to 5 ms.

The original low contrast ratio of IPA has also improved significantly in the modern IPS generation by using innovative electrode geometries, such as herringbone patterned electrodes and novel pixel structures. The obscuration from the co-planar electrodes initially limited the effective transmittance and has been reduced through development. Compared to the non-planar structures of MVA, the IPS structure is relatively simple, but the IPS electrode is more complex.

The high color saturation, sufficient black levels, touch insensitivity, and the large viewing angle allowed IPS to initially penetrate the high-end markets, such as medical imaging. IPS then rapidly expanded market share in consumer applications, such as tablets, smartphones, and Apple's Retina displays. IPS provides good color performance at high pixel densities and works well for touch

Figure 24.27 (Left) The transmittance through an IPS cell is not affected by applying pressure to an IPS, (right) whereas with a VAN cell, pressure is easily visible.

screens. Compared to TN displays, IPS requires more power and is more expensive to manufacture, so TN has maintained some market share.

Other common defects of the IPS display include persistence image (a faint remnant of the old image stays after a new image replaces it) after the display has been left on for a long period of time. The IPS monitor also shows slight color and brightness shifts from edge to edge even with the high pixel resolution.

24.4.6 Liquid Crystal on Silicon Cells

Liquid crystal on silicon (LCoS) cells are pixelated liquid crystal cells fabricated over a mirrored surface on a silicon integrated circuit. LCoS are reflective displays; all the other LC cell configurations described here are transmissive. As shown in Figure 24.28, the chip provides a compact electronic means of addressing the LCoS panel, delivering the pixilated electric field to the liquid crystal. The optical components are a transparent conductive electrode, such as ITO, a thin layer of liquid crystal about 5 μm thick, and a reflective surface over the silicon structure.

Figure 24.28 shows an LCoS panel schematic with a cover glass with an ITO layer, a liquid crystal layer, and a mirror over Si integrated circuit. LCoS are used for many compact systems, such as head up displays where weight and volume are at a premium.

Figure 24.29 shows the configuration for a typical projector. Light from the lamp is shaped by illumination optics and the color is modulated by a spinning color wheel. Since the LCoS is a reflective device, it is generally illuminated on-axis using a polarizing beam splitter (PBS). The

Figure 24.28 Schematic of an LCoS cell. Light enters from the top, transmitting through the LC layer, reflecting from a mirror coated directly on a circuit addressing the electrodes. The light double passes the LC layer and exits out the top.

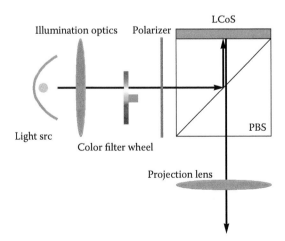

Figure 24.29 For a typical LCoS projector, polarized light reflecting from a PBS is incident on the LCoS panel. As the retardance of the pixels is modulated, the polarization state from each pixel changes. The light coupled into the orthogonal polarization transmits through the PBS and is projected onto a screen. Color images can be generated by rotating a color wheel with three color filters in the incident beam, and creating red, green, and blue frames at three times the overall frame rate.

24.4 Configurations of Liquid Crystal Cells

illuminating light passes through a pre-polarizer and is reflected from the PBS. The LCoS is illuminated with linear polarized light, which transmits through the ITO and liquid crystal layer, reflects, and then makes a second pass through the liquid crystal and ITO before exiting the device. The change in polarization state introduced by the LCoS panel is transformed into intensity variations by an analyzer following the LCoS. In the *dark state*, the light reflects without polarization change and is reflected back into the illumination system by the PBS. In the on state, the LCoS rotates the polarization 90°, and this reflected light transmits through the PBS and onto the viewing screen. The PBS is typically a MacNeille-type cube beam splitter or a wire grid beam polarizing splitter. Other projector configurations may utilize one, two, or three panels for different wavelengths in a variety of configurations.

24.4.7 Blue Phase LC Cells

Another configuration for LCD uses the *blue phase* (BP) mode, a highly twisted cholesteric LC configuration with a regular cubic structure. The BP can switch in approximately 0.1 ms. BP has been developed for 3D TV, which, due to the need to project alternating images to the left and right eyes, must project twice the number of images and operate at twice the speed. The blue phase is an example of the very complex director configurations that can now be fabricated.

In BP, the directors of a cholesteric LC are arranged in a double twisted structure, the *double twist cylinder*, where the molecules twist in two dimensions simultaneously. A view of the orientation distribution of the directors is shown in Figure 24.30a.[7] At the rim of a slice of the cylinder, all directors lie tangent to the cylinder and 45° from the circumference, as shown in Figure 24.30b. Along each radial axis on that slice, the directors have a 90° twist, as shown in Figure 24.30c. The directors rotate from one end of the slice, at 45° relative to the circumference, to the center axis of the cylinder, where all directors are aligned vertically, and then they

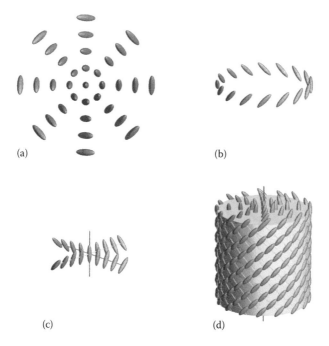

Figure 24.30 (a) Top view of a blue phase mode double twist cylinder showing the orientation variation of the directors on one of the layers. (b) 3D view of the director orientation around the circumference. (c) The orientations of the directors are shown in two orthogonal axes on a slice of the cylinder. (d) All directors are rotating about the central axis of the double twist cylinder.

Figure 24.31 (Left) Basic double twist cylinders stacking on each other, where disclination occurs at where they are in contact. (Middle) BP II has a simple cubic structure. (Right) BP I has a face-centered cubic structure.

continue to rotate in the same direction and reach the other side of the circumference with a total rotation of 90°. The structure of the slice continues down the cylinder as shown in Figure 24.30d. The cylinder is stable up till the 45° twist, which has a distance in the order of 100 nm from the central axis.

Different configurations of double twist cylinders are constructed by stacking the cylinders orthogonally with interlaced disclinations. These defects appear at regular distances throughout the 3D structure. Two types of the BP are shown in Figure 24.31. The periodic cubic structure induces Bragg reflection since the BP structure contains periodic disclination defects on the order of the visible wavelength. Light of a particular color diffracts with the color controlled by an applied electric field. The applied voltage induces birefringence in the liquid crystal through the Kerr effect.

The BP mode has been used in IPS structures, which is modeled with the Kerr effect in a macroscopic scale. Research has focused on lowering the driving voltage of BP LCs, by using materials with high Kerr constants or with optimized structure design.

24.5 Polarization Models

The design and analysis of liquid crystal cells require the calculation of color and polarization. Liquid crystal cells are spatially varying anisotropic materials. Simulation of the LC cells is done to design and tolerance the cells and understand their optical properties. Computer simulation of LC cells involves two stages:

1. Determining the distribution of LC directors given the boundary conditions and electric and magnetic fields
2. Given the director distribution, to determine the optical properties for light transmitted through the cell, such as the Jones matrix as a function of angle of incidence and wavelength

For many of the most common LC cells, such as the twisted nematic and VAN mode cells, the cell has a planar structure and can be divided into layers parallel to the faces with a constant director in each layer, similar to a multi layer thin film. The optical properties of such cells can be analyzed by the generalization of multi layer films to anisotropic layers. The IPS cell is not a plane parallel structure and cannot be analyzed so simply as a multilayer film.

24.5.1 Extended Jones Matrix Model

The *extended Jones matrix method* of cell simulation divides the LC into many thin retarder layers with varying optic axis direction (Figure 24.32) and calculates a Jones matrix for each layer.[8,9] Consider an arbitrary ray propagating straight through a cell, with an x and y defined in its transverse

24.5 Polarization Models

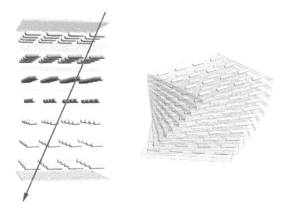

Figure 24.32 (Left) A light ray passes through an LC cell that is modeled as a series of retarders. (Right) Model of an LC cell as a stack of thin birefringent plates with twisting and tilting directors, that is, as a series of discrete layers of retarders.

plane for Jones matrix definition. When the director for a layer is perpendicular to the ray, the birefringence is maximum, $n_E - n_O$. For a thickness Δt and director twisted by ϕ from the x-axis, the layer's Jones matrix is

$$\mathbf{J}(\phi) = \mathbf{R}(\phi) \cdot \mathbf{LR} \cdot \mathbf{R}(-\phi) = \begin{pmatrix} \cos\phi & -\sin\phi \\ \sin\phi & \cos\phi \end{pmatrix} \cdot \begin{pmatrix} e^{-i\Theta_O} & 0 \\ 0 & e^{-i\Theta_E} \end{pmatrix} \cdot \begin{pmatrix} \cos\phi & \sin\phi \\ -\sin\phi & \cos\phi \end{pmatrix}. \quad (24.2)$$

LR is a linear retarder Jones matrix where the ordinary component's phase changes by $\Theta_O = 2\pi n_O \Delta t/\lambda$ across the layer, and the extraordinary phase changes by $\Theta_E = 2\pi n_E \Delta t/\lambda$. If the director is tilted by θ from the ray ($\pi/2 - \theta$ from the transverse plane), the effective refractive index of the extraordinary mode, $n_e(\theta)$, polarized in the plane of the director, is reduced from its maximum value n_E to n_e,

$$\frac{1}{n_e(\theta)^2} = \frac{\cos^2\theta}{n_O^2} + \frac{\sin^2\theta}{n_E^2}. \quad (24.3)$$

The birefringence, $n_O - n_e(\theta)$, encountered by the ray is projected onto the transverse plane of the ray and an equivalent retarder Jones matrix is calculated using the thickness for each layer. Light propagating along a director, $\theta = 0$, experiences no birefringence and no retardance; light propagating perpendicular to a director, $\theta = \pi/2$, experiences the full birefringence, $n_O - n_E$. The overall cell Jones matrix is calculated by multiplying the Jones matrices for the individual layers.

The extended Jones matrix methods can use Jones matrices since the light is propagating in nearly a straight line through the cell.

24.5.2 Single Pass with Polarization Ray Tracing Matrices

The single pass calculation can also be performed in three dimensions using the **P** polarization ray tracing matrices two ways: (1) propagating a single ray through the cell, or (2) propagating two rays, ordinary and extraordinary. Method (1) works best for cell designs with significant twist, rotation about the z-axis, and parallels the extended Jones matrix method, producing an equivalent result. Method (2) is more accurate for cells without twist, such as Fréedericksz cells and VAN-mode cells, where most of the light remains in either the E-mode throughout the cell or the O-mode.

Consider Method (1). As light enters the cell, the **P** for refraction into the glass cell windows is calculated based on the dielectric tensor ε_1 for the first layer. Next, the light refracts into the first LC layer, $q = 1$, where the high and low index modes have slightly different propagation directions, $\mathbf{k}_{1,H}$ and $\mathbf{k}_{1,L}$, and refractive indices, $n_{1,H}$ and $n_{1,L}$. An average **k** is calculated weighted by the fraction of light in each mode. The birefringence in the transverse plane Δn_1 is calculated and a corresponding pure retarder polarization ray tracing matrix $\mathbf{P}_1$ is constructed as well as an average optical path length, OPL_1. The ray continues on a straight path through the LC cell and is divided into Q segments. The dielectric tensors are calculated at the centers of each ray segment ε_q from the director distribution. Then, a pure retarder **P** matrix, $\mathbf{P}_q$, is written for each short ray segment. The **P**'s are multiplied together to describe the polarization change through the cell as a sequence of retarders, yielding a net $\mathbf{P}_{Total}$ for propagation through the LC cell,

$$\mathbf{P}_{Total} = \mathbf{P}_Q \mathbf{P}_{Q-1} \cdots \mathbf{P}_q \cdots \mathbf{P}_2 \mathbf{P}_1 = \prod_{q=1}^{Q} \mathbf{P}_{Q-q+1}, \qquad (24.4)$$

from which the retardance is calculated. The average optical path length is found from the sum

$$OPL_{Total} = \prod_{q=1}^{Q} OPL_q. \qquad (24.5)$$

Sets of OPL_{Total} for ray grids are used to calculate wavefront aberration functions for focusing through cells.

Method (2). As light refracts into the first LC layer, $q = 1$, the refractive indices, $n_{1,H}$ and $n_{1,L}$, and two slightly different propagation directions, $\mathbf{k}_{1,H}$ and $\mathbf{k}_{1,L}$, are used to generate two rays, one for each of the high and low index modes. Two polarization ray tracing matrices are constructed for the first modeled layer, $\mathbf{P}_{1,H}$ and $\mathbf{P}_{1,L}$. These two rays are continued through the cell and two polarization ray tracing matrices are constructed for each additional layer $\mathbf{P}_{q,H}$ and $\mathbf{P}_{q,L}$. Finally, the layer matrices are multiplied together to yield two end-to-end matrices, one for the low index mode $\mathbf{P}_{L,Total}$ and one for the high index mode $\mathbf{P}_{H,Total}$,

$$\mathbf{P}_{L,Total} = \mathbf{P}_{L,Q} \mathbf{P}_{L,Q-1} \cdots \mathbf{P}_{L,q} \cdots \mathbf{P}_{L,2} \mathbf{P}_{L,1} = \prod_{q=1}^{Q} \mathbf{P}_{L,Q-q+1} \qquad (24.6)$$

and

$$\mathbf{P}_{H,Total} = \mathbf{P}_{H,Q} \mathbf{P}_{H,Q-1} \cdots \mathbf{P}_{H,q} \cdots \mathbf{P}_{H,2} \mathbf{P}_{H,1} = \prod_{q=1}^{Q} \mathbf{P}_{H,Q-q+1}. \qquad (24.7)$$

Because these are single mode **P**, each has the properties of polarizers. The differences in phase yield the retardance and the differences in transmission yield the diattenuation for the ray path. This two-mode method is most applicable to untwisted cells, such as the Fréedericksz cell and IPS cell, where, once started, little coupling occurs between the low and high modes.

The extended Jones matrix method ignored the difference between $\mathbf{k}_{1,H}$ and $\mathbf{k}_{1,L}$ and used the average propagation direction for the ray direction within the LC.

24.5.3 Multilayer Interference Models

The previous two methods analyze the LC cell as a series of retarders in a single a pass. Because of the refractive index variations along the path, some light is reflected into the reverse direction continuously along the path. The cell can be considered as a multilayer structure similar to a thin film with interference occurring between different layers, layers that are anisotropic. The first full multilayer interference model was attributed to Berreman in 1972 based on a four-element vector (E_x, E_y, B_x, B_y) containing the electric and magnetic fields.[10,11] A 4 × 4 matrix is developed for each layer, and matrix multiplication yields the relationship between the amplitude, phase, and polarizations of the incident and exiting plane waves. A different multilayer interference model was developed by Mansuripur.[12] For most cells, the single pass methods and multilayer interference methods produce almost the same values, and since LC cells are relatively thin, the differential refractive index is not changing very rapidly, and only multiple low energy reflected light is present.

These models use practical simplifications of the simulation, (1) dividing the continuously varying dielectric tensor into a set of separate layers, and (2) assuming the modes propagate along the same ray path from the LC entrance to exit. When light refracts from glass into the LC, the light refracts at different angles into the high and low index modes. The next layer has a slightly rotated dielectric tensor, so most of the high mode refracts into the high mode, but a tiny bit couples into the low mode. Similarly, a tiny fraction of the low mode refracts into the high mode; these tiny fractions have a third and fourth propagation direction. This continues at each of the layers generating new small beams at additional angles. LC display cells are thin. The result is a large set of rays that exit from almost the same ray intercept with almost the same propagation direction, but a small amount of spreading in location and angle occurs. This spreading is not accounted for in the models presented here.

When setting up an LC model, it is not initially clear how many layers the model should use to divide the cell. As the number of layers is increased, the LC's polarization matrix changes until it asymptotically approaches its final value. One common technique is to start with a small number of layers and keep doubling the number of layers until the calculation converges.

24.5.4 Calculation for Liquid Crystal Cell ZLI-1646

An example of a *twisted nematic liquid crystal cell* simulated with a multilayer interference model is performed using *Polaris-M* (Airy Optics Inc.) by modeling the cell as a stack of interfering biaxial thin films. The cell is formed from the liquid crystal mixture ZLI-1646 at a wavelength at 589 nm. This replicates the configuration from the reference *Optics of Liquid Crystal Displays*, 2nd edition by Pochi Yeh and Claire Gu. The directors are anchored to the top and bottom surfaces of the cell. The orientation of the directors through the LC cell is defined by the twist angle ϕ, rotation in the plane parallel to the glass plates, and the tilt angle θ (rotation on the plane orthogonal to the glass plates). The total twist angle is defined by the angles of the grooves on the two alignment layers, Φ = 90°. The maximum tilt angle θ_{max} is a function of the applied voltage and occurs halfway through the cell. The cell gap has a thickness of d, and the position in the cell is z/d. The cell is assumed isotropic in x and y; it is a planar structure. Figure 24.33 shows the director orientation for three applied voltages.

By increasing the applied voltage from 0 V (right), the LC cell switches from behaving as a stack of thin *A-plates* with a 90° twist to a stack of *C-plates* (left). The polarization properties of this LC cell are calculated over an angle of incidence range of 0.08 NA to study the LC performance variation with incident angle. Figure 24.34 maps the magnitude and orientation of the angle of incidence across the beam. Figure 24.35 maps the variation of the retardance versus angle of incidence with increasing voltage. Ellipses indicate the extraordinary eigenpolarization (slow state) and the size of the ellipse indicates the magnitude of the retardance. Ideally, in each figure, all the ellipses would be identical, so the behavior of the cell would be independent of angle. As the voltage increases from 0 V,

860 Chapter 24: Liquid Crystal Cells

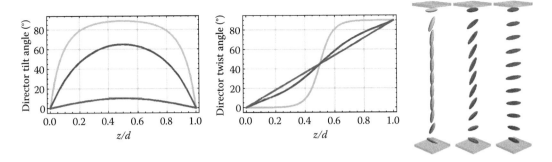

Figure 24.33 (Left) The tilt angle and (middle) the twist angle of a 90° TN cell for three different applied voltages are shown for 5.03 V (turquoise), 2.2 V (pink), and 1.3 V (purple). (Right) The corresponding director orientations through the LC cell.

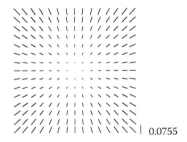

Figure 24.34 AOI map for 0.08 NA with a key for the scale at the lower right.

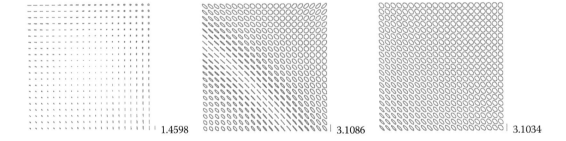

Figure 24.35 Retardance maps versus field of view for twisted nematic LC for 5.03 V, 2.2 V, and 1.3 V. The label at the bottom right of each map shows the maximum retardance in radians over the field.

the on-axis retardance changes from about one-half wave (3.14 radians) to almost zero, while the ellipticity of the fast state increases toward circular polarization. At 1.3 V, the fast axis rotates 180° around normal incidence with some ellipticity present.

A small amount of diattenuation is also observed in the LC cell model in Figure 24.36. One source of diattenuation occurs refracting into and out of the glass windows due to the Fresnel equations. Since the polarization state evolves from the front, through the cell, to the back window, these two outer surface diattenuation contributions are no longer aligned and may substantially add or cancel depending on the LC setting. Interference between LC layers also generates additional diattenuation. As is seen in Figure 24.36, LC cell diattenuation is generally minor, less than 0.02 in this case.

24.6 Issues in the Construction of LC Cells   861

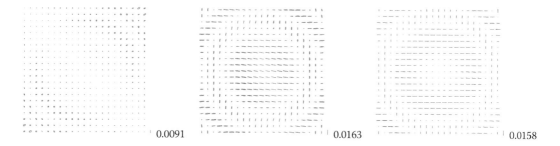

0.0091   0.0163   0.0158

Figure 24.36 Diattenuation map for 5.03 V, 2.2 V, and 1.3 V. The label at the bottom right of each map shows the maximum diattenuation over the field.

24.6 Issues in the Construction of LC Cells

This section examines the construction of LC cells. Following their invention, LC displays took a long time to come to market due to their complexity and difficulty of manufacture. Among the earliest commercial LC applications were digital watches and calculator displays, applications where speed and accurate color were not requirements. As the technology steadily solved the many technical challenges, LCs moved into more and more demanding applications. Laptops and conference room projectors, for example, needed more pixels than calculators, but these applications still did not require great speed or fine color control. By the early 2000s, LC displays were finally competitive as high-end TVs.

24.6.1 Spacers

Fabricating and maintaining a constant thickness of the thin liquid crystal layer is challenging. To maintain proper distance between the glass plates, *spacers*, such as precision glass beads or plastic spheres, are placed between the glass plates to fix the separation, as seen in Figure 24.37. When a voltage is applied across the LC cell, the two electrodes and plates are electrostatically attracted toward each other, squeezing the cell, attempting to reduce its volume. The spacers hold the plates apart, but between the spacers, the glass plates are pulled toward each other and the cell thickness varies. This variation of cell thickness can be seen in the retardance map of the untwisted nematic cell in Figure 24.38. More recent LC cell designs use pillars or pyramids micro-lithographically fabricated onto one of the glass faces as spacers.

24.6.2 Disclinations

Disclinations are locations in the LC where the director abruptly changes its orientation; they are defects in the positional order of the directors. Consider the schematic in Figure 24.39 (right), which shows a disclination in the Fréedericksz cell. In Figure 24.12 (left) at 0 V, the directors lie parallel

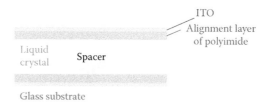

Figure 24.37 Typically, the glass substrate is 0.7–1.1 mm thick, the alignment layers are 10–20 nm thick, the ITO electrodes are about 100 nm thick, and the spacer is about 1–7 µm.

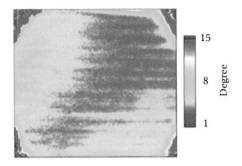

Figure 24.38 (Left) The retardance of a single-pixel polarization controller, a Fréedericksz cell, shows non-uniformities in horizontal stripes due to cell thickness variations. These periodic variations of about one degree of retardance result from the periodic spacing between rows of spacers. In between rows of spacers, the glass plate bows due to the attractive force between the electrodes. The resulting thickness variations cause these retardance variations.

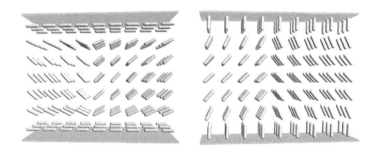

Figure 24.39 The disclination between two volumes of (left) a parallel aligned cell and (right) a VAN LC cell occurs where the director abruptly changes direction. In both figures, the left four columns of directors have rotated clockwise and the right five columns have rotated counterclockwise. Thus, the disclinations occur between columns four and five. Such discontinuous changes of director cause visible lines within the corresponding pixels when used with pairs of polarizers.

to the glass surfaces. When a voltage is applied, the molecules have an equal probability of rotating clockwise or counterclockwise. Ideally, the whole pixel's directors should rotate together either clockwise or counterclockwise. If one region of a pixel rotates clockwise, and another rotates counterclockwise as seen in Figure 24.39 (left), a disclination will form at the boundary, and the pixel will appear non-uniform from different angles. Figure 24.39 (right) shows a disclination in a VAN mode cell. In nematic LCs, disclinations usually appear as a line defect, but a variety of topological forms can occur. Disclinations are undesirable, causing visible defects and reducing specifications. Chiral dopants, which are helical molecules, can be added to ensure a uniform rotation direction, either clockwise or counterclockwise.

24.6.3 Pretilt

To avoid disclinations, most LC cells introduce a *pretilt* in the alignment surfaces as shown in Figure 24.40. Small alignment grooves are placed in a thin layer of soft material, typically polyimide (PI layer).[13] Polyimide can be buffed unidirectionally with velvet-like cloth to create small alignment grooves. The LC molecules at the boundary layer fit into the grooves. When the surface layer of molecules has a small pretilt of 2°–4°, the LC molecules have already started their rotation at 0 V; this breaks the balance between rotating clockwise and counterclockwise when a voltage is

24.6 Issues in the Construction of LC Cells

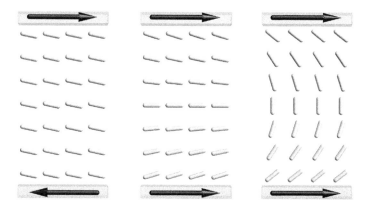

Figure 24.40 Pretilt examples set the tilt of the boundary layer of the LC axes. The pretilt is created in the alignment layers by rubbing. The rubbing directions for different cell designs are indicated by the arrows. (Left) Parallel alignment, (middle) splay cell, and (right) bend cell.

applied. Now, when a small field is applied, the directors continue to rotate in the direction of the pretilt, avoiding the disclination, and keeping the pixel from breaking up into different domains. Buffed polyimide has been steadily replaced by UV photo aligned layers with grooves created interferometrically.[14]

24.6.4 Oscillating Square Wave Voltage

The LC material filling a cell always contains some free ions, such as sodium and calcium ions. When a DC voltage is applied across an LC cell for a long time, these ions drift through the LC and accumulate near one of the electrodes. Groups of positive ions form at one side of the cell while the negative ions collect on the other side of the cell. Their concentration buildup degrades the liquid crystal performance and frequently leads to cell failure.

To avoid this ion buildup problem, LC cells are driven with square wave voltages, as shown in Figure 24.41. The exact frequency is not of importance since the LC works for either DC or square wave voltage; the oscillation is what is necessary to prevent ion buildup. Typical square wave frequencies lie between 500 and 5000 Hz. The LC directors cannot move far in the brief time between positive and negative voltages, so the directors remain nearly fixed.

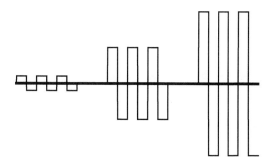

Figure 24.41 The driving voltages to a liquid crystal cell should be square waves centered on 0 V, not DC voltages.

24.7 Limitations on LC Cell Performance

The ideal LCD has many target specifications. It should be very bright for high settings and have a very *dark state* for low settings. When all the pixels are set to the same level, the color and brightness should be constant across the display. The color should also be constant over viewing angle. The colors should be saturated. The display should switch fast, and modulation at 120 Hz would be ideal. Performance should be consistent over a large range of temperatures. For smartphones, the display should be insensitive to touch and to forces on the front surface. For a mass market, the cell should be inexpensive to manufacture in high volume with excellent yield.

Real liquid crystal cells used in displays must find a balance between these many often conflicting objectives. For LC televisions, two of the driving specifications are the quality of the *dark state*, and the contrast ratio, the ratio between the bright and *dark state* levels. Some of the limitations on the *dark state* and the contrast ratio arise from the following:

- Dispersion of retardance and spectral bandwidth
- Variation of retardance with angle and angular bandwidth
- Scattering and depolarization
- Misaligned polarizers
- Variation of the retardance axis

The retardance magnitude uniformity is limited by the spatial uniformity of the applied electric field and temperature. The retardance orientation uniformity is affected by the alignment layer. Many LC retarders are slightly elliptical retarders, weak diattenuators, and weak depolarizers. Figure 24.42 shows the polarization properties of a single-pixel Fréedericksz cell, an electrically adjustable LC retarder, where less uniformity is observed at low voltage.

The issues and problems of this small LC component include the following: a slow response time of around 50–80 ms, a fairly high temperature variation about 0.5% per °C, and depolarization of around 1%–10%, which produces scattering. It is very challenging to maintain a uniform thickness and temperature of the LC cell. The non-uniformity shown in Figure 24.13 has a magnitude retardance variation of 14°.

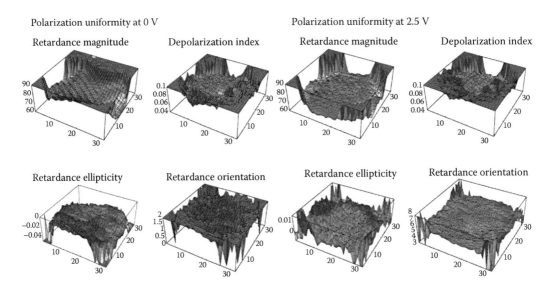

Figure 24.42 Retardance magnitude, retarder ellipticity, depolarization index, and retardance orientation of an LC retarder with 0 V (left) and 2.5 V (right) applied voltage.

24.7 Limitations on LC Cell Performance

24.7.1 LC Cell Speed

The LC cell speed is characterized by the response time measured in milliseconds that a pixel takes to transform from one value to another (typically from 10% to 90% brightness) and then back to the original value, as shown in Figure 24.43. It includes the rise and fall times as the pixel changes from state to state. Since liquid crystal cells modulate retardance by rotating molecules, this is an intrinsically slow modulation mechanism, particularly compared to electro-optical modulators, such as $LiNbO_3$ modulators used in fiber optics, which only need to move electrons within molecules.

When an object is moving across the screen, if the trailing pixels behind the object have a slow response time and do not shift their color quickly, a shadow-trail artifact is observed following the moving object, as shown in Figure 24.44. A slow response time is not acceptable for motion pictures, sports, and video games. The typical response times for current LCD are 8–16 ms for a transition from black-to-white-to-black, and 2–6 ms for a transition from gray-to-gray. Human eyes can perceive differences in response times greater than 5 ms. Because of limitation of the eye (for a 60-Hz frame rate), response times below 10 ms are hard to perceive.

Several technologies can reduce the response time, motion blurring, and trailing. In the response time compensation (RTC) and the over driving circuit (ODC) technologies, an over-voltage is applied to force the LC molecules to rotate into position faster as shown in Figure 24.45, producing artifacts like Figure 24.46. Overdrive significantly reduces pixel transition time. Double overdrive technology improves the transition times by applying overdrive to both the rise time and the fall time. However, if overdrive is applied aggressively or is controlled poorly, an overdrive color trailing (a pale or dark halo) appears behind the moving object due to the intervening state (Figure 24.47). Without overdrive, typical

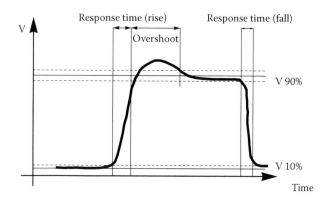

Figure 24.43 The response time of a liquid crystal can be defined as the time the cell takes to change from 10% to 90% brightness. To increase the response time, the voltage can be increased briefly above the target voltage to accelerate the change. This may result in an overshoot in the target retardance.

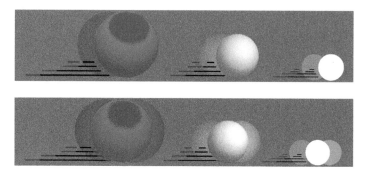

Figure 24.44 Example of motion blur due to slow response time.

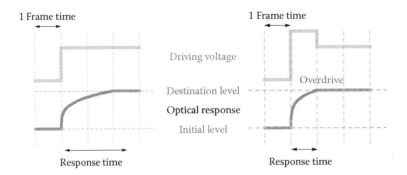

Figure 24.45 (Left) Drive voltage and response times without overdrive and (right) with overdrive.

Figure 24.46 (top) A frame from a video of a moving car, and (bottom) the corresponding view from a display with a poor response time blurs the moving objects.

Figure 24.47 An example of the artifacts due to a large amount of RTC overshoot. In this example, the RTC overshoot causes a pale halo behind the red moving object, and a dark halo behind the white and yellow moving objects.

Polarized Light and Optical Systems

24.7 Limitations on LC Cell Performance

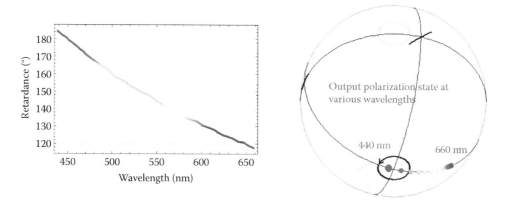

Figure 24.48 (Left) Dispersion measurement of an example Fréedericksz cell. (Right) The output polarization state expressed as a Stokes spectrum is shown as a function of wavelengths on a Poincaré sphere.

black-white-black response times are 5 ms for the TN mode, 12 ms for the VA mode, and 16 ms for the IPS mode, where the gray-to-gray transition times are higher. With overdrive, the typical times are 2 ms for the TN mode, 6 ms for the VA, and 5 ms for the IPS.

24.7.2 Spectral Variation of Exiting Polarization State

The refractive index of a liquid crystal is a function of wavelength and polarization state. Figure 24.48 shows the dispersion of a common liquid crystal cell. Each material has a unique dispersion characteristic. Consider an example modulator constructed from a variable linear retarder oriented at 45° to the polarization state generator. On the Poincaré sphere, the polarization state is stretched out along the trajectory and so a polarizer cannot extinguish the whole spectrum at once.

For maximum polarization state variation, the incident polarization state is 45° from the generator. In this situation, each degree of retardance variation moves the polarization state 1° around a great circle about the retardance axis at 45°. The polarization of the exiting wavelengths will be spread out along the great circle. Thus, each wavelength has a different extinction, the net extinction being an integral over the weighted spectrum. Therefore, for contrast ratio of 100, the dispersion of the retardance cannot be much greater than about twice 5°. Similarly, for a contrast ratio of 10,000, the retardance dispersion cannot be much larger than about 1°, a very tough specification.

For an LC cell, the dispersion is likely to be less at the end of the operating range with the smaller retardance. For a Fréedericksz cell, the dispersion is greater at the low voltage end of its operating range. In contrast, a VAN mode cell is intrinsically near zero retardance at 0 V, so the low voltage setting will be preferable for lowest retardance dispersion.

24.7.3 Variation of Retardance with Angle of Incidence

Most liquid crystal devices have substantial angular polarization aberrations. This angular response can be measured by focusing light through the LC cell in an imaging Mueller matrix polarimeter. Consider a monochromatic spherical wave illuminating an LC cell. The LC layer in an LC display is illuminated through a polarizer. If the polarizer's absorption axis is along y, then the x- and z-components are transmitted, so the incident light has a linear polarization state like latitude vectors in a region about the equator, since latitude vectors have no component along the y-axis. The LC layer behaves as a series of thin birefringent layers, and each angle sees different projections of the director. The retardance varies as a function of angle of incidence, so the exiting polarization state will also show variation with angle. These LC retardance variations can be as large as several

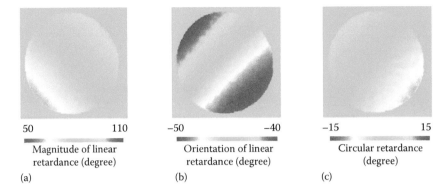

Figure 24.49 Variation of retardance of a Fréedericksz cell as a function of angle of incidence over a 30° cone, (a) linear retardance magnitude, (b) linear retardance orientation, and (c) circular retardance magnitude. The cell is illuminated with an NA = 0.55 numerical aperture microscope objective and the transmitted light is collected with another objective. The whole data set is measured simultaneously as a Mueller matrix image.

degrees of retardance per degree of angle of incidence. This resulting spatially varying polarization state is incident on the analyzing polarizer leading to transmission variations with angle. Because of the LC's variation of retardance with wavelength, the transmission of colors also varies with wavelength. The underlying variation of retardance with angle and wavelength becomes visible as a variation of color with angle, which is very noticeable to the eye.

The variation of retardance with angle is also observed as a variation of contrast with angle. Hence, contrast is commonly measured as a function of angle of incidence after integrating over red, green, and blue spectral bands. Also, if the transmission axes of the front and back polarizers are not parallel (they are most commonly crossed), the relative angle of the polarizers will vary as the incident beams move diagonally in angle with respect to the polarizers, as shown in Figure 1.16.

For example, the retardance variation measured for a Fréedericksz LC shown in Figure 24.49 is rather large, a 60° linear retardance across the 30° field of view. Each of the three retardance parameters shown has a linear variation with angle. The linear retardance magnitude varies most rapidly in the 45° plane and the retardance orientation rotates most rapidly in the 135° plane. At normal incidence and in the 45° plane, the retardance is linear (the circular retardance is near zero) and the orientation varies in two different directions.

Figure 24.50 shows the linear retardance magnitude measured for two LC single-pixel untwisted cells at three different applied voltages. The mean values and the peak-to-peak linear retardance of the measurement are shown in Table 24.2, where the first LC is a little less uniform than the second one.

24.7.4 Compensating LC Cells' Polarization Aberrations with Biaxial Films

To address the variation of retardance with angle, the technology of discotic *field-widening films* was developed. Discotic LC molecules are disk-like or pancake-like negative uniaxial molecules. They are manufactured into multilayer films and used as compensating films to compensate for angle of incidence and wavelength variations of LC cells. The structure of typical discotic molecules is shown in Figure 24.51. In multilayer films, they can be engineered to twist and tilt from layer to layer, as in the example of Figure 24.52, representing a twisted discotic film.

When used as a compensating film, the discotic layers are oriented to cancel the retardance aberration of the LCs. A discotic LC layer paired with a nematic LC layer with opposite uniaxial properties can cancel their polarization, yielding a polarization matrix which is an identity matrix at all angles. Typically, the eigenpolarization of the discotic compensating film has oriented orthogonal to the eigenpolarization of the corresponding LC layer. Figure 24.53 (left) shows a layer of the

24.7 Limitations on LC Cell Performance

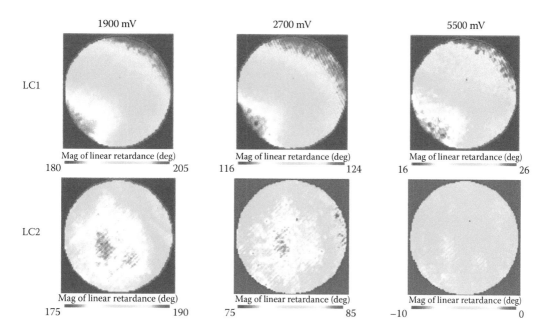

Figure 24.50 Linear retardance magnitude plot of two LCs.

Table 24.2 The Mean Linear Retardance Magnitude and Peak-to-Peak Values of the Measurement Shown in Figure 24.50

	1900 mV	2700 mV	5500 mV
LC1	Mean value = 193.6°	Mean value = 116.4°	Mean value = 21.7°
	22.0° peak-to-peak	17.2° peak-to-peak	9.7° peak-to-peak
LC2	Mean value = 180.8°	Mean value = 79.4°	Mean value = −4.7°
	7.8° peak-to-peak	9.1° peak-to-peak	5.0° peak-to-peak

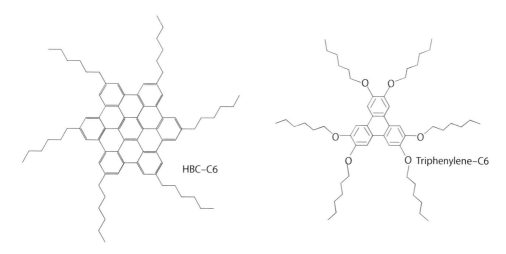

Figure 24.51 Typical discotic molecules as a derivative of a hexabenzocoronene and 2,3,6,7,10,11-hexakishexyloxytriphenylene.

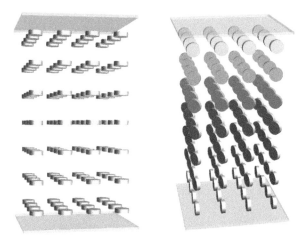

Figure 24.52 (Left) Schematic of a uniform film of discotic molecules, represented as cylinders. (Right) Schematic of multilayer wide-field film with discotic molecules rotating through 90° from the top to the bottom of the film.

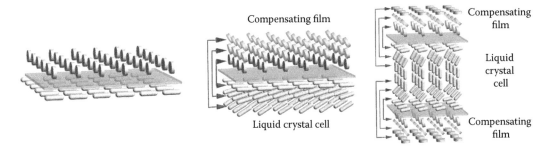

Figure 24.53 (Left) Compensating film layer of discotic molecules is shown compensating the adjacent nematic LC layer. (Middle) Each layer of the compensating film compensates on the layer of the LC, reducing the combination to an identity matrix. (Right) Compensating films can be applied to both sides of the LC cell, with each film correcting the aberration of half the cell.

compensating discotic film (above blue plane) canceling the retardance aberration for a layer of the nematic LC. By pairing the layers as shown by the red arrows (Figure 24.53, middle), near zero retardance is obtained for each pair of layers, yielding near zero retardance for the compensating discotic film plus LC cell pair. Thus, the polarization aberrations can be compensated throughout the entire thickness of an LC cell.

24.7.5 Polarizer Leakage

The contrast ratio is also limited by the non-ideal contrast of polarizers. An LC device using polarizers with a contrast ratio of 100:1 cannot exceed a contrast ratio of 100:1. Dichroic polarizers are absorptive, so the extinction depends on the thickness. Polarizing films constructed have several relevant defects. The polarizer material may have wrinkles and pinholes that leak unpolarized light. The polarizer may have areas of thinner dichroic material and reduced contrast ratio. The orientation of the front and rear polarizing films is never parallel. The alignment between the generator and analyzer polarizers will have small spatial variations contributing to a loss of extinction ratio. Figure 24.54 shows the Mueller matrix image of a low-quality polarizing film. The red and blue streaks in the M_{03}, M_{13}, M_{30}, and M_{31} images indicate variation of the transmission axis.

24.7 Limitations on LC Cell Performance

Figure 24.54 Mueller matrix image of poor-quality polarizing film, showing spatial variations of the polarizer's orientation in vertical streaks.

24.7.6 Depolarization

Depolarization is the reduction of the *degree of polarization* when polarized light interacts with a sample. Depolarization is particularly critical to minimize in LC cells because half of any depolarized light will leak through the analyzer and thus has a particular impact on the *dark state* and the contrast ratio specification.

Depolarization occurs in all materials, including liquid crystals, due to *scattering*. Liquid crystals are moderate sized molecules that naturally have some scattering. This scattering is minimized by optimizing the mixtures of LCs and solvents used. Liquid crystals have an additional rather unique contribution to depolarization. Since cells are driven with square wave voltages (Section 24.6.4), time-varying fluctuations in the electric field's magnitude allow the molecules to vibrate. If the directors vibrate, the retardance vibrates, allowing a time-dependent retardance. A time-averaged retardance and a time-averaged *Mueller matrix* manifest themselves as depolarization. Figure 24.55 shows measured depolarization data as a *degree of polarization map* for a Fréedericksz cell. Such levels of depolarization in the 1% to 3% range are unacceptable in cell phone and LC television displays. For more on depolarization, see Chapter 6.

Depolarization adversely affects LC display performance differently from incorrect retardance or retardance non-uniformity. When some depolarization is present, a fraction of the exiting light, the

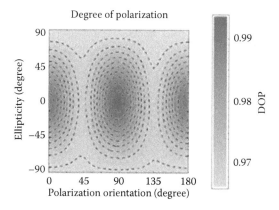

Figure 24.55 A degree of polarization map for a Fréedericksz cell on a flattened Poincaré sphere shows the exiting degree of polarization varies from 0.97 (3% depolarization) for 45°, right, 135°, and left circularly polarized light to 0.99 (1% depolarization) for 0° and 90° light.

depolarized component, can be treated as *unpolarized light*. Fifty percent of the depolarized light will pass through the analyzer and 50% will be blocked. In the *dark state*, the leaked depolarized light increases the *dark state* intensity and, if significant, has a severe effect on the contrast ratio. Thus, for *high contrast* to be achieved, the LC displays must have very low levels of depolarization. In the white state, half of the depolarized light is blocked by the analyzer decreasing the white state brightness, a less critical problem than *dark state* leakage. Scattering is a common cause of depolarization in liquid crystals. Liquid crystal depolarization also arises from spatial averaging; micron-scale retardance variations cause adjacent parts of the beam to emerge with different polarization states that average at the polarimeter, resulting in a depolarized component in the measurement. An imaging polarimeter measures the average retardance within each of the polarimeters' pixels and any subpixel retardance variations are measured as depolarization. Temperature variations, electric field variations, edge effects in pixels, and *disclinations* in the LC all contribute to depolarization.

24.8 Testing Liquid Crystal Cells

Mueller matrix polarimeter testing has become widespread in the LC industry. Polarimeters are used to test the glass for LC cells for the presence of any small birefringence; small birefringences can significantly alter the cell performance, appearing as defects in color and uniformity. Polarimeters and ellipsometers verify that the LC cell's retardance and orientation are within specifications, so that when the polarizers are attached, the intensity and color modulate properly.[15] Mueller matrix polarimeters are used when setting up production, to verify that the production processes stay within specification and to assist in failure analysis when problems do occur.

The cell gap, twist angle, pretilt, and rubbing direction for liquid crystal cells can be measured by measuring retardance as a function of angle of incidence and then fitting these parameters using optimization algorithms. Mueller matrix polarimeters are used to measure the retardance magnitude and retardance eigenstates. The diattenuation and depolarization are also determined and can be helpful in understanding cell performance.

Figure 24.56 shows commercial Mueller matrix polarimeters used for liquid crystal cell testing (left, center) of angle of incidence variations and (right) of spatial uniformity. On the production

  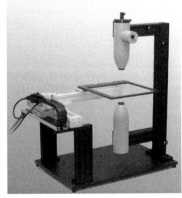

Figure 24.56 (Left) A Mueller matrix spectropolarimeter for testing LC cells requires a capability to adjust angle of incidence. The top blue component is the polarization state generator. A fiber enters the top, bringing light from a monochromator. The bottom blue component is the polarization state analyzer. The sample is placed on the black tray, which tilts to vary angle of incidence. (Center) A closeup of the pair of rotation stages for angle of incidence testing. The silver rotary stage on the right adjusts the angle of incidence. The black round rotary stage on the left side varies the azimuth of the angle of incidence, allowing tests to be performed in different planes. (Right) A Mueller matrix spectropolarimeter configured for spatial uniformity mapping has *xy*- translation stages (silver) to move the sample holder, the black rectangle). (Courtesy of Axometrics Inc., Huntsville, Alabama.)

24.8 Testing Liquid Crystal Cells

Figure 24.57 (Left) A Mueller matrix polarimeter for production testing of large-screen LCD TVs and for the stress birefringence testing of the associated glass. The head can scan over 1.5 m while the part under test is carried through the fixture from front to back. (Right) Closeup of a three-head polarimeter, for high-speed testing of large-screen LCD TVs. Testing is necessary on- and off-axis, so using three heads is far faster than scanning a single head in angle of incidence. Rotary stages above and below the polarimeter heads allow the fixtures to rotate about the vertical axis to vary the azimuth. (Courtesy of Axometrics Inc., Huntsville, Alabama.)

lines for LC displays and LC TVs, entire LC TV panels, large pieces of films, and large glass plates must be tested. Figure 24.57 shows a Mueller matrix polarimeter for integration into a production line with a capability of scanning over 1.5 m while simultaneously measuring Mueller matrices at three angles of incidence.

24.8.1 Twisted Nematic Cell Example

An example of LC cell testing is shown for "cell A" in Figure 24.58. The *Mueller matrix spectrum* of the cell at normal incidence has been measured (not shown) and the retardance properties have been calculated. The retardance eigenstates are plotted on the *Poincarè sphere*; for best cell performance, ideally the eigenstates would all be located at left circularly polarized light, but cells typically have some spectral variation. To measure the cell gap and director orientations through

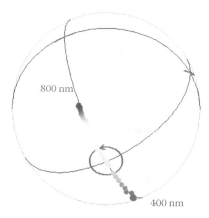

Figure 24.58 Retardance eigenstates of cell A, a twisted nematic (TN) cell as a function of wavelength, are designed to pass through left circularly polarized light unchanged. Green circularly polarized light is unchanged, but other wavelengths vary linearly with wavelength. (Courtesy of Axometrics Inc., Huntsville, Alabama.)

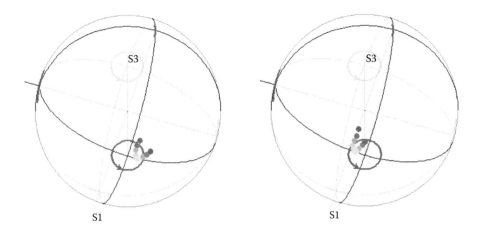

Figure 24.59 Tilt test where color indicates the variation of angle of incidence as the cell rotates in a fixture in the polarimeter (see Figure 24.56 left and center). The retardance eigenstates for a TN cell (left) at 500 nm and (right) at 550 nm shows quadratic and very small variation of their retardance eigenstates. (Courtesy of Axometrics Inc., Huntsville, Alabama.)

the cell, the Mueller matrix is measured as the cell is tilted through normal incidence in multiple planes. The retardance eigenstates of "cell B" shown in Figure 24.59 have a quadratic variation with superior spectral performance. These functions can be fitted by optimizing the cell gap and twist angles in an optimization program to measure the cell's gap and twist angle.

24.8.2 IPS Tests

An example of a Mueller matrix spectropolarimetric test of an IPS cell at normal incidence is shown in Figure 24.60. The configuration of the cell is shown in Figure 24.26 (left). All the LC directors are parallel, so the retardance spectrum as a function of wavelength is constant at low voltage. Figure 24.61 shows tilt scans of the retardance for a cell without pretilt, compared to Figure 24.62 for a cell with pretilt, indicating how such tests are used to set up and maintain production machinery.

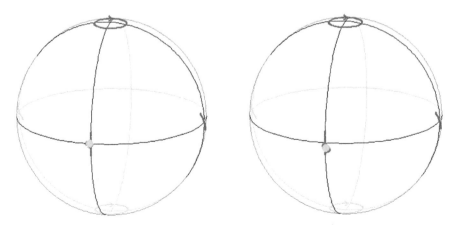

Figure 24.60 (Left) Retardance eigenstates of a well-aligned IPS cell at normal incidence are constant with wavelength. (Right) Introducing a small tilt off normal incidence introduces a slight ellipticity at all wavelengths. (Courtesy of Axometrics Inc., Huntsville, Alabama.)

24.8 Testing Liquid Crystal Cells

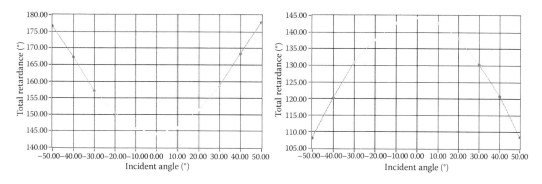

Figure 24.61 Retardance measurements of tilt scans of an IPS cell with photo aligned alignment layers (left) along the alignment direction and (right) perpendicular to the rubbing direction are centered at 0°; this symmetry indicates no pretilt in the polyimide alignment layer. (Courtesy of Axometrics Inc., Huntsville, Alabama.)

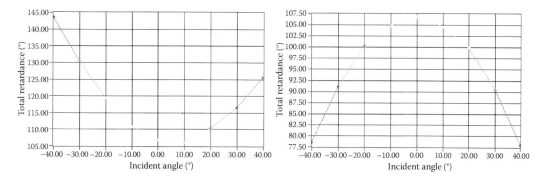

Figure 24.62 Retardance measured from tilt scans of an IPS cell with rubbed alignment layers (left) along the rubbing direction and (right) perpendicular to the rubbing direction is not centered at 0°; the asymmetry indicates a pretilt in the polyimide alignment layer in this case of about 2°. (Courtesy of Axometrics Inc., Huntsville, Alabama.)

24.8.3 VAN Cell

Another example of pretilt determination using polarimetry is shown in Figure 24.63 for a vertically aligned cell, similar to Figure 24.20 (left). The cell's retardance is measured as a function of angle of incidence. The cell of Figure 24.61 has a pretilt of 90°; thus, the retardance is symmetric. The cell of Figure 24.62 has a pretilt of 89° and an asymmetry in the retardance about the origin is observed.

24.8.4 MVA Cell Test

By simulating the measurement using methods shown in Section 24.5 and converting the result to Mueller matrices, the measured data can be fitted to the simulation result. The best-fit model parameters (cell gap, twist angle, etc.) are calculated iteratively for the smallest RMS difference between the simulated and the measured Mueller matrices.

The multi-angle Mueller matrix imaging technique can determine the 3D structure of the LC directors' arrangement in the LC cell. There is fundamentally no limit to the cell gap and pretilt ranges that can be measured. The cell can be positioned in any orientation. By scanning the LC cell with a high-resolution imaging polarimeter, the polarization behavior within an LCD pixel can be studied. It is possible to design a multi-domain pixel, investigate the LC behavior near patterned electrodes, and analyze bad and damaged pixels. Figure 24.64 shows the parameter map of an MVA LCD pixel.

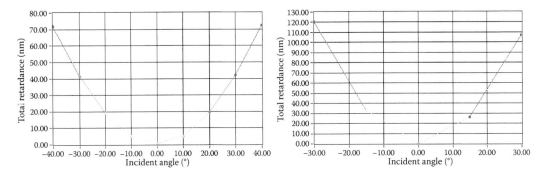

Figure 24.63 Retardance versus tilt angle for a VAN cell (left) with a pretilt of 90° is symmetric, while (right) a pretilt of 89° yields an asymmetric tilt scan. (Courtesy of Axometrics Inc., Huntsville, Alabama.)

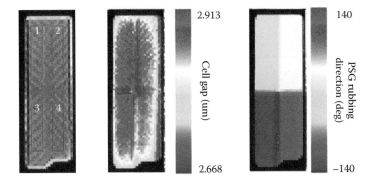

Figure 24.64 The measured parameters of an MVA LCD pixel. (Left) The intensity image with four subpixel regions indicated. (Center) An LC layer thickness map calculated from multi-angle Mueller matrix images. (Right) The orientation map for the alignment layers. (Courtesy of Axometrics Inc., Huntsville, Alabama.)

24.8.5 Sheet Retarder Defect

The retarders incorporated into LC displays are usually produced at high speed in wide sheets by stretching plastic or by spraying self-aligning molecules onto a moving substrate. For quality control, retardance must be monitored for defects. Figure 24.65 shows an example of how a microscopic defect on a compensating film can appear in a false color retardance image due to non-uniform deposition of the retarding layer.

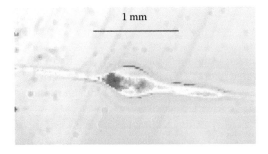

Figure 24.65 Appearance of a defect in an LC compensating film produced on a continuous web process shown in false color retardance. (Courtesy of Axometrics Inc., Huntsville, Alabama.)

24.8.6 Misalignment between Analyzer and Exiting Polarization State

Extinction of a polarized state is described by a generalization of Malus' law. On the Poincaré sphere, the transmission of a state through an ideal polarizer is given by $\sin^2\psi$, where ψ is the angle between the polarization state and the analyzer's extinction axis. As the output state moves past the extinction axis, it is unlikely to pass exactly through the extinction axis but is likely to have some minimum ψ. Thus, getting a good *dark state* and consistently high contrast ratio involves keeping the polarization trajectory close to the extinction axis. For a contrast ratio of 100, ψ must be less than 0.1 radians or 5.7°, which is not difficult to achieve. A contrast ratio of 10^{-4} requires a ψ less than 0.01 radians.

24.9 Problem Sets

24.1 Address the following performance factors of LC cells.
   a. What is the principal factor that limits the speed of liquid crystal cells?
   b. Why do liquid crystal cells have some depolarization?
   c. Why are liquid crystal cells cheaper than other retardance modulators?
   d. How can liquid crystal cells be configured as phase-only modulators?
   e. List five disadvantages of liquid crystal cells.
   f. List five advantages of liquid crystal cells.

24.2 A certain twisted nematic liquid crystal cell can be modeled in its *off state* by 91 parallel layers, each with a retardance of 0.060463 radians, with the fast axis of the first layer at 0°, the second layer at 1°, the third layer at 2°, and so on, until the 91st layer is oriented at 90°.
   a. Calculate the Jones matrix **J** for transmission through the cell.
   b. What are the eigenpolarizations of the cell?
   c. How can this LC cell and two linear polarizers be oriented to produce a dark state?
   d. How can this LC cell and two linear polarizers be oriented to produce a bright, fully transmitting state?
   e. Find the Jones matrix $\mathbf{J}_1$, eigenpolarizations, and retardance for the first half of the cell, layers 1 to 45. Draw the ellipse for at least one of the eigenpolarizations.
   f. Find the Jones matrix $\mathbf{J}_2$, eigenpolarizations, and retardance for the second half of the cell, layers 46 to 91. Draw the ellipse for at least one of the eigenpolarizations.
   g. Compare the polarization of the first and second half of the cell. Explain with Pauli matrices how the Jones matrices $\mathbf{J}_1$ and $\mathbf{J}_2$ combine to form **J**.

24.3 Compare the expression for a series of weak polarization elements with the exact equation for the following liquid crystal structure. Light is propagating in the +z-direction. A twisted nematic cell is modeled as 60 layers of linear retarders. Each layer is a linear retarder with retardance $\pi/300$. The retarder axes (the directors) rotate in the x–y plane through 360° such that the axis is along the line $(\cos\theta_q, \sin\theta_q, 0)$, $\theta_q = q \times 6°$ for $q = 1$ to 60.
   a. List the Jones matrices in the order they will be multiplied. Then, multiply the linear retarders in the correct order and calculate the Jones matrix $\mathbf{J}_{LC}$ for the entire sequence.
   b. Express $\mathbf{J}_{LC}$ as in normalized Pauli matrix form.
   c. Express the first two retarders $\mathbf{J}_1$ and $\mathbf{J}_2$ in weak polarization element form. These have 6° and 12° directors.

d. Make a table of the σ_1 components for each of the 60 retarders. Sum the components.
e. Make a table of the σ_2 components for each of the 60 retarders. Sum the components.
f. Obtain a weak polarization element expression for the sequence of weak Jones matrices. Compute the polarization properties of the sequence.

Acknowledgment

The authors wish to thank Jon Herlocker, David Serrano, and Matt Smith, who provided materials and assistance in the preparation of this chapter.

References

1. J. A. Castellano, *Liquid Gold: The Story of Liquid Crystal Displays and the Creation of an Industry*, World Scientific (2005).
2. C. H. Gooch and H. A. Tarry, The optical properties of twisted nematic liquid crystal structures with twist angles ≤ 90 degrees, *J. Phys. D* 8.13 (1975): 1575.
3. L. Vicari, (ed.), *Optical Applications of Liquid Crystals*, CRC Press (2016).
4. N. Konishi, K. Kondo, and H. Mano, 34-cm Super TFT-LCD with wide viewing angle, *Hitachi Rev.* 45.4 (1996): 165–172.
5. S. Aratani et al., Complete suppression of color shift in in-plane switching mode liquid crystal displays with a multidomain structure obtained by unidirectional rubbing. *Jpn. J. Appl. Phys.* 36.1A (1997): L27.
6. Y. Momoi, O. Sato, T. Koda, A. Nishioka, O. Haba, and K. Yonetake, Surface rheology of rubbed polyimide film in liquid crystal display, *Opt. Mater. Express* 4 (2014): 1057–1066.
7. S. He, J.-H. Lee, H.-C. Cheng, J. Yan, and S.-T., Wu, Fast-response blue-phase liquid crystal for color-sequential projection displays, *J. Disp. Technol.* 8 (2012): 352–356, 10.1109/JDT.2012.2189434.
8. P. Yeh, and C. Gu, *Optics of Liquid Crystal Displays*, Chapter 8, 2nd edition, John Wiley & Sons (2010).
9. D.-K. Yang and S.-T. Wu, *Fundamentals of Liquid Crystal Devices*, 2nd Edition, John Wiley & Sons (2015).
10. D. W. Berreman, Optics in stratified and anisotropic media: 4 × 4-matrix formulation, *J. Opt. Soc. Am.* 62(4), 502–510 (1972).
11. I. J. Hodgkinson and Q. H. Wu. *Birefringent Thin Films and Polarizing Elements*, World Scientific, Singapore (1997).
12. M. Mansuripur, Effects of high-numerical-aperture focusing on the state of polarization in optical and magneto-optic data storage systems, *Appl. Opt.* 30.22 (1991): 3154–3162.
13. V. G. Chigrinov, V. M. Kozenkov, and H.-S. Kwok, *Photoalignment of Liquid Crystalline Materials: Physics and Applications*, Vol. 17, John Wiley & Sons (2008).
14. F. S. Yeung, J. Y. Ho, Y. W. Li, F. C. Xie, O. K. Tsui, P. Sheng, and H. S. Kwok, Variable liquid crystal pretilt angles by nanostructured surfaces, *Appl. Phys. Lett.* 88(5) (2006): 051910.
15. S. T. Tang and H. S. Kwok, Transmissive liquid crystal cell parameters measurement by spectroscopic ellipsometry, *J. Appl. Phys.* 89.1 (2001): 80–85.

25

Stress-Induced Birefringence

25.1 Introduction to Stress Birefringence

Stress is the distribution of forces inside a body that the neighboring parts of its materials exert on each other. *Strain* is the resulting deformation of the material, the atoms moving closer together or further apart in response to the forces, with a corresponding change in dimensions and volume. Such changes of volume change the refractive index and birefringence of the material, a change labeled *stress-induced birefringence*. This chapter presents algorithms for polarization ray tracing through lenses with stress birefringence and has examples showing the impact of stress on image quality. For such analyses, the stress distributions are usually described by arrays of stress tensors in files generated by computer-aided design (CAD) programs performing finite element analyses of stress and strain.

Consider some examples where stress is generated in optical elements. Each optical element has a distribution of forces throughout its volume. For example, gravity pulls a lens downward against its mount with the forces between the lens and mount squeezing the lens and generating internal stresses. Tightening screws in the lens mount further increases the forces against the lens, thus increasing the stresses inside. Heating the lens and the mount causes the lens and mount materials to expand at different rates, based on their coefficients of thermal expansion, modifying the stress distribution. Many lens coatings are applied in vacuum to hot lenses because the coating material will be deposited more uniformly and densely so that the resulting coatings are stronger. When the lens is removed from the chamber and the lens and coating cool and contract at different rates, the surface of the lens often ends up in compression, with a coating-induced stress near the lens surfaces.

880 Chapter 25: Stress-Induced Birefringence

Stress-induced birefringence is common and often unavoidable in optical systems. Optical materials undergo strain at the molecular level due to various environmental conditions, such as external forces and pressures, vibration, or temperature change. The microscopic strains induce birefringence and affect the wavefront and point spread function of optical systems. The associated stress-induced retardance is generally undesirable, changing the wavefront aberration and polarization aberration in complex patterns. Thus, when dealing with optical elements with stress birefringence, it is useful to be able to ray trace the effects of stress birefringence to assess its impact. Further, the optical system can be ray traced with different levels of stress birefringence to *tolerance* the maximum amount of stress that is acceptable based on the system's image quality specifications.

Scottish physicist David Brewster[1] discovered stress-induced birefringence in isotropic substances in 1816; this phenomenon is also called mechanical birefringence, photoelasticity, and stress birefringence. Stress presents in optical systems in two different forms, mechanically-induced or residual. *Mechanical stress* results from physical pressure, vibration, or thermal expansion and contraction. It often arises from mounts squeezing or applying force to elements. Such a property of an optical material is referred to as *photoelasticity*.[2] *Residual stress* is a permanent stress inside an element, independent of external forces. It commonly occurs during fabrication of injection-molded lenses or when glass is poorly annealed. As the material cools from liquid to solid, stress can easily become frozen into the material, particularly when the outer surface solidifies before the inner material. Most optical glass is annealed, heated to its annealing temperature, a temperature where the molecules have enough thermal energy to rearrange themselves slightly to reduce stress, but where the glass is not hot enough to deform. The glass is then cooled slowly to avoid the introduction of additional stress. Either mechanical stress or residual stress alters the material's molecular structure slightly, some molecules are closer than equilibrium, others further apart, changing the optical properties in the direction of the stress and inducing birefringence.

Isotropic materials with stress-induced birefringence behave as spatially varying weakly uniaxial or biaxial materials. This induced birefringence is readily observed with *interferometers*[3,4] and *polariscopes*,[5,6] as shown in Figure 25.1. The colorful patterns of materials with stress birefringence between crossed polarizers is due to the wavelength-dependent variation of the retardance, with more birefringence for blue light and less for red light.

Transparent isotropic optical elements can exhibit temporary birefringence when subjected to stress and revert to isotropic when the stress is released. This stress can be induced by overtightening a knob of an optical mount as shown in Figure 25.2.

Stress birefringence can be intentionally induced in tempered glass for strength. In photoelastic modulators, an enormous sound wave is resonantly built up in the crystal to create a sinusoidally varying retardance for use as a polarization modulator. In general, stress birefringence is undesirable. For example, the stress that is common in injection-molded lenses induces wavefront aberrations and causes polarization aberrations, thus increasing the size of the point spread function and degrading the image.

 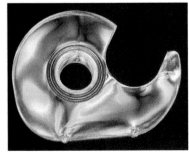 

Figure 25.1 The colors in this plastic cup, plastic tape dispenser, and prescription eyeglasses placed between crossed polarizers indicate large amounts of spatially varying stress.

25.3 Theory of Stress-Induced Birefringence

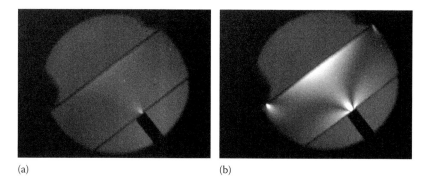

(a)   (b)

Figure 25.2 (a) An isotropic glass plate placed in an optical mount and viewed between crossed polarizers. (b) The same glass plate as a screw is tightened against the glass on the bottom. Because of the stress created, induced birefringence has caused significant light leakage between the crossed polarizers.

25.2 Stress Birefringence in Optical Systems

In most optical designs, stress must be nearly eliminated to ensure that the image quality of a high-performance optical system is maintained. A quick analysis can be performed by understanding the allowable induced birefringence. Some typical birefringence tolerances on lenses and glass blanks are 2 nm/cm for critical applications such as photolithography systems, polarimeters, and interferometers; 5 nm/cm in precision optics; 10 nm/cm for microscope objectives; and 20 nm/cm for eyepieces, viewfinder, and magnifying glasses.[7,8] Many lenses are fabricated by injection molding plastics these days due to its low cost and ease of fabrication. However, plastic optics, particularly the harder plastics such as polycarbonate, suffer from significant stress birefringence resulting from the manufacturing process, including molding, cooling, and mounting. Molding parameters such as the pressure applied to the resin in a mold, the time spent in the mold, and the rate the molded lens cool are often adjusted to minimize stress birefringence, but completely eliminating it is difficult.

When significant amount of stress presents in an optical system, it can be important to incorporate the stress birefringence into a polarization ray trace to simulate its effect on image formation, fringe visibility, and other optical metrics. This chapter presents the mathematical description of stress and its relationship to optical birefringence. Standard methods to translate the non-uniform stress distribution in optical components to retardance are presented and algorithms for polarization ray tracing through components with stress are shown. The finite element model of example plastic lenses will be analyzed using retardance maps, Jones pupil images, and polarization point spread matrices to visually demonstrate the induced birefringence and its effects.

25.3 Theory of Stress-Induced Birefringence

The stress birefringence can be described mathematically. When stress is applied to an object, the material deforms slightly as atoms reposition themselves in response to applied forces, as shown in Figure 25.3. The magnitudes of stress discussed in this chapter only induces small changes in atomic and molecular positions but causes negligible change in the material's physical shape. This applied stress, contact stress, or principal stress is characterized by the force applied in kg·m·s^{-2} or Newton (N) to the relevant cross-sectional area; the resulting stress is given in units of N/m^2 or Pascals (Pa). For a material such as glass or clear plastic, compression increases the refractive index in the compressive stress direction; atoms move closer together in this direction but further apart in the perpendicular plane. Conversely, tension decreases refractive index in the tensile stress direction as atoms expand along this axis. For small stress, after releasing the force, the glass returns to an isotropic state. This is referred to as elastic deformation. Above some threshold, irreversible

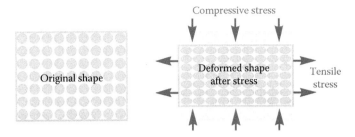

Figure 25.3 The forces on an object generate compressive stress and tensile stress. (Left) The original shape and atomic position of an optic. (Right) The changes when stress is applied, the shape of the object slightly deformed; for the object shown, the change is exaggerated to visualize the effect of compressive and tensile stress.

changes to the molecular arrangements occur, and the object cannot return to its original shape. For example, glass may shatter or a plastic lens may become dented.

Stress typically induces small changes in a material's refractive index with Δn less than 1. Typically, a 15-MPa stress yields a Δn in the order of 0.0001 for glass, as shown in Figure 25.4. When an external force is applied along one direction to an isotropic piece of glass, it becomes a uniaxial material with its optic axis along the direction of external force. Application of a second force along a different direction causes the glass to become a biaxial material. These stresses alter the dielectric tensor material. Therefore, the stressed optical element is simulated as an anisotropic material. A real optical component generally has varying stress throughout its volume, such as Figures 25.1 and 25.2; hence, stress birefringence should be simulated as a spatially varying birefringent material.

The typical stress-induced birefringence causes a very small amount of ray splitting. Figure 25.5 shows the very small change in refractive index (Δn) due to typical stress-induced birefringence for N-BK7 and polycarbonate (PC), which results in a negligibly small amount of ray doubling ($\Delta \theta°$); thus, the two split modes propagate very close to each other, typically within a few micrometers. Since a few wavelengths of ray separation in a ray trace is insignificant and would usually not affect the accuracy of many calculations, this ray doubling can be safely ignored. The induced birefringence of the two orthogonally polarized modes can be modeled as retardance along a single ray path. In the ray trace analysis, a polarization ray tracing **P** matrix will be constructed to keep track of such stress-induced retardance. The following sections present the algorithm to translate the

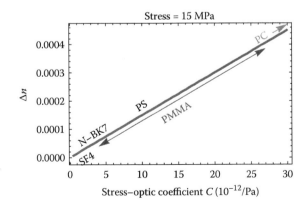

Figure 25.4 The change of refractive index Δn plotted as a function of material stress-optic coefficient C for 15 MPa applied stress, where $\Delta n = C \cdot \text{stress}$. Glass such as N-BK7 and SF4 have $C < 5$. Polymers such as polymethylmethacrylate (PMMA), polystyrene (PS), and polycarbonate (PC) have larger C, which also vary significantly with temperature.

25.4 Ray Tracing in Stress Birefringent Components

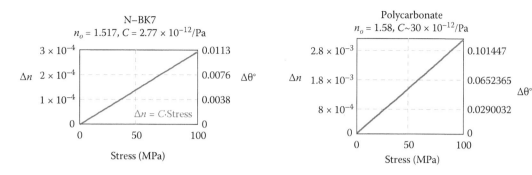

Figure 25.5 Refractive index change Δn and its corresponding ray split $\Delta\theta$ as a function of applied stress for glass N-BK7 (left) and polymer polycarbonate (right). The ray split is calculated with algorithms presented in Chapter 19.

stress of an optical material into the resultant retardance, and the refraction is typically modeled based on Snell's law using the unstressed refractive index.

25.4 Ray Tracing in Stress Birefringent Components

Algorithms will now be developed for simulating stress birefringence in optical systems by polarization ray tracing to allow its effects on image formation, fringe visibility and other optical metrics to be calculated. The applied stress and its induced retardance have a linear relationship. Therefore, the stress-induced birefringence is represented as a spatially varying linear retarder. **P** matrices keep track of stress-induced retardance along ray paths. The algorithms take the stress of an optical material as input and calculate the resultant retardance experienced by light rays. The calculation of the stress distributions involves methods from mechanical engineering that are not treated here.[9]

When multiple stresses are applied to an isotropic material in different directions, the stressed component becomes biaxial. The applied stress is represented as a 3 × 3 *strain tensor* **S**, defined in terms of *normal stresses* σ and *shear stresses* τ in an (x, y, z) coordinate system,

$$\mathbf{S} = \begin{pmatrix} \sigma_{xx} & \tau_{xy} & \tau_{xz} \\ \tau_{xy} & \sigma_{yy} & \tau_{yz} \\ \tau_{xz} & \tau_{yz} & \sigma_{zz} \end{pmatrix}. \tag{25.1}$$

S contains six *stress tensor coefficients* σ_{xx}, σ_{yy}, σ_{zz}, τ_{xy}, τ_{xz}, and τ_{yz},[10–14] which represent the force per unit area from different directions in 3D. The amount of deformation in the optic relative to the original shape due to **S** is characterized by the *strain tensor* Γ,

$$\Gamma = \begin{pmatrix} \gamma_{xx} & \gamma_{xy} & \gamma_{xz} \\ \gamma_{xy} & \gamma_{yy} & \gamma_{yz} \\ \gamma_{xz} & \gamma_{yz} & \gamma_{zz} \end{pmatrix}, \tag{25.2}$$

where γ's are the *strain tensor coefficients*. The material refractive index is represented as a 3 × 3 dielectric tensor ε. Isotropic materials have a refractive index n_0, a diagonal ε_0 (n_0^2 × identity matrix), which changes and becomes non-diagonal $\boldsymbol{\varepsilon}$ when external stress is applied in arbitrary directions,

$$\boldsymbol{\varepsilon}_0 = \begin{pmatrix} n_0^2 & 0 & 0 \\ 0 & n_0^2 & 0 \\ 0 & 0 & n_0^2 \end{pmatrix} \rightarrow \boldsymbol{\varepsilon} = \begin{pmatrix} n_{xx}^2 & n_{xy}^2 & n_{xz}^2 \\ n_{xy}^2 & n_{yy}^2 & n_{yz}^2 \\ n_{xz}^2 & n_{yz}^2 & n_{zz}^2 \end{pmatrix}, \tag{25.3}$$

where n_0 is the unstressed refractive index. Similarly, the impermeability tensor, the inverse of $\boldsymbol{\varepsilon}_0$, also changes from a diagonal tensor to a non-diagonal tensor with external stress,

$$\boldsymbol{\eta}_0 = \boldsymbol{\varepsilon}_0^{-1} = \begin{pmatrix} 1/n_0^2 & 0 & 0 \\ 0 & 1/n_0^2 & 0 \\ 0 & 0 & 1/n_0^2 \end{pmatrix} = \begin{pmatrix} \eta_0 & 0 & 0 \\ 0 & \eta_0 & 0 \\ 0 & 0 & \eta_0 \end{pmatrix} \rightarrow \boldsymbol{\eta} = \begin{pmatrix} \eta_{xx} & \eta_{xy} & \eta_{xz} \\ \eta_{xy} & \eta_{yy} & \eta_{yz} \\ \eta_{xz} & \eta_{yz} & \eta_{zz} \end{pmatrix}. \tag{25.4}$$

Thus, the stress alters the index ellipsoid from spherical to ellipsoidal,

$$\frac{x^2}{n_0^2} + \frac{y^2}{n_0^2} + \frac{z^2}{n_0^2} = x^2 \eta_0 + y^2 \eta_0 + z^2 \eta_0 = 1$$
$$\downarrow \tag{25.5}$$
$$\eta_{xx} x^2 + \eta_{yy} y^2 + \eta_{zz} z^2 + 2\eta_{xy} xy + 2\eta_{xz} xz + 2\eta_{yz} yz = 1,$$

where $x = \left(\dfrac{D_x}{\sqrt{2u}}\right)$, $y = \left(\dfrac{D_y}{\sqrt{2u}}\right)$, and $z = \left(\dfrac{D_z}{\sqrt{2u}}\right)$, and $u = \dfrac{1}{2}\mathbf{E}\cdot\mathbf{D} = \dfrac{1}{2}\left(\dfrac{D_x^2}{n_x^2} + \dfrac{D_y^2}{n_y^2} + \dfrac{D_z^2}{n_z^2}\right)$ is the energy density of the light field.

The link between the stress/strain and their optical effect on isotropic, non-magnetic, and non-absorbing material are the 6 × 6 *stress optic tensor* $\mathbf{C}$ and *strain optic tensor* $\boldsymbol{\Omega}$,

$$\mathbf{C} = \begin{pmatrix} C_1 & C_2 & C_2 & 0 & 0 & 0 \\ C_2 & C_1 & C_2 & 0 & 0 & 0 \\ C_2 & C_2 & C_1 & 0 & 0 & 0 \\ 0 & 0 & 0 & C_3 & 0 & 0 \\ 0 & 0 & 0 & 0 & C_3 & 0 \\ 0 & 0 & 0 & 0 & 0 & C_3 \end{pmatrix} \text{ and } \boldsymbol{\Omega} = \begin{pmatrix} p_1 & p_2 & p_2 & 0 & 0 & 0 \\ p_2 & p_1 & p_2 & 0 & 0 & 0 \\ p_2 & p_2 & p_1 & 0 & 0 & 0 \\ 0 & 0 & 0 & p_3 & 0 & 0 \\ 0 & 0 & 0 & 0 & p_3 & 0 \\ 0 & 0 & 0 & 0 & 0 & p_3 \end{pmatrix} \tag{25.6}$$

for isotropic materials and polymers with unidirectional symmetric structure. $\mathbf{C}$ is a function of the stress optic coefficients, C_1 and $C_3 = C_1 - C_2$ with units of inverse Pascals (1/Pa). $\boldsymbol{\Omega}$ is a function of

25.4 Ray Tracing in Stress Birefringent Components

Table 25.1 Refractive Index n_0, Strain Optic Coefficients (p_1 and p_2), Young's Modulus E, Poisson's Ratio ν, and Stress Optic Tensor Coefficients (C_1 and C_2) for Glasses and Plastics at 633 nm

Materials	n_o	p_1	p_2	E (GPa)	ν	C_1 (10^{-12}/Pa)	C_2 (10^{-12}/Pa)
Corning 7940 fused silica[12,15]	1.46	0.121	0.270	70.4	0.17	0.65	4.50
Corning 7070 glass[15]	1.469	0.113	0.23	51.0	0.22	0.37	4.80
Corning 8363 glass[15]	1.97	0.196	0.185	62.7	0.29	5.41	4.54
Al_2O_3[12]	1.76	−0.23	−0.03	367	0.22	−1.61	0.202
As_2S_3[15]	2.60	0.24	0.22	16.3	0.24	72.5	59.1
Polystyrene[16]	1.57	0.30	0.31	3.2	0.34	53.9	62.0
Lucite[17]	1.491	0.30	0.28	4.35	0.37	35.4	24.9
Lexan[17]	1.582	0.252	0.321	2.2	0.37	13.0	98.1

the strain optic coefficients, p_1 and p_2 with $p_3 = (p_1 - p_2)/2$. These stress/strain optic coefficients are directly related to each other, the Young's modulus (E) and Poisson's ratio (ν) as

$$C_1 = \frac{n_o^3}{2E}(p_1 - 2\nu p_2) \quad \text{and} \quad C_2 = \frac{n_o^3}{2E}\left[p_2 - \nu(p_1 + p_2)\right]. \tag{25.7}$$

Then, the matrices Ω and $\mathbf{C}$ relate the strain and stress to the material's refractive index as*

$$\Delta\eta = \begin{pmatrix} 1/n_{xx}^2 - 1/n_0^2 \\ 1/n_{yy}^2 - 1/n_0^2 \\ 1/n_{zz}^2 - 1/n_0^2 \\ 1/n_{xy}^2 \\ 1/n_{yz}^2 \\ 1/n_{xz}^2 \end{pmatrix} = \Omega \begin{pmatrix} \gamma_{xx} \\ \gamma_{yy} \\ \gamma_{zz} \\ \gamma_{xy} \\ \gamma_{yz} \\ \gamma_{xz} \end{pmatrix} \text{ and } \Delta\eta = \mathbf{C}_0 \begin{pmatrix} \sigma_{xx} \\ \sigma_{yy} \\ \sigma_{zz} \\ \tau_{xy} \\ \tau_{yz} \\ \tau_{xz} \end{pmatrix} \text{ or } \begin{pmatrix} n_{xx} - n_0 \\ n_{yy} - n_0 \\ n_{zz} - n_0 \\ n_{xy} \\ n_{yz} \\ n_{xz} \end{pmatrix} \approx -\mathbf{C} \begin{pmatrix} \sigma_{xx} \\ \sigma_{yy} \\ \sigma_{zz} \\ \tau_{xy} \\ \tau_{yz} \\ \tau_{xz} \end{pmatrix}.$$

(25.8)

Table 25.1 contains a set of strain and stress optic tensor coefficients C_1 and C_2. The stress optic tensor coefficients for plastic are typically larger than glass, which means an equivalent stress applied on plastic yields a larger change of refractive index than glass. More compressible materials have larger response in general. PC is a very strong polymer, but generally has high stress-induced birefringence. Thus, PC is avoided in application where stress is a problem, but is popular in eyeglasses for its strength and the eye protection provided.

---

* See Equation 25.9 for the relationship between C_0 and C.

Example 25.1 Uniaxial Stress Optic Effect

An isotropic material under a stress in one direction becomes a uniaxial material with its optic axis along the direction of the applied stress. From Equation 25.8, when there is only a stress acting along the x-direction,[18–20]

$$C_0 \sigma = \Delta \eta = \frac{1}{n_{xx}^2} - \frac{1}{n_0^2}$$

$$= \frac{n_0^2 - n_{xx}^2}{n_0^2 n_{xx}^2} = \frac{(n_0 - n_{xx})(n_0 + n_{xx})}{n_0^2 n_{xx}^2}, \text{ where } n_0 \approx n_{xx}$$

$$\approx \frac{(n_0 - n_{xx})(2n_0)}{n_0^4}$$

$$\approx -\frac{2}{n_0^3} \Delta n.$$

Thus,

$$\Delta n \approx -\frac{n_0^3}{2} C_0 \sigma = -C \sigma. \tag{25.9}$$

For common polymers, such as polymethyl methacrylate (PMMA), C_0 is in the order of 10^{-12} Pa^{-1}, which is equivalent to a *Brewster*. Several stress optic coefficients C of glass and polymers are listed in Table 25.2.

Table 25.2 Stress Optic Coefficients C

Materials	C (10^{-12}/Pa)
N-BK7 (Schott glass)[21]	2.77
F2 (Schott glass)[21]	2.81
SF4 (Schott glass)[21]	1.36
Poly methyl methacrylate (PMMA)[22–24]	~3.3, 4.5[a]
Polycarbonate[25,27]	30 to 72[a]
LBC3N (HOYA)[27]	0.43
FF5 (HOYA)[27]	3.31

Schott glasses were measured at 589.3 nm,[21] and polymers and HOYA glasses were measured at 632.8 nm.[27]

[a] Parameter varies with modified material formulations for different applications.

25.4 Ray Tracing in Stress Birefringent Components

From Equation 25.8, the stressed impermeability tensor is $\boldsymbol{\eta} = \boldsymbol{\eta}_0 + \Delta\boldsymbol{\eta}$, which can be converted to a dielectric tensor in the principal coordinate system. When η is expressed in its principal coordinates, it is a diagonal matrix with its eigenvalues (L_1, L_2, L_3) along the diagonal,

$$\boldsymbol{\eta}_{principle} = \begin{pmatrix} L_1 & 0 & 0 \\ 0 & L_2 & 0 \\ 0 & 0 & L_3 \end{pmatrix}, \text{ so } \boldsymbol{\varepsilon}_{principle} = \begin{pmatrix} 1/L_1 & 0 & 0 \\ 0 & 1/L_2 & 0 \\ 0 & 0 & 1/L_3 \end{pmatrix}. \tag{25.10}$$

The dielectric tensor in principal coordinates is rotated to the global coordinate system as

$$\boldsymbol{\varepsilon} = \boldsymbol{R}^{-1} \cdot \boldsymbol{\varepsilon}_{principle} \cdot \boldsymbol{R}, \tag{25.11}$$

where $\boldsymbol{R}$ is $(\mathbf{v}_1\ \mathbf{v}_2\ \mathbf{v}_3)$ whose columns are the eigenvectors of η.

Stress applied to uniaxial or biaxial crystals requires additional stress optic coefficients to the coefficients shown in Equation 25.6; these are related to other strain coefficients. Their stress optic tensor require more non-zero components to account for interaction between off-diagonal ε components in both shear and normal stress.[10,28,29]

Using the results in Equations 25.8, 25.10, and 25.11, we obtain the stressed dielectric tensor that alters the polarization properties of a ray in a way similar to a biaxial dielectric tensor. As described in Chapter 19, when a ray passes through a stressed biaxial material, it splits into two modes with orthogonal electric field vectors $\mathbf{E}_1$ and $\mathbf{E}_2$, which carry two different refractive indices n_1 and n_2. The following assumptions are made to simplify the calculation of the stressed $\mathbf{P}$ matrix while retaining the necessary accuracy for useful optics calculations (generally better than $\lambda/100$ for phases):

1. The applied stress only causes small changes to the refractive index. Hence, refraction into and out of an element is accurately calculated by Snell's law using the unstressed refractive index.
2. Ray doubling is ignored since two modes exiting the stress birefringent material are essentially on top of each other, as shown in Figure 25.5, and the path is well modeled by a single ray segment traced from the entrance to exiting face. Therefore, $\hat{\mathbf{k}} \approx \hat{\mathbf{S}}$ and $\hat{\mathbf{D}} \approx \hat{\mathbf{E}}$.

Using these assumptions, the refracted propagation vector is $\mathbf{k}$,

$$n_i \sin\theta_i = n_0 \sin\theta_0 \text{ and } \mathbf{k} = \frac{n_i}{n_0}\mathbf{k}_i + \left(\frac{n_i}{n_0}\cos\theta_i - \cos\theta_0\right)\hat{\boldsymbol{\eta}}, \tag{25.12}$$

where n_i and n_0 are the incident and unstressed refractive indices, θ_i and θ_0 are the incident and refraction angles, $\mathbf{k}_i$ is the incident propagation vector, and $\hat{\boldsymbol{\eta}}$ is the surface normal. The refractive index and electric field vector for the two modes are calculated with the method of Section 19.5.

By combining the anisotropic constitutive relation in Equation 19.7 and the Maxwell equations, the eigenvalue equation for **E** is

$$\left[\boldsymbol{\varepsilon} + (n\mathbf{K})^2\right]\mathbf{E} = 0, \tag{25.13}$$

where $\mathbf{K} = \begin{pmatrix} 0 & -k_z & k_y \\ k_z & 0 & -k_x \\ -k_y & k_x & 0 \end{pmatrix}$ and $\hat{\mathbf{k}} = (k_x, k_y, k_z)$ is the propagation direction within the stressed material. For non-zero **E**,

$$\left|\boldsymbol{\varepsilon} + (n\mathbf{K})^2\right| = 0 \tag{25.14}$$

with two refractive index solutions, n_1 and n_2 correspond to the two eigenmodes. Then, $\mathbf{E}_1$ and $\mathbf{E}_2$ are obtained through the singular value decomposition of $[\varepsilon + (n\mathbf{K})^2]$ with n_1 and n_2. Their Poynting vectors are the cross product of the electric and magnetic vectors,

$$\mathbf{S} = \frac{\mathcal{R}e[\mathbf{E} \times \mathbf{H}^*]}{\left|\mathcal{R}e[\mathbf{E} \times \mathbf{H}^*]\right|}, \tag{25.15}$$

where

$$\mathbf{H} = n\,\mathbf{K} \cdot \mathbf{E}. \tag{25.16}$$

The **P** matrix of the ray within the stressed material is a linear retarder. It maps vectors $(\hat{\mathbf{E}}_1, \hat{\mathbf{E}}_2, \hat{\mathbf{S}})$ to $\left(\hat{\mathbf{E}}_1 e^{i\frac{2\pi}{\lambda}n_1 d}, \hat{\mathbf{E}}_2 e^{i\frac{2\pi}{\lambda}n_2 d}, \hat{\mathbf{S}}\right)$. Thus,

$$\mathbf{P}_{stressed} = \left(\hat{\mathbf{E}}_1 e^{i\frac{2\pi}{\lambda}n_1 d} \quad \hat{\mathbf{E}}_2 e^{i\frac{2\pi}{\lambda}n_2 d} \quad \hat{\mathbf{S}}\right) \left(\hat{\mathbf{E}}_1 \quad \hat{\mathbf{E}}_2 \quad \hat{\mathbf{S}}\right)^T, \tag{25.17}$$

where λ is the wavelength of the light and d is the distance the light travels inside the material.

Example 25.2 Uniaxial Stress of Compression and Tension

A compressive stress increases refractive index and produces a positive uniaxial effect; an isotropic medium changes into a weakly positive uniaxial medium. The direction of compression is the slow axis. The compressive stress σ has a negative sign.

A tensile stress or tension produces a negative uniaxial effect altering an isotropic material into a negative uniaxial material. Tension decreases refractive index; hence, the direction of such a stress denotes the fast axis. The tensile stress σ has a positive sign.

The phase of the light polarized along the fast axis is ahead of the mode polarized along the slow axis. For compression stress along y, retardance $\delta = 2\pi \frac{\Delta n}{\lambda} d = \frac{2\pi}{\lambda}(n_y - n_x)d > 0$. For tensile stress along y, retardance $\delta = 2\pi \frac{\Delta n}{\lambda} d = \frac{2\pi}{\lambda}(n_y - n_x)d < 0$. The linear retarder **P** matrix with retardance δ for this ray propagating in the z-direction is

$$\begin{pmatrix} e^{i\frac{2\pi}{\lambda}n_x d} & 0 & 0 \\ 0 & e^{i\frac{2\pi}{\lambda}n_y d} & 0 \\ 0 & 0 & 1 \end{pmatrix} \quad \text{or} \quad \begin{pmatrix} e^{-i\delta/2} & 0 & 0 \\ 0 & e^{i\delta/2} & 0 \\ 0 & 0 & 1 \end{pmatrix}, \qquad (25.18)$$

and the two eigenmodes are $e^{-i\delta/2}\mathbf{E}_x = e^{-i\delta/2}(1, 0, 0)$ and $e^{i\delta/2}\mathbf{E}_y = e^{i\delta/2}(0, 1, 0)$.

25.5 Ray Tracing through Stress Birefringence Components with Spatially Varying Stress

In optical elements with stress birefringence, the stress varies in a complex way. For example, when plastic optical elements are manufactured by injection molding, stress from lens mounts and vacuum windows create complex spatially varying stress birefringence, like Figure 25.2. Mechanical engineering software packages, such as *SigFit*,[30] calculate the stress distributions within mechanical parts due to forces such as bolting items together, welding, and gravity. This *finite element modeling* (FEM) of stresses and many other parameters such as vibration characteristics are part of a mechanical design process. The resultant stress distribution is calculated at a discrete set of data points, thus the designation finite element. For injection molding, other software packages such as *Moldflow* and *Timon3D*[31–33] analyze the complex physical process of injection molding by simulating the flow of the viscous melted plastic resin into a heated mold, the non-uniform cooling of the mold and part, the solidification under high compression, and the separation of the lens from the mold. Modeling has improved efficiency and quality in plastic molding to incorporate new polymers and meet the high demand in high-quality electronics, consumer products, and automobile industry.[34] In both cases, mechanical stresses and molding stresses, the distribution of stress inside a 3D object is expressed as an array of 3 × 3 stress tensors. Figure 25.6 shows the 3 × 3 symmetric stress tensor distribution on the surface of an injection-molded lens, where high stress variations are observed in the diagonal elements.

Simulating polarization change through an optical element with spatially varying stress involves four general steps listed below. These steps calculate *OPL*, retardance, and **P** matrices for a finite element stress model:

1. Extract the element's optical shape from the finite element program's files.
2. Calculate the optical path of a ray inside the element, and divide it into short segments.
3. Extract the stress information along that optical path.
4. Convert the stress optical distribution into retarder **P** matrices for the segments.
5. Multiply the **P** matrices to obtain an overall **P** matrix from entrance to exit.

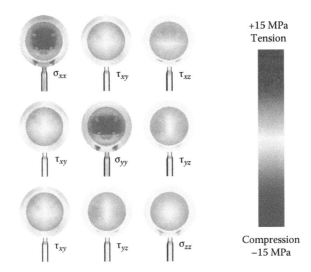

Figure 25.6 The 3 × 3 stress tensor distribution of an injection-molded lens. Red indicates tension, blue indicates compression, and gray indicates zero stress. The plastic entered the lens mold from the bottom at the region called the gate. The cylinder of plastic extending from the bottom of the lens is for handling the lens, which is moved by robot to a saw, where the cylinder is removed and the lens falls into packaging for delivery.

25.5.1 Storage of System Shape

Although the storage details of a 3D object vary between CAD systems, the idea is similar; the goal is to discretize a continuous region to a finite number of sub-domains. In general, an object stored in a CAD file is using numerous simple building blocks or finite *elements*, such as cuboids or tetrahedrons. Each element is specified by its vertices or *nodes*. A cuboid requires eight nodes, and a tetrahedron requires four nodes as shown in Figure 25.7b and c.

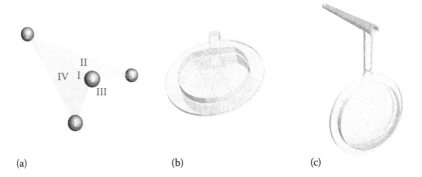

Figure 25.7 (a) A tetrahedron element has four vertices (1, 2, 3, and 4) and four surfaces (I, II, III, and IV). (b) and (c) are surface plots of two different injection-molded lens structures. The surfaces are represented by surface triangles.

25.5 Ray Tracing through Stress Birefringence Components with Spatially Varying Stress

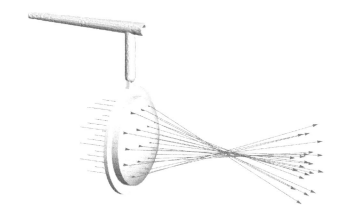

Figure 25.8 Refraction through an injection-molded lens uses surface triangles and Snell's law to calculate ray intercepts. The figure shows a simulation of collimated rays converging through the lens.

25.5.2 Refraction and Reflections

Refraction and reflection are performed using the unstressed refractive indices and the surface triangles at the object surface extracted from a CAD file. Each surface triangle has its surface normal specified in global coordinate systems. For a given triangle, two vectors are calculated by taking the differences between the triangle vertices. The cross product of these two vectors approximates the surface normal for the triangular face. For a given ray, a ray intercept is calculated using the algorithm in Ref. 35. Figure 25.8 shows a ray grid refracting through an injection-molded lens. Using the ray intercept routine and Snell's law, a set of incident collimated rays converge to an image after refracting through the lens surfaces.

25.5.3 Stress Data Format

Stress information is contained in CAD files as an array of stress tensors, each with six stress coefficients, σ_{xx}, σ_{yy}, σ_{zz}, τ_{xy}, τ_{xz}, and τ_{yz} in Equation 25.8. A stress tensor is assigned to each element building block. When a ray propagates through a 3D object, the ray passes through multiple elements and experiences varying stress tensors along the ray segment. Usually, each stress tensor is associated with the center of one element block as shown in Figure 25.9. Thus, for an object composed of N element blocks, the stress file provides N stress data points distributed throughout the object's volume. The light ray propagates through this cloud of points and interpolation can estimate the stress at arbitrary locations along the ray.

The stress component σ_{xx} of an example injection-molded lens is shown in Figure 25.10. It is common that most of the stress is concentrated in a thin layer at the surface. A stress tensor map with all nine components of another example lens is shown in Figure 25.11 with the magnitude of the stress represented by the color scale. Higher stress is observed around the *gate* (where the plastic melt flows into the mold) and the *flange* (an annular structure around the outside of the lens for mounting) of the plastic lens.

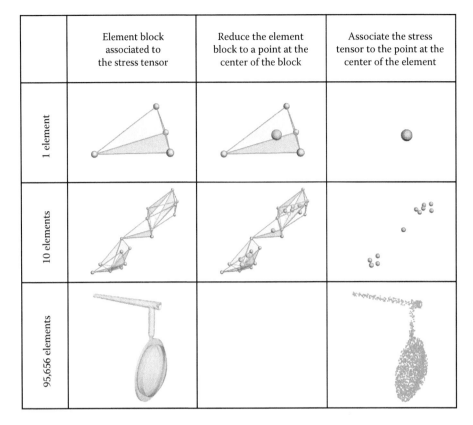

Figure 25.9 A tetrahedron element defined by four corners (top row, left column) is reduced to one data point, shown as a red point at the center of the element. Ten tetrahedrons shown in the middle row are reduced to 11 data locations. The last row shows that all 95,656 tetrahedrons of an object are reduced to 95,656 data locations, each of them are associated to a stress tensor.

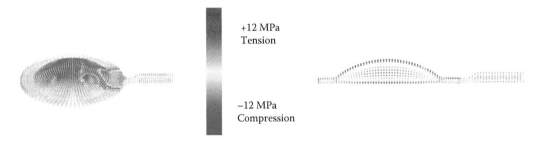

Figure 25.10 Two views of the stress tensor coefficient σ_{xx} from the CAD file of an injection-molded lens.

25.5.4 Polarization Ray Tracing Matrix for Spatially Varying Biaxial Stress

When a ray propagates through material with spatially varying stress, it experiences retardance change throughout the ray path. This is similar to a ray propagating through a spatially varying birefringent material; hence, the ray path is modeled as a stack of constant biaxial materials as shown in Figure 25.12. Figure 25.12 (left) shows a spatially varying birefringent material with a rotating optic axis and changing retardance magnitude. Such spatially varying behavior is simulated by dividing the material into thin slabs along the ray path as shown in Figure 25.12 (right). Each of the

25.5 Ray Tracing through Stress Birefringence Components with Spatially Varying Stress

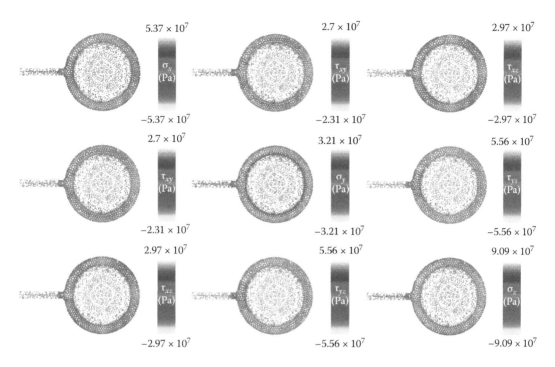

Figure 25.11 Stress tensor maps for the nine components are plotted across an object cross section in color scale.

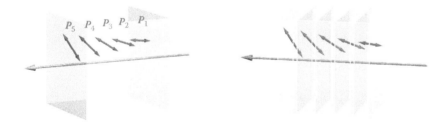

Figure 25.12 For a single ray, a spatially varying biaxial material (left) can be simulated as a stack of constant biaxial slabs (right). A different stack is calculated for each ray.

slabs represents a constant biaxial material with its unique optic axis orientation and retardance magnitude.

The concept of slicing a ray path into segments through a material is applied to the stress data shown in Figure 25.13. (a) Along a ray path, a ray passes many stress data points. (b) The ray path is then divided into steps. (c) The midpoint of a step is unlikely to land exactly on top of a data point. (d) The stress tensor at the center of each step is interpolated from the data; for example, the weighted average of the three closest data points can be used. (e) A **P** matrix is calculated for the step from the stress tensor. The data point closest to the step contributes more than the data point data further away from the step. The number of steps should be chosen to be sufficiently large to model spatial variation of the stress birefringence and ensure accurate results.

The stress grid is intended to sample the spatial stress variation densely enough that the change between nearby modes is small. In this limit, the stress can be interpolated. However, for more rapidly varying stresses, higher-order fitting equations could be needed, and interpolation would be performed on the retardance magnitude and retardance orientation separately.

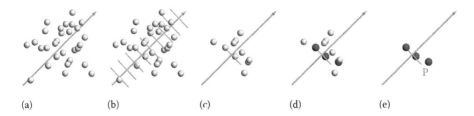

Figure 25.13 The ray path inside the spatially varying biaxial material is divided into steps. Interpolation is used to obtain stress information at each step. (a) A ray passes near many stress data points shown in blue along its ray path. (b) The ray path is then divided into steps shown as orange parallel lines along the ray path. (c) It is unlikely that the step exactly intersects any data point. (d) The N closest data points, three are highlighted in purple, are used to interpolate the stress for the step. (e) A P matrix is calculated for the step from the stress tensor, the ray direction, and the step length.

One straightforward method to calculate the weighted average stress is similar to the interpolation algorithm shown in Chapter 20. Consider a set of 3×3 stress tensors ($\mathbf{S}_1, \mathbf{S}_2, \ldots \mathbf{S}_N$) at N data locations ($\mathbf{r}_{s1}, \mathbf{r}_{s2}, \ldots, \mathbf{r}_{sn}, \ldots \mathbf{r}_{sN}$), describing the stress distribution of an object. A ray propagates through the object and its ray path within the stressed object is evenly divided into M steps ($\mathbf{r}_1, \mathbf{r}_2, \ldots \mathbf{r}_m, \ldots \mathbf{r}_M$). The distance between each step is d. For step m, the Q closest data points are ($\mathbf{r}_{m1}, \mathbf{r}_{m2}, \ldots \mathbf{r}_{mq}, \ldots \mathbf{r}_{mQ}$) with stress tensors ($\mathbf{S}_{m1}, \mathbf{S}_{m2}, \ldots, \mathbf{S}_{mq}, \ldots, \mathbf{S}_{mQ}$). Q is chosen to produce a reasonable stress estimation at a location other than the N data points. $Q = 3$ is sufficient for the examples shown in this chapter. These data points are distance $\left(\left| \bar{r}_{m1} \right|, \left| \bar{r}_{m2} \right|, \ldots, \left| \bar{r}_{mQ} \right| \right) = (|\mathbf{r}_{m1} - \mathbf{r}_m|, |\mathbf{r}_{m2} - \mathbf{r}_m|, \ldots, |\mathbf{r}_{mQ} - \mathbf{r}_m|)$ away from step m. The stress at step m is interpolated from this data. Using

$$\mathbf{S_m} = \sum_{q=1}^{Q} A_q \mathbf{S_q}, \tag{25.19}$$

where

$$A_q = \frac{\left(\left| \bar{r}_{mq} \right| + \varepsilon^{-17} \right)^{-2}}{\sum_{q=1}^{Q} \left[\left(\left| \bar{r}_{mq} \right| + \varepsilon^{-17} \right)^{-2} \right]} \tag{25.20}$$

is a scaling factor based on distance $|\bar{r}_{mq}|$, which has a maximum of 1 and $\varepsilon \approx 10^{-17}$. A_q approaches $\mathbf{S}_q$ when it is near point r_m and deemphasizes points further away. If $\mathbf{r}_m$ is exactly on top of $\mathbf{r}_{mq}$, $|\bar{r}_{mq}| = 0$ and $A_q \approx 1$. Between data points, A_q averages effects from the closest Q data points.

Each step behaves as a linear retarder whose effect is represented by a **P** matrix. The stressed $\mathbf{P}_m$ matrix corresponding to step m is calculated using methods described in Section 25.3 and Equation 25.17 with average $\mathbf{S}_m$ in Equation 25.19. The net effect of the sequences of retardances along the ray path through the stressed element is obtained by multiplying the **P** matrices,

$$\mathbf{P}_{stress} = \prod_{m=1}^{M} \mathbf{P}_{M-m+1}. \tag{25.21}$$

25.5 Ray Tracing through Stress Birefringence Components with Spatially Varying Stress

To represent a ray refracting into the stress material, propagating through the stressed material, and refracting out of the material, the total **P** matrix is

$$\mathbf{P}_{out} \cdot \mathbf{P}_{stress} \cdot \mathbf{P}_{in}, \tag{25.22}$$

where $\mathbf{P}_{in}$ and $\mathbf{P}_{out}$ are isotropic **P** matrices calculated as described in Chapter 9 with the unstressed refractive index for the incident and exiting surfaces.

25.5.5 Examples of Spatially Varying Stress Function

Consider the example of light propagating through tempered glass, such as in an automobile windshield. As illustrated in Example 25.3, the parabolic stress distribution in a piece of tempered glass provides strength and stability through equalizing tension and compression. An imbalance in stress can cause unexpected weakness leading to spontaneous breakage.

Example 25.3 Parabolic Stress in Tempered Glass

Tempered glass uses a parabolic stress distribution produced in a glass plate by controlled cooling. The example stress distribution is shown in Figure 25.14, where positive stress is tension and negative stress is compression applied along the x-direction. Having compression on the surface results in stronger molecular bonds and a stronger material. The tension in the center (a less dense region) balances the forces across the glass thickness.

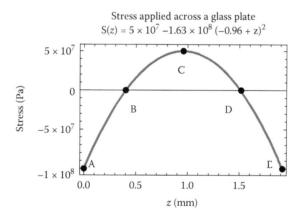

Figure 25.14 Parabolic stress across to a glass plate thickness.

Consider 500 nm light normal incident onto the plate with index 1.52. The ray experiences the retardance induced by compression stress from A to B in Figure 25.14. With the method described in Section 25.5.4, the ray path is divided into 30 slices of linear retarders, and the resultant $\mathbf{P}_{AB}$ is

$$\begin{pmatrix} 0.707 + 0.707i & 0 & 0 \\ 0 & 0.707 - 0.707i & 0 \\ 0 & 0 & 1 \end{pmatrix},$$

which has 90° retardance magnitude with a y fast axis. Then, it travels from B to C with an increasing tension stress. The resultant $\mathbf{P}_{BC}$ for that segment is

$$\begin{pmatrix} 0.707 - 0.707i & 0 & 0 \\ 0 & 0.707 + 0.707i & 0 \\ 0 & 0 & 1 \end{pmatrix},$$

which has 90° retardance magnitude with an x fast axis. The overall $\mathbf{P}$ from A to E is

$$\begin{pmatrix} 1 & 0 & 0 \\ 0 & 1 & 0 \\ 0 & 0 & 1 \end{pmatrix},$$

where retardance is cancelled to zero by the balance between tension and compression. The stress is purposely set up to distribute tensions and compressions through the component. Each segment does not need to have a quarter wave of retardance, but the important point is to balance the resultant retardance to zero.

Another example of stress tensor as a function of location on the component cross section is shown in Example 25.4. A 2 mm × 2 mm glass plate under a stress with the stress tensor function of Equation 25.23 becomes a rotationally symmetric retarder. Figure 25.16 shows the simulation results of the stressed plate under a linear polariscope. About 10 waves of retardance are observed across the 2 mm width. The orientation of the zero intensity cross rotates with the polariscope, since the stress axis of the plate is radially orientated. The simulated 2 × 2 Jones pupil (Figure 25.17) shows the coupling between the vertical (y) and horizontal (x) polarization through the stress plate. The coupling of xx and yy has the same pattern, which is the opposite pattern of the cross-coupling leakage of xy and yx. In the phase components, the xx and yy coupling shows the radial cycles from $-\pi$ to π, while the cross couplings have jumps from $-\pi/2$ to $\pi/2$. When the amplitude and the phase images are considered together, the cross-coupling Jones pupils are real valued and have a saddle shape varying from +1 to −1.

Example 25.4 Simulation of a Rotationally Stressed Plate

A stressed plane parallel plate under polariscope is simulated using $\mathbf{P}$ matrix multiplication. Consider of 0.1844 mm thick plate made of N-BK7 with a stress tensor:

$$\mathbf{S}(x,y) = \begin{pmatrix} x^2 & x \times y & 0 \\ x \times y & y^2 & 0 \\ 0 & 0 & 0 \end{pmatrix} \cdot 2000 \text{ MPa}. \tag{25.23}$$

The stress magnitude and orientation is shown in Figure 25.15; the magnitude is zero along $(x, y) = (0, 0)$ axis and increases quadratically from the center.

25.5 Ray Tracing through Stress Birefringence Components with Spatially Varying Stress 897

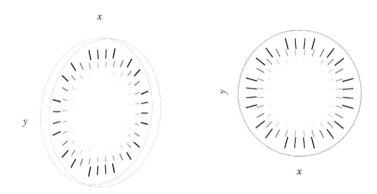

Figure 25.15 Stress distribution of Equation 25.23. The stress is applied along the line and the stress magnitude is proportional to the length of the line. The larger the stress, the darker the line is, so larger stress is located toward the end of the plate.

Consider placing the stressed plate in a polariscope with two crossed polarizers. The stressed plate is a spatially varying A-plate. The polariscope's **P** matrix is obtained from the 0° and 90° linear polarizer **P** matrices as $\mathbf{P}_{0°} \cdot \mathbf{P}_{stress} \cdot \mathbf{P}_{90°}$, and the resultant intensity image is shown in Figure 25.16.

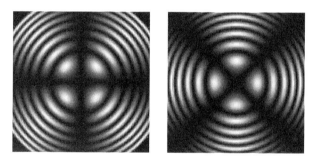

Figure 25.16 Linear polariscope images of a stressed plate; the stress plate is in between horizontal and vertical polarizers shown on the left and in between 45° and 135° polarizers shown on the right. Black denotes zero intensity and white denotes high intensity.

The Jones pupil corresponding to $\mathbf{P}_{stress}$ is plotted in Figure 25.17.

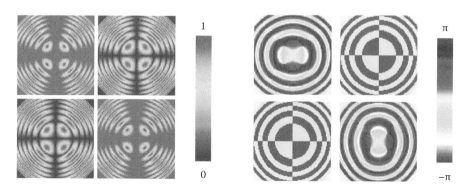

Figure 25.17 Jones pupil amplitude (left) and phase (right) of the stress waveplate.

25.6 Effects of Stress Birefringence on Optical System Performance

There are many ways to analyze ray tracing results for systems containing stress birefringence, such as polarization state change or birefringence change. The most common figure of merit is retardance. Retardance magnitude and fast axis orientation across the pupil give insight into the stress location, orientation, and magnitude. In most optical systems, a quarter of a wave of retardance yields a significant amount of aberration, causing noticeable image degradation in the point spread function. In this section, simulations of various injection-molded lenses and images taken by polariscope for stress-induced plastic and glass elements are presented.

25.6.1 Observing Stress Birefringence Using Polariscope

A quick way to observe birefringence is to use the polariscope. Since the polariscope measures the integrated retardance along the ray path, it is commonly used to analyze stress birefringence of transparent samples. Different configurations of polariscope reveal the magnitude and orientation of induced birefringent. Polariscopes have been used for quality inspections during glass and clear plastic lens manufacturing processes to identify defects and stress. Polariscopes are discussed in Section 7.7.1.

The polariscope images of some plastic samples (Figures 25.18 through 25.21) are taken from different polariscope configurations. The light leakage through the sample is a sign of stress birefringence in polariscope with cross polarizers. A plastic tape dispenser in Figure 25.18 and the CD substrate in Figure 25.19 show large variation of birefringence near the edges where the shape changes abruptly. Since the tape dispenser and the CD substrate are not typically used as optical elements in polarization-sensitive optical systems, large stress birefringence is not a problem. A pair

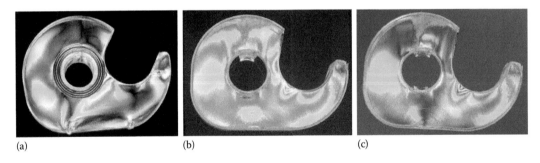

Figure 25.18 A plastic tape dispenser is placed under (a) crossed linear polarizers, (b) crossed circular polarizers, and (c) sensitive tint plate polariscope.

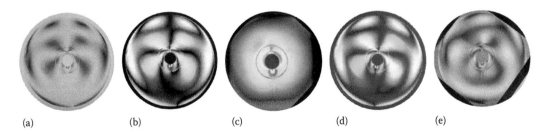

Figure 25.19 A CD substrate is placed under (a) parallel linear polarizers, (b) crossed linear polarizers, (c) crossed circular polarizers, (d) sensitive tint plate polariscope, and (e) crossed polarizers with a quarter waveplate.

25.6 Effects of Stress Birefringence on Optical System Performance

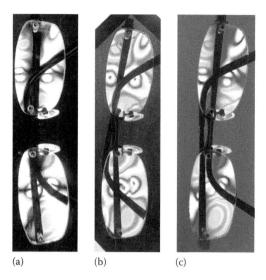

Figure 25.20 A pair of plastic glasses is placed under (a) crossed linear polarizers, (b) crossed circular polarizers, and (c) sensitive tint plate polariscope.

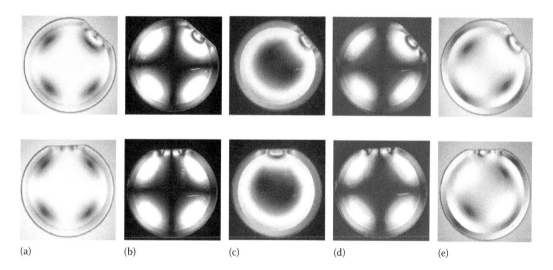

Figure 25.21 A plastic lens is placed under (a) parallel linear polarizers, (b) crossed linear polarizers, (c) crossed circular polarizers, (d) sensitive tint plate polariscope, and (e) crossed polarizers with a quarter waveplate. The plastic lens is oriented to show stress. The gate for plastic injection appears different for the two orientations, except in circular polariscope, which produces intensity independent of fast axis orientation.

of prescription eyeglasses, however, shows unexpectedly high birefringence. Since human eyes are insensitive to polarization, the birefringence has little effect on the performance of the eyeglasses.

The injection-molded convex lens in Figure 25.21 shows the typical *Maltese cross pattern* of lenses between crossed polarizers. The Maltese cross mostly comes from thin compressive layers on the surfaces of the lens as shown in Figure 25.10. The gate of the injection mold, where the plastic flows into the mold, is located at the side of the lens and typically has high stress birefringence relative to the rest of the lens.[36] The two sets of images in Figure 25.21 show the polariscope images of the lens in five configurations of polariscope; the bottom row is rotated by 45°. By changing the orientation of the lens with different polariscope configurations, orientation of the stressed induced retardance can be qualitatively calculated. This top row of Figure 25.21 provides information on the linear vertical/horizontal and circular retardance; the bottom row provides information on the linear 45°/135° and circular retardance. In between linear cross polarizers, the main pattern is invariant with the lens orientation because the stress is predominantly radially oriented, similar to Figure 25.15. From the circular polariscope configuration, no pattern change is detected by rotating the lens; the retardance magnitude is revealed in one snapshot independent from the lens orientation. The color shift between blue and yellow when rotating the lens for 45° in the sensitive tint plate polariscope reveals that the birefringent axis is 45° with respect to two orthogonal directions of the Maltese cross pattern.

Mounting optical components can apply external forces that induce birefringence. Figures 25.22 through 25.24 show polariscope images of a rectangular plane parallel glass plate with increasing

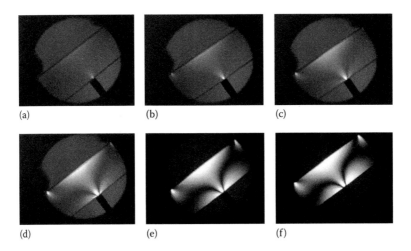

Figure 25.22 A plane parallel glass plate in a linear polariscope with increasing force from (a) to (f).

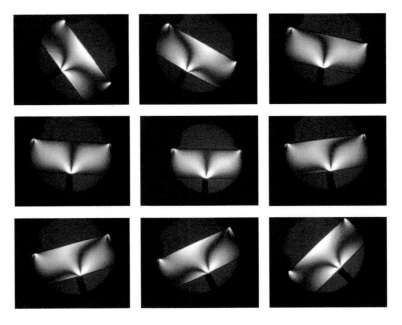

Figure 25.23 Linear polariscope images as the plane parallel glass plate of Figure 25.22f is rotated. The dark band indicates when its fast axis is aligned with the polariscope axes.

Polarized Light and Optical Systems

25.6 Effects of Stress Birefringence on Optical System Performance

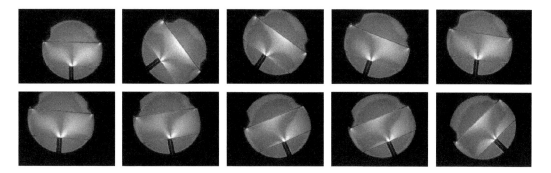

Figure 25.24 With the addition of a tint plate retarder, the resultant color pattern changes. The pinkish background represents near zero retardance or nearly isotropic conditions.

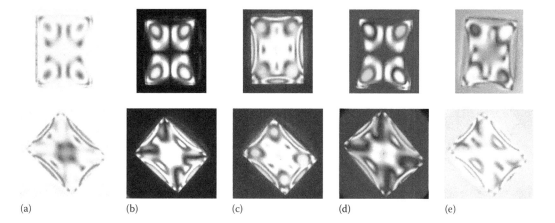

(a)   (b)   (c)   (d)   (e)

Figure 25.25 *Glas Spannungen* with intrinsic stress is placed under (a) parallel linear polarizers, (b) crossed linear polarizers, (c) crossed circular polarizers, (d) sensitive tint plate polariscope, and (e) crossed polarizers with a quarter waveplate. The stressed glass is viewed in 0° and 45° orientations in the two rows that reveal all the stress. The birefringence pattern appears different for the two orientations, except in circular polariscope, which produces intensity independent of fast axis orientation.

forces applied by a screw in the middle of one edge. The plate is supported by two pins on the opposite edge. It shows that unnecessary stress applied to optical elements causes stress that adds to the wavefront aberration. As shown later, this undesired stress induces polarization aberrations.

A glass sample formed with intentionally high stress, as observed under a polariscope in Figure 25.25. The view of the stress pattern changes in different polariscope configurations. Testing is usually performed at 0° and 45° to observe both components of stress, associated with Stokes parameters S_1 and S_2, with different glass orientations providing kaleidoscope-like images.

25.6.2 Simulations of Injection-Molded Lens

The simulation methods explained in Section 25.5 are applied to calculate the retardance induced by an injection-molded lens shown in Figure 25.26 with stress shown in Figure 25.27. The complete 3 × 3 stress tensor of this lens is also shown in Figure 25.11.

A collimated grid of rays is traced through this stressed lens. The retardance variation across the lens is calculated from the ray tracing **P** matrices and plotted in Figure 25.28. Rapid variation of retardance appears on the flange and the gate of the lens. The retardance is lower at the center

902 Chapter 25: Stress-Induced Birefringence

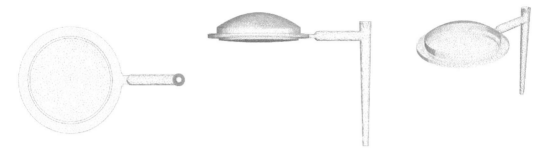

Figure 25.26 The CAD image of an injection-molded lens in three different views.

Figure 25.27 The stress images of the injection-molded lens (the same lens as in Figure 25.26) with a cut-away through the CAD variation file.

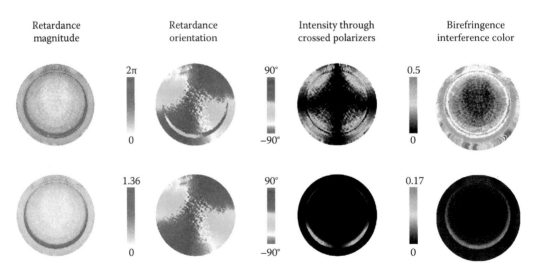

Figure 25.28 Retardance magnitude, retardance orientation, polariscope images, and birefringence color calculated from a polarization ray trace of an injection-molded lens with two sets of stress optic coefficients. (Top row) $\{C_1, C_2\}$ is $\{5 \times 10^{-14}\,\text{Pa}^{-1}, 5 \times 10^{-13}\,\text{Pa}^{-1}\}$. (Bottom row) $\{C_1, C_2\}$ is $\{6.5 \times 10^{-13}\,\text{Pa}^{-1}, 4.5 \times 10^{-12}\,\text{Pa}^{-1}\}$. The gate of the lens is located at the top of the lens shown.

Polarized Light and Optical Systems

25.6 Effects of Stress Birefringence on Optical System Performance

and slightly higher toward the edge. This lens is also simulated in a polariscope with cross polarizers. The light leakage through the polariscope is due to the shape of the lens as well as the stress-induced retardance. Different stress optic coefficients result in different induced retardance across the object. The set of simulated figures with the higher stress coefficients gives higher retardance magnitude and induces more leakage through the polariscope.

25.6.3 Simulation of a Plastic DVD Lens

DVD and CD systems are sensitive to stress-induced polarization from injection-molded lenses since the DVD signal is routed twice through a polarization beam splitter. The DVD signal is degraded by too much stress-induced aberrations. Too much degradation causes the bit error rate to increase to unacceptable levels. Without stress, the uncoated lens has zero retardance and performs as a conventional ray trace would predict for isotropic lenses; with stress, the lens has stress-induced retardance. The stress-induced retardance of an example DVD pickup lens (Figure 25.29) across the pupil is shown in Figure 25.30.

The DVD pickup lens configuration in Figure 25.29 (left) is illuminated through and then images through a polarizing beam splitter. The image quality is evaluated by its amplitude response matrix **ARM**, as shown in Figure 25.31. The main **ARM** components xx and yy are nearly Airy disks for the unstressed lens (Figure 25.31, left) with a small amount of light coupled into the off-diagonal elements xy and yx. When the polarization ray trace analysis is repeated with the stress data, the

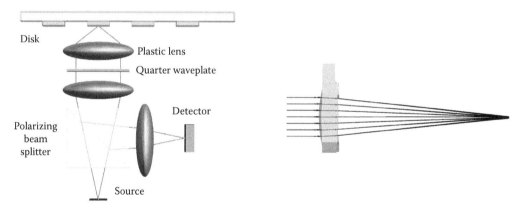

Figure 25.29 (Left) Optical layout of an optical pickups system in a CD player.[37] (Right) A model of an injection-molded lens focuses on a collimated incident beam.

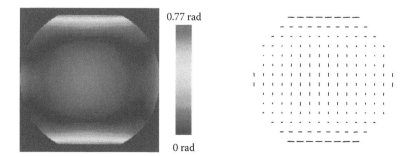

Figure 25.30 Induced retardance magnitude (left) and retardance orientation (right) of a stress lens across the pupil.

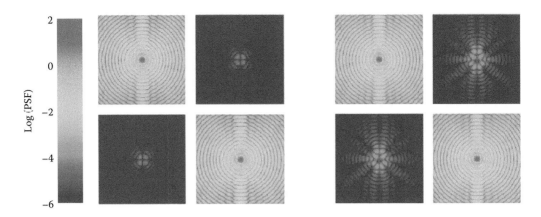

Figure 25.31 The **ARM** of an unstressed (left) and stressed (right) injection-molded lens represented in log scale.

main *xx* and *yy* components are only slightly affected, but much more light is now coupled into the orthogonal polarization states, as shown by the off-diagonal elements *xy* and *yx*. The corresponding polarization modulation transfer functions (MTFs) are shown in Figure 25.32 with an overall lower modulation.

As illustrated in Figure 25.33, the cross-coupling components in the resultant **ARM** depart more and more from an Airy disk as the stress increases. At extreme levels of stress (Figure 25.33e), the lens loses its ability to form an image.

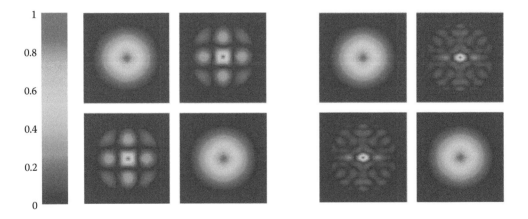

Figure 25.32 The corresponding MTFs of an unstressed (left) and stressed (right) injection-molded lens for Figure 25.31.

25.7 Conclusion

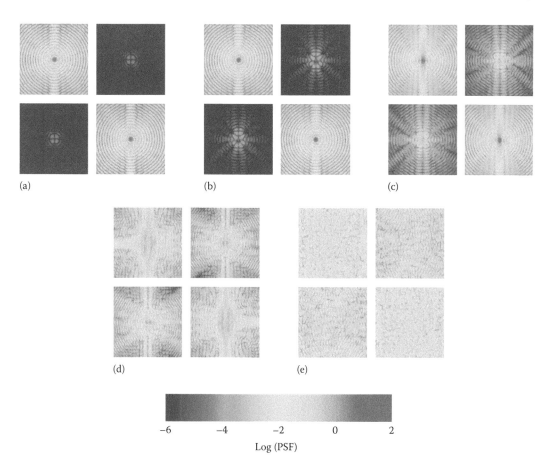

Figure 25.33 Point spread functions of polarized light refracting through a lens with (a) no stress, (b) some modeled stress, (c) 10 times the stress than (b), (d) 100 times the stress, and (e) 1000 times the stress.

25.7 Conclusion

Stress in optical elements affects performance. In this chapter, complex stress distributions are modeled as a varying anisotropic material with spatially varying dielectric tensor. Stress in isotropic materials was treated here. Anisotropic material requires more complex stress optic tensors and strain optic tensors to describe its directional-dependent change induced by external stress. Similar algorithms with more elements than Equation 25.6 can still model the optical performance. This ray tracing method, modeling optical components with spatially varying optical properties, can be applied to simulate gradient index optics, liquid crystal, and other similar optical elements.

Being able to analyze and simulate the effect of stress in optical elements is important for industrial inspection, product control, and tolerance analysis for precision optical system. Stress birefringence can be measured by Mueller matrix polarimeters, polariscopes, the sensitive tint plate method, and the Senarmont method. Other birefringence measurement methods are also developed for higher accuracy and shorter processing time using photoelastic modulators, optical heterodyne interferometry,[38] phase shifting technique,[39] and near-field optical microscopy.[40] By comparing measurement and simulation, improvements can be made in manufacturing procedures for molded lenses and other elements. The stress simulation has been used to predict the product performance and the trade-off between the processing time and the product quality in injection molding and to reduce the expensive cost of trial and error.

25.8 Problem Sets

25.1 What are the differences between ray tracing through a sample with stress birefringence and a uniaxial sample?

25.2 Images in Figures 25.1 and 25.2 are both taken with crossed polarizers. Why does Figure 25.2 contain only black and white? And why does Figure 25.1 contain color?

25.3
a. Is the polarizer in polarizing sunglasses on the outer surface, the inner surface, or spread throughout?
b. If the stress birefringence of a pair of polarizing sunglasses was located in the middle, but not in the outer surfaces, what would be the effect in the three cases in (a)?
c. Perform measurements with a polariscope or linear polarizers. What do your measurements show?

25.4 For the change in refractive index Δn of 0.0001, what is the corresponding stress for the materials in Table 25.2?

25.5 In the weak stress limit for isotropic materials, the change in refractive index Δn is related to the principal stress difference $\sigma_{xx} - \sigma_{yy}$ through the stress-optic coefficients C_1 and C_2:

$$\Delta n = C(\sigma_{xx} - \sigma_{yy}) = \frac{1}{2}n_0^3(C_1 - C_2)\sigma,^*$$

where n_o is the refractive index of unstressed material. At a wavelength of 633 nm:

Lucite	$n_0 = 1.491$	$C_1 = 35.4 \times 10^{-12}$/Pa	$C_2 = 24.9 \times 10^{-12}$/Pa
Fused silica	$n_0 = 1.46$	$C_1 = 0.65 \times 10^{-12}$/Pa	$C_2 = 4.5 \times 10^{-12}$/Pa

a. For principal stress difference $\sigma_{xx} - \sigma_{yy} = 100$ MPa, what is the change in refractive index for lucite plastic? And for fused silica glass?
b. Given the above stress, what is the retardance magnitude (expressed in waves) for a 1-mm-thick piece of lucite? And for a 1-mm-thick piece of fused silica?
c. Assume 1-mm thickness for both fused silica and lucite. What is the principal stress difference needed in the glass to get the same retardance as 100 MPa stress in the plastic?

---

* A negative sign might apply for another convention of compression and tension.

25.6 Using Figures 25.23 and 25.24, locate the birefringent axis of the stressed glass.

25.7 Explain what information can be concluded about the sample in Figure 25.25.

Acknowledgments

This chapter incorporates the work of Greg Smith, who developed the spatially varying stress algorithms used in many of the examples. Some of this work resulted from collaborations with Nalux Co. Ltd and Emhart Glass.

References

1. D. Brewster, On the effects of simple pressure in producing that species of crystallization which forms two oppositely polarized images, and exhibits the complementary colors by polarized light, *Philos. Trans. R. Soc. Lond.* 105 (1815): 60–64.
2. E. G. Coker and L. N. G. Filon, *Treatise on Photoelasticity*, London: Cambridge University Press (1931).
3. G. Birnbaum, E. Cory, and K. Gow, Interferometric null method for measuring stress-induced birefringence, *Appl. Opt.* 13 (1974): 1660–1669.
4. E. R. Cochran and C. Ai, Interferometric stress birefringence measurement, *Appl. Opt.* 31 (1992): 6702–6706.
5. A. V. Appel, H. T. Betz, and D. A. Pontarelli, Infrared polariscope for photoelastic measurement of semiconductors, *Appl. Opt.* 4 (1965): 1475–1478.
6. W. Su and J. A. Gilbert, Birefringent properties of diametrically loaded gradient-index lenses, *Appl. Opt.* 35 (1996): 4772–4781.
7. R. K. Kimmel and R. E. Park, *ISO 10110 Optics and Optical Instruments—Preparation of Drawings for Optical Elements and Systems: A User's Guide*, Washington, DC: Optical Society of America (1995).
8. ISO/DIS 10110—Preparation of drawings for optical elements and systems. Part 2: Material imperfections—Stress birefringence (1996).
9. A. Y. Yi and A. Jain. Compression molding of aspherical glass lenses—A combined experimental and numerical analysis, *J. Am. Cer. Soc.* 88.3 (2005): 579–586.
10. A. Yariv and P. Yeh, *Optical Waves in Crystals: Propagation and Control of Laser Radiation*, John Wiley & Sons (1984), pp. 319–329.
11. D. A. Pinnow, Elastooptical materials, in *CRC Handbook of Lasers with Selected Data on Optical Technology*, ed. R. J. Pressley, Cleveland, OH: The Chemical Rubber Company (1971).
12. M. Huang, Stress effects on the performance of optical waveguides, *Int. J. Solids Struct.* 40 (2003): 1615–1632.
13. K. Doyle, V. Genberg, and G. Michaels, Numerical methods to compute optical errors due to stress birefringence, *Proc. of SPIE* 4769 (2002): 34–42.
14. S. He, T. Zheng, and S. Danyluk, Analysis and determination of the stress-optic coefficients of thin single crystal silicon samples, *J. Appl. Phys.* 96(6) (2004): 3103–3109.
15. N. F. Borrelli and R. A. Miller, Determination of the individual strain-optic coefficients of glass by an ultrasonic technique, *Appl. Opt.* 7 (1968): 745–750.
16. R. E. Newnham, *Properties of Materials: Anisotropy, Symmetry, Structure*, Oxford University Press (2004).
17. R. M. Waxler, D. Horowitz, and A. Feldman, Optical and physical parameters of Plexiglas 55 and Lexan, *Appl. Opt.* 18 (1979): 101–104.
18. M. G. Wertheim, Mémoire sur la double refraction temporairement produite dans les corps isotropes, et sur la relation entre l'élasticité mécanique et entre l'élasticité optique, *Ann. Chim. Phys.* 40 (1854): 156–221.
19. A. Kuske and G. Robertson, *Photoelastic Stress Analysis*, Wiley (1974).
20. M. Born and E. Wolf, *Principles of Optics: Electromagnetic Theory of Propagation, Interference and Diffraction of Light*, 6th edition, Pergamon Press (1980), pp. 703–705.

21. Schott Optical Glass Catalogue (http://www.us.schott.com).
22. V. N. Tsvetkov and N. N. Boitsova, *Vysokomol. Soedin.* 2 (1960): 1176.
23. B. E. Read, Dynamic birefringence of poly(methyl methacrylate), *J. Polym. Sci. C* 16(4) (1967): 1887–1902.
24. D. W. van Krevelen, *Properties of Polymers*, 2nd edition, Amsterdam: Elsevier (1976).
25. D. L. Keyes, R. R. Lamonte, D. McNally, and M. Bitritto, Polymers for photonics, *Opt. Polym.* (2001): 131–134.
26. S. Shirouzu et al., Stress-optical coefficients in polycarbonates, *Jpn. J. Appl. Phys.* 29 (1990): 898.
27. HOYA Cooperation, *Optical Glass Master Datasheet* (http://www.hoyaoptics.com).
28. M. Zgonik, P. Bernasconi, M. Duelli, R. Schlesser, P. Günter, M. H. Garrett, D. Rytz, Y. Zhu, and X. Wu, Dielectric, elastic, piezoelectric, electro-optic, and elasto-optic tensors of $BaTiO_3$ crystals, *Phys. Rev. B* 50(9) (1994): 5941.
29. R. B. Pipes and J. L. Rose, Strain-optic law for a certain class of birefringent composites, *Exp. Mech.*, 14(9) (1974): 355–360.
30. *SigFit* is a product of Sigmadyne, Inc., Rochester, NY.
31. R. Y. Chang and W. H. Yang, Numerical simulation of mold filling in injection molding using a three-dimensional finite volume approach, *Int. J. Numer. Methods Fluids* 37 (2001): 125–148.
32. H. E. Lai and P. J. Wang, Study of process parameters on optical qualities for injection-molded plastic lenses, *Appl. Opt.* 47 (2008): 2017–2027.
33. Y. Maekawa, M. Onishi, A. Ando, S. Matsushima, and F. Lai, Prediction of birefringence in plastics optical elements using 3D CAE for injection molding, *Proc. SPIE* 3944 (2000): 935–943.
34. L. Manzione, *Applications of Computer Aided Engineering in Injection Molding*, Oxford University Press (1988).
35. T. Möller and B. Trumbore, Fast, minimum storage ray-triangle intersection, *J. Graph. Tools*, 2(1) (1997): 21–28.
36. Y.-J. Chang et al., Stimulations and verifications of true 3D optical parts by injection molding process, *Proc. ANTEC* 22(24) (2009).
37. What is Light? Chapter 3: Applications of Light: CDs and DVDs, Canon Science Lab (http://www.canon.com/technology/s_labo/light/003/06.html, accessed January 2015).
38. R. Paschotta, Optical Heterodyne Detection, *Encyclopedia of Laser Physics and Technology* (http://www.rp-photonics.com/optical_heterodyne_detection.html, accessed January 2015).
39. E. Hecht, *Optics*, Addison-Wesley (2002).
40. Y. Oshikane et al., Observation of nanostructure by scanning near-field optical microscope with small sphere probe, *Sci. Technol. Adv. Mater.* 8(3) (2007): 181.

26

Multi-Order Retarders and the Mystery of Discontinuities

26.1 Introduction

Throughout this book, retardance has proven to be one of the more difficult concepts to define. This chapter examines the spectra of high-order retarders and how the retardance varies with wavelength. The retardance spectra of these compound multi-order retarders appear to "turn around" and avoid integer numbers of waves of retardance, when it may be obvious that the retardance must be continuously increasing. Such issues that occur in measurement and modeling of retarders are addressed in this chapter. The operation of *retardance unwrapping* of the principal retardance is developed to explain the behavior of retardance of the compound retarders. This allows us to generalize the concept of retarder order for what at first appears to be mysterious jumps in the retardance.

Two approaches to retardance discontinuities are developed. One approach uses a dispersion model to describe the retardance behavior. Another approach considers multiple wavefronts exiting the compound retarder system, that is, a *multivalued optical path length* (*OPL*) view.

The *common definition* of a retarder is a device that divides a beam into two orthogonal modes and introduces a relative phase difference δ.[1] Another view of retarders is provided by the Mueller calculus and Poincaré sphere. A retarder rotates polarization states on the Poincaré sphere about an axis by the retardance δ; as light propagates through a retarder, the incident state on the Poincaré sphere is rotated about a retardance axis to another state. In this Mueller/Poincaré picture, cascading retarders is equivalent to cascading rotations on the Poincaré sphere. This view of retarders and the associated Mueller matrices have ambiguities of retardance order; all retarders with retardance $2n\pi \pm \delta$, with integer n, have the same Mueller matrix. In this Mueller retardance picture, the final polarization state always ends up in the right place but the retardance is only known to modulo 2π

910　Chapter 26: Multi-Order Retarders and the Mystery of Discontinuities

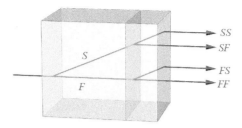

Figure 26.1 A misaligned two-element compound linear retarder has an arbitrary angle between the two fast axes, different from 0° or 90°. Four output beams exit a misaligned retarder with four different combinations of the fast and slow *OPL*s. For an aligned retarder, only the *FF* and *SS* beams exit.

and the retardance fast and slow states may be either of two orthogonal states along an axis through the Poincaré sphere.

A *multi-order retarder* is a retarder with more than a half wave of retardance. A *compound retarder* is a combination of two or more retarders in sequence. A *misaligned retarder* is defined here as a compound retarder where the fast axes of the components are not at 0° or 90° to each other as shown in Figure 26.1. For misaligned compound multi-order retarders, more than two beams are involved and the common definition of retarders is inadequate. Using the *retarder space* and algorithms first introduced in Chapter 14 (Jones Matrix Data Reduction with Pauli Matrices), a retardance unwrapping algorithm is derived. An example compound retarder system is studied to demonstrate discontinuities in the retardance space. These discontinuities correspond to the retardance spectra missing the identity matrix as the retardance transitions from less than an integer number of waves of retardance to greater than that integer number of waves. The explanation for this breakdown to the conventional definition of retardance arises from the interference of the four or more beams as shown in Figure 26.1 such that the simple definition of a retarder does not apply.

In this chapter, the retardance spectrum of an example Quartz and Sapphire misaligned retarder is modeled and compared to Mueller matrix imaging polarimeter measurements.

26.2 Mystery of Retardance Discontinuity

Polarimeters measure *principal retardance*, which varies between 0 and π. As shown in Figure 26.2, the orientation of the retarder's fast axis flips by 90° at a retardance of 0° and 180° while the retardance

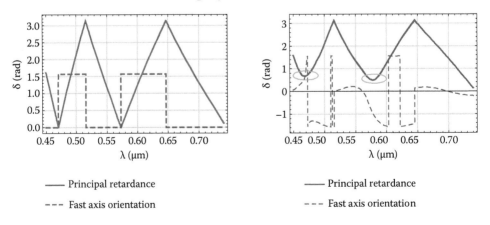

Figure 26.2 (Left) Principal retardance (solid line) and fast axis orientation (dashed line) of a non-compound retarder as a function of wavelength. (Right) Principal retardance (solid line) and fast axis orientation (dashed line) of a misaligned retarder with non-parallel or non-perpendicular fast axes as a function of wavelength. The principal retardance has a tendency to avoid zero retardance (green ovals).

26.2 Mystery of Retardance Discontinuity

abruptly increases or decreases. Similar behavior can be observed in interferometers. Most interferometers measure phase between 0 and 2π and often use phase unwrapping to extend the range of phase beyond one wave.[2-4] For non-compound retarders (Figure 26.2, left), the principal retardance can be unwrapped to a smooth curve with no discontinuity as shown in the red curve of Figure 26.3. On the other hand, the principal retardance of a compound retarder does not reach zero, as shown in Figure 26.2 (right), and thus the unwrapped retardance has discontinuity as the blue curve shown in Figure 26.3. In this chapter, this mystery of *retardance discontinuity* will be explained.

In Section 17.2.1, the effect of rotating the polarization state analyzer (PSA) in a Jones matrix polarimeter is analyzed. The resultant family of Jones matrices forms a trajectory in the retarder space. The concept of the retarder space will be used to understand the unwrapped retardance in later sections. Figure 26.4 shows a trajectory of a half wave horizontal fast axis linear retarder (HLR) as the PSA rotates from 0 to π; fast axis orientation θ rotates from 0 to π. Figure 26.5 shows two different views of the trajectories of HLR with retardance δ_0 (written in blue) as the PSA rotates.

Since the Mueller matrices for retarders repeat every $n\pi$, the retardance space trajectory repeats every π rotation of θ. For the half wave retarder ($\delta_0 = \pi$), the trajectory stays in $\delta_R - \delta_{45}$ plane and moves around the π retardance sphere in a semicircle. When $\delta_0 = 0$, the retardance trajectory remains along the δ_R-axis; the Jones matrix for an empty sample compartment appears as a circular retarder as the PSA rotates. As δ_0 increases from 0, the trajectory starts to curve and form a spiral. All the initial points for each δ_0 start from the δ_H-axis. When the trajectory reaches a distance π from the origin, the retardance components *jump* to the opposite point with respect to the origin and continue to move until it returns to the starting point as θ reaches π. Unwrapping retardance is equivalent to understanding the trajectory of retarders in the retarder space.

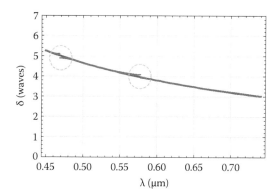

Figure 26.3 The unwrapped retardance of a single retarder of Figure 26.2 (left) in red and the compound retarder of Figure 26.2 (right) in blue showing retardance discontinuities.

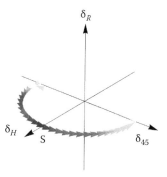

Figure 26.4 A trajectory of HLR in the retarder space as the PSA rotates from 0 to π. The trajectory starts at (π, 0, 0), at where the letter "S" is, moves to (0, π, 0), and to (0, $-\pi$, 0), and comes back to the starting point.

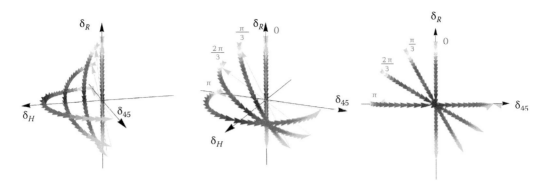

Figure 26.5 Trajectories of HLR ($\delta = 0$, $\pi/3$, $2\pi/3$, and π) in the retarder space as the PSA rotates from 0 to π are shown from the side views (left two figures) and the $\delta_{45} - \delta_R$ view (right). Each trajectory starts on a point along δ_H and then follows the same color scheme as Figure 26.4.

26.3 Retardance Unwrapping for Homogeneous Retarder Systems Using a Simple Dispersion Model

A dispersion model of retardance and the use of the model for unwrapping the continuous retardance of a homogeneous retarder system are shown in this section.

26.3.1 Dispersion Model

The Jones matrices for monochromatic beams with different absolute phase terms cannot be distinguished; for example,

$$\mathbf{J}_1 = \begin{pmatrix} 1 & 0 \\ 0 & 1 \end{pmatrix} = \mathbf{J}_2 = \begin{pmatrix} 1 & 0 \\ 0 & e^{-2\pi i} \end{pmatrix} = \mathbf{J}_3 = \begin{pmatrix} 1 & 0 \\ 0 & e^{-4\pi i} \end{pmatrix}, \quad (26.1)$$

where $\mathbf{J}_1$, $\mathbf{J}_2$, and $\mathbf{J}_3$ are Jones matrices for 0, 1, and 2 waves of retardance. Thus, the absolute phase term is often ignored for monochromatic light sources, and all three matrices are treated as an identity matrix. In the Michelson interferometer with polychromatic light, the location of the zero optical path difference (*OPD*) can be found, the so-called white light or zero order fringe. Thus, with polychromatic light, it may be possible to distinguish absolute phases in cases such as Equation 26.1.

A simple model for the wavelength-dependent retardance of a waveplate assumes the retardance varies as $1/\lambda$,

$$\delta(\lambda) = \frac{\delta_0 \lambda_0}{\lambda} \quad (26.2)$$

where δ_0 is the retardance for a reference wavelength λ_0. This model follows from the assumption that the ordinary and extraordinary refractive indices are not dependent on the wavelength. Equation 26.2 is the dispersion model of the retardance. This model is used for the retardance unwrapping of the principal retardance for homogeneous and inhomogeneous retarder systems.

26.3.2 Retardance of the Homogeneous Retarder System

In Section 26.3.3, a retardance unwrapping algorithm using the dispersion model equation is introduced. The Quartz and the Sapphire retarder models and their *principal retardance* and fast axes

26.3 Retardance Unwrapping for Homogeneous Retarder Systems Using a Simple Dispersion Model

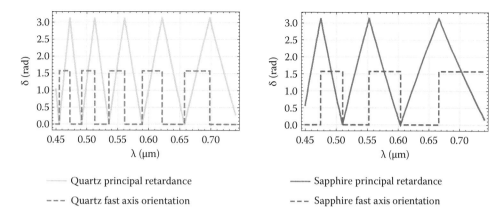

Figure 26.6 The principal retardance spectra measured by Mueller matrix polarimetry (solid lines) and fast axis orientation (dashed lines). (Left) Quartz retarder and (right) Sapphire retarder oscillate between 0 and π. Each time the retardance changes direction, the fast axis changes by 90°.

orientation for 0.45 μm < λ < 0.74 μm are shown in Figure 26.6. These models are created to compare with experimental results. The principal retardance and the fast axis orientation are calculated by the algorithms in Chapter 14 (Jones Matrix Data Reduction with Pauli Matrices). For this example, both retarders' ordinary axes are along the horizontal axis. For Quartz, a positive uniaxial material, horizontal is the fast axis and its thickness is 0.5831 mm. For Sapphire, a negative uniaxial material, horizontal is the slow axis and its thickness is 0.37427 mm. The fast axis orientations measured by a Mueller matrix polarimeter as a function of wavelength switches between 0° and 90° because the value of the principal retardance is bounded between 0 and π although the retardance is steadily increasing from longer to shorter wavelengths; hence, the measured fast axis orientation changes from 0° to 90° each time the principal retardance reaches 0 or π. The Mueller matrix of the retarders varies continuously as the wavelength changes from 0.74 to 0.45 μm.

From a Jones or Mueller matrix spectrum, the retarder order at each λ is determined using the dispersion model by rearranging sections of the principal retardance spectrum. Retardance unwrapping is an algorithm to arrange the pieces of the principal retardance to generate a plausible overall retardance spectrum. Our example of a compound retarder, formed from a Sapphire retarder followed by a Quartz retarder, is used to understand the retardance unwrapping algorithm. When two horizontal linear retarders made from Sapphire and Quartz are aligned so that ordinary axes of each crystal are aligned or perpendicular to each other, the combination forms another linear retarder system since the eigenpolarizations are parallel or orthogonal to each other. Here, the two ordinary axes are aligned and the retarder's fast axes are perpendicular to each other. Thus, the total retardance is the difference of the individual retarders, the Quartz retardance subtracted from the Sapphire retardance,

$$\delta_{Total}(\lambda) = \delta_{Quartz}(\lambda) - \delta_{Sapphire}(\lambda), \tag{26.3}$$

where $\delta_{Quartz}(\lambda) > 0$ and $\delta_{Sapphire}(\lambda) > 0$.

Figure 26.7 shows the continuous principal retardance of the Quartz retarder and the Sapphire retarder and the total retardance $\delta_{Total}(\lambda)$ of the homogeneous retarder system.

Figure 26.8 compares the simulation and an actual Mueller matrix polarimeter measurement of the principal retardance magnitude and orientation of the compound retarder. The measurement is taken every 0.01 μm between 0.45 and 0.74 μm. The simulation closely matches the measured retardance data.

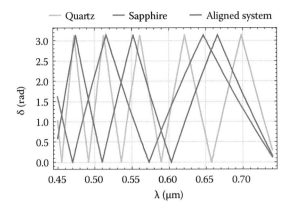

Figure 26.7 The principal retardance is plotted as a function of wavelength for the Quartz and Sapphire (green and blue) retarders and the combination with the fast axes crossed so the retardances subtract (red).

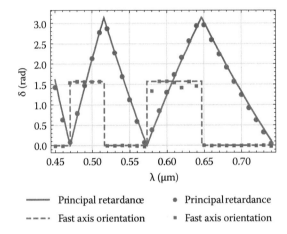

Figure 26.8 Principal retardance (red) and orientation (dark red) of the Sapphire and Quartz compound retarder as a function of wavelength for the simulation (lines) and measurement (dots).

Examining the curves from longer toward shorter wavelengths, the principal retardance oscillates between 0 and π more rapidly as the wavelength becomes shorter. Coming from the right side of Figure 26.8, the retardance increases to π with a horizontal fast axis and then decreases with a vertical fast axis. When the retardance reaches zero, it increases again with a horizontal fast axis, and so on. To unwrap the principal retardance, a *mode number* $q = \gamma$ is assigned to each segment of the principal retardance with different fast axis orientation; odd q's correspond to a horizontal fast axis, and even q's correspond to a vertical fast axis. Figure 26.9 shows the mode numbers with an odd value γ, odd mode numbers in blue, and even in red.

26.3.3 Homogeneous Retarder's Trajectory and Retardance Unwrapping in Retarder Space

Figure 26.10 shows the retarder space trajectory (see Section 14.5 for more details) for the principal retardance, which remains within a sphere of radius π, jumping from one side of the sphere to the other at a radius $\delta_H = \pi$. The left figure corresponds to the $q = \gamma$, and the red color indicates the principal retardance vector $(\delta_H, \delta_{45}, \delta_R)$ for the longest wavelength. In each plot, arrow color changes from red → orange → green → blue → purple as the wavelength becomes shorter. The δ_R-axis

26.3 Retardance Unwrapping for Homogeneous Retarder Systems Using a Simple Dispersion Model

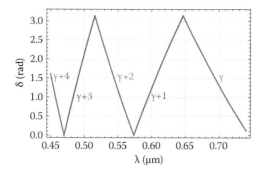

Figure 26.9 Each segment of the principal retardance is assigned a mode number q in units of half wavelengths. Starting from the right side of the graph, blue segments have odd mode numbers and the red segments have even mode numbers.

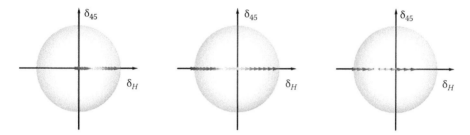

Figure 26.10 The principal retardance vector trajectories are shown in the retarder space as the wavelength changes. Each figure corresponds to a different mode number starting from the longest wavelength to the shortest wavelength: (left) $q = \gamma$, (middle) $q = \gamma + 1, \gamma + 2$, (right) $q = \gamma + 3, \gamma + 4$.

points out of the page, and each segment shows the points along the same fast axis. The left figure has the horizontal fast axis (along $+\delta_H$-axis) and the retardance is increasing. Once the retardance reaches the π sphere at $(\pi, 0, 0)$, its fast axis changes to the vertical direction (along the $-\delta_H$-axis) and jumps to the symmetric point $(-\pi, 0, 0)$. The next trajectory is continued in the middle figure, and so on. The origin $(0, 0, 0)$ is equivalent to the identity matrix, which represents no retardance, one wave retarder (2π), or a $2n\pi$ retarder for integer n. As the wavelength reduces, the fast axis orientation changes four times alternating along the horizontal and vertical directions.

The retardance unwrapping algorithm maintains the fast axis orientation as the wavelength reduces and estimates the true retardance at each wavelength. Thus, there should be no more jumps from one side to the other of the sphere. One assumption is that if one starts measuring a multi-order retarder at a long enough wavelength, the retardance will be less than a half wave of retardance. This gives a known starting point for the unwrapping, which would be the true retardance of that retarder. When this is the case, then when $q = 1$, the *principal retardance is the true retardance*. For even q's, the true retardance is $q\pi - \delta_{principal}$ and the fast axis orientation remains along the horizontal axis. For odd q's, the true retardance is $(q - 1)\pi + \delta_{principal}$ with the fast axis along the horizontal axis, that is,

$$\delta_{unwrapped} = \begin{cases} \delta_{principal} & \text{when } q = 1 \\ q\pi - \delta_{principal} & \text{when } q \in even \\ (q-1)\pi + \delta_{principal} & \text{when } q \in odd \end{cases} \quad (26.4)$$

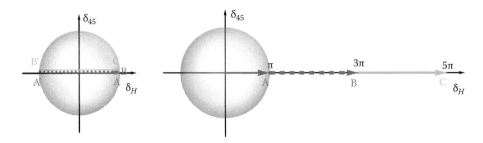

Figure 26.11 (Left) Principal retardance trajectory of the aligned retarder in the retarder space as the wavelength decreases. (Right) Retardance trajectory after retardance unwrapping.

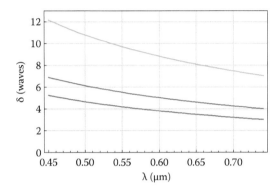

Figure 26.12 The unwrapped retardance as a function of wavelength for the two linear retarders individually (green and blue) and the combined system with their fast axes aligned (red).

Thus, the range of the unwrapped retardance has no upper limit.

Figure 26.11 (left) shows the principal retardance trajectory in the retarder space for the aligned example system as the wavelength reduces; when the principal retardance reaches the boundary value π, the trajectory moves to the origin symmetric point on the sphere and changes its fast axis to the orthogonal direction. Figure 26.11 (right) shows the retardance trajectory of the same system after the retardance is unwrapped; the horizontal retardance increases continuously, keeping the fast axis orientation along the $+\delta_H$ horizontal direction. As shown in Section 14.5, the distance from the origin to a point in the retarder space is the total retardance.

Retardance plots in Figure 26.7 are unwrapped in Figure 26.12. In this figure, the total retardance (red) value is always $\delta_{Quartz}(\lambda) - \delta_{Sapphire}(\lambda)$ (Equation 26.3). There is no discontinuity in the unwrapped retardance for homogeneous retarder systems that consist of two or more linear retarders with parallel or perpendicular fast axes.

26.4 Discontinuities in Unwrapped Retardance Values for Compound Retarder Systems with Arbitrary Alignment

Things become interesting with misaligned retarders, compound linear retarders whose fast axes are neither parallel nor perpendicular. Such retarders may result from misalignment or be intentionally aligned at arbitrary angles, as with *Pancharatnam-design retarders*.[5–7] Sequences of linear retarders whose axes are neither parallel nor perpendicular are in general *elliptical retarders* as is seen in the next section and have discontinuous retardance spectra. In practice, the axes of compound retarders would always be expected to have a small misalignment, within some alignment tolerance.

26.4.1 Compound Retarder Jones Matrix Decomposition

If the fast axis of the second retarder is slightly misaligned from that of the first retarder (it always will be in practice), the net retardance shows a different behavior from the aligned system. For example, if the two horizontal fast axis linear retarders with retardance $\delta_1(\lambda)$ and $\delta_2(\lambda)$ are misaligned by θ, the Jones matrix of the system is

$$\mathbf{J}_{Total} = \mathbf{J}_1 \mathbf{J}_2$$
$$= e^{\frac{-i(\delta_1+\delta_2)}{2}} \begin{pmatrix} \cos^2(\theta) + e^{i\delta_2}\sin^2(\theta) & (1-e^{i\delta_2})\cos(\theta)\sin(\theta) \\ e^{i\delta_1}(1-e^{i\delta_2})\cos(\theta)\sin(\theta) & e^{i\delta_1}(e^{i\delta_2}\cos^2(\theta) + \sin^2(\theta)) \end{pmatrix}. \quad (26.5)$$

Using Equation 14.77, the principal retardance is

$$\delta = 2\arccos\left[\cos\left(\frac{\delta_1+\delta_2}{2}\right)\cos^2(\theta) + \cos\left(\frac{\delta_1-\delta_2}{2}\right)\sin^2(\theta)\right], \quad (26.6)$$

which reduces to $\delta = \delta_1 + \delta_2$ if $\theta \to 0$ and $\delta = \delta_1 - \delta_2$ if $\theta \to \pi/2$. Equation 26.6 shows that the retardance of the system depends on θ. If both retarders are half wave retarders with horizontal fast axes, the total retardance will be 2π (one wave) when the two fast axes are aligned, and the total retardance will be 0 when the fast axes are orthogonal. Figure 26.13 shows how the total retardance of the system changes as the second retarder's fast axis orientation (θ) varies.

Note that in Figure 26.13, zero principal retardance occurs three times but they imply different total retardance; at $\theta = \pi/2$, zero principal retardance means zero total retardance but at $\theta = 0$ and π, the total retardance is 2π.

$\mathbf{J}_1$ and $\mathbf{J}_2$ in Equation 26.5 can be written as the sum of *Pauli matrices*,

$$\mathbf{J}_1 = \mathbf{LR}_1(\delta_1, 0) = \cos\left(\frac{\delta_1}{2}\right)\sigma_0 - i\sin\left(\frac{\delta_1}{2}\right)\sigma_1 \quad \text{and}$$
$$\mathbf{J}_2 = \mathbf{LR}_2(\delta_2, \theta) = \cos\left(\frac{\delta_2}{2}\right)\sigma_0 - i\cos(2\theta)\sin\left(\frac{\delta_2}{2}\right)\sigma_1 - i\sin(2\theta)\sin\left(\frac{\delta_2}{2}\right)\sigma_2, \quad (26.7)$$

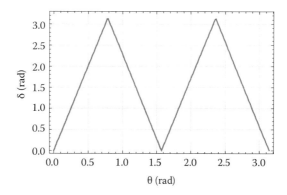

Figure 26.13 The principal retardance for two combined half wave linear retarders as a function of the fast axis orientation (θ) of the second retarder with respect to the first retarder.

where $\mathbf{LR}(\delta, \theta)$ is a linear retarder with a principal retardance δ and its fast axis along θ. The apparent retardance discontinuities (Figure 26.3) are explained by separating the $\mathbf{J}_{Total}$ in Equation 26.5 into two parts, a θ-independent Jones matrix $\mathbf{J}_{Major}$ and θ-dependent Jones matrix $\mathbf{J}_{Minor}$,

$$\mathbf{J}_{Total} = \mathbf{LR}_1(\delta_1, 0)\mathbf{LR}_2(\delta_2, \theta) = \mathbf{J}_{Major}\mathbf{J}_{Minor}. \quad (26.8)$$

$\mathbf{J}_{Major}$ matrix is a θ-independent part, a horizontal fast axis linear retarder with retardance $\delta_{Major} = \delta_1 + \delta_2$,

$$\mathbf{J}_{Major} = \mathbf{LR}(\delta_1 + \delta_2, 0) \quad (26.9)$$

and $\mathbf{J}_{Minor}$ is a θ-dependent part,

$$\begin{aligned}\mathbf{J}_{Minor} &= c_0\left[\boldsymbol{\sigma}_0 + d_1\boldsymbol{\sigma}_1 + d_2\boldsymbol{\sigma}_2 + d_3\boldsymbol{\sigma}_3\right] \\ &= \left[\cos^2(\theta) + \cos(\delta_2)\sin^2(\theta)\right]\left[\boldsymbol{\sigma}_0 + \frac{i}{\cot(\delta_2) + \cot^2(\theta)\csc(\delta_2)}\boldsymbol{\sigma}_1 \right.\\ &\quad \left. - \frac{i\cot(\theta)}{\cot(\delta_2) + \cot^2(\theta)\csc(\delta_2)}\boldsymbol{\sigma}_2 + \frac{i\sin^2(\delta_2/2)\sin(2\theta)}{\cos^2(\theta) + \sin^2(\theta)\cos(\delta_2)}\boldsymbol{\sigma}_3\right].\end{aligned} \quad (26.10)$$

$\mathbf{J}_{Minor}$ is a retarder since the coefficients d_1, d_2, and d_3 are pure imaginary (see Chapter 14 [Jones Matrix Data Reduction with Pauli Matrices]). Using Equation 14.77, the retardance of $\mathbf{J}_{Minor}$ is

$$\delta_{Minor} = 2\arctan\left[\sqrt{\csc^2\left(\frac{\delta_1}{2}\right)\csc^2(\theta) - 1}\right]. \quad (26.11)$$

Note that $\mathbf{J}_{Major}$ contains the phase difference between the *FF* and *SS* modes in Figure 26.1 while $\mathbf{J}_{Minor}$ contains the phase difference between the *FS* and *SF* modes. As shown in the previous section, $\mathbf{J}_{Major}$ always has continuous unwrapped retardance. Thus, the discontinuities in the unwrapped retardance of $\mathbf{J}_{Total}$ come from $\mathbf{J}_{Minor}$. Note that the amplitude of $\mathbf{J}_{Minor}$ has θ dependence; the discontinuity becomes maximum at $\theta = \pi/4$. Figure 26.14 shows the principal retardance of $\mathbf{J}_{Major}$ in red and $\mathbf{J}_{Minor}$ in blue when $\theta = 10°$.

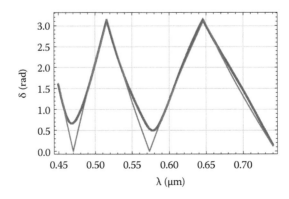

Figure 26.14 The principal retardance of $\mathbf{J}_{Major}$ (red) and $\mathbf{J}_{Minor}$ (blue) is plotted for 10° fast axis misalignment.

26.4 Discontinuities in Unwrapped Retardance Values for Compound Retarder Systems

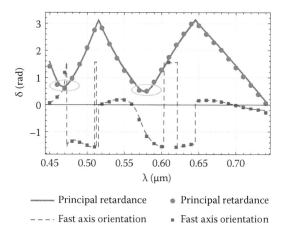

Figure 26.15 (Lines) Simulated principal retardance (blue) and the fast axis orientation (dark blue) of a system of $\mathbf{J}_{Total}$, a Sapphire retarder with fast axis at 10° followed by a Quartz retarder with horizontal fast axis are plotted as a function of wavelength. Green circles indicate regions where the principal retardance changes its slope without going to zero, thus avoiding one wave, two waves, and so on, retardance values. (Dots) Measured principal retardance (blue) and the fast axis orientation (dark blue) of the system are also plotted as a function of wavelength.

As an example, consider our two laboratory linear retarders in sequence—a Sapphire retarder with retardance δ_2 at 10° followed by a Quartz HLR with retardance δ_1 at 0°. The compound retarder system has a Jones matrix

$$\mathbf{J}_{Total} = \mathbf{LR}_1(\delta_1, 0)\mathbf{LR}_2(\delta_2, 10°)$$

$$= e^{\frac{-i(\delta_1+\delta_2)}{2}} \begin{pmatrix} \cos^2(10°) + e^{i\delta_2}\sin^2(10°) & (1-e^{i\delta_2})\cos(10°)\sin(10°) \\ e^{i\delta_1}(1-e^{i\delta_2})\cos(10°)\sin(10°) & e^{i\delta_1}\left[e^{i\delta_2}\cos^2(10°) + \sin^2(10°)\right] \end{pmatrix}. \quad (26.12)$$

Using the simple dispersion model for retardance, the simulated principal retardance and the fast axis orientation of the combination ($\mathbf{J}_{Total}$) are plotted in Figure 26.15, which also shows the measured principal retardance and the fast axis orientation yielding a close match between simulated values and measurements. Coming from the right side, the principal retardance increases to π (half wave), and then the fast axis changes to vertical and the principal retardance decreases. The principal retardance only decreases to 0.7 before turning around and increases to π a second time, the second maximum from the right. At this point, the retardance (δ) is expected to be 3π, corresponding to the second maxima in the top figure in Figure 26.8; however, the retardance never reaches 0, which would correspond to $\delta = 2\pi$. Thus, it appears as if the retardance has passed from π to 3π without passing through 2π!

26.4.2 Compound Retarder's Trajectory in Retarder Space

Full wave retarders ($2n\pi$ retardance) have Jones matrices, which are the identity matrix, corresponding to the origin (0, 0, 0) of the *retarder space*. The trajectory of the principal retardance of the example compound system repeatedly misses the origin of the retarder space as the wavelength scans, as shown in Figure 26.16. The trajectories' segments from π to π leaving one side of the π sphere jump to the other side as the wavelength decreases. Each segment shows the trajectory as it approaches the π sphere boundary. The left figure's trajectory starts at $\lambda = 0.74$ μm and moves as λ becomes shorter, and the right figure's trajectory ends at $\lambda = 0.45$ μm. Unlike the aligned system's retarder space trajectory in Figure 26.10, the compound system's retarder space trajectory

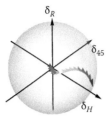 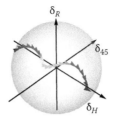 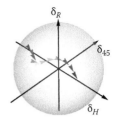

Figure 26.16 The retarder space inside the π sphere (purple sphere) and the principal retardance component trajectories are plotted as the wavelength decreases (left) λ = 0.74 to 0.64 μm, (middle) λ = 0.64 to 0.515 μm, and (right) λ = 0.515 to 0.45 μm. All figures have the same color scheme as Figure 26.10; red → yellow → green → blue → purple. The left figure's ending point is where the retardance is π. The middle figure starts from the symmetric point of the ending point of the left figure, and so on. Note that the trajectories for the second and third figures miss the origin.

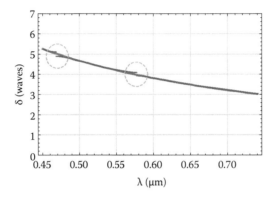

Figure 26.17 The unwrapped retardance plotted as a function of wavelength for the aligned (red) and misaligned (blue) two-retarder system.

repeatedly misses the origin, which indicates discontinuities in unwrapped retardance; points closest to the origin in Figure 26.16 middle and right are where the discontinuity happens, although the retarder space trajectory is not discontinuous. When unwrapped in retarder space, the discontinuities at integer numbers of waves become obvious.

Using the principal retardance of $\mathbf{J}_{Total}$ and the fast axis orientation, a retardance can be assigned by using the retardance unwrapping algorithm in Equation 26.4. Discontinuities are clearly visible when the unwrapped retardance for the misaligned (blue) and aligned (red) systems are plotted together in Figure 26.17. The blue plot has similar values as the red plot since the misalignment is small. However, the blue plot has discontinuities whenever the retardance value crosses $2n\pi$ boundaries; that is, the unwrapped retardance of the compound system increases from π to 3π without passing through 2π, the point of origin. The discontinuities come from non-zero amplitude in $\mathbf{J}_{Minor}$, that is, the existence of multiple modes exiting the compound system, which will be discussed in the next section.

26.4.3 Multiple Modes Exit the Compound Retarder System

Another view of the multi-order retarders considers each of the four ray paths as a different polarizer, with the output beams interfering upon exit. Thus, a ray propagating through a retarder can be modeled as coherent addition of two rays exiting from two polarizers with corresponding *OPL*s

26.4 Discontinuities in Unwrapped Retardance Values for Compound Retarder Systems

due to birefringence. A retarder has two modes, the fast and slow modes. For a misaligned system of two retarders, four modes exit, $F_1 F_2$, $F_1 S_2$, $S_1 F_2$, and $S_1 S_2$, where F and S stand for the fast and slow modes, and 1 and 2 stand for the first and the second retarders. At normal incidence, $F_1 F_2$, $F_1 S_2$, $S_1 F_2$, and $S_1 S_2$, lie on top of each other. When θ is close to zero, two of the modes ($F_1 F_2$ and $S_1 S_2$) have most of the intensity and the other two modes ($F_1 S_2$ and $S_1 F_2$) are weak. The $F_1 F_2$ and $S_1 S_2$ modes for $\mathbf{J}_{Total}$ in Equation 26.12, that is, θ = 10°, together have $\cos^2(10°) \approx 97\%$ of the intensity and the other two modes have $\sin^2(10°) \approx 3\%$ of the intensity. Therefore, the unwrapped retardance is similar to the aligned system. However, whenever the phase difference between $F_1 F_2$ and $S_1 S_2$ modes is a multiple of 2π (full waves of retardance), effects from the other two modes become substantial, creating the discontinuities as shown in Figure 26.17.

The Jones matrix for an HLR can be represented as a sum of horizontal and vertical polarizers with OPLs along the fast and slow axes, OPL_{f1} and OPL_{s1}, respectively,

$$\mathbf{J}_{HLR} = e^{+i\frac{2\pi}{\lambda}OPL_{f1}} \begin{pmatrix} 1 & 0 \\ 0 & 0 \end{pmatrix} + e^{+i\frac{2\pi}{\lambda}OPL_{s1}} \begin{pmatrix} 0 & 0 \\ 0 & 1 \end{pmatrix}. \tag{26.13}$$

Thus, a linear retarder at θ is equal to a sum of linear polarizers at θ and θ + π/2 with absolute phases equal to the OPLs along the fast and slow axes, respectively,

$$\mathbf{LR}(\theta, OPL_{f2}, OPL_{s2})$$
$$= \frac{e^{+i\frac{2\pi}{\lambda}OPL_{f2}}}{2} \begin{pmatrix} 1+\cos(2\theta) & \sin(2\theta) \\ \sin(2\theta) & 1-\cos(2\theta) \end{pmatrix} + \frac{e^{+i\frac{2\pi}{\lambda}OPL_{s2}}}{2} \begin{pmatrix} 1-\cos(2\theta) & -\sin(2\theta) \\ -\sin(2\theta) & 1+\cos(2\theta) \end{pmatrix}. \tag{26.14}$$

Therefore, the Jones vector for each mode can be calculated from the matrix multiplication of two linear polarizers with associated OPLs as absolute phases. For example, the first mode ($F_1 F_2$) is a horizontal linear polarizer followed by a linear polarizer at θ with associated absolute phases, yielding the Jones vector

$$\mathbf{X}_1 = F_1 F_2 = e^{+i(OPL_{f1}+OPL_{f2})} \begin{pmatrix} \cos(\theta) \\ \sin(\theta) \end{pmatrix}. \tag{26.15}$$

Similarly, the other three modes' Jones vectors are

$$\mathbf{X}_2 = S_1 F_2 = e^{+i(OPL_{s1}+OPL_{f2})} \begin{pmatrix} \cos(\theta) \\ \sin(\theta) \end{pmatrix}$$
$$\mathbf{X}_3 = F_1 S_2 = e^{+i(OPL_{f1}+OPL_{s2})} \begin{pmatrix} -\sin(\theta) \\ \cos(\theta) \end{pmatrix} \tag{26.16}$$
$$\mathbf{X}_4 = S_1 S_2 = e^{+i(OPL_{s1}+OPL_{s2})} \begin{pmatrix} -\sin(\theta) \\ \cos(\theta) \end{pmatrix}.$$

For simplicity, one OPL or absolute phase can be set to zero; here, OPL_s is chosen. This does not affect the retardance calculation. Using Equation 26.2, each mode's phase is

$$\arg(\mathbf{X}_1) = OPL_{f1} + OPL_{f2} = \frac{\delta_1 \lambda_0}{\lambda} + \frac{\delta_2 \lambda_0}{\lambda}$$

$$\arg(\mathbf{X}_2) = OPL_{s1} + OPL_{f2} = 0 + \frac{\delta_2 \lambda_0}{\lambda}$$

$$\arg(\mathbf{X}_3) = OPL_{f1} + OPL_{s2} = \frac{\delta_1 \lambda_0}{\lambda} + 0 \tag{26.17}$$

$$\arg(\mathbf{X}_4) = OPL_{s1} + OPL_{s2} = 0 + 0 = 0,$$

where δ_i is the retardance of the ith retarder at λ_0.

The retardance is a function of the amplitude and OPD between the four modes. When the misalignment is small, most of the intensity of the exiting light has the phases in $\mathbf{X}_1$ and $\mathbf{X}_4$. Therefore, the retardance of the compound system as a function of wavelength follows the curve of the OPD between $\mathbf{X}_1$ and $\mathbf{X}_4$,

$$\begin{aligned}\delta_{major} &= \arg(\mathbf{X}_1) - \arg(\mathbf{X}_4) \\ &= OPL_{f1} + OPL_{f2} - OPL_{s1} - OPL_{s2} = \frac{\lambda_0(\delta_1 + \delta_2)}{\lambda}.\end{aligned} \tag{26.18}$$

However, whenever δ_{major} becomes $2n\pi$, effects from the OPLs of the other two modes ($\mathbf{X}_2$ and $\mathbf{X}_3$) increase, and the retardance of the compound system deviates from δ_{major}.

26.4.4 Compound Retarder Example at 45°

In this section, the properties of a Sapphire retarder with the fast axis at 45° followed by a Quartz retarder with the fast axis at 0° is simulated and measured. Section 26.6 (Appendix) shows the principal retardance for a Sapphire retarder with the slow axis at θ followed by a Quartz retarder with the fast axis at 0° as θ varies from 0° to 90°. Figure 26.18 compares a simulation and measured principal retardance and fast axis orientation as a function of wavelength of the compound retarder. Both simulated and measured principal retardance plots have minima far from zero, which indicates that this system will have large discontinuities in unwrapped retardance values near $2n\pi$.

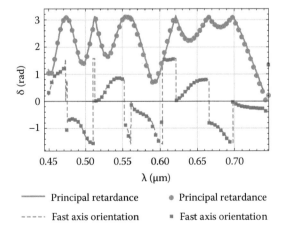

Figure 26.18 Simulated (lines) and measured (dots) principal retardance and the fast axis orientation of a system of a Sapphire retarder with fast axis at 45° followed by a Quartz retarder with horizontal fast axis are plotted as a function of wavelength.

26.4 Discontinuities in Unwrapped Retardance Values for Compound Retarder Systems

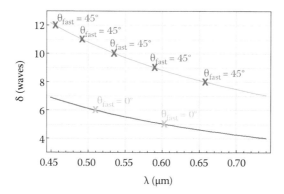

Figure 26.19 The unwrapped retardance of the Quartz (green) and Sapphire (blue) retarders with the fast axis orientation (θ_{fast}) of the compound system are plotted. The × symbols mark wavelengths where individual plates have integer waves of retardance and don't contribute to the axis of the retarder.

To better understand the behavior of the fast axis orientation of the compound system, Figure 26.19 shows the Sapphire (blue) and Quartz (green) retarders' retardance as a function of wavelength along with the fast axis orientation of the compound system, θ_{fast}. When the retardance of the Sapphire retarder becomes $2n\pi$, the fast axis of the system is aligned with the Quartz retarder, that is, $\theta_{fast} = 0$. When the retardance of the Quartz retarder becomes $2n\pi$, the system's fast axis is aligned with the Sapphire retarder, that is, $\theta_{fast} = \pi/4$.

Using Equation 26.4, the principal retardance can be unwrapped assuming that the overall behavior of the retardance is $1/\lambda$. Figure 26.20 shows the unwrapped retardance of the Quartz HLR (green), a 45° fast axis Sapphire retarder (blue), and the combined system (orange). Each mode in this system has 25% of the total intensity, and thus the discontinuities in the unwrapped retardance are much larger than Figure 26.15.

Note that the discontinuities in the orange line occur whenever the retardance of one of the retarders is $2n\pi$; when one of the retarders has multiple waves of retardance, interferences occur between the two modes in the same polarization, that is, interference between $S_1 F_2$ and $F_1 F_2$, as well as interference between $F_1 S_2$ and $S_1 S_2$.

The retardance space trajectory for the principal retardance of the example system is shown in Figure 26.21. For clarity, the top and side views reveal the 3D character of the space curve. The trajectory starts at point A (0.675, 1.463, 0.675) and crosses the inside of the π sphere; once the

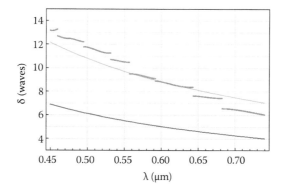

Figure 26.20 The green line shows the retardance of the Quartz retarder, $\delta_1(\lambda)$ as a function of wavelength. The blue line is the retardance of the 45° fast axis Sapphire retarder, $\delta_2(\lambda)$. The orange line shows the retardance of the combined system, $\delta_{Total}(\lambda)$.

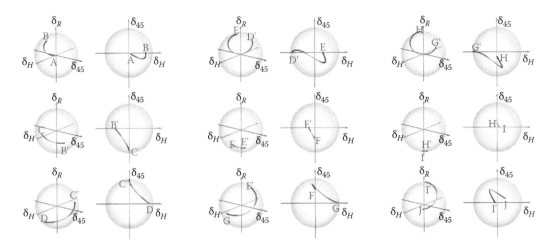

Figure 26.21 A principal retardance trajectory of the compound retarder. The two left columns show the trajectories in one viewpoint and the two right columns show the trajectories viewed from δ_R. When this trajectory reaches the boundary of π, the trajectory jumps to the opposite point on the π sphere and the fast axis changes to the orthogonal state.

trajectory reaches the opposite boundary of the sphere (point B), it jumps to the opposite point (point B') on the π sphere and the fast axis changes to the orthogonal state. Thus, points B and B' correspond to the same Jones matrix and retarder components. The trajectory follows A → B → B' → C → C' → D → D' → E → E' → F → F' → G → G' → H → H' → I → I' → J. Point A corresponds to the retardance for the longest wavelength and point J corresponds to the retardance for the shortest wavelength.

26.5 Conclusion

The mysterious behavior of compound retarders and their discontinuity is explained by the interference of the multiple modes exiting the system. Using the wavelength dependence of the retarders and the fast axis orientation, an algorithm to unwrap the principal retardance is derived. Discontinuities occur when the unwrapped retardance of compound retarders avoids the values around $2n\pi$. This is because the trajectory of the principal retardance in the retardance space can easily avoid the origin, which is just a single point representing the identity matrix. But for the unwrapped retardance, this point at the origin becomes the entire 2π sphere, 4π sphere, and so on. As the retardance increases, the trajectory appears to jump discontinuously across the $2n\pi$ sphere in the retarder space.

For small misalignment of fast axis orientation in a compound retarder, the trajectory passes close to the origin; thus, the discontinuities are small. These compound multi-order retarders have a single Jones and Mueller matrix and perform a single rotation on the Poincaré sphere. However, because of the multiple exiting beams, they don't exist in a single state of "retardance."

Although this chapter only showed examples of two waveplates, systems with multiple waveplates can be understood in a similar logic; for N waveplates, there will be 2^N different *OPLs* and the mathematics is analogous. Also, understanding the multi-order nature of the retardance provides greater insights on understanding liquid crystals, biaxial films, and optical fiber.

26.7 Problem Sets

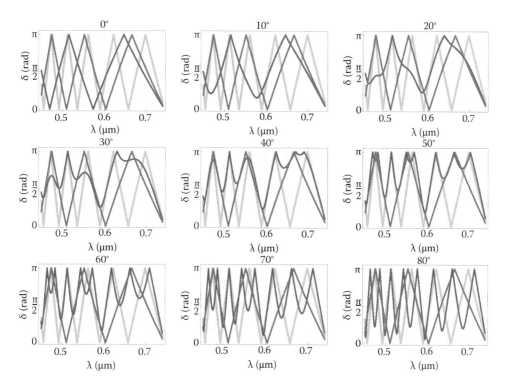

Figure 26.22 The principal retardance is plotted as a function of wavelength for Quartz and Sapphire retarders (green and blue) and a compound retarder (red), a Sapphire retarder with the slow axis at θ followed by a Quartz retarder with the fast axis at 0°, as the angle between the two fast axes θ varies.

26.6 Appendix

This section shows Figure 26.22, the principal retardance for a set of compound retarder systems.

Whenever the principal retardance of the misaligned HLR system has minima other than zero, the unwrapped retardance has discontinuity. At θ = 90°, the fast axes of two HLRs are aligned to each other since Quartz is a positive uniaxial material and Sapphire is a negative uniaxial material. Thus, the total retardance at θ = 90° is

$$\delta_{Total}(\lambda) = \delta_{Quartz}(\lambda) + \delta_{Sapphire}(\lambda), \qquad (26.19)$$

where $\delta_{Quartz}(\lambda) > 0$ and $\delta_{Sapphire}(\lambda) > 0$.

26.7 Problem Sets

26.1 Consider a compound retarder system consisting of two retarders with slight misalignment. What happens to the unwrapped retardance when one of the retarders is an integer number of retardance?

26.2 What if we have two birefringent wedges instead of two retarders? Now you are using the thickness instead of a wavelength to unwrap the retardance. How does the retardance unwrap?

26.3 Given the Mueller matrix spectrum in Figure 26.23, find the retardance magnitude and orientation through wavelength and then unwrap its retardance.

26.4 Given the Mueller matrix spectrum in Figure 26.24, plot the retardance trajectory in retarder space.

26.5 What are the retardance and fast axis orientation when two one-wave retarders are aligned at θ = 0°, 45°, 90°?

26.6 Plot the trajectory of a horizontal fast axis one-wave retarder in retarder space as the fast axis of the retarder rotates.

26.7 How does the gap, shown in Figure 26.17, change as you rotate the second retarder?

26.8 Why do multi-wave retarders avoid $2n\pi$?

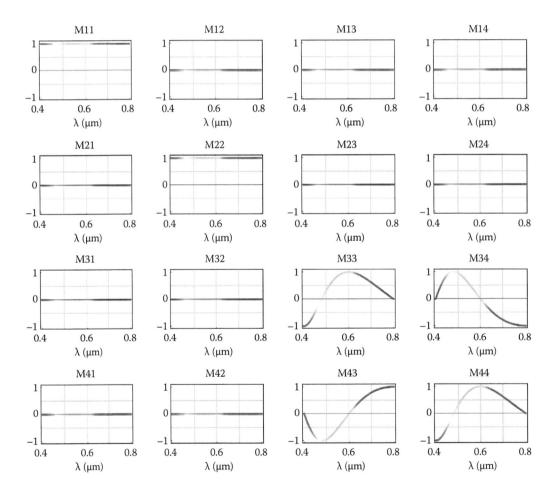

Figure 26.23 Mueller matrix spectrum.

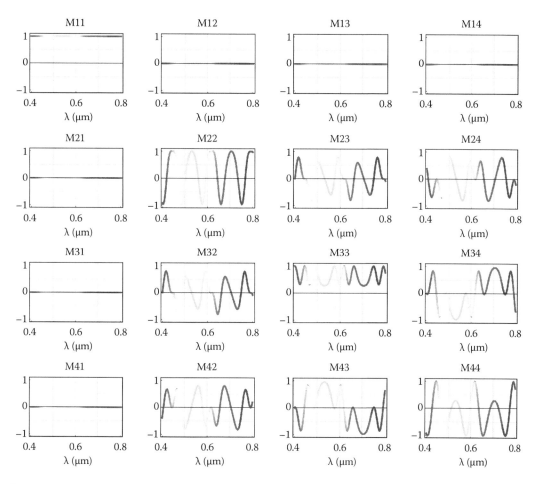

Figure 26.24 Mueller matrix spectrum.

References

1. R. C. Jones, A new calculus for the treatment of optical systems, *J. Opt. Soc. Am.* 31 (1941): 488–493, 493–499, 500–503; 32 (1942): 486–493; 37 (1947): 107–110, 110–112; 38 (1948): 671–685; 46 (1956): 126–131.
2. T. R. Judge and P. J. Bryanston. A review of phase unwrapping techniques in fringe analysis, *Opt. Lasers Eng.* 21(40) (1994): 199–239.
3. J. M. Huntley and H. Saldner, Temporal phase-unwrapping algorithm for automated interferogram analysis, *Appl. Opt.* 32(17) (1993): 3047–3052.
4. E. P. Goodwin and J. C. Wyant, Field guide to interferometric optical testing, *SPIE* 37 (2006).
5. P. Hariharan and D. Malacara, A simple achromatic half-wave retarder, *J. Mod. Opt.* 41.1 (1994): 15–18.
6. R. Bhandari and G. D. Love, Polarization eigenmodes of a QHQ retarder—Some new features, *Opt. Commun.* 110.5 (1994): 479–484.
7. S. Pancharatnam, Achromatic combinations of birefringent plates, *Proc. Indian Acad. Sci. A* 41 (1955): 130–136.

27

Summary and Conclusions

27.1 Difficult Issues

Our final chapter examines the big picture of polarized light and optical systems, highlighting several slippery and difficult topics such as coherence, scattering, and depolarization. To tie together key concepts, an example telescope analysis is included, which shows how the polarization ray tracing and polarization aberration approaches are complementary.

Polarization engineering is a rapidly evolving area of optics. In such a changing environment, how can polarization engineers communicate better, make fewer mistakes, and complete projects faster? The following sections summarize concepts that have been particularly difficult to define or describe.

For polarization engineers to communicate better, the language must be clear and understood through the whole cycle of development and production, from those who specify optical systems, the optical designers who perform the analysis, the engineers who perform detailed design of all the parts and subsystems, the subcontractors and vendors making components, metrology and quality control testing these parts, to the system integration and test team.

Problems occur with the introduction of new technologies when technical terms are not understood across these boundaries. As a field matures, terms and methods of communication become standardized. A good example of the standardization of terms and methods is seen in the optics' industries implementation of *interferograms*. In the 1950s, interferograms were not common and interferometers were not standard equipment in optics laboratories since coherent light sources (lasers) were not available, partially coherent sources, lightbulbs with pinholes, were not bright, and imaging sensors were primitive. *Foucault tests*, *Ronchi tests*, and similar geometric optical tests were standard. After the introduction of lasers and the spread of custom research interferometers in the 1960s and early 1970s, vendors still could not easily produce optics to interferometric specifications. The technology was new and expensive. Interferometers were custom instruments with interferograms still recorded onto film, developed chemically, and finally printed. Customers and vendors did not always agree on their interferometric tests. Because interferometry was a new technology, the language had not become standardized and technical communication remained an issue. By the 1980s, commercial interferometer technology with charge-coupled devices became affordable and widespread. Interferograms could be readily passed back and forth between departments, customers, and vendors. Measurements and terminology had become standardized and interferograms could serve as an efficient means of communication. Lens designers could predict interferograms for their optical designs; later metrology departments could confirm the performance of optics using standardized interferometers.

Mueller matrix images and Mueller matrix spectra have recently made a similar transition. In the 1970s, *ellipsometers* became standardized for measuring surfaces and films. But in the 1980s and 1990s, polarimeters for remote sensing and for optical system measurement were all custom instruments. In the 2000s, *Mueller matrix polarimeter* measurement technology was commercialized and, in the 2010s, became widespread and affordable. Now, a system integrator, such as an aerospace company or display company, can specify parts in terms of Mueller matrices, and vendors understand the specifications and can perform the tests in-house or outsource the metrology. A common language for polarization engineering has evolved, but is still not fully standardized.

Some of the more complex and difficult topics in polarization analysis are considered next.

27.2 Polarization Ray Tracing Complications

27.2.1 Optical System Description Complications

For an accurate polarization ray trace of an optical system, anything that makes a significant contribution to the polarization properties needs to be included. In conventional ray tracing, optical systems are defined only by the materials and shape of the optical elements. Polarization analysis requires much more information, much of which is not commonly used in conventional analysis.

Consider the complications associated with thin films, gratings, stress birefringence, and liquid crystal cells in the polarization ray trace. First, the *thin films* prescriptions, thicknesses, and indices must be specified and assigned to surfaces. Since many thin film designs are proprietary to the coating vendors, the prescriptions are not readily available to the optical designer; hence, measured thin film data or encrypted coating prescriptions need to be incorporated in ray tracing codes. Diffraction gratings have strong polarization effects. To include *diffraction gratings* in a polarization ray trace, the profile of the diffraction grating rulings is needed, and then rigorous coupled wave analysis can be performed at each ray intercept; unfortunately, the complex RCWA algorithm slows the calculation by typically a factor of one thousand per ray intercept. Grating polarization data can also be imported and interpolated, but grating vendors provide only a limited amount of measured data, angles and wavelengths, in-plane and out-of-plane, for commercial gratings; thus, vendor information is insufficient for accurately analyzing the polarization properties of most spectrometers and similar systems. *Stress birefringence* limits the imaging performance of many plastic lenses. For accurate analysis, the three-dimensional distribution of stress birefringence must be calculated with molding simulations or measured data. *Liquid crystal cells* are complex elements where the three-dimensional distribution of the director needs to be used for calculating the complex amplitude transmittance.

Often, such component information is not available, but this information shortage should not bring the polarization analysis and design process to a halt. For thin films with unknown designs, the designs for similar thin films can be used or a witness sample of the film can be measured in a polarimeter or ellipsometer. During the ray trace, the properties can be interpolated on a ray-by-ray basis. Similarly, grating and liquid crystal measurements can be used in lookup tables for polarization ray tracing. Often, a *reasonable* distribution of stress birefringence can be used to put an upper limit on the allowable stress and calculations can perform tolerance analysis without the actual stress distributions. For liquid crystal cells, since they are such critical optical components manufactured in enormous quantities in semiconductor fabs costing billions of dollars, imaging Muller matrix polarimetric metrology is widely deployed to measure the cell's director distributions and to obtain accurate experimental data for polarization ray tracing models.

Thus, a successful polarization analysis requires the integration of many science models for the physics of thin films, reflection, refraction, gratings, crystals, stress, scattering, diffraction, and more.

27.2.2 Elliptical Polarization Properties of Ray Paths

In propagating through optical systems, light rays encounter a sequence of polarization effects. The accumulated effect is described in the exit pupil where it is desirable to describe the ray path's polarization properties in terms of an equivalent diattenuation and retardance. In systems of isotropic lenses and mirrors, each ray intercept acts as a homogeneous linear polarization element; the diattenuation and retardance are linear and aligned with the *s*- and *p*-planes. For skew ray paths, the ray's *s*- and *p*-planes rotate from surface to surface. Sequences of "misaligned" linear diattenuators and linear retarders become *elliptical diattenuators* and *elliptical retarders*, thus complicating the polarization description of ray paths and wavefronts (see Section 14.3). Thus, for a general-purpose polarization ray trace, elliptical polarization must be calculated, but in lenses and other systems, the ellipticity of ray paths may be a low priority.

27.2.3 Optical Path Length and Phase

The optical design of imaging systems involves calculating phase, amplitude, and polarization. Small variations of phase, in the order of half a wavelength or so, have a large impact on image quality. Small variations of amplitude or polarization state, in the order of 10%, have a much smaller impact on image quality. Hence, optical design is principally concerned with phase since it is the top priority.

To ensure that the wavefronts exiting an imaging optical system remain nearly spherical, optical design methods calculate the phase on a reference sphere in the exit pupil by calculating the *optical path length* (*OPL*) along ray paths through the optical system. To improve the transmission, *antireflection coatings* are added to lenses. Multiple reflections occur within these thin film structures and *partial waves* are generated in the forward and reverse directions. For the films to improve transmission, these partial waves must substantially *constructively interfere* in the forward direction and *destructively interfere* in the reverse direction. Thus, for an antireflection-coated optical system, a substantial number of partial waves are propagating in the forward direction. Each of these partial waves has a different *OPL*, amplitude, and polarization. Therefore, in a system with coatings, the *OPL* becomes multivalued. The *OPL* calculated by the ray tracing program is only the "direct" ray, without any multiple reflections, which is typically the first and usually the brightest of the partial waves. Figure 27.1 (left) shows an example of a single wavefront exiting from an uncoated lens. Figure 27.1 (middle) shows a set of partial waves exiting a lens with an ideal *quarter wave coating*, where each partial wave is separated by exactly one wavelength; thus, all the partial waves constructively interfere, suppressing the reflections from the lens and boosting the transmission. Each of these partial waves has a different *OPL*. When the wavelength changes (Figure 27.1, right), the partial waves are no longer separated by an integer number of wavelengths; thus, many more waves are seen in the figure. The lenses' transmittance suffers since the exiting wavefronts do not completely constructively interfere.

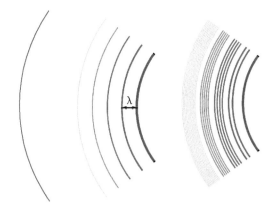

Figure 27.1 (Left) The spherical wavefront from the direct light through the system. (Middle) Ideal quarter wave antireflection coatings yield an additional series of partial waves separated by one wavelength with diminishing amplitude. (Right) If the wavelength changes, multiple partial waves are generated with non-integer wavelength separations.

When an *ultra-short laser pulse* enters an optical system with coatings, a sequence of ultra-short pulses exit spread out in time, corresponding to the partial waves created by the coatings.

Because of these multiple reflections, our bedrock concept of optical path length becomes more complicated, blurred by a legion of smaller partial waves; *OPL* has become a multivalued function. These partial waves cannot be physically separated from each other. An entering photon has a *probability amplitude* to be in any of the partial waves. The resulting electromagnetic wave is the vector sum of the electric field over all the partial waves and has a definite phase, amplitude, and polarization state. Thus, although *the optical path length is no longer well defined, the phase is well defined* and can be uniquely measured by an interferometer.

27.2.4 Definition of Retardance

For similar reasons, the definition of *retardance* does not generalize well to systems with complex crystal assemblies and ray paths through such optical systems. When light transmits through a sequence of *anisotropic materials*, a set of two, four, eight, or more *partial waves* can exit, due to ray doubling, each with a different *OPL*. Again, the *OPL* becomes multivalued, but the amplitude, phase, and polarization state are still uniquely defined. Retardance is simple to define for a single crystal *waveplate* but more difficult to define in such general circumstances such as multiple birefringent plates. For a single birefringent plate, since only two waves exit, the retardance is simply the difference between the optical path lengths of the two modes. With a sequence of n anisotropic materials, 2^n beams exit, each with different optical path lengths, amplitudes, and polarization states. No single "retardance" number can describe a complex set of optical path differences. A different approach to defining retardance is to consider a *retarder* as an element that rotates polarization states around the *Poincaré sphere* about an axis by an angle equal to the retardance. With this definition, the number of exiting beams does not matter and the retardance and fast axis can be specified.

27.2.5 Retardance and Skew Aberration

Extending the concept of retardance to ray paths through optical systems is also complicated. These ray paths can be characterized by *polarization ray tracing matrices*, **P**. Since the entering and exiting rays need not be collinear, the *eigenvectors* of the **P** usually do not represent actual polarization states. The polarization-dependent phase changes can be divided into two categories, *geometrical*

27.2 Polarization Ray Tracing Complications

Figure 27.2 The linear skew aberration transforms a uniform incident linearly polarized state into a skew aberrated state. This linear form of skew aberration occurs in most lenses and other rotationally symmetric systems.

transformation and *proper retardance*. Coordinate system rotations, such as occur after sequences of *refractions*, can masquerade as *circular retardance* and are an example of geometrical transformation. Polarization changes on *reflection*, which masquerade as a half wave of linear retardance, are another example. The *parallel transport* of transverse vectors along ray paths through optical systems calculates the geometrical transformation, and the collection of geometrical transformations across a wavefront is the *skew aberration*.

Skew aberration is the polarization aberration of *non-polarizing optical systems*; skew aberration is the systematic polarization changes that remain when all the *diattenuation* and *retardance* are removed from an optical system. Skew aberration depends only on the sequence of *propagation vectors*. *Skew tilt* is the linear variation of skew aberration across the pupil orthogonal to the *meridional plane* as shown in Figure 27.2. A constant skew would uniformly change the polarization and would not affect the *point spread function*. It is the variation of skew (i.e., skew aberration) that affects the *Mueller point spread matrix* and thus degrades image quality. Skew aberration is typically a small effect in lenses but is a large effect in *corner cubes*.

The *proper retardance* refers to a physical property by which optical path length accumulation depends on the incident polarization state. **P** describes the polarization state changes due to diattenuation, retardance, and geometric transformations. The parallel transport matrix **Q** describes the associated non-polarizing optical system and thus keeps track of just the geometric transformation. To calculate the proper retardance, the geometric transformation needs to be removed, $\mathbf{M} = \mathbf{Q}^{-1}\mathbf{P}$. The difference in eigenvalue arguments of $\mathbf{M}_R$, the unitary part of the polar decomposed **M**, gives the proper retardance; the proper retardance cannot be assigned to a ray through an optical system in a black box whose internal ray path is unknown.

27.2.6 Multi-Order Retardance

A *multi-order retarder* is a retarder with more than a half wave of retardance and a *compound retarder* is a combination of two or more retarders in sequence. When a compound retarder has *fast axes* that are misaligned with respect to one another, we call it a *misaligned retarder*. For misaligned compound multi-order retarders, more than two beams are involved and the common definition of retarders is inadequate. For these retarders, the optical path length can be multivalued as shown in Figure 27.3.

Compound retarders have what at first appears to be *mysterious behavior*; the *unwrapped retardance* of compound retarders avoids the values around $2n\pi$ and the retardance appears to jump, implying *retardance discontinuities*. These discontinuities are better understood in *retardance space* where the trajectory of the principal retardance can easily avoid the *origin*, which is just a single point representing the identity matrix (and the only point with zero retardance) as shown in Figure 27.4. For the unwrapped retardance, this point at the origin becomes the entire 2π sphere,

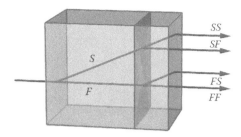

Figure 27.3 A misaligned two-element compound linear retarder has four output beams exiting with four different combinations of the fast and slow *OPL*s. For an aligned retarder, only the *FF* and *SS* beams exit.

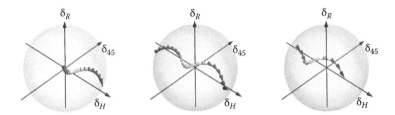

Figure 27.4 The retarder space inside the π sphere (purple sphere) and the principal retardance component trajectories are plotted as the wavelength decreases (from the left figure to the right figure). The middle figure starts from the symmetric point of the ending point of the left figure, and so on. Note that the trajectories for the second and third figures miss the origin.

4π sphere, and so on. As the retardance increases, the trajectory appears to jump discontinuously across the 2nπ sphere in the retarder space.

For small misalignment of fast axis orientation in a compound retarder, the trajectory passes close to the origin; thus, the discontinuities are small. These compound multi-order retarders have a single Jones and Mueller matrix and perform a single rotation on the Poincaré sphere. However, because of the multiple exiting beams, they do not exist in a single state of "retardance."

27.2.7 Birefringent Ray Tracing Complications

Anisotropic materials provide *retardance, ray doubling, polarization separation*, and other phenomena that cannot be generated by *isotropic materials*. The associated light–matter interactions are different from glass and other isotropic materials. An anisotropic material's directional properties are described by the *dielectric tensor* and *gyrotropic tensor* rather than a single *refractive index* as in isotropic material. The two modes' reflected and refracted rays propagate in different directions rather than in the same direction as with *s*- and *p*-polarizations. *Wavefronts* in anisotropic materials have more complex shapes, orthogonal to **k** but not orthogonal to **S**, and different residual aberrations compared to wavefronts in isotropic material. Propagation within the anisotropic material induces a phase difference between polarization states, the *retardance*.

At a *birefringent interface*, light divides into two eigenpolarizations, mutually orthogonal polarizations propagating in different directions with distinct refractive indices. This *ray doubling* characteristic of birefringent materials, which produces two new ray segments in both transmission and reflection, complicates the calculations of *refracted* and *reflected amplitude coefficients* in polarization ray tracing. The *OPL* associated with each ray becomes a function of polarization state.

The stress-induced birefringence in optical components is modeled as a *spatially varying birefringence medium* in Chapter 25. Such modeling predicts the spatial retardance variation throughout the

27.2 Polarization Ray Tracing Complications

stressed optical component and its effects on the PSF and MTF for different incident polarizations. Stress birefringence modeling is a powerful tool to predict the performance of components with residual stress to tolerance allowable stress, and optimize the molding process parameters, for example the tradeoff between processing time (time in the mold and molding temperatures) and product quality.

27.2.8 Coherence Simulation

The standard ray tracing algorithms trace rays one wavelength at a time to determine wavefronts on a wavelength-by-wavelength basis; thus, these algorithms simulate *coherent monochromatic light* through the optical system. Often, the optical system properties are needed for *partially coherent* or *incoherent light*. The coherent, partially coherent, and incoherent simulation cases can be compared with a simple example configuration, light reflecting from the front and back surfaces of a glass plate or a retarder. In monochromatic light, the principal and reflected beams produce *fringes*. When the wavelength is swept, the fringes move. Thus, for light with a small spectral bandwidth, the fringe motion is small and the fringes are slightly blurred; the *fringe visibility* is reduced but the fringes are visible. This is the partially coherent case. For a larger bandwidth, so many different fringe patterns are superposed that in the large spectral bandwidth limit, fringes are completely lost. This is the incoherent case.

The monochromatic ray trace, either a conventional or polarization ray trace, calculates the coherent configuration; it will predict fringes from the glass plate. For an LED with a smaller spectral bandwidth, fringes may or may not be visible, but the monochromatic ray trace is always predicting fringes. Thus, for proper simulation of partial coherence (finite spectral bandwidth), the calculation is generally repeated for many wavelengths and the fringe patterns are combined to evaluate the effect of the spectral bandwidth on the reduction of fringe visibility. This is the essence of *partial coherence calculations*, repetition of the ray trace for a large number of wavelengths, fields, or configurations. Sometimes, shortcuts can reduce the amount of calculation, but usually the simplest procedure is just to ray trace a system a hundred times or more. Thus, individual wavelengths are ray traced to calculate *optical path lengths*, *Jones matrices*, and *polarization ray tracing matrices*, but light from different wavelengths, or light from different spatially incoherent sources, are added incoherently as *Mueller matrices* or the equivalent.

27.2.9 Scattering

Evaluating the effects of *scattering* in an optical system is an important optical analysis task. For example, *telescopes* have *baffles* to prevent light from outside the field of view from reaching the focal plane. Before building expensive telescopes, for example, to be placed in orbit, it is common to calculate the out-of-field rejection of the baffle design. If the sun or moon is, for example, 3°, 10°, or 30° from the telescope's axis, how much light scatters from the baffles and telescope structures and reach the focal plane? This involves tracing rays to the baffles and other scattering surfaces and then tracing many rays from each ray intercept to other scattering or optical surfaces and finding the flux along the ray paths that do reach the focal plane. At each ray intercept at each surface, a calculation of the *bidirectional reflectance distribution function* is commonly performed. Millions or billions of rays may need to be calculated with *Monte Carlo* routines to get accurate stray light estimates because of the vast number of potential ray paths.

Scattering effects can be very polarization dependent, for example, when scattering near *Brewster's angle* or from *diffraction gratings*. In such cases, a scattering calculation needs the *polarization bidirectional reflectance distribution function* (PBRDF) for the surfaces of interest, a Mueller matrix function of the incident and scattering angles.* This may be measured with

---

* Also known as the Mueller matrix bidirectional reflectance distribution function, MMBRDF.

a *Mueller matrix polarimeter* illuminating and collecting scatter from a large number of angles. Alternatively, a suitable analytic PBRDF model may exist, such as the *ScatMech* library from the National Institute of Standards and Technology.[1,2] Scattering ray trace simulations are an example of an incoherent ray trace since the path lengths vary by thousands of wavelengths.

27.2.10 Depolarization

Some optical systems have measurable *depolarization*, where incident polarized light exits with a reduced *degree of polarization*. Some optical elements, such as *diffraction gratings, holographic optical elements, multimode waveguides*, and *liquid crystal cells*, are more likely to contribute depolarization. Depolarization is a statistical process, a randomization of the polarization state often due to roughness, scattering, and variations of a material at a microscopic scale. With *laser illumination* through an optical system, in the absence of depolarization, the wavefront is smooth with a slowly varying amplitude, phase, and polarization. Depolarization causes the wavefront to become noisy with some high-frequency content. A rough surface or a coating that is cracked or peeling imposes many small phase variations to the wavefront, which cause such noise. For a large amount of depolarization, such as scattering from a rough surface or light transiting an *integrating sphere*, the emerging wavefront becomes a *speckle pattern* with a rapidly varying polarization state.

The phase changes from randomly scattering surfaces, integrating spheres, and the like cannot be exactly simulated with a ray tracing program. A surface with roughness of the scale of 0.1 μm would require more than $(10^5)^2$ points or more than 10^{10} points per square centimeter to characterize for ray tracing, an impossible task.

Depolarization is not described by *Jones matrices* and *polarization ray tracing matrices*; it is a statistical process. Depolarizing optical components can be ray traced with *Mueller matrices*. In general, when light is incident on a depolarizing surface, the light scatters into many directions. Scattered light ray tracing programs are programmed to calculate many scattered rays, leaving an interface for each incident ray, and can be programmed to calculate with Mueller matrices. *Polarization BRDF* models are available for many depolarizing surfaces and can be used for each scattering calculation. Such calculations are usually performed by choosing rays, leaving a ray intercept using *Monte Carlo* methods, and not by following regular ray grids through an optical system.

A ray tracing calculation with *Mueller matrices* is an example of an *"incoherent" ray trace*, where incoherent refers to the ray trace, not the *coherence* of the light. A Mueller matrix does not track the *absolute phase* of the light. Mueller matrices can be added to simulate the combination of polarized or partially polarized beams that do not have a phase relationship with each other; they are mutually incoherent.

Such a calculation is appropriate for an *integrating sphere* illuminated with laser light. The output is a *speckle pattern* that cannot be exactly simulated because the description of the surface (with 10^{10} points per square centimeter for example) is unreasonable. Light reaches a surface following the integrating sphere from paths leaving many different areas of the sphere. The optical path lengths of rays incident on a particular point are randomly distributed over hundreds of waves of *optical path length*; thus, the rays are incoherently combined. A Mueller matrix ray trace can calculate the Mueller matrix, which would be measured by averaging over many many speckles, even though the exact distribution of speckles cannot be calculated. For this *coherent light*, an *incoherent ray trace* is appropriate due to the scattering nature of the system.

For analyzing an optical system with some depolarization, such as from a diffraction grating, a *coherent ray trace* can be performed for the majority of the light, the image-forming part. The scattered light part of the light can be represented with Mueller matrices, and a separate *incoherent ray trace* can be performed for this depolarized fraction.

27.3 Polarization Ray Tracing Concepts and Methods

27.3.1 Jones Matrices and Jones Pupil

The *Jones matrix* operates on *Jones vectors*, which describe the *polarization ellipses* with respect to an *x–y local coordinate system* in the *transverse plane*. To use Jones vectors and matrices in optical design for the ray tracing of highly curved beams, different local coordinate systems are required for each ray segment to define the direction of the Jones vector's x- and y-components in space. These local coordinate systems lead to complications due to the intrinsic singularities of local coordinates.

The full description of the wavefront and polarization at the exit pupil is divided into a combination of four functions: the *wavefront aberration function*, *amplitude function*, *aperture function*, and *polarization aberration function* at the *exit pupil*. This combination is called the *Jones pupil*,

$$JonesPupil(x,y) = aperture(x,y) \cdot a(x,y) \cdot \mathbf{J}(x,y) \cdot e^{-2\pi i W(x,y)}. \tag{27.1}$$

aperture(x,y) is an aperture function that is one inside the aperture and zero outside the aperture. *a(x,y)* is an amplitude function that describes the transmission along the ray path, also known as the *apodization*. $W(x,y)$ is the wavefront aberration function that characterizes the optical path difference between each ray and the *chief ray*. $\mathbf{J}(x,y)$ is the Jones matrix along ray paths from the *entrance pupil* to the *exit pupil*. (x,y) is the pupil coordinate. The *JonesPupil(x,y)* is spatially varying and is the starting point for the polarization image formation calculations.

27.3.2 P Matrix and Local Coordinates

The 3 × 3 *polarization ray tracing matrices*, **P** matrices, enable ray tracing in *global coordinates*, which provide an easy basis to interpret polarization properties. *Diattenuation* can be calculated by the *singular value decomposition* (SVD) of **P** since its singular values are the maximum and minimum *amplitude transmittances*.

Given the distribution of polarization states on a *spherical wavefront*, representing this information on a computer screen or page involves choices on how to flatten the information. Two most commonly used *basis vector* functions for this transformation are the *dipole coordinates*, a *latitude* and *longitude* system, and the *double pole coordinates*, a coordinate system that matches the polarization and direction changes when lenses focus collimated wavefronts.

Chapter 11 presented methods for converting the polarization representation from 3D to 2D, and vice versa, using the dipole and double pole coordinates. Figure 27.5 shows Jones pupil's dependence on the choice of local coordinates when converting the **P** matrix pupil.

27.3.3 Generalization of PSF and OTF

A *point spread function* (PSF) is a fundamental metric for optical imaging systems and describes the image of a point object. For systems with *polarization aberration*, the PSF depends on the incident polarization state. Thus, the concept of a PSF is generalized so that it describes imaging of arbitrary polarization states by introducing an *amplitude response matrix* (**ARM**). Given a wavefront aberration function and Jones pupil function, the *Mueller point spread matrix* (**MPSM**) is calculated, which shows how the PSF varies with the incident polarization state. Similarly, the *optical transfer function* (OTF) of conventional optics can be extended to an *optical transfer matrix* (**OTM**) to describe how the spatial filtering of an object during imaging depends on the incident polarization state. Figure 27.6 shows the relationship between Jones pupil, **ARM**, **MPSM**, and **OTM**. Off-diagonal elements in **ARM**, **MPSM**, or **OTM** describe the coupling between polarization components in the image. The relative magnitude of the off-diagonal elements with respect to

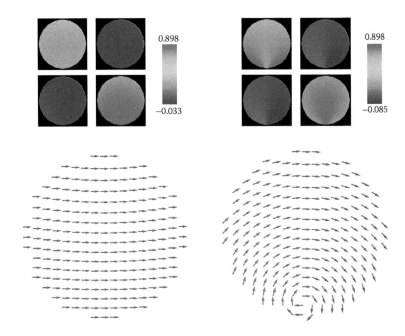

Figure 27.5 (Top) Jones pupil of the cell phone lens system using (left) double pole coordinates and (right) s–p coordinates to convert from the **P** matrix pupil. (Bottom) Local x coordinates across the exit pupil from the (left) double pole coordinates and (right) s–p coordinates.

$$\langle \text{Imaging polarized light} \rangle$$
$$\textit{Jones pupil} = \mathbf{J}_{\text{pupil}}$$
$$\downarrow \Im$$
$$\mathbf{ARM} = \Im[\mathbf{J}_{\text{pupil}}] \rightarrow \mathbf{MPSM}$$
$$\downarrow \Im$$
$$\mathbf{OTM} = \mathbf{MTM}\,\exp(i\,\mathbf{PTM})$$

Figure 27.6 Relationships between the 2 × 2 amplitude response matrix (**ARM**), the 4 × 4 Mueller point spread matrix (**MPSM**), the 4 × 4 optical transfer matrix (**OTM**), the 4 × 4 modulation transfer matrix (**MTM**), and the 4 × 4 phase transfer matrix (**PTM**) are shown.

the diagonal elements tells the amount of polarization mixing occurring in the image plane. Thus, for the analysis of optical systems with polarization aberrations, matrix representations of the PSF and OTF are necessary.

27.3.4 Ray Doubling, Ray Trees, and Data Structures

At an anisotropic element, one incident wavefront yields two exiting wavefronts in transmission, a system with N anisotropic elements can have 2^N transmitted wavefronts. Each of the 2^N wavefronts focuses at different locations, has different polarization state variation, and has differing amounts of wavefront aberrations. One incident ray exits with 2^N optical path lengths. *Retardance*, which is defined as the optical path difference between two overlapping rays, needs to be redefined for the case of three or more overlapping wavefronts.

27.3 Polarization Ray Tracing Concepts and Methods

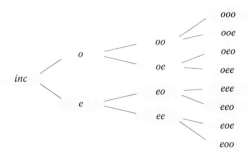

Figure 27.7 A *ray tree* for one incident ray refracting through three uniaxial interfaces, where the *ordinary mode* is *o* and the *extraordinary mode* is *e*.

To simulate the propagation of wavefronts through optical systems, a large number of rays are traced to accurately sample the shape, amplitude, and polarization over the wavefront. We extend the *polarization ray tracing calculation* to optical systems with anisotropic materials. The ray tracing results of these wavefronts are stored in the **P** matrices. Then, the properties of these wavefronts, such as the *wavefront aberrations*, *polarization aberrations* including *diattenuation* and *retardance*, and the *image quality* are calculated from the **P** matrices. Because of the *ray doubling* behavior of the anisotropic materials, the algorithm for polarization ray tracing is further generalized to handle the exponentially increasing number of ray segments as the number of anisotropic elements increased.

To characterize wavefronts transmitted through anisotropic materials, *families of rays* need to be traced. When tracing large grids of rays to accurately sample the *multiple exiting wavefronts*, accessing the ray properties along these ray paths through a specific surface or a series of surfaces is an important step in optical design. K incident rays refracting through N anisotropic intercepts result in $K \times 2^N$ exiting rays and $K\left(\sum_{n=1}^{N} 2^n\right)$ ray segments. These branches of ray segments can be organized by the *cumulative mode label* (introduced in Table 19.1) into a *ray tree* such as is shown in Figure 27.7.

Such a *ray tree* keeps track of the *multiplicity of rays* generated by sequences of anisotropic elements. The *cumulative mode label* identifies the individual polarized wavefront of a specific sequence of *ray splittings* as the wavefront propagates through the series of optical surfaces. These mode labels accommodate the *automation of ray doubling* in a computer ray trace without imposing any assumptions and assist in sorting the exiting rays for each polarized wavefront. The data structure derived from the mode label systematically manages the multiplicity of rays generated by anisotropic elements. These structures are useful to handle many special cases unique to anisotropic interactions: *inhibited refraction*, *total internal reflection*, and *conical refraction*.

27.3.5 Mode Combination

After tracing all the ray splitting through an optical system, parts of these exiting wavefronts are usually overlapping. A method to analyze the resultant wavefront represented by these ray trees is as follows. The eigen-wavefront produced from each separate mode sequence is first reconstructed by *interpolation*. As illustrated in Chapter 20, each of these reconstructed wavefronts has different aberrations, but a smooth distribution at the exit pupil. Therefore, they can be interpolated with the electric field of the sample rays.

When the eigen-wavefronts overlap at the image plane, they interfere. Each of these wavefronts produces an image that can be calculated using methods presented in Chapter 16. Then, the overall image is the *summation of these eigen-images*.

27.3.6 Alternative Simulation Methods

This book's emphasis has been on polarization, but other methods are necessary for related polarization problems in optical systems with *nonlinear optics*, *waveguides*, *photonic devices*, *metamaterials*, and *nanostructured materials*. Another class of simulation methods, *wave propagation algorithms*, propagates the fields through devices and optical systems, diffracting the light from the first to second surface, and so on. Wave propagation algorithms are necessary at longer wavelengths, that is, radio, where the *ray approximations* break down, and in waveguides where structures have dimensions on the order of the wavelength. For subwavelength structures, periodic structures can be simulated with *rigorous coupled wave analysis* (Chapter 23). Non-periodic structures are often simulated with *finite difference time domain algorithms* (FDTD), which solve the time-dependent *Maxwell's equations* on a space–time grid taking small steps in space and time. First the electric field is propagated in step, then the magnetic field, and then the process is repeated. FDTD is a very general light field propagation method, but because of the amount of computation involved, the region of solution is generally limited to volumes less than 0.1 by 0.1 by 0.1 mm.[3]

27.4 Polarization Aberration Mitigation

Optical designers have a variety of methods to change the *polarization aberrations* of any particular optical system.[4] *Fresnel aberrations* and coating-induced polarization aberrations tend to be of small magnitude with low-order functional variation (constant, linear, quadratic, etc.).[5] The following summarizes several mitigation approaches:

A. Reducing *angles of incidence*: Since the diattenuation and retardance increase quadratically for small angles of incidence, reducing the largest angles of incidence can significantly reduce polarization aberration. Reducing the range of angles at mirrors and lenses reduces the variation of retardance and diattenuation.

B. Reducing *coating polarization*: The optical coating prescriptions for antireflection coatings of lenses and reflection-enhancing coatings of mirrors provide design degrees of freedom (thicknesses and materials) to adjust the diattenuation and retardance. In our experience, these coating prescriptions can be adjusted to moderately reduce the polarization properties, but cannot zero out diattenuation or retardance for substantial angle and wavelength ranges. The surfaces of antireflection-coated lenses typically have one-third or less the diattenuation of uncoated lens surfaces, providing great benefit. The reflection-enhancing coatings for mirrors often increase the retardance and diattenuation of metal mirrors in some wavebands.

C. Compensating *polarization elements*: Polarization aberrations can be introduced in several ways. Simply placing a (spatially uniform) weak polarizer (diattenuator) and a weak retarder in the system could zero out the polarization aberration at one point in the pupil, leaving overall polarization aberrations smaller. A spatially varying diattenuator and retarder with polarization magnitude approximately equal to the cumulative diattenuation and retardance but orthogonally oriented would nearly eliminate the polarization aberration. Such a polarization plate could be considered as the matrix inverse of the *Jones pupil*. Such correction plates might be fabricated from *liquid crystal polymers* with spatially varying magnitude and orientation of diattenuation or retardance, similar to the *vortex retarders* used in coronagraphy.[6–8] Wedged, spherical, and aspheric

27.4 Polarization Aberration Mitigation

crystalline elements or element assemblies can provide a wide variety of compensating polarization aberrations.[9] Since polarization aberrations of telescopes and fold mirrors tend to be small, spatially varying anisotropic thin films, which can only provide small retardances, could provide another path toward compensation.[10]

D. Crossed *fold mirrors*: Fold mirrors tilted about opposite axes, such that the *p*-polarized light exiting one mirror is *s*-polarized on the second, have a compensating effect for both diattenuation and retardance.[11,12] A linear variation of polarization about zero will still remain across the pupil.

E. Compensating *optical elements*: The diattenuation of lenses has the opposite sign (greater *p*-transmission) compared to the diattenuation of mirrors. Thus, including lenses would reduce the diattenuation from the primary and secondary mirrors in the example system in Section 27.5. Similarly, sets of coatings might be selected to have opposite retardance contributions. Despite several concerted attempts, the author (Chipman) has not been able to change the sign of the diattenuation or retardance of an antireflection- or reflection-enhancing coating over a useful spectral bandwidth. In practice, this approach has never been very successful.

Considering these mitigation approaches, novelty, fabrication issues, scattering, tolerances, and risk must be balanced against the magnitude of the polarization aberration. For critical systems, with low levels of polarization aberration, the imperfections associated with these mitigation approaches, in particular irregularity, absorption, and scattering, can easily become worse than the problem.

To design optical systems, typically a *merit function* is defined to characterize the wavefront and image quality, and an *optimization* program adjusts the system's constructional parameters to find acceptable configurations.

1. If polarization ray tracing parameters such as *diattenuation* and *retardance* are included in the merit function, the *optimizer* can balance the *polarization aberrations* against the *wavefront aberrations* and other constraints, pushing the solutions toward reduced polarization aberration.
2. Similarly, if the coating and polarization element constructional parameters are included in the optimization, the *optimizer* can explore the coating design space and polarization element configuration to find compensation schemes. For example, overcoated layers on aluminum will modify the polarization.

These two steps are complicated, but advanced users can apply these methods, often through the use of the optical design program's macro languages, to evaluate polarization mitigation strategies listed above.

27.4.1 Analyzing Polarization Ray Tracing Output

Polarization ray tracing is complicated! Conventional optical design produces a *wavefront aberration function*, a scalar function. Incorporating the effects of coatings and isotropic elements, *polarization ray tracing* produces arrays of *polarization ray tracing matrices*. These polarization ray tracing matrices can then be transformed into *Jones matrices* defined between spherical surfaces in the entrance pupil and exit pupil. The ray tracing problem has moved from scalar wavefront functions with one degree of freedom into an eight-dimensional representation using Jones matrices. Three dimensions are hard to visualize. Four-dimensional spaces are very difficult to visualize. Eight dimensions are truly daunting. To understand and operate on this eight-dimensional data, the designer needs to extract the most critical information. To effectively communicate the information, the quantity of information needs to be reduced, priorities need to be established, and the total information needs to be simplified without oversimplifying. Only certain polarization properties are likely to be important, but which are the most important properties vary from system to system. Buried in all the data, only a few aberration patterns may be important. Finding such features can often summarize the important information from a vast mass of calculated data.

The polarization ray tracing algorithm needs to be general purpose. The algorithm needs to make accurate calculations for a broad range of optical systems. To do this, it needs to calculate all eight degrees of freedom in the Jones calculus to handle a general set of problems. But many systems have no significant *circular retardance* or *circular diattenuation* (see Section 27.2.2). Hence, these two degrees of circular freedom can often be set aside, typically after the calculation, as having lower priority. Similarly, some systems have no significant retardance, like uncoated lenses. Other systems have no significant polarization aberration, so that only the optical path length and wavefront aberration are needed, but the polarization ray tracing calculations may be needed to verify, for example, that a set of coatings have minimal effect on the image formation of a particular lens.

The polarization analysis challenge remains. The optical designer is confronting an eight-dimensional Jones pupil function and needs effective tools in the polarization ray tracing program to quickly understand the data and prioritize the optical effects. In the next section, the application of polarization aberration expansions to polarization ray tracing data is shown to provide a method for reducing the polarization ray tracing output to a small number of parameters and simple functions. Further, if the coating design is changed or fold mirror angles changed, the system doesn't necessarily need to be retraced; in this example, design rules were derived from the aberrations to show how the performance scales with such changes.

27.5 Comparison of Polarization Ray Tracing and Polarization Aberrations

The polarization ray tracing method and polarization aberration method will be compared for an example system. Ray tracing methods treat system performance and aberrations numerically as grids of *OPL*s, Jones matrices, and other values. Aberration theory describes system aberrations in terms of simple closed-form functions. Both methods have their strengths and weaknesses and complement each other. Aberration theory has strengths in the simplicity of the aberration representation, representation of the aberration with a small number of parameters, the ability to describe how performance scales with numerical aperture and object size, and the ability to pinpoint at which surfaces aberration is arising and suggest methods for aberration balancing. Aberration theory becomes complicated as systems lose symmetry, become tilted or decentered, or incorporate free-form optical surface shapes. The strength of ray tracing is its ability to analyze arbitrary systems and to provide (nearly) exact answers. The aberration approach can lead to powerful design rules to assist in making optical design trade-offs.[13]

An example Cassegrain telescope (Figure 27.8) consisting of a primary, secondary, and fold mirror is analyzed, comparing the ray tracing and aberration methods with the goal of relating the constructional parameters of the telescope to the coating effects on the PSF. A closed-form expression for the Jones pupil is found in terms of the second-order polarization aberrations: constant (piston), linear (tilt), and quadratic (defocus) terms for the diattenuation, retardance, amplitude, and phase. With these aberrations, the entire Jones pupil can be reduced to 12 parameters, its second-order polarization aberration coefficients. These coefficients can be directly related to the mirror coatings and the angles of incidence through the system. The following are two of the aberrations to be considered: (1) The *XX* and *YY* PSF components, the two bright co-polarized components, are shifted in opposite directions. This shear PSF is related to the fold mirror coatings and image-space numerical aperture. (2) The *XX* and *YY* PSFs have different magnitudes and orientations of on-axis coating-induced astigmatism, which are related to the primary and secondary mirror angles of incidence and coatings. This example combines the polarization aberration analysis of the fold mirror of Section 12.3 with the Cassegrain telescope analysis of Section 12.5. This example was initially analyzed for coronagraphs for the imaging of exoplanets and debris disks around stars where the

27.5 Comparison of Polarization Ray Tracing and Polarization Aberrations

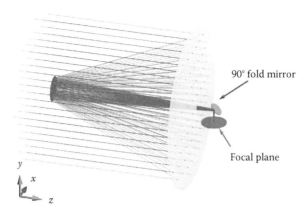

Figure 27.8 An example Cassegrain telescope system with a primary mirror at $F/1.2$, a Cassegrain focus of $F/8$, and a 90° fold mirror in the $F/8$ converging beam. The 90° fold mirror is folded about the x-axis. The primary mirror has a clear aperture of 2.4 m. All three mirrors are coated with aluminum with index $n = 2.80 + 8.45i$ at 800 nm. Y-polarized light refers to the polarization in the entrance pupil in the plane of incidence of the axial ray at the fold mirror, or vertical in the picture. X-polarized light is s-polarized at the fold mirror.

control of the PSF needs to be exquisite to image planets with expected brightnesses of $<10^{-8}$ of the star within several Airy disk radii.

Figure 27.8 shows the Cassegrain telescope and fold mirror illuminated with an on-axis collimated beam. The fold mirror tilts about the x-axis reflecting light propagating toward $+z$ into the $-y$-axis. The primary mirror is parabolic. The conic constant of the hyperbolic secondary mirror is chosen to eliminate spherical aberration; there are no on-axis wavefront aberrations. Therefore, from a conventional ray trace, the on-axis system is perfect, diffraction limited. During the polarization ray trace, any deviations from ideal imaging are due to the mirrors coating polarization and are not mixed with the effects of wavefront aberration.

27.5.1 Aluminum Coating and Polarization Aberration Expression

An aluminum coating has been chosen for this analysis. The amplitude and phase coefficients for aluminum at 800 nm are plotted in Figure 27.9. The diattenuation is shown in Figure 27.10 (left) where it is zero at normal incidence and increases approximately quadratically to about 0.05 at the 45° central angle of the fold mirror. Similarly, the retardance, shown in Figure 27.10 (right),

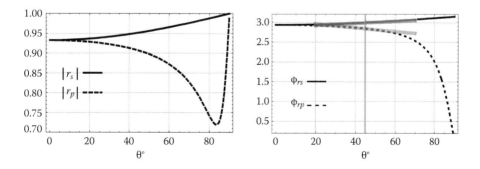

Figure 27.9 Reflection coefficients for the amplitude (left) and the phase (right) for angles of incidence θ between 0° and 90° for a bare aluminum mirror. ϕ_{rs} and ϕ_{rp} are the reflected phase for s- and p-polarized light. The green vertical line highlights the phase changes at 45°. The different slopes or linear phase shifts (red ϕ_{rs} and blue ϕ_{rp} lines) at the 45° incident angle cause different XX and YY image shifts.

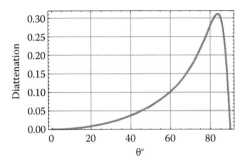
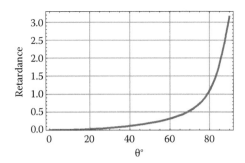

Figure 27.10 (Left) Diattenuation and (right) retardance of bare aluminum mirror at 800 nm.

increases quadratically from 0 to about 0.15 at 45°. Hence, the retardance is expected to generate about 0.15/0.05 ≈ 3 times more polarization aberration.

For the polarization aberration analysis, the Fresnel equations can be replaced with simpler linear and quadratic polynomial fits valid over the range of angles in use. For the on-axis mirrors, the coating diattenuation $D(\theta)$ and retardance $\delta(\theta)$ functions are expanded about normal incidence in quadratic functions as

$$D(\theta) \approx a_2\,\theta^2 + O(\theta^4),\ \delta(\theta) \approx b_2\,\theta^2 + O(\theta^4), \tag{27.2}$$

with second-order coefficients a_2 for diattenuation and b_2 for retardance. These quadratic fits are shown in Figure 27.11 (left) and Figure 27.12 (left). For the fold mirror, a first-order expansion about the axial ray angle of incidence $\theta_0 = 45°$ suffices,

$$D(\theta) \approx a_0 + a_1\,(\theta - \theta_0) + O(\theta^2),\ \delta(\theta) \approx b_0 + b_1\,(\theta - \theta_0) + O(\theta^2) \tag{27.3}$$

and the fits are shown in Figure 27.11 (right) and Figure 27.12 (right). The fit coefficients are tabulated in Table 27.1. These Taylor series coefficients can be evaluated at other wavelengths to describe the polarization aberration as a function of wavelength. Fits to coating design program output can provide the coefficients for other metals and for arbitrary multilayer coatings by the curve fitting method of Math Tip 13.1.

Since the fold mirror is in a converging beam, the non-zero slopes of the *s*- and *p*-phases are important and have been highlighted in Figure 27.9 (right). These slopes indicate linear phase shifts

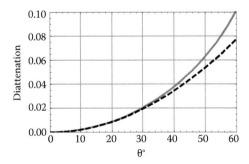
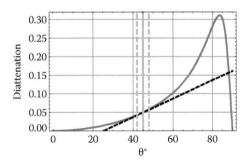

Figure 27.11 (Left) Aluminum diattenuation versus angle of incidence with quadratic fit about 0° and (right) linear fit about 45°. Solid brown lines are the exact diattenuation. Dashed black lines are the quadratic and linear fit of diattenuation (see Table 27.1). The solid green line in the right figure indicates the angle of incidence of the axial ray and the dashed green lines indicate the range of the NA = 0.06 beams.

27.5 Comparison of Polarization Ray Tracing and Polarization Aberrations

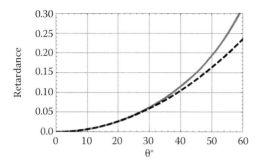 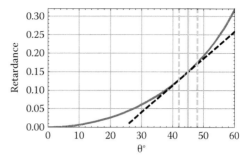

Figure 27.12 (Left) Aluminum retardance in radians versus angle of incidence with quadratic fit about 0° and (right) linear fit about 45°. Solid magenta lines are the exact retardance. Dashed black lines are the quadratic and linear fit of retardance. The solid green line in the right figure indicates the angle of incidence of the axial ray and the dashed green lines indicate the range of the NA = 0.06 beams.

Table 27.1 Aluminum Mirror Coating Fitting Coefficients

$a_0 = 0.049$
$a_1 = 0.0026$
$a_2 = 0.000024$
$b_0 = 0.150$
$b_1 = 0.0079$
$b_2 = 0.000070$

that move the locations of the X- and Y-polarized PSF components from the geometrical image location. Since the slopes are different with opposite sign, the corresponding image components move in different directions by a small fraction of the Airy disk radius.

27.5.2 Polarization Ray Trace and the Jones Pupil

A polarization ray trace of the telescope and fold mirror was performed with Polaris-M (Airy Optics Inc.). The Jones pupil calculation is displayed in Figure 27.13. From the amplitude maps (Figure 27.13, left), it is seen that this Jones pupil is very close to the identity matrix times a constant (~0.806); the 0.806 accounts for average amplitude reflection losses from three aluminum reflections. Since the Jones pupil is close to the identity matrix, the coating-induced polarization aberrations are small. Deviations from the identity matrix are due to the mirror's diattenuation and retardance.

The diagonal elements, J_{XX} and J_{YY}, contain different amplitude variations. The overall J_{XX} amplitude is about 5% larger than the overall J_{YY} amplitude because of the s- and p-reflection difference at 45° as shown in Figure 27.11 (left). This will cause the x-polarized image to be about 9% brighter than the y-polarized image. The amplitude images of the Jones pupil in Figure 27.13 (left) are close to the identity matrix, only a small fraction of the light has its polarization changed. The off-diagonal J_{XY} and J_{YX} elements show the polarization coupling between orthogonal polarizations. This polarization cross-talk has relatively low amplitude compared to the diagonal elements. The amplitudes A_{XX} and A_{YY} are nearly constant (<2% variation), but the A_{XY} and A_{YX} are highly apodized, showing a Maltese cross pattern (dark along x- and y-axes) shifted downward.

The phases of the four elements in the Jones pupil, shown in Figure 27.13 (right), represent contributions to the wavefront aberration function from the aluminum mirrors. The Fresnel phase changes are different for the s- and p-components leading to different wavefronts for these two components. Along the y-axis and along a horizontal line below the x-axis, the amplitudes of the A_{XY}

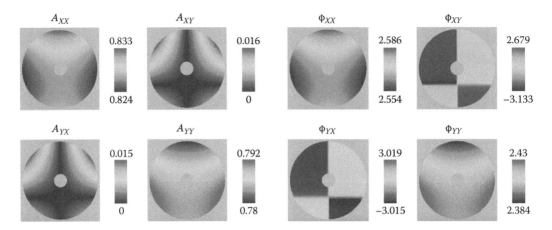

Figure 27.13 Values of the Jones pupil elements, given by Equation 27.4, are displayed as color-coded images to show variation across the exit pupil. The four images on the left show the amplitude as a function of pupil position and the four images on the right show the phase. Amplitudes A_{XX} and A_{YY} are constant to within 2%. Since A_{XY} and A_{XY} are highly apodized, their diffraction patterns are significantly larger than images associated with the diagonal terms. A scale appears to the right of each box: units are amplitude (left) and phase in radians (right). The phase of a complex number changes by π when the amplitude passes through zero. This causes the phase discontinuities in ϕ_{XY} and ϕ_{YX}.

and A_{YX} terms change sign, so the phases ϕ_{XY} and ϕ_{YX} change by π along these lines. Since the on-axis geometrical wavefront aberration of the telescope is zero, the phase variation across the pupil is entirely from the mirror coatings. ϕ_{XX}, the wavefront aberration function of the telescope illuminated with x-polarized light and analyzed with an x-analyzer, has an overall linear variation of about 0.008 waves with a small additional deviation, which primarily has the form of astigmatism. The diagonal elements ϕ_{XX} and ϕ_{YY} have a different linear variation indicating a wavefront tilt between the XX and YY wavefronts. The source of this tilt difference is seen in Figure 27.12 (right), where the fold mirror at 45° angle of incidence has different slopes of the phase change for s-polarized light and p-polarized light. If the Fresnel phases were linear about 45°, only tilt would be introduced. The deviations from linear introduce higher-order aberrations including small amounts of astigmatism (from quadratic deviation), coma (from cubic deviation), and other aberrations.

Thus, polarization ray tracing can provide a detailed picture of the polarization aberrations. If the object moves off-axis, the coatings are modified, or the configuration of the system changed, the polarization ray trace can readily recalculate the polarization aberrations. A large number of rays have been calculated to generate this Jones pupil.

27.5.3 Aberration Expression for the Jones Pupil

To develop a simpler polarization aberration model for an optical system, the Jones pupil (Figure 27.13) can be fit to a Jones matrix aberration function containing aberration terms appropriate for that particular optical system. For wavefront aberrations, defocus, spherical aberration, coma, astigmatism, and so on, are appropriate functions for a large fraction of optical systems. Similarly, for polarization aberrations, constant, linear, and quadratic functions for diattenuation, retardance, amplitude, and phase provide a good fit to many systems including this Cassegrain telescope with a fold mirror.

The Jones pupil of Figure 27.13 is a smooth function. Each element's amplitude (left) and phase (right) has low-order variations that are well described by simple constant, linear, and quadratic (piston, tilt, and defocus) terms as functions of polar pupil coordinates ρ, the normalized radial

27.5 Comparison of Polarization Ray Tracing and Polarization Aberrations

distance, and ϕ, the azimuth measured from the x-axis. The Jones pupil description has four components: the wavefront aberration $W(\rho, \phi)$ from the coatings, amplitude transmission $A(\rho, \phi)$, and the diattenuation and retardance shown here combined into the Jones pupil function $\mathbf{JP}(\rho, \phi)$

$$\mathbf{JP}(\rho,\phi) = A(\rho,\phi)e^{i2\pi W(\rho,\phi)/\lambda}\mathbf{J}(\rho,\phi). \tag{27.4}$$

The wavefront aberration and amplitude transmission are scalar terms. Examining the aluminum mirror phase changes (Figure 27.9, right), the difference $(\phi_s - \phi_p)$ is the retardance and the average of the variation, $(\phi_s + \phi_p)/2$, is the wavefront aberration. The coatings introduce small amounts of the wavefront aberrations piston w_0, tilt w_1, and defocus w_2 with the functional forms

$$W(\rho,\phi) = w_0 + w_1 \rho \sin\phi + w_2 \rho^2. \tag{27.5}$$

Higher-order wavefront aberrations, spherical aberration, coma, and so on, are not generated in significant amounts by the aluminum coatings in this example. Similarly, small polarization-independent amplitude variations (variations of the s- and p-average reflectance) are generated from the Fresnel aberrations, and the associated piston, tilt, and defocus are expressed with coefficients a_0, a_1, and a_2 as

$$A(\rho,\phi) = a_0 + a_1 \rho \sin\phi + a_2 \rho^2. \tag{27.6}$$

The diattenuation and retardance aberrations are each described by three terms yielding the six polarization aberration terms (Equation 27.7) $\mathbf{J}_1$, $\mathbf{J}_2$, ... $\mathbf{J}_6$, listed in Equation 27.8, defined in terms of the Pauli matrices, $\boldsymbol{\sigma}_1$, $\boldsymbol{\sigma}_2$, and $\boldsymbol{\sigma}_3$, and the identity matrix $\boldsymbol{\sigma}_0$ (see Chapter 14).

Each aberration term of the Jones matrix function is defined over the pupil using polar coordinates ρ and ϕ. The magnitude of each term is specified by an aberration coefficient, d_0, d_1, and d_2, for the diattenuation terms, and Δ_0, Δ_1, and Δ_2 for the retardance terms. All six aberration coefficients are much less than one, so when these six terms are cascaded, only the first-order terms in $\boldsymbol{\sigma}_1$ and $\boldsymbol{\sigma}_2$ are significant, as described in Section 15.3. The result describes the Jones pupil as the sum of the six aberration terms,

$$\begin{aligned}\mathbf{J} &= \mathbf{J}_6\mathbf{J}_5\mathbf{J}_4\mathbf{J}_3\mathbf{J}_2\mathbf{J}_1 \\ &= \boldsymbol{\sigma}_0 + \boldsymbol{\sigma}_1 \left[\frac{(d_0 + i\Delta_0) - (d_1 + i\Delta_1)\rho\sin\phi + (d_2 + i\Delta_2)\rho^2\cos 2\phi}{2}\right] \\ &\quad + \boldsymbol{\sigma}_2 \frac{(d_1 + i\Delta_1)\rho\cos\phi + (d_2 + i\Delta_2)\rho^2\sin 2\phi}{2},\end{aligned} \tag{27.7}$$

where $\mathbf{J}_1$ is diattenuation piston, $\mathbf{J}_2$ is retardance piston, $\mathbf{J}_3$ and $\mathbf{J}_4$ are tilt, and $\mathbf{J}_5$ and $\mathbf{J}_6$ are defocus terms:

$$\begin{aligned}\mathbf{J}_1 &= \boldsymbol{\sigma}_0 + \frac{d_0}{2}\boldsymbol{\sigma}_1, & \mathbf{J}_2 &= \boldsymbol{\sigma}_0 + \frac{i\Delta_0}{2}\boldsymbol{\sigma}_1, \\ \mathbf{J}_3 &= \boldsymbol{\sigma}_0 + \frac{d_1\rho}{2}(-\boldsymbol{\sigma}_1\sin\phi + \boldsymbol{\sigma}_2\cos\phi), & \mathbf{J}_4 &= \boldsymbol{\sigma}_0 + \frac{i\Delta_1\rho}{2}(-\boldsymbol{\sigma}_1\sin\phi + \boldsymbol{\sigma}_2\cos\phi), \\ \mathbf{J}_5 &= \boldsymbol{\sigma}_0 + \frac{d_2\rho^2}{2}(\boldsymbol{\sigma}_1\cos 2\phi + \boldsymbol{\sigma}_2\sin 2\phi), & \mathbf{J}_6 &= \boldsymbol{\sigma}_0 + \frac{i\Delta_2\rho^2}{2}(\boldsymbol{\sigma}_1\cos 2\phi + \boldsymbol{\sigma}_2\sin 2\phi).\end{aligned} \tag{27.8}$$

Equation 27.4 incorporating Equations 27.5, 27.6, and 27.7 provides an accurate expression for the example telescope's Jones pupil.

Table 27.2 lists the aberration coefficients for the retardance, diattenuation, amplitude, and wavefront for the Jones pupil of the telescope end to end determined by curve fitting the coefficients of Equation 27.4 incorporating Equations 27.5, 27.6, and 27.7 to the Jones pupil polarization ray trace data (Figure 27.13). Each of the aberration coefficients, d_0, d_1, d_2, Δ_0, Δ_1, and Δ_2, is the value of the diattenuation or retardance at the edge of the pupil for that specific term.

All of the polarization aberration coefficients are much less than one, so that the combination of terms in Equation 27.7 is an accurate representation for the cascaded polarization of the three elements, and the overall matrix depends little on the order of the six terms chosen. Figure 27.14 shows the Jones pupil of Figure 27.13 as approximated by the polarization aberration function. Figure 27.15 shows the small residual difference between the polarization ray trace result shown in Figure 27.13 and the polarization aberration fit shown in Figure 27.14; the fit matches the exact polarization ray tracing within 0.002 in amplitude, which is better than 0.2% of the average amplitude ~0.8.

The phase and retardance values of Table 27.2 are provided in radians. They can be divided by 2π to express the aberration in waves. For example, the aluminum coatings have contributed $\Delta_1/2\pi \sim 0.004$ waves, or 4 milliwaves of polarization dependent tilt. It is seen from the last line in Table 27.2 that the wavefront aberration contributions w_1 and w_2 contribute less than 5 milliwaves of aberration in the present example.

Equation 27.7 is a general-purpose diattenuation and retardance aberration equation which with different coefficient values will closely approximate the Jones pupils of many camera lenses,

Table 27.2 Polarization Aberration Coefficients for Telescope's Jones Pupil Shown in Figure 27.13

	Polarization Aberration Coefficients		
Diattenuation	$d_0 = 0.050$	$d_1 = -0.008$	$d_2 = -0.007$
Retardance	$\Delta_0 = -0.151$	$\Delta_1 = -0.023$	$\Delta_2 = -0.022$
Amplitude	$a_0 = 0.806$	$a_1 = -0.002$	$a_2 = 0.0000$
Wavefront	$w_0 = 2.492$	$w_1 = -0.004$	$w_2 = 0.000$

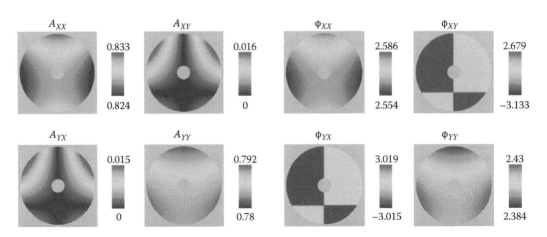

Figure 27.14 The Jones pupil calculated using Equation 27.7 with the aberration coefficients of Table 27.2 provides an accurate representation of the Jones pupil (left) for Jones pupil amplitude and (right) for Jones pupil phase in radians.

27.5 Comparison of Polarization Ray Tracing and Polarization Aberrations

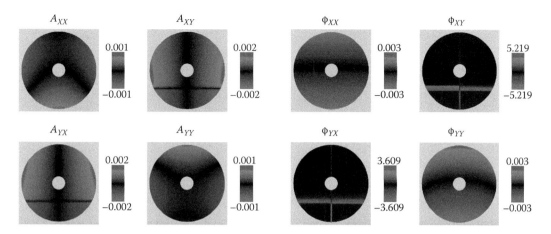

Figure 27.15 The differences between the Jones pupil obtained by polarization ray tracing in Figure 27.13 and the aberration expansion fit in Figure 27.14 are small. In all of the plots, black represents a difference of zero; the fit generated by the aberration expansion has the same value as the ray tracing data. This residual contains small contributions from polarization aberration terms of higher order than Equation 27.7.

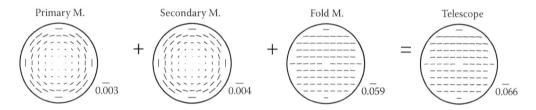

Figure 27.16 Diattenuation maps for each mirror element (the first three panels) and the cumulative diattenuation for the entire telescope (the last panel). The key in the lower right corner of each panel shows the scale of the largest diattenuation. For this example telescope, the dominant source of diattenuation is the 45° fold mirror.

microscope objectives, telescopes, and many other optical systems. The coefficient fitting helps determine if higher-order terms are needed to describe the system's polarization aberrations.

27.5.4 Diattenuation and Retardance Contributions

The individual diattenuation contributions from the three mirror elements were calculated during the polarization ray trace for a grid of rays. The surface-by-surface contributions are shown in the first three panels of Figure 27.16. The fourth panel in Figure 27.16 shows the cumulative diattenuation for the entire telescope as viewed looking into the exit pupil from the image plane. The primary and secondary mirrors produce *diattenuation defocus*, $\mathbf{J}_5 = \boldsymbol{\sigma}_0 + d_2 \, \rho^2 \, (\boldsymbol{\sigma}_1 \cos 2\phi + \boldsymbol{\sigma}_2 \sin 2\phi)/2$, a rotationally symmetric, tangentially oriented diattenuation with a magnitude that increases quadratically* from the center of the pupil. The fold mirror introduces *diattenuation tilt*, $\mathbf{J}_3 = \boldsymbol{\sigma}_0 + d_1 \, \rho \, (-\boldsymbol{\sigma}_1 \sin \phi + \boldsymbol{\sigma}_2 \cos \phi)/2$, a horizontally oriented diattenuation with a linear variation along the vertical axis, as well as having a constant offset at the center described by the term, *diattenuation piston*, $\mathbf{J}_1 = \boldsymbol{\sigma}_0 + d_0 \boldsymbol{\sigma}_1/2$. Since the fold mirror makes the largest contribution, the entire telescope's diattenuation is similar to the fold mirror.

---

* Throughout, *linear* implies *approximately linear* and *quadratic* implies *approximately quadratic*.

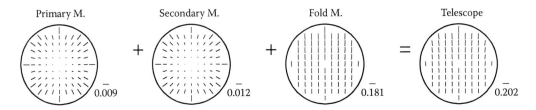

Figure 27.17 Retardance maps for each mirror element (the first three panels) and the cumulative retardance (the last panel). The key in the lower right corner of each of the four panels shows the scale of the largest retardance in radians. This figure shows that the dominant source of retardance is the 90° fold mirror (third panel).

Figure 27.17 shows the individual surface contributions to the retardance aberration in the first three panels and the cumulative retardance aberration in the last panel. The primary and secondary mirrors produce *retardance defocus*, $\mathbf{J}_6 = \mathbf{\sigma}_0 + i\Delta_2\, \rho_2\, (\mathbf{\sigma}_1 \cos 2\phi + \mathbf{\sigma}_2 \sin 2\phi)/2$, a rotationally symmetric tangentially oriented fast axis, which increases quadratically from the center, while the fold mirror introduces *retardance tilt*, $\mathbf{J}_4 = \mathbf{\sigma}_0 + i\Delta_1 \rho\, (-\mathbf{\sigma}_1 \sin\phi + \mathbf{\sigma}_2 \cos\phi)/2$, and *retardance piston*, $\mathbf{J}_2 = \mathbf{\sigma}_0 + i\Delta_0\, \mathbf{\sigma}_1/2$, with a vertically oriented fast axis. Since the fold mirror has the largest retardance, the resultant retardance for the entire telescope is similar to the fold mirror retardance. The cumulative linear retardance map (the fourth panel of Figure 27.17) is primarily a constant retardance with a linear variation from bottom to top, and a smaller variation of retardance orientation from left to right. Constant retardance is a constant difference in the wavefront aberration between *XX* and *YY*, a "piston" between polarization states; it changes polarization states, but this piston does not degrade image quality. The linear variation of retardance indicates a difference in the wavefront aberration tilt; *X*- and *Y*-polarizations get different linear phases, and so their images are shifted from the nominal image location by different amounts. This effect is very interesting and will be analyzed later.

The retardance defocus from the primary and secondary mirrors causes *astigmatism*. For *X*-polarized light, the relative phase is *advanced quadratically moving along the x-axis* from the center to the edge of the field and is retarded quadratically moving to the edge of the field along the *y*-axis. This causes astigmatism arising from the different quadratic variations of ϕ_{rs} and ϕ_{rp} about the origin in Figure 27.9 (right). Hence, the *X*-polarized image, being astigmatic by 0.022 radians (0.012 + 0.010 or 3.4 milliwaves), becomes slightly elongated in opposite directions on either side of the best focus. Similarly for *Y*-polarized light, the relative phase is *advanced moving along the y-axis* from the center to the edge of the field and is retarded moving to the edge of the field along the *x*-axis. Therefore, the *Y*-polarized image is astigmatic with the opposite sign. For unpolarized light, the coating-induced astigmatic image is the average over the PSF of all polarization components, which is also the sum of the PSF for any two orthogonal components. Thus, when the astigmatism of *X*-polarized light is added to the astigmatism for *Y*-polarized light where the astigmatism is rotated by 90°, the combination forms a radially symmetric PSF, which is slightly larger than the unaberrated image. Inserting a polarizer will reveal the astigmatism at any particular polarization orientation. More information on retardance defocus and the associated astigmatism in Cassegrain telescopes is found in Reiley.[14]

These polarization aberrations affect the PSF. A detailed discussion of the image defects will be delayed until after the design rules are introduced.

27.5.5 Design Rules Based on Polarization Aberrations

The analysis above was for aluminum coatings at one wavelength. With the aberration tools, the telescope's polarization aberrations for many coatings can be accurately estimated by combining two of our expressions: (A) The Jones pupil is described in terms of six simple polarization

27.5 Comparison of Polarization Ray Tracing and Polarization Aberrations

aberration terms: constant ($\mathbf{J}_1$ and $\mathbf{J}_2$), linear ($\mathbf{J}_3$ and $\mathbf{J}_4$), and quadratic terms ($\mathbf{J}_5$ and $\mathbf{J}_6$) in Equation 27.7, and (B) the coating polarization has been described in terms of simple constant (a_0 and b_0), linear (a_1 and b_1), and quadratic terms (a_2 and b_2) in Equations 27.2 and 27.3. If the coating wavelength is changed, or if the coating design is changed, then a_0, a_1, a_2, b_0, b_1, and b_2 in Equations 27.2 and 27.3 also change. Hence, changes in (A) or (B) can be simply related to the Jones pupil polarization aberration coefficients: d_0, d_1, d_2, Δ_0, Δ_1, and Δ_2.

A list of *design rules* will be considered based on the behaviors of the aberrations of Figure 27.14 for the following list of the example telescope's image defects:

1. Diattenuation at the center of the pupil.
2. Retardance at the center of the pupil.
3. The PSF shears between the *XX*- and *YY*-components, the two bright co-polarized components. These PSFs shift in opposite directions because of the fold mirror, thus stretching point images!
4. The polarization-dependent astigmatism, whose orientation rotates with the incident polarization state.
5. The fraction of light in the ghost PSF in components *XY* and *YX*. These dim cross-polarized components have PSFs about twice as large as the *XX*- and *YY*-components.

By analyzing a simple three-element system with aberration theory, each image defect can be related to the fold mirror angle, numerical aperture, and coating choices.

27.5.5.1 Diattenuation at the Center of the Pupil

The diattenuation at the center of the pupil corresponds to the diattenuation piston term $\mathbf{J}_1$ with magnitude d_0. This arises only from the fold mirror; the primary and secondary mirrors have zero diattenuation at the center of the pupil. The coefficient d_1 is close to the value of the average diattenuation, averaging over the pupil. $\mathbf{J}_1$ is primarily responsible for the 9% difference in the flux transmitted to the focal plane between the incident *X*- and *Y*-polarized components. Unpolarized light has a ~4% *DoP* when incident at the focal plane. This is typically calibrated out when using focal plane Stokes polarimeters. The magnitude of d_0 only depends on the diattenuation at the fold mirror evaluated at the axial ray's angle of incidence θ_3 (for surface 3, 45° in this case). The angle θ_0 in the Taylor series expansion of Equation 27.3 is θ_3 for the telescope's fold mirror. This leads to two design rules for the average diattenuation.

Design rule 1 The average diattenuation (diattenuation piston value) characterized by d_0 is quadratic in the fold mirror angle θ_3.

Design rule 2 The diattenuation piston value, d_0, is linear in the coating parameter a_0 defined in Equation 27.3. If $a_0(\lambda)$ is calculated for a coating as a function of wavelength, the spectral variation of the average diattenuation will be proportional to $a_0(\lambda)$. If the aluminum is overcoated with silicon monoxide or some other material, $a_0(\lambda)$ can be recalculated to compare the resulting average diattenuation between coatings.

27.5.5.2 Retardance at the Center of the Pupil

The retardance at the center of the pupil corresponds to the retardance piston term $\mathbf{J}_2$ with value Δ_0. $\mathbf{J}_2$ also arises only from the fold mirror. The coefficient Δ_0 is close to the value of the average retardance, averaging over the pupil. $\mathbf{J}_2$ does not change the polarization (Stokes parameters) of unpolarized light transiting the system; retardance only modifies polarized and partially polarized light. $\mathbf{J}_2$ introduces a constant phase difference between the *X*- and *Y*-incident components; thus, $\mathbf{J}_2$ does not degrade the PSF. The value of Δ_0 only depends on the retardance at the fold mirror evaluated at the axial ray's angle of incidence θ_3. This leads to two more design rules.

Design rule 3   The retardance piston characterized by Δ_0 is quadratic in the fold mirror's tilt angle θ_3.

Design rule 4 Δ_0 is linear in the coating parameter b_0 the retardance at the fold mirror's nominal angle of incidence (Equation 27.3).

27.5.5.3   Linear Variation of Diattenuation

The diattenuation tilt $\mathbf{J}_3$ with magnitude d_1 describes a linear variation of diattenuation across the pupil, corresponding to unpolarized light exiting with a smaller *DoP* at the bottom of the pupil and a larger *DoP* at the top of the pupil. $\mathbf{J}_3$ arises only from the fold mirror; the primary and secondary mirrors do not contribute. $\mathbf{J}_3$ causes the *X*-polarized input to be brighter at the top of the pupil and be linearly dimmer toward the bottom of the pupil. *Y*-polarized input has the opposite variation. This apodization has a small effect on the shape and structure of the *XX*- and *YY*-PSFs, much smaller than the other polarization imaging defects. $\mathbf{J}_3$ does contribute significantly to the off-diagonal Jones pupil elements J_{XY} and J_{YX} and thus to the brightness of the ghost PSFs, I_{XY} and I_{YX}. The value of d_1 depends on the slope of the diattenuation a_1 at the fold mirror evaluated at θ_3 and on the range of angles at the fold mirror, characterized by the *F/#* in image space, *F*/8, or alternatively the numerical aperture, *NA* = 0.06. This leads to the design rules for the diattenuation tilt $\mathbf{J}_3$.

Design rule 5   The diattenuation tilt characterized by d_1 is linear in the angle θ_3. This follows from the slope of a quadratic function (Equation 27.2) being linear.

Design rule 6 d_1 is linear in the coating diattenuation slope parameter a_1.

Design rule 7 d_1 is also linear in the coating diattenuation quadratic parameter angle a_2, which follows from the slope of a quadratic function being linear.

27.5.5.4   Linear Variation of Retardance, the PSF Shear between the XX- and YY-Components

The retardance tilt $\mathbf{J}_4$ with value Δ_1 describes a linear phase variation for the *XX*-polarized component and an opposite linear phase variation for the *YY*-polarized component. This term arises only from the fold mirror; the primary and secondary mirrors do not contribute. $\mathbf{J}_4$ is a very important term because it shifts the images of the *XX*- and *YY*-components in opposite directions, causing the overall PSF to become elliptical. $\mathbf{J}_4$ also contributes to the off-diagonal Jones pupil elements J_{XY} and J_{YX} and thus to the brightness of the ghost PSFs, I_{XY} and I_{YX}. The value of Δ_1 depends on the slope of the retardance at the fold mirror evaluated at θ_3 and on the range of angles at the fold mirror, characterized by the numerical aperture. This leads to the design rules for the retardance tilt $\mathbf{J}_4$.

Design rule 8   The retardance tilt characterized by Δ_1 is linear in the fold mirror angle θ_3, which follows from the slope of a quadratic function being linear.

Design rule 9   The retardance tilt magnitude Δ_1 is linear in the coating retardance slope parameter b_1.

Design rule 10   The retardance tilt magnitude Δ_1 is also linear in the coating retardance quadratic parameter b_2, which follows from the fact that the slope of a quadratic function, b_1, is linear in the quadratic parameter.

27.5.5.5   The Polarization-Dependent Astigmatism

The retardance defocus term $\mathbf{J}_6$ with magnitude d_2 describes a quadratic variation of retardance from the center of the pupil, which is tangentially oriented. This arises primarily from the primary and secondary mirrors with a small contribution from the fold mirror. $\mathbf{J}_6$ causes the *X*-polarized input exiting into *X*-polarized output to become astigmatic. From the center, the phase is advanced quadratically along the x-axis and delayed quadratically along the y-axis. For YY, this astigmatism is rotated by 90°. The effect on unpolarized light is like spinning the astigmatic PSF; the PSF is rotationally symmetric, but enlarged by the astigmatism. $\mathbf{J}_6$ also contributes to the off-diagonal Jones pupil elements J_{XX} and J_{YY} and thus to the brightness of the ghost PSFs, I_{XY} and I_{YX}. The magnitude of d_2 depends on the quadratic variation of the retardance about normal incidence b_2 and on the angle of incidence of the marginal ray at the edge of the primary θ_1 and secondary θ_2 mirrors. This leads to the design rules for the retardance defocus $\mathbf{J}_6$.

> **Design rule 11** The polarization-dependent astigmatism term, retardance defocus, is characterized by Δ_2, and thus is quadratic in the sum of the angles $\theta_1 + \theta_2$ assuming identical coatings. Therefore, Δ_2 is quadratic in the NA if the design F/# is scaled by just changing the entrance pupil diameter.
>
> **Design rule 12** The retardance defocus magnitude Δ_2 is quadratic in the coating retardance quadratic parameter b_2.
>
> **Design rule 13** The maximum fraction of flux coupled into the orthogonal state for weak linear polarization elements is quadratic in the diattenuation and quadratic in the retardance, and occurs for incident light polarized at 45° to the diattenuation axis or retardance fast axis.

27.5.5.6 The Fraction of Light in the Ghost PSF in XY- and YX-Components

The fraction F_{YX} of the X-polarized incident light coupled into the ghost PSF image I_{YX} depends on the tilt coefficients d_1 and Δ_1 and the defocus coefficients d_2 and Δ_2, but not the piston coefficients. This is also equal to the fraction F_{XY} of the Y-polarized incident light coupled into the ghost PSF image I_{XY}. The fraction of incident flux incident in X- or Y-polarization coupled into the orthogonal polarization state is found by integrating the off-diagonal element's (the $\boldsymbol{\sigma}_2$ term) magnitude squared $|\mathbf{J}_{YX}|^2$ or $|\mathbf{J}_{XY}|^2$ over the pupil and normalizing by π, the area of the pupil,

$$F_{XY} = \frac{\int_0^{2\pi}\int_0^1 |\mathbf{J}_{XY}|^2 \rho d\rho d\phi}{\int_0^{2\pi}\int_0^1 \rho d\rho d\phi} = \frac{d_1^2 + \Delta_1^2}{16} + \frac{d_2^2 + \Delta_2^2}{24}. \qquad (27.9)$$

F_{XY} is quadratic in the tilt and defocus coefficients; hence, an order of magnitude reduction in these polarization aberrations reduces the ghost brightness by two orders of magnitude.

> **Design rule 14** The fraction of the flux incident in either the X-polarization or the Y-polarization coupled into the orthogonal ghost images I_{YX} (fraction F_{YX}) or I_{XY} (fraction F_{XY}) depends quadratically on d_1 and Δ_1. Since d_1 and Δ_1 are linear in the angle θ_3, F_{YX} and F_{XY} are quadratic in the fold mirror angle θ_3 (see Design rules 5 and 8). F_{YX} and F_{XY} are also quadratic in the coating retardance slope parameters a_1 and b_1 (see Design rules 6 and 9). F_{XY} and F_{YX} are also quadratic in the coating diattenuation quadratic parameter a_2 (see Design rules 7 and 10).

Analysis of the defocus aberrations leads to additional design rules for the fractions F_{YX} and F_{XY}. The diattenuation defocus, d_2, and the retardance defocus, Δ_2, are quadratic in the sum of the marginal ray angles $\theta_1 + \theta_2$ assuming identical coatings.

Design rule 15 The fractions F_{YX} and F_{XY} are fourth order in the sum of the marginal ray angles $\theta_1 + \theta_2$ assuming identical coatings, and thus fourth order in the *NA*, assuming the design F/# is scaled by just changing the entrance pupil diameter. Thus, small decreases in F/# can yield large increases in ghost PSF brightness (see Design rule 11).

Design rule 16 The fractions F_{YX} and F_{XY} are fourth order in the coating diattenuation quadratic parameter a_2 (see Design rule 12).

Thus, these design rules describe the scaling of the polarization aberration as the pupil size changes, F/# changes, coating prescription changes, and fold mirror angle changes for a telescope of the configuration in Figure 27.8. As presented, these relations only apply to the on-axis image. The off-axis equations are more complicated, but coronagraphs and other astronomical systems usually have small enough fields that the polarization aberration variation over the field is not significant and therefore these design rules do not change over this system's practical field of view. As more fold mirrors or other components are added to this system, these polarization aberration equations need to be generalized to relate the polarization aberrations to the coating prescriptions and optical prescription. Lam and Chipman discussed polarization aberration reduction with two- and four-mirror combinations.[15] Discussions of higher-order polarization aberration terms can be found in McGuire and Chipman,[16,17] Ruoff and Totzeck,[18] and Sasián.[19,20]

27.5.6 Amplitude Response Matrix

In conventional scalar image formation calculations, the amplitude response function is calculated as the Fourier transform of the exit pupil function. This electric field distribution is then squared to obtain the PSF.[21] To evaluate the image formed by systems with polarization aberration, McGuire and Chipman introduced a Jones calculus version of the amplitude response function named the *Amplitude Response Matrix*, **ARM**,[22,23]

$$\mathbf{ARM} = \begin{pmatrix} \mathfrak{F}[J_{XX}(x,y)] & \mathfrak{F}[J_{XY}(x,y)] \\ \mathfrak{F}[J_{YX}(x,y)] & \mathfrak{F}[J_{YY}(x,y)] \end{pmatrix}, \tag{27.10}$$

where $\mathfrak{F}$ is a spatial Fourier transform over each of the Jones pupil elements (see Section 16.4). For a plane wave incident on the telescope with Jones vector **E**, the amplitude and phase of the image are given by the matrix multiplication, **ARM·E**. The **ARM** for the three-mirror telescope of Figure 27.8 is shown in Figure 27.18. Table 27.3 summarizes the system and the parameters most relevant to this imaging calculation.

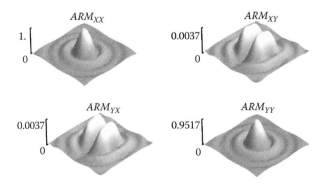

Figure 27.18 The absolute value of the amplitude of the 2 × 2 **ARM** at an on-axis field point is shown for the example telescope of Figure 27.8 normalized by the peak of the *XX*-component.

27.5 Comparison of Polarization Ray Tracing and Polarization Aberrations

Table 27.3 Parameters Associated with the Imaging Calculation

Wavelength	800 nm
Image space F/#	8
Entrance pupil diameter	2.4 m
Effective focal length	19.236 m
Number of rays across entrance pupil	65
Number of rays across the Jones pupil array	513
Spacing in the **ARM** and **PSM** viewing from object space	9.0 milliarcsec

The **ARM**'s diagonal elements are close to the well-known Airy disk pattern but are slightly larger due to the aberrations in ϕ_{XX} and ϕ_{YY}. Each is slightly astigmatic. Their centroids are slightly shifted due to the differences in their tilt. The off-diagonal elements have much lower amplitudes and contain interesting structure, mostly due to the fold mirror. We refer to these off-diagonal PSF images as the ghost PSFs.

For unpolarized illumination, the incident X- and Y-polarizations are incoherent with respect to each other. Thus, the output components ARM_{XX} (X in X out) and ARM_{YX} (X in Y out) are coherent with each other but incoherent with ARM_{XY} and ARM_{YY}. Hence, for unpolarized illumination, the two output X-components in the **ARM** are incoherent with respect to each other, as are the two output Y-components. Thus, the PSF for an unpolarized source has four additive components $I = I_X + I_Y = (|ARM_{XX}|^2 + |ARM_{XY}|^2) + (|ARM_{YX}|^2 + |ARM_{YY}|^2)$.

27.5.7 Mueller Matrix Point Spread Matrices

The distribution of flux and polarization in the image of an incoherent point source, such as a star, can be described with a 4 × 4 *Mueller matrix point spread matrix* (**PSM**), the Mueller matrix generalization of the PSF (Section 16.5). This **PSM** is calculated by the transformation of the **ARM**'s Jones matrices into Mueller matrix functions using Equation 6.102 or 6.107. The Mueller matrix representation of polarization properties is familiar to most astronomers who make or work with astrophysical measurements of the four Stokes parameters.[24–27] The example telescope's **PSM**, calculated from the **ARM** (Figure 27.18), is shown in Figure 27.19. The contribution of each of the 16 elements varies across the **PSM** and changes depend on the incident Stokes parameters. Hence, each element of the matrix is shown with its contribution and appears as miniature PSFs with different shapes. An example of **PSM** measurements is found in McEldowney.[7]

The PSF for unpolarized illumination is described by the Stokes parameter image in the first column ($m00, m10, m20, m30$) inside the red rectangle. Since $m10$, $m20$, and $m30$ are not zero, the PSF of an unpolarized star is not unpolarized. In this example, the Q-component's 4.7×10^{-2} contribution mostly arises from the diattenuation of the fold mirror, which is reflecting more 0° (s-polarized) light than 90° (p-polarized) polarized light. The U-component (at 4.36×10^{-3}) is mostly due to the diattenuation contributions at 45° and 135° from the primary and secondary seen in the first two panels in Figure 27.16. The ellipticity (from the V-component) arises when weakly polarized light reflected from the primary and secondary interacts with the retardance from the fold mirror. The spatial variations of Q, U, and V introduce polarization fluctuations in the region of the diffraction rings. Figure 27.20 maps the *DoP* in these zones. Such polarization fluctuations in the PSF of a star are clearly a concern when measuring the polarization of exoplanets and debris disks.

Figure 27.19 contains a graphical equation describing the 4 × 4 **PSM** operating on an X-polarized incident beam (represented by the 4 × 1 matrix), which yields a 4 × 1 Stokes image (I_X, Q_X, U_X, V_X), as represented by the rightmost term in the equation. For an unpolarized collimated incident beam, the resulting Stokes image is contained in the first column of the **PSM**, shown inside the red box. The $m10$ element describes the ~9% *DoP* for the image of the unpolarized source. The $m20$ element

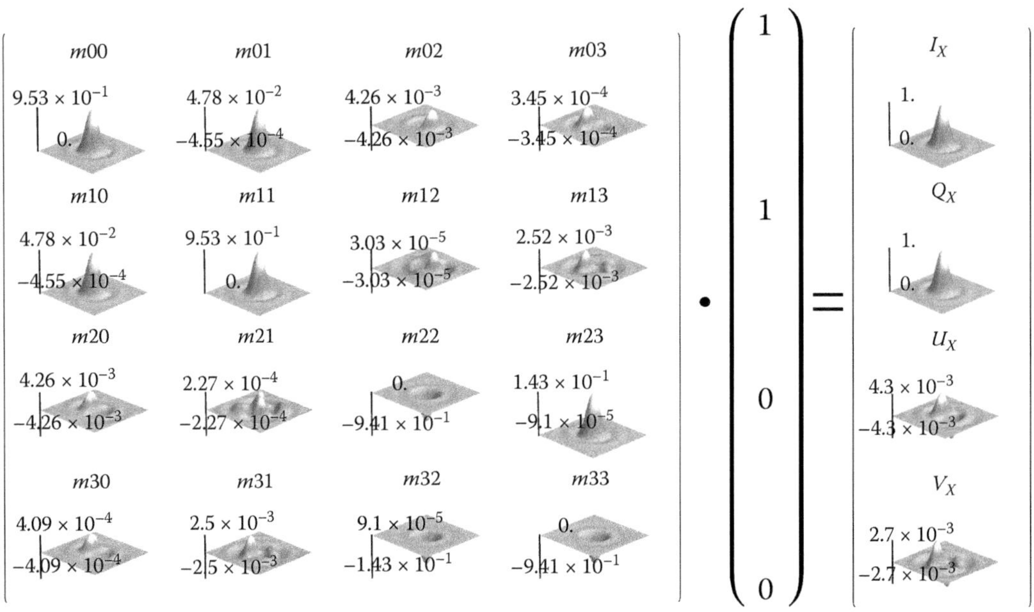

$$\begin{pmatrix} m00 & m01 & m02 & m03 \\ 9.53\times10^{-1} & 4.78\times10^{-2} & 4.26\times10^{-3} & 3.45\times10^{-4} \\ m10 & m11 & m12 & m13 \\ 4.78\times10^{-2} & 9.53\times10^{-1} & 3.03\times10^{-5} & 2.52\times10^{-3} \\ m20 & m21 & m22 & m23 \\ 4.26\times10^{-3} & 2.27\times10^{-4} & 0. & 1.43\times10^{-1} \\ m30 & m31 & m32 & m33 \\ 4.09\times10^{-4} & 2.5\times10^{-3} & 9.1\times10^{-5} & 0. \end{pmatrix} \cdot \begin{pmatrix} 1 \\ 1 \\ 0 \\ 0 \end{pmatrix} = \begin{pmatrix} I_X \\ Q_X \\ U_X \\ V_X \end{pmatrix}$$

Figure 27.19 The 4 × 4 point spread matrix (**PSM**) (left) operates on an example X-polarized incident beam with Stokes parameters (1, 1, 0, 0) (middle) to calculate the point spread function. The resultant polarization distribution is a 4 × 1 Stokes image (right) with components (I_X, Q_X, U_X, V_X). The subscript X represents Stokes parameters resultant from X-polarized incident light. The normalized magnitude of each matrix element is shown on each vertical scale. The U_X image indicates a small variation of polarization orientation, while V_X indicates small variations of ellipticity. The red box on the left (m00, m10, m20, m30) is the resultant Stokes parameter image (I, Q, U, V) for a collimated beam of unpolarized incident light, such as an unpolarized star.

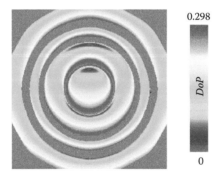

Figure 27.20 The *DoP* variation throughout the PSF of an unpolarized object. Regions with intensity below 0.0008 of the peak have been removed due to noise and are shown in gray.

describes small variations of linear polarization orientation within the PSF while m30 characterizes even smaller ellipticity variations.

The light coupled into orthogonal components has a significant impact on the outer portions of the PSF because they arise from the highly apodized Jones pupil components A_{XY} and A_{YX} as shown in Figure 27.14 (left). To see this, compare the PSF arising from J_{XX} and J_{YX}. These PSF terms are calculated using the resultant Stokes image components on the right side of Figure 27.19 as

$$I_{XX} \propto \frac{I_X + Q_X}{2} \quad \text{and} \quad I_{YX} \propto \frac{I_X - Q_X}{2}. \tag{27.11}$$

27.5 Comparison of Polarization Ray Tracing and Polarization Aberrations

The two terms in Equation 27.11, I_{XX} and I_{YX}, are compared in Figure 27.21, where it is seen that the peak of I_{YX} is about 10^{-5} of I_{XX}. This "ghost PSF" should be very important in imaging applications that require contrast ratios of 10^{-8} or greater.

Figure 27.21 compares the I_{XX} PSF component with the ghost component I_{YX}. Figure 27.22 shows the irradiance along an x-axis cross section, indicated by the vertical plane, through the two PSFs in $\log_{10}$ scale at the plane drawn through the two images shown in Figure 27.21. This ghost PSF has its light spread away from the center. In Figure 27.22, the Airy disk's zeros of I_{XX} are not at the same location as the zeros for the cross-coupled term I_{YX}. Thus, the zeros of I_{XX} are washed out by the light leakage from the non-zero I_{YX}. The PSF I_{YX} cannot be corrected by wavefront compensation for either the XX- or YY-components alone because most of the image spread is due to I_{YX}'s apodization (Figure 27.13). A linear Polaroid placed at the image plane can pass I_{XX} and remove I_{YX}, but will still pass the other ghost I_{XY}, and thus will not correct for this polarization aberration.

The shape of I_{YX} shows that the I_X and Q_X Airy disks are not exactly on top of each other. The image plane irradiance distribution for the I_{YX} term sits beneath the Airy diffraction pattern characteristic of the I_{XX} term. Figure 27.22 (left) shows a slice normal to the axis at the RMS best focus through the PSF for I_{XX} and for I_{YX} in Figure 27.21. Figure 27.22 (right) shows a high-dynamic range image of the irradiance across the focal plane in the vicinity of the core of the PSF for I_{YX}. The concentric pink circles superposed on Figure 27.22 (right) shows the first and second zeros of the Airy diffraction pattern of I_{XX}. These dark rings overlay regions with non-zero I_{YX}.

The right column in Figure 27.19 (I_X, Q_X, U_X, V_X) is the Stokes parameter PSF for the X-polarized component of an incident beam. The flux of this component is $I_X = I_{XX} + I_{YX}$. Similarly, the PSF I_Y

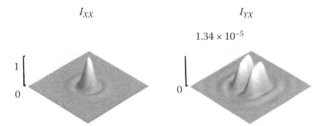

Figure 27.21 PSF of I_{XX} and I_{YX} calculated from Equation 27.11 is shown normalized to the peak of I_{XX}.

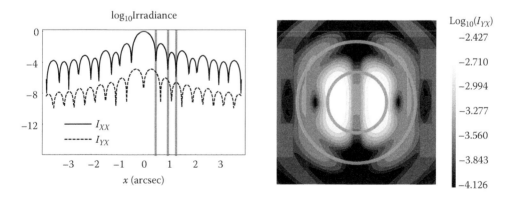

Figure 27.22 (Left) Cross sections through the $\log_{10}$PSF image for I_{XX} and I_{YX} between −1 and +1 arc second along the x-direction. The solid and dashed black curves show I_{XX} and I_{YX}, respectively. Note that the first three minima of I_{XX} (dark rings in the Airy disk) are close to the angles for the local maxima of the I_{YX} curve. On average, the polarization-coupled I_{YX} flux is about 10^{-4} below I_{XX}. (Right) PSF for I_{YX} shown in $\log_{10}$ contour with its scale on the right ranging from −2.4 to −4.1, normalized to the peak intensity of I_{XX}. The superposed pink circles show the location of the first and second Airy dark ring of I_{XX}.

for a Y-polarized incident beam is calculated by multiplying the Stokes parameters (1, −1, 0, 0) to the **PSM**, and $I_Y = I_{YY} + I_{XY}$. Finally, the PSF for unpolarized incident light is $(I_X + I_Y)/2$, which can also be calculated by multiplying the unpolarized Stokes parameters (1, 0, 0, 0) to the **PSM**.

This demonstrates that for unpolarized starlight passing into a "generic" optical system like that shown in Figure 27.8, the PSF is the sum of two nearly Airy diffraction patterns, I_{XX} and I_{YY}, plus two secondary or "ghost" PSFs, I_{YX} and I_{YX}, which originates from the system's polarization crosstalk, the off-diagonal elements in the Jones pupil.

The Jones pupil, **ARM**, and **PSM** can be calculated in basis sets other than x and y. Here, x and y are aligned parallel and perpendicular to the fold mirror's s-state. The resulting Jones pupil and **ARM** for other bases are found by Cartesian rotation of the matrices of Figures 27.13 and 27.18. The overall flux distributions, $I = I_X + I_Y$, for an unpolarized source or point source of arbitrary polarization are unchanged by such a change of the basis. Similarly, the **PSM** is rotated by the same rotation operation as Mueller matrices, and again, the net flux for any source polarization is unchanged; the corresponding Stokes images are just rotated versions of the ones presented above. The advantage of the x- and y-basis chosen here is that the off-diagonal elements I_{XY} and I_{YX} have their smallest values in this basis. As the basis set rotates or becomes elliptical, the scale of these off-diagonal image components increases rapidly and quickly approaches Airy disks due to the coupling between bright diagonal elements and weak off-diagonal elements from the rotation operation. Thus, the fold mirror's s- and p-basis (our x and y) is the best basis for viewing the value and functional form of the ghost PSF components for the example system of Figure 27.8.

27.5.8 Location of the PSF Image Components

For the telescope of Figure 27.8, consider a Stokes imaging polarimeter measuring the PSF of an unpolarized star as a Stokes image. The PSFs for the X-polarized, $I_X = I_{XX} + I_{YX}$, and Y-polarized light, $I_Y = I_{XY} + I_{YY}$, at the focal plane are very close in form to the classical Airy diffraction pattern because the polarization-induced wavefront aberration, ϕ_{XX} and ϕ_{YY} in Figure 27.13, is less than 8 milliwaves and the amplitude apodization is less than 0.015. But these two PSF images are not exactly superposed; the peaks of I_X and I_Y are displaced from each other by 0.625 milliarcsec. The PSF cross sections through the maxima of I_X, I_Y, and $I_X - I_Y$ (the star's Stokes Q image) are shown in Figure 27.23. The shift between the I_X and I_Y PSFs arises from the difference in slopes of the s- and p-phases in the Fresnel coefficients (red and blue tangent lines in Figure 27.9 (right), which is the cause of the overall linear variations in ϕ_{XX} and ϕ_{YY}. Their difference $Q = I_X - I_Y$ is sheared from I_X and I_Y by 5.8 milliarcsec, as shown in Figure 27.23, and is due to the shift between I_X and I_Y. These PSF shifts and PSF ellipticities are listed in Table 27.4 for a single 45° fold mirror before

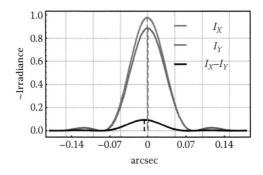

Figure 27.23 The cross-section profiles of the I_X (red) and I_Y (blue) PSF images in arc seconds from the center of the PSF. The black line shows Stokes Q image, the difference between the two PSFs.

Table 27.4 The Shape of the PSF Calculated from PSM in Figure 27.19 Is Described by the Following Parameters: The PSF's Flux, the Radius of Encircled Energy, the PSF Shears, and the PSF Ellipticity for X- and Y-Polarized Incident Light

Characterize the shape of PSF	
PSF shear in object space:	
Between I_X and I_Y	0.625 milliarcsec
Between I_X and $(Q = I_X - I_Y)$	5.820 milliarcsec
Flux in PSF:	
$\dfrac{\text{Flux of } I_{YX}}{\text{Flux of } I_{XX}}$	0.0048%
$\dfrac{\text{Flux of } I_{YY}}{\text{Flux of } I_{XX}}$	90.6%
$\dfrac{\text{Flux of } I_{YX}}{\text{Flux of } I_{XX}}$	0.0046%
$\dfrac{\text{Peak of } I_Y}{\text{Peak of } I_X}$	90.6%
$\dfrac{\text{Peak of } (I_X - I_Y)}{\text{Peak of } I_X}$	$\dfrac{\text{Peak of } Q}{\text{Peak of } I_X} = 9.6\%$
Radius of 90% encircled energy in object space:	
$r_{XX} = r_{YY}$	0.15 arcsec
$r_{YX} = r_{XY}$	0.36 arcsec
Ellipticity of PSF:	
Unpolarized incident light	7.502×10^{-6}
X-polarized incident light	0.00199
Y-polarized incident light	0.00208

the focal plane. The ellipticity of the PSF image was calculated by fitting an ellipse to the PSF at the half power points.

In astronomical applications involving the precise measurement of the location of the centroid of the PSF, distortions of the shape of the PSF are important. Most systems incorporate multiple folds. These relay optics with multiple folds may increase the shear between the PSF's polarization components. The variation of linear phase across the pupil, as seen in Figure 27.8, is approximately linear; thus, the shear between polarization components is linear in the $F/\#$.

References

1. https://www.nist.gov/services-resources/software/scatmech-polarized-light-scattering-c-class-library (accessed June 30, 2017).
2. T. A. Germer, Polarized light diffusely scattered under smooth and rough interfaces, *Proc. SPIE* (2003): 5158.
3. R. M. Hao, *FDTD Modeling of Metamaterials: Theory and Applications*, Artech House Publishers (2009).

4. P. W. Maymon and R. A. Chipman, Linear polarization sensitivity specifications for space-borne instruments, *Proc. SPIE* 1746, Polarization Analysis and Measurement, 148 (1992).
5. R. A. Chipman, Polarization analysis of optical systems, *Opt. Eng.* 28(2) (1989): 280290.
6. N. Clark and J. B. Breckinridge, Polarization compensation of Fresnel aberrations in telescopes, in *SPIE Optical Engineering and Applications, International Society for Optics and Photonics* (2011).
7. S. McEldowney, D. Shemo, and R. Chipman, Vortex retarders produced from photo-aligned liquid crystal polymers, *Opt. Express* 16 (2008): 7295–7308.
8. D. Mawet, E. Serabyn, K. Liewer, Ch. Hanot, S. McEldowney, D. Shemo, and N. O'Brien, Optical vectorial vortex coronagraphs using liquid crystal polymers: Theory, manufacturing and laboratory demonstration, *Opt. Express* 17 (2009): 1902–1918.
9. D. R. Chowdhury, K. Bhattacharya, A. K. Chakraborty, and R. Ghosh, Polarization-based compensation of astigmatism, *Appl. Opt.* 43 (2004): 750–755.
10. I. J. Hodgkinson and Q. Wu, *Birefringent Thin Films and Polarizing Elements*, World Scientific Publishing Company (1998).
11. P. W. Maymon and R. A. Chipman, Linear polarization sensitivity specifications for space borne instruments, *Proc. SPIE* 1746 (1992): 148–156.
12. S. C. McClain, P. W. Maymon, and R. A. Chipman, Design and analysis of a depolarizer for the Moderate resolution Imaging Spectrometer Tilt (MODIS T), *Proc. SPIE* 1746 (1992): 375–385.
13. J. B. Breckinridge, W. S. T. Lam, and R. A. Chipman, Polarization aberrations in astronomical telescopes: The point spread function, *Publ. Astron. Soc. Pacific* 127(951) (2015): 445–468.
14. D. J. Reiley and R. A. Chipman, Coating-induced wave-front aberrations: On-axis astigmatism and chromatic aberration in all-reflecting systems, *Appl. Opt.* 33(10) (1994): 2002–2012.
15. W. S. T. Lam and R. Chipman, Balancing polarization aberrations in crossed fold mirrors, *Appl. Opt.* 54.11 (2015): 3236–3245.
16. J. P. McGuire and R. A. Chipman, Polarization aberrations. 1. Rotationally symmetric optical systems, *Appl. Opt.* 33.22 (1994): 5080–5100.
17. J. P. McGuire and R. A. Chipman, Polarization aberrations. 2. Tilted and decentered optical systems, *Appl. Opt.* 33.22 (1994): 5101–5107.
18. J. Ruoff and M. Totzeck, Orientation Zernike polynomials: A useful way to describe the polarization effects of optical imaging systems, *J. Micro/Nanolithogr. MEMS MOEMS* 8.3 (2009): 031404.
19. J. Sasián, *Introduction to Aberrations in Optical Imaging Systems*, Cambridge University Press (2013).
20. J. Sasián, Polarization fields and wavefronts of two sheets for understanding polarization aberrations in optical imaging systems, *Opt. Eng.* 53.3 (2014): 035102.
21. J. W. Goodman, *Introduction to Fourier Optics*, Roberts and Company Publishers (2005).
22. J. P. McGuire and R. A. Chipman, Diffraction image formation in optical systems with polarization aberrations. I: Formulation and example, *JOSA A* 7.9 (1990): 1614–1626.
23. J. P. McGuire and R. A. Chipman, Diffraction image formation in optical systems with polarization aberrations. II: Amplitude response matrices for rotationally symmetric systems, *JOSA A* 8.6 (1991): 833–840.
24. T. Gehrels (ed.), *Planets, Stars and Nebulae: Studied with Photopolarimetry*, Vol. 23, University of Arizona Press (1974).
25. C. U. Keller, Instrumentation for astrophysical spectropolarimetry, *Astrophysical Spectropolarimetry*, Proceedings of the XII Canary Islands Winter School of Astrophysics, Puerto de la Cruz, Tenerife, Spain, November 13–24, 2000, eds. J. Trujillo-Bueno, F. Moreno-Insertis, and F. Sánchez, Cambridge, UK: Cambridge University Press (2002), pp. 30–354.
26. J. Tinbergen, *Astronomical Polarimetry*, Cambridge University Press (2005).
27. F. Snik and C. U. Keller, Astronomical polarimetry: Polarized views of stars and planets, *Planets, Stars and Stellar Systems*, Netherlands: Springer (2013), pp. 175–221.

Index

Page numbers followed by f and t indicate figures and tables, respectively.

Aberration, 17; *see also* Polarization aberrations
    amplitude, 545, 653
    defined, 543, 545, 653
    depolarization, 582–583
    Fresnel, 545
    paraxial, *see* Paraxial polarization aberrations
    polarization, 543, 653
    Seidel, 544, 550
    skew, *see* Skew aberration
    wavefront, 543, 572, 580, 653, 663
Aberration-free system
    ARM of, 602
    MPSM of, 604
Absentee layer, 488
Absent mode, 702
Absolute phase, 580
    about, 166
    change, 133
    of Jones matrix, 129
    of Jones vector, 77, 78
    of light, 33
Absorption, 365
    axis, 187
    factor, 148
Absorption coefficient, 365
Achromatic retarders, 715
Active materials, optically, 169
Addition form, of polarization ray tracing matrix, 344–347
Addition of circularly polarized beams (example), 50
Addition of Mueller matrices, 201–202
Addition of polarized beams; *see also* Interference of polarized light
    Gaussian wave packet example, 109–112, 112t
    of polarized light of two different frequencies, 103–105
    of polychromatic beams, 105–109
Adjoint of vector, 36
Aerosol, 260
    polarimeter, 8
Aerosol polarimeter sensor (APS) instrument, 261
Airborne multi-angle spectro polarimetric imager (AirMSPI), 258, 261f, 262f
Air passing, collimated plane wave, 724
Air to calcite, ray parameters refract and reflect, 757
Airy disk, 19, 543
    patterns, 773
Algorithm
    multilayer thin films, 487–488
    proper retardance, 643–545
    skew aberration, 655–657
Alignment layer, 843
Aluminum beam splitters, 485–486
Aluminum-coated three-fold mirror system, 647–649

Aluminum diattenuation *vs.* angle of incidence, 944
Aluminum oxide, 500
Aluminum retardance
    in radians *vs.* angle of incidence, 945
Aluminum wire grid polarizer, 822
Amplitude aberration, 382, 545, 653
Amplitude coefficients, 298, 299–300, 383–385, 544, 550; *see also* Fresnel equations
    coating and, 479
    multilayer coatings, 487, 490, 499–500
    thin films, 479, 490, 499–500
Amplitude depolarization, 200
Amplitude distribution, 92
Amplitude function, 598, 937
Amplitude grating, 812
    structure, 812
Amplitude response matrix (**ARM**), 593, 594, 623, 774, 937, 938, 954, 958
    of aberration-free system, 602–603
    absolute value of, 954
    amplitude of, 775
    of Cassegrain telescopes, 472, 473f, 602–603
    components, 903
    critical angle CCR and, 619–621
    encircled energy (EE), 775
    overview, 601
    parameters, imaging calculation, 955
    polarization structure of images and, 607
    polarized pupil with unpolarized object, 610–611
    scale of, 605–607
    skew aberration and, 664
    solid CCR and, 615, 617
    of uncoated single-element lens, 452–453, 453f
Amplitude transmission, 382
Amplitude transmittances, 141
Analytical manipulation, 494
Analyzer vector
    about, 178
    calculation, 251
Angle dependence; *see also* Polarization problems in optical systems
    of polarizers, 12
    of retarders, 13
Angle of incidence (AOI), 614
Angle of incidence, paraxial, 544, 550–553, 562, 572, 586
    variation of, 551
Angle of incidence maps, 562–563
    cell phone lens, 362–363, 362f
Angle of linear polarization (AoLP), 262
Angle of major axis, 68
Angle of polarization (AoP), 68
Angles of incidence, 940
Angular dispersion, 815

Angular momentum, 137
Anisotropic devices, 718
Anisotropic interfaces, 480, 725
Anisotropic material, 11, 36, 297
Anisotropic media; see also Birefringent materials
  classes of, 670f
  electromagnetic waves in, 672–673
  optical properties of, 669
Anisotropic plate assembly, 726
Annealing, 15
Anti-commutative, Pauli matrices as, 510
Anti-pole, 430
Antireflection-coated interfaces, 21
Antireflection coatings, 8, 20, 23, 482–485, 546, 931
  single-layer, 485
Antireflection grating's groove orientation, 826
Antireflection subwavelength antireflection coatings (SWG), 825–827
  intensity transmission of, 827
AOI, see Angle of incidence
AOI map, 860
Aperture function, 598–599, 937
Apertures, 366–367, 366f
Aperture stop, 17, 369, 369f
Apodization, 382, 545, 599, 653, 786, 937
Approximately linear and quadratic function, 449
Arbitrary shape gratings, 813
Arctan function, 68, 499
ARM, see Amplitude response matrix
ARMOS model, 733
Artifacts in polarimetric images; see also Polarimetry
  about, 280–281
  pixel misalignment, 281
Associative rule for matrix multiplication, 165, 167
Astigmatism, 362, 558, 572, 583, 786, 950
  into on-axis beam, 25
Atmospheric polarization
  about, 259
  images (example), 259–262
Average degree of polarization, 195
Average transmission, 179
Axometrics AxoStep Muller matrix imaging polarimeter, 580

Baker–Hausdorff–Campbell formula, 531–532, 533
Bandpass filters, 491
Bare aluminum mirror
  diattenuation, 944
  reflection coefficients, 943
Basis functions, 567
Basis polarization states, 38, 51t, 65t, 75t
Basis vectors, 656
Beam combination
  illumination pattern, 718
  with polarization ray tracing matrices, 715–736
  principles, 716
  wavefronts, pupil of imaging systems, 718
Beam overlapping, 725
Beam splitters, 93, 715
  coatings, 479
  metal, 485–486
  polarizing, 492–496

Berry phase, 635, 636, 654
Biaxial crystal, 271, 272f
  blocks, ray splitting, 706–707
Biaxial cube, reflection inside, 707–710
Biaxial material, 673, 674
Biaxial multilayer films, 17, 843
Bidirectional reflectance distribution function, 935
Binodal astigmatism, 544, 558
Binodal polarization, 558
Birefringence, 13, 118, 122, 675–676
  of a ray, 745
Birefringent components, 716
Birefringent filters, 345
Birefringent interface
  configurations of, 682–683
  examples, 706–710
  mode labeling for rays, 671t
  parameters, 671t
  polarization ray tracing matrices for, 695–705
  reflections and refractions at, 681–693
Birefringent materials
  defining, 673–679
  eigenmodes of, 679–681
  interface, see Birefringent interface
  KTP crystal, ray tracing, 670, 672, 686–693
  optical path length, 684–685
  optical properties of, 669
  optic axis of, 744
  ray doubling, data structure for, 694–695
  ray tracing in, 669–672
  refractive index, 675, 683
Birefringent plate
  collimated wavefront passes, 717
Birefringent ray tracing complications, 934
  anisotropic materials, 934
  birefringent interface, 934
  ray doubling, 934
  spatially varying birefringence medium, 934
  wavefronts, 934
Birefringent retarders, 168
Birefringent-to-birefringent interface, 703
Birefringent-to-isotropic interface, 701–703
Birefringent waveplates, 13
Blaze angle, 819
Blazed grating, 813
Blaze wavelength, 818
Blazing, 818
Blue phase (BP) mode, 855
  double twist cylinder, 855
  double twist cylinders stacking, 856
Brewster's angle, 68f, 305–306, 636n, 787, 935; see also Fresnel equations
Brightness
  response time of liquid crystal, 865

CAD file, 891
Calcite, 11
  A-plate, 271
Calcite $CaCO_3$ crystal structure
  with vertical optic axis, 743
Calcite quarter waveplate
  polarization aberrations of, 726

Calibration, 279
Canonical basis, for polarization states, 336
Canonical forms, 509
    Jones matrix, 546
Canonical local coordinates, 640–642
Canonical summation form, Pauli matrices, 514–515
Cartesian form, 500
Cartesian rotation matrix, 42, 137, 511
Cassegrain telescopes, 24, 25, 469–473, 470f, 598, 942, 943
    angle of incidence, 470, 471f
    ARM, 472, 473f, 602–603
    astigmatism, 470, 471–472
    diattenuation pupil maps, 470, 471f
    MPSM, 604–605
    OTM, 608–609
    PSFs, 472, 473f
    specifications, 470t
Cavity Q-factor, 111
CCD detector, 715
c-Coefficients, Pauli, 512, 512f
CCRs, *see* Corner cube retroreflectors
Cell phone lens, 17, 361–364
    aberrations, 439–440
    angle of incidence maps, 362–363, 362f
    angle of incidence (*AoI*) variation, 362
    diattenuation, 363–364
    Jones pupil, 364, 364f
    optical path length (*OPL*) variation, 361–362
    prescription, 416–417, 417t–418t
    wavefront aberration of, 404–405
Characterized matrix, 487
Chief ray angle, 660
    of incidence, 552
Chief rays, 372, 372f–373f, 585, 658
Chromatic aberration, 394, 499
    defocus, 497–499
Circular diattenuation, 942
Circular polariscope, 266–267; *see also* Polariscopes
Circular polarized light, 636
Circular retardance, 265, 629, 636, 663, 666
    tilt, 659
Circular retarder, 122, 137, 169, 184, 632
    Jones matrix, 619
Circularly polarized basis, 49, 50
Circularly polarized light, 22, 35, 43–45, 71f, 104
Coating polarization, 940
Coatings, 667
Co-cross-polarized systems
    extinction ratio map, 804
CODE V's U.S. patent library, 658, 666–667; *see also* Skew aberration
Coherence simulation, 935
    partially coherent/incoherent light, 935
Coherent combination, of polarization ray tracing matrices, 344–347
Coherent monochromatic light, 935
Coherent ray tracing, 407–409, 936
Color displays, 843
Coma, 572, 583
Combining **P** matrices for interferometers, 346–347
Combining wavefronts process, 716

Commutator, 149
Commuting matrix, 123
Complete analyzer, 254
Complete polarimeters, 222
Complex Pauli coefficients, 508, 510–511, 548–550
Complex pulses, 109
Complex vector, 34
Complex wavefront aberrations, 786
Compound retarder, 185, 186f, 267n
    Jones matrix decomposition, 917–919
    principal retardance trajectory, 924
Condition number, 282, 284
Conical refraction, 692–693
Conoscope, 13, 270–271, 693, 694; *see also* Polariscopes
Conservation of momentum, 297
Constant degree of circular polarization, 83
Constitutive relation, 676
Contrast ratio, 845
Conventional optical design, 17, 20
Conversion into Jones vectors (example), 52
Convolution, **MPSM**, 604
Coordinate vectors, 640
Co-propagating overlapped wavefronts, 722
Co-propagating wavefront combination, 718
    beam combining configuration, 718
    incident beam yields, 718
Corner cube retroreflectors (CCRs), 275, 614–618
    critical angle, 618–622
Corner cubes, 347–351, 667
    system, 653
C-plate, 14
    crystal waveplates, aberrations of, 769–771
    twisted nematic (TN cell), 848
Critical angle, 306–307; *see also* Fresnel equations
    corner cube retroreflector, 618–622, 655
Crossed linear polariscope, 264, 265
Crossed mirror configuration, 464–469
Crossed polarizers, 14, 15f, 22
Crystal, anisotropic, 11
Crystal axes, 673
Crystalline quartz, 677
Crystalline solid structure, 838
Crystal polarizers, 715, 785, 786
    field of view (FOV), 785
    Glan–Taylor polarizer, *see* Glan–Taylor polarizer
    materials for, 786
        anisotropic, 785
Crystal prisms, 778
    uniaxial crystal blocks, 778
Crystal's index ellipsoid, 751
Crystal's principal indices, 754
Crystal waveplates, aberrations of, 765–771
    birefringent material, 765
    C-plate, 769–771
    field of view (FOV)
    refracts into uniaxial plane parallel plate, 766
    A-plate, 767–769
        converging beam, 768
        negative uniaxial, 767
        y-oriented calcite, 767
    principal section with incident angles, 768
Cumulative mode label, 939

Cumulative **P** matrix, for reference and test paths, 342–344
Curvature, 585
Cutoff frequency, **OTM**, 609–610
Czerny–Turner monochromator, 818, 819

Dark fringes, 271
Dark state, 846
Data reduction, *see* Mueller matrices
Data structure for ray doubling, 694–695
Daughter rays, 695
Decreasing phase convention, 379
Defocus, chromatic aberration, 497–499
Defocus (quadratic), polarization aberrations, 543, 544f, 557
    diattenuation, 553–556, 557, 564–566
    one wave of, with a circular aperture, 599
    retardance, 557, 563–564
Degree of circular polarization (*DoCP*), 68
Degree of linear polarization (*DoLP*), 67, 244, 259, 262f, 280
Degree of polarization (*DoP*), 66–70, 80, 105, 193, 493, 871, 936
    maps, 196–198
    surfaces, 196–198
Degrees of freedom (DoF)
    about, 200
    in Jones matrix, 507, 508t
    Jones vector, 38
    in polarization matrices, 194
Density matrices, 109
Depolarization, 871, 936; *see also* Mueller matrix
    aberration, 582–583
    about, 6, 20, 117, 193–195
    addition of Mueller matrices, 201–202
    defined, 193
    degree of polarization surfaces/degree of polarization maps, 196–198
    depolarization index/average degree of polarization, 195
    non-depolarizing Mueller matrices, 192
    physically realizable Mueller matrices, testing for, 198–200
    weak depolarizing elements, 200–201
Depolarization index (DI), 195
Depolarized beam, 193
Depolarizers, 6
DFT, *see* Discrete Fourier transform
Diagonal depolarization, 200
Diagonal depolarizer Mueller matrix, 194
Diagonal elements, of **ARM**, 602
Diagonal Jones matrix, 143
Diagonal matrix, 335
Diattenuation, 937
    aberration, 653
    about, 6
    components, 140
    defined, 119
    higher-order polarization aberrations, 578–580
    linear, 545, 572, 573
    matrix properties of Jones matrices, 144–145
    of metal at non-normal incidence, 315–316
    microscope, 581
    for multi-element lens, 23
    parameter, defined, 178
    paraxial, 553

    polarization-dependent transmittance, 130
    as polarization properties, 117, 119–120, 630, 649
    in polarization ray tracing, 20, 21
    seven-element lens system, 561–562, 564–567
    transmittance, 177–180
    vector, 178
    weakly polarizing ray intercepts, 549
Diattenuation calculation of **P** matrix, 334–337
    singular value decomposition (SVD), 335–337
Diattenuation defocus, 949
    paraxial, 553–556, 557, 562, 564–566; *see also* Defocus (quadratic), polarization aberrations
Diattenuation maps
    cell phone lens, 363, 363f
    crossed mirror configurations, 465–469, 466f, 467t, 469f
    singlet lens, 449–451, 450f
Diattenuation piston, 557, 558f, 562, 564–566, 949; *see also* Piston (constant), polarization aberrations
Diattenuation space
    inhomogeneous Jones matrices, 533–534, 534f–535f
Diattenuation tilt, 557, 562, 949; *see also* Tilt (linear), polarization aberrations
Diattenuators
    Jones matrix, 139
    matrices, 521–523, 528–529
    Mueller matrix, 182; *see also* Polarizer/diattenuator Mueller matrices
    operation of, 187
    polarization element/polarization properties, 5–6
    optical element, 120
    polarizer, 180–182; *see also* Poincaré sphere operations
Dichroic and birefringent materials, 118; *see also* Jones matrices
Dichroic dyes, 243
Dichroic polarizer, 12, 118, 179
Dichroism, 118, 118f, 120
Dielectric anisotropy, 840
Dielectric refraction, 308–309; *see also* Fresnel refraction/reflection
Dielectric tensors, 669, 673–674
Diffracted wavevectors, 816
Diffraction gratings, 252, 935, 936
Diffraction-limited **MPSM**, 654
Diffraction limited wavefront, 405
Diffractive optical elements (DOEs), 93, 811
    application of, 811
Dipole coordinates, 424, 426–430
    $\hat{a}_{LOC}$, 429
    defined, 426
    local to global polarization state, 427–428
    singularities, 428–429
    wavefront aberration function, 440–411
Dipole electromagnetic wave, 56
Dipole model, 56
Dipole moment, 16, 840
Dipole oscillators, 306
Dipole radiation, 56
Dipole radiator model, 424
Dipoles, 423
Direction cosines, 32
Direction of energy propagation, 752
Director distributions, with splay, 844

Index

Disclinations, 862
Discotic LC molecules, 868
Discotic molecules
    compensating film layer, 870
    schematics for, 840
    uniform film of, 870
Discontinuities, see Retardance discontinuities
Discrete Fourier transform (DFT), 238, 594–597
    1D DFT, 595–596
    2D DFT, 596
    MPSM of patent lens system, 666
Discrete ray data yielding, 731
Disk-shaped molecules, 17
Dispersion, 13
Division of amplitude, 93
Division-of-amplitude polarimetry (DOAP), 241, 246–248; see also Simultaneous polarimetric measurement
Division-of-aperture polarimeter, 241
Division-of-aperture polarimetry, 242; see also Simultaneous polarimetric measurement
Division-of-focal-plane polarimetry, 242–245; see also Simultaneous polarimetric measurement
Division of wavefront, 92
Double-crossed mirror system, 464–469
Double image, 664
Double pole coordinates, 424, 430–436, 937
    $â_{LOC}$, 436
    defined, 430
    parabolic reflector and, 431, 434
    wavefront aberration function, 440–411
Double pole local coordinates, 598
Double pole model, 56, 424
Double pole polarization, 656
Double pole spherical wave, 57
Double refraction, 670, 679, 741
    through a calcite rhomb, 680f
    through a ulexite biaxial block, 680f
    through three biaxial crystal blocks, 706–707
Double singularity, 431–432
Double twist cylinder, 855
Dove prism
    parallel transport matrix for, 639–640
Dual rotating retarder Mueller matrix polarimeter, 274 275; see also Mueller polarimetry configurations
Dual tetrahedron Mueller polarimeter (example), 284
Dummy surfaces, optical systems, 367–368

Earth's energy balance, 260
Earth Venture Instrument program, 258
Eccentricity, 74
Effective depth, multilayer film, 491, 492f
$eiei$-mode, 695
eie-mode, 695
Eigenmodes
    birefringent materials, 669, 679–681
    subaperture, 616f
    unitary matrix and, 630
Eigenpolarizations, 480, 841; see also Eigenmodes
    about, 124–126, 168, 185
    defined, 120
Eigenvalues
    and eigenvectors, 124, 125
    for Pauli matrix sum, 513

Eigenvectors, 630, 932
    for Pauli matrix sum, 513
eio-mode, 695
Elastic constant, 843
Electric field aberrations, 568–572
Electric field amplitude, 727, 728, 728
Electric field intensity, 802
Electric field of unpolarized light, 65
Electric field vectors, 33f
Electromagnetic waves in anisotropic media, 672–673
Ellipsoid of wave normal, 750
Ellipsometers, 930
    in microlithography industry, 12
    sample-measuring polarimeter, 8
Ellipsometry, 8–10
Elliptical diattenuators, 931
    about, 140–141
    example, 141, 142
Elliptically polarized light, 35, 45–48
Elliptical polarization parameters, 73–74
Elliptical polarizer, 130
    Mueller matrix, 176, 177
Elliptical retarders, 122, 141–143, 142, 916, 931; see also Jones matrices
    Jones matrices, 529–531
    Mueller matrix, 171
Ellipticity, 4, 45, 74
Energy density, 749
Entrance pupil, 370, 370f, 654
    local coordinates, 423–424
Entrance Pupil Diameter (EPD), 370
EPD, see Entrance Pupil Diameter
Equi-rectangular projection, 83
Etendué, 368, 374–375
Even permutation, Pauli matrices, 510
Exit pupil, 370–372, 371f, 500, 550, 551, 564, 580, 581, 582, 594, 598–601, 654, 663
    local coordinates, 423–424
Exponential form of polarization components, 525
Exponential Pauli coefficients, 509
Exponentiation of Jones matrices, 518–519
Extended Jones matrix method, 856–857
    birefringent plates with twisting and tilting directors, 857
External reflection, 306, 309–310; see also Fresnel refraction/reflection
Extinction ratio, 120
    distribution, 802
Extraordinary modes, 670, 680
Extraordinary principal axis, 743
Extraordinary ray, 742

Fabry–Perot interferometers, 717
Faraday effect, 677
Faraday rotation, 169
Fast axis, 122, 168
    unchanged convention, 134, 135t, 136
Fast-mode, birefringent interfaces, 681–682
Fast single-element lens, see Uncoated single-element lens
Fast–slow–fast-mode, ray trace parameter, 707t
Fiber optic pulse measurement, 718
Field curvature, 583
Field distribution, 717

Polarized Light and Optical Systems

Field Guide to Geometrical Optics (Greivenkamp), 583
Field of view (FOV)
    diattenuation aberrations across, 556–557
    retardance aberrations across, 556–557
    skew aberration and, 654, 658, 667
Finite differences, 498
Finite difference time domain algorithms (FDTD), 940
Finite element modeling (FEM), 889
Fixed analyzer polarimeter, 251–253
Fixed polarizer polarimeter, 254–255
Flange, 892
Floquet's condition, 830
Flowcharts, coherent/incoherent imaging, 594
Flux, 37
Flux components, finding (example), 77
Flux vector, 224
Focusing grating, 813
Fold mirrors, 455–460, 567, 941
    combinations of, 461–469
    parallel transport matrix for, 639
Forward problem, 117
Foucault tests, 929
Four-dimensional phase space, 374
Fourier coefficients, 252
    function, 830
Fourier optics, 18
Fourier transform, 18, 71, 664
Fourier transform algorithms, 719
Fourier transformation operation, 500
Fourier transform spectrometers, 718
Fourth-order wavefront aberrations, 583
FOV, *see* Field of view
Fréedericksz cell, 841, 846, 858
    degree of polarization map, 871
    director orientations, 847
    dispersion measurement, 867
    retardance modulator for adjustable linear retardance, 847
    retardance of, 868
    single-pixel, 864
    variation of director orientation, 841
Frequency domain analysis, 238
Fresnel aberrations, 447–475, 545, 940
    Cassegrain telescopes, 469–473
    fold mirrors, 455–460
        combinations of, 461–469
    overview, 447–448
    uncoated single-element lens, 448–455
Fresnel amplitude coefficients, 496
Fresnel coefficients, 811
    about, 300–301
    for aluminum (example), 313
    approximate representations of
        about, 316–317
        Taylor series for Fresnel coefficients, 317–318
Fresnel equations, 21, 21f, 22, 447, 448, 544, 550
    amplitude coefficients, 299–300
    Brewster's angle, 305–306
    critical angle, 306–307
    Fresnel coefficients, 300–301
    Fresnel coefficients, approximate representations of
        about, 316–317
        Taylor series for Fresnel coefficients, 317–318

Fresnel refraction/reflection
    dielectric refraction, 308–309
    external reflection, 309–310
    internal reflection, 311–312
    metal reflection, 313–316
    intensity coefficients, 301–303
    intensity/phase change with incident angle, 307–308
    Jones matrices with Fresnel coefficients, 308
    normal incidence, 304–305
    overview, 295–296
    propagation of light
        homogeneous/isotropic interfaces, 297
        light propagation in media, 297–298
        plane of incidence, 296
        plane waves/rays, 296
    s- and p-polarization components, 298–299
Fresnel refraction/reflection
    dielectric refraction, 308–309
    external reflection, 309–310
    internal reflection, 311–312
    metal reflection, 313–316
Fresnel rhomb, 474–475, 474f
Fresnel transmission, 727
    coefficients, 727
Fresnel zone patterns, 813
Fringes, 935
Fringe visibility, 91, 97, 935
Full sky polarimeter, 259

Gaussian function, 594
Gaussian wave packet example, 109–112, 112t
Geometrical ray tracing, 359; *see also* Polarization ray tracing (PRT)
    cell phone lens, 361–364
Geometrical transformation, 618, 629, 649
    non-polarizing optical systems, 633–634
    parallel transport matrix, 636–640
    parallel transport of vectors, 634–636
        with reflection, 636, 637f
    polarization ray tracing matrix, separating from, 642–645
    rotation of local coordinates, 631–633
    skew aberration and, 654, 665
Glan–Foucault prisms, 778
Glan–Taylor polarizer, 669, 742, 785, 786, 787
    aberrations of, 797
    configurations for calculating extinction, 800
    co-polarized system to cross-polarized system, 799
    FOV of, 800
    light incident on hypotenuse calcite/air interface, 787
    limited FOV, 787–789
    multiple polarized wavefronts, 793–795
    multiple potential ray paths, 789–793
    pairs of, 799
        extinction ratio, 799
    performance of, 800
    polarized wavefronts exiting, 796–797
    Zernike polynomial fitting, 797
Glan–Taylor prisms, 778, 801
Glan–Thompson prism polarizers, 179, 280, 742, 786
Glan–Thompson prisms, 778
Glan-type polarizers, 694, 778
Global coordinates, 325, 631, 674

Index

Gold
    phase discontinuities in, 500, 501f
Grating
    equation, 814
        in-plane, 814
        out-of-plane diffraction, 814
    free spectral range, 818
    parameters and notation definitions, 814
    resolving power, 818
    vector, 816
Greivenkamp, John, 583
Grid tracing, of rays emerging, 716
Gyrotropic tensors, 669, 673, 676

Hafnium oxide (HfO_2), 489–490
Half linear wave retarders, 134
Half wave circular retarder, 184
Half wave elliptical retarders (HWR), 172, 173
Half wave film, 488–489
Half wave linear retarder (HWLR), 6, 121, 136, 169, 170t, 333–334; *see also* Linear retarders
    Jones matrix for, 337
    P matrix for, 333–334, 338
Half wave retarders, 169
Helicity of light, 4, 35
Hermitian adjoint, 335
    for Pauli matrix sum, 513
Hermitian matrices, 144–145, 148, 150, 182
Hermitian matrix, 523–525, 643
Hexabenzocoronene
    discotic molecules, 869
Higher-order aberrations, 498
Higher-order basis functions, 567
Higher-order polarization aberrations
    diattenuation, 578–580
    electric field aberrations, 568–572
    Fourier transform and, 594
    orientors, 567, 572–578
    retardance, 578–580
    vector Zernike polynomials, 568–572, 569t–570t, 571f
High numerical aperture wavefronts, 436–437
Hollow aluminum corner cube, 347–351
Hologram, 102
Holography, defined, 102
Homogeneous diattenuator, 130
Homogeneous interfaces, 297, 480
Homogeneous Jones matrices, 523–527, 632
    components of, 526–527
    examples, 532
Homogeneous matrices, 631
Homogeneous retarder system, retardance of, 912–914
Horizontal fast axis linear retarder (HLR), 911
    trajectories of, 912
    trajectory of, 911
Horizontal linear polarizer, 128
Hyperspectral imaging, 260

Ideal depolarizer, 195
    Mueller matrix, 194
Ideal diattenuators, 181
Ideal horizontal linear polarizer, 182
Ideal polarization element matrices, notation, 127t
Ideal polarizer, 5, 119, 174

Ideal reflection at normal incident, 631
Ideal retarder, 120
Identities, Pauli matrices, 509–510, 510t
Identity matrix, 165, 167, 201
Illumination
    extinction ratio map, 803
Illumination pattern, 718
Illumination systems, 17, 717
Image formation, through A-plate, 771–777
Image formation, with polarization aberrations
    amplitude response matrix, 593, 594, 601–603, 605–607, 610–611, 615, 617, 619–621
    critical angle corner cube retroreflector, 618–622
    discrete fourier transformation, 594–597
    Jones exit pupil, 598–601
    Jones pupil function, 598–601
    Mueller Point Spread Matrix, 582, 593, 594, 602, 603–607, 611–614, 620–622
    optical transfer matrix, 593, 608–610, 621–622
    overview, 593–594
    polarization structure of images, 607
    polarized pupil with unpolarized object, 610–614
    skew aberration and, 663
    solid CCRs, 614–618
Image quality of lens, 17
Imaging polarimeters, 241
Imaging systems, 17
Impermeability tensor, 750
Incident, 716
Incident angle
    radian and diattenuation, 728
Incident beam
    diffracts into multiple orders, 816
Incident light descriptions, 831
Incident plane wave, 299
Incoherent imaging systems, 608
Incoherent light, 91
Incoherent ray tracing, 407–409, 936
Incomplete polarimeters, 222
Index ellipse, 751
Index ellipsoid, 749, 750
Index indicatrix, 750
Indium tin oxide (ITO), 843
    glass substrate, 861
Inhibited reflection, 692
Inhomogeneous interface, 297, 480
Inhomogeneous Jones matrices, 531–532
    diattenuation space, 533–534, 534f–535f
    example, 534
    retardance spaces, 534, 534f–535f
Inhomogeneous matrices, 126, 149, 631
Inhomogeneous polarizer, 188
Injection-molded lens, 890
    stress tensor coefficient, 892
In-plane switching (IPS), 846
    LC cell, 852
        advanced S-IPS (AS-IPS), 853
        electrodes, 853
        Mueller matrix spectropolarimetric test, 874
        transmittance through, 853
        well-aligned, retardance eigenstates of, 874
Instrumental polarization, *see* Polarization aberrations
Integrating spheres, 6, 193

Intensity, in radiometry, 37
Intensity distribution, 96
Intensity gradients, 280
Intensity interference pattern, 102
Intensity modulation, 843
Intensity modulator, 16, 842
Intensity/phase change with incident angle, 307–308
Intensity reflection coefficients, 311
Intensity transmission coefficients, 21, 311
Intensity transmittances, 130
Interference, defined, 91, 92
Interference colors, 267–269; *see also* Polariscopes
Interference fringes, 91
Interference of horizontally/vertically polarized light (example), 98–99
Interference of polarized light; *see also* Polarized light
    addition of polarized beams
        addition of polarized light of two different frequencies, 103–105
        addition of polychromatic beams, 105–109
        Gaussian wave packet example, 109–112, 112t
    holography, polarization in, 102
    interferometers, 93–95
    light waves, combining, 92–93
    of nearly parallel monochromatic plane waves, 95–100
    overview, 91–92
    plane waves at large angles, interference of, 100–101
Interference of right/left circularly polarized light (example), 99–100
Interferograms, 24, 91, 929
Interferometer(s), 93–95, 630, 655, 715, 880; *see also* Interference of polarized light
    beam splitter coatings in, 479
    combining P matrices for, 346–347
Interferometer with PBS, 337–344
    analyzer, 341
    reference path, 338–340
        cumulative **P** matrix for, 342
    test path, 338, 340–341
        cumulative **P** matrix for, 342–344
Internal reflection, 306, 311–312; *see also* Fresnel refraction/reflection
Interpolation method, 716, 732
Interrupted sinusoidal projection, 83
Inverse-distance weighted interpolation algorithm, 733
Inverse problem, 117
IPA technology, 853
Irradiance, 37, 164
Irregular pupil shapes, 594
Isoplanatic patch, 594
Isotropic interfaces, 297, 480
Isotropic material, 297
Isotropic media, 669, 674t; *see also* Birefringent materials
    classes of, 670f
    propagation in, 36; *see also* Polarized light
Isotropic modes, 670
Isotropic-to-biaxial interface, 705
Isotropic-to-birefringent interface, 700–701
Isotropic-to-isotropic intercept, 698–700

Jones calculus, 425
Jones exit pupil, 550, 551, 564, 580, 581, 582, 594, 598–601
    critical angle CCR, 618–619

Jones matrices, 323, 545, 630, 832, 911, 918, 924, 935, 936, 941; *see also* Jones vectors
    about, 21
    canonical form, 546
    circular retarder, 619
    coherent ray tracing, 407–409
    component calculation, 207
    decomposition
        compound retarder, 917–919
    in decreasing/increasing phase conventions, 154t–155t
    defined, 123, 857
    degrees of freedom, 507, 508t
    diattenuation, 119–120
    diattenuation space, 533–535
    diattenuator, 521–523, 528–529
    dichroic and birefringent materials, 118
    eigenpolarizations, 124–126
    elliptical diattenuator, 140–141
    elliptical retarder, 141–143
    elliptical retarders and retarder space, 529–531
    exponentiation of, 518–519
    with Fresnel coefficients, 308
    for HLR, 921
    homogeneous, 523–527, 632
    increasing phase sign convention, 153
    inhomogeneous, 531–532
    interface, 383–385
    Jones matrix notation, 127, 127t
    for linear diattenuator (example), 131
    for linear diattenuator, derivation (example), 131
    linear diattenuator Jones matrices, 130–132
    linear diattenuators, 139–140
    logarithms of, 519–520
    matrix properties of
        Hermitian matrices, diattenuation, 144–145
        polar decomposition, separating retardance from diattenuation, 149–152
        unitary matrices/unitary transformations, retarder, 145–149
    for monochromatic beams, 912
    to Mueller matrices (conversion) using Pauli matrices, 207
    into Mueller matrices (transformation) using tensor product, 202–206
    to Mueller matrix, converting (numerical example), 205–206
    non-polarizing Jones matrices for amplitude/phase change, 143–144
    notation, 127, 127t
    objectives, 123
    overview, 117–118, 507–509
    Pauli matrices and, *see* Pauli matrices
    polarizer Jones matrices, 128–130, 129t
    for ray intercept, 546
    for reflection and refraction, 329
    retardance, 120–123, 121t
    retarder, 520–521, 527
    retarder Jones matrices
        circular retarder Jones matrices, 137
        linear retarder Jones matrices, 133–137, 135t
        vortex retarders, 137–139
    rotation of Jones matrices, 127
    sequences of polarization elements, 515–518

3D generalization of, *see* Polarization ray tracing matrix (**P**)
   weak polarization elements, 508, 535
   weak polarization interactions, 546–548
Jones pupils, 21, 404, 424, 545, 550, 567, 580, 594, 598–599, 937, 940, 948; *see also* Local coordinates
   artifacts, 438
   converting **P** pupils to, 437–439
   elements, value of, 946
   function, 593, 598–601
   padding, 604–605
   polarization ray tracing, 949
   polarized pupil with unpolarized object, 610–614
   skew aberration and, 664, 665
Jones rotation matrix, 127, 146
Jones vectors, 323–324, 568, 631, 921; *see also* Polarized light
   about, 37–41, 39t
   addition of, 49–50
   elements, 124
   elliptically polarized (example), 47
   with local coordinates, *see* Local coordinates
   method for polarized light calculations, 32t
   for normalized linearly polarized light, 43
   orthogonal, 48–49
   for polarization states, 5
   rotation of, 42
   rotation operation, 43
   sign conventions, 75
   and Stokes parameters, conversions between, 75–78
   into Stokes parameters, converting (example), 78

Kriging interpolation, 730, 734
K-sphere, 634, 636
KTP (potassium titanyl phosphate) crystal, ray tracing, 670, 672, 686–693

Lagrange invariant (*H*), 373–374, 661
Laser beams, 104
Laser light, 70
Lasers, high-power, 257
Laser systems
   high-power, 786
Latitude, 423, 424; *see also* Local coordinates; Longitude
Least square fit, 495
Left circularly polarized beam, 44
Left-handed coordinate system, 641
Left hand rule, 43
Lens design, 17
Light
   defined, 2
   monochromatic, 2
Light beams, 368
Light distribution, 728
Light frequency
   in radians per second, 751
Light intensity, 686
Light-measuring polarimeter, 8, 221, 222
Light polarization, 745
Light propagation, 787; *see also* Fresnel equations
   homogeneous/isotropic interfaces, 297
   in media, 297–298
   plane of incidence, 296
   plane waves/rays, 296

Light rays, 368, 368f
   four-dimensional phase space of, 374
Light reflection
   diattenuation of, 823
Light–surface interaction, 719
Light transmission
   diattenuation of, 823
Light waves, 92–93; *see also* Interference of polarized light
$LiNbO_3$ modulators
   in fiber optics, 865
Linear diattenuation (LD), 139–140, 179, 180, 545, 572, 573; *see also* Jones matrices
   followed by retarder (example), 189
   Jones matrices, 130–132
Linear diattenuators, 546
   followed by retarder (example), 189
Linearly polarized light, 34, 42–43, 67
Linearly polarized Stokes parameters, 73
Linearly polarized wavefront, 423
Linear polariscope, 263–266; *see also* Polariscopes
Linear polarization sensitivity, 179
Linear polarization state, 122, 168
Linear polarizer
   defined, 120
   on eigenpolarization states (example), 129
   on eigenpolarization states, operation of (example), 129
   polarization elements, 5
   uncoated lens surface, 22
Linear retardance, 545, 572, 573
Linear retarders, 546
   defined, 122
   with horizontally polarized fast axis, 134, 169
   and linear polarized components, 6
Linear retarders, **P** matrix for, 331–332, 332t
   HWLR, 333–334, 338
   QWLR, 330–333, 339–340
Linear skew aberration, 654, 655, 659, 660
Line spread function, 818
Liquid crystal display (LCD), 12, 16–17, 843; *see also* Polarization problems in optical systems
   configuration, 843
   device, 717
   image quality, 837
   monitors, 847
   TVs, 873
Liquid crystal on silicon (LCoS), 275
Liquid crystal on silicon (LCoS) cells, 854
   schematic of, 854
Liquid crystals (LCs), 273
   polymers, 940
Liquid crystals (LCs) cell, 837, 838, 839, 841, 845
   cell configuration, 843
   cell diattenuation, 860, 861
   cell performance, limitations, 864
      cell speed, 865–867
      depolarization, 871–872
      exiting polarization state, spectral variation of, 867
      polarization aberrations with biaxial films, 868–870
      polarizer leakage, 870–871
      retardance with angle of incidence, 867–868
   cell technologies, 846
   compensating film, 876

configurations of
 blue phase LC cells, 855–856
 Fréedericksz cell, 846–847
 in-plane switching (IPS), 852–854
 liquid crystal on silicon (LCoS) cells, 854–855
 super twisted nematic cell (STN), 849–850
 90° twisted nematic cell, 847–849
 vertically aligned nematic (VAN), 850–852
construction of, 842–843, 861
 disclinations, 861–862
 oscillating square wave voltage, 863
 pretilt, 862–863
 spacers, 861
disclinations, 872
display, high contrast ratio intensity modulation, 845
distributions of directors, 841, 842
driving voltages, 863
limitations, 864
linear retardance magnitude plot, 869
mean linear retardance magnitude and peak-to-peak values, 869
molecules, temperature influences, 839
Mueller matrix spectropolarimeter, 872
projector, 845
restoring forces, 843–845
retardance and orientation, 872
retardance magnitude, 864
TV sets, 845
types of, 838
Liquid crystal (LC) variable linear retarders, 254, 254f
Liquid crystal (LC) variable retarders, 259
Lissajous figure, 103
Littrow monochromator, 819, 821
Local basis vectors, 636
Local coordinates, 323–324, 325, 629, 649
 converting **P** pupils to Jones pupils, 437–439
 dipole coordinates, *see* Dipole coordinates
 double pole coordinates, *see* Double pole coordinates
 for entrance and exit pupils, 423–424
 high numerical aperture wavefronts, 436–437
 Jones matrices, 665
 objective of, 424
 orthogonal transformations between, 325–331
 right-handed, 632
 rotation of, 631–633
 singularities, 323–324
Logarithms of Jones matrices, 519–520
Longitude, 423, 424; *see also* Latitude; Local coordinates
Lu–Chipman decomposition, 630
Luneburg lens, 17n
Lyot filters, 715, 717, 742

Mach–Zehnder interferometer, 93, 94, 94f, 95f
MacNeille, S. M., 492
MacNeille beam splitter, 492–493, 493f–494f
MacNeille-type cube beam splitter, 855
Magnesium fluoride (MgfO$_2$), 483–484, 485, 489–490, 493, 500
Magnetic field, 4, 36–37
 vectors, 33f
Magnetic induction, 676
Magnetic permeability tensor, 676
Major axis, orientation of, 45, 47, 83

Maltese cross, 22
 Cassegrain telescopes, 472
 pattern, 803, 899, 900
 uncoated single-element lens, 451–452, 451f
Marginal rays, 372, 372f–373f, 585, 658
 angle, 660
Mathematica, 734
Mathematics of polarimetric measurement/data reduction
 Mueller data reduction matrix, 226–230
 Mueller matrix elements, measuring, 225–226
 null space/pseudoinverse, 230–240
 Stokes polarimetry, 222–225
MATLAB, 734
Matrix cube, for Pauli matrix sum, 514
Matrix exponential, 518; *see also* Exponentiation of Jones matrices
Matrix inverse, for Pauli matrix sum, 513
Matrix logarithm, 518; *see also* Logarithms of Jones matrices
Matrix multiplication, 123
Matrix product of Mueller matrices, 201
Matrix properties of Jones matrices
 Hermitian matrices, diattenuation, 144–145
 polar decomposition, separating retardance from diattenuation, 149–152
 unitary matrices/unitary transformations, retarder, 145–149
Matrix square roots, 151
 for Pauli matrix sum, 514
Matrix transpose, for Pauli matrix sum, 513
Matrix vector multiplication, 41
Maxwell's equations, 296, 672, 830, 940
 for periodic boundary conditions, 818
Maxwell's fisheye lens, 17n
Measurements, polarization aberrations, 580–583
Mechanical quality factors, 255
Media, light propagation in, 297–298
Media refractive index, 815
Meridional plane, 667
Meridional ray, 548, 654, 658
Merit function, 941
Metal beam splitters, 485–486
Metal reflection; *see also* Fresnel equations
 about, 313–314
 normal incidence reflectance of metals, 315
 retardance/diattenuation of metal at non-normal incidence, 315–316
Metal reflectors, 490
Metric thicknesses, 486
Michael–Levy interference color chart, 268f, 271
Michelson interferometers, 718
Microlithography, 26, 814
 optics, 667
Micro-polarizer array polarimeter method, 242–243
Mie theory, 260
Mirror element
 retardance maps, 950
Mode combination surface
 coordinate system, 731
Mode locked lasers, 111
Modulated polarimeters, 240
 example, 234
Modulation transfer functions (MTFs)
 polarization, 904
Modulation transfer matrix (**MTM**), 594

Mollweide projection, 83
Monochromatic light, 2
Monochromator/spectrometer
  linear dispersion of, 818
Monte Carlo methods, 936
Monte Carlo routines, 935
Motion blur, due to slow response time, 865
MOTM, *see* Mueller matrix optical transfer matrix
MPSM, *see* Mueller point spread matrix
MSPI/MAIA imaging polarimeters, 258–259
MTM, *see* Modulation transfer matrix; Mueller modulation transfer matrix
Mueller coherence matrix, 198
Mueller–Jones matrices, 192
Mueller matrices, 21, 774, 832, 871, 874, 875, 909, 935, 936
  about, 164–165
  Cassegrain telescope, 472, 473f
  columns, meaning of, 164
  conversion of Jones matrices to Mueller matrices using Pauli matrices, 207
  depolarization
    about, 193–195
    addition of Mueller matrices, 201–202
    degree of polarization surfaces/degree of polarization maps, 196–198
    depolarization index/average degree of polarization, 195
    physically realizable Mueller matrices, testing for, 198–200
    weak depolarizing elements, 200–201
  for horizontal linear polarizer (example), 174
  image interpretation, 276–279
  incoherent ray tracing, 407–409
  Jones matrices into Mueller matrices using tensor product, transforming, 202–206
  for linear polarizer, 175
  multiplication (example), 174
  non-depolarizing Mueller matrices, 192
  non-polarizing Mueller matrix, 165–166
  origins of, 213–214
  overview, 163
  for physicality, testing (example), 199
  Poincaré sphere operations
    indicating polarization properties, 188
    of polarizers and diattenuators, 187
    of retarders on Poincaré sphere, 182–185
    of rotating linear retarder, 186
  point spread matrix (PSM), 955
  polarimeter, 8, 10, 10f, 24f
  polarization elements, sequences of, 165
  polarizer/diattenuator Mueller Matrices
    basic polarizers, 174–177, 176t
    diattenuators, 180–182
    polarizance, 180
    transmittance and diattenuation, 177–180
  ray tracing with
    about, 211
    Mueller matrices for reflection, 212–213
    Mueller matrices for refraction, 212
  retarder Mueller matrices, 168–174, 170t
  rotating polarization elements about light direction, 166–167
  rotation matrix for, 72

  transforming into Jones matrices, 207–210
  uncoated lens, 453–455, 454f
  weakly polarized (example), 191
  weak polarization elements, 190–192
Mueller matrix image, 868
  of poor-quality polarizing film, 871
Mueller matrix imaging polarimeter, 580–582
  measurements, 910
Mueller matrix optical transfer matrix (MOTM), 617
Mueller matrix polarimeters, 867, 872, 913, 930, 936
  testing, 872
Mueller matrix polarimetry
  principal retardance spectra, 913
Mueller matrix spectrum, 873
Mueller modulation transfer matrix (MTM), 608
Mueller optical transfer matrix, 608
Mueller point spread matrix (**MPSM**), 582, 593, 594, 602, 623, 654, 774, 937
  of aberration-free system, 604
  of Cassegrain telescope, 604–605
  critical angle CCR and, 620–622
  diffraction-limited, 654
  o- and *e*-wavefronts, 775
  overview, 603
  polarization structure of images and, 607
  polarized pupil with unpolarized object, 611–614
  scale of, 605–607
  skew aberration and, 664–665
  for U.S. Patent 2,896,506, 665–666
Mueller polarimetry configurations, 272–276; *see also* Polarimetry
  about, 272–273
  dual rotating retarder Mueller matrix polarimeter, 274–275
  polarimetry near retroreflection, 275–276
Mueller rotation matrix, 166
Multi-angle imager for aerosols (MAIA), 257, 258, 259
  imaging polarimeters, 258–259
Multi-angle spectro polarimetric imager (MSPI), 258
Multi-domain vertically aligned (MVA) cell, 851
  configuration for, 852
  LCD pixel, 876
Multilayer thin films, 486–496
  algorithms, 487–488
  polarizing beam splitters, 492–496
  quarter and half wave films, 488–489
  reflection-enhancing coatings, 489–492
Multi-order retardance, 933; *see also* Retardance
  compound retarder, 933
  misaligned two-element compound linear retarder, 934
  principal retardance component, 934
  retardance discontinuities, 933
  retardance space, 933
  unwrapped retardance, 933
Multiple birefringent plates
  wavefront exiting systems of, 723
Multiple scattering, 260
Multivalued optical path length (OPL), 711, 909
  misaligned two-element compound linear retarder, 910

NA, *see* Numerical aperture
Nearly parallel monochromatic plane waves, interference of, 95–100
Nearly singular Mueller polarimeter (example), 285

Negative uniaxial crystals, 674
  birefringence spectra of, 745
Nematic liquid crystals, 839
Nematic molecules
  schematics for, 840
Nicol prism, 779, 778
  FOV for, 778
Nilpotent matrix (example), 144
Nodal aberration theory, 400
Non-co-propagating wavefront combination, 728
Non-depolarizing Mueller matrices, 192, 195, 202
Non-ideal polarization elements, 221
Non-orthogonal coordinate system of Stokes parameters, 78–79, 166
Nonorthogonal igenpolarizations, 149
Non-periodic grating, 813
Non-polarizing beam splitter, 93, 486
Non-polarizing Mueller matrix, 165–166
Non-polarizing optical element, 165
Non-polarizing optical systems, 633–634
  skew aberration, 654, 667
Non-polarizing reflection, 639
Non-polarizing rotations, 629
Non-sequential ray tracing, 407
Normal (ray intercept), 586
Normal incidence, 304–305; see also Fresnel equations
  reflectance of metals, 315
  reflection, 645–647
Normalized coordinates, wavefront analysis, 393
Normalized flux, 41
Normalized Jones vector, 38, 39t
Normalized Pauli summation, 549
Normalized propagation vectors, 95
Normalized Stokes parameters
  about, 64, 80
  defined, 178
North pole, 4
Null space, 220, 225, 230–240, 232, 282
Numerical aperture (NA), 23, 25, 26f, 373
  skew aberration and, 654, 658, 667
  wavefronts, high, 25–26; see also Polarization problems in optical systems

Object plane, 369
Odd permutations, Pauli matrices, 510
Off-axis telescopes, 567
oie-mode, 695
oio-mode, 695
Open set, Jones matrices, 523
Optical activity, 137
Optical axis, 25, 583
Optical coherence tomography, 718
Optical design
  about, 17–20
  polarization ray tracing, 20–21
  programs, 20
Optical detectors, 37
Optical elements, 941
Optical indicatrix, 749
Optical isolators, 715
Optically active materials, 676
Optical path difference (OPD), 182, 307, 642, 645, 724, 766, 912
  contribution, 724

Optical path length (OPL), 17, 20, 120, 143, 166, 359, 545, 630, 716, 858, 931
  A-plate, 768
    retardance magnitude through the quarter wave, 769
  birefringent materials, 684–685
  within birefringent plate, 724
  cell phone lens, 361–362
  compound retarder, 910
  decreasing phase convention, 379
  eigenmode propagating, in anisotropic material, 718
  eigenstates, 722
  grid locations, on spherical surface, 734
  intermediate matrices, 720
  interpolated function, 733, 735
  misaligned retarder, 910
  misaligned two-element compound linear retarder, 910
  multi-order retarder, 910
  multivalued, 711
  o- and e-wavefronts, 770, 769
  off-axis collimated wavefront
    birefringent plate, 724
  polarization state distributions, 730
  polarization states, 723
  ray grids, 729
  rays and ray segments, 718
  ray tracing, 378–379, 379f
  retarder space, 910
  tangential and sagittal rays, 771
  waveplate, 724
Optical power, 585
Optical rotatory power, 676
Optical systems
  apertures, 366–367, 366f
  aperture stop, 369, 369f
  defined, 364
  dummy surfaces, 367–368
  entrance pupil, 370, 370f
  etendué, see Etendué
  exit pupil, 370–372, 371f
  interfaces, 367
  Lagrange invariant (H), 373–374
  marginal and chief rays, 372, 372f–373f
  numerical aperture (NA), 373
  object plane, 369
  polarized light, 375–376
  specifications of, 364–368, 365f
  surface equations, 366
  surface parameters, 365t
  wavelength, 365
Optical thicknesses, 486
Optical transfer function (OTF), 500, 593, 937
Optical transfer matrix (**OTM**), 593, 608–610, 623, 937
  of Cassegrain telescope, 608–609
  critical angle CCR and, 621–622
  cutoff frequency, 609–610
  polarized pupil with unpolarized object, 611–614
  scale of, 609–610
Optical tweezers, 139
Optical waveplates
  wavefront aberrations, 742
Optic axis, 11, 673, 674
Order-dependent terms, 546

Order of the retarder, 645
Ordinary modes, 670, 680
Ordinary refractive index, 741
Orientation, plane of incidence, 551, 552–553
Orientor basis set, 573
Orientors, 545, 567, 572–578
    defined, 573
Orthogonality of two Jones vectors (example), 48
Orthogonal Jones vectors, 48–49, 48f
Orthogonal matrices, 168, 326
Orthogonal polarization states, 74–75, 75t, 336
Orthogonal unit vectors, 168
Orthonormal polarization states, 50
OSA Handbook of Optics, 785
Oscillating electric dipole, 430
Oscillating element polarimetry, 220
OTF, *see* Optical transfer function
OTM, *see* Optical transfer matrix
Out-of-plane diffraction grating equation, 816
Out-of-plane grating equation, 817
    graphic depiction, 817
Over driving circuit (ODC) technologies, 865
Overlapping wavefronts, *see* Wavefronts, overlapping

Padding, Jones pupil, 604–605
Pancharatnam-design retarders, 916
Pancharatnam phase, 635, 654
Parallel polarizer linear polariscope, 265
Parallel transport, 660, 667
    matrix, 631, 636–640, 642, 654
Parallel transport of vectors, 634–636
    with reflection, 636, 637f
Paraxial angle of incidence, 544, 550–553, 562
Paraxial calculation, 562
Paraxial chief ray, 586, 660
Paraxial limit, 660
Paraxial marginal ray, 586, 660
Paraxial optics, 497, 543, 583–585, 657, 660, 664
Paraxial polarization aberrations
    angle of incidence, 550–553
    binodal polarization, 558
    diattenuation, 553
        across field of view, 556–557
        defocus, 553–556
    over series of surfaces, 558–560
    overview, 550
    piston, 557, 558f
    plane of incidence, 550–553
    retardance, 553
        across field of view, 556–557
        defocus, 555–556
    seven-element lens system, 560–567
    tilt, 557
Paraxial ray trace, 550, 586–587, 633
    equations, 544
    form, 585, 585t
    skew aberration in, 660–661
Paraxial region, 544, 583–584
    polarization aberrations, 550–560
Paraxial skew aberration, 662–663
Paraxial skew rays, 588
Partial coherence calculations, 935
Partial depolarizer Mueller matrix, 194

Partially linearly polarized light, 67
Partially polarized light, 66–70, 73
Partial polarizers, 5, 130
    Mueller matrix for, 180
Partial waves, 11, 491, 670
Patterned vertical-aligned cell (PVA), 851
    director distribution, 852
Pauli *c*-coefficients, 512, 512f
Pauli coefficients
    combination of weak diattenuations, 550f
    complex, 548–550
Pauli logarithm coefficients, 523–525
Pauli matrices, 543, 547, 573, 917; *see also* Jones matrices
    as anti-commutative, 510
    canonical summation form, 514–515
    Cartesian rotation matrix, 511
    eigenvalues, 513
    eigenvectors, 513
    even permutation, 510
    expansion in sum of, 510–511
    functions, 513–514
    identities, 509–510, 510t
    odd permutations, 510
    overview, 508–509
    polarization element rotated about optical axis, 511–512
    sign convention, 511
Pauli vectors, 549–550
Periodic grating, 812
Phase change, 143
    with incident angle, 307–308
Phase convention, 133
Phase depolarization, 200
Phase discontinuities, in thin films, 499–501
Phase distribution, 92
Phase grating, 812
    structure, 813
Phase matching, 297
Phase modulation polarimetry, 220
Phase modulators, 842, 843
Phase of light, 76
Phase retardation (optical path difference), 631
Phase sign convention
    decreasing, 54–55
    increasing, 55–56, 153, 154t–155t
Phase transfer matrix (PTM), 594, 609
Phase unwrapping operation, 500
Photoelasticity, 880
Photoelastic modulator (PEM), 254, 259
    polarimeters, 255–258
Photon, polarization of, 65
Physically realizable Mueller matrices, testing for, 198–200
Piezoelectric transducer, 255
Piston (constant), polarization aberrations, 543, 544f, 557, 558f
    diattenuation, 557, 562, 564–566
    retardance, 557, 564
Pixel gain, variation in, 259
Pixel misalignment, 281; *see also* Artifacts in polarimetric images
Planar aligned nematic cell, 846
Planar wavefront incident
    birefringent plate, 721

Plane of incidence
    paraxial, 550–553
        orientation, 551, 552–553
    ray intercepts, 547
Plane of polarization, 25
Plane waves at large angles, interference of, 100–101
Plate birefringence, 777
Poincarè sphere, 4–5, 252, 630, 845, 867, 873, 909, 932
    about, 80–83
    flat mappings of, 83–85
Poincaré sphere operations
    indicating polarization properties, 188
    of polarizers and diattenuators, 187
    of retarders on Poincaré sphere, 182–185
    of rotating linear retarder, 186
Point spread function (PSF), 18–20, 500, 582, 593, 603, 611, 654, 818, 937, 958
    of Cassegrain telescope, 472, 473f
    cross-section profiles, 958
    DoP variation, 956
    $\log_{10}$ scale, 957
    shape of, 959
    skew aberration effect on, 663–665
    of uncoated single-element lens, 453–455, 454f
Point spread matrix (PSM), 453–455, 956
    Mueller matrix, 955
    polarization fluctuations, 955
    shape of, 959
Polarcor, 180
Polar Decomposition Example, 151
Polar decompositions, 149–152, 643
Polarimeters, 630, 631–633, 910
    calibration, 279–280
    classes of
        division-of-amplitude polarimeter, 241
        division-of-aperture polarimeters, 241
        imaging polarimeters, 241
        modulated polarimeters, 240
        time-sequential polarimeter, 240
    complete/incomplete, 222
    with four polarizers (example), 224
    function of, 2, 8
    light-measuring, 221
    optimization, 282–286
    polarization generators/analyzers, 222
    remote sensing, 259
    sample-measuring, 221
Polarimetric data reduction equation, 223, 228
Polarimetric data reduction matrix, 223, 224, 253
Polarimetric measurement equation, 223
Polarimetric measurement matrix, 223, 228, 236, 252
Polarimetric measurement vector, 227
Polarimetry, 580
    artifacts in polarimetric images
        about, 280–281
        pixel misalignment, 281
    and ellipsometry, 8–10
    mathematics of polarimetric measurement/data reduction
        Mueller data reduction matrix, 226–230
        Mueller matrix elements, measuring, 225–226
        null space/pseudoinverse, 230–240
        Stokes polarimetry, 222–225

Mueller matrix image interpretation, 276–279
overview, 219–220
polarimeter
    complete/incomplete, 222
    light-measuring, 221
    polarization generators/analyzers, 222
    sample-measuring, 221
polarimeter calibration, 279–280
polarimeter optimization, 282–286
polarimeters, classes of
    division-of-amplitude polarimeters, 241
    division-of-aperture polarimeters, 241
    imaging polarimeters, 241
    modulated polarimeters, 240
    time-sequential polarimeter, 240
polarization images, 220
sample-measuring polarimeter
    about, 262
    Mueller polarimetry configurations, 272–276
    polariscopes, 263–272
Stokes polarimeter configurations
    atmospheric polarization images (example), 259–262
    MSPI/MAIA imaging polarimeters, 258–259
    photoelastic modulator (PEM) polarimeters, 255–258
    rotating element polarimetry, 249–253
    simultaneous polarimetric measurement, 242–248
    variable retarder/fixed polarizer polarimeter, 254–255
Polariscopes, 880, 898; *see also* Polarimetry; Sample measuring polarimeter
    CD substrate, 898
    circular, 266–267
    conoscope, 270–271
    defined, 14
    interference colors, 267–269
    linear, 263–266
    plastic glasses, 899
    plastic lens, 899
    plastic tape dispenser, 898
    with tint plate, 269–270
Polaris-M, 17, 17n
Polaris-M polarization analysis program, 560, 615, 670
Polaris-M polarization ray tracing program, 827
Polarizance, 180
Polarizance vector, defined, 180
Polarization
    about, 2
    analyzer, 221, 222
    components, exponential form of, 151
    critical optical systems, 1
Polarization aberration function, 21, 323, 394–395, 395f, 397–398, 452, 455, 599
Polarization aberrations, 323, 447, 448, 786; *see also* Aberration
    about, 1, 12
    aluminum coating, 943–945
    amplitude response matrix, 954–955
    analysis, 944
    conventional optical design, 20
    defined, 543
    design rules, 950–951

Index

diattenuation at center of pupil, 951
ghost PSF in XY- and YX-components, 953–954
linear variation of diattenuation, 952
linear variation of retardance, 952
polarization-dependent astigmatism, 952–953
PSF shear between XX- and YY-components, 952
retardance at center of pupil, 951–952
diattenuation aberration, 653
diattenuation/retardance contributions, 949–950
division, 653
functions, 773, 774, 775, 827, 937
higher-order, see Higher-order polarization aberrations
image formation with, 593–623
Jones pupil, 945–946
aberration expression, 946–949
and Jones vector, 318
of lenses, 22–25; see also Polarization problems in optical systems
liquid crystal displays/projectors, 16
measurements, 580–583
mitigation, 940
analyzing polarization ray tracing output, 941–942
Mueller matrix point spread matrices, 955–958
overview, 545–546
paraxial, see Paraxial polarization aberrations
polarization aberration expression, 943–945
polarization ray trace, 945–946
polarization ray tracing, comparison of, 942
PSF image components, location of, 958–959
retardance aberration, 653
second-order, 557
skew aberration, 653, 667; see also Skew aberration
understanding of, 630
weakly polarizing ray intercepts, 548–550
weak polarization interactions, 546–548
Polarization and Directionality of Earth's Reflectances (POLDER), 260
Polarization beam splitter (PBS), 95f
cubes, 492
Polarization bidirectional reflectance distribution function (PBRDF), 935
models, 936
Polarization dependent loss (PDL), 179
Polarization economy, 837
Polarization elements, 940
about, 2
about light direction, rotating, 166–167
defined, 117
depolarizers, 6
optical properties, 118
in polarization ray tracing, 20
polarizers, 5–6
retarders, 6
sequences of, 165; see also Mueller matrix
Polarization ellipse, 2, 3, 22, 34, 38, 40, 74, 103, 572
rotation, 72–73
Polarization engineering, 1, 929
Polarization fringes, 91
Polarization generator, 221, 222f
Polarization images, 220
Polarization-independent amplitude change, 525
Polarization-independent phase change, 525
Polarization independent transmission, 120

Polarization leakage, 15
Polarization matrix propagation method, 21
Polarization models, 856
extended Jones matrix method, 856–857
liquid crystal cell ZLI-1646, 859–861
multilayer interference models, 859
retardance of single-pixel polarization controller, 862
single pass, with polarization ray tracing matrices, 857–858
Polarization piston, 557, 558f
Polarization problems in optical systems
angle dependence of polarizers, 12
high numerical aperture wavefronts, 25–26
liquid crystal displays/projectors, 16–17
polarization aberrations of lenses, 22–25
stress birefringence in lenses, 14–15
wavelength and angle dependence of retarders, 13
Polarization properties; see also Jones matrices
homogeneous Jones matrices, 523–527
inhomogeneous Jones matrices, 531–532
polarization elements and, 5–7
of polarization elements and optical elements, 117
Polarization quality, wavefront analysis, 406
Polarization ray tracing (PRT), 1, 20, 118, 323–324, 550, 615, 629, 642, 649, 665, 723, 892–895
for addition, 716, 721
additional information obtained from, 381–382, 381t
amplitude transmission, 382
for birefringent interface, 695–705
brief history of, 411–412
cell phone lens, see Cell phone lens
hollow aluminum corner cube, 347–351
interferometer with PBS, 337–344
with P matrix, see Polarization ray tracing matrix (P)
proper retardance algorithm for, 643–545
use of, 410–411
Polarization ray tracing complications, 930
elliptical polarization properties, of ray paths, 931
geometrical transformation, 933
linear skew aberration transforms, 933
optical path length/phase, 931
antireflection coatings, 931
constructively interfere, 931
spherical wavefront, from direct light through the system, 932
optical system description complications, 930
diffraction gratings, 930
liquid crystal cells, 930
reasonable distribution, of stress birefringence, 931
stress birefringence, 930
thin films, 930
parallel transport, 933
proper retardance, 933
retardance, 932
skew aberration, 932, 933
Polarization ray tracing (PRT) matrices, 324–334, 385–392, 932, 935, 936
coherent combination, 344–347
definition, 324–325
diattenuation calculation, 334–337
Jones pupils, converting into, 437–439
at normal incidence, 386–387
orthogonal transformations, 325–331
for refraction and reflection, 330

retarder, 331–334
two gold fold mirrors, 387–390
Polarization ray tracing methods, 937
   alternative simulation methods, 940
   cell phone lens system, 938
   data structures, 938–939
   Jones matrices, 937
   Jones pupil, 937
   mode combination, 939–940
   optical transfer matrix (**OTM**), 937
   P matrix, 937
   point spread function (PSF), 937
   ray doubling, 938–939
   ray trees, 938–939
Polarization rotation, 655
Polarization state analyzer (PSA), 263
   in Jones matrix, 911
   rotates, 911
Polarization state generator (PSG), 263f
Polarization states
   exiting, 164
   and Poincaré sphere, 2–5
   of sources, 56–59
   time-varying, 103
   variations of, 17
Polarization structure, of images, 607
Polarization transformations, 184
Polarization vector, 672
   about, 5, 32–33
   defined, 32
   to Jones vectors, converting, 51–54
   method for polarized light calculations, 32t
   properties of, 34–36
Polarization vortices, 138
Polarized flux, 36–37, 66, 75–78
   components, 50–51, 51t
Polarized incident beam, 722
Polarized incident light, 220
Polarized light, 1
   change of basis, 49
   circularly polarized light, 43–45
   description of, 31, 32t
   elliptically polarized light, 45–48
   Jones vectors
      about, 37–41, 39t
      addition of, 49–50
      orthogonal, 48–49
      rotation of, 42
   linearly polarized light, 42–43
   magnetic field/flux/polarized flux, 36–37
   methods for calculations, 32t
   and optical systems, 721
   overall phase, evolution of, 41–42
   phase sign convention
      decreasing, 54–55
      increasing, 55–56
   polarization state of sources, 56–59
   polarization vector
      about, 32–33
      to Jones vectors, converting, 51–54
      properties of, 34–36
   polarized flux components, 50–51, 51t
   propagation in isotropic media, 36

Polarized light of two different frequencies, addition of, 103–105
Polarized part of flux, 76
Polarized sunglasses, 219
Polarizer; *see also* Poincaré sphere operations
   about, 2
   angle dependence of, 12; *see also* Polarization problems in optical systems
   basic, 174–177, 176t
   characterized by diattenuation, 177
   defined, 119
   ideal, 174
   operation of, 187
   over aperture (example), 201
   tests, 132
Polarizer/diattenuator Mueller matrices
   basic polarizers, 174–177, 176t
   diattenuators, 180–182
   polarizance, 180
   transmittance and diattenuation, 177–180
Polarizer Jones matrices, 129t
Polarizer Mueller matrices for basis polarization states, 176t
Polarizing beam splitters (PBS), 6, 6f, 93, 246, 247, 247f, 479, 729; *see also* Beam splitters coatings
   collimated wavefronts, 729
   multilayer thin films, 492–496
   two collimated wavefronts, 729
Polarizing films, 870
Polycarbonate (PC), 882
Polychromatic beams, addition of, 105–109
Polychromatic light
   defined, 63
   description of, 63–64
Polychromatic unpolarized beam, measurement of, 105
Polyimide, 843, 862
Polymethylmethacrylate (PMMA), 882
Polynomial curve fitting, 495–497
Polynomials, 494–496
Polystyrene (PS), 882
Positive uniaxial crystals, 674
   birefringence spectra of, 744
Potassium titanyl phosphate (KTP) crystal, ray tracing, 670, 672, 686–693
Poynting vectors, 37, 368, 716, 720, 752, 777
   mapping, 720
Primary cloudbow, 261
Principal axes, 673, 674
Principal refractive indices, 673
Probability amplitude, 932
Product notation, 124
Projection screens, 6
Propagation sphere, 425, 425f; *see also* **k**-sphere
Propagation vectors, 17, 32, 324, 325f, 655, 660, 662, 667, 672, *see* Snell's law
Propagation vector sphere, 663
Proper retardance
   algorithm for, 643–545
   calculation of, 629, 631, 649
   definition of, 642
   identifying, 633
   parallel transports for reflection and, 636

Index

4'-propyl-bicyclohexyl-4-carboxylic acid 3,4,5-trifluoro-phenyl ester ($C_{22}H_{29}F_3O_2$)
   chemical structure of, 839
PRT, see Polarization ray tracing
Pseudo electric field, 702
Pseudoinverse, 230–240, 233, 240, 496
PSF, see Point spread function
PTM, see Phase transfer matrix
Pulse trains, 111
Pure diattenuator, 130
Pure retarder, 168

Quadratic diattenuation coefficients, 561
Quadratic radially oriented retardance, 600–601
Quadratic retardance coefficients, 562
Quality, wavefront analysis
   polarization, 406
   wavefront, 405
Quantum optics, 109
Quarter wave coating, 931
Quarter wave film, 488–489
Quarter wave layers, 489
Quarter wave linear retarder (QWLR), 331, 331f
   about, 6, 121
   Jones matrices for, 134, 338–339, 516
   Mueller matrices for, 169, 170t, 171
   P matrix, 330–333, 339–340
   Poincaré sphere, 183
   sequence, 516
   spinning (example), 202
Quarter waveplates, 766
Quarter wave retarder (QWR), 169, 247f
Quarter wave thick magnesium fluoride coatings, 8
Quartz HLR, 919
Quartz retardance, 913
   principal retardance, 914
Quartz retarder, 913, 922
   function of wavelength, 923
   unwrapped retardance, 923
Quasi-monochromatic light, 70

Radially symmetric optical system, 658
Radially symmetric systems, 550
Radiometry with polarization elements, 221
Ray diffracts
   into multiple diffraction, 812
Ray doubling, 745, 939
   data structure for, 694–695
Ray ellipsoid, 749, 750
Ray grids, 716–718
   angular variation of retardance, 727
   birefringent plate, 725
   corresponding modes, 731
   cross- and co-polarized systems, 801
   interpolation functions, 733
   irregular grids, combination, 730
      inverse-distance weighted interpolation, 732–734
      misaligned ray data, 730–731
   at locations, 732
   point source on spherical wavefronts, 717
   positions of ray, 730
   propagating, 728
   quarter wave calcite A-plate with optic axis, 772

Ray intercepts, 17, 360
   algorithm for, 378
   calculation, 584–585
   finding, 377–378
   Jones matrix for, 546
   multiplicity with surface, 378
   planes of incidence, 547
   s- and p-components, 382–383
   weakly polarizing, 548–550
Rayleigh scattering, 260
Ray paths, 642
Ray propagation, 719
   through co-polarized system, 800
Ray splitting, see Double refraction
Ray stop criteria, 695
Ray tracing
   algorithms, 729
   amplitude coefficients, 383–385
   in birefringent materials, see Birefringent materials
   coherent and incoherent, 407–410
   defined, 1
   description, 359
   geometrical, 359; see also Polarization ray tracing
   goals for, 360–361, 376–377
   interface Jones matrix, 383–385
   light beams, 368
   with Mueller matrix; see also Mueller matrix
      about, 211
      Mueller matrices for reflection, 212–213
      Mueller matrices for refraction, 212
   non-sequential, 407
   in optical design, 17
   optical path length, see Optical path length
   optical system and, see Optical systems
   polarization, see Polarization ray tracing
   ray intercept, see Ray intercepts
   reflection and refraction, 380–381
   s- and p-components, 382–383
   steps in process of, 360
   through birefringent components, 742
   wavefront analysis, see Wavefront analysis
Ray tracing DOEs, 818
   diffractive retarders, 825
   diffractive subwavelength antireflection coatings, 825–829
   eigenpolarizations, 818
   reflection diffractive gratings, 818–821
   wire grid polarizers, 822–825
Real unitary matrices, 168
Reciprocal ellipsoid, 750
Rectangle function, 594
Reduced angles, 587–588
Reduced thicknesses, 587–588
Reference beam, 93, 102
Reference path, ray tracing, 338–340
   cumulative **P** matrix for, 342
Reference sphere, 370
Reflection, 636, 637f, 638, 646, 649
   at birefringent interface, 681–693
   diffractive grating, 819
   inside a biaxial cube, 707–710
   Mueller matrices for, 212–213
   non-polarizing, 639

normal incidence, 645–647
and refraction, 380–381
Reflection-enhancing coatings, 489–492, 497
effective depth, 491, 492f
Reflection gratings
diffractive, 818
Reflection/transmission at glass (example), 302
Reflective coatings, 489; *see also* Reflection-enhancing coatings
Reflective diffraction gratings, 818
Refraction, 636, 638; *see also* Reflection and refraction
at birefringent interface, 681–693
Mueller matrices for, 212
Refractive grating
in-plane incident light ray, 815
Refractive index, 20, 36, 297, 365
birefringent material, 675, 683
KTP, aragonite, and mica, 706t
reduced thicknesses and angles, 587–588
skew aberration and, 660
Remote sensing polarimeters, 259
Research Scanning Polarimeter (RSP), 260
Resolution, DFT, 597
Response time
drive voltage and response times, 866
motion blur, due to slow, 865
Response time compensation (RTC), 865
overshoot, 866
Retardance, 932
aberration, 653
with angle, 868
angular variation of, 727
circular, 629, 636, 663, 666
common units for, 121
components, 171
critical angle CCR, 618
defined, 307, 630, 642, 649
from diattenuation, separating, 149–152
higher-order polarization aberrations, 578–580
importance of, 629
linear, 545, 572, 573
measurement technique, 265
of metal at non-normal incidence, 315–316
microscope, 581
for multi-element lens, 23
and optical path lengths, 6
parameters, 173
paraxial, 553
polarization ray tracing, 20, 21, 117
proper, *see* Proper retardance
of pure retarder Mueller matrix, 168
range, 645
seven-element lens system, 562, 566–567
three-dimensional polarization ray trace matrices, 631
Retardance defocus, 950; *see also* Defocus (quadratic), polarization aberrations
paraxial, 555–556, 557
quadratic radially oriented, 600–601
seven-element lens system, 563–564
Retardance discontinuities, 500
mystery of, 910–912
retarder, unwrapped retardance of, 911

Retardance maps, 457
crossed mirror configurations, 465–469, 466f, 467t, 468f, 469f
vs. field of view, 860
Retardance modulation, 845
Retardance modulators, 841, 843
Retardance of P matrices, 337
Retardance piston, 557, 564, 950; *see also* Piston (constant), polarization aberrations
Retardance space, 933
inhomogeneous Jones matrices, 534, 534f–535f
Retardance tilt, 557, 564, 950; *see also* Tilt (linear), polarization aberrations
Retardance trajectory, 916
Retardance unwrapping, 645
Retardation plate, 122
Retarders, 6, 118, 120, 121f, 145–149, 168, 630, 909
angle dependence of, 13; *see also* Polarization problems in optical systems
axis, 183
Jones matrices, 133–137, 520–521, 527
Mueller matrices, 168–174, 170t
multi-order, 909
order of, 645
P matrix, 331–334
linear retarders, 331–332, 332t
on Poincaré sphere, operation of, 182–185
Retarder space
Compound retarder at 45°, 922–924
compound retarder's trajectory, 919–920
half wave linear retarders, 917
homogeneous retarder's trajectory and retardance unwrapping, 914–916
Jones matrices, 529–531
multiple modes, exit compound retarder system, 920–922
principal retardance, 915, 916, 918, 919, 920
principal retardance trajectory, 916
principal retardance vector trajectories, 915
Quartz HLR, 919
unwrapped retardance, 916, 920
Retroreflection
polarimetry near, 275–276; *see also* Mueller polarimetry configurations
testing, 275
Retroreflectors, corner cubes as, 347–351
Right circularly polarized light, 43
Right circularly polarized time helix, 44
Right-handed local coordinates, 632
Right hand rule, 43
Rigorous coupled wave analysis (RCWA)
algorithm, 811, 815, 829–832, 830
diffraction efficiency, 820
TE and TM's, 821
divides, 830, 832
grating/diffractive optical element, 830
simulation, 828
Rochon prisms, 778
Ronchi tests, 929
Rotating element polarimetry, 220; *see also* Polarimetry
rotating analyzer plus fixed analyzer polarimeter, 250
rotating analyzer polarimeters, 249–250
rotating retarder/fixed analyzer polarimeters, 251–253

Index

Rotating linear retarder, 186
　operation of, 186
Rotating quarter wave retarder polarimeter (example), 253
Rotating retarders, 137
　imaging polarimeter, 281
Rotation axis, 168
Rotation matrix, 168, 327–328, 632
Ruoff, J., 404, 412

Sag/saggita, 366
Sampled functions, 594
Sample measuring polarimeters, 221, 222
　about, 262
　Mueller polarimetry configurations, 272–276
　polariscopes, 263–272
　polarization properties and, 8
s- and p-components, at ray intercepts, 382–383
Sapphire, aluminum oxide in, 500
Sapphire retarder, 923
Scale
　ARM, 605–607
　MPSM, 605–607
　OTM, 609–610
Scattered light, 102
Scattering, 6, 871, 935
Second-order polarization aberrations, 557
Seidel aberrations, 544, 550
Seidel wavefront aberration expansion, 398–401
Semi-major axis, 47
Semi-minor axis, 47
Sensitive tint plate polariscope, 270
Sequence of polarizers (example), 188
Sequence of retarders, 184
Sequence of rotations of sphere, 184
Sequence of three quarter wave retarders (example), 185
Sequences of polarization elements, 515–518
Seven-element lens system, 560–567
Sheet polarizers, 12, 179
Shifting functions, DFT, 596–597
Simultaneous polarimetric measurement; see also Polarimetry
　division-of-amplitude polarimetry (DOAP), 246–248
　division-of-aperture polarimetry, 242
　division-of-focal-plane polarimetry, 242–245
Simultaneous polarimetric measurement, 242–248
sinc function, 594
Single-element lens, see Uncoated single-element lens
Single-layer antireflection coatings, 485; see also Antireflection coatings
Single-layer thin films, 480–486
　antireflection coatings, 482–485
　metal beam splitters, 485–486
Singularities, of local coordinates, 323–324
Singular matrices, 523
Singular value decomposition (SVD), 150, 282, 335–337
S-IPS mode panels
　motion picture image, 851
Skew aberrated local coordinate, 665
Skew aberration, 618, 629; see also Polarization aberrations
　algorithm, 655–657
　CODE V's U.S. patent library, 658, 666–667
　definition of, 654–655
　effect on PSF, 663–665
　ith ray's, 657
　linear, 654, 655, 659, 660
　overview, 653–654
　paraxial, 662–663
　in paraxial ray trace, 660–661
　polarization aberration and, 653, 667
　polarization state rotation and, 653–654
　U.S. Patent 2,896,506, 658–659
　　PSM for, 665–666
Skew rays, 588, 654, 658
　parallel transport, 660
Skew tilt, 663, 664, 667; see also Tilt (linear), polarization aberrations
Slow axis, 122, 168
　unchanged convention, 134, 135t
Slow-mode, birefringent interfaces, 681–682
Snell's law, 17, 754, 306, 360, 480, 583–584, 683, 891
Solid corner cube retroreflectors (CCRs), 614–618
South pole, 4
Space helix, 44, 45f
Spatial filtering operation, 608
Spatially modulated polarimeters, 240
Speckle pattern, 193, 193f, 936
Spectral bandwidth, 2, 70–71; see also Stokes parameters
Spectrally modulated polarimeters, 240
Spectral resolution, 818
Spherical aberration, 498, 572, 583
Spherical interface, 551
Spherical polygon, 660, 663
Spherical wave, 22, 375
Spinning quarter wave linear retarder
　about, 186
　example, 202
s-/p-polarization components, 298–299
Statistics, skew aberration, 666–667
Stokes image of building (example), 69
Stokes parameter image, 607, 608, 614
Stokes parameter rotation by 45° (example), 72
Stokes parameters, 5, 776
　elliptical polarization parameters, 73–74
　example, 65
　for Gaussian wave packet spectral components, 112t
　and Jones vector, conversions between, 75–78
　and Jones vector sign conventions, 75
　linearly polarized Stokes parameters, 73
　measurement, 104
　method for polarized light calculations, 32t
　non-orthogonal coordinate system of, 78–79
　orthogonal polarization states, 74–75, 75t
　partially polarized light/degree of polarization (DoP), 66–70
　phenomenological definition of, 64–65, 65t
　Poincaré sphere
　　about, 80–83
　　flat mappings of, 83–85
　polarization ellipse, rotation, 72–73
　polarized flux, 75–78
　polychromatic light, description of, 63–64
　spectral bandwidth, 70–71
　unpolarized light, 65
Stokes polarimeter
　configurations; see also Polarimetry
　　atmospheric polarization images (example), 259–262
　　MSPI/MAIA imaging polarimeters, 258–259

photoelastic modulator (PEM) polarimeters, 255–258
rotating element polarimetry, 249–253
simultaneous polarimetric measurement, 242–248
variable retarder/fixed polarizer polarimeter, 254–255
with six measurements, 230
Stokes polarimetry, 222–225
Stokes vectors, 64
image, 582
Strehl ratio, 19
Stress birefringence, 880
in lenses, 14–15; *see also* Polarization problems in optical systems
Stress data format, 891–892
flange, 892
Stress-induced birefringence, 255, 879; *see also* Birefringence
computer-aided design (CAD) programs, 879
isotropic glass plate, 881
material's refractive index, 882, 883
mechanical stress, 880
N-BK7 and polycarbonate (PC), 882
object generate compressive stress and tensile stress, 882
optical materials, 880
on optical system performance, 898
glas spannungen, 901
induced retardance magnitude, 903
injection-molded lens, 904
injection-molded lens, CAD image, 902
injection-molded lens, simulations of, 901–903
injection-molded lens, stress images, 902
linear polariscope images, 900
pickups system, optical layout, 903
plane parallel glass plate, 900
plastic DVD lens, simulations of, 903–905
polarized light refracting, point spread functions of, 905
retardance magnitude, 902
tint plate retarder, 901
using polariscope, 898–901
in optical systems, 881
molding parameters, 881
plastic tape dispenser, 880
ray tracing, 883–889
diagonal tensor to non-diagonal tensor, 884
finite element modeling (FEM), 889
Jones pupil amplitude, 897
parabolic stress in tempered glass, 895–896
polarization ray tracing matrix, 892–895
reflections, 891
refraction, 891
refraction, through an injection-molded lens, 891
refractive index, 885
rotationally stressed plate, 896–897
spatially varying stress, 889–897
storage of system shape, 890
strain tensor coefficients, 883
stress data format, 891–892
stressed plate, linear polariscope images, 897
stress optic tensor coefficients, 885
stress/strain and optical effect, 884
tetrahedron element, 890
varying stress function, 895

residual stress, 880
retardance magnitude, 898
stress distribution, 897
theory of, 881
tolerance, 880
uniaxial stress
compression and tension, 888–889
optic effect, 886
Subwavelength grating (SWG), 813
intensity transmission of, 827
Jones pupil
amplitude of, 828
wavefront aberration, 829
phase shift of, 828
polarization-dependent aberrations, 828
retardance dispersions of, 825
retardance of fused silica, 826
transmitted phase and retardance, 827
Supernumerary bows, 261
Super twisted nematic cell (STN), 849
director orientations, 849
liquid crystal cell's directors, 850
Surface admittance, 489
Surface equations, optical systems, 366
Surface grating, 812
Surface parameters, optical systems, 365t
K/S-Surfaces, 749
SVD, *see* Singular value decomposition
Symmetrical phase convention, 134, 135t, 153
Systematic errors, 280

Taylor series
approximations, 494–495
for Fresnel coefficients, 317–318
Tensor product, 203
Test beam, 93, 102
Testing liquid crystal cells
analyzer and exiting polarization state, 877
IPS tests, 874–875
Mueller matrix polarimeter testing, 872, 873
MVA cell test, 875–876
sheet retarder defect, 876
twisted nematic cell example, 873–874
VAN cell, 875
Test path, ray tracing, 338, 340–341
cumulative P matrix for, 342–344
Tetrahedron orientation, 282
Thin film equations, 22
Thin films, 479–501
multilayer, 486–496
phase discontinuities in, 499–501
single-layer, 480–486
wavefront aberrations and, 497–499
3D polarization ray tracing, 324
3D TV, 855
Three-fold mirror system, 636, 637f
aluminum-coated, 647–649
Throughput, *see* Etendué
Tilt (linear), polarization aberrations, 543, 544f, 557
diattenuation, 557
retardance tilt, 557, 564
Time helix, 44, 45f
Time-modulated polarimeters, 240

Time-sequential polarimeter, 240
Time-varying polarization state, 103
Time-varying Stokes parameters, 104
Tint plate, polariscope with, 269–270; *see also* Polariscopes
Total internally reflected mode, 692
Total internal reflection (TIR), 306–307, 778, 785, 939
    o-mode, 778
Totzeck, M., 404, 412
Transmission axis, 5, 187
Transmittance, 177–180, 187
Transverse electric mode, 299
Transverse electric (TE) modes, 811
    diffraction efficiencies, 820
Transverse magnetic (TM) modes, 299, 811
    diffraction efficiency, 820
    polarization
        wire grid polarizer transmission, 824
Transverse plane, 34, 323
Trim retarder, 849
Twisted nematic (TN cell), 847
    C-plate, 848
    dark state, 849
    different voltages, 860
    director distributions, 848
    helix distribution of the liquid crystal directors, 848
    polarization ellipses of normal modes, 849
    retardance as function applied voltage, 848
    retardance eigenstates, 873
    tilt test, 874
Twisted nematic liquid crystal cell, 16, 859
Two linear diattenuators with rotation, cascading, 208
Two-retarder polarization controller, 94
Twyman–Green interferometer, 93, 94, 94f

Ultra-short laser pulse, 932
Uncoated single-element lens, 448–455
    amplitude response matrix (**ARM**), 452–453, 453f
    angle of incidence maps (AoI map), 449, 449f
    diattenuation map, 449–451, 450f
    Maltese cross, 451–452, 451f
    point spread matrix (PSM), 453–455
Uniaxial crystal, ray normally incident, 750
Uniaxial interfaces
    incident ray refracting, 939
    reflections/refractions, 746–749
Uniaxial materials, 674, 744
    amplitude coefficients, of transmitted mode, 764
    anisotropic materials, 741
    birefringence, 744
    converging beam, 743
    converging beam, polarization ray trace of, 771
    descriptions of, 743
    dielectric tensor for uniaxial birefringent, 743
    double refraction/ray splitting, 745, 746
    eigenmodes of, 745–746, 751
    eigenmodes/ordinary *o*-mode/extraordinary e-mode, 743
    e-mode transmission, 760
    exiting polarization ellipses, 771
    extraordinary e-mode, 745
    extraordinary mode in transmission, 764
    imaging doubling through calcite, 742
    incident rays with incident angles, 770
    index ellipsoid, 755
    index ellipsoid of, 751
    K-surface, 753, 754
        extraordinary mode in positive uniaxial material, 757
        isotropic material, 755
        poynting vectors, 754
    negative uniaxial crystals, 744
        birefringence spectra of, 745
    normal incident beam, polarization ray trace of, 777
    normally incident ray, 751
    normal surface, 753
    o- and *e*-wavefronts, 773
    off-axis ray trace from air, 755
    o-mode's **P** matrices, 772
    o-mode transmittion, 759
        sagittal and tangential ray bundles, 768
    ordinary *o*-mode, 745
    polarization states, 771
    polarization structure, 772
    positive uniaxial crystals, 744
    p-polarized incident light, 763
    principal section, 745
    ray propagating, 753, 757
    ray refracting, 745, 746
    ray trace through blocks, 742
    ray with propagation vector, 752
    reflection matrix, 765
    refraction angle, 756
    refractive index, 745
    refractive index of exiting *e*-mode, 770
    retardance magnitude of spherical wavefront, 771
    several possible index ellipses, 752
    s-polarized incident light, 762
    S-surface, 756
Uniaxial medium, 744
Uniaxial-to-isotropic interface, 704
Unitary change of basis, 49, 127
Unitary Jones matrices, 145
Unitary matrices, 630, 643; *see also* Orthogonal matrices
Unitary transformation, 146, 147, 149, 168, 181
Unit propagation sphere, 425
Unpolarized generator, 226
Unpolarized incident, 664
Unpolarized light, 65, 71f, 104, 164, 178, 666, 872;
    see also Stokes parameters
Untwisted nematic cell, 846
Unwrapped retardance, 916
    compound retarder Jones matrix decomposition, 917–919
    compound retarder systems with arbitrary alignment, 916
U.S. Patent 2,896,506, skew aberration, 658–659
    PSM for, 665–666
UV photo, 863

Values estimation, by interpolation, 730
VA panels
    motion picture image, 851
Variable beam splitter, 102
Variable retarder/fixed polarizer polarimeter, 254–255
Vectors, 572
    basis, 656
    coordinate, 640

parallel transport of, 634–636
polarization, 672
propagation, 655, 660, 662, 667, 672
Vector Zernike polynomials, 567, 568–572, 569t–570t, 571f, 575–578
Vertically aligned nematic (VAN), 846, 850
  cell's directors at several voltages, 850
  dark state, 851
  eigenpolarizations in cell without pretilt, 851
  LC cell, 844, 862
    with negative dielectric anisotropy, 844
    with negative dielectric anisotropy sandwich, 844
  mode cell, 844, 867
  retardance *versus* tilt angle, 876
Vertical quarter wave linear retarder (VQWLR), 170t, 183, 183f
Volume grating, 812
Vortex retarder, 582

Walk-off angle, 777
Walk-off plate, 777
  mode index for calcite *versus* optic axis angle, 777
  vs. wavelength for uniaxial materials, 778
Wavefronts, 17, 716–718, 725
  aberration function, 719, 937, 941
  combination, 730
  combination procedures, 716
  converges through birefringent plate, 717
  eieieiei-wavefront, 801
  eie-mode
    with zero piston, tilt, and defocus, 799
  eie-mode, residual wavefront, 797
  encounter beam splitters, 716
  incident collimated beam with planar, 721
  individual, 726
  interfering, 717
  local polarization ellipses of, 823
  o- and *e*-modes for uniaxial material, 721
  off-axis beam, 724
  off-axis collimated, 724
  overlapping, 715
  overlapping region, polarization of, 726
  piston, 667
  point source, propagating toward lens, 717
  polarization amplitudes, 801
  polarization amplitudes
    through co-polarized system, 802
  polarization variations of, 734–736
  quality, 405
  quarter wave calcite A-plate, 772
  reconstruct, 730
  reconstruction, 731
  resultant, 725
  shear, 723
  sheared non-planar, 726

Wavefront aberrations, 17, 18, 19f, 448, 543, 572, 580, 653, 663
  thin films contributions to, 497–499
Wavefront aberration function, 11, 360, 545, 593, 598
  cell phone lens, 404–405
  description, 393–394
  dipole *vs.* double pole coordinates, 440–441
  evaluation of, 395–398
Wavefront analysis, 393–406
  aberration function
    polarization, 394–395
    wavefront, *see* Wavefront aberration function
  normalized coordinates, 393
  quality
    polarization, 406
    wavefront, 405
  Seidel wavefront aberration expansion, 398–401
  Zernike polynomials, 401–405
Wavelength and angle dependence of retarders, 13
Wave nature of light, 92
Waveplates, 122, 630
  achromatic, 766
  first-order quarter, 766
  half, 766
  polarization ellipse distribution, 736
  polarization ellipse on plane, 735
  second-order quarter, 766
  true zero-order quarter, 766
Waves, retardance magnitude, 829
Weak depolarizing elements, 200–201
Weak diattenuator Mueller matrix, 190
Weakly linearly polarizing interfaces, 548
Weakly polarizing optical elements, 543, 545
Weakly polarizing ray intercepts, 548–550
Weak polarization aberration, surfaces, 558–560
Weak polarization elements, 190–192, 508, 535
Weak polarization interactions, Jones matrix, 546–548
Weak retarder Mueller matrix, 190, 191
Wedges
  wavefront exiting systems of, 723
White light beam, 106
Winding number theorem, 324, 428, 429, 430–431
Wire grid polarizers, 12, 242, 786, 822
  angle of incidence variation, 823
  optimization, 823
  transmission
    performance comparison of, 825
    of TM polarization, 824
Wollaston prisms, 246, 742, 778, 785, 786

Young's double slit interferometer, 92–93, 92f

Zernike coefficients, for *eie*-Wavefront, 798–799
Zernike polynomials, 401–405, 544, 545, 568–572, 575–578
  fitting, 797